UTILISATION OF ELECTRICAL POWER and ELECTRIC TRACTION

For

B.Sc. (Engineering); B.E., B.Tech; (Electrical Engineering);

A.M.I.E. (Elect.) Section (B); Diploma in Elect. Engg. Competitive and other equivalent Engineering Examinations

G.C. GARG

F.I.E., (Elec. Engg.); C. Engg. (I)

T.T.T.I. (Holland-Collaboration), Chandigarh

Ex-Registrar, Technical Education, Haryana

Formerly, Head of Electrical Engineering Deptt.

Govt. Polytechnic, Ambala City (Haryana)

KHANNA BOOK PUBLISHING CO. (P) LTD.

Publisher of Engineering and Computer Books

4C/4344, Ansari Road, Darya Ganj, New Delhi-110002

Phone : 011-23244447-48 **Mobile:** +91-9910909320

E-mail : contact@khannabooks.com

Website : www.khannabooks.com

Dear Readers,

To prevent the piracy, this book is secured with HIGH SECURITY HOLOGRAM on the front title cover. In case you don't find the hologram on the front cover title, please write us to at contact@khannabooks.com or whatsapp us at +91-99109 09320 and avail special gift voucher for yourself.

Specimen of Hologram on front Cover title:

Moreover, there is a SPECIAL DISCOUNT COUPON for you with EVERY HOLOGRAM.

How to avail this SPECIAL DISCOUNT:

Step 1: Scratch the hologram

Step 2: Under the scratch area, your "coupon code" is available

Step 3: Logon to www.khannabooks.com

Step 4: Use your "coupon code" in the shopping cart and get your copy at a special discount

Step 5: Enjoy your reading!

Thanks,
Khanna Publishing House

ISBN: 978-93-86173-35-5

Utilisation of Electrical Power and Electric Traction
by **G.C. Garg**

First Edition: 2019
Reprint: 2026

To view complete list of books, please scan the QR Code:

Published by:
Khanna Book Publishing Co. (P) Ltd.
CIN: U22110DL1998PTC095547

Visit us at: www.khannabooks.com
Write us at: contact@khannabooks.com

Printed in Bharat.

In Loving Memory of
My Dear Daughter
BABITA

PREFACE

The Author feels great pleasure in presenting the book **'Utilisation of Electrical Power and Electric Traction'** which is prepared on the basis of engineering scientific study covering the syllabi of all the universities running electrical engineering courses, Engineering colleges and Technical Education Boards of all the states in India. This book is also meets the requirement of Foreign Universities where this subject is taught.

In this book total there are 10 chapters i.e. (i) Industrial utilization, (ii) Control of Industrial Motors, (iii) Electric Heating, (iv) Illumination, (v) Electrolytic Process, (vi) Electric Welding, (vii) Refrigeration & Air Conditioning, (viii) Electric Traction, (ix) Economic Aspects, (x) Objective Types of Questions and Answers.

All the above chapters have been arranged in such a manner that at the end of each chapter sufficient number of short questions with answers, objective type Questions such as Fill in the blanks, True/False Questions and matching type questions have been given. Number of solved examples in each chapter has been given to make the subject more illustrative and clear.

Another added feature of this book is that at the end of all the chapters, large number of classified objective type of Questions of various states of India with their answers have been given to make the book more useful from the examination point of view.

The author sincerly thanks all the Scientists, Authors & Engineers whose Works have been the source of inspiration to the in preparing this Edition of Text Book.

The Author wishes to thank Sh. Mukesh Seth Managing Director, Khanna Book Publishing Co. (P) Ltd. for bringing out this book in such a short duration of time maintaining Excellent quality with dedication.

Any constructive suggestions for the improvement of this book will be greatfully acknowledged.

Lastly but not the least I wish to express my gratitude to my wife Kanta Rani for her patience and encouragement during the preparation of this book.

—G.C. GARG

CONTENTS

International System (SI) OF UNITS

The System International d'Unites originally proposed by the Comite International des Poides et Mesures (CIPM) and subsequently recommended by the General Conference on Weights and Measures by the International Standards Organisation (ISO) and by the International Electrotechnical Commission (IEC) are known commonly as the SI units (not S.I. unit) It is very essential that all concerned adhere to the style recommended for SI units. Students of engineering should, taken care in using and writing the SI units correctly. Some such units are given below.

1. This system employs the following basic units :

	Units	*Symbol*
Length	metre	m
Mass	kilogramme	kg
Time	second	s
Thermodynamic temperature	kelvin	K
Electric current	ampere	A
Luminous intensity	candela	cd
Supplementary Units are :		
Plane angle	radian	rad
Solid angle	steradian	sr

2. Derived units are in general power products or quotients of basic units with factors 1 (coherent units). Some of them have standardized names and often also standardized symbols.

Examples are the following :

	Units	*Symbol*
Velocity	metre per second	m/s
Area	square metre	m^2
Volume	cubic metre	m^3
Density	kilogramme per cubic metre	kg/m^3
Force	newton	$N = kgm/s^2$
Pressure	newton per square metre	N/m^2
		$J = Nm = kgm^2/s^3$
Power	watt	W = J/s
Voltage	volt	V
Resistance	onm	$\Omega = V/A$
Frequency	hertz	$Hz = s^{-1}$

3. Non-coherent units are connected to basic or derived units by a *constant factor* preferably a power 10^3. In using these powers, some standardized prefixes are the following :

Power	*Prefix*	*Symbol*
10^{12}	tera	T
10^{9}	giga	G
10^{6}	mega	M
10^{3}	kilo	k
10^{-2}	centi	c
10^{-3}	milli	m
10^{-6}	micro	μ
10^{-9}	nano	n
10^{-12}	pico	p

The use of double prefixes should be avoided. Example : not kMW, but GW

4. Requirements of common usage permit in special cases the use of non-coherent units derived by other factors, such as

Name	*Symbol*	*Equal to*
Hour	h	3600 s
Day	24 h	
Bar	b	10^5 N/m²

5. There are many other units and names in legal or common use in various countries such as inch, pound, gallon kilogramme force (= kilopond), atmosphere (of varying definitions), hectopieze (= bar), Pascal (Pa = N/m² = 1000 N), thermie (th = 10^6 cal), kilocalorie (kcal = 4.1868 KJ), degree Fahrenheit etc. If a person desires to use such units each quantity should be given first in SI units and followed in brackets by the unit of common usage as in the above example.

6. Correct way of writing full names of units in lower cases : kelvin, newton, joules etc.

1 Industrial Utilisation

Introduction

The development of modern industry has revolutionised the conventional driving systems available earlier, namely chain drive and rope drive. A modern production unit for its efficient working may be considered to consist of three different important parts: Engine or motor (Prime-mover), Transmission system and the machine that conducts the assigned work. The first two components, the prime-mover with its control system and the transmission system which may comprise a set of shafts, pulleys, belts and gears etc., is to impart motion and to operate the given working machine. This combination is called the *electric drive*, if the prime-mover used is an electric motor.

1.1 General

For obtaining electric drives, both A.C. and D.C. motors are used. However, A.C. system is preferred. The utilization of electric energy is always advantageous as it is cheaper, it can be easily transmitted at comparatively low line losses, it is easy to maintain the voltage at consumer premises within the prescribed limits and it is possible to increase or decrease the voltage without appreciable loss of power. Inspite of advantages of A.C. system sometimes it becomes essential to use D.C. energy as industrial drive. The following are the applications of D.C. :

(*i*) D.C. motors are used in industry where high accuracy of speed control is required.

(*ii*) For traction purpose, as in such application a very high starting torque is required. The high starting torque can be obtained from D.C. series motor at low operating cost.

(*iii*) Battery charging and electrolysis.

(*iv*) The speed of A.C. motors is almost constant, whereas it can be varied easily in case of D.C. motors. Thus for variable speed applications such as lift and Ward-Leonard system etc., the D.C. motors are preferred. Under electric drive, it will be studied the characteristics of a few common types of motors with a view to be able to select a motor to suit a particular drive. Before this, however, the names of various types of motors are listed in Article 1.11.

1.1.A. Transmission drives used for motors.

Transmission drive means to transmitt the mechanical power developed by the motor to the driven machines. The different methods of transmission drive in general use are:

1. **Direct Coupling** : In this type of drive, driving member is connected directly to the driven member. In this system, metal flanges are provided on both the members on the shaft ends and are held together by means of bolts. In case of flexible coupling, the belt ends are covered by means of leather bushes or rubber which gives a cushioning effect to the drive so that vibrations and sudden shocks etc. are not transmitted to the motor.

2. **Gear drive** : Gear drive is used for low speed operations. In this type of drive, the coupling between motor and drive is through a gear box. In this normally the motors are manufactured with gearbox incorporated in the motor frame itself. Such type of method is used in D.C. series motors and other big size motors.

3. **Vertical drive** : In this type of drive, the shaft of the motor is kept vertical and load is arranged over it as in flour mills.

4. **Belt drive** : This method of drive is very simple and less expensive. In the transmission of power where a speed change is desired this type of drive is employed. In this the leather belt gives the most economical form of drive. The maximum power of about 200 to 300 H.P. can be transmitted by means of belts.

5. **Rope drive** : In this type, drive is done by means of ropes which are running over the pulleys, having a number of grooves, is used for power ranges beyond the limit of V-belt drive. The advantages of this type of drive is:

(*i*) ability of taking sudden loads and
(*ii*) negligible slip

Rope drive is capable of transmitting more power than the belt drive. This type of drive is smooth and noiseless.

6. **Chain drive** : Although this drive is more costlier than belt and rope drive, but is more efficient and no slip chains are capable of transmitting large power and can he used for more greater speed ratios.

1.2 Electric Drive

An electric drive is defined as a form of machine equipment designed to convert electric energy into mechanical energy and provide electrical control of these processes.

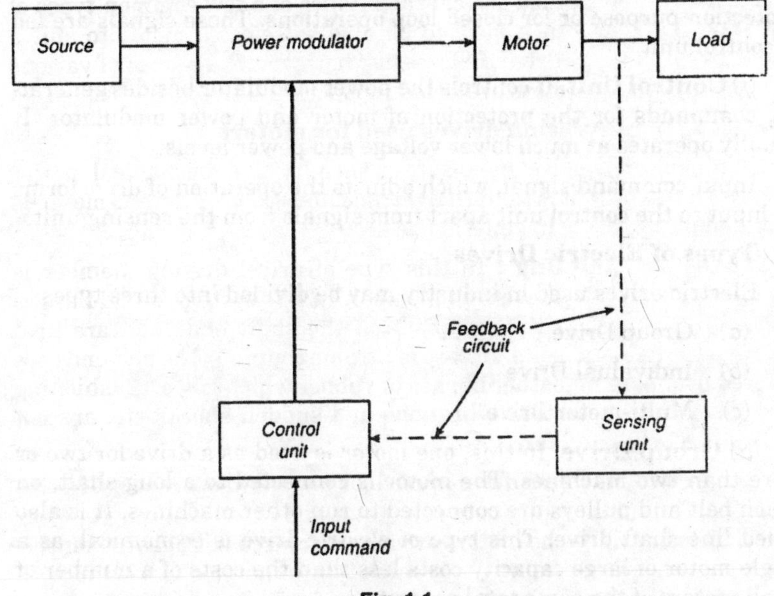

Fig. 1.1

Block diagram of an electric drive as given in fig 1.1 :

(*a*) **Source.** It is either type of electrical power d.c. or a.c. supply.

(*b*) **Power Modulator.** It performs the following functions :

(*i*) It converts electrical energy received from the source in the form suitable to the motor *i.e.* if the source is d.c. and an induction motor is to be employed then it converts d.c. into a variable frequency a.c.

(*ii*) During transient operations, such as starting, braking and speed reversal it restricts source and motor currents within permissible limits.

(*iii*) It selects the mode of operation of motor *i.e.* motoring or braking.

(*c*) **Load.** is usually a machinery designed to perform a given operation *i.e.* fans, machine tools, domestic appliances, trains, pumps etc.

(*d*) **Motors.** Motors commonly used in electrical drives are D : C motors–shunt, series, compound and permanent magnet. Induction motors–squirrel cage and wound rotor ; synchronous motors– wound field and permanent magnet ; Brushless dc motors ; stepper motors ; and switched reluctance motors.

(*e*) **Sensing unit.** It is employed for sensing the drive parameters, such as speed, motor current etc. which may be required either for protection purpose or for closed loop operations. These signals are fed to control unit.

(*f*) **Control Unit.** It controls the power modulator besides generating commands for the protection of motor and power modulator. It usually operates at much lower voltage and power levels.

Input command signal, which adjusts the operation of drive forms an input to the control unit apart from signals from the sensing unit.

Types of Electric Drives

Electric drives used in industry may be divided into three types:

(*a*) Group Drive

(*b*) Individual Drive

(*c*) Multi-motor Drive.

(*a*) **Group Drive.** In this, one motor is used as a drive for two or more than two machines. The motor is connected to a long shaft, on which belt and pulleys are connected to run other machines. It is also called line shaft drive. This type of electric drive is economical, as a single motor of large capacity costs less than the costs of a number of small motors of the same total capacity.

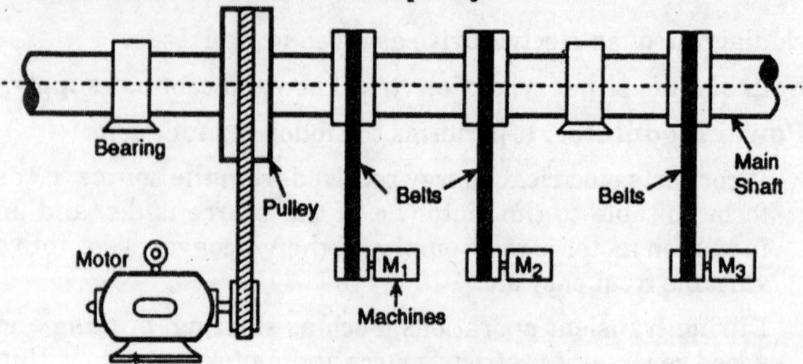

Fig. 1.1 (*b*) Group Drive.

The use of this kind of drive is restricted due to the below mentioned reasons :

(*i*) In case of fault in the motor, all the connected machines to this motor will cease to operate.

(*ii*) If at certain instance all the machines are not in operation, then the motor will be working at low capacity.

(*iii*) It is not possible to install a new machine at a far away distance.

(*iv*) Speed control of different machines using belts and pulley is cumbersome.

(*b*) **Individual Drive.** In this type of electric drive a single electric motor is used to drive one individual machine. Though it costs more than Group drive but each operator has complete control on his machine, which enables him to either increase the speed of motor or to stop it while not in operation. Machines can be located at convenient places. If there is a fault in one motor, this will not effect the production of the industry appreciably.

(*c*) **Multi-motor Drive.** It consists of several individual drives, each performs different functions and are a part of big, complicated machinery. This type of drive have their application in paper-making machines, cable manufacturing units, metal-cutting machines, rolling mills and similar other types of machinery.

Now-a-days electric drive is being extensively used in the industry because of its certain inherent advantages over the other types of drives Some of the advantages are listed below.

1.3 Advantages of Electric Drive

The electric drive is being employed in all industries due to its following advantages over the mechanical drive :

1. Cost is too low as compared to other systems of drive.
2. The system is more simple and clean.
3. The control is very easy and smooth.
4. Flexible in layout.
5. Facility for remote control.
6. Transmission of power from one place to other can be done with the help of cables instead of long shafts etc.
7. Its maintenance cost is quite low.
8. It can be started at any time without delay in time.

1.4 Disadvantages of Electric Drive

No doubt electric drive is more advantageous but it has several disadvantages also such as :

1. Electric drive system is tied only up to the electrified area.
2. The condition arising under the short circuits, leakage from conductors and break down of over head conductor may lead to fatal accidents.
3. Failure in supply for a few minutes may paralyse the whole system.

1.5 Group *vs* Individual Drive

During selecting a drive, first of all it should be determined whether the drive should be individual or group drive. The advantages of individual drive are :

(*i*) Disconnection of Individual drive or its failure will not affect the smooth working of different machines.

(*ii*) For transmission of drive there is no necessity of a shaft.

(*iii*) The cost of several machines is more than the cost of a single motor but the failure of single motor will bring the whole plant to a stand still.

The following factors can be considered for the broad classification :

1. Type of electric supply : A.C. or D.C.
2. Nature of drive: individual or group drive.
3. Nature of load : wheather the load demands less or high starting torque.
4. Electrical characteristics of the motor :

   (*i*) Starting characteristics (*ii*) Running characteristics

   (*iii*) speed control (*iv*) Braking.
5. Rating and size of motors ; whether the motor is for intermittent, continuous or variable load cycle. Whether the overload capacity or pull out torque meet the requirements.
6. Mechanical characteristics :

   (*i*) Nature of bearings (*ii*) Type of enclosure

   (*iii*) Transmission of drive and (*iv*) Noise.
7. Economic Consideration

   (*i*) Running expenditure (*ii*) Capital expenditure.

There are generally large number of machines to be driven in any industry. In individual drive each machine and tool is driven by a separate motor. Although in this type initial cost is increased. In group drive a single motor of large capacity is to drive number of machines and tools in industry.

1.6 Advantages of Group Drive

Advantages of group drive will be the disadvantages of individual drive.

(*i*) *Initial Cost.* Initial cost of Group drive is less as compared to that of the individual drive.

(*ii*) *Sequence of Operation.* Group drive system is useful because all the operations are stopped simultaneously.

(*iii*) *Space Requirement.* Less space is required in Group drive as compared to individual drive.

(*iv*) *Low Maintenance Cost.* It requires little maintenance as compared to individual drive.

1.7 Disadvantages of Group Drive

Although the initial cost of group drive is less but yet this system is not used extensively because of following disadvantages :

(*i*) *Power factor.* Group drive has low power factor.

(*ii*) *Efficiency.* Group drive system when used, if all the machines are not working together the main motor shall work at very much reduced load.

(*iii*) *Reliability.* In group drive, if the main motor fails then whole industry will come to stand-still.

(*iv*) *Flexibility.* Such arrangement is not possible in group drive *i.e.* this arrangement is not suitable for the place where flexibility is the prime factor.

(*v*) *Speed.* Group drive does not provide constant speed.

(*vi*) *Types of machines.* Group drive is not suitable for driving heavy machines such as cranes, lifts and hoists etc.

1.8 Advantages of Individual Drive

1. Individual drive give desired operation as each machine is driven by its own individual motor.
2. Individual motor works at good power factor.
3. Efficiency of the system is high.
4. Individual drive is more reliable.
5. Machine may be fitted wherever convenient.
6. More useful where constant speed is required.
7. Individual drive is most suitable for driving heavy machines such as lifts, hoists and cranes etc.

1.9 Comparison Between two drives.

Points	*Individual Drive*	*Group Drive*
1. Initial cost	More	Less
2. Power factor	Works at good p.f.	Low p.f.
3. Efficiency	High.	Low, when motor works at light load.
4. Reliability	More reliable.	If main motor fail whole industry will come to stand still.
5. Flexibility	Machines may be fitted wherever convenient.	Such arrangement is not possible in group drive.
6. Speed	More useful where constant speed is required.	Does not give constant speed.
7. Sequence of operation	Useless, where sequence of operation is required.	Useful, because all the operations are stopped simultaneously

8. Types of machines	This is must for driving heavy machines *e.g.* Cranes, lifts etc.	This system not employed in such cases.
9. Space utility	Space can be fully & conveniently utilized	More space is required in group drive.

1.9 (*a*) Choice of Electrical Drives

Some of the important factors to choose an electrical drive are :

(*i*) **Requirements related to the source :** Type of source, and its capacity, magnitude of voltage, voltage fluctuations, power factor, harmonics and their effect on other loads and ability to accept regenerated power.

(*ii*) **Steady state operation requirements :** Nature of speed torque characteristics, speed regulation, speed range, efficiency, duty cycle, quadrants of operation, speed fluctuations and ratings.

(*iii*) **Transient requirements :** Starting, breaking, values of acceleration and de-aceleration, reversing performance.

(*iv*) Capital and running cost, maintenance.

(*v*) Environment and location.

(*vi*) Reliability.

(*vii*) Space and weight restrictions.

1.9. (b). Specifications of Commonly used motors

For a particular application for the selection of motors, the following features have to be considered:

(*i*) Horse power rating

(*ii*) Whether the rating is continuous or intermittant or short time.

(*iii*) Type of Supply : D.C. or A.C.

(*iv*) In case of A.C. whether single-phase or three phase.

(*v*) Frequency (*vi*) Speed of motor

(*vii*) Enclosure

(*viii*) Whether the drive is individual or group

(*ix*) Ambient temperature (*x*) starting method

(*xi*) Braking method to be used.

1.9 (c) Choice of an electric motor

For a particular drive the selection of an electric motor depends upon the following factors :

1. Rating of the motor

2. Type of motor : whether it is A.c. or D.c. motor with necessary starting and operating characteristics.
3. Type of bearings and enclosure to suit working conditions. Along with the nature of the mechanical load, starting conditions of load and also matching of the speed-torque characteristics of motor with load are to be examined.
4. The drive is for one or more than one machines.
5. The type of drive or transmission *i.e.* by direct or through gears, a belt or chain.

1.10. Steady State Stability of an Electric Drive

It is quite important to investigate the conditions of stable operation of an electric drive, when its equilibrium state is disturbed. In practice, it is necessary to consider two types of disturbance :

1. Changes from the state of equilibrium take place slowly and the effect of either the inertia of the rotating masses or that of inductance is in-significant.
2. Sudden and fast changes from the equilibrium state, as a result of which the effect of neither the inertia nor the inductance can be ignored. Study of stability under conditions enunciated for first type of disturbance relates to the field of steady state stability, while for the second type of disturbance pertain to the field of dynamic or transient stability.

The drive is said to be in equilibrium, if at a given speed the torque developed by the motor is exactly equal to the load torque. However, this state of equilibrium may be stable, unstable or neutral, as observed in other physical processes. If the drive comes out of the state of equilibrium under the influence of any unexpected external or internal factors, in the case of stable equilibrium, the drive again returns to the steady state but in case of unstable equilibrium the speed of drive either increases tremendously or decreases and come to rest. When the drive, after coming out the state of equilibrium due to disturbances, preserves its steady state at different speeds it is said to be in natural state.

The stability of the motor-load combination is defined as the capacity of the system which enables it to develop forces of such a nature as to restore equilibrium after any small departure therefrom.

Not all drives possess steady state stability. Let us understand the conditions, which ensure steady state stability and thereby determine the criteria for steady state stability of an electric drive.

1.10.1. Criteria for Steady State Stability. An accurate and complete study of the problem of stability of drives requires the solution of the differential equations of the drive system and interpretation of the results obtained. In general, it may be said that if the transient portion of the solution to these equations approaches zero with increasing time, the effects of any disturbance die away, and the system is stable.

However in the absence of any such complete analysis, considerable insight to the problem may be gained by a study of steady state speed torque curves, of both motor and load, and assuming that all departures from equilibrium will be along these curves.

Let the equilibrium values of the torque and speed be denoted by T_m, T_L and ω and let small deviation be denoted by ΔT_m, ΔT_L and $\Delta\omega$. After a small displacement from the equilibrium, the torque equation becomes,

$$J\frac{d\omega}{dt} + J \cdot \frac{d(\Delta\omega)}{dt} + T_L + \Delta T_L - T_m - \Delta T_m = 0 \quad \text{...(1)}$$

But $$J \cdot \frac{d\omega}{dt} + T_L - T_m = 0 \quad \text{...(2)}$$

Hence, $$J \cdot \frac{d(\Delta\omega)}{dt} + T_L - \Delta T_m = 0 \quad \text{...(3)}$$

The above equation is to be interpreted as the torque equation in which all quantities are expressed in terms of their deviations from equilibrium values. If we assume that these increments are so small that they may be expressed as linear function of the change in speed then,

$$\Delta T_m = \frac{dT_m}{d\omega} \cdot \Delta\omega \quad \text{...(4)}$$

$$\Delta T_L = \frac{dT_L}{d\omega} \cdot \Delta\omega \quad \text{...(5)}$$

where $dT_m/d\omega$ and $dT_L/d\omega$ indicate derivatives at the point of equilibrium. Substituting these relations in equation (3) and re-arranging, we have,

$$J \cdot \frac{d(\Delta\omega)}{dt} + \left[\frac{dT_L}{d\omega} - \frac{dT_M}{d\omega}\right]\Delta\omega = 0 \quad \text{...(6)}$$

whose solution is,

$$\Delta\omega = (\Delta\omega)_0\, e^{-1/J\left[\frac{dT_L}{d\omega} - \frac{dT_m}{d\omega}\right]t} \quad \text{...(7)}$$

The quantity $(\Delta\omega)_0$ is initial values of the deviation in speed. In order that the system be stable, it is necessary that the exponent be negative, so that the speed increment will disappear with time. Under such conditions, the system will return to its equilibrium speed. Whereas if the exponent is greater than zero the speed deviation will increase with time and the system will move away from equilibrium. If the exponent is exactly equal to zero, the equation is insufficient to discuss about stability. The exponent will always be negative if,

$$\frac{dT_L}{d\omega} - \frac{dT_m}{d\omega} > 0 \quad \text{...(8)}$$

This is equivalent to saying that for a decrease in speed the motor torque must exceed the load torque and for increase in speed the motor torque must be less than the load torque.

This relationship is illustrated in Fig. 1.2 in which the load torque T_L' results in a stable operating point and the load torque curve T_L' results in an unstable situation.

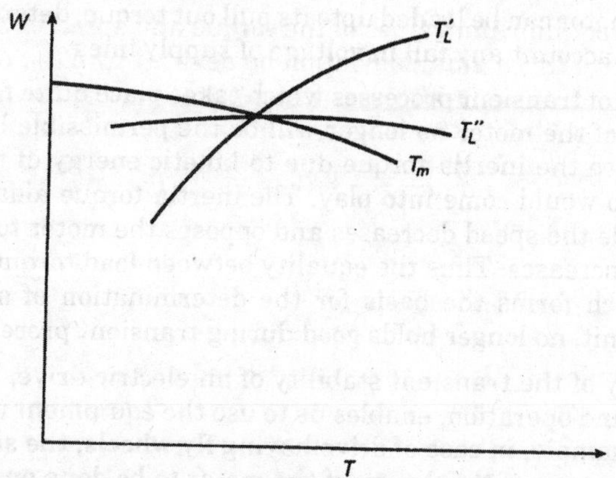

Fig. 1.2 Speed-Torque Curves for motor-load combination.

The stability of the operating point is most easily evaluated by assuming a small change in speed away from equilibrium. With this speed increment assumed, the relative values of motor and load torque will determine whether the speed will return to its previous value, that is, if an increase in speed bring about a greater increase in load torque than motor torque the speed will tend to decrease and return to its original value, which is then a stable point. For the converse case, the speed continue to increase, and the system is obviously unstable.

It should be emphasized that this criterion for stability is an approximation in the sense that it assumes that the steady state speed torque curves are transversed when small disturbance cause a departure in speed from some operating point. This assumption may or may not be valid, as there are many instances in which the transient speed-torque relations are quite different from the steady state ones. Despite this fact, the above criterion is quite useful.

1.10.2. Transient Stability of an electric drive

Concept of Transient stability. While analysing the steady state stability of an electric drive, only its initial and final conditions are examined on the basis of the speed-torque characteristics of the motor and load, without paying any attention to the inertia torques and to the time taken to change over from the initial condition to the final one. The nature of the motion of the drive during this period is also not considered. The essence of the study of transient stability of an electric drive or, in general, any moving system is to take into account the influence of the above mentioned factors, such a study enables us to estimate more accurately the performance of the drive with respect to its stability of motion.

With slow and steady changes in load, the driving motor may be loaded upto its maximum capacity of torque or power. For example an induction motor can be loaded upto its pull out torque, determined after taking into account any fall in voltage of supply lines.

In case of transient processes which takes place quite fast, the pull out torque of the motor no longer will be the permissible limit of load torque. Since the inertia torque due to kinetic energy of the rotating masses also would come into play. The inertia torque aids the motor torque while the speed decreases and opposes the motor torque, while the speed increases. Thus the equality between load torque and motor torque which forms the basis for the determination of steady state stability limit, no longer holds good during transient processes.

A study of the transient stability of an electric drive, both during its design and operation, enables us to use the equipment more rationally. For example, in case of drive having fly wheels, the size of lattice can be decreased, if the design of the motor to be done on the basis of transient stability limit. However, in practice, this is done only partially, allowing the transient stability limit to be used as a reserve capacity, when sudden changes in load occur.

1.10.3. Transient Stability of a Synchronous Motor.

Let us consider a drive system consisting of a synchronous motor driving a constant torque load. The speed torque characteristics of the motor and load, it is obvious that $\frac{dT_L}{d\omega} = 0$ and $\frac{dT_m}{d\omega} = +\alpha$. Hence by applying the criterion of steady state stability, it can be seen that the operation of the drive system under consideration will be stable if $\left(\frac{dT_L}{d\omega} - \frac{dT_M}{d\omega}\right)$ is equal to $+\alpha$ and unstable if it is equal to $-\alpha$. That is for the given drive system, the criterion of steady state stability does not give any useful information. Under such circumstances, to investigate stability using only the steady state characteristics of motor is quite insufficient. It become necessary to study the essence of the processes involved during transition of synchronous motor from one operating condition to the other. In other words it is imperative to investigate the transient stability of such a drive system.

If a large load is suddenly applied to the shaft of a synchronous motor, the motor must slow down momentarily at least to have a larger value of the torque angle required to supply the added load. In fact until the new angle is attained, a considerable portion of the energy supplied

to load comes from the stored energy in the rotating mass as it reduces its speed. As the required value of torque angle is reached, the rotor does not attain its equilibrium since the rotor speed is less than the synchronous speed. The torque angle, hence, must increase further in order to allow replenishment of the deficit of stored energy in the rotating mass. This process involves a series of oscillations of the rotor about its final position even when equilibrium is finally restored.

Similar oscillations accompanied with torque and current pulsations occur in synchronous motors driving loads whose torque requirements very cyclically at a fairly rapid frequency, like in motors driving reciprocating air or ammonia compressors. If the natural frequency of mechanical oscillations of the rotor of synchronous motor becomes equal or close to the frequency of a significant harmonic of the load. Cycle variations oscillations of very high magnitude are produced. Exact description of such process can only be given in terms of the related electromechanical differential equation and information about the restorating of equilibrium about the rotor can be observed only by solving the equation.

The equation of motion in terms of power can be written as,

$$P_M = P_{dyn} + P_L \qquad \text{...(1)}$$

where, P_m, P_{dyn} and P_L denote the electromagnetic power developed by the motor, dynamic power and load power at the shaft respectively.

The dynamic power is determined from the angular acceleration. The angular position of the shaft at any instant is taken as the electrical angle 'δ' between a point on it and a reference which is rotating at synchronous speed. Often the angle δ is assumed to be the same as the torque or power angle. With a sudden application of load, since the rotor slows down, the angular acceleration will be negative and hence the dynamic power will be given by :

$$P_{dyn} = -P_j \frac{d^2 \delta}{dt^2} \qquad \text{...(2)}$$

where

$$P_j = J \cdot w \cdot \frac{2}{\text{Poles}} \qquad \text{...(3)}$$

The electromagnetic power P_m usually has two components

(*i*) damping power which is assumed, to vary linearly with the departure '$d\delta/dt$' from synchronous speed

(*ii*) Synchronous power produced by synchronous motor action, which is a function of load angle, δ.

Thus the electromechanical equation becomes,

$$P_j \frac{d\delta^1}{dt^2} + P\,d\,\frac{d\delta}{dt} + P\,(\delta) = P_L \qquad \text{...(4)}$$

where P_d is the damping power/unit departure in speed.

Neglecting damping and assuming a cylindrical rotor synchronous machine. Eq. (4) becomes

$$P_j \frac{d^2\delta}{dt^2} + P_m \sin\delta = P_L \qquad ...(5)$$

where $P_m = \frac{VE}{X_S}$ in which V, E and X_S denotes the applied voltage and e.m.f. due to excitation and synchronous reactance of the motor. From Eq. (5) we get,

$$\frac{d^2\delta}{dt^2} = \frac{P_L - P_m \sin\delta}{P_j}$$

Multiplying both sides by $\frac{d\delta}{dt}$ we have,

$$\frac{d^2\delta}{dt^2}\left(\frac{d\delta}{dt}\right) = \left(\frac{P_L - P_M \sin\delta}{P_j}\right)\frac{d\delta}{dt}$$

so, $$\frac{1}{2}\cdot\frac{d}{dt}\left(\frac{d\delta}{dt}\right)^2 = \left(\frac{P_L - P_M \sin\delta}{P_j}\right)\frac{d\delta}{dt}$$

so, $$\frac{d\delta}{dt} = \sqrt{\int_{\delta_0}^{\delta} \frac{2\,(P_L - P_M \sin\delta)}{P_j}\, d\delta}$$

where δ_0 is the load angle before the disturbance *i.e.* at time, $t = 0$. Also since the motor was running at synchronous speed at time, $t = 0$, $\frac{d\delta}{dt} = 0$.

Ultimately for the machine to be stable, δ must stop changing as the synchronous speed is reached and at this final equilibrium point also $\frac{d\delta}{dt} = 0$.

Therefore the criterion for the stability is that $\frac{d\delta}{dt} = 0$. Hence,

$$\sqrt{\int_{\delta_0}^{\delta} \frac{2\,(P_L - P_M \sin\delta)}{P_j}\, d\delta} = 0.$$

i.e. $$\int_{\delta_0}^{\delta} (P_L - P_m \sin\delta)\, d\delta = 0 \qquad ...(6)$$

Let us consider a synchronous motor having the power-angle curve of Fig. 1.3 (*a*). With the motor initially loaded with a load of power P_{L_1} and the operating point is at A corresponding to a power angle δ_0. As the load on the shaft is suddenly increased to P_{L_2} the power angle swings to a value of δ_f at which instant the speed is again synchronous. From Eq. (6) it follows that this system will be stable if,

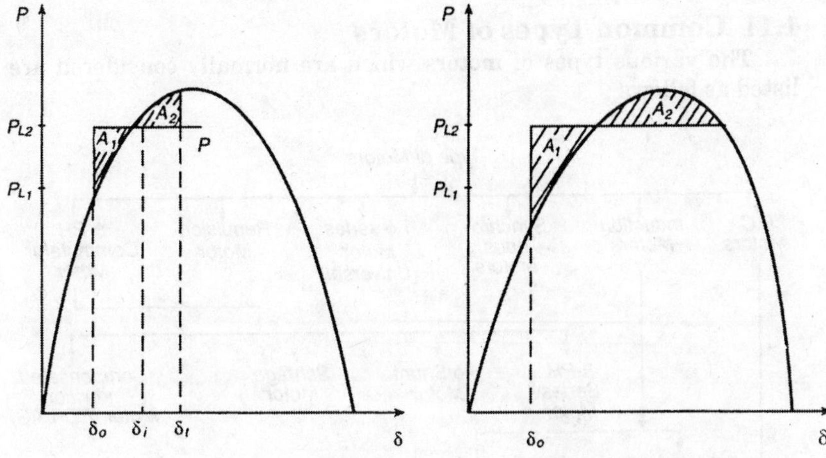

(a) Power-Angle curve of synchronous-motor and variation of load

(b) Equal area criterion for transient stability.

Fig. 1.3

$$\int_{\delta_o}^{\delta_i} (P_{L_2} - P_M \sin \delta)\, d\delta + \int_{\delta_i}^{\delta_f} (P_{L2} - P_M \sin \delta)\, d\delta = 0 \quad ...(7)$$

where δ_i is the power angle corresponding to the new load P_{L_2}.

It can be seen that the term $(P_{L_2} - P_M \sin \delta)$ is positive for value of power angle between δ_o and δ_i and negative for those between δ_i and δ_f. Therefore Eq. (7) can be rewritten as,

$$\int_{\delta_o}^{\delta_i} (P_{L_2} - P_M \sin \delta)\, d\delta = \int_{\delta_i}^{\delta_f} (P_M \sin \delta - P_{L_2})\, d\delta$$

or Area, A_1 = Area, A_2

This method of determining the transient stability of a drive system is called the equal area criterion of stability.

The equal area method gives thus a simple indication of whether synchronism is maintained or not. In Fig. 1.3 (b) it can be seen that,

(i) If Area, $A_2 > A_1$ motor remain in synchronism and stability is maintained

(ii) If area, A_2 = Area A_1, the point of operation is just stable and

(iii) If area, $A_2 < A_1$, the motor loses its synchronism.

It may be noted that while driving the criterion for transient stability of the synchronous motor drive, the damping power term has been neglected. Hence equal area criterion method gives slightly pessimistic results as regards Transient stability.

1.11 Common Types of Motors

The various types of motors which are normally considered are listed as follows :

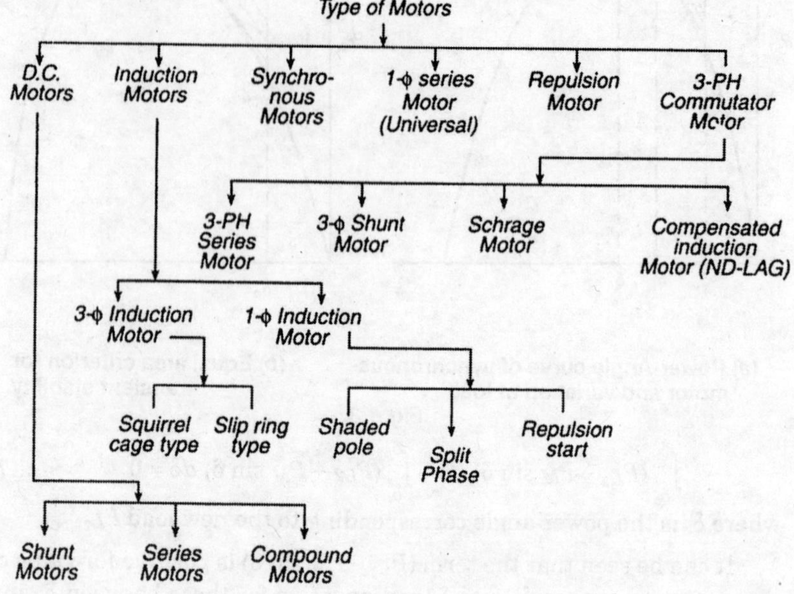

Fig. 1.4

1.12 Selection of Motors

The selection of the motor is based upon :

(*i*) Condition under which motor is supposed to work.

(*ii*) Type of load the motor is supposed to handle. Accordingly the factors which are considered for selection of motor for a particular service are as under :

1. **Electrical characteristics :**

   (*a*) Running characteristics

   (*b*) Starting characteristics

   (*c*) Speed control

   (*d*) Braking characteristics.

2. **Mechanical characteristics :**

   (*a*) Type of enclosure

   (*b*) Type of bearings

   (*c*) Transmission of drive

(*d*) Noise level.

3. **Size and rating of motor :**

(*a*) Rating of motor

(*b*) Whether the motor is to be used continuously, intermittently or for variable loads.

4. **Cost**

(*a*) Capital cost

(*b*) Running cost.

1.12 (*a*) Starting Conditions of the Load :

The size of the motor is not only determined by the full load torque requirement but by the accelerating and braking torque also. The acceleration of the moving masses, both the load and motor itself calls for motor torque in excess of load torque by an amount which depend upon the time available for acceleration of the masses involve. In case of induction motor which is the work horse of industry power transferred from the stator to rotor is through the rotating magnetic field. Mechanical power developed by the rotor on the other hand depends upon the actual speed of the rotor. The difference of two powers is therefore the rotor heating which is proportional to the slip of the rotor slip of Induction motor at starting is one, therefore all the rotor input is lost as rotor copper losses. These losses go on reducing progressively as the speed increases till full load speed is reached. Thus rotor heating primarily depends upon the starting time. In the case of aluminium. die. cast rotor, the weight of rotor being very small it has low moment of inertia and therefore require less time to attain full speed. Such motors besides free from common troubles of opening of rotor cage make saving on copper, so scarce in our country. Allowable temp. rise is also depends upon the type of its construction.

For instance integrally cast rotors are very rugged and withstand heavy temperatures whereas in rotors having rotor bars brazed to end rings, the safe temperature is limited specially due to radial expansion and contraction of the rings which flexes the end bar and cause their fatigue failure. A prolonged accelerating time causing starting losses to continue for too long is likely to damage the motor as a result of high temp. Thus starting duty after determines the rotor heating which in turn decides the size of motor.

Losses, power factor, maintenance and depreciation etc.

It can thus be seen that the choice of an electric motor for a given drive includes a large number of factors and although in any particular case some of these may be definitely fixed.

(*a*) ***Running characteristics.*** Operating characteristics of various types of loads are different. Some machines require constant speed irrespective of the torque characteristics and some require a variable speed control. Therefore, it is desired to know about the operating or running characteristics of various types of motors. The electrical behaviour of a motor when running under normal load conditions can be explained by considering following factors :

(*a*) Speed-torque characteristics

(*b*) Speed-current characteristics

(*c*) Torque-current characteristics

(*d*) Losses and efficiency

(*e*) Magnetising current and power factor at various loads in case of a.c. motors only.

1.13 Types of Loads Torque

Load can be classified under two categories :

(*a*) **Active or potential load :** Loads which are due to the forces exerted on account of gravity, compression or tension in a spring or in any elastic body are termed as Active loads. These loads are also termed as potential load as they are related with change in the potential energy of various elements of an electrical drive system. The active or potential load may be positive or negative depending upon whether they aid or oppose the motion of the drive. Active torque come into play for hoists, elevators, lifts and railway locomotives on gradients etc.

(*b*) **Passive load/Torque.** These are due to friction, cutting and deformation of inelastic body. These torques always oppose the motion of drive and change their sign when the direction of rotation of the drive is changed, for example frictional torque always acts in a direction opposite to that of driving torque.

1.13.1. Components of Load Torques

Load torque can be divided into following components :

(*a*) **Frictional torque (T_F).** Friction will be present at the motor shaft and also in various moving parts of the load. T_F is equivalent value of various friction torques referred to the motor shaft.

(*b*) **Windage torque (T_W).** When a motor runs, wind generates torque, opposing the motion. This opposing torque is termed as windage torque.

(*c*) **Torque required to do some useful mechanical work (T_L):** Nature of this torque depends upon particular application. It may be:

(*i*) Constant and independent of speed ;

(*ii*) Some function of speed ;

(*iii*) Depend on the position or path followed by load ;

(*iv*) Time variant or time-invariant ;

(*v*) Vary cyclically and its nature may also change with the load's mode of operation.

1.13.2. Nature and Classification of Load Torques

Broadly load torque can be classified into torque varying with speed and with time.

(A) **Variation with speed**

Load torque, depending upon the speed can be classified as :

(*i*) **Constant load torque.** The torque is independent of the speed and is constant. A low speed hoist is an example of load where torque is constant and independent of speed. At low speeds, windage torque is negligible, therefore net torque is mainly due to gravity which is constant and independent of speed *e.g.* paper mill drive.

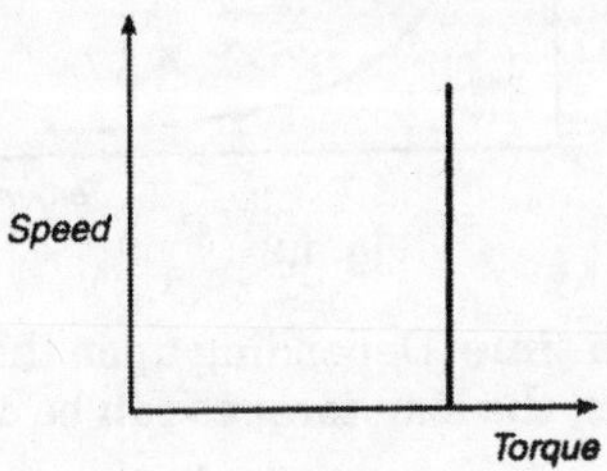

Fig. 1.5

(*ii*) **Load torque v/s speed.** The torque increases linearly with speed as shown below. This occurs in case of fluid friction where lubricant is used.

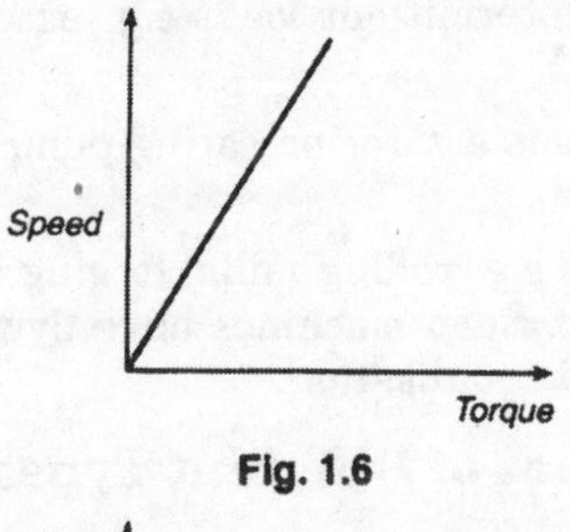

Fig. 1.6

Fig. 1.7

(*iii*) **Load torque v/s speed^2.** The torque increases as a square of speed as shown in fig 1.7. This occurs in case of air and fluid friction such as in fans, compressors, aeroplanes, centrifugal pumps, ship-propellers, coilers, high speed hoists and traction etc.

(*iv*) **Lead torque v/s $\frac{1}{\text{speed}}$** This occurs where deformation of material takes place for example in grinding and metal drawing etc.

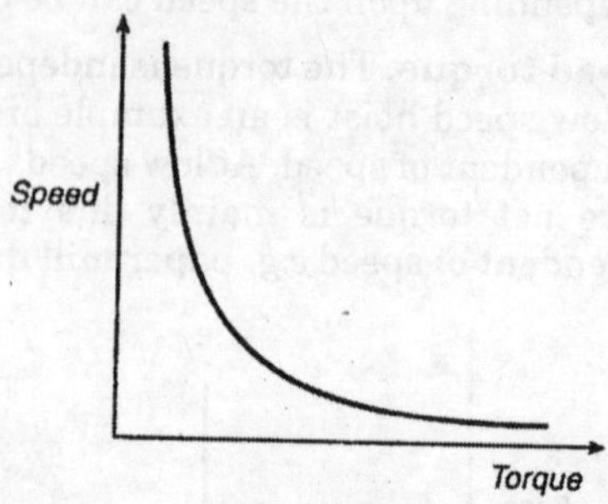

Fig. 1.8

(B) **Variation with time.** Depending upon the time for which load exists on a electric drive, the load torques can be classified as :

(*i*) Continuous and constant load, for example in centrifugal pumps operating under same condition for a long time.

(*ii*) Continuous but variable loads *e.g.* hoisting winches, conveyors etc.

(*iii*) Short time intermittent loads *e.g.* excavators, cranes and hoists etc.

(*iv*) Pulsating loads *e.g.* reciprocating pumps and textile looms etc.

(*v*) Impact loads *e.g.* rolling mills, forging hammers, shearing machines etc. Such machines have flywheel attached with them for load equalisation.

1.14. Characteristics of Different Types of Mechanical Loads.

1. **Lifting load.** This involves lifting of some weight by a crane or hoist etc. In lifting of loads the torque required independent of speed. Load torque remains constant as shown in Fig. 1.9.

2. **Air and fluid friction.** In these types of loads the torque changes with the square of speed such as in blowers, fans and water wheels etc. In this case Torque $\propto$ (speed)^2 as shown in Fig. 1.10.

3, **Load due to friction.** Whenever two surfaces move one over the other on a shaft, rotates in a bearing or bush, an opposite face always

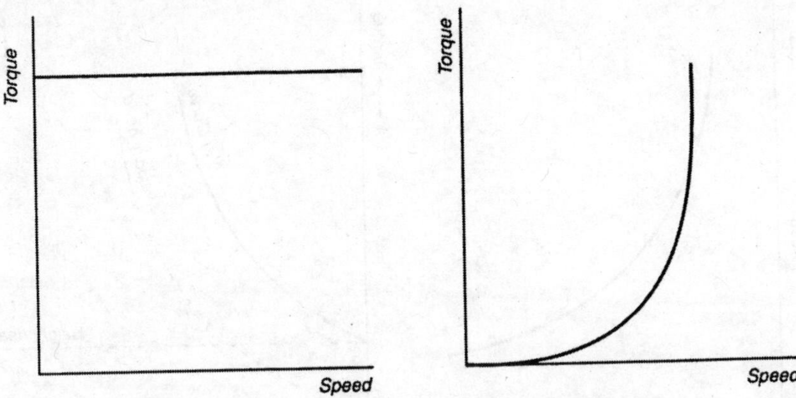

Fig. 1.9

Fig. 1.10

act at the surface. This force is called the frictional force. However this force can be decreases by the using a lubricant. When no lubricant is used friction is called dry friction, torque in that case is constant with speed. In fluid friction where a lubricant is used the torque is linearly proportional to speed as shown in Fig. 1.11.

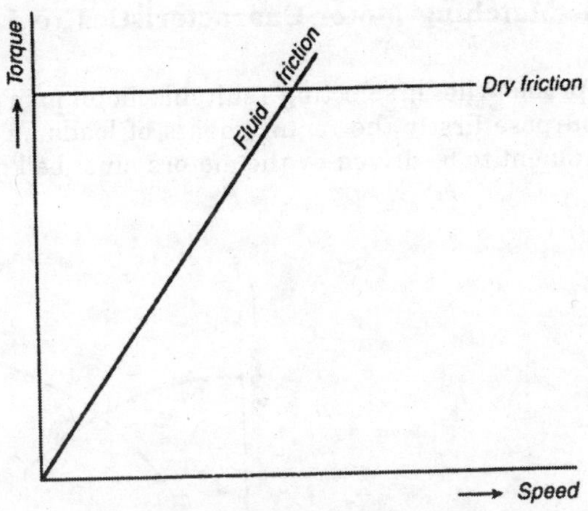

Fig. 1.11

4. **Deformation loads.** Such types of loads occur in crushing, grinding and metal drawing etc. In such cases torque is inversely proportional to the speed as shown in Fig. 1.12 (*a*).

In most cases we have a combination of two or more types of mechanical loads *e.g.* in a centrifugal pump the mechanical load consist mainly of a height head and velocity head. Height head means pump

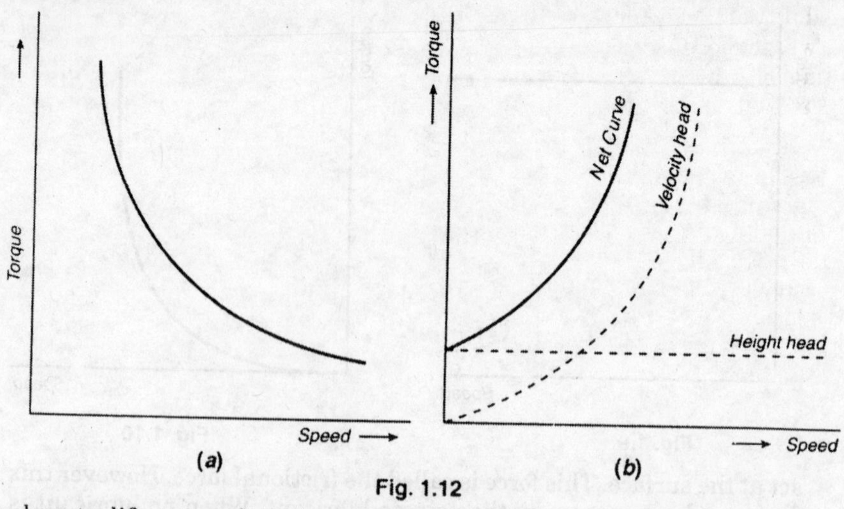

Fig. 1.12

has to lift water from lower level to higher level against gravity. Therefore the torque required for this load will be constant and independent at speed as shown in Fig. 1.12 (*b*). Velocity head means the load due to fluid friction and therefore the torque required to overcome this load will be proportional to square of speed.

1.14.1. Matching Motor-Characteristics to Load Requirement

Basic problem lies in selecting a suitable motor for a particular load. For this purpose firstly the requirements of loads *i.e.* machines and other equipment to be driven by the motors must be known. Then the

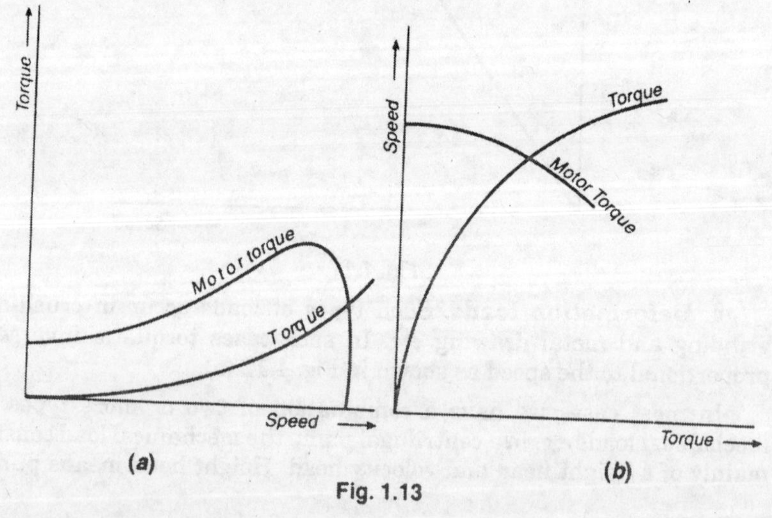

Fig. 1.13

ability of the motor to carry such load is seen. This ability depends upon two factors : thermal and mechanical. Thermal consideration means the capacity to withstand the temp. rise and mechanical is for ensuring that the peak load is safely carried by the motor.

Suppose it is desired to select a motor for driving a fan the load torque varies as the square of speed. Among A.C. motors, induction motor is most suitable. The stable point S is shown on the curve in the Fig. 1.13 (*a*). If the supply is D.C. then a shunt motor is suited to drive the fan. The point of stable operation is the point where the motor torque and fan torque curve intersect each other as shown in Fig. 1.13 (*b*).

On the opposite in a shearing and bunching machine, the load is fluctuating. In this case either a d.c. compound motor or an induction motor with a flywheel can be used.

1.15 D.C. Motors

In all d.c. motors, the torque is given by the equation,

$$T = \frac{\phi \times Z \times P \times I_a}{2\pi A}$$

or $\quad T \propto \phi I_a \because 2\pi, Z, P, A$ are constant

or $\quad T \propto (I_a)^2$ in case of series motor

$\because \phi \propto I_a$ in that case

Speed, $\quad N = \frac{E}{\phi Z} \times \frac{60A}{P}$ r.p.m. $\left[\because E = \frac{\phi Z N P}{60 A}\right]$

or $\quad N = \frac{V - I_a R_a}{\phi Z} \times \frac{60A}{P}$ [Also, $E = V - I_a R_a$]

where $\quad V$ = Supply Voltage

I_a = Armature Current

R_a = Armature resistance

or $\quad N \propto \frac{V - I_a R_a}{\phi}$

all other quantities being constant.

(*i*) **D.C. Shunt Motor.** As we know in shunt motors the field is connected in parallel with armature.

(*a*) ***Speed–current characteristics.*** From the expression for speed

i.e. $\quad N = \frac{V - I_a R_a}{\phi}$,

it is clear that the speed is directly proportional to the back e.m.f,. E_b and inversely proportional to flux, ϕ. But in this case if the applied voltage, V is kept constant, the flux ϕ will be also constant. Now when

load on the motor is increased the current in armature increases thus increasing the voltage drop in armature and hence decreasing the speed of motor. As the drop in armature at full load is very small as compared to applied voltage V, therefore, drop in speed from no load to full load is again very small and the motor can be used as compared to a constant speed motor as shown in Fig. 1.14.

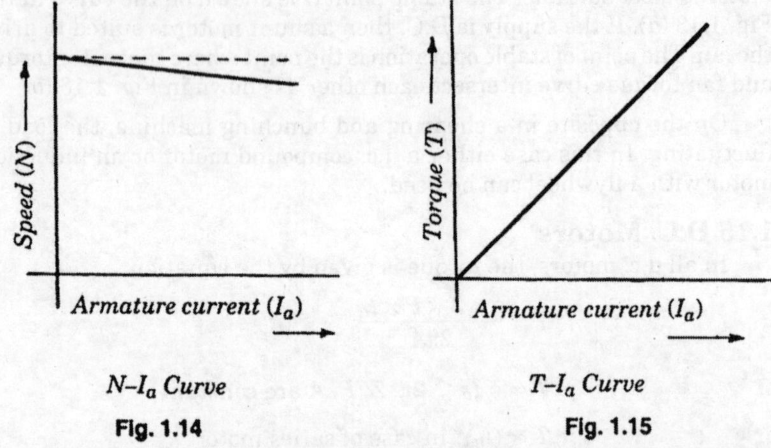

N–I_a Curve

Fig. 1.14

T–I_a Curve

Fig. 1.15

(*b*) Torque–Current Characteristics. **From the expression for torque $T \propto \phi\, I_a \propto I_a = KI_a$ in case of shunt motor, it is observed that the torque varies directly as the load current, hence the curve follows a linear law as shown in Fig. 1.15.**

(*c*) Speed Torque Characteristics. **In a d.c. shunt motor, when the supply voltage V is constant the field flux ϕ is also constant and hence speed of motor depends mainly on armature current. The speed decreases with the increase in armature current. But torque also increases with the increase in current. Hence the torque speed characteristics for a d.c. shunt motor will be a straight line with a slight drop as shown in Fig. 1.16.**

(*d*) Performance curves. **The curves for efficiency, speed torque and current plotted against B.H.P. of a motor are called the performance curves. The performance curves for a shunt motor are shown in Fig. 1.17 from the speed B.H.P. curve it is seen that the motor has definite speed at no load and hence it cannot run away when the load is suddenly thrown off. Also there is a slight fall in speed from no load to full load and hence the motor can be used when a substantially constant speed is required. It is noted from the B.H.P. efficiency curve that the efficiency of the shunt motor increases rapidly in the beginning and reaches its maximum value and then falls nearly at full load. The efficiency of a shunt motor is maximum when the variable losses are equal to constant losses. The curve for armature current, B.H.P. shows that even**

at zero output a certain value of current is required to meet the losses occurring in the motor.

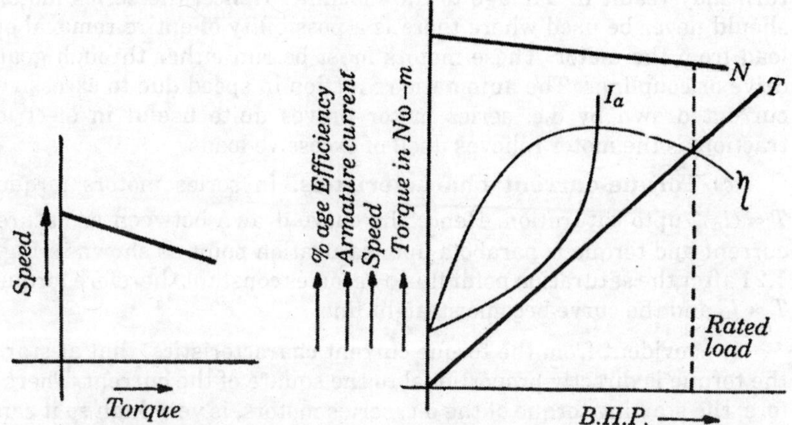

Fig. 1.16 *N-T* Curve. Fig. 1.17 Performance curves.

(*ii*) **D.C. Series Motors.**

In this motor, the field winding is connected in series with the armature winding. It means as the load on the motor increases the current through the series winding also increases or in other words the flux varies according to the load on the motor. The flux ϕ first increases with the increase in load current and then become maximum at saturation point and finally become constant as shown in Fig. 1.18.

(*a*) **Speed torque characteristics.** Speed of the motor decreases with the increases in torque as shown in Fig. 1.19. Hence series motors are best suited for services where the motor is directly coupled to the load such as in electric trains.

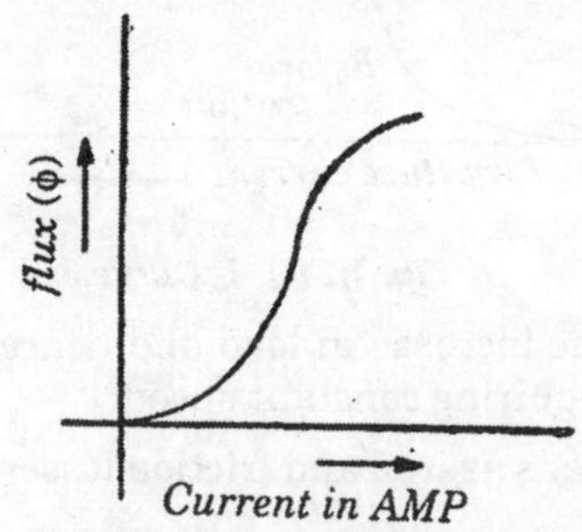

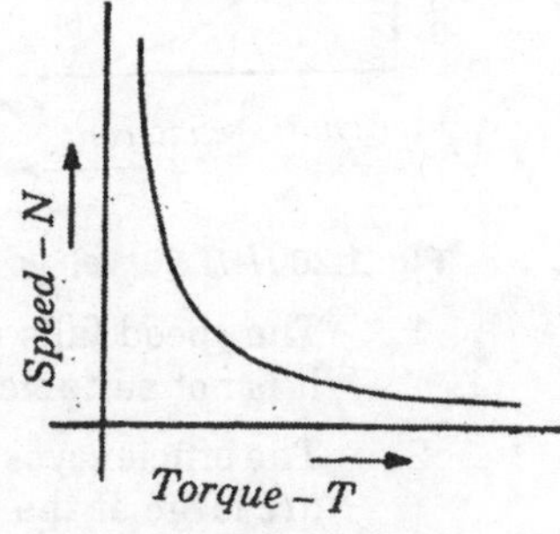

Fig. 1.18. Magnetising curve. Fig. 1.19. *N-T* curve.

(*b*) **Speed-current characteristics.** Speed in a series motor is inversely proportional to flux/pole and directly proportional to the applied voltage. If the applied voltage is kept constant then speed at no load $N \propto 1/\phi \propto 1/I_a$ because $\phi \propto I_a$. Hence at light loads the speed is dangerously high and comes down sharply with the increase in load as shown in Fig. 1.20. A series motor is a variable speed motor. Dangerously high speed at no load may produce heavy centrifugal forces which in turn may result in damage to the machine. Hence, the series motor should never be used where there is a possibility of entire removal of load from the motor. These motors must be run either through gear drive or couplings. The automatic reduction in speed due to excessive current drawn by d.c. series motor proves quite useful in electric traction as the motor relieves itself of excessive loads.

(*c*) **Torque-current characteristics.** In series motors torque $T \propto (I_a)^2$ upto saturation. Hence the curve drawn between armature current and torque is parabola upto saturation point as shown in Fig. 1.21 after the saturation point flux ϕ becomes constant therefore torque $T \propto I_a$ and the curve become straight line.

It is evident from the torque current characteristics, that at start the torque is directly proportional to the square of the current. Therefore, the starting torque of the d.c. series motors, is very high so it can be used where large starting torque is required such as in electric trains, cranes, trolleys and other electric vehicles.

(*d*) **Performance Curve.** The curve drawn between B.H.P. and torque, speed current, efficiency of a d.c. series motor are known as the performance curves. The following points should be noted from these curves :

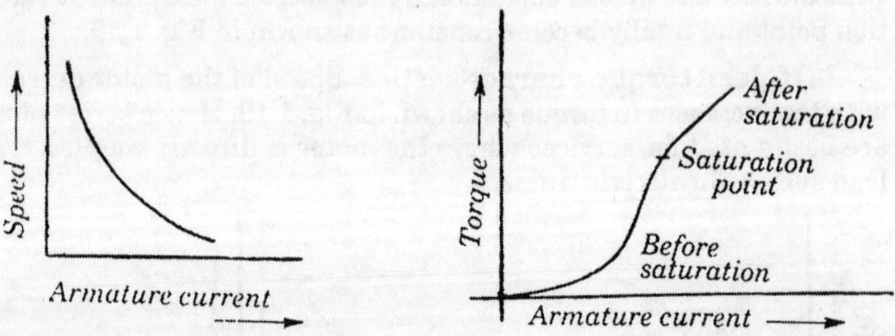

Fig. 1.20 *N–I*a *Curve.* *Fig. 1.21 T–I*a *Curve.*

1. The speed falls rapidly with the increase in load and hence it is not suitable for services requiring constant speed.
2. The efficiency is less at light loads as iron and friction losses are large at the starting.
3. It develops a greater starting torque as compared to shunt motor and hence suitable for services where large starting torque is required.

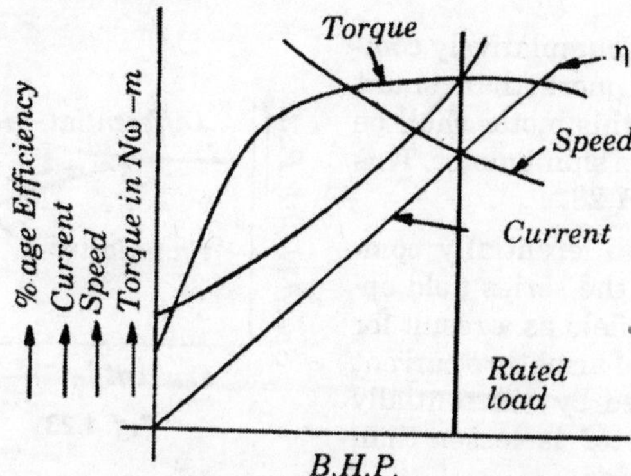

Fig. 1.22 Performance Characteristics.

(*iii*) **Compound Motor**

There are two types of compound motors namely :

(*a*) Cumulative compound motor.

(*b*) Differentially compound motor.

(*a*) **Cumulative Compound.** In these motors series field winding is connected to assist the shunt field. The characteristics of such motors are the combination of series and shunt motors.

Cumulatively compound motors are used in driving machines which are subject to sudden application of heavy loads *i.e.,* in rolling mills, shears and punches etc. This type of motor is also used where a large starting torque is required.

(*b*) **Differential Compound.** In this motor series field opposes the shunt field. Speed remains constant and sometimes increases with the increase in load. This motor may rotate in opposite direction at high loads. This type of motor is seldom used practically.

1.16 Speed current characteristics of cumulatively compounded motor and differentially compounded motor

In case of compound motor, field has both shunt winding as well as series winding. In case of cumulatively compounded motor series field helps the shunt field with the result that for the same value of armature current, the field of compound winding is more than that of shunt or series winding.

Speed N of shunt motor drops down with increase in current.

$$N \propto \frac{E_b}{\phi}$$

Since flux of cumulatively compound motor is more than shunt motor, speed of this motor shall be lesser than that of shunt motor. This is shown in Fig. 1.23.

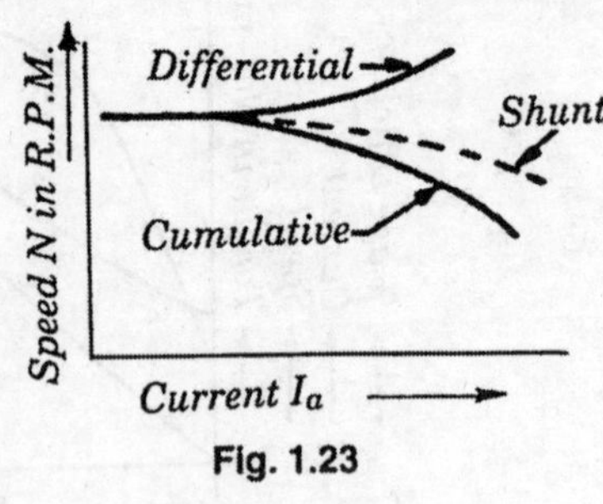

Fig. 1.23

In case of differentially compounded motor, the series field opposes the shunt field as a result for the same value of armature current the flux produced by differentially compounded motor is lesser than that of shunt motor.

$$N \propto \frac{E_b}{\phi}$$

Since flux is lesser than that of shunt motor, speed is more than that of shunt motor, for same value of armature current. This is also shown in Fig. 1.23.

1.16.1. Torque-current characteristics of cumulatively compounded motor and differentially compounded motor.

Torque $\quad T \propto \phi I_a$

Flux in case of cumulatively compounded motor is more than that of shunt motor so torque developed in cumulatively compound motor is more than that of shunt motor for same armature current. This is shown in Fig. 1.24.

In case of differentially compounded motor, flux is less than that of shunt motor so torque developed is lesser than that of shunt motor as is shown in Fig. 1.24.

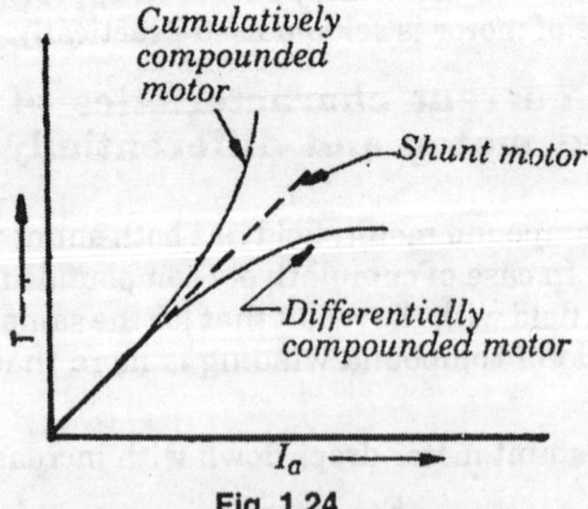

Fig. 1.24

Thus cumulatively compounded motors are used in driving machines which are subject to sudden application of heavy loads such as occur in Rolling Mills, Shear and Punches.

This type of motor is also used when high starting torque is required and series motor cannot be used conveniently such as in cranes and elevators.

1.17 Starters of shunt and compound wound motors.

As we know that starters are used to limit the starting current to a safe value and obtain the required starting torque.

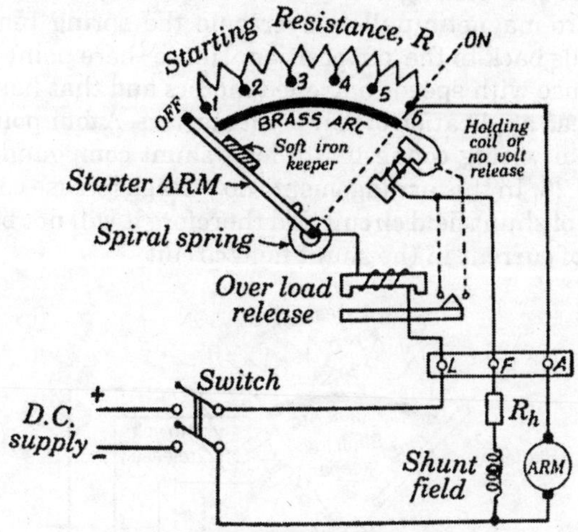

Fig. 1.25 Three point starter.

The starter of a shunt motor (three point starter) is shown in Fig. 1.25. At the start, all resistances are in series with the armature, the holding coil (through the arm) is across the line. As the speed builds up and back e.m.f. increases, the arm. is moved to the right. This decreases the armature resistance and the armature has a greater e.m.f.

At normal speed when the back e.m.f. is large enough to limit the armature current, the holding coil and arm. make contact. Both the field and armature are directly across the line. Current through the holding coil keeps the arm in this position. If for any reason the line voltage is fail, the magnetic field of the holding coil ceases and under the action of spring the arm goes back to its off position. This will keep the line voltage open. Because of this action the starter protects the motor. Restarting after a line voltage failure can only be done by using the protective series resistance in the armature circuit. The starter is also provided with over load releases coil to protect the motor against the flow of excessive current due to over load. This coil is connected in series with motor and so carries full load current. When motor is over loaded it draws heavy current, which also flows through the coil and magnetises it to such an extent, that it pulls, its armature upwards and so short circuits the no voltage release coil, as shown in Fig. 1.25. The no voltage release coil being short circuited demagnetizes and releases the starting arm which goes back to the off position and the motor is automatically disconnected from the supply mains.

In three point starter, as discussed above no volt release coil is connected in series with the shunt field rheostat and therefore the current flowing through the field is the same current that flows through the holding coil. Thus if sufficient resistance is cut in by the field rheostat so that the holding coil current is no longer able to create sufficient electro-magnetic pull to overcome the spring tension, the starter arm pulls back to the 'off' position. Hence there point starter is unsuitable for use with speed controlled motors and that has resulted in the wide spread application of four point starters. A four point starter with its internal wiring connected to long shunt compound motor is shown in Fig. 1.26. In this arrangement "no voltage release coil" circuit is independent of shunt field circuit and therefore it will not be effected by the change of current in the shunt field circuit.

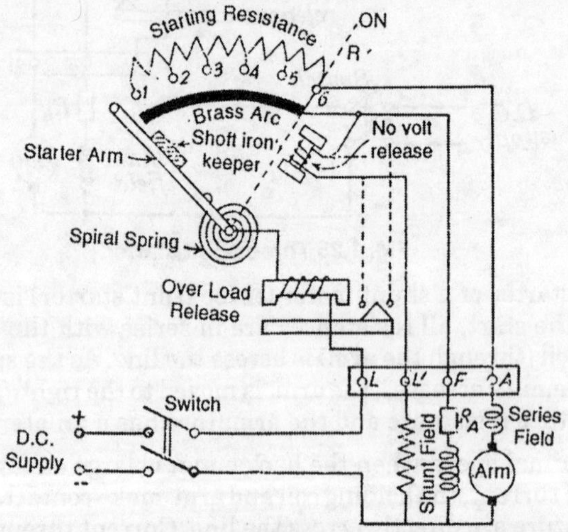

Fig. 1.26 Four point starter.

1.17.1. D.C. Series motor starter and speed controller. Series motor starter is not only used for starting of d.c. series motor but is also used for controlling its speed and that is why, it is usually known as good controller. The controller is in the form of rotating drum having segments which makes contact with the fixed points. Speed controller of the type is shown in Fig. 1.27 in which *K* is an arc breaking coil, *EM* is a breaking electromagnet ; *FLS* is switch for rotation in one direction and *RLS* is switch for rotation in reverse direction. The controller has six positions for forward and six positions for reverse rotation of motor. The working positions of the controller are shown by vertical dotted lines. The electromagnet E.M. is connected in parallel with the motor and releases the motor at starting. When the motor is disconnected it is breaked mechanically.

Now when the knife blade switch is closed and the controller is set in forward position, the connections are along vertical line one. In this position segments make contacts with fixed points 6 and 7, 8 and 9. In this position currents flows from +ve bus bar through armature winding of motor, arc braking coil *K*, all the starting resistances 1-6 fixed points 6 and 7 through controller segments, series field winding, the forward limit switch *FLS*, fixed points 9 and 8 through controller segments and returns back to the negative bus bar.

In the second forward position, the segments of the controller come in contact with fixed points 5 and 6 thereby bringing some of the starting resistance out of circuit, so speed of series motor increases. In the subsequent forward position 3rd, 4th and 5th additional steps of starting resistances are brought out of the circuit and finally in 6th position all the starting resistances are short circuited and motor attains maximum speed.

In all the six forward positions the direction of current in armature as well as in series field winding is same (from right to left) as shown in Fig. 1.27.

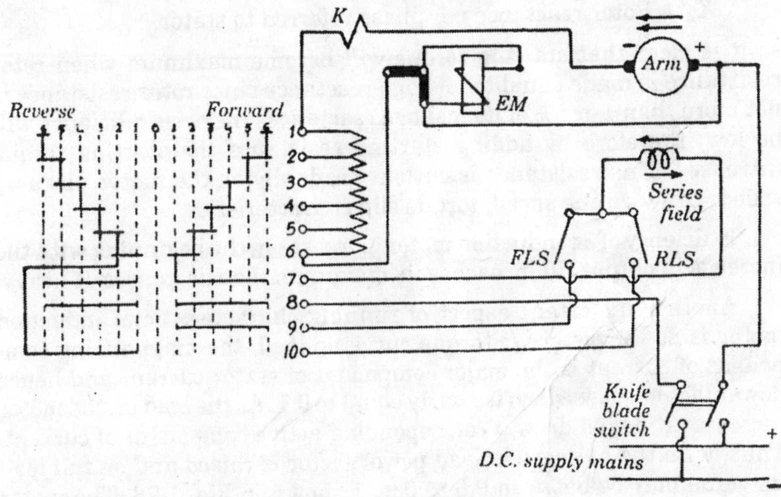

Fig. 1.27 Drum type series motor speed controller.

But in first reverse position the current flows from the bus bar through armature winding, arc braking coil *K*, all the starting resistance 1-6, fixed points 6 and 10 through controller segments, the reverse limit switch *RLS*, the series field winding, fixed points 7 and 8 through controller segments and returns back to the –ve bus bar. Hence it is obvious that in reversing positions direction of flow of current in armature winding remains unchanged while reverses through the series field winding, there by reversing the direction of rotation of motor.

1.18 A.C. Motors

(*i*) **Three-phase Induction Motor.** While in running condition there is no basic difference between squirrel cage and slipring induction motor. The only drawback of squirrel cage motor is that speed control can not be achieved by inserting resistance in rotor circuit as in case of slip ring motor. So starting torque developed by slip ring motor is more than that of squirrel cage motor.

It is because of the reason that starting torque per phase in case of 3ϕ. Induction motor is given by the equation :

$$\text{Starting torque per phase or } T_s = \frac{kV^2 R_2'}{(R_1 + R_2')^2 + (X_1 + X_2')^2}$$

where V = Applied voltage per phase to stator

R_1 = Stator resistance per phase

R_2' = Rotor resistance per phase referred to stator

X_1 = Stator reactance per phase

X_2' = Rotor reactance per phase referred to stator.

so it is clear that starting torque will become maximum when rotor resistance is made equal to leakage reactance since rotor resistance is not more than 1 or 2% of its leakage reactance otherwise efficiency will be low, therefore by adding starting resistance the starting torque increases. This resistance is cut out gradually as the motor picks up speed. Draw torque speed, torque slip characteristics.

Efficiency of 3ϕ induction motor increases in the beginning with the increase in output till it reaches its maximum then decreases at heavy.

Another important aspect of running characteristic of induction motor is power factor v/s torque curve no load, the magnetising component of current is the major component of stator current and hence power factor is low which is nearly equal to 0.1. As the load on the motor increases it has to draw a corresponding active component of current. Thus, with the increase in load power factor is raised and on full load its value may be between 0.8 to 0.9 as shown in Fig. 1.28. The motor always works on low power factor. The starting torque of slipring Induction motor is high and hence these are suitable for heavy duty starting squirrel cage motor is the most commonly used motor. It is used to drive pumping sets, machine tools and the other operations where fairly constant speed is desired.

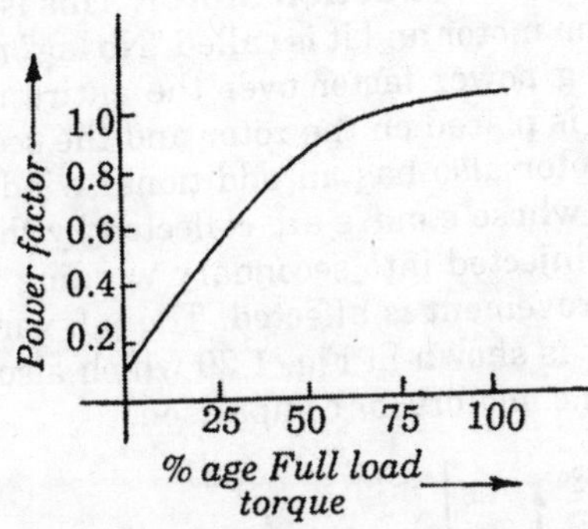

Fig. 1.28 Torque speed characteristics.

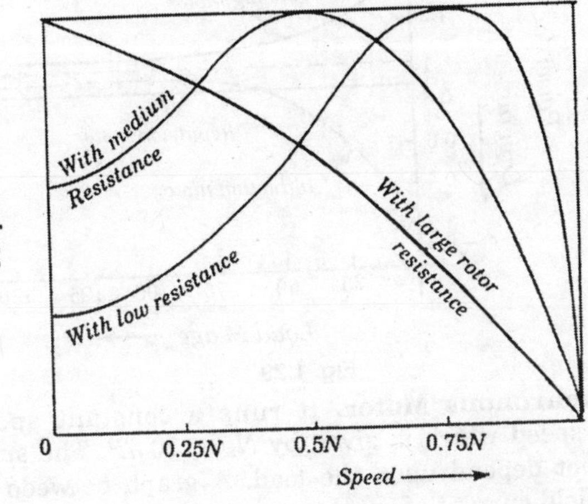

Fig. 1.28 (*a*) Torque speed curves of Induction Motor.

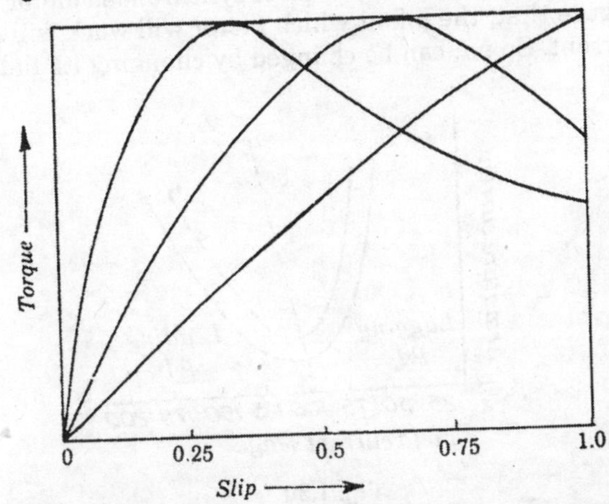

Fig. 1.28 (*b*) Torque Slip Curves of 3 φ Induction Motor.

(*ii*) **Compensated Induction Motor.** This is an improvement of the plain induction motor and it is called "No lag" motor. It gives unity or slightly leading power factor over the entire range of loads. The primary winding is placed on the rotor and the secondary winding on the stator. The rotor also has an additional winding known as commutator winding whose e.m.f.'s are collected by the brushes from the commutator and injected into secondary winding in such a way that power factor improvement is affected. The p.f. variation according to load on the motor is shown in Fig. 1.29 which also shows p.f. characteristics of other a.c. motors for comparison.

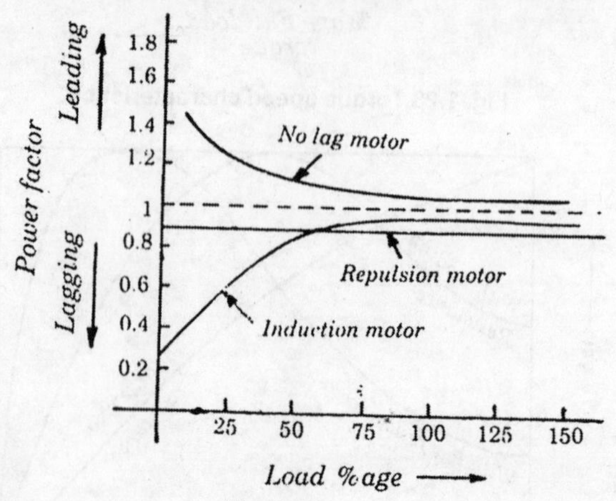

Fig. 1.29

(*iii*) **Synchronous Motor.** It runs a constant speed called synchronous speed which is given by $N_s = 120\,f/P$. The speed of the motor does not depend upon the load. A graph between armature current and field current of synchronous motor is shown in Fig. 1.31. This graph is also called the V-curves of synchronous motor. It is clear from this figure that, the p.f. at which motor will work will depend on the field current. So p.f. can be changed by changing its field current.

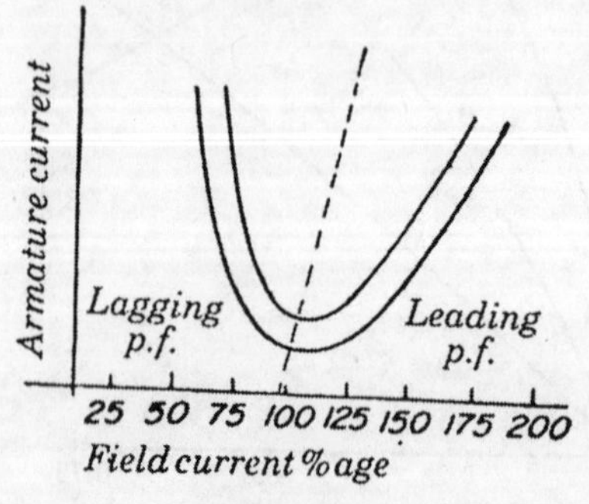

Fig. 1.30

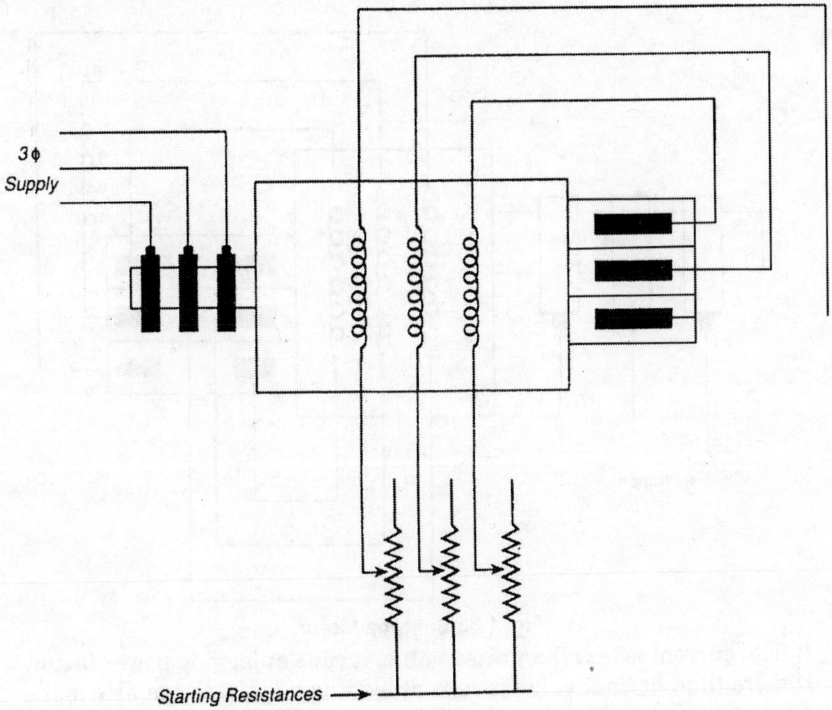

3ϕ
Supply
Starting Resistances

Fig. 1.31

3ϕ
A.C Supply
Starting Rheostat
Slip rings
Exciter
Motor
D P switch

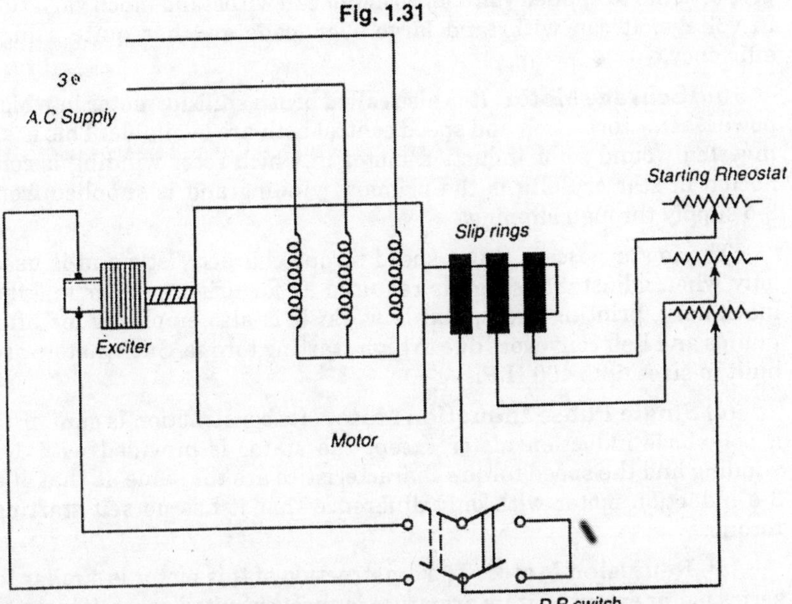

Fig. 1.32 Synchronous Motor.

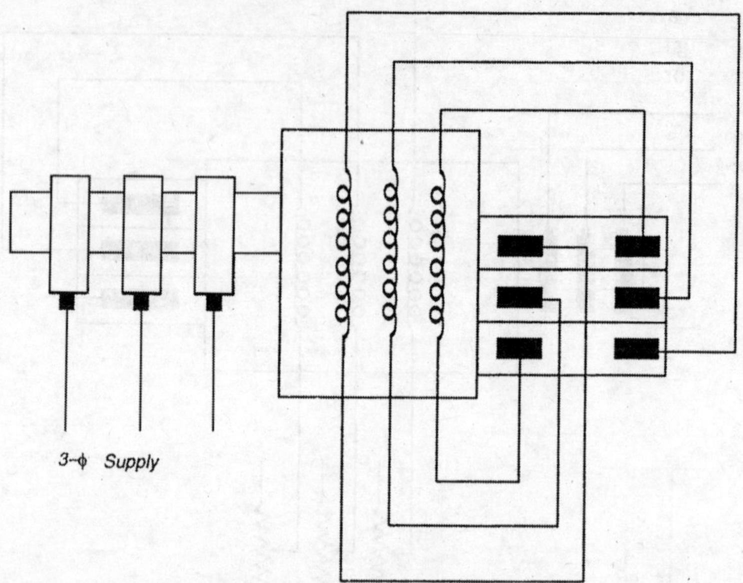

Fig. 1.33 Schrage Motor.

If field current is less than rated value, it runs at lagging power factor. If more than normal value it runs at leading p.f. Motor can also made at unity p.f. by changing the field excitation. Torque developed is proportional to applied voltage. So motor can withstand much variation in voltage. It can withstand large over loads and has quite a high efficiency.

(*iv*) **Schrage Motor.** It is also called brush shifting motor in which power factor correction and speed control both are available. This is an inverted wound rotor induction motor in which rotor winding is connected in star or delta is the primary winding and is supplied from 3-ϕ supply through slipping.

The motor possess shunt speed torque characteristic and is used only when adjustable speed is required as for driving paper making machinery, printing press, textile work. It is also employed for lifts, pumps and Belt conveyors due to high starting torque. Such motors are built in sizes upto 400 H.P.

(*v*) **Single Phase Induction Motor.** Its construction is similar to a polyphase induction motor except the stator is provided with 1-ϕ winding and the speed torque characteristics are the same as that of a 3-ϕ induction motor with only difference that it has no self starting torque.

(*iv*) **Repulsion Motor.** The construction of this motor is similar to series motor except that its armature is short circuited on itself instead of being connected in series with stator. The operating characteristics of repulsion motor are similar to those of series motor *i.e.* high starting torque and high speeds at light loads. Its speed can be adjusted by shifting the brushes with the help of a special lever mechanism provided on motor.

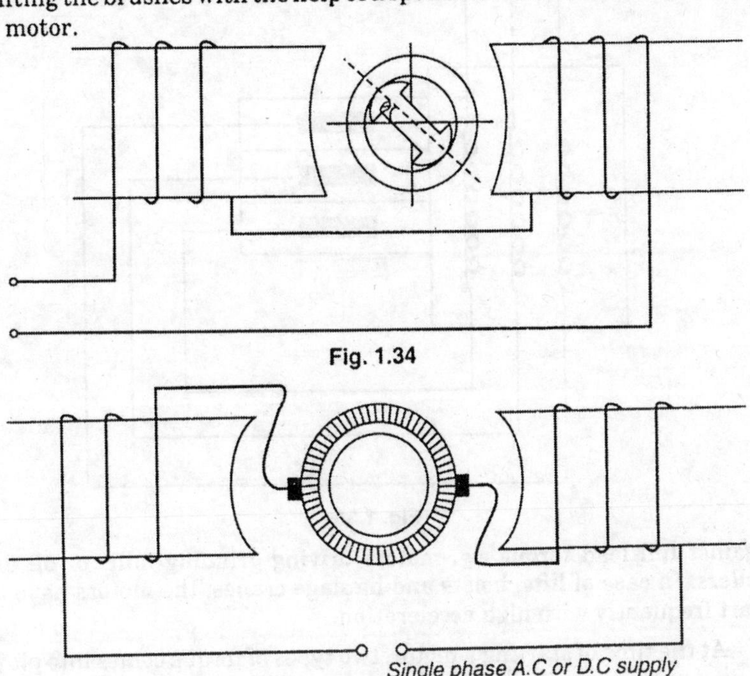

Fig. 1.34

Fig. 1.35

(*vii*) **Universal Motor.** It is a series wound motor, which operates at approximately the same speed and output on either d.c. or a.c. of approximately same voltage. The characteristics of motor is similar to those of d.c. series motors. The motor built up in fractional horse power upto about 1/4 H.P. These are suitable for sewing machines, table fans. Vacuum cleaners, portable drills, hair dryer, cooler pumps and small blowers etc.

(*viii*) **Three Phase Series Motor.** The stator of such a motor is similar to the plain induction motor in construction except that all the six ends of the three windings are brought out. The rotor accommodates a commutator winding with three sets of brushes per pole pair on the commutator. The brushes are connected to the supply in series with the stator winding as shown in Fig. 1.36.

The speed-torque characteristics are just similar to the d.c. series motor characteristics. This is used for hoisting work and for electric traction when only a.c. supply is available.

(*b*) **Starting Characteristics.** The starting torque exerted by a motor should be large enough to accelerate the motor and load to the rated speed in reasonably short time. Some motors may have to start

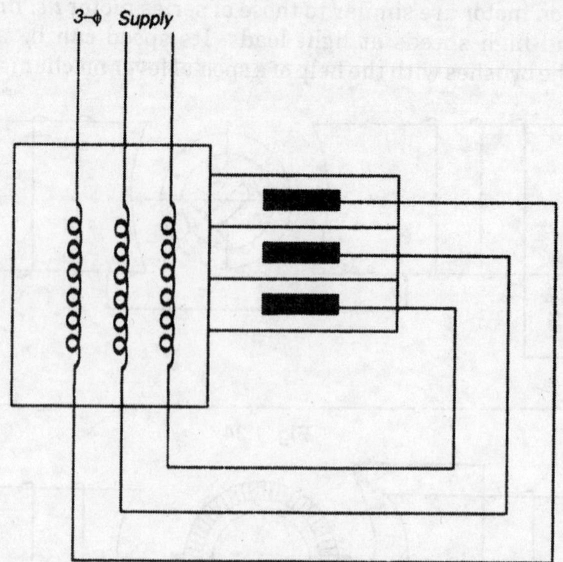

Fig. 1.36.

against full load torque *e.g.* motors driving grinding mills or oil expellers. In case of lifts, hoists and haulage cranes, the motors have to start frequently with high acceleration.

At the time of starting a motor, two types of torque comes into play:

(*i*) The torque required to overcome the static friction.

(*ii*) The torque necessary to accelerate the motor and its loads to the desired speed.

Starters are used for starting purpose to limit the starting current within reasonable limit and also that requisite of starting torque is developed.

1.19 D.C. Motors

In a D.C. motor armature current is given by:

$$I_a = \frac{V - E}{R_a}$$

where,
V = Supply voltage
E = Induced e.m.f. in armature
R_a = Armature resistance

when motor is at rest, the induced e.m.f. E in the armature is zero [$\because E \propto N$]. Consequently, if full voltage is applied across the motor terminals, the armature will draw heavy current [$I_a = V/R_a$] because armature resistance is relatively small. The armature current is relatively 20 to 30 times to that of full load current. This heavy current has the following effects :

(*i*) It will blow out the fuses and prior to that it may damage the armature winding due to excessive heating effect :

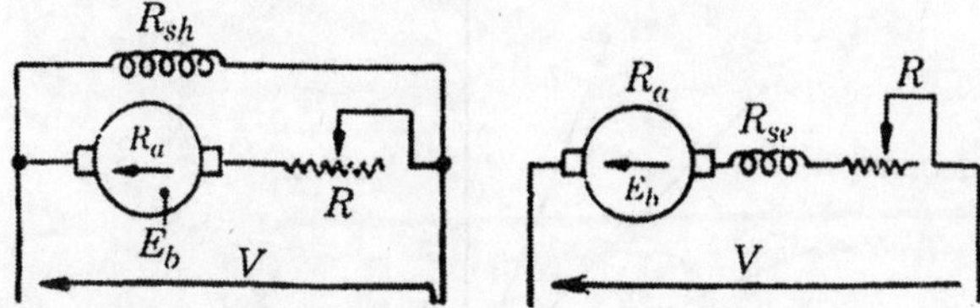

Fig. 1.37

(*ii*) Excessive voltage drop will occur in the lines to which the motor is connected. Thus to avoid this heavy current at start a variable resistance is connected in series with the armature called starting resistance or starter. In case of shunt motor torque developed at the time of starting is proportional to full load torque. But in case of D.C. series motor since ϕ is also proportional to I_a, so starting torque is proportional to square of full load current, before saturation and proportional to armature current after saturation.

This is reason why series motor is more suitable for heavy duty starting than shunt motor.

1.19.1. Braking of dc motors during lowering of loads.

DC Shunt Motor. When a dc shunt motor used in a hoisting mechanism, is switched on, for lowering a load, the torque developed by the motor and that due to load torque act in unison to accelerate the motor with increase in speed, the e.m.f. induced in armature also increases and attains a value equal to the applied voltage, when the speed becomes the ideal no load speed at this moment, the armature current, and hence the electromagnetic torque becomes zero, so that the downward motion of the hoist is sustained by the downward moving load. When the speed becomes greater than the ideal no load speed $E > V$ and therefore the armature current becomes negative. The drive then acts as a generator and provides the braking torque. The drive attains a steady state speed when the braking torque developed by the motor is equal to the load torque. During this operation, the e.m.f. induced in the armature is directed against the applied voltage as it happens during motoring operation. The type of braking that occurs is regenerative braking, since power is fed back to the supply.

The speed torque characteristics of the motor during motoring and braking operations are shown in Fig. 1.38. There are actually three characteristics– one being the inherent one, with no additional resistance in the armature circuit and the other two being the modified ones, corresponding to increased values of resistance in the armature circuit. It is clear, from the Fig. 1.38 that braking can occur only at speeds greater than the ideal no load speed of the motor.

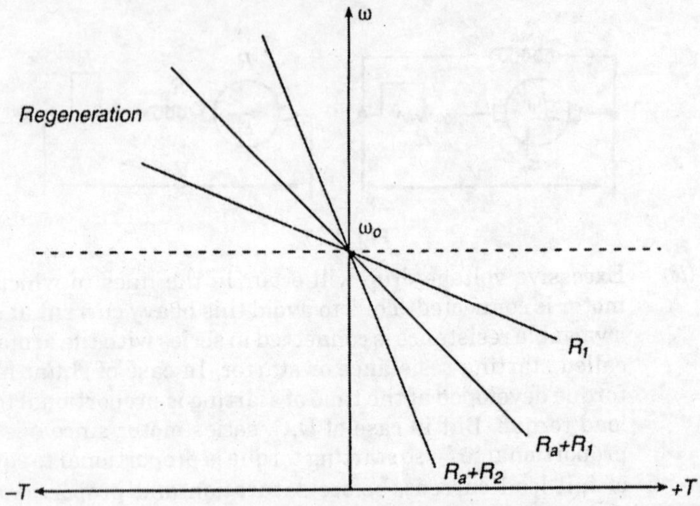

Fig. 1.38 Motoring and Regenerative braking characteristics of a D.C. Shunt Motor.

At speeds lower than the ideal no load speed, retarding torque can be provided only by means of rheostatic braking. During this type of braking the motor acts as a generator, whose armature terminals are short circuited through an external resistance R_{ex}. The voltage and speed equation corresponding to this operation are :

$$E = -I\,(R_a + R_{ex}) \quad \text{...(1)}$$

and

$$w = -T\,\frac{(R_a + R_{ex})}{K + K_e\,\phi^2} \quad \text{...(2)}$$

since $V = 0$.

The speed-torque characteristics during rheostatic braking are as shown in Fig. 1.39 (*b*). These are approximately straight lines passing through the origin. The slope of the characteristics depend upon the total resistance in the armature circuit. Braking torque can be obtained at very low speeds, *i.e.* 0.1 to 0.01 of the rated speed. Obviously complete stoppage of the downward moving load is not possible because the current required for producing the braking torque is caused only by the movement of the load.

Instead of the separate excitation, shown in Fig. 1.39 (*a*) shunt excitation can also be provided with. But as soon as the motor reaches near about one half of the rated speed, the process of self excitation itself becomes ineffective and hence no braking torque will be developed.

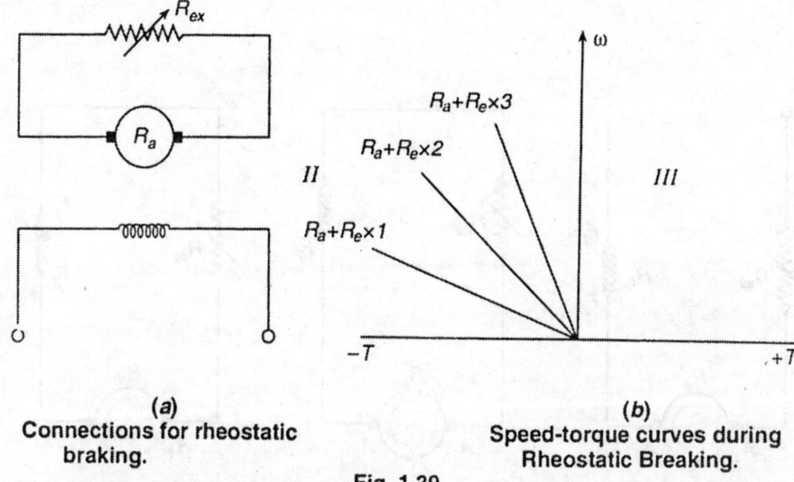

(*a*)
Connections for rheostatic braking.

(*b*)
Speed-torque curves during Rheostatic Breaking.

Fig. 1.39

1.19.2. DC Series Motor. If the speed of a series motor is increased, the current and hence the flux reduces. Therefore, it is not possible to get an emf. greater than the terminal voltage by this means. It is quite known that the speed is extremely high even before the motor reaches actual no load. Also, since there is no means of making the field current greater than the armature current in a simple series motor, regeneration is not practicable with series motors. This can be confirmed by observing the speed-torque characteristics which does not cross into the second quadrant of the quadrantal diagram. In electric traction, where regenerative braking of series motors is used the motors are actually reconnected as separately excited machines.

Rheostatic braking, of course, is possible with series motors as shown in Fig. 1.40 but care should be taken to interconnect the armature winding and the field winding in such a way as to ensure that the direction of the current in the field remains the same, in spite of its change in armature winding. Then only, the self excitation of the series generator will take place. Also the external resistance connected should not be of such a value to make the total resistance of the motor circuit greater than the critical value. Due to the presence of residual magnetism and the speed of rotation of the rotor, an e.m.f. is induced in the armature. Under its influence, current, flux and e.m.f. increased until they reach their steady state values, given by the equation,

$$E = I\,(R_a + R_f + R_{ex})$$

where R_f is the resistance of the series field winding.

Rheostatic braking of series motors can also be effected by feeding their field winding from independent sources and connecting a resistance across the armature terminals *i.e.* by reconnecting them as separately excited motors.

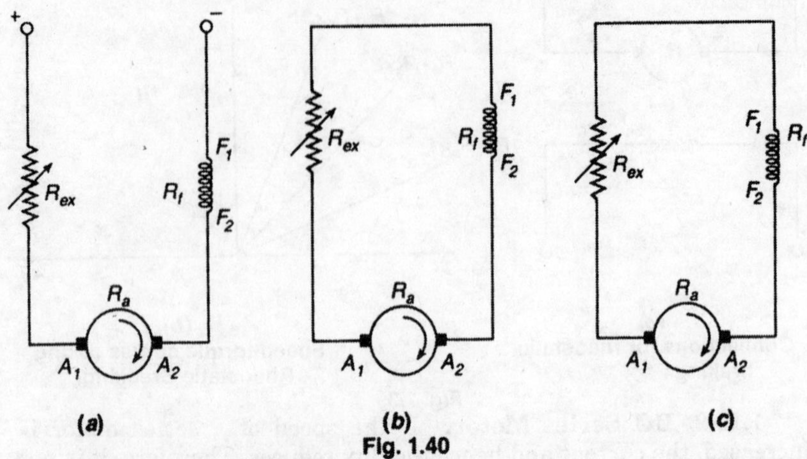

Fig. 1.40

1.19.3. Braking While Stopping

(*a*) DC Shunt Motor. Regenerative braking is possible, when an adjustable speed motor is used. Before bringing the motor to rest, the field excitation is increased to the permissible maximum value, as a result of which the speed of the motor falls to the minimum value and the kinetic energy released from the rotor is fed back to the supply. If the speed of the motor were to reduce in the ratio 4 : 1, the energy during braking would reduce in the ratio 16 : 1. Since the motor operates with a weak field at rated conditions, armature has to be designed to carry a large current to produce the rated torque and hence the motor will be larger in size, poor in efficiency and costlier.

Plugging of dc motor involves reconnecting the motor to the line with reversed polarity the motor now produces a torque in opposite direction to that of the rotation. The rotor speed decreases until it become zero and then the rotor accelerates in the opposite direction. Therefore plugging is used to get either a quick reversal or get a rapid stop for the latter it is necessary to employ some means which will disconnect the motor at the time when it passes through zero speed.

The connections corresponding to the normal working and plugging of dc shunt motors are shown in Fig. 1.41 (*a*) and 1.41 (*b*).

In order to reverse the torque developed, it is necessary to reverse the magnetic field or the armature current. Since the field winding, usually has a large time constant due to large inductance armature current is reversed. When the armature current is reversed, the back emf of the armature will no longer oppose the applied voltage, but will aid it. This condition is actually worse than starting motor direct on line, since nearly double the supply voltage is applied across the armature having a low resistance. Therefore additional resistance R_{ex} is inserted in the armature circuit, simultaneously with the reversing of connections of the armature.

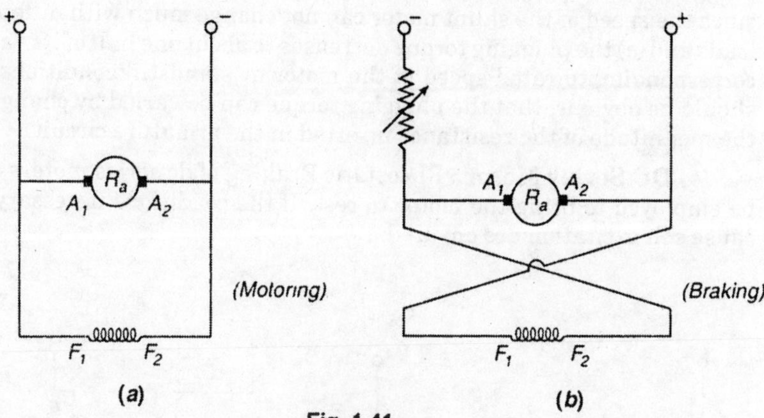

Fig. 1.41

The plugging operation of a dc shunt motor can be explained with the help of the quadrantal diagram shown in Fig. 1.41 (c) the characteristic A is exactly of the same shape as the dc shunt motor with a

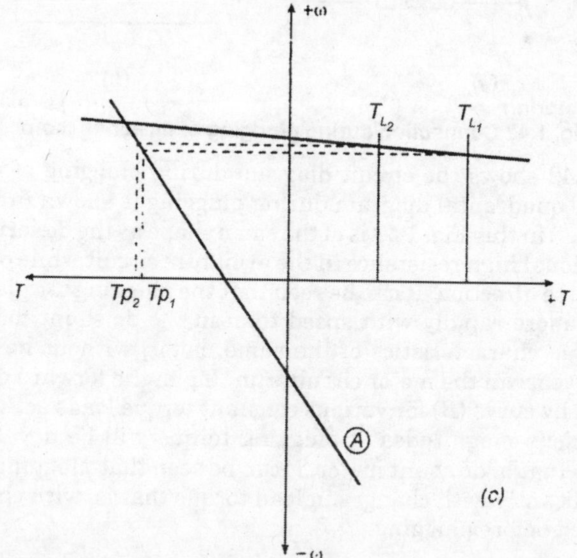

Fig. 1.41 (c)
Quadrantal Diagram during plugging of d.c. shunt motor.

sufficiently high series resistance in the armature running in the reverse direction. The curve (B) corresponds to the operation of the motor in the forward direction, with no series resistance in the armature. For different values of constant load torques acting on the motor, different plugging torques are obtained. It can be observed from the quadrantal diagram than (*a*) the value of plugging torque developed will be corresponding to the speed of operation of the motor before plugging, (*b*) the plugging torque vary only slightly with the motor load, since the speed of the shunt motor can not change much with different loads and (*c*) the plugging torque decreases to about one half of its value corresponding to rated speed of the motor at standstill conditions. It should be obvious, that the plugging torque can be varied by changing the magnitude of the resistance inserted in the armature circuit.

(*b*) **DC Series Motor.** Rheostatic Braking of dc series motors can be employed to bring the motor to rest. If the conditions necessary to cause self excitation are ensured.

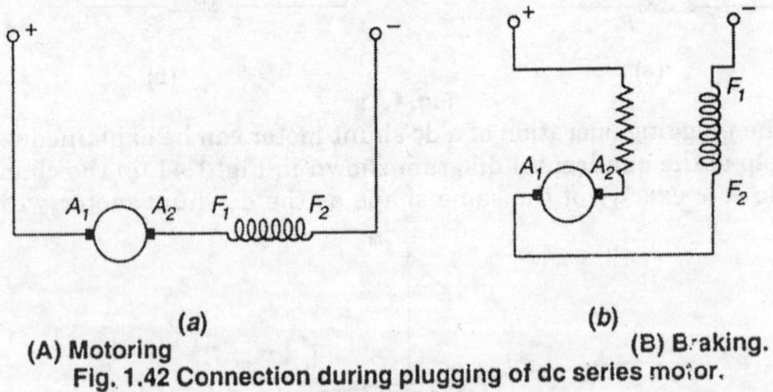

(A) Motoring (B) Braking.

Fig. 1.42 Connection during plugging of dc series motor.

Fig. 1.42 shows the circuit diagram during plugging of dc series motor. The quadrantal diagram during plugging is shown in Fig. 1.41 (*c*). Curve (*A*) in this Fig. 1.43 is of the same shape as the dc series motor with additional high resistance in the armature circuit while operating in the reverse direction. It can be seen that the plugging torque reduces somewhat more rapidly with speed than in the dc shunt motor. The space torque characteristics of the same motor, without having any extra resistance in the motor circuit, running in the forward direction, is depicted by curve (*B*) for various constant torque loads acting on the motor, various magnitudes of plugging torque will be developed. In contrast to that of dc shunt motor, it can be seen that plugging torques vary significantly with changes in load torque that is, with changes in motor speed before plugging.

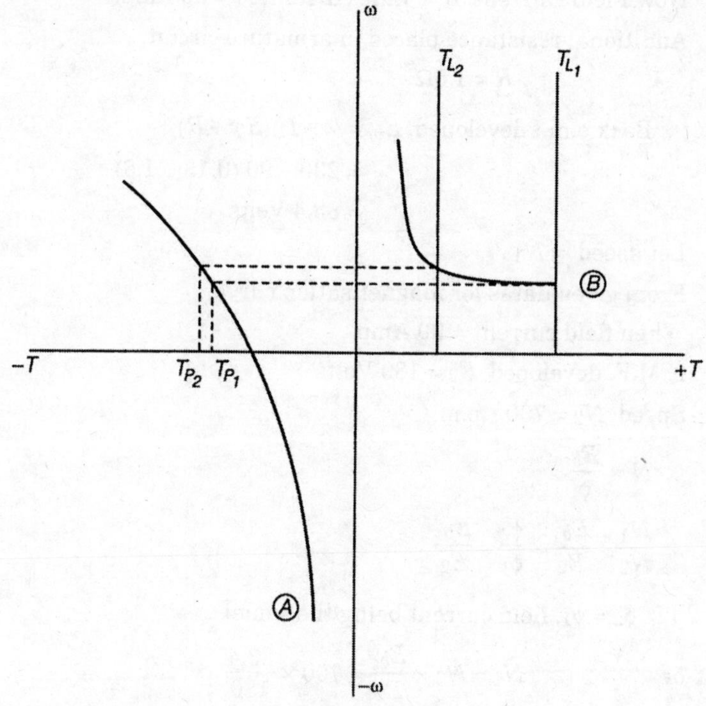

Fig. 1.43

Example1.1 *A 230 V, d.c. series motor has an armature resistance of* 0.08Ω *and field resistance of* 0.05Ω*. The magnetisation characteristic of the machine at 700 r.p.m. is as follows :*

Field current	30	60	90	120	150 A
E.M.F.	70	125	180	210	230 V

The motor is run on a 230 V supply. An additional resistance of 1.5Ω is connected in series with the armature. Determine the torque and speed when the motor draws a current of 90 A.

[A.M.I.E. Section B. Elect, Drives and Their Control, Winter, 1994]

Solution: Given : $R_a = 0.08\Omega$, $R_{se} = 0.05\Omega$

Between motor terminals, the total resistance,

$$R_T(motor) = R_a + R_{se}$$

$$= 0.08 + 0.05 = 0.13\Omega$$

Now Field Current, I_f = Line current, I_1 = 90 Amp.

Additional resistance placed in armature circuit,

$$R = 1.5\Omega$$

$\therefore$ Back e.m.f developed, $E_{b_f} = V - I_1 (R_T + R)$

$$= 230 - 90\,(0.13 + 1.5)$$

$$= 83.4 \text{ Volts.}$$

Let speed $= N_1$

From given datas for magnetisation curve,

When field current = 90 Amp.

E.M.F. developed, E_2 = 180 Volts

Speed, N_2 = 700 r.p.m.

$$\because N \propto \frac{E_b}{\phi}$$

$$\therefore \frac{N_1}{N_2} = \frac{E_{b_1}}{E_2} \times \frac{\phi_2}{\phi_1} = \frac{E_{b_1}}{E_2}$$

[$\because$ $\phi_2 = \phi_1$, field current being the same]

or $$N_1 = N_2 \times \frac{E_{b_1}}{E_2} = 700 \times \frac{83.4}{180}$$

$$= 325 \text{ r.p.m.} \quad \textbf{Ans.}$$

Torque developed, $$T = \frac{9.55 \times E_b\, I_a}{N}$$

$$= \frac{9.55 \times 83.4 \times 90}{324}$$

$$= 222 \text{ N-m} \quad \textbf{Ans.}$$

Example 1.2 *A 239 V D.C. shunt motor with constant field drives a load whose torque is proportional to its speed. When running at 750 r.p.m. it takes 30 A. Find the speed at which it will run if a 10Ω resistance is connected in series with its armature. The resistance of armature may be neglected.*

Solution : Neglecting shunt field current,

Armature current, I_a = Line current, I_L = 30 Amps.

Speed, N_1 = 750 r.p.m.

Suppose the speed is reduced to N_2 r.p.m, then $N_2 = KN_1$ after inserting additional resistance of 10Ωs in the circuit of armature.

As with field excitation constant, torque developed is proportional to the armature current *i.e.*

$$T \propto I_a \qquad ...(1)$$

Load torque, $T \propto N$ ...(2) given

From (1) and (2) equations we get,

$$I_a \propto T \propto N$$

or $$I_{a_2} = I_{a_1} \times \left[\frac{N_2}{N_1}\right] = KI_{a_1} = 30\,K \qquad ...(3)$$

As $N \propto [V - I_a(R_a + R)]$ Since flux, ϕ remaining the same

or $$\frac{N_2}{N_1} = \frac{V - I_{a_2}R}{V}$$ Keeping Armature resistance negligible

Substituting, $\frac{N_2}{N_1} = K$ and from equation (3)

$I_{a_2} = 30K$, and $R = 10\Omega$ in the equation derived above, we get

$$K = \frac{230 - 30\,K \times 10}{230}$$

or $$K = \frac{230}{230 + 300} = 0.433$$

$\therefore$ Speed desired $= KN_1$

$= 0.433 \times 750$

$= 325$ r.p.m. **Ans.**

Example 1.3 *A 200 V d.c. series motor runs at 700 r.p.m. when operating at its full-load current of 20 A. The motor resistance is* 0.5Ω *and the magnetic circuit can be assumed unsatuated. What will be the speed if (i)* the load torque is increased by 44% *(ii)* the motor current is 10 A ?

Solution : Developed back e.m.f.,

$$E_{b_1} = V - I_1(R_a + R_{se})$$

$$= 200 - 20\,(0.5) = 190 \text{ Volts}$$

(*i*) When the load torque is increased by 44% that means

$$T_2 = 1.44T_1$$

or $$\phi_2 I_a = 1.44\,\phi_1 I_1 \quad [\because T \propto \phi I_a]$$

[As $\phi \propto I$]

So putting the values in the above equation

$$I_2^2 = 1.44 I_1^2$$

or $$I_2 = \sqrt{1.44}.I_1 = 1.2 \times 20 = 24 \text{ Amp.}$$

Developed Back e.m.f, $E_{b_2} = V - I_2 (R_a + R_{se})$

$$= 200 - 24 \times 0.5 = \textbf{188 Volts}$$

Since $E_b \propto \phi N$

$$\therefore \text{ Speed, } N_2 = \frac{E_{b_2}}{E_{b_1}} \times \frac{\phi_1}{\phi} \times N_1$$

$$= \frac{188}{190} \times \frac{20}{24} \times 700$$

$$= 577.2 \text{ r.p.m.} \qquad \textbf{Ans.}$$

(*ii*) When the motor current is 10 Amps.

Then developed back e.m.f.,

$$E_{b_3} = V - I_3 (R_a + R_{se})$$

$$= 200 - 10 \times 0.5 = 195 \text{ volts.}$$

$$\text{and speed, } N_3 = \frac{E_{b_3}}{E_{b_1}} \times \frac{\phi_1}{\phi_3} \times N_1$$

$$= \frac{195}{190} \times \frac{20}{10} \times 700$$

$$= 1{,}436 \text{ r.p.m.} \qquad \textbf{Ans.}$$

1.20 Induction Motors

(*i*) **Slipring Induction Motors.** Induction motor have their torque maximum when resistance of the rotor is same as that of its reactance. Normally the reactance of the rotor is more than that of resistance. Thus to produce maximum torque in the motor at the time of starting, external resistance in each phase of rotor circuit is added through their slip rings. This serves two purposes : one increases the torque at the time of starting and secondly limits the rotor current, within reasonable limits. This in turn limits the stator current/motor current.

As the motor picks up more and more speed, slip goes on reducing, the reactance of rotor also goes down.

Thus to produce the maximum torque, the resistance is to be reduced and, therefore, cut-off. In the mean time voltage induced in the

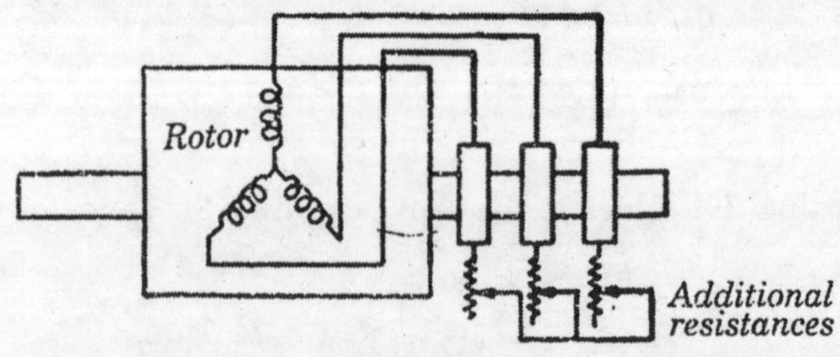

Fig. 1.44

rotor also goes down and even with reduced resistance also, the current is within the reasonable limits. Thus as soon as the motor picks up full load speed, the whole of resistance is cut off. These motors on account of their robust construction are more reliable, cheaper and easier to maintain as compared to squirrel cage motors. These motors are employed where constant torque is required *e.g.* in paper mills, textile mills compressors and conveyors.

In case of squirrel cage induction motor, it is not possible to add external resistance in the rotor circuit for starting because of cage type rotor construction, so such motors shall have low starting torque and hence cannot be started by this method.

(*ii*) **Starting of squirrel cage induction motors.** Since external resistance cannot be added into rotor of squirrel cage induction motor, so squirrel cage induction motor is started by applying reduced voltage to the stator of the motor. Reduced voltage applied would reduce the starting current while producing the desired starting torque. Induction motors up to 3 H.P. can be started directly on line as the current drawn is not very high to create high voltage drops to disturb the other working points on the same line and in addition small motors are so light that they pick up rated speed without any loss of time thus giving no opportunity to high currents flowing in motor to produce heat to damage the motor. Motors above 3 H.P. need starting device as the rotor of induction motor acts just like a short circuited secondary of transformer at the time of starting. There are four methods of starting 3 – ϕ squirrel cage induction motor :

(*a*) D.O.L. starter

(*b*) Stator resistance starter.

(*c*) Auto transformer starter.

(*d*) Star delta starter.

(*a*) **Direct on line starter.** In this method of starting motor is directly connected to the supply in on step and no provision is there for reducing the in-rush of starting current. Schematic diagram of D.O.L. starter employing electromagnetic contractors is shown in Fig. 1.45. Electromagnetic contractors consists of a coil ⊢⁄\⁄⊣ . Normally open (NO) contacts ⊣ ⊢ and normally closed (NC) contacts ⊣⁄⊢ on Pressing.

The start push button which is normally held open by a spring operating coil of the main contactor is energized. As a result of this all the three main (NO) contactors and one seal in (NO) interlock contactor closed. Now even it start push button is open. Main contactor coil remains energized through the seal in contactor. As a result of closing of main contactors motor is connected to the supply. Stop button is pressed, contactor coil gets de-energized and seal in contactors open out thereby preventing restart of motor on the resumption of supply. In case of overload, overload coils open the normally closed contactors, thereby

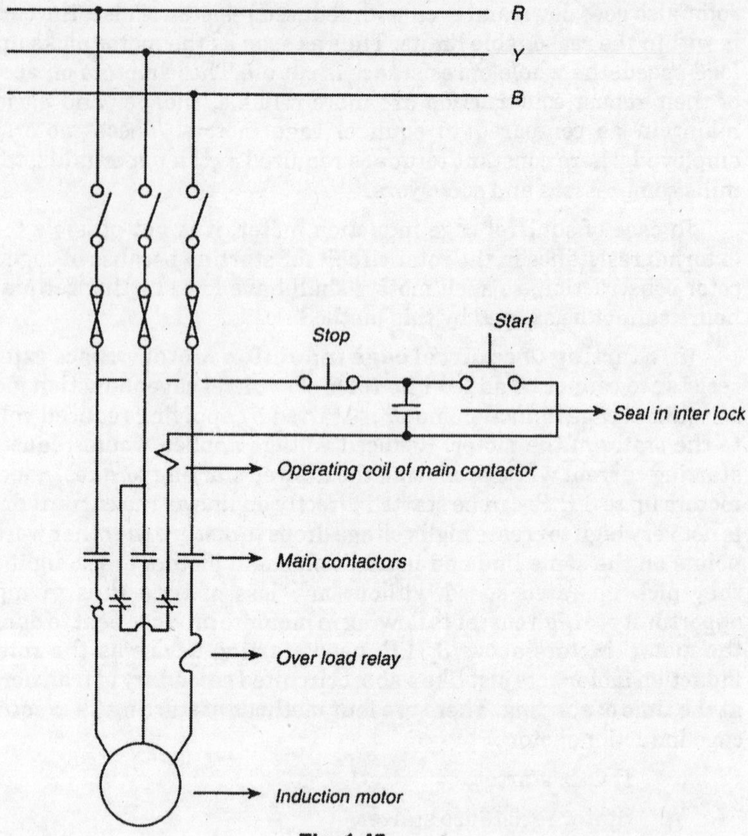

Fig. 1.45

de-energizing the operating coil of main contactors and disconnecting the supply from the motor. Push button station can be installed at any place thus making the remote control of motor possible.

Advantages of D.O.L. starters : Highest starting torque, lowest cost and greatest simplicity. This type of starter can be used for High H.P. motors also provided power system supplying the motor is strong and motor is designed to withstand the mechanical forces due to the inrush of current associated with full voltage starting.

Disadvantages : excessive voltage drop in the weak power system and that the torque may be limited to protect certain types of loads.

Starters are manufactured in two designs viz. oil immersed type and air break type. AIR break type starters are preferred to oil break starters on account of the following disadvantages of the later are:

1. High maintenance cost on account of carbonizing and sludging of oil brought about by repeated switching off load currents.
2. Oil immersed starters are bulky.
3. Presence of oil has danger of fire hazard.
4. Oil break starters can not be assembled in multizier formation and therefore not amenable to group motor control centre.

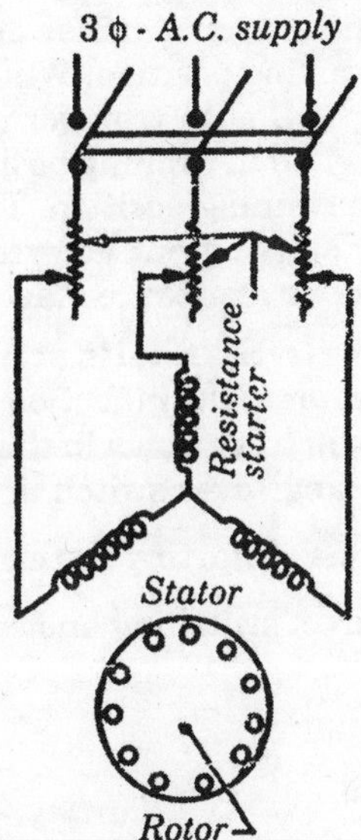

Fig. 1.46 Stator resistance starter.

(*a*) **Stator resistance starter.** Stator resistance starter is shown in Fig. 1.46. In this type of starter, resistance is connected in series with each phase of stator winding. This reduces the voltage fed to stator at the time of starting. Reduced voltages are induced in the rotor also. Since rotor is short circuited on itself, so reduced current flow. As soon as motor picks up more and more speed, more and more resistance is cut-off from the stator currents as the voltage induced in the rotor automatically falls down due to reduced slip. When rated speed is achieved, whole of resistance is cut off. This method is not very useful as lot of power is wasted in additional resistance at the time of starting.

(*b*) **Auto transformer starting.** Auto transformer starter has been shown in Fig. 1.47.

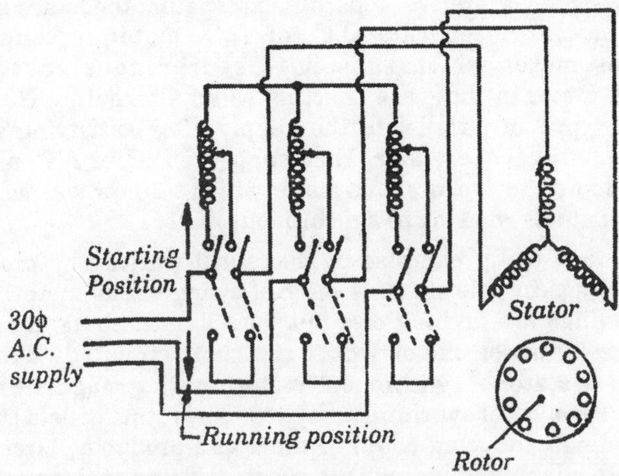

Fig. 1.47 Auto transformer starting.

In this type of starting device reduced voltage is fed to the stator of the induction motor through auto transformer by putting the switch in starting position. When the motor attains about 80 to 85% of its rated speed, auto transformer is removed from the circuit by putting the switch in running position. Full voltage is fed to the stator of the motor in running position. This method develops reduced torque at the time of starting but no excessive copper losses take place as in case of the stator resistance starter.

(*c*) **Star delta starter.** Star-delta starter is shown in Fig. 1.48. The stator of the induction motor is connected in star when the change over switch is thrown in the starting position and is connected in delta when change over switch is thrown in the running position. The current at time of starting by this method gets reduced to $\frac{1}{3}$rd of the one drawn by direct switching and starting torque is also accordingly reduced.

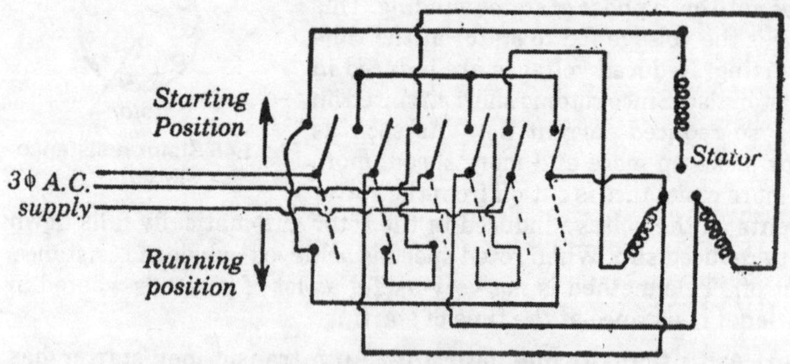

Fig. 1.48 Star delta starter.

(*d*) **Starting of Synchronous Motor.** Synchronous motor is not self-starting motor, therefore, separate external methods are used to start this motor. A separate D.C. or A.C. motor is coupled to synchronous motor which runs it to synchronous speed. The synchronous motor in that case is acting as an alternator. The alternator is then put in parallel to the supply after satisfying all the conditions of parallel operations. Then supply to D.C. or A.C. motor is disconnected and the synchronous motor which till now was acting as an alternator starts working as synchronous motor.

Synchronous motor can also be made self-starting by providing damper windings in pole shoes of rotor of synchronous motor. These damper windings are nothing else but bars like the bars of rotor of squirrel cage induction motor which are short circuited. When 3-ϕ supply is fed to stator of synchronous motor rotating magnetic field is set up which revolves at synchronous speed. This rotating field cuts the damper windings, inducing e.m.f. in them and producing circulating currents as the damper windings are shorted. These current carrying conductors placed in magnetic field, develop torque and as a result, the motor starts running. As soon as nearing synchronous speed is achieved, the D.C. current to the field windings of the rotor of synchronous motor is made to flow forming poles on the rotor. These rotor, poles then magnetically get locked up with poles of rotating magnetic field, and as a result the motor continues running at synchronous speed. The resistance of the damper windings can be adjusted in such a way that starting torque between 50 to 100% full load torque is developed. The starting current can be limited by the use of auto-transformer.

1.21 Synchronous Motors.

Synchronous motors can have only two speeds-synchronous speed or zero speed. Since it takes no magnetising current from a.c. side as in case of induction motor ; its power factor can be made unity or leading by over exciting the field. This enables the motor to perform phase compensation action in addition to driving the load. Torque developed by synchronous motor varies directly as voltage. Unlike in Induction motor where it varies as square of the applied voltage. Therefore synchronous motors are better suited to withstand large voltage variations. In case of induction motor, in order to keep down the magnetizing current and therefore lagging component of stator current, air gap is kept small. This is not in the case of synchronous motor. Vector diagram of synchronous motor on active and reactive power axes is shown in Fig. 1.49 (*a*). Therefrom it will be observed that by making generated voltage E or excitation of motor more and more, power or load torque develop to pull it out becomes large. Synchronous motor can, therefore, be made to take large over loads. Synchronous motor is a motor, having constant speed, capable of improving p.f., capable of stable operation inspite of

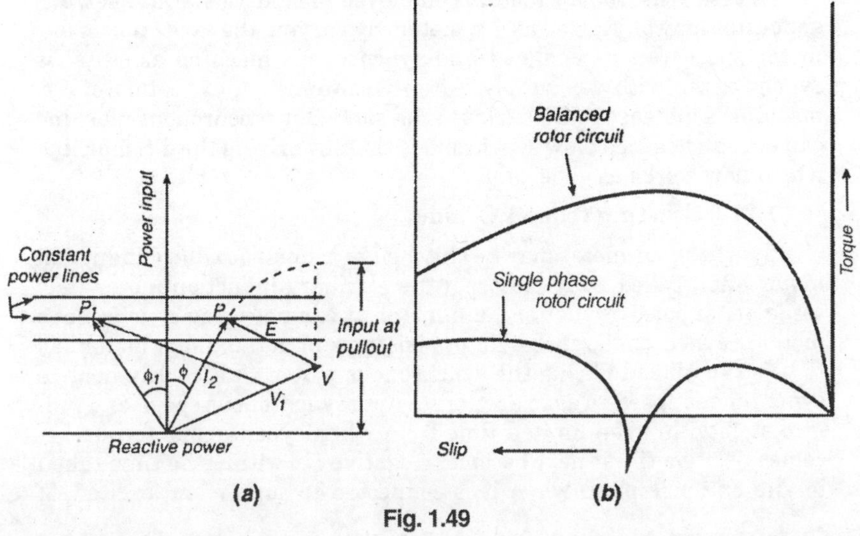

Fig. 1.49

voltage variations, capable of taking large overloads and having high operating efficiency.

Uses. These motors are used where constant speed operation is not a disadvantage, such as for fans, blowers, dc generators, line shafting, compressors, p.f. correction, frequency changer drive and voltage stabilizer.

Voltage stabilizing action of synchronous motor is explained as follows. Synchronous motor is installed at the receiving end of a transmission line. Due to line drawing more and more lagging load, voltage received at substation will decrease. This decrease in applied voltage from V to V_1 which will shift the operating point on constant power line from P to P_1, the motor will now take power at leading p.f. This will relieve the line of carrying lagging component of current and voltage drop in the line will reduce. This therefore, helps in maintaining the voltage at substation constant.

1.21.1 Starting of Synchronous Motor.

All the advantages of synchronous motor are equally balanced by the drawback of being non-self starting. Motor has, therefore to be started and brought up to synchronous speed by some external means before it is capable of operating with load angle, δ. Starting methods may use d.c. or a.c. supply.

1.21.2 Starting from D.C. Supply.

If exciter of synchronous motor is on the same shaft and d.c. supply is available, exciter can be worked as motor and synchronous machine synchronized to a.c. supply as an ordinary alternator and then exciter is switched off from d.c. supply.

In case synchronous motor is employed in motor generator set, d.c. generator may be started as d.c. motor. By varying the excitation of d.c. motor speed can be changed and synchronous machine as usual is synchronized with a.c. supply. After synchronizing, excitation of d.c. machine is increased which tries to run slow. But synchronous machine can not run at speed blow synchronous thereby driving the d.c. machine which now works as generator.

1.21.3 Starting from A.C. side.

Synchronous motor may be clutched to a small auxiliary induction motor often called pony motor. If the auxiliary induction motor has same no. of poles as that of main motor, it can not bring synchronous motor to synchronous speed. Rotor resistance of the auxiliary motor has to be so small as to bring the synchronous motor. Sufficiently near to synchronous speed so as to accelerate upto synchronous speed and pull into step within the limited time from when synchronous torque becomes positive till, it next becomes negative *i.e.* within the time taken to slip one pole pitch when it is connected to supply and excited. If auxiliary induction motor has one pole pair less than main synchronous motor, the set can be taken above synchronous speed and synchronized after disconnecting the auxiliary motor from supply in the same manner as alternator is synchronized.

Damper winding embedded in the pole faces may be used to start the motor as squirrel cage induction motor. In that case some type of starter will be required for induction motor. At starting rotating magnetic field has maximum speed with respect to field winding. If the d.c. field winding circuit is closed, it will reduce the effective rotor resistance of induction motor which will result in low starting torque. If on the other hand it is left open, voltage induced in it at stand still will be very high. Field circuit is therefore closed through resistance during acceleration of the machine as induction motor. Unless starting torque required is high, damper winding should have low resistance. This enables the motor to come very close to synchronous speed and also damping action during its synchronous motor action is very effective. When full induction motor speed is reached, it may be possible that motor may pull into synchronism due to reluctance motor action even before d.c. excitation is applied. This may result in locking up of rotor poles with wrong "would be polarity" so that when d.c. excitation is actually given, slipping of poles through one pole pitch may occur. This state of affairs should be avoided as it causes hunting and consequent power surges. This is achieved by applying weak d.c. excitation before full induction motor speed is reached so that locking of poles should be with right polarity. Another difficulty with this method is that the induction motor secondary winding on the rotor does not form a balanced three phase secondary winding of induction motor. An extreme case of rotor unbalanced circuit is a single phase rotor. Speed torque curve of balanced rotor and single phase rotor is shown in Fig. 1.49 (*b*). As rotor unbalance is reduced dip in speed torque characteristic in vicinity of half speed will become smaller and curve will approach more and more to that of the balanced rotor induction motor. The load torque, in no case should be more than the lowest dip torque, otherwise motor will operate at half the synchronous speed and can not bring the speed of motor to synchronous value.

Where high value of synchronous speed is required, 3-ϕ pole face winding is brought out to three extra slip rings and motor started by inserting resistance in the rotor circuit. There will be five slip rings in that machine two for d.c. excitation and three for induction motor secondary.

1.21.4. Self Synchronizing method of starting of Synchronous Motor. Auxiliary motor used is 3-ϕ star connected squirrel cage induction motor. Supply to the stator of this motor is given through the stator of main synchronous motor as shown in Fig. 1.50 which can be regarded as reactors in series with supply. When motor picks up speed, star point connection of the main motor can be closed by circuit breaker thereby shorting out the auxiliary motor.

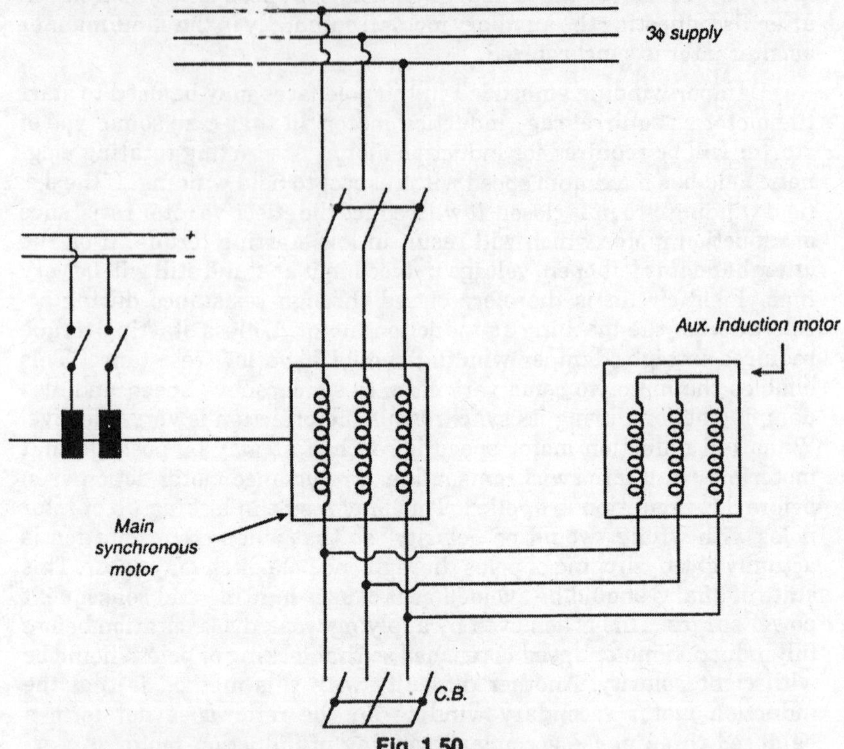

Fig. 1.50

1.22 Single Phase Induction Motor

As already discussed single phase induction motors are not self-starting and some external means are used for starting it. The various methods for producing starting torque are given below :

1. **Split Phase Method.** In this method an additional winding known as auxiliary winding having low inductive reactance but connected in series with high resistance R is used in parallel to the main winding but placed 90° from this winding in space as shown in Fig. 1.51 (*a*). The main winding is a low resistance and high inductance winding. As soon as both these windings which form stator are connected across the single-phase supply, current start flowing. Auxiliary winding being high resistance has the current I_s flowing through it which is almost in phase with applied voltage, whereas current I_m flowing through main winding which is highly inductive lags behind the voltage by approximately 90°. Thus current I_s and I_m differ in phase by approximately 90° as shown in Fig. 1.51 (*b*). They produce a rotating magnetic field as a result torque is developed in the rotor and it starts running. As soon as the motor picks up about 75 to 80% of rated speed, auxiliary

winding gets disconnected by means of centrifugal switch and the motor starts running as single-phase induction motor without auxiliary winding. Auxiliary winding being highly resistive ought to be disconnected from the circuit failing which lot amount of losses would continue taking place. The split phase motor runs at almost constant speed even at varying load but where moderate torque is required and is as such used in fans, blowers, duplicating machines, refrigerators etc. They are available in sizes from 1/20 to 1/30 H.P. with speeds of 3450 to 1725 R.P.M., 1140 R.P.M. and 865 R.P.M.

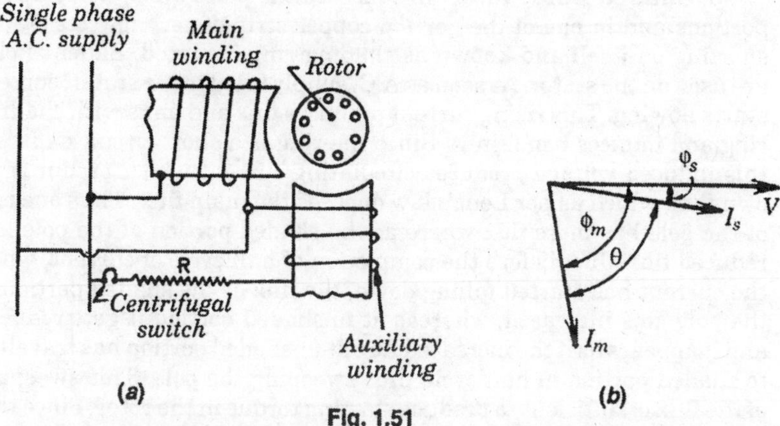

Fig. 1.51

2. **Capacitor starting.** This type of starting differs from the previous method in a way that a capacitor is connected in series with auxiliary winding instead of resistance as discussed above and shown in Fig. 1.52 (*a*).

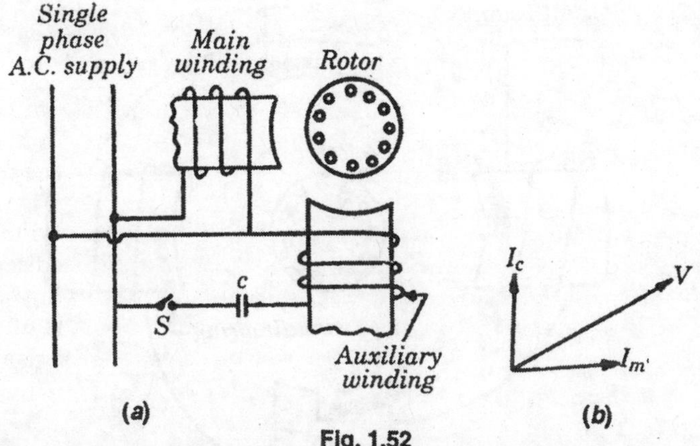

Fig. 1.52

Main winding as such is highly inductive whereas auxiliary winding has low inductance. Current I_m flowing in the main winding lags behind voltage V, whereas current, I_c through auxiliary winding leads the voltage V by some angle. The capacitance C can be adjusted so that, current I_c and I_m differ in phase by 90° as shown in Fig. 1.52 (*b*) thus producing rotating magnetic field which gives the starting torque to the motor. The motor starts running but as soon as it picks up speed 75 to 80% of rated speed, auxiliary winding is disconnected from supply side. Since the two fields can be made exactly 90° apart, so high starting torque is produced. Capacitor can be made to remain in the circuit if so desired as it shall improve the power factor.

3. **Shaded Pole Motor.** In this case the poles are split up in two portions and in one of the portion copper strip of very low resistance shorting on itself and known as shading ring is placed. Salient Poles are used on the stator. As soon as A.C. supply is fed to the stator, current starts flowing. This rising current produces flux and links with shading ring and induces e.m.f. in it. Since the ring is a short circuit on itself, this induced voltage produces circulating current which produces its own flux, which as per Lenz's law opposes the main flux. Thus one end of the pole has more flux where as the shaded portion of the pole has reduced flux. Just before the completion of half cycle of current, when the current has started falling down, the flux at the shaded portion of the pole gets increased, whereas at unshaded portion it gets reduced and it appears as if the increased flux at unshaded portion has travelled to shaded portion in half cycle thus sweeping the pole. This sweeping of the flux is sufficient to produce starting torque in the rotor. Since this method of starting produce very small amount of torque, hence this method is used where starting torque required is less such as hair dryers and small fans.

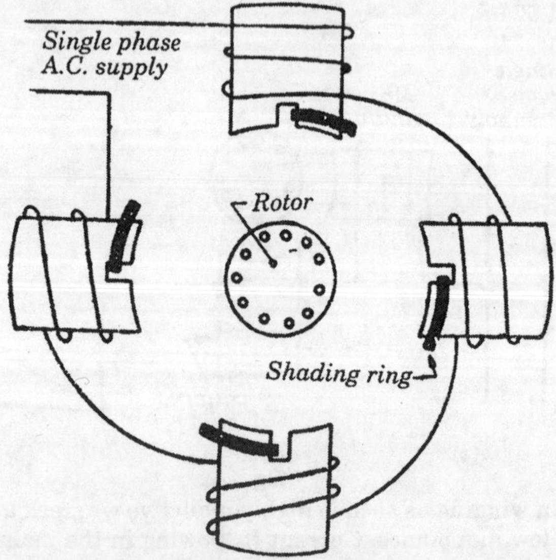

Fig. 1.53 Shaded Pole Motor.

4. **Capacitor start capacitor run method.** As the name indicates and as shown in the figure, the motor is started by capacitor C_S and then it is cut off by centrifugal switch or transfer switch. Capacitor C_R remains always in circuit. This method gives high starting torque and high power factor.

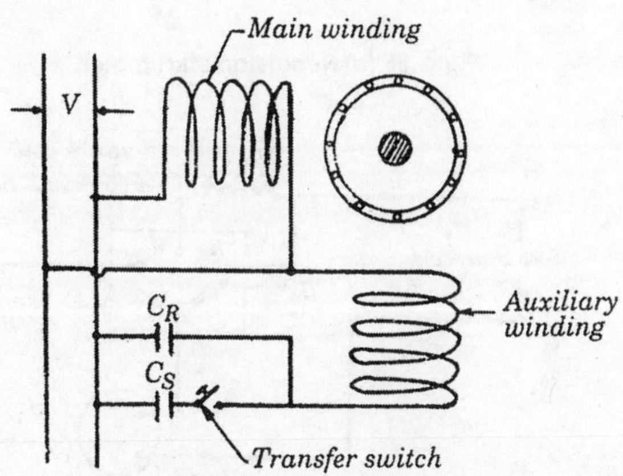

Fig. 1.54. Capacitor start capacitor run motor.

These motors are used for constant speeds under varying loads such as blowers, drilling machines, refrigerators and grinders etc.

5. **Repulsion start method.** In this method, the rotor has in addition to the squirrel cage winding, a repulsion winding similar to a d.c. motor armature and commutator with a centrifugal device which short circuits all the commutator segments and raises the brushes as the motor attains nearly synchronous speed.

This type of motor has high starting torque and runs as an induction motor having constant speed characteristics. The direction of the motor can also be reversed by changing the position of the brushes. This method is suitable for compressors and refrigerators.

In all the above types of motors, the speed control can be had for the drives within the range ± 20%. The nature of speed control requirement for an industrial drive depends upon its type. The speed is regulated with the insertion or removal of line reactance, by adjusting the voltage, or by unique winding interconnections.

Where a motor must be started or stopped frequently, it is sometimes desirable to bring it and its load to rest very quickly. This often necessitates a positively timed 'dynamic braking' involving the use of d.c. source.

The Figure 1.55 (*b*) shows the speed control of a shaded pole motor by a tapped reactance coil connected in series with the motor to obtain three speeds on the single phase motor.

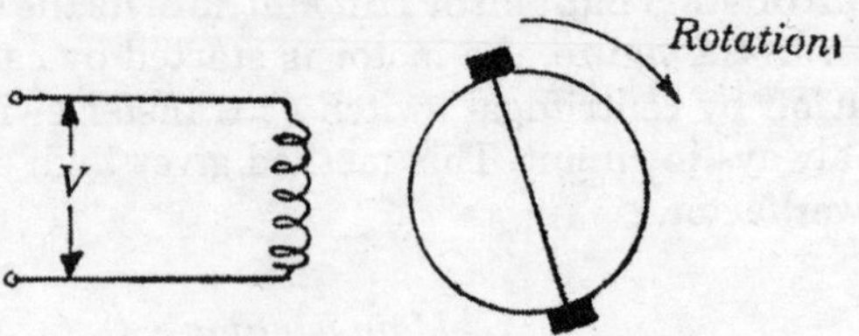

Fig. 1.55 (*a*) Repulsion start motor.

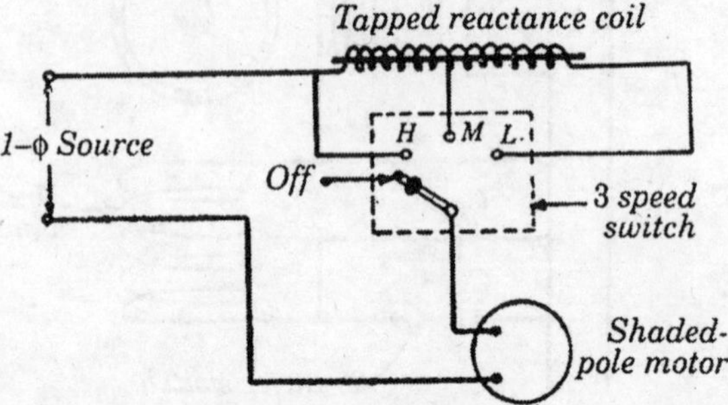

Fig. 1.55 (*b*) Tapped reactance-coil method of controlling the speed of a shaded-pole motor.

A tapped auto transformer may be connected in series with the motor to obtain many speeds. Even on the winding may be tapped to have many speeds on the motor. In case of capacitor start and capacitor run motors the winding interconnections may be in L shaped or T-shaped.

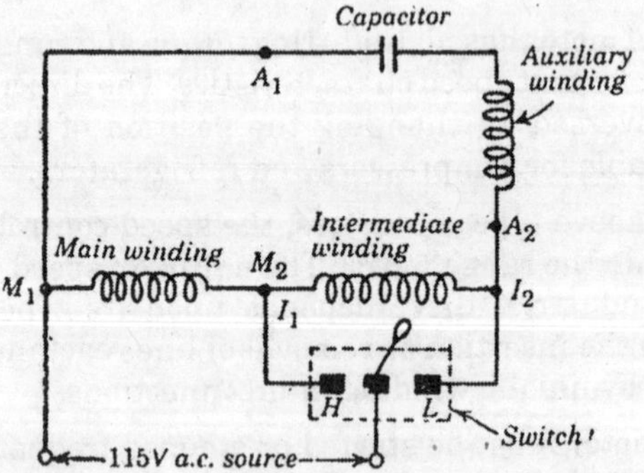

Fig. 1.56 (*a*) L-connection of a tapped winding capacitor run motor to control speed.

1.23 Effect of starting on Power Supply, motor and load

While studying starting of electric drive system, it is necessary to consider three factors (*i*) effect of starting upon the Power Supply, (*ii*)

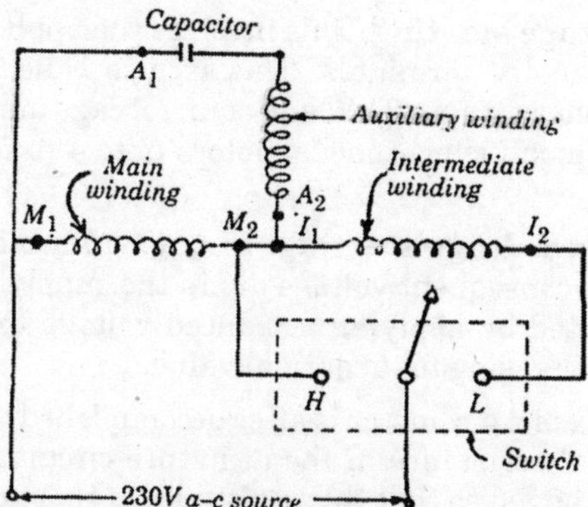

Fig. 1.56 (*b*) T-connection.

effect of starting upon the driving motor itself and (*iii*) effect of starting upon the connected mechanical load.

The supply network to which the motor is connected may affect the selection of the starting device from the following view point. The excessive voltage drop due to the peak starting current may interfere with the supply in such a way that it can not be tolerated by other equipment or other consumers connected to the same power supply network.

Since starting is associated with excessive currents, the effect of starting upon the motor itself must be carefully considered. The starting currents will add to the motor heating by an amount that depends upon their R.m.s. values and upon the frequency of starting. In a d.c. motor the limitation may be good commutation rather than heating, as dc machines have a certain maximum limit for the current dictated by the commutation process.

The equipment connected to the driving motor may impose strict constraints upon the type of accelerating cycle and upon the maximum permissible acceleration. Discomfort to passengers in lifts and trains may very well set an upper limit to the rate of acceleration. Cranes, excavators and similar material handling equipment must be so equipped that the opèrator has precise control over the load, whether it be light or heavy. In case of loads which exhibit friction the motor must develop sufficient torque to start the rotation, after which the torque must perhaps be reduced for limiting the acceleration to a safe value.

1.24 Methods of Starting Electric Motors

The different methods of starting of the various types of electric motors would have been dealt with in detail in a course on "Rotating electrical machines" and hence, they have been just listed here for the sake of continuity of thought.

1. **Full voltage starting.** This involves the application of full line voltage to the motor terminals. This is also called "direct on line starting". D.C. motors upto 2 kW and squirrel cage induction motors as well as certain small synchronous motors upto 4 to 5 kW are usually line started.

2. **Reduced voltage starting.** In order to avoid heavy starting current and the consequent voltage dip in the supply lines majority of motors are started by applying a reduced voltage to their terminals subsequently increasing to its normal value.

The starting of a d.c. motor is after accomplished by the addition of suitable external resistance in the armature circuit and the starting controller is arranged so that this resistance is short-circuited in steps as the motor comes upto speed.

Reduced voltage starting of induction motors is achieved by (*i*) Stator resistance starting, (*ii*) Stator reactor starting, (*iii*) Star-delta starting and (*iii*) auto transformer starting.

The above methods apply equally to the synchronous motors, with reduced voltage starting the transition to full voltage may be made either before or after synchronization although the former method is usually preferred.

It may be noted that the torque at starting, in all the above methods gets reduced.

3. **Increased torque starting.** With a wound rotor induction motor, resistance can be added in rotor circuit so as to decrease the starting current while increasing the starting torque, even upto the value of maximum torque that can be developed by motor.

4. **Starting by means of smooth variation of voltage or frequency :** With a.c. motor-dc generator set, dc motors can be started by smooth variation of applied voltage and with variable frequency sources both induction and synchronous motors can be started by smooth variation of supply frequency simultaneously varying proportionally the applied voltage to the motors.

1.24.1. Acceleration Time.

The time required to attain a change in speed from ω_1 to ω_2 may be determined from the equation of motion,

$T_M = T_L + j\,\dfrac{d\omega}{dt}$ where both T_M and T_L are functions of ω.

i.e. $$T_M - T_L = J\,.\,\frac{d\omega}{dt}$$

i.e. $$dt = J\,.\,\frac{d\omega}{T_M - T_L} \text{ therefore } T = J\int_{\omega_1}^{\omega_2} \frac{d\omega}{T_M - T_L} \quad ...(1)$$

In the above equation the inertia J is assumed to be constant. In case J varies with time or speed the equation can be rewritten suitably. For the case, where ω_1 is zero at 't' equal to zero and ω_2 is equal to a steady state speed ω_0 attained by the drive ; the accelerating time is given by,

$$t = J\int_0^{\omega_0} \frac{1}{T_M - T_L} \, . \, d\omega \qquad \text{...(2)}$$

The value of the integral in equation (2) is the area under the curve relating $\left(\frac{1}{T_M - T_L}\right)$ to ω. For a typical motor load combination, whose speed torque characteristics are shown in Fig. 1.57 (a) the reciprocal of the accelerating torque $1/(T_M - T_L)$ plotted as a function of speed is given in Fig. 1.57 (b). The shaded area is the value of the integral in

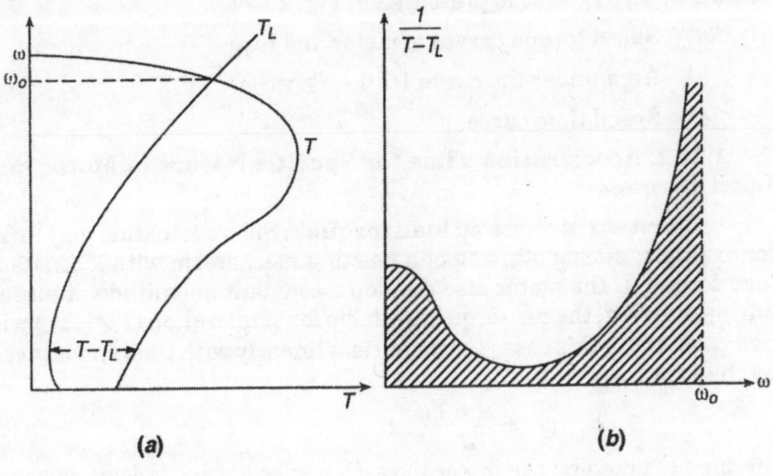

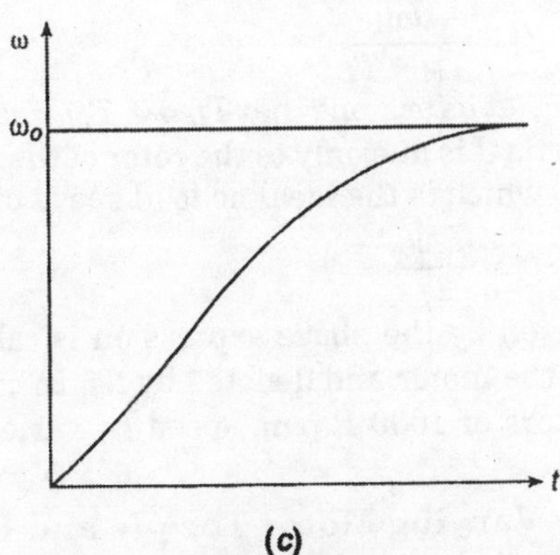

Fig. 1.57

question and when multiplied by total inertia J gives the accelerating time. It should be noted that the reciprocal of $(T_M - T_L)$ would become infinitely large as the speed approaches its steady state value and this would lead to an infinite accelerating time. This difficulty is removed in practice by computing the time required to reach say 95 to 98 percent of the final steady state speed. The desired value of accelerating time for any one of the methods studied earlier.

It should also be observed that the speed versus time curve can be constructed from the relationship between the reciprocal of the accelerating torque and speed. The time required to reach any speed ω_x is given by the expression

$$t_x = J \int_0^{\omega_x} \frac{1}{(T_M - T_L)} \, . \, d\omega \qquad \text{...(3)}$$

The resulting motor speed-time curve for the drive system characterised in Fig. 1.57 (*a*) is presented in Fig. 1.57 (*c*).

(*a*) Speed torque curves of motor and load

(*b*) Area under the curve $1/T_M - T_L$ vs. ω

(*c*) Speed time curve.

1.24.2. Acceleration Time for Specific Nature of Motor and Load Torques

1. **Constant motor and load torques :** Such a situation may arise for example, during starting of a hoisting mechanism with a constant load torque. If the motor also develop a constant magnitude of torque during starting, the net torque available for acceleration $(T_M - T_L)$ will be a constant in this case ; *i.e.* speed rises linearly with time and is given by the expression.

$$\omega = \frac{T_M - T_L}{J} \, . \, t \qquad \text{...(4)}$$

the time t_1 taken by the motor to reach a speed of ω_1, rad/sec. is given by,

$$t_1 = \frac{J\omega_1}{T_M - T_L} \qquad \text{...(5)}$$

During starting let us assume that $T_L = 0$, T_M = rated torque T_f and the moment of inertia J is due only to the rotor of the motor. The motor reaches a speed ω_o which is the ideal no load speed of the motor. Then

$$t_1 = \frac{J\,\omega_0}{T_f} \qquad \text{...(6)}$$

The time t_1 given by the above expression is called **mechanical time constant** of the motor and denoted by t_m. In case of d.c. motors and induction motors of 1000 R.p.m. speed t_m varies from 0.4 to 0.6 seconds.

(*ii*) **Linearly Varying Motor Torque and Constant Load Torque :** As explained earlier, a linearly varying motor torque, druing starting, can be exhibited as a d.c. series motor with shunted armature or shunt motor connection or a slipring induction motor whose rotor circuit resistance is quite high. At $\omega = 0$ the motor torque corresponds to the starting torque T_{st} and at $\omega = \omega_s$, the synchronous speed of the induction motor $T_M = 0$ Fig. 1.58 shows the relationship between motor torque and slip.

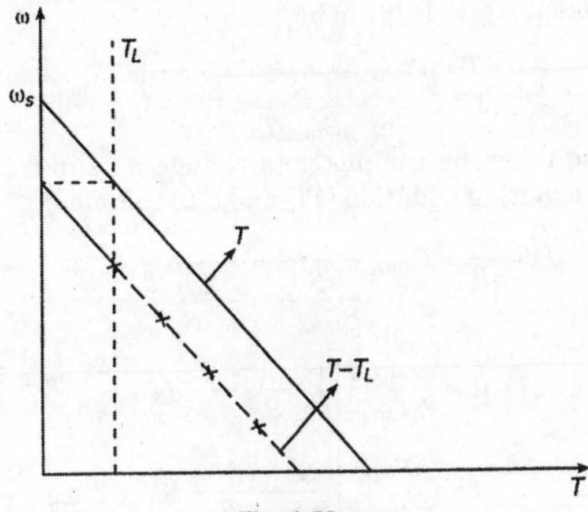

Fig. 1.58

Linear motor speed -Torque and constant load curves.

$$\frac{T_{st}}{1} = \frac{T_M}{T_S} \text{ or } S = \frac{T_m}{T_{st}} \qquad ...(7)$$

The accelerating torque, $J\dfrac{d\omega}{dt} = T_M - T_L$ ...(8)

Using equation (7) and (8) we have,

$$-\omega_s \frac{ds}{dt} = \frac{s\,T_{st} - T_L}{J}$$

Since $\omega = \omega_s(1 - s)$ and hence $d\omega = -\omega_s\, ds$

Therefore $\dfrac{ds}{dt} + \dfrac{T_{st}}{J\omega_s} S - \dfrac{T_L}{J\,\omega_s} = 0$ ...(9)

The quantity $\dfrac{J\omega_s}{T_{st}}$ must have a dimension of seconds and is termed as the starting time constant, t_{st} at the drive. Here J represents the total moment of inertia of the drive system, inducing the inertia at any external rotating masses other than the rotor of the driving motor equation (9) then becomes

$$\frac{ds}{dt} + \frac{1}{t_{st}} . S - \frac{1}{t_{st}} . \frac{T_L}{T_{st}} = 0 \qquad ...(10)$$

1.24.3 Optimum Value of S_{max} for minimum Accelerating Time : If load and friction torques were negligible the equation of motion could be written as,

$$T_M = J\frac{d\omega}{dt} = J \cdot \frac{d}{dt}[\omega_s(1-s)]$$

$$= -\omega_s J \cdot \frac{ds}{dt} \quad ...(11)$$

Neglecting stator resistance, the electromagnetic torque developed by the induction motor is given by,

$$T = T_{max} \cdot \frac{2}{\frac{S}{S_{max}T} + \frac{S_{max}T}{S}} \quad ...(12)$$

The time taken by the motor to change its slip from S_1 to S_2 is obtained by equating equation (11) and (12) therefore,

$$-J\omega_s\frac{ds}{dt} = T_{max}\frac{2}{\frac{S}{S_{max}T} + \frac{S_{max}T}{S}}$$

i.e.

$$\int_o^t dt = \frac{-J\omega_s}{2T_{max}}\left[\int_{S_1}^{S_2}\frac{S}{S_{max}T}ds + \int_{S_1}^{S_2}\frac{S_{max}T}{S}\cdot ds\right]$$

$$t = \frac{J\omega_s}{2T_{max}}\left[\frac{S_1^2 - S_2^2}{2S_{max}T} + S_{max}T\log_e\frac{S_1}{S_2}\right] \quad ...(13)$$

It is clear from the above equation that the time required for the slip to change from S_1 to S_2 is a function of the slip at which the maximum torque of the motor occurs. The value of $S_{max\,T}$ that would give minimum time to change the slip of the motor from S_1 to S_2 can be obtained by differentiating equation (13) w.r.t. $S_{max\,T}$ and equating the result to zero.

$$(S_{max}T)_{OPT} = \left(\frac{S_1^2 - S_2^2}{2\log_e(S_1/S_2)}\right)^{1/2} \quad ...(14)$$

The minimum time can be determined by substituting equation (14) in equation (13). The optimum value of $S_{max\,T}$ to accelerate the motor from $S_1 = 1$ (stand still condition) to any arbitrary slip, S in the minimum time is given by :

$$(S_{max}T)_{OPT} = \left[\frac{1-S^2}{2\log_e(1/S)}\right]^{1/2} \quad ...(15)$$

The optimum value of rotor resistance to change the slip of motor from S_1 to S_2 in minimum time can be obtained from equation (14) and the relation,

$$\frac{R_2}{S_{max}T} = [X_1 + X_2]$$

Hence $[R_2]_{OPT} = (X_1 + X_2)\,(S_{\max T})_{OPT}$

$$= [X_1 + X_2]\left[\frac{S_1^2 - S_2^2}{2 \log_e S_1/S_2}\right] \quad ...(16)$$

1.25 Energy Relations During Starting

Certain useful expressions regarding the energy losses in a motor during periods of changing speed like starting, braking and sudden changes in applied voltage, load etc. can be obtained as described below:

(*i*) **D.C. Shunt Motor.** An expression of the form given below may be obtained for the energy lost in the armature of a d.c. shunt rotor.

The voltage equation of d.c. motor is,

$$V = E + I_a R_a \text{ or } I_a R_a = V - E$$

For shunt motor

$$I_a R_a = V - K\omega. \quad ...(17)$$

the above equation (17) is valid for a separately excited motor also. The equation of motion on no load will be,

$$T_M = KI_a = J\frac{d\omega}{dt} \quad ...(18)$$

Multiplying equation (17) by equation (18), we get

$$I_a^2 R_a = \frac{J.V}{K}\frac{d\omega}{dt} - J\,\omega\frac{d\omega}{dt} \quad ...(19)$$

On no load the $I_a R_a$ in the motor will be negligibly small and hence $V = K\omega_o$ where ω_o is the no load speed of shunt motor.

Substituting $\frac{V}{K} = \omega_o$ in equation (19) we get,

$$I_a^2 R_a \,.\, dt = J\omega_o \,.\, d\omega - J_\omega \,.\, d\omega \quad ...(20)$$

If the motor speed were to change from ω_1 to ω_2 from time t_1 to t_2 energy dissipated in the armature circuit is given by,

$$W = \int_{t_1}^{t_2} I_a^2 R_a\, dt = J\,\omega_o \int_{\omega_1}^{\omega_2} d\omega - J\int_{\omega_1}^{\omega_2} \omega \,.\, d\omega$$

$$= J\,\omega_o\,(\omega_2 - \omega_1) - \frac{J}{2}(\omega_2^2 - \omega_1^2)$$

Hence energy lost during starting, when the motor changes its speed from zero to no load speed ωo will be,

$$W_{st} = J\omega_2^2 - \frac{J}{2}\,\omega_2^2 = \frac{J\,\omega_0^2}{2} \text{ Joules.}$$

Thus energy lost in armature of a d.c. motor during starting on no-load will be equal to the kinetic energy absorbed by the armature, while accelerating from stand still to no-load speed. Further this energy is independent of the armature circuit resistance.

If the motor were started with a constant load torque T_L and energy lost during starting could be determined as follows :

The equation of motion will now be of the form

$$T_M = KI_a = T_L + J\frac{d\omega}{dt} \qquad ...(22)$$

Multiplying equation (17) and by equation (22) we get,

$$I_a^2 R_a = J\frac{V}{K} \cdot \frac{d\omega}{dt} - J \cdot \omega \cdot \frac{d\omega}{dt} + \frac{V}{K} T_L - T_L \cdot \omega \qquad ...(23)$$

If the motor speed were to change from ω_1 to ω_2 the energy lost would be given by,

$$W = \int_{t_1}^{t_2} I_a^2 R_a \, dt$$

$$= J \cdot \frac{V}{K}\int_{\omega_1}^{\omega_2} d\omega - J\int_{\omega_1}^{\omega_2} \omega \, d\,\omega + \frac{V}{K} T_L \int_{t_1}^{t_2} dt - T_L \int_{t_1}^{t_2} \omega\,(t)\, dt$$

$$= J \cdot \frac{V}{K}(\omega_2 - \omega_1) - \frac{J}{2}(\omega_2^2 - \omega_1^2) + \frac{V}{K} T_L (t_2 - t_1) - T_L \int_{t_1}^{t_2} \omega\,(t)\,\omega\, dt \qquad ...(24)$$

V/K in the above equation can be replaced by ω_o the no load speed. Therefore, the energy lost during starting on load, when the speed changes from zero to ω_r will be,

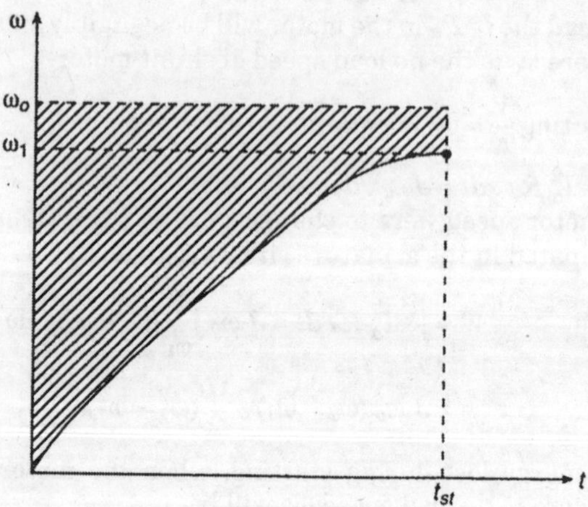

Fig. 1.59 Speed-time curve during starting of a d.c. shunt motor.

$$W_{st} = J\,\omega_o\,\omega_r - \frac{J}{2}\omega_r^2 + T_L\,\omega_o\,t_{st} - T_L\int_o^t \omega\,(t)\,dt$$

$$= J\left(\omega_o\,\omega_r - \frac{\omega_r^2}{2}\right) + T_L\left(\omega_o\,t_{st} - \int_o^t \omega\,(t)\,dt\right) \qquad ...(25)$$

where t_{st} is the accelerating time.

The second term in equation (25) shows that the energy lost during starting depends both on the accelerating time and on the nature of speed variation with time during acceleration. In fact the expression

$(\omega_o t_{st} - \int_o^t \omega(t)\, dt)$ can be determined as the shaded area shown in Fig. 1.59.

(*ii*) **D.C. Series Motor.** Since a d.c. series motor should not be started on no-load, let us assume a constant load torque T_L to be present, while determining the energy loss during starting.

The equation of motion will be,

$$T_M = K I_a^2 = J \frac{d\omega}{dt} + T_L \quad ...(26)$$

Hence the total energy dissipated in the armature circuit resistance is given by the following expression,

$$W = \int_{t_1}^{t_2} I_a^2 R_a \, dt = J \frac{R_a}{K} \int_{\omega_1}^{\omega_2} d\omega + \frac{T_L}{K} R_a \int_{t_1}^{t_2} dt$$

$$= \frac{J R_a}{K} (\omega_2 - \omega_1) + \frac{T_L R_a}{K} (t_2 - t_1) \quad ...(27)$$

Thus the energy dissipated while the motor speed changes from ω_1 to ω_2 is proportional to R_a, the armature circuit resistance, rather than independent of this quantity as in the case of dc shunt motor.

(*iii*) **Three Phase Induction Motor**

Let us consider, now a 3-ϕ induction motor. The torque developed is given by the expression,

$$T = \frac{3}{\omega_s} I_2^2 \frac{R_2}{S} \quad ...(28)$$

Neglecting load and friction torques in the motor, we have

$$T = J \frac{d\omega}{dt} = -J \omega_s \frac{ds}{dt} \text{ since } \omega = \omega_s (1 - s) \quad ...(29)$$

The total energy loss in the rotor resistance of induction motor when its slip changes from S_1 to S_2 is given as,

$$W = 3 \int_{t_1}^{t_2} I_2^2 R_2 \, dt \quad ...(30)$$

where t_1 and t_2 are the instants corresponding to S_1 and S_2 respectively.

Equating equation (28) and (29) we have,

$$3 I_2^2 R_2 = -J\omega_s^2 . S . \frac{ds}{dt}$$

Substituting this in equation (30) we get,

$$W = -J\omega_s^2 \int_{S_1}^{S_2} S . ds$$

$$= \frac{1}{2} J \omega_s^2 (S_2^2 - S_1^2) \text{ Joules} \qquad ...(31)$$

During starting the slip of induction motor changes from 1 to 0 and hence the energy lost in the rotor is given as,

$$W_{st} = \frac{1}{2} J \,.\, \omega_s^2 \text{ Joules} \qquad ...(32)$$

It may be noted that the energy lost in the rotor circuit does not depend on the rotor circuit resistance, but only on the moment of inertia of the rotating masses and the initial and final speeds. From equation (32), it can be seen that the energy lost in the rotor circuit, during starting, is equal to the kinetic energy of the rotor at its final speed.

The above observation that the total energy loss in the rotor circuit during a change in speed is independent of the magnitude of the rotor resistance may be explained as follows. An increase in rotor resistance will reduce the magnitude of current at a given speed and hence reduce the magnitude of accelerating torque. This, in turn, will mean that the time taken to accelerate will be lengthened.

The fact that the total energy loss in the rotor circuit does not depend on the rotor circuit resistance doesn't imply that the total energy loss in the motor is independent of rotor circuit resistance. Neglecting the magnetizing current, both the stator current and rotor current referred to stator will be equal and therefore,

$$\frac{\text{Stator copper losses}}{\text{Rotor copper losses}} = \frac{R_1}{R_2}$$

Hence, although the energy lost in the rotor circuit is independent of R_2, the energy lost in the stator will be decreased by an increase in R_2. So that the total energy lost in the motor can be decreased by an increase in R_2.

An expression for the total energy lost in the motor during a change in speed can be easily obtained by modifying eqn. (30) as

$$W_m = 3 \int_{t_1}^{t_2} I_2^2 (R_1 + R_2)\, dt \qquad ...(33)$$

Since $I_2 \approx I_1$

going through the same steps as from Eqn. (28) to Eqn. (31) we obtain

$$W_m = \frac{1}{2} J \omega_s^2 \left(1 + \frac{R_1}{R_2}\right) (S_1^2 - S_2^2) \qquad ...(34)$$

During starting, the energy lost in the motor will be,

$$(W_m)_{st} = \frac{1}{2} J \omega_s^2 \left(1 + \frac{R_1}{R_2}\right) \qquad ...(35)$$

It is obvious that the first term in the right hand side of the above equations represents the energy lost in the rotor circuit, which is equal to the kinetic energy of the rotor and the second term denotes the energy lost in the stator.

The energy relations derived, so far, were under the assumption that the load torque on the motor was absent. Let us now, consider the presence of a load torque T_L. Under this condition,

$$T_M - T_L = J\frac{d\omega}{dt}$$

i.e. $$dt = J\frac{d\omega}{T_M - T_L} \qquad ...(36)$$

Also $$d\omega = -\omega_s \,.\, ds \qquad ...(37)$$

Energy lost in the rotor circuit, $W = 3\int_{t_1}^{t_2} I_2^2 R_2 \, dt.$

$$= \omega_s \int S\, T_M \, dt$$

Substituting equation (36) and (37) in equation (38) we get,

$$W = -J\,\omega^2 \int_{S_1}^{S_2} \frac{T_M}{T_M - T_L} J \,.\, ds$$

$$= -J\,\omega_s^2 \int_{S_1}^{S_2} \left[1 + \frac{T_L}{T_M - T_L}\right] J\, ds \qquad ...(39)$$

Comparing Equations (32) and (39) it can be seen that the energy lost in rotor during starting of an induction motor on load will be greater than that on no load. In the above equation both T_M and T_L are functions of speed. If these functional relations are known, it is possible to integrate the expression in equation (39) analytically or graphically in order to determine the energy loss.

It may be recalled that equations (32) and (39) were drived irrespective of the nature of the speed-torque characteristics of the motor and load as well as the method of starting the motor, while the energy loss on rotor during starting of the motor on no load is independent of the method by which it is started. The energy loss during starting on load will be minimum with direct on line start, since T_M the torque developed by the motor will be higher in magnitude at rated voltage. During reduced voltage starting ($V < V_{rated}$) the energy loss increases, inspite of the reduction in starting current. This can be clearly seen from equation (39) if we substitute the new motor developed torque T_M instead of T_M' bearing in mind that,

$$T_M' = \left(\frac{V}{V_{rated}}\right)^2 \,.\quad T_M' < T_M$$

The entire energy loss during starting is dissipated as heat in the rotor and stator winding of squirrel cage motors and mostly in external resistances introduced in the rotor circuit of slip ring induction motors. Larger the magnitude of this external resistance, in relation to both the rotor and stator resistance, the cooler the motor will operate. But, as observed earlier an increased resistance in the rotor circuit will increase the starting time.

1.26 Methods to Reduce the Energy loss During Starting.

The following methods are commonly used to reduce the loss in energy during starting of motors :

(*i*) Reducing the moment of inertia of the rotor

(*ii*) Starting of d.c. shunt motors by smooth adjustment of applied voltage

(*iii*) Starting of multi-speed induction motors in discrete steps of speed and

(*iv*) Starting of induction motors by smooth variation of supply frequency (*V/f* control)

(*i*) **Reducing the Moment of Inertia of Rotor**

It is obvious from equations (21), (25), (27), (32) and (39) that the total energy loss in motors during transient operations can be reduced by reducing the moment of inertia of the drive system. In order to achieve such a reduction, a single motor of certain power rating can be replaced by two motors of one half of the rating. As a consequence of such a change, motors with a smaller diameter and hene reduced moment of inertia will be employed.

Yet another method is to use specially designed motors having large axial length for the same power rating and speed of rotation, these motors will have smaller diameter than the general purpose motors.

(*ii*) **Starting of D.C. Shunt Motors by Smooth Variation of applied voltage.** This method necessitates the presence of a variable direct voltage source. Smooth adjustment of applied voltage is equivalent to applying the voltage in a large number of small voltage steps. Let us assume, for easiness sake, that the motor is on no load and develops a constant torque so as to ensure uniform acceleration of the rotor. Starting in steps assumes the presence of a number of discrete steps in speed like $\omega_{01}, \omega_{02}, \omega_{03}$... and the starting process to take place from 0 to ω_{01}, from ω_{01} to ω_{02} and from ω_{02} to ω_{03} and so on upto the no load speed of rotor ω_0. Since the speed of d.c. shunt motor is approximately directly proportional to applied voltage, the steps in speed will be of equal magnitude, if the voltage is varied in equal steps. Therefore the loss in energy during starting with m equal steps of voltage can be expressed as,

$$W_{st} = m\left[\frac{1}{2}J\left(\frac{\omega_0}{m}\right)^2\right] = \left[\frac{1}{2}J\,\omega_0^2\right]\frac{1}{m} \qquad ...(40)$$

from equation (40) it is obvious that larger the steps in voltage, less will be the energy loss during starting.

(*iii*) **Starting of Multispeed Induction Motors in Discrete steps.** Pole changing motors of two or more ratio of speeds can be started in discrete steps of speed. Since there is no change in the expression for energy loss during starting of an induction motor from that of a d.c. shunt motor, it can be easily argued that this method of starting of multispeed induction motor will also involve only less energy loss during starting.

(*iv*) **Starting of Induction Motors by Smooth Variation of Supply Frequency.** Just as in the case of d.c. shunt motors being started by smooth adjustment of applied voltage, this method will involve changes in speed in a very large no. of steps. If the speed steps are equal in magnitude and a large number of steps in frequency are effected. The loss in energy during starting will be given by the expression.

$$W_{st} = \frac{1}{m}\left[\frac{1}{2} J\omega_0^2\right]$$

where ω_0 is the final speed achieved by the motor.

1.27 Speed Control

While selecting a motor for a particular drive due attention has to be given to the required speed variation. Different industrial drives requires variation of speed between different limits. Control for an industrial drive depends upon the nature of work being carried out. Under this section, the speed control of different type of motors is discussed.

(*A*) **D.C. Motors.** The speed of a d.c. motor and its torque are given by equation,

$$N \propto \frac{V - I_a R_a}{\phi} \qquad \text{...}(i)$$

$$T \propto \phi I_a \qquad \text{...}(ii)$$

where V = Terminal voltage

I_a = Armature current

R_a = Resistance of armature circuit

ϕ = flux/pole.

Thus equation (*i*) Suggest that there can be following methods of controlling the speed of d.c. motor:

(*a*) By varying the field flux.
(*b*) By varying the resistance in armature circuit.
(*c*) By varying the applied voltage.

(*a*) **Field control method.** The base or rated speed of a motor is that which is obtained with rated voltage applied to the armature and rated current flowing through the field winding. Since the field flux can be reduced by reducing some resistance in the field circuit, therefore, speed above normal can be obtained by this method.

(*i*) **Shunt Motors.** When a variable resistance R is connected in series with the shunt field winding as shown in Fig. 1.60. The shunt field current is reduced and hence flux ϕ.

Consequently, the motor runs at a speed higher than the normal speed. The amount of increase in the speed depends upon the value of variable resistance.

There is a limit to the maximum speed obtainable with this method because of poor commutation at weaker flux. A ratio of maximum to minimum speed of 6 : 1 is fairly common.

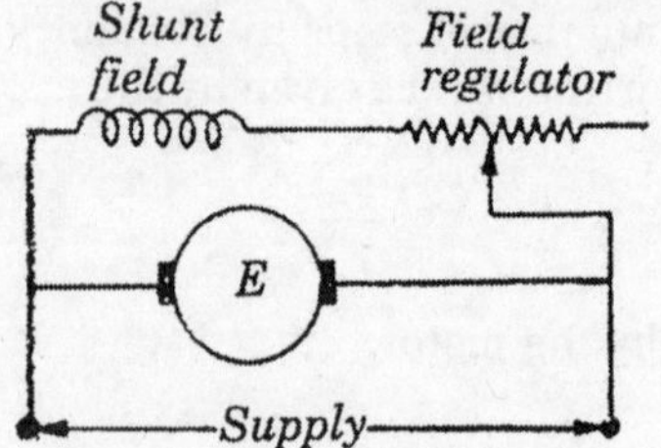

Fig. 1.60 Field control of shunt motor.

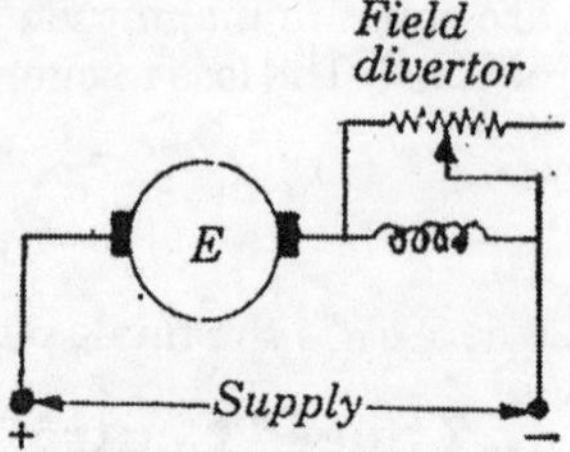

Fig. 1.61 Field diver-

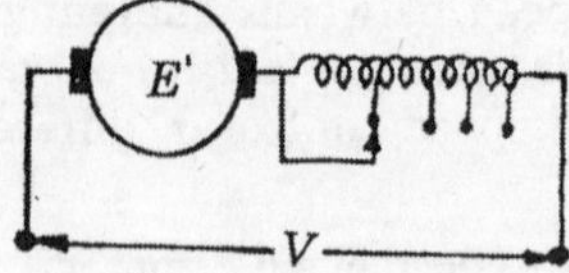

Fig. 1.62 Tapped field.

(*ii*) **Series Motor.** The variation of flux can be brought out by any one of the following methods :

Field diverter. In this method, a variable resistance R is connected in parallel with the series field as shown in Fig. 1.61. Its effect is that it diverts the path of the line current. A part of this current flows through the diverter and the current flowing through the series field winding is reduced. Consequently, the speed of motor is increased.

Tapped field control. In this method the number of turns of the series field winding can be changed by short circuiting a part of it since flux produced by the winding depends upon the ampere turns as the number of turns are reduced, therefore, the speed of the motor is increased.

(*b*) **Armature Control Method.** In this method the speed below the normal can be obtained by inserting external resistance in the armature circuit. This can be employed in all three types of d.c. motors.

(*i*) **Shunt Motor.** In this shunt motor flux is constant when applied terminal voltage and shunt field resistance are constant. Therefore, speed of motor is directly proportional to induced e.m.f., E. The value of E depends upon the drop in the armature circuit. When a variable resistance is connected in series with the armature as shown in Fig. 1.63 the induced e.m.f. is reduced and hence the speed. The decrease in speed depends upon the value of resistance. The method is neither economical nor efficient as a large amount of power is wasted in the controlled circuit.

(*ii*) **Series Motor.** In this case also a resistance is connected in series with the armature as shown in Fig. 1.64. If the current drawn by the motor is constant then flux is constant and the speed depends upon the induced e.m.f. (E). With the addition of variable resistance induced e.m.f. reduces and hence the speed. By changing the resistance, the speed below normal can be obtained. But this method is not economical.

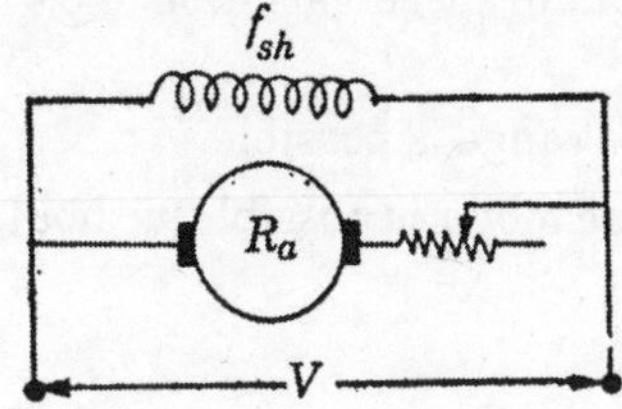

Fig. 1.63 Armature Control.

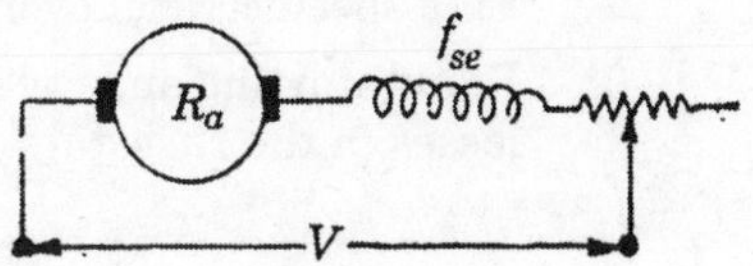

Fig. 1.64 Armature Control.

Voltage Control Method. Here the voltage across the motors is changed by different methods :

Series Parallel Control. This method is generally applied to series motor. The voltage across the motor is changed by connecting them in series or in parallel or in combination of both. This is widely used in electric traction. Two identical series motors are shown connected in series and parallel in Fig. 1.65 & 1.66. When the motors are

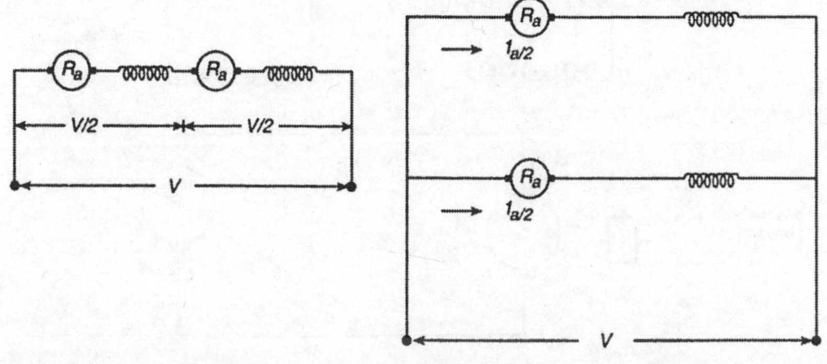

Fig. 1.65 **Fig. 1.66**

connected in series low speeds are obtained and when connected in parallel high speeds are obtained. Uniform speed control is not possible in this case.

Ward Leonard System. This method is used to supply variable voltage to the motor. A d.c. generator *G* is connected mechanically with the prime mover which rotates the generator at a constant speed as shown in Fig. 1.67. The field winding of d.c. generator is connected to a constant voltage a.c. supply line through a field regulator and reversing switch. The d.c. motor is fed from the generator *G* and its field winding is connected directly to a constant d.c. supply line. The voltage of generator fed to the motor can be varied from zero to maximum value. By reversing switch the direction of rotation of motor can also be changed. This system is initially very costly but is suited where unlimited speed control and reversing of direction is required.

Advantages:

1. It is possible to achieve lowest limit of speed variation of 5 to 7% of its base speed.
2. Fine speed control over very wide range is possible.
3. Rapid starting and reversing of the motor is possible without losses in the rheostats.

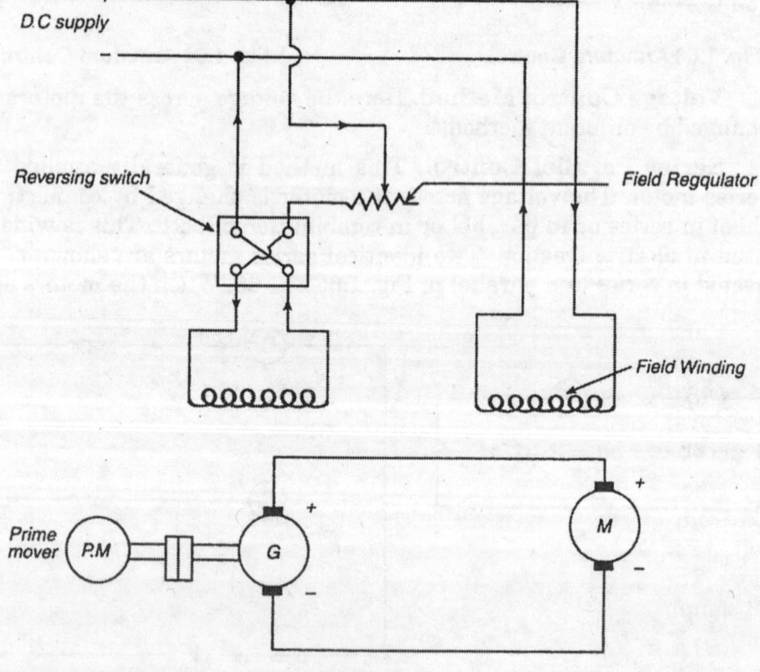

Fig. 1.67

4. Speed control is not by a cumbersome rheostat as in case of rheostatic method of speed control. Since controlling rheostat is in the field circuit of generator where very small current flows. It is very small in size which can be handled very easily.
5. Regenerative braking is also possible. This is achieved by suddenly decreasing the generated voltage of the generator to a value less than the back e.m.f. of the main motor. Now main motor will fed from the generator set will work as generator and power will be supplied back to the network.
6. For any setting of field regulator speed of the motor is independent of the load on the motor.

Main disadvantage of this system is that we require 3 machines and as such it becomes very costly in capital investment since there are two conversion stages of power *i.e.* electrical to mechanical and then to electrical and back to mechanical. The efficiency of speed control is low.

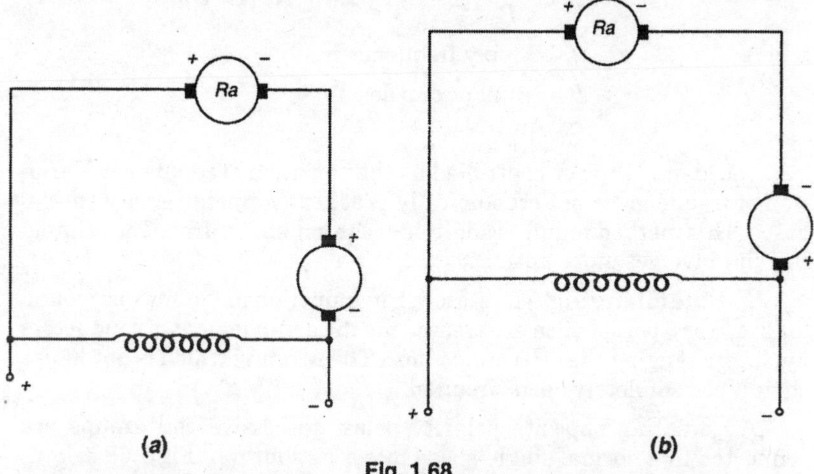

Fig. 1.68

Voltage Booster. The voltage applied to motor is changed above or below normal by using a Booster in series with the armature of motor as shown in Fig. 1.68 for increasing speed of motor the polarity of Booster is shown in Fig. 1.68 (*a*) and for decreasing the speed polarity is shown in Fig. 1.68 (*b*) where the Booster acts as motor.

Rectifier Control. In this method the d.c. supply to the motor is fed from mercury arc rectifiers or silicon controlled rectifier whose voltage can be adjusted. These being static devices ; their efficiency is more. These also occupy less space but their control apparatus is little bit complicated.

(*B*) **A.C. Motors.** In the broad sense it is difficult and inconvenient to get variable speed from A.C. motors. Synchronous motor is a constant speed motor which runs only at one speed.

A.C. commutator motors can be run at variable speeds by either applying a variable voltage or by the movement of brushes around the commutator.

In 1–ϕ A.C. motors, the speed can be controlled by voltage variation from a transformer. Transformer tapping gives wide variation of speed between zero and maximum.

In three phase a.c. motor brush shifting method is most suitable for speed control.

In repulsion motor the speed can be varied over a wide range by brush shifting.

A three phase induction motor is practically a constant speed motor but its speed can be varied by any one of the following methods:

(*a*) **Frequency control.** The speed of an induction motor is given by ;

$$N = \frac{120 f}{P}(1 - S) \text{ r.p.m.} = N_s (1 - S)$$

where f = Supply frequency

P = Number of poles

S = Slip

Speed may thus be controlled by the variation of frequency. Variation of frequency is not economically practically except in very special cases. This method requires additional equipment for frequency changing and involves more initial cost.

(*b*) **Pole changing.** The stator of an induction motor may be wound for 2, 4, 6 or 8 poles which are created my the arrangement of conductors by, interchanging the coil connections. The winding which is not in use should be completely open circuited.

For creating opposite polarity poles, successive coil groups are connected in a normal single-speed motor as shown in Fig 1.69 (*a*).

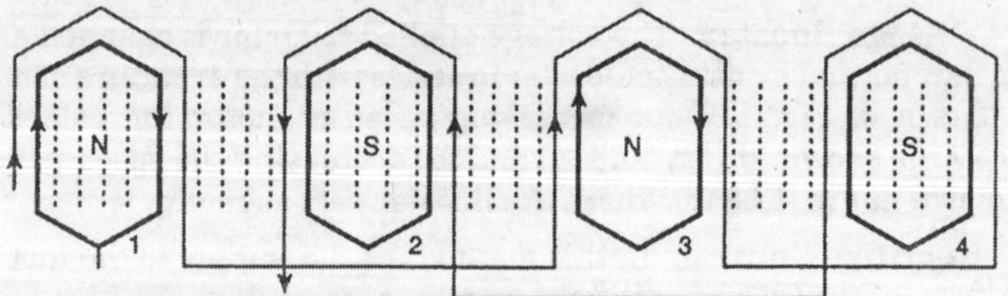

Fig.1.69 (*a*) Pole changing Method for Induction Motor.

It may be noted that all coils are in series, but the end connections are in the alternate groups. From the Fig. 1.69 (*a*) it is clear that group 1 is connected to group 3, and group 4 is connected to group 2. Now change the end connections, so that successive coils may have the same polarity as shown in Fig 1.69 (*b*).

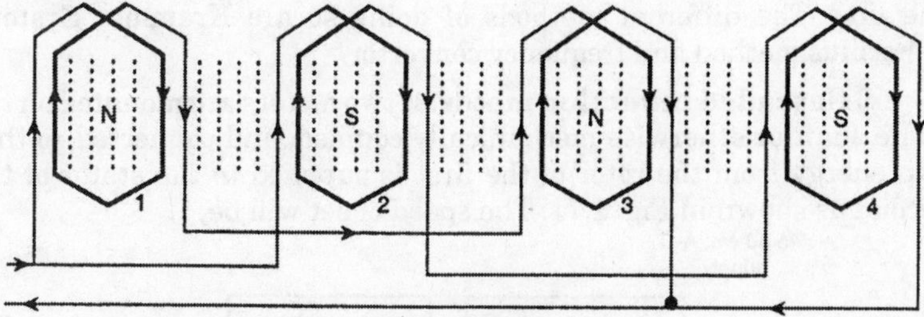

Fig. 1.69 (*b*)

Both series groups 2-4 and 1-3 are connected in parallel, so that four successive north poles are formed. Hence, south poles are created between the North poles and the motor is now connected for eight poles. Thus, for eight poles a pole winding can be reconnected. Normally this method is used for squirrel cage Induction Motors.

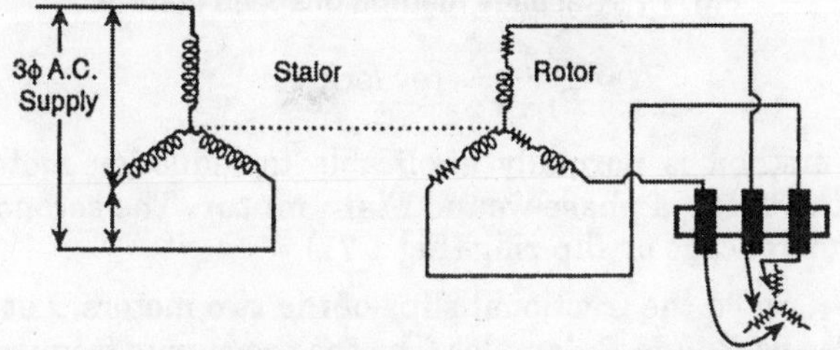

Fig. 1.70 Rotor Resistance Control of Induction Motors

Two separate primary windings can be employed wound for different number of poles and thus giving two different speeds. Squirrel cage motors designed for pole changing are called multi speed induction motors. Owing to complicacy of connections it is not advisable to design the motor having more than four different speeds.

(*c*) **Rotor resistance control.** Wound rotor motors are started by connecting starting resistance in the secondary circuit which are shorted in steps. If these resistances are designed for continuous operation these can serve dual purpose of starting and speed control which is obtained by inserting resistance into the armature as similar to a d.c. shunt motor.

In the rotor circuit resistance is added, thus varying the speed. This is because the percentage rotor copper loss is equal to the percentage slip. Moreover, the speed control by this process results in power loss in the additional resistance (see Fig.1.70)

So long as the load torque is consant, the speed is a function of external resistance. In case of large motors liquid-type resistors are used, while in medium sized motors metallic resistances are used.

(*d*) **Injected e.m.f. in rotor circuit.** The energy at slip frequency is converted to electrical energy at supply frequency and returned to the line. The different methods of doing so are Krammer System, Scherbius method and frequency convertor.

(*e*) **Cascaded induction motor.** If two motors are mounted on the same shaft or otherwise mechanically coupled, and connected so that slip energy from the rotor of the first is supplied to the stator of the second as shown in Fig. 1.71. The speed of set will be,

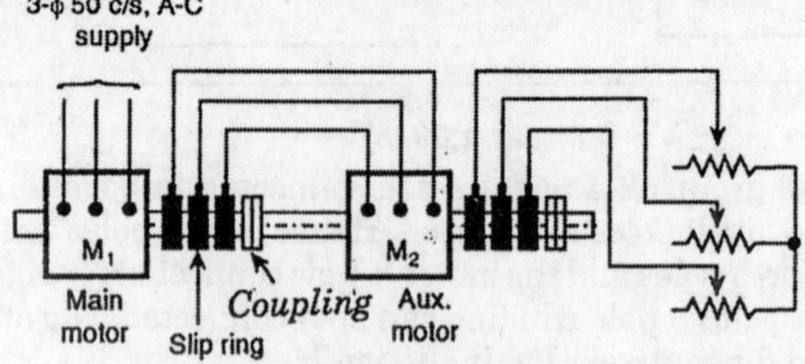

Fig. 1.71 Cascade method of speed control.

$$N = \frac{f}{P_1 + P_2} \text{ rev./sec}$$

This method is normally applicable to induction motors which requires the use of a phase wound **Main motor.** The second machine can be squirrel cage or slip-ring (Fig 1.71).

Let S_1, S_2 be the fractional slips of the two motors. Let the main motor poles be P_1 and P_2 and let f, be the the supply frequency.

Let Synchronous speed of main motor $= \dfrac{120 f}{P_1}$

Slip speed of main motor $= \dfrac{120 f}{P_1}(1 - S_1)$

and Synchronous speed of second motor $= \dfrac{120 S_1 f}{P_2}(1 - S_2)$

$$= \frac{120 S_1 f}{P_2}$$

This is because S_2 is very small as it is working normally. If the motors are coupled solidly in that case both the machines working at the same speed.

$$\therefore \quad \frac{120 f(1 - S_1)}{P_1} = \frac{120 S_1 f}{P_2}$$

or $$\frac{120 f}{P_1} - \frac{120 S_1 f}{P_1}$$

$$\therefore \quad = \frac{120 S_1 f}{P_2}$$

$$\therefore \quad S_1 = \frac{P_2}{P_1 + P_2}$$

In case more than normal speed is required, in that case the leads to the stator of the second motor are interchanged to have the torque reversed. This is called **Differential cascading** and the speed of the set is given as : $\frac{120f}{P_1 - P_2}$ under this condition.

1.28 Starting and speed control of a.c. commutator motors

The speed control and starting of a.c. commutator motor is achieved by applying reduced voltage or by shifting its brushes.

1.28.1 Starting and speed control of single phase induction motors

Since single phase induction motors are not self starting, some means of their starting are required. The methods employed are discussed below :

(*i*) **Split phase method.** In this method two windings, one main winding and the other auxiliary or starting winding is provided on the stator. The main winding is of high inductance and low resistance. When this winding is connected across the mains, it draws current I_m lagging behind the supply voltage V, by a large phase angle ϕ. The starting winding is of low inductance and is connected in series with the high resistance. It draws current I_s very approximately in phase with the supply voltage, V. The phase angle between I_m and I_s thus is very large. Therefore, the two currents produce a rotating field thereby starting of motor.

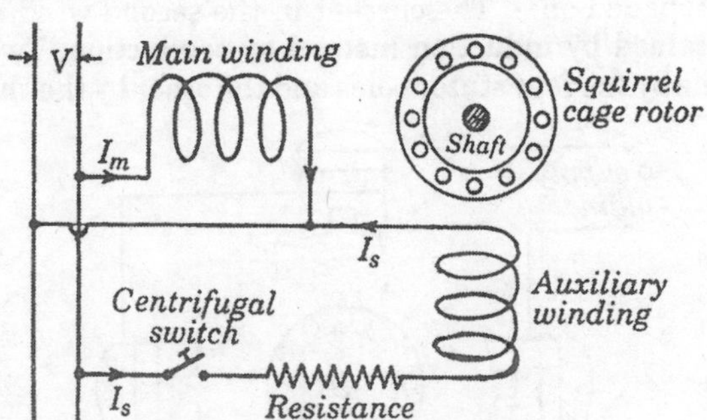

Fig. 1.72 Split phase motor.

When the motor picks up the speed to nearly 75 to 80% of synchronous speed the centrifugal switch cuts off the starting winding.

The general purpose split phase motor is used for constant speed operation where moderate torque is desirable. This is usually employed in blowers, duplicating machines, fans, centrifugal pumps, refrigerators etc. and available in sizes varying from 1/20 h.p. to 1/30 h.p. with speeds, 2880, 1440, 960 and 720 r.p.m.

(*ii*) **Capacitor Starting.** In this method a capacitor is used in place of resistance in the starting winding for splitting up the phase at starting. Such an arrangement provides a high starting torque because the two magnetic fields can be displaced in phase by 90°.

The capacitor can be switched off at will. However, if capacitor, C remains in circuit it will improve the power factor of the system to which it is connected. No centrifugal switch is thus provided.

These types of motors are commonly used where low starting torque is required such as table and *ceiling fans.*

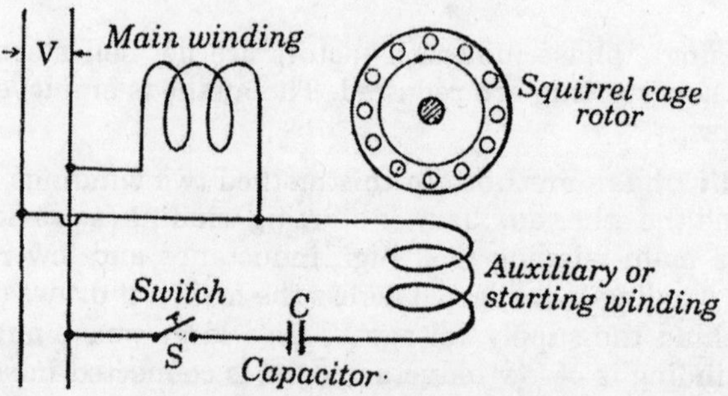

Fig. 1.73 Capacitor start motor.

(*iii*) **Shaded pole method.** In this method the stator has four shaded poles, each pole provided with its own exciting coils. A short circuited turn of low resistance is placed about a pole tip on each of the main poles (shaded one). The current in the second winding (shading rings) is obtained by induction instead by conduction. Two fields are produced, one by the four stator poles and the other by the shading rings

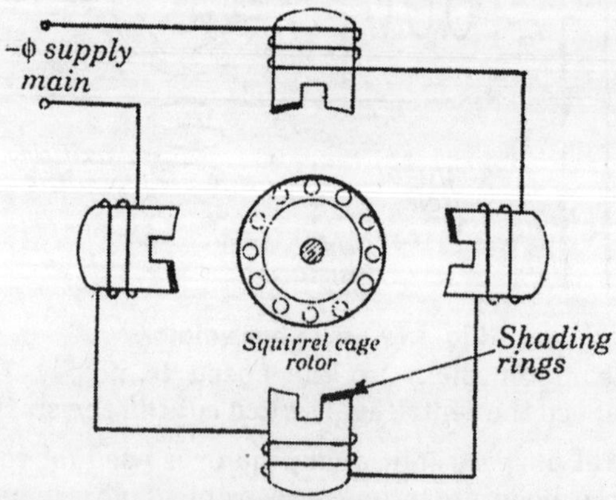

Fig. 1.74 Shaded pole motor.

which on interaction produce starting torque. The shaded pole motor produces a weak starting torque, so this method is only suitable for small motors such as hair dryers, small fans etc.

Example 1.4 *A 4-pole 50 c/s Induction motor has a full load speed of 1200 r.p.m. How fast it will run at half the load?*

Solution : Let, $1200 = \frac{1200 \times 50\,(1-S)}{4}$ where S, is the slip

$$\therefore \quad 1200 = 1500(1-S)$$

or $$1 - S = \frac{12}{15}$$

or $$S = 1 - \frac{12}{15} = \frac{15-12}{15} = \frac{3}{15} = \frac{1}{5}$$

As slip is proportional to load at half load,

$$\therefore \quad \text{Slip} = \frac{1}{5} \times \frac{1}{2} = \frac{1}{10}$$

$$\therefore \text{ Half load speed } = 1500\left(1 - \frac{1}{10}\right) = 1500\left(\frac{9}{10}\right)$$

$$= 1350 \text{ r.p.m.} \quad \textbf{Ans.}$$

Example 1.5 *At stand-still condition of a 3–φ, 4-pole, 440 V induction motor, the rotor resistance and reactance are 0.5 ohm and 0.8 ohm, respectively per phase. Calculate the starting current, when the speed is 1440 r.p.m, and the supply frequency is 50 c/s.*

Solution : Rotor resistance/phase $= 0.5\Omega$

Rotor resistance/phase $= 0.8\Omega$

Speed, $$N_2 = \frac{120 \times f}{P} = \frac{120 \times 50}{4} = 1500 \text{ r.p.m.}$$

$$\therefore \quad \text{Slip} = \frac{1500 - 1400}{1500} = \frac{60}{1500} = \frac{1}{25} = 0.04$$

At the time of starting, slip $S_1 = 1$

$$\therefore \text{ Rotor current / phase } = \frac{440/\sqrt{3}}{\sqrt{(0.5)^2 + (0.8)^2}}$$

$$= \frac{254.04}{\sqrt{0.89}} = \frac{254.04}{0.9434} = \textbf{269.28 Amp.} \quad \textbf{Ans.}$$

Now Rotor current at 1440 r.p.m is given as below :

Rotor current/phase, $$I = \frac{440/\sqrt{3}}{\sqrt{\left(\frac{0.5}{0.04}\right)^2 + (0.8)^2}}$$

$$= \frac{254.04}{\sqrt{(12.5)^2 + 0.64}}$$

$$= \frac{254.04}{\sqrt{56.89}} = \frac{254.04}{12.526}$$

$$= \mathbf{20.28\ Amp.} \quad \textbf{Ans.}$$

Example 1.6 : *The maximum torque of a 440 V, 3-φ, 4- pole. 60 c/s induction motor is 440 N.m at a slip of 0.1. If the motor works at 50 c/s, 400 V supply, find the maximum torque, slip and the speed at which it occurs. Neglect stator impedance.*

Solution: The angular speed, w_s of a motor is given by :

$$w_s = \frac{2\pi N_s}{60} \text{ and } T_{max} = \frac{3(V)^2}{3\omega_s . X_T}$$

or
$$T_{max} = \frac{3(V)^2}{\left(\frac{2\pi}{60}\right)\left(\frac{120f}{P}\right) \times 2\pi f(L_T)} = \frac{3(V)^2 P}{8\pi^2 f^2 L_T}$$

∴
$$\frac{T_{max_2}}{T_{max_1}} = \frac{(V_2)^2 f_1^2}{(V_1)^2 f_2^2}$$

or
$$T_{max_2} = 400\left[\frac{400}{440} \times \frac{60}{50}\right]^2$$

$$= \mathbf{475.65\ Nm.} \quad \textbf{Ans.}$$

Slip at max. frequency is, $\frac{R_2}{X_T}$

thus
$$\frac{S_2}{S_1} = \frac{f_1}{f_2}$$

∴
$$S_2 = 0.1 \times \frac{60}{50} = \mathbf{0.12} \quad \textbf{Ans.}$$

The synchronous speed $= \frac{120 \times 50}{4} = 1500$ r.p.m.

∴ Speed at Max. torque $= 1500(1 - 0.12)$
$= 1500 \times 0.88 =$ **1320 r.p.m. Ans.**

1.29 Breaking Characteristics

In many industrial drives the provision of apparatus for stopping a motor and its driven machine is as important as that for starting it. A planer machine for instant must be quickly stopped at the end of its stoke in order to get high rate of production, in other cases rapid stops must be possible to prevent danger to operator or damage to the manufactured product. For lifts and hoists, effective braking must be provided.

Braking. A good braking system must possess following features:

(*i*) It should be fast and reliable.

(*ii*) The breaking force must be capable of being controlled.

(*iii*) Kinetic energy of rotating parts of the motor and its driven machines must be suitably dissipated.

Types of braking system. Following are two types of braking system which may be employed :

1. **Friction Braking or Mechanical Braking.** In this the stored energy is dissipated as heat by a brake shoe or brake lining rubbing on a wheel or brake drum.

2. **Electric or Electrodynamic Braking.** In this the stored energy is converted into electrical energy and dissipated in resistance as heat or in certain cases returned to the supply.

1.30 Advantages of Electric Braking over Mechanical Braking

(*i*) It is more smooth than mechanical braking.

(*ii*) It is quite fast therefore higher speeds can be maintained.

(*iii*) It is quite cheap from maintenance point of view since no replacement of breaking lining is needed as in case of mechanical breaking system.

(*iv*) Other advantage is in regenerative braking the electric energy produced during breaking system can feed back to supply system.

(*v*) Heat produced during the period of electric braking is in no way harmful but heat produced on the lining etc. during the application of brakes may result in the failure of brakes.

1.31 Disadvantages of Electric Braking

(*i*) It cannot be applied to all types of motors.

(*ii*) Its initial cost is very high.

(*iii*) It is, however, complicated than mechanical braking.

(*iv*) Electric braking can stop the motor but it cannot hold it stationary. To hold it stationary mechanical braking is required.

1.32 Methods of Electric Braking

Here the kinetic energy of the motor is converted into electrical energy and is dissipated as heat in a resistance or returned to supply system.

Electric braking is superior to mechanical braking since it is much quicker and eliminates the cost of maintenance of mechanical breaks.

However, in order to finally bring the motor to a stand still position and hold it there, friction breakers are also essential. There are three types of electric braking :

(*i*) Plugging
(*ii*) Rheostatic braking
(*iii*) Regenerative braking.

(*i*) **Plugging.** It is the simplest type of braking and involves the reversing of the connections to the armature of the motor so that it tends to run in reverse direction. In order to stop the machine a special device must be installed to cut off the supply as soon as it comes to rest this method can be applied to d.c. motors, induction motors and synchronous motors. It is particularly suitable where the drive has to be rapidly reversed such as in planner machines etc.

1.33 Plugging as applied to D.C. Shunt Motor

The direction of rotation of d.c. shunt motor can be reversed by reversing the current through the armature. The expression for the armature current, $I_a = \dfrac{V - E_b}{R_a}$. During normal runnings the back e.m.f. E_b is opposite to the applied voltage V but at the instant of reversal, a voltage equal to $V + E_b$ is impressed across armature. This will cause a great rush of current in armature circuit, to prevent this the starting resistance is re-inserted in circuit as shown in Fig. 1.75.

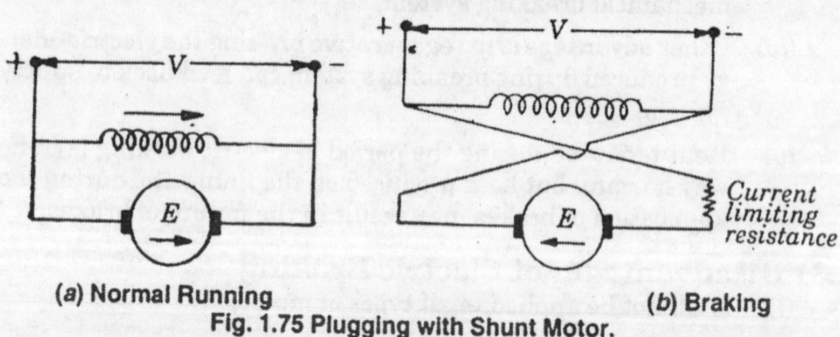

Fig. 1.75 Plugging with Shunt Motor.

The plugging operation of a d.c. shunt motor can be explained with the help of quadrantal diagram [Refer Fig. 1.76 (*a*)].

Curve *A* corresponds to operation of motor in reverse direction and curve *B* corresponds to operation of motor in forward direction. Supposing that motor is rotating in forward direction with point of operation represented by '*a*'. As soon as operation of plugging is done, the point of operation will shift to '*b*' on the curve *A*. Torque T_p corresponding to point '*b*' is called plugging torque. Motor speed will become zero as soon as point of operation shifts to '*c*' along the path *bdc*.

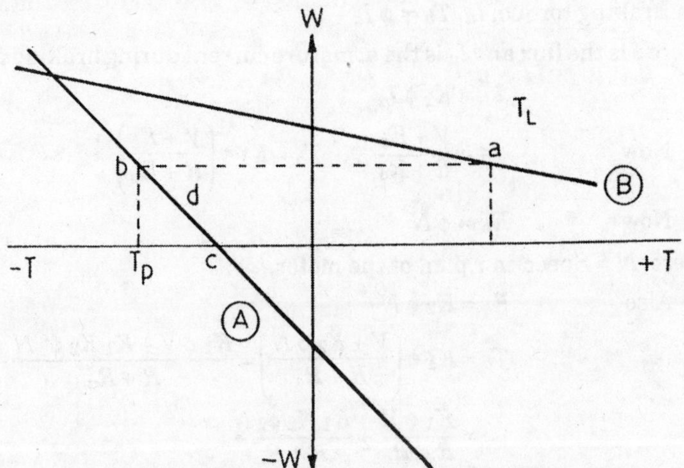

Fig. 1.76 (*a*) Plugging as Applied to D.C. Shunt Motor.

1.34 Plugging as Applied to D.C. Series Motor

The method is same as in case of d.c. shunt motor. The change over of connections is shown in Fig. 1.77. In both cases, the motor must be disconnected from the supply when it occurs to rest otherwise it may rotate in the opposite direction.

For that the connections have been shown in Fig. 1.77 (*b*). Fig. 1.77 (*a*) shows that connections of D.C. series motor before plugging is applied. As shown in Fig. 1.77 (*b*), the direction of field current remains unchanged, where as direction of armature current, I_a gets reversed and is given by as in the previous case,

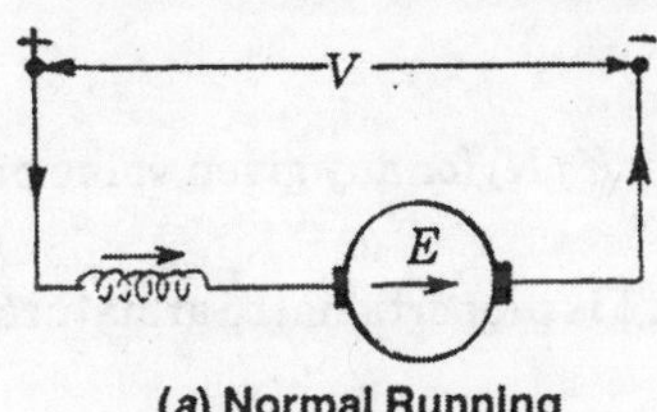

(*a*) Normal Running

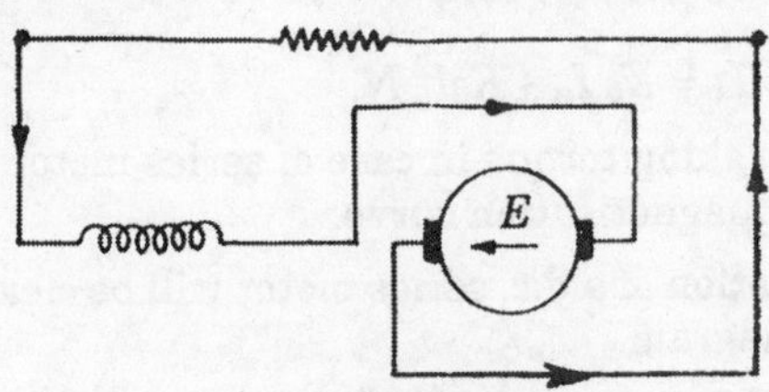

(*b*) Braking

Fig. 1.77 Plugging with Series Motor.

$$I_a = \frac{V + E_b}{R + R_a}$$

where R is the additional resistance connected in series to limit the current.

Braking torque, I_a. $T_b \propto \phi I_a$

where ϕ is the flux an ! I_a is the armature current during braking period.

$$T_b = K_1 \phi I_a$$

Now $$I_a = \frac{V + E_b}{R + R_a} \therefore T_b = K_1 \phi \left(\frac{V + E_b}{R + R_a}\right)$$

Now $$E_b \propto \phi N$$

where, N = Speed in r.p.m. of the motor.

Also $$E_b = K_2 \phi N$$

$$\therefore \quad T_b = K_1 \phi \left(\frac{V + K_2 \phi N}{R + R_a}\right) = \frac{K_1 \phi V + K_1 K_2 \phi^2 N}{R + R_a}$$

$$= \frac{K_1 \phi V}{R + R_a} + \frac{K_1 K_2 \phi_2 N}{R + R_a}$$

$$\frac{K_1 V}{R + R_a} = K_3 = \text{Constant}$$

and $$\frac{K_1 K_2}{R + R_a} = K_4 = \text{Constant}$$

$$\therefore \quad T_b = K_3 \phi + K_4 \phi^2 N.$$

Now in case of shunt motors flux, ϕ is constant.

$$K_3 \phi = K_5$$

and $$K_4 \phi^2 = K_6$$

Braking torque in shunt motor = $(K_5 + K_6 N)$ for any given value of resistance, R.

In case of D.C. series motor, since flux ϕ is proportional to armature current, therefore, braking torque,

$$T_b = K_3 \phi + K_4 \phi^2 N \quad \text{as} \quad \phi \propto I_a$$

$$\therefore \quad T_b = K_5 I_a + K_6 I_a^2 N.$$

The valve of braking torque in case of series motor can, therefore, be found from the magnetisation curve.

Plugging operation of a d.c. series motor will be clear from the Fig. 1.76 quadrantal diagram.

Curve A in this figure represents the operation of motor in reverse direction and curve B in forward direction. In case of d.c. series motor plugging torque varies considerably with changes in load torque that is, with changes in motor speed before plugging.

1.35 Plugging as Applied to Induction Motor

If any of the two supply phases are interchanged with each other the direction of rotation of the magnetic field reverses and, therefore the torque on the rotor also reverses providing the braking, effect supply, however, it has to be cut off when the motor comes to rest, otherwise the motor will start rotating in the reverse direction.

The normal working of 3-ϕ induction motor is shown in Fig. 1.78 (*a*). The rotor rotates at speed N in the direction of rotation of the rotating magnetic field (rotating at speed N_s). When breaking by the method of plugging is applied to the 3-ϕ Induction motor, the two phases to the stator are interchanged. This is shown in Fig. 1.78 (*b*).

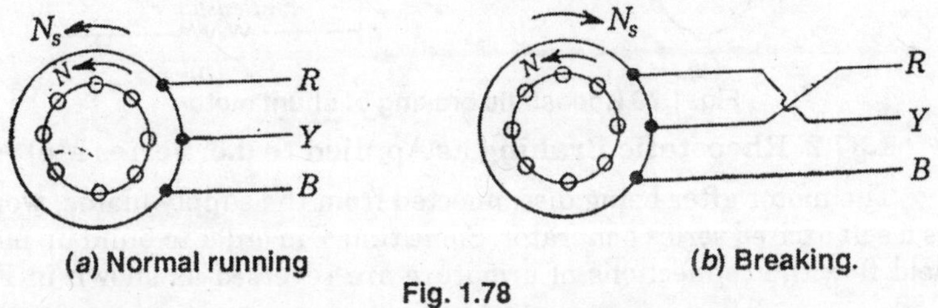

(*a*) Normal running (*b*) Breaking.

Fig. 1.78

1.36 Plugging as Applied to Synchronous Motor

If the d.c. excitation to the synchronous motor is reversed the d.c. and a.c. fields will rotate in the opposite direction and there can be no braking effect. But in case of motors filled with damper windings, the eddy currents induced in them may provide braking. By interchanging the two phases of stator supply the braking can be provided due to induction motor action.

1.37 Rheostatic Braking

With this method of braking the motor is disconnected from the supply and used as a generator driven by the momentum of the equipment to be braked and supplying current to resistance. D.C. or synchronous motors can be braked in this way. The induction motors however, requires a separate source of d.c. excitation during this type of braking. This type of braking as applied to different motors is explained below.

1.37.1 Rheostatic Braking as Applied to the d.c. Shunt Motor

The armature is disconnected from the supply and connected across a resistance. The motor then works as a separately excited generator and the braking torque is applied by the current flowing in resistance as shown in Fig. 1.79.

The magnitude of the braking effect can be controlled by varying either the resistance in armature circuit or the field current. The braking is not effective if the supply fails during the period. Armature current during breaking period is equal to

$$I_a = \frac{E}{R + R_a}$$

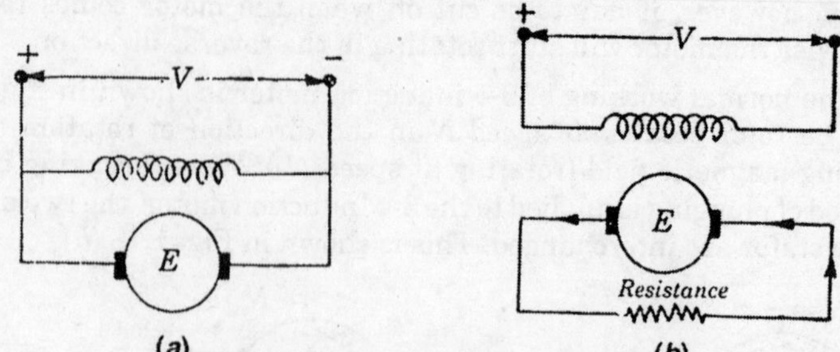

Fig. 1.79 Rheostatic braking of shunt motor.

1.37.2. Rheostatic Braking as Applied to d.c. Series Motor

The motor after being disconnected from the supply mains, works as a self excited series generator. Some times, in order to build up more field flux the connections of armature are reversed as shown in Fig. 1.80.

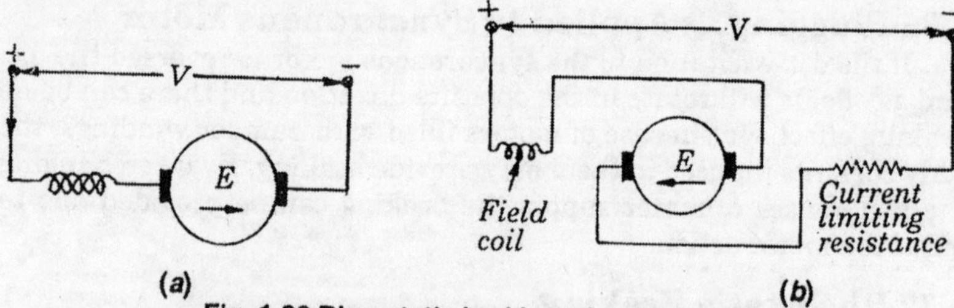

Fig. 1.80 Rheostatic braking of series motor.

In such cases if the resistance is above a certain critical value equal to the initial slope of current voltage characteristics, the machines will not self excited, therefore, resistance should be low enough.

The armature current during the period of braking is given by,

$$I_a = \frac{E_b}{R + R_a}$$

The resistance inserted in the circuit should be less than the critical resistance, otherwise the generator shall not be self-exciting.

1.37.3. Rheostatic Braking as Applied to D.C. Compound Motors

Rheostatic braking to D.C. compound motors is applied in a similar way as to D.C. shunt motors.

Relation between the braking torque and speed for D.C. motor during rheostatic braking is determined as below :

$$T_b = \text{Braking torque} \propto \phi I_a = K_1 \phi I_a.$$

But $$I_a = \frac{E_b}{R + R_b}$$

$$\therefore \quad T_b = K_1 \phi . \frac{E_b}{R + R_a}$$

also $\quad E_b \propto \phi N = K_2 \phi N$

$$\therefore \quad T_b = K_1 \phi . \frac{K_2 \phi N}{R + R_a} = \frac{K_1 K_2 \phi^2}{R + R_a} . N$$

$$R + R_a = \text{Constant.}$$

$$\therefore \quad \frac{K_1 + K_2}{R + R_a} = \text{Constant} = K_2$$

$$\therefore \quad T_b = K_3 \phi^2 N.$$

In case of shunt motors, since flux, ϕ is constant, therefore, torque is proportional to the speed but in case of D.C. series motor, since flux is proportional to the armature current, therefore, torque can be determined graphically from the magnetization curve.

1.37.4. Rheostatic Braking as Applied to Induction Motor

The stator is disconnected from the supply and direct steady current is passed through its windings. A flux produced. When the short circuited rotor conductors cut this steady flux, e.m.f. is induced in them provide the necessary braking effect. If the rotor is wound, the braking torque can be controlled by insertion of suitable resistance in the rotor circuit.

1.37.5. Rheostatic Braking as Applied to Synchronous Motors

The field excitation is maintained and the motor after being disconnected from the supply is connected to resistance in star or delta. It now works as an alternator and the kinetic energy is dissipated in the form of losses in the resistances.

1.38 Regenerative Braking

In the first two methods, the mechanical energy of the revolving parts was converted into electrical energy which was further dissipated in the external resistance. In case of plugging, even energy was drawn during the braking period also and was wasted. In case of regenerates braking, the machine is made to work in a way that it generates energy and feeds it back to the supply system.

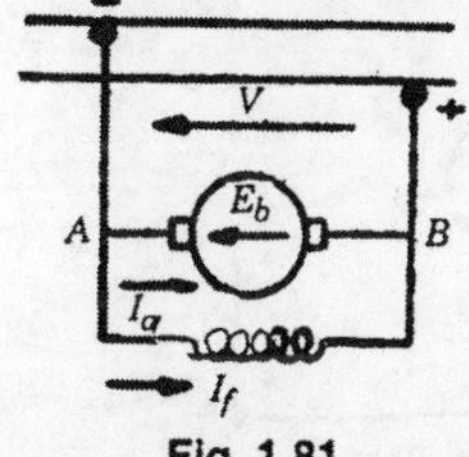

Fig. 1.81

As shown in Fig. 1.81, the shunt machine draws power from supply if applied voltage V, is greater than induced voltage, E_b.

If induced voltage, E_b becomes greater than the applied voltage, the machine instead of taking power from the supply (motoring action) sends back power to supply (generating action). This can be achieved if,

(i) Supply voltage V is reduced.

(ii) The field excitation is increased so that E_b becomes greater than the applied voltage, V.

(iii) The speed of the motor becomes greater than the no-load speed of the motor, in that case again the value of E_b becomes greater than the applied voltage, V.

Thus whenever the value of E_b becomes greater than V, the direction of current in the armature gets reversed, the direction of field remaining same, the torque produced is in the opposite direction. Thus braking effect is produced. When the torque is reversed, the motor tends to stop thus reducing the speed and thus reducing E_b. In order that braking torque is still applied, the field excitation is increased so that E_b still remains greater than applied voltage. Thus it is not possible to obtain regenerative braking down to very low speeds as E_b falls tremendously down and it cannot be increased by the increase in excitation because too much increase in field cannot be obtained.

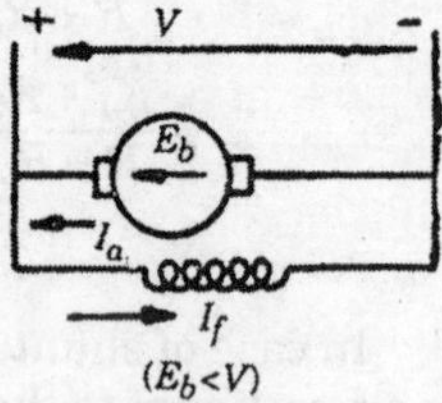

Fig. 1.82

Thus regenerative braking cannot stop the motor. It can only be applied to overhauling loads which end to drive the motor at speeds greater than the no-load speed of the motor. This can take place in a crane motor due to action of lowering load or in a motor of an electric locomotive moving down gradient.

This type of braking is mostly employed in electric traction. The kinetic energy of motor is converted in electric traction in which it is fed back to the supply lines instead of wasting it in rheostats. This method can be applied to D.C. motors and induction motors.

1.38.1. Regenerative Braking Applied to D.C. Shunt Motors

If the e.m.f. generated by the motor is greater than the supply voltage, power will be fed back to the supply. The e.m.f. in a shunt motor depends upon the field flux and speed. During this type of braking the field is disconnected from the armature and connected to a separate supply as shown in Fig. 1.83. The field current is so adjusted that

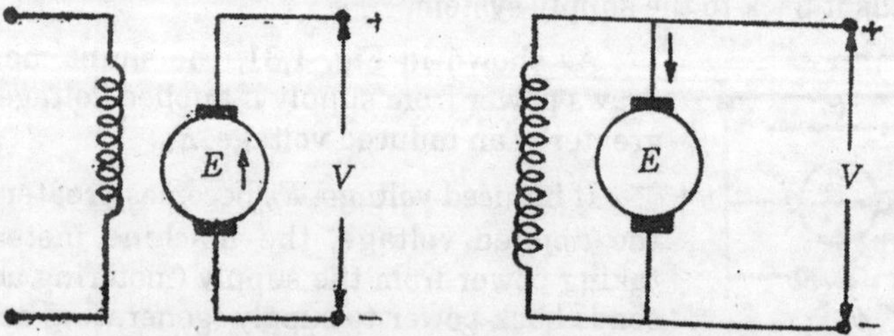

Fig. 1.83 Regenerative braking with shunt-motor.

$E > V$ and this is returned to the mains. Alternatively, if the field excitation, does not change. But the load causes the speed to increase, the induced e.m.f. may become more than the supply voltage and power will be fed back to the supply.

1.38.2. Regenerative Braking as Applied to D.C. Series Motors

The conditions with a series motors are not so simple as shunt motor because here the reversal of current necessary to produce the regeneration would cause a reversal of the field and therefore of the back e.m.f. The complications necessary to obtain satisfactory operation with series motors are not normally warranted for industrial purposes, although in the traction work this type of braking is frequently employed.

One method of regenerative braking with series motors is french method. If there is a single series motor it is provided with one main series field winding and 2 or 3 auxiliary field winding. During normal running the main and auxiliary field windings are placed in parallel with each other as shown in motoring action in Fig. 1.84 (*a*). During regenerative braking, the auxiliary winding are placed in series with each other and combination is connected in parallel across the armature the main series field winding as shown in generating action in Fig. 1.84 (*b*). In this case the machine works as a differentially compound generator and sends back voltage to supply lines.

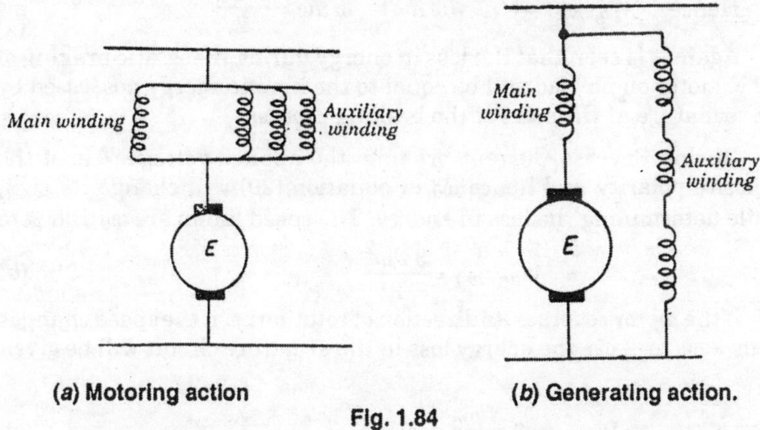

(*a*) Motoring action (*b*) Generating action.

Fig. 1.84

In locomotives there are four to six motors and therefore, no auxiliary winding are required. The regenerative braking can be applied in this case as shown in Fig. 1.85. An alternative method of regenerative braking is by using an exciter for controlling the excitation of the field winding during regeneration.

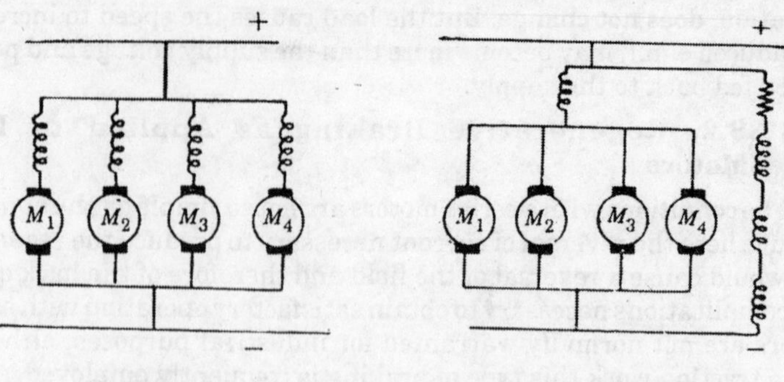

Fig. 1.85

1.39 Energy Relations During Braking

As during starting of motors, certain expressions can be obtained for energy lost by the motors during braking.

DC motors with separate and shunt excitation : Let us suppose that the braking takes place with no load on the motor. The energy dissipated in the armature circuit during rheostatic braking is determined using equation (20). It must be noted that during rheostatic braking, $V = 0$ and the braking takes place from speed ω_0 to stand still.

Hence $$W_{br(rhe)} = \int_{\omega_0}^{0} - J\omega\, d\omega \int_{0}^{\omega_0} \omega\, d\omega = \frac{J\,\omega_0^2}{2} \quad \text{...(a)}$$

Again it is seen that the loss in energy during rheostatic braking of a d.c. motor on no load will be equal to the kinetic energy possessed by the armature at the start of the braking process.

During Reverse Current braking, the applied voltage V is of the opposite polarity and hence ω_0 in equation (20) will change its sign, while determining the loss in energy. The speed limits are ω_0 and zero

$$W_{br(rev.)} = \frac{3\,J\omega_0^2}{2} \quad \text{...(b)}$$

If the motor reverses its direction of rotation *i.e.* if the speed changes from $+\omega_0$ to $-\omega_0$, the energy loss in the armature circuit will be given by,

$$W_{(rev.)} = 4\,\frac{J\,\omega_0^2}{2} \quad \text{...(c)}$$

1.40 Induction Motors

Let us assume that braking is performed with motor on no load.

During reverse current braking or plugging the slip changes from 2 to 1 substituting these limits of slip in equation (31) we get,

$$W_{br\,(rev.)} = \frac{J\,\omega_s^2}{2}\,(2^2 - 1^2) = \frac{3\,J\,\omega_s^2}{2} \quad \text{...(d)}$$

The heat dissipated in the rotor circuit is, therefore, three times the kinetic energy at no load speed. Thus this type of braking draws from the supply twice as much energy as there already exists in the drive and dissipates the whole lot in the Rotor Circuit resistances.

Incidentally, if the motor were reversed to run nearly at ω_s, the slip would vary from 2 to 0.

The loss during reversal of speed will then be given by,

$$W_{rev.} = \frac{J\,\omega_s^2}{2}(2^2 - 0^2) = 4\,\frac{J\omega_s^2}{2} \qquad ...(e)$$

Both equation (*d*) and (*e*) represent the loss in rotor circuit alone. The total energy dissipated by the motor during braking will be, $\frac{3\,J\omega_s^2}{2}\left(1 + \frac{R_1}{R_2}\right)$ Joules and that during reversal of speed will be,

$$\frac{4J\omega_s^2}{2}\left(1 + \frac{R_1}{R_2}\right) \text{ Joules.}$$

During Rheostatic braking, the energy dissipated in the rotor circuit will be given by the stored kinetic energy available at the beginning of the braking process *i.e.*

$$W_{b(rhe.)} = \frac{J\,\omega_s^2}{2} \qquad ...(f)$$

Loss of energy in the stator, during rheostatic braking is given by,

$$W_{br\,(s)} = 3\,I_{ac}^2\,(eq.)\,R_1\,t_{br} \qquad ...(g)$$

where t_{br} denotes the braking time during which the motor speed falls from ω_s to zero.

Incidentally, the time during braking can be determined by substituting proper value of slip in equation,

$$t = \frac{J\,\omega_s}{2\,T_{max}}\left[\frac{S_1^2 - S_1^2}{2\,S_{max}\,T} + S_{max}\,T\,\log_e \frac{S_1}{S_2}\right]$$

1.41 Mechanical Characteristics

1. **Types of Enclosures.** Some form of protection of live and moving parts is usually desirable to prevent the ingress (incoming) of dust and dirt or larger foreign bodies such as spanners, tools and even careless human beings. If however some types of enclosure is adopted to give this protection it will also prevent the free ingress of cooling air. Hence an enclosed motor will not be able to give such a large output for a given temperature rise as an unprotected motor of some dimensions. The various types of enclosures are defined in the Indian standard specifications and the reputed manufacturers, manufacture the motor enclosure as per I.S.I. specification. Some of the enclosure fall into the following classes :

(*i*) **Open Type.** The bearings are supported on simple pedestals or end brackets so that there is no restriction to ventilation and also no protection. Many old type of motors were of this type. But modern practice confines its use chiefly to a large d.c. motors which are installed in a room separate from the machine shop.

(*ii*) **Protected Type.** The access to live and rotating parts is limited to prevent accidental contact and ingress of foreign material. Ventilation is not obstructed. Such motors are used mostly for industrial purposes.

(*iii*) **Screen Protected Type.** A better protection is afforded if the openings of the protected type are covered with wire mesh screen. The degree to which ventilation achieved depends on the fineness of the screen and the possibility of its becoming clogged with dirt.

(*iv*) **Drip Proof Type.** Where motors are to be installed in very damp situations, a drip proof enclosure can be used in which the openings are protected by overhanging cowls to prevent falling moisture from entering the machine.

(*v*) **Splash Proof Type.** In such machines, the ventilating openings are so constructed that liquid drops or dust particles reaching the machine at any angle between the vertical and 100° from it cannot enter the machine.

(*vi*) **Pipe Ventilated Type.** Another method of overcoming the heating difficulty of the totally enclosed motor, particularly with large sizes, employs a duct or pipe through which clean air is led to motor from out side the dust laden building or areas.

(*vii*) **Totally Enclosed Type.** In some situations where the atmosphere is very dusty such as in coal handling plants, saw mills or in oil or chemical plants, a totally enclosed type motor is used. Such machine have solid frames and end shields without any opening. These motors are cooled by the surface radiation only with the help of cooling fins. The output of such motors is about 50 to 75% of that obtainable from a protected motor of same dimensions. Motors of these types are of small ratings.

(*viii*) **Flame Proof Type.** For the use in coal mines or other situations where an explosive atmosphere may arise, a special form of totally enclosed motor must be used which is capable of withstanding any explosion of gas inside the motor without transmitting the flame to the outside. The ventilating openings in the machines are made long enough for the flame to be contained within the frame.

1.42 Bearings

Bearing are those parts of machines which house and supports the main shaft. These provide friction free movement of rotating parts and keep the moving part in a fixed physical position. The two types of bearings are generally employed for motors :

(*i*) Ball or roller bearing.

(*ii*) Sleeve or bush bearing.

(*i*) **In ball bearing.** The shaft is supported on the walls or roller which are movable. Ball bearing are used for motors upto 100 H.P. It is expensive but lower frictionless, requires less space and low maintenance. This bearing gives more noise so it is not recommended where noise is to be avoided.

(*ii*) **Sleeve or Bush Bearings.** These are also known as plain bearings and friction in such cases is reduced by the use of lubricating film. For fractional horse power motors, it is in the form of a self-aligning porous bronze bush known as sintered bush. For large horse powers it is in the form of journal bearings. When properly designed, the complete load is transferred to a hydrodynamic oil film developed in between the shaft and sleeve. The advantage of this type of bearing is that it gives noiseless operation and has longer life.

1.43 Transmission of Power

Mechanical power available at the motor shaft has to be transmitted to the driven machines. The various methods of transmission of drive in general use are :

1. **Direct Drive.** The motor is coupled directly with machine to be driven by solid or flexible coupling such as motor generator set. The drive is 100% efficient and requires very less space but can be used when speed of driven member required is same as that of driving member.

2. **Belt Drive.** This method make use of a belt from motor to driven machine on pulley system. Thus method is quite sufficient and give satisfactory service. The general rules which are used in belt drive system are :

(*i*) The diameter of the driven pulley should not be more than six times the diameter of motor pulley.

(*ii*) The distance between pulleys centres should not be less than four times the diameter of larger pulley.

3. **Rope or V Belt Drive.** Drive by means of ropes running over pulleys with a number of V-grooves can be used where the power to be transmitted is beyond the limit for belt drive. This type of drive provides flexibility on sudden loads and the slip is negligible. They are capable of transmitting greater power than flat belts and are noiseless and smooth in operation.

4. **Gear Drive.** For low speed drives the coupling between the motor and the drive is through a gear box. Motors are usually manufactured with gear box incorporated in the motor frame itself for reasons of economy.

5. **Chain Drive.** In the belt and rope drives there is about 4% slip where as the chain drive has the advantage that there is no slip. Chains are capable of transmitting large horse power and also the operation is salient compact and difficult.

6. **Vertical Drive.** In this type the motor shaft is kept vertical and load is arranged over it such as in flour mills.

1.44 Noise Level

This is another important features to be considered while making the selection of a motor. Noise should be kept as low as possible in the workshops, hospitals and other domestic purposes. The noise may be due to bearing, vibrations magnetic pulsations and faulty foundations.

To reduce the noise, journal bearing may be used in place of ball bearings. The motor should be mounted on a heavy concrete or cast iron block. The electrical connections should be made through the flexible conduits.

1.45 Size and Rating

The size of a motor depends upon firstly, the temperature rise which in turn depends upon whether the motor is to operate a continuous, intermittent or variable load and secondly, the maximum torque required of the motor. Keeping in view the torque requirements of load, the rating of a motor will be decided by the load conditions. The different rating for electrical motors are :

1. **Continuous Rating.** Drives like pumps, fans etc. require a constant power and the choice of a suitable rating of motor is a simple affair. It is that output which a motor can give continuously for long time without exceeding the given temperature rise and the motor should be able to give 20% over load for two hours.

2. **Continuous Maximum Rating.** It is similar to continuous rating but it does not allow overload. It is only used for big motors larger than 2.5 H.P. per r.p.m.

3. **Intermittent Rating.** The intermittent rating of motor is in accordance with the intermittent load. The intermittent loads may be of two types :

(*a*) The motor may be loaded for a short time and then shut off for a sufficient long time allowing the motor to cool down. In such a case motor with a suitable short time rating is adopted.

(*b*) The motor may be loaded for sometime and then shut off for sometime. Before the motor can cool down completely the load cycle repeats. In such a case a suitable continuous or short time rated motor is chosen which when operating on the particular. load cycle will not exceed the specified temperature limits.

1.46 Cost of the Motor

It is assumed that the motor selected should be one which is the cheapest but a motor with initial cost being low may not necessarily be the cheapest. The motor selected should prove to be the cheapest overall till the end of its life. It is assumed that high speed motors are relatively less costly, have higher efficiency and are smaller in size.

Example 1.7 *A squirrel cage induction motor takes twice full load current and develops half full load torque when switched on with star-delta starter. Find starting current and starting torque in terms of full load values when started by auto-transformer with 40% tapping.*

[A.M.I.E., (Sec. B), Utilization of Elect. Power 1977]

Solution. Starting current with star-delta starter,

$I_{st} = 2\, I_f$ where I_f = full load current

Short circuit current,

$$I_{sc} = 3 \times I_{st} \text{ with star-delta starter}$$

$$= 3 \times 2I_f = 6I_f$$

or $$\frac{I_{sc}}{I_f} = 6$$

Starting torque with star delta starter,

$$T_{st} = \frac{1}{3}\left(\frac{I_{sc}}{I_f}\right)^2 . S_f . T_f$$

or $$\frac{1}{2} T_f = \frac{1}{3} (6)^2 S_f . T_f$$

or $$S_f = \frac{1}{24} \quad \text{where } S_f = \text{Full load slip.}$$

Starting current when started by auto-transformer with 40% tapping.

$$I_{st} = K^2 I_{sc} = (0.4)^2 \times 6\, I_f = 0.96\, I_f. \qquad \textbf{Ans.}$$

Starting torque when started by auto-transformer with 40% tapping

$$= K^2 \left(\frac{I_{sc}}{I_f}\right)^2 S_f . T_f = (0.4)^2 \times (6)^2 \times \frac{1}{24} \times T_f = 0.24\, T_f$$

or 24% of full load torque. **Ans.**

Example 1.8 *A delta connected, 400 V, 6 pole, squirrel cage induction motor takes a full load current of 40 A at slip of 4%, the power factor and efficiency being 0.9 and 0.88 respectively. The impedance per phase is 4 Ω. Determine the starting current from supply and starting torque if the motor is started by :*

(*i*) *Star-delta starter*

(*ii*) *An auto-transformer with 70% tapping.*

Solution. Full load current, $I_f = 40$ A

Full load slip, $s_f = 0.04$

Short circuit current/phase when normal voltage of 40 V is applied

$$= \frac{V}{Z_p} = \frac{400}{4} = 100 \text{ A.}$$

Short-circuit line current when normal voltage of 400 V is applied.

$I_{sc} = 100\sqrt{3}$ A (Since motor is δ connected)

(*a*) **With a star-delta starter**

(*i*) Starting current,

$$I_{st} = \frac{1}{3} I_{sc} = \frac{1}{3} \times 100\sqrt{3} = \textbf{57.8 Amp. Ans.}$$

(*ii*) Starting torque,

$$T_{st} = \frac{1}{3}\left(\frac{I_{sc}}{I_f}\right)^2 S_f T_f = \frac{1}{3}\left(\frac{100\sqrt{3}}{40}\right)^2 \times 0.04 \times T_f$$

$$= 25\% \text{ of full load torque. } \textbf{Ans.}$$

(*b*) **With an Auto-transformer starter with 70% tapping,**

$K = 0.7$

(*i*) Starting current,

$$I_{st} = K^2 I_{sc} = (0.7)^2 \times 100\sqrt{3}$$

$$= 84.8 \text{ amp. } \textbf{Ans.}$$

(*ii*) Starting torque,

$$T_{st} = (0.7)^2 \left(\frac{100\sqrt{3}}{40}\right)^2 \times 0.04\, T_f$$

$$= 36.8\% \text{ of full load torque. } \textbf{Ans.}$$

1.47 Types of Loads

Loads can be of two types-those which provide active torque and those which provide passive torque.

Active torques are due to either gravitational force or deformation in elastic bodies. The active torque due to gravitational pull are obtained in case of hoists, lifts and elevators and railway locomotives operating on gradients. Such torques are also developed during compression or release of spring. Since the functioning of hoisting mechanism, operation of locomotives on gradients and compression or release of springs are all closely associated with a change in potential energy of drive, active torques are also closely connected to the potential energy. When a load is moved upward or a spring is compressed the stored potential energy increases and active torques develop opposes the action that take place *i.e.* the torque is directed against the upward movement or compression. On the other hand when a load is brought downwards the stored potential energy decreases and torque associated with it aids the action. Thus it is seen than active torques continue to act in the same direction even after the direction on drive is reversed.

Passive torques are those due to friction or due to shear and deformation in inelastic bodies. They always oppose motion, retarding the rotation of the driven machine. Moreover with change in direction of motion, the sense of torque also changes. For example when a weight is being lifted up, the friction torque adds to useful torque, but when lowered down it subtracts from the latter.

1.48 Different Types of Industrial Loads

There are three different types of loads under which the motor is required to work, these are :

(*i*) Continuous loads,
(*ii*) Intermittent loads,
(*iii*) Variable or Fluctuating loads.

The size of motor depends upon, firstly the temperature rise, which in turn depends whether the motor is to operate a continuous, intermittent or variable load and secondly, the maximum torque required of the motor. Thus keeping in mind the torque requirements of the load, the rating of a motor will be decided by the load conditions as under :

(*i*) **Continuous load.** Drives like pumps and fans etc. require a constant power as these keep on operating continuously and the choice of a suitable rating of motor is a simple affair.

Manufacturers charts are referred.

It is essential to calculate the exact amount of power needed by the drive so as to make the correct choice of the motor 'H.P.' If the H.P. of the motor is less than what is needed, the motor will overheat and subsequently burn out. On the other hand, if the H.P. of the motor is more than what is demanded by the load, more will be operating at less than full load value, and hence lower efficiency and power factor.

(*ii*) **Intermittent loads.** Intermittent loads may be of two types :

(*a*) The motor may be loaded for a short time and then shut-off for a sufficiency long time allowing the motor to cool down [Refer Fig. 1.86 (*a*)]. In such a case a motor with a suitable short time rating is adopted such as in mixies etc.

(*b*) The motor may be loaded for some time and then shut off for some times. Before the motor can cool down completely the load cycle repeats. [Refer Fig. 1.86 *b*]. In such a case a suitable continuous or short time rated motor is chosen which when operating on the particular load cycle, will not exceed the specified temperature limits.

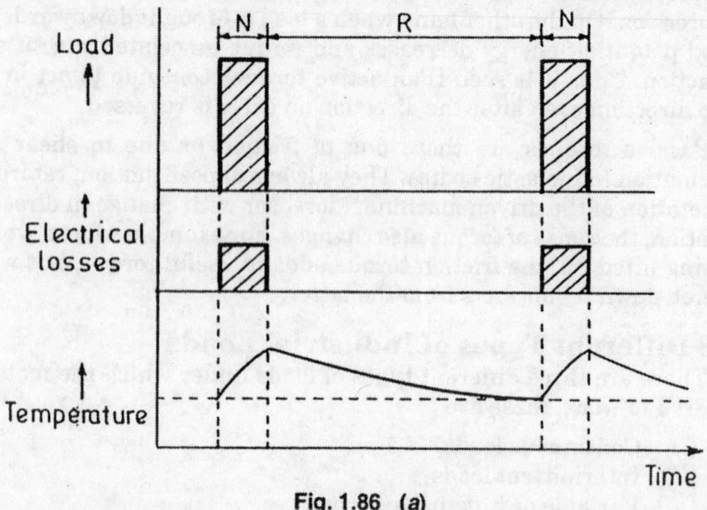

Fig. 1.86 (*a*)

(*iii*) **Variable or Fluctuating loads.** In case of fluctuating loads the most accurate method of selection of a suitable rating of motor is to draw the heating and cooling curves (as per load fluctuations) for a number of motors. The smallest motor, which does not exceed the permitted temperature rise when operating on the particular load cycle, is then chosen for the purpose.

However, an approximate and a simpler method of selection of a suitable rating of motor is to assume that heating is proportional to

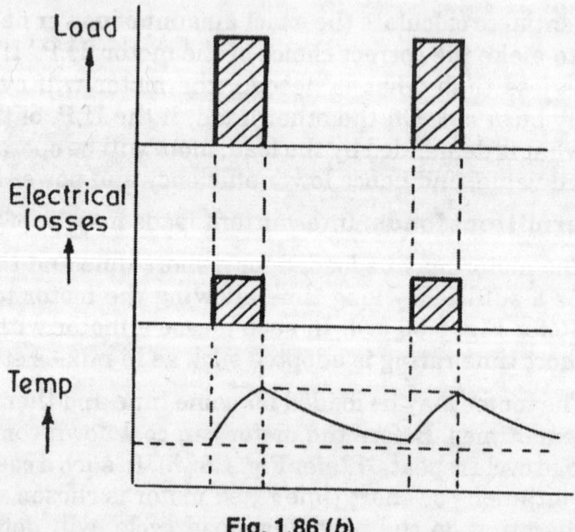

Fig. 1.86 (*b*)

square of the current and hence square of the load. The suitable continuous rating of the motor would be the r.m.s. value of the load curve. The procedure will be clear by the following examples.

Example 1.9. *A motor driving a colliery winder has the following load cycle :*

Acceleration period, 0 to 15 sec., load rising from 0 to 1050 H.P.

Full speed period, 15-85 sec., load constant at 600 H.P.

Deaccelerating period 85 to 95 sec., regenerative braking, the H.P. returned falling uniformly from 300 to 0 H.P. Decking 95 to 120 sec., motor is stationary. Estimate the size of continuous rated motor suitable for the load.

Solution. In the Example 1.3, the load varies uniformly during the accelerating period. Hence the $(H.P.)^2$ *Vs* time curve will be a parabola. The area of the shaded portion of a parabola is shown in Fig. 1.85 is $\frac{2}{3}$ times the area of the circumscribing rectangle, *i.e.* shaded Area $OAB = \frac{2}{3}$ (Area of rectangle $OABC$) and Area $OBC = \frac{1}{3}$ (Area of rectangle $OABC$).

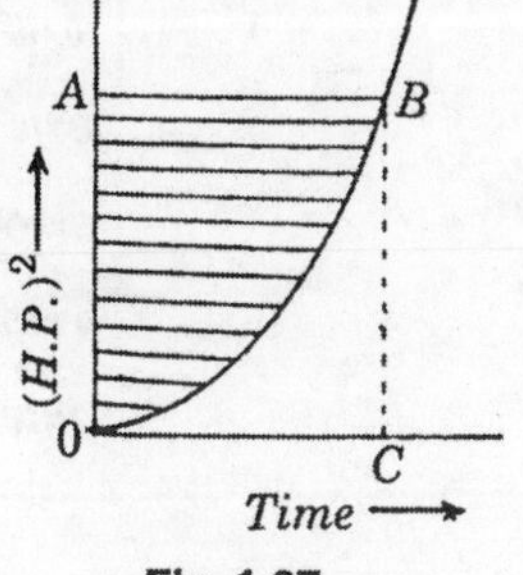

Fig. 1.87

The curve between H.P. *Vs.* time and $(H.P.)^2$ *Vs.* time are shown in Fig. 1.87.

Total area under the $(H.P.)^2$ curve is,

$$= \frac{1}{3} \times (1050)^2 \times 15 + (600)^2 \times 70 + \frac{1}{3} (300)^2 \times 10$$

$$= 55.125 \times 10^5 + 252 \times 10^5 + 3 \times 10^5$$

$$= 310.125 \times 10^5 \text{ H.P. sec.}$$

The above area should be equal to $(H.P._{eq.})^2 \times 120$, since the load cycle period is 120 sec.

i.e. $(H.P._{eq.})^2 \times 120 = 310.125 \times 10^5$

$\therefore$ $H.P._{eq.} = 507$ H.P.

The nearest rating of motor available in the market might be chosen.

Example 1.10. *The load cycle of a motor operating a lift for 9 minutes is as follows :*

Loading period at the bottom	*...4 min.*	*: 1 H.P.*
Load going up	*... 1 min.*	*: 20 H.P.*
Loading period at the top	*...3 min.*	*: 1 H.P.*
Load going down	*...1 min.*	*: – 15 H.P.*

Regenerative braking takes place when the load is descending. The cycle is repeated continuously. Estimate a suitable Horse Power rating for the motor.

Solution. The load curve is shown by the full time and the $(H.P.)^2$ curve is shown in dotted in Fig. 1.87. Between the 8th and 9th minute power is fed back into the lines. Still, however, heating of the motor takes place.

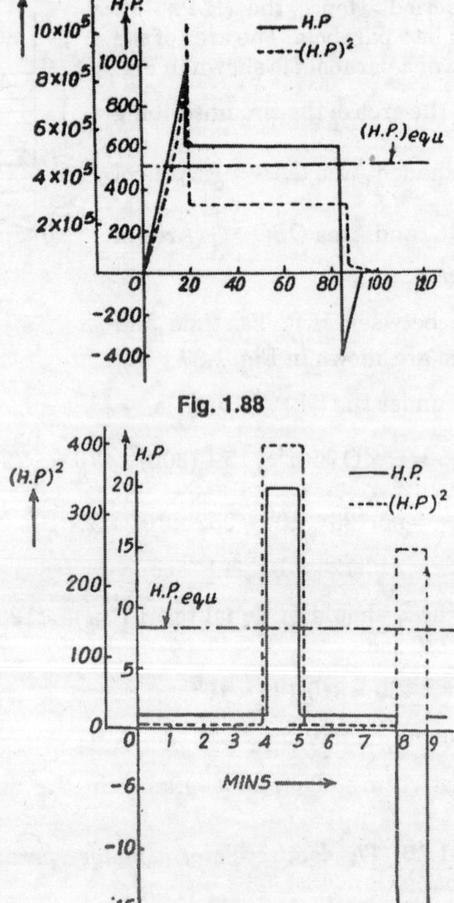

Fig. 1.88

Fig. 1.89

The total area under the $(H.P.)^2$ curve,

$$= (1)^2 \times 4 + (20)^2 \times 1 + (1)^2 \times 3 + (-15)^2 \times 1$$
$$= 4 + 400 + 3 + 225 = 632 \text{ H.P. min.}$$

The above area should be equal to $[(H.P._{eq.})^2 \times 9]$ H.P. min.

i.e. $(H.P._{eq.})^2 \times 9 = 632$

$\therefore$ $H.P._{eq.} = 8.4$ H.P.

Thus on the assumption that heat produced is proportional to the square of the load, the equivalent continuous rating of the motor works out as 8.4 H.P. The nearest standard rating of the motor will be 10 H.P.

1.49 Load Equalisation

The process of smoothing out the fluctuating load is known as load equalisation. In this process, energy is stored during the interval of light load and given out during the interval of load thus power from the supply remains approximately constant. The most common method of load equalisation is by use of flywheel.

In industry there are many applications, where load variations are pretty large. A very particular example is the steel rolling mill operation in which the load is very large during the rolling operation and very little as the charge passes by. A typical example of the load cycle of this operation is shown in Fig. 1.90.

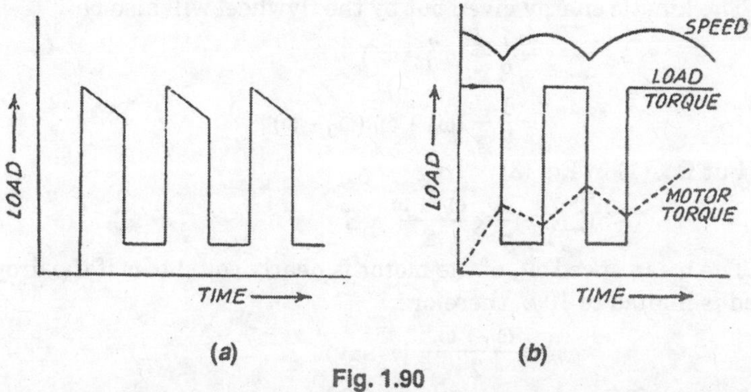

Fig. 1.90

It would require a motor of large output, though on the average, the load on the motor is not large. It could well be a short time rated motor. As the load cycle is quite regular, it is quite possible to devise ways to cater for the large load torque for a short while, thus effecting a considerable economy in motor size. The best way doing this is to increase the moment of inertia of the motor by means of a flywheel, which stores energy during the lean period of the cycle (when the load is small) ; and give it up during its peak period, when the load is large. The flywheel stores energy when the load is small and the speed picks up ; and gives it up to the system in time of demand, when the load is large. Thus, it brings about an equalization of load to a ascertain extent. It is thus practical to have a motor, whose torque is not as large as the load torque required, the gap being adjusted by the flywheel.

1.50 Flywheel Calculations

Let T_L = Load torque

T_m = Torque exerted by the motor

T_0 = No load torque

T_F = Torque supplied by the flywheel

ω = Angular speed of the motor at time, t in radians/sec.

ω_0 = Angular speed at no load in radians/sec.

J = Moment of inertia of flywheel in kg/m^1

t = Time in seconds

g = Acceleration due to gravity in metres/sec.^{2}

The slip of the motors, $S = \dfrac{\omega_0 - \omega}{\omega_0}$ ...(1)

Case I : Load increasing

Now consider that the load on the motor is increasing. As explained earlier during this period the flywheel will decelerate and impart its stored kinetic energy to the load. Therefore, the torque on the motor will be,

$$T_m = T_L - T_F \quad ...(2)$$

The kinetic energy given out by the flywheel will also be,

$$= \frac{1}{2}\frac{J}{g}(\omega_0^2 - \omega^2)$$

$$= \frac{1}{2}\frac{J}{g}(\omega_0 + \omega)(\omega_0 - \omega) \quad ...(3)$$

Put Eq. (1) in Eq. (3)

$$\text{K.E.} = \frac{J}{g} \times \frac{\omega_0 + \omega}{2} \times S \quad ...(4)$$

The mean speed ω_m of the motor is nearly equal to ω if the drop in speed is limited to 10%, therefore,

$$\omega_m = \frac{\omega_0 + \omega}{2} = \omega \text{ (say)} \quad ...(5)$$

In view of Eqs. (4) and (5)

$$\text{Energy given out by flywheel} = \frac{J}{g}\omega S \quad ...(6)$$

Power given out by the flywheel = Rate of change of energy

$$= \frac{d}{dt}\left(\frac{J}{g}\omega s\right) = \frac{J}{g} \cdot \omega \cdot \frac{ds}{dt} \quad ...(7)$$

$$\text{Flywheel torque, } T_F = \frac{\text{Power given out by the flywheel}}{\omega}$$

$$= \frac{\frac{J}{g} \cdot \omega \cdot \frac{ds}{dt}}{\omega} = \frac{J}{g} \cdot \frac{ds}{dt} \quad ...(8)$$

In view of Eq. (2) and (8), the motor torque,

$$T_m = T_L - \frac{J}{g} \cdot \frac{ds}{dt} \quad ...(9)$$

If the drop of speed, *i.e.* slip is limited to 10%, then,

$$S \propto T_m \quad ...(10)$$

$$\frac{ds}{dt} = K \cdot \frac{dT_m}{dt} \quad ...(11)$$

where K is a constant.

Set Eq. (11) in Eq. (9),

$$T_m = T_L - \frac{JK}{g} \frac{dT_m}{dt} \quad ...(12)$$

or
$$T_L - T_m = \frac{JK}{g} \times \frac{dT_m}{dt} \quad ...(13)$$

or
$$\frac{dT_m}{T_L - T_m} = \frac{g \cdot dt}{JK} \quad ...(14)$$

Integrating Eq. (14)

$$-\log_e (T_L - T_m) = \frac{g}{KJ} t + C_1 \quad ...(15)$$

where C_1 is the constant of proportionality and can be determined from the initial condition, when

$$\begin{Bmatrix} J = 0 \\ T_m = T_0 \end{Bmatrix} \quad ...(16)$$

From Eq. (15) and (16), $-\log_e (T_L - T_0) = C_1$ ...(17)

Substituting Eq. (17) in Eq. (15)

$$-\log_e (T_L - T_m) = \frac{g}{KJ} \cdot t - \log_e (T_L - T_o) \quad ...(18)$$

or
$$-\frac{g}{KJ} \cdot t = \log_e \frac{(T_L - T_m)}{(T_L - T_0)} \quad ...(19)$$

or
$$e^{-\frac{g}{KJ} t} = \frac{T_L - T_m}{T_L - T_0} \quad ...(20)$$

$\therefore$
$$T_L - T_m = (T_L - T_0)\, e^{-\frac{g}{KJ} t} \quad ...(21)$$

or
$$T_m = T_L - (T_L - T_0)\, e^{-\frac{g}{KJ} t} \quad ...(22)$$

[**Note.** $T_m = T_L - (T_L - T_0)\, e^{-\frac{t}{JK}}$. If the values of torques are given in NW-m.

Case II. Load Decreasing

While load decreases or is totally removed, the motor starts accelerating and so the kinetic energy is stored by the flywheel. The speed of the flywheel will increase to normal and thereby decreasing the slip. Under these conditions, $\frac{ds}{dt}$ is – ve.

Now motor torque will be, $T_m = T_o + T_F$ ...(23)

But $$T_F = -\frac{J}{g}\frac{ds}{dt}$$ ...(24)

The minus sign is used as $\frac{ds}{dt}$ is negative in view of Eqs. (23) and (24)

$$T_m = T_o - \frac{J}{g}\frac{ds}{dt}$$ ...(25)

If the above Eq. (25) is solved in the similar way as Eq. (9) was solved then the expression, we get,

$$T_m = T_o + (T_m' + T_o)\, e^{-\frac{g}{KJ}t}$$ ...(26)

T_m' is the torque of the motor when the load is decreased or removed.

1.51 Quadrantal Diagram of Speed Torque Characteristics

In view of the fact that both active and passive torques can be present in general, in a drive system, the motor driving the load may operate in different regions, not only as a motor, but for specific periods, also as a generator and as a brake. Further in many applications, the motor may be required to run in both directions. Therefore in sketching the speed torque characteristics of either the load or motor, it is preferable to use all four quadrants of speed torque plane for plotting, rather than to confine the characteristics to the first quadrant alone. When drawn in this manner, the diagram is referred to as quadrantal diagram.

The conventions used for + ve and negative values of speed, motor torque and load torque in a diagram of this type must be understood very clearly. The speed is assumed to have a positive sign, if the direction of rotation is anticlockwise or is in such a way to cause an upward or forward motion of the drive. In the case of reversible drive, the positive sign for speed may have to be signed arbitrarily either to anticlockwise or clockwise direction of rotation.

The motor torque is said to be positive if it produce an increase in speed in the positive sense. The load torque is assigned a positive sign when it is directed against the motor torque.

Fig. 1.91 shows the four quadrant operation of a motor driving a hoist, consisting of a cage with or without load, a rope wound onto a drum to hoist the cage and a balance weight of magnitude greater than that of the empty cage but less than that of the loaded one. The arrows in this Fig. 1.91 indicates the actual direction of motor torque, load torque and motion in the four quadrants. It can be easily seen that they

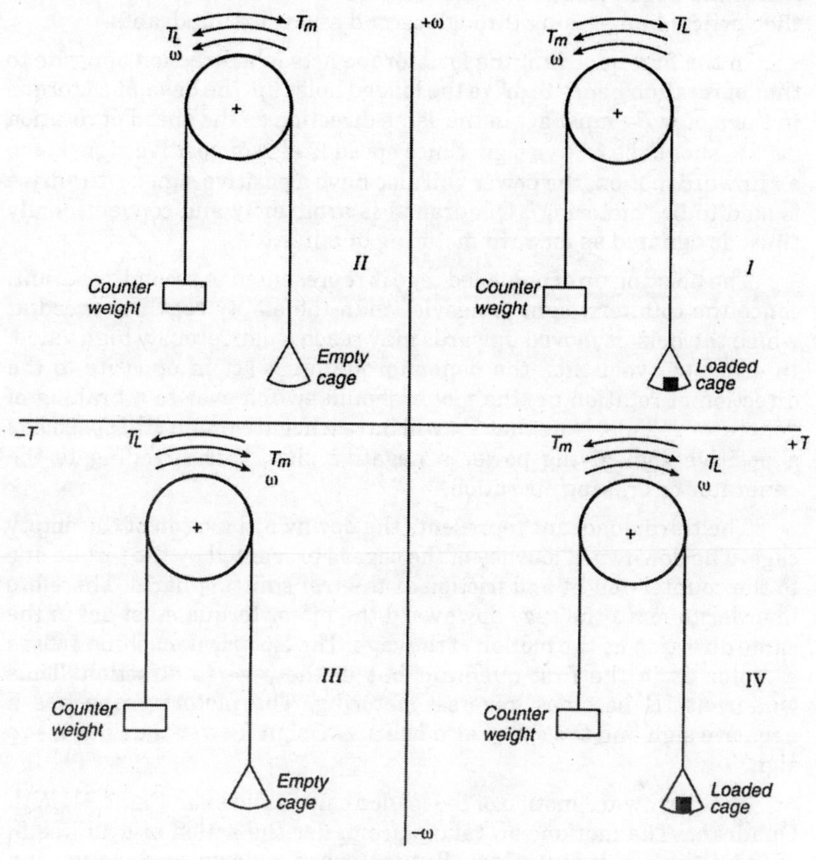

Fig. 1.91

correspond to the sign conventions stated earlier for speed, motor torque and load torque.

The load torque of the hoisting mechanism may be assumed to be constant *i.e.* independent of speed, since the force due to friction and windage are small in case of low speed hoists and the torque is primarily due to gravitational pull on the cage. This torque being an active, load torque doesn't change its sign even when the direction of rotation of the driving motor is reversed. Therefore the speed torque curves of a hoist load can be represented by means of vertical lines passing through two quadrants. The speed torque characteristics of the loaded hoist is shown in Fig. 1.92 by means of vertical lines passing the Ist and fourth quadrants. Since the counter weight is assumed to be heavier than the empty cage. The inherent tendency of load viz. the empty cage is to move in an opposite direction to that of load presented by the loaded cage and hence the speed torque curve of the unloaded hoist is represented by the vertical line passing through second and third quadrants.

In the first quadrant the load torque acts in a direction opposite to that of rotation hence to drive the loaded hoist up, the developed torque in the motor T_m must act in the same direction as the speed of rotation *i.e.* T_m should be of + ve sign. Since speed is also of positive sign, being an upward motion, the power will also have a positive sign *i.e.* the drive is said to be "motoring". Quadrant I is arbitrarily and conventionally thus, designated as forward motoring quadrant.

The hoisting up of unloaded cage is represented in second quadrant. Since the counterweight is heavier than the empty cage the speed at which the hoist is moved upwards may reach a dangerously high value. In order to avoid this, the motor torque must act in opposite to the direction of rotation *i.e.* the motor should switch over to a braking of generator regime. Note that T_m will have a negative sign still speed has a positive sign giving power a negative sign, corresponding to the generator or braking operation.

The third quadrant represents the downward motion of the empty cage. The downward journey of the cage is prevented by the torque due to the counterweight and friction at the transmitting parts. Therefore in order to move the cage downward the motor torque must act in the same direction as the motion of the cage. The electrical machine acts as a motor as in the first quadrant but in the reverse direction. Thus Quadrant III becomes 'reverse' motoring. The motor torque has a negative sign and the speed also has a – ve sign. Power thus has a +ve sign.

The downward motion of the loaded cage is shown in Fig. 1.91, IVth Quadrant. The motion can take place under the action of load itself ; without the use of any motor. But the speed of downward motion can be dangerously high. Therefore in this case the electrical machine must act as a brake, limiting the speed of downward motion of the hoist. The motor torque has a positive sign. Since it causes a decrease in speed in the downward motion. The speed of course has a negative sign, being a downward journey. The power thus acquire a negative sign, corresponding to the braking operation of the motor. A second basic type of loading that occur is the one characterized by dry friction. This type of the load presents to the motor a passive torque which is essentially independent of the speed. It is characterized also by requirement of an extra torque at very near zero speed. In power applications, it is often called as the break away torque and in control system, it is referred to as striction. The speed torque curves for this type of load are shown in Fig. 1.92 (*a*).

Another type of friction loading is used by control system engineers is known as viscous friction. It is a torque whose magnitude is directly proportional to the speed. The viscous friction T-N characteristics are shown in Fig. 1.92 (*b*). Yet another basic type of load torque is one whose

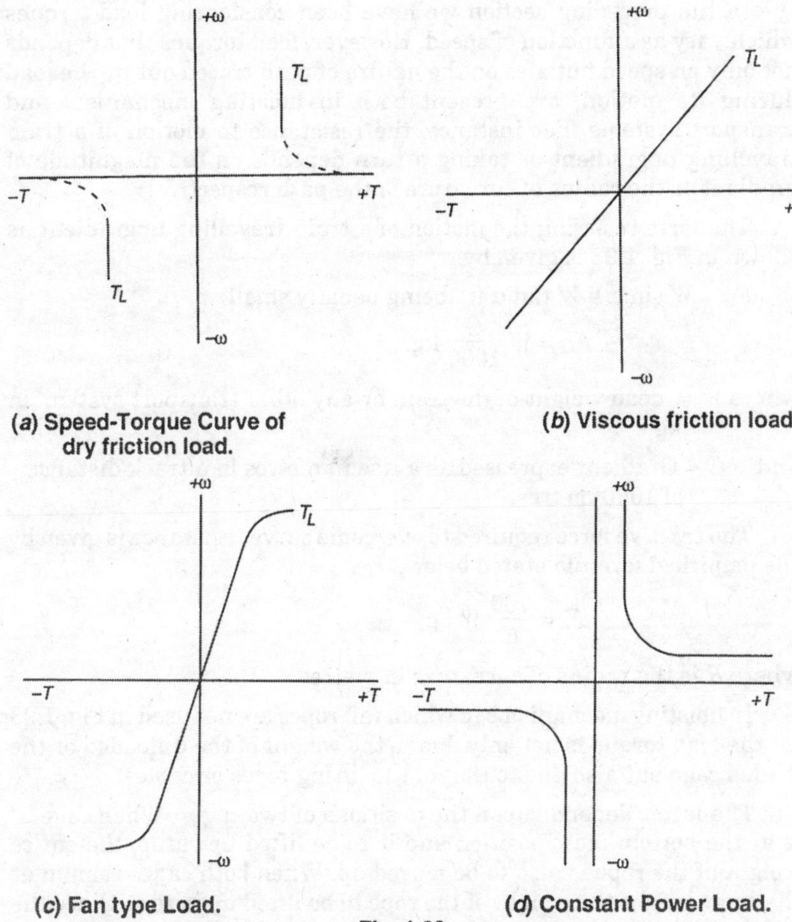

(*a*) Speed-Torque Curve of dry friction load.

(*b*) Viscous friction load

(*c*) Fan type Load.

(*d*) Constant Power Load.

Fig. 1.92

magnitude is proportional to some power of the speed. Such a load is best illustrated by a fan or blower. The torque produced by the fan is directly proportional to the square of the speed throughout the range of usable fan speeds. The speed-torque curves for the fan type of load are presented in Fig. 1.92 (*c*). Centrifugal pumps, propellers in ships or aeroplanes also have the same type of speed-torque characteristic.

Hyberbolic speed-torque characteristics as shown in Fig. 1.92 (*d*) is associated with certain type of lathes, boring machines and milling machines and steel mill coilers etc.

In general, the load torque in any specified application may consist of any of the above mentioned loads in varying proportions.

1.52 Load Torques that depends on the path or position taken by load during motion

In the preceding section we have been considering load-torques which vary as a function of speed. However, load torques that depends not only on speed but also on the nature of path traced out by the load during its motion, are present both in hoisting mechanism and transport systems. For instance, the resistance to motion of a train travelling upgradient or taking a turn depends on the magnitude of gradient or the radius of curvature of the path respectively.

The force resisting the motion of a train travelling upgradient as shown in Fig. 1.93 is given by

$$F_G = W \sin \alpha \equiv W \tan \alpha \text{ } (\alpha \text{, being usually small})$$

$$F_G = W \frac{G}{1000} \text{ kg.}$$

where W = dead weight of the train or any other transport system in kg.

and G = Gradient expressed as a rise in metres in a track distance of 1000 metres.

The tractive force required to overcome curve resistance is given by the empirical formula stated below,

$$F_c = \frac{700}{R} W \text{ kg.}$$

where R is the radius of curvature in metres.

In hoisting mechanisms in which tail ropes are not used in Fig. 1.93 (*b*) the load torque is not only due to the weight of the unloaded or the loaded cage but also due to that of the lifting ropes or cables.

The latter depends upon the positions of two cages. When cage '1' is at the bottom most position and is to be lifted upwards, the entire weight of the rope is also to be moved up. When both cages remain at the same height the weight of the rope to be lifted up is zero. Since the weight of the rope at both sides balance each other. When cage 1 is at the higher position than cage II, a portion of the weight of the rope acts in such a way to aid the upward motion of cage I. In fact when cage I occupies the top most position the whole weight of the rope aid the upward movement. The force that resist the upward motion of the load F_r due to the varying weight of rope depending on the position of the load is given as,

$$F_r = W_r \left(1 - \frac{2x}{h}\right) \text{ kg.}$$

where W_r = Total weight of the rope in kg

x = Height of the cage at any arbitrary position from the bottom most position in meters and

h = The desired maximum height to which the cage is to be moved upwards in meters.

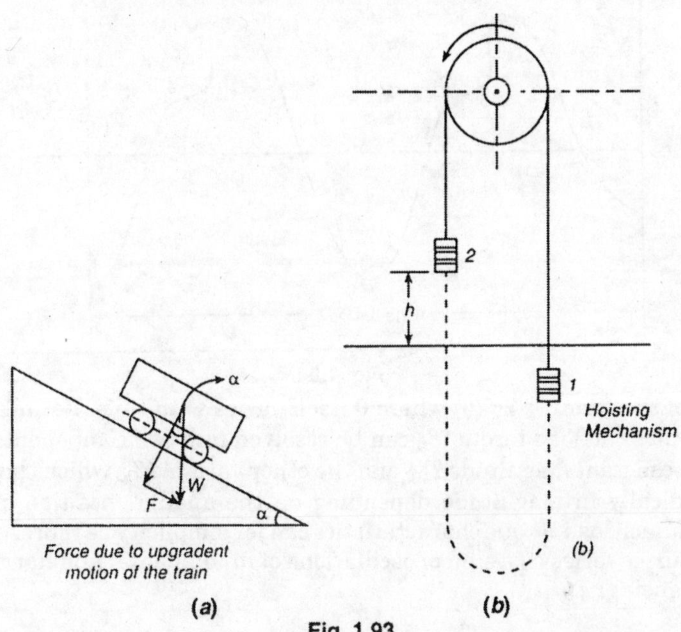

Fig. 1.93

Since for high values of 'h' the weight of rope may be considerably greater than that of the load to be lifted upwards, the force F_r affects to a large extent the performance of the drive to be used in hoisting mechanism. By using tail ropes, as shown in Fig. 1.93 by dotted lines in Fig. 1.93 (*b*) the weight of the connecting rope can be balanced and more or less smooth movement of the cage can be ensured.

Another example of a load torque which depends on path traced out during motion is that of a planer machine. At a particular position of the moving table containing the workpiece the load torque comes in the form of a sudden blow ; in a different position, after the cutter has come out of the job ; the magnitude of the load torque decreases sharply.

1.53 Load Torque that vary with Angle of Displacement of shaft

In all machines having crank shafts for example, in reciprocating pumps and compressors, frame saws, weaving looms, rocking pumps, used in petroleum industry etc. load torque is function of the position of the crank *i.e.* the angular displacement of the shaft or the rotor of the motor. Load torque in the drives used for steering ships also belongs to this category.

Fig. 1.94 shows the approximate relationship between load torque and angular displacement of the shaft"0' for a reciprocation compressor

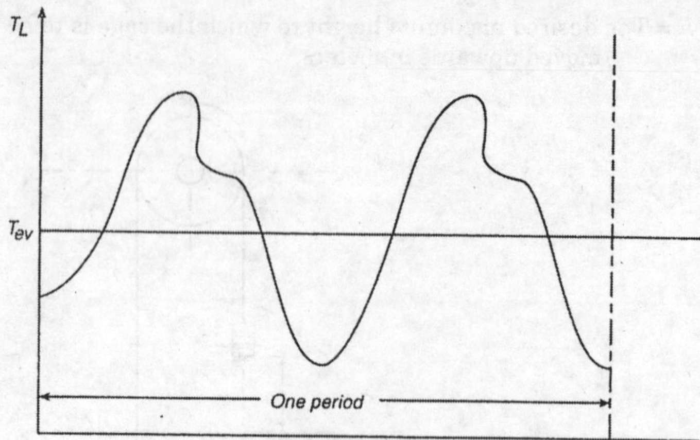

Fig. 1.94

It is of the form $T_L = f(\theta)$ where θ itself varies with time. For all such machines the load torque T_L can be resolved into two components one of the constant magnitude T_{av} and the other variable T_L which changes periodically in magnitude depending on the angular position of the shaft. Such load torque characteristics can for simplicity be represented by Fourier series as a sum of oscillations of fundamental and harmonic frequencies *i.e.*

$$T_L' = \sum_{r=0}^{m} T'_{L_1} \sin (r\theta + \phi_r)$$

$\theta = \omega t$ where ω represents the angular speed of the shaft of the motor driving the compressor.

During changes in speed, since only small deviations from a fixed value of speed ω_a occur. The angular displacement can be represented by $\theta = (\omega_a + \Delta\omega)\, t$ then the variable portion of the load torque may be expressed as,

$$T_L' = \sum_{0}^{m} T_{Lr}' \sin [(r\omega_a t + \phi_0) + r\Delta\omega t]$$

the term $r.\Delta\omega t$ being of very small magnitude can be neglected. Thus restricting to small deviation in angle from equilibrium position, a load torque which varies with the angular displacement of the shaft can be transformed to one which varies periodically with respect to time.

1.54 Load Torque that vary with Time

Of equal or perhaps greater importance in motor selection is variation of load torque with time. This variation in certain applications can be periodic and repetitive. One cycle of variation being called a duty cycle. It is convenient to classify different types of loads under the following groups :

(*a*) **Continuous Constant Loads :** Centrifugal pumps or fans operating for a long time under the same conditions ; paper making machines etc.

(*b*) **Continuous Variable Loads :** Metal cutting lathes ; hoisting winches, conveyors.

(*c*) **Pulsating Loads :** Reciprocating pumps and compressors, frame saws, textile looms and generally all the machines having crank shaft.

(*d*) **Impact Load :** Apparent, regular and repetitive load peaks or pulses which occur in rolling mills, presses, shearing machines, forging, hammers etc. Drives for such machines are characterized by heavy flywheels.

(*e*) **Short time Intermittent Loads :** Almost all forms of cranes and hoisting mechanisms, excavators, roll trains etc.

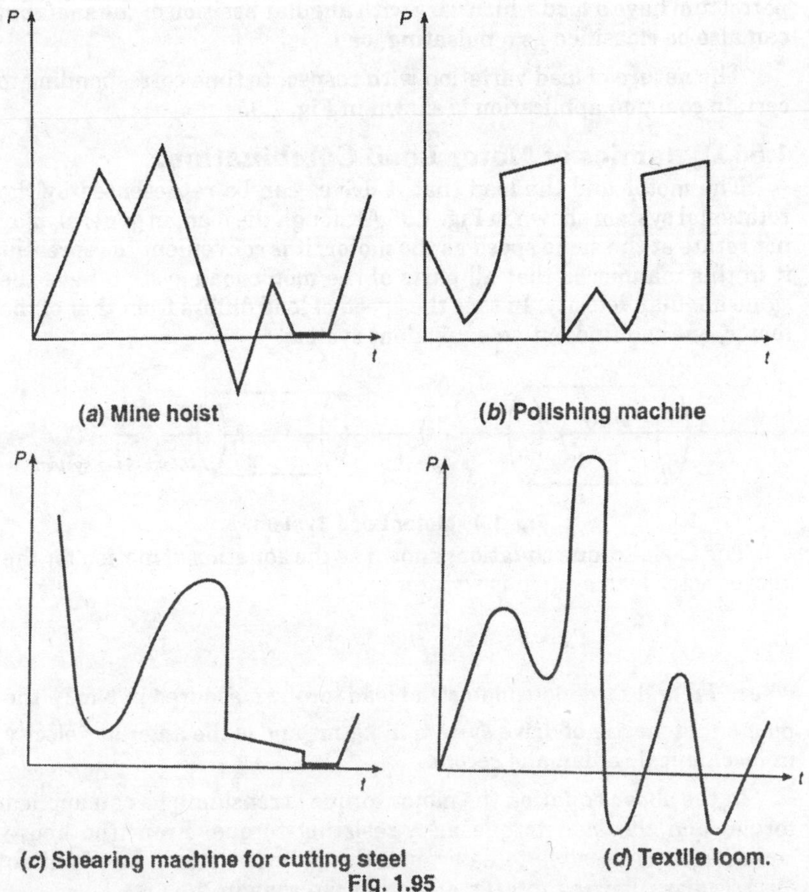

(*a*) Mine hoist (*b*) Polishing machine

(*c*) Shearing machine for cutting steel (*d*) Textile loom.

Fig. 1.95

(*f*) **Short time loads :** Motor-generator sets for charging batteries. Servo motors used for remote control of clamping rods of drilling machines.

Certain machines like stone crushers and ball mills do not strictly fall under any of the above groups. If these loads were characterized by frequent impacts of comparatively small peaks, it would be more appropriate to classify them under continuous variable loads rather than under impact loads, since both of them are periodic in nature and hence may be expressed as a sum of sinusoidal waves of different amplitude frequency and phase.

One and the same machine can be represented by a load torque which varies either with speed or with time. For example, a fan load whose load torque is proportional to the square of speed is also a continuous constant load. Load torque of a crane is independent of speed and also short time intermittent in nature. Rocking pumps for petroleum have a load which vary with angular position of the shaft but can also be classified as a pulsating load.

The nature of load variation with respect to time corresponding to certain common application is shown in Fig. 1.95.

1.55 Dynamics of Motor Load Combination.

The motor and the load that it drives can be represented by the rotational system shown in Fig. 1.95 Although the load, in general, may not rotate at the same speed as the motor, it is convenient to represent it in this manner so that all parts of the motor-load system have the same angular velocity. In case the speed of load differs from that of the motor, one can find out an equivalent system.

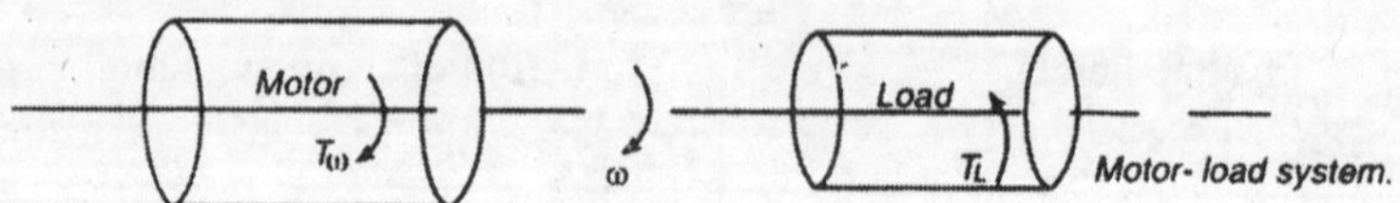

Fig. 1.96 Motor Load System.

The basic-torque equation, known as the equation of motion for the above motor load system is written as,

$$T_M = T_L + J\frac{d\omega}{dt}$$

where T_M and T_L denote motor and load torque measured in Nm ; J the moment of inertia of drive system in kg-m^2 and ω the angular velocity in mechanical radian and second.

In the above equation the motor torque is considered as an applied torque and the load torque as a resisting torque. From the above equation, it is possible to determine the different states at which an electric drive causing rotational motion can remain.

(i) $T_M > T_L$ *i.e.* $d\omega/dt > 0$ *i.e.* the drive will be accelerating, in particular picking up speed to reach rated speed.

(ii) $T_M < T_L$ *i.e.* $d\omega/dt < 0$ *i.e.* the drive will be decelerating and particularly coming to rest.

(iii) $T_M = T_L$ *i.e.* $d\omega/dt = 0$ *i.e.* the motor will continue to run at the same speed if it were running or will continue to be at rest if it were not running.

The above statements namely that when $T_M > T_L$ the drive accelerates and that when $T_M < T_L$ the drive accelerates are valid only when T_L happens to be a passive load. The reverse may occur with active loads. For example, if we were to switch on the motor for hoisting up a winch. While it is coming down on its own weight, until the direction of rotation changes deceleration of the drive and not acceleration takes place, when $T_M > T_L$. In case $T_M < T_L$ in the above situation when the motor has been switched on for moving the winch up, the load will continue to come down and motor will accelerate instead of decelerating.

The term $J\, d\omega/dt$ which represent the inertia torque, is also known as dynamic torque, since it is present only during transient condition *i.e.* when the speed of drive varies. During acceleration of the drive, the inertia torque is directed against motion but during braking, it maintains the motion of the drive. Thus inertia torque is determined both in magnitude and sign as the algebraic sum of the motor and load torques.

In the view of the above, the signs for T_M and T_L in equation correspond to motoring operation of the driving machine and to passive load torque or to a braking torque caused by active loads, respectively. The equation of motion can, in general, be written as :

$$\pm T_M = \pm T_L + J\frac{d\omega}{dt} \quad ...(1)$$

The signs to be associated with T_M and T_L in eq. (1) depends as indicated earlier, on the regions of operation of the driving motor and the nature of load torque. The equation of motion enables us to determine the variation of torque current and speed with respect to time during transient operation of the drive.

1.55.1 Equivalent system.

Seldom a motor shaft directly coupled to load shafts. In general, the different loads connected to the motor will have different speed requirements. Speed changing mechanism such as gears, V-belts etc. will be used to obtain different speeds. Since the ultimate objective is to select a motor suitable for the application, it is desirable to refer all mechanical quantities such as load torque, inertia torque etc. to one single axis of rotation, conveniently, the output shaft of the motor. The principle of conservation of energy will be used for this purpose.

1.56 Determination of Referred Load Torque

Let the speed of motor shaft be W_m and that of the load be W_h. Equating power, we have,

$$T_L \cdot W_L \frac{1}{\eta} = T_L' W_M$$

i.e.
$$T'_L = T_L \frac{W_L}{W_M} \times \frac{1}{\eta} = \frac{T_L'}{i\eta} \quad ...(2)$$

where T_L = Load torque

T_L' = Load torque referred to the motor shaft

$i = \frac{W_m}{W_L}$ = Speed transmission ratio (gear ratio) and

η = Efficiency of transmission.

When there are several stages in transmission between the driving motor and the driven machine, as shown in Fig. 1.97 given below with gear ratio $i_1, i_2, i_3, ... i_n$ and the respective efficiencies $\eta_1, \eta_2, ... \eta_n$, the load torque referred to the motor shaft is given as,

$$T_L' = T_L \times \frac{1}{i_1 \times i_2 .... i_n} \times \frac{1}{\eta_1 \cdot \eta_2 ... \eta_n} \quad ...(3)$$

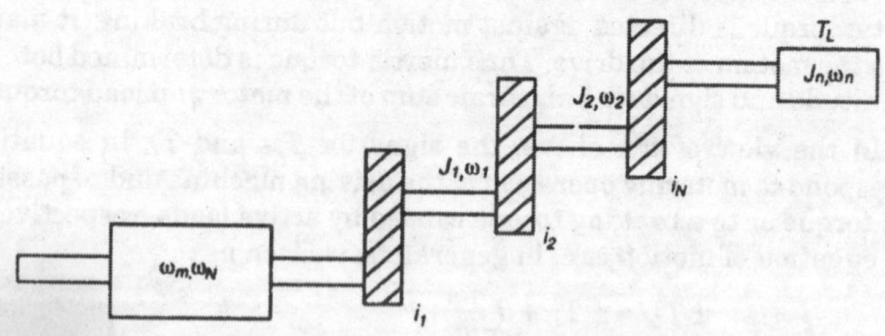

Fig. 1.97 Motor load system with gear.

1.57 Determination of Referred Moment of Inertia

Moment of inertia are referred to a given shaft on the basis that the total amount of kinetic energy stored in the moving parts and referred to the given shaft remains unchanged. With the rotating parts having moments of inertia $J_m, J_1, J_2 ,..., J_n$ and angular speeds $\omega_1, \omega_2 ...., \omega_n$. The equivalent moment of inertia referred to the shaft may be found as follows :

$$\frac{J' \omega_n^2}{2} = J_M \frac{\omega_M^2}{2} + J_1 \frac{\omega_1^2}{2} + J_2 \frac{\omega_2^2}{2} + ... + J_n \frac{\omega_n^2}{2}.$$

where J', the moment of inertia referred to the motor shaft

$$= J_M + J_1 \left(\frac{\omega_1}{\omega_m}\right)^2 + J_2 \left(\frac{\omega_2}{\omega_m}\right)^2 + ... + J_n \left(\frac{\omega_n}{\omega_m}\right)^2$$

$$= J_M + \frac{J_1}{i_1^2} + \frac{J_2}{(i_1 . i_2)^2} + \dots \dots + \frac{J_n}{(i_1 . i_2 \dots . i_n)^2} \quad \dots(4)$$

1.58 Referring forces and Masses having Translational Motion to a Rotating Shaft.

In many machines, for example in hoists, cranes and shaping machines, some of the moving parts rotate, while other go through a translation motion.

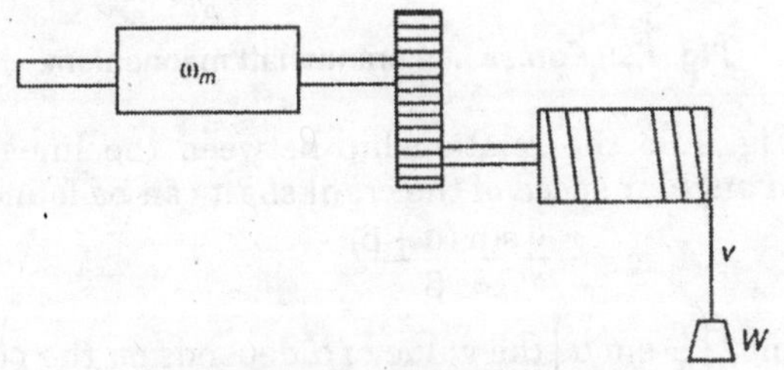

Fig. 1.98 Motor-hoist load system.

Fig. 1.97 shows the forces and torques acting on a hoist drive. If the moving mass has a velocity of v_m/sec and the motor shaft has an angular velocity ω_m rad/sec.

$$f_r \times v . \frac{1}{\eta} = T'_L \, \omega_M.$$

where f_r is the resisting force developed by the load due to the gravitational pull of the moving weight, ω and η, the efficiency of transmission.

$$\text{Referred Load torque, } T_L' = \frac{f_r . v}{\omega_m . \eta} \quad \dots(5)$$

A mass having translational motion is referred to a rotating one on the basis of constant kinetic energy

$$\frac{mv^2}{2} = J' \frac{\omega_m^2}{2}$$

The M.I. referred to the motor shaft

$$J' = m \, (v/\omega_m)^2 = W/g \, . \, (v/\omega_m)^2 \quad \dots(6)$$

1.59 Referring torques and masses which undergo translational motion at variable speeds.

In certain type of machines which convert the rotational motion into a translational one with the help of crank-shaft, the speed and acceleration of the moving masses vary both in magnitude and in sign during each revolution of the crankshaft. The kinetic energy stored in these masses will change from zero to a maximum. The moment of inertia of these masses referred to the crankshaft is

$$J' = \frac{mv^2}{\omega^2} \quad \dots(7)$$

where m = mass of the given body which undergoes the translational motion with a veloc ty v and ω, the angular velocity of crankshaft.

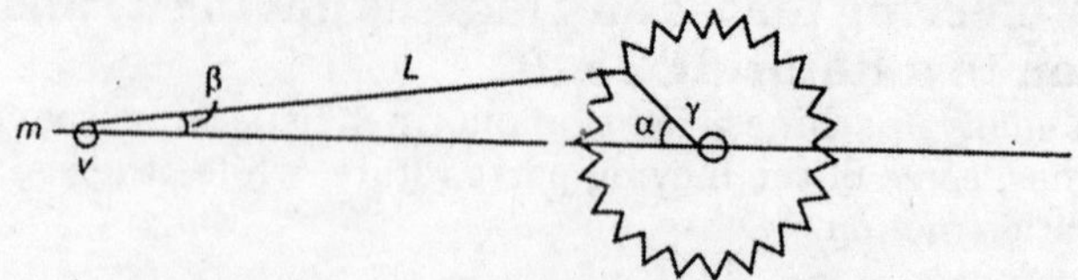

Fig. 1.99 Forces in a crankshaft mechanism.

Using Fig. 1.98 the relationship between the linear velocity of piston, v and angular speed of the crankshaft can be found as,

$$v = \frac{r\,\omega \sin(\alpha + \beta)}{\cos \beta} \quad ...(8)$$

where $\beta = \sin^{-1}\left(\frac{r}{L}\sin\alpha\right)$ the value of α depends on the position of the crank pin. Substituting, v from equation. The moment of inertia referred to the crankshaft

$$J' = \frac{mr^2 \sin^2(\alpha + \beta)}{\cos^2 \beta} \quad ...(9)$$

The total moment of inertia will be the sum of the moment of inertia J' and that of all other moving parts referred to the crankshaft. If there is some intermediate stage of transmission between the crankshaft and the driving motor, the moment of inertia referred to the crankshaft should then be referred to the motor shaft, using equation (5).

Hence the load torque referred to the motor shaft,

$$T_L' = \frac{f \,.\, r \sin(\alpha + \beta)}{\eta\, i \,.\, \cos \beta} \quad ...(10)$$

where f is the resisting force offered by the part, which undergoes translational motion, i the gear ratio ω_M/ω. If any and η the efficiency of transmission.

In mechanisms using crankshaft, the moment of inertia varies as a function of α and hence the equation of motion has a more complex form.

Kinetic energy stored in crankshaft $(K.E.)_c = \frac{J\omega^2}{2}$ ...(11)

The dynamic power, $P_{dyn.} = \frac{d\,(K.E)C}{dt} = J\omega\frac{d\omega}{dt} + \frac{\omega^2}{2}\frac{dJ}{d\alpha}\cdot\frac{d\alpha}{dt}$

$$= J\omega\frac{d\omega}{dt} + \frac{\omega^3}{2}\cdot\frac{dJ}{d\alpha} \qquad \left(\because \frac{d\alpha}{dt} = \omega\right) \quad ...(12)$$

The inertia torque, $T_{dyn.} = P_{dyn}/\omega$

$$= J\frac{d\omega}{dt} + \frac{\omega^2}{2} \cdot \frac{dJ}{d\alpha} \qquad ...(13)$$

The equation of motion becomes

$$T_M - T_L = J \cdot \frac{d\omega}{dt} + \frac{\omega^2}{2}\frac{dJ}{d\alpha} \qquad ...(14)$$

The additional term that has appeared is obviously due to the variable moment of inertia.

Example 1.11. *A 10-pole 50 Hz induction motor has to supply a load torque of 1000 N-m for 10 seconds followed by a non-load period during which it regains its original speed. The motor has a full load slip of 4% ; and can exert a maximum torque of 300 N-m. The flywheel connected to the motor has an inertia of 1400 kg-m² while the motor has a moment of inertia of 100 kg-m² only. Determine the torque exerted by the motor, at the end of the decelerating period and the time taken for the motor torque to fall to 300 N-m after the peak is over.*

Solution. Moment of inertia of the rotating parts,

$$J = 1400 + 100 = 1500 \text{ kg-m}^2$$

Load torque, T_L = 1,000 N-m

Load period, t = 10 seconds

No load torque, T_o = 0.

Now synchronous speed,

$$N_s = \frac{120f}{P} = \frac{120 \times 50}{10} = 60 \text{ r.p.m.}$$

Slip = 4% of 600 = 600 × 0.04 = 24 r.p.m.

$$= 24 \times \frac{2\pi}{60} = 0.8\,\pi \text{ rad./sec.}$$

$$K = \frac{\text{Slip}}{\text{Max. Torque}} = \frac{0.8\,\pi}{\frac{300}{9.81}} = \frac{0.8\,\pi}{300} \times 9.81 = 0.026\,\pi$$

[Please note that torque is given in Newton-metres]

$$\frac{g}{JK} \cdot t = \frac{9.81 \times 10}{1500 \times 0.026\,\pi} = 0.795$$

[1 kg-m = 9.81 N-m]

Motor torque at the end of de-acceleration period,

$$T_m = T_L - (T_L - T_o)\, e^{-\frac{g}{KJ}t} = 1000 - (1000 - 0)\, e^{-0.795}$$

$$= 1000 - 1000\, e^{-0.795} = 1000\,(1 - e^{-0.795})$$

$$= 1000\,(1 - 0.452) = 1000 \times 0.548$$

$$= 548 \text{ N-m. } \textbf{Ans.}$$

After the peak is over, the motor accelerates the flywheel to its normal speed and t :e expression for the motor torque is given as :

$$T_m = T_o + (T'_m - T_o)\, e^{-\frac{g}{KJ}t}$$

where T'_m is the motor torque at the peak instant.

Since $T_o = 0$, $T'_m = 548$ N-m $= \frac{548}{9.81}$ kg-m

[∵ 1 kg-m = 9.81 N-m]

$$\therefore \quad \frac{300}{9.81} = 0 + \left(\frac{548}{9.81} - 0\right) e^{-\frac{g}{KJ}t}$$

or
$$e^{-\frac{g}{KJ}t} = \frac{300}{9.81} \times \frac{9.81}{548} = \frac{300}{548} = 0.548$$

or
$$-\frac{gt}{JK} = \log_e 0.548 = -0.604$$

or
$$t = \frac{0.604\,.\,JK}{g} = \frac{0.604 \times 1500 \times 0.8\pi \times 9.81}{9.81 \times 300}$$

$$= 0.604 \times 1500 \times 0.00267\,\pi$$

$$= \mathbf{7.6\ Seconds.}\quad \textbf{Ans.}$$

Example 1.12. *A 4 pole, 50 Hz induction motion has a flywheel on its shaft. Total inertia at the motor shaft is 1,000 kg-m². Load torque is 100 kg-m for 10 seconds followed by a no-load period long enough for the flywheel by regain its full speed. Motor has a slip of 6% at a torque of 50 kg-m. Calculate the speed at the end of deceleration period. Assume motor speed-torque characteristic to be a straight line in the region of interest and neglect friction and windage.*

(Utilization of Elect. Power, A.M.I.E., Section B, Winter 1987)

Solution. Load torque,

$T_L = 100$ kg-m $= 100 \times 9.81 = 981$ N-m

Load period, $t = 10$ seconds

No. load torque, $T_0 = 0$

Synchronous speeds, $N_s = \frac{120f}{P} = \frac{120 \times 50}{4} = 1{,}500$ r.p.m.

Slip, $S = 0.06 \times 1500 = 99$ r.p.m.

or $\frac{2\pi \times 90}{60} = 9.42$ radian/sec.

$$K = \frac{\text{Slip}}{\text{Torque}} = \frac{9.42}{50 \times 9.81} = 0.0192$$

$$\therefore \quad T_m = T_L - (T_L - T_m)\, e^{-t/KJ}$$

$$= 981 - (981 - 0)\, e^{-\frac{10}{0.0192 \times 1,000}}$$

$$= 981\,(1 - e^{-0.52}) = 981\,(1 - 0.594) = 398 \text{ N-m.}$$

Slip speed, $S = KT_m = 0.0192398$ rad./sec.

$$= \frac{0.0192 \times 398}{2\pi} \times 60 = 73 \text{ r.p.m.}$$

Speed at the end of deceleration period

$$= 1,500 - 73 = 1437 \text{ r.p.m.}$$

Example 1.13. *An 8-pole, 50 Hz, 3-phase cage induction motor equipped with a flywheel is driving a mill, which requires a load toque of 1400 Kg-m during the rolling operation lasting 10 seconds. The torque then drops to 200 Km-m for 30 seconds, the cycle being repeated. The motor is capable delivering a maximum torque of 1050 Kg-m running at a slip of 10%. The moment of inertia of moving parts 50,000 kg-m^2. Determine the speed of the motor after 5 seconds during the rolling operation.*

Solution. Slip = 10%

$$= 750 \times 0.1 = 75 \text{ r.p.m.}$$

$$= \frac{75 \times \pi}{60} \text{ r.p.s.} = 7.85 \text{ rad./sec.}$$

As $S = KT_m$

$\therefore$ $K = \frac{S}{T_m} = \frac{7.85}{1050} = 0.0075$

with flywheel accelerating,

$$T_m = T_L - [T_L - T_O]\, e^{\frac{g}{JK}t}$$

Now $\frac{g}{JK} = \frac{9.81}{50,000 \times 0.0075} = 0.0026$

when $t = 5$

$\therefore$ $T_m = 1400 - [1400 - 200]\, e^{-0.0026 \times 5}$

$$= 1400 - 1200 \times e^{-0.013} \cong 420 \text{ kg-m.}$$

At this torque, the slip = 0.0075 × 490 ≅ 36.7 r.p.m.

Speed of the motor would be = 750 − 36.7 = 713 r.p.m.

Example 1.14. *A motor fitted with a flywheel supplies a load torque of 1000 kg-m for 10 seconds, after which it is at no load and regions, its original speed of 500 r.p.m. The motor can exert a maximum torque of 600 kg-m only. Determine the moment of inertia of the flywheel, assuming the full load slip of motor to be 10%.*

Solution. Slip = 10%

$$= 500 \times 0.1 = 50 \text{ r.p.m.}$$

$$= \frac{50 \times 2\pi}{60} = \frac{5\pi}{3} \text{ radian/sec.}$$

$$K = \frac{S}{T_m} = \frac{5\pi/3}{600} = 0.0087$$

$$T_m = T_L - (T_L - T_0)\, e^{-\frac{g}{Jk}t}$$

Substituting the values,

$$600 = 1000 - [1000 - 0]\, e^{-\frac{g}{Jk}t}$$

or
$$\frac{g}{JK}t = -\log_e \frac{T_2 - T_m}{T_L - T_0}$$

$$= -\log_e \frac{1000 - 600}{1000 - 0} = -\log_e \frac{400}{1000}$$

$$= -2.3 \log_{10} \frac{4}{10} = -2.3 \log_{10} 0.4$$

$$= -2.3 \times 1.6021 = -2.3\,(-1 + 0.6021)$$

$$= -2.3 \times -0.3979 = 0.915$$

$$J = \frac{g}{k \times 0.915} \times t = \frac{9.81 \times 10}{0.0087 \times 0.915}$$

$$= \frac{98.1}{0.0079605}$$

$$= \frac{981 \times 1{,}000{,}000}{79605} = 11607 \text{ kg-m}^2 \textbf{ Ans.}$$

Example 1.15. *A 3-ϕ, 8-pole 50 c/s induction motor equipped with a flywheel supplies a constant load torque of 100 Nw-m and at wide intervals an additional load torque of 300 Nw-m for 6 sec. The motor runs at 735 r.p.m. at 100 Nw-m and torque speed curve may be assumed to be linear over the working range.*

Find the minimum moment of inertia of the flywheel if the motor torque is not to exceed 250 Nw-m. *(Elect. Power-1988)*

Solution. Motor torque, $T_m = 250 \text{ Nw-m} = \frac{250}{9.81}$ kg-m.

Load torque, $T_L = 100 + 300 = 400 \text{ Nw-m} = \frac{400}{9.81}$ kg-m

$$T_0 = 100 \text{ Nw-m} = \frac{100}{9.81} \text{ kg-m}$$

Synchronous speed of the motor,

$$N_s = \frac{120f}{P} = \frac{120 \times 50}{8} = 750 \text{ r.p.m.}$$

$\therefore$ Slip at 120 Nw-m

$$= 750 - 735 = 15 \text{ r.p.m.}$$

$$= \frac{2\pi \times 15}{60} = 1.572 \text{ rad./sec.}$$

As slip at 100 Nw-m $\left(\frac{100}{9.81} \text{ kg m}\right)$ should be proportional to the torque at this moment,

$$K = \frac{1.572 \times 9.81}{100} = 0.154$$

Now we know,

$$-\frac{g}{KJ} t = \log_e \frac{T_L - T_m}{T_L - T_o}$$

or

$$J = \frac{g \times t}{K \log_e \left(\frac{T_L - T_o}{T_L - T_m}\right)} = \frac{9.81 \times 6}{0.154 \times \log_e \left(\frac{400 - 100}{400 - 250}\right)}$$

$$= \frac{9.81 \times 6}{0.154 \log_e 2} = \frac{9.81 \times 6}{0.154 \times 2.3 \log_{10} 2}$$

$$= \frac{9.81 \times 6}{0.154 \times 2.3 \times 0.3010} = \frac{9.81 \times 6}{0.154 \times 0.6923}$$

$$= \mathbf{551.6 \text{ kg-m}^2}. \textbf{ Ans.}$$

Example 1.16. *A motor fitted with a flywheel has to supply the load torque of 200 Kg-m for 10 sec. followed by a no-load period. During the no load period, the motor regains its speed. If it is desired to limit the motor torque to 100 Kg-m, what should be the moment of inertia of the flywheel ? The no load speed of the motor is 500 r.p.m. and has a slip of 10% at a torque of 100 Kg-m.*

Solution. T_m = 100 kg-m, T_L = 200 kg-m, T_0 = 0 kg-m

Load period, t = 10 sec.

$\therefore$ Slip, $S = 0.01 \times 500 = 50$ r.p.m.

$$= \frac{2\pi \times 50}{60} = \frac{10\pi}{60} \text{ radian/sec.}$$

Now $$K = \frac{S}{T_m} = \frac{10\pi}{6 \times 100} = \frac{\pi}{60}$$

Now $$T_m = T_L - (T_L - T_0)\, e^{-\frac{g}{KJ}}$$

or $$(T_L - T_0)\, e^{-\frac{g}{KJ}t} = T_L - T_m$$

or $$-\frac{g}{KJ}t = \log_e \frac{T_L - T_m}{T_L - T_0}$$

or $$J = \frac{g \times t}{K \log_e\left(\frac{T_L - T_0}{T_L - T_m}\right)} = \frac{9.81 \times 10}{\frac{\pi}{60}\log_e\left(\frac{200}{100}\right)}$$

$$= \frac{9.81 \times 10 \times 60}{\pi \log_e 2} = \frac{9.81 \times 600}{\pi \times 2.3 \times 0.3010}$$

$$= \frac{9.81 \times 600}{3.14 \times 2.3 \times 0.3010} = \mathbf{2.704\ kg\ m^2\ Ans.}$$

Example 1.17. *A motor equipped with the flywheel has to supply a load torque of 600 N-m for 10 sec. followed by no load period long enough for flywheel to regain its full speed. It is desired to limit the motor torque to 450 N-m. What should be the moment of inertia of the motor and flywheel ? The no load speed of the motor is 600 r.p.m. and has 8% slip at a torque of 450 N-m. The speed / torque characteristic of the motor can be assumed to be a straight line in the region of interest.*

(A.M.I.E. Sec. B. Util. of Elect. Power Summer 1986)

Solution. Motor Torque, $T_m = 450$ N-m

Load torque, $T_L = 600$ N-m

No. load torque, $T_0 = 0$

Load period, $t = 10$ sec.

Slip speed, $S = 0.08 \times 600 = 48$ r.p.m.

or $$= \frac{2\pi \times 48}{60} = 5.02 \text{ rad./sec.}$$

$$K = \frac{S}{T_m} = \frac{5.02}{450} = 0.01117$$

$$T_m = T_L - (T_L - T_0)\, e^{\frac{t}{KJ}} \quad [\because \text{Slip} \propto \text{Torque}]$$

$$= T_L - T_L \cdot e^{\frac{t}{KJ}} \quad [\because T_0 = 0]$$

or $$= T_L\left(1 - e^{\frac{t}{KJ}}\right)$$

or $$1 - e^{\frac{t}{KJ}} = \frac{T_m}{T_L} = \frac{450}{600} = 0.75$$

or $$e^{t/KJ} = 1 - 0.75 = 0.25$$

or $$J = \frac{-t}{K \log_e 0.25} = \frac{-10}{0.01117 \times (-1.3863)}$$

$$= 646 \text{ kg-m}^2 \ \textbf{Ans.}$$

1.60 Speed-Time Relationship

It is often necessary to find out the time taken by a motor to accelerate to a particular speed, or falling from a particular speed, for which a study of speed time relationship is necessary.

These relationships are based on the fundamental laws of dynamics, determining the relationships between force, velocity and acceleration.

(*i*) **Linear Motion**

If a is the acceleration in metres/sec^2.

W is the weight in kg.

g is the acceleration due to gravity in metres/sec^2.

Then tractive force,

$$F_t = \frac{W}{g} \times a \qquad ...(i)$$

or $$a = \frac{F_t \times g}{W} \text{ m/sec}^2 \qquad ...(ii)$$

But acceleration is also given as,

$$a = \frac{dv}{dt} \qquad ...(iii)$$

or $$dt = \frac{1}{a} dv \text{ sec.} \qquad ...(iv)$$

Integrating both sides, we get,

$$t = \int \frac{dv}{a} \qquad ...(v)$$

Further velocity, $v = \frac{ds}{dt}$ ...(*vi*)

or $$ds = v \,.\, dt$$

or $$S = \int v \, dt \text{ metres} \qquad ...(vii)$$

where S = distance travelled in metres.

(*ii*) **Angular Motion.** Let J be the moment of inertia of the rotating parts in kg-m^2, α be the angular velocity of the motor in radians/sec^2.

Then the torque exerted by the motor is given as :

$$T_m = \frac{J\,\alpha}{g} \text{ kg-metre} \qquad ...(viii)$$

or $$\alpha = \frac{T_m\, g}{J} \text{ radians/sec}^2 \qquad ...(ix)$$

If w is the angular velocity, then

$$\alpha = \frac{dw}{dt} \text{ radians/sec.}^2 \qquad ...(x)$$

where w is the angular velocity in radians/sec.

or $$dt = \frac{dw}{\alpha} \quad ...(xi)$$

or $$t = \int \frac{dw}{\alpha} \text{ sec.} \quad ...(xii)$$

Angular Motion, $w = 2\pi \frac{dn}{dt}$ ...(xiii)

where n is the number of revolutions

or $$dn = \frac{w\,dt}{2\pi} \quad ...(xiv)$$

or $$n = \int \frac{w}{2\pi}\, dt \quad ...(xv)$$

$\therefore$ $$n = \frac{1}{2\pi} \int w\, dt.$$

1.61 Determination of Time Required and Revolutions Made in Bringing a Motor to Standstill Point

Let the brakes be applied to a particular motor and assume that,

T_E = The electric braking torque,

T_M = The mechanical braking torque,

Then total braking torque,

$$T_B = T_E + T_M \quad ...(xvi)$$

we know that for constant excitation the electric braking torque is proportional to speed. So, it is assumed that the excitation remains constant, then,

$$T_E = K \,.\, w \quad ...(xvii)$$

where K is a constant and w is the angular velocity.

From equations (*xvi*) and (*xvii*)

$$T_B = T_M + K \,.\, w \quad ...(xviii)$$

Time Calculations

From equation (*ix*) angular retardation, a_r is given as :

$$a_r = \frac{T_B \,.\, g}{J} \quad ...(xix)$$

In view of equations (*xviii*) and (*xix*)

$$a_r = \frac{(T_M + K \,.\, w) \times g}{J} \quad ...(xx)$$

Time taken to slow down is given as :

$$t = \int \frac{-1}{a_r}\, dw \quad ...(xxi)$$

or $$t = \frac{-J}{(T_M + K \,.\, w) \times g}\, dw$$

or $$= \frac{-J}{kg} \log_e (T_M + Kw) + C_1 \quad ...(xxii)$$

The minus sign is used to indicate that a is – ve

when $t = 0$,

then w = initial speed = w_i

$$\therefore \quad 0 = -\frac{J}{Kg} \log_e (T_M + Kw_i) + C_1 \quad \text{...(xxiii)}$$

or
$$C_1 = \frac{J}{Kg} \log_e (T_M + Kw_i) \quad \text{...(xxiv)}$$

Adjust equation (*xxiv*) in equation (*xxii*)

$$t = -\frac{J}{Kg} \log_e (T_m + Kw) + \frac{J}{kg} \log_e (T_M + kw_i)$$

$$= \frac{J}{kg} \log_e \left(\frac{kw_i + T_M}{kw + T_M} \right) \quad \text{...(xxv)}$$

The time taken by the motor to attain the rest position can be calculated if in equation (*xxv*), the value of w is substituted as zero thus.

Time taken to bring the motor to standstill

$$= \frac{J}{kg} \log_e \frac{kw_i + T_M}{T_M} \quad \text{...(xxvi)}$$

Speed Calculations

In view of equation (*xxv*)

$$\log_e \frac{kw_i + T_M}{kw + T_M} = \frac{kg}{J} t \quad \text{...(xxvii)}$$

$$\frac{kw_i + T_M}{kw + T_M} = e^{+\frac{kg}{J} t} \quad \text{...(xxviii)}$$

or
$$kw_i + T_M = e^{\frac{kg}{j} t} (kw + T_M) \quad \text{...(xxix)}$$

or
$$kw = (kw_i + T_M) e^{-\frac{kg}{J} t} - T_M \quad \text{...(xxx)}$$

or
$$w = \left[\left(\frac{kw_i + T_M}{K} \right) e^{-\frac{kg}{J} t} - \frac{T_M}{K} \right] \quad \text{...(xxxi)}$$

Substituting the value of w in equation (*xv*)

$$n = \frac{1}{2\pi} \int \left[\left(\frac{Kw_i + T_M}{K} \right) e^{-\frac{kg}{J}} - \frac{T_M}{K} \right] dt$$

$$= \frac{1}{2\pi K} \left[\frac{(Kw_i + T_m)\, e^{-\frac{kg}{J} t}}{-Kg/J} T_{Mt} \right] + C_2 \quad \text{...(xxxii)}$$

where C_2 is a constant of integration.

Since from initial conditions, when the speed is w_i, $t = 0$; $n = 0$

$$\therefore \quad 0 = \frac{1}{2\pi K} \left[(T_M + Kw_i) \times \frac{(-J)}{kg} \right] + C_2$$

or
$$C_2 = \frac{J}{Kg} (Kw_i + T_M)$$

or $$n = \frac{1}{2\pi K}\left[\frac{-J}{kg}(Kw_i \times T_M)\, e^{-\frac{kg}{J}t} - T_M\, t + \frac{J}{kg}(Kw_i + T_M)\right]$$

or $$n = \frac{1}{2\pi K}\left[\frac{J}{gk}(Kw_i + T_M)\left(1 - e^{-\frac{kg}{J}t}\right) - T_M \times t\right] \qquad ...(xxxiii)$$

Example 1.18. *A 500 r.p.m. synchronous motor and its load have a total moment of inertia of 42 kg-m^2 and a frictional torque of 140 Nw-m. The motor is braked by a rheostatic method, the field current and the braking resistance being kept constant at a value which gives an initial braking torque of 5,400 Nw-m. Find the time taken to bring the motor to rest.* *[A.M.I.E. (Section B) May 1986]*

Solution. $T_M = 140 \text{ Nw-m} = \frac{140}{9.81} = 14.3 \text{ kg-m}$

Initial braking torque,

$$T_B = Kw_i = 5{,}400 \text{ Nw-m} = \frac{5400}{9.81} = 550 \text{ kg-m}$$

$$w_i = \frac{2\pi \times 500}{60} = 52.34 \text{ radians/sec.}$$

$$K = \frac{T_E}{w_i} = \frac{550}{52.34} = 10.5$$

Taking the equation (*xxvi*),

Time taken to bring the motor to rest,

$$= \frac{J}{gK}\log_e \frac{Kw_i + T_M}{T_M}$$

$$= \frac{420}{9.81 \times 10.5}\log_e \frac{550 + 14.3}{14.3} = 4.077 \log_e 39.47$$

$$= 4.077 \times 3.6756$$

$$= 14.98 \text{ or say 15 sec.} \quad \textbf{Ans.}$$

Example 1.19. *A 400 V, 8 pole, 3-phase synchronous motor and its load have a total moment of inertia of 630 kg-m^2. Determine the time taken and the number of revolutions made for it to come to standstill if (i) it is braked by plugging, which produces a constant braking torque of 165 kg-m. (ii) Rheostatic braking is employed which gives an initial electric braking torque of 690 kg-m. Assume a frictional torque 1.4 kg-m and supply frequency 50 Hz.*

[A.M.I.E. (Sec. B) Util. of Elect. Power Nov. 1982]

Solution.

Total moment of inertia of rotating parts, $J = 630 \text{ kg-m}^2$

Frictional Torque, $T_M = 1.4 \times 9.81 = 13.73$ N-m

(i) Electrical torque produced by plugging, $T_E = 165 \times 9.81$

$$= 1{,}618 \text{ N-m}$$

Total braking torque,

$$T_B = T_E + T_M = 1618 + 13.73 = 1631.73 \text{ N-m}$$

Angular retardation, $a_r = \frac{T_B}{J} = \frac{1631.73}{630} = 2.59$ rad./sec.^{2}

Synchronous speed of motor, $N_s = \frac{120 f}{P} = \frac{120 \times 50}{8} = 750$ r.p.m.

Time taken to stop the motor, $t = \frac{w_i}{a_r} = \frac{78.5}{2.59} = 30.3$ sec. **Ans.**

Number of revolutions made before coming to standstill,

$$= \frac{1}{2} w_i t = \frac{1}{2} \times \frac{750}{60} \times 30.3$$

$$= 189.4 \text{ revolutions.} \quad \textbf{Ans.}$$

(ii) Initial braking torque, $Kw_i = 690 \times 9.81 = 6{,}769$ N-m

and $$K = \frac{6769}{78.5} = 86$$

Time taken to bring the motor to rest,

$$t = \frac{J}{K} \log_e \frac{Kw_i + T_M}{T_M}$$

$$= \frac{630}{86} \log_e \frac{6{,}769 + 13.73}{13.73} = 45.5 \text{ seconds.} \quad \textbf{Ans.}$$

Number of revolutions made,

$$n = \frac{.1}{2\pi \times K}\left[\frac{J}{K}(Kw_i + T_M)(1 - e^{-Kt/J}) - T_M t\right]$$

$$= \frac{1}{2\pi \times 86}\left[\frac{630}{86}(6{,}769 + 13.73)(1 - e^{-86 \times 45.5/630}) - 13.73 \times 45.5\right]$$

$$= \frac{1}{172\,\pi}[7.3 \times 6782.73\,(-e^{-45.5/73}) - 625]$$

$$= 90 \text{ revolutions.} \quad \textbf{Ans.}$$

Example 1.20. *Determine the time taken by a D.C. shunt motor to fall in speed from 1500 r.p.m. to 500 r.p.m. at constant excitation. Given that a resistance of 20 ohm is connected across the armature e.m.f. induced in the armature at 1500 r.p.m. is 250 V and the moment of inertia of the armature is 20 kg-m^2. Neglect the effect of friction.*

Also determine the time taken if the frictional torque is equivalent to 2 kg-m^2.

Solution. Motor power supplied to the resistance connected across the armature,

$$= \frac{E^2}{R} = \frac{250 \times 250}{20} = 3125 \text{ watts.}$$

Electric braking torque,

$$T_E = \frac{3125 \times 60}{2\pi \times 1500 \times 9.81} = 2.028 \text{ kg-m}$$

We know as studied earlier that,

$$T_E = Kw_i$$

where, w_i = initial speed $= \frac{2\pi \times 1500}{60} = 157$ rad./sec.

$$K = \frac{T_E}{w_i} = \frac{2.028}{157} = 0.0129$$

Applying the equation,

$$t = \frac{J}{kg} \log_e \left(\frac{Kw_i + T_M}{Kw + T_M} \right)$$

(*a*) when $T_M = 0$

$$\therefore \quad t = \frac{J}{kg} \log_e \frac{w_i}{w} = \frac{20}{0.0192 \times 9.81} \log_e \frac{1500}{500}$$

$$= 157.8 \log_e 3 = 173.3 \text{ sec.} \quad \text{Ans.}$$

(*b*) $Kw_i = 2.028$

$$\therefore \quad Kw = \frac{2.028 \times 500}{1500} = 0.676$$

The mechanical braking torque is same as frictional torque,

$$\therefore \quad t = \frac{J}{kg} \left(\log_e \frac{Kw_i + T_M}{Kw + T_M} \right) = 157.8 \log_e \left(\frac{2.028 + 2}{0.676 + 2} \right)$$

$$= 157.8 \log_e \frac{4.028}{2.676} = 157.8 \log_e 1.505$$

$$= 157.8 \times 0.4087 = 64.5 \text{ sec.} \quad \textbf{Ans.}$$

1.62 Service Capacity and Rating

There are two factors which govern the size of a motor for a particular service :

(*a*) Maximum temperature rise when operating normally under the required load conditions.

(*b*) The maximum torque which the motor is required to give.

Of the two factors, if the first is satisfactory, the second one is also generally satisfactory unless the motor has to operate a load of heavy peaks followed by relatively long intervals of no load.

1.63 Temperature Rise

Electrical machines are often in continuous operation, for long periods. Hence one of the most severe constraints for their operation is their temperature rise. Not only the resistance of the windings increases with the temperature causing greater copper losses and consequently decrease in operating efficiency, the insulation of the windings also gets damaged ; and may not be able to withstand increase in temperature to that extent. Hence the capacity (output) of the machine is primarily limited on that count ; and in fact the temperature rise is one of the most important limiting factors for machine operation.

The machines are designed such that the rate at which heat is generated in the machine is equal to the rate of heat dissipation from it, so that a final steady temperature is reached, within the permissible limits for the machine. The increase in temperature is dependent on the following factors :

(*i*) Amount of heat produced.

(*ii*) Amount of heat dissipated from the surface of the motor.

The rate of rise in temperature will be high in the beginning but will decrease gradually with the passage of time and finally the temperature of the motor attains a steady value. At this point the heat produced and dissipated will be equal

Following assumptions are made for calculating the temperature rise :

(*i*) The losses, *i.e.* heat produced remains constant during the temperature rise.

(*ii*) The heat dissipated is directly proportional to the difference in temperature of the motor and cooling medium.

(*iii*) Temperature of the cooling medium remains unchanged.

1.63.1. Temperature Rise Calculations.

For finding the expression for temperature rise of an electrical machine after time t seconds from the instant of switching on. Let,

Power converted in to heat = W Joules/sec. or Watts

Mass of the Active parts of the machine = M kg

Specific heat of material = S Joules/kg/ ^oC

Surface area of cooling = A metres^2

Co-efficient of cooling = α in watts per metre^2 of surface per ^oC of difference between surface and ambient cooling temperature.

(i) **Heating Curve.**

Suppose a machine attains a temperature rise of θ^{o}C above ambient temperature in ^{o}C after t seconds of switching on the machine and further rise of temperature by $d\theta$ in very small time dt.

Energy converted into heat = $W \cdot dt$ Joules

Heat absorbed = $MSd\,\theta$ Joules.

Heat dissipated = $A\,\theta\,\alpha\,dt$ Joules.

Since energy converted into heat

= Heat absorbed + Heat dissipated

$$\therefore \quad W\,dt = M.S.\,d\theta + A.\,\theta.\,\alpha\,dt \qquad ...(i)$$

or $$(W - A.\,\alpha.\,\theta)\,dt = MS\,d\theta$$

or $$\frac{dt}{MS} = \frac{d\theta}{W - A\alpha\theta}$$

or $$\frac{dt}{MS/A\alpha} = \frac{d\theta}{W/A\alpha - \theta} \qquad ...(ii)$$

When final temperature is reached, there is no absorption of heat. Whatever heat is generated, has to be dissipated.

$$W\,dt = A\,\theta_m\,\alpha\,dt$$

or steady state temperature, $$\theta_m = \frac{W}{A\,\alpha} \qquad ...(iii)$$

Substituting $\dfrac{W}{A\,\alpha} = \theta_m$ in eq. *(ii)*, we get,

$$\frac{A\,\alpha}{MS}\cdot t = \int \frac{d\,\theta}{(\theta_m - \theta)}$$

Integrating both sides, we get,

$$\frac{A\,\alpha}{MS} t = -\log_e(\theta_m - \theta) = K_1 \qquad ...(iv)$$

where K_1 is a constant of integration.

Substituting, $t = 0$, $\theta = \theta_1$, ambient temperature from initial conditions, we get,

$$0 = -\log_e(\theta_m - \theta_1) + K_1$$

or $$K_1 = \log_e(\theta_m - \theta_1)$$

Substituting $K_1 = \log_e(\theta_m - \theta_1)\,r$ in eq. *(iv)*, we get,

$$\frac{A\alpha}{MS} t = \log_e(\theta_m - \theta) + \log_e(\theta_m - \theta_1) = \log_e\left[\frac{(\theta_m - \theta_1)}{(\theta_m - \theta)}\right]$$

or $$e^{\frac{A\propto}{MS} t} = \frac{\theta_m - \theta_1}{\theta_m - \theta}$$

or $$\theta_m - \theta = (\theta_m - \theta_1)\,e^{-\frac{A\alpha}{MS} t}$$

or $$\theta = \theta_m - (\theta_m + \theta_1)\, e^{-\frac{A\alpha}{MS}t}$$
$$= \theta_m - (\theta_m - \theta_1)\, e^{-t/\lambda} \quad ...(v)$$

where $\lambda = \frac{MS}{A\alpha}$ and is known as heating time constant.

If motor is started from ambient temperature, $\theta_1 = 0$, we get,

$$\theta = \theta_m - \theta_m\, e^{-t/\lambda} = \theta_m\,(1 - e^{-t/\lambda}) \quad ...(vi)$$

Substituting $t = \lambda$ in above expression, we get,

$$\theta = \theta_m\,(1 - e^{-1}) = 0.632\,\theta_m$$

1.63.2 Heating time Constant.

It is defined as the time taken in attaining the final steady temperature if the initial rate of increase of temperature were maintained. It is also defined as the time required to heat it, up to 0.632 times its final temperature rise.

The usual value of heating time constant is up to 90 minutes for motors of output up to 15 kW and up to 5 hours for large rating motors.

(ii) **Cooling Curve.** Let the machine be switched off after reaching steady state temperature rise of θ_m. When the machine is switched off, no heat is produced, therefore,

$$O = \text{Heat absorbed} + \text{Heat dissipated}$$

or $$O = MSd\,\theta - A\theta\,\alpha'\,dt$$

where α' is the rate of heat dissipation during cooling.

or $$dt = \frac{-MSd\theta}{A\,\alpha'\,\theta}$$

integrating both sides, we get,

$$t = \frac{-MS}{A\,\alpha'} - \log_e \theta + K_2 \quad ...(vii)$$

where K_2 is a constant of integration.

To get the value of K_2 substitute, $t = 0, \theta = \theta_m$ from initial conditions in the above expression,

$\therefore$ $$O = \frac{-MS}{A\,\alpha'} \log_e \theta_m + K_2$$

or $$K_2 = \frac{-MS}{A\,\alpha'} \log_e \theta_m$$

Substituting the value of K_2 in expression *(vii)*, we get,

$$t = \frac{-MS}{A\,\alpha'} \log_e \theta + \frac{MS}{A\,\alpha'} \log_e \theta_m$$

or $$t = \frac{-MS}{A\,\alpha'} \log_e \frac{\theta}{\Phi_m}$$

or $$\theta = \theta_m\, e^{-A'/MS} \quad ...(viii)$$

or $$\theta = \theta_m e^{-t/\lambda}$$

where $$\lambda' = \frac{MS}{A\,\alpha'}$$

The heating and cooling curve follow an exponential law. The heating and cooling curves for a motor which attains final temperature, the maximum permissible value and then cools down to ambient temperature are shown Fig. 1.100. Such curves are obtained in case of motors continuously operated on full load.

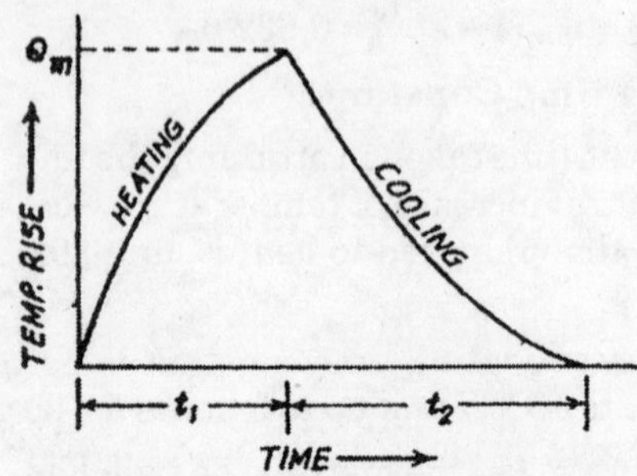

Fig. 1.100 Heating and cooling curves.

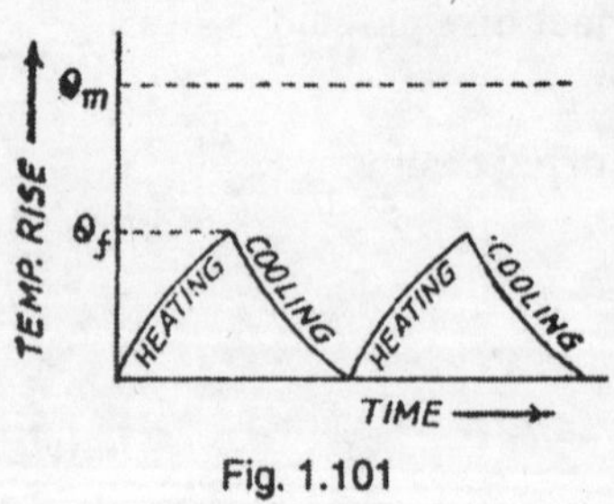

Fig. 1.101

The heating and cooling curves for an intermittently loaded motors are shown in Fig. 1.101.

A motor of smaller capacity can be used on intermittent load without exceeding the safe temperature rise as obvious from the curves.

Again, if there is an intermittent load on the motor such that the interval between the full load periods is not sufficient to allow the motor to cool to ambient temperature the shape of the curve will be as under shown in Fig. 1.102.

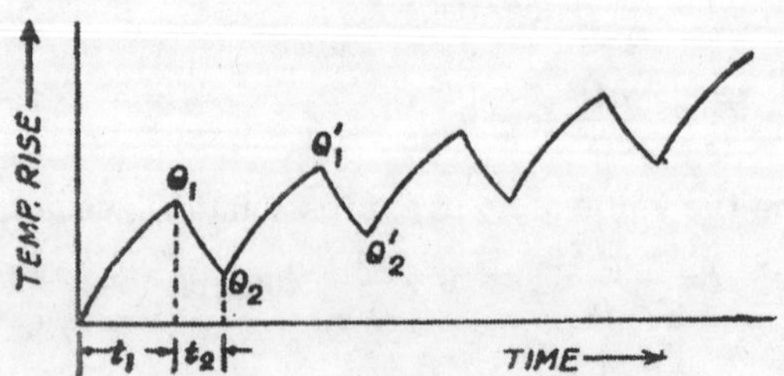

Fig. 1.102

1.64 Standard Ratings for Motors

The rating of electric machine indicates its load capacity, limited by the considerations of temperature rise. According to I.S.I. specifications, three kinds of ratings are generally specified as follows :

1. **Continuous Rating.** (Often indicated on the name plate of the machine as *Rating* : Cont. 10 H.P.)

It is the maximum load which the motor is capable of delivering continuously for an indefinite period, without the temperature exceeding specified limits ; and also capable of undertaking 25% overload for a period not exceeding 2-hours under the same conditions.

Thus if a motor is rated continues 10 H.P., it means that it is capable of giving an output of 10 H.P. continuously for an indefinite period ; and 12.5 H.P. for a period of two hours, without the temperature exceeding specified limits.

2. **Continuous Maximum Rating.** The load capacity as above but no overload capacity. Thus they are little inferior to the continuous rated motors.

This is the output which the motor can give continuously for a long time without exceeding the temperature rise. The motor cannot give any over load. It is used for motors of sizes larger than 2.5 H.P. per r.p.m.

3. **Short Time or Intermittent Rating.** This is the output, which a motor can deliver for a specified period say 1 hr. or $\frac{1}{2}$ hr. or $\frac{1}{4}$ hr. etc. without exceeding the specified temperature rise.

Many motors are not meant for continuous operation ; and could be designed accordingly, effecting an overall economy. *Short Time Rating* indicates the maximum load of the motor for the specified time followed by a no load period, during which the machine regains its original temperature.

Intermittent rating of a machine is defined as that load which is applied during a certain fraction of time of a load cycle and the temperature rise limit is not exceeded.

1.65 Estimation of Rating of a Motor

As the primary limitation for the operation of an electric motor is the temperature rise, its rating is calculated on the basis of its average temperature rise. The average temperature rise is calculated on the basis of its average heating during the period, which is proportional to the square of the current and the time for which the load persists.

Thus if a motor has load, L_1 for time t_1 and load L_2 for time t_2 and so on.

The average heating $\propto L_1^2 \times t_1 + L_2^2 \times t_2 + \ldots\ldots\ldots$

(Actually the heating is proportional to the square of the current, but as the load can also be expressed in terms of the current drawn, the proportionality could be taken for the load as well).

$$\text{Size of the motor} = \sqrt{\frac{L_1^2 \times t_1 + L_2^2 \times t_2 + ... L_n^2 \times t_n}{t_1 + t_2 + ... + t_n}}$$

In fact the load on the machine is generally expressed by its load cycle. There may be periods of no load as well in it. As a matter of fact, when the machine is running on no load, the heat generated is rather small, but heat dissipation continues at the same rate, as long as the machine is running. Thus there is a difference in the heating up of a motor running at no load and that while at rest. For that reason, the American practice is to consider the period at rest as one-third in the calculation of the size of the motor. It actually gives a higher value, which is advantageous for calculating the size being on the safe side.

Example 1.21. *A motor operates at full load of 100 H.P. for 10 minutes at 3/4 load for the next 10 minutes, at half load for the next 20 minutes, after which there is no load period for the next 20 minutes, the cycle is repeated. Calculate a proper size of a continuously rate motor for the purpose.*

Solution. Size of the motor,

$$= \sqrt{\frac{(100)^2 \times 10 + (75)^2 \times 10 + (50)^2 \times 20 + 0 \times 20}{10 + 10 + 20 + 20}}$$

$$\cong 61 \text{ H.P.}$$

However, considering the no load period as $\frac{1}{3}$rd according to American practice. The size

$$= \sqrt{\frac{(100)^2 \times 10 + (75)^2 \times 10 + (50)^2 \times 20 + 0 \times 20}{10 + 10 + 20 + 20/3}}$$

$$\cong 66 \text{ H.P.}$$

Example 1.22. *A motor has to be selected for a load rising uniformly from zero to 200 H.P. in 10 minutes, after which it is constant at 200 H.P. for next 10 minutes, followed by a no load period of 15 minutes, before the cycle repeats itself. Estimate a suitable size of continuous rated motor.*

Solution. Size of motor,

$$= \sqrt{\frac{(1/2 \times 200)^2 \times 10 + (200)^2 + 10 + 0}{10 + 10 + 15 \times \frac{1}{2}}}$$

$$= \sqrt{\frac{(100)^2 \times 10 + (200)^2 \times 10}{25}}$$

$$= 100\sqrt{2} \cong 140 \text{ H.P.} \quad \textbf{Ans.}$$

Example 1.22 *(a) load drops down to half the valve and remains constant at that for 40 seconds. The motor is at no load for the next one minute after which the cycle of operation is repeated. Estimate a suitable size of the motor for the purpose.*

Solution. The average load during the accelerating period

$$= \frac{1}{2} \times 2000 = 1000 \text{ H.P.}$$

Thus, the ineffect motor could be supposed to be working at 1000 H.P. for the 1st, 20 + 40 seconds, *i.e.* 1 minute.

It is at no load for the next one minute before the cycle is repeated.

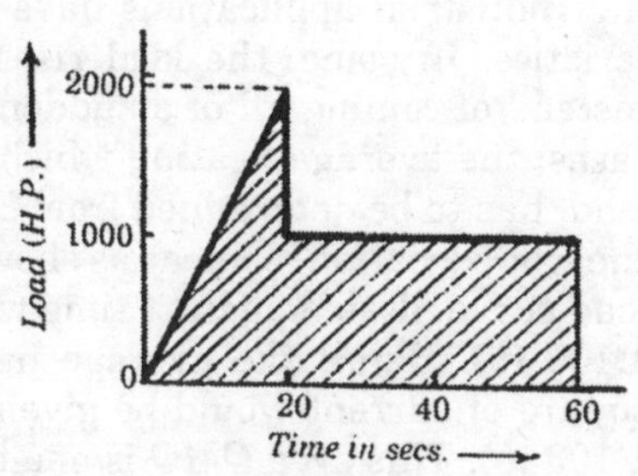

Fig. 1.103

Size of the motor,

$$= \sqrt{\frac{(1/2 \times 2000)^2 \times 20 + (1000)^2 + 40 + 0}{20 + 40 + 60 \times 1/3}}$$

$$= \sqrt{\frac{(1000)^2 \times 60}{80}} = 866 \text{ H.P.}$$

It should however be seen in all such cases, that the nearest standard size to be calculated, value has to be specified. Also, the size has to be such that the maximum load is not more than 125% of the rated value. Thus in example 2.17, we shall have to specify the nearest size, *i.e.*, 150 H.P. The maximum load being 200 H.P. is not above its over load capacity ; and hence, that would be an appropriate size.

However for the motor in example 1.22 *(a)*, the nearest size would be 900 H.P. Though the maximum load is 2000 H.P. yet it comes only during the accelerating period, rather for a short time. The actual maximum load is only 1000 H.P. which is well within the overload limits of 900 H.P. motor.

Example 1.23. *The driving motor of a machine has to work under the following load cycle :*

Light load for 5 minutes, 20 H.P. for the next 20 minutes, no load for next 5 minutes when the motor is at rest. The load cycle is then repeated indefinitely. Would it be proper to have a smaller sized continuous rated motor for the purpose instead of the maximum load size of 20 H.P. What would be the appropriate size ?

Solution. A continuous rated motor of a size less than the maximum load of 20 kW could be used. Appropriate size,

$$\sqrt{\frac{0 \times 5 + (20)^2 \times 20 + 0 \times 5}{5 + 20 + 5/3}} \cong 17 \text{ H.P.}$$

[**Note.** Only the no load period rest is counted as one-third, not the light load periods when the motor is running.]

Industrial applications have often different types of load characteristics. In some, the load rises uniformly from zero to full value, instead of coming all of a sudden, or in other different ways. In such cases, the average heating which is proportional to the square of the load, has to be determined from the actual nature of load variation. If the load variation is linear as shown in Fig. 1.104 (*a*). The square of the load curve plotted against time would be a parabola, as shown in Fig. 1.103 (*b*). Hence the average heating, which is proportional to the square of current would be given by the shaded portion *OPQ* in Fig. 1.103 (*b*). This area *OMQ* is one-third of the rectangular area *OMPQ*, while the unshaded part *OMP* = One-third of area *OMPQ*.

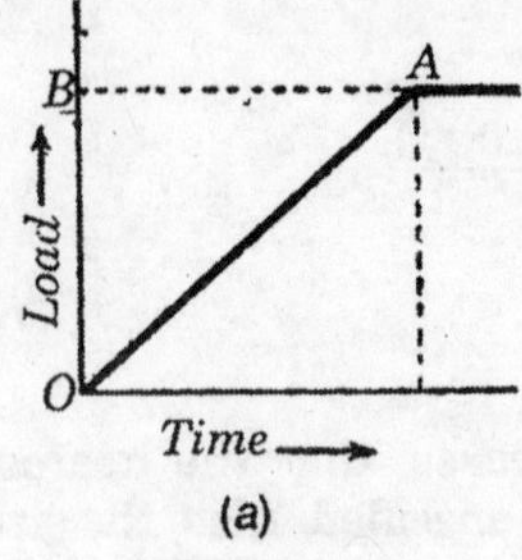

(a)

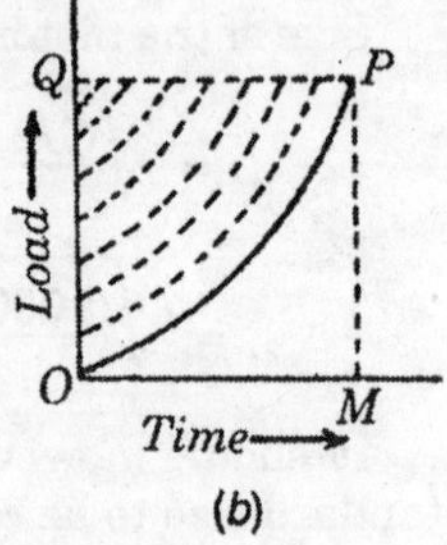

(b)

Fig. 1.104

Thus, the average heating is actually $\propto \frac{1}{3}(OB \times \text{Time})$

If the variation of load is not linear, but given by a curve (as a function of time) the average heating has to be found by computing the square of the function and taking its average accordingly.

However in practice, in case of uniform load variations as shown in Fig. 1.102 (*a*), average heating is just computed on the basis of half the load for the period for simplicity in calculations. Thus, the load rising uniformly from zero to *OB* in time *t* is computed as $\frac{1}{3} OB \times t$. This actually gives a size on the safer side ; and is generally accepted.

Example 1.24. *A motor has a heating time constant of 70 minutes and cooling time constant of 90 minutes. When run continuously on full load of 200 KW, the final temperature rise is 40° C. The motor is driving an intermittent load having 10 minutes on load period followed by 15 minutes shut down. What is the maximum value of load it can supply during the on load period without the maximum temperature rise* *exceeding 40° C ? Assume losses to be proportional to the square of the load.*

Solution. Cooling time constant, $\lambda' = 90$ minutes

Heating time constant, $\lambda = 70$ minutes

Load period, $t_1 = 10$ minutes

Off period, $t_2 = 15$ minutes.

Maximum temperature rise, $\theta_m' = 40^\circ C$

Continuous rating of motor, $W = 200$ kW

From the expression already discussed,

$$\theta_m' = \frac{1 - e^{-t_1/\lambda}}{1 - e^{-\left(\frac{t_1}{\lambda} + \frac{t_2}{\lambda'}\right)}}$$

or

$$\theta_m = \theta_m' \frac{\left[1 - e^{-\frac{t_1}{\lambda} + \frac{t_2}{\lambda}}\right]}{1 - e^{-t_1/\lambda}}$$

$$= 40 \frac{\left[1 - e^{-\left(\frac{10}{70} + \frac{15}{90}\right)}\right]}{1 - e^{-10/70}} = 40 \frac{(1 - 0.07338)}{(1 - 0.8669)} = 80^\circ C.$$

Let maximum value of load that can be supplied be W' kW, then

$$\left(\frac{W'}{W}\right)^2 = \frac{80}{40}$$

or

$$W' = W\sqrt{\frac{80}{40}} = 200\sqrt{2}$$

$$= 283 \text{ kW.} \quad \textbf{Ans.}$$

Example 1.25. *The heating time constant of a 30 H.P. motor is 90 minutes. If the motor is to be run continuously on full load, it attains a maximum steady temperature rise of 40° C. Calculate the half hour rating of the motor for this temperature rise. Assume that losses are proportional to the square of the load and the motor cools down completely between each load cycle.*

Solution. Rating of motor = 30 H.P. = 30 × 0.7355 = 22 kW

Heating time constant, λ = 90 minutes

$$\theta_m' = \frac{\theta}{1 - e^{-t/\lambda}} = \frac{40}{1 - e^{-30/90}} = \frac{40}{1 - e^{-1/3}} = 141^\circ C$$

Since losses are proportional to the square of load

$$\therefore \quad \frac{\text{Total losses at } W \text{ kW}}{\text{Total losses at 22 kW}} = \left(\frac{W}{22}\right)^2 \qquad [\because \theta_m \propto \text{losses}] \qquad ...(i)$$

and also
$$\frac{\text{Total losses at } W \text{ kW}}{\text{Total losses at 22 kW}} = \frac{\theta_m'}{\theta_m} \qquad ...(ii)$$

Comparing equations (*i*) and (*ii*)

$$\left(\frac{W}{22}\right)^2 = \frac{\theta_m'}{\theta_m} = \frac{141}{40}$$

or
$$W = 22\sqrt{\frac{141}{40}} = 13.2 \text{ kW. } \textbf{Ans.}$$

Example 1.26. *A 25 H.P. motor has a heating time constant of 90 minutes. When run continuously on full load, it attains a final steady temperature of 45°C. Calculate the temperature of the motor after 30 minutes, the initial temperature being 10°C.*

Solution. Final steady temperature = 45°C

Initial temperature = 10°C

Final temperature = 35°C

Now $\theta = \theta_m (1 - e^{-t/\lambda})$

When $t = 30$ minutes and $\lambda = 90$ minutes

Then
$$\theta = 35 (1 - e^{-30/90})$$
$$= 35 (1 - e^{-1/3}) = 12.8° \quad \textbf{Ans.}$$

Example 1.27. *A 15 H.P. motor has a heating time constant of 2 hours. Determine the time in which it attains a temperature of 40° C from its value of 10°C while working continuously on full load, if its final steady temperature is 45° C.*

Solution. Given, $\theta = 30°\text{C}$, $\theta_m = 35°\text{C}$, $\lambda = 2$ hours

Now $\theta = \theta_m (1 - e^{-t/\lambda})$

or
$$30 = 35 (1 - e^{-t/2})$$

or
$$1 - e^{-t/2} = \frac{30}{35}$$

or
$$e^{-t/2} = 1 - \frac{30}{35} = \frac{1}{7}$$

or
$$t = -2 \log_e \frac{1}{7}$$
$$\approx 1.7 \text{ or 1 hr. 40 minutes. } \textbf{Ans.}$$

Example 1.28. *A motor running continuously on full load has a temperature rise of 40° C. The heating time constant is 120 minutes. How long should be motor be run at twice the continuously rated output without overheating ? The motor has maximum efficiency at full load.*

Solution. Heating time constant, $\lambda = 120$ minutes

Final temperature rise, $\theta_m = 40^\circ C$

Let the losses at full load be W watts.

Since maximum efficiency occurs at full load, therefore, at full load copper loss is equal to iron loss and each is equal to half the total losses at full load, *e.g.* $W/2$ Watts.

Iron loss at twice the full load $= \dfrac{W}{2}$

[Since iron loss remains constant]

Copper loss as twice the full load $= \dfrac{W}{2}(2)^2$

Since copper loss various as the square of the load.

Total loss at twice full load = 2.5 W

Let the maximum temperature rise be θ_m', then,

$$\frac{\theta_m'}{\theta_m} = \frac{2.5\,W}{W}$$

or $$\theta_m' = 2.5\,\theta_m = 2.5 \times 40 = 100^\circ C.$$

Let the motor be run for t minutes at twice the continuously rated output without overheating (*i.e.* attaining temperature θ of 40°C).

Since $$\theta = \theta_m'(1 - e^{-t/\lambda})$$

or $$40 = 100\,(1 - e^{-t/120})$$

or $$e^{-t/120} = 0.6$$

or $$t = -120 \log_e 0.6 = 61.3 \text{ minutes}$$

or ≈ 1 hr. **Ans.**

Example 1.29. *Calculate the maximum over load that can be carried by a 25 KW motor, if the temperature rise is not exceed 50°C after one hour on over load. The temperature rise on the full load after one hour is 30°C and after 2 hours is 40°C. The losses vary as square of the load.*

Solution. The expression for temperature rise is given as,

$$\theta = \theta_m (1 - e^{-t/\lambda}) \qquad ...(i)$$

From given values,

when $t = 1$ hour, $\theta = 30^\circ$C

when $t = 2$ hour, $\theta = 40^\circ$C

Substituting these values of t and θ, we get,

$$30 = \theta_m (1 - e^{-1/\lambda}) \quad ...(ii)$$

and

$$40 = \theta_m (1 - e^{-2/\lambda}) \quad ...(iii)$$

Dividing expression (*iii*) by expression (*ii*), we get,

$$\frac{1 - e^{-2/\lambda}}{1 - e^{-1/\lambda}} = \frac{4}{3}$$

or

$$1 + e^{-1/\lambda} = \frac{4}{3}$$

or

$$e^{-1/\lambda} = \frac{1}{3}$$

Substituting $e^{-1/\lambda} = \frac{1}{3}$ in expression (*ii*), we get

$$\theta_m = \frac{30}{1 - \frac{1}{3}} = 45^\circ\text{C.}$$

If temperature rise after one hour is not to exceed 50°C, final temperature rise θ_m' will be given as,

$$50 = \theta'_m (1 - e^{-t/\lambda})$$

or

$$\theta'_m = \frac{50}{1 - e^{-1/\lambda}} \text{ Since } t = 1 \text{ hour}$$

$$= \frac{50}{1 - \frac{1}{3}} = 75^\circ\text{C Since } e^{-1/\lambda} = \frac{1}{3}$$

Let the maximum load that can be carried by a 25 kW motor be P kW, since losses vary as the square of the load,

$$\therefore \quad \frac{\theta'_m}{\theta_m} = \frac{\text{Losses at } P \text{ kW}}{\text{Original losses}}$$

$$\frac{75}{45} = \left(\frac{P}{25}\right)^2$$

or

$$P = 25\sqrt{1.667} = 32.25 \text{ kW.} \quad \textbf{Ans.}$$

Example 1.30. *At full load of 10 H.P., temperature rise of a motor is 25°C after 1 hour and 40°C after 2-hours. Find :*

(*a*) *The final temperature rise on full load.*

(*b*) *Heating time constant of motor.*

(c) *Half hour rating if iron losses, which remain constant are 80% of copper losses at full load.* *(P.U.)*

Solution. Applying equations given below for 25°C and 40°C temperature conditions, we get

$$\theta = \theta_m (1 - e^{-t/\lambda})$$

i.e. $$25 = \theta_m (1 - e^{-1/\lambda}) \quad ...(i)$$

and $$40 = \theta_m (1 - e^{2/\lambda}) \quad ...(ii)$$

Dividing (ii) by (i) $\dfrac{40}{25} = \dfrac{1 - e^{-2/\lambda}}{1 - e^{-1/\lambda}}$

If $e^{-1/\lambda} = x$ then, we get

$$\frac{40}{25} = \frac{1 - x^2}{1 - x}$$

$$\therefore \quad \frac{8}{5} = 1 + x$$

or $$8 = 5 + 5x \text{ or } 5x = 3 \;\because x = \frac{3}{5}$$

$$\therefore \quad e^{-1/\lambda} = \frac{3}{5}$$

$$\therefore \quad e^{1/\lambda} = \frac{5}{3}$$

$$\therefore \quad \frac{1}{\lambda} = \log_e \frac{5}{3} = 0.51$$

$$\therefore \quad \lambda = \frac{1}{0.51} = 1.96 \text{ hr.}$$

$$\therefore \quad 25 = \theta_m (1 - e^{-1/\lambda}) = \theta_m \left(1 - \frac{3}{5}\right) = \frac{2}{5}\theta_m$$

$$\therefore \quad \theta_m = \frac{5 \times 25}{2} = 62.5^\circ\text{C.}$$

If temperature rise after half an hour is 62.5 °C, final temperature rise θ_m will be given as,

$$62.5 = \theta_m' (1 - e^{-t/\lambda}) = \theta_m' \left(1 - \frac{1}{e^{t/\lambda}}\right) = \theta_m' \left[1 - \frac{1}{\left(\frac{5}{3}\right)^{0.5}}\right]$$

$$= \theta'_m \left(1 - \frac{1}{1.29}\right)$$

$$\therefore \quad \theta_m' = \frac{62.5}{0.29} \times 1.19 = 278^\circ\text{C}$$

Let copper losses at full load P $= 1$

Iron losses at full load will be $= 0.8$

Total losses at full load will be $= 1.8$

Copper losses at P' load $= \left(\frac{P'}{P}\right)^2 \times 1$

Iron losses at P' load $= 0.8$

Total losses at P' load $= 0.8 + \left(\frac{P'}{P}\right)^2$

$$\frac{\text{Total losses at } P}{\text{Total losses at } P'} = \frac{1.8}{0.8 + \left(\frac{P'}{P}\right)^2} = \frac{\theta_m}{\theta_m'} = \frac{62.5}{278}$$

or $$0.8 + \left(\frac{P'}{P}\right)^2 = \frac{278}{6.25} \times 1.8 = 8$$

$\therefore$ $$\left(\frac{P'}{P}\right)^2 = 8 - 0.8 = 7.2$$

$\therefore$ $$\frac{P'}{P} = \sqrt{7.2} = 2.68$$

$\therefore$ $$P' = 2.68 \times 10 = 26.8 \text{ H.P.} \quad \textbf{Ans.}$$

Example 1.31. *A 220 V d.c. series motor runs at 1000 r.p.m. and takes 20 A. Combined resistance of armature and series field is 0.4 ohms. Calculate the resistance to be inserted in series so as to reduce the speed to 800 r.p.m. assuming torque to vary as square of the speed and linear magnetisation curve.*

Solution. Motor current, $I_1 = 20$ A

Back e.m.f. $E_{b1} = V - I_1 (R_a + R_{se})$

$= 200 - 20 \times 0.4 = 192$ V

Speed, $N_1 = 1{,}000$ r.p.m.

Speed $N_2 = 800$ r.p.m.

Since $T \propto N^2$

$\therefore$ $$\frac{T_2}{T_1} = \left(\frac{N_2}{N_1}\right)^2$$

or $$\frac{I_2 \phi_2}{I_1 \phi_1} = \left(\frac{N_2}{N_1}\right)^2 = \frac{800}{1000} = 0.64 \qquad [\because T \propto \phi I_a]$$

or $$\frac{I_2}{I_1} \times \frac{I_2}{I_1} = 0.64 \qquad [\because T \propto \phi I]$$

or $$I_2 = \sqrt{0.64} \times I_1 = 0.8 \times 20 = 16 \text{ A}$$

Let the resistance of R ohms be connected in series with the motor circuit.

Back e.m.f. $E_{b_2} = V - I_2 (R + R_a + R_{se})$

$$= 200 - 16 (R + 0.4) = (193.6 - 16 R) \text{ volts}$$

Since $$\frac{E_{b2}}{E_{b1}} = \frac{N_2}{N_1} \times \frac{\phi_2}{\phi_1}$$

$\therefore$ $$\frac{193.6 - 16 R}{192} = \frac{800}{1{,}000} \times \frac{16}{20}$$

or $$R = \frac{193.6 - 0.64 \times 192}{16} = 4.42\ \Omega \textbf{ Ans.}$$

Example 1.32. *A 250 V d.c. shunt motor with constant field excitation drives a load whose torque varies as the square of the speed. The armature current is 20 A. When the motor is running at 500 r.p.m. Find the speed of the motor when running with a 25 Ω resistor connected in series with the armature. Neglect motor losses.*

Solution. Supply voltage, $V = 250$ volts

Armature current, $I_1 = 20$ A

Speed, $N_1 = 500$ r.p.m.

Let the speed be reduced to N_2, where $N_2 = KN_1$ after inserting additional resistance of 25 Ω in the armature circuit.

Since torque, which is proportional to armature current in case of shunt motor, varies as the square of the speed.

So new armature current, $I_2 = I_1 \left(\frac{N_2}{N_1}\right)^2 = K^2 I_1 = 20 K^2$

Since $N \propto V - IR$ flux remaining the same neglecting armature resistance.

$\therefore$ $$\frac{N_2}{N_1} = \frac{V - 20 K^2 \times 25}{V}$$

or $$K = \frac{250 - 500 K^2}{250}$$

or $2K^2 + K - 1 = 0$ or $K = \frac{1}{2}$

$\therefore$ New speed, $N_2 = KN_1 = \frac{1}{2} \times 500 = $ **250 r.p.m. Ans.**

Example 1.33. *A 250 V d.c. shunt motor, taking an armature current of 150 A and running at 550 r.p.m. is braked by reversing the connections to the armature and inserting additional resistance in series with it. Calculate :*

(*a*) *The value of series resistance required to limit the initial current to 240 A.*

(*b*) *The initial value of braking torque.*

(*c*) *The value of braking torque when the speed has fallen to 200 r.p.m.*

The armature resistance is 0.09 Ω. Neglect winding, friction and iron losses.

Solution. Full load armature current, $I = 150$ A

Terminal voltage, $V = 250$ V

Induced e.m.f., $E = V - IR = 250 - 150 \times 0.09 = 236.5$ V

Voltage across the armature at braking instant,

$$= V + E = 250 + 236.5 = 486.5 \text{ V}$$

(*a*) Resistance required in the armature circuit to limit the current to

$$240 \text{ A} = \frac{486.5}{240} = 2.027 \ \Omega$$

Resistance to be inserted in the armature ckt.

$$= 2.027 - 0.09 = 1.937 \ \Omega \ \textbf{Ans.}$$

Full load torque, $T_f = \dfrac{V \times I}{2\pi N/60} = \dfrac{250 \times 100}{2\pi \times 550} \times 60 = 650$ N-m

(*b*) Initial braking torque $= T_f \times \dfrac{\text{Initial braking current}}{\text{Full load current}}$

$$= 650 \times \frac{240}{150} = 1{,}040 \text{ N-m. } \textbf{Ans.}$$

(*c*) When the speed has fallen to 200 r.p.m. since the induced e.m.f. falls in the same proportional as the speed

$$\therefore \qquad E' = E \times \frac{200}{550} = 236.5 \times \frac{200}{500} = 94.6 \text{ V}$$

$$\therefore \text{ Current, } I' = \frac{250 + 94.6}{2.027} = 170 \text{ A}$$

$$\text{Electric braking torque} = 650 \times \frac{170}{150} = \textbf{737 N-m. Ans.}$$

Example 1.34. *A 30 KW, 400 V, 3-phase, 4 pole 50 Hz induction motor has full load slip of 5%. If the ratio of standstill reactance to esistance per rotor phase is 4. Estimate the plugging torque at full speed.*

Solution. Synchronous speed,

$$N_s = \frac{120\,f}{P} = \frac{120 \times 50}{4} = 1{,}500 \text{ r.p.m.}$$

Full load slip, $S_f = 0.05$

Speed at full load,

$$N_f = N_s\,(1 - s) = 1500\,(1 - 0.05) = 1425 \text{ r.p.m.}$$

Full load torque,

$$T_f = \frac{\text{Power developed in kW} \times 100}{2\pi N/60} = \frac{30 \times 1000}{2\pi \times 1425/60} = 200 \text{ N-m}$$

Since torque developed,

$$T \propto \frac{5\,R_2\,E_2^2}{R_2^2 + S_2^2 X_2^2} \therefore \frac{T_2}{T_1} = \frac{S_2\,R_2\,E_2^2/R_2^2 + S_2^2 X_2^2}{S_1\,R_2\,E_2^2/R_2^2 + S_2^2 X_2^2}$$

$$= \frac{S_2\,[R_2^2 + S_1^2 X_2^2]}{S_1\,[R_2^2 + S_2^2 X_2^2]} = \frac{S_2}{S_1} \cdot \frac{1 + S_1^2 \left(\frac{X_2}{R_2}\right)^2}{1 + S_2^2 \left(\frac{X_2}{R_2}\right)^2} = \frac{S_2}{S_1} \cdot \frac{1 + 16\,S_1^2}{1 + 16\,S_2^2} \left[\because \frac{X_2}{R_2} = 4\right]$$

Slip, $S_p = 2 - 0.05 = 1.95$

Plugging torque,

$$T_p = \frac{1.95}{0.05} = \frac{1 + 16\,(0.05)^2}{1 + 16\,(1.95)^2} \times T_f = 39 \times \frac{1.04}{61.84} \times 200$$

$$= \mathbf{131\ N = m.} \quad \textbf{Ans.}$$

Example 1.35. *A 25 h.p., 220 V d.c. shunt motor with a full load speed of 600 r.p.m. is to be braked by plugging. Estimate the value of the resistance which should be placed in series with it to limit the current to 130 A. What should be the initial value of the electric braking torque and value when speed has fallen to half of its full load value ? Armature resistance of motor is 0.1 Ω. Full load armature current is 95 A.*

Solution. Full load armature current, $I = 95$ A

Terminal voltage, $V = 220$ Volts

Induced e.m.f. of the motor $E = V - IR$

$$E = 220 - 90 \times 0.1 = 210.5 \text{ V}$$

Voltage across the armature at the instance of braking, $V + E$;

$$= 220 + 210.5 = 430.5 \text{ Volts}$$

Resistance required in the armature circuit to limit the current to 130 A,

$$= \frac{430.5}{130} = 3.3115$$

External resistance required in armature circuit

$$= 3.3115 - 0.1 = 3.211\ \Omega\ \textbf{Ans.}$$

Full load torque, $T_f = \dfrac{\text{h.p. output} \times 735.5}{2\pi N/60}$ N-m

$$= \frac{25 \times 735.5}{2\pi \times 600/60} = 292.65 \text{ N-m.}$$

As in case of d.c. shunt motor the flux remain unchanged, therefore, torque is directly proportional to current.

∴ Initial braking torque

$$= T_F \times \frac{\text{initial breaking Current}}{\text{full load current}} = 292.65 \times 130/95 = 400.5 \text{ N-m.}$$

Ans.

When the speed has fallen to half its full load value, the back e.m.f. also falls to half of its original value *i.e.* $\dfrac{210.5}{2} = 105.25$ V

$$\text{Current} = \frac{220 + 105.25}{3.3115} = 98.22 \text{ A}$$

Electric breaking torque at half full load speed,

$$= 292.65 \times \frac{98.22}{95} = 302.57 \text{ N-m. } \textbf{Ans.}$$

Example 2.36. *A 50 h.p., 440 d.c. shunt motor is braked by plugging. Calculate the value of the resistance to be placed in series with the armature circuit to limit the initial braking current to a double of the full load current. Armature resistance = 0.1 Ω, full load armature current = 100 A.*

Solution. Full load armature current, $I = 100$ A

induced e.m.f. of the motor $E = V - IR$

$$E = 440 - 100 \times 0.1 = 430 \text{ V.}$$

Voltage across the armature at instant of braking

$$= V + E = 440 + 430 = 870 \text{ V}$$

Initial braking current $= 2 \times$ full load current $= 2 \times 100 = 200$ A.

Resistance required to be placed in series with armature circuit

$$= \frac{870}{200} - 0.1 = 4.25\ \Omega\ \textbf{Ans.}$$

Example 1.37. *The outside of a 12.5 KW d.c. motor is equivalent to a cylinder of 100 cm diameter and 140 cm in length. The machine may* *be considered to be of homogeneous material weighting 650 Kg and of specific heat 700 J/Kg/°C. Assuming that the outer cylindrical surface dissipates heat at 0.0012 W/cm²/°C. Estimate the thermal time constant of the machine and its final temperature rise when operating on full load with an efficiency of 85%.*

Solution. Output of machine = 12.5 kW or 12500 Watts

Power loss in the motor, $P = \text{Power output} \times \dfrac{1-\eta}{\eta}$

$$= 12500 \times \frac{1-0.85}{0.85} = 2206 \text{ watts.}$$

Mass of active parts of machine, m = 650 kg

Specific heat of material, $C_p = 700$ J/kg/°C

Coefficient of cooling,

$$\alpha = 0.0012 \times 10{,}000 \text{ W/m}^2/\,^\circ\text{C} = 12 \text{ W/m}^2/\,^\circ\text{C}$$

Surface area of cooling, $S = \pi\, dl = \pi \times 1 \times 1.40 = 4.4 \text{ m}^2$

Thermal time constant,

$$\lambda = \frac{m\,C_P}{S\alpha} = \frac{650 \times 700}{4.4 \times 12} = 8617 \text{ seconds. } \textbf{Ans.}$$

Final temperature rise, $\theta_m = \dfrac{P}{S\alpha}$

$$= \frac{2206}{4.4 \times 12} = 41.8\,^\circ\text{C. } \textbf{Ans,}$$

Example 1.38. *A certain motor has to perform the following duty cycle :*

100 h.p. for 10 minutes   *No load of 5 minutes*

50 h.p. for 8 minutes   *No load for 4 minutes.*

The duty cycle is repeated indefinitely. Draw the curve for the load cycle. Assuming that the heating is proportional to the square of the load, determine the suitable size of a continuously rated motor.

Solution.

Since heating is proportional to the square of load.

Rating of continuously rated motor,

= R.M.S. horse power

$$= \sqrt{\frac{(100)^2 \times 10 + 50^2 \times 8}{27}} = 66.67 \text{ h.p.}$$

i.e. a motor of 70 h.p. will be adequate. **Ans.**

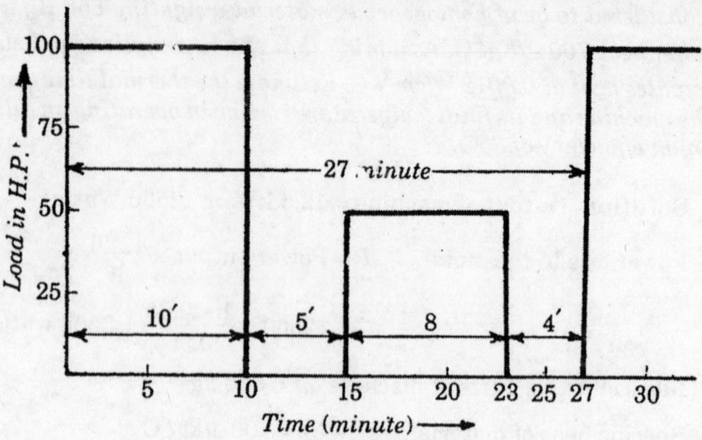

Fig. 1.105

Example 1.39. *A motor has to exert power starting from zero and rising uniformly to 150 h.p. in 8 minutes after which it works at a constant rate of 100 h.p. for 5 minutes. The motor remains on no load for the next 5 minutes. The cycle again starts and is repeated infinitely. Determine the suitable size of the motor.*

Solution. The load cycle is shown in Fig. 1.106 given below.

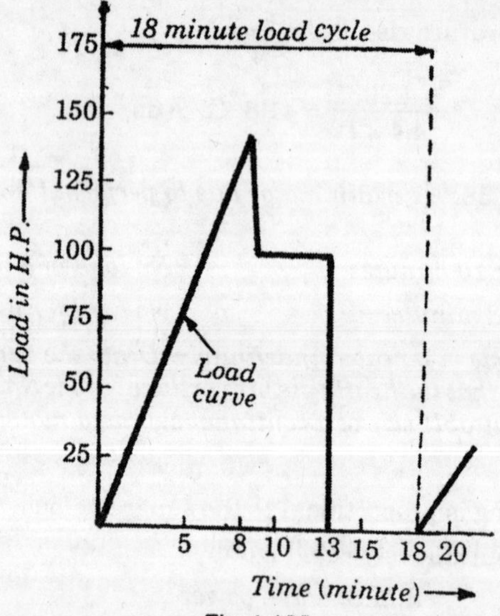

Fig. 1.106

Assuming heating proportional to the square of the load.
(Rating of motor)^2 × time of load cycle

$= \int (\text{H.P.})^2\, dt$ over the complete cycle.

$$\text{and} \int \text{HP}^2\, dt = \int_0^8 \left(\frac{150}{8} t\right)^2 dt + \int_8^{13} (100)^2\, dt + \int_{13}^{18} (0)^2\, dt$$

$$= \frac{22500}{64}\left[\frac{t^3}{3}\right]_0^8 + 10{,}000\, [t]_8^{13} + 0$$

$$= \frac{22500}{64} \times \frac{512}{27} + 10{,}000 \times 5 = 56667$$

$$\text{Rating of motor} = \frac{56667}{18} = 3148$$

or Rating of motor $= \sqrt{3148} = 56$ h.p.

i.e. a motor of 60 h.p. will be adequate. **Ans.**

Example 1.40. *A motor has a heating time constant of 70 minutes and cooling time constant when stationary of 90 minutes. When run continuously on full load of 25 h.p. the final temperature rise is 40°C.*

(*i*) *What load could the motor deliver for 10 minutes. If it is followed by a shut down period long enough for its cooling, without temperature rise exceeding 40°C.*

(*ii*) *If it is on an intermittent load of 10 minutes or load followed by 15 minutes shut down. What is the maximum value of load, which it could supply during the on period without the maximum temperature rise exceeding 40°C.*

Assuming losses to be proportional to square of the load.

Solution. Final temperature $= \theta_m = 40°\text{C}$

Heating time constant $\lambda = 70$ minutes

Cooling time constant $\lambda' = 90$ minutes

Continuous rating of motor $= 25$ h.p.

Load period $= t_1 = 10$ minutes

No load period $t_2 = 15$ minutes

(*i*) When load period is followed by a shut down period long enough for its cooling, let the load supplied by the motor for 10 minutes without temperature rise exceeding 40°C be *P,H.P.*

Final temperature rise,

$$\theta'_m = \frac{\theta_m}{1 - e^{-t/\lambda}} = \frac{40}{1 - e^{-10/70}} = 300$$

Sine losses are proportional to the square of load,

$$\therefore \quad \frac{\text{Total losses at } P \text{ h.p.}}{\text{Total losses at 25 h.p.}} = \left(\frac{P}{25}\right)^2 = \frac{\theta_m'}{\theta_m}$$

$$\left(\frac{P}{25}\right)^2 = \frac{\theta_m'}{\theta_m} = \frac{300}{40}$$

$$P = 25\sqrt{7.52} = 68.5 \text{ h.p. } \textbf{Ans.}$$

(*ii*) When the motor is on intermittent load of 10 minutes on load followed by 15 minutes shut down.

$$\theta_m = \theta_m' \frac{1 - e^{-t_1/\lambda}}{1 - e^{-(t_1/\lambda + t_2/\lambda')}}$$

or

$$\theta'_m = 1 - e^{-(t_1/(\lambda + t_2\lambda')}$$

$$\theta_m \; 1 - e^{-t_1/\lambda}$$

$$= 40 \frac{1 - e^{-(10/70 + 15/90)}}{1 - e^{-10/70}} = 40 \frac{1 - 0.7338}{1 - 0.8669} = 80^\circ\text{C}$$

Let maximum value of load that can be supplied be P' h.p.

$$\therefore \quad \left(\frac{P'}{25}\right)^2 = \frac{\theta_m'}{\theta_m} = \frac{80}{40}$$

or

$$P' = 25\sqrt{\frac{80}{40}} = 25\sqrt{2}$$

$$P' = 35.35 \text{ h.p. } \textbf{Ans.}$$

Example 1.41. *A 25 h.p., 3ϕ, 10 pole, 50 Hz induction motor provided with a flywheel has to supply a load torque of 800 N-m for 10 second followed by a no-load period during which flywheel regains the full speed. The full load slip of motor is 4% and the torque-speed curve may be assumed linear over working range. Find the moment of inertia of the flywheel. If motor torque is not to exceed twice the full load torque.*

Solution. Output of motor $= 25$ h.p.

Syn. speed of motor $N_s = \dfrac{120f}{P} = \dfrac{120 \times 50}{10} = 600$ r.p.m.

Full load slip $= S_f = 0.04$

Full load speed $N_f = N_0 (1 - S_f) = 600 (1 - 0.04)$

$= 576$ r.p.m.

Full load torque, $T_f = \dfrac{25 \times 735.5}{2\pi \times \dfrac{576}{60}} = 304.8$ N-m

Motor torque, $T_m = 2 \times T_f = 2 \times 304.8 = 609.6 \text{ N} - m$

Load torque, $T_L = 800 \text{ N} - \text{m}$

No-load torque, $T_0 = 0$

Slip speed, $S = 0.04 \times 600 = 24$ r.p.m.

$$= 2\pi \times \frac{24}{60} = 2.5133 \text{ rad./S.}$$

As the slip at full load is proportional to full load torque.

$\therefore$ Proportionality constant ;

$$k = \frac{\text{Slip}}{\text{Full load torque}}$$

$$= \frac{2.5133}{304.8} = 0.008246$$

Since $T_m = T_L - (T_L - T_O)\, e^{-t/kJ}$

$$= T_L\,(1 - e^{-t/kJ}) \qquad T_0 = 0$$

or $$1 - e^{-t/kJ} = \frac{T_m}{T_L} = \frac{609.6}{800} = 0.762$$

or $$e^{-t/kJ} = 1 - 0.762 = 0.238$$

or $$J = \frac{-t}{K \log_e 0.238}$$

$$= \frac{-10}{0.008246 \log_e 0.238}$$

$$J = 844.8 \text{ kg-m}^2. \quad \textbf{Ans.}$$

Example 1.42. *Determine the ratio of starting to full-load current of a 10 H.P, 400 V, 3-phase induction motor with star-delta starter. The motor has a full-load power factor of 0.8 lagging and an efficiency of 0.9. The short ckt. current of motor is 7.2 A at 160 V. Neglect the effect of magnetising current.*

Solution: Output = 10 H.P. = $10 \times 735.5 = 7355W = 7.355KW$

Full load line current, $I_f = \dfrac{\text{Output in KW} \times 1000}{\sqrt{3}\,.\,V_L\,.\,\cos\phi \times \eta}$

$$= \frac{7.355 \times 1000}{\sqrt{3} \times 400 \times 0.8 \times 0.9} = 14.75 \text{ Amp.}$$

Short ckt. current at $160V$ = 7.2 Amp.

Short ckt. current at normal voltage of 400 V,

$$I_{sc} = 7.2 \times \frac{400}{160} = 18.0 \text{ Amps.}$$

Starting current with star Delta starter,

$$I_{st} = \frac{I_{sc}}{3} = \frac{18.0}{3} = 6.0 \text{ Amps.}$$

Ratio of starting current to full load current

$$= \frac{I_{st}}{I_{fl}} = \frac{6.0}{14.75} = \mathbf{0.407\ Ans.}$$

Example 1.43. *A 50 KVA, 400 V, 3 phase, 50 Hz squirrel cage induction motor has full load slip of 5%. Its standstill impedance is* 0.866 Ω/phase. *It is started using a tapped auto-transformer. If the maximum allowable supply current at the time of starting is 100 Amp. Calculate the tap position and the ratio of starting torque to full-load.*

Solution. Full load current drawn by a 3-ϕ induction-motor,

$$I_f = \frac{\text{Output in KVA} \times 1000}{\sqrt{3} \times \text{line voltage}} = \frac{50 \times 1000}{\sqrt{3} \times 400} = 72.2 \text{ Amp.}$$

$$\text{Short Ckt. current, } I_{sc} = \frac{\text{Supply line voltage}/\sqrt{3}}{\text{standstill impedence per phase}}$$

$$= \frac{400/\sqrt{3}}{0.866} = 266.7 \text{ Amp.}$$

Supply current at start, I_{st} = 100 Amp.

Top position of Auto-transformer,

$$K = \sqrt{\frac{I_{st}}{I_{sc}}} = \sqrt{\frac{100}{266.7}}$$

$$= 0.6124 \text{ or } 61.24\% \ \mathbf{Ans.}$$

Full load slip, S = 5% or 0.05

Ratio of sarting torque to full load torque,

$$\frac{T_{st}}{T_f} = K^2 \left[\frac{I_{sc}}{I_f}\right]^2 \times S_f$$

$$= 0.6124^2 \left[\frac{266.7}{72.2}\right]^2 \times 0.05 = 0.256 \ \mathbf{Ans.}$$

It must be noted that the 3–ϕ induction motor has been assumed to be star-connected and losses have been neglected.

Example 1.44. *A 220 V, D.C. shunt motor takes 22 Amp. at rated voltage and runs at 1000 r.p.m. Its field resistance is* 100Ω *and armature circuit resistance is* 0.1Ω. *Compute the value of additional resistance required in the armature circuit to reduce the speed to 800 r.p.m. when:*

(*i*) *The load torque is proportional to speed.*

(*ii*) *The load torque varies as the square of the speed.*

Solution. Under Normal Conditions,

Line current, $I_L = 22$ Amp.

Shunt field current, $I_{sh} = \frac{V}{R_{sh}} = \frac{220}{100} = 2.2$ Amp.

Armature current, $I_{a_1} = I_{L_1} - I_{sh} = 22 - 2.2 = 19.8$ Amp.

Back e.m.f., $E_{b_1} = V - I_{a_1}R_a = 220 - 19.8 \times 0.1 = 218.02$ Volt

Speed, $N_1 = 1000$ r.p.m.

(*i*) Let the additional resistance required in the armature circuit be of R ohms to reduce the speed to 800 r.p.m. When the load torque is proportional to speed.

$$T_2 = T_1 \times \frac{N_2}{N_1} = T_1 \times \frac{800}{1000} = 0.8T_1$$

or $I_{a_2}\phi_2 = 0.8 I_{a_1}\phi_1$ $\quad \because T \propto \phi\, I_a$ and $\phi_2 = \phi_1$

or $I_{a_2} = 0.8 \times 19.8 = 15.84$ Amp.

$$E_{b_2} = V - I_{a_2}(R + R_a) = 220 - 15.84(R + 0.1)$$

$$= 218.416 - 15.84R$$

Since $\frac{E_{b_2}}{E_{b_1}} = \frac{N_2}{N_1}$ (Flux Remáins Constant)

$$\therefore \quad \frac{218.416 - 15.84R}{218.02} = \frac{800}{1000}$$

or $R = 2.778\ \Omega$ **Ans.**

(*ii*) Let the additional resistance required in the armature circuit be of R′ohms to reduced the speed to 800 r.p.m. when no load torque varies as the square of the speed.

$$\therefore \quad T_3 = T_1 \times \left(\frac{N_2}{N_1}\right)^2 = T_1 \times \left(\frac{800}{1000}\right)^2 = 0.64T_1$$

or $I_{a_3}\,\phi_3 = 0.64 I_{a_1}\phi_1$

[$\because$ $T \alpha\, \phi\, I_a$ and $\phi_3 = \phi_1$]

or $I_{a_3} = 0.64 \times 19.8 = 12.672$ Amp.

$$E_{b_3} = V - I_{a_3}(R' + R_a) = 220 - 12.672\,(R' + 0.1)$$

$$= 218.7328 - 12.672\,R'$$

Since $\frac{E_{b_3}}{E_{b_1}} = \frac{N_3}{N_1}$ (Flux remians Constant)

$$\therefore \quad \frac{218.7328 - 12.672R'}{218.02} = \frac{800}{1000}$$

or $R' = 3.497\Omega$ **Ans.**

Example 1.45. *The following data refers to a 500 h.p. rolling-mill induction motor equipped with a flywheel. No-load speed : 40 r.p.m., slip* *at full load : 12%, load torque during actual rolling : 41,500 kg-m. Duration of each rolling period : 10 seconds. Determine the inertia of the flywheel required in the above case to limit the motor torque to twice its full load value. Neglect the no-load losses and assume that the rolling mill torque falls to zero between each rolling period. Assume that the motor slip is directly proportional to the torque.*

Solution : Output of motor = 500 h.p. $= 500 \times 735.5W$

No load speed, $N_0 = 40$ r.p.m.

Full load slip, $S_f = 12\%$ or 0.12

Full load speed, $N_f = 40(1 - 0.12) = 35.2$ r.p.m.

$$\text{Full load torque, } T_f = \frac{500 \times 735.5}{2\pi \times \frac{35.2}{60}} = 99{,}766 \text{ N-m}$$

Motor torque, $T_m = 2 \times T_f = 2 \times 99{,}766 = 1{,}99{,}532$ N-m

Load torque, $T_L = 41500 \times 9.81 = 4{,}07{,}115$ N-m

No load torque, $T_0 = 0$

Load period, $t = 10$ sec .

Slip-speed = Full load slip × No load speed

= 0.12 × 40 = 4.8 r.p.m.

or $\frac{2\pi \times 4.8}{60} = 0.5026$ rad./ sec

$$K = \frac{\text{Slip - speed}}{\text{Torque}} = \frac{0.5026}{99{,}776} = 5.0383 \times 10^{-6}$$

$\because = S \propto$ Torque

$$T_m = T_L - (T_L - T_o)e^{-t/kJ} = T_L(1 - e - t/kJ)$$

$(\because T_0 = 0)$

$$J = \frac{-t}{K \log_e\left(1 - \frac{T_m}{T_L}\right)}$$

$$= \frac{-10}{5.0383 \times 10^{-6} \log_e\left(1 - \frac{199532}{4{,}07{,}115}\right)}$$

$$= 2.947 \times 10^6 \text{ kg-m}^2 \textbf{ Ans.}$$

Example 1.46. *An electric motor has load variation as given below: Torque 240 Nm for 20 minutes, 140 Nm for 10 minutes, 300 Nm for 10 minutes, 200 Nm for 20 minutes. If the speed of the motor is 720 r.p.m., find the power rating of motor.*

Solution : Equivalent load torque,

$$T_{eq} = \sqrt{\frac{T_1^2 t_1 + T_2^2 t_2 + T_3^2 t_3 + T_n^2 t_n}{t_1 + t_2 + t_3 + t_n}}$$

$$= \sqrt{\frac{(240)^2 \times 20 + (140)^2 \times 10 + (300)^2 \times 10 + (200)^2 \times 20}{20 + 10 + 10 + 20}}$$

$= 225.39$ Nm

Speed of Motor, $N = 720$ r.p.m.

$\therefore$ Power rating of motor, $P = T_{eq} \times \dfrac{2\pi N}{60}$

$$= 225.39 \times 2\pi \times \frac{720}{60}$$

$= 16{,}994$ watts or **17 KW Ans.**

(**Note :** Equivalent torque method is based upon the assumption that the motor current is proportional to the torque and heating is proportional to the square of the current (*i.e.* heating is proportional to the square of torque) so the equivalent torque is determined in the following manner *i.e.*

$$T_{eq} = \sqrt{\frac{T_1^2 t_1 + T_2^2 t_2 + T_3^2 t_3 + ... T_n^2 . t_n}{t_1 + t_2 + t_3 + ... + t_n}}$$

A motor having torque nearby equal to T_{eq} is selected.)

Example 1.47. *Based on the r.m.s. torque, estimate the kW rating of a 750 r.p.m. motor used for driving an equipment having the following load torque curve :*

(*i*) *For first 10 seconds, the torque is constant at 40 kg-m*

(*ii*) *For next 30 seconds the torque varies linearly with time from 35 kg-m to 15 kg-m*

(*iii*) *For the last 50 seconds, the torque is constant and equal to 10 kg-m.*

Solution. The load cycle is shown in the figure 1.107

$(T_{eq})^2 \times$ Time of cycle $= \int T^2 dt$ over the cycle

$$= \int_0^{10} 40^2 dt + \int_{10}^{40} \left[35 - \frac{35-15}{30}(t-10)\right]^2 dt + \int_{40}^{90} 10^2 . dt$$

$$= \int_0^{10} 1{,}600 dt + \int_{10}^{40} \left[1{,}225 - \frac{140}{3}(t-10) + \frac{4}{9}(t-10)^2\right] dt + \int_{40}^{90} 100\, dt$$

$$= [1600]_0^{10} + \left[1225\,t - \frac{140}{6}(t-10)^2 + \frac{4}{27}(t-10)^3\right]_{10}^{40} + (100)_{40}^{90}$$

$$= 16{,}000 + 1225 \times 30 - \frac{70}{3} \times 900 + \frac{4}{27} \times 27000 + 5{,}000$$

$$= 16000 + 36750 - 21000 + 4000 + 5000 = 40{,}750$$

or $$T_{eq} = \sqrt{\frac{40{,}750}{90}} = 21.278 \text{ kg-m}$$

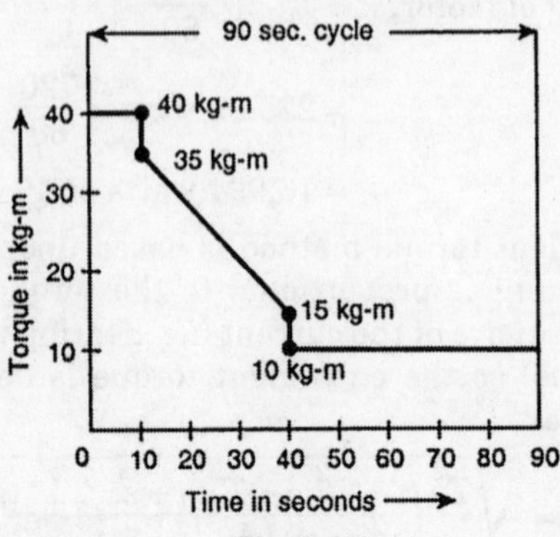

Fig. 1.107

Power rating of motor,

$$P = T_{eq} \text{ in kg.m} \times 98.1 \times \frac{2\,\pi\,N}{60}$$

$$= 21.278 \times 9.81 \times \frac{2\,\pi \times 750}{60}$$

$$= 16{,}400W \text{ or } \mathbf{16.4\ kW\ Ans.}$$

Example 1.48. *A 400 volts, 4 Pole, 50 Hz induction motor develops 25 H.P. at 4% slip. If rotor resistance to standstill reactance is 1 : 4, Calculate the initial plugging torque.*

Solution. Synchronous speed, $N_s = \frac{120\,f}{P} = \frac{120 \times 50}{4} = 1500$ r.p.m.

Full load slip = 0.04

Full load speed, $N_f = N_s\,(1 - s) = 1500(1 - 0.04)$

$= 1440$ r.p.m.

Full load Torque, $T_f = \frac{\text{Power developed} \times 1000}{2\pi N/60}$ N-m

$$= \frac{25 \times 735.5 \times 1000 \times 60}{2\,\pi \times 1440} = 121.99 \text{ N-m}$$

Since Torque required, $T \propto \frac{SR_2E_2^2}{R_2^2 + S_2^2 X_2^2}$

$$\therefore \quad \frac{T_2}{T_1} = \frac{\dfrac{S_2R_2E_2^2}{R_2^2 + S_2^2X_2^2}}{\dfrac{S_1R_2E_2^2}{R_2^2 + S_1^2X_2^2}} = \frac{S_2}{S_1}\left[\frac{R_2^2 + S_1^2X_2^2}{R_2^2 + S_2^2X_2^2}\right]$$

or

$$\frac{T_2}{T_1} = \frac{S_2}{S_1} \cdot \frac{(1 + S_1^2\left(\dfrac{X_2}{R_2}\right)^2}{1 + S_2^2\left(\dfrac{X_2}{R_2}\right)^2} = \frac{S_2}{S_1} \cdot \frac{1 + 16S_1^2}{1 + 16S_2^2}$$

$$\text{Slip} = 2 - 0.04 = 1.96$$

$$\text{Plugging Torque} = \frac{1.96}{0.04} \times \frac{1 + 16(0.04)^2}{1 + 16(1.96)^2} \times T_f$$

$$= 49 \times \frac{1.0256}{62.46} \times 122$$

$$= \mathbf{98.15\ N\text{-}m\ \ Ans.}$$

Example 1.49. *A 3-phase, 50 Hz, 8 pole, 400 volt synchronous motor and its driven machine have a total moment of inertia of 300 Kg–m^2. Determine the time taken and the number of revolutions made by it to come to stand still if (i) the reheostatic braking is used which gives an initial braking torque of 450 kg or alternatively (ii) braked by plugging which produces a constant braking torque of 120 kg. Assume a frictioinal torque of 1.0 kg.*

Solution. Total moment of inertia of rotating parts,

$$J = 300 \text{ kg-m}^2$$

Frictional torque $T_f = 1 \times 9.81 = 9.81$ N-m

(*i*) Electric torque produced by plugging, $T_E = 120 \times 9.81 = 1177.2$ N-m

Total braking torque, $T_B = T_E + T_F = 1177 + 9.81 = 1187$ N-m

$$\text{Angular retardation,} \quad \beta = \frac{T_B}{J} = \frac{1187}{300} = 3.95 \text{ rad./s}^2$$

$$N_s = \frac{120f}{P} = \frac{120 \times 50}{8} = 750 \text{ r.p.m.}$$

$$w_1 = \frac{2\pi \times 750}{60} = 78.5 \text{ rad./s}$$

Time taken to stop the motor, $t = \dfrac{w}{\beta} = \dfrac{78.5}{3.95} = 19.87$ seconds

$$\text{Now } t = -\int \frac{dw}{\beta}$$

or $$dt = \frac{-dw}{\beta}$$

No. of revolutions made,

$$n = \frac{1}{2\pi}\int w dt = -\frac{1}{2\pi}\int_{w_1}^{0} w\frac{dw}{\beta} = \frac{-1}{2\pi\beta}\left[\frac{w^2}{2}\right]_{w_1}^{0}$$

$$= \frac{1}{2\,\pi \times 3.95}\left[\frac{(78.5)^2}{2}\right] = 124 \text{ revolutions}$$

Initial Braking torque $= Kw_1 = 450 \times 9.81$

$$= 4414 \text{ N-m}$$

$$K = \frac{4414}{78.5} = 56.23$$

Time taken to bring the motor to rest,

$$t = \frac{J}{K}\log_e \frac{Kw_1 + T_F}{T_F}$$

$$= \frac{300}{56.23}\log_e \frac{4414 + 9.81}{9.81} = 5.33 \log_e 145.13$$

$$= 5.33 \times 4.97 = 26.5 \text{ seconds}$$

Number of revolutions made

$$n = \frac{1}{2\pi K}\left[\frac{J}{K}(Kw_1 + T_F)(1 - e^{-kt/J}) - T_F t\right]$$

$$= \frac{1}{2\pi \times 56.23}\left[\frac{300}{56.23}(4414 + 9.81)(1 - e^{-56.23 \times 26.5/300}) - 9.81 \times 26.5\right]$$

$$= \frac{1}{353.12}[7588.9(1 - 0.00697) - 260]$$

$$= \frac{1}{353.12}[7276] = 21 \text{ revolutions} \quad \textbf{Ans.}$$

1.66 Selection of Motor for a Particular Service

The following factors play an important part in selecting the size and suitable type of motor for a particular drive :

(*i*) **Load Characteristics.** It includes the load starting torque and speed requirements etc.

(*ii*) Surroundings.

(*iii*) Thermal characteristics of the motor.

(*iv*) Noise.

(*v*) Annual running charges of the motor.

Initially first four factors are thoroughly studied and the annual charges for the type of motors selected are assessed. The final selection made should be economical.

Following table indicates the load characteristics of motors suitable for various services :

Load characteristics *(1)*	***Type of industry equipment*** *(2)*	***Suitable D.C. motor*** *(3)*	***Suitable A.C. motors*** *(4)*
(*i*) Low torque nearly constant speed (*ii*) Frequent start (*iii*) Load duration comparatively small	Vacuum cleaners Portable tools Office appliances, Hair dryer, kitchen appliances etc.	Universal motor	(*i*) Single phase universal motor for constant load (*ii*) Capacitor start single phase induction motor for carrying loads
(*i*) Small load (*ii*) High torque (*iii*) High output	Small compressors Garage equipment Shoe machinery, wood working machinery. Refrigerator		(*i*) Replusion type single phase motor (*ii*) Repulsion start induction motor
(*i*) Low load requirement between the range of 10-15 watts. (*ii*) Intermittent on continuous loads (*iii*) Constant speed (*iv*) Noiseless requirement	Electric clocks, Photography Instruments Hair dryers		Shaded pole single phase motor
(*i*) Constant speed (*ii*) Load requirement up to 1 kW (*iii*) Starting torque between 1.5 to 2 times full load torque	Electric fans Blowers pumps.	Shunt motor	(*i*) Squirrel cage induction motor (*ii*) Synchronous motor with starting arrangement
(*i*) Unloaded or loaded start (*ii*) Nearly constant speed	Line shaft conveyors, Grinders (Pulp)	Series motor	(*i*) Squirrel cage induction motor (*ii*) Slip ring induction motor (*iii*) Synchronous motor for line shaft.
(*i*) Variable speed (*ii*) High starting torque.	Conveyors, shear Floor mills	Compound motor	Squirrel cage induction motor
Note. Motors coupled with flywheel are used.			

(*i*) Constant speed (*ii*) High starting torque	Rolling mill, paper mill, Rubber mill	Compound motor	(*i*) Squirrel cage induction motor (*ii*) Synchronous motor
(*i*) Frequently starting and reversing (*ii*) Starting at full load (*iii*) Intermittent service	Elevators, small crane and hoists Large cranes and hoists	Series motor Series motor Series motor	Squirrel cage induction motor, Slip ring induction motor, Squirrel cage, induction motor
(*i*) High load (*ii*) High starting torque (*iii*) Motor should be capable of withstanding heavy duty of rolling	Iron and steel works	Specially designed shunt motor	Specially designed induction motor
(*i*) Flammable gases surrounding the motor (*ii*) Variable load	Mining work	Compound motor	(*i*) For small load up to 10 H.P. squirrel cage induction motor (*ii*) For high load slipring induction motor
Note. D.C. motors is preferred for high loads.			
(*i*) Variation in speed to a large extent (*ii*) High starting torque	Electric-trains	Series motor	(*i*) Single phase series motor (*ii*) Three phase induction motor
(*i*) Accuracy (*ii*) Constant speed (*iii*) Low load (*iv*) Low starting torque.	Lathe Milling machines	Shunt motor	Induction motor with pole changing arrangement.

1.67 Characteristics and Applications of various d.c. and AC 1 – ϕ and 3 – ϕ motors

D.C. Shunt Motor. This motor has nearly constant speed and torque varies proportional to the current. Its starting torque and maintenance cost is medium, continuous speed range of 4 : 1 is possible by usual ways.

It can be used for driving constant speed line shafts, lathes, vacuum cleaners, pressure blowers, centrifugal pumps, washing machines and printing press etc.

D.C. Series Motors. This has a very high starting torque at it varies as the square of the current. The speed is varying and is dangerously high at light or no loads. Its maintenance cost is medium and speed can be regulated.

These motors are used for heavy duty applications such as electric traction, rolling mills and cranes etc.

Commulative Compound Motors. This motor has high starting torque of the order of 2 to 3 times of the full load torque. Requires medium maintenance and speed can also be adjusted within limits.

These are used for driving compressors, stamping machines, door lifts, shearing machines and passenger elevators etc.

Differential Compound Motor. Its starting torque is low and has unstable operation. Constant but dangerously high speed at no load.

It is seldom used in practice.

Synchronous Induction Motor. This is also a constant speed motor and its starting torque is about twice the final load torque. This can also work on varying p.f. both lagging and leading. Speed control is not possible.

This is used for driving large fans, compressors, line shaft and pumps where fairly constant and starting torque both are required.

3-ϕ Squirrel Cage Induction Motor. This motor is a nearly constant speed motor with a poor starting torque. It has high over load capacity and operates always at lagging p.f. varying from 0.7 to 0.9.

The squirrel cage induction motors can be used for driving low and medium power drives where speed control is not required. These are used in tube wells, lathes, drilling machines, saws, grinders, blowers, and line shafts etc.

Slip Ring Induction Motor. The speed of slipring induction motor is also constant. Its starting torque is about twice the full load torque. It has high overload capacity. Its speed can control over a small range and operates at about 0.75 to 0.9 lagging P.F.

It is used in drives requiring high power at high starting torque and adjustable speed *e.g.* line shafts, lifts, pumps winding machines and compressors etc.

3-ϕ Series Motor. It has high starting torque of about 2.5 times the full load torque. P.F. varies rom 0.9 to 1 lagging. Speed control of this motor is possible over full range but maintenance cost is high.

It is used in drives where load increases with speed such as fans, centfrifugal pumps, and where high starting torque is required such as haulage and hoist work.

Scharge Motor. It has nearly constant speed for a particular setting and high starting torque speed can be controlled over a wide range and p.f. is also nearly unity.

This motor is used where adjustable but constant speed is required such as for driving paper making machines, calendar and textile works etc.

1-φ Synchronous Motor. It operates at constant speed and speed control is not possible. Its initial cost is quite more than the comparable squirrel cage induction motor.

It is used in teleprinters, clocks, control apparatus of all timing devices, sound recording and reproducing devices.

Single Phase Series Motor. This motor possess high starting torque of about 3 times the full load torque. The speed can be controlled on full range with the help of a transformer. Its maintenance cost is high and P.F. is about 0.8 – 0.9 lagging.

In small sizes it is used for driving domestic appliances like vacuum cleaners and mixers etc.

Repulsion Motor :- A repulsion motor also has high starting torque like a series motor. Speed control over a wide range is possible by shifting of brushes.

It is used for services requiring large starting torque and adjustable but constant speed just as in coil winding machines.

Capacitor start Single Phase Induction Motor. This has high starting torque of 1.5 times the full load torque. Speed control is not possible and power factor is 0.7 to 0.9 lagging.

Mainly used for compressors, Refrigerators and portable hoists.

Capacitor Start-Capacitor run Type Motor :- Its efficiency is high and P.F varies from 0.9 to 1 (lagging). Operation is quiet and speed is almost constant.

Used in water cooler and large air conditioners.

1.68 Particular Applications of Electric Drives

Industrial applications are far too numerous ; and it is not practical to decide the suitability of a particular electric motor, except by way of broad guidelines for the purpose. It would be difficult even categorize them for this purpose, but the broad fields of applications are discussed here under :

We will now suggest possible choice of motor for particular service keeping in view all external forces.

1. **Domestic Applications.** Small universal motor, usually series type is used for various domestic appliances such as for domestic refrigerators shavers, vacuum cleaners, mixi, and cloth washing machines etc.

2. **Cranes.** D.C. series motors or D.C. compound motors are preferred on account of their high starting torque. Easy speed control for lifts; hoists also for D.C. series motor and D.C. compound motors is used.

3. **Lifts.** D.C. compound wound or A.C. slip ring induction motors are used. To get speed control, shunt field regular method or Ward-Leonard method is used.

4. **Lathes, Milling and Grinding Machines.** Lathes are usually driven by constant speed. Squirrel cage induction motor for milling and grinding machines also, A.C. squirrel cage induction motor is used.

5. **Planer.** D.C shunt or compound motor is used, if D.C. supply is available otherwise slip ring induction motor is used.

6. **Shapers and Cranes.** Constant speed squirrel cage induction motors are used.

7. **Punches and Shears.** Punches and shears are usually equipped with flywheel to supply momentary demand of power. High slip squirrel cage induction motor is used for punches and shears.

8. **Sugar Mills.** In sugar mills, a centrifugal is used to separate out sugar crystals from the syrup by the action of centrifugal forces. The duty cycle of a centrifugal motor is as shown in Fig. 1.105. It consists of starting the motor to get a speed of about 200 r.p.m, charging the syrup with the motor disconnected from the supply, an intermediate spinning of the centrifugal at about 500 r.p.m spinning at a speed of about 1000 r.p.m and braking of the motor successively at different speeds till a ploughing speed of about 50 r.p.m is obtained. In order to reduce the energy lost during acceleration and retardation of the motor the centrifugal action is performed at different speeds.

Four speed pole changing motors having two distinct stator windings that enables us to obtain synchronous speeds 1500/750/214/107 r.p.m. or 1000/500/214/107 r.p.m. are commonly employed as drives for sugar centrifuge. They are not only capable of providing the desired fixed speeds of operation but also of returning a portion of the energy back to the supply during braking by means of regenerative braking accomplished by switching over to a higher pole operation from a lower

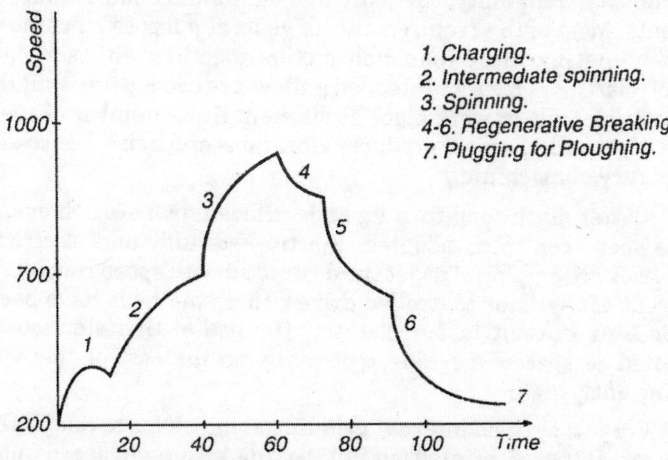

Fig. 1.108 Duty Cycle of a Sugar Centrifuge.

pole one. In order to satisfy the duty cycle as shown in Fig. 2.108 first the 28 pole winding is switched on to get a speed of about 200 r.p.m. The supply is cut off as the charging starts and as soon as it is over the intermediate spin speed of about 450 r.p.m. is obtained by energizing the 12 pole winding. The final spin speed near to 950 r.p.m. is attained by switching over to the 6-pole winding. Once the final spin speed of the centrifuge has been reached, super synchronous regenerative braking is done by connecting the supply to 12, 28 and 56 pole windings successively to bring the speed down to near about 500, 214 and 107 r.p.m. respectively. Ploughing speed of about 50 r.p.m. is accomplished by plugging the 56 pole motor. Automatic control of the entire duty cycle is achieved by feeding the control equipment from the output of the technogenerator which is directly coupled to the motor driving the centrifuge.

Motors used for driving centrifuge have their own special features in construction. They are vertically mounted so as to be coupled with the shaft of the centrifuge. The motors may have larger air gap to take care of the possible rotor oscillation about the vertical axis. Since the motors operate in a humid environment, the insulation used must be humid proof. Due to frequent switching of the motor at different windings wound for different poles wide fluctuations in motor current occur during the duty cycle of a centrifuge. To protect the motor from overheating, thermal elements which operates a few degrees below the maximum permissible temperature of the winding are embedded in the windings. These elements, often give visual and audible warning signals so that the particular duty cycle may be completed.

9. **Machine Tool Applications.** Squirrel cage induction motors are the most widely used ones for most machine tool drives due to its simplicity, reliability, low cost and minimum maintenance requirements. Most of the requirements of general purpose machine tools are easily met by these induction motors together with simple manual controls. Gear boxes and stepped pulleys are used to transmit the power to cutting tools or work place at different finite number of speeds. The use of gear mechanism produces vibrations and noise, hence affects the accuracy of machining.

Better machining turning and surface finish are achieved, if stepless speed control is adopted. Electro-hydraulic and electromagnetic controls are employed to get the desired smooth speed control. With the advent of thyristor controlled drives, these methods have been superseded. At present in our country, the use of thyristorized drives is limited to special machine tools only, as the cost of these drives is sufficiently high.

For use such as internal grinding spindles for horological applications drilling of printed circuit boards etc. high speed motors are generally employed. These are usually high frequency, 3-ϕ induction motors operating upto speeds of 180000 r.p.m. and giving fairly high output powers in small sizes. Typical ratings are 1.2 kW at 120000 r.p.m. for high speed grinder drives and 0.5 kW at 150000 r.p.m. for high precision horological applications.

Certain machine tool applications require two or more fixed speeds. Pole changing motors with single or two stator windings and pole amplitude modulated motors are used to achieve the desired speeds.

Drives for Numerical Controlled Machines. With the advances made in the digital processing area, numerical controlled machines are slowly replacing conventional machine tools. With conventional machine tools the machine utility is quite low, whereas with the numerical controlled machines the utility could be five to six times more. Taking into account the operational inefficiencies, co-ordinations required high rate of rejections, number of stage inspections etc. One numerical controlled machine possibly can do the job of perhaps 4 to 6 conventional machines. As regards cost, with the development of indigenous drives and controls, numerical controlled machines may become economical soon.

Essentially the motors for numerical controlled machines are required for spindle drive and number of axes drives. The number of motors in a specific numerical controlled machine could be from 2 to 6 depending on the type and sophistication of the machine. The principal requirements for these drives are very fast response, wide speed range, very low inertia, precise positioning accuracies, severe duty cycles, low vibrations, better thermal capacity, low maintenance cost etc. Depending upon the control system the drive motors for axes are required to be provided with a resolver or a digital encoder, coupled to the motor shaft for position feed back along with a techogenerator integral with the motor, for velocity feed back. The motor ratings for the spindle drive will be usually upto 20 kW and 3000 r.p.m. and for axes drive upto 10 kW and 5000 r.p.m.

Electronically commutated d.c. motors or inverter fed induction motors are used for spindle drives whereas, stepper motors or high energy rare earth permanent magnet d.c. servo motors are employed for axes drives.

10. Petro chemical Industry

Pump drives. The petroleum and chemical industries use fluid handling equipment extensively. Pumps have been driven by induction motor with control of flow, when desired, accomplished by throttling valves. It has been estimated that 30% of more of electrical energy used by a typical refinery is wasted as a result of the throttling effect of control valves. However adjustable-speed pumping utilizing multispeed induction motors or inverter fed induction motors offer substantial economic benefits based on reduced energy and maintenance cost. Since conservation of energy has become a must for all industries and variable flow applications are commonly present in many industries, the energy saving capability of adjustable speed drives has been discussed in detail below.

Fig. 1.109 (*a*) shows a typical pump characteristic describing the head versus flow variation of a centrifugal pump. This curve indicates that the pump will produce limited flow if applied to a piping system in which a large change in pressure is required across the pump to lift the liquid and overcome resistance to flow at, *P*. Higher flow rates can be obtained as the required pressure differential is reduced (as at point Q).

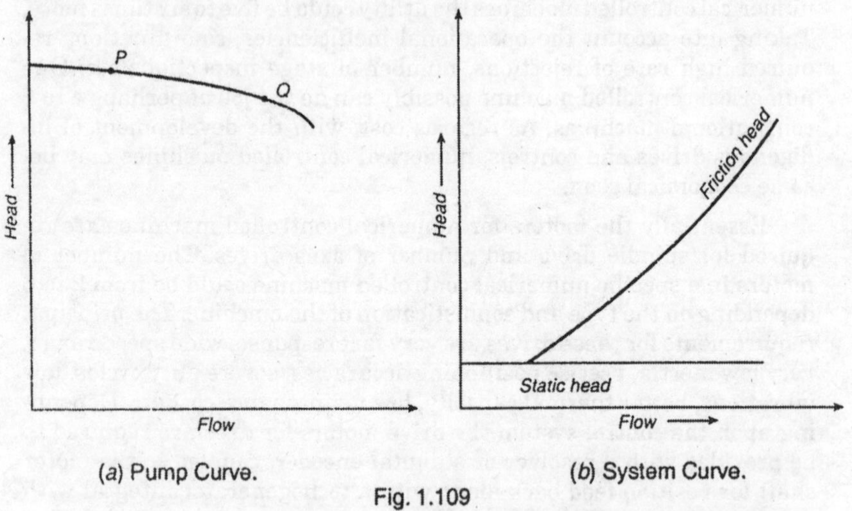

(*a*) Pump Curve. (*b*) System Curve.

Fig. 1.109

The system curve in Fig. 1.109 (*b*) which represents the characteristics of the piping system to which the pump is applied, is required to determine where along the pump characteristics it will operate in a given application. The head required at zero flow is known as static head or lift. This shows how many metres of elevation the pump must lift the fluid irrespective of the flow rate. The other component of head, called the friction head, increases with increasing flow. It is a measure of the resistance to flow offered by the pipe and its accompanying valves, elbows and other system elements. The intersection of pump and system characteristic denotes the natural operating point for the system without flow control as illustrated in Fig. 1.110 given below.

If a control valve is introduced into the piping system at the pump outlet and partially closed, it reduces flow by increasing the friction in the system. The modified system curve and the new operating point are also shown in Fig. 1.110. It may be observed that the desired reduction in flow has been achieved but the head has been increased.

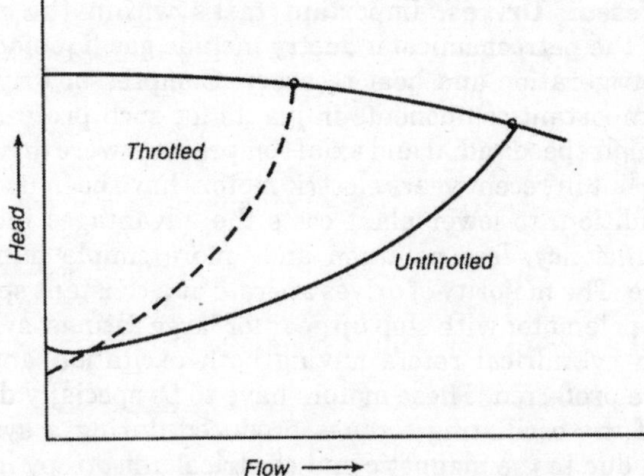

Fig. 1.110 Throtting valve control of flow.

If the desired flow could be accomplished by modifying the pump curve rather than the system curve as shown in Fig. 1.111, both flow and head could be reduced together. Actually the head required across the pump would be reduced rapidly relative to flow reduction. Reducing the speed of the pump results in change of the pump characteristic as shown. Since the pump output power is proportional to the product of head and flow for any fluid, it is obvious that there is a substantial reduction in required pump output power when the speed of the pump is reduced.

Variable speed drives using solid state devices have much lower losses than others and hence substantial energy savings are achieved by using, them for pumping application.

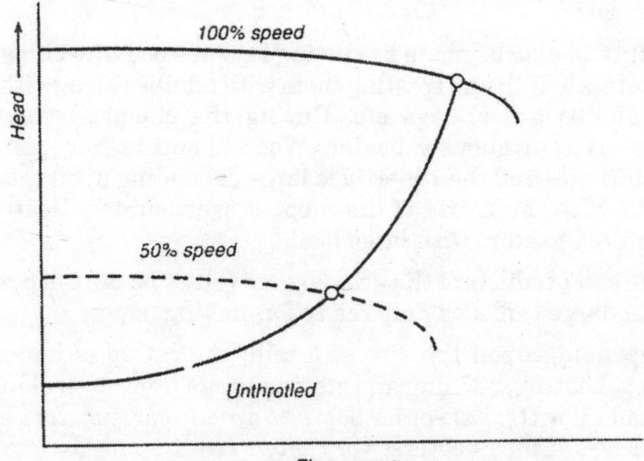

Fig. 1.111 Variable speed control of flow.

Compressor Drives. Important tasks within the production processes of the petrochemical industry include gas liquefaction, compression, refrigeration and heat recovery. Compressor drive systems constitute important components in plants for such processes. In the past, large high speed radial and axial compressors were driven mainly by turbines, but in recent years electric motors have been used increasingly. In addition, to lower plant costs the advantages include high operating efficiency, less pollution and more simple handling and maintenance. The majority of drives aperate at a constant speed, using a four or six polemotor with step up gear for large ratings, synchronous motors with cylindrical rotors having both excitation and starting windings are preferred. These motors have to be specially designed to take care of the oscillating torques produced during a synchronous acceleration due to the magnetic and electrical anisotropy in the rotor and special cooling conditions.

11. **Paper Mills.** Pulp making and the paper making are the two main processes that take place in the paper mill and the drives required are quite different for each one of them.

Pulp Making. Pulp is made in two ways—purely by mechanical means and by both mechanical and chemical processing. The former involves grinding logs of wood of about a meter length on large grind stones. Grinders operate at almost constant speed and can be started under light load conditions. Hence synchronous motors are considered as most suitable for grinder drives since they usually work with speeds of 200-300 r.p.m., Geared drives are used, especially when the motors are installed in a separate room for protection from humid atmosphere. Usually making pulp by purely mechanical means consumes more than 50% of the total power requirement of a paper mill ; hence large size grinders driven by 3000-4000 kW motors are normally considered as economical.

Pulp can also be made by cutting logs of wood into chips of several centimeters length and treating them with alkalies alongwith other raw material like grass, rags etc. During the chemical treatment the material is continuously beater. Wood choppers have random load characteristics and their inertia is large depending upon the size of the disc, on which the knives of the chopper are mounted. Beaters usually are required to start with large load.

The end products of the grinders as well as beaters are refined and stored in large tanks as pulp ready for making paper.

Depending upon the size of a mill, the rating of motor used for chipping, beating, refining and storing range from several hundreds to thousand kilowtts. Except for beaters, synchronous motors are used in these drives. Since beaters, very often require speeds less than 200 r.p.m and large starting torque, slip ring induction motor drives are found to be more suitable.

Paper Making. The machine that makes paper in a paper mill has to perform the job of forming sheets, removing water from sheets, drying of sheets, pressing of sheets and reeling up of sheets. Therefore, the paper is made in following five sections (*i*) Couch section (*ii*) Press Section (*iii*) Dryer Section (*iv*) Calendar Section and (*v*) Reel Section.

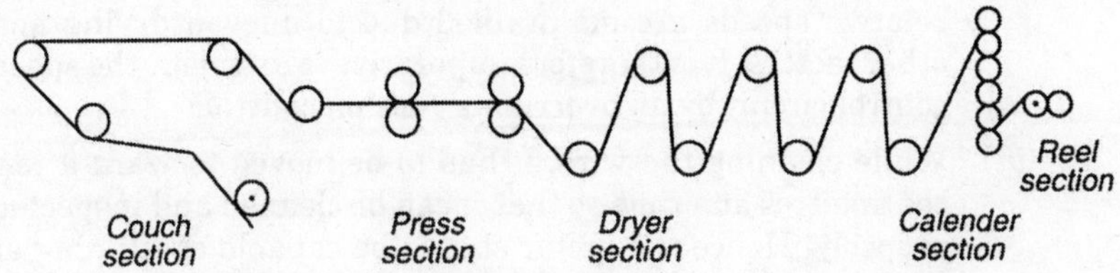

Fig. 1.112

Fig. 1.112 shows a schematic layout of the different sections that make paper. The paper pulp suspension with a moisture content of 98% to 99% is transferred uniformly to the wire. Most of the quantity of water drips through the wire mesh and the rest is removed by suction. At the end of the wire section, the moisture content would have reduced to about 80%. in the press section, which follows the wire section. Sheet is pressed between woollen felts so as to squeeze out water from the wet sheet and web leaves the press section with a moisture content of 65 to 60%. In dryer section, the sheet is further dried by passing it over and under the heated cylinders until the desired dryness, usually 6% of moisture content is obtained. In calendar section sheet is subjected to pressure and friction so that a compact and smooth surface of sheet results. In reel section the sheet is wound up on a mandrel.

Requirements of Paper Machine Drive

(*i*) For the actual formation and production of sheets and from the point of view of economy, it is necessary to maintain the speed of paper machine constant.

(*ii*) For a paper machine to be multipurpose, its speed should be adjustable over a range as large as 10 : 1.

(*iii*) In the wet end of the paper machine, the web becomes elongated by about 5%. Care should be taken to allow for this elongation by varying the speeds of individual sections. As this elongation is a function of the quality of the paper, speeds of sections should be independently adjustable.

In dry end of paper machine, the web hangs freely between sections without any support. Therefore a definite amount of tension must be set and regulated. This also necessitates a definite difference between the speeds of successive sections.

The relative speeds of two sections have a direct bearing on the pull on the sheet. Hence steady state accuracy of 0.1% or better is specified for the speed control system so as to avoid learing and folding of sheets.

(*iv*) As web is manually introduced into the calendar, an arrangement to take up sag is must.

(*v*) In the last two sections viz. the calendar and the reel, variations in tension in sheet can occur even if the correct relative speeds are maintained due to uneven drying and other factors. It is therefore imperative to augment the speed control circuit by an overriding tension control.

(*vi*) While cleaning the wire, it has to be moved forward a few centimetres at a time so that it can be cleaned and inspected properly. Hence the motor should be capable of running at inching speeds of 10-25 m/min. as long as particular button is pressed.

(*vii*) Each section should be able to run at the crawling speed of 10-25 m/min. for running in felts, wire and heating up of dryer cylinders.

(*viii*) Smooth and quick starting of the sections, without causing excessive starting current are required.

(*xi*) Control system employed should be flexible in nature.

Types of Drives Used for Paper Machine

There are two types of drives employed for making paper from pulp *i.e.* line shaft drive and sectional drive.

Line Shaft Drive. In this form of drive, the various sections of the paper machines are driven from a line shaft running the full length of paper machine. Cone pulleys and belt combinations, drive the various sections of the machine from line shaft through right angled gear reductions.

Normally electric motors are employed to drive the transmission shaft. Both a.c. and d.c. drives can be used for practically loss-less speed control.

In A.C. drive only the a.c. commutator motor with shunt characteristic is of use for obtaining an economic speed control system. Its advantage is in the possibility of connecting the motor directly to a 3-ϕ supply thereby eliminating an a.c.-d.c. converter, which would be required if a d.c. drive were used. However, the speed of an a.c. commutator motor depends on load and, hence its use as a paper machine drive with strigent requirement of constant speed is no longer advisable. Also its speed range as well as the power required greatly affects the motor size. The open loop speed control of the a.c. commutator motor is sluggish in comparison with a d.c. drive, as speed is varied by adjusting an induction regulator and shifting the brush rocker.

In d.c. drives, the speed of the paper machine is controlled by varying the armature voltage of a separately excited d.c. motor. The variable d.c. voltage is obtained from alternating voltage supply by means of either rotary converters or static converters.

Sectional Drives. In sectional drive each section of the paper machine has its own electrical motor. All the motors are operated from a common supply bus. By varying bus bar voltage, the speed of the paper machine can be controlled. By adjusting the field excitation of any motor it is possible to vary the speed of that particular motor with respect to other motors.

Comparison Between Line Shaft Drive and Sectional Drive. The requirements of a line shaft drive are not as stringent as those of the sectional drive. It has a number of major limitations as compared to the sectional drive system. The lower cost of the electrical equipment of a line shaft drive system is offset significantly by the additional cost of the mechanical equipment required and its maintenance.

Draw. In a line shaft drive, the draw (speed difference between sections) is generally achieved as follows. The individual sections are coupled to the line shaft by means of cone pulleys and belts. By shifting the belt position the section speed can be trimmed with respect to the master speed. The greatest problem in setting the draw is the slipping of belts. If the belt starts slipping the draw of the particular section gets affected and there is no automatic corrective action.

In sectional drives the draw is obtained by trimming the speed with reference to the sections and any variation in draw due to any reason is automatically corrected by means of closed loop feed back control systems.

Repeatability of Draw Settings. In a sectional drive the draw settings are done by calibrated accurate potentiometers. Therefore, the draw settings, which are quite critical can be accurately recorded and reproduced when the particular grade of paper is produced again.

In line shaft drive, it is difficult to achieve the above.

Loop Removal. In sectional drive, slight increase in speed of a section is easily obtained, without disturbing the original draw settings for removal of loop.

In line shaft drives the loop removal has to be done by changing the draw settings. This invariably results in over correction. For example assume that the paper has started forming a loop between dryer and calendar. If the operator speeds up the calendar to get the correct value, whatever loop that has already formed will continue to remain without reducing or increasing. The operator now speeds up more and more until the loop is fully taken up and finds that the paper has become tight after some time. Then he has to reduce the speed. This process goes on until the correct draw setting is obtained.

Rate of Loop Removal. In sectional drives, the rate of loop removal can be made constant at all operating speeds by suitable circuitary.

In the line shaft system, for a constant rate of loop removal the belt shifting will be different at different operating speeds. At lower speeds the shifting will be more as compared to that at higher speeds.

Independent operation of sections. With sectional drives, any section can be started and run at any required speed. For instance during starting the wire section will be switched on for 'crawl' in order to do washing, whereas the dryers will be running at a higher speed for initial warming up.

In line shaft system, all the sections will necessarily have to run at line shaft speed.

Crawl. While paper is being made one of the sections may develop a minor trouble and it may have to be stopped. After the fault has been rectified, it may have to be put on 'crawl' for check up. With sectional drive this is possible but not so with line shaft.

Inch Reverse. During threading paper can get jammed in the dryer sections. To remove the paper it is essential to have "inch reverse" facility. Line shaft system can not offer this feature.

Electrical Interlocks for safety. With sectional drives, each section can have its own set of devices that permit starting and safety interlocks. The former do not allow the sections to be started until all the necessary safety conditions for that particular section are fulfilled. Also if any fault develops during running the safety drive will trip only that particular section.

With line shaft system, human negligence can lead to major accidents. An operator can engage the clutch of the drive under the most dangerous conditions.

In case of sectional drives, overloading in a particular section can be sensed and the drive can be shut off if overloading is beyond the safe limits. Whereas with the line shaft system overloading in a particular section has negligible effect on the drive motor and hence the overloaded section will be continued to be driven until the machine breaks down.

12. **Machine Tools Shears.** To drive small portable tools such as drill etc. high speed motors with gear arrangements are often used. D.C. series motor and slipring induction motor is used for punches and shears.

13. **Electric Traction.** D.C. series motor which is simple and robust is used for suburban service where high rate acceleration is essential for main line works 1-ϕ A.C. compensated series motors are used. Due to poor starting torque these motors are not suitable for suburban service.

14. **Cement Works.** 3-ϕ 11 kV slipring induction motors are used as the size of motors used for rotating the current kilns is very large.

Cement Mills

Cement Making. The raw materials for producing cement contain lime and silica as main components and aluming and ferric oxide as fluxing components. The limestone mixed from the quarries is crushed and transported to the plant by dumpers wagons, trucks or ropeways depending on the area and distance involved. In fact if the quarry is within 1-2 km from the plant the crusher might be located right next to the plant and in the line of supply of lime stone. The crushed limestone together with the required proportion of corrective additives like clay bauxite, iron ore etc. is grounded in grinding mills. The fine dry powder coming out is homogenised in silos by passage of air from bottom and through the medium. It is then fed into kiln, which is the heart of the cement plant for producing cement clinker at high temperature. If the kiln receives finely ground and precisely composed dry feed as mentioned above, the cement plant is called as a dry process one. In wet process, the raw materials are ground with water to produce a slurry before entering the kiln feed tank. Dry process is preferred to wet process because less fuel is required by such kilns. Wet process is necessitated sometimes, since certain materials contain so much water that adding a little more water and using wet process is better than trying to dry the raw materials. The clinker coming out of the kilns is air cooled in special types of coolers and then transported to the storage. After aging in storage for atleast three days the clinker mixed with the right amount of gypsum is fed to cement grinding mills ground to required fineness. The cement is stored in silos drawn for packing in gunny bags and despatched by wagons or trucks to the dealers.

Types of Drives. The driving motors used in the cement industry can be broadly classified as follows :

(*i*) Raw mill and cement mill drives

(*ii*) Kiln drives

(*iii*) Crusher drives

(*iv*) Waste gas fan drives

(*v*) Compressor drives etc.

Raw Mills and Cement Mill Drives. Slipring induction motors of 6.6 kV are widely used. In order to improve the power factor of the line, current drawn, high voltage capacitors of adequate reliability and automatic capacitor control switchgear and circuit breakers are to be used. Even after adding the price of the capacitor and the control gear, the slipring motor is cheaper than the synchronous motor of same rating. Liquid resistance starters are usually employed to start the motor and to bring it upto full speeds. Gear boxes are also attached in order to get the desired mill speed of about 15 r.p.m.

From the point of view of voltage dips during starting the starting current of mill drives for large cement plants is normally restricted to 1.75 times the full load current.

The starting torque for the mill motor for large cement plant is limited to 125% of the rated torque and the pull out torque is restricted to nearly 240% of the rated torque.

Normally the motors should be able to withstand 50% overload for one minute occuring four times per hour at equal intervals. The motors for such drives are generally designed for a duty cycle of three consecutive starts from cold condition and two consecutive starts from hot conditions per hour against full load.

Twin Drives. Due to large ratings required for the raw and cement mill drives and due to the limitations in the availability of large size gear boxes and motors, twin drives are employed in these mills. The two motors have to be more or less identical to each other and so also their liquid resistance starters.

Gear less Drives. In developed countries gearless drives are being increasingly used for large mills. The rotor is shrunk on to the mill and the air gap between the rotor and the overlapping stator is maintained by levitation using a sophisticated electric closed loop control. The supply frequency is rectified into d.c., which is then converted to a.c. to a much lower frequency so as to provide a mill speed of approximately 15 r.p.m. This arrangement completely dispenses with the gear box, which is normally the source of maintenance problems. These types of drives would become economically viable in a few years. When power diodes are thyristors would be available in plenty at a much lower cost than those prevalent today.

Kiln Drives. The rotary kiln is an indispensable part of a cement plant. There are different types of rotary kilns depending on whether the cement is manufactured by means of wet or dry process. But in general they are tubular. Slightly tilted from the horizontal and have a ring gear fitted around them, which engages with one or two pinions. Each pinion drive shaft is driven by a variable speed motor.

In early years variable speed a.c. commutator motors were employed for kiln drives. However due to requirements of higher output ratings and of speed range in excess of 1 : 2, the commutator motor become an expensive proposition in addition to maintenance problems. This was followed by Ward-Leonard drives. On account of higher capital cost, lower efficiencies and greater maintenance problems as compared to thyristor controlled d.c. drives, the Ward-Leonard drives has been superseded by the d.c. motor with static supply.

The rating of the motors used for driving the kilns vary from 100-1000 KW The maximum speed of the kiln is about 1 r.p.m. and the kiln motor has to be designed for a speed range of the order of 1 : 10.

The starting torque required may be between 200% to 250% of full load torque. The motors are also specially designed to pick up speed at full load within the normal time of 15 seconds. Quite often kiln motors have to cater to over loads to the tune of 200% to 250% for small periods of time. The motor and control equipment have also to be specially designed for inching and spotting of the kiln during maintenance and routine check.

Twin Motor D.C. Drives. To cope up with increasing kiln capacities the modern trend is to use twin motor d.c. drives for kiln application. In this case two d.c. motors with separate pinions drive the same gear wheel at the periphery of the kiln drum. Although this arrangement has certain advantages for the designer ofelectric motors, it does involve extra expenditure on the electrical side.

The twin motor drive system must be designed such that motive power is supplied in equal parts by the two motors to prevent over-loading of either one of them or its mechanical transmission system. The speed of the two motors must necessarily be the same because they are coupled through the gear system. This can be achieved either with series connection of two d.c. motors or with parallel connection by means of a closed loop control system.

Crusher Drives. The motors used in crusher are of the slipring type. Stalling considerations play a very important part in the design of these motors. Normally the motors are designed to withstand locked rotor current during running without any external resistance introduced in the rotor circuit for one minute. This is quite important, since very often the crusher tend to get jammed, when a big sized boulder gets trapped between the jas of the crusher. Generally the starting torque for such drives is limited to 160% of the full load torque and pull out torque is limited to 200% to 250% of the full load torque. The motors are also normally designed for 15% overload for 15 seconds and 20% overload for 10 seconds taking into consideration the adverse loading conditions encounter practice.

Fan Drives. The motors used are of the slipring type with a speed variation, generally between 1000 and 750 r.p.m. The cast iron grid resistance controllers are normally used for starting and controlling the speed of these drives. As the motors are located out door or in semi out door location, totally enclosed motors of TEFC are employed. The starting torque is around 120% of the full load torque and pull out torque is between 200% to 250% of the rated torque. The sliprings and brush gears are totally enclosed and kept external to the motor enclosures for ease in maintenance. Due to higher frequency and finer control, thyristorized d.c drives are replacing slipring motors in fan drives too.

Compressor Drives. The air compressor motors are normally of smaller rating varying between 300 kW to 450 kW Generally the compressors are started on no load. Depending on specific system studies either a squirrel cage or a slipring motor can be used. The enclosures are usually of TEFC type and the speeds vary between 1000 and 750 rpm.

Textile Mills. From the stage, cotton is picked in the field to the stage it leaves the mills as finished cloth, it undergoes different processes viz. cotton to slivers, spinning, weaving and finishing.

Cotton to slivers. The process of separating the seeds from cotton is known as ginning. Ginning mills are usually located in the cotton growing areas. Bales of ginned cotton are first transported to the textile mills. There they are opened and the impurities are picked up and removed in the blow room. After further opening and cleaning, cotton is transformed into laps and fed to cording section. Here it is opened completely and is converted into slivers. The slivers are gathered in cans and then processed on a drawing machine, which makes them uniform by straightening the fibre. The slivers are then changed into laps from before feeding them for combing which parallels the fibre and upgrades it.

Spinning. The sliver at this stage is in a fiagele condition and is also bulky. After reducing the diameter in two or three stages it is processed on 'speed frame', which makes it suitable for final spinning. Due to twisting a continuous yarn of sufficient strength is produced during spinning. This yarn is wound on bobbins located in cone winding machines.

Weaving. Before the yarn is actually woven, it is warped *i.e.* made into a uniform layer. Weaving consists of joining two sets of threads one which extends throughout the length of the fabric and the other whose threads go across. This process is done in a loom.

Finishing. This consists of a number of processes like bleaching, dyeing, printing, calendering, stamping and packing. The impurities like oil and greases are removed and the fabric made white during bleaching. Dyeing involves giving a colour or shade to the cloth. Printing produces designs and patterns in multi colours.

Motors used for Different Textile Processes. All machines used in accomplishing the different processes described above require electric motors as their drives, special environmental, operating and drive conditions demand specially designed motors for textile industry.

Loom Motor. In order to accomplish the 'pick up' process in a short time the starting torque of the loom motor should be high. The loom being essentially a reciprocating mechanism causes both torque and current pulsations. Also loom motors are subjected to frequent starts and stops. These result in a higher temperature rise and is taken care by having good thermal dissipation capacity of the motor.

Loom motors are either totally enclosed or totally enclosed fan cooled, 3-ϕ high toque squirrel cage induction motors. Presence of lot of fluff in the atmosphere requires a smooth surface finish of the housing and end shields so that the fluff does not get collected on the surface of the motor. The insulation of the motor must be able to withstand high moisture content.

The ratings of the motors used for driving looms for light fabrics such as cotton, silk, rayon, nylon etc. are 0.37, 0.55, 0.75, 1.1 and 1.5 kW white those of the motors used for making heavy fabrics (wool and canvas) are 2.2 and 3.7 kW. They are usually 6 or 8 poles.

Card Motors. The general requirement of card motors is almost similar to that of loom motors except that the former is required to have a very high starting torque and must be able to withstand a prolonged starting period. Both the above requirements for the card motors are due to the very high inertia of card drum. Once the drum is started the operating is continuous and uninterrupting, upto like that of a loom where frequent starts and stops are involved.

The commonly used drives for card motors are again totally enclosed and totally enclosed fan cooled 3-ϕ high torque squirrel cage Induction motors. The usual ratings of motors for cards of light fabrics are 1.1 and 1.5 kW and those for cards of heavy fabrics are 2.2, 3, 3.7 and 5.5 kW. Here again the preferred synchronous speeds are 750 and 1000 r.p.m.

Spinning Motors. For good quality spinning, it is essential that the starting torque of spinning motors should be moderate and the acceleration should be smooth. If the starting torque were low, the tension of the yarn would be insufficient and hence the yarn would get entangled and break. If the starting torque were high, the acceleration would be high and the yarn would snap.

In general three types of drives are used for spinning frame operation. Single speed motor, 2-speed motor and two motor drive. Normally a 4 pole or 6 pole squirrel cage induction motor is used as single speed drive.

In order to maximize production with minimum breakage, two speed motors (4/6 or 6/8 poles) are used. Although these motors would be larger in size and costlier, the increased production may compensate for the additional initial outlay.

In case of two motor drive, two separate motors are used to drive the common pulley of the ring frame. Although this drive is costlier and requires more space, it has the following advantages :

(*i*) Any desired speed differential can be got by aqjusting the speed ratios.

(*ii*) The tension of the yarn can be adjusted independently.

(*iii*) Production can be continued even when one of the motors fail.

Totally enclosed fan cooled motors ranging from 5-30 kW are used as spinning motors. From what has been discussed above it is clear that some form of controlled torque starting of textile machinery drives is imperative. Proper starting will minimize yarn breakage, improve the quality of product and increase the life of the machine. Less down time obtained will also increase production.

The most common electrical method of controlled torque starting involve the use of standard squirrel cage motors and different methods of applying reduced voltage to the motors during a selected starting period. One of the most effective methods of controlling the stator voltage of motors used in textile industry is that of using series reactors. This method gives stepless, closed transition increased to approximately full speed. Since the motor terminal voltage is a function of the current drawn from the line, during acceleration the motor voltage will increase as the line current decreases, resulting in greater accelerating energy at the higher speeds and no significant pull on the yarn being processed. Reactor with fixed tapping may be used to get variety of starting torques. Variable iron core inductors offer almost infinite choice of starting torques. Since the normal starting time is limited to a few seconds, the effect of the poor p.f. during starting on the overall plant P.F. is not appreciable. Although primary resistor starting has all the advantages of reactor starting, it has not proved practicable for textile plants because of the possible fire hazards, when the resistors get overheated during frequent starting.

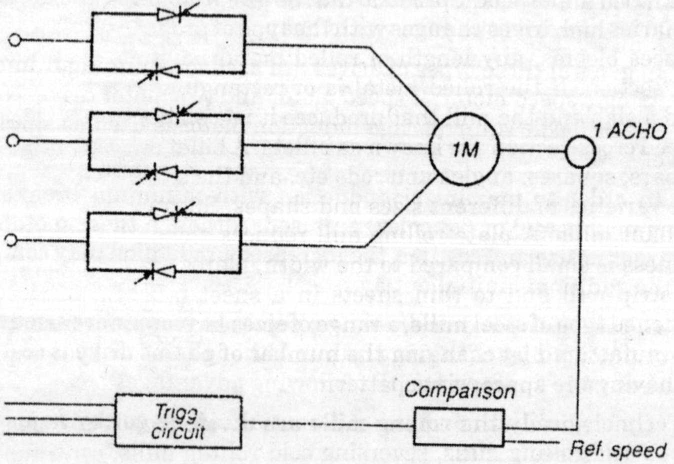

Fig. 1.113 Solid state stator voltage control of textile motor.

With the ever increasing use of solid state devices, nowadays, a.c regulators are being used to provide control of starting torque. Fig. 2.112 shows the commonly used circuit. Through a tachometer speed feed back signal, machine starting can be automatically controlled by presenting the reference input signal from a potentiometer. Thus a large variety of torque time machine starts is possible from the same controller without any tap changing or adjustment of airgap of reactors.

15. **Belt Conveyers.** High torque squirrel cage or slipring induction motors are used for this purpose.

16. **Refrigeration and Air Conditioning.** 1-ϕ, 220 V capacitor type induction motors with D.O.L. starters are usually employed.

17. **Pumps.** Drip proof, totally enclosed polyphase induction motor. Squirrel cage type in case of centrifugal pumps and slipring type in case of reciprocating pumps are usually employed.

18. **Textile Mills.** Three phase A.C. motors are used, as their speed remains constant due to dependence on supply frequency. D.C. motors cannot be used because their speed varies with voltage.

19. **Iron and Steel Industry.** Motors used in mill shops are D.C. compound motors or induction motors. Speed control in case of D.C. motors is obtained by shunt field regulator or Ward-Leonard method.

20. **Steel Mills.** Rolling of steel, the process during which the cross-section of the metal gets reduced, while increasing its length proportionately is the major function of steel mills.

Steel mills usually produce blooms, slabs, rails, rods, sheets, strips, beams, bars or angles. Rolling mills are generally classified according to their products, since, very often a particular mill is capable of manufacturing a specific product and the design of millstand, millihead, auxiliaries and drives changes with the type of product. A blooming mill produces blooms, any length of rolled metal having almost a square cross section. If the rolled metal is of rectangular cross section, it is called a slab and the mill that produces it is a slabbing mill. Blooms of reduced cross-section are known as billets. A billet can be further rolled into bars, squares, angles and rods etc. and the same mill can produce these varieties of different sizes and shapes. Such mills are termed as merchant mills. A plate rolling mill reduces slabs into plates, whose thickness is small compared to the width, plates are changed to strips in a strip mill and to thin sheets in a sheet mill. In all the above mentioned type of steel mills, a range of sizes having similar shapes can be manufactured by changing the number of passes or by using stand rolls having the appropriate pattern.

Technologically the rolling mills are divided into four types continuous cold rolling mills, reversing cold rolling mills, continuous hot rolling mills, and reversing hot rolling mills.

Sheet steel having high mechanical properties, uniform gauge and good surface quality can be produced only by cold rolling. Blooms, slabs, billets, strips etc. are manufactured by hot rolling mills.

In a reversing mill there is only one stand carrying the rolls that press the metal and the metal is passed through this stand alternately forward and backward several times in order to reduce it to desired size. Each motion or travel is known as pass. A continuous mill or tantem mill consists of several stands each one of them carrying pressing rolls. The metal passes through all stands in only one direction and gets rolled simultaneously by some or all of them.

Reversing cold rolling mills produce sheets with a minimum thickness of 0.15 mm and a maximum width of 2 m and more, black plates with a minimum thickness of 0.07 mm and a maximum width upto 1.3m and bands with a minimum thickness upto 0.0015 mm and a maximum width upto 1 m. The turnover of a reversing cold rolling mill is limited by the low sheet speed owing to the continuous acceleration and retardation and time taken, for adjusting the roll gap during each pass. This disadvantage overcome in continuous cold rolling mills, in which the metal passes in one direction, through several successive stands.

Continuous hot rolling mills, usually, manufacture billets, strips and the products of a merchant mill. Blooming and slabbing mills, which produce blooms and slabs respectively are generally of reversing hot type.

Reversing Hot Rolling Mills

Process. Reversing hot rolling mills roll out blooms or slabs from hot steel ingots, which come directly either from steel making shops or from heating chambers called socking pits. Red hot ingots are taken out one by one by special crane having tongs and placed on a car known as ingot buggy. The ingot buggy travels a short distance on rails along the same line as the mill bed and takes the ingot to the front end of the mill bed. The mill bed consists of a large number of rolls.

Fig. 1.114 given below shows the schematic diagram of the various stages in the hot reversing mills. The ingots brought upto the front edge of the mill bed are tilted and placed on the receiving table such that they lie with its longitudinal axis horizontally on the bed. Then they move over to the weighing table where it is possible to weigh the passing ingots. The travel of ingots over the roll table is facilitated by the rotation of the rolls on which they are laid from the weighing table, the ingots move over an approach table, several intermediate tables and a few front work tables before reaching the main working roll of the millstand. On the other side of the stand the mill bed has a few back work tables, several intermediate tables and a runout table in succession ending with a shear table, where the rolled metal is cut into standard lengths. The length of the mill bed on either side of the stand will be determined by the length of the product of the mill before shearing.

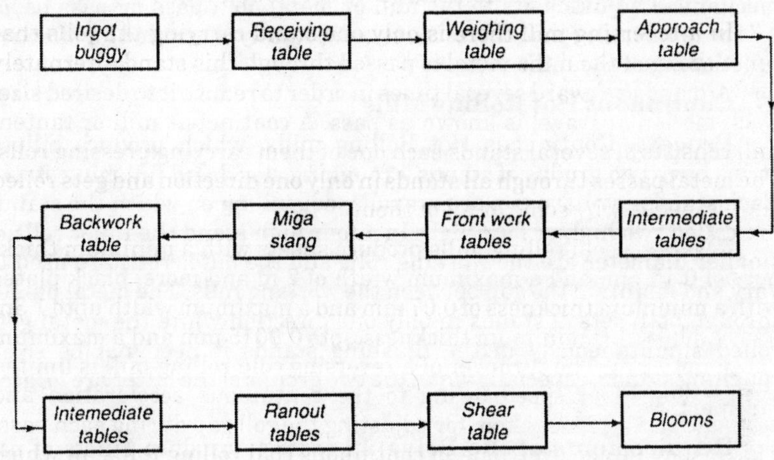

Fig. 1.114

The ingots are rolled by the wörking rolls into blooms, as they pass through the mill stand several times in forward and reverse directions. The gap length between the main rolls is reduced gradually in successive passes by bringing down the top roll of the main rolls by a screw down mechanism. This adjustment of gap length is made during the time when the forward ingots is stopped for reversal of its direction of motion.

Alignment of the metal to required entry position in mill rolls is done by devices called manipulator side guides. Tilt fingers, in conjunction with manipulators turn metal over to permit rolling on all four sides.

Requirements of the Drive. The main drive of the hot reversing null consists of either a motor driving both rolls through pinions or a twin drive arrangement in which each roll is separately driven. A single armature motor may be used for mill of medium size out of double armature motor is necessary for large mills. In very large mills a double armature twin drive may be used. Main requirements of the drive for the main working rolls are :

(*i*) Wide range of speed variation.

(*ii*) Ability of the motor and its control equipment to permit frequent starts and reversals.

(*iii*) Reversals of the direction of rotation of the motor without causing serious disturbance in the power handing circuits.

(*iv*) Automatic control of operation.

(*v*) High reliability.

(*vi*) Accuracy.

In addition, there are a number of auxiliary drives which are used to convey the ingot, bloom or slab to or from the mill and in making mechanical adjustments to the mill or metal on a pass by pass basis. All these variable speed drives must be integrally controlled for effective functioning of the mill.

Continuous Hot Rolling Mills

Process. Continuous Hot Rolling mills, which produce billets, strips have two groups of stands – roughing stand and finishing stand. Each stand may have two or four rolls, depending on which the stands are called two high or four high. In a four high stand the inner rolls of smaller diameter are the working rolls and the outer rolls are used to vary and maintain the gap between the working rolls. The metal passes through both sets of stands in only one direction white the metal gets rolled simultaneously in the finishing stands it need not be so in roughing stands, especially when the two groups of the latter are located far apart.

Requirements of the Drive. In order to produce billets of different dimensions, different gap lengths between the working rolls are to be set for a specific gap setting, the speeds of the stand motors are maintained at values differing slightly from each other to take care of the gradual reduction in thickness and lengthening of the metals *i.e.* the speed of the motors must be capable of variation over a range of 1.5–2. In order to avoid sag of metal between stands, the speed of the stand motors should be controlled to a high degree of accuracy. Also, the metal comes into contact with the working rolls at its speed of operation. Since this gives rise to a sudden application of load on the stand motor, its speed may drop and cause sag of the metal between stands. Hence by means of an automatic control scheme the speed of the motor must be restored to its set value as quickly as possible.

Reversing Cold Rolling Mills

Process. The strip to be rolled is received by the mill in the form of a real, wound on a mandrel, mounted on one of the reel stand as shown in Fig. 1.115. Another stand carries an empty mandrel. The end of the strip is threaded manually through the stand rolls and round the empty mandrel. The mill is then run with slow speed and the strip is wound on the coiling drum for a few turns with low tension and roll pressure. The speed is then increased with a uniform acceleration and the metal is rolled at the required pressure and tension. During this one reel gets coiled and the other is uncoiled. When only a few turns are left to be rolled, the speed is reduced and the direction is reversed, thus ensuring that the strip does not come out of the uncoiled reel. The same procedure is repeated while rolling in the opposite direction. A number of such reversed rolling with gradual gap length reduction produces the desired strip.

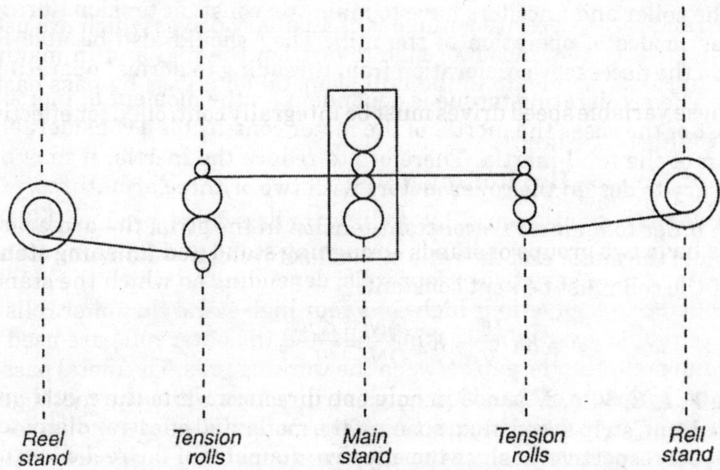

Fig. 1.115

Requirement of the Drive. The mill consists of a reversing stand, two reversible coilers in most cases an additional uncoiler for the first pass. The mill stand is, often of four high type, *i.e.* it has two work rolls with a diameter between 200 mm to 500 mm and two back up rolls whose diameters are about two to four times greater than that of the work rolls. The back up rolls are required to prevent bending of the work rolls. The torque required for rolling is provided by either one or two stand motors. The drive is provided to the back up rolls through gears and the work rolls are friction driven. Sometimes the work rolls are themselves driven by individual motors. The coiler and uncoiler are driven by separate motors. In addition to coiling and uncoiling the strip, the coilers also have the job of developing a defined strip tension between themselves and the stand during the individual passes. This tension is not only required for avoiding either looping or breaking of strips, but also to prevent variation in the gauge of the material being rolled. The roll gap is adjusted by means of a screw down system.

One of the characteristics of a rolling mill is that of the armature intertia of the mill motor would be larger than that of the rolls. Hence in large capacity mills, to keep the ratio of motor inertia to load inertia low, motors with multiple armatures are chosen.

In principle, the mill stand requires a four quadrant drive at different speeds. Hence, the control system must provide both armature voltage control and field weakening. Field is kept at its rated value until the maximum back e.m.f. is reached and subsequently reduced depends upon the speed requirements. When twin drives are used, separate controls are provided for the two motors with additional circuitry to ensure equal load sharing.

The coiler and uncoilers have to maintain constant tension during various modes of operation of the mill. They should also be able to provide the necessary acceleration from threading to normal operating speed. The accelerating torque is dependent on the moment of inertia. In most of the cases the inertia of the motor constitutes a considerable portion of the total inertia. Therefore to reduce the inertia, it may be necessary to design the coiler motors with two or three armatures.

In order to maintain a constant tension in the strip, the armature current of the motor and the ratio of the flux to effective diameter of the reel of the coil must be kept constant.

$$F = K_1 \frac{IE}{V} = K_2 \frac{I\phi N}{ON} = \frac{K_2 I\phi}{O},$$

where F, I, E, V, ϕ, N and O denotes strip tension, armature current, induced emf. strip speed, flux, speed of the motor and effective diameter of the reel respectively : since the effective diameter of the reel changes as the strip is unreeled at the input end and reeled up at the output end. The flux also to be varied linearly with the coil diameter. If the maximum strip tension is required at maximum speed from the point of view of rating of the motor, the optimum condition is to make the field weakening range identical with the diameter range. However this may not always be possible, since it would be very difficult to get a field weakening range better than 1 : 3. In such cases it is necessary to operate the system both in armature and field control ranges, which demands a motor of higher power rating.

Continuous Cold Rolling Mills. The metal in such mills passes in only one direction through several successive stands. The stands are often of four high type.

The mill consists of an uncoiler and coiler reel, the latter demands accurate torque and speed control. The most important features of continuous cold rolling mills are given below :

The strips are subjected to a large interstand tension, which is kept at constant value depending on the desired thickness of the sheet to be produced. The threading of the metal through the stands is done at a low speed. As soon as the threading is over, the speed of the mill is raised to leaves the first stand. Depending on the products to be obtained, a wide variation in speed of the mill drive is called for.

Drives Used in Different Mills. Use of d.c. motors is universally accepted for choosing the main drives in both reversing and undirectional continuous mills. Motors for reversing mills must have high starting torque, wide speed range, precise speed control be capable to withstand overload and pull out torque which may be as high as three times the rated value and have good commutation throughout. Acceleration from zero to base speed and then to top speed and subsequent reversal from top speed forward to top speed backward must be as small as possible. Therefore the motors used for steel mill duty are of totally enclosed, forced ventilated and higher class of insulation.

As regards controlling the speed of the drive, Ward Leonard system with a flywheel is unsurpassed for the purpose. A liquid rheostat known as slip induction motor provided for the starting and regulation of the slip of the induction motor prime mover of the Ward Leonard Igner set. This is also used as a loading resistance for the induction motor while applying d.c. rheostatic braking to bring the Ilgner set to a quick stop at a fast controlled rate. D.C. supply required for rheostatic braking is obtained from the exciter of the variable voltage d.c. generator and the stored energy of the Ilgner set is dissipated in the slip regulator. In addition to this electrical braking, a mechanical braking arrangement is also provided to bring the Ilgner set to rest in case of emergency, of course, nowadays, the Ward-Leonard Ilgner system is rapidly losing its foothold against the thyristor converter, which has brought phenomenal changes in the design of control systems for the mill drives.

1.69 Conclusion

For selecting a particular motor for a particular drive various factors have to be taken into consideration as already discussed. All these factors may not favour the same choice and some of the factors may be conflicting. For example in a particular drive a high starting torque is needed. The choice for a motor should be naturally a D.C. series motor but if the motor can not be mounted in line with the drive, the choice is that of belt drive. Now belt drive cannot be used with D.C. series motor for fear of belt getting broken and the motor running on on-load thus developing infinitely high speed. Thus other aspects require the choice of D.C. series motor but it cannot be recommended as the drive used is belt drive.

Thus due consideration has to be given to each and every aspect before a motor is recommended for a use for any particular drive.

1.70 Specifications for Commonly used Motors.

Commonly used motors are :

- Squirrel cage induction motor
- Slipring induction motor
- Single phase induction motor
- Synchronous motor
- D.C. series motors
- D.C. shunt motors

Important specifications of these motors will include the following particulars:

- Rating in H.P.
- Type of supply *viz.* 1–ϕ, 3–ϕ or D.C. supply
- Voltage

- Speed
- Frequency
- Rating *i.e.* continuous, intermittent or short time rating
- Type of enclosure
- Type of drive
- Ambient temperature
- Type of starting
- Type of braking [if used]

However it should be noted that in general, high speed motors are relatively less costly having higher efficiency and are smaller in size.

1.71 Steady State Stability of an electric Drive.

Equilibrium speed of a motor-load system is obtained when motor torque equals the load torque. Drive will operate in steady-state at this speed, provided the speed is in a state of stable equilibrium.

Concept of steady-state stability has been developed to readily evaluate the stability of an equilibrium point from the steady state speed-torque curves of the motor-load system, thus avoiding solution of differential equations valid for transient operation of the drive.

For example let us examine the steady state stability of equilibrium point A in figure 1.116 (*a*) shown below.

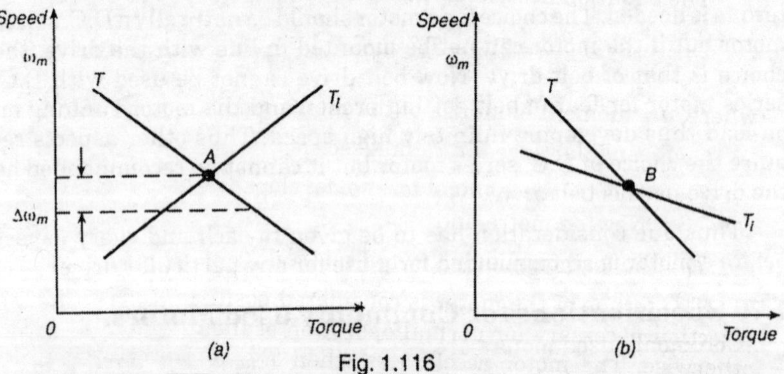

Fig. 1.116

The equilibrium point will be termed as stable when the operation will be restored to it after a small departure from it due to a disturbance in the motor or load. Let the disturbance causes a reduction of $\Delta\omega_m$ in speed. At new speed, motor torque is greater than the load torque, consequently, motor will accelerate and operation will be restored to A. Similarly an increase of $\Delta\omega_m$ in speed caused by a disturbance will make load torque greater than the motor torque, resulting into deceleration and restoration of operation to point A. Hence the drive is steady-state stable at point A.

Let us now examine equilibrium point B as shown in figure 1.116 (*b*) above, which is obtained when the same motor drives another load.

A decrease in speed causes the load torque to become greater than the motor torque, resulting in the deceleration of drive and operating point moves away from B. Similarly, when working at B an increase in speed will make motor torque greater than the load torque, which will move the operating point away from B. Thus B is an unstable point of equilibrium.

1.72 Transient Operations Characteristics.

Starting, braking, speed change and speed reversal are transient operation. It may also be expressed as the operation of a drive between two steady state conditions. During this period, for an electric drive, the speed, torque and current undergo change.

This may be caused due to sudden change of load, braking of drive, starting of drive, reversal of direction of rotation of the drive, sudden change in supply voltage or frequency or both. The study of transient operation characteristics helps in determining the requisite control equipment for starting, braking etc. so to achieve minimum energy loss during such operation and to ensure that the machine operates stably after the transients die over.

In relation to it, a brief description about the dynamics of electric drive is given here.

The equation describing the rotary motion of a motor is given by,

$$J\frac{d\omega}{dt} = T_{em} - T_l$$

where, J = moment of inertia of the rotating part in kg-m^2

ω = angular speed in radian/sec.

T_{em} = electro mechanical torque developed by the motor in Newton metre and

T_l = the load torque in Newton-metre.

The torque of the motor is considered positive when it acts in the direction necessary for normal operation of the drive and is negative otherwise. The motor accelerates when T_{em} is greater than T_l and decelerates when $T_l > T_{em}$. It operates at steady state torque when the net torque $(T_{em} - T_l)$ is zero.

Short Questions with Answers

Question 1. *Define electric drive.*

Ans. An electric drive is defined as a form of equipment designed to convert the electrical energy in to mechanical energy and is used for providing electrical control to this process.

Question 2. *What is a group drive ?*

Ans. Group drive means a drive in which a single electric motor drives a line shaft by means of which an entire group of working machines may be operated. It is also many times called the line shaft drive.

Question 3. *Why the d.c. motors are mostly gear motors ?*

Ans. The reason is that because, the low speed motors are much expensive than the high speed motors for the same output ratings.

Question 4. *What will be the effect on direction of rotation of a d.c. motor, if the terminals of the supply are reversed.*

Ans. There will be no effect. The motor will rotate in the same direction, as before the reversal of the supply line terminals.

Question 5. *For traction purpose why d.c. motor is preferred ?*

Ans. Because of the reason that d.c. series motor develops large starting torque and slows down with the increase in load and therefore gets automatically relieved from heavy excessive load.

Question 6. *What are the main disadvantages of a cage motor ?*

Ans. Cage motor is starting torque with large starting current and there is no possibility of speed control.

Question 7. *Due to which reasons the speed-torque characteristic curves are commonly used in utilisation studies ?*

Ans. In engineering applications and utilisation studies, it is normally the load torque which is known to us and the speed at which the motor can drive this load is unknown and it is to be determined. Thus the torque is independent variable and speed is dependent variable. Consequently, speed-torque characteristic curves with torque is abcisssa are commonly used in studies of utilisation.

Question 8. *What is multi-motor arive ?*

Ans. This type of drive consists of several individual drives each of which serves to operate one of many working members or mechanism in many production units.

Question 9. *Does a shaded pole motor have large starting torque?*

Ans. No, rather it has very little starting torque.

Question 10. *Why usually single phase induction motors are set on rubber spring mountings ?*

Ans. In a single phase induction motor, pulsating torque is developed as a direct consequences of the pulsating power in a single phase circuit.

Question 11. *Is the torque developed by a single phase induction motor at synchronous speed is zero?*

Ans. No, it is always negative.

Question 12. *How the power factor of a sysnchronous motor is changed keeping the shaft load undisturbed?*

Ans. It is done by varying the field excitation.

Question 13. *Define individual drive.*

Ans. When each machine has its own prime mover, it is an individual drive.

Question 14. *How selection of motor is done ?*

Ans. Selection of motor is done on the basis of their electrical and mechanical characteristics. The electrical characteristics is: starting and running characteristics. The mechanical characteristics is : enclosure and transmission system used.

Question 15. *Why D.C. Series motor should not be run without applying the load ?*

Ans. Due to the reason that on no load D.C. Series motor would attain very high speed and the motor may get damaged due to very high centrifugal forces set up in the rotating parts.

Question 16. *Why induction motors are called asynchronous motors ?*

Ans. Due to the reason that the rotor does not turn in synchronous with the rotating field developed by the stator currents.

Question 17. *Is the maximum torque of a three phase induction motor dependent on the rotor resistance ?*

Ans. No, it is always independent of the rotor resistance.

Question 18. *While controlling the speed of a 3-phase induction motor, why the V/f is kept constant ?*

Ans. In order to kept flux level constant.

Question 19. *What are the main disadvantages of a squirrel cage motor ?*

Ans. With large starting current there is a low starting torque and no possibility of speed control.

Question 20. *By which method the power factor of a synchronous motor is changed by keeping the shaft load undisturbed ?*

Ans. By variation of the field excitation.

Question 21. *How can the speed of a linear induction motor be controlled ?*

Ans. By varying both the frequency and voltage simultaneously.

Question 22. *Is torque developed by a single phase induction motor at synchronous speed zero ?*

Ans. No, it is negative.

Question 23. *Is a shaded-pole motor have High Starting Current?*

Ans. No, It has very small staring current.

Question 24. *Write whether repulsion motor has shunt speed torque characteristics or series speed-torque characteristics ?*

Ans. It has series speed-torque characteristics.

Question 25. *What is a stepper motor ?*

Ans. The motor which usually turns in discrete movement is called the stepper motor.

Question 26. *What is the cheapest method of starting a 3-phase induction motor ?*

Ans. By the use of Direct on Line (D.O.L.) starter or simply D.O.L. starting.

Question 27. *Which type of protection is used in the starters meant for 3-phase induction motors.*

Ans. Over load and under-voltage type protection.

Question 28. *While controlling the speed of a D.C. shunt motor, what should be done to achieve a constant torque drive.*

Ans. The applied voltage should be maintained constant so as to maintain field strength constant.

Question 29. *How speed variation is affected in a repulsion motor?*

Ans. In a repulsion motor speed variation is affected either by varying the impressed voltage or by changing the brushes position.

Question 30. *What do you know about plugging ?*

Ans. Plugging means, stopping of an electric motor by reversing its torque instantaneously till it stops.

Question 31. *What is the most cheap method of electric braking ?*

Ans. The Regenerative braking.

Question 32. *Electric breaking is accomplished in an induction motor by interchanging two stator terminals. Name the method.*

Ans. Plugging or Counter-current braking.

Question 33. *When it is required to control the induction motor speed during deceleration, what kind of braking is preferable to employ?*

Ans. D.C. dynamic braking is most suited for controlling the speed of an induction motor during deceleration by controlling the D.C. excitation.

Question 34. *What are the factors on which the rate of rise of temperature depends ?*

Ans. The rate of temperature rise depends upon :

(*i*) The amount of heat dissipated per °C rise of the surface of the machine.

(*ii*) Amount of heat produced.

Question 35. *Why the cooling time constant of a rotating machine is usually larger than its heating time constant ?*

Ans. The cooling time constant of a rotating machine is normally bigger than its heating time constant owing to poor ventilation conditions when the machine cools.

Question 36. *What is meant by the load equalization ?*

Ans. Load equalization is the process of smoothing out of fluctuations in load.

Fill in the blanks.

1. In individual drive, the operator has ------ control of his machine.
2. In group drive, number of machines are rotated through belts and pulleys by a big ------.
3. Starting torque of a d.c. series motor is ------.
4. In a centrifugal pump, the mechanical load is ------.
5. The characteristics of a d.c. compound motor are ------ the series and shunt motor.
6. The torque of a D.C. shunt motor ------ with the increase in load.
7. Speed of the Scharage motor can be controlled by ------.
8. If the field of a synchronous motor is over excited, the power factor will be ------.
9. Starting torque of a three phase induction motor can be increased by ------.
10. Direction of rotation of split phase motor can be reversed by reversing either of ------ windings.
11. For small domestic appliances ------ motors are generally used.
12. The reversal of rotation of motor for electric braking is known as ------.
13. For traction work, the best suited motor is ------.
14. Textile industry requires ------ motor
15. Generally in rolling mills, the type of braking applied is -----.
16. For punch and shear machines ----- motor is used.
17. In group drive the p.f. is ------ improved.
18. In lifting, torque is ------ of speed.
19. A d.c. series motor can safely started ----- load.
20. Starting torque of repulsion motor is ------.

Ans.

1. complete	2. main motor
3. Very high	4. lifting
5. in between	6. increases
7. shifting the brushes	8. leading
9. increasing the rotor resistance	10. two
11. universal	12. Plugging
13. D.C. series motor	14. Schrage
15. Plugging	16. D.C. Compound Motor
17. not	18. independent
19. without	20. high.

True or False Questions

1. In group drive initial cost is high.
2. In a shunt motor, torque is proportional to the armature current.
3. In case of blowers and centrifugal pumps, torque varies as the square of the speed.
4. In case of light loads, the power factor of a 3-ϕ induction motor is very low.
5. At no load, d.c. series motor can be started safely.
6. We treat a d.c. shunt motor is a constant speed motor.
7. Starting torque of a shaded pole motor is very poor.
8. Starting torque of a repulsion motor is low.
9. Plugging can be applied to an induction motor.
10. The characteristics of a D.C. compound motor remains in between those of D.C. Series and shunt motor.

Ans.

1. False	2. True	3. True	
4. True.	5. False	6. True	
7. True	8. False	9. False	10. True.

Choose the Correct Answer.

1. A traction drive which does not require a specific warming-up time, has not to carry a bulk of fuel and has the highest efficiency *i.e.* apporximately 40% is :

   (*a*) Diesel driven electric locomotive.

   (*b*) Gas turbine electric locomotive

   (*c*) Electric locomotive operated through a thermal power plant (*d*) Electric locomotive energised through a hydro-electric power plant

(*e*) None as above.

2. In case of individual drive :

(*a*) each operator has complete control of his machine.

(*b*) no load losses are eliminated

(*c*) the initial cost is less than that of group drive.

(*d*) both (*a*) and (*b*).

3. The basic elements of an electric drive are :

(*a*) electric motor, transmission and control system

(*b*) electric motor and the transmission system

(*c*) the control and transmission system

(*d*) conversion equipment and electric motor or only prime mover.

4. The important drawback of electric drive is that :

(*a*) it is difficult drive

(*b*) costlier in initial cost and as well as in maintenance cost

(*c*) failure in electric supply makes the drive stand-still

(*d*) all of the above.

5. The selection of an electric motor is governed by :

(*a*) nature of the load to be handled.

(*b*) conditions of the environment.

(*c*) availability of nature of electric supply

(*d*) all of the above.

6. D.C. motors are more useful for use in :

(*a*) steel mills, electric excavators and cranes

(*b*) machines and lathe tools

(*c*) jaw crushers and flour mills

(*d*) paper industry

7. At all speeds, load torque constant is represented by a

(*a*) compresor (*b*) fan

(*c*) centrifugal pump load.

8. For having heavy fluctuation of load, the machine used is :

(*a*) planer (*b*) lathe

(*c*) punching machine (*d*) printing machine

9. Normally variable speed operation is preferred in :

(*a*) ceiling fan (*b*) refrigerator

(*c*) exhaust fan (*d*) water pump.

10. If d.c. motor flux approaches zero then its speed will :

(*a*) approach to zero

(*b*) approach to infinity

(*c*) remains unchanged

(*d*) between zero and infinity (∞)

11. A fly wheel is usually fitted to :

(*a*) a d.c. series motor driving a constant torque load.

(*b*) a separately excited d.c. motor driving pulsed torque load

(*c*) a comulatively compound motor driving pulsed torque load

(*d*) a differentially compound motor driving pulsed torque load.

12. D.C. Shunt motors are used for driving of :

(*a*) lathes (*b*) machine tools

(*c*) line shafts (*d*) all of these

13. The induction motor has normally lagging power factor during,

(*a*) operation only (*b*) starting only

(*c*) both starting and operation (*d*) none of the above.

14. For reversal of speed quickly, the motor preferred is :

(*a*) Squirrel cage inducitón motor

(*b*) Synchronous motor

(*c*) d.c. motor

(*d*) slip ring induction motor.

Ans.

1. (*d*)	2. (*a*)	3. (*a*)	4. (*d*)	5. (*d*)
6. (*a*)	7. (*a*)	8. (*c*)	9. (*a*)	10. (*b*)
11. (*c*)	12. (*d*)	13. (*c*)	14. (*c*).	

Choose the Correct Answer

1. The Basic elements of an electric drive are :-

(*a*) Electric motor, control and the transmission system.

(*b*) Electric motor and the transmission system

(*c*) The control and transmission system

(*d*) Electric motor and conversion equipment.

2. Old engineering workshop is to be changed over from engine drive to an electric drive :

(*a*) Individual drive

(*b*) Group drive

(*c*) Multi-motor drive.

3. These days electric drive is becoming more popular because of the reason :

(*a*) It is clean, simple, reliable and compact

(*b*) In layout it is flexible, easy starting and availability of facility for remote control. Also it provides smooth and easy control.

(*c*) Its maintenance and initial cost is cheaper.

(*d*) All of the above.

4. For a paper mill requiring constant speed operation and flexibility of control, the type of drive used is :

(*a*) Individual drive

(*b*) Group drive

(*c*) Individual or Multi-motor drive

(*d*) Multi-motor drive.

5. In production industries, introduction of automation has become possible only because of the use of :

(*a*) Individual drive

(*b*) Group drive

(*c*) Electric drive

(*d*) Individual and Multi-motor drives.

6. For a particular drive, the selection of an electric motor is governed by :

(*a*) Environmental conditions

(*b*) Nature of the load to be handled

(*c*) Nature of electric supply available

(*d*) All of the above

7. Still d.c. motors are preferred to be used in :

(*a*) Lathes and machine tools

(*b*) Jaw crushers and flour mills

(*c*) Electric excavators, cranes and steel mills

(*d*) Paper industry

8. A.C. in to D.C. Conversion can be affected with the use of :

(*a*) silicon diodes (*b*) Motor generator set

(*c*) Mercury arc rectifier (*d*) Any of the above

9. For a flour mill which is the significant electrical characteristic in selection of the electric motor :

(*a*) Running characteristics (*b*) Braking charteristics

(*c*) Starting characteristics (*d*) Efficiency

10. A typical active load is :

(*a*) Hoist (*b*) Blower

(*c*) Pump (*d*) Lathe

11. An elevator drive is required to operate in :

(*a*) One quadrant only (*b*) Two quadrants

(*c*) Three quadrants (*d*) Four quadrants

12. Load torque constant at all speeds is represented by a

(*a*) FAN (*b*) Compressor

(*c*) Centrifugal Pump load (*d*) All of the above

13. An example of a motor having short time duty is found in :

(*a*) Centrifugal Pumps (*b*) Crane drives

(*c*) Drilling machines (*d*) All of the above

14. The load cycle for a motor driving a power press will be :

(*a*) Continuous (*b*) Variable

(*c*) intermittent and variable (*d*) periodical but continuous

15. Travelling speed of a crane changes from :

(*a*) 20 to 25 m/s (*b*) 10 to 20 m/s

(*c*) 1.0 to 2.5 m/s (*d*) 0.2 to 1.0 m/s

16. A d.c. shunt motor runs at rated speed. If its field circuit gets open-circuited, then soon after this the motor speed would tend to :

(*a*) decrease

(*b*) remains unchanged

(*c*) increase

(*d*) fluctuate around its previous speed.

17. An unsaturated D.C. Shunt motor drives a constant torque load at a speed of 1,000 r.p.m. taking an armature current of

20 A when 230 V is applied to the motor. If the applied voltage is reduced to 115 V, the motor runs at :

(*a*) 500 r.p.m. taking an armature current of 20 A

(*b*) 1,000 r.p.m. taking an armature current of 20 A

(*c*) 500 r.p.m. taking an armature current of 40 A

(*d*) 1,000 r.p.m. taking an armature current of 40 A

18. A D.C. shunt motor drives a load at rated speed and supply voltage. If both the load and voltage are halved, the speed of the motor will be :

(*a*) Almost doubled (*b*) Almost halved

(*c*) The rated speed

(*d*) Slightly less than the rated speed.

19. In an unsaturated D.C. series motor :

(*a*) Torque is proportional to armature current

(*b*) Torque is proportional to the square root of the armature current

(*c*) Torque is proportional to the square of armature current

20. The demand for a large increase in torque of a D.C. series motor is met by a :

(*a*) Large decrease in current

(*b*) Large decrease in speed.

(*c*) Large increase in speed

(*d*) Small decrease in speed

21. T, is the load torque of a D.C. series motor having linear magnetization and negligible armature resistance. Speed of the motor is :

(*a*) Inversely proportinal to $\sqrt{T}$

(*b*) Directly proportional to $\sqrt{T}$

(*c*) Inversely proportional to T^2

(*d*) Directly proportional to T^2

22. An electric train employing a D.C. series motor is running at a fixed speed, when a sudden slight drop in the mains voltage occurs. This would result in :

(*a*) Drop in speed and rise in current

(*b*) Rise in speed and drop in current

(*c*) Rise in speed and rise in current

(*d*) Drop in speed with current unaltered.

23. In case of a D.C. series motor it is possible to have finite no-load speed if a resistance is connected across its :

(*a*) Field terminals

(*b*) Armature terminals

(*c*) Field and armature together

(*d*) It is always very high at no load

24. A, D.C. series motor is running at rated speed without any additional resistance in series. If an additional resistance is placed in series, the speed of motor :

(*a*) Increases (*b*) Decreases

(*c*) Remains unchanged

(*d*) Oscillates around the rated speed.

25. A D.C. series motor is used for an overhauling load. It can work stably if :

(*a*) The armature is shunted by a resistor

(*b*) The field winding is reversed

(*c*) A resistor in put is series with the machine

(*d*) A divertor is put across the field

26. A fly wheel is normally fitted to :

(*a*) D.C. series motor driving a constant torque load

(*b*) Separately excited D.C. motor driving pulsed torque load

(*c*) Commulatively compound motor driving pulsed torque load.

(*d*) Differentially compound motor driving pulsed torque load.

27. Which motor should not be used for centrifugal pumps ?

(*a*) Shunt (*b*) Series

(*c*) Comulatively compound

(*d*) Differentially compound

28. For continuously running rolling mills with intermittant loading, the most suitable D.C. drive is :

(*a*) D.C. series motor

(*b*) D.C. shunt motor

(*c*) D.C. differentially compounded motor

(*d*) D.C. commulatively compounded motor

29. Consider the following statements: If the terminal voltage of a D.C. shunt motor is halved with the load torque varying as the square of speed, then :

1. Speed is halved
2. Speed remains unaltered
3. Armature current is doubled
4. Armature current remains unaltered of these statements :

(*a*) 1 and 3 are correct (*b*) 1 and 4 are correct

(*c*) 2 and 3 are correct (*d*) 2 and 4 are correct

30. Breakdown torque of a 3-ϕ induction motor of negligible stator impedence is :

(*a*) Directly proportional to the rotor resistance

(*b*) Inversely proportional to the rotor resistance

(*c*) Directly proportional to the reactance

(*d*) Inversely proportional to the rotor leakage reactance.

31. The torque-slip characteristic of a polyphase induction motor becomes almost linear at small values of slip, because in this range of slips :

(*a*) The effective rotor circuit resistance is very large compared to the rotor reactance

(*b*) The rotor resistance is equal to the stator resistance

(*c*) The rotor resistance is equal to the rotor reactance

(*d*) The rotor resistance is equal to the stator reactance

32. Unbalanced supply voltage given to a 3-phase delta-connected induction motor will cause :

(*a*) Zero sequence currents

(*b*) Less heating of motor

(*c*) Negative Sequence Componet current

(*d*) All the above

33. Consider the following statements:

In a 3-phase induction motor connected to a 3-phase supply; If one of the lines suddenly gets disconnected, then the :

1. Motor will come to stand still.
2. Motor will continue to run at the same speed with line current unchanged.
3. Motor will continue to run at a slightly reduced speed with increase in line current
4. Rotor current will have both of *s-f* and (2-*s*) *f* component frequencies, where, *S* is the slip and *f* is the supply frequency.

Of these statements :

(*a*) 1 and 4 are correct (*b*) 1 and 2 are correct

(*c*) 3 and 4 are correct (*d*) 2 and 3 are correct

34. An induction motor having full-load torque of 60 N-m when delta connected develops a starting torque of 120 N-m. For the same supply voltage, if the motor is changed to star-connection, the starting torque developed will be :

(*a*) 40 N-m (*b*) 60 N-m

(*c*) 90 N-m (*d*) 120 N-m

35. In case of 3-phase slip-ring induction motor, as the rotor resistance is increased, the starting torque:

(*a*) Increases

(*b*) Decreases

(*c*) Increases up to a certain value of resistance and then remains constant.

(*d*) Increases up to a certain value of resistance and then decreases.

36. One slip-ring connection of a 3-phase induction motor gets open circuited, the motor will,

(*a*) Continue to run at the same speed

(*b*) Run at a lower speed

(*c*) Run at a higher speed

(*d*) Stop

37. A 3-phase induction motor is fed from a balanced 3-phase supply on the rotor side, short-circuiting the stator terminals. The speed of the rotating magnetic field is

(*a*) zero (*b*) Equal to rotor speed.

(*c*) Equal to synchronous speed (*d*) Equal to slip speed

38. A synchronous motor is supplying a load at unity p.f. If the load on the motor is increased keeping its excitation and terminal voltage constant, the power factor:

(*a*) will remain the same (*b*) will become leading

(*c*) will become lagging (*d*) None of the above

39. A 3-phase synchronous motor connected to an infinite bus is operating at half-full load with normal excitation. When the load on the synchronous motor is suddenly increased :

(*a*) Its speed will first decrease and then become synchronous.

(*b*) Its speed will first increase and then become synchronous.

(c) Its speed will fluctuate around synchrounous speed and then become synchronous.

(d) Its speed will remain unchanged.

40. A synchronous motor is running clockwise. If the direction of field current is reversed, the motor would :

(a) Come to stop

(b) Run in reversed direction

(c) Run as before

(d) Run as before but at less speed.

41. For the same rating, the size of the single phase induction motor is about --- that of the corresponding three phase induction motor.

(a) Three times (b) the same as

(c) 1.5 times (d) One-third

42. A universal motor runs at :

(a) Higher speed with D.C. supply and with less sparking

(b) Higher speed with A.C. supply and with less sparking

(c) Same speed with both A.C. & D.C. supplies

(d) Higher speed with A.C. supply but with increased sparking at the brushes.

43. Speed of the universal motor is :

(a) Dependent on frequency of supply.

(b) Proportional to frequency of supply.

(c) Independent of frequency of supply.

(d) None of the above.

44. The torque-speed characteristic of a repulsion motor resembles which of the following D.C. motor characteristic ?

(a) Separately excited

(b) Shunt

(c) Series

(d) Compound

45. The main reason for using a hysteresis motor for high quality tape recorders and record players is that

(a) Its speed is constant (synchronous)

(b) It develops extremely steady torque

(c) It requries no centrifugal switch

(d) Its operation is not affected by mechanical vibrations

46. A fluctuating voltage supply is deterimental to a referigrator motor but not to a ceiling fan, although both are single phase induction motors because, the refrigerator motor :

(*a*) is made more robust than the fan motor

(*b*) is subjected to short duty cycle but the fan motor is subjected to continous duty.

(*c*) is enclosed in a sealed unit while the fan motor is open to environment.

(*d*) load is constant, but the fan motor load is voltage dependent.

47. Consider the following single phase motors

I. Capacitor start motor

II. Capacitor start and run motor

III. Permanent split capacitor motor

IV. shaded pole motor

The correct sequence of the increasing order of their cost is:

(*a*) IV, III, II, I (*b*) IV, III, I, II

(*c*) III, IV, II, I (*d*) III, IV, I, II

48. For speeds higher than say 3000 r.p.m., the machine used is:

(*a*) Induction Motor (*b*) Synchronous motor

(*c*) Universal motor (*d*) None of the above

49. Consider the follwoing statements regarding the starting of a d.c. shunt motor :

1. External field resistance should be minimum.

2. External field resistance should be maximum.

3. Reduced voltage should be applied to the armature

4. Increased voltage should be applied to the armature

Of these statements

(*a*) 1 and 3 are correct (*b*) 1 and 4 are correct

(*c*) 2 and 3 are correct (*d*) 2 and 4 are correct

50. A d.c. shunt motor has external resistances of R_a and R_f in the armature and field circuits respectively. Armature current at starting can be reduced by keeping :

(*a*) R_f maximum and R_a minimum

(*b*) R_a maximum and R_f maximum

(*c*) R_a minimum and R_f maximum

(*d*) R_a maximum and R_f minimum

51. A starting torque of 80 Nm is developed in an induction motor by an auto-transformer starter with a tapping of 30%. If the tapping of auto-transformer is 60%, then the starting torque will be :

(*a*) 40 Nm (*b*) 160 Nm

(*c*) 240 Nm (*d*) 320 Nm

52. Following controls are considered for d.c. motors :

I. Control flux.

II. Armature resistance control

III. Supply voltage control

which of the above controls play significant role in the speed control of d.c. motors.

(*a*) I only (*b*) II and III only

(*c*) I and III only (*d*) I, II and III

53. In a coal mine, Ward Leonard method of speed control, as shown in the given figure below, is used to lift coal from the pit through a crane and bucket arrangement. After lifting the coal to desired level and emptying at the dumping ground, the bucket has to be lowered in to the pit. The action to be taken as this stage is to reverse the :

(*a*) Direction of Current in F_2 and slowly increase it.

(*b*) Current in F_3 and shlowly increase it.

(*c*) Supply terminal of M_1

(*d*) Direction of current in F_1 and slowly increase it.

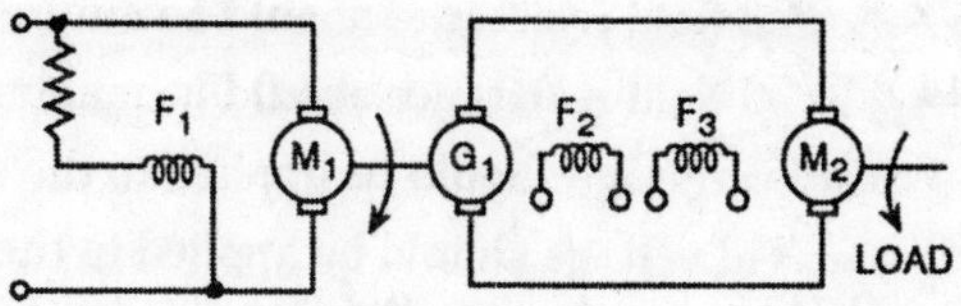

Fig. 1.117

54. While operating on variable frequency supplies, the a.c. motor requires variable voltage as well in order to :

(*a*) Protect the insulation

(*b*) Avoid the effect of saturation

(*c*) Improve the capabilities of the inverter

(*d*) Protect the thyristor from $\frac{dv}{dt}$

55. Consider the following statements regarding speed control induction motors by means of external rotor resistors :

1. Regulation in speed is accompanied by reduced frequency.
2. With a arge resistance in the rotor circuit, the speed would vary considerably with variation in torque.
3. The method is very complicated.

The disadvantages of such a method of speed control would include:

(*a*) 1 and 2 (*b*) 2 and 3
(*c*) 1 and 3 (*d*) 1, 2, 3

56. In rotor resistance control of induction motor, the hardness of speed torque characteristic.

(*a*) increases (*b*) decreases
(*c*) remains same.

57. Out of the following, which of the motors has least range of speed control ?

(*a*) D.C. shunt motor
(*b*) Wound rotor induction motor
(*c*) Schrage motor
(*d*) Synchronous motor

58. A d.c. separately excited motor runs from 220 V supply. If the motor is braked by plugging with field excitation kept constant and a resistance of 10Ω is inserted in the armature circuit, the initial braking current is (neglecting all losses):

(*a*) 22 A (*b*) 11 A.
(*c*) 44 A (*d*) 0.

Answers

1. (*a*),	2. (*b*),	3. (*d*),	4. (*c*),	5. (*d*),
6. (*d*),	7. (*c*),	8. (*d*),	9. (*b*),	10. (*a*),
11. (*d*),	12. (*b*),	13. (*b*),	14. (*d*),	15. (*c*),
16. (*c*),	17. (*d*),	18. (*d*),	19. (*d*),	20. (*b*),
21. (*a*),	22. (*a*),	23. (*b*),	24. (*b*),	25. (*a*),
26. (*c*),	27. (*b*),	28. (*d*),	29. (*c*),	30. (*d*),
31. (*a*),	32. (*c*),	33. (*c*),	34. (*a*),	35. (*d*),
36. (*b*),	37. (*d*),	38. (*c*),	39. (*c*),	40. (*c*),
41. (*c*),	42. (*a*),	43. (*c*),	44. (*c*),	45. (*b*),
46. (*d*),	47. (*b*),	48. (*c*),	49. (*a*),	50. (*d*),
51. (*d*),	52. (*d*),	53. (*a*),	54. (*b*),	55. (*a*),
56. (*b*),	57. (*d*),	58. (*c*).		

Questions of Matching Type with Answers

Note : Each given question consists of lists. Match the items given in the list No.I and those given in the list No. II. By using the codes given in each of the Questions, select your answers to these Questions.

1. *List - I* (*D.C. Motors*) — *List - II* (*Characteristics*)

List - I (D.C. Motors)		List - II (Characteristics)	
A.	Commulatively Compounded motor	1.	Fairly constant speed irrespective of the load
B.	Differentially Compounded motor	2.	It may start in reverse direction
C.	Series motor	3.	Definite no-load speed
D.	Shunt motor	4.	Never started without load

Codes

	A	B	C	D
(*a*)	2	3	1	4
(*b*)	3	2	4	1
(*c*)	1	3	2	4
(*d*)	4	1	3	2

Ans. (*b*)

2. Typical characteristics of a d.c. series motor, operating under loaded condition, are shown in the figure given below. List -I gives the identification of these characteristics while list -II gives the description of important load characteristics of d.c. motors working on load.

2. *List - I* (*see figure curves labelled*) — *List - II*

List - I	List - II
A.	1 Efficiency *vs* load
B.	2 Torque *vs* load
C.	3 Current *vs* load
	4 Speed *vs* load

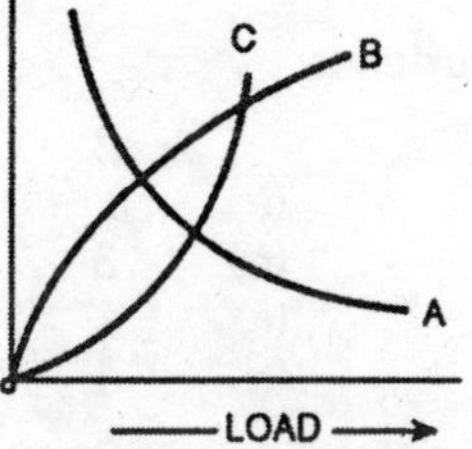

Codes :

	A	B	C
(*a*)	4	2	3
(*b*)	2	3	1
(*c*)	3	2	1
(*d*)	4	1	2

Ans. (*d*)

3.

	List - I (*Motors*)		*List - II* (*Applications*)
A.	Permanent magnet d.c. motor	1.	Cassette tape recorder
B.	Stepper motor	2.	Ceiling fan
C.	Single phase induction motor	3.	Hand drill
D.	Universal motor	4.	Digital control

Codes :

	A	B	C	D
(*a*)	1	4	2	3
(*b*)	1	3	2	4
(*c*)	4	1	3	2
(*d*)	1	2	4	3

Ans. (*a*)

4.

	List - I (*Motors*)		*List - II* (*Applications*)
A.	Shaded pole motor.	1.	Record players and tape recorders
B.	Universal motor.	2.	Electric clocks signalling and timing devices
C.	Reluctance motor.	3.	Small fans
D.	Hystereisis motor	4.	Electric shavers

Codes :

	A	B	C	D
(*a*)	3	4	1	2
(*b*)	3	4	2	1
(*c*)	4	3	1	2
(*d*)	1	2	4	3

Ans. (*b*)

5. *List - 1* *(Type of Motors)* — *List - II* *(Torque-speed characteristics)*

A. **Hysteresis motor.**

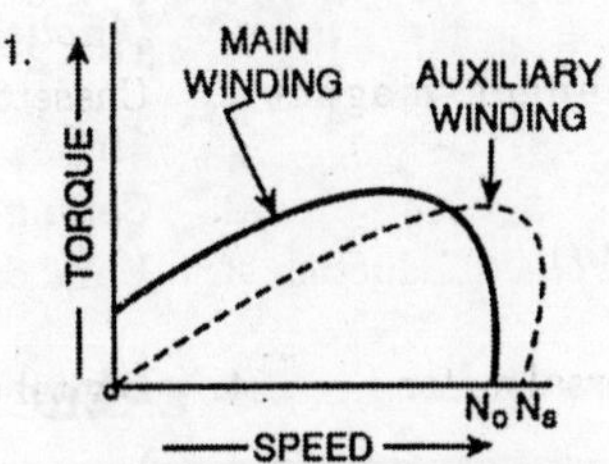

B. **Capacitor start split phase motor.**

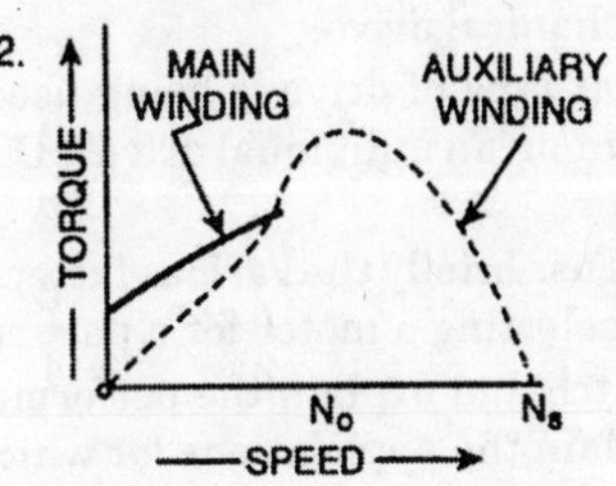

C. **Reluctance motor.**

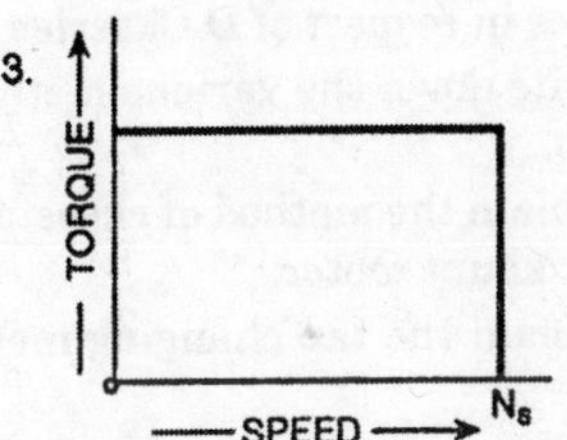

Codes :

	A	B	C
(*a*)	3	1	2
(*b*)	2	3	1
(*c*)	2	1	3
(*d*)	1	3	2

Ans. (*a*).

6.

List - I (Type of 1-ϕ)	*List - II* (Type of appliance)
A. Permanent magnet type	1. Rocket.
B. Shaded pole..	2. Refrigerator Compresor.
C. Universal	3. Sowing machine
D. Capacitor start capacitior-run induction motor	4. Photo-Copying machine.

Codes :	*A*	*B*	*C*	*D*
(*a*)	1	2	3	4
(*b*)	1	2	4	3
(*c*)	4	1	3	2
(*d*)	1	4	3	2

Ans. (*d*)

EXERCISES

1. State the advantages and disadvantages of electrical drive over mechanical drive.
2. What type of drive is being used in modern day industry. 'A group drive' or 'an individual drive'? Discuss the advantages of one over the other.
3. Discuss briefly the various factors which are required to be considered for selecting a motor for a particular industry.
4. Sketch and explain the performance of D.C. series motor and hence explain the applications for which it is most suitable.
5. Draw and explain the speed current and torque current characteristics in respect of D.C. series motor and D.C. shunt motor.
6. (*a*) Write down the various methods of controlling the speed of D.C. motors.

   (*b*) Explain the method of rheostatic control for controlling the speed of D.C. shunt motor.

   (*c*) Explain the tap changing method of speed control of D.C. series motor.

   (*d*) Explain how the speed of D.C. shunt motor can be brought above or below the rated speed.
7. State the various methods of speed control of three-phase induction motor. Explain any one method in detail.
8. (*a*) What is the necessity of starter in induction motor ?

   (*b*) Explain any one method of starting :

   (*i*) 3-phase squirrel cage induction motor.

   (*ii*) 3-phase slipring induction motor.

   (*iii*) Single phase motor.
9. (*a*) Compare electric braking system to that of mechanical braking system.

   (*b*) What are the various method of braking D.C. motor ?

   (*c*) Explain the process of 'plugging' as applied to D.C. series motor.
10. Distinguish between rheostatic and regeneration braking as applied to electric traction. Under what circumstances the mechanical or regenerative braking applied ?

11. Explain how induction motor is stopped by,

(*i*) plugging

(*ii*) rheostatic braking.

12. State giving reasons whether you would choose plugging or rheostatic braking for quickly braking to rest :

(*i*) 3-phase induction motor,

(*ii*) Synchronous motor.

13. Describe the following features pertaining to the relation for motor :

(*a*) Types of enclosures

(*b*) Bearings

(*c*) Transmission of drive and noise.

14. Why D.C. series motor is preferred over other types of D.C. motors for use in electric tractions ?

15. What do you understand by 'series' and 'shunt' speed-torque characteristics of electric motors used in industrial drives ? Name (at least one of each type) d.c. motors and a.c. motors with such characteristics.

16. How does a 3-phase synchronous motor differ from a 3- ϕ induction motor as far as the starting and running features are concerned ? With the help of suitable diagrams, explain the different methods used in industrial drives to start a 3-phase synchronous motor.

17. Discuss the merits and demerits of using :

(*a*) Star-delta starter and

(*b*) Auto-transformer starter with a 3-phase cage induction motor,

18. Describe in brief speed control of induction motor by pole changing.

19. A motor is to be selected for driving a load having large torque of short duration followed by a long no-load period. A flywheel of suitable inertia is already mounted on the load shaft. Out of the three types of D.C. motors (series, shunt and compound) which will you recommend for this application and why ?

20. State giving reasons what type of motor be suitable for the following drives :

- Propeller shaft of a ship
- Rotary printing press
- Spinning machine in textile mills
- Rotary compressors.

Would you suggest any load equaliser for the above drives ? Justify.

21. What do you mean by load equalisation ? How is it affected?

22. Explain the following :

(*i*) Why the D.C. series motor is used for traction application ?

(*ii*) Why is a series D.C. motor not to be run on no load ?

(*iii*) Why electric drive is preferred over mechanical drive ?

(*iv*) Why individual drive is preferred over group drive now-a-days ?

(*v*) Why it is objectionable to start a large squirrel cage induction motor by directly switching it on the lines ?

23. Determine approximately the starting torque of an induction motor in terms of full load torque when started by means of :
(*a*) star delta switch (*b*) an auto-transformer with 50% tapping.

Ignore magnetising current. The motor draws 5 times the full load current when switched on direct and its full load slip is 4%.

24. A machine on a short test run gave the following mean temperature rise :

Time, hr.	0	0.25	0.50	0.75	1.0	1.25	1.50
Temp. rise °C	42.5	45.0	47.4	49.5	51.4	53.0	54.4

Estimate graphically the final temperature and the time constant. Justify the procedure adopted by you.

[**Ans.** 62°C, 1.6 hr.]

25. A motor driving a mining equipment has to deliver a load rising uniformly from zero to maximum of 1,500 kW for 20 seconds during acceleration period, 1000 kW for 50 seconds during the full load period and during deceleration period of 10 seconds when regenerative braking is taking place the kW returned to the supply falls from an initial value of 500 kW to zero uniformly. The interval for decking before the next load cycle starts to 20 seconds. Estimate a suitable kW rating for the motor.

[**Ans.** 815 kW]

26. A motor driving a colliery winder equipment has to deliver a load rising uniformly from zero to maximum of 2,000 H.P. in 20 seconds during the acceleration period, 1,000 H.P. for 40 seconds during the full load period and during the deceleration period of 10 seconds when regenerative braking is taking place, the horse power returned to supply falls from 330 to zero. The interval for decking before the next load cycle starts is 20 seconds. Estimate the H.P. rating of the motor.

[**Ans.** 865 H.P.]

27. Assuming an exponential law of temperature rise, calculate the final steady temperature on full load and the time constant for an electric motor whose temperature rise after one hour is 25°C and after 2 hours it is 45°C.

[**Ans.** 125°C, 4.5 hr.]

28. A 500 r.p.m. synchronous motor and its load have a total movement of inertia of 420 kg-m^2 and a frictional torque of 140 N-m. The motor is braked by rheostatic method, the field current and the breaking resistance being kept constant at a value which gives an initial electric braking torque of 5400 N-m. Find the time taken to bring the motor to rest.

[**Ans.** 15 seconds]

29. A 10 pole, 50 Hz synchronous motor is coupled to a load having a moment of inertia of 500 kg-m^2. The motor is braked by rheostat braking, the field circuit working resistance being kept constant so as to give an initial braking torque of 500 kg-m. If the total frictional torque is 50 kg-m ; Calculate the time taken by the motor to come to rest.

[**Ans.** 15.4 seconds]

30. A 3-phase, 8-pole, 50 Hz, induction motor, equipped with a flywheel, supplies a constant load torque of 100 N-m and at wide intervals, an additional load torque of 300 N-m for 6 seconds. The motor runs at 735 r.p.m. at 100 N-m and torque-speed curve may be assumed to be linear over the working range.
Find the minimum moment of inertia of the flywheel if the motor torque is not be exceed 250 N-m.

[**Ans.** 500 kg-m^2]

31. A motor fitted with a flywheel is required to supply a load torque of 50 kg-m for a period of 12 sec. followed by a no load period long enough for the motor to regain its full speed. Calculate the moment of inertia of flywheel needed to limit the motor torque to 20 kg-m. The no load speed of the motor is 1000 r.p.m. and it has a slip of 8% at a torque of 25 kg-m., state assumptions made and prove the formulae you use.

[**Ans.** 620 kg-m^2]

32. (*a*) Distinguish amongst :

(*i*) Continuous rating (*ii*) Short time rating and (*iii*) Intermittent rating.

(*b*) The 15-minute rating of a domestic mixer is 500 watts. If the heating time constant is 45 minutes, determine the continuous rating of the motor used in the mixer. The maximum efficiency of the motor occurs at 80% of full load. Drive the formulate employed.

33. (*a*) Give an account of torque-current characteristics of a D.C. shunt and series motors.

(*b*) A squirrel cage type induction motor when started by means of a star-delta starter takes 180% of full load line current and develops 35% of full load torque at starting. Calculate the starting torque and current in terms of full load values, if an auto-transformer with 70% tapping were employed.

34. (*a*) Explain how an induction motor is brought to stop by dynamic braking ?

(*b*) The load cycle for a motor which is required to drive haulage is as follows :

0—15 minutes	40 H.P.
10—10 minutes	No load
15—25 minutes	45 H.P.
25—30 minutes	35 H.P.

The above load cycle is repeated indefinitely. What would be the size of the motor suitable for this load ?

Short Answer Type Questions

Q.1. What is an Electric drive?

Ans. Arrangement to run a machine is called drive. Drive consists of prime mover and transmission system. If the prime mover in a drive is electrical motor, the drive is called electrical drive.

Q.2. Mention at least two advantages of electrical drives.

Ans. (i) Electrical drive is cheaper and economical,

(ii) Remote control is possible

(iii) It is simple.

Q.3. List different types of motors used in electrical drives.

Ans. (i) DC motors :

(a) DC series motors

(b) DC shunt motors

(c) Compound motor

(ii) 3 Phase AC motors :

(a) Induction motors (i) Squirrel cage (ii) Slip rings

(b) Synchronous motors

(iii) Single phase motors :

(a) Split phase induction motors

(b) Capacitor start motors (c) Reluctance motors

(d) Hysteresis motors (e) Universal motors

Q.4. Define Braking.

Ans. The process of stopping a drive immediately is called braking.

Q.5. State the requirements of a good braking system. *[Dec 2008]*

Ans. (i) It should be reliable,

(ii) It should have quick action

(iii) The heat produced during the process should be properly dissipated,

(iv) It should have full control in the hands of the operator

Q.6. What is difference between electric braking and machanical braking?

Ans. In mechanical braking, some machanical system is used and the energy spent is dissipated in the form of heat. In electrical braking, some electrical system is used and the energy is spent in the form of electrical energy through resistance or returned to the source of supply.

Q.7. Name different types of electric braking.

Ans. (i) Plugging

(ii) Rheostatic braking

(iii) Regenerative braking

Q.8. What is plugging?

Ans. In plugging, the connections of the armature of the motor are reversed and as soon as the motor starts running in opposite direction, it is stopped by applying mechanical brakes.

Q.9. What is Rheostatic braking?

Ans. In Rheostatic braking, the motor is disconnected from supply and is run as a motor due to kinetic energy of motor and energy *so* produced is dissipated in a Rheostat.

Q.10. What is Regenerative braking?

Ans. In Regenerative braking, the motor is disconnected from supply and is made to run as a generator. Power so generated is fed to supply. This is the most efficient form of braking.

Q.11. Write two applications of DC motors.

Ans. (i) Electric Traction.

(ii) Lift, crane and flour mills

Q.12. What is FKW motor?

Ans. The motors of power rating in fraction of kilowatt is called FKW (fraction kilowatt hour motor).

Q.13. Motor used in Mixer/grinder is ....

Ans. (i) Universal motor (ii) DC series motor.

Q 14. Motor used in domestic referigerator is ....

Ans. Repulsion motor.

Q.15. Which motor is used in lift?

Ans. DC series motor.

Q.16. Give one application of DC series motor.

Ans. Electric traction.

Q.17. What is a group drive?

Ans. The drive which is used to run a number of machines simultaneously is called group drive.

Q.18. What are different means of power transfer?

Ans. (i) Belt (ii) Rope

(iii) Chain (iv) Gear

Q.19. Which is the most efficient form of braking?

(a) Mechanical (b) Plugging

(c) Rheostatic (d) Regenerative.

Ans. (d) Regenerative.

Q.20. Load/Torque is constant in a .... load. [lifting/frictional].

Ans. Lifting.

Q.21. What is fly wheel?

Ans. It is a wheel of large mass attached with a rotating machine such as motor/engine to damp out vibrations.

Questions Bank with Answer

Q.1. What are different types of mechanical loads? Draw their characteristics. ***[June 2009]***

Ans. The different mechanical loads and their characteristics are following.

(i) Lifting loads : Examples of lifitng loads are to lift some load by crane, lift etc. In this load, torque is constant and independent of speed.

This fig. 1.117 shows Torque speed characteristics of lifting load.

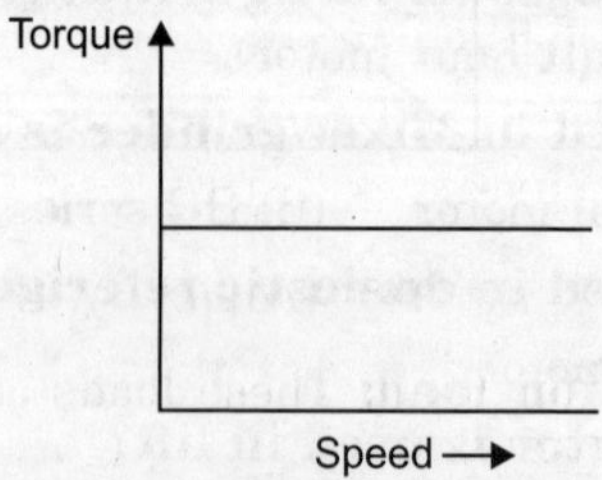

Fig. 1.117

(ii) Air and Fluid loads: Examples of these loads are found in air blowers and in water turbines. In these air and water are the loads. The torque varies as square of the speed. The fig. 1.118 shows T/N characteristic of these loads.

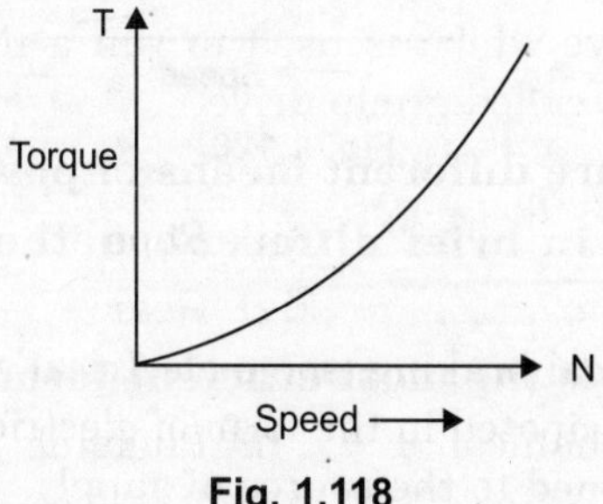

Fig. 1.118

(iii) Frictional Load: (a) When no lubricant is used, T is constant and example of frictional load is ceiling fan where load is air friction in frictional load.

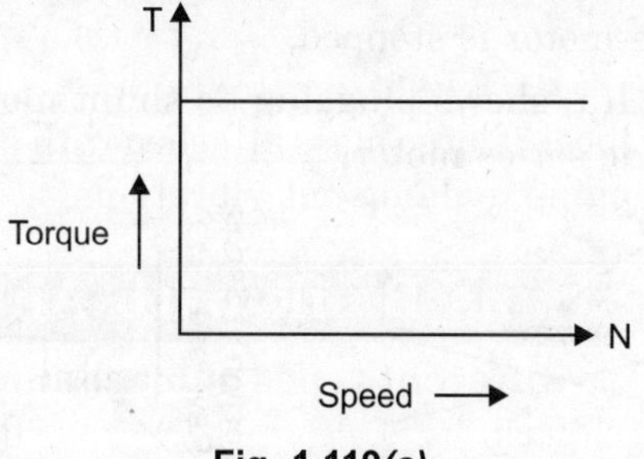

Fig. 1.119*(a)*

(b) When a lubricant is used $T \alpha N$ as shown in fig. 1.119*(b)*.

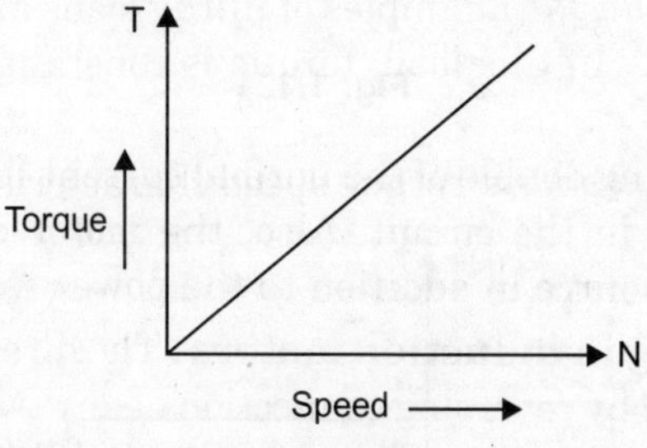

Fig. 1.119*(b)*

(iv) Deformation load: These loads occur in grinding and crushing processes. In this case, $T \alpha \frac{1}{N}$ as shown in fig. 1.120.

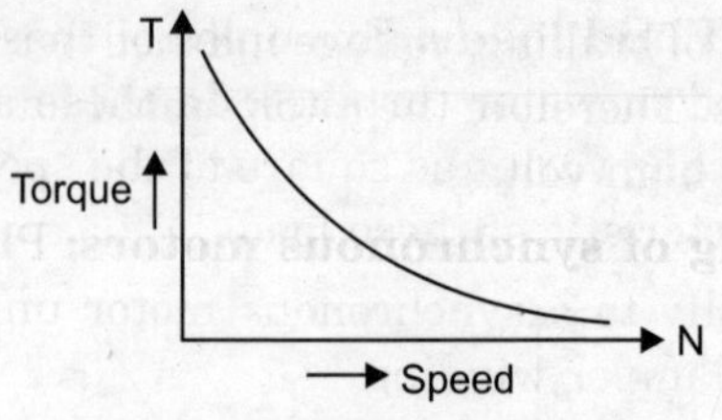

Fig. 1.120

Q.2. Discuss in brief different methods of electrical braking.

Ans. In electrical braking, some electrical system is used and energy spent is dissipated in the form of electrical energy through resistance or returned to the source of supply.

Electrical braking is of three types

(1) Plugging (2) Rheostatic braking (3) Regenerative braking

1. Plugging : (i) Plugging in DC motors: Connections of the armature are reversed. The motor runs in the reverse direction. At this stage the motor is stopped.

The fig. 1.121(a) shows plugging dc shunt motor and 1.121(b) shows plugging dc series motor.

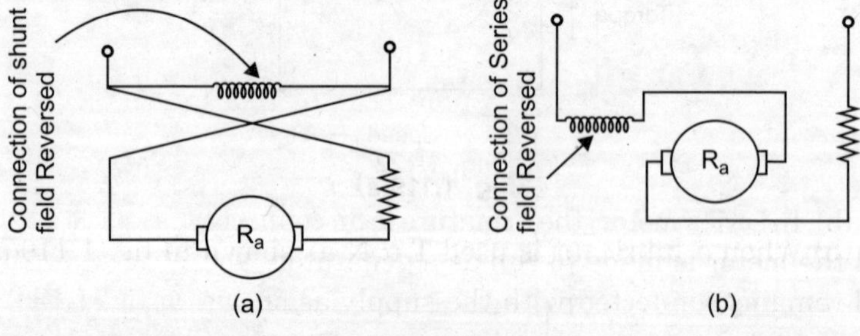

Fig. 1.121

During braking double of the normal current flows, so a limiting resistor r is put in the circuit. Also, the motor draws additional power from the source in addition to the power wasted in braking.

(ii) Plugging in induction motors : The direction of induction motor is reversed by reversing connection of any two phases. Brakes are applied when the motor becomes stand still before starting in reverse direction, and the motor is switched off from the supply to stop it from moving in reverse direction, and the motor stops instantly.

At the instant of braking, voltage induced in the rotor is twice of the normal values. Therefore the rotor should have good insulation to withstand this high voltage.

(iii) Plugging of synchronous motors: Plugging cannot be applied succes fully to a synchronous motor unless the motor is provided with a damper winding.

2. Rheostatic braking: In this type of braking, the motor is disconnected from the supply and is made to run as generator due to kinetic energy of the motor and electrical energy so produced is dissipated as heat in a Rheostat. In this respect this method is superior to plugging as no energy from the supply is required.

(i) Rheostatic braking of d.c. motors

(a) In case of shunt motor, the armature is disconnected (but field remains connected) from the supply. The armature is connected through a resistor r that can be varied to adjust torque of the armature as shown in fig. 1.122.

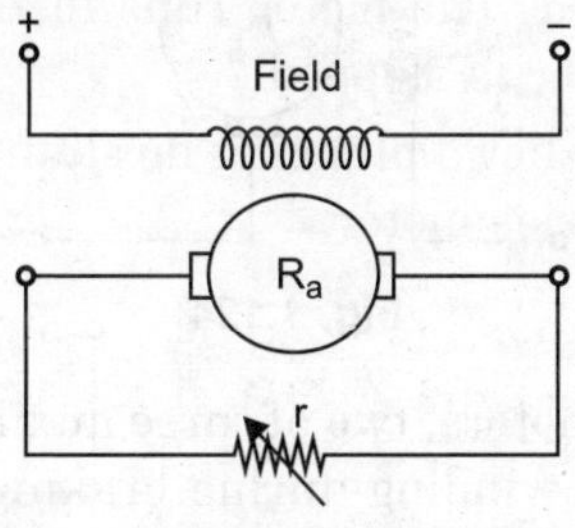

Fig. 1.122

(b) In series motor, the armature lc disconnected from the supply and the motor is made to run as a generator. The field is reversed and remains connected with the supply as shown in fig. 1.123.

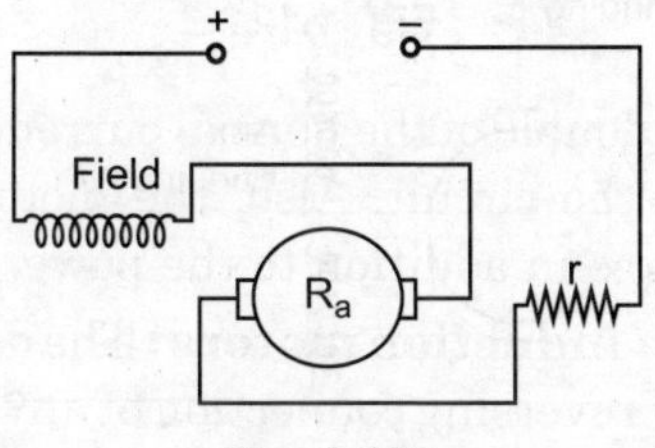

Fig. 1.123

The value of r should be such that the motor remains self excited.

(ii) Rheostatic braking in induction motors : The induction motor is disconnected from the supply and direct current is passed through the stator. The dc required is 3 to 5 times the full load current of the motor.

(iii) Rheostatic braking in synchronous motors: The synchronous motor is disconnected from the supply and is made to run as an alternator. Note that field excitation must be maintained.

3. Regenerative braking: This is the most superior method of braking as some part of kinetic energy of the motor is returned to the supply mains.

(i) Regenerative braking with dc motors: In case of shunt motor, the field is disconnected from the armataure and the motor remains connected to the supply through the field. If the back emf is more than the supply voltage *i.e.,* ($E_b > V$), current will be reversed and a part of the energy is returned to the source.

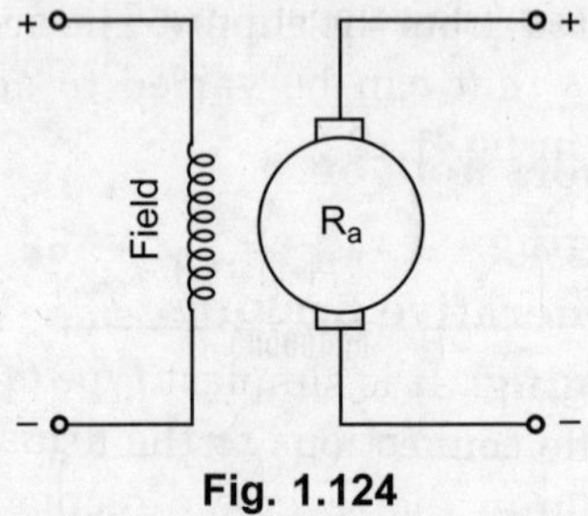

Fig. 1.124

In case of series motors, two of three auxiliary winding are put in series to the main winding during braking process. The motor works as a differentially compound dc motor and sends energy back to supply.

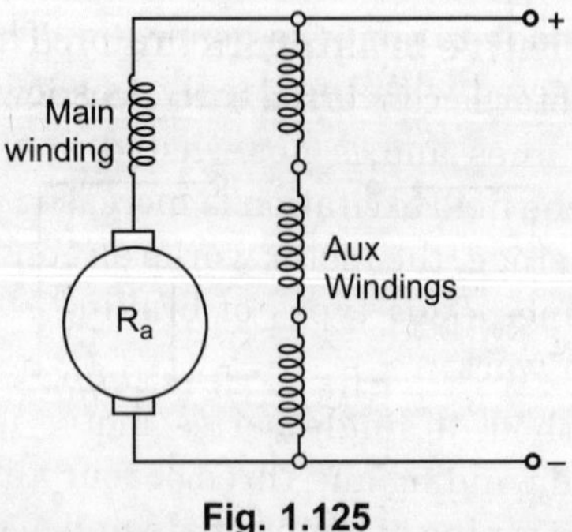

Fig. 1.125

The series motors are mostly used in electric trains and usually regenerative braking is adopted on these motors.

Q 3. What is difference between plugging and Rheostatic braking?

Ans.

	Plugging	Rheostatic braking
1.	Motor remains connected to supply.	Motor is disconnected from supply.
2.	Motor is made to run in the reverse direction by interchaning the supply terminals.	Motor continues to run in the same direction.
3.	Power is drawn from the source during braking.	No power is drawn from the source during braking.
4.	Plugging cannot be applied to synch onous motors unless they are equipped with damper windings	Rheostatic braking can be successfully applied to synchronous motors even if they are not supplied with damper windings.

Q.4. write short note on

(i) Plugging

(ii) Regenerative braking

Ans. (i) Plugging: It is simplest type of braking and involves the reversing of the connections to the armature of the motors so that it tends to run in the reverse direction. In order to stop the machine, a special device must be installed to cut off the supply as soon as it comes to rest. This method can be applied to DC: motors, induction motors and synchronous motors. It is particularly suitable where the drive has to be rapidly reversed such as in planning machines etc.

(ii) Regenerative braking: In this type of braking, the kinetic energy of the motor is converted into electrical energy which is fed back to supply lines and is not wasted in the Rheostat. During braking period the field excitation is increased more than the supply voltage. At this stage, the motor works as generator and feeds back to the supply lines. This types of braking is mostly employed in electric traction work.

In this method, a single series motor is provided with one main series field winding and three or four auxiliary field winding. During normal running of motor, main and auxiliary field windings are connected in parallel with each other but are in series with the armature of the machine. When brakes are applied, the auxiliary windings are connected in series with each other and the combination is connected in parallel across the main series field as shown in fig.1.126.

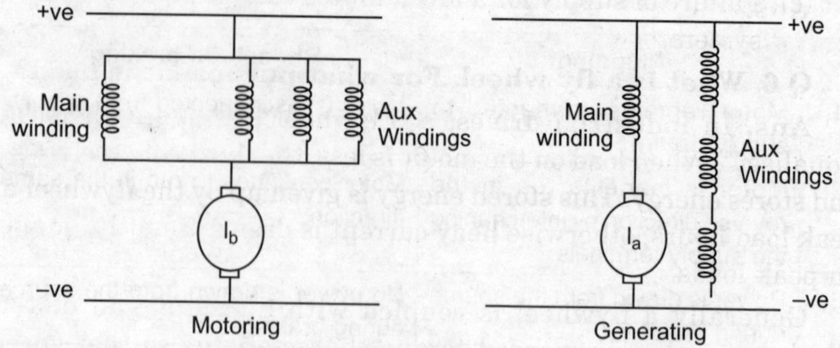

Fig.1.126

The machine acts as a differentially compound generator and sends back electrical energy to supply lines.

This type of braking is very useful in hilly area where trains return about 20% of total energy to the supply lines while moving down a gradient. This also reduces wear and tear on brake shoes.

Q 5. What are the advantages and disadvantages of electrical drives over mechanical drives?

Ans. Advantages of electric drive: In the industries, two types of drives *i.e.* (i) Mechanical drives and (ii) Electric drives can be employed. But the electric drives are being mostly used due to their following advantages over the mechanical drives.

(a) Their initial as well as maintenance cost is low as compared to other mechanical drives

(b) Their operation is more simple and clean.

(c) Their control is very easy and smooth.

(d) Their layout is very flexible

(e) They can have remote controls.

(f) Transmission of power from one place to the other can be done with the help of cables instead of long shafts etc.

Disadvantages of electric drive: Although, electric drive is more advantageous but is has some serious disadvantages as mentioned below :

(a) Electric drive system can only be employed if the area is electrified

(b) There is every possibility of short circuit, leakage from conductors and breakdown of over head conductors. This may lead to fatal accident.

(c) Failure of supply for a few minutes can paralyse the whole system.

Q 6. What is a fly wheel. For what purpose it is used?

Ans. In industrial drives, a 'Flywheel' is used as a "Load Equaliser". When load on the motor is less, the flywheel accelerates and stores energy. This stored energy is given up by the flywheel at peak load hours, otherwise heay current is drawn from the supply on peak loads.

Generally a flywheel is coupled with dc compound motor, which has a dropping speed torque characteristic, so that speed may fall with increase in load and enables the flywheel to give up its stored energy. For a load, for which the motor is to run in the same direction for a longer time, the flywheel may be mounted on the shaft of the motor generator set.

Q 7. Explain Rheostatic braking of series motors.

Ans. In this method of braking, motor is disconnected from supply and is used as a generator driven by momentum of the equipment to be braked and the power so generated is dissipated as heat in a Rheostat.

In series motor, the armature is disconnected from the supply and motor is run as a generator. The field is reversed and remains connected to supply as shown in fig.1.127.

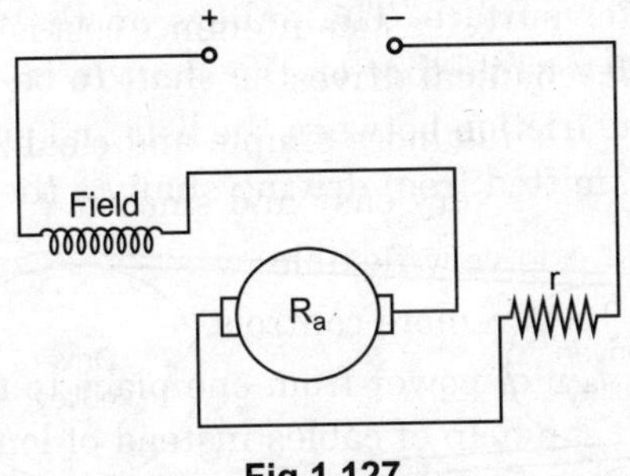

Fig.1.127

The value of r should be such that the motor remains self excited.

Q 8. Write short note on

(i) Industrial drives

(ii) Power transfer by belts, gears and pulleys.

Ans. Industrial drives: Usually large number of machines are required to run in an industry. Each machine may be driven by providing a separate prime mover or by providing a common shaft using belts and pulleys. Accordingly there are two type of drives,

(i) Individual drive (ii) Group drive

In **Individual drive** each machine tool is driven by a separate motor. In most cases the motor and its controlling unit are placed on the machine which results in good appearance, safety and cleanliness. Individual drive is required where consistency of speed and flexibility of control is required.

In **Group drive** a single motor of large capacity is used to drive a number of machines and tools in an industry through a long shaft. Its initial cost is small as compared to individual drive but it suffers from many draw backs due to which group drive does not suit to modern industry. However this system is adopted when some factories previously running on diesel engine drive are changed to electric drive simply by replacing the engine by motor.

(ii) Power transmission by belts, rope gear and chain: Power transmitted from one shaft to another by means of

(i) Belts (ii) Ropes (iii) Chains (iv) Gears.

Belt drive is the most common method for transmission of power when distance between two shafts is large. Both the shafts are provided with pulleys on them. The pulleys accommodate the belt on their outer surface. The pulleys on the rotating shaft is called driver pulley while that on the shaft to be rotated is called driven pulley. The friction between the belt and pulley enables the power to be transmitted from driving shaft to the driven shaft.

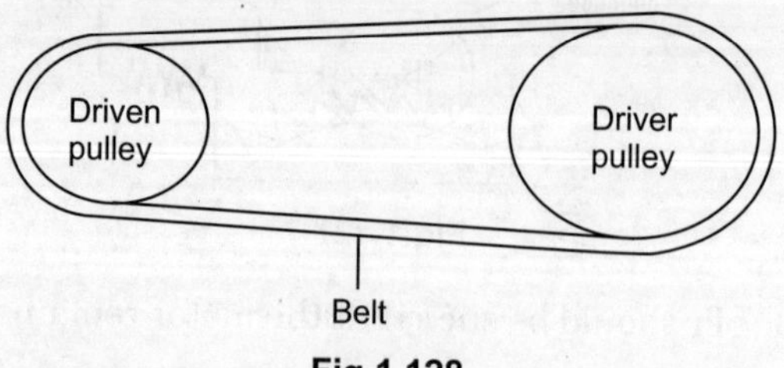

Fig.1.128

Rope drives are used if power is to be transmitted over large distances usually more than 10 meters and upto 30 meters.

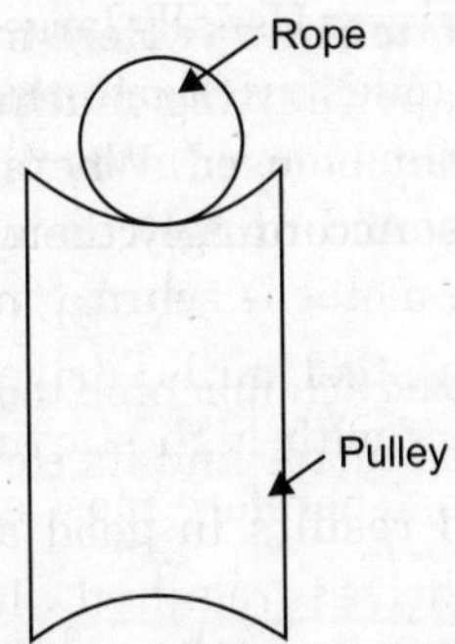

Chain drives are used for power transmission in motor cycles, cycles, printing machines, textile mills, paper mills when distance between centres of shafts is short.

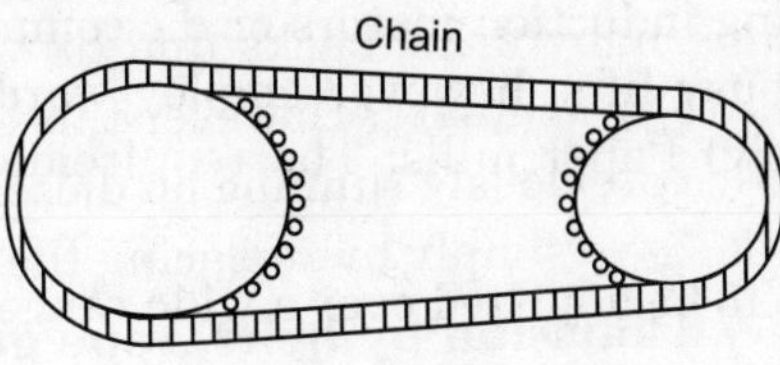

Fig.1.129

Gears are used for transmission of power in scooters, cars, trucks etc. Gear is a body of circular or frustum of a cone shape with identical teeth on its circumferential surface. Gears work in pairs. Teeth of one gear fitted on a shaft, mesh into a corresponding space on the other mating gear fitted on another shaft and vice-versa.

Fig.1.130

Q .9. Select drives for the following types of applications. Give reasons for your choice.

(i) Printing machinery **(ii) Steel mills**

(iii) Lift **(iv) Paper mills**

Ans. (i) Printing machinery: Here motors are employed for printing presses and paper cutting machines. Induction motors using rotor resistance may be used. Where large speed variations are requrired, d.c compound or a.c commutator motors may be used. The most suitable motor is schrage motor.

(ii) Steel mills: In steel mills, drives are used for rolling where heavy duty motors with high starting torque are required for continuous operation. Therefore most suitable drives for such applications are:

(a) DC series motors (b) Slip ring induction motor

(c) Plain synchronous motor.

(iii) Lifts: Lifts require easy starting and braking, and very smooth acceleration to avoid jerks to passengers. Suitable short time rated slip ring induction motors or d.c compound motors are employed. Very busy lifts, however employ ward-leonard control with d.c supply (iv) Paper mills: The requirement of paper mill drive are:

(a) Speed has to be adjusted over a wide range to suit different types of paper.

(b) It should be possible to control the speed within about 0.1% of the set value otherwise there may be variation in the quality of paper.

(c) Crawling speeds are required for inspection on some occasions. It is usual to provide d.c motor with Ward-Leonard speed control. In case of a.c motors, schrage motors are used with electric interlocking.

Q.10. What are the advantages and disadvantages of electrical braking over mechanical braking?

Ans. (i) It is cheaper and economical.

(ii) In regenerating type of electric braking, a part of the energy is returned to the source of supply thus, the cost of operation is lesser.

(iii) The heat produced in mechanical braking may spoil the braking system whereas in electric braking the heat produced is dissipated in resistance and is not so harmful.

(iv) The mechanical braking produces jerk and so not comfortable to the driver as well as to passengers, whereas the electric braking is smooth.

Disadvantages of electric braking:

(i) During electric braking the motor runs as generator, so the motor should possess the required characteristic.

(ii) Additional equipment is required for electric braking. Generally for reliability, mechanical as well as electric brakings both are provided in vehicles.

Q 11. Give comparison between Individual and Group drive.

Ans. The following table gives a comparison of the two drives.

	Individual drive	Group drive
1.	Initial cost is high	Initial cost is low
2.	Operating cost is low because different machines are operated only when there is necessity.	Operating cost is high because the main motor has to be operated even when only one machine is required to be operated.
3.	Individual motor works at good p.f.	Group drive motor works at poor power factor.
4.	System is more reliable as each machine has individual drive.	Reliability is poor, if main motor fails whole industry will come to stand still.
5.	Machines may be fitted where ever convenient.	Machines have to be fitted near the main shaft.
6.	Speed can be controlled as per the need.	Speed control is not easy.
7.	Only individual drive can be em ployed for driving heavy machines such as cranes, lifts and hoists etc.	This system cannot be employed in such cases.
8.	Space can be fully and conveniently utilised.	Space cannot be used fully and conveniently.

Q 12. Explain how drive is selected for

(i) Textile mills **(ii) Paper mills**

Ans. (i) Textile mills: Motors used should be totally enclosed type and moisture proof and they should have constant speed of operation. Therefore DC shunt motors and synchronous motors are best suited for textile mills.

(ii) Paper mills: Refer to Q 9 (iv)

Q 13. Discuss in brief various FKW motor and their applications.

Ans. FKW motors are single phase motors of rating less than one KW. Various FKW motors and their applications are given below.

(i) Single phase synchronous motor: It is a constant speed motor which operates at a power factor 0.9 to 1.0. Speed control of such motors is not possible. Its initial cost is comparatively higher than squirrel cage induction motors. These are used in such applications requiring cosntant speed but very low torque such as teleprinters, clocks, all kinds of timing devices, recording instruments, sound recording and reproducing apparatus.

(ii) Single phase series motor: This motor posseses a high starting torque (3 times full load torque with a starting current of twice the full load current). The speed of the motor varies inversely as the current. The speed can be controlled over a wide range with a transformer. Motor normally works at *a* power factor 0.8 to 0.9. Its maintenance cost is high. It is used in domestic appliances such as vacuum cleaners, mixers, refrigerators etc.

(iii) Repulsion Motor: The motor also has a high starting torque like a series motor. At light load, it has high speed. Speed control is possible by shifting of brushes. The motor operates at 0.8 to 0.9 power factor lagging. Its maintenance cost is high. It is used for services requiring large starting torque and adjustable but constant speed such as in coil winding machines.

(iv) Capacitor start induction motor: This motor has high starting torque of about 1.5 times the full load torque. Speed control of such motors is not possible and power factor is 0.7 to 0.9 lagging.

(v) Capacitor start capacitor run induction motor: The motor's operating characteristics are similar to capacitor start induction motor. It operates at a better power factor and efficiency. It is used in water cooler and large size air conditioners.

Q.14. What are main applications of AC series motors, explain?

Ans. AC series motor are either three phase or single phase.

(i) Three phase AC series motor: This motor has speed torque characteristics similar to that of d.c series motor. It has high starting torque of about 2.5 times the full load torque. Power factor varies from 0.9 to 1 lagging. Speed control of this motor is possible over full range but its maintenance cost is very high.

This is used as drive where load increases with speed such as fan and centrifigual pumps. This is also employed where high starting torque is required such as for haulage and hoist work.

(ii) Single phase AC series motor : This motor possesses high starting torque of about 3 times the full load torque. The speed can be controlled on full range with the help of a transformer. Its maintenance cost is high and power factor is about 0.8 to 0.9 lagging.

In small sizes, it is used for driving domestic appliances like vacuum cleaners, mixers, hair driers etc. In large sizes it is used in traction.

2 Control of Industrial Motors

2.1 Introduction

Necessary apparatus required for opening and closing the circuits for operating, reversing, speed control and braking of different types of electric motors along with protective equipment for saving against faults is known as the **Control Device**.

Manual control is the simplest method of motor control which is altogether unsuitable for different kinds of modern automatic equipment. In order to eliminate the short-comings of manual control the need arises to the development of semi-automatic and fully automatic control equipment. Semi-automatic control system makes use of control apparatus which can be energised by pressure on a push button. In the case of automatic control, the driven machinery may start up the drive motor itself or it can be done by suitable controllers.

In the national economy automatic electric control is also of greater importance because of the season that it permits the productivity of labour to be raised, makes for good working conditions, improves the quality of the products which are being produced, improves the realiability of operation of the equipment and lowers the expenditure of electrical energy.

By automation of the drive the operation of the equipment simplifies and allows it to be remotely controlled. Keeping in view above advantages of fully automatic control equipment it, applications are increased very much inspite of the fact that it cost four times as much as the corresponding manual control system.

Electric motors provide one of the principal sources of power for driving modern machine tools and other industrial equipment. The motor has become so closely related with the driving of machine elements that in most cases the motor is included as an integral part of the design of the machine.

This relationship of motor and machine through *direct* drive has focused attention on the design, construction, installation and maintenance of equipment to control the motor.

The term **motor control** in modern communication refers to speed control, reversing, methods of acceleration and declaration and many other functions of motor controllers in addition to merely starting and stopping of motors.

The motor control system may be manual or automatic. The term 'automatic control systems' will apply to an assembly of such devices which are not protective relays, although functioning in conjunction with them and performing the operations in order to :

(*a*). Prevent, locate and eliminate faults affecting the whole or part of a power system.

(*b*) Distribute electric power to the individual parts of a power system.

(*c*) Restore normal connections in the systems and the continuity of power supply to loads after faults.

The primary aim of this chapter is to familiarize the reader with the general principles of operation and construction of control elements based on electrical machines.

2.2 Starting and Speed Control of Motors

The motor may be started by connecting it directly across the line. However, the driven machine may be damaged if started with such sudden torque. Starting may have to be slow and gradual.

This will not only protect the machine but the line current in-rush on starting may not be too great. Thus there arises the necessity of motor controllers for the starting of the motors.

Speed control can also be obtained by installing controllers at suitable one or two places with the motor. These controllers can maintain very precise speeds for purposes of industrial processing. Other controllers are required to change the speeds of motors in steps, or gradually.

One properly installed and adjusted, motor starters will maintain reliable starting time, voltages, current, and torques for the benefit of the driven machine and power system. Proper-sized starting fuses, circuit-breakers, and disconnect switches (isolators) and good wiring practices that are governed by electrical code of practice.

2.3 Principles of Starting and Speed Control of D.C. Motors

When a d.c. electric motor is connected to an external source of supply so that current is sent through the armature and field from this external source, then if the excitation of the field and the directions of

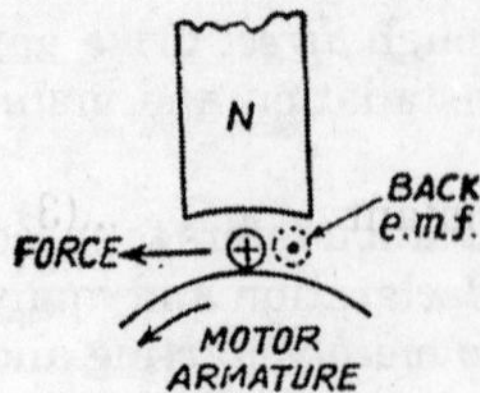

Fig. 2.1. Principle of operation of d.c. motor.

the armature currents are the same, the magnetic drag will be set up in the direction of force as shown in Fig. 2.1 and under its influence the armature will rotate. When the motor armature rotates the armature conductors cut the field flux, and as a result, they have e.m.f. induced in them. The direction of such induced e.m.f. in an individual conductor is in the opposite direction to the direction of the current. The direction of this e.m.f. is given by the Fleming's right-hand rule. This induced e.m.f. is called the "back e.m.f.". The supply voltage has to do work against this back e.m.f. in forcing the current through the armature. The magnitude of this back e.m.f. is given by the expression of the generated, e.m.f.

Back e.m.f. $$E_b = \frac{\phi ZN}{60} \times \frac{P}{A} \text{ volts} \quad \text{...(1)}$$

where ϕ = Flux per pole in webers

Z = No. of armature conductors

N = Speed in r.p.m.

P = No. of poles

A = No. of parallel paths through the armature winding

The speed of a d.c. motor automatically adjusts itself to the load so that the electric power required to drive the current through the armature is equal to the mechanical power required to drive the load. The load here includes all those losses in the motor which are produced by rotation, namely, the magnetic and mechanical losses.

Hence $E_b . I$ is proportional to Speed × (Total retarding torque)

Thus resultant e.m.f. acting in the armature circuit is,

$$E - E_b = E\left(\frac{\phi ZN}{60} \times \frac{P}{A}\right)$$

where E is the e.m.f. applied to be armature terminals.

Therefore, $$I_a = \frac{E - \left(\frac{\phi ZN}{60}\frac{P}{A}\right)}{R_a},$$

where R_a is armature resistance,

Hence, speed $$N = \frac{E - R_a I_a}{\phi Z} \times \frac{60A}{P} \quad \text{...(2)}$$

In modern machines the drop of voltage in the armature, $I_a R_a$ is small compared with E, so that we have, approximately,

$$N = \frac{1}{\phi} \times \left(\frac{E}{Z} \times \frac{60A}{P} \right)$$

or $\qquad N \propto \frac{1}{\phi}$ so long as E is constant ...(3)

and $\qquad N \propto E$, in case of constant excitation ...(4)

Thus the speed of a d.c. motor is inversely proportional to the flux per pole.

The speed control of d.c. motors can be achieved by the following methods :

(*a*) Rheostatic control of d.c. motors

(*b*) Field control of d.c. motors

(*c*) Series parallel control of d.c. series motors

(*d*) Ward-Leonard control

(*e*) Booster control

(*f*) Metadyne control

(*g*) Series parallel control

(*a*) **Rheostatic Control of D.C. Motors.** When an adjustable resistance R is placed in series with the armature making the total resistance in the armature circuit $R + R_a$, then the back e.m.f. for any armature current I_a is given by, $E_b = E - (R + R_a) I_a$.

At no load, when there is no external resistance R in series with the armature, then the back e.m.f., E_b is approximately equal to applied e.m.f., E.

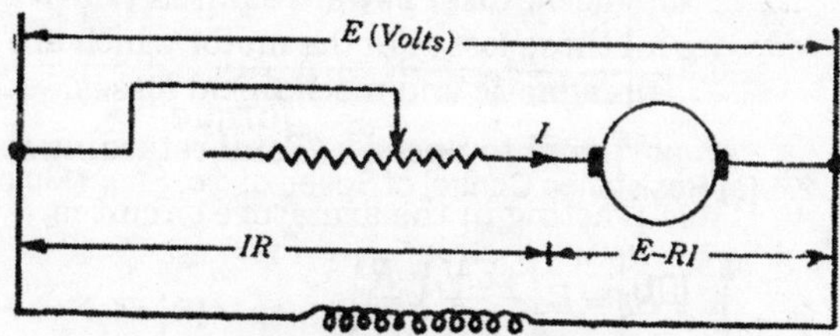

Fig. 2.2. Rheostatic control.

Since, for constant excitation, the speed is proportional to the back e.m.f. [equation 4] then denoting the no-load speed by N_0,

$$\frac{N}{N_0} = \frac{E - (R + R_a) I_a}{E}$$

Therefore, $\qquad N = N_0 \left(1 - \frac{R + R_a}{E} . I_a \right)$ ...(5)

Thus, for a given resistance $(R + R_a)$ the speed is thus a linear function of the armature current, I_a. The graph of N against I_a being a dropping straight line as given in Fig. 2.3.

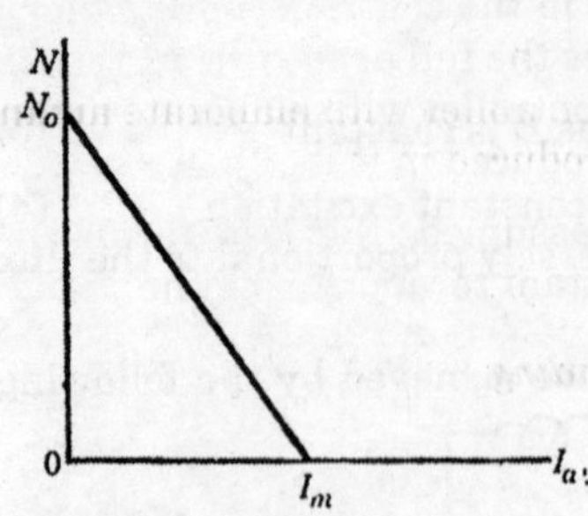

Fig. 2.3. Speed-armature current curve of d.c. motor.

From the above graph (Fig. 2.3), it is seen that for each value of R, there will be a certain value of armature current (I_m) which just stalls the motor. This value is given by,

$$O = N_0\left(1 - \frac{R + R_a}{E} \cdot I_m\right)$$

Here

$E_b = 0$. Therefore, $I_m = \dfrac{E}{R + R_a}$.

From Eq. (5) above it can be seen, that for a given value of the resistance, R the speed is not a constant, but is a function of the load current. This is one of the disadvantages of this method of speed control of d.c. motors. Where the speed is to be kept sensibly constant on a rapidly-changing load, the Ward-Leonard control

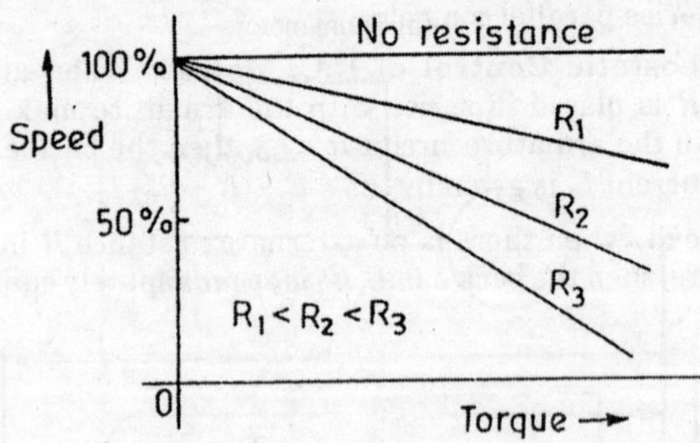

Fig. 2.4 (*a*) Resistance Control of Speed of d.c. Shunt Motor.

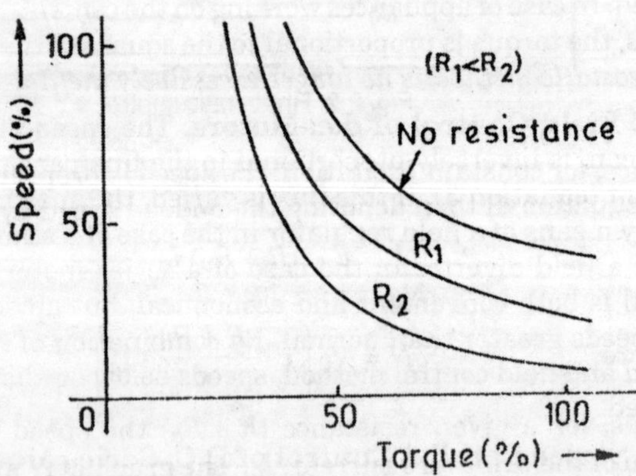

Fig. 2.4 (*b*) Resistance Control of Speed of d.c. series motor.

described later is applied. Another disadvantage of the method is that there will be considerable loss of energy in the controller, particularly at low speeds, since the resistance carries the full armature current. It also necessitates a large and expensive controller with elaborate arrangement for the dissipation of the heat produced in it.

In the above discussion, we have assumed, the load (armature current, I_a) to be constant, *i.e.* one of constant torque during the change

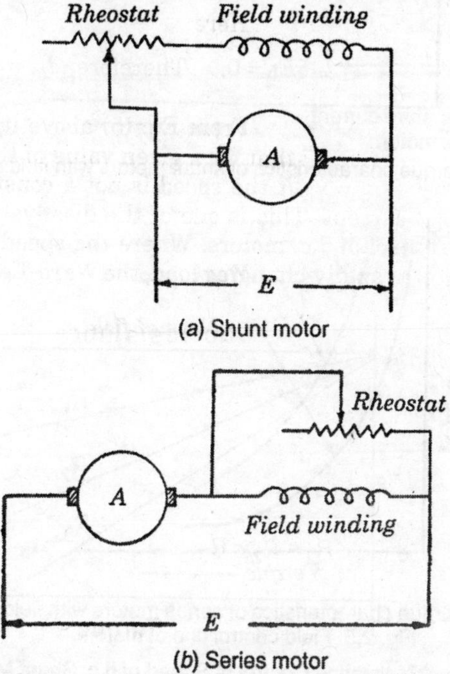

(*a*) Shunt motor

(*b*) Series motor

of speed. In case of appliances working on the centrifugal principle, such as fans, the torque is proportional to the square of the speed, and *there, the rheostatic method is no longer excessively inefficient.*

(*b*) **Field Control of d.c. Motors.** The speed of a d.c. motor, we have seen, is inversely proportional to the flux per pole so long as E is constant (equation 3). If this flux is varied, the speed will vary. This is done by means of a field regulator in the case of a shunt motor (Fig. 2.5 *a*) and a field diverter in the case of a series motor (Fig. 2.5*b*). This method is both convenient and economical, but obviously it will only give speeds greater than normal. By combination of rheostatic control method and field control method, speeds below or above normal can be obtained.

(*c*) **Series Parallel Control of D.C. Series Motor.** This method in conjunction with auxiliary rheostatic control is used for electric

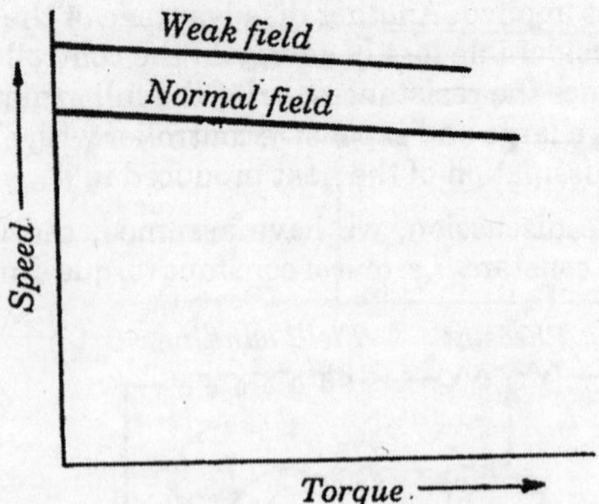

(c) Speed-torque characteristics of shunt motors with field control.

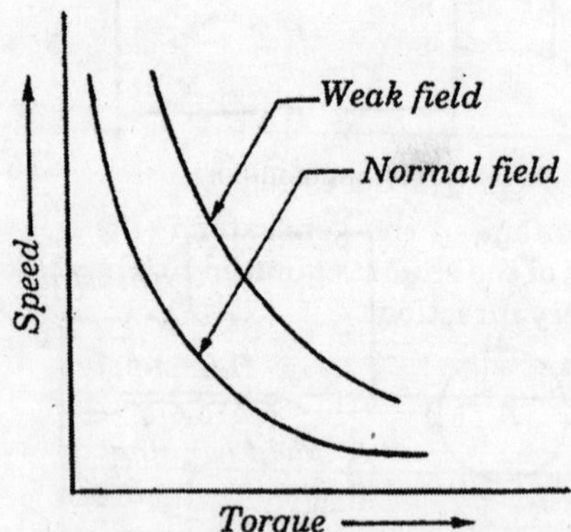

(d) Speed-torque characteristics of series motors with field control.
Fig. 2.5. Field control of d.c. motors.

traction. It requires two motors mechanically coupled. When the two motors are coupled in series, the torque produced is four times the torque produced by the motors in parallel. The various speeds can be attained through the various changes in connections as the controller handle is moved up to the full speed position.

(d) **Ward-Leonard Control.** This method is commonly used where a very delicate speed control over the whole range from zero to full speed is required, as, for example, with colliery winders and elevators. The method consists simply in working the motor with a constant excitation and applying to its armature sufficient voltage to give the speed required. A variable voltage supply is, therefore, required and it is obtained from a motor-generator set or converter set. The scheme is shown in Fig. 2.7. The speeds in the reversed direction are also obtained with this method by reversing the generator excitation by means of the reversing switch.

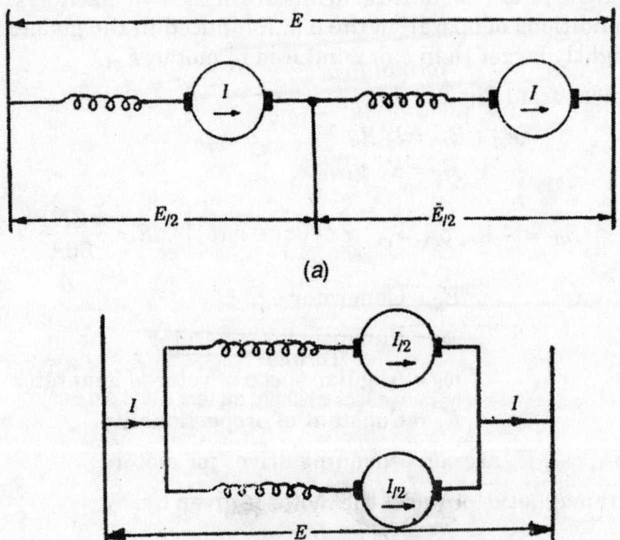

Fig. 2.6 (*b*) Series-parallel control.

The variable voltage of the generator is obtained by varying its excitation by means of the shunt regulator, *R*. The converter set always runs in the same very direction.

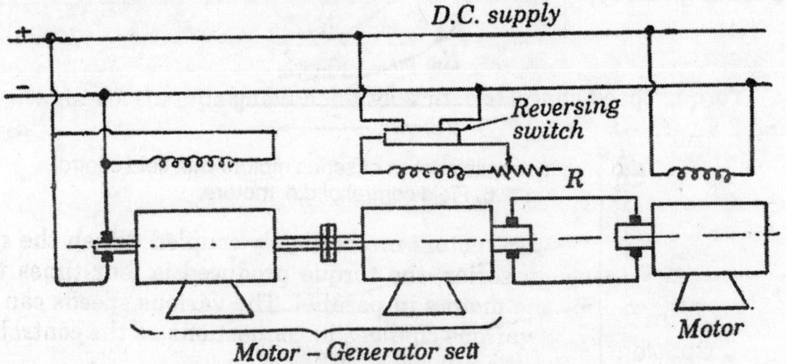

Fig. 2.7. Ward-Leonard control.

The steady state speed torque characteristics of a Ward-Leonard system can be derived from the equivalent ckt. shown in Fig. 2.8, R_a

Fig. 2.8

represents the total resistance of armature of the two machines. Under normal conditions of operation the e.m.f. induced in the generator, E_g will be slightly larger than e.m.f. induced in motor, E_m.

It is clear from Fig. 2.8 that,

$$E_g + E_m = I_a R_a$$

but $$E_g = K_g \phi_g \omega_g. \quad ...(a)$$

and $$E_m = -K_m \phi_m W_m \quad ...(b) \quad \left[\because E = \frac{\phi ZNP}{60A} + \frac{\phi ZP\omega}{2\pi A}\right]$$

where E_g = Generator e.m.f.

ϕ_g = Flux/pole for generator

ω_g = Angular speed of rotor of generator

K_g = Constant of proportionality.

E_m, ϕ_m, ω_m, K_g are corresponding drives for motor.

Electromagnetic torque of the motor is given by,

$$T = K_m \phi_m I_a \quad ...(c)$$

where K_m = Constant of proportionality

and I_a = Armature Current

Eliminating E_g, E_m and I_a from equations *a*, *b* and *c*, the motor speed is given by,

$$\omega_m = \frac{K_g \phi_g \omega}{K_m \phi_m} - \frac{R_a T}{K_m \phi_m^2} \quad ...(d)$$

Torque speed characteristics obtained using eqn. (*d*) are shown in Fig. 2.9.

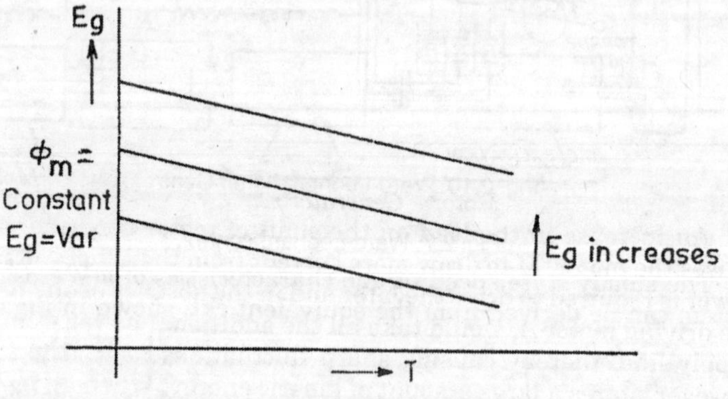

Fig. 2.9

Where this voltage control method is applied to very large motors such as colliery winders (mine hoists), reversing rolling mills etc., the supply is always an alternating one and a load equilizing arrangement, *i.e.* flywheel is incorporated in order to cope with the violent peak loads, which occur in such cases. The necessary modification is shown in Fig. 2.10 and this method is known as **Ward-Leonard Ilgner-Control.**

This supply being alternating one the motor A of the convertor set is a three-phase a.c. motor. Because direct current supply is required for excitation of the variable speed motor M and variable voltage generator B, the convertor set has an excitor, E directly coupled to it. A flywheel for load equilizing is also coupled to the converter set.

The function of the flywheel is to kept off the fluctuations in load from the mains. Its speed is reduced when the peak load comes on and is raised again when there is a light load on it. The necessary variations in the speed of the convertor set are obtained automatically by means of a 'slip regulator', R_g which puts resistance in the rotor circuit of the motor A when the load increases and cuts it out again when the load decreases. This resistance is generally liquid resistance R controlled through a torque motor (T.M.) fed through CT's from a.c. supply.

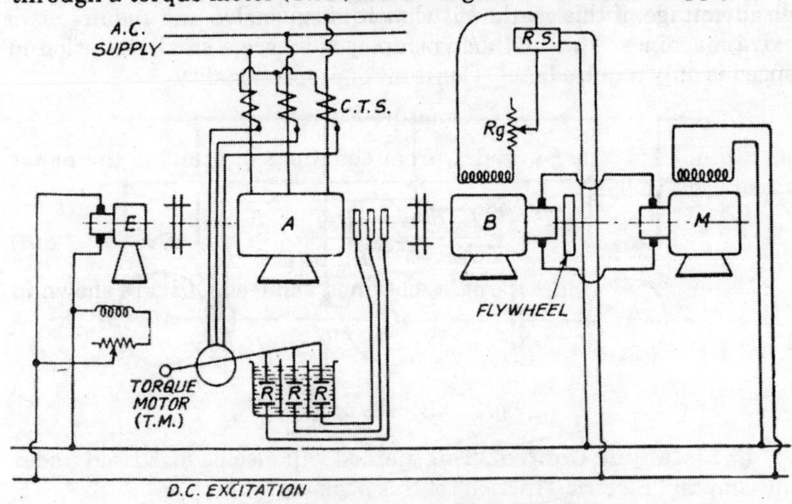

Fig. 2.10. Ward-Leonard Ilgner Control.

An increase in the load on the shaft of motor-generator set, A-B causes the motor, M to draw more current from the generator, so more power is required to drive the later and in the absence of any fly-wheel, the driving motor A, would take all the additional power from the a.c. supply line, thereby causing sharp fluctuations in it. The flywheel, however, stores a large amount of kinetic energy. When an increase in the generator load causes the motor, A to slow down, some of the kinetic energy of the flywheel is spent to restain the peak load on the shaft of the motor M. When the load on motor M decreases, the motor A picks up the speed and the flywheel stores up the kinetic energy.

(*e*) **Booster Control.** When the Ward-Leonard control is modified such that the armature of the motor M is connected in series with that of generator, G and with the supply, as shown in Fig. 2.11, then the voltage applied to motor M is the resultant of the supply voltage and the voltage in generator G. When the excitation of generator G is adjusted such that its voltage is equal and opposite to that of the supply; then the resultant voltage will be zero and the motor M will not rotate. As the excitation of generator G is reduced, the resultant voltage will increase and thus the speed of motor M will increase. With zero excitation of the generator G, the motor speed will be one-half of full speed. If now the excitation of generator G is reversed the generator will "boost" instead of "bulk" the supply voltage, the motor speed will increase still further. On full excitation when the generator voltage is equal to the supply voltage the motor M will be running at its full speed. By this method the speed can be varied over a wide range depending upon the size of the generator (booster) G. The exact size of the booster generator G depends upon the range of speed control needed. The main disadvantage of this method is that it is expensive and requires two extra machines. This method is employed where a small variation in speed is only required.

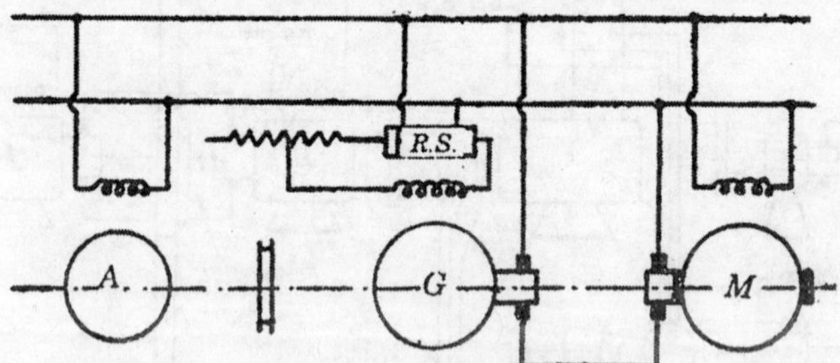

Fig. 2.11 Booster control.

(*f*) **Metadyne Control.** This method will also be discussed under the chapter "Electric Traction" of this book.

(*g*) **Series Parallel Control.** This method will also be dealt under the chapter 'Electric Traction' of this book.

Example 2.1. *A 240 volts shunt motor has a field resistance of 400 ohm and armature resistance of 0.1 ohm. The armature current is 50 amperes and the speed is 900 r.p.m. Assuming a straight line magnetisation curve, calculate the additional resistance in the field to increase the speed to 1000 r.p.m. for the same current.*

(*A.M.I.E., Sec. B. Utilization of Electric Power, Nov., 1988*)

Solution. When the motor running under normal working working conditions, then Back e.m.f.,

$$E_{b_1} = E - I_a R_a$$

$= 240 - 50 \times 0.1 = 235$ volts

Also shunt field current,

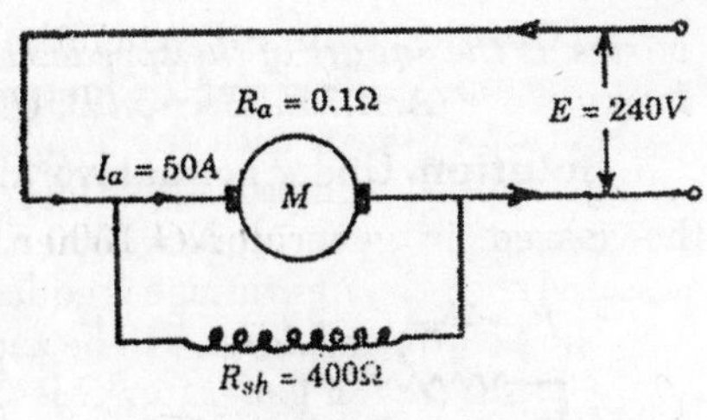

Fig. 2.12

$$I_{sh_1} = \frac{V}{R_{sh}} = \frac{240}{400}$$

$= 0.6$ ampere.

Speed, $N_1 = 900$ r.p.m.

Since, speed,

$$N \propto \frac{E_b}{\phi} \text{ where } \phi \text{ is the field flux}$$

Then $$N_a \propto \frac{E_{b_1}}{\phi_1}$$

Let a resistance R ohm be connected in series with the field to make the speed, $N_2 = 1000$ r.p.m.

then $$\frac{N_2}{N_1} = \frac{E_{b2}}{E_{b1}} \times \frac{\phi_1}{\phi_2}$$

As armature current and applied voltage E is same, so back e.m.f. remains the same and

$$\frac{N_2}{N_1} = \frac{\phi_1}{\phi_2} \quad \text{or} \quad \phi_2 = \frac{N_1}{N_2}\phi_1 = 0.9\ \phi_1.$$

As magnetisation curve is a straight line one, then $\phi \propto I_{sh}$

Thus $$\frac{\phi_2}{\phi_1} = \frac{I_{sh_2}}{I_{sh_1}}$$

or $$I_{sh_2} = \frac{\phi_2}{\phi_1} I_{sh_1} = 0.9\ I_{sh_1} \qquad \left[\because \frac{\phi_2}{\phi_1} = 0.9\right]$$

$$= 0.9 \times 0.6 = 0.54 \text{ ampere.}$$

Thus total resistance, $R + R_{sh}$ in the field circuit will be,

$$R + R_{sh} = \frac{E}{I_{sh_2}} = \frac{240}{0.54} = 444.4 \text{ ohms.}$$

∴ Additional resistance in the field circuit,

$$R = 444.4 - 440 = 40.4 \text{ ohms.}$$ **Ans.**

Example 2.2. *Explain briefly how the speed of d.c. series motor is controlled ?*

A d.c. series motor runs at 1000 r.p.m. taking 100 A at 400 volts. A diverter having double the resistance of the field winding is then connected in parallel with it. Estimate the change in speed if the torque varies as the square of the speed. Assume unsaturated field and neglect losses. (*A.M.I.E., Section B, Utilization of Elect. Power, May 1988*)

Solution. Under normal working conditions :

Speed, $N = 1000$ r.p.m.

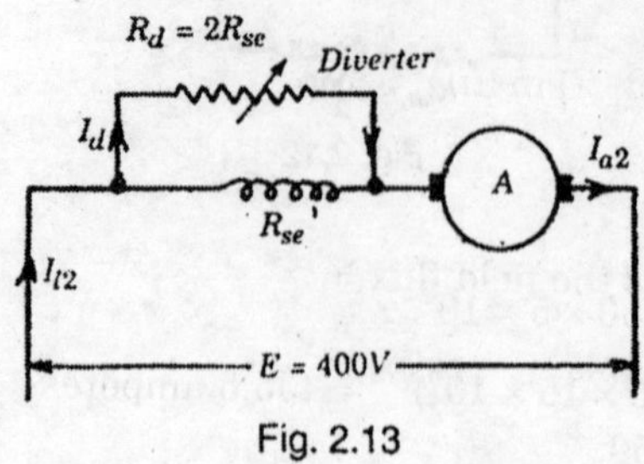

Fig. 2.13

Applied voltage, $E = 400$ V. And armature current and series field current and line current are the same

i.e. $I_{a_1} = I_{se_1} = I_{l_1} = 100$ A

After the connection of field diverter, let the speed be N_2 and line current I_{12}

Then, armature current,

$$I_{a_2} = I_{l2}$$

Series field current,

$$I_{se_2} = I_{l2} \cdot \frac{R_d}{R_{se} + R_d} = \frac{2}{3} I_{l2}$$

Since losses are negligible, therefore, back e.m.f. E_b may be taken as same as the supply voltage which remains constant,

So $$\frac{N_2}{N_1} = \frac{\phi_1}{\phi_2} \text{ (Since back e.m.f., } E_b \text{ is constant)}$$

As field flux, ϕ is proportional to the series field current, I_{se}.

Then $$\frac{N_2}{N_1} = \frac{I_{se_1}}{I_{se_2}} \text{ [where } \phi_1 \propto I_{se_1} \text{ and } \phi_2 \propto I_{se_2}]$$

Substituting the values,

$$I_{se_1} = 100 \text{ amperes and } I_{se_2} = \frac{2}{3} I_{l2}$$

$$\frac{N_2}{N_1} = \frac{100}{\frac{2}{3} I_{l2}} = \frac{150}{I_{l2}} \qquad ...(i)$$

Now, torque, $T_1 \propto$ armature current,

$I_{a_1} \times$ flux, $\phi_1 \infty I_{a_1}, \phi_1$ and torque, $T_2 \propto I_{a_2} \phi_2$

Because $I_{a_1} = I_{l1}$ and $I_{d2} = I_{l2}$

Therefore $$\frac{T_2}{T_1} = \frac{I_{l1} \phi_2}{I_{l1} \phi_1} = \frac{I_{l2} I_{se2}}{I_{l1} I_{se_1}}$$

or $$\frac{T_2}{T_1} = \frac{I_{l2} \times \frac{2}{3} I_{l2}}{100 \times 100} = \frac{\frac{2}{3} I_{l2}^2}{100 \times 100}$$

or $$\frac{T_2}{T_1} = \frac{I_{l2}^2}{15 \times 10^3} \qquad ...(ii)$$

It is given that torque varies as the square of speed.

Hence, $$\frac{T_2}{T_1} = \left(\frac{N_2}{N_1}\right)^2 \qquad ...(iii)$$

Substituting values from (*i*) and (*ii*) in (*iii*)

$$\frac{I_{12}^2}{15 \times 10^3} = \left(\frac{150}{I_{l2}}\right)^2$$

or $$I_{l1}^2 \times I_{l2}^2 = 150 \times 150 \times 5 \times 10^3$$

$$I_{l2} = (15 \times 15 \times 15 \times 10^5)^{1/4} = 135.5 \text{ amperes.}$$

And speed, $$N_2 = N_1 \times \frac{150}{I_{l2}} \qquad \text{[from } (i)]$$

$$= 1000 \times \frac{150}{135.5} \equiv 1110 \text{ r.p.m.}$$

Hence change in speed, $N_2 - N_1 =$ **110 r.p.m. Ans.**

Example 2.3. *A 250 volt d.c. series motor takes 40 amperes of current when developing in full load torque at 1500 r.p.m. Its resistance is 0.5 ohm. If the load torque varies as the square of the speed determine the resistance to be connected in series with the armature to reduce the speed to 1200 r.p.m. Assume that the flux is proportional to the field current.* (*A.M.I.E., Section B, Utilisation of Elect. Power, May 1986*)

Solution. Let R_h be the resistance to be added in series with the armature of the motor.

Given : I_1 = Motor armature current = 40 Amperes.

V = Rated voltage = 250 volts.

E_b = Back e.m.f. = $V - I_1 (R_h + R_{se}) = 250 - 40 (R_h + 0.5)$

Initially when no additional resistance is connected in series with the armature.

Then $$E_{b_1} = 250 - 40\,(0.5) \qquad [\text{Since } R_h = 0]$$
$$= 230 \text{ volts} \qquad ...(i)$$

It is given that the load torque varies as the square of the speed, thus

$$\frac{T_2}{T_1} = \left(\frac{N_2}{N_1}\right)^2 = \left(\frac{1200}{1500}\right)^2 \qquad ...(ii)$$

$$\left[\begin{array}{l}\text{Since } N_1 = 1500 \text{ r.p.m.} \\ \text{and } N_2 = 1200 \text{ r.p.m.}\end{array}\right]$$

Also Torque $T \propto$ flux $\phi \times$ Armature current,
$I_a = (I_l$ series motor)

Therefore, $T \propto \phi\, I_l$ or $\dfrac{T_2}{T_1} = \dfrac{\phi_2\, I_{l2}}{\phi_1\, I_{l1}}$

Assuming flux proportional to the field current.

$$\phi_2 \propto I_{l2} \text{ and } \phi_2 \propto I_{l1}$$

[$\because I_{l2} = I_{se2}$ and $I_{l1} = I_{se_1}$ in case of series motor]

$$\frac{T_2}{T_1} = \frac{I_{l2}^2}{I_{l1}^2} = \frac{I_{l2}^2}{(40)^2} \qquad ...(iii)$$

Substituting (*ii*) in (*iii*)

$$\frac{I_{l2}^2}{1600} = \left(\frac{1200}{1500}\right)^2$$

or $$I_{l2} = \frac{1200 \times 1200}{1500 \times 1500} \times 1600 = 32 \text{ amperes.}$$

When the motor current is 32 amperes and resistance R_h ohm in addition is connected in series with the motor armature, then

Back e.m.f. $$E_{12} = V - I_{l2}\,(R_h + 0.5)$$
$$= 250 - 32\,(R_h + 0.5) = 250 - 16 - 32\,R_h$$
$$= 234 - 32\,R_h \qquad ...(iv)$$

Since in a d.c. motor, Back e.m.f. $E_b \propto N\,\phi$

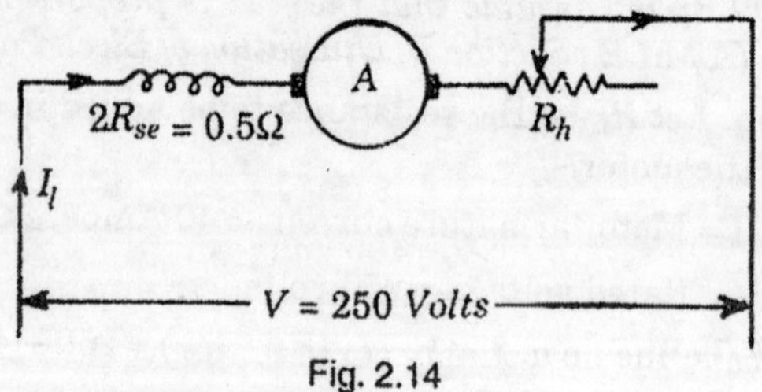

Fig. 2.14

Then $$\frac{E_{b2}}{E_{b1}} = \frac{N_2\,\phi_2}{N_1\,\phi_1}$$

Therefore, $$\frac{234 - 32\,R_h}{230} = \frac{1200 \times 32}{1500 \times 40} \qquad \begin{bmatrix} \phi_2 \propto I_{l1} \\ \text{and } \phi_1 \propto I_{l1} \end{bmatrix}$$

or

$$R_h = \left[234 - \frac{1200 \times 32 \times 230}{1500 \times 40}\right] \times \frac{1}{32} = \frac{234 - 147.2}{32} = \mathbf{2.7125 \text{ ohms.}} \ \textbf{Ans.}$$

Hence a resistance, 2.7125 ohms should be added in series armature circuit.

Example 2.4. *A four pole, d.c. series motor runs at 500 r.p.m., at 400 volts supply taking 20 amperes of current, all the four field coils* *being normally connected in series. Estimate the current and speed of the motor if the field coils are connected in 2 parallel groups of 2 in series. The load torque varies as the square of the speed. Assume the flux to the proportional to current. Ignore losses.*

(*A.M.I.E., Section B, Utilization of Elect. Power, May 1987*)

Solution. The given values are,

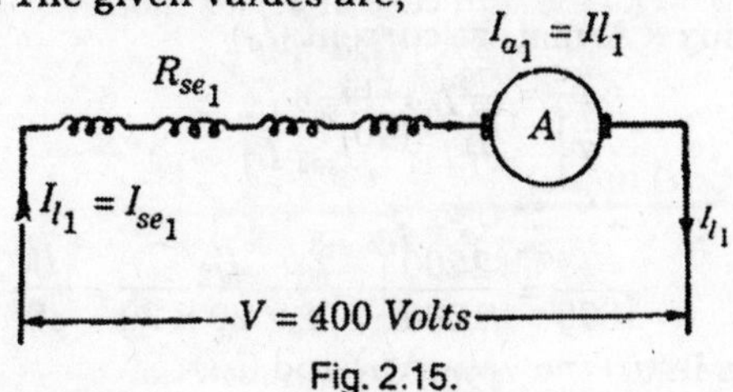

Fig. 2.15.

$I_{l1} = 20$ amperes $\qquad V = 400$ volts

$N_1 = 500$ r.p.m. and Torque $\propto$ (speed)^2, flux, $\phi \propto$ current, I_{se}

Also Losses are negligible, hence, $E_b = V$.

In case II when the field coils are connected in two parallel groups, each group having 2 coils in series, then the field current is halved.

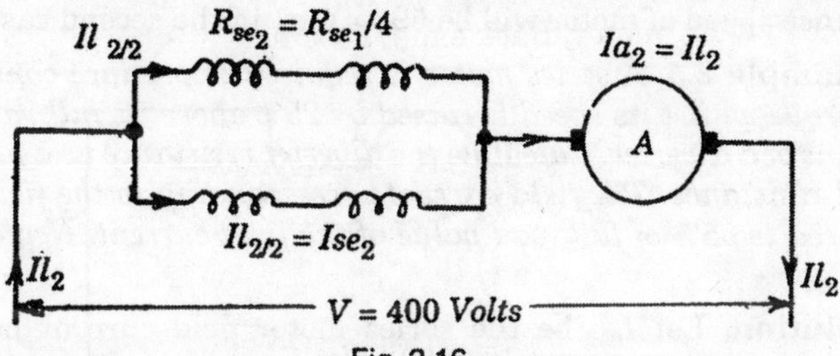

Fig. 2.16

Now set I_{l2} be the now supply current and N_2 the speed.

Then $\qquad I_{se2} = \dfrac{I_{l2}}{2}$ and $E_b = V$

Since speed is proportional to the back e.m.f. and inversely proportional to the flux.

Then $\qquad \dfrac{N_2}{N_1} = \dfrac{E_{b2}}{\phi_2} \times \dfrac{\phi_1}{E_{b1}}$ As $E_{b1} = E_{b2} = V$.

Therefore, $\qquad \dfrac{N_2}{N_1} = \dfrac{\phi_1}{\phi_2}$

or $\qquad \dfrac{N_2}{N_1} = \dfrac{I_{se_1}}{I_{se_2}}$ [Since $\phi \propto I_{se}$ (given)]

or $\qquad \dfrac{N_2}{N_1} = \dfrac{I_{l1}}{I_{l2/2}}$

or $$\frac{N_2}{N_1} = \frac{20 \times 2}{I_{l2}} = \frac{40}{I_{l2}} \quad ...(i)$$

As torque proportional to the square of speed

$$\frac{T_2}{T_1} = \left(\frac{N_2}{N_1}\right)^2 \quad ...(ii)$$

and Torque ∝ Flux × Armature current (I_a)

$$\therefore \quad \frac{T_2}{T_1} = \frac{\phi_2 I_{a2}}{\phi_1 I_{a1}} = \frac{I_{se2} I_{l2}}{I_{se1} I_{l1}}$$

or $$\frac{T_2}{T_1} = \frac{\frac{I_{l2}}{2} \times I_{l2}}{I_{l1} \times I_{l1}} = \frac{I_{l2}^2}{2 \times 20 \times 20} = \frac{I_{l2}^2}{800} \quad ...(iii)$$

Substituting in (ii) the values (i) and (iii)

$$\frac{I_{l2}^2}{800} = \left(\frac{40}{I_{l2}}\right)^2 \quad \therefore \quad I_{l2}^4 = 1600 \times 800$$

or $$I_{l2} = (1600 \times 800)^{1/4} = 33.69 \text{ amperes. } \textbf{Ans.}$$

Hence speed, $N_2 = N_1 \times \frac{40}{I_{l2}}$ (from i) $= 500 \times \frac{40}{33.62} =$ **594 r.p.m.**

Hence speed of motor will be 594 r.p.m. in the second case. **Ans.**

Example 2.5. *A series motor is required to produce constant full load torque whilst its speed is raised by 20% above its full rated value by means of a diverter. Calculate the diverter resistance as a percentage of field resistance. The field current corresponding to the field flux at new speed is 65% of full load value of the field current. Neglect motor losses.*

Solution. Let I_{se_1} be the series motor field current before the diverter is connected and I_{a_1} be the armature current, then,

$$I_{se_1} = I_{a_1}$$

When diverter is connected across the field resistance, then,

$$I_{se_2} = 0.65\, I_{se_1} = 0.65\, I_{a_1} \quad ...(i)$$

Because load torque is constant $I_{a_1} \phi_1 = I_{a2} \phi_2$

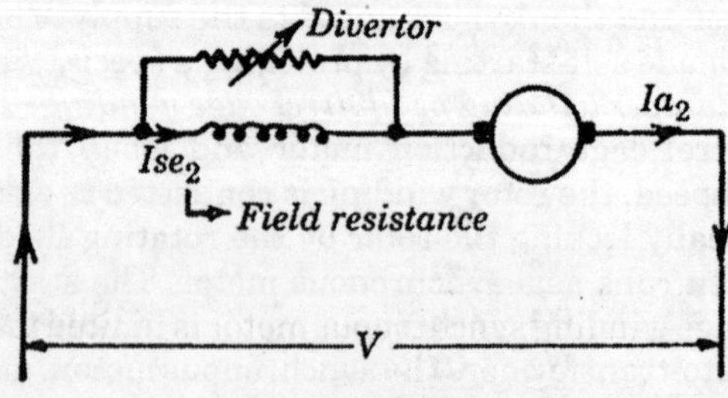

Fig. 2.17

or $$I_{a2} = \frac{\phi_1}{\phi_2} \times I_{a_1} \qquad ...(ii)$$

Since speed $N \propto \frac{E_b}{\phi} \propto \frac{V - IR}{\phi} \propto \frac{V}{\phi}$ [∵ Losses are negligible]

When the diverter is connected across the field resistance the speed N_2 becomes 1.2 N_1.

i.e. $$\frac{N_2}{N_1} = 1.2.$$

Also $$\frac{N_2}{N_1} = \frac{V/\phi_2}{V/\phi_1} = \frac{\phi_1}{\phi_2} \text{ therefore, } \frac{\phi_1}{\phi_2} = 1.2.$$

Now $I_{a2} = 1.2\, I_{a_1}$ [putting value in (*ii*)...(*iii*)]

Let $R_{div.}$ be the diverter resistance and R_{se} be the field resistance.

The series field current, $I_{se2} = \frac{R_{div}}{R_{se} + R_{div}} \cdot I_{a2}$

Substituting values from (*i*) and (*iii*)

$$0.65\, I_{a1} = \frac{R_{div.}}{R_{se} + R_{div.}}\, 1.2\, I_{a1} \text{ or } \frac{R_{se} + R_{div.}}{R_{div.}} = \frac{1.2}{0.65}$$

or $$\frac{R_{se}}{R_{div.}} = \frac{1.2}{0.65} - 1 = 0.55 \text{ or } R_{div.} = \frac{1}{0.55} R_{se}$$

i.e. Resistance of diverter is 118% of series field resistance for raising the speed by 20%. **Ans.**

2.4. Principles of Speed Control of A.C. Motors

A.C. motors are two types :

(*a*) Synchronous motors and (*b*) Induction motors

The induction motors are three-phase type used for industrial and agricultural pump drives and also single phase type used in domestic and commercial appliances.

(*a*) **Synchronous Motors.** Since the synchronous motor has got no self-starting torque, therefore, it requires some external source for starting purpose. It runs at a fixed speed, called the *synchronous speed.* The synchronous motor is first run upto synchronous speed by an auxiliary motor and then synchronised to the supply. The synchronous motor can be made self-starting by providing a special winding on rotor poles, called *damper winding* or *squirrel cage winding.* The motor first runs as squirrel cage induction motor and when the rotor attains synchronous speed, the rotor winding is connected to exciter terminals thus magnetically locking the rotor by the rotating field of the stator. The motor then runs as a synchronous motor. The starting current of the squirrel cage winding synchronous motor is limited to the prescribed value by an auto-transformer. The synchronous motors may be provided with a cylindrical rotor having slip ring induction motor winding to start as a slip ring induction motor with a good starting torque. When such a motor attains synchronous speed, d.c. supply is switched on to the rotor and the rotor comes into synchronism and the motor continues to run as synchronous motor. No speed control can be achieved with synchronous motor because this type of motor always runs at 'Synchronous Speed'.

Synchronous speed is given by $\frac{120f}{P}$, where f is the frequency of the a.c. supply and P, the number of poles of the motor.

(*b*) **Polyphase Induction Motors**

(*i*) **Starting.** The polyphase (three-phase) induction motor is used more extensively than any other form of a.c. motor. As indicative from its name the current in its rotor is not drawn from the a.c. supply, but is induced by the relative motion of the rotor conductors and the magnetic field produced by the stator currents. The polyphase stator winding on connection with the polyphase a.c. supply, a magnetic field is produced which rotates at synchronous speed. The stator flux cuts the rotor conductors and induce currents in them, hence by Lenz's law, the rotor follows after the stator field. Now it is on the relative velocity of the rotor and the stator field that the magnitude of the induced rotor current, and therefore, of the torque, depends. Hence even on no load the rotor speed must be less than *synchronous speed,* because of friction and iron losses. When extra load is put on the motor, the torque must increase in order to cope with it ; hence induced current must increase; and therefore, the relative velocity of rotor to stator field must increase. But the speed of the stator field is fixed by the frequency of the supply, *i.e.* by the synchronous speed, therefore, as the load on the motor increases, the actual speed of the rotor decreases. The speed torque characteristics of polyphase induction motor is, therefore, somewhat similar to that of a d.c. shunt motor.

There are two types of rotor winding, the "squirrel-cage" and the "phase-wound". The squirrel cage rotor consists of a number of copper bars permanently short circuited through copper end rings and therefore no external resistance can be connected in series for starting purposes. The squirrel cage motor is self-starting and are used as small motors because it takes a large current from the line with small starting torque.

In order to limit the starting current from the mains, the auto-transformer or star-delta starting of squirrel cage induction motor can be adopted.

The phase wound rotor is provided with a distributed winding, it is nearly always a two-layer wave winding in large motors. The separate phases are generally connected in star, and the three free ends connected each to one of the three insulated slip rings mounted on the motor shaft. When running normally these slip-rings are short circuited, but for starting purposes they are connected to three-phase star connected starting resistance. The starting resistance is included so as to limit the starting current from the mains as the large slip-ring motors may draw large value of current from the mains in the absence of starting resistance in the rotor circuit.

(ii) **Speed Control of Induction Motors.** A polyphase (three-phase) induction motor is practically a constant speed motor just as a d.c. shunt motor. The speed of an induction motor cannot be changed without the expense of its efficiency and good speed regulation. The speed of an induction motor is given by the expression,

$$N = \frac{120f}{P}(1 - S)$$

where f is the supply frequency, P the number of poles and S the slip.

For changing the speed of the induction motor, it is essential to change at least one of the above three elements : f, P or S. All a.c. motors including induction motors, without commutators, are fundamentally single-speed machines because the magnetic field rotates at synchronous speed. The economical control of the speed of induction motors is, in consequence, much more difficult than that of d.c. motors.

The speed control can be achieved by the following methods :

(a) By changing the supply frequency

(b) By changing the number of poles

(c). By changing the slip of the motor

(d) By cascading the two motors.

(a) **By changing the supply frequency.** Though this method of speed control provides wide range of gradual variations in the speed throughout its range but it is difficult to get the variable frequency supply. Installation of auxiliary equipment required for this purpose results in high initial cost excessive maintenance cost and lowering of the efficiency. This type of speed control is applied where wide, continuously variable speed range and good regulation of speed makes its use highly desirable. An example of this kind is the testing of scale-model aeroplanes in a wind tunnel. The supply frequency to the motor under speed control can be varied by varying the speed of the prime motor of the generator. Even in such case the speed variation is limited because the efficiency of the prime mover falls rapidly with the change in speed from that for which they are designed.

In this method it is essential that the line voltage be varied directly with the frequency in order to maintain nearly constant flux density so as to avoid the overheating of the machine.

Typical speed-torque curves for four different frequencies are shown in Fig. 2.18.

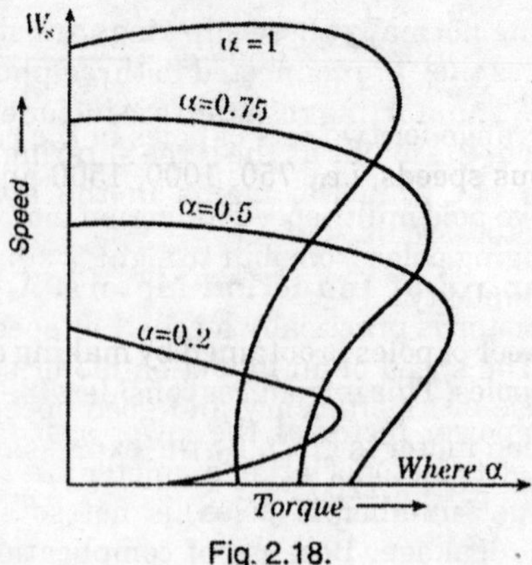

Fig. 2.18.

The slip at which maximum torque occurs becomes large as the operating frequency decreases and the max. torque gets reduced slightly. The starting torque increases for small reduction in frequency, but attains a maximum and then decreases with further reduction in frequency.

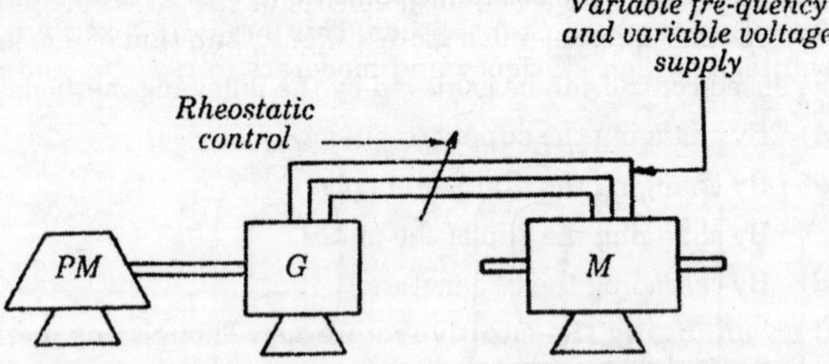

Fig. 2.19. Speed control of induction motor by changing the supply frequency.

(*b*) **By changing the number of poles.** The stator of the induction motors is wound so that the number of poles can be varied by changing the inter-connections of the coils of the stator winding by means of multiple control switches. The rotor of these motors is usually of the squirrel cage type, since such a rotor is suitable for any number of poles. If the rotor is phase wound, then its number of poles must also be changed. Thus this method of speed control is not practicable with wound rotor motors because it involves considerable complications in design of motors and switching as same numbers of pole are required for both the windings. Squirrel cage motors designed for pole changing control of speed are called *multi-speed induction motors.*

If P_1 and P_2 are the number of poles, then there are two synchronous speeds, given by

$$n_1 = \frac{120f}{P_1} \quad \text{and} \quad n_2 = \frac{120f}{P_2}$$

It is easy to arrange for two sets of poles in the ratio 1 : 2. Thus in all four synchronous speeds, *i.e.*, 750, 1000, 1500 and 3000 r.p.m. can be obtained with two pole multi-speed induction motor. Fig. 2.20 shows the method of changing poles from four to eight, a single phase winding being shown for simplicity.

The large number of poles is obtained by making all the poles of one name, consequent poles. This introduces considerable magnetic leakage resulting in lower power factor at the lower speeds. The method is, therefore, best suited to motor of larger diameter in which the pole pitch, even with the larger number of poles, is not so small as to cause excessive magnetic leakage. Because of complications in design and switching of the interconnections of the stator winding. It is not feasible to obtain more than four different synchronous speeds from any motor by this method of speed control. It is because of this reason, this method is not employed for general purpose speed control applications. This method is mainly employed in applications such as ventilating fans, conveyors, machine tools or other applications which require only two or four approximately constant speeds. This method is simple and has an advantage of high efficiency and moderate initial cost and maintenance.

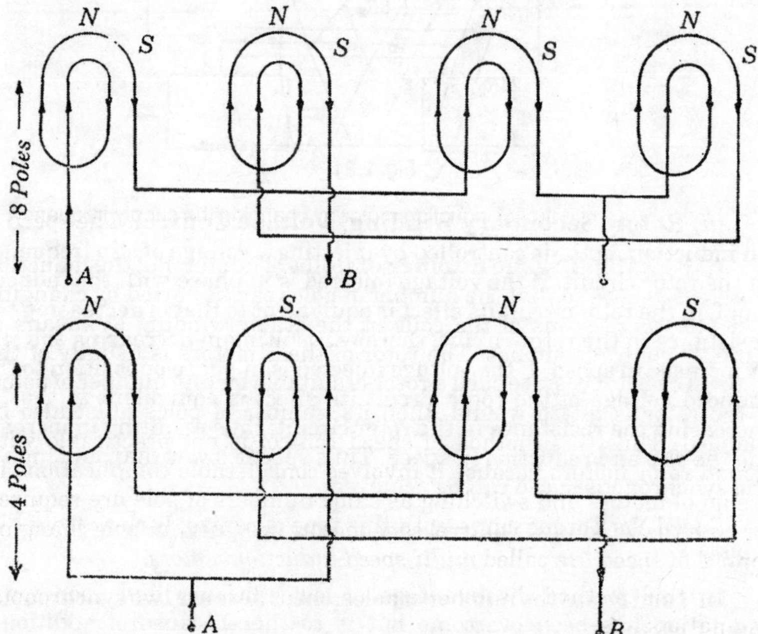

Fig. 2.20 Portions of single-phase winding for 8 and 4 poles.

(*c*) **By changing the slip of the motor**

(*i*) **Rheostatic Control.** This method can be applied to motors with wound rotors only. Because the speed depends upon the resistance of rotor circuit, resistance (additional resistance per phase) is included in the rotor circuit. In this method drop in speed is proportional to the power dissipated in rheostat. Hence it is a very wasteful method of speed control if speeds much below synchronism are required. A further disadvantage of this method is that the speed is only a function of the internal resistance so long as the load torque remains constant. For very large motors, resistance controllers are almost invariably of the liquid type, but for small and medium-sized machines metallic resistance with a controller of the barrel type are common. This method is often employed, because of its simplicity and convenience, where speed is to be reduced for a shorter period only.

The introduction of external resistance in the rotor circuit of a slip ring induction motor will modify the speed torque characteristics as shown in Fig. 2.21.

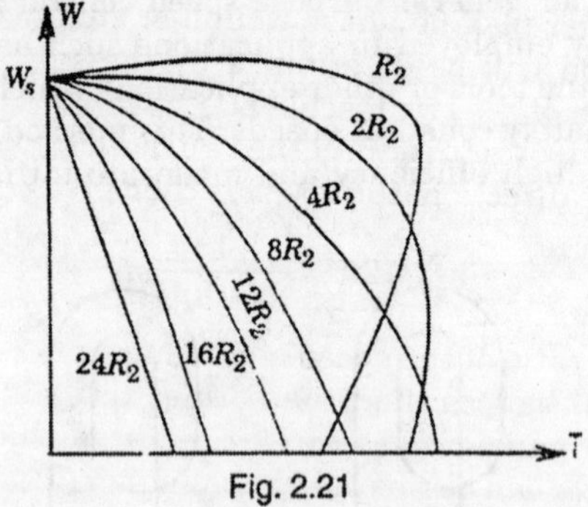

Fig. 2.21

(*ii*) **Rotor (Secondary Winding) Voltage Control.** The speed of an induction motor is controlled by injecting a voltage of slip frequency in the rotor circuit. If the voltage injected is in phase with the induced e.m.f. in the rotor circuit its effect is equivalent to that of decreasing the resistance in the rotor circuit, thereby resulting in decreasing the slip or increase in speed. If the voltage injected is in phase opposition to the induced voltage of the rotor circuit its affect is equivalent to that of increasing the resistance in the rotor circuit, thus resulting in increasing the slip and reduction in speed. Thus speeds lower than and higher the synchronous speed can be achieved.

Speed V/s Torque curve with this type of control is clear from Fig. 2.22.

In this method disadvantage of low efficiency and poor speed regulation has been overcome but is costlier because of additional auxiliary equipment for injecting the voltage of slip frequency in the rotor circuit. The method is employed with motors of very large rating, such as for motors in *steel mills*.

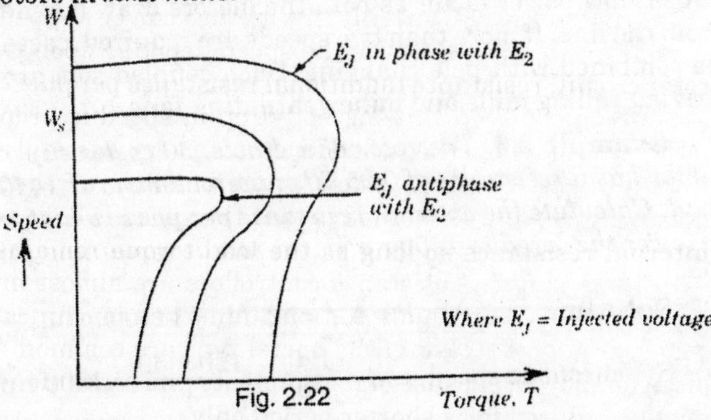

Fig. 2.22

(*d*) Speed Control by Cascading the Two Motors

In this method two induction motors are mechanically coupled as shown in Fig. 2.23. The rotor of the main motor is phase wound and the rotor of the other may be phase wound or squirrel cage. The speed of the combination will be determined by the sum or difference of the number of poles of the two machines. The common method of applying the cascade control is to connect the two motors such that their torques act in the same direction, the speed is then given by,

$$120f \times \frac{1}{p_1 + np_2}$$

where n is the ratio of the speed of auxiliary motor to the main motor. This speed will be less than the normal speed of the main motor. If speeds greater than normal speed are required, the torque of the second (auxiliary) motor can be reversed by changing the two stator leads. This is called *differential cascading* and the speed is given by

$$N = 120f \times \frac{1}{p_1 - p_2}$$

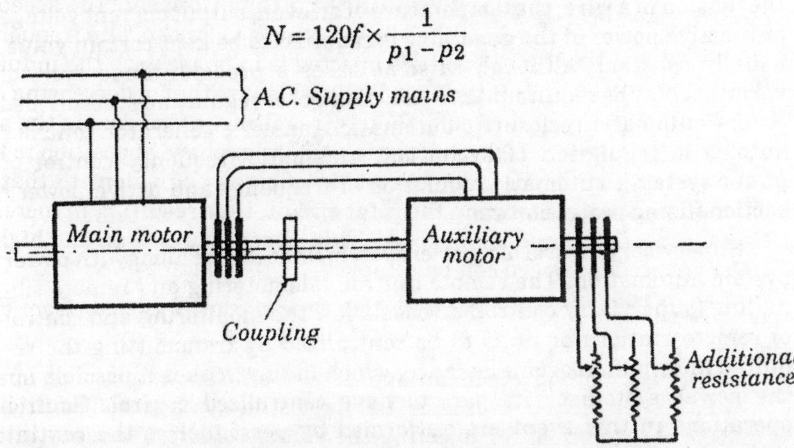

Fig. 2.23. Speed control by cascading two motors.

The method is costly because of necessity of second method, it causes low power factor as both the motors draw magnetising current from the line. If more than two speeds are required, cascade control can be combined with pole changing. Such coupled sets are employed for driving rolling mills and mine ventilating fans.

Example 2.6. *The rotor of a 4-pole, 50 cycles slip ring induction motor has a resistance of 0.25 Ω/phase and runs at 1440 r.p.m. at full load. Calculate the external resistance per phase which must be added to lower the speed to 1200 r.p.m. the torque remaining same as before.*

Solution. $P = 4, f = 50, N_1 = 1440$ r.p.m. (Given)

Synchronous speed, $N_s = \frac{120f}{P} = \frac{120 \times 50}{4} = 1500$ r.p.m.

Slip, $S_1 = \frac{N_s - N_1}{N_s} = \frac{1500 - 1440}{1500} = 0.04$

Let r be the resistance per phase added to the rotor circuit

$N_2 = 1200$ r.p.m.

after adding resistance in rotor circuit.

Then slip, $S_2 = \frac{1500 - 1200}{1500} = 0.2$

For constant input power to the rotor circuit.

$S \propto R_2$ where R_2, is the rotor circuit resistance.

$\therefore \frac{S_2}{S_1} = \frac{0.25 + r}{0.25}$ or $\frac{0.2}{0.04} = \frac{0.25 + r}{0.25}$ or $r = 1$ ohm. **Ans.**

2.5 Automatic Regulating System

In industrial processes a certain quantity such as speed of a motor, the tension in a wire, the temperature of an oven, output current voltage or reactive power of the generator is required to be kept certain value, usually constant, although in some cases a predetermined cycle of variation may be required. Automatic control or regulating systems also covers automatic reclosure ; automatic transfer ; generator control ; automatic regulation of excitation, automatic frequency control of power system ; automatic regulation of frequency and active power ; sactionalizing protection etc.

Supervisory control of a power system finds its use alongwith power system automation. The remote control, telemetering and remote signalling (supervisory control) devices allow the monitoring and control of remote stations or units to be centralized by transmitting the required information over a distance, which in turn makes it possible for the power system or its part to have centralized control. Control operations in this event are performed by personnel at the central station.

It should be noted that if an operation can be performed automatically or manually by supervisory control with much the same technical and economic results preference should be given to automation. In this event, the possible error of operators are prevented, their work is made easier and reliability is improved.

2.6 Elements of Automatic Control Systems

Any automatic control device including those used in power systems has in one form or another the following essentials :

(*i*) an element to sense the effect of an external factor (actuating device) ;

(*ii*) an element to convert it into an output signal in a predetermined manner (this is an operating device) ;

(*iii*) and an element which carries into effect the action of the output signal so that the variable may be under control (this is a connecting link, which may be thermionic valve, rotating amplifier or magnetic amplifier).

These elements may use combinations of devices performing amplification, signal delay, logical and mathematical operations (summation, inversion, multiplication, division etc.) and also transducers, relays and regulators or controllers.

2.7 Rotary Amplifiers (Amplidyne)

An amplifier is a unit meant to control rather heavy power (output value) by means of a low-power signal (input value). The output value is a function of the input signal and amplification is afforded by the energy of an external source. The electriçal amplifiers are classified as electronic, thyratron, transistorized, magnetic, magneto-electric and rotary amplifiers.

The output (controlled) power of the rotary amplifiers is produced due to mechanical power of the drive motor. The rotary amplifiers are essentially d.c. commutator motors. Depending upon the method of excitation, the amplifiers are classified as direct magnetized amplifiers and cross-magnetized amplifiers or amplifying.

The cross-magnetized amplifiers (amplifying), in which the main flux is along the quadrature axis of the machine, are as follows :

1. full-pitch winding amplifying ;
2. half-pitch winding amplifying ;
3. segregated magnetic system amplifying.

This section converse full-pitch amplifying. The amplifying are cross-magnetized machines where the main-Field flux is the quadrature-armature reaction flux. Fig. 2.24 (*a*) illustrates the amplidyne schematic diagram. The amplidyne is essentially a d.c. generator with an additional set of short-circuited brushes occupying a position along the quadrature axis of the machine p-q. The amplidyne stator carries a number of windings *viz.* control windings CW (usually two or four) arranged along the axis d-d, and a compensating winding C mounted coaxially with the former. The degree of compensation afforded by the compensating winding C is controlled by means of a shunt resistor R_{sh}. The same circuit comprises a commutating winding C.C. to improve commutation.

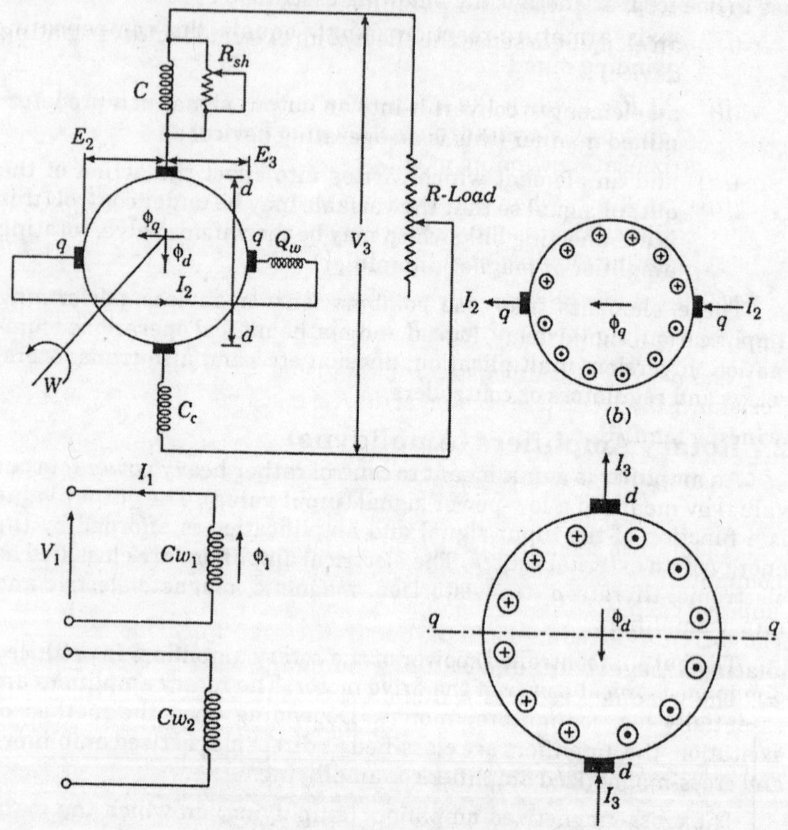

Fig. 2.24 Amplidyne schematic diagram.

The principle of operation of amplidyne is as follows :

Assume that the drive motor runs at a rated speed, *i.e.*, $\omega = \omega_r$ and a d.c. voltage V_1 is applied to one of the control windings, then the direct armature reaction flux ϕ_d is proportional to the m.m.f.

$$F_d = I_3 \, W_A \quad \text{...(i)}$$

where W_A is the number of turns in the parallel circuit of the armature winding. Also the compensating winding m.m.f. is,

$$F_c = I_3 \, W_c \quad \text{...(ii)}$$

where W_C is the number of turns in the compensating winding.

The degree of amplidyne compensation is characterized by the compensation factor.

$$F = \frac{F_c}{F_d}$$

The amplidyne may run under one of the following three conditions:

(*i*) at $K = 1$, when the machine is compensated, *i.e.*, the direct axis armature-reaction m.m.f. equals the compensating winding m.m.f ;

(*ii*) at $K < 1$, when the machine is under compensated, *i.e.*, the direct-axis armature-reaction m.m.f is higher than the compensating winding m.m.f ;

(*iii*) at $K > 1$, when the machine is over compensated, *i.e.*, the direct-axis armature-reaction mmf is lower than the compensating winding m.m.f ;

The amplifying are usually produced slightly F_d, *i.e.*, F_d being 8 per cent higher than the F_d or with $K = 1$.

The degree of compensation is adjusted by means of the shunt resistance R_{sh}. With due account of the R_{sh} resistance the compensating winding mmf.

$$F_e = I_3 W_c \, \frac{R_{sh}}{R_c + R_{sh}}$$

To improve commutation in the direct-axis circuit of the armature, commutating poles are arranged on the stator along the *d-d* axis. To improve commutation in the quadrature-axis circuit of the armature, the current I_2 is reduced and the quadrature-axis flux is maintained by placing a magnetising quadrature winding Qw on the stator (Fig. 2.24 *a*). The winding produces a flux $\phi\, Q_w$ that is commutative with the quadrature armature reaction flux.

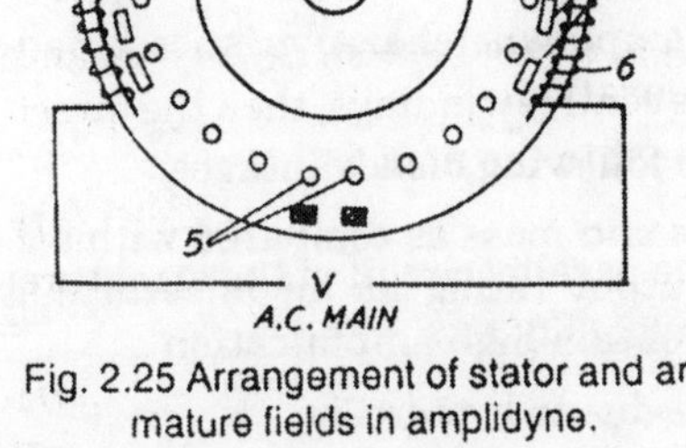

Fig. 2.25 illustrates the arrangement of stator and armature windings in a two-pole cross-magnetised amplidyne. The demagnetizing winding 6 serves to demagnetized the stator by a.c. so as to eliminate residual voltage due to hysteresis.

Fig. 2.25 Arrangement of stator and armature fields in amplidyne.

Legend

1 = Control windings.

2 = Distributed compensating winding.

3 = Distributed winding.

4 = Concentrated commutating winding.

5 = Magnetizing quadrature winding.

6 = Demagnetizing winding.

Such a mechanical design of the stator and armature is most characteristic of amplifying rated upto 20 kW.

The amplidyne is rotated by means of a d.c. or a.c. motor that can be either built integral in a common frame or placed separately.

The most important characteristic of the amplidyne is its power gain. A high power gain is afforded due to the fact that the amplidyne functions as a two-state amplifier. One stage is from the control winding to the quadrature field and the other stage, from the quadrature field back to the direct-axis output circuit. Therefore, the total power gain $k_p = k_{p1} \,.\, k_{p2}$.

where k_{p_1} is the power gain of the first stage, k_{p_2} is the power gain of the second stage.

In this case $k_{p_1} = \dfrac{V_2 I_2}{V_1 I_1}$; $k_{p_2} = \dfrac{V_3 I_3}{V_2 I_2}$

or
$$k_p = k_{p1}\, k_{p2} = \frac{V_3 I_3}{V_1 I_1} = \frac{I_3^2 R_{load}}{I_2^2 R_1}$$

where R_1 is the control field resistance.

The rotary amplifiers are machines of serial production widely used in automatic control systems and automated electric drives. The rotary amplifier which are most popular have the following advantages :

(*i*) high power gain (upto 0.5×10^4 for amplidyne rated below 750 W and upto 10^5 for amplidyne of higher output ratings);

(*ii*) low input power which makes it possible to feed the control windings from electronic valve and semiconductor amplifiers;

(*iii*) rather fast response, *i.e.,* short time constants of the amplidyne circuit. The equivalent electromagnetic time constant is of the order 0.1 to 0.2 S ;

(*iv*) sufficiently high reliability, long service life and wide range of power control ;

(*v*) provision for varying the amplidyne characteristics by adjusting the degree of compensation.

The amplidyne, however, has the following disadvantages :

(*i*) relatively large dimensions and mass as compared with d.c. generators of the same output rating as an unsaturated magnetic circuit is used to obtain high amplification ;

(*ii*) Presence of residual voltage due to hysteresis.

The amplidyne motor generator sets of upto 10 kW output rating are in great favour in radio electronic machines and installations. The amplifying are also employed in radio factories for tuning TV sets and radio receivers, where the sine voltage is maintained constant in amplitude and varying in shape as a function of time.

The amplifying are employed in great variety of automatic control systems, where these function as power amplifiers. To perform this function, several control fields are provided to afford comparison of signals and to employ feedback circuits. The amplifying are also employed in controlling reversing mills, cold rolling mills, blast furnaces and mine hoists, metal cutting and paper industry etc. Now-a-days there is a trend in the industry to replace amplifying by static devices, such as magnetic amplifiers operating in conjunction with transistor amplifiers because of inherent disadvantages of rotary amplifiers (amplifying).

2.8 Magnetic Amplifiers

A magnetic amplifier is a steel-cored choke coil inserted into an a.c. circuit. By magnetizing this choke, coil with a direct current it is possible to vary the alternating current in the controlled circuit within a wide range. There is a great variety of magnetic amplifiers, however, the so called "self-magnetizing" system of magnetic amplifiers enjoys the widest amplification factor and high rate of response.

An elementary circuit diagram of such a magnetic amplifier is shown in Fig. 2.26. The magnetic amplifier consists of two identical cores 1 and 1′ made of electrical sheet steel. It is desirable that the core steel has high permeability and magnetic saturation. Coldrolled steel grades are used for high-power magnetic amplifiers and permalloys are used for low-power amplifiers.

Each core is provided with working winding 2 and 2′ with a number of turns equal to N_w. Both windings are connected opposite and parallel

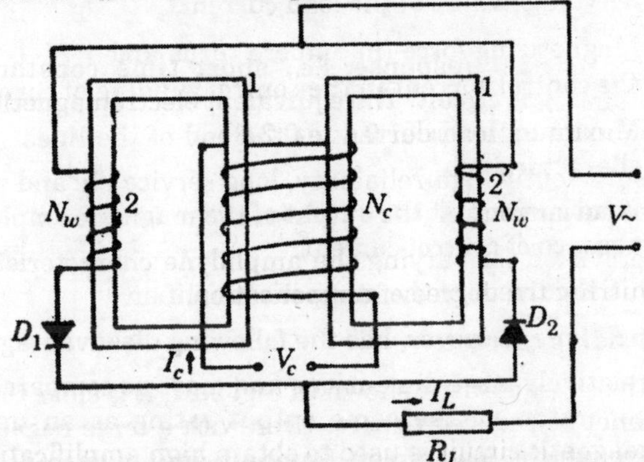

Fig. 2.26. Circuit diagram of a self-magnetizing magnetic amplifier with an a.c. output.

to each other. Each winding is connected in series with a corresponding rectifier D_1 and D_2. These rectifiers provide amplification factor and rate of response. Therefore, all modern magnetic amplifiers are usually based on this circuit. Load R_L is connected in series with the magnetic amplifier in an a.c. circuit fed with an a.c. voltage, $V_\sim$.

The control winding 3 with the number of terms N_c linkages with both cores. The control signal V_c is fed to this winding. If the system is designed to respond to several control signals, the cores are provided with several control windings.

The magnetic amplifier works on the principle of working of choke coil. Thus, by varying the initial magnetization of the choke coil, one can control the current flowing through the load. The dependence of the average value of load current on the magnetizing force of the control winding of a real amplifier is illustrated in Fig. 2.27.

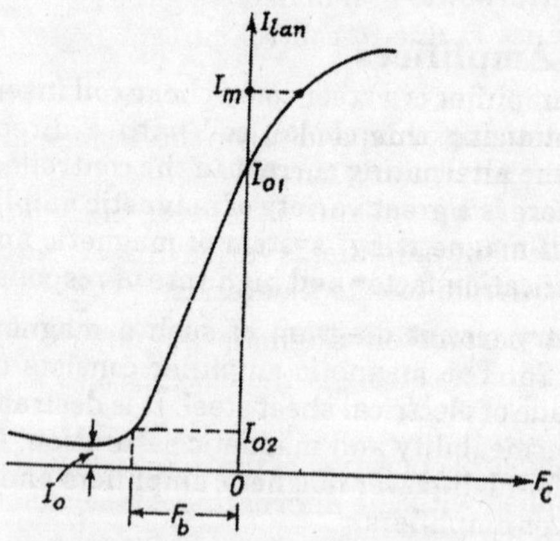

Fig. 2.27 Characteristic of a magnetic amplifier with internal feedback.

where I_{av} = Average value of the load current.

F_c = Magnetizing force (in ampere turns and $= N_c I_c$ where I_c is the control current in the control winding of turns N_c).

I_m = Maximum load current at the end of the linear part of the characteristic.

I_{a_1} = Load current, at the output of the magnetic amplifier, in the absence of control current.

I_{a_2} = Initial current of the magnetic amplifier.

I_o = Standing current.

From the characteristic shown in Fig. 2.27, it is clear that the load current cannot be reduced to zero. Even with a large negative control current, through the load circuit flows a standing current defined by the magnetizing force required for remagnetization of the cores. Therefore, a magnetic amplifier is characterized by the ratio :

$$K_l = \frac{I_m}{I_0}$$

In commercial amplifiers this ratio range from 10 to 30.

The amplifying properties of magnetic amplifiers are defined by their amplification factor.

$$K_a = \frac{P}{P_c}$$

The load power, P_l is formed by the formula, $P_l = V_r\, I_r$.

where V_r, and I_r are the rated voltage and load current of the amplifier.

Under the control power P_c if understood, the I^2R loss in the control winding of resistance r_c due to control current I_{ol} for the rated load current, $P_o = R_c\, I_{ol}^2$.

The amplification factor depends upon the design parameters of the magnetic, amplifier, particularly, on the amount of copper in the control winding. The greater the amount of copper, the higher is K_a. The amplification factor of the amplifier is varied in a wide range, depending on their designation, from tens to thousands.

An important parameter magnetic amplifiers is their rate of response, *i.e.*, the rate of setting of the output voltage when applying the control signal.

From the principle of operation of the magnetic amplifier it follows that it can be supplied with the alternating current only. However, it can be loaded with d.c. consumers, such as field windings of electric machines, an armature of a d.c. motor, a storage battery to be charged, etc. For this purpose the amplifier output is connected to a rectifier providing a rectified current for supplying the load ; therefore, the load current is unidirectional because the current of the reverse polarity cannot pass through the rectifier. Where the load current must be reversed, there are employed more complicated circuits, so-called, push-pull magnetic amplifiers.

The advantages of magnetic amplifiers are simple design, absence of moving parts and simple servicing ; long service life, high amplification factor, possibility for simply summing up input signals. These are used in electrical control systems, for measurement and automatic regulation systems. As such it is used as *Power Amplifier*. For example, in automatic control of a.c. and d.c. motors or in smooth regulation of lighting of some area.

The other amplifiers used are Electronic and Ionic Amplifiers, Semiconductor (Transistorized) Amplifiers and Thyristors.. These devices are featured by high mechanical strength, long service life, small size and weight, absence of moving parts, simple design, unlimited number of allowable switching operations etc.

2.9 Pilot Devices and Accessories

Different types of pilot devices are employed for switching the control circuit in electric drives. Push buttons, float switches, master switches, thermostats, pressure regulators, limit switches and plugging switches are most important which shall be described here.

2.10 Push Buttons

A push button is a device that provides control of a motor or drive by pressing a button. By using more than one push button it is possible to control a drive from as many places as there are push buttons through the same magnetic controller. The rated voltages are upto 500 volts a.c. or d.c.

The contacts of push buttons are usually double break, as shown in Fig. 2.28, actuated by plastic buttons. Two sets of contacts are provided so that when the button is pressed, one set is opened and the other set is closed. Thus by connecting to the proper set of contacts, either a normally open or a normally closed contact is obtained.

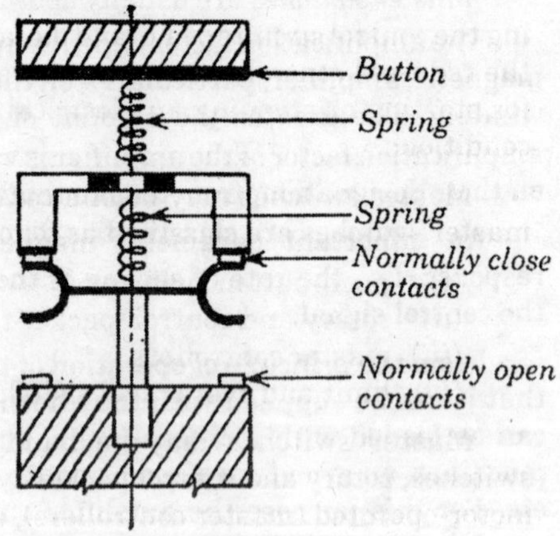

Fig. 2.28. Push button.

Push buttons are made of two types of services ; standard-duty type for the usual application, and heavy-duty type for frequent use of push buttons.

Since control buttons are subject to high momentary voltages, caused by the inductive effect of the coils to which they are connected, good clearance between contacts and insulation to ground (earth) is provided. The current broken by a push button is usually small.

The main operating conditions of a motor are controlled by push buttons, such as start, stop, forward, reverse, fast and slow. Auxiliary push buttons may be installed at different points to control any or all of these functions. They may also be used as remote stop buttons with manual controllers equipped with potential trip or low voltage protection. The breaking capacity of push-button contact ranges from 80 to 100 W when used in d.c. circuits and is upto 1500 VA in a.c. circuits. The contact elements have an electrical service life of at least 200,000 opening operations, and a mechanical service life of atleast 1,000,000 switching cycles.

Various enclosures for push buttons may be obtained in addition to the general type to provide for unusual conditions. They are termed as water tight, dust-tight, explosion-proof and submersible.

Provisions for pad locking stop buttons in the open position (for safety purposes) are also available and used extensively.

2.11 Controllers and Master Switches

Controllers are devices used for manually starting, regulating the speed, reversing and stopping the electric motors. They are employed for controlling a.c. and d.c. motors, particularly, in load handling and hoisting mechanisms. Controllers may be of a *face-plate type, drum, or cam type.* Cam type controllers can perform upto 600 closures per hour. They can be made to handle much greater currents than drum-type controllers.

Master switches are usually operated by hand and serve for switching the control circuits of a.c. and d.c. apparatus. By making or breaking the required control circuit, the operator can remotely supply a signal for starting or stopping an electrical drive, and for changing its duty conditions.

Master switches may be of a contact or non-contact type. Contact master switches are classified as follows :

(*i*) push-button switches ;
(*ii*) rotary and control packet-type switches ;
(*iii*) master controllers ;
(*iv*) limit and over travel switches.

Master switches may be hand or foot operated (push-button switches, rotary and control packet-type switches, master controllers), motor-operated (master controllers), and operated by the moving part of the controlled machinery (limit and over travel switches). They may be of a fixed-position type whereby the switch remains in the selected position after the actuating force has been removed (the operating handle is released), and of a return action type, in which case the switch returns automatically to the initial (zero) position when released by the operator.

2.12 Rotary Selector Switches

These are intended for manually switching direct and alternating-current circuits rated at upto 500 volts. They serve for making infrequent changes in the connections of control circuit, are used as voltmeter and ammeter selector switches, servomotor control switches, for switching non-automatically connected and disconnected electrical equipment on and off and, finally, for changing the number of pole pairs of small multi-speed induction motors.

The rotary switch is made up of a set up of sections assembled into one or two stacks, through which passes central operating shaft. The operating shafts are coupled with handle, either directly or *via* gear wheels and intermediate shaft.

The rotary switch contacts are rated for a continuous current upto 20 amperes. Their breaking capacity ranges from 1 to 120 amperes a.c. and 0.1 to 20 amperes d.c. depending on the number of breaks, the supply mains voltage and the load inductance.

2.13 Rotary Control Switches

Rotary control switches consist of a set of sections carrying the fixed contacts and accommodating the moving bridge-type contacts, the latter being actuated by the central operating shaft and handle of the switch. The sequence of contact closure with and without interruption of the circuit current is varied by changing the shape of the bridge contacts.

These switches are used for controlling a.c. and d.c. circuits rated at upto 220 volt. Their operating handle may be of a fixed position or spring return type.

2.14 Over Travel and Limit Switches

Over-travel and limit switches serve for switching control circuit when a moving parts under control passes beyond its normal travel (over-travel switches), or when the moving part reaches the end of its normal travel (limit switches). Limit switches are used, for example, to limit the travel of load hoisting and handling mechanisms, to limit the travel of carriages of metal working machines and many other mechanisms and, also, to start and stop electric motors according to the path covered by the processed workpiece (for example, to start, stop, and reverse the movement of the table of a rolling mill according to the position occupied by the ingot). The development and implementation of automatic production lines calls for the use of a wide variety of overtravel and limit switches.

The limit switch may be of a pusher, lever, spindle or rotary type.

Rotary limit switches are used in case when the switch is to be operated by a rotating part of the controlled mechanism. The contacts of these switches are actuated by cam discs.

Over-travel and limit switches may be adjustable or non-adjustable. Their switching capacity depends upon the constructional features of the contact system.

2.15 Float Switches

Float switches are designed and used for automatic control of a.c. and d.c. pump motor magnetic starters or for automatic direct control of light motor loads. These are employed for maintaining a certain level of liquid in the tanks. The raising or lowering of a float mechanically attached (*i.e.,* rod operated or chain operated float switch) to the electrical contacts of the motor may start the pump to empty or to fill the tank, whatsoever is required. These switches are also required for opening or closing of piping valves for the control of fluids. The switches may not be submerged in water but are used for tank or sump pumping operations or tank filling.

2.16 Pressure Switches and Regulators

There is a place for the pressure switch in virtually any industrial application, where a pressure sensing element exists. Industrial pressure switches are designed to cover the wide range of requirements encountered in the control of pneumatic or hydraulic machines. These controls are most commonly employed in the control circuits of welding equipment, machine tools, high pressure lubricating systems, and motor-driven pumps and compressors.

Because of the diversity of application, a wide selection of pressure ranges is available. A diaphragm-actuated device is designed for applications requiring sensitive response to small pressure ranges. A seamless metal-bellows-actuated control is used for pressures as high as 140 kg/sq. cm and the piston operated hydraulic switches are suitable for pressures upto 1050 kg/sq. cm.

The range and differential (difference between the "cut-in" and "cut-out" pressures) can be changed in the field. Most commonly used are single-pole switches, but two-pole switches are also used.

Gauge type pressure regulators provide accurate control of pressure or vacuum conditions on the systems on which they are installed. Used as pilot control devices with magnetic starters, they govern operation of pump or compressor motors similar to pressure switches. Reverse action regulators may also be used for pressure systems interlocks to prevent an operation until the pressure in that system has increased to the point desired.

Pressure regulators consist of a pressure gauge and control relay. Delicate contacts on the gauge energize the relay, causing it to be opened or closed, as the case may be. Thus, the relay contacts are used to control a large motor starter, avoiding possible burning of the gauge contacts. Regulators for standard operation open the circuit at high pressure and close the circuit at low pressure. Special regulators operate on reverse operation to close the circuit at high pressure and to open the circuit at low pressure.

2.17 Thermostats or Temperature Switches

Thermostats are designed for automatic control of temperature maintaining equipment. General industrial temperature controllers are recommended for heating applications where the temperature to be controlled is higher than the normal or ambient temperature.

Gas vapour or liquid pressure expands a metal bellows against the force of a spring to give a motion to an operating pin, which snaps a precision switch to its operating position, where the setting is reached. This is very similar to industrial pressure switches. The operating pressure is generated in proportion to the temperature of a liquid in a closed bulb. The precision switch shape back to its normal position when the pressure in the bellows drops enough to allow the main spring to compress the bellows.

There are many different designs of the more familiar 'thermostat' for automatic control of space heating and cooling equipment, such as air conditioners, refrigerators and cold storages. Electrically, the thermostat is a two or three (or more) wire control pilot device (switch) that is temperature actuated. Temperature actuated switches are used in control circuits to operate heaters, blowers, fans, solenoid valves, pumps and other devices.

2.18 Speed Governors

A tachogenerator is employed as a speed regulation system element. Tachogenerators are fractional horse power electrical machines running as generators that function to convert a rotational speed into a proportional electrical signal.

A general output characteristic equation for an ideal tachnogenerator is as follows :

$$V_{output} = K_g \, \omega = K_g \frac{d\theta}{dt}$$

where θ is an angle of turn of the tachogenerator rotor ; K_g is steadystate gain (steepness of output characteristic), ω is circular motion of the rotor (*i.e.*, speed).

Discussed below in the operation of a radar aerial spinning drive running at constant speed.

The desired speed is governed by the respective value of the reference voltage V_r which is delivered to the *EA* input via a comparison unit *C*. A control voltage V_{cont} passed through the E_A and the amplidyne is applied to the control winding of the servo-motor whose field winding is permanently connected to a voltage V_g. The motor rotor starts running and turns the aerial, A and the ITG (induction tachogenerator) rotor. The tachogenerator output voltage V_g is compared in the *C* with the reference voltage V_r and a constant value of V_{cont}. determined by the error voltage V_e is set across the amplifier stage output. The d.c. servomotor rotor and the aerial revolve continuously at a constant speed.

The reference and tachogenerator voltages (V_r and V_g respectively) are phased out so that any increase in the tachogenerator voltage causes a decrease of the error voltage V_e. Hence, any variation in the ser-

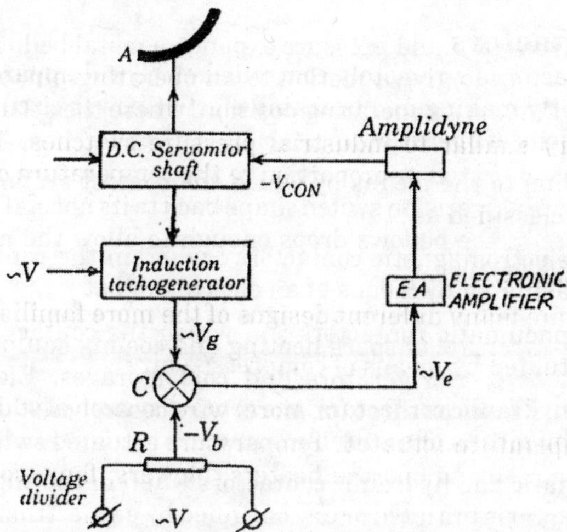

Fig. 2.29. Block-diagram of radar antenna drive.

vomotor rotor and the aerial speed will cause such a change of voltage V_g and V_e, and consequently V_{const}, that the speed will tend to retain the original value. So, a tachogenerator inserted in a feedback circuit has made it possible to stabilize the aerial speed.

2.19 Plugging Switches

Plugging switches, also called zero-speed switches, were preliminary designed for incorporation into control circuits as pilot devices to provide quick automatic stopping of machines driven by squirrel cage motors. If properly adjusted, plugging switches will prevent reversal of rotation of the controlled drive after it has been brought to a standstill by the reversal of motor connections. Plugging switches are often installed on machine tools that must come to a sudden stop at some point in their cycle of operation to prevent inaccuracies in the work or damage to the machine. Another use is found in processes where the machine must come to a complete stop before the next step of work is started ; here the reduced stopping time saves time for a greater total output.

The shaft of the plugging switch, is mechanically connected to the motor shaft or a suitable shaft on the driven machine. Motion of the motor is transmitted to the switch contacts by a centrifugal mechanism or a magnetic induction (eddy current disc) arrangement within the switch. The contacts are wired to the reversing starter, which controls the motor. Acting as a link between the motor and reversing starter, the switch allows the starter to apply just enough "reverse" power to bring the motor to a quick stop.

2.20 Contactors

A contactor is a remotely controlled electrical apparatus designed for frequently making and breaking electric power circuits under normal conditions of operation.

According to the means by which the contact system is actuated, contactors classified as :

1. electromagnetic contactors : wherein the contact system is actuated by means of an electromagnet ;
2. pneumatic contactors : in which the contact system is actuated by means of compressed air ;
3. hydraulic contactors : the contact system of which is actuated by means of a suitable liquid.

Pneumatic and hydraulic contactors wherein the supply of operating air or liquid is initiated or discontinued by an electromagnet or some other remotely controlled operating mechanism are not dealt with in this book. Only electromagnetic contactors will be described here.

Electromagnetic contactors have found wide application and are the main power control elements of modern automatically controlled electric drive systems. Contactors are also classified as :

(*i*) d.c. contactors for use in direct current circuits ;
(*ii*) d.c. contactors for use in industrial frequency a.c. circuits ;
(*iii*) high-frequency contactors for use in high frequency circuits (upto 1000 Hz).

The magnetic system (operating circuit) of the contactor may be designed for a current differing from that of the power circuit (main contacts). For example, the operating coil of an a.c. contactor may be designed for d.c. supply, and that of a high frequency contactor may be designed for d.c. or 50 Hz. a.c. supply.

According to their circuit-breaking function, contactors may be classified as :

(*i*) line (power circuit) contactors used for making and breaking all kinds of electric circuits (for example, contactors *L*, *1F* and *2F*, *1E* and *2E* as shown in Fig. 2.30).

(*ii*) accelerating contactors (say, contactors 1A, 2A and 3A shown in Fig. 2.30) used for switching over starting resistor steps R_1, R_2 and R_3.

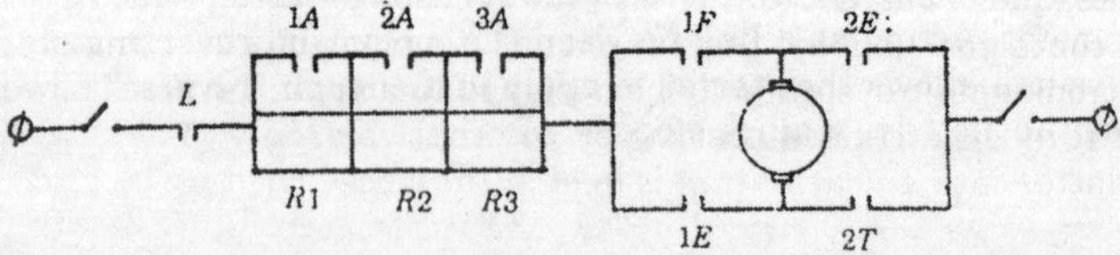

Fig. 2.30 Elementary circuit diagram for starting and reversing a d.c. motor.

Accelerating contactors have only to change the circuit resistance and, therefore, operate under higher duty conditions then power circuit (line) contactors.

Contactors are made to meet the following general requirements :

(*i*) high making and breaking capacity—not below ten times the rated current and, in individual cases upto 20 times the rated current ;

(*ii*) continuous duty at high frequency of opening ;

(*iii*) long electrical service life—from 3 million cycles of operation (including interruption of starting currents) ;

(*iv*) long mechanical service life—from 10 to 20 million cycles ;

(*v*) easiness in manufacture, small weight and size ;

(*vi*) high service reliability.

Contactors are designed in the main for currents of upto 600 amperes (certain series are rated upto 300 A), voltages of 220, 440, 650 and 750 volt d.c. ; 380, 500 and 660 V a.c. of industrial frequency (50 or 60 Hz), and of upto 1500 V high frequency current. The most typical frequency of operation is 600 and 1200 to 1500 closures per hour. Some series of contactors allow upto 3000 closures per hour, and welding contactors even upto 14000 closures per hour.

Contactors consist of a system of main contacts, an arc-quenching and electromagnetic systems, and auxiliary contacts. The main contacts serve for making and breaking the power circuit. Depending upon the normal position of the main contacts, contactors may be with *making, breaking* and *mixed contacts,* also referred to as *normally open, normally closed* and *normally open and closed contacts.* The contacts are considered to be in their normal position when the operating coil of the contactor is de-energized and all its mechanical latches are released.

The main contacts may be of lever or bridge type. Lever-type contacts are used in rotary moving system, and bridge type contacts are used in straight-travel moving contact systems.

The arc-quenching system serves for exiting fishing the arc burning between the main contacts when they separate.

The electromagnetic system provides for remote control (opening and closing) of the contactor.

The design features of the electromagnetic system are determined by the kind of current (a.c. or d.c.) that is used for the control circuit of the contactor and by the functional diagram of the system. The electromagnetic system may be designed to attract the armature and hold it in the attracted position or, alternatively, only to attract it. In the latter case, the armature is held in the closed position by latch.

The auxiliary contacts are used to switch over the control, interlocking and signalling circuits of the contactor. They are designed to carry a continuous current of not more than 20 amperes and to interrupt current of not over 50 amperes. Auxiliary contacts may be making and breaking and are mostly of a bridge type, though they may also be of a lever type.

The principal difference between a contactor and motor starter is that the contactor does not contain overload relays. Contactors are used to switch, with pilot control devices, such as lighting, heating and controlling a.c. motors where overload protection is provided separately. The larger sizes are used extensively for remote control of relatively high-current circuits where expense prohibits running the power leads to the remote controlled location. This is one of the main advantages of electromagnetic control over manual control. Pilot devices, such as push buttons, float switches, pressure switches, limit switches, or thermostats, are used to provide the necessary control for operating contactors. Automatic control with the use of some of these pilot devices is a decided advantage of magnetic controllers.

2.21 Electromagnetic Relays

A relay is an apparatus which is used to actuate or control a high power system by the action of a relatively small power on the relay. The functions performed by relays and their constructional details art of a very wide variety.

The relays can be classified according to various parameters. The most reasonable is to classify them according to the parameter which actuates the relay. Relays are accordingly classified as : current relays, voltage relays, power relays, power directional relays, reactance relays, impedance relays, thermal relays, gas pressure relays, acoustic relays, optical relays and relays responsible to various chemical reactions, etc.

Generally, the relay comprises three basic functioning elements: sensitive, intermediate and actuating members. The sensitive (supervising) member responds to the controlled parameter X and converts it into a physical quantity required for the operation of the relay. The sensitive members is a moving element is relays having movable parts. The intermediate member (a spring in contact relays) compares the common primary parameter with the reference (set) value and whenever the controlled parameter exceeds the preset quantity it makes the relay operate, *i.e.*, transfers the primary action to the actuating members.

The actuating members acts upon the controlled circuit, changing the parameter Y. A moving contact system in relays having moving parts (contact relays) serves as the actuating member.

Relays may also have a fourth element which establishes a time delay between the moment when the sensitive element has to actuate the relay and the moment the actuating element operates. The time delay can be produced by electromagnetic, mechanical and other means.

Electromagnetic relays are of relatively unsophisticated design and allow one to obtain different enough characteristics that match well with the characteristics of the contact systems. While relatively small in size, the electromagnets develop comparatively great pull and due to this find extensive application in different types of relays operating both with d.c. and a.c. current relays (under current and over current relays), voltage relays (over voltage, under voltage, auxiliary, signal relays, etc.), time relays, frequency relays and other which are used as protective relays, as well as motor control relays and automation relays, are all of the electromagnetic type.

Electromagnetic relays are manufactured in output power rating ranging from one to several thousand watts. The power in the control circuit ranges from fractions to several scores of watts or to several hundred volt-amperes (V.A. burden of instrument transformers). The operating time ranges from 1-2 to 20 ms, depending upon the purpose, construction and connection diagram. Electromagnetic time relays provide a time delay of upto 5-10 seconds. Electromagnetic relays have a permissible frequency of operation of upto 1500-4000 cycles per hour with a mechanical service life of 10 to 20 million cycles and an electric service life of several million closures. Typical relays are described below.

2.22 Protective Relays

Protective relays such as instantaneous overcurrent relays are used in protective circuits of electrical installations. According to the method the sensitive member is connected to the current transformer secondary and the action of the actuating member energizes the trip circuit of the circuit breaker to trip.

2.23 Voltage Relays

These type of relays differ from current relays basically in the number of the coil turns and the coil wire diameter. They are manufactured as over-voltage and under voltage relays. The power consumption is about 1 VA.

2.24 Electromagnetic Time Relay (Direct and alternating current time relay)

It produces a time lag with the aid of a clock work mechanism. The relay is manufactured for D.C. and A.C. service in various modifications ensuring different time lags within the range from 0.1 to 0 seconds. Power consumed by the D.C. relays is 15 to 30 W and that consumed by the A.C. relays is 10 to 20 VA.

2.25 Control and Automation Relays

These type of relays are manufactured for currents from 2 to 1500 A with one opening contact or with one opening and one closing contacts with self-resetting and with manual or electromagnetic resetting. The relays with self resetting return into the open position automatically after the protected circuit is tripped. With the manual or electromagnetic resetting the armature of the relay after picking up is brought under a latch and does not return into the initial position after tripping the circuit. In this case the resetting is performed by operating personnel either manually or remotely.

2.26 Polarized Electromagnetic Relays

The sensitive member of polarized relays is acted upon by a polarizing flux in addition to the working flux. Due to this the relay acquires a property to respond to the direction of the current in its windings.

2.27 Basic Construction and Operation of an Electromagnetic Relay

Fig. 2.31 shows the basic construction of a typical electromagnetic relay. The relay essentially consists of a coil to which the supply voltage is applied, a solenoid containing a core upon which the coil is wound, relay contacts, and a movable steel armature which is held against one of the contacts (b_1 or b_2 as shown in Fig. 2.31) by means of a spring. The relay action depends on the interaction of the magnetic field set up by the coil carrying the current and the movable steel armature.

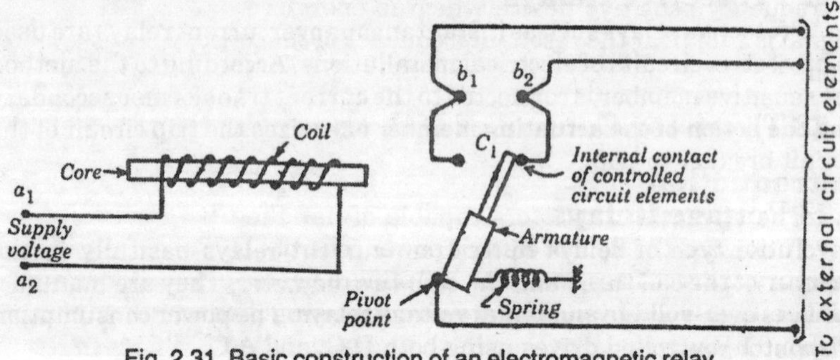

Fig. 2.31. Basic construction of an electromagnetic relay.

Before the voltage is applied to the coil the relay is in *normal* or deenergized position. The working of the relay depends on the interaction of the magnetic field set up by the coil carrying the current and the movable steel armature. The relay remains in the working (energized) position till the coil remains energized. Although relay contacts are controllable manually, but these are remotely controlled.

Relay contacts are made of metals and alloys having low resistance in order to prevent overheating. Also case is taken in selecting the metal having properties against welding action of the contacts that occurs when the contacts close and the arcing phenomenon that forms when the contacts open, the design and shape of the relay contacts should be such so as to resist their pitting action. Some common shapes of relay contacts are shown in the Fig. 2.32. The metals used in the manufacture of relay contacts are silver, tungsten, palladium and certain alloys etc. The current flowing through the relay contacts is generally smaller (of the order of a 5 Amperes) for D.C. and A.C. voltages.

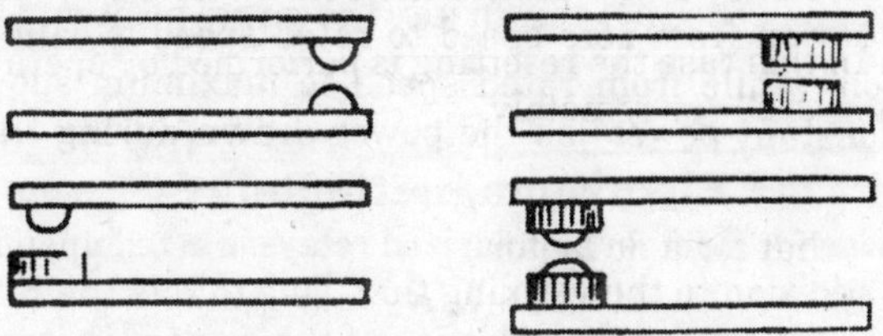

Fig. 2.32. Some common relay contacts.

A.C. relays in addition have a shading coil placed over a portion of the pole face at the air gap to reduce vibration and chattering of the contacts. The function of the shading coil which has a single turn of wire fastened over approximately half of the pole face area is to shift the phase of the flux in the shaded portion of the pole face. This action prevents the total force of the flux (taken over the entire pole) from decreasing to zero at any point in the cycle. Thus, this results in minimizing the chattering phenomenon of the contacts. In certain A.C. relays a capacitor is connected in series with the relay coil, producing a frequency sensitive device which is energized only when the A.C. signal is at or near the resonant frequency of the coil capacitor combination.

2.28 Thyristor Control of D.C. and A.C. Motors

Introduction

The advent of thyristors capable of handling large currents has revolutionized the field of electric power control. Thyratrons, ignitrons, mercury arc rectifiers, magnetic amplifiers, motor-alternator sets have all been replaced by solid state circuits using diodes and thyristors. Thyristor controlled drives using both D.C. and A.C. motors find wide application in industry as variable speed drives. Changing from electro-mechanical and other form of control to electronic control, generally, has the advantages of higher accuracy; better reliability and response and higher efficiency.

2.29 Thyristorised Speed Control of DC Motors

2.29.1. Introduction

The dc motors are undisputable electrical equipment in any application. Their usefulness mainly results from the fact speed of dc motors can be easily controlled.

Before the advent of thyristors other methods were used for the speed control of these motors. Well known among them was Ward-Leonard system. The introduction of thyristor has completely changed the picture and now thyristor controlled dc motors are dominating the field.

2.29.2. Characteristics of dc shunt motors

Figure 2.33 shows at a glance the characteristics of the dc shunt motor. The region from zero speed to rated speed is called "constant torque region" while from rated speed to maximum speed is called "region of constant power" as the power drawn during this region is constant.

We know that for a dc motor :

$$V = E_b + I_a R_a \qquad ...(2.1)$$

$$E_b \propto \phi N \qquad ...(2.2)$$

where E_b = back emf of motor; ϕ = flux; N = speed; T = torque; R_a = armature resistance; I_a = armature current; and V = applied armature voltage.

If armature drop is neglected then

$$V \propto \phi N \qquad ...(2.3)$$

Or if the field flux held constant, varying armature voltage can vary motor speed. During this region if the armature current is constant, the torque developed by the motor will be constant. Hence this is called constant torque region. Varying the armature voltage, controls normally motor speed upto the rated speed.

Speed above rated speed can be obtained by weakning the field flux as can be seen from equation (2.3). Mechanically a dc motor can be run at 5 to 6 times its rated speed but electrically it is not possible to go beyond 3.5 times of rated speed. Beyond this value effect of armature reaction is very strong compared to already weak field flux.

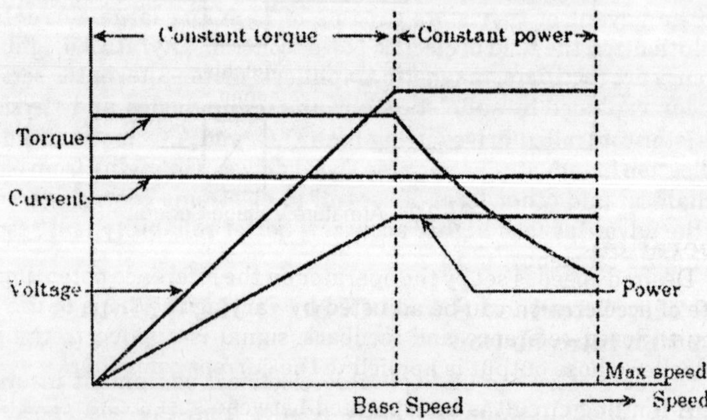

Fig. 2.33. Characteristics of dc Shunt Motor.

During field weakening range voltage and current drawn by the machine are constant and hence the region is called 'constant power control' and the power drawn from the main will be constant. But the torque will get reduced as the field flux is being reduced. This is clear from the relation

$$N \propto \frac{1}{\phi} \propto \frac{1}{T} \qquad ...(2.4)$$

for region of constant power.

Based on the application sometimes it may only be necessary to control speed upto the rated speed or beyond rated speed. A few applications may even demand both. We will discuss each of these cases in detail under the following headings:

(*a*) Armature voltage control.

(*b*) Field control.

(*c*) Combined armature and field control.

2.29.3. Armature Voltage Control

Figure 2.34 shows the circuit for controlling motors speed from zero upto rated speed. Most of the applications require speed control in either direction with the facility of regenerative braking. Hence a dual converter has been used to give four quadrant operation.

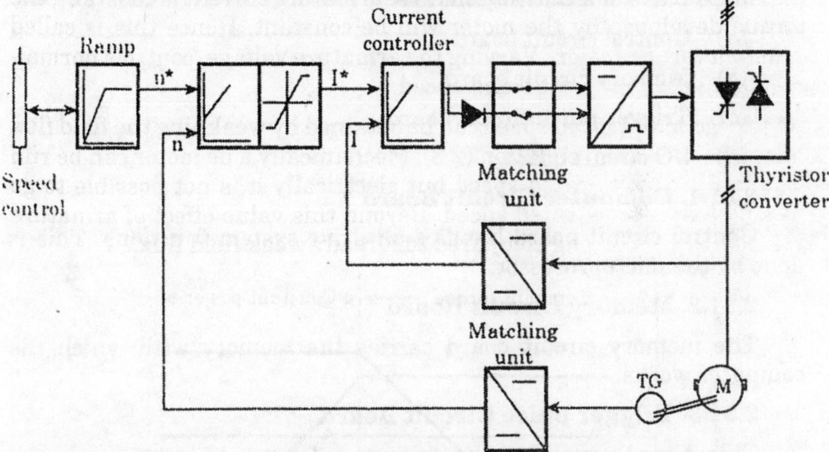

Fig. 2.34. Armature Voltage Control.

Operation

Desired speed is set by the operator on the reference potentiometer. Rate of acceleration can be adjusted by varying the slope of the ramp circuit. Speed reference and feedback signal is applied to the speed controller whose output is applied to the current controller.

A limiting circuit is incorporated between speed controller & current controller to limit the armature current to maximum allowable value.

Output of current controller is fed to the firing circuit which adjusts the firing angle to give the required armature voltage for desired motor speed.

2.30 Microprocessor Based Drives

Now a days microprocessor based drives are most commonly used in the industry. In these drive all functions are controlled by a microprocessors. The main control circuits in this drives are described here.

2.31 Control circuit in a micro processor based DC drive

All functions in a microprocessor based drive are performed in digital technology. These drives are capable of controlling all drive functions such as :

(*a*) Torque control

(*b*) Speed control

(*c*) Protection & supervision of drive

(*d*) Protection of motor from overload

(*e*) Reference handling

Main parts of a control unit are :

(*a*) Control circuit board

(*b*) Memory circuit board

(*c*) Trigger pulse circuit board

(*d*) I/O circuit board

2.31.1. Computer Circuit Board

Control circuit board handles all drive system functions. This is done by the microprocessor.

2.31.2. Memory Circuit Board

The memory circuit board carries the memory with which the computer works.

2.31.3. Trigger pulse Circuit Board

This board amplifies the trigger pulse to the thyristors, and provides galvanic isolation between control & isolation circuit.

2.31.4. I/O Circuit Board

This circuit board is the input/output board for control signals to & from the drives.

2.32 Field Controller

Figure 2.35 shows the schematic diagram for controlling the motor speed beyond rated speed by weekning the field flux. As the motor speed

is to be increased, field reference is reduced. This vary the firing angle of the thyristor convertor in such a way that the field current is reduced. Steady state condition is reached when the field reference is equal to the feedback value.

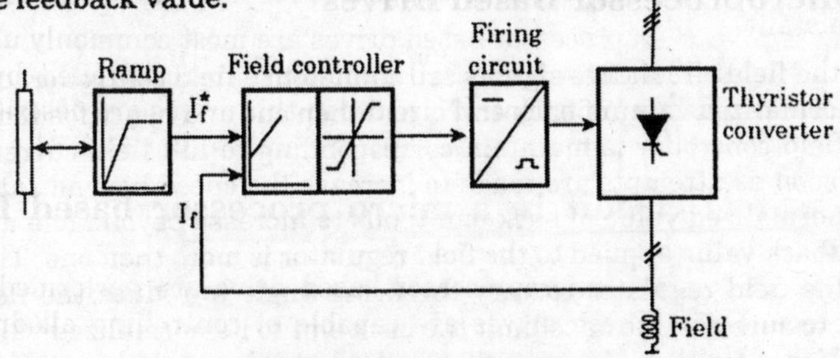

Fig. 2.35. Field Current Control.

2.33 Combined Armature & Field Current

Figure 2.36 shows the combined armature & field control scheme. With this, the speed can be controlled from zero to maximum value with automatic changeover from armature control to field control and vice

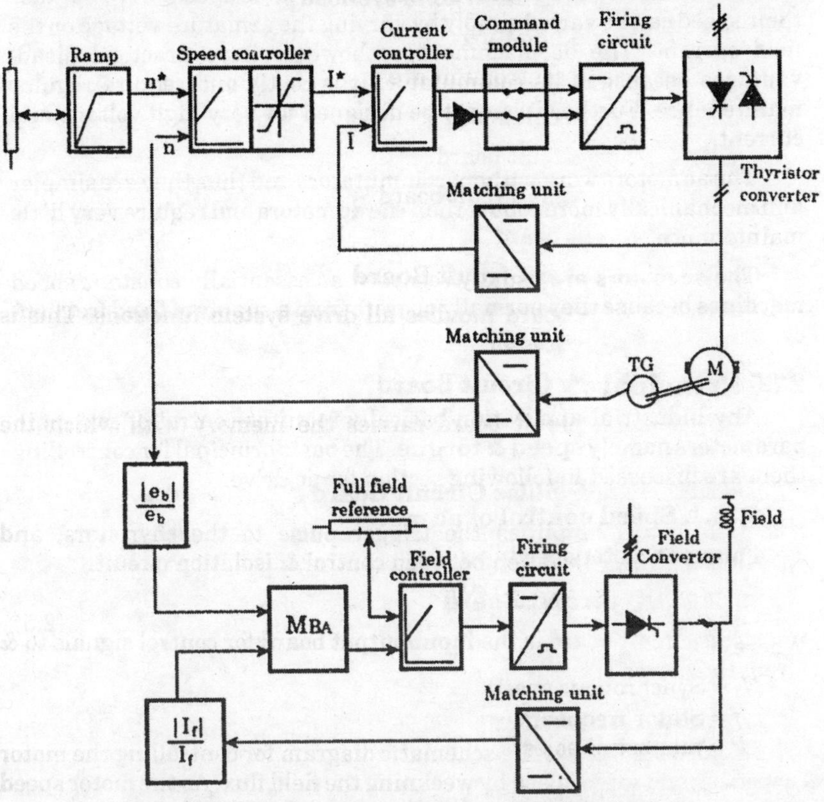

Fig. 2.36. Combined Armature Voltage and Field Control.

versa. Speed from zero to rated value is obtained as usual from armature voltage control. If now attempts are made to increase the speed beyond rated value, motor back emf will tend to increase beyond its rated value because field circuit is sluggish in response due to its large time constant.

In the field circuit two signals are applied to a field controller, one is the normalized value of back emf is less then one and hence feedback to the field controller is maintain corresponding to full field current. But as soon as attempts are made to increase the speed beyond rated value, normalized value of back emf tends to increase beyond one and the feedback value applied to the field regulator is more then one. This forces the field regulator to vary the firing angle to reduce the field current to such a value as to limit the back emf to its normalized value of one. Hence for obtaining speed beyond rated value, field is automatically weakened.

2.34 Thyristorised speed control of AC motors

2.34.1. Introduction

The dc motors are commonly used in variable speed drive because their speed can be varied simply by varying the armature voltage or the field current. The dc machine does however have practical disadvantages because of its commutator. It is costly and require regular maintenance. Further it cannot be designed for very high voltage and current.

The ac motors work without commutators and thus they are simpler and mechanically more robust than the dc motors and require very little maintenance.

The ac motors are usually regarded as essentially constant speed machines because they normally operate from a supply of fixed frequency.

2.35 Principal

Any industrial application basically requires control of two main parameters namely **speed & torque**. The basic principal for controlling them are discussed in following section for ac drive.

2.35.1. Speed control of ac motor

For an ac motor speed

$$N_s = \frac{120\,f}{P} \qquad \text{...(2.5)}$$

Where

N_s = Synchronous speed

f = Stator frequency

P = number of poles

Thus the speed of an ac motor can be controlled by either varying number of poles or stator frequency or both. However it is not possible to have too many different stator winding combinations for having different number of poles. Further the speed can not be varied in a step less manner.

The most commonly and attractive method available is to vary stator frequency. From equation 10.5 it is clear that

$$N_s \propto f \quad ...(2.6)$$

This equation shows that speed can be widely varied by changing stator frequency. DC link inverters & cycto converters are based on this principle.

Another method is by varying slip at which the motor is operating as can be seen from following equation

$$N = N_s (1 - S) \quad ...(2.7)$$

where

N = Speed of ac motor

N_s = Synchronous speed

S = Slip

Stator voltage control, slip power recovery & rotor resistance control methods are based on this principle.

Based on this we can broadly classify various methods of speed control of ac drives under following main areas :

(*a*) Frequency control
(*b*) Slip control.

2.36 Frequency Control

There are two methods for controlling frequency of ac drives.

2.36.1. Cyclo converters

In cyclo converters ac voltage of one frequency is directly converted into a variable voltage variable frequency supply.

2.36.2. DC link inverters

In this type of arrangement, power from main circuit is first converted into dc and then again to ac through an inverter circuit. The name **DC Link Inverter** derived because of presence of the dc link in the circuit. This class of inverters are either of voltage source or of current source type depending on the method of torque control.

In voltage source type the arrangement may generate stepped wave or pulse width output depending on the tyre of design. In the former type frequency is controlled by the inverter circuit, while the voltage is controlled by the rectifier circuit. Several different circuits are possible depending on the type of the commutation circuitary used.

In pulse width modulated circuits, voltage and frequency both are controlled at the same Place.

In current source inverter, current rather than voltage is used as a parameter for the torque control. Current control is effected through rectifier while frequency is controlled through inverter circuit. Generally all type of inverter circuits use forced commutation principle, but load commutation can be used along with synchronous motor. The back emf of motor providing commutation voltage.

The pulse width modulation can also be used with advantage along with current source inverter to reduce torque pulsation. Combination of pulse width modulation and stepped wave CSI is also used with PWM operation restricted to low frequency region.

2.37 Slip control

Slip control is possible using any of the following methods :

2.37.1. Stator voltage control

Figure 2.37 shows basic circuit making use of thyristors for controlling the stator voltage. The method is not only reliable but simple also. Here thyristors are connected back to back and fired symmetrically at identical points of ac wave cycle. The effective voltage applied to motor can be varied from zero to maximum value by controlling the firing angle.

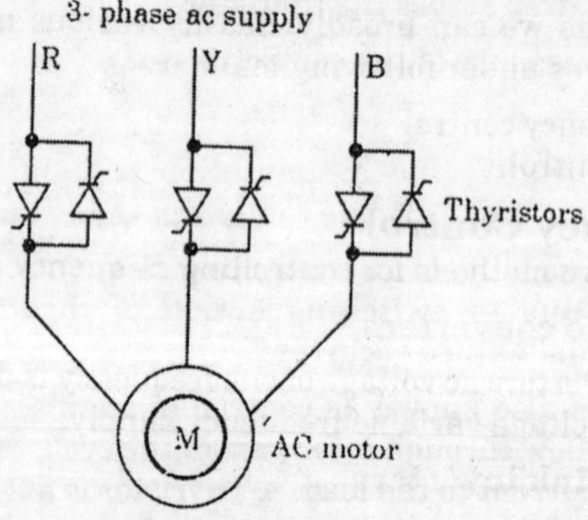

Fig. 2.37

2.37.2. Rotor Voltage Control

Rotor voltage control can be done either through rotor resistance control or through slip power recovery. Rotor resistance control is occasionally used alone because of number of demerits associated with the method.

2.38 Superiority of Static methods of Control

Static methods of speed control of ac motors have tremendous advantage over conventional methods of speed control as discussed below :

1. They provide smooth control of speed.
2. High starting torque can be produced by controlling v/f ratio.
3. Response is very accurate and fast.
4. Static control systems provide better stability.
5. Static control systems are maintenance free.
6. Static control systems are compact.

2.39 Thyristors for speed control

In the field of electric power control, the advent of thyristors capable of handling large currents has been very much appreciated. For variable speed drives in industry, the thyristors controlled drives employing both d.c. and a.c. motors find large applications. Previously saturable reactors were employed in conjunction with high power silicon rectifiers for d.c. drives. Now-a-days, thyristors are largely used for a.c. to d.c. conversion.

Thyristor control has the following advantages and disadvantages.

Advantages

1. Reliable and simple operation
2. The control device requires less space with minimum maintenance, smaller in size, lighter in weight and cheaper in cost.
3. The efficiency of the control system is very high due to low voltage drop across the thyristor.
4. Faster response of the control device.

Disadvantages

1. Motor heating and commutation problems are very high because of higher ripple content in the converter output.
2. More possibility of interference with the communication networks due to switching action of thyristors and non-sinusoidal nature of current.

A Thyristor is also known as controlled rectifier, a device which allows current to flow through it for part of the cycle only and there by controls average current to the load. A Thyristor is a 3-terminal device

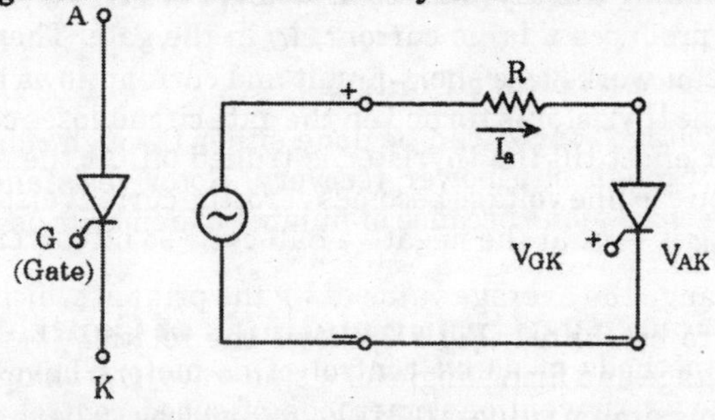

Fig. 2.38

similar to a diode with a third terminal called the **'Gate'** which allows a thyristor to be turned on under separate control as shown in Fig 2.38(*a*).

Thyristor which is connected in a circuit is shown in Fig 2.38 (*b*). If V_{GK} becomes slightly $+ve$ such that the gate current exceeds a little value of the order of several milli-amperes and λV_{AK} is $+ve$, the thyristor will turn on suddenly and remains as long as V_{AK} is $+ve$ and current flows from *A* to *K*. Now if V_{GK} is $-ve$, thyristor works in open circuit condition in all conditions.

For making the thyristor off, a reverse voltage is to be applied from *K* to *A* or allows the current to fall to zero naturally. Fig. 2.39 (*a*) shows the action in a thyristor controlled circuit.

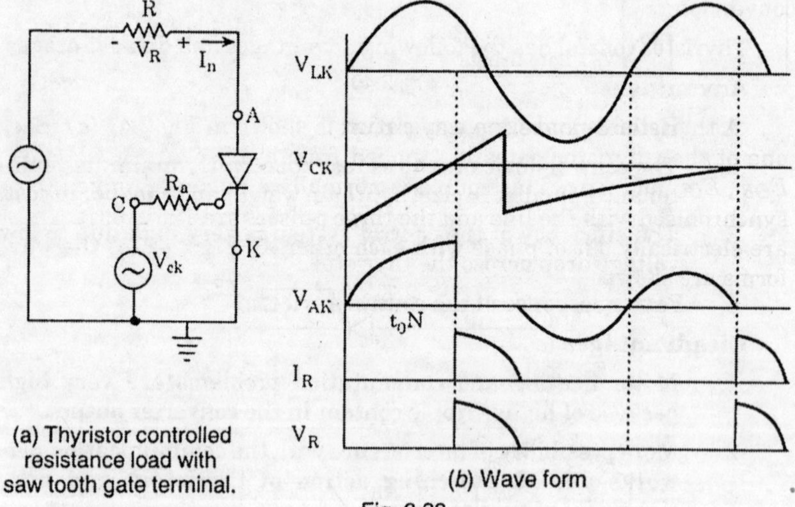

(a) Thyristor controlled resistance load with saw tooth gate terminal.

(*b*) Wave form

Fig. 2.39

For producing I_G, V_{CK} is applied to the gate. No current flows in the thyristor when V_{CK} is $-ve$ although V_{AK} is $+ve$.

For making the thyristor conducting, at to *N*, V_{CK} will be slightly $+ve$ and produces a large current, I_G in the gate. Then the cathode-anode circuit works as a short circuit and current flows in the resistor. Suppose the thyristor is turned on the gate circuit loses control and has no further effect till the thyristor is truned off. At the end of the half cycle, when the line voltage becomes zero, the current also becomes zero. As the V_{AK} is $-ve$ at the negative half cycle so no current will flow.

To change the average value of V_R the print at which V_{CK} becomes $+ve$ has to be shifted. Fig 2.3 shows the relation between average output voltage and firing angle for a sine wave. By taking the half wave form, the maximum value is $0.314V_{max}$. which can be decreased to zero by delaying the angle of firing.

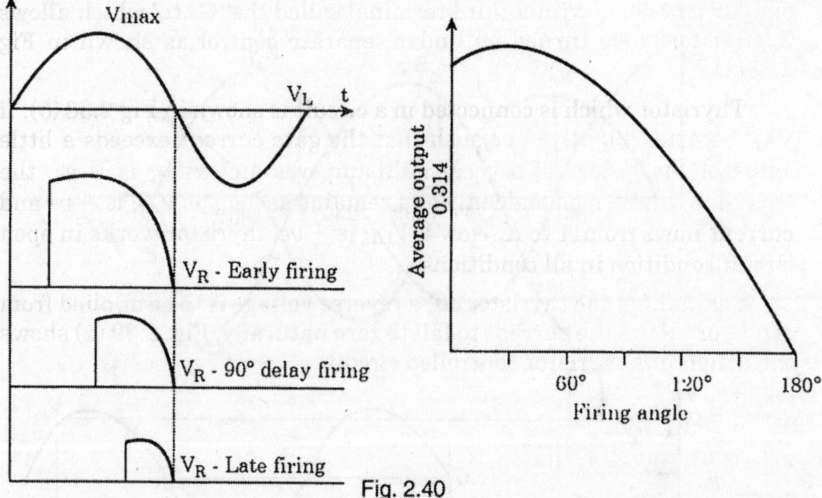

Fig. 2.40

A thyristor 3-phase one way circuit is shown in Fig 2.41 (*a*). Each one of the thyristor gates is supplied from a different signal source E_{GB}, E_{GC} and E_{GA}. This source is normally a pulser arranged to be synchronised with the line and the three pulsers are 120° apart. These are electrically out of phase with each other. In Fig 2.41 (*b*) the wave forms are shown.

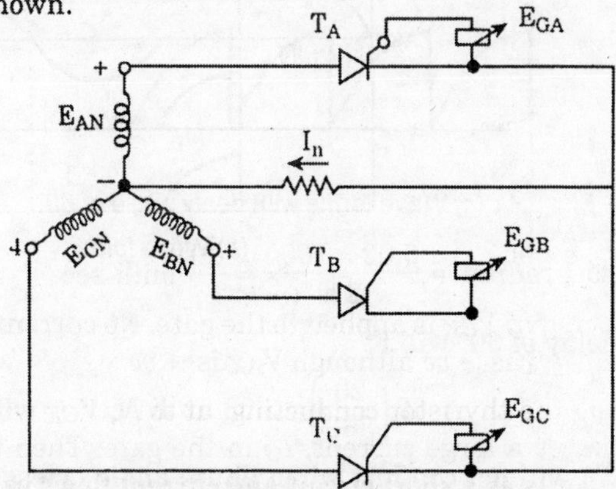

Fig. 2.41 (*a*) Thyristor 3 phase one way out

On cycle is $\frac{1}{50}$ sec. for 50Hz circuits.

$$\therefore \quad 1 \text{ cycle} = 2\pi \text{ radians}$$

$$\pi = 360° = \frac{1}{50} = 0.02 \text{ sec.}$$

Hence $$1° = \frac{0.02}{360} \text{ sec} = \frac{1}{18} \text{ milli-sec}$$

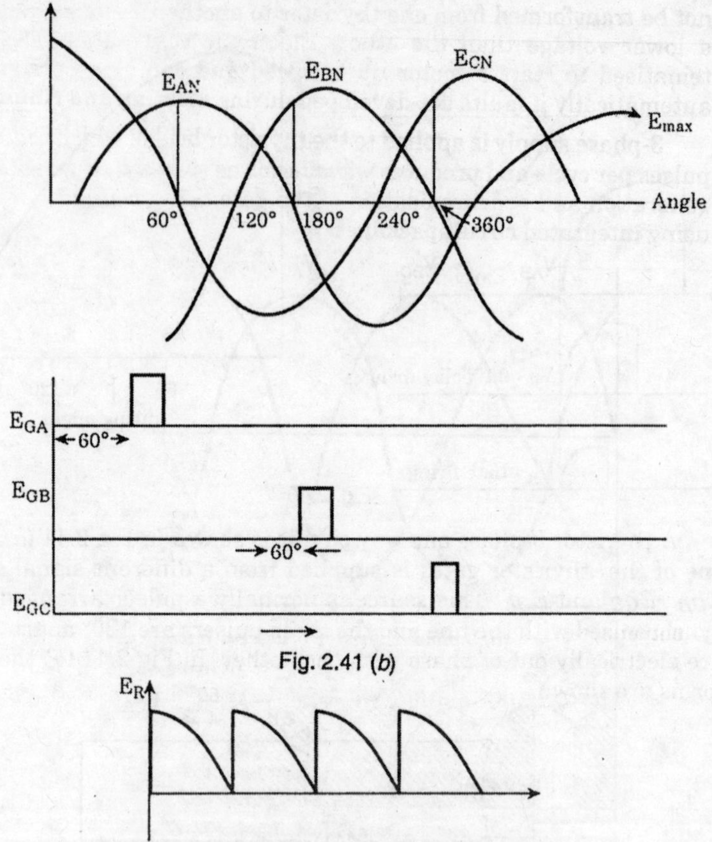

Fig. 2.41 (*b*)

Fig. 2.41 (*c*) Wave forms with dealy angle of 60°

$$\text{Also 1 radian} = \frac{0.02 \times 1000}{2\,\pi} = \frac{20}{2\,\pi} \text{ milli-sec}$$

Hence a delay of 60° will be

$$= 60\left(\frac{20}{360}\right) = 3.33 \text{ milli-sec}$$

Tyristor T_A will not conduct when phase angle is less than 60° as it is not fired. E_{GA} produces a pulse at 60° which turns T_A off; when it starts conducting voltage E_{AN} which applies itself to these resistor. After that conduction extends to 180° as there is no voltage present to turn E_{AN} off. Second time T_B is truned on till the phase angle is 180°. Across the load the voltage wave form is a set of sine waves with different amplitude but with a reduced d.c. component. The minimum phase angle can be reduced to 30° using a pulse to trigger the ckt. For example we apply a pulse to trigger the phase A say at 20°, $E_{CN} > E_{AN}$ and hence E_{AN} will not turn on. When $E_{AN} > E_{CN}$ the pulse has gone and E_{AN} can not be applied to the load. But the current can not be transformed from one thyristor to another if one thyristor is at a lower voltage than the other. Electronic controllers can be systematised to start a motor up to speed and can also be shut down automatically if faults are developed during starting and running.

3-phase supply is applied to the thyristor bridge which requires six pulses per cycle and produces wave forms as shown in Fig 2.42 (*b*). The generation and synchronisation of the gate firing signals can be done using integrated circuit packages.

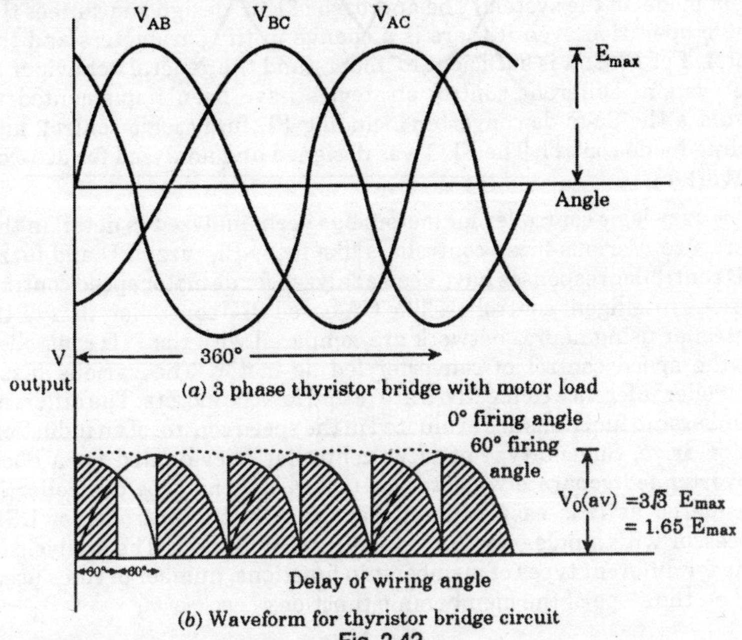

(*a*) 3 phase thyristor bridge with motor load

(*b*) Waveform for thyristor bridge circuit

Fig. 2.42

2.40 Fuzzy Controlled Chopper Fed d.c. Drives

Introduction : A sugeno type fuzzy controller for closed loop control of d.c. drive fed by buck type d.c. to d.c. converter is designed and considered for evaluation. The controller is used to change the duty cycle of the converter and thereby, the voltage fed to the armature of the separately excited d.c. motor to regulate the speed.

Here let, B : friciton coefficient of the motor; i_a : armature current; J: moment of inertia of the motor; K : torque constant of the motor or motor back emf. constant; L_a: armature inductance; R_a: armature resistance. T_L: load torque applied; V_s : the supply voltge; ω : motor angular speed

The dc motor drives have the advantage of high controllability and are used in many applications, such as, robotic manipulators, position control, steel mining, paper and textile industries. In some industrial applications, the dynamic response of drives is bounded by certain limitations, such as, transient time and steady state error. General PI and PID controllers are widely used for motor control applications. But it does not give satisfactory results when control parameters, loading conditions and the motor itself are changed.

In recent years, the AI techniques, such as, fuzzy logic controller (FLC) have shown high potential for motor control applications. Fuzzy logic controller has the advantage that it can be designed without the exact model of the system. The approach of FLC design guarantees the stable operation even if there is a change in the parameters and the motor. For FLC, it is sufficient to understand the general behaviour of the system. Different control strategies have been implemented to regulate the dc-to-dc converter including PI, fuzzy logic control, and sliding mode control. The FLC was designed and analyzed for dc-to dc converters.

Fuzzy logic controller for motors has been analyzed in detail in the literature. Various fuzzy controllers like fuzzy PI, fuzzy PD and fuzzy PID controller responses have been analyzed for dc motor speed control. Various intelligent controllers like GA based PID controller, fuzzy PID controller using neural network are compared with the PID controller for the speed control of converter fed dc motor. The various fuzzy controller inference connectives are tested for a dc motors. The different membership funcitons are evaluated in the speed control of an induction motor drive. Sugeno type of fuzzy controller is evaluated for a buck converter fed separately excited dc motor. Sugeno type controller is considered as it is easy to implement in a micro-controller of DSP processor with a pulse width modulating (PWM) unit. The analysis is done for different types of membership functions, number of rules used and on the slope of the membership functions.

Proposed System

Figure 2.43 shows the block diagram of the complete proposed system. The system consists of buck converter type dc-to-dc converter for driving the separately excited motor. Any processor with an in-built PWM unit can be used to implement the fuzzy logic. The PWM waveform generated is used to switch the dc-to-dc converter.

The designed closed loop control has two loops. One is outer fuzzy speed control loop and another one is inner current control. In outer speed control loop, the speed is fed back and is compared with set speed. After comparison, error signal and the change in error are calculated and are given as input to fuzzy controller. In this work, the error is normalized to per unit value with respect to the set speed. This helps in using the fuzzy controller for any set speed. The fuzzy controller will attempt to reduce the error to zero by changing duty cycle of switching signal.

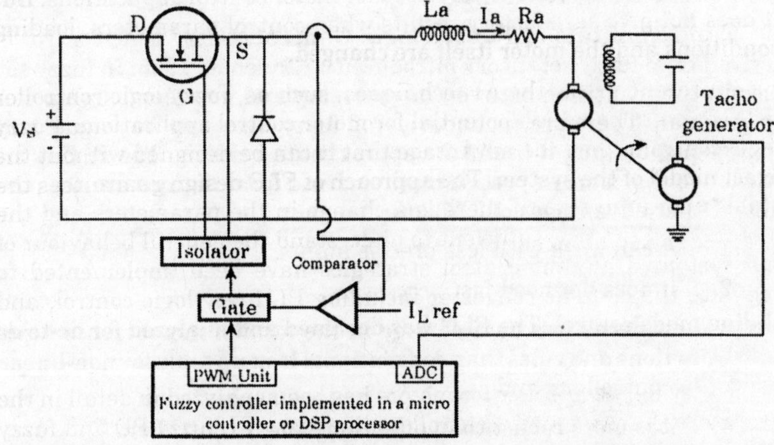

Fig. 2.43 Block diagram of the proposed system.

The designed fuzzy controller with an inner current control has the advantage of being a universal controller so that it can be used for any motor with any specification. Changing $I_{L\ ref}$ can change the controller to work for motor of any current rating. The controller is designed to work with buck converter type dc/dc converter and so the speed can be controlled over a wide range from zero to the rated speed.

MATLAB/Simulink model of the dc motor and the dc-to-dc converter was developed and simulated. The fuzzy controller was designed by using the fuzzy logic toolbox and then the closed loop operation was simulated and evaluated.

Mathematical Model Of Dc Motor And Buck Converter

This simulation of the entire set up was done using equation models of the motor and buck converter. The dc motor powered from dc-to-dc converter has been modelled with the following modelling equations. Mode 1 is when the MOSEFET switch is ON and Mode 2 is when the MOSFET switch is OFF.

When the MOSEFET switch is ON,

$$V_s = R_a\, i_a + L_a \frac{di_a}{dt} + K\,\omega \qquad ...(1)$$

$$Ki_a = J\,\frac{d\omega}{dt} + B\omega + T_L \qquad ...(2)$$

When the MOSFET switch is OFF.

$$0 = R_a\, i_a + L_a \frac{di_a}{dt} + K\,\omega \qquad ...(3)$$

$$Ki_a = J\frac{d\omega}{dt} + B\omega + T_L \quad ...(4)$$

FUZZY CONTROLLER

Fundamentals of Fuzzy Logic Control. Fuzzy logic control is derived from fuzzy set theory introduced by Zadeh in 1965. In fuzzy set theory, the transition between membership and non-membership can be gradual. Therefore, boundaries of fuzzy sets can be vague and ambiguous, making it useful for approximate systems. The FLC is an attractive choice when precise mathematical formulations are not possible. Other advantages of FLC are

1. it can work with less precise inputs;
2. it does not need fast processors;
3. it needs less data storage in the form of membership functions and rules than conventional look up table for non-linear controllers; and
4. it is more robust than other non-linear controllers.

Fuzzification

Fuzzy logic uses linguistic variables instead of numerical variables. The process of converting a numerical variable (real number or crisp variable) into a linguistic variable (fuzzy number) is called fuzzification. The simplesft form of membership function is triangular membership function and it is used here as the reference. As Sugeno type of implication is considered, the singleton membership function is used for the output variable namely the change in duty cycle. Error is normalized by dividing the actual error in speed with the reference speed. The normalization is useful for using the fuzzy controller for any speed reference.

Defuzzification

The reverse of fuzzification is called defuzzification. The use of FLC inference engine produces requied output in a linguistic form. According to real world requirements, the linguistic variables have to be transformed to crisp output. Weighted average method is the best well-known defuzzification method for Sugeno type fuzzy controller and is used in this work. This method has the advantage of easy implementation in micro-controllers and toher embedded processors.

COMPUTER SIMULATION AND RESULTS

The simulation of buck converter fed dc motor is done based on equation modelling technique in MATLAB/ Simulink toolbox. The Simulink model developed is given in Figure 2.44.

The fuzzy toolbox is used to test in evaluating the fuzzy controller proposed. The parameters of the dc motor used are given in the Table 2.1. The computer simulation is run for a step change in motor reference speed and the actual change in speed is recorded. The step change in load torque is also applied and the corresponding change in the speed is recorded. The set speed is initially 750 rpm and changed to rated speed of 1500 rpm at 2 *s* with no load. The load is initially 0 and then changed to full load at 4 *s* and removed at 6 *s*. The simulated results are given in Figure 2.45.

Table 2.1. The dc motor parameters

dc Motor Parameters	*Value*
The dc supply voltage, *V*	220
Armature resistance (R_a), ω	2.25
Armature inductance (L_a) , *m H*	46.5
Inertia constant (*J*), Nm^2	0.07
Damping constant (*B*)	0.002
Torque constant (*Kt*), *Nm/rad/s*	1.1
Back emf constant	1.1
Speed, rpm	1500

The membership functions used for error and change in error are changed and simulated. The simulated response of the fuzzy controller for set reference with a triangular membership funtion is given in Figure 2.45 (*a*). The results of using seven linguistic variables and 49 rules are given in solid line. The dashed lines correspond to the five linguistic variables and 25 rules. The transient response is satisfactory without any oscillations and minimum settling time. The steady state error is also at its minimum in using the triangular membership functions. It can be readily seen that the error is minimum if seven linguistic sets are used compared to five linguistic sets.

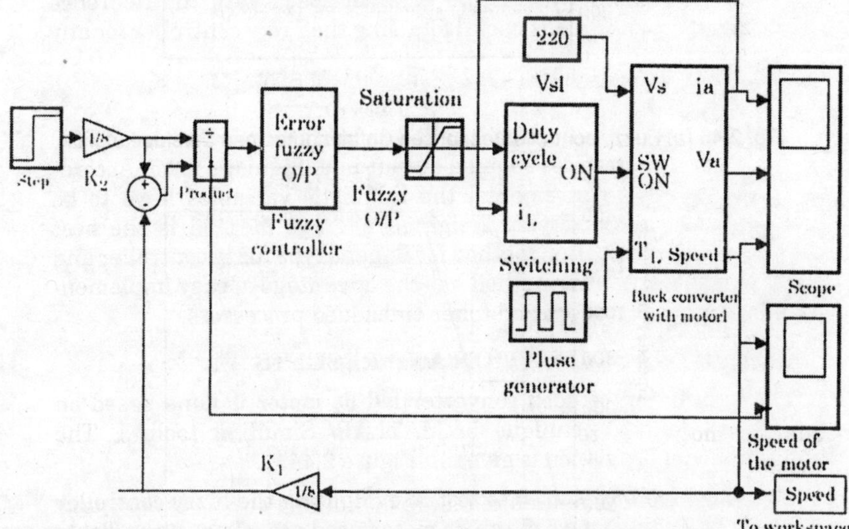

Fig. 2.44 Simulink model of the proposed system.

Figures 2.45 (*b*) to 2.45 (*e*) show the results of using trapezoidal, Gaussian, bell shaped and PI membership functions, respectively. The transient response is almost the same for the given set of rules irrespective of the membership functions used. But the steady state error is increased for other membership functions compared to the triangular membership function. It can be seen from the response curves that the steady state error is reduced for memebreship funcions with the reduced value of the nucleus. For triangular membership function, the nucle us is a single point and is the height of the function. For membership functions with a wide nucleus, such as, PI and trapezoidal membership functions, the steady state error is very high.

The triangular membership function is found to be more suitable for chopper fed dc drive in reducing the steady state error. The error is reduced if seven linguistic sets aree used compared to five linguistic sets.

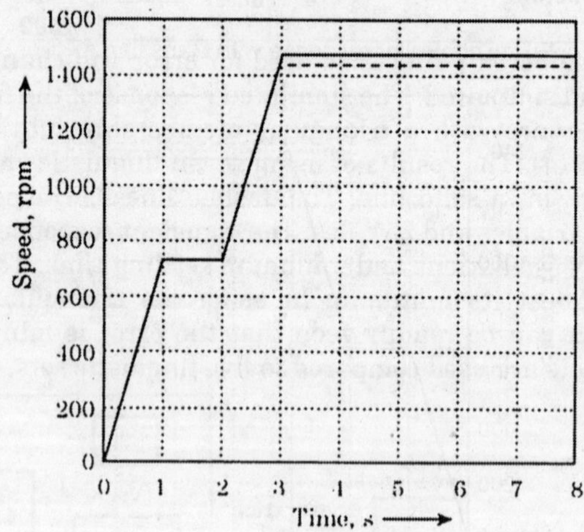

Fig. 2.45 (*a*) Fuzzy controller response with triangular membership function

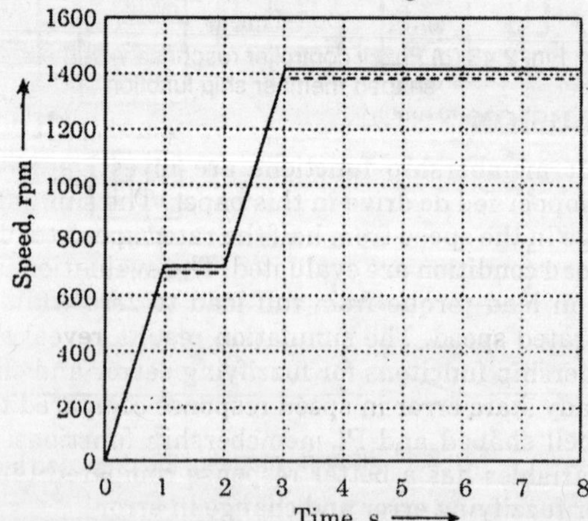

Fig. 2.45 (*b*) Fuzzy controller response with trapezoidal membership function.

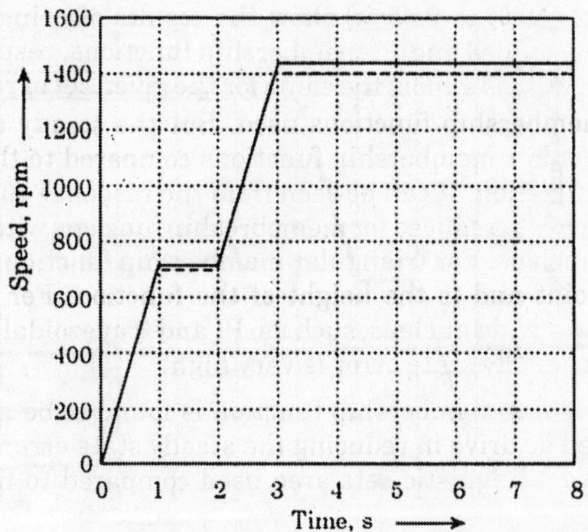

Fig. 2.45 (*c*) Fuzzy controller response with Gaussian membeship function.

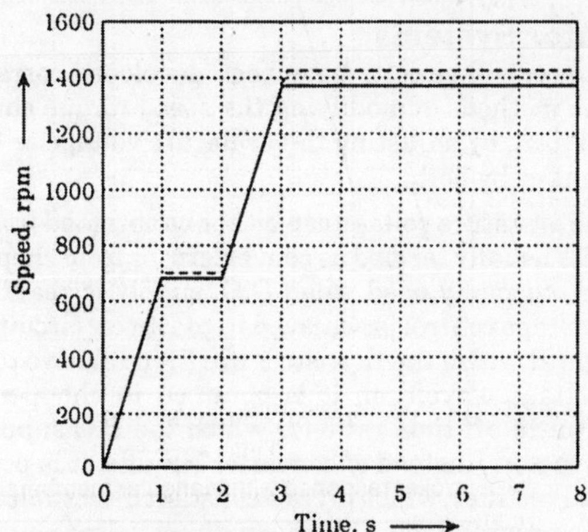

Fig. 2.45 (*d*) Fuzzy controller response with the bell shaped member ship function

CONCLUSION

Different membership functions are investigated for the fuzzy control of chopper fed dc drive in this paper. The simulated results for a step change in the speed from half the rated speed to full rated speed nunder no load condition are evaluated. The evaluation is also done for the change in load torque from full load to zero while the motor is running at rated speed. The simulation results reveal that the triangular membership funcitons for fuzzifying eerror and change in error reduces steady state error in speed response compared to trapezoidal, Gaussian, bell shaped and PI memebership functions. Use of seven linguistic variables has a better response compared to five linguistic variables for fuzzifying error and change in error.

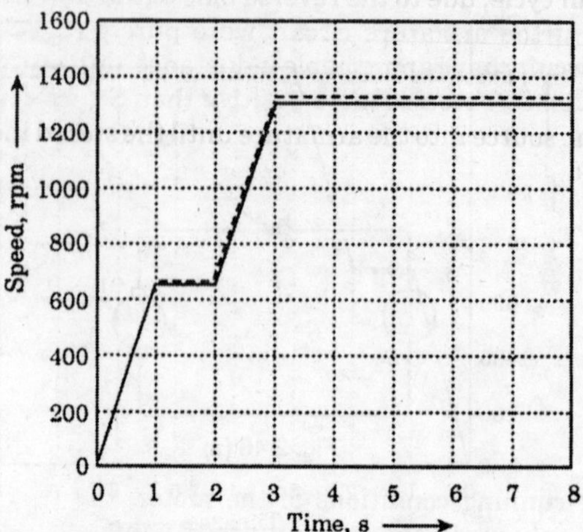

Fig. 2.45 (*e*) Fuzzy controller response with the PI member ship function

2.41 DC Motor Systems

Solid state control circuits have been developed corresponding to each one of the methods of modifying the speed torque characteristics of a D.C motor, *viz.*, by adjusting the armature voltage or by adjusting the field current or by both.

Adjustable armature voltage can either be obtained from controlled rectifier circuits usually termed as converters] or from chopper circuits. The latter are, normally used when D.C. supply is readily available. White motor voltage control is achieved in converter circuits by varying the phase angle at which the thyristors are fired relative to the applied alternating voltage waveform, it is obtained in chopper circuits by changing the **on** to **off** time ratio for which the D.C supply voltage is applied to the motor. Instead of converter circuits, it is possible to use an uncontrolled rectifier, which gives a constant direct voltage, followed by a chopper to provide a variable mean direct voltage output.

Variation in field current can be obtained by supplying the field winding from an auxiliary controlled rectifier circuit.

2.42 Controlled Rectifier Circuits

Rectifier circuits are classified according to the number of output voltage pulses in a period of the A.C. line power frequency. The simplest converter is of $m = 1$, where m equals the number of voltage pulses per cycle of the supply voltage. The basic forms of these converters are shown in Figs. 2.46 - 2.48. Fig. 2.46 (*a*) shows a dc motor fed by such a converter. During standstill conditions of the motor, when S_1 is more positive than S_2, the thyristor TH may be gated on producing the output voltage and current waveforms shown in Fig. 2.46(*b*). That is during the positive half cycle of the supply voltage waveform, the current flows from the source into the motor armature circuit. At the start of the negative half cycle, due to the reverse bias to the thyristor, the current would stop if the armature circuit were purely resistive. Since the armature circuits have appreciable inductance, during the negative half cycle although S_2 becomes more positive than S_1, current continues to flow from the source into the armature until the stored inductive energy is dissipated.

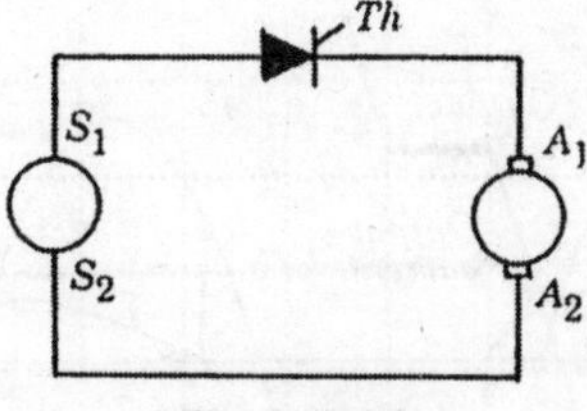

Fig. 2.46 (*a*)

During running conditions of the motor, the thyristor can start conducting only when the instantaneous value of the source voltage becomes equal to or exceeds the induced e.m.f. in the armature and when it receives a gating pulse. The thyristor reaches its blocking state when the armature inductance, has discharged its stored energy and the current falls to zero, as during standstill conditions of the motor. The presence of the induced e.m.f, however, causes a different output voltage waveform [Fig. 2.47 (*c*)].

If the motor were driven by a load connected to its shaft such that terminal A_2 is positive with respect to terminal A_1, power would flow from the motor armature to the **ac** source. This mode of operation is known as inverting. If during operation the **ac** supply should fail [indicated by opening the switch *S* in Fig. 2.47(*d*), there would be no **ac** source to commutate the **dc** current. That is, due to the small circuit impedance; when *S* opens, the armature current continues to flow and increase until the fuse or the thyristor opens or the motor fails. Another possibility of failure during inverting operation exists when the magnitude of the induced e.m.f of the motor becomes too large with respect to the **ac** line voltage as shown in Fig. 2.47 (*e*). The thyristor begins to conduct at point *A*. At point *B* the ac line voltage and the induced e.m.f impress a negative anode voltage on the thyristor, but due to the inductance of the motor armature the current continues to flow. At point *C* the **ac** line voltage and the induced e.m.f produce a positive anode voltage and the thyristor continues to conduct. The armature current continues to increase during successive half-cycle periods, until a circuit failure occurs.

Fig. 2.48 (*a*) shows another form of a single pulse converter, having a free-wheeling diode. This circuit, because of the presence of the freewheeling diode, cannot feed energy back into the **ac** line. The corresponding output voltage and current waveforms are shown in Fig. 2.48 (*b*). During the negative half cycle, the energy stored in the armature inductance flows from the armature through the freewheeling diode at negligibly small voltage. Hence, the conduction of current during the negative half-cycle is for a much longer period than in Fig. 2.47(*c*). The form factor (ratio of r.m.s to average value) of the armature current is lower than in Fig. 2.47(*c*).

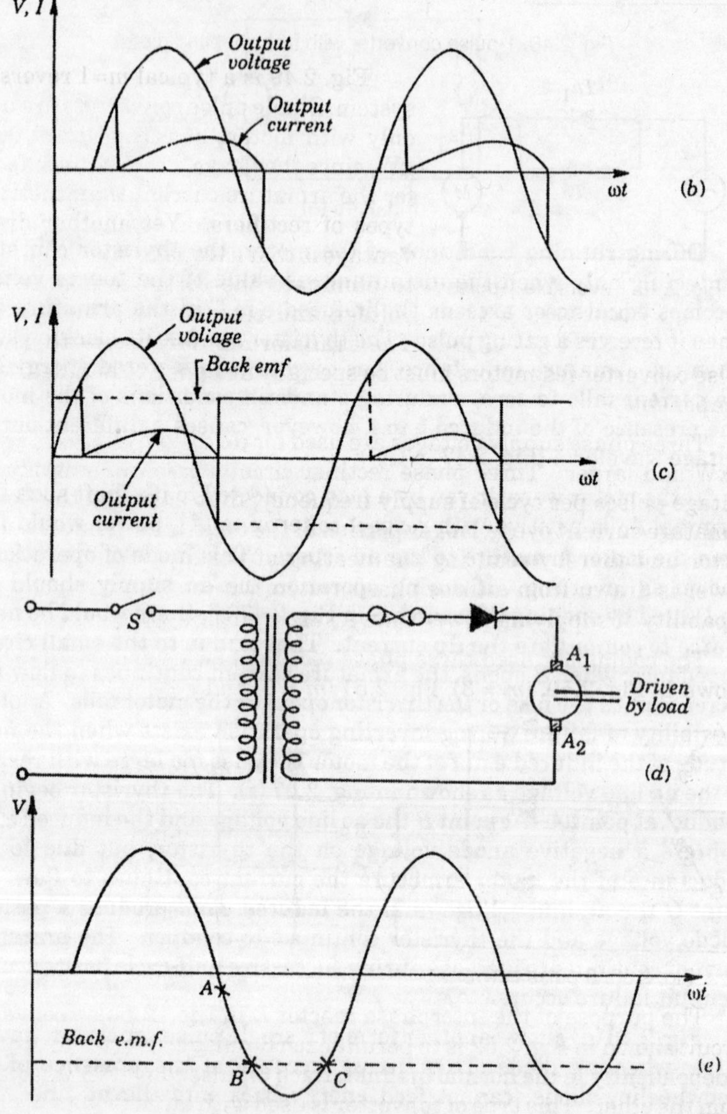

Fig. 2.47. Operation of 1-pulse Converter.

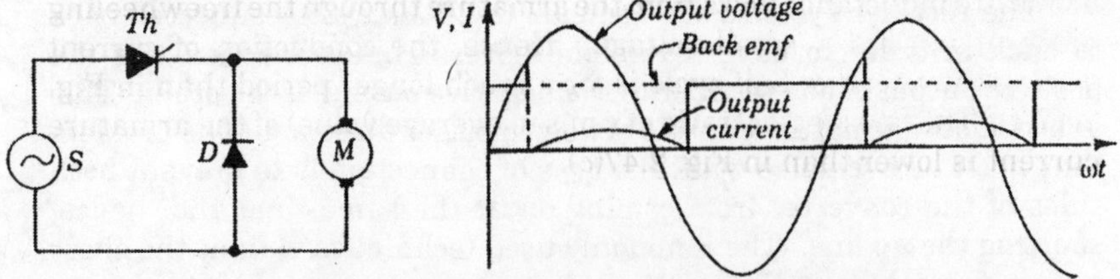

Fig. 2.48. 1-pulse converter with freewheeling diode.

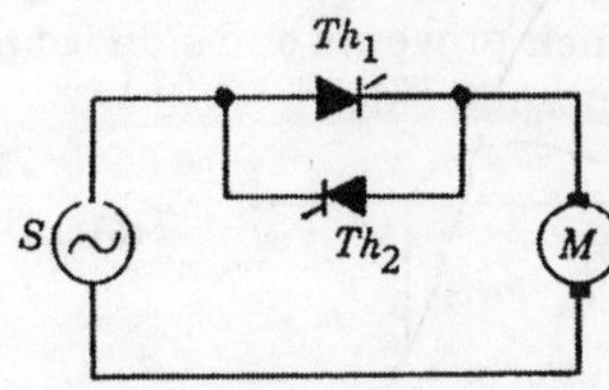

Fig. 2.49. Feversible 1-pulse converter

Fig. 2.49 is a typical m=1 reversing system. Single pulse converters are used only with motors of rating less than 3 kW, since they have a higher form factor for the armature current than the other types of rectifiers. Yet another disadvantage is that a transformer is required to eliminate the direct components of voltage and current from the **ac** line. The transformer for use with single pulse converter fed motors must be specially designed to handle the **dc** component.

Three phase supply voltages are used for **dc** drive systems of about 5 kW and larger. Three phase rectifier circuits give more number of voltage pulses per cycle of supply frequency, thus ensuring the flow of armature current over a longer portion of the cycle, increasing the form factor and thereby reducing the heating of the armature. Also, the power is drawn from a three phase system which generally has more capability of supplying power than a single phase system.

The commonly used three phase rectifiers for **dc** motor drives are shown in Fig. 2.50. (m = 3), Fig. 2.51 (m = 6) and Fig. 2.52 (m = 12).

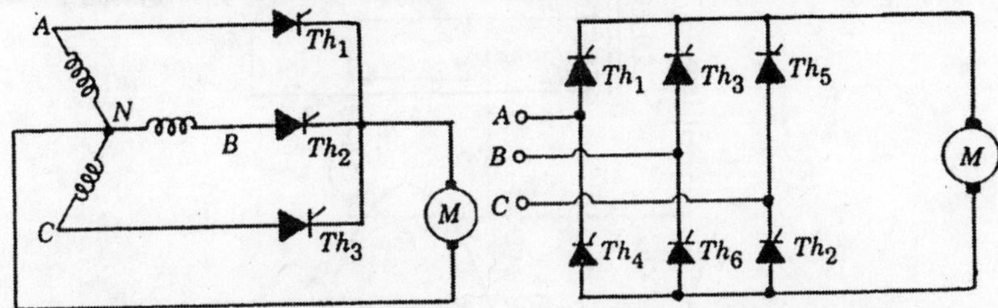

Fig. 2.50. 3-pulse converter.

Fig. 2.51. 6-pulse converter.

The purpose of the interphase reactor L in the 12 pulse converter circuit shown in Fig. 2.52 is to permit each 6-pulse converter to operate independently in the normal manner. Each 6-pulse converter is isolated from the other. This type of converter is used with large motors in order to reduce the current ripple to a value less than that obtained with a 6-pulse converter.

The rectifier systems of Figs. 2.50 to 2.52 may be connected back to back in order to have reversible drives. This connection ties the positive output of one rectifier to the negative output of the other. This allows motor armature current to flow in either direction. The problem that must be overcome with this type of connection is to prevent both sides of the converter from coming on at the same time and, hence, shorting the **ac** line. The commonly used techniques to limit the short circuit current are : (*i*) insertion of a reactor in the armature circuit, (*ii*) provide time delay before turning on the opposite rectifier and (*iii*) introduction of a current sensing device which prevents one side from turning on, if current were flowing in the opposite side.

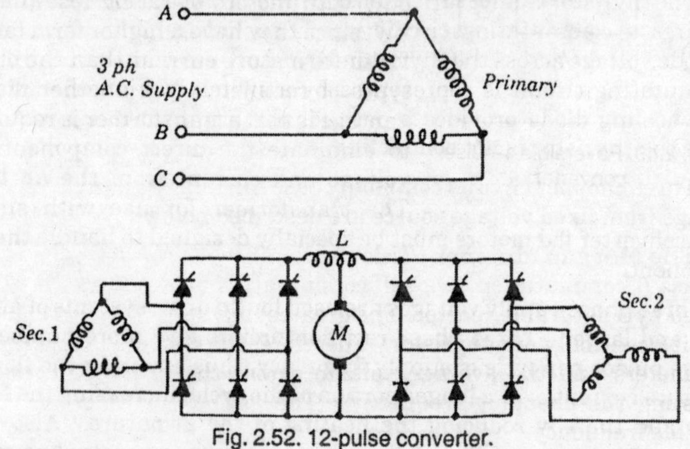

Fig. 2.52. 12-pulse converter.

2.43 Chopper Circuits

Chopper circuits are used to control the speed of **dc** motors which are fed from fixed voltage sources like batteries or uncontrolled rectifiers. Fig. 2.53 (*a*) shows the basic chopper circuit that supplies

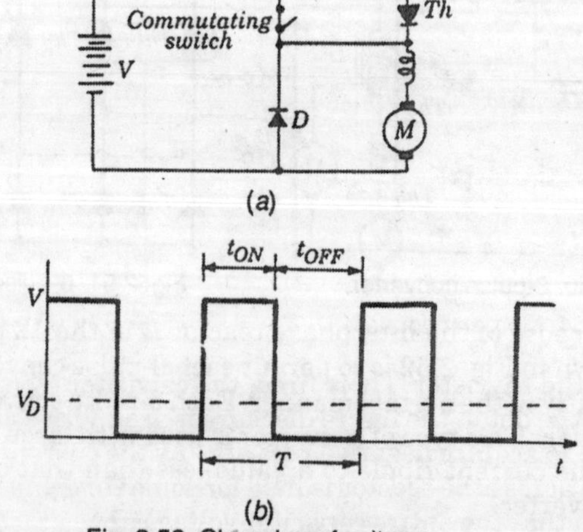

Fig. 2.53. Chopping of voltage.

variable direct voltage to a series motor. The input to the chopper is a fixed direct voltage, V. The thyristor acts as a switch which is made on and off at the rate of several hundred hertz. The relative on-to-off time ratio of the thyristor determines the mean output voltage. Variable output voltage can be obtained by varying the on-to-off time ratio (sometimes called mark-space ratio) either by adjusting the frequency of switching, keeping on-time constant or by adjusting on-time, keeping the frequency of switching constant or by adjusting both on-time and frequency of switching. The average output voltage V_0 is equal to $V - t_{on} / (t_{on} + t_{off})$ as indicated in Fig. 2.53(*b*).

The thyristor cannot turn itself off while carrying current and hence requires a commutating circuit which applies a negative anode to cathode voltage across the thyristor for a short-period to turn it off. The commutating circuit is represented by a switch in Fig. 2.53 (*a*). The freewheeling diode provides a path for the armature current to flow while the thyristor is not conducting.

Three chopper circuits commonly used for obtaining variable direct voltage from fixed voltage source are described below :

The Morgan chopper : This circuit, shown in Fig. 2.54, is typical of class *B* commutation, *i.e.*, self-commutation by a resonating circuit helped by a saturable reactor. The greatest advantage of this circuit is that it needs only one thyristor. Therefore, the on-time is fixed by *LC* parameters and the average voltage across the motor is varied by adjusting the chopping frequency. The gating oscillator generates variable frequency.

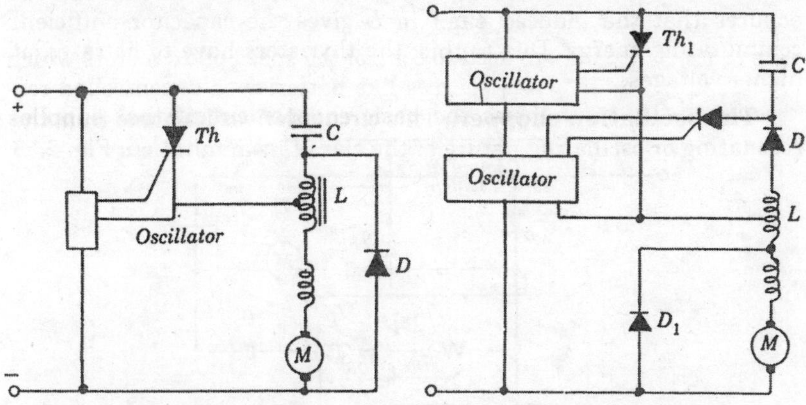

Fig. 2.54. Morgan chopper.

Fig. 2.55. Jones chopper.

When the thyristor 'Th' is fired, the capacitor *C*, having a positive polarity at the dot in the figure discharges around the circuit formed by *C*-Th and *L* to acquire negative polarity. As the current again reverses, the voltage across the reactor is held for some time and then saturation occurs such that the entire capacitor voltage appears across the thyristor. The thyristor gets reverse biased and if the discharge current is more than the thyristor current, it turns off. The capacitor continues to carry load current until it charges up fully with the dot positive again. The freewheeling diode offers a path to dissipate further stored energy and if the energy is dissipated before the thyristor is turned on again, the motor will coast.

The Jones chopper : Fig. 2.55 shows the circuit, which is characterized by class *D* commutation. A charged capacitor *C*, switched by an auxiliary thyristor Th_2 and the auto transformer *T* constitute the commutating circuit. Although, in principle, both the on-time and off-time can be varied due to the presence of Th_2, usually, as indicated in the figure, the off-time commutating period *T* is varied by means of the gating oscillator of Th_1, keeping on-time fixed by means of the oscillator of Th_2.

When thyristor Th_1 is fired, the capacitor charged positively at the dot discharges around the circuit formed by *C*-Th_1-*L* and *D* and reverses polarity. The diode *D* prevents further oscillation of the resonating *LC* circuit. Hence, the capacitor retains its charge until Th_2 switched on. Then, the discharge causes Th_1 to be reverse biased and turns it off. The capacitor charges up with the dot positive again and Th_2 turns off, since the current through it falls below the holding value when the capacitor is recharged. The process repeats itself when, Th_1 is switched on again.

Even if the capacitor had not charged fully by the time Th_1 was switched on again, no damage would be done, since the load current ensures that the induced e.m.f in *L* gives the capacitor sufficient commutating energy. Due to this, the thyristors have to be rated at higher voltages.

The oscillation chopper : This circuit derives its name from the resonating or oscillating nature of the class *D* commutation. Fig. 2.56

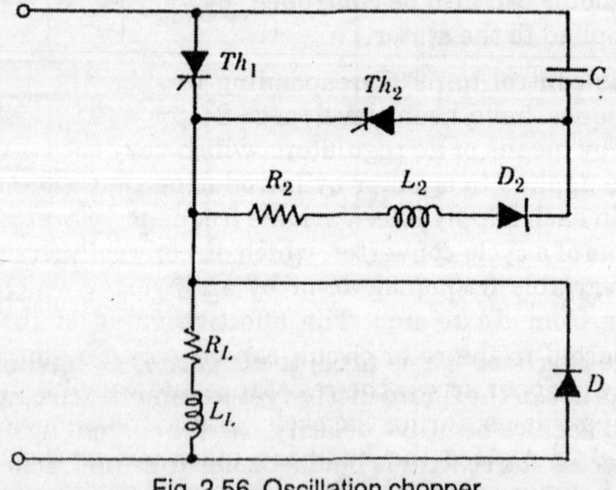

Fig. 2.56. Oscillation chopper.

shows the basic circuit, which differs from Morgan and Jones Chopper in that there is neither a saturable reactor nor an auto-transformer in the load circuit.

Th_2 must be triggered first so that C may get fully charged with the dot positive. Then, when Th_1 is switched on, so that load current can flow, C reverses polarity through the resonating circuit $C - Th_1 - R_2 - L_2$ and D_2. The capacitor remains in that charged condition because of D_2 until Th_2 is fired. This discharges C, causes Th_1 to be reverse biased and turns it off.

In choosing one power control circuit over the other, the following points should be borne in mind :

1. Power supply : (*a*) **dc** (*b*) **ac**-single phase or three phase.
2. Form factor for the motor : (*a*) rectifier type (*b*) need of a choke.
3. Need for inversion.
4. Need for reversed.
5. Necessity of a transformer either because of **ac** line voltage available or rectifier used.
6. System cost : (*a*) transformer (*b*) choke (*c*) rectifier (*d*) motor.
7. Electrical energy costs : (*a*) rectifier power factor (*b*) rectifier efficiency.
8. Amount of **ac** line pollution caused by the rectifier.

2.44 AC Motor Systems

The speed-torque characteristics of an induction motor can be controlled by adjusting the stator voltage, by adjusting the frequency of the source, by adjusting the resistance of the rotor circuit and by injecting slip frequency e.m.f's to the rotor circuit. The speed of a synchronous motor can also be controlled by varying the frequency of the voltage applied to the stator.

Solid state control units corresponding to each one of the above control techniques have been developed. Stator voltage variation is accomplished by means of **ac** regulators, which vary the r.m.s value of the **ac** voltage applied to a motor by introducing thyristors connected back-to-back in each supply line. Variable frequency power is obtained either by means of a cyclo-converter, which directly converts **ac** of fixed frequency to variable frequency **ac** or by an inverter which converts electric power from **dc** to **ac**. The effective value of the external resistance inserted in the rotor circuit can be varied by introducing a high frequency chopper across the resistance and varying the time for which the chopper is on during the cycle. Static frequency converters are being used to replace the auxiliary machines in the Scherbius system. The Kramer scheme has also been altered by using a diode bridge rectifier in place of the rotary converter, but a **dc** motor is still required to convert the rectified slip power to mechanical power.

2.45 AC Regulators

AC regulators are used to control the r.m.s value of the **ac** voltage applied to motor. There are two methods of control : on off control and phase control.

In **on-off control**, the thyristors are employed as switches to connect the motor to the source for a few cycles of the source voltage and then to disconnect it for a comparable period. The thyristor, thus, acts as a high speed contactor. This method is known as integral cycle control.

In **phase control**, the thyristors are used as switches to connect the motor to the supply for a chosen portion of each cycle of the supply voltage. The power circuit configurations for on-off control and phase control do not differ in any manner.

Smooth variation of three phase **ac** voltage can be realized by means of different configurations of the power circuit. Circuits of three phase half wave and full wave **ac** regulators for delta connected motors or star connected motors in which the neutral point is not inaccessible are shown in Fig. 2.57 (*a*) and (*b*). Either regulator shown in the figure

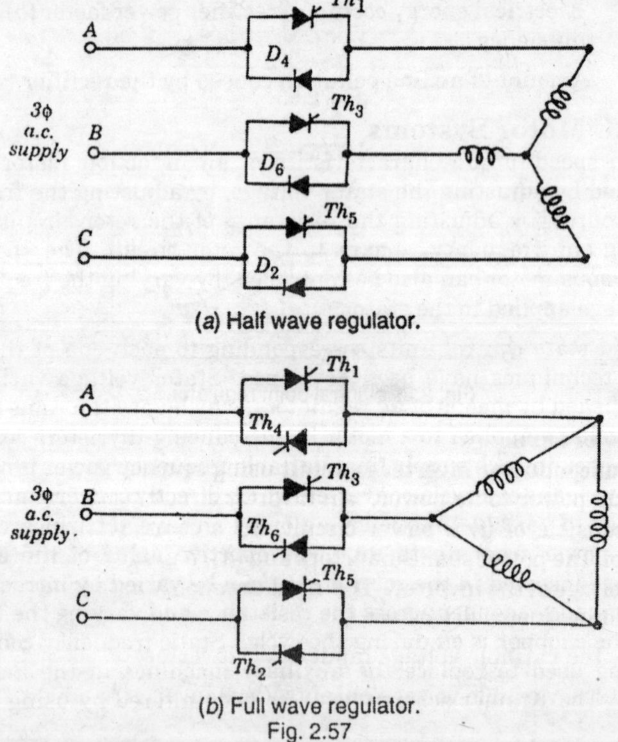

(*a*) Half wave regulator.

(*b*) Full wave regulator.

Fig. 2.57

could, of course, be used with either connection of the motor. The half wave controller, shown in Fig. 2.57 (*a*) effects a saving in the cost of semiconductor devices and does not give rise to **dc** components in any part of the system. However, it introduces more harmonics into the line current than does the full wave regulator. As regards the heating of the motor windings are concerned, star connected motor fed through a full wave **ac** regulator is preferred to a delta connected motor fed from a full wave **ac** regulator. This is so, because any third harmonic voltages developed by the motor back e.m.f. can cause circulating currents, when the motor is delta connected.

For delta connected load circuits in which each end of each phase is accessible, the arrangement shown in Fig. 2.58 may be used and has the advantage of reducing the current of the devices. For star connected load circuits in which the neutral point is accessible and can be opened, the regulator circuit shown in Fig. 2.59 reduces the number of thyristors required to three and considerably simplifies the control circuitry.

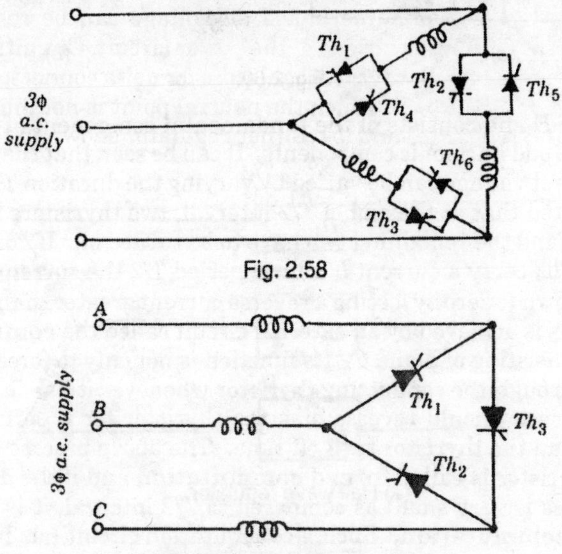

Fig. 2.58

Fig. 2.59 Neutral point regulator.

2.46 Inverters

A semiconductor inverter is a unit using semiconductor devices to convert **dc** power to **ac** power. An inverter circuit used for driving **ac** motors consists of (*i*) a power circuit, (*ii*) a commutating circuit for turning off the power semiconductors and (*iii*) a unit for controlling the output voltage of the inverter. The most commonly used inverter power circuits for **ac** motor drives are :

1. the variable voltage input (VVI) inverter,
2. the variable voltage output (VVO) inverter,

3. the pulse width modulated (PWM) inverter, and
4. the current controlled inverter.

Fig. 2.60 (*a*) illustrates the principle of **dc** to **ac** conversion. Thyristors Th_1 and Th_4 are switched on for a period $T/2$. During this interval Th_2 and Th_3 are off. For the next $T/2$ interval Th_2 and Th_3 are switched on. Th_1 and Th_4 being made off. Repetition of such switching on and off of the relevant thyristors leads to the output voltage wave shown in Fig. 2.60 (*b*). The output voltage waveform is a square wave of

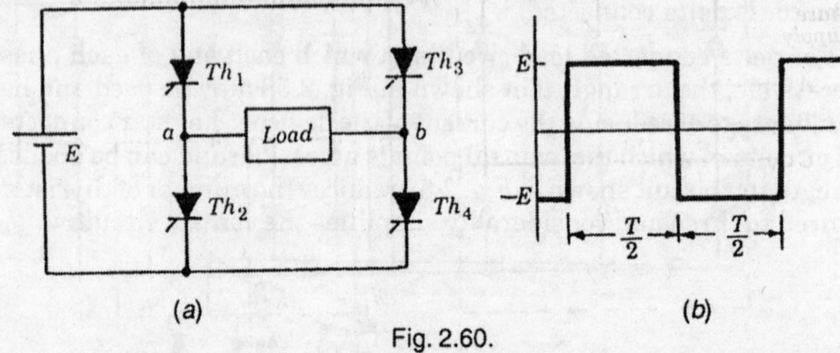

(*a*) (*b*)

Fig. 2.60.

amplitude E and consists of the fundamental component of frequency $1/T$ and all odd harmonic components. It can be seen that the frequency of the output voltage can be varied by varying the duration $T/2$. It may also be noted that at the end of $T/2$ interval, two thyristors have to be turned off and the remaining two have to be turned on. If, for example, Th_1 and Th_4 carry a current I, after a period $T/2$ the current has to be brought down to zero by forcing a reverse current greater than I through them. This is achieved by an external circuit called the **commutation circuit** consisting of L and C. Its function is not only to force a reverse current through the conducting thyristor whenever it has to be turned off, but also it should reverse bias the thyristor for a period of time greater than the thyristor turn off time. The above process of turning off the thyristor is called **forced commutation** and if the duration of this process is very small as compared to $T/2$ interval, it is known as **impulse commutation.** Such a commutation circuit has hardly any effect on the load voltage waveform.

(*a*) Fuse bridge single phase inverter

(*b*) Waveform of load voltage.

Fig. 2.61 and 2.62 show two simplified versions of the *VVI* three phase inverter. In Fig. 2.61 a 6-pulse bridge converter is employed to convert the normal three phase **ac** power to **dc** power at variable direct voltages through the *LC* filter, Fig. 2.63 shows the voltage waveforms V_{AN}, V_{BN}, V_{CN} and the line-to line voltages V_{AB}, V_{BC} and V_{CA}. Each line voltages waveform is displaced in time phase by 120° electrical degrees from one another and is of quasi-square wave with 120° width.

The *VVI* power circuit shown in Fig. 2.62 is similar to that shown in Fig. 2.61, the 6-pulse controlled bridge rectifier has been replaced by a 6-pulse uncontrolled bridge recitifier and a chopper performs the function of adjusting the output voltage from the diode bridge. This circuit has the ability of feeding from the same **dc** bus several inverters using one main uncontrolled rectifier.

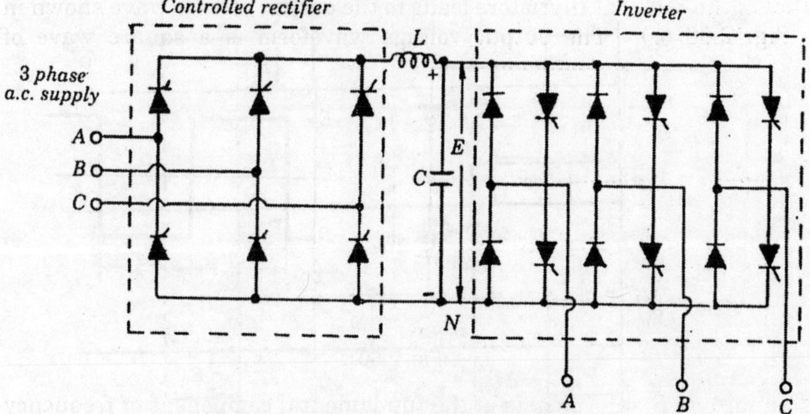

Fig. 2.61. Variable voltage input converter.

The inverters mentioned above cannot return power back to the **ac** supply lines unless another phase controlled rectifier is added to form a reversing system.

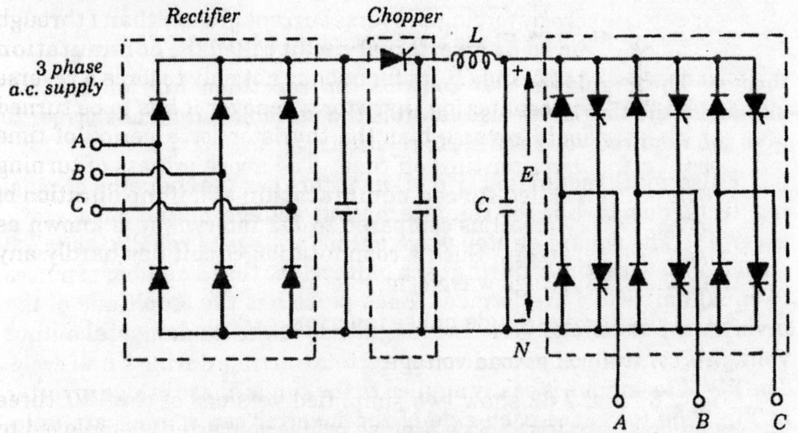

Fig. 2.62. Variable voltage input inverter.

Fig. 2.64 shows a VV0 type inverter. The output voltage waveform of the inverter is the same as that of inverters shows in Figs. 2.61 and 2.62, but its amplitude is fixed by the constant **dc** bus voltage. The input voltage to the **ac** motor is obtained through a 3-phase auto-trans-

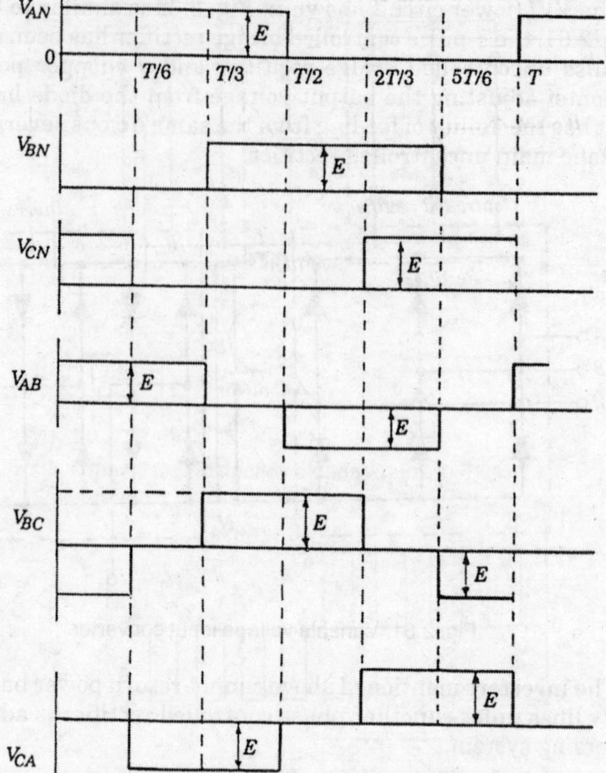

Fig. 2.63. Phase and line voltage waveforms.

former. The regulator that controls the repetition rate of the gate signals to the thyristors also controls the variable auto-transformer to give the required volts per hertz relationship.

Pulse width modulated (PWM) inverters use the chopping or pulsing technique to control the alternating voltage output of a static inverter. The square or step wave output voltage is rapidly switched on and off several times during each half cycle so that a number of pulses of equal amplitude are formed. Each pulse has the amplitude of the inverter input voltage V_{dc}. The magnitude of the fundamental output voltage is controlled by variation of the total on time during a half cycle. The PWM technique is easily applied to the single phase bridge inverter in Fig. 2.60 (*a*), since each side of the inverter can comm' ate independently and an interval of zero voltage is obtained whenever the two load terminals are connected to the same dc line. By commutating one side of the bridge several times during a half-cycle, the output voltage waveform of Fig. 2.65 (*a*) can be obtained. Simple PWM inverters are easily fabricated to generate a waveform with only two pulses per half cycle within a six step envelope, as shown in Fig. 2.65 (*b*). Simplified

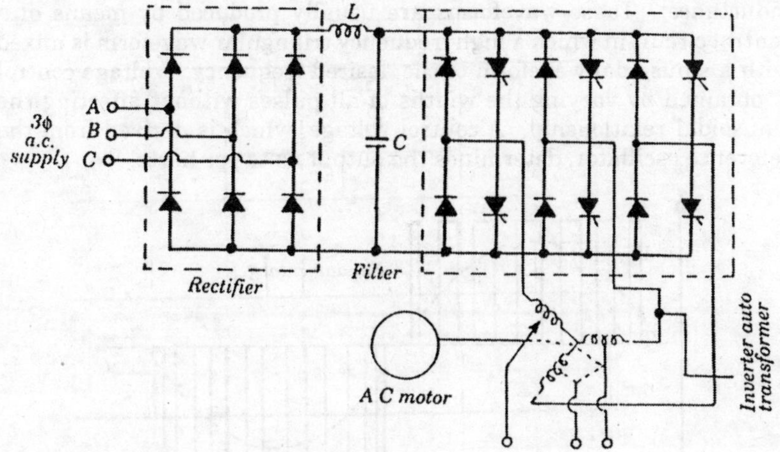

Fig. 2.64. Variable voltage output inverter.

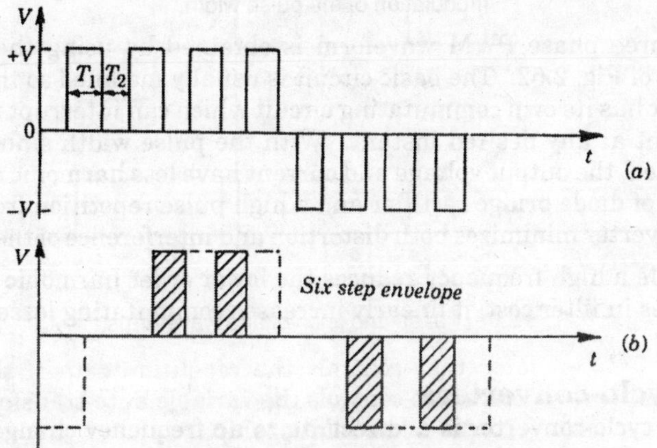

Fig. 2.65. Basic waveforms of PWM.

fifth and seventh harmonics are present in this waveform and they cause appreciable deterioration in the low speed performance of the ac motor. In order to eliminate the low order harmonics, more refined techniques of PWM are used, in which high frequency pulsing occurs throughout the half-cycle.

In sophisticated PWM systems, the pulse width is varied throughout the half-cycle in a sinusoidal manner (Fig. 2.52). Actually, the pulses should be regularly spaced and the pulse width at a particular position should be proportional to the area under the sine wave at that position. In the PWM waveform, the lowest harmonic frequency is at the pulse repetition frequency and, if this is much higher than the fundamental frequency, adequate filtering is provided by the machine inductance. These waveforms are usually produced by means of a control circuit in which a high frequency triangular waveform is mixed with a sinusoidal waveform of the desired frequency. Voltage control is obtained by varying the widths of all pulses without affecting the sinusoidal relationship. A control voltage, which is derived from the reference oscillator, determines the output volts per hertz.

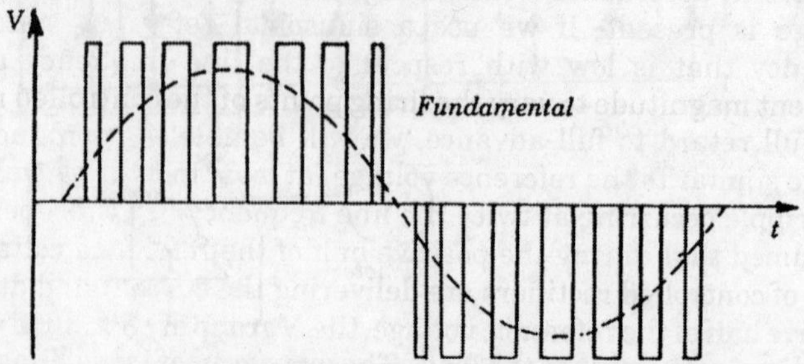

Fig. 2.66. Output voltage waveform of a PWM Inverter with sinusoidal modulation of the pulse width.

A three phase PWM waveform is obtained by using the bridge inverter of Fig. 2.62. The basic circuit is usually modified so that each thyristor has its own commutating circuit which can interrupt the flow of current at any desired instant. With the pulse width sinusoidally modulated, the output voltage and current have less harmonic content. The use of diode bridge rectifier and a high pulse repetition frequency in the inverter minimizes both distortion and interference of the mains.

While a high frequency reduces the lower order harmonic content and saves in filter cost, it linearly increases commutating losses in the inverter.

2.47 Cyclo-converters

The cyclo-converter is a 'direct' **ac** to **ac** frequency changer. The term 'direct' means that the energy does not apear in any other form than **ac** input or **ac** output contrary to an inverter which converts **dc** to **ac**.

While both the cyclo-converter and inverter have as their output **ac** power of variable voltage and frequency, methods of attaining these outputs are quite different for the two circuits.

The single phase centre tapped circuit is shown in Fig. 2.67. This circuit is similar to a single phase centre tapped rectifier except that

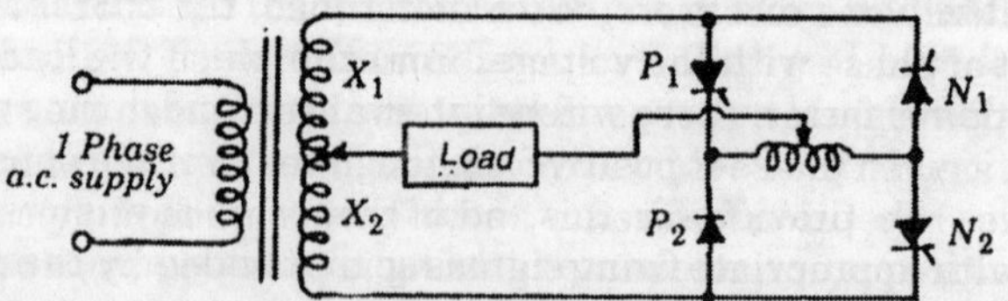

Fig. 2.67. Single phase cyclo-converter.

two groups of controlled rectifiers are provided to permit current reversal through the load.

The thyristors P_1 and P_2 will provide positive current to the load and the thyristors N_1 and N_2 provide negative current. Let us assume that a firing circuit that can modulate the load points of the controlled rectifiers in accordance with a control voltage, called the **reference voltage** is present. If we use a sinusoidal reference voltage of a frequency that is low with respect to the line frequency and of a sufficient magnitude to vary the firing points of the controlled rectifiers from full retard to full advance, we will be able to reproduce a load voltage similar to the reference voltage (at least in average value) with some ripple occurring at twice the line frequency. In this operation, it is assumed that during the positive half of the reference voltage the *P* group of controlled rectifiers are delivering the current and during the negative half of the reference voltage, the *N* group of controlled rectifiers are delivering current to the load. The waveform of load voltage during this type of operation is illustrated in Fig. 2.68. From this waveform it can be seen that the low frequency output voltage is synthesized from the higher frequency line voltage. The description made above is a somewhat simplified picture of cycloconverter operation as it gives the voltage waveform produced by having no more than one rectifier in conduction at any particular instant of time. The load is assumed to be resistive although most of the loads encountered in practice will be a combination of resistance and inductance. The voltage waveform and the firing control with such loads will be slightly different and some of these differences will be explained below :

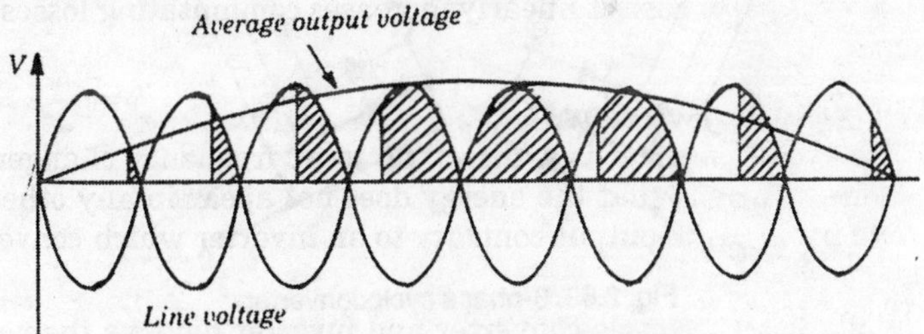

Fig. 2.68. Output voltage waveform.

It must be noted that either group (*P* or *N*) of controlled rectifiers mentioned in the above paragraph is capable of providing output voltage of either polarity, but that a given group is capable of conducting current in only one direction. With motor load, the current normally will be out of phase with the voltage. For example, if the load presents a lagging power factor, there will be intervals of time during which the current in a given phase is positive, although the average phase voltage is negative. To provide for this, both groups of thyristors must be supplied with appropriate firing signals so that both groups are producing the same average voltage, allowing current to flow in either direction. As the P group of tyhristors is advanced in firing angle to produce more positive voltage, the N group firing angles are retarded to produce less negative voltage and therefore both groups are producing the same average load voltage.

Using the thyristor firing pattern described in the preceding paragraph, there will be intervals in time when point X_1 of the supply transformer (Fig. 2.67) is connected to point X_2 by virtue of thyristor P_1 and N_2 both being triggered. The purpose of the reactor shown in the same figure is to limit the magnitude of the current in this loop. In modern cycloconverters different methods are employed to suppress this circulating current between groups and thereby reduce the size of the reactors or eliminate them completely.

The basic circuit can be extended to polyphase operation by simply providing additional controlled rectifier elements in the positive and negative groups. The basic power circuit scheme of a three phase cycloconverter is shown in Fig. 2.69.

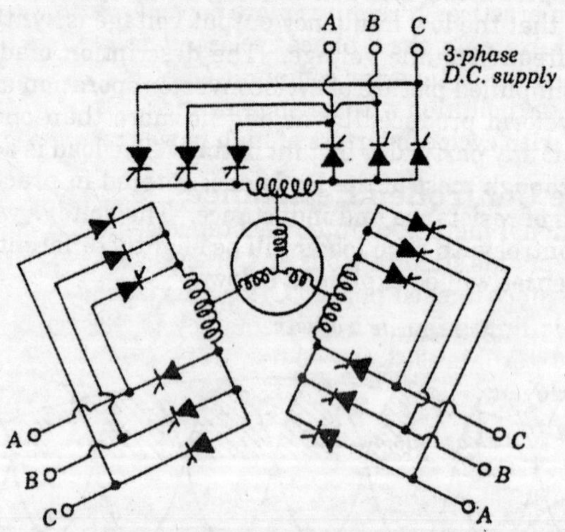

Fig. 2.69. 3-phase cycloconverter.

It should be pointed out that independent control of output frequency and voltage is obtained with only one parameter variation, *viz.*, the firing points of the controlled rectifiers to be varied. The frequency of the output voltage is controlled by the rate at which the firing points are varied about the quiescent point and the output voltage is controlled by the maximum excursion of the firing points from the quiescent point. The cycloconverter, with its associated firing circuit, produces an output voltage that is a replica of the reference voltage.

The operation of a cycloconverter is characterized by several features. They are, generally, used as step down frequency converters. This follows from the fact that the low frequency ac voltage is synthesized from sections of the higher frequency input voltages. There is no fixed minimum ratio of input to output frequency; however, a 3:1 ratio is often found to be the limit for practical applications. Below this ratio, the efficiencies of both the cycloconverter and the motors fed by them start falling significantly.

Reversibility is another feature of cycloconverter drive systems. A cycloconverter fed ac motor drive will respond to a change in polarity of the input signals by changing direction of rotation of the motor without the use of contractors to reverse phase sequence.

The ability of the cycloconverter to handle power flow in either direction is another important feature. This, together with the above mentioned reversibility feature, provides an induction motor drive capable of operating in any of the four quadrants of the motor's speed torque curve.

While the cycloconverter has many attractive characteristics from a theoretical point of view, there are several limitations for its wide application in practice. It requires more power semiconductors than an inverter. For example, the 3-phase cycloconverter needs 18 thyristors, whereas the rectifier-inverter combination of Fig. 2.61 needs only 12 thyristors. Line pollution with harmonics and low power factor can also be problems with cycloconverters of high power rating.

2.48 Pulse Controlled Resistance

Conventional method of rotor resistance control of slipring induction motor demands simultaneous and precise variation of all the three balanced resistors in each phase. Often this is difficult to achieve. To overcome this limitation, a high frequency thyristor chopper, which enables the external resistance to be varied simultaneously and steplessly is employed.

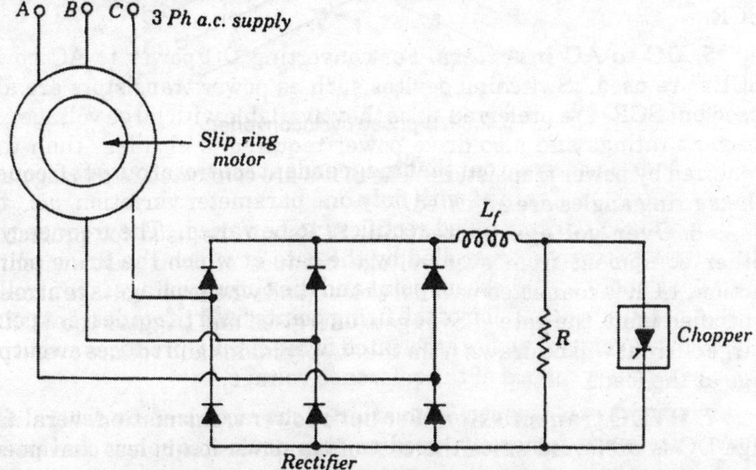

Fig 2.70. Pulsed rotor resistance control.

Fig. 2.70 shows a circuit in which the rotor slip power is rectified in a diode bridge rectifier and fed through a filtering choke to an external resistance. The thyristor in the chopper connected across the resistor is switched on and off at a high frequency. The ratio of on-time to off-time determines the effective value of rotor circuit resistance and thus controls the motor speed by changing its speed torque characteristics.

2.49 Applications of SCRs in the field of Control Engineering

SCRs are used in industry for a variety of purposes. The property of SCR by which it can be switched on and off and the efficiency of gate control to trigger the SCR has made the SCR ideally suitably for many industrial applications. Since the SCRs are compact, reliable, have low losses and fast turn-on and turn-off times, they are more suitable as compared to the conventional saturable-core reactors. The SCRs are used in a wide range of applications, most important of which are :

1. **Speed Control of Motors.** By varying the firing angle we can control the armature voltage by keeping the input voltage constant and hence vary the speed. This is called phase control.

2. **AC and DC switches.** These acts as circuit breakers for making and breaking a circuit. When the SCR is turned off the circuit breaks.

3. **DC to DC or AC to AC converters.** SCRs are used for converting voltage of one frequency into a voltage of some other frequency. They are also used for obtaining variable DC voltage from a source of constant DC voltage. This uses on-off control or chopper control.

4. **AC to DC rectifiers.** SCRs are used for obtaining variable DC voltage from AC supply. This is achieved by using full wave control circuits. Variable DC is obtained by controlling the firing angle of the SCR.

5. **DC to AC inverters.** For converting DC power to AC power SCRs are used. Switching devices such as power transistors are also used but SCRs are preferred since they available with large voltage and current ratings and also drive power required is also less than that required by power transistors. The SCRs are connected in a bridge and their firing angles are adjusted.

6. **Over voltage protection.** SCRs can be used for protecting other equipment from over voltage because of their fast switching action. SCR is connected parallel to load and whenever voltage exceeds specified limit, the gate of SCR gets energised and triggers the SCR. A large current will be drawn from the supply mains and reduce over-voltage at the load.

7. **HVDC transmission.** For bulk power transmission, high voltage DC is preferred since the cost of transmission is less and power conductors are used.

SCRs have taken the place of mercury arc converters in HVDC transmission since there is reduction of space by about 80% reduction in main tenance costs, no arc backs, no warming up time and very short turn-off time.

The SCRs are connected in bridge form and the firing angles are adjusted to get rectifier and inverter operation.

8. **Illumination controllers.** Illustration controllers are mainly used in the stage lighting systems. As the music sound varies so does the illumination level of the stage lights. This is achieved by using SCR. The frequency waves of sound are converted into a voltage signal with the help of a transducers and this signal acts as input to the SCR gate.

When firing angle of SCR changes, the output voltage of SCR changes too and so the intensity of the lamps varies. This is an automatic control and is preferred to the manual rheostat control which does not have high accuracy and fast response.

2.50. Speed Control of DC Motors

From the relation $N = \left[\frac{V - I_a R_a}{K\phi}\right]$

where V = applied stator voltage

ϕ = flux

I_a, R_a = armature current and resistance.

the speed can be controlled by varying the applied stator voltage or flux if voltage drop across the armature is considered to be negligible.

2.51. Phase Control of dc Shunt Motor

A schematic diagram for DC motor control using field control method is given in Fig. 2.71.

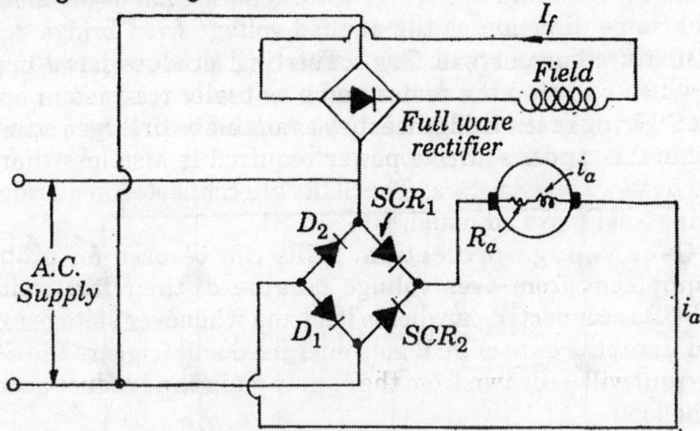

Fig. 2.71. DC shunt motor control.

The field is supplied from a full wave rectifier bridge. A symmetrical half controlled bridge is used for varying the armature voltage.

SCRs 1 and 2 are fired in each half cycle. In the positive half cycle, SCR_1 and D_1 will conduct. If motor speed is low and armature inductance is high, current i_a will continue to flow even after the supply voltage polarity is reversed. D_2 will get forward biased and i_a will free-wheel through D_2 and SCR_1. However counter e.m.f in a separately excited DC motor is very steady therefore i_a will quickly be reduced to zero. Similarly in the negative half cycle SCR_1 and D_2 will conduct. Current and voltage waveforms are given in Fig. 2.72. Natural commutation will occur at P and Q instead of P', Q'. The voltage exerted to the motor is given by the hatched portion. Slight extension of hatched portion beyond P and Q is due to L_a.

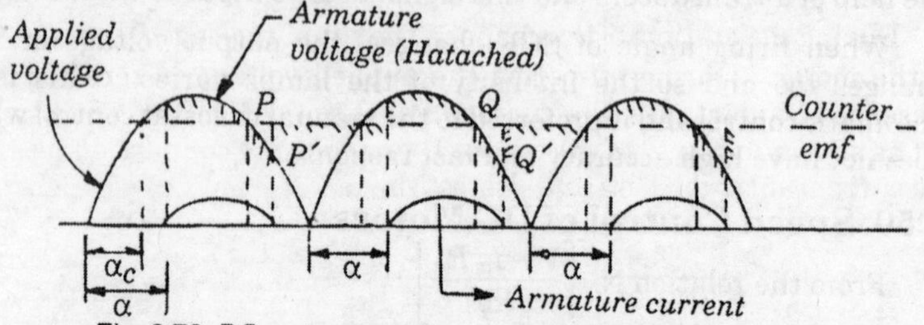

Fig. 2.72. DC motor control current and voltage converter.

2.52. Reversible Speed Control of DC Motor Using Dual Converter

The reversible speed of **dc** motor along with plugging and regenerative breaking is obtained by using dual converter. The scheme is explained in Fig. 2.73. If bridge 1 is operative, the armature current will be i'_a. In case bridge-2 is operative, the armature current will be i'_a. For plugging the sequence of operation is as follows. Let bridge 1 be operative and motor be run in clockwise direction at full speed. Now stop firing bridge 1 and fire bridge 2. The back e.m.f of the motor will be in the same direction as the applied voltage from bridge 2. Thus large armature current will flow. This will produce large breaking torque which will stop the motor and eventually rotate it in opposite direction. Firing in the bridge 2 is to be stopped before the motor starts running in the opposite direction.

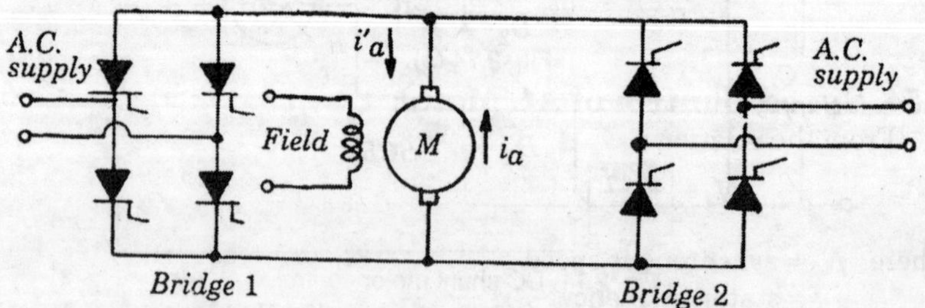

Fig. 2.73. Speed control of using dual converter.

In regenerative braking, after the gating pulses to converter 1 have stopped converter 2 is operated in the inverter mode. The counter e.m.f. in the armature will drive the current through converter 2 and power is fed back into the AC mains. Thus kinetic energy of the motor is converted into electrical power. This will slow down the motor. When the motor stops, the firing angle has come down to less than 90 degrees, the circuit will operate as a rectifier and apply DC voltage across the armature. This will cause the motor to rotate in opposite direction.

2.53. Chopper Control of DC Series Motor

The use of chopper circuits for the speed control of d.c. series motor is given in Fig. 2.74 (*a*).

During the on-time of the chopper, the supply voltage E_{DC} is applied to the motor. During off-period the applied voltage is zero. Due to presence of inductance, the current waveform will not be rectangular but as shown in Fig. 2.74 (*b*). Diode D provides the conduction path for the current interval T-off and protect the motor from voltage peaks.

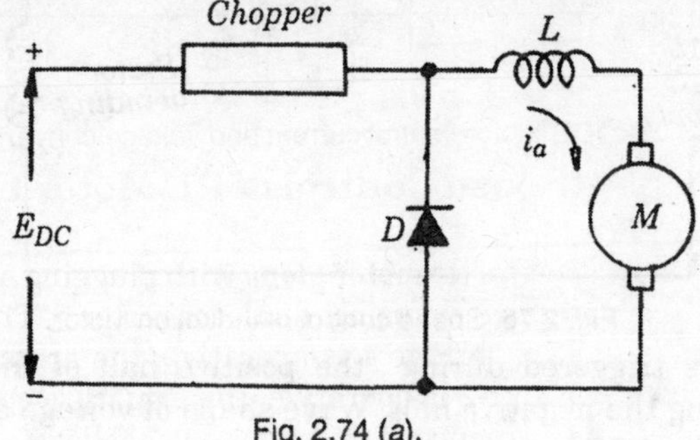

Fig. 2.74 (*a*).

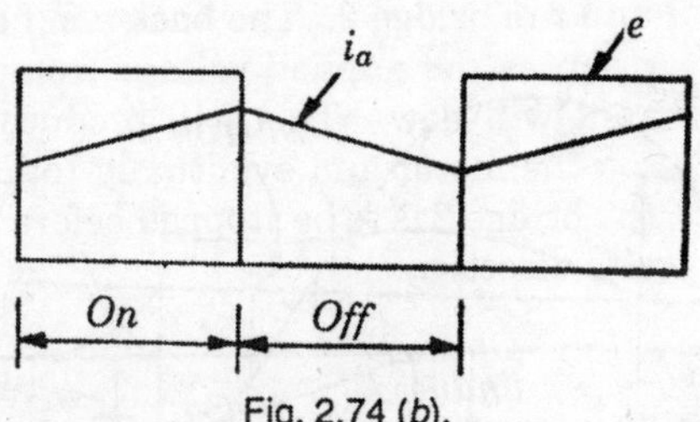

Fig. 2.74 (*b*).

2.54. Speed Control of AC Motor

From the relation :

$$N_s = \left[\frac{120f}{P}\right]$$

where N_s = synchronous speed

f = stator frequency

P = number of poles

the speed of an a.c. motor can be controlled by varying number of poles or stator frequency or both. Also from the relation,

$$N = N_s (1 - S)$$

where N = speed of motor

S = slip.

speed can also be controlled by varying slip.

2.55 Slip Control

1. **Stator Voltage Control.** Speed of an induction motor can be varied by applying a variable voltage to the motor. The arrangement for 1-ϕ induction motor is given in Fig. 2.75.

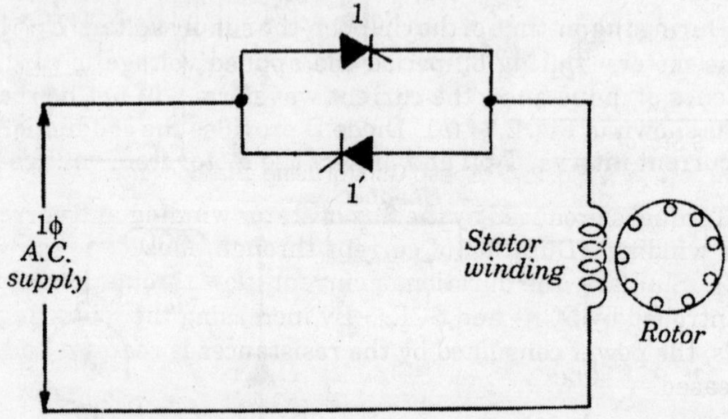

Fig. 2.75. Speed control of Induction Motor.

SCR_1 is triggered during the positive half of the supply and SCR_1' during the negative half. Wave shape of voltage and current is shown in Fig. 2.76.

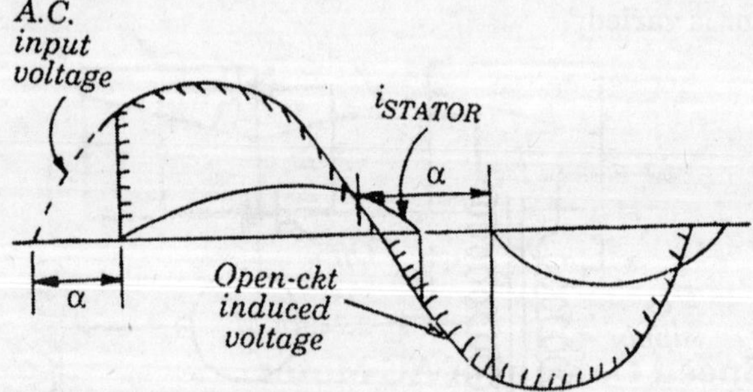

Fig. 2.76. Current and voltage waveforms for 1-ϕ induction motor control.

SCR_1 is triggered at a firing angle α and the stator gets the supply. Current continues to flow even after the reversal of the supply voltage due to the inductance of the winding.

SCR_1' conducts during the negative half of the cycle. After SCR_1 is extinguished and before SCR_1' 'is fired' the stator winding remains open circuited. Some voltage is induced in it due to the current flow in the rotor winding. By varying the firing angle α, the r.m.s value of the supply voltage is changed and thereby speed is changed. The speed of 3-ϕ, I.M. can similarly be controlled using 6 SCRs, 2 for each phase.

2. **Rotor Resistance Control.** Speed of the slip ring induction motor can be varied by varying the rotor resistance. The scheme is given in Fig. 2.77.

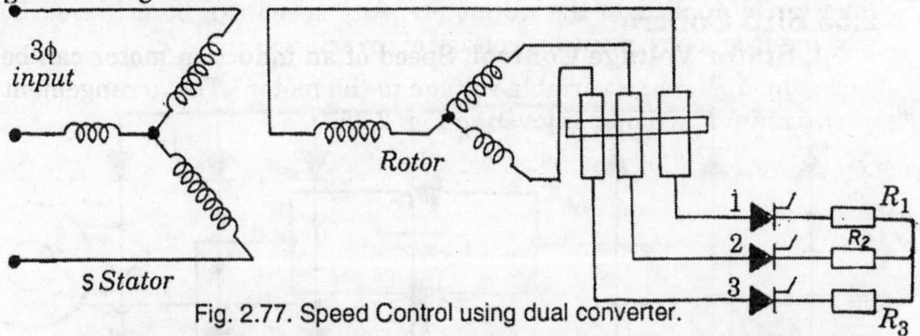

Fig. 2.77. Speed Control using dual converter.

Torque is produced by the flux of stator winding and current of the rotor winding. Duration of current through phase 1 is controlled by SCR_1. Similarly the duration of current flow through phases 2 and 3 is controlled by SCR_2 and SCR_3. By increasing the firing angle of the SCRs, the power consumed by the resistances is reduced and speed is increased.

2.56. Frequency Control

1. **Cyclo Converter Control.** With cycloconverter **ac** output is obtained from **ac** input but at different frequencies. As shown in Fig. 2.78 by controlling the firing of the SCRs, the frequency of the **ac** output voltage which is applied to the motor is varied and hence the speed of the motor is varied.

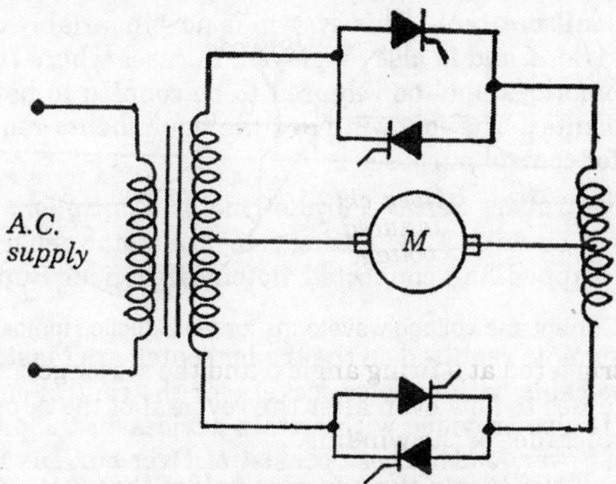

Fig. 2.78. Cyclo converter control.

2. **Inverter Control.** The principle of parallel and series inverters has already been discussed previously. By controlling the firing of SCRs in an inverter, the frequency of output **ac** is varied. Inverter is also used for speed control of induction motors. Fig. 2.79 shows the schematic arrangement of the control scheme for an inverter fed induction motor. The **ac** supply is first rectified and then **dc** voltage is applied through a filter to the inverter. Controlled output **ac** voltage from the inverter is applied to the **ac** motor. Any speed can be achieved by varying frequency of output voltage of inverter.

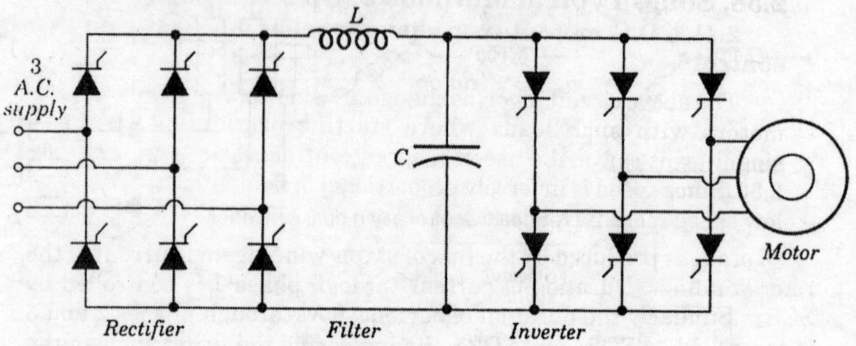

Fig. 2.79. Invertor control.

2.57. Multiple Unit Control

The coatches where electric motors are installed are known as motors coatches. For city and suburban services, electric trains are made up of a number of motor and trailer coatches may be reqd whereas during off peak periods one motor coatch and one trailer may be good enough. Each motor coatch may have two or four motors and it is desirable that it should be possible to control the operation of all the motors in train from a single point and this control system is known as **multiple unit control.** This system is now invariably used for all motor coatch trains and is also employed, in cases where two or more than two locomotives may be required to be coupled to haul a heavy train on gradients. The following are the equipments required on a motor coach for control purposes :

Master controller, Series Parallel motor Controller, multi-core cable for control circuits, Accelerating relays, a motor-generator Set or a Battery or Tapped-line connected Potential-Divider Supply to the control ckt.

The equipments required on the trailer coatch are Master Controller, multi-care cable with coupler sockets for the control circuits. The motor coatch is also provided with certain devices for the protection of motors against over loads. These consist of Over current Relay with Driver's Control Switch, Isolating switch for Master Controller,

Switches with fuses for isolating the control ckt. of the motor controller from the control Bus Line.

With all types of controllers, a Dead Man's Handle Device is provided in order to stop the main automatically in case the driver fails and/or is not in a position to control the operation. This is in the form a contact attached to the knob of the controller handle. If the driver is not able to operate the handle properly, the knob rises which causes the contact to close and operate the main ckt. breaker and apply the brakes.

2.58. Some Typical Methods of Speed Control

2.41.1. D.C. motor to run above or below normal through A.C. control

The above circuit involves change of field flux only. Thus small D.C. motors with such loads where starting or running torques are singnificant can make use of a speed control circuit as shown in Fig. 2.80. Since speed is inversely proportional to field flux, more flux means low speed more torque and *vice versa*.

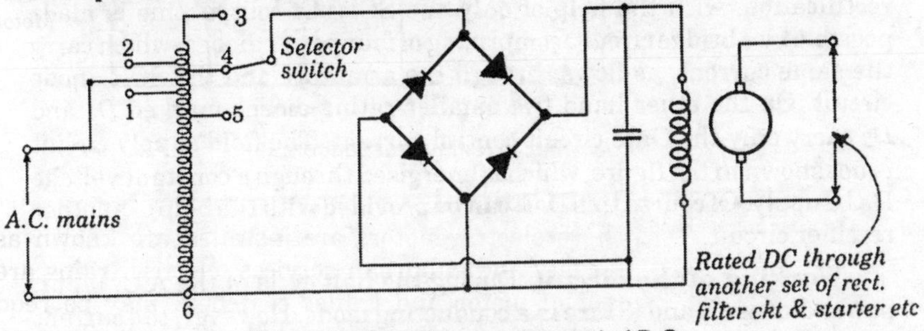

Fig. 2.80. Circuit for speed control of D.C. motor across A.C. mains.

The A.C. mains feed an auto transformer which will give highest A.C. voltage across the rectifier circuit, when the input from mains is given between terminals 1 and 6, but output is taken from across terminals 3 and 6. The A.C. is rectified by a bridge rectifier, filtered and fed across the motor field. The armature is getting normal D.C. through a second set of rectifier/filter energised directly by the mains or a suitable transformer. The flux being highest at this stage the speed will be lowest as per required circuit design, but more flux for a given armature current will produce larger torque.

When the selector switch is thrown to terminal 5 on the output of the auto-transformer and to terminal 2 on the input side, the voltage across the bridge/motor field will be lowest. Obviously the field flux will go down and speed will increase at the cost of torque. Thus the upper speed has to be limited to stability of operation.

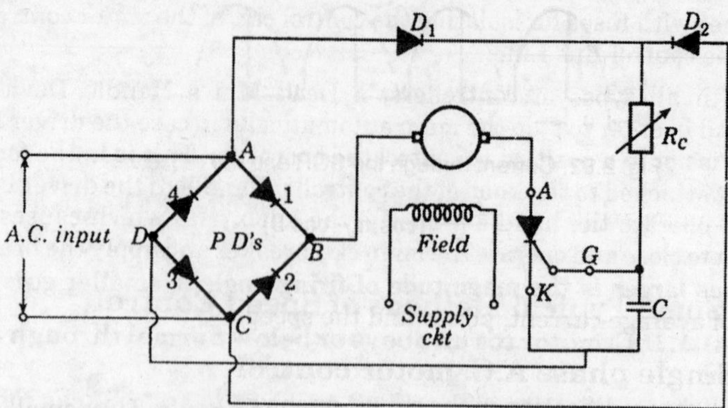

Fig. 2.81. A controlled rectifier circuit feeding the armature to give desired speed control below normal.

2.59 D.C. shunt motor to run below normal

The interesting feature of this control circuit is full wave controlled rectification, with the help of only one SCR. Of course, this is made possible by a bridge circuit, comprising of four power diodes, which carry the same current, as flows through the armature and the SCR anode circuit. On the other hand two smaller rating diodes marked D_1 and D_2 carry only the Gate circuit control current. The field supply circuit is not shown in the figure, which is energised through a constant voltage D.C. supply. Of course this D.C. can be provided with the help of another rectifier circuit.

Working of the circuit. During the half cycle of the A.C. input, power diodes (1) and (3) are in a conducting mode. However, the current through the armature and SCR can flow only after firing of the SCR, which otherwise offers such a high resistance so as to permit only leakage current to flow through the armature. The firing gate potential is now decided by the magnitude of control resistor R_C and the charging voltage across capacitor C.

The charging current flows through the control circuit diode D_1. As soon as the Gate potential and corresponding anode voltage touch a break over level, the SCR fires and the armature gets the desired current.

The process is repeated by power diodes (2) and (4) along with control diode D_2 during the negative half cycles. Thus the average power received by the armature is decided by the firing angle θ which is varied through resistor R_C and so varies the speed of the drive. Obviously the mechanical power as well as the torque undergo reduction along with the speed in the given case.

$$I_{av} = \frac{1}{\pi}\int_{\theta}^{\pi} i\,d\theta \quad \text{amps.}$$

Fig. 2.82. Current through the SCR and the armature.

$$= \frac{I_{max}}{\pi} (\cos \pi - \cos \theta) \qquad \text{amps.}$$

Thus larger is the magnitude of firing angle θ, smaller goes the value of average current, power and the speed.

2.60 Single phase A.C. motor control

A slight modification in the circuit as shown in Fig. 2.83 can control the speed of a single phase A.C. drive.

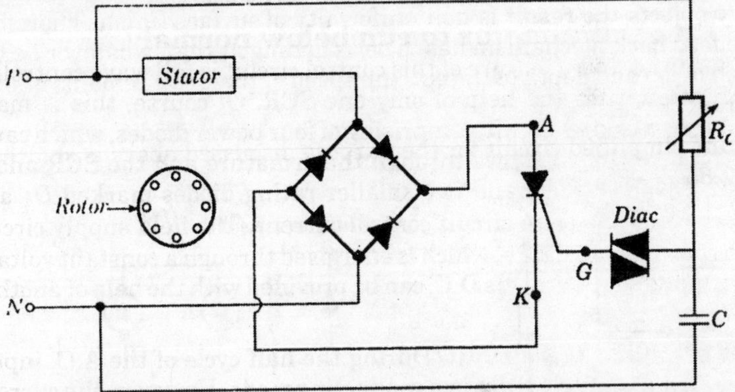

Fig. 2.83. Circuit for single phase A.C. drive speed control with one SCR.

As evident from the circuit diagram, the current circulating through the stator winding to the machine is alternating. This current is negligible as long as the SCR remains in a non-conducting stage. The firing of the device is now controlled through a bilateral conducting thyristor *viz.* a diac. During either of the half cycles when the capacitor charges to approx. twenty volts, the diac conducts. It sends a triggering current to the gate circuit and the SCR fires. It reduced resistance makes the rectifier power diodes conduct in pairs alternately. Although this keeps the flow of current through the SCR undirectional, but the stator current reverses in each of the half cycles. The current is delayed by and angle θ which is controlled by R_C and *C*. The stator current is shown in Fig. 2.84 which speaks of a reduced average power supplied to the machine. The results in speed variation dependent upon the firing angle θ.

Note. Three such configurations as shown in Fig. 2.83, with ganged operation of three control resistors as R_C, can result in power/speed control of three phase cage induction motor. Obviously it is at the cost of torque. Therefore the circuit has limited applications where/requisite starting torque is not high, but starting to running speed has to increase gradually.

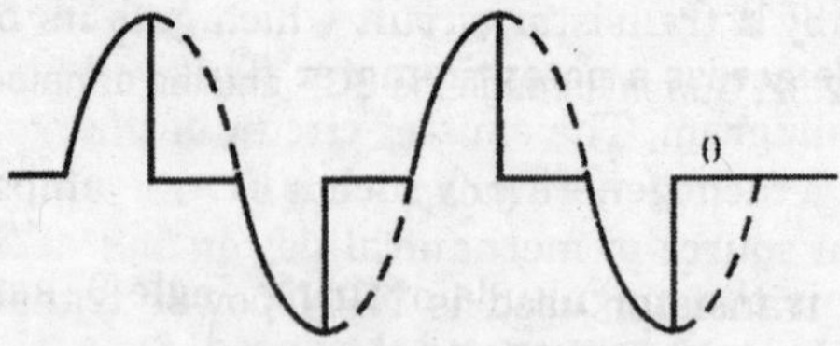

Fig. 2.84. Bilateral current through the stator.

2.61 Constant Speed D.C. Drive

It is established fact that when mechanical load on any motor is increased, its speed tends, to go down. In mechanical mechnining of heavy objects the result is non uniformity of surface finish. Therefore some feed back mechanism has to be as arranged that, whatever be the length of job and depth of the cut, the speed of the drive remains almost constant.

One simplified circuit for the purpose discussed above is shown in Fig. 2.85.

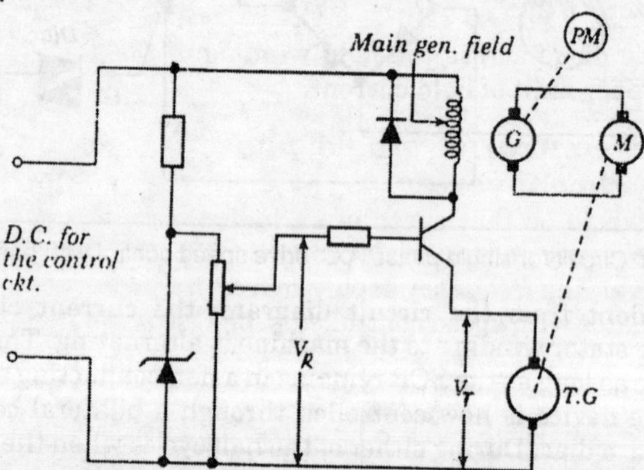

PM = Prime Mover ; *G* = Main D.C. Generator; *M* = D.C. motor armature driving the Mechanical Load ; *TG* = Tachogenerator coupled to the motor shaft.

Reference voltage = V_R.

Voltage generated by the Tachogenerator = V_T.

Fig. 2.85. Constant speed D.C. drive control circuit. (Field supply for the motor and Tachogenerator not shown in the Fig. It is from a rated D.C. supply source).

2.62 Working of the circuit

From the circuit diagram it is clear that the speed of the motor on no load will depend upon its armature voltage being supplied by the main Generator. Its field is maintained at the rated voltage by a source not shown in the circuit diagram. The generator voltage further depends upon its field flux taking all other parameters to be constant. This field flux is controlled by a transistor circuit which gets its base potential from a zener diode across a potentio-meter. This voltage is marked as V_R in the circuit diagram. The emitter circuit of the transistor gets a potential V_T from a tachogenerator which is driven by the same motor, which is the main source of mechanical power and needs a constant speed. Since the transistor used is NPN power transistor its base emitter voltage $V_{BE} = V_R - V_T$ volts. Its base current $I_B \propto V_{BE}$ and collector current $I_C = \beta I_B$ (neglecting leakage) and is practically the field current for the main generator. By adjusting V_R, I_C can be adjusted to a desired level deciding the field current/generated voltage of the main generator. This is turn decides the motor r.p.m. and V_T.

Now let us consider effect of mechanical load on the motor. As the motor r.p.m. tends to fall the voltage V_T generated by the tachogenerator also falls. But V_{BE} of the transistor increases, its I_B increases same is the case of I_C or the current through the field of the generator. More field current means more flux, more voltage and improved motor r.p.m.

Note. The diode across the field winding of the main generator decides the upper limit of field current.

2.62 Speed control of single phase induction motor through cyclo-converter

It is known to us that speed of an induction motor is directly a function of frequency and inversely proportional to the number of poles. To arrange a varying frequency supply source in the conventional form *viz.*, an alternator with speed controlled prime mover is a costly affair.

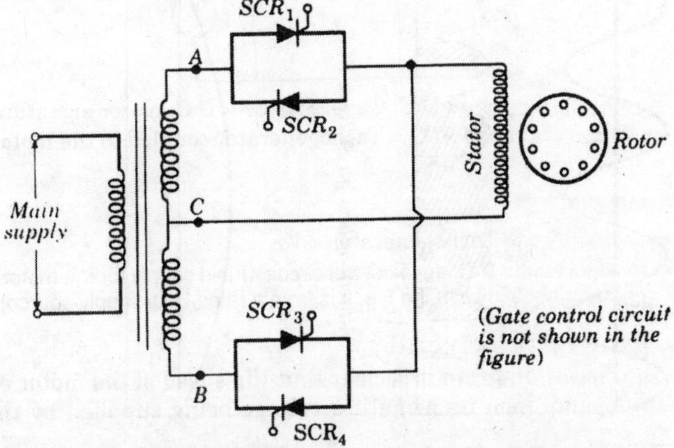

Fig. 2.86. A simplified single phase cyclo-converter circuit.

A solid state circuit as shown as Fig. 2.86 gives an idea about frequency variation at the output, the input being taken from supply mains.

Working of the circuit. During positive half cycle of the A.C., input across the centre tapped secondary of the input transformer. SCR_1 and SCR_4 are in conducting mode whereas SCR_2 and SCR_3 are in a blocking mode. Since the peak of the A.C. input is kept lower than break over voltage, a triggering gate pulse is essential to cause firing of the SCR's in conducting mode. This pulse is provided only to the gate of SCR_1, that too with a delayed angle θ, which is larger than 90° and at a required magnitude.

The negative half cycle brings SCR_2 and SCR_3 in a conducting mode. Now SCR_3 is fired but at a reduced delay. This firing is continued

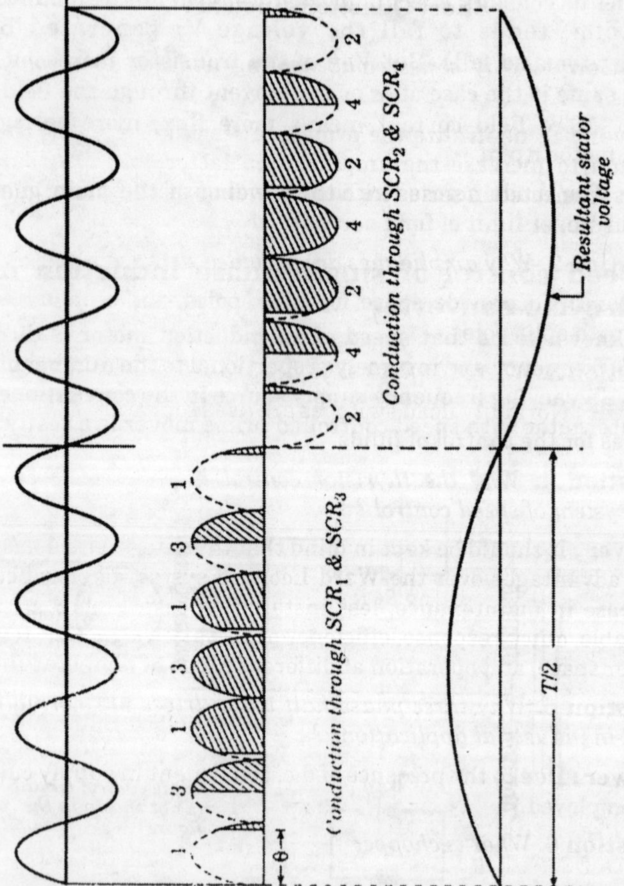

Fig. 2.87. Change of time period/frequency established through sequential firing of different SCR pairs at varying firing angle θ.

alternately for SCR_1 and SCR_3 (with gradual variation of θ as well as gate current) as shown in Fig. 2.87 for a period *T*/2. The process is repeated with firing of SCR_2 and SCR_4 alternately for an other period of *T*/2 seconds

The variation of period of conduction of different *SCR's* through the motor stator virtually appears to be an A.C. applied across if with a time period *T* which is practically seven times the time period of A.C. mains in the given example. It is nothing but equivalent to exciting the stator at a reduced frequency resulting in speed variation of the rotor.

This way it is possible to have gradual variation of speed which can be a very small fraction of the rated r.p.m. (1/20 to 1/30). Of course the gate control circuitry involves a higher technological design and skill. The control panel is quite costly (Around Rs. 60,000 for a 5 H.P. drive).

Short Question-Answers

Question 1. *Why series compensating winding is employed in amplidyne ?*

Answer : To neutralize the armature reaction set up by the load current and to increase the amplification factor, the compensating winding is connected in series with the power output brish terminals of the amplidyne.

Question 2. *Why amplidynes are provided with split poles.*

Answer : To provide space for inter poles, the amplidynes are provided with split-poles.

Question 3. *Where the float switches are being used ?*

Answer : Normally fload switches are used for opening and closing pipe values for the control of fluids.

Question 4. *Why the thyristor control is preferred over Ward Leonard system of speed control ?*

Answer : It should be kept in mind that thyristor control has some different advantages over the Ward Leonard system of speed control such as ease in maintenance, less installation cost, higher accuracy, more reliable, quick response, efficiency of operation is higher, requires lesser floor space, and operation at different range of temperatures etc.

Question 5. *Why three-phase half wave drives are normally not employed in industrial applications ?*

Answer : Due to the presence of d.c. component in supply currents it is not employed.

Question 6. *What is chopper ?*

Answer : Chopper is a kind of circuit in which thyristor is used to switch on the supply to load for a certain interval and after that switch off the supply for another certain interval.

Question 7. *What is meant by duty ratio of a d.c. chopper ?*

Answer : Duty ratio is the ratio of on period to the time period *i.e.*,

Duty ratio, $\lambda = \frac{T_{on}}{T}$

or $T_{on} \times \text{frequency}$

or $\frac{T_{on}}{T_{on} + T_{off}}$

Question 8. *Mention the advantages of chopper fed drive over converter fed drive ?*

Answer : The following are the advantages of chopper fed drive over converter fed drive :

(*i*) The chances discontinuous operation in chopper drive are tower than that in case of converter drive. So speed regulation and transient response of chopper drive are better than that in converter drive.

(*ii*) Ripple in the output of the chopper is much less that in the output of a converter. Lesser ripple means lesser losses.

Question 9. *The commutators of the thyristor controlled d.c. motors made larger than those for ordinary d.c. motors. Reason ?*

Answer : In order to provide extra insulation to withstand larger and rapid voltage fluctuations, the commutators of the thyristor controlled d.c. motors are made.

Question 10. *What is the difference between on-off control and phase control in the case of a.c. voltage controller ?*

Answer : The thyristor are used to connect the motor to source of supply for a certain number of cycles and then to disconnect it for another certain number of cycles; each of the on and off items consists of an integral number of cycles. In on-off control, also known as integral cycle control. On the other hand in the case of phase control, the thyristors are employed as switches for cnnecting the motor to the supply for a little portion of each cycle of the supply voltage.

Question 11. *State two advantages of current source inverter.*

Answer : The main advantages of current source inverter are mentioned below :-

(*i*) short-circuits and misfiring of devices, because of constant input current, do not pose any difficulty.

(*ii*) Without free wheeling diodes, it can handle reactive or regenerative loads.

Question 12. *Mention any two drawbacks of cycloconverters.*

Answer : The two draw backs are:

(*i*) They can produce a sub-frequency output only.

(*ii*) It needs more power than in case of an inverter.

Question 13. *Give some applications of cyclo converter.*

Answer : Normally cycloconverter drives are used for motors of big size. They have also been employed in gearless cement mills or ball mill drives.

Question 14. *What is meant by GTO ?*

Answer : It (Gate-turn-off switch) is a special thyristor which can be turned on by a positive gate signal and can be turned-off by a negative gate signal.

Choose the Correct Answers

1. Thyristors may also be used for the control of d.c.:

(*a*) separately excited motor (*b*) series motor

(*c*) shunt motor (*d*) all of the above.

2. For speed control of d.c. motors using controlled rectifiers, armature voltage control gives speed :

(*a*) above rated speed

(*b*) below rated speed

(*c*) above as well as below rated speed.

3. The armature current of a d.c. motor fed from a thyristor power convertor contains ripple. The ripple in the armature current affects :

(*a*) commutating capability of the motor.

(*b*) over load capacity of the motor

(*c*) torque capability of the motor

(*d*) control ability of the speed of the motor

4. An SCR chopper circuit supplies power to a d.c. motor. What will be the nature of motor armature current.

(*a*) Sinusoidal (*b*) Exponential rise and decay

(*c*) Constant (*d*) none of the above

5. For speed control of motors using choppers, it is desirable to employ :

(*a*) frequency modulation method

(*b*) pulse width modulation method

(*c*) phase control mehtod

(*d*) any one of (*a*), (*b*) and (*c*)

6. A power chopper converts :

(*a*) a.c. to d.c. (*b*) d.c. to d.c.

(*c*) d.c. to a.c. (*d*) a.c. to a.c.

7. d.c to d.c. convertor or chopper can :

(*a*) only step down d.c. voltage

(*b*) only step up d.c. voltage

(*c*) step up or step down d.c. voltage

8. In a two-quadrant chopper the load voltage is varied from positive maximum to negative maximum by varying the time ratio of the chopper.

(*a*) from 0 to 1. (*b*) from 1 to 0

(*c*) from 0 to 0.5 (*d*) from 0.5 to 1

9. In a chopper fed d.c. drive, the chopping frequency is approximately equal to :

(*a*) 50 Hz (*b*) 300 Hz

(*c*) 1,000 Hz (*d*) 5,000 Hz

10. A 1000 r.p.m., 180 V, 25 A separately excited d.c. motor has an armature resistance of 2π. The motor is connected to a 230 V d.c. supply thorough a chopper. To run the motor at half rated speed and at rated torque, the chopper duty cycle should be :

(*a*) between 0 to 0.5 (*b*) between 0.5 to 1.0

(*c*) 1.0 (*d*) 0.5

11. In an inverter, the commutation method, is :

(*a*) Line commutation (*b*) Forced commutation

(*c*) Either (*a*) or (*b*) (*d*) None of the above.

12. In integral cycle control of an A.C. regulator, the input power factor is equal to ----where α is the duty cycle,

(*a*) α (*b*) α^2

(*c*) $\sqrt{\alpha}$ (*d*) $\frac{1}{\alpha}$

13. A cyclo-converter uses :

(*a*) Natural commutation

(*b*) Forced Commutation

(*c*) Either (*a*) or (*b*)

(*d*) Both (*a*) and (*b*) combine

14. In case of a cyclo-converter, it can be varied :

(*a*) Output voltage only

(*b*) Output frequency only

(*c*) Both output voltage and output frequency at a time

(*d*) None of the above.

15. At the input terminals of a cyclo-converter generally the power factor is :

(*a*) Low and lagging (*b*) Low and leading

(*c*) High and lagging (*e*) High and leading

16. A thyristor can be termed as:

(*a*) d.c. switch (*b*) a.c. switch

(*c*) either (*a*) or (*b*) (*d*) square wave switch

17. A 3-phase a.c. to d.c. converter which requires neutral point connection is :

(*a*) 3-phase semiconverter

(*b*) 3-phase full converter

(*c*) 3-phase half-wave converter

(*d*) 3-phase full converter with diodes.

18. The number of thyristors required for single phase cyclo-converter of the midpoint type and for three-phase three-pulse type cyclo-converter are respectively :

(*a*) 4, 6 (*b*) 8, 18

(*c*) 4, 18 (*d*) 4, 36

19. To change the stator volatage, a 3-ϕ induction motor is supplied by a 3-ϕ a.c. regulator. The change in speed will be more if,

(*a*) load torque is constant

(*b*) with the increase in speed the load torque increases

(*c*) with the increase in speed the load torque decreases

(*d*) either (*a*) or (*b*)

20. For maintaining the ratio V/f constant, A 3-ϕ induction motor is controlled by voltage and frequency control. At 50 c/s, the starting torque is 150 Nm. At 50 c/s , the starting torque,

(*a*) remains less than 150 Nm

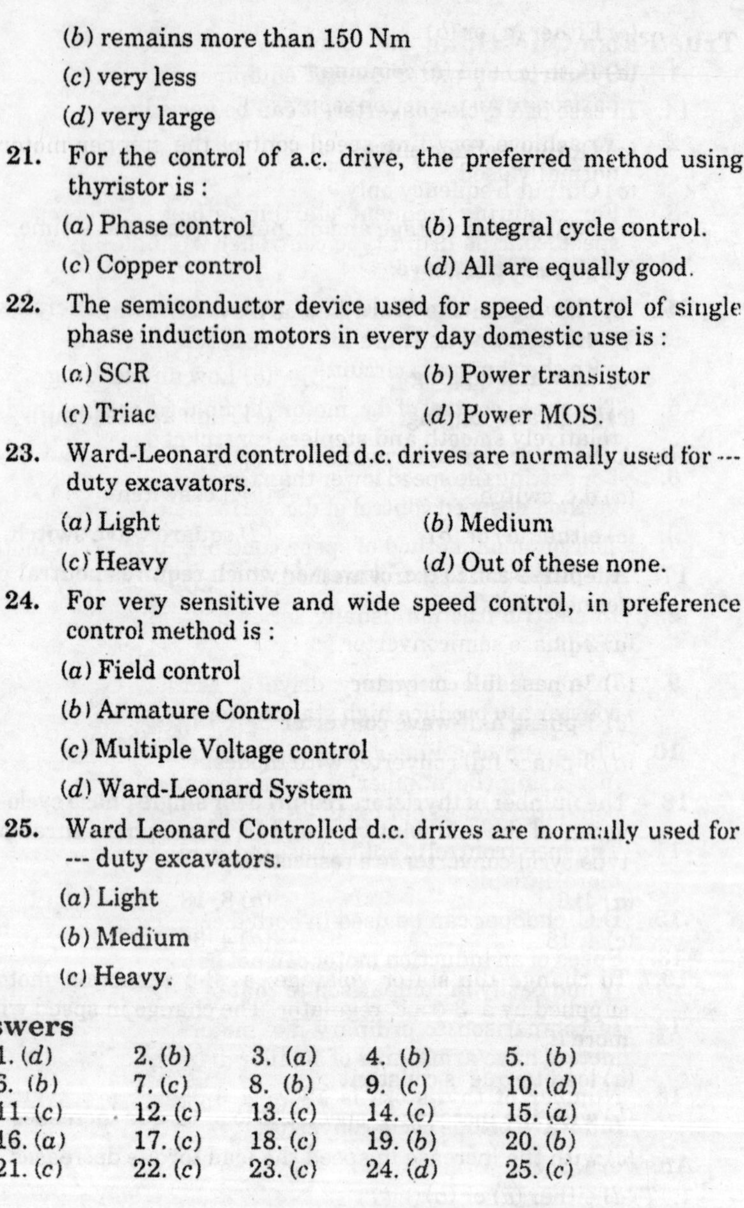

(*b*) remains more than 150 Nm

(*c*) very less

(*d*) very large

21. For the control of a.c. drive, the preferred method using thyristor is :

(*a*) Phase control (*b*) Integral cycle control.

(*c*) Copper control (*d*) All are equally good.

22. The semiconductor device used for speed control of single phase induction motors in every day domestic use is :

(*a*) SCR (*b*) Power transistor

(*c*) Triac (*d*) Power MOS.

23. Ward-Leonard controlled d.c. drives are normally used for --- duty excavators.

(*a*) Light (*b*) Medium

(*c*) Heavy (*d*) Out of these none.

24. For very sensitive and wide speed control, in preference control method is :

(*a*) Field control

(*b*) Armature Control

(*c*) Multiple Voltage control

(*d*) Ward-Leonard System

25. Ward Leonard Controlled d.c. drives are normally used for --- duty excavators.

(*a*) Light

(*b*) Medium

(*c*) Heavy.

Answers

1. (*d*)	2. (*b*)	3. (*a*)	4. (*b*)	5. (*b*)
6. (*b*)	7. (*c*)	8. (*b*)	9. (*c*)	10. (*d*)
11. (*c*)	12. (*c*)	13. (*c*)	14. (*c*)	15. (*a*)
16. (*a*)	17. (*c*)	18. (*c*)	19. (*b*)	20. (*b*)
21. (*c*)	22. (*c*)	23. (*c*)	24. (*d*)	25. (*c*)

True/False Questions

1. Synchronous motor control equipment is complex in comparison to an induction motor.
2. To achieve very fine speed control the stepper motors are normally used.
3. For requiring frequent starting, stopping, reversing and speed control, drum type controller is employed with a d.c. motor of 10 KW.
4. At low speed the Ward-Leonard system of speed control is more efficient as there are no resistors connected in series with the armature circuit.
5. The speed control of d.c. motors by field control method gives relatively smooth and stepless control of speed.
6. For getting the speed lower than the normal the field divertor method of speed control of d.c. series motors is used.
7. The common method of speed control of d.c. series motors is the Rheostatic control method.
8. In electric traction usually series-parallel control is widely used.
9. In a variable frequency drive of induction motor *V*/ is constant to produce high starting torque.
10. The speed of a linear induction motor can be increased by increasing the number of poles, the supply frequency and pole-pitch being kept fixed. [A.M.I.E. Sect. B. Summer 1996]
11. The speed control of split-phase induction motors is relatively more difficult.
12. D.C. chopper can be used in both d.c. and a.c. drives.
13. Speed of an induction motor can not be controlled as efficiently and easily in comparison to that of a d.c. shunt motor.
14. In comparison to ordinary d.c. motors, the tyristors drive motors have armatures of smaller diameter.
15. Number of thyristors is 12 in a 3-phase to single phase full-wave bridge cyclo-convertor.

Answers

1. (*T*)	2. (*T*)	3. (*T*)	4. (*T*)	5. (*T*)
6. (*F*)	7. (*T*)	8. (*T*)	9. (*F*)	10. (*F*)
11. (*T*)	12. (*T*)	3. (*T*)	14. (*F*)	15. (*T*).

Fill in the Blanks

1. In a single phase fully controlled bridge rectifier, the number of thyristors used are ---
2. In a single phase half bridge inverter the number of thyristor used are ---
3. Three-phase inverter can operate in --- mode or --- mode.
4. A single phase half-wave a.c. regulator has normally one thyristor and ---.
5. A cyclo-converter is a ---------- converter.
6. A bridge type 1-phase cyclo-converter always changes its frequency from f to $f/4$, then one half-wave of output contains ----- of input.
7. D.C. motor automatic speed control system uses ----------.
8. In a full-wave thyristor controlled rectifier, conduction angle and firing delay angle is always --------.
9. A chopper is also a -------- converter.
10. From a chopper circuit output voltage is ---------- on the duty cycle and ---------- of load current.

Answers

1. four 2. Two
3. 120° ; 180° 4. one diode
5. a.c. to a.c. 6. four half-waves.
7. negative feedback 8. 180°
9. d.c. to d.c.
10. only dependent; independent.

EXERCISES

1. How do you describe control gear ?
2. What are the advantages of automatic control over the manual control ?
3. Describe briefly the starting and speed control of a.c. and d.c. motors.
4. Describe the Ward-Leonard method of speed control and illustrates its applications.
5. Describe the principle, construction and working of the amplidyne illustrating its applications.
6. Describe the speed control and starting principles of a.c. single phase motors.
7. Describe the magnetic amplifier explaining its field of application.

8. What is the difference between a contactor and a motor starter ?
9. Differentiate between the push button and the master switch ?
10. Where do we use float switches and plugging switches ?
11. What is Relay ? Where do we use the Relays ? Give the basic construction and field of application of a typical electromagnetic relay.
12. Write short notes on the following items of control equipment :
    (*a*) Rotating amplifiers
    (*b*) Magnetic amplifier
    (*c*) Master switches
    (*d*) Push Buttons
    (*e*) Thermostats
    (f) Plugging switches
    (*g*) Float switches
    (*h*) Electromagnetic
    (*i*) Pressure regulators
    (*j*) Electromagnetic contactors
    (*k*) Rotary switches
    (*l*) Speed Governors
    (*m*) Limit Switches.

3 Electric Heating

3.1. Introduction

Electric heating is preferred over other types of heating methods *i.e.* by wood, coal oil and gas. Practically all heating requirements can be fulfilled by some method of electric heating.

We are very much familiar with the fact that whenever current, 'I' is made to flow through any ckt. having a resistance of R ohms, power dissipated in that circuit is I^2R watts. If the current flows for t seconds, energy consumed is $I^2R\,t$ Joules/watt-secs. This energy is being converted into heat and can be worked out as,

$$H = \frac{I^2Rt}{4.2} \text{ Calories or} = 0.24\, I^2Rt \text{ Calories}$$

where 4.2 is a constant called **mechanical equivalent of heat.**

Electric heating is based on the principle that when electric current passes through a medium (Solid, liquid and gas) heat is produced.

There are three modes of transmission of heat :

(*i*) Conduction

(*ii*) Convection

(*iii*) Radiation

Liquids are heated by convection method, Solids by conduction method and distant objects are heated from the source of heat by radiation. The heat of sun reaches to us by radiation.

Electric heating is extensively used for domestic applications as well as for industrial applications. The examples of domestic/Industrial appliances making use of heating effect of current are quoted as below:

A. Domestic Applications of Electric Heating :

1. Room heater for heating the building.

2. Immersion heaters for water heating.
3. Hot plates for cooking
4. Gysers
5. Electric kettles
6. Hot air driers
7. Electric Iron
8. Cofee percolators
9. Pope corn plant
10. Electric ovens for baking products
11. Electric Toasters

B. Industrial Applications of Electric Heating :

(*i*) Melting of metals

(*ii*) Heat treatment of metals like annealing, tempering, Soldering, Brazing and Case hardening etc.

(*iii*) Electric welding

(*iv*) Moulding of glass for making glass appliances

(*v*) Baking of insulators

(*vi*) Moulding of plastic components

(*vii*) Enamelling of copper conductors

(*viii*) Heat treatment of pointed surfaces

(*ix*) Making of plywood

(*x*) Enamelling of copper conductor

(*xi*) Viterous enamelling of wire wound resistors, danger plates and other metallic components.

If we go into the details of generation, transmission and distribution of electric power and finally its application for heating purposes, it is revealed that electric power is generated by burning of coal, *i.e.* by generating heat and ultimately heat can be produced by electric power.

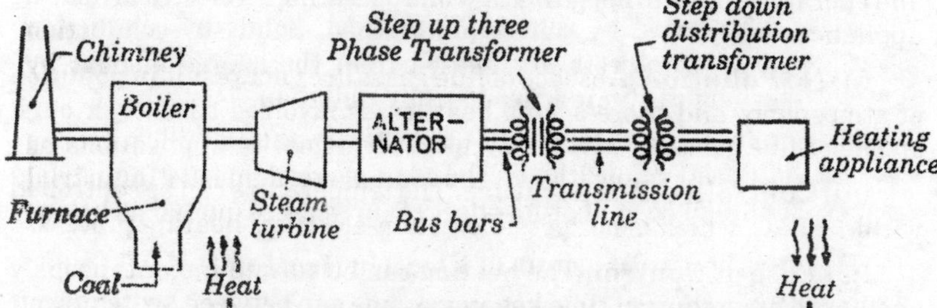

Fig. 3.1. Simple scheme of heat energy being converted into electical energy and vice versa.

The above diagram shows production of steam in a boiler with burning of coal in a thermal power plant. The steam turbine converts the heat energy into mechanical energy necessary for driving the alternator. The alternator converts this mechanical energy into electrical energy which is transmitted to far off places by stepping up the voltage and taken through the transmission lines to be distributed through step down transformers. This electrical energy is again converted into heat energy by using different heating appliances. This cycle involves a lot of losses right from the (*i*) heat losses in the boiler and turbine unit, (*ii*) efficiency of the turbine, (*iii*) losses of the generation, (*iv*) Line losses in transmission and distribution, (*v*) transformer losses etc. When we analyse the situation further it is found that if,

(*i*) Instead of distributing electricity, coal is distributed to produce heat at different places, the losses still do take place. At each place of utilisation of coal instead of electricity, involves heat losses identical to those in boiler/furnace.

(*ii*) It involves transportation to each place of utility.

(*iii*) It involves storing cost.

(*iv*) It involves dust, smoke, pollution of atmosphere.

(*v*) It provides no controlling facilities.

(*vi*) It does not provide flexibility in utilisation.

(*vii*) Can we warm up an operation theatre straight way with coal?

(*viii*) Can the beauty of place be maintained by directly using the coal for heating purposes ?

(*ix*) Can bad conductors of heat be uniformly heated with heat from coal ?

The above few considerations give an idea about the importance of ignoring the losses involved in the cycle shown on Fig. 3.1.

3.2. Advantages of Electric Heating over Other Methods of Heating

(*i*) **Clean and Neat Atmosphere.** There is no coal dust or smoke and operators hand do not go black while operating an electrical heating appliance.

(*ii*) **No Pollution.** Absence of flue gases does not result in pollution of atmosphere and there are no heat losses involved as that through smoke or flue gases.

(*iii*) **Controlled Temperature.** The temperature can be controlled within ± 5° C, which is not possible in non-electrical heating process.

(*iv*) **Ease of Control.** The heating can be started instantaneously or stopped at a required time keeping a time gap between switching off and cooling of the heating circuit. Automatic switching controls are possible.

(*v*) **Localised Application.** A workpiece can be heated up to particular depth for heat treatment whereas the piece as a whole receives heat in non-electrical heating process.

(*vi*) **Low Ambient Temperature.** The temperature around an electrical furnace is much lower as compared to that around non-electrical furnace which is troublesome for the workers.

(*vii*) **Uniform Heating.** The heating can be generated from within the workpiece resulting in uniform heating through induction heating.

(*viii*) **Heating of Bad Conductor of Heat and Electricity.** Wood, plastic and bakery items can be uniformly and suitably heated with dielectric heating process.

(*ix*) **Highest Efficiency of Utilisation.** Heat produced electrically does not go waste through chimney and other by-products. Most of the heat produced is utilised by the material being heated electrically. This results in the highest efficiency as compared to other heating processes.

(*x*) **Cheap Furnaces.** The electrical furnaces do not require big space for installation, no storage of coal and fire wood is necessary, no chimney to be constructed, no extra heat insulation necessary. This brings the cost involved to a lower value as compared to non-electrical furnaces.

(*xi*) **Mobility of job.** Pieces undergoing heat treatment can be mounted on a conveyer passing through the heating cabinets making use of electric heaters.

3.2.1. Classification of Heating Methods.

1. Low temperature up to 400°C.
2. Medium Temperature 400 – 1150°C.
3. High temperature - temperatures above 1150°C are classified as high temperatures.

3.3. Heating Methods

Basically heat will be produced due to circulation of current through a resistance. The current may be circulating directly due to application of potential difference or it may be due to induced eddy currents. Similarly in magnetic materials, hysteresis losses can cause heating. Inter molecular fraction produced by repeated changes in alignment can also be source of heat. This will be evident from dielectric heating. Production of an arc between an electrode and the material to be heated can also be a method of heating, or striking an arc between two electrodes in the vicinity of some object can heat the object. Bombardment by some high energy particles like α, β, γ rays or X-rays or accelerating ions can produce heating of a surface.

The various methods of producing heat for general industrial work, heating of building, baking, drying and for welding may be conveniently classified as follows :

Resistance Heating. This method is based upon the I^2R loss. Whenever current is passed through a resistive material heat is produced because of I^2R loss. There are two methods of resistance heating: As mentioned below :

(*i*) **Direct Resistance Heating.** In this method of heating the material or charge to be heated is taken as a resistance and current is

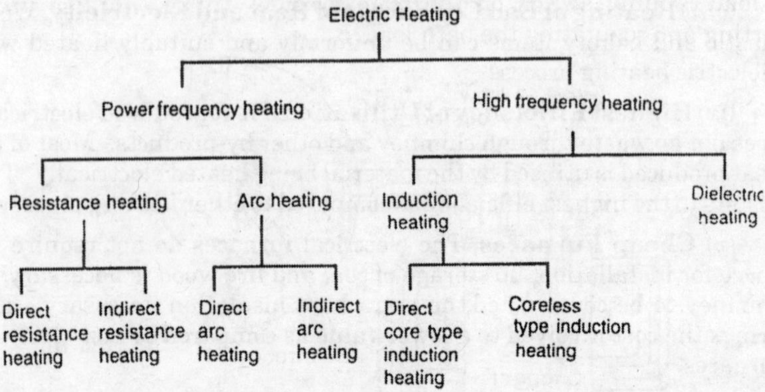

passed through it. The charge may be in the form of powder, pieces or liquid. The two electrodes are immersed in the charge and connected to the supply. In case of D.C. or single phase A.C., two electrodes are required but there will be three electrodes in case of three phase supply. When metal pieces are to be heated a powder of high resistivity material is sprinkled over the surface of the charge to avoid direct short circuit. The current flows through the charge and heat is produced. This method has high efficiency since heat is produced in the charge itself.

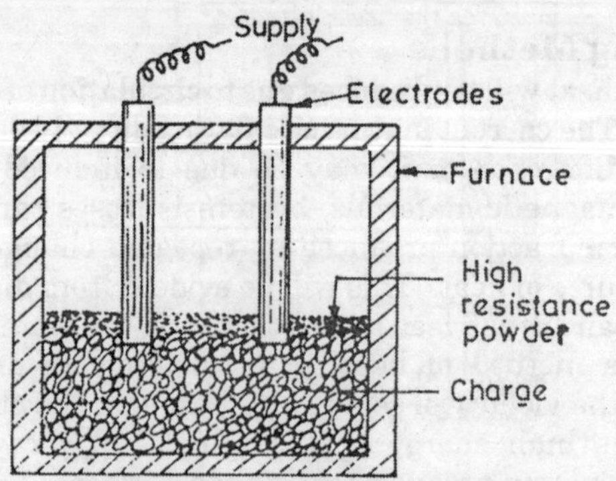

Fig. 3.2 Direct Resistance Heating.

Though automatic temperature control is not possible in this method but it gives uniform heat and high temperature. One of the major application of the process is salt bath furnances having an operating temperature between 500°C to 1400°C.

An immersed electrode type medium temperature salt bath furnance is shown in Fig. 3.2 (a). The bath makes use of a supply voltage across two electrodes varying between 5 to 20 volts. For this purpose a special double wound transformer is required which makes use of 3-ϕ primary and a 1-ϕ secondary. This speaks of an unbalanced load.

The variation in the secondary voltage is done with the help of an off load tapping switch on the primary side. This is necessary for starting and regulating the bath load.

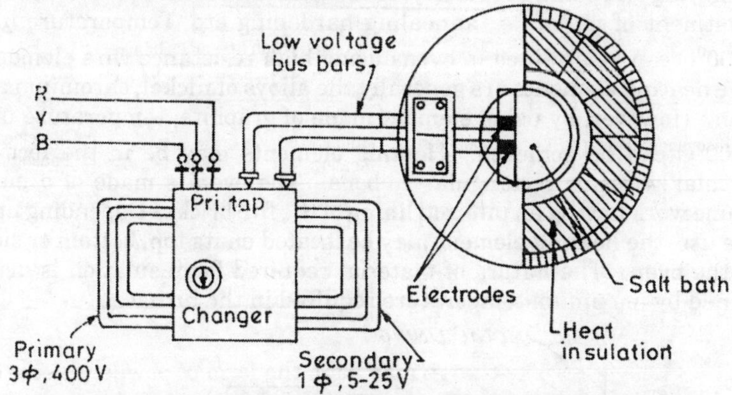

Fig. 3.2. (a) A simplified view of an electrode type salt bath furnace.

Indirect Resistance Heating. In this method the current is passed through a highly resistance element which is either placed above or below the oven depending upon the nature of the job to be performed.

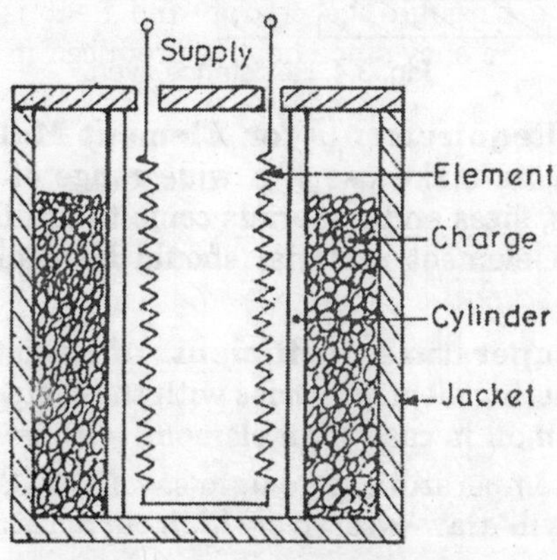

Fig. 3.3. Indirect Resistance Heating.

The heat proportional to I^2R losses produced in heating element delivered to the charge either by radiation or by convection.

Sometimes in case of industrial heating the resistance is placed in a cylinder which is surrounded by the charge placed in the jacket as shown in Fig. 3.3. The arrangement provides an uniform temperature.

Automatic temperature control can be provided in this case. Both A.C. & D.C. supplies can be used for this purpose at full mains voltage depending upon the design of heating element, the common example of this type of heating is resistance oven.

3.4. Resistance Furnace or Oven

This is an enclosed chamber with a provision for ventilation used for drying and baking of pottery, commercial and domestic cooking, heat treatment of metal *i.e.,* annealing hardening etc. Temperature upto 1000°C can be attained in ovens using high resistance wire elements. The heating elements are generally the alloys of nickel, chromium and iron. However, by using element made of graphite, temperature upto 3000°C can be achieved. Heating elements may be in the form of circular wires or rectangular ribbons. The oven is made of a metal frame work having an internal lining of the fire bricks. Depending upon the use, the heating element may be located on its top, bottom or sides of the oven. The nature of material required for insulation is determined by maximum temperature required in the furnace.

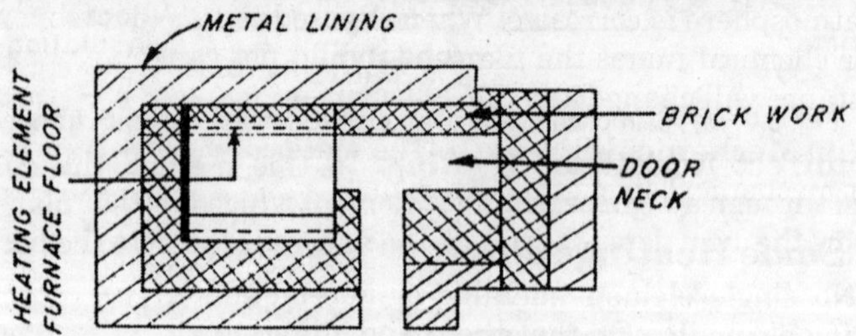

Fig. 3.4. Resistance oven.

Material Requirements for Element Making. The indirect resistance furnaces make use of a wide range of temperatures and different shapes, sizes and materials come to fulfill different requirements. A good element material should have following important properties *viz :*

(*i*) **Low Temperature Coefficients of Resistance.** Since resistance of a conducting element varies with the temperature, this variation should be small in case of an element. Otherwise when switched ON from room temperature to go upto say 1200°C, the low resistance at initial stage will draw excessively high currents at the same operating voltage.

(*ii*) **Resistance Coefficient Positive.** If temperature coefficient is negative the element will draw more current when hot. A higher current means more wattage, a higher temperature or a still lowest resistance, which can cause instability of operation.

(*iii*) **High Melting Point.** Its melting point should be sufficiently higher than its operating temperature. Otherwise a small rise in the operating voltage will destroy the element.

(*iv*) **High Specific Resistance.** The resistivity of the material used for making element should be high. This will require small lengths and shall give convenient size.

(*v*) **High Oxidizing Temperature.** Its oxidizing temperature should be higher than its operating temperature. Otherwise oxidised layers from the surface will flake off changing the resistance of the filament and giving it a smaller life.

(*vi*) **Ductile.** To have convenient shapes and sizes, the material used should have high ductility and flexibility. It should not be *brittle and fragile*.

(*vii*) **Should Withstand Vibrations.** In most industrial processes quite strong vibrations are produced. Some furnaces have to open or rock while hot. The element material should withstand the vibrations while hot and should not break open.

(*viii*) **Anti-corrosive.** In many heat treatment processes the heating atmosphere is corrosive. When exposed to this atmosphere or any other chemical fumes the filament should not corrode. Otherwise its resistance will change and life will shorten.

(*ix*) **Mechanical Strength.** The material used should have sufficient mechanical strength of its own.

3.5. Some Heating Alloys

No single element metallic or non-metallic is used for making heating elements. On the other hand, different alloys have been tried to give different operating temperature. The following list gives the names, composition and safe operating temperature for different alloys used as heating elements.

S.No.	*Name of alloy*	*Composition*	*Safe operating temperature*
1.	Constantan	Nickel 45% Copper 55%	400°C
2.	Nichrome	Nickel 80% Chromium 20%	1500°C
3.	Kanthal	Chromium Aluminium Cobalt, Iron	1300°C
4.	Silicon carbide	1450°C	

3.6. Causes of Failure of Heating Elements

Sooner or later, with the passage of time each heating element breaks open some day, rendering it unserviceable. Some of the factors responsible for its failure can be quoted as follows :

(*i*) Formation of hot spots. (*ii*) Oxidation.

(*iii*) Corrosion. (*iv*) Mechanical failure.

(*i*) **Formation of Hot Spots.** In this case the point of failure is the place where the element shines brightest during operation. This speaks of a comparatively higher operating temperature at the particular spot as compared to the rest of the filament. The formation of hot spots can be further attributed to different causes *viz.*

(*a*) *Unequal spacing.* If the spacing of different turns/parts of the element is non-uniform, the temperature will be highest, where spacing is minimum. That will result in formation of a hot spot.

(*b*) *Shielding by supporting network.* If the supporting structure is a bad conductor of heat and obstructs the transfer of heat the temperature of the filament will remain comparatively higher near the supports. This will result into formation of hot spots near the supports.

(*c*) *Excessive oxidation.* The portion of the filament undergoing excessive oxidation with respect to the rest of the element offers a higher resistance to the same current. The heat produced in this portion is thus higher resulting into a hot spot.

(*ii*) **Oxidation.** The outer/exposed surface of the element gets oxidised at high temperatures. During switching operations the oxide layer gets flaked off, exposing inner surface to atmosphere. During further operation the exposed inner surface gets oxidised. The working cross-section at such spots becomes comparatively smaller resulting into formation of hot spots and ultimate break down.

(*iii*) **Corrosion.** If the filaments are directly exposed to chemical fumes during industrial operations, the reactions will corrode the contact region where fumes actually touch the element. From this portion the element will ultimately fail.

(*iv*) **Mechanical Failure.** During alloying a portion may have a higher content of the higher resistivity metal/constituent. The portion will produce more heat and fail ultimately. If one of the constituent is iron the size of the element will increase wherever iron content is relatively high. The portion becomes fragile when cold and the element breaks mechanically from such portions.

3.7. Temperature Control of Resistance Furnaces

If a constant voltage is continuously energising an electric furnace, the temperature will gradually be increasing with the passage of time. It will cross the desired level of heating at one stage. It is at this stage when some controlling mechanism should come into action and should stop further rise of temperature. In doing so the heat generated by the heating elements has either to be cut off momentarily or reduced. In either case the temperature will try to fall from a predetermined level. At this stage the procedure has to be reversed, *i.e.* the supply if it has been reduced or stopped has to be fully restored. In either case we can easily guess that the temperature will vary about a mean value but it will not be exactly constant. How closely it remains near a mean value depends upon the *response* of the furnace to the controlling action and the *mode* of control.

The control exercised on larger furnaces is automatic which can be electromagnetic or electronic involving servo-mechanism operations. The modes of control can be :

(*i*) Two step

(*ii*) Three step

(*iii*) Continuous.

The two steps can be (*a*) ON - OFF, (*b*) Low-High.

As indicated by the name ON - OFF mode, when the temperature reaches the desired level the supply to the heating elements should be switched off. In case of low and high mode the circuit has not to be switched off but the heat generated should be reduced. In three step control the levels of temperature involved are three *viz.* low - medium - high. In case of continuous control, the control mechanism continuously varies the energy input depending upon the temperature inside the furnace so that the input is reduced when temperature rise and *vice versa*.

The basic control involves following adjustments which can vary the temperature inside a resistance furnace :

(*i*) Intermittent switching, *i.e.* ON - OFF control.

(*ii*) Variation of number of heating elements.

(*iii*) Variation in circuit configuration.

(*iv*) Change of applied voltage.

(*i*) **Intermittent Switching.** As mentioned earlier by switching the heating supply OFF for some time the temperature rise can be stopped beyond a desired limit. Of course the switching has to be done before the temperature actually reaches the desired limit. Even after switching off the supply, the temperature rises to some extent in totally enclosed furnaces. After some time the temperature starts falling off. When the fall reaches again a predetermined value the supply is again switched ON. Since entire furnace is switched ON and OFF in this process there is lot of arcing at the switching contacts. As such the method is suitable for small furnaces only. Air circuit breakers are ideal

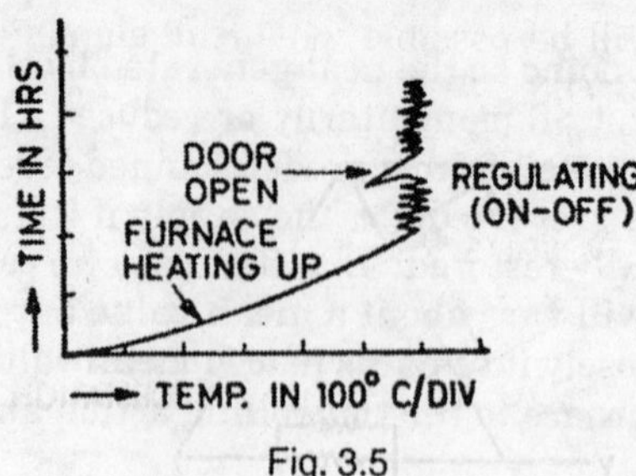

Fig. 3.5

if switching action has to be frequent to keep the temperature within close working limits.

The figure 3.5 shows the results of ON-OFF control on temperature as a function of time.

(ii) **Variation of Number of Heating Elements.** In this case instead of cutting off the supply to the entire furnace, the number of heating elements are changed in the supply circuit. Obviously with smaller number of heating elements, supply of heat is reduced and *vice versa*. Lesser the supply of heat smaller should be the temperature and so on. This is applicable for continuous mode and multistep modes of temperature control. High/low, two step mode can also work on this practice.

Note. In case of a three-phase circuit, case should be taken that while removing from heating circuit, equal number of elements are switched off from each phase. This is necessary to keep balanced load conditions. Secondly from a row of heating elements the switched off elements should be equally spaced e.g. If sex elements are connected per phase and while regulating, three have to be removed out of circuit if either of the switch is opened three elements will be switched off. If S_1 is opened, elements *A, C* and *E* will come out. These are alternately spaced and will keep a uniformity of operating temperature of inside the furnace. If on the other hand, *ABC* are removed, one side of the furnace will cool down and other shall remain hot if special arrangements do not exist for forced circulation of heat.

(iii) **Variation in Circuit Configuration.** In case of three-phase secondary load, the heating elements will give different heat when in star formation and more heat when in delta formation. When say *X* number of elements connected in parallel come per phase in star the voltage across these elements is $\frac{V_L}{\sqrt{3}}$ volts. If suppose line voltage is 400 volts the voltage in star formation per phase will be 231 volts. Since power is $\frac{V^2}{R}$ watts it will be *one-third* of the power when the same

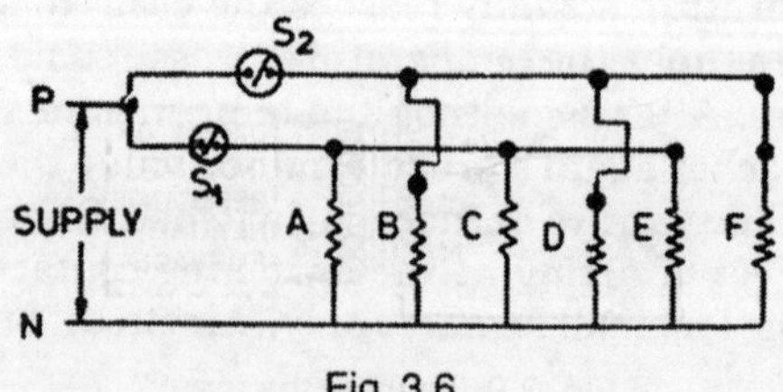

Fig. 3.6.

elements get full 400 volts. This will be possible when the elements appear in delta formation (See Fig. 3.7).

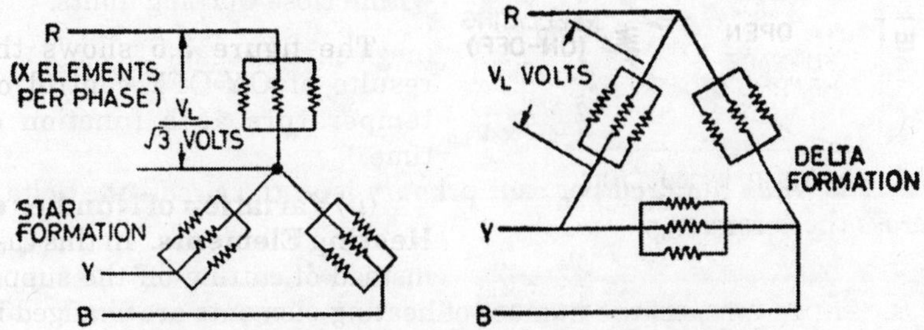

Fig. 3.7. Star/Delta configuration of heating elements.

In single phase circuits series parallel grouping can cause a change in power dissipation resulting in change of temperature.

In Fig. 3.8, when all four elements are in parallel, voltage across each element of the same resistance is V volts. In the series parallel formation the voltage across each element is reduced to $V/2$ volts. Obviously power dissipated in the second case will be *one-fourth* of the previous value. (Neglecting temperature coefficient and assuming constant resistance of the elements in either connections.)

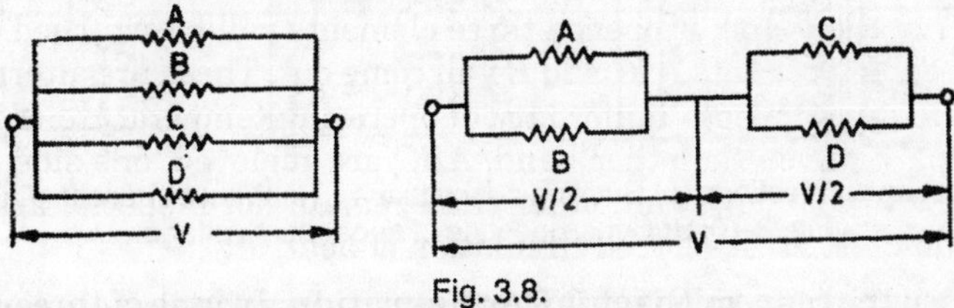

Fig. 3.8.

(*iv*) **Change of Applied Voltage.** (*a*) This is of course very simple to guess that lesser is the magnitude of applied voltage smaller will be the power dissipated and lesser will be the temperature produced. In case of furnace transformer having high voltage primary; it is advisable to have a tapping control on the high voltage side. It is because of *lesser* magnitude of *current* on this side. For simplicity let us take into consideration a single phase circuit involving a multi-tap step down transformer.

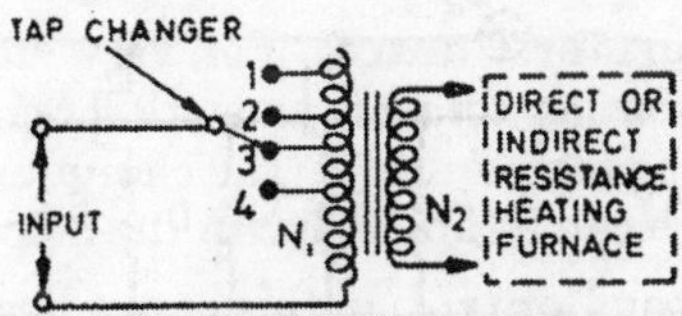

Fig. 3.9. Tapping the input.

Let there be four tappings on the primary for say 100%, 80%, 60% and 50% turns. When input is fed across 50% turns the voltage across the secondary,

$$= \frac{N_2}{N_1/2} \times V_{in} \text{ (neglecting drops)}$$

$$= \frac{2 N_2}{N_1} \times V_{in} \quad \text{...(i)}$$

whereas when hundred per cent primary is in the circuit the voltage across the secondary

$$= \frac{N_2}{N_1} \times V_{in} \quad \text{...(ii)}$$

which is just half that of the previous value. So by selecting a suitable primary tapping, voltage across the secondary can be reduced or increased causing a change of temperature in the heating furnace.

(*b*) **Bucking-boosting the Secondary Voltage.** In this case the secondary may have two sections one with a *lesser* number of *turns*. If the direction of current in the smaller section is same as that of remaining secondary, its induced e.m.f. helps or adds to the voltage induced in the rest of the secondary. This action is boosting action, and results in increased secondary voltage. The effect is shown in Fig. 3.10.

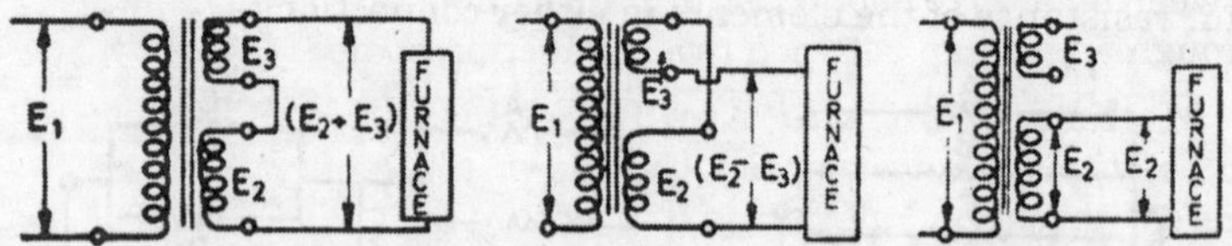

(*a*) Boosting connections (*b*) Bucking connections (*c*) Without boosting/bucking
Fig. 3.10. Bucking/Boosting of secondary voltage.

When the connections of this extra secondary winding are reversed the induced e.m.f. in this winding is in opposition to the main secondary. Thus the second effect is *Bucking* effect and the voltage across the furnace is reduced. The third possibility is to neglect the voltage developed across the extra secondary thus the three different voltages possible against the furnace can be,

(*i*) $E_2 + E_3$ ⇒ boosting

(*ii*) $E_2 - E_3$ ⇒ Bucking

(*iii*) E_2 ⇒ no bucking-no boosting.

(*c*) **Auto-transformer Control.** For very small resistance furnaces, the auto transformer control can work for changing the applied voltage and as such the temperature. The circuit used is shown in Fig. 3.11. Required voltage can be applied with the help of voltage selector.

(*d*) **Series Reactor Control.** In this case a heavy duty coil wound on a core is placed in series with the furnace as and when designed. A

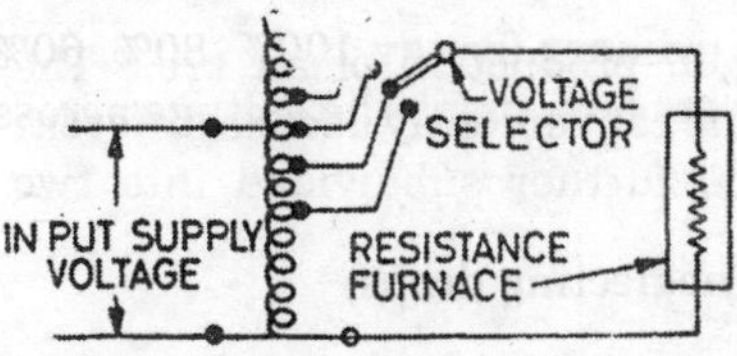

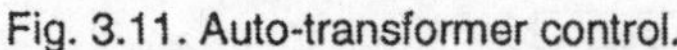
Fig. 3.11. Auto-transformer control.

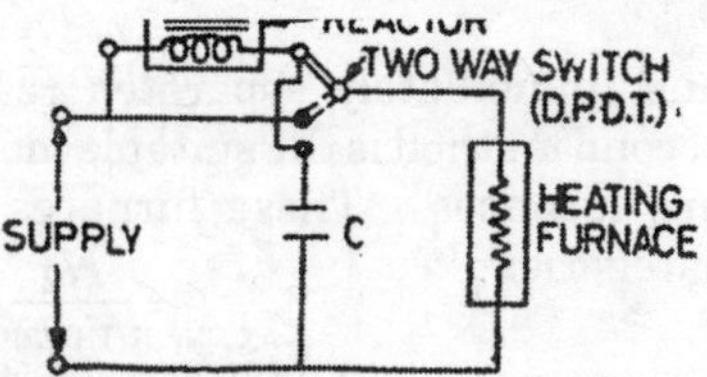

Fig. 3.12. Series Reactor Control.

part of applied voltage drop across the impedance produced by the coil and the rest appears across the furnace, thereby reducing the temperature inside the furnace. The circuit involved is shown in Fig. 3.12. The D.P.D.T. switch provides Low, High, two step made of temperature control. Since the introduction of inductive reactance in series with the furnace can reduce the power factor, a power capacitor is simultaneously introduced in the circuit for keeping the p.f. near unity. The inductor comes in series whereas the capacitor comes in parallel with the furnace load.

3.8. Arc Furnaces

The furnaces used for melting/extraction of ferrous and non-ferrous metals need a high temperature operation. One of the methods is through production of electric arc which gives an Arc temperature between 3000°C to 3500°C on L.T. operation. Arc is the flow of cur 'ent through an air gap between two conducting bodies.

Methods of Striking Arc. There are two methods of striking an arc between two electrodes. On one of the methods i **H.T. strike** whereas the second method is **L.T. strike method.**

In the first method a constant gap is maintained across a pair of electrodes say made of carbon. The electrodes are connected across the H.T. secondary of a step-up transformer. The primary is fed with variable A.C. voltage as shown in the circuit. To strike an arc the primary input voltage is gradually increased, thus increasing the high tension voltage across the secondary/electrodes. A stage comes when the medium between the two electrodes is ionised and becomes conducting. It is at this stage when an arc is struck between the two electrodes. After the striking of the arc the p.d. across the electrodes has to be reduced from H.T. to L.T. since the conducting gap between the two electrodes has now negligibly low resistance as compared to the gap resistance before striking of the arc. The H.T. required is dependent upon the size and nature of the gap*, but runs into thousands of volts.

In the second case a low voltage of the order of mains voltage enough to strike the arc. In this case the electrodes arc connected to the lower voltage side of the transformer. These are momentarily short circuited

*A few mm. gap of a sparking plug used in auto-engines needs 10-20 kV, H.T. for producing spark, which is struck momentarily and not maintaines like an arc of the furnace.

and immediately separated resulting in production of an arc. The second method is the suitable method and conveniently used in different arc furnaces. These furnaces can be further subdivided into two categories :

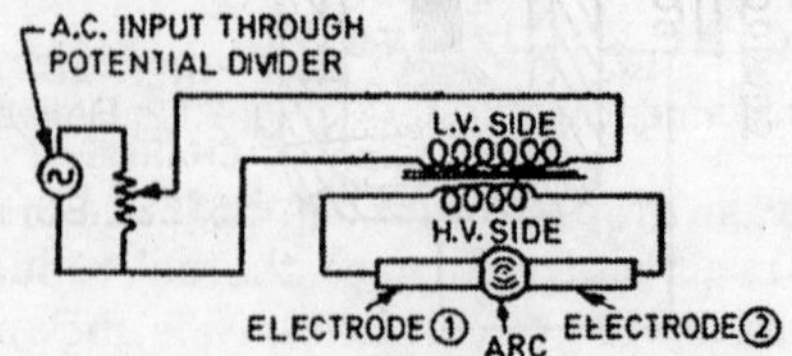

Fig. 3.13. H.T. method of striking an arc.

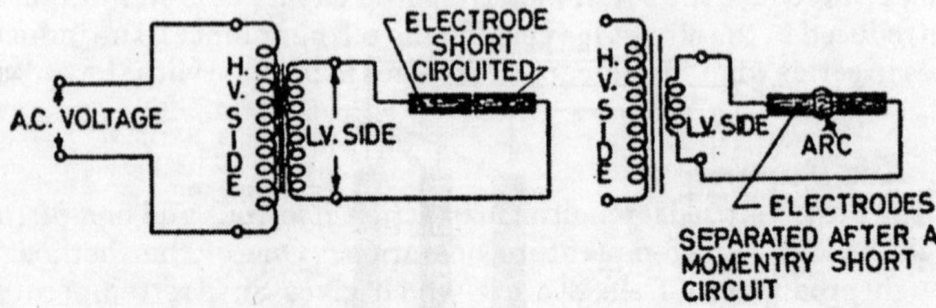

Fig. 3.14.

(*i*) Direct-arc furnaces. (*ii*) Indirect-arc furnaces.

When the arc is struck between electrode/electrodes and the charge to be heated, so that the arc current flows through the charge and there is a direct contact between the arc and the charge, it is the case of a direct-arc furnace. The heating process is faster in this case, heating efficiency is too high and in addition to arc heat, additional heat is produced by the current passing through the charge undergoing heating.

In the second case heat produced by the arc set up between two electrodes does the heating indirectly. The arc does not come in contact with the charge but heat is transferred through radiation from arc to top layer of charge and through conduction from top layer to the bottom layers in the charge.

Direct-Arc Furnaces. Electrically these furnaces can be further sub-divided into two categories *viz.*

(*i*) Conducting bottom. (*ii*) Non-conducting bottom.

The former has the bottom of the furnace as a part of the electrical circuit whereas no current flows through the body of the furnace in the second case. It will be clear from the study of the Fig. 3.15.

Most of the furnaces used are non-conducting bottom type due to insulation problem faced in case of conducting bottom.

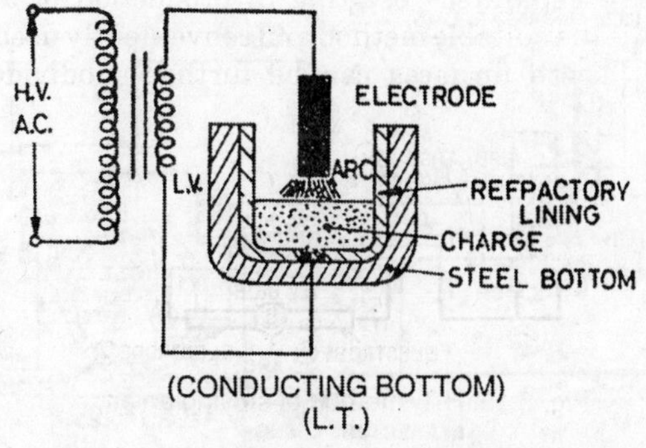

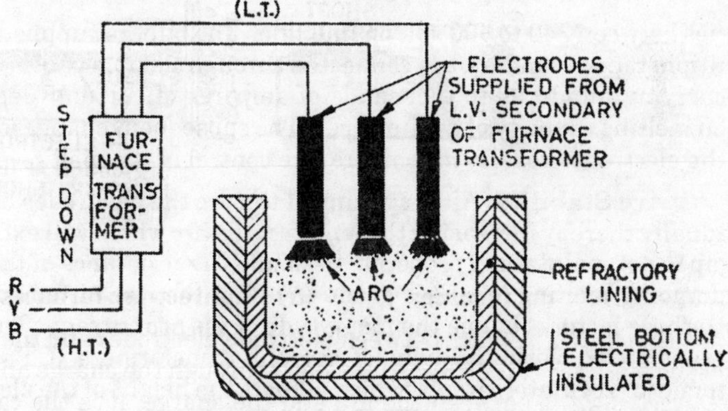

Fig. 3.15. Conducting/Non-conducting bottom type direct-arc furnaces.

Basic Electrical Requirements. The diagram shown in Fig. 3.16 meets the fundamental requirement of an electric arc furnace. These requirements can be emunerated as:

(*i*) Supply to the electrodes.

(*ii*) Arc stability.

(*iii*) Over-current protection.

(*iv*) Isolation of the furnace from supply.

(*v*) Selection of electrode material.

(*i*) **Supply to the electrodes.** The electrodes have to be short circuited momentarily and them separated from the charge inside the furnace, which requires a low voltage on the secondary and a high reactance transformer winding. For this purpose the furnace transformer has as high a reactance as 35% and multiple tappings on the secondary. The primary has a line voltage of 11 kV whereas secondary

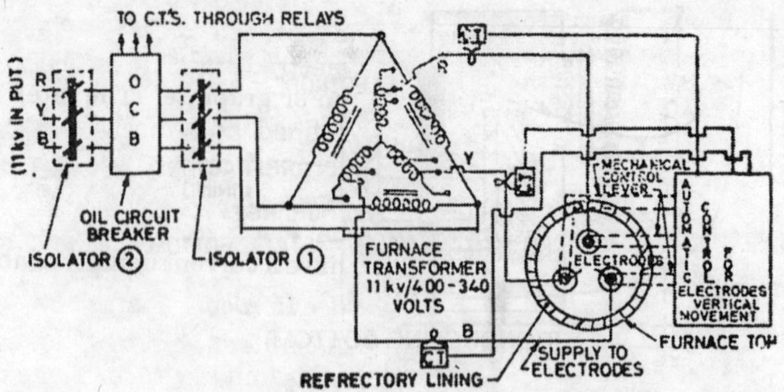

Fig. 3.16 Basic circuit for electric arc furnace.

varies between 340 to 400 volt on line side. The bubars supplied by this multiple tape secondary are connected through flexible cables capable of carrying hundreds to thousands of amperes of current (depending upon melting capacity of the furnace). The subsequent change of supply to the electrodes results in temperature control of the melt.

(*ii*) **Arc Stability.** Alter striking of the arc the electrodes burn out gradually thereby increasing the length of the arc which will extinguish if gap is not maintained properly. Secondly the resistance of the metal undergoing melting, changes gradually with the rise of temperature. Both these factors change the magnitude of the arc current. The effect is transmitted with the help of current transformers (CT's) to an automatic* regulator circuit, which adjusts the height of the electrodes during the melting process. The stability** of arc for a fixed *given* gap is further due to *high reactance* of the furnace transformer.

(*iii*) **Over current/ground or leakage fault protection.** With the exception of momentary short circuit, the furnace is not supposed to draw an over-current (*i.e.* more than the furnace is rating) continuously. For this purpose the current transformers operate a relay/relays circuit tripping the oil circuit breaker. The tripping can be due to 11 kV mains voltage fluctuation as well.

(*iv*) **Isolation of The furnace from the supply.** When metal scrap has to be charged into the furnace or melt/slag removed from the furnace it has to be disconnected from the supply. For this purpose an isolator switch is connected between the oil circuit breaker and furnace transformer. The O.C.B. is tripped off and then line isolator opened.

A second isolator is provided between mains and the O.C.B., this is necessary to disconnect the O.C.B. from the line during maintenance and repairs. Frequent trippint carbonises the oil of the breaker and burns the contacts which have to be taken care of periodically.

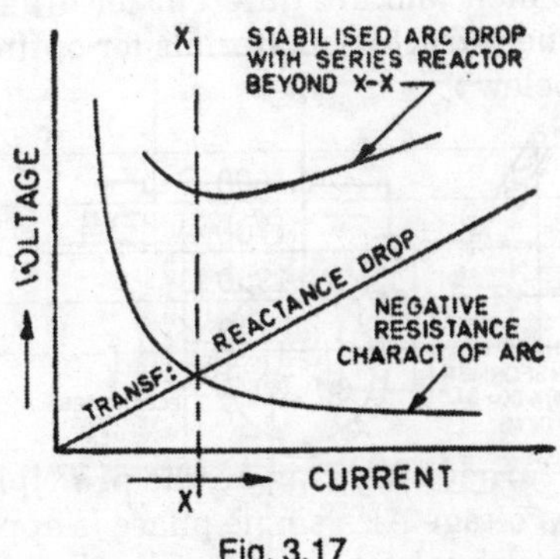

Fig. 3.17

(*v*) **Selection of electrode material.** The electrodes used are either carbon or graphite. The later is a refined and specially treated form of carbon and has four times conductivity. The carbon electrodes are cheap and have a current density of about 8 – 15 A/cm^2 against 32 to 60 A/cm^2 for graphite. Thus carbon is suitable for low capacity arc furnaces only.

3.9. Basic Mechanical Requirements

1. **Charging.** For charging purposes the roof as a whole along with electrode supporting column shifts to one side with furnace for 4 tonne capacity and above. The iron or other metallic scrap is crane loaded into the furnace and hammered with hanging loads. Door charging is possible for small furnaces.

2. **Removal of Melt and Slag.** When the melt is ready the front door called tapping spout is opened and the furnace (steel tank) is tilted forward by 45° as a whole, throwing the melt into a *ladel pit* provided for the purpose from where it is distributed for the specific operations. For removal of slag the furnace is tilted backwards by 15°, and the slag door opened. This results into discharge of slag into a slag pit.

3. **Shell and Lining Structure.** The shell is made of steel and cylindrical in shape. It has two doors for tapping and slag removal opposite each other where roof is detatchable otherwise charging door is an extra provision. It is desirable if the working furnace is air leak-proof. This avoids contamination of atmosphere from furnace fumes and provides controlled atmosphere heating facilities.

The inner surface of the steel shell has to be lined with something heat resistant and which prevents leakage of heat from the melt to the steel shell or chamber. The linings can be basic or acidic in nature

* The electric controls make use of amplidyne generators controlling D.C. motors which maintain the gap of the electrodes and the melt to a desired length. The circuit used is quite complicated and beyond the scope of this book.

** When arc current is increased say due to more supply voltage, its dia. increases. Which means reduced arc resistance and tendency to draw more current. Thus this negative resistance characteristic of the arc needs a high series reactance which drops more voltage when current tries tc rise and keep the arc current stable (Fig. 3.17).

* It is never opened on load otherwise a severe arching damages the isolator contacts permanently.

depending upon the nature of the melt and are different for different working temperatures. Some of the refractory materials for optimum working temperatures are given below :

1. Mullite	1600°C
2. Claybonded fused alumina	1700°C
3. Impervious Alumina	1950°C
4. Zirconia	2400°C

3.10. Indirect Arc Furnace

Fig. 3.18 shows a rocking arc furnace working on the principle of indirect arc. In the given case low voltage A.C. single phase is applied across the electrodes. The arc is struck by short circuiting the electrodes manually or automatically for a moment and with drawing these apart. The heat from the arc is transferred to the charge top layer and refractory lining, through radiation and from top layer of the charge to other portion of the charge through conduction. To distribute the heat properly, the furnace has to be rocked continuously, exposing further layers not only to the heat transferred from the arc, but from the exposed lining. This results in prolonged life of the lining as well since during rocking a portion of the lining is not continuously exposed to heat as in the case of non-rocking furnaces.

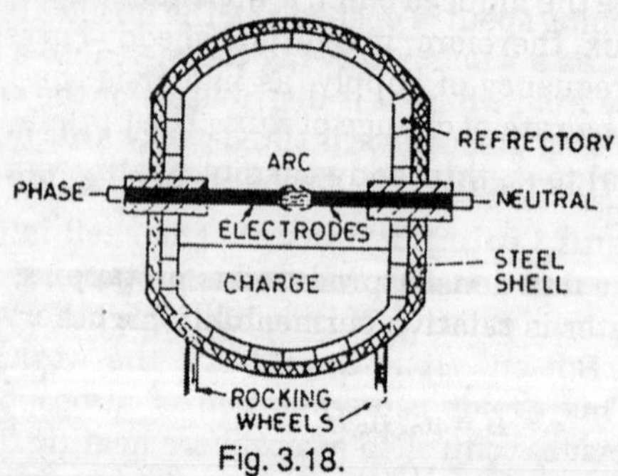

Fig. 3.18.

Since the unit is working on a single phase, a great unbalance will result if a furnace transformer without three-phase primary is not used. The furnaces are not supposed to handle melting beyond one tonne, for the same reason. The unit works quite efficiently due to totally enclosed construction, the carbon electrodes further improve the quality of ferrous melts. Special motors with reduction gears and reversing directions are used for rocking.

INDUCTION HEATING

3.11. Introduction

(*i*) **Eddy Currents.** From our previous knowledge it is known to use that whenever flux linking with any conduction body changes, the result is an induced e.m.f. which is a function of rate of change of flux, or rate of change of current.

Consider a coil of N turns wound on one limb of a magnetic core. Let the coil be supplied with alternating current from a source of A.C. supply. Obviously the flux produced by this current will be changing in magnitude as well as direction. This changing flux links with the coil as well as with the magnetic core resulting in an induced e.m.f. as per $e = -N\dfrac{d\Phi}{dt}$ volts or $e = -L\dfrac{di}{dt}$ volts. Since the core is a complete circuit single coil turn, any induced e.m.f. will result in circulation of current through this core. This current is known as **eddy current.** Obviously the magnitude of this current depends upon the magnitude of induced e.m.f. and the resistance offered by the core material to the flow of this current. If, i amperes is the magnitude of current and R ohms the resistance of the core material, then $i^2 R$ watts will be the power dissipated by the core. This power will be utilised in *heating** the core. It is this effect of induced eddy currents which provides the basis for induction heating.

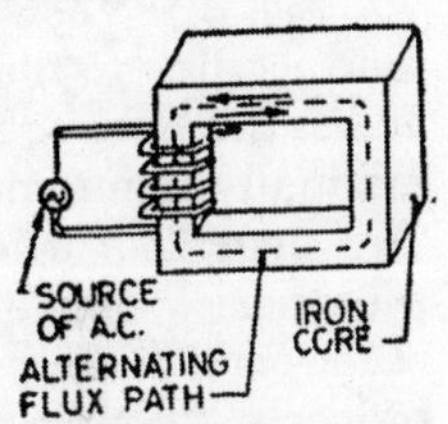

Fig. 3.19

(*a*) Since the induced e.m.f. is dependent upon the rate of change of flux, therefore, magnitude of eddy current is proportional to frequency of supply, as higher is the frequency, greater will be rate of change of flux. Heat produced is thus proportional to i^2, thus eddy current heating effect is proportional to f^2.

(*b*) Since flux density produced is proportional to reluctance, so greater is relative permeability higher will be the flux density, B.

where, $B = \mu_o . \mu_r . H$

and $H = \dfrac{NI}{l}$

where N = No. of turns on the coil.

I = Current through the coil.

l = Length of the magnetic circuit.

or heating is a function of μ_r or relative permeability.

---

*In order to minimise the heating effect and the power lost in heating the core, the cores of transformers and other electromagnetic machines are laminated. The laminations are insulation coated from one side to increase the effective resistance and reduce eddy current losses.

(c) Greater is the number of turns on the coil greater will be the magnitude of flux. Thus eddy current heating is a function of N.

(d) Larger is the current flowing through the coil greater will be the flux produced and higher will be the eddy currents. So we can say that eddy current heating is proportional to the magnitude of current exciting the induction coil.

(e) Lesser is the resistivity of the core material, greater will be the magnitude of eddy currents or eddy current heating is proportional to resistivity/conductivity of the heating material.

(f) $$W_e = KB_{max}^2 f^2$$

where W_e = Power dissipated through eddy current

K = Constant of proportionality

B_{max} = Flux density (Maximum)

and f = Frequency.

(ii) **Hysteresis.** We also know that if a piece of magnetic material is magnetised, de-magnetised and re-magnetised time and again some energy is lost in the process. The losses so occurring in any electromagnetic device are named as *hysteresis losses.* The energy lost is converted into *heat.* This heat is over and above the heat produced by the eddy currents.

During positive half cycle the core is magnetised to maximum at positive peak. When the sinusoidal current becomes zero, the magnetic flux density is not reduced to zero due to residual flux. During negative

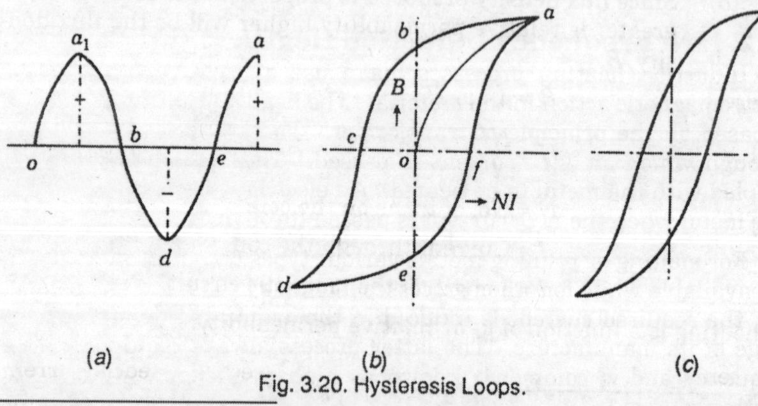

Fig. 3.20. Hysteresis Loops.

*In case of electromagnetic machine the stampings used for core/armature, material is silicon steel having a small area of hysteresis loop as shown in Fig. 3.20 (c), this ensures low magnitude of hysterisis losses.

f cycle some coercive force is necessary to reduce the residual flux to zero. This happens at C midway somewhere between b and d. If d be the peak of negative half cycle, the core is having *maximum reversed flux*, to be demagnetised and remagnetised by the coming positive half cycle. The path traced by the flux density corresponding to NI at different instantaneous values of the exciting current, is shown in Fig. 3.20 (*b*) and (*c*). The configurations are termed as hysteresis loops. Greater is the area* covered by one such loop, larger is the magnitude of hysteresis loss and more is the heat produced.

(*a*) Greater is the frequency of a.c. supply, larger will be the number of such loops traced/sec, and more will be the heat produced.

(*b*) The experimental verification has established the hysteresis losses to be proportional to $B_{max}^{1.6}$.

(*c*) Since B is further linked with μ_r, the hysteresis losses are proportional to relative permeability of the material. As a matter of fact a permanent magnetic material has a broader hysteresis loop as shown in Fig. 3.20 (*b*). Which speaks of more power dissipation and more heat generation.

(*d*) Above certain temperature of the magnetic material under going induction heating, the hysteresis effect ceases to exist. The temperature at which it occurs is known as **'Curie temperature'**. After that the behaviour of the material is like any non-magnetic material, being induction heated.

(*e*) $W_h = KB_{max}^{1.6} f$, where W_h = Power dissipated in a core due to hysteresis. B_{max} = Maximum flux density and f is the frequency of induction.

3.12. Low Frequency Induction Heating

Induction heating processes make use of currents induced by electromagnetic action in the material to be heated. Induction heating is based on the principle of transformer. There is primary winding through which an A.C. current is passed. The coil is magnetically coupled with the metal to be heated. An electric current is induced in this metal when the A.C. current is passed through the primary coil.

For melting or refining metals various types of induction furnaces are available while for other processes like case hardening and soldering, the required current is induced in the article at the appropriate stage of its manufacture. The latter process always requires a high frequency and is commonly known as high frequency eddy current heating.

The following are different types of induction furnaces employed for melting purposes :

(1) **Core type**

(*a*) Direct core type, (*b*) Vertical core type (*c*) Indirect core type

(2) **Coreless type**

(*a*) **Direct Core type induction furnace.** It is like a transformer. The change forms the secondary winding and consists of one turn only formed by the metal to be melted. The current in the charge is very high, say of the order of several thousand amperes. The charge is magnetically coupled to the primary winding. Electromagnetic forces are set-up by the high current in the melten metal. When there is no molten metal, no current will flow in the secondary. To start the furnace molten metal is poured in the annular hearth or a sufficient quantity is left over from the previous charge. There are few drawbacks in the furnace which are given below:

(*i*) Leakage reactance is high, power factor is low due to poor magnetic coupling.

(*ii*) Low frequency supply is required and additional motor generator set or a frequency converter is required.

(*iii*) If current density increases about $5A/\text{mm}^2$ the pinch effect is there and that causes a complete interruption of secondary.

(*iv*) Odd shape of crucibles are not convenient from metallurgical point of view.

(*v*) Furnace cannot function if the secondary circuit is open. It must be closed.

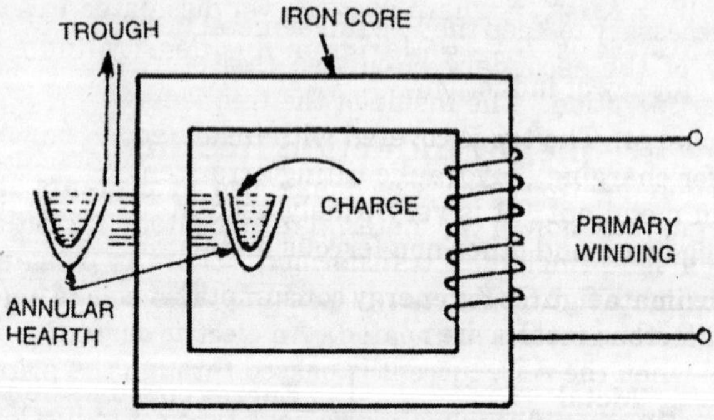

Fig. 3.21. Direct type induction furnace.

(*b*) **Vertical core-type induction furnace.** This furnace is an improved over core type furnace. It has vertical channel for the charge, thus the crucible used is also vertical which is suitable from metallurgical point of view. The modified form of this furnace is shown in Fig. 3.22 is known as **Ajax Wyatt vertical core type furnace.**

The magnetic coupling in this furnace is better than core type. Hence leakage reactance is comparatively low and power factor is high,

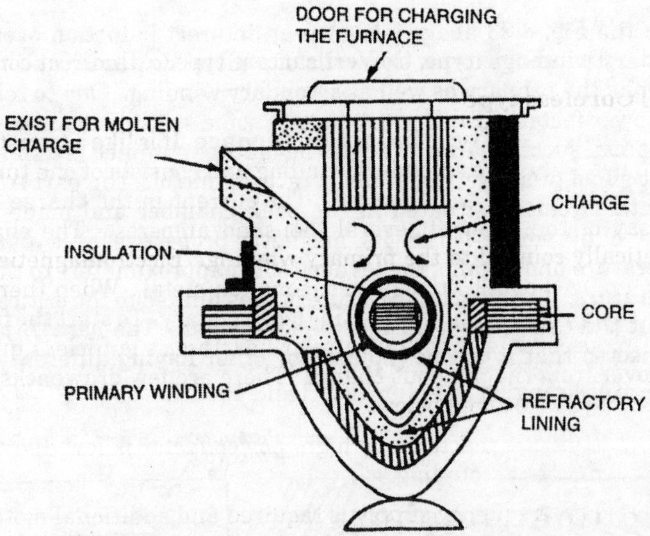

Fig. 3.22. Ajax Wyatt Vertical Core Furnace.

so it can operate from normal frequency supply. The tendency of the circuit to rupture due to the pinch effect, is counteracted by the weight of the charge in the main body of the crucible. The circulation of molten metal is kept up round the Vee portion by convection currents, as indicated and by the electromagnetic forces between the current in the lower halves of Vee.

It is necessary to keep the Vee full of metal in order to maintain the continuity of the secondary circuit so this furnace is suitable for continuous operation. The inside of the frequency is lined depending upon the charge. The top is covered with insulated cover which can be removed for charging. Hydraulic tilting arrangement is there to take the molten metal out. It is very widely used in foundries for melting the refining brass and other non ferrous metals.

Approximate figures for energy consumption for melting and refining of non-ferrous metals are :

Metal	Energy consumption/tonne
Copper and bronze	300-500 kWh
Brass	200-300 kWh
Zinc	100-150 kWh

(c) **Indirect core type induction furnance.** The induction principle can also be used for general heat treatment of metallic and other charges if an inductively heated element is employed to transmit the heat to charge by radiation. So far as the charge is concerned the conditions are similar to those as in the resistance oven.

In the Fig. 3.23 above is shown an indirect induction oven. The secondary windings forms the walls of a metal container and the iron core links the primary as well as secondary winding. Due to relatively poor power factor (0.8). It must possess some special advantage for its commercial success. The particular advantage obtained is a simple and almost a fool proof method of temperature control. The part *AB* of the magnetic circuit is situated in the oven chamber and made from a special alloy which looses its magnetic properties at a particular temperature and regains then when cooled approximately to the same temperature. The oven temperature is thus limited to the critical value without the use of external control equipment. The bar *AB* is made detached so that it can be replaced by other having different critical temperature ranging between 400°C and 1000°C.

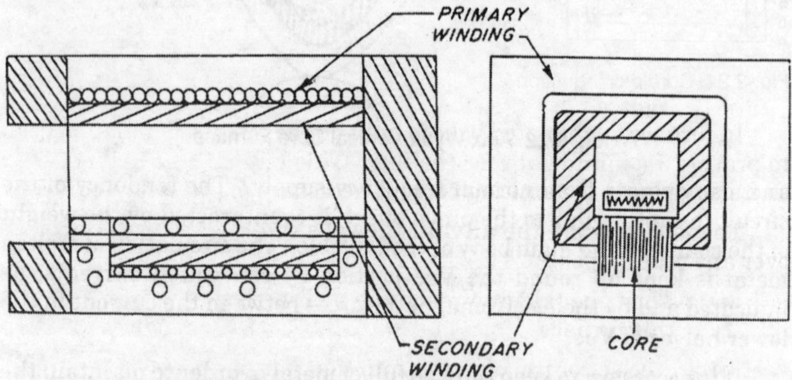

Fig. 3.23. Indirect Core Type Induction Furnace.

(*d*) **Coreless Induction Furnace.** The general construction of this type of furnace is shown in Fig. 3.24 which indicates that a convenient shape of crucible can be employed. The flux produced by the primary winding sets up eddy currents in the charge. The eddy currents developed in any magnetic current is given as :

$$\text{Eddy current} \propto B^2 \times f^2.$$

where *B* is flux density, *f* is frequency. These eddy currents are sufficient to heat the metal to melting point and also sets up electromagnetic forces which produce a striking action. Construction of this type of furnace is shown in Fig. 3.25 below. The furnace has a refractory lining. The primary coils are wound around it. The alternating flux produced by the primary winding induces eddy currents in the charge. The direction of resultant eddy current will be in opposite direction to the current in primary coil. The artificial cooling of primary coil is necessary because of high amount of copper losses. The temperature of the primary winding increases considerably.

The coil is constructed in the form of hollow tube through which cold water is circulated.

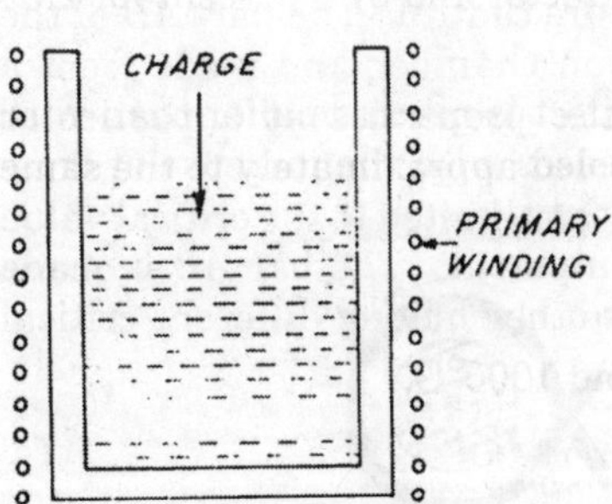

Fig. 3.24. Coreless induction furnace.

Steel former
Fritted lining
Earthing pole
Magnetic yoke
Water cooled induction coil

Fig. 3.25. High frequency coreless induction furnance

In furnace a laminated yoke is built on the outside of induction coil to provide a definite path for the flux. This minimises the stray field and also reduces the exciting current necessary to produce the flux.

The following are the **advantages** of a coreless induction furnace over any other type :

(*i*) The time taken to reach the melting temperature is less in this furnace

(*ii*) Precise control of power into the charge can be employed

(*iii*) Crucible of any shape can be used

(*iv*) Intermittant operation possible without any loss in time for warming

(*v*) Charging and pouring is simple

(*vi*) There is no dust, smoke and noise etc.

(*vii*) Automatic stirring in the charge due to eddy currents

(*viii*) Eraction cost of furnace is low

(*ix*) Opening cost is also low.

3.13. Skin Effect

The steady direct current when flowing through a conductor distributes itself uniformly over the whole X-section of a conductor. But in case of alternating current the distribution is not uniform. The alternating current tends to concentrate near the surface of the conductor and no current flows through the core of conductor. This phenomenon is called the **skin effect** and it causes the resistance of the conductor to increase slightly particularly when the diameter of the conductor is large. At low frequencies the skin effect is very small but its effect becomes significantly at higher frequency. At supply frequency of 50 Hz or less the increase in ohmic resistance is negligible for solid copper conductor upto 1 cm dia. The resistance increases by about 2.5 percent for 1.5 cm. dia solid copper conductor and by 8 percent for 2.54 cm. dia. conductor.

In stranded conductors the skin effect is much smaller than with solid conductors. The skin effect increases with the increase of cross section, supply frequency and permeability.

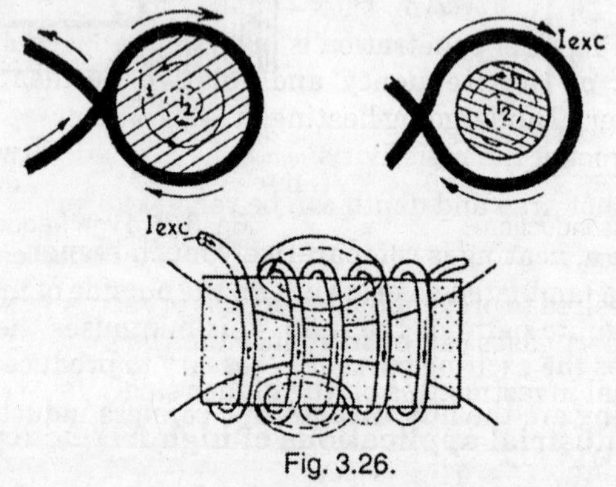

Fig. 3.26.

Consider a conducting core material enclosed in a coil carrying alternating current. The flux produced by the coil will induce concentric eddy currents circulating through the conductor shown as i_1 and i_2 in Fig. 3.26.

From Lenz's law it is known to us that the tendency of the induced eddy currents is to oppose the very cause producing these currents. Obviously in the given case the eddy currents, are induced due to the exciting current, which is alternating and produces alternating flux, while flowing through the coil. So the flux produced by the main coil has to be neutralised by the flux due to the induced eddy currents. Obviously the demagnetising effect of eddy current i_1, does not permit the main flux to penetrate the core and induce current i_2. This in other words is the shielding produced by the outer layer eddy current i_1 which keeps the effective resultant flux to remain on the outer surface or skin of the core. Hence the name *skin effect.* Moreover, higher is the frequency of exciting current (I_{exc}), smaller is the depth of penetration. This effect is being utilised for heat treatment of surfaces of steel parts of machine tools. Same effect allows the use of hollow conductor for making coils for the induction furnaces and water cooling of such conductors.

3.13.1. Characteristics of high frequency eddy current/induction heating.

(*i*) The heat generated is closer to the outer surface and no heat is generated at the core as shown in Fig. 3.27.

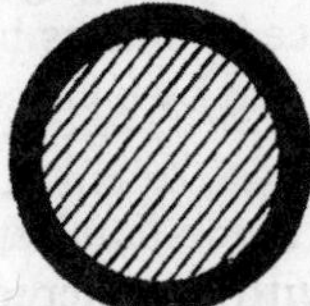

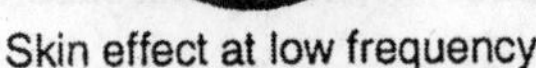

Skin effect at low frequency   Skin effect at high frequency

Fig. 3.27

(*ii*) The depth of penetration is inversely proportional to square root of the frequency and relative permeability of the material undergoing heating. It is directly proportional to the sq. root of the resistivity.

(*iii*) The hot area and depth can be remarkably controlled.

(*iv*) Rate of heating is comparatively much higher.

(*v*) If desired to prevent oxidation, use of vacuum can be provided or inert gases can be introduced in the heating chamber.

(*vi*) Initial investment is on the higher side.

3.13.2. Industrial applications of high frequency induction heating

(*i*) **Quick surface hardening of steel parts.** The portion of the steel component to be hardened is covered with the induction coil. The coil is composed of hollow water cooled copper tube. Necessary insulation is provided to prevent short circuit. The coil is now subjected to high frequency A.C. supply source. The process of hardening/heating involves few seconds. The job is now cooled by dipping in plain water or salt bath or oil as per mechanical requirements.

(*ii*) **Brazing.** Two pieces to be soldered with non-conventional soldering compounds *viz.* silver, copper alloys can be jointed together by placing the soldering alloy, between suitably flux coated surfaces. The pieces are placed close enough. The induction heating is applied which not only heats the ends to be soldered, but simultaneously melts the solder in a few seconds. Immediate cooling of the joint results in good soldering.

(*iii*) **Annealing.** Brass, bronze etc. can be annealed efficiently without having a brittle surface if the metal is pre-heated to a specific temperature. This momentary working temperature can be quickly reached through induction heating at high frequency.

(*vi*) **Butt welding.** Moderate sized members can be butt welded by momentary induction heating of the ends to be welded together. The process is identical to brazing, the desired, fusing temperature is reached, however, in bit longer time.

(*v*) **Sterilisation of medical instruments.** Heating of such instruments, to a certain temperature for a specific time makes the instruments germs free.

(*vi*) **Drying of metallic painted surfaces.** Spray painted surfaces which are metallic in nature gives a lasting shine when subjected to uniform heating to a desirable temperature. The process becomes very fast when heating chambers make use of high frequency induction heating.

3.14. Dielectric Heating

This is also sometimes called high frequency capacitive heating and is employed for heating of insulating materials like wood, plastic and ceramic etc. The supply frequency applied for this type of heating is between 10-30 mega-cycles per second and the applied voltage up to 20 kV.

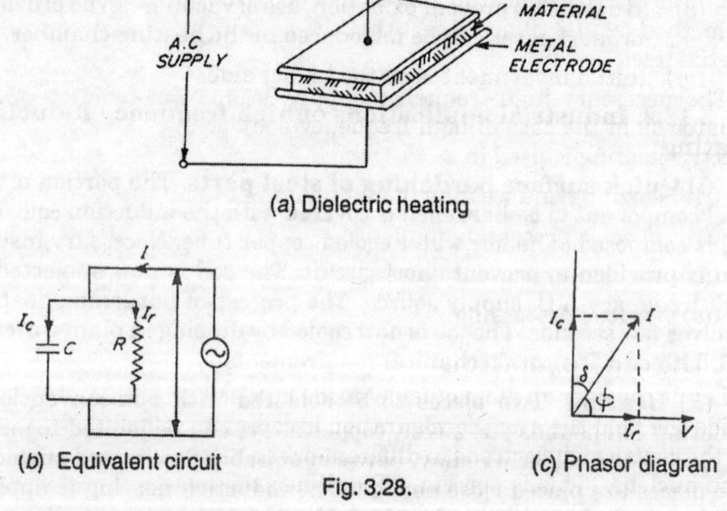

(*a*) Dielectric heating

(*b*) Equivalent circuit (*c*) Phasor diagram

Fig. 3.28.

The principle of operation of dielectric heating is that when a capacitor is subjected to a sinusoidal voltage, the current drawn by it is never leading the voltage exactly by 90°. The angle between voltage and current is slightly less than 90° (say angle, δ) as shown in Fig. 3.28 above, where δ is called the loss angle.

$$(\delta = \frac{1}{2\pi} \frac{\sqrt{\rho \times 10^7}}{\mu_r \cdot f} m)$$

where ρ = Resistivity

μ_r = Relative permeability

and f = Frequency

This clearly shows that there is a small component of current which is in phase with the applied voltage and in turn produces a power loss in the dielectric. At normal supply frequency of 50 c/s this loss may be small enough but at high frequencies of the order of 10 - 30 megacycles/second the loss becomes so large that it is sufficient to heat the dielectric. So the material to be heated is placed between two sheet type electrodes which forms a capacitor as shown in Fig. 3.28 and its equivalent circuit. The value of the capacitance is given by :

$$C = K K_0 A/d \text{ farads.}$$

where K = Relative permittivity

K_0 = Permittivity of vacuum and its value is, 8.854×10^{-12}

A = Surface area of the electrode in m^2

d = distance between the two electrodes in metres.

The advantage of heating material of poor thermal conductivity by this method results from the fact that the heat is produced within the material itself.

The necessary high frequency supply is obtained from a valve oscillator as in the case of high frequency eddy current heating. The dielectric heating is used in :

(*i*) seaming and welding in manufacture of synthetics

(*ii*) in wood processing industry

(*iii*) for baking foundary cores

(*iv*) for food processing.

3.15. Dipole Formation

Let us consider an atom outside/inside an electrostatic field. When outside the field the atom configuration is concentric. That is nucleus is at the centre and electrons in different orbits having the same centre as the nucleus. This is per Fig. 3.29. When the same atom is placed between a pair of positively and negatively charged plates (respectively), the positive plate repels the positively charged nucleus which is being attracted by the negative plate. This atom, construction is no

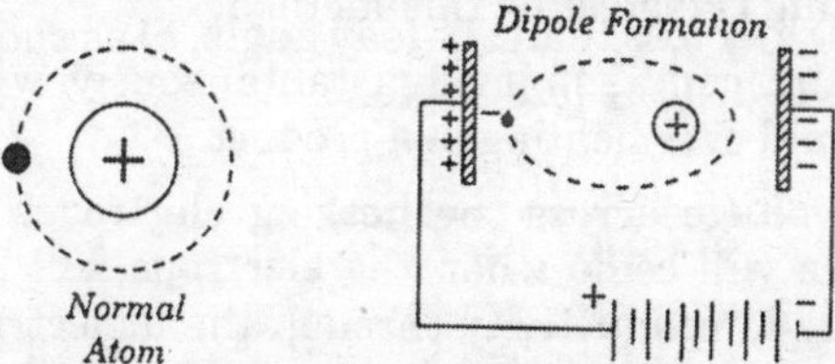

Fig. 3.29. Atom stretched under the influence of electrostatic field.

longer concentric but eccentric. Such a stretched atom is known as **dipole**.

Now let us consider the atoms/molecules of a dielectric medium placed in between two strongly charged plates. Fig. 3.30 (*a*) the atoms of dielectric behave as dipoles, so that the +ve charged nuclei face the negative plate and the negative ends of the dipoles towards the positive plate.

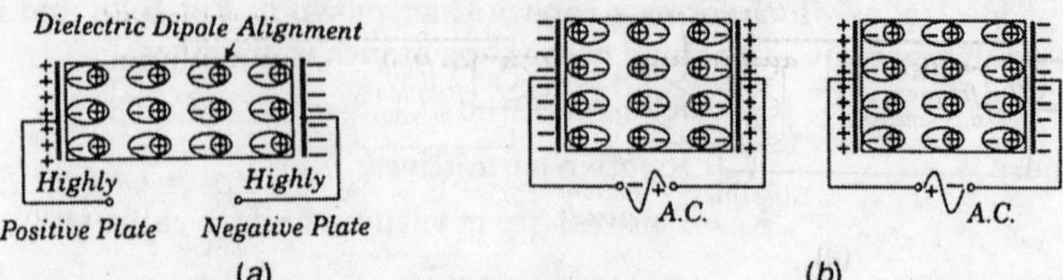

Fig. 3.30. Reversing of dipole alignment with reversal of polarities on the plates in different half cycles.

3.16. Generation of Dielectric Heat

Suppose the D.C. source of above circuit is replaced with a source of alternating current as shown in Fig. 3.30 (*b*). Obviously during positive half cycle the positive ends of the dipoles will be towards one plate and negative towards the other. During negative half cycle this alignment will reverse because the polarities of the plates will reverse. During this change of alignment of the dipoles, intermolecular friction results in production of heat. Higher the applied frequency faster will be the changes taking place inside the dielectric medium and greater will be the heat produced.

From the above discussion it can be concluded that,

(*i*) Heat will be generated from within the dielectric medium resulting in uniform heating of the entire structure unlike external heating.

(*ii*) Heating will be faster with increase in frequency.

(*iii*) Since no naked flame appears in the process even flammable articles like plastics or wooden products can be safely heated by this process.

(*iv*) Bad conductors of heat can only be heated by the above process.

(*v*) Heating is fastest by this method.

(*vi*) Heating can be stopped instantaneously, which is necessary to avoid over heating of a product.

Note. If voltage across the heating electrodes is gradually increased, a stage will come when the electrons will leave the parent atoms, resulting in conductivity through the dielectric. At this stage the dielectric is said to be raptured.

3.17. Some Applications of Dielectric Heating

(1) **Plywood Industry.** Figs. 3.31 and 3.32 shows the dielectric heating of wood assemblies with glue lines for joining. It is known to us that under normal conditions the glue lines may take a day *or* more to dry and setting of the joint. However this setting takes a few minutes when heated through high frequency di-electric heating. Making of multilayer plywood needs faster glueing and dielectric heating is a necessary for uniform heating in plywood industry.

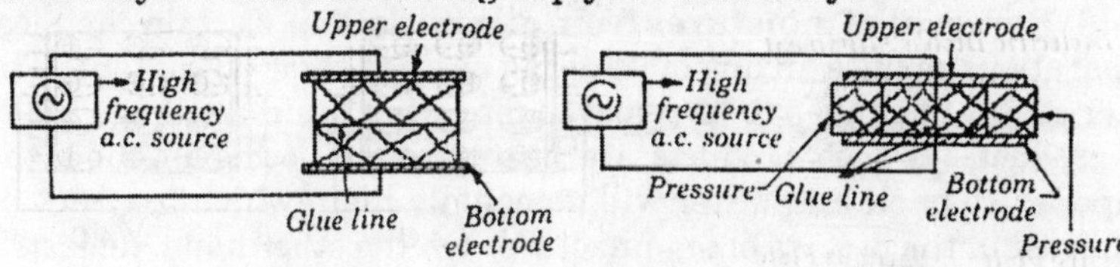

Fig. 3.31. Glueing of wood assemblies through dielectric heating.

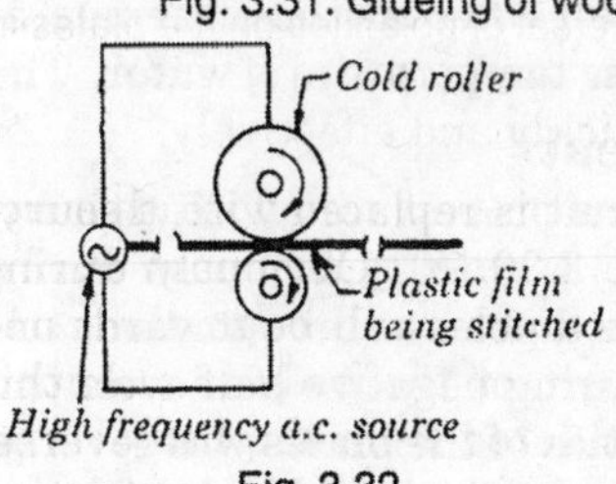

Fig. 3.32.

(2) **Sand core baking.** Sand cores made of dry sand, water and some resin have to be baked before being used in moulding processes. The drying is done when the cores placed on a conveyor belt are made to pass through an alternating dielectric field is shown in Fig. 3.31. The cores are placed on a core plate which is further carried by a conveyor belt moving between a pair of electrodes connected to a high voltage, high frequency power source.

(3) **Plastic Industry.** Most of the domestic appliances have some part made of plastic. Plastic moulding becomes faster and fault free if pre-heating is uniform. This pre-heating is done through dielectric heating and takes fraction of a minute to give the desired heat to the plastic compounds before sending into the moulding section.

(4) **Tombacco Industry.** Drying of tombacco in cigarette industry after mixing with glycerine etc. is done with the help of dielectric heating.

(5) **Bakeries.** Modern bakeries making use of automatic machines make use of dielectric heating for baking of biscuits, cakes and other products.

(6) **Electronic Sewing.** Plastic garments like raincoats cannot be stitched with ordinary sewing machines, otherwise rain water can enter the stitching holes. Hot rollers used for making polythene covers are also not suitable as heating will not be uniform and faster on thick material. Therefore, cold rollers fed with high frequency supply do the work satisfactorily as shown in Fig. 3.32.

(7) **Dehydration of Food.** Food articles required to be preserved for a period covering months, have to be de-hydrated and sealed in air tight containers. The food can be used after soaking in water and cooking will not be necessary if dehydration is done for a cooked stuff. Dehydrated peas, packed sweets like Gulabjamuns etc. available in the market are an example of this process which involves passing the food through alternating electrostatic field resulting in dehydration through dielectric heating of the product.

(8) **Removal of moistures from oil emulsions.** Certain chemical preparations require mixing of watering paste in some oil content. The water content has then to be removed for giving desired storage life to the product. If such products are heated by any other method, the temperature of oil and water will go equally high, which may not be permitted in the process of manufacture. On the other hand, dielectric constants for oil and water being much different, if heating is done through dielectric heating water will boil off whereas the oil temperature will remain quite lower than the boiling temperature of water. The emulsions may as such be dried safely, quickly and effectively.

(9) **Electro-medical application.** In a process called 'diathermy' was part of the human body suffering from certain type of pain can be cured by heating the inner parts of the effective area through dielectric heating. For applying diathermy, the part of the body is clamped between two electrodes which remain insulated from the body. The electrodes are then energised from high frequency high voltage source thus generating dielectric heat in the tissues and bones coming between the electrodes.

(10) **Book Binding.** High quality foreign books are passed through alternating electrostatic fields between tubular electrodes. This results in drying of glue on the cover bindings between 5 - 10 seconds, through dielectric heating. The books can be packed and despatched without waiting for the covers to dry, which cannot be done otherwise.

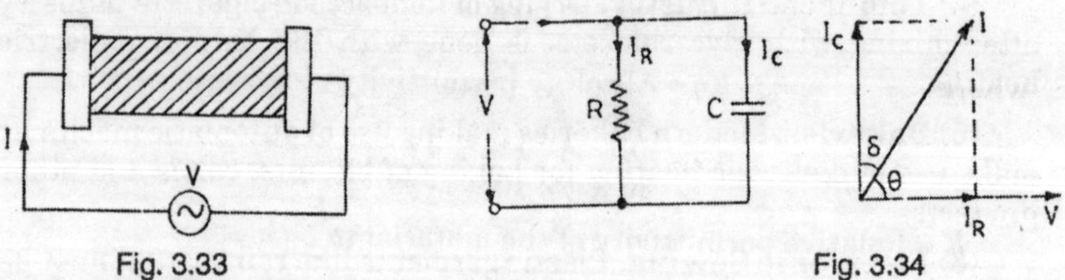

Fig. 3.33 Fig. 3.34

3.18. Dielectric Heating, Principle

As has already been told that this type of heating process is used only to heat non-conducting materials. The materials to be heated is placed between two conducting electrodes across which alternating voltage of high frequency is applied.

Two electrodes, separated by a dielectric medium and across which some potential difference is applied from a capacitor. The equivalent circuit of which is shown in Fig. 3.34. The vector diagram for the same circuit is also shown separately. Since the capacitor formed as a result of above circuit may not be pure capacitor, so a resistor R is shown in parallel. The resistor R is very high so that current flowing through it is very small so much so that capacitor current, I_c can be considered to be the same as the total current, I

$$P = \text{Power consumed} = VI \cos\theta$$

$$I = I_c - I_R$$

but I_R is so small that it can be neglected,

$$\therefore \quad I = I_c$$

$$\therefore \quad P = V.I_c \cos\theta$$

$$I_c = \frac{V}{X_c}$$

$$X_c = \frac{1}{2\pi fC}$$

$$\therefore \quad I_c = V.2\pi fC$$

$$\therefore \quad P = V(V\,.\,2\pi fC)\cos\theta = V^2 2\pi fC \cos\theta$$

$$\theta = (90^\circ - \delta)$$

$$\cos\theta = \cos(90^\circ - \delta) = \sin\delta = \tan\delta = \delta$$

when δ is small and it should be in radians.

$$\therefore \quad P = 2\pi f V^2 C\delta \text{ watts.}$$

This power consumed is converted into heat. The capacitance of parallel plate condenser is given by,

$$C = \frac{K_0 KA}{t} \text{ farads}$$

where K_0 = Absolute permittivity of air and is equal to

$$= \frac{1}{36 \times \pi \times 10^9} \approx 8.854 \times 10^{-12} \text{ Farads/metre}$$

K = Relative permittivity of the material to be heated.

A = Surface area of plates in m^2.

t = Thickness of the material to be heated in metres.

thus it is clear that heat produced/power consumed,

$$= 2\pi f V^2 C\delta.$$

Capacitance C and loss angle δ shall be constant for a particular material to be heated. Thus heating produced is proportional to square of voltage and frequency. This is why high voltage at high frequency is used for dielectric heating. For sufficient heating, voltage of about 20 kV at a frequency of 10 to 30 Mc/s is used.

Example 3.1. *A piece of plywood is to be heated by dielectric heating. The area of cross-section of the piece is 0.5 m² and the thickness is 2.5 cm. If the frequency of 25 mega cycle per sec. is used and the power obsorbed is 1000 watt. Find the voltage employed, necessary for heating. The relative permittivity of wood is 2.5 and power factor is 0.046.*

Solution. Capacitance of capacitor formed,

$$C = \frac{K_0 \,.\, KA}{t}$$

$$K_0 = 8.854 \times 10^{-12}$$

$$K = 3.6$$

$$A = 0.5 \text{ m}^2$$

$$t = 0.025 \text{ m}$$

$$C = \frac{8.854 \times 10^{-12} \times 3.6 \times 0.5}{0.025}$$

$$= 637.488 \times 10^{-12} \text{ farads}$$

$$\cos \Phi = 0.046$$

$$\Phi = 17.36°$$

$$\delta = (90° - \Phi) = (90° - 87.36) = 2.64°$$

or, $$\delta = \frac{\pi}{180} \times 2.64 = 0.046 \text{ radian}$$

$$P = 2\pi f V^2 C \delta$$

$$1000 = 2\pi \times 2.5 \times 10^6 \times V^2 \times 637.488 \times 10^{-12} \times 0.046$$

or $$V^2 = \frac{1000}{2\pi \times 25 \times 10^{-6} \times 637.488 \times 0.046}$$

$$= \frac{10^9}{2\pi \times 25 \times 637.488 \times 0.046} = 217{,}095{,}023$$

∴ $$V = 465.93 \text{ volts.} \quad \textbf{Ans.}$$

3.19. Design of Heating Element

Assume that the whole of the heat energy is dissipated solely by radiation.

Heat dissipated according to Stefan's Law,

$$H = 5.72K \cdot e\left[\left(\frac{T_1}{100}\right)^4 - \left(\frac{T_2}{100}\right)^4\right] \text{Watts/m}^2 \quad ...(1)$$

Electrical input,

$$P = \frac{V^2}{R}$$

where V is the supply voltage and R is the resistance of heating element and is given by the expression,

$$R = \rho\frac{l}{a} = \frac{\rho l}{\pi d^2/4} = 4\frac{\rho l}{\pi d^2}$$

For a circular wire of diameter d, length, l and resistivity, ρ

Electrical input,

$$P = \frac{V^2}{R} = \frac{V^2}{4\rho l/\pi d^2} = \frac{\pi d^2 V^2}{4\rho l}$$

or

$$\frac{l}{d^2} = \frac{\pi V^2}{4\rho . P} \quad ...(2)$$

Surface Area, $S = \pi.d.$

Heat dissipated $= \pi dl.H$

At steady temperature we know,

Power input = Heat dissipated

$\therefore$

$$P = \pi dl \times H$$

or

$$\frac{\pi d^2 V^2}{4\rho l} = \pi dl \times H$$

or

$$\frac{d}{l^2} = \frac{4\rho H}{V^2} \quad ...(3)$$

Solving expression (2) and (3) length and diameter of wire can be determined.

For ribbon type of conductor, let w be the width and t the thickness.

Electrical input,

$$P = \frac{V^2}{R} = \frac{V^2}{\rho l/a} = \frac{V^2}{\rho l/wt} \quad [\because a = wt]$$

$$= \frac{V^2.wt}{\rho l}$$

or

$$\frac{l}{wt} = \frac{V^2}{\rho P} \quad ...(4)$$

Since at steady temperature,

Power input = Heat dissipated

$$\therefore \qquad P = 2wl.H$$

(neglecting the side area $2tl$ as the thickness is negligible)

or $$\frac{V^2\, wt}{\rho l} = 2wlH$$

or $$\frac{t}{l^2} = \frac{2H\rho}{V^2} \qquad ...(5)$$

By solving equations (4) and (5) length l and width w for a ribbon of thickness t will be evaluated.

3.20. Efficiency and Losses

The efficiency of the oven

$$= \frac{\text{Heat reqd. to rise the temp. of the charge to the reqd. value}}{\text{Heat reqd. to rise the temp. of the charge to the reqd. value + losses}}$$

The heat required to rise the temperature of the charge to the required value,

W = Mass of charge in kg × Specific heat of charge in J /*kg* °*C* × Temp. rise.

= m s t joules.

The heat used for rising temperature of the oven for furnace can be calculated in a similar way by knowing the weight of refractory material and its specific heat. In case the oven is used continuously this loss becomes negligible.

Heat used for raising the temperature of container is calculated in exactly the same way as for the oven of furnace. The container usually has to be heated up fresh for each charge.

Since the heat is continuously conducted through the walls so this source of loss is most important. It can be determined knowing the mean value of the inside and outside surface areas A in square metres. The thickness of walls t in metre; inside and outside temperatures T_1 and T_2 in °C and thermal conductivity of walls K in MJ per hour per square metre, per metre, from the following expression :

Heat loss by conduction through walls

$$= \frac{KA}{t}(T_1 - T_2) \text{ MJ/hour.}$$

Though there is no specific formulae for determination of loss occurring due to opening of door for inspection of the charge however, the loss may be taken as 0.575 to 1.15 MJ/m^2 of the door area if the door is opened for a period of 20 to 30 seconds.

Note. The efficiency described above lies between 60 to 80 percent.

3.21. Radiant Heating or Infra-red Heating

In the ordinary resistance furnaces the heat is transferred from heating elements to the charge partly by radiation and partly by convection. The latter predominating at low and medium temperatures but in this form of heating suitable for low and medium temp. heating purely radiant or infra-red heating is used. In this method, of heating, elements consist or tungston filament lamps together with reflectors to direct the whole of the heat emitted on to the charge. The lamps are operated at 2,300°C instead of 3,000°C giving greater proportion infra-red radiations and a longer life. The reflectors are plated with rhodium. The lamps employed are usually of rating between 250 W to 11000 watts operating at 115 volts. The operation at low voltage results in a robust filament. With this arrangement charge temp. between 200°C and 300°C can be obtained. Heat emission intensities up to 7000 watts/square metre of the chamber surface can be obtained. These are much higher than those obtained with ordinary resistance furnaces (1,500 watt/m^2). In radiant heating heat absorption remains practically constant whatever the charge temp. where as it falls rapidly as the temp. of charge rises in the ordinary resistance furnace. Paint drying is a common application of this method. It is used for drying foundry moulds, low temp. heating of plastics and for various dehydration and other processes.

Infra-red heating posses the following advantages :

(*i*) Rapid heating

(*ii*) Compactness of heating units

(*iii*) Flexibility

(*iv*) Safety

3.21.1. Infra-red heating Furnace

This machine is useful for large scale production. The simplified view of infrared furnace had been shown in fig. 3.35. The heaters are installed on the upper and low areas to provide uniform heating. The infara-red sources are quartz lamps whose range of temperature is 400° – 1000°C. The machine has a max. temperature of 300° C. when is divided into different heating zones (called preheating & reflow zones).

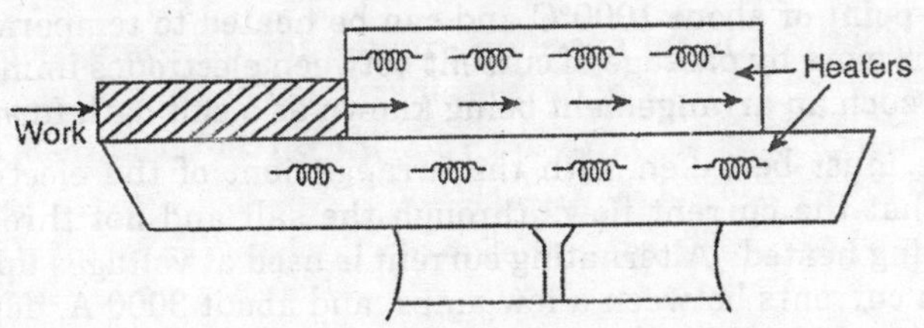

Fig. 3.35.

In this the lamps with their reflectors are grouped close together to form the wall and if possible to form the top of the heating chamber. These lamps are mounted on an angle iron frame work. The best results can be obtained by keeping the distance of 25-30 cm between the infra-red lamps and the object to be heated. Continuous process of heating can also be arranged by using conveyor belt on which articles to be heated are placed. Fig. 3.36 (*a*) shows a typical infra-red furnace and Fig. 3.36 (*b*) shows infra-red reflow time verses temperature graph.

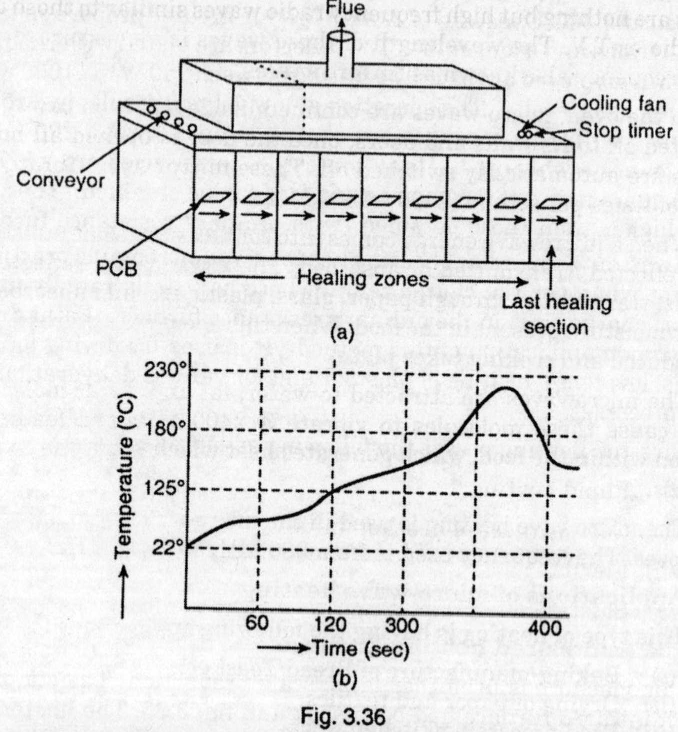

Fig. 3.36

3.22 Salt Bath Heating or Furnance

In order to prevent oxidation during hardening, a process applied to tool steels employs a method in which they are heated by immersion in molten sodium chloride or some order suitable salt. Such salts have a fusing-point of about 1000°C and can be heated to temperatures of 1500°C or more by passage of current between electrodes immersed in the salt, such an arrangement being known as a salt bath furnace.

Care must be taken with the arrangement of the electrodes to ensure that the current flows through the salt and not through the metal being heated. Alternating current is used at voltages up to 20 V and with currents between a few amps. and about 3000 A, depending on the type of furnace.

A step-down transformer is thus needed for the supply to such a furnace and means must be provided for varying the voltage, by tappings or an induction regulator, so that a constant power input can be maintained during the heating process when the resistance of the salt is continuously decreasing.

3.23 Microwave Heating

In this system electricity is converted into electromagnetic waves, which generates energy and this energy is used to cook the food. These waves are nothing but high frequency radio waves similar to those used by radio or T.V. The wavelength of these waves is very short of very high frequency also known as **"microwaves"**.

In the oven, micro-waves are confined inside the oven cavity and reflected off to its walls and doors, once the door is opened all microwaves are automatically switched off. These microwaves vibrate millians of times per second (2400—2500 MHz)

When a microwave energy comes into contact with some substance it is reflected, transmitted or absorbed. These waves are reflected by metals, transmitted through paper, glass, plastic etc. and absorbed by water/moisture present in the food. When this energy is absorbed, heat is produced and cooking takes place.

The microwaves are attracted to water, fat and sugar molecules. They cause these molecules to vibrate @ 2400 MHz/sec. leading to friction within the food, which generates heat which begin the cooking process.

The microwave heating is used in the micro-wave oven for baking purposes. The frequency used is from 900 MHz to 2400 MHz.

Applications of micro-wave heating

This type of heating is having the following applications :

(*i*) Baking-manufacture of Bread/Toast etc.
(*ii*) Drying of paper and textiles.
(*iii*) Food processing/kitchen work.
(*iv*) Treatment of diseases like cancer etc.
(*v*) Manufacture of plastics.
(*vi*) Processing of cement and timber etc.

Advantages of Microwaves Heating over Conventional Heating

It has many advantages as mentioned below :

(*i*) It has Neat and Clean system
(*ii*) It provides uniform heating to the substance.
(*iii*) The system provides quick heating.
(*iv*) The depth of penetration of heat in to the material is much more.

(v) Within the material, the heat is generated directly which gives much faster temperature rise.

Microwave oven

This oven basically converts the microwave energy in to thermal energy and supplies to the substance which is to be heated. Domestic ovens are generally of about 1 KW rating whereas the industrial ovens can be of several kilowatts rating capacity.

A microwave oven basically comprises of two blocks or chambers.

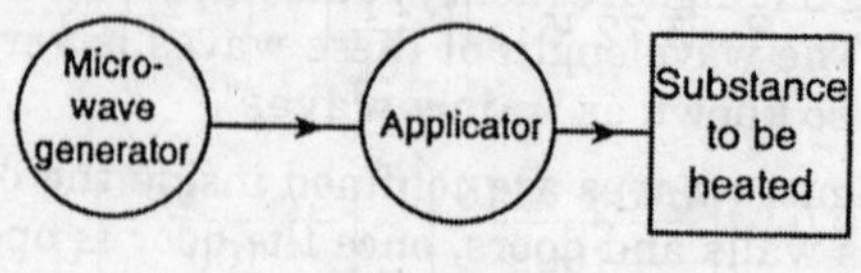

Fig. 3.37

(i) **Microwave Generator :** It generates microwaves energy to be supplied to the substance to be heated.

(ii) **Applicator :** This a link between microwave generator and the substance to be heated. It should be kept in mind that there should be proper impedence matching between the two for maximum power transfer to the substance and that too without any leakage.

The applicator may be of the following types :

1. Multimode Cavity Type
2. Monomode Cavity Type.
3. Wave Guide (hollow Conducting-tube) type.

Example 3.2. *A resistance oven employing nichrome wire is to be operated from 220 V single phase supply and is to be rated at 16 KW. If the temperature of the element is to be limited to 1.170°C and average temperature of the charge is 500°C. Find the dia. and length of the element wire.*

Radiating efficiency = 0.57,

Emissivity = 0.9,

Specific resistance of nichrome = 109×10^{-6} ohm–cm.

Solution. Power supplied,

$$P = 16 \text{ kW} = 16{,}000 \text{ watts}$$

Supply voltage, $V = 220$ V

Radiant efficiency, $K = 0.57$

Emissivity, $e = 0.9$

Resistivity, $\rho = 109 \times 10^{-6}\ \Omega\text{–cm}$

$$= 1.09 \times 10^{-6} \ \Omega\text{-m.}$$

From the equation (2) as discussed previously

$$\frac{l}{d^2} = \frac{\pi V^2}{4\rho P}$$

or $$\frac{l}{d^2} = \frac{\pi \times (220)^2}{4 \times 1.09 \times 10^{-6} \times 16{,}000} = 2{,}175{,}000 \qquad ...(i)$$

Now $$H = 5.72\, K \,.\, e \left[\left(\frac{T_1}{100}\right)^4 - \left(\frac{T_2}{100}\right)^4\right] \text{W/m}^2$$

$$= 5.72 \times 0.57 \times 0.9 \left[\left(\frac{273 + 1170}{100}\right)^4 - \left(\frac{273 + 500}{100}\right)^4\right] \text{W/m}^2$$

$$= 2.93436[(14.43)^4 - (7.73)^3] \text{ W/m}^2 = 11.65 \times 10^4 \text{ W/m}^2$$

We know heat dissipated = Electrical power input

So, $\pi dl \times 116{,}500 = 16{,}000$

or $$dl = \frac{16{,}000}{\pi \times 1{,}16{,}500} = 0.0438$$

or $$d^2 l^2 = 0.0019 \qquad ...(ii)$$

Multiplying expressions (*i*) and (*ii*), we get

$$l^3 = 2{,}175{,}000 \times 0.0019 = 4{,}132$$

or $$l = \sqrt[3]{4132} = 16 \text{ metre} \qquad \textbf{Ans.}$$

and $$d = \frac{0.0438}{16} \text{ metre} = \frac{43.8}{16} = \mathbf{2.74\ mm} \qquad \textbf{Ans.}$$

Example 3.3. *A 30 KW, 3Φ, 400 V resistance oven is to employ nickel-chrome strip 0.254 mm thick for the three star-connected heating elements. If the wire temperature is to be 1,100°C and that of the charge to be 700°C. Estimate a suitable width for the strip. Assume emissivity = 0.9 and radiating efficiency to be 0.5. What would be the temperature of the wire if the charge were cold ?*

Solution. Power per phase,

$$P = \frac{30 \times 1000}{\sqrt{3}} = 10{,}000 \text{ watts.}$$

Phase voltage, $$V_P = \frac{400}{3} = 231 \text{ volts}$$

Resistance of strip, $$R = \frac{(V_p)^2}{P} = \frac{(231)^2}{10{,}000} = 5.34\ \Omega$$

If l is the length of strip and w is the width of strip then, Resistance of strip,

$$R = \frac{\rho l}{w \times t} = \frac{101.6 \times 10^{-8} \times l}{w \times 0.254 \times 10^{-3}}$$

or $$\frac{l}{w} = \frac{5.34 \times 0.245 \times 10^{-3}}{101.6 \times 10^{-8}} = 1{,}335 \qquad ...(i)$$

Heat dissipated from surface,

$$H = 5.72\ \text{K.e} \left[\left(\frac{T_1}{100}\right)^4 - \left(\frac{T_2}{100}\right)^4\right]$$

$$= 5.72 \times 0.5 \times 0.9 \left[\left(\frac{1373}{100}\right)^4 - \left(\frac{973}{100}\right)^4\right]$$

$$\doteq 68{,}600 \text{ watts/m}^2.$$

Surface area available for heat radiation,

$$= 2\,wl \text{ metres}^2$$

Since heat dissipated = Power input

$$68{,}600 \times 2 \times wl = 10{,}000$$

or $$wl = \frac{10{,}000}{137{,}200} = 0.0782 \qquad ...(ii)$$

Dividing expression (*ii*) by expression (*i*), we get

$$w^2 = \frac{0.0728}{1335}$$

or $$w = \frac{\sqrt{0.0728}}{1335} = 7.38 \text{ mm.}$$ **Ans.**

Example 3.4. *A cubic water tank has surface area of 6.0 m² and is filled to 90% capacity six times daily. The water is heated from 20°C to 65°C. The losses per square metre of tank surface per 1°C temperature difference is 6.3 watts. Find the loading in KW and the efficiency of the tank. Assume specific heat of water = 4,200 J/kg/°C and one KWh = 3.6 MJ.*

Solution. Let the one side of the tank be l metres.

Surface area of the tank,

$$6\,l^2 = 6.0 \text{ m}^2$$

or $$l = 1.0 \text{ m}$$

Volume of the tank = $l^3 = 1.0 \text{ m}^3$

Water to be heated daily,

$$= 6 \times 0.9 \text{ m}^3 = 5.4 \text{ m}^3 = 5400 \text{ kg.}$$

[∵ 1 cubic metre water weighs 1000 kg]

Heat required to raise the temperature of water

$$= ms(t_2 - t_1) = 5400 \times 4200(65 - 20)$$

$$= 1020 \text{ MJ} = 283.5 \text{ kWh}$$

Losses from the surface of the tank,

$$= \frac{6.3 \times 6.0 \times 45 \times 24}{1000} \text{ kWh} = 40.8 \text{ kWh}$$

Energy supplied $= 283.5 + 40.8 = 324.3$ kWh.

Loading in kW $= \frac{324.3}{24} = 13.5$ kW. **Ans.**

Efficiency of the tank

$$= \frac{\text{Output}}{\text{Input}} \times 100 = \frac{283.5}{324.5} \times 100 = 87.4\%. \quad \textbf{Ans.}$$

Example 3.5. *An electric furnace consuming 5 KW takes 15 minutes to just melt 2.5 Kg of aluminium, the initial temperature being 15°C. Find the efficiency of the furnace when the specific heat of aluminium is 0.212 per gram per °C, melting point 658°C and latent heat of fusion 320 joules per gram.*

Solution. Quantity of aluminium to be melt,

$$m = 2.5 \text{ kg}$$

Initial temp., $t_1 = 15°\text{C}$

Melting temp., $t_2 = 658°\text{C}$

Sp. heat of aluminium,

$$S = 0.212 \text{ calorie/gm/°}C$$

$$= 0.212 \times 1000 \times 4.18 \text{ J/kg/°C}$$

$$= 886.16 \text{ J/kg/°}C$$

Latent heat of aluminium,

$$L = 320{,}000 \text{ J/kg}$$

Heat required to melt 2.5 kg of aluminium,

$$W = ms(t_2 - t_1) + mL$$

$$= 2.5 \times 886.16\ (658 - 15) + 2.5 \times 320{,}000 \text{ Joules}$$

$$= 2.2245 \text{ MJ} = \frac{2.2245}{3.6} \text{ kWh.}$$

$\because$ (1 kWh = 3.6 MJ) = 0.618 kWh

Energy input $= 5 \times \frac{15}{60} = 1.25$ kWh.

Efficiency of the furnace,

$$= \frac{\text{Output}}{\text{Input}} = \frac{0.618}{1.25} \times 100 = 49.5\% \quad \textbf{Ans.}$$

Example 3.6. *A slab of insulating material 150 cm² in area and 1 cm. thick is to be heated by dielectric heating. The power required is 400 watts at 30MHz. Material has permittivity of 5 and p. f. 0.05. Determine the voltage necessary. Absolute permittivity = 8.854 × 10⁻¹² F/m.*

Solution. Power absorbed,

$$P = 400 \text{ watts}$$

Power factor, $\cos \Phi = 0.05$

Frequency of supply,

$$f = 30 \times 10^6 \text{ Hz}$$

Capacitance of the capacitor formed, $= \dfrac{\varepsilon_r . \varepsilon_o . A}{t}$

$$= \frac{5 \times 8.854 \times 10^{-12} \times 150 \times 10^{-4}}{1 \times 10^{-2}}$$

$$= 8.854 \times 7.5 \times 10^{-12} \text{ F} = 66.405 \times 10^{-12} \text{ F}$$

Since $P = 2\pi f C V^2 \cos \phi$

$$\therefore \quad V = \sqrt{\frac{P}{2\pi f C \cos \phi}}$$

$$= \sqrt{\frac{400}{2\,\pi \times 30 \times 10^6 \times 66.405 \times 10^{-12} \times 0.05}}$$

$$= 800 \text{ volts} \quad \textbf{Ans.}$$

Example 3.7. *An insulating material 2 cm thick and 200 cm² in area is to be heated by dielectric heating. The material has permittivity of 5 and power factor 0.05. Power required is 400 watts and frequency of 40 MHz is to be used. Determine the voltage necessary and the current that will flow through the material. If the voltage were to be limited to 700 V, will be the frequency to get the same loss.*

Solution. Power absorbed,

$$P = 400 \text{ watts}$$

Power factor, $\cos \Phi = 0.05$

Supply frequency, $f = 40 \times 10^6$ Hz

Capacitance of the capacitance formed,

$$C = \frac{\varepsilon_o . \varepsilon_r . A}{t} = \frac{8.854 \times 10^{-12} \times 5 \times 200 \times 10^{-4}}{0.02}$$

$$= 44.27 \times 10^{-12} \text{ F}$$

Since $P = 2\pi f C V^2 \cos \Phi$

$$\therefore \quad V = \sqrt{\frac{P}{2\pi f C \cos \Phi}}$$

$$= \sqrt{\frac{400}{2\pi \times 40 \times 10^6 \times 44.27 \times 10^{-12} \times 0.05}}$$

$= 848$ volts **Ans.**

Current flowing through the material,

$$I = \frac{P}{V \cos \Phi} = \frac{400}{848 \times 0.05} = 9.434 \text{ Amp.} \quad \textbf{Ans.}$$

Since heat produced $\propto f V^2$

$$\therefore \quad f_2 V_2{}^2 = f_1 V_1{}^2$$

$$\therefore \quad f_2 = f_1 \left(\frac{V_1}{V_2}\right)^2 = 40 \times 10^6 \left(\frac{848}{700}\right)^2 = 58.7 \times 10^6 \text{ Hz.} \quad \textbf{Ans.}$$

Example 3.8. *A 3-phase electric arc furnace has the following data:*

Current drawn = 5,000 A

Arc voltage = 50 V

Resistance of transformer referred to secondary = 0.002 Ω

Resistance of transformer referred to secondary = 0.004 Ω

(*i*) *Calculate the power factor and KW drawn from the supply.*

(*ii*) *If the overall efficiency of the furnace is 65%, find the time required to melt 2 tonnes of steel when latent heat of steel = 8.89 K. cal. / kg, specific heat of steel = 0.12, melting point of steel = 1370°C and initial temperature of steel = 20°C.*

Solution. Arc voltage = 50 volts

Current drawn, I =5,000 A

Voltage drop due to resistance of transformer

$= 5{,}000 \times 0.002 = 10\text{V}$

Voltage drop due to resistance of transformer

$= 5{,}000 \times 0.004 = 20 \text{ V}$

Since arc drop is resistive in nature, therefore, from Fig. 3.35 open circuit phase voltage of transformer secondary,

$$= \sqrt{(50 + 10)^2 + (20)^2}$$

$$= \sqrt{3600 + 400} = \sqrt{4000} = 63.25 \text{ V}$$

(*i*) P.F. of the supply,

$$= \frac{50 + 10}{63.25} = 0.9487. \quad \textbf{Ans.}$$

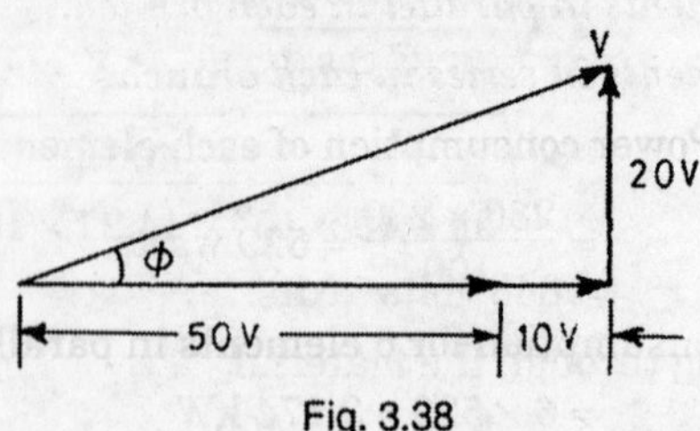

Fig. 3.38

Power drawn per phase,

= Current drawn per phase × Secondary phase voltage × Power factor

$$= 5000 \times 63.25 \times 0.9487 = 300{,}000 \text{ W or } 300 \text{ kW}$$

Total power drawn from the supply,

$$= 3 \times 300 = 900 \text{ kW}. \quad \textbf{Ans.}$$

(*ii*) Energy required to melt 2 tonnes of steel,

$$= ms\,(t_2 - t_1) + mL = m\,[s\,(t_2 - t_1) + L]$$

$$= 2{,}000[0.12 \times (1370 - 20) + 8.89]$$

$$= 3{,}41{,}780 \text{ kilo-calories}$$

$$= \frac{3{,}41{,}780}{860} \text{kWh} = 397.42 \text{ kWh}$$

Power utilised = Power drawn from supply mains × Overall efficiency

$$= 900 \times 0.65 = 585 \text{ kW}.$$

Time required for melting

$$= \frac{\text{Energy required in kWh}}{\text{Power required in kW}}$$

$$= \frac{397.42}{585} = \textbf{40 minutes 46 seconds.} \quad \textbf{Ans.}$$

Example 3.9. *Six resistances each of 100 ohms are used in resistance furnace. How much power is drawn with following connections and supply voltage :*

(*i*) *250 volt single phase (a) Six groups in parallel, each of the resistance (b) Two groups in parallel, each of three resistances in series.*

(*ii*) *400 volt three phase star connections*

(*a*) *Two heating elements in parallel in each phase.*

(*b*) *Two heating elements in series in each phase.*

(*iii*) *400 volt three phase delta connections*

(*a*) *Two elements in parallel in each branch.*

(*b*) *Two elements in series in each branch.*

Solution. (*i*) Power consumption of each element,

$$= \frac{230 \times 230}{100} = 529 \text{ watts.}$$

(*a*) Power consumption for 6 elements in parallel,

$$= 6 \times 529 = 3.174 \text{ kW}$$

(*b*) Power consumption each group of three resistances in series

$$= \frac{230 \times 230}{(100 + 100 + 100)} = \frac{529}{3} \text{ watts}$$

Power consumption of two groups in parallel

$$= \frac{529}{3} + \frac{529}{3} = 353 \text{ watts.}$$

(*ii*) Phase voltage $= \frac{400}{\sqrt{3}}$

(*a*) Equivalent resistance of two heating elements in parallel

$$= \frac{100}{2} = 50\Omega$$

Power consumed by two heating elements in parallel,

$$= \frac{400}{\sqrt{3}} \times \frac{400}{\sqrt{3}} \times \frac{1}{50} = 1.06 \text{ kW.}$$

(*b*) Equivalent resistance of two heat elements in series

$$= 100 + 100 = 200 \text{ ohms.}$$

Power consumed $= \frac{400}{\sqrt{3}} \times \frac{403}{\sqrt{3}} \times \frac{1}{200} = 0.625 \text{ kW.}$

(*iii*) Phase voltage or voltage across each branch = 400 volts

Equivalent resistance of heating elements in parallel

$$= \frac{100}{2} = 50 \text{ ohms.}$$

(*a*) Power consumed by two heating elements in parallel

$$= \frac{400 \times 400}{50} = 3.2 \text{ kW.}$$

(*b*) Equivalent resistance of two heating elements in series

$$= 100 + 100 = 200 \text{ ohms.}$$

Power consumed $= \frac{400 \times 400}{200} = 800 \text{ watts.}$

Example 3.10. *A 20 KW, 1-phase 220 volt resistance oven employs circular nickel chrome wire for its heating elements. If the wire temperature is not exceed 1170°C and the temperature of the charge is to be 500°C. Calculate the length and size of wire required. Assume a radiating efficiency of 0.6 and the specific resistance of nickel chrome. 100×10^{-6} Ω cm and emissivity 0.9.*

Solution. Power supplied

$$P = 20 \times 1000 = 20{,}000 \text{ watts}$$

Supply voltage $V = 220$ volts

Emissivity, $e = 0.9$

Radiant efficiency, $K = 0.6$

Specific resistance, $\rho = 100 \times 10^{-6}$ ohm cm.

We know that $$\frac{l}{d^2} = \frac{\pi V^2}{4 \rho P}$$

or $$\frac{l}{d^2} = \frac{3.14 \times 220 \times 220}{4 \times 100 \times 10^{-6} \times 2000}$$

or $$\frac{l}{d^2} = 18{,}700 \qquad \text{...(1)}$$

According to 'Stefan's Law,

$$H = 5.72\, K \times e \left[\left(\frac{T_1}{1000}\right)^4 - \left(\frac{T_2}{1000}\right)^4\right] \text{ watt/cm}^2$$

$$H = 5.72 \times 0.6 \times 0.9 \left[\left(\frac{1170 + 273}{1000}\right)^4 \left(\frac{500 + 273}{1000}\right)^4\right]$$

$$= 12.3 \text{ watt/cm}^2$$

Surface area $= \pi dl \text{ cm}^2$

Total heat dissipated $= \pi.d.l.(12.3)$ watts

Total heat dissipated is equal to electric power input,

So $\pi dl(12.3) = 20{,}000$

or $$dl = \frac{20{,}000}{3.14 \times 12.3} = 5.8$$

or $$d^2 l^2 = 266{,}000 \qquad \text{...(2)}$$

From equation (1) and (2)

$$l^2 = 266{,}000 \times 18{,}700$$

or $l = 1710$ cm. or 17.1 metres. **Ans.**

and $\qquad d = \dfrac{518}{1710} = 0.3$ cm. **Ans.**

Example 3.11. *Estimate the energy required to melt $\frac{1}{2}$ tonne of brass in a single phase induction furnace. If the melt is to be carried out in $\frac{1}{2}$ hour, what must be the average power input to furnace ?*

Specific heat of brass = 0.094

Latent heat of fusion of brass = 30 kilo-calories / kg

Melting point of brass = 920°C

Furnace efficiency = 60.2%

The temperature of the cold charge may be taken as 20°C.

Solution. Quantity of brass to be melted = 500 kg.

Heat reqd. to raise the temperature of brass from 20° to 920°C

$$H_1 = \text{Mass} \times \text{Specific heat} \times (\text{Rise in temperature})$$

$$= 500 \times 0.094 \times (920 - 20) = 42{,}300 \text{ kcals.}$$

Heat reqd. to heat the metal at 920°C (which is melting point)

$$H_2 = 500 \times 39 = 19{,}500 \text{ k-cals.}$$

Total heat required,

$$H = H_1 + H_2 = 42{,}300 + 19{,}500 = 61{,}800 \text{ k-cals.}$$

Energy required

$$= \frac{\text{Heat required (Kcal)} \times 4180}{3600 \times 1000 \times \text{Furnace efficiency}}$$

$$= \frac{61800 \times 4180}{3600 \times 1000 \times 0.602} = 119.5 \text{ kWh.} \quad \textbf{Ans.}$$

Example 3.12. *An electric toaster consists of two resistance elements each of 190 ohms. Calculate the power drawn from 250 volts a.c. single phase supply, when the elements are connected in (i) parallel and (ii) series.*

Solution.

(*i*) When the elements are connected in parallel equivalent resistance,

$$R_1 = \frac{1}{\frac{1}{190} + \frac{1}{190}} = 95\ \Omega$$

Power drawn, $\quad P_1 = \dfrac{V^2}{R_1} = \dfrac{(250)^2}{95} = 657.47$ watts $\quad$ **Ans.**

(*ii*) When the elements are connected in series, equivalent resistance,

$$R_2 = 190 + 190 = 380\Omega$$

Power drawn, $P_2 = \frac{V^2}{R_2} = \frac{(250)^2}{380} = 164.89$ watts **Ans.**

Example 3.13. *Six resistances each of 100 ohms are used in resistance furnace. How much power is drawn with following connections and supply voltage :*

(*i*) *250 volt single phase*

(*a*) *Six groups in parallel, each of the resistance.*

(*b*) *Two groups in parallel, each of three resistances in series.*

(*ii*) *400 volt three phase star connections :*

(*a*) *Two heating elements in parallel in each phase.*

(*b*) *Two heating elements in series in each phase.*

(*iii*) *400 volt three phase delta connections :*

(*a*) *Two elements in parallel in each branch.*

(*b*) *Two elements in series in each branch.*

Solution. (*i*) Power consumption of each element

$$= \frac{230 \times 230}{100} = 529 \text{ watts.}$$

(*a*) Power consumption for 6 elements in parallel

$$= 6 \times 529 = 3.174 \text{ KW.}$$

(*b*) Power consumption each group of three resistances in series

$$= \frac{230 \times 230}{(100 + 100 + 100)} = \frac{529}{3} \text{ watts}$$

power consumption of two groups in parallel

$$= \frac{529}{3} + \frac{529}{3} = 353 \text{ watts}$$

(*ii*) Phase voltage $= \frac{400}{\sqrt{3}}$ Volts

(*a*) Equivalent resistance of two heating elements in parallel

$$= \frac{100}{2} = 50\ \Omega$$

Power consumed by two heating elements in parallel

$$= \frac{400}{\sqrt{3}} \times \frac{400}{\sqrt{3}} \times \frac{1}{50} = 1.06 \text{ kW}$$

(*b*) Equivalent resistance of two heating elements in series

$$= 100 + 100 = 200\ \Omega$$

Power consumed

$$= \frac{400}{\sqrt{3}} \times \frac{40}{\sqrt{3}} \times \frac{1}{200} = 0.625 \text{ kW}$$

(*iii*) Phase voltage or voltage across each branch = 400 Volts.

Equivalent resistance of heating elements in parallel

$$= \frac{100}{2} = 50\ \Omega$$

(*a*) Power consumed by two heating elements in parallel

$$= \frac{400 \times 400}{50} = 3.2 \text{ kW}$$

(*b*) Equivalent resistance of two heating elements in series

$$= 100 + 100 = 200\ \Omega$$

Power consumed $= \dfrac{400 \times 400}{200} = 800$ Watts.

Example 3.14. *A 15 KW ; 220 V, single phase resistance oven employs nickel-chrome wire for its heating elements. If the wire temperature is not to exceed 1,000°C and the temperature of the charge is to be 600°C, calculate the diameter and length of the wire. Assume radiating efficiency to be 0.6 and emissivity as 0.9. For nickel chrome resistivity is 1.016 × 10^{-6} ohm-metre.*

Solution. Power supplied

$$P = 15 \text{ kW} = 15{,}000 \text{ Watts}$$

Supply voltage, $V = 220$ Volts

$K = 0.6$

Emissivity, $e = 0.9$

Resistivity, $\rho = 1.016 \times 10^{-5}\ \Omega\text{-m}$

We know that $\dfrac{l}{d^2} = \dfrac{\pi V^2}{4\rho P} = \dfrac{\pi(220)^2}{4 \times 1.016 \times 10^{-6} \times 15{,}000}$

$$= 24{,}94{,}309 \qquad \text{...}(i)$$

Also $H = 5.72\, K.e \left[\left(\dfrac{T_1}{100}\right)^4 - \left(\dfrac{T_2}{100}\right)^4\right] \text{W/m}^2$

$$= 5.72 \times 0.6 \times 0.9 \left[\left(\frac{1000 + 273}{100}\right)^4 - \left(\frac{600 + 273}{100}\right)^4\right] \text{W/m}^2$$

$$= 63{,}174 \text{ W/m}^2$$

Since heat dissipated = Electrical power input

$$\therefore \quad \pi d.l. \times 63{,}174 = 15{,}000$$

or $$d^2 l^2 = \left(\frac{15{,}000}{\pi \times 63{,}174}\right)^2 = 0.00571216 \qquad ...(ii)$$

Multiplying expressions (*i*) and (*ii*) we get,

$$l^3 = 24{,}94{,}309 \times 0.00571216 = 14{,}247.8 \text{ m}^3$$

or $$l = 24.24 \text{ meters. } \textbf{Ans.}$$

and diameter of wire,

$$d = \frac{\sqrt{0.00571216}}{24.24} = 0.00311 \text{ m} = 3.11 \text{ mm. } \textbf{Ans.}$$

Example 3.15. *A 30 KW, 3-phase, 400 V resistance oven is to employ nickel-chrome strip 0.025 cm thick for the 3-phase star connected heating elements. If the wire temperature is to be 1100°C and that of charge is to be 700°C, estimate a suitable width for the strip. Assume radiating efficiency as 0.6 and emissivity as 0.9. The specific resistance of the nichrome-alloy is* 1.03×10^{-6} Ω-m. *State any assumption made.*

Solution. Power per phase,

$$P = \frac{30 \times 1000}{3} = 10{,}000 \text{ Watts}$$

Phase voltage, $$V_P = \frac{400}{\sqrt{3}} = 230.94 \text{ V}$$

Resistance of strip,

$$R = \frac{(V_P)^2}{P} = \frac{(230.94)^2}{10{,}000} = 5.333\ \Omega$$

If 'l' is the length of the strip and 'w' is the width of the strip then resistance of strip,

$$R = \frac{\rho l}{w \times t} = \frac{1.03 \times 10^{-6} \times l}{w \times 0.025 \times 10^{-2}}$$

or $$\frac{l}{w} = \frac{0.00025\, R}{1.03 \times 10^{-6}}$$

$$= \frac{250 \times 5.333}{1.03} = 1{,}295 \qquad ...(i)$$

Heat dissipated from the surface,

$$H = 5.72\, K.e. \left[\left(\frac{T_1}{100}\right)^4 - \left(\frac{T_2}{100}\right)^4\right] \text{ W/m}^2$$

$$= 5.72 \times 0.6 \times 0.9 \left[\left(\frac{273 + 1100}{100} \right)^4 - \left(\frac{273 + 700}{100} \right)^4 \right] \text{ W/m}^2$$

$$= 82{,}082 \text{ W/m}^2.$$

Surface area available for heat radiation,

$$= 2\, wl \text{ metres}^2$$

Since, heat dissipated = Power input

$$\therefore \quad 82{,}082 \times 2\, wl = 10{,}000$$

$$\therefore \quad wl = \frac{10{,}000}{2 \times 82{,}082} = 0.060915 \text{ m}^2 \qquad \ldots(ii)$$

Dividing expression (*ii*) by expression (*i*) we get

$$w^2 = \frac{0.060915}{1{,}295} = 0.000047 \text{m}^2$$

or $$w = \sqrt{0.000047} = 0.00686 \text{ m} = 6.86 \text{ mm.} \qquad \textbf{Ans.}$$

Example 3.16. *A piece of plywood is to be heated by dielectric heating. The area of cross-section of the piece is 0.5 m² and the thickness is 2.5 cm. If the frequency of 25 Mega cycle per sec. is used and the power absorbed is 1000 watt. Find the voltage employed necessary for heating. The relative permittivity of wood is 2.5 and power factor is 0.046.*

Solution. **Capacitance of capacitor formed,**

$$C = \frac{K_0 \,.\, K \,.\, A}{t}$$

$$K_0 = 8.854 \times 10^{-12}$$

$$K = 3.6$$

$$A = 0.5 \text{ m}^2$$

$$t = 0.025 \text{ m}$$

$$\therefore \quad C = \frac{8.854 \times 10^{-12} \times 0.5 \times 3.6}{0.025}$$

$$= 637.488 \times 10^{-12} \text{ farads}$$

$$\cos \Phi = 0.046$$

$$\Phi = 87.36°$$

$$\delta = (90 - \Phi)°$$

$$= (90 - 87.36)° = 2.64°$$

$$\delta = 2.64° = \frac{\pi}{180} \times 2.64 = 0.046 \text{ radian}$$

$$P = 2\pi f V^2 C \delta$$

$$1000 = 2\pi \times 2.5 \times 10^6 \times V^2 \times 637.488 \times 10^{-12} \times 0.046$$

or $$V^2 = \frac{1000}{2\pi \times 2.5 \times 10^{-6} \times 637.488 \times 0.046}$$

$$= \frac{10^9}{2\pi \times 25 \times 637.488 \times 0.046} = 217{,}095{,}023$$

or $V = 465.93$ Volts. **Ans.**

Example 3.17. *In a 3-phase arc furnace 10 tonne steel is to melt in 2 hours, estimate the average input to the furnace, if overall efficiency is 50%. If the current input is 9,000 A with the above KW input and the resistance and reactance of furnace leads (including transformer) are 0.003 Ω and 0.005 Ω respectively, estimate the arc voltage and total KVA taken from the supply.*

Specific heat of steel $= 444\ J\ Kg^{-1}\ {}^{\circ}C^{-1}$

Latent heat of fusion of steel = 37.25 KJ/Kg.

Melting point of steel = 1,370°C

Solution. Energy required to melt 10 tonne steel,

$$= mS(t_2 - t_1) + mL = m[S(t_2 - t_1) + L]$$

$$= 10{,}000\ [444(1{,}370 - 20) + 37{,}250]\ \text{Joules}$$

Assuming initial temperature of steel as 20°C

$$= 6{,}366.5 \times 10^6 J = \frac{6{,}366.5 \times 10^6}{60 \times 60 \times 1{,}000}\ \text{kWh}$$

$$= 1{,}768.5\ \text{kWh}$$

Time taken in melting of steel = 2 hours

Average output

$$= \frac{\text{Total energy required in kWh}}{\text{Time of melting in hours}}$$

$$= \frac{1{,}768.5}{2} = 884.25\ \text{kW}$$

Average input $$= \frac{\text{Average output}}{\text{Overall efficiency}} = \frac{884.25}{0.5}$$

$$= 1{,}768.5\ \text{kW}$$ **Ans.**

Current input, $I = 9{,}000$ A

Voltage drop due to resistance of furnace leads (including transformer)

$$= 9{,}000 \times 0.003 = 27\ \text{V}$$

Voltage drop due to reactance of furnace leads (including transformer)

$$= 9{,}000 \times 0.005 = 45 \text{ V}$$

Let the arc drop be V_A volts, resistive in nature from the Fig. 3.37.

Open circuit phase voltage of transformer secondary

$$= \sqrt{(V_A + 27)^2 + (45)^2} \text{ volts}$$

and power-factor,

$$\cos \Phi = \frac{V_A + 27}{(V_A + 27)^2 + (45)^2}$$

Total power input $= 3 \times$ power drawn per phase

$= 3 \times$ current drawn per phase $\times$ secondary phase voltage $\times$ p.f.

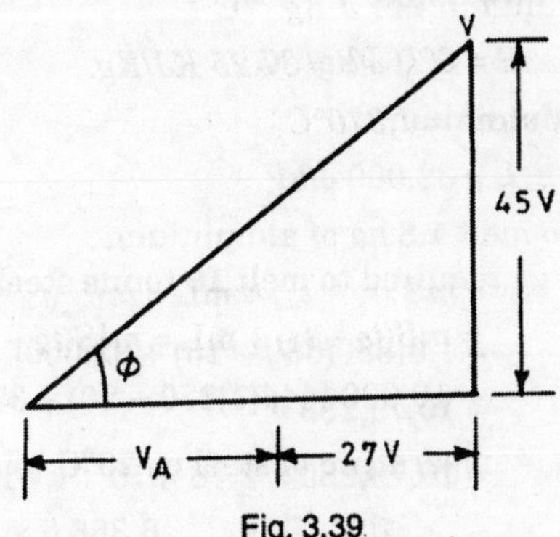

Fig. 3.39

$$\therefore \quad 1{,}768.5 \times 1{,}000 = 3 \times 9{,}000 \times \sqrt{(V_A + 27)^2 + (45)^2} \times \frac{V_A + 27}{\sqrt{(V_A + 27)^2 + (45)^2}}$$

or $$V_A + 27 = \frac{1{,}768.5 \times 1{,}000}{3 \times 9{,}000} = 65.5 \text{ V}$$

or Arc voltage, $V_A = 65.5 - 27 = 38.5$ V. **Ans.**

Open circuit phase voltage (secondary)

$$= \sqrt{(V_A + 27)^2 + (45)^2} = \sqrt{(38.5)^2 + (45)^2} = 79.5 \text{ V}$$

Total kVA taken from supply $= 3 \times$ current drawn per phase $\times$ secondary phase voltage $\times 10^{-3}$

$$= 3 \times 9{,}000 \times 79.5 \times 10^{-3} = 2{,}146 \text{ kVA.} \quad \textbf{Ans.}$$

Example 3.18. *Estimate the efficiency of a high frequency induction furnace which takes 10 minutes to melt 1.8 Kg of aluminium. The input to the furnace being 5 KW and initial temperature 15°C. Given :*

Specific heat of aluminium = 880 J/Kg/°C

Melting point of aluminium = 660°C

Latent heat of fusion of aluminium = 32 KJ/Kg and 1 J = 2.78 × 10^{-7} KWh.

Solution. Quantity of aluminium to be melt,

$$m = 1.8 \text{ kg}$$

Initial temperature, $t_1 = 15°C$

Melting temperature,

$$t_2 = 660°C$$

Specific heat of aluminium

$$S = 880 \text{ J/kg/ °C}$$

Latent heat of aluminium,

$$L = 32{,}000 \text{ J/kg}$$

Heat required to melt 1.8 kg of aluminium,

$$= mS(t_2 - t_1) + mL = m[S(t_2 - t_1) + L]$$

$$= 1.8[880(660 - 15) + 32{,}000]$$

$$= 10{,}79{,}280 \text{ J}$$

$$= 10{,}79{,}280 \times 2.78 \times 10^{-7} \text{ kWh} = 0.3 \text{ kWh}$$

Energy input $= 5 \times \frac{10}{60} = 0.8333$ kWh

Efficiency of furnace,

$$\eta = \frac{\text{Output}}{\text{Input}} \times 100 = \frac{0.3}{0.8333} \times 100 = 36\% \quad \textbf{Ans.}$$

Example 3.19. *A 3-phase electric arc furnace has the following data:*

Current drawn = 5,000 A

Arc voltage = 50 V.

Resistance of transformer referred to secondary = 0.002 Ω

Reactance of transformer referred to secondary = 0.004 Ω.

(*i*) *Calculate the power factor and KW drawn from the supply.*

(*ii*) *If the overall efficiency of furnace is 65%, find the time required to melt 2 tonnes of steel when latent heat of steel = 8.89 Kcal / Kg. specified heat of steel = 0.12, melting point of steel = 1370°C and initial temperature of steel= 20°C.*

Solution. Arc voltage = 50 volts

Current drawn, $I = 5{,}000$ A

Voltage drop due to resistance of transformer,

$$= 5{,}000 \times 0.002 = 10 \text{ V}$$

Voltage drop due to reactance of transformer,

$$= 5{,}000 \times 0.004 = 20 \text{ V}$$

Since arc drop is resistive in nature, therefore, from the Fig. 3.40.

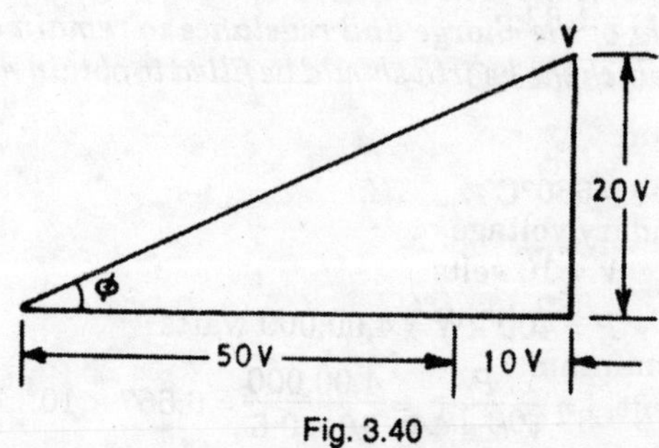

Fig. 3.40

Open circuit phase voltage of transformer secondary

$$= \sqrt{(50 + 10)^2 + (20)^2} = \sqrt{4000} = 63.25 \text{ V}$$

(*i*) P.F. of the supply $= \dfrac{50 + 10}{63.25} = 0.9487.$ **Ans.**

Power drawn per phase

= Current drawn per phase × Secondary phase voltage × Power factor

$$= 5000 \times 63.25 \times 0.9487$$

$$= 304{,}000 \text{ W or } 300 \text{ kW}$$

Total power drawn from the supply

$$= 3 \times 300 = 900 \text{ kW} \quad \textbf{Ans.}$$

(*ii*) Energy required to melt 2 tonnes of steel,

$$= mS\,(t_2 - t_1) + \text{mL} = m[S\,(t_2 - t_1) + \text{L}]$$

$$= 20{,}00[0.12 \times (1370 - 20) + 8.89]$$

$$= 341{,}780 \text{ kilo-calories}$$

$$= \frac{341{,}780}{869} = 397.42 \text{ kWh}$$

Power utilised = Power drawn from supply mains × overall efficiency

$$= 900 \times 0.65 = 535 \text{ kW}$$

Time required for Melting

$$= \frac{\text{Energy required in kWh}}{\text{Power required in kW}}$$

$$= \frac{397.42}{585} = 40 \text{ minutes, } 46 \text{ seconds.} \quad \textbf{Ans.}$$

Example 3.20. *A low frequency induction furnace, whose secondary voltage is maintained constant at 10 volts, takes 400 KW at 0.6 p.f. when the hearth is full. Assume the resistance of the secondary circuit to vary inversely as the height of the charge and reactance to remain constant, find the height upto which the hearth should be filled to obtain maximum heat.*

Solution. Secondary voltage

$$V = 10 \text{ volts}$$

Power drawn, $P = 400 \text{ kW} = 4{,}00{,}000 \text{ watts}$

Secondary current, $I = \frac{P}{V \cos \Phi} = \frac{4{,}00{,}000}{10 \times 0.6} = 6.667 \times 10^4 \text{ A}$

Impedance of secondary circuit when hearth is full,

$$Z = \frac{V}{I}$$

$$Z = \frac{10}{6.667 \times 10^4} = 1.5 \times 10^{-4} \ \Omega$$

Resistance of secondary circuit when hearth is full,

$$R = Z \cos \Phi = 1.5 \times 10^{-4} \times 0.6 = 0.9 \times 10^{-4} \ \Omega$$

Reactance of secondary circuit

$$X = Z \sin \Phi = 1.5 \times 10^{-4} \times 0.8 = 1.2 \times 10^{-4} \ \Omega$$

Let the height of charge be x times of full hearth *i.e.*

$$h = xH$$

Since resistance varies inversely as the height of charge,

∴ Resistance of the charge with height of charge as xH

$$= \frac{R}{x} = \frac{0.9 \times 10^{-4}}{x} \ \Omega$$

Power drawn or heat produced will be maximum when resistance of secondary circuit will be equal to reactance of secondary circuit.

i.e.
$$\frac{0.9 \times 10^{-4}}{x} = 1.2 \times 10^{-4}$$

or
$$x = \frac{0.9}{1.2} = \frac{3}{4}$$

i.e. Maximum heat will be obtained with the height of charge as three fourth of height of hearth.

Example 3.21. *A plywood board of 0.5 × 0.25 × 0.02 metre is to be heated from 25°C to 125°C in 10 minutes by dielectric heating employing a frequency of 30 MHz. Determine the power required in this heating process. Assume specific heat of wood 1500/J/Kg/°C; weight of wood 600 $^{kg}/_{m^3}$ and efficiency of process 50%.*

Solution. Volume of plywood to be heated

$$= 0.5 \times 0.25 \times 0.02 = 0.0025 \text{ m}^3$$

Weight of plywood $= 0.0025 \times 600 = 1.5$ kg

Heat required to raise the temperature of plywood board from 25°C to 125°C

$$= 1500 \times 1.5(125 - 25) = 2{,}25{,}000 \text{ J or Watt-Sec.}$$

$$= \frac{2{,}25{,}000}{60 \times 60} = 62.5 \text{ Watt-Hr.}$$

Since it is to be done in 10 minutes, so power required,

$$= \frac{62.5}{10/60} = 375 \text{ watts}$$

Power input required

$$P = \frac{\text{Power required}}{\text{Efficiency of process}} = \frac{375}{0.5} = 750 \text{ watts.} \quad \textbf{Ans.}$$

Example 3.22. *Dielectric heating is to be employed to heat a slab of insulating material 20 mm thick and 1530 mm^2 in area. Power required is 200 watts and a frequency of 3 MHz is to be used. The material has a permittivity of 5 and p.f. of 0.05. Determine the voltage necessary and the current which will flow through the material.*

Solution. Power absorbed,

$$P = 200 \text{ watts}$$

Power factor, $\cos \Phi = 0.05$

Supply frequency, $f = 3 \times 10^6$ Hz.

Capacitance of the capacitor formed

$$C = \frac{K_o K A}{t}$$

or

$$\left(\frac{\varepsilon_0\, \varepsilon_r A}{t}\right) = \frac{8.854 \times 10^{-12} \times 5 \times 1500 \times 10^{-6}}{20 \times 10^{-3}}$$

$$= 3.32 \times 10^{-12} \text{ F}$$

Since

$$P = 2\,\pi f\, C V^2 \cos \Phi$$

∴ Voltage required

$$V = \sqrt{\frac{P}{2\,\pi f C \,.\, \cos \Phi}}$$

$$= \sqrt{\frac{200}{2\,\pi \times 3 \times 10^{6} \times 3.32 \times 10^{-12} \times 0.05}}$$

$= 8000$ V. **Ans.**

Current flowing through the material

$$I = \frac{P}{V \cos \Phi} = \frac{200}{8000 \times 0.05} = 0.5 \text{ A} \quad \textbf{Ans.}$$

Example 3.23. *A small piece of plywood having dimensions : length 5 cm, width 2 cm, and thickness 1 cm, is placed in between two electrodes having dimensions, length 25 cm width 2 cm and width 2 cm distance between them. Electric current at frequency of 10 MHz is passed through electrodes for heating of wood by the process of dielectric heating. If the power consumption for heating the piece of wood is 1 KW, determine (i) Voltage across the electrode and (ii) Current through it during the process of heating. Assume that the plywood is an imperfect dielectric and has relative permittivity = 5 and power factor 0.04.*

Solution. The arrangement of heating material is shown in Fig. 3.41. With this configuration the capacitance of the parallel plate condenser formed with two dielectrics, plywood, and air is given by expression.

$$C = \varepsilon_0 \left[\frac{A_1\, \varepsilon_{r2}}{t} + \frac{A_2}{\dfrac{t_1}{\varepsilon_{r1}} + \dfrac{t_2}{\varepsilon_{r2}}} \right] \text{Farads}$$

where $A_1 = (25 - 5) \times 2 = 40 \text{ cm}^2 = 0.004 \text{ m}^2$

$A_2 = 5 \times 2 = 10 \text{ cm}^2 = 0.001 \text{ m}^2$

$t = 2 \text{ cm} = 0.02 \text{ m}$

$t_1 = 1 \text{ cm} = 0.01 \text{ m}$

$t_2 = 2 - 1 = 1 \text{ cm} = 0.01 \text{ m}$

$\varepsilon_0 = 8.854 \times 10^{-12}$ F/m

ε_{r_1} = Relative permittivity of plywood = 5

ε_{r_2} = Relative permittivity of air = 1

∴ Capacitance of the capacitor,

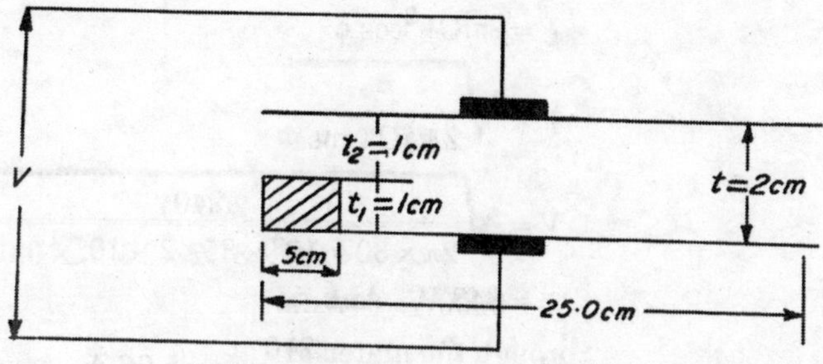

Fig. 3.41

$$C = 8.854 \times 10^{-12} \left[\frac{0.004 \times 1}{0.02} + \frac{0.001}{\frac{0.01}{5} + \frac{0.01}{1}} \right]$$

$$= 2.509 \times 10^{-12} \text{ Farads}$$

(*i*) Voltage across the electrode,

$$= \sqrt{\frac{P}{2\pi f C \cos \Phi}}$$

$$= \sqrt{\frac{1000}{2\pi \times 10 \times 10^6 \times 2.509 \times 10^{-12} \times 0.04}}$$

$$= 12{,}594 \text{ V.} \quad \textbf{Ans.}$$

(*ii*) Current through plywood,

$$I = \frac{P}{V \cos \Phi} = \frac{1{,}000}{12{,}594 \times 0.04} = 1.985 \text{ A.} \quad \textbf{Ans.}$$

Example 3.24. *Dielectric heating is to be employed to heat a slab of insulating material 2 cm thick and 160 sq-cm in area. The power required is 240 watts and frequency of 30 mega cycles per second is to be used. The material has a relative permitivity of 5 and power factor of 0.05. Determine the voltage necessary and current which will flow through the material. If the voltage were to be limited to 640 V; to what value would the frequency have to be used ?*

Solution. Power absorbed = 240 watts

Power factor = 0.05

frequency = 30×10^6 Hz

$$\text{Capacitance of capacitor} = \frac{\varepsilon_0 \varepsilon_r A}{t}$$

$$= \frac{8.854 \times 10^{-12} \times 160 \times 10^{-4} \times 5}{0.02}$$

$$= 35.42 \times 10^{-12} F$$

$$P = 2\pi f C V^2 \cos\phi$$

or $$V = \sqrt{\frac{P}{2\pi f C \cos\phi}}$$

or $$V = \sqrt{\frac{240}{2\pi \times 30 \times 10^6 \times 35.42 \times 10^{-12} \times 0.05}}$$

$$= \mathbf{848\ V} \quad \textbf{Ans.}$$

$$I = \frac{P}{V\cos\phi} = \frac{240}{848 \times 0.05} = \mathbf{5.66\ A} \quad \textbf{Ans.}$$

Heat produced $\propto f V^2$

$$f_2 V_2^2 = f_1 V_1^2$$

or $$f_2 = f_1 \left(\frac{V_1}{V_2}\right)^2$$

$$= 30 \times 10^6 \left(\frac{848}{640}\right)^2$$

$$= \mathbf{52.67 \times 10^6\ Hz} \quad \textbf{Ans.}$$

Example 3.25. *Determine the diameter and length of the nichrome wire to be used as heating element in a 10 KW, 220 volts single phase resistance furnace. The temperature of wire is not to be exceed 1000°C and charge is to be at 500°C. Assume radiating efficiency as 0.7, emissivity 0.9 and resistivity as 1.016 micro-ohm-m.*

Ans. Power supplied, $P = 10 \times 1000$ W

Supply Voltage, $V = 220$ V

Maximum Temperature, $T_1 = 1000$°C

Temp. of charge, $T_2 = 500$°C

Radiating efficiency, $k = 0.7$

Emissivity, $e = 0.9$

Heat dissipated (H)

$$H = 5.72 \times 10^4\, ke \left[\left(\frac{T_1}{1000}\right)^4 - \left(\frac{T_2}{1000}\right)^4\right] \text{Watts/sqm}$$

$$= 5.72 \times 0.7 \times 0.9 \times 10^4 \left[\left(\frac{1000}{1000} + 273\right)^4 - \left(\frac{773}{1000}\right)^4\right] \text{Watts/sqm}$$

$$= 5.72 \times 0.7 \times 0.9 \times 10^4\ [2.626 - 0.357]$$

$$= 8.176 \times 10^4 \text{ Watt/sqm}$$

Circular wire element, $P = \dfrac{V^2}{\dfrac{4\rho R}{\pi d^2}}$ or $\dfrac{d^2}{L} = \dfrac{4\rho}{\pi V^2}$

$$\text{or} \qquad \frac{d^2}{l} = \frac{4 \times 1.016 \times 10 \times 1000}{10^6 \times \pi \times 220 \times 220} = \frac{40640}{152114.28 \times 10^6}$$

$$= \frac{152114.28 \times 10^6}{40640} \qquad \text{...}(i)$$

Area of radiating surface = πdl

$$\pi dl \times H = P$$

$$\pi dl \times 8.176 \times 10^4 = 10000$$

$$dl = \frac{10000}{\pi \times 8.176 \times 10^4} = \frac{1}{8.176\pi}$$

$$\text{or} \qquad d^2 l^2 = \left(\frac{1}{8.176\pi}\right)^2 \qquad \text{...}(ii)$$

Multiplying (*i*) and (*ii*)

$$\frac{l}{d^2} \times d^2 l^2 = \frac{152114.28 \times 10^6}{40640} \times \frac{1}{8.176} \times \frac{1}{8.176} \times \frac{1}{\pi} \times \frac{1}{\pi}$$

$$l^3 = 5679$$

$$l = 17.84 \text{ m} \quad \textbf{Ans}$$

Put l in eq. (*i*)

$$d = \frac{1}{8.176\pi \times 17.84} = 0.00218 \text{ m.} \quad \textbf{Ans}$$

Example 3.26. *A state of insulating material 15 sq.cm in area and 1 cm thick is to be heated by dielectric heating. The power required is 400 watts at 30 MHz. The material has a permittivity of 5 and power factor 0.05. Calculate the voltage necessary. Absolute permittivity 8×10^{-12} F/m.*

Solution.

Power Required = 400 watts

$$f = 30 \times 10^6 \text{ Hz}$$

Power factor = 0.05

Absolute permitivity = 8×10^{-12}

Permittivity of material = 5

Capacitance of a Capacitor = $\dfrac{\varepsilon_o \, \varepsilon_r \, A}{t}$

$$= \frac{8 \times 10^{-12} \times 15 \times 10^{-4} \times 5}{0.01} = 6 \times 10^{-12} \text{ F}$$

$$P = 2\pi f C V^2 \cos\phi$$

$$V = \sqrt{\frac{P}{2\pi f C \cos\phi}}$$

$$= \sqrt{\frac{400}{2 \times \pi \times 30 \times 10^6 \times 6 \times 10^{-12} \times 0.05}}$$

$$= \sqrt{\frac{400}{2 \times \pi \times 180 \times 10^{-6} \times 0.05}}$$

$$= \sqrt{7077140.8} = \mathbf{2660V} \quad \textbf{Ans.}$$

Short Questions Answers

Q.1. What are the different modes of heat transfer ?

Ans. Conduction, convention and radiations are the three modes of transfer of heat.

Q.2. Write the names of the materials which are normally used as heating element material.

Ans. Normally used materials as heating element are Nickle-chromium (Nichrome) alloys, Nickle-Chromium-iron alloys, Nickle copper alloys, (Eureka or Constantun), and iron-chromium aluminum alloy (Kantal).

Q.3. What properties are considered for selecting material for heating element ?

Ans. For selection of material for heating elements, the properties to be considered are, resistivity, melting point, temperature co-efficient of resistance and free from oxidization.

Q.4. What are the causes of failure of heating elements ?

Ans. The main causes of failure of heating elements are contamination and corrosion, formation of hot spots, oxidation and intermitancy of operation etc.

Q.5. What is an oven ?

Ans. Having low temperature heating chamber with provision for ventilation also, is termed as oven.

Q.6. Name the few applications of infra-red heating.

Ans. Infra-red heating is mainly used for :

(*i*) Drying of furniture

(*ii*) Drying of pottery and paper textile.

(*iii*) Paint storing and

(*iv*) Prior to moulding, pre-heating of plastic.

Q.7. Name the type of electrodes used in arc furnaces.

Ans. In arc furnaces the electrodes used are of three types namely; Carbon, graphite and self baking electrodes.

Q.8. What are the applications of high frequency eddy current heating ?

Ans. The main applications of high frequency eddy current heating are : Surface hardening, sodering, annealing, welding, drying of paints, melting of precious metals, sterilization of surgical instruments etc.

Q.9. What type of supply is used for direct core type induction furnace in our country ?

Ans. In our country 1 - ϕ, 50 Hz, 600 V supply is used for direct core type induction furnace.

Q.10. How amount of heating is controlled in high frequency eddy current heating ?

Ans. The amount of heating is controlled by controlling the supply frequency and the flux density in high frequency eddy current heating.

Q.11. At what power factor would you like to operate the electric arc furnace ?

Ans. Generally at power factor of about 0.9, but not below than 0.8 p.f. in any case.

Q.12. How can the rate of di-electric heating be varied ?

Ans. Rate of di-electric heating can be varied by varying either supply voltage or frequency of supply.

Q.13. Why dielectric heating is employed only where other methods are impracticable or very slow ?

Ans. Because of the reason that very high cost of equipment required for di-electric heating.

Q.14. How is control of power affected in electric arc furnaces ?

Ans. The power input to arc furnaces can be controlled by raising or lowering the electrodes or by varying the voltage applied to the arc furnace.

Preferably both voltage and electrode controls (voltage can be changed by changing of tappings on supply transformer).

Choose the Correct Answer

1. Dielectric heating is more effective when :

(a) Voltage is low, but frequency is high

(b) Voltage is high, but frequency is low

(*c*) Both voltage and frequency are low

(*d*) Both the voltage and frequency are high.

2. Heating of some electrolyte through Immersion type heaters is:
   (*a*) A direct resistance heating process
   (*b*) A indirect resistance heating process.
3. Rocking furnaces can be :
   (*a*) Indirect resistance heating type
   (*b*) Indirect are type
   (*c*) Both as above
   (*d*) None as above.
4. Butt welding is :
   (*a*) Possible through momentary high frequency eddy current heating for small members
   (*b*) Welding is entirely a different process the concept is meaningless
   (*c*) None of the above statements are correct.
5. Formation of hot spots along the length of a heating element indicates:
   (*a*) Higher voltage application
   (*b*) Efficient performance of the device
   (*c*) Chances of approaching failure
   (*d*) Characteristic of the material composing the element
6. Ajax-Wyatt vertical furnace is most suitable for :
   (*a*) Ferrous melts
   (*b*) All non-ferrous metals
   (*c*) Only non-conducting materials
   (*d*) Ideal for brass melting.
7. Capacitor bank used across ferrous material melting furnaces should have :
   (*a*) Continuous constant rating
   (*b*) Varying reactive component from instant to instant
   (*c*) Different values only before and after the melting
8. In indirect resistance heating, if the operating temperature exceeds 1200°C the element material should consist of :
   (*a*) Nichrome
   (*b*) Tungsten
   (*c*) Plantinum
   (*d*) Silicon carbide/Molybdenum.
9. In high frequency induction furnace the heating is due to :
   (*a*) Circulation of eddy currents through the charge
   (*b*) The hysteresis effect pre-domination

(c) Skin effect penetration

(d) None out of above, both eddy currents and hysteresis effect remain away from the charge.

10. The conducting bottom type furnances are meant for :

(a) Indirect arc heating (b) Direct arc heating

(c) Direct resistance heating (d) Non out of the above.

11. Intermittent switching for temperature control is suitable for :

(a) Indirect resistance heating (b) Direct resistance heating

(c) Arc heating (d) Induction heating.

12. **Fill in the blanks** by suitable words selected from the three words given in brackets.

(a) The overall efficiency of electric heating is ____.
(low, high, very low)

(b) The material used for the heating element of ovens should have ____ resistivity. (low, medium, high)

(c) The temperature coefficient of resistance of the material of heating element should be ____. (low, high, negative)

(d) The electrodes used for direct resistance heating one made of _____. (mild steel, copper, carbon)

(e) The electrodes used in furnaces are of ____.
(copper, carbon, graphite)

(f) Induction furnaces require ________ frequency supply.
(low, high, power)

(g) Dielectric heating can be only done at _____ frequency.
(low, power, high)

(h) Dielectric heating is used for heating of ____materials.
(ferrous, non-ferrous, insulating)

13. **Write 'Yes' or 'No' for** the following sentences:-

(i) Electric heating is more efficient as compared to other forms of heating.

(ii) Resistance heating can be done at d.c. supply.

(iii) Arc heating is done at high frequency supply.

(iv) Induction heating utilises a.c. supply at high frequency.

(v) Dielectric heating is employed for heating of metal alloys.

(vi) Insulating materials like wood plaster are only heated by dielectric heating.

(vii) Ajax-Wyatt furnace is only suitable for melting non ferrous metals.

(viii) In direct core type induction furnace, pinch effect can be removed by using d.c. supply.

14. In case of direct arc heating the suitability of arc is established through:

(*a*) Controlling reactors (*b*) Voltage control

(*c*) Current control (*d*) Electrode movement.

Answers

1. (*d*)	2. (*b*)	3. (*c*)	4. (*b*)	5. (*c*)
6. (*d*)	7. (*b*)	8. (*d*)	9. (*d*)	10. (*b*)

11. (*a*) 12. (*i*) high (*ii*) high (*iii*) low (*iv*) mind steel (*v*) carbon or graphic (*vi*) high (*vii*) high (*viii*) Insulating

13. (*i*) Yes (*ii*) No (*iii*) No (*iv*) yes (*v*) No (*vi*) Yes (*vii*) Yes (*viii*) No.

14. (*d*.)

1. Fill in the blanks

(*i*) The electric heating is ______.

(*ii*) The material used for the heating element of an oven should have ______ resistivity.

(*iii*) The insulating materials can be heated only by ______.

(*iv*) The temperature co-efficient of resistance of the material of heating element should be ______.

(*v*) From Sun, we get energy through ______.

(*vi*) The electrodes used in arc furnaces are of ______.

(*vii*) Induction furnaces may be core type or ______ type.

(*viii*) The heating method using infra-red radiation is called the ______ heating.

(*ix*) The microwave heating used radiations of very short ______.

(*x*) The frequency of microwaves is ______ high.

(*xi*) Die-electric heating can be only done at ______ frequency.

(*xii*) Induction furnaces can work only on ______ supply.

(*xiii*) Heating element used in household appliances are made of ______.

(*xiv*) In flammable articles like plastics can safely be heated by using ______ heating processes.

(*xv*) Indirect furnaces are not built in ______ sizes.

(*xvi*) Di-electric heating works on ______ supply and not on ______ supply.

(*xvii*) Induction heating is used for heating ______.

(*xviii*) The different modes by which heat is transferred are conduction, ______ and Radiation.

(*xix*) Thermal conductivity is measured in ______.

(*xx*) Large furnace installations are often equipped with ______ charging.

(*xxi*) The di-electric loss depends upon ______.

Answers

(i) Economical	*(ii)* High
(iii) Electricity	*(iv)* Low
(v) Radiations	*(vi)* Carbon
(vii) Coreless	*(viii)* Radiant
(ix) Penetrates	*(x)* very
(xi) High	*(xii)* A.C.
(xiii) Nichrome	*(xiv)* di-electric
(xv) large	*(xvi)* A.C., D.C.
(xvii) Magnetic or Non-magnetic material	
(xviii) Convection	*(xix)* MJ/m/°C/hour
(xx) Side door	*(xxi)* Voltage & frequency.

Choose the correct Answer

(i) For arc healing, the electrodes are made of

(*a*) Copper (*b*) Aluminium

(*c*) Graphite (*d*) ACSR Conductor

(ii) The method of heating used in an electric room heater convector is :

(*a*) Resistance heating (*b*) Drooping Characteristics

(*c*) Straight Characteristics (*d*) Exponential characteristics

(iii) The ideal method of heating plastics is

(*a*) Coal/oil fired furnace (*b*) di-electric-heating

(*c*) Induction heating (*d*) Resistance heating

(iv) The material used as a heating element for a furnace should have :

(*a*) High resistivity (*b*) High melting point

(*c*) Low temp. coefficient (*d*) All of the above

(v) The heating element to be used in a furnace, used for heating around 1600°C should be of material,

(*a*) Nichrome (*b*) Eureka

(*c*) Molybdenum (*d*) Silicon-Carbide.

(vi) For radiant heating around 2000°C, the heating elements used should be of material,

(*a*) Tungsten alloy (*b*) Copper alloy

(*c*) Carbon (*d*) Stainless steel alloy

(vii) In a resistance furnace, the temperature is controlled by :

(*a*) Variation of operating voltage

(*b*) Variation of resistance of heating ckt.

(*c*) Switching on and off the supply periodically

(*d*) All of the above.

(*viii*) The device necessarily used for automatic temperature control in a furnace is :

(*a*) Thermo-stat (*b*) Auto-transformer

(*c*) Thermo-couple (*d*) None of the above

(*ix*) Radiant heating is used for:

(*a*) Melting of ferrous metals (*b*) Annealing of metals

(*c*) Drying of points and varnishes

(*d*) None of the above.

(*x*) In direct arc furnace, which of the following is of high value :

(*a*) Current (*b*) Voltage

(*c*) Power (*d*) All of the above.

(*xi*) The main application of indirect arc furnace is to melt :

(*a*) Iron (*b*) Steel

(*c*) Non Ferrous metals (*d*) None of the above.

(*xii*) High frequency induction heating is used for

(*a*) Ferrous metals only (*b*) Non-ferrrous metals only

(*c*) Both ferrous & Non -ferrous metals.

(*d*) None of the above

(*xiii*) The di-electric loss in a di-electric is proportional to :

(*a*) Voltage impressed on the di-electric

(*b*) The square of the voltage impressed on the di-electric.

(*c*) The square root of the voltage impressed on the di-electric.

(*d*) None of the above.

(*xiv*) The power factor will be maximum in case of :

(*a*) Electric arc heating (*b*) Resistance heating

(*c*) Induction heating (*d*) Di-Electric arc heating

(*xv*) The method heating used for non-conducting material is :

(*a*) Induction heating (*b*) Die-electric heating

(*c*) Electric resistance heating (*d*) D-electric arc heating

(*xvi*) The most modern method for food processing is

(*a*) Induction heating (*b*) Resistance heating

(*c*) Di-electric heating (*d*) Eddy current heating

(*xvii*) Hysteresis loss and eddy current loss are used in :

(*a*) Resistance heating

(*b*) Di-electric heating

(*c*) Induction heating of steel

(*d*) Induction heating of brass.

(*xviii*) The method suitable for heating of conducting medium is :

(*a*) Induction heating (*b*) Indirect arc heating

(*c*) Eddy current heating (*d*) Radiant heating

(*xix*) Which of the following methods of heating is independent of supply frequency:

(*a*) Electric arc heating (*b*) Induction heating

(*c*) Electric resistance heating

(*d*) Di-electric heating

(*xx*) The main application of indirect arc furnace is to melt:

(*a*) Iron (*b*) Steel

(*c*) Non-ferrous metals (*d*) None of the above

(*xxi*) Ni-Cr-Fe alloy wires can be safely used for temperatures upto ....

(*a*) 2,500°C (*b*) 2,000°C

(*c*) 1,150°C (*d*) 850°C

(*xxii*) The highest value of thermal conductivity is for :

(*a*) Aluminium (*b*) Brass

(*c*) Copper (*d*) Iron

(*xxiii*) A body reflecting entire radiations incidenting on it, is called the:

(*a*) White body (*b*) Gray body

(*c*) Black body (*d*) Transparent body

(*xxiv*) The material used as heating element for a furnace should have :

(*a*) High resistivity (*b*) High melting point

(*c*) Low temp. co-efficient (*d*) All of the above

(*xxv*) The simplest and most commonly used method for temp. control is :

(*a*) External series resistance in the heating ckt.

(*b*) Change of connections in the heating ckt.

(*c*) Use of variable number of heating elements

(*d*) Transformer tappings.

Answers

1. (*c*)	2. (*a*)	3. (*b*)	4. (*d*)	5. (*c*)
6. (*a*)	7. (*d*)	8. (*a*)	9. (*c*)	10. (*a*)
11. (*c*)	12. (*c*)	13. (*b*)	14. (*b*)	15. (*b*)
16. (*c*)	17. (*c*)	18. (*a*)	19. (*c*)	20. (*c*)
21. (*d*)	22. (*c*)	23. (*a*)	24. (*d*)	25. (*b*).

TRUE/FALSE QUESTIONS

Q.1. The body which reflects all the incident radiations is called the perfect black body.

Q.2. The electric furnaces, nickel-chrome-iron alloy is the most useful material as heating element.

Q.3. For transfer of heat by conduction in the indirect resistance heating, an enclosure is required.

Q.4. Direct arc furnaces are normally of 3 - ϕ type, whereas indirect arc furnaces are generally of 1 - ϕ type.

Q.5. For direct core type induction furnaces high frequency supply is very much required.

Answers.

1. False 2. True 3. False 4. True 5. False.

MULTIPLE CHOICE QUESTIONS

1. Induction heating process is based on

(*a*) electro-magnetic induction principle

(*b*) resistance heating principle

(*c*) thermal ion release principle

(*d*) none of these

2. Which heating method has maximum power factor

(*a*) resistance heating

(*b*) dielectric heating

(*c*) induction heating

(*d*) arc heating

3. Which heating method has leading power factor

(*a*) resistance heating

(*b*) dielectric heating

(*c*) induction heating

(*d*) arc heating

4. In electric press, amount of heat is controlled by

(*a*) tapping the heating element

(*b*) thermostat control

(*c*) changing the supply voltage

(*d*) none of the above

5. Ajax Wyatt furnace is started when

(*a*) it is fully empty

(*b*) it is filled below core level

(*c*) it is filled above core level

(*d*) series parallel control

6. For heating non-conducting materials we will use

(*a*) induction heating

(*b*) arc heating

(*c*) dielectric heating

(*d*) eddy current heating

7. In electric press, mica is used because it is

(*a*) bad conductor of heat but good conductor ofelectricity

(*b*) bad conductor of heat but bad conductor of electricity

(*c*) good conductor of heat but bad conductor of electricity

(*d*) good conductor of heat but good conductor of electricity

8. Which of the following methods of heatings is not dependent on the frequency of supply

(*a*) electric resistance heating

(*b*) induction heating

(*c*) dielectric heating

(*d*) all of above

9. In heating the ferro-magnetic material by induction heating, heat is produced.

(*a*) due to induced current flow through the charge

(*b*) due to hysteresis loss taking place below curie temperature

(*c*) due to hysteresis loss as well as eddy current loss taking place in the charge

(*d*) due to none of the above factors

10. In induction heating

(*a*) greater the size of the object will be the frequency employed

(*b*) greater the thermal conductivity of the charge more will be the frequency

(*c*) greater the depth of penetration of heat required less will be the frequency of supply (*d*) all above are correct

11. Dielectric heating is due to

(*a*) friction between polarized atoms of insulating material subjected to reversing electric field

(*b*) friction between polarized atoms of insulating material subjected to reversing magnetic field.

(*c*) friction between the magnetic molecules in insulating material subjected to reversing magnetic field

(*d*) friction between the magnetic molecules in the insulating material subjected to reversing electric field

12. Induction heating takes place

(*a*) in insulating material

(*b*) in conducting material which are magnetic

(*c*) in conducting material which are non-magnetic

(*d*) in conducting material which may or may not be magnetic

13. High resistivity mateiral is chosen for heating element to

(*a*) produce large amount of heat

(*b*) reduce the length of heating element

(*c*) to increase the life of the heating element

(*d*) to reduce the effect of oxidation

14. Property of low temperature co-efficient of heating element is desired

(*a*) to avoid initial rush of current

(*b*) to avoid change in kW rating with temperature

(*c*) due to above two reasons

(*d*) due to none of above reasons

15. Temperature control in resistance furnace is done by

(*a*) voltage variation

(*b*) resistance variation

(*c*) periodical switching on and off of the supply

(*d*) all of above methods

16. Temperature control by voltage variation is done with the help of

(*a*) auto-transformer

(*b*) induction regulator

(*c*) series imepedance

(*d*) all of above methods

17. Resistance variation method of temeprature control is done by connecting resistance elements in

(*a*) series

(*b*) parallel

(*c*) series parallel connections

(*d*) star delta connections

(*e*) all above ways.

18. In the indirect resistance heating method, maximum heat transfer takes place by

(*a*) convection

(*b*) conduction

(*c*) radiation

(*d*) equally by all above

19. Function of choke in arc furnace is

(*a*) to improve P.F.

(*b*) to stabilize the arc

(*c*) to reduce the severity of the surge

(*d*) case of starting.

20. It is desirable to operate the arc at p.f. of

(*a*) unity

(*b*) 0.707 leading

(*c*) 0.707 lagging

(*d*) zero

21. Dielectric heting used to heat

(*a*) insulating material

(*b*) magnetic material

(*c*) conducting material

(*d*) non magnetic material

22. Hysteresis los and eddy current loss are used in

(*a*) dielectric heating

(*b*) resistance heating

(*c*) induction heating of brass

(*d*) induction heating of steel

Answers

1. (*a*)	2. (*a*)	3. (*b*)	4. (*b*)	5. (*c*)
6. (*c*)	7. (*c*)	8. (*a*)	9. (*c*)	10. (*d*)
11. (*a*)	12. (*d*)	13. (*b*)	14. (*c*)	15. (*d*)
16. (*d*)	17. (*e*)	18. (*d*)	19. (*b*)	20. (*c*)
21. (*a*)	22. (*d*).			

EXERCISES

1. State a few advantages of electric heating over other forms of heating.

2. What are the specific advantages of dielectric heating ?

3. (*a*) Make a neat sketch of Ajax-Wyatt core type induction furnace.
   (*b*) Explain its principle of working.
   (*c*) State how the pinch effect is overcome.

4. Describe the construction and working of any type of induction furnace.

5. Discuss the factors which determine the choice of frequency for core-less induction furnace. Enumerate the various methods of providing power supply for such a furnace.

6. (*a*) Explain the process of dielectric heating.
   (*b*) What is the range of voltage and frequency used in dielectric heating and explain the reason of your answer.
   (*c*) Is dielectric heating process used on conducting materials or on non-conducting materials. Explain your answer.

7. A plywood board 50 cm × 20 cm × 2 cm is to be heated to 150°C in 15 minutes by dielectric heating employing a frequency of 30 Mega c/s. The wood has a specific heat of 0.35, a weight of 0.56 gm/cm^3, the ambient temperature 30°C, relative permittivity of 5, absolute permittivity 8.854×10^{-12} (in RMKS units) and a power factor of 0.05. Determine the power required, the voltage across the board and current through it during heating process.

   [Ans. 440 W, 456 V, 19 A**]**

8. (*a*) Describe with neat sketches (*i*) the direct arc furnace (*ii*) the indirect arc furnace.
   (*b*) Why are indirect arc furnaces not built in large sizes ?
   (*c*) What is the purpose of using reactors in electric arc furnaces ?
   (*d*) At what power factor would you like to operate the arc furnaces?

9. **Four resistances each of 60 ohms are used in a resistance oven. Calculăte the power drawn from 240 V single phase a.c. supply when the elements are connected in (*i*) Series (*ii*) Parallel.**

[Ans. (*i*) $P_1 = 240$ watts.(*ii*)$P_2 = 3840$ watts]

10. **A 4.5 kW, 400 V, 3 Φ resistance oven is to have 3-star connected nichrome wire heating elements. If the wire temperature is to be 1000°C and that of the charge 500°C. Estimate the dia and length of wire. The resistivity of nichrome alloy is 42.5×10^{-8} Ω m. Assume the radiation efficiency and the emissivity of the element as 1 and 0.9 respectively.**

[Ans. Diagonal = 0.375 mm, length = 9.23 metres]

Questions Bank with Answer

Short Answer Type Questions.

Q.1. Define electrical heating.

Ans. Heating by means of converting electricity into heat is called electrical heating.

Q.2. Give atleast two advantages of electrical heating.

Ans. (i) Cleanliness (ii) Ease of control

(iii) Quick operation (iv) Absence of flue gases

(v) High efficiency of utilization

Q.3. What are various types of electrical heating?

Ans. Electrical heating is of two types

(i) Power frequency heating

(ii) High frequency heating

In power frequency heating, alternating current is applied at 50 Hz. Power frequency heating is further divided into

(i) Resistance heating

(ii) Arc heating.

Resistance heating may be direct resistance heating or indirect resistance heating. Infrared heating is also a form of direct resistance heating. Similarly arc heating may also direct electrical arc heating or indirect arc heating. In high frequency heating, alternating current is supplied at very high frequency. It is further subdivided into (i) Induction heating (ii) Dielectric heating

Q.4. Define resistance heating.

Ans. Electrical heating based on heating effect of electric current is called resistance heating. In resistance heating, electric current of power frequency is passed through a heating element. Power lost in resistance is converted into heat. If a current is passed through a resistance R for t second, the heat produced is given by

$$H = \frac{I^2Rt}{J} \text{ calories}$$

Where J = Joul's equivalent of heat

Q.5. State desirable properties of resistance heating element.

Ans. Resistance heating element should possess.

(i) High specific resistance

(ii) Low temperature co-efficient of resistance

(iii) High melting point.

Q 6. The resistance heating element is commonly made of ....

Ans. Tungsten, Molybdenum or an alloy of Nickel and Chromium (nichrome).

Q.7. What are the causes of failure of heating element. ***[Dec 2007]***

Ans. Chief causes of failure of heating element are

(i) Formation of hot spots (ii) Oxidation

(iii) Intermittancy of operation (iv) Corrosion of element

Q.8. What is infrared heating?

Ans. This is also called radiant heating and is used where low and medium temperatures are required. In this method of heating special tungsten filament lamps together with their reflectors are used. Lamps operate at 2300° C. The low operating temperature ensures longer life and greater amount of infrared radiation.

Q.9. Name few applications of infrared heating.

Ans. (i) Drying of furniture, pottery, paper and textile,

(ii) Pre-heating of plastic prior to moulding,

(iii) Curing of joint pains.

Q.10. What is a thermostat?

Ans. It is a bimetallic strip which bends and opens the contact when temperature rises beyond preset conditions.

Q.11. What is principle of Arc production? ***[Dec 2006]***

Ans. When a high voltage is applied across two electrodes, separated by an air gap, the air in the gap gets ionised and becomes a conducting medium. Current starts flowing between the two electrodes in the form of a continuous spark known as an Arc.

Q.12. What is direct Arc heating?

Ans. In direct arc heating, heat of arc between two electrodes is directly used to heat the charge. Direct arc furnaces are used to melt the steel and other metals

Q.13. What is indirect arc heating?

Ans. In this method of heating, heat produced due to arc struck between two electrodes is used to heat up the surrounding charge.

Q 14. Electrodes used in Arc furnaces **are made of ....**

[June 2009]

Ans. Graphite.

Q.15. What is induction heating?

Ans. Heating produced due to eddy currents induced in a inductive core by electromagnetic induction is called induction heating.

Q.16. Give two applications of induction heating.

Ans. (i) Surface hardening of axles of automobiles.

(ii) Melting and refining of non-ferrous metals.

(iii) Production of steel

Q.17. Induction furnace requires .... frequency of supply. Low/high.

Ans. High.

Q.18. State applications of Ajax watt furnaces.

Ans. The furnace is widely used in foundries for melting and refining of brass and other non-ferrous metals.

Q.19. Define power factor.

Ans. Ratio of active power to apparent power of an AC circuit is called power factor

Power factor, $$\text{P.F.} = \frac{\text{Active power}}{\text{Apparent power}}$$

$$\text{P.F.} = \frac{VI \cos f}{VI} = \cos \phi$$

Q.20. Ajax watt furnace operates at power factor of ......

Ans. 0.8 to 0.85.

Q.21. What is the advantage of high power factor.

Ans. High power factor means more heating for the same amount of current at low power factor.

Q.22. Define kilowatt hour.

Ans. It is commercial unit of electricity consumed. It is the amount of energy consumed in one hour by an electrical appliance of power rating one KW.

Q.23. What is dielectric heating?

Ans. In this method of heating, dielectric loss in the dielectric is utilised for heating of insulating materials.

Q.24. Dielectric heating is used for heating of .... materials. (Ferrous/Non-Ferrous/Insulating)

Ans. Insulating.

Q.25. Name two applications of dielectric heating.

[June 2009/08}

Ans. (i) Drying of textiles (ii) Dehydration of tobacco
(iii) Manufacturing of plywood (iv) Bakery

Medium/Long Answer Type Questions

Q.1. What are the advantages and applications of electrical heating?

Ans. Following are the advantages of electrical heating.

(i) It is a simple, clean and pollution free method of heating

(ii) It is a highly efficient method and requires very little maintenance.

(iii) It is economical and provides uniform heating.

(iv) Automatic control is possible.

(v) Heating of insulating materials is also possible which is not possible by any other method.

Applications of Electrical heating:

(i) **Domestic applications** : Cooking, space heating, water heating, ironing and heating of buildings.

(ii) **Industrial applications** : Refining of metals, manufacture of high grade steel, manufacture of rubber, plastic, foundary, plastic moulding, drying of paints and potteries.

Q.2. What is resistance heating? Discuss its types.

or

Explain direct and indirect resistance heating. In what applications they are used.

Ans. When current flows through a wire, heat is produced. This is called resistance heating. The wire is known as "Heating element" and this heat is transferred to the material (called charge) to be heated.

Resistance heating may be of two types.

(i) Direct resistance heating : In this method the charge itself is used as an element. All the heat is directly given to the charge through electrodes, so heat loss is zero. This method is used in geysers.

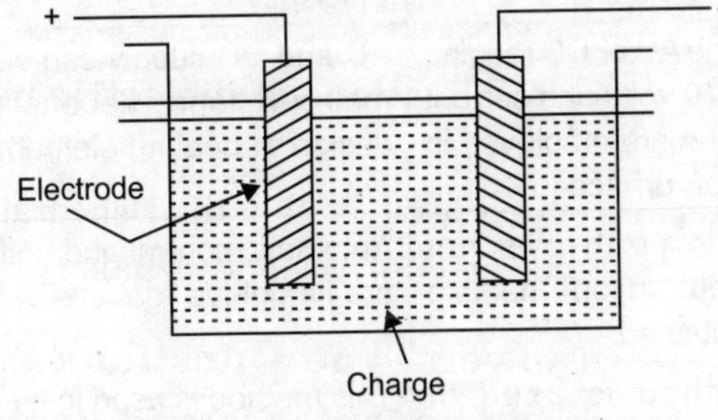

Fig. 3.42

When the electrodes are connected with the supply, they are heated and entire heat is transferred to the charge. Automatic control is not possible however, uniform heating can be obtained.

(ii) Indirect resistance heating : In this method the heat produced by resistance is not transferred directly but indirectly to the charge (*i.e.,* load) Therefore heat is not fully utilized to heat loss and efficiency is low

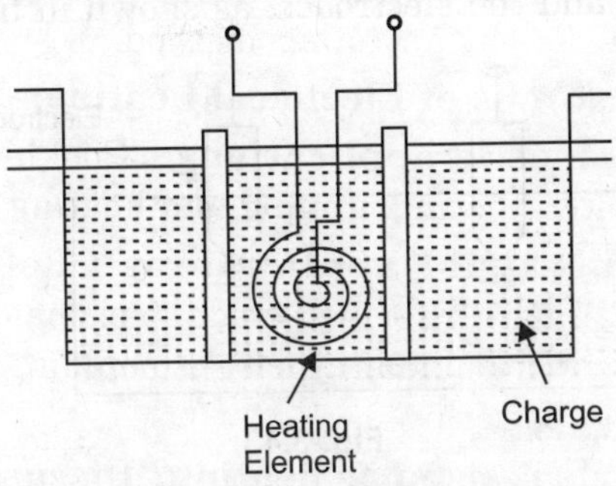

Automatic temperature control and uniform heating are its unique features. This method is used in electric iron. In an electric iron temperature is controlled by thermostat.

Q.3. Differentiate between direct and indirect resistance heating.

Ans.

	Direct resistance heating	Indirect resistance heating
1.	In this type of heating material to be heated acts as a resistance and current is passed through it	In this method of heating, current is passed through a wire of high resistance and heat is transferred to the charge by convection or radiation.
2.	Efficiency is high	Efficiency is low
3.	Low voltage AC or DC ranging from 2 to 20 volts and current upto 3000 A are employed in this type of furnace	AC and DC supply can be used at full mains voltage depending upon the design of heating element
4.	Automatic control is not possible but uniform heating can be obtained.	Automatic control and uniform heating are its unique features
5.	This method is used in geysers.	This method is used in electrical irons, cookers and milk boilers etc

Q.4. What is arc heating? Discuss its types.

Ans. When two electrodes are placed nearby separated by an air gap and subjected to a high voltage, the air in the gap is ionised and an arc is produced. This arc has a very high temperature and can be used for heating. This is called arc heating. The arc heating is of two types.

(i) Direct arc heating : In this method the arc is struck between charge and the electrodes, as shown in fig 3.43.

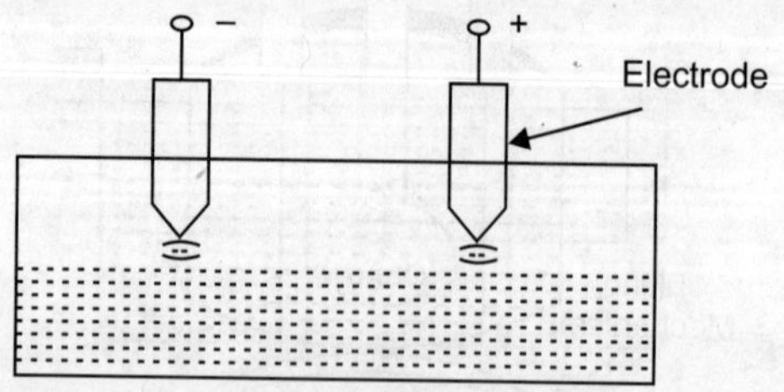

Fig. 3.43

Heat of arc is utilized directly to heat the charge.

(ii) Indirect heating: In this method arc is struck between the two electrodes, which are projected from the sides above the charge. The heat of the arc reaches the charge through radiation.

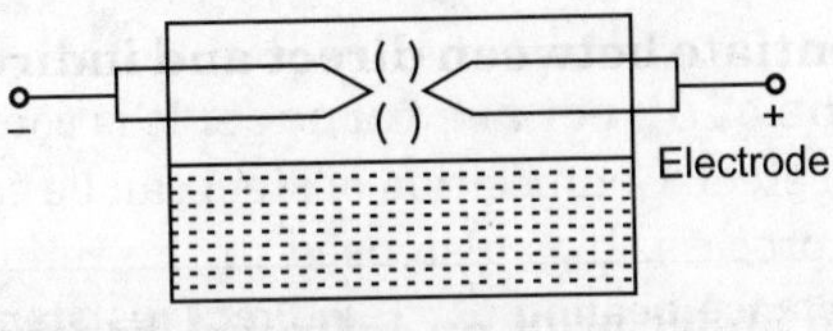

Fig. 3.44

Q 5. Describe construction and working of an arc furnace. What are its uses.

OR

What is direct arc heating? What are its applications?

Ans. When voltage across two electrodes separated by an air gap is increased, a stage is reached when the air in the gap gets ionised and provides a conducting path. Current starts flowing between the two electrodes in the form of a continuous spark which is called arc. Are between the two electrodes produces heat and has a temperature in between 1000°C to 3500°C depending upon the material of electrodes used.

Heat produced in the arc can be utilized directly for heating the charge in a direct arc furnace or indirectly in an indirect arc furnace by transferring the heat to the charge by radiation. Direct arc furnaces are commonly used due to their high efficiency. The furnace gives a very high temperature.

Fig. 3.45 shows a three phase direct arc furnace.

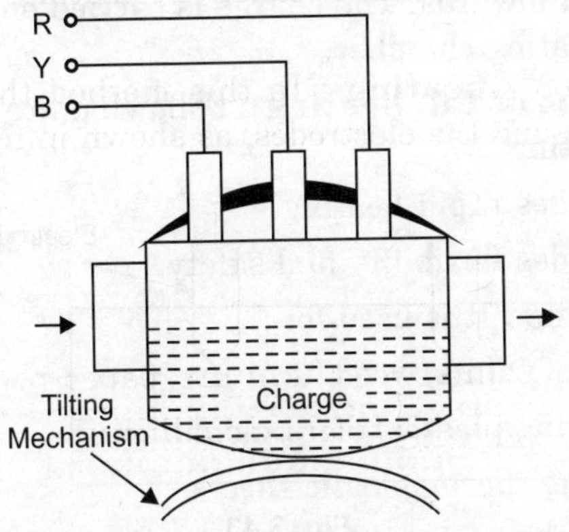

Fig. 3.45

The chamber of this furnace is lined with suitable refractory material. Three electrodes put at the corners of an equilateral triangle project on the charge through the roof and three arcs are formed.

Applications of direct arc furnace: It is commonly used for the production of steel. Composition of steel can be easily controllled during refining process. It is also used in the refining of metals.

Q.6. Write a short note on infrared heating.

Ans. If a heating element is operated at a temperature of about 2000°C, it emits high amount of infrared (I.R) radiations. The heating at this temperature is known as Infrared or I.R. heating.

Basically, heat energy from an incandscent lamp is focussed on the body to be heated in the form of electromagnetic radiations.

In ordinary resistance furnaces, heat is transferred from heating element to the charge partly by radiation and partly by convection but in infrared or radiant heating radiation is dominant. In this method the heating elements consist of tungsten filament lamps with reflectors to direct the whole heat on the charge. The lamps are operated at 2300°C instead of 3000°C. At this temperature, infrared radiations are more than light radiation. The reflectors are plated with Rhodium. The lamps of rating of 250 W to 1000 W are used at 115 Volts. There may be a number of such lamps in an I.R furnace. A temperature upto 300°C can be obtained. The radiant efficiency of infrared furnace is 7000 W per m^2 whereas in the case of resistance furnace, this value is 1500 W/m^2.

In Infrared Furnaces, no heat insulation is required, as all the heat is directed inwards. The charge is carried on a conveyor belt through the heating chamber.

Advantages of I.R. heating : Following are the advantages of infrared heating

(i) It provides rapid heating.

(ii) It provides flexibility and safety.

Applications : It is used for

(i) Drying of paints, wood furniture, paper, pottery, textile etc.

(ii) Preheating plastic before moulding.

(iii) Softening thermoplastic sheets.

Q.7. Describe principle and applications of induction heating.

Ans. Consider a piece of metal forming the core of a solenoid or coil as shown in fig.

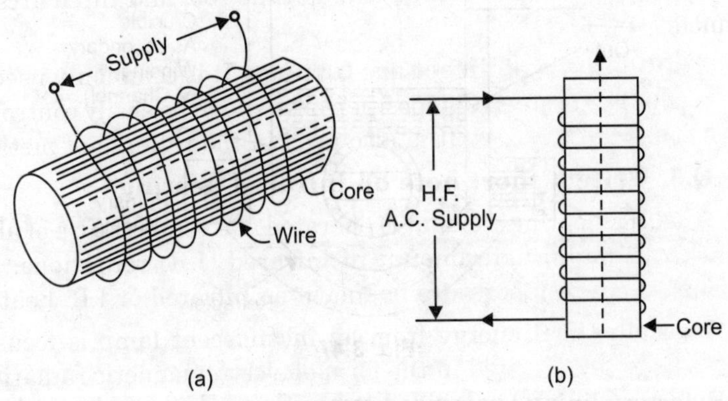

Fig. 3.46

When an alternating voltage is applied across the coil, an alternating flux is set up in the core and an e.m.f, is induced. The eddy currents are induced in the metal piece. The metal piece provides closed paths for flow of eddy currents. These closed paths are like the short circuited secondaries having resistance and reactance. The power loss due to the eddy currents in the core produces heat. The eddy current loss is directly proportional to square of supply frequency $(f)^2$: square of maximum flux density (B_{max}) and inversely proportional to the resistivity of the material of core material (P).

When a core is heated by passing high frequency current around it, the amount of heat produced on the outer periphery will be very large and reduces considerably when we move towards the centre. This is because the opposition to current elements on the outer periphery is less. The above phenomenon is known as skin effect.

Applications: In the industry, this method of heating is employed for surface hardening of axles of various automobiles, and in the production of steel.

Q.8. Describe construction and working of ajax watt furnace.

OR

Write a short note on Ajax watt furnace.

Ans. Fig.3.47 shows construction of the furnace. It has a vertical core for heating the charge.

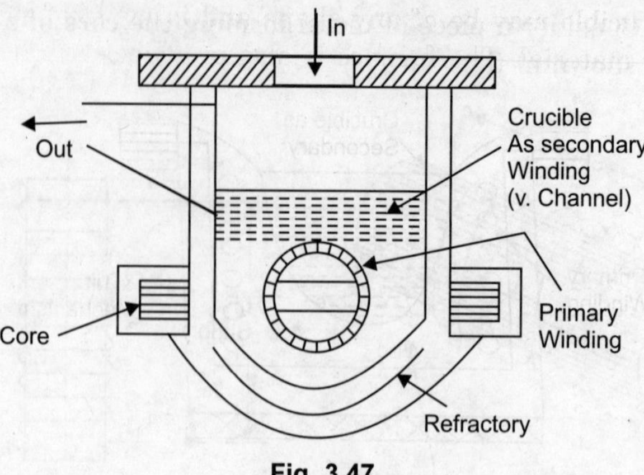

Fig. 3.47

Refactory lining is provided to avoid any damage due to heat. The separated doors are provided for charging and discharging the metal to be heated. When high frequency AC supply is given to the primary winding, heat is produced in the core due to eddy currents induced in the core.

The furnance is suitable for continuous operation. The crucible should be kept full of metal. The furnace is available up to 1 ton capacity and power consumption is about 200 KW.

The furnace is used for melting brass and other non=ferrous alloys and metals.

Q 9. Write a short note on coreless induction furnace.

or

Give construction and working of coreless induction furnace.

Ans. This Furnace has no core. Any shape of crucible can be used. A winding may be placed around the crucible full of charge. The crucible acts as short circuited secondary and eddy currents induced heat up the crucible. The fig shows complete construction of a coreless furnace. Its construction is in 3 parts.

(i) Primary winding.

(ii) Crucible as secondary winding

(iii) Supporting and tilting mechanism.

Primary winding coils are made hollow and are cooled by circulating water through them *so* that they can carry high frequency currents upto 10^6 hertz. This is wound around the crucible.

The crucible may be of any shape and is made of some heat resistance material. The furnace is also provided with a thermal insulation.

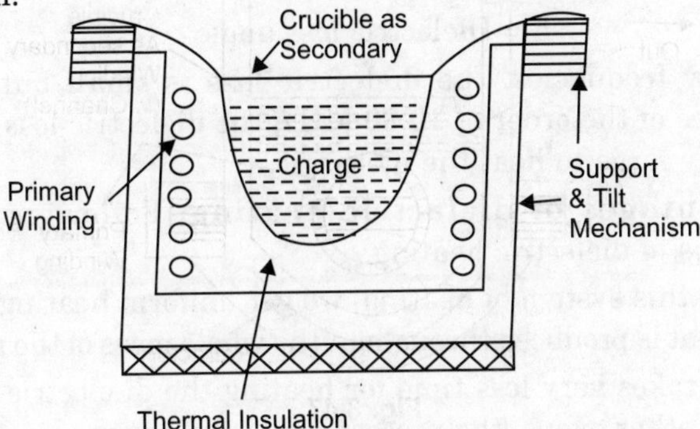

Fig. 3.48

The power factor of the furnace is very low. For improvement of power factor a capacitor bank may be used.

The furnace is used to manufacture high grade steel.

Q 10. Describe the principle, advantages and applications of dielectric heating.

Ans. In dielectric heating, dielectric to be heated is placed between two metallic sheet type electrodes across which alternating voltage of high frequency is applied as shown in fig. 3.49.

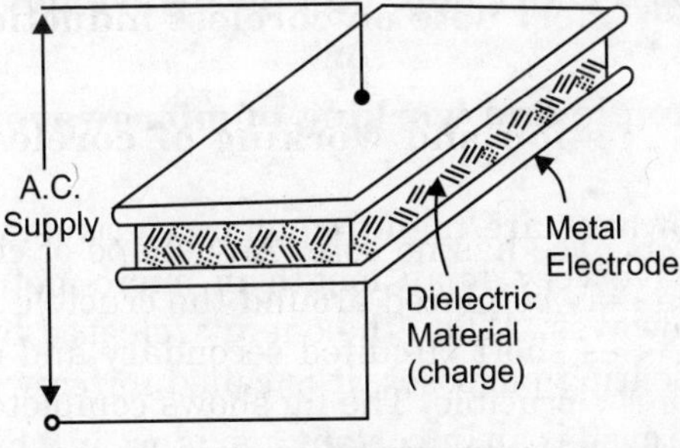

Fig. 3.49

As two metallic plates separated by an insulating material (dielectric) form a capacitor, dielectric loss takes place within the dielectric which is given by—

$$\rho = V^2 \, 2 \pi f C \sin \delta$$

where v = voltage applied

f = Frequency of AC supply

C = Capacitance of capacitor formed as result of dielectric placed in between the two electrodes.

δ = Dielectric loss angle

At low frequecies, the dielectric loss is small but at high frequencies of the order of 10-30 MHz, the dielectric loss becomes sufficiently large to heat the dielectric.

Advantages of dielectric heating: Following are the advantages of dielectric heating.

(i) In this system of heating, we get uniform heating because heat is produced throughout the whole mass of the material.

(ii) It takes very less time for heating the di-electric whereas in other ovens, the process may take hours.

(iii) Normally, materials heated by this method are combustible which cannot be heated by other methods.

Applications: (i) Dielectric heating is used for heating of insulating materials like wood, plastic and ceramics,

(ii) Food processing.

(iii) Dehydration of fruits,

(iv) Drying of textiles,

(v) Manufacture of paper and plywood,

(vi) Vulcanization of rubber.

Q.11. Write a short note on microwave heating.

or

State principle and working of microwave oven.

[June 2009]

Ans. Microwaves are high-frequency electromagnetic waves. These waves have very small length in micro-metres and hence named as microwaves., When the energy released by these waves is utilised for heating purposes, it is called mircrowave heating.

Mircrowave oven is an important appliance in which microwave heating is utilised for cooking, defrosting or reheating the food items.

In microwave ovens, the microwaves are generated by magnetron. These microwaves are attracted and absorbed by water, fat and suger contents of food where their energy is released and converted into heat energy. The microwaves are distributed and thus the food is cooked uniformly.

The microwaves are absorbed by the food upto a depth of about

2.5 cm. Cooking then continues as the heat is dissipated within the food. Cooking times vary according to the recipient used and the properties of the food

Q.12. What is induction heating? Describe principle and construction of core type induction furnace.

Ans. Induction heating is also called eddy current heating. The metal to be heated is wound with a winding. When a.c supply is given to the winding, by induction, voltage is induced in the core and as it offers a closed path, it produces eddy currents and causes power loss in the core due to the resistance of the core. This power lost is utilized to heat a metal.

It is high frequency heating and heat produced due to eddy currents is proportional to the square of frequency. The frequencies used are upto 2 kilo hertz. The depth of penetration of heat into the metal is inversely proportional to the square root of the frequency of applied voltage

Core type furnace is basically a transformer. The charge to be heated is used as short circuited seocndary winding. The fig shows the furnace with crucible of odd shape. The crucible with charge is put as secondary winding. The primary is given supply, eddy currents are induced in the charge which is heated.

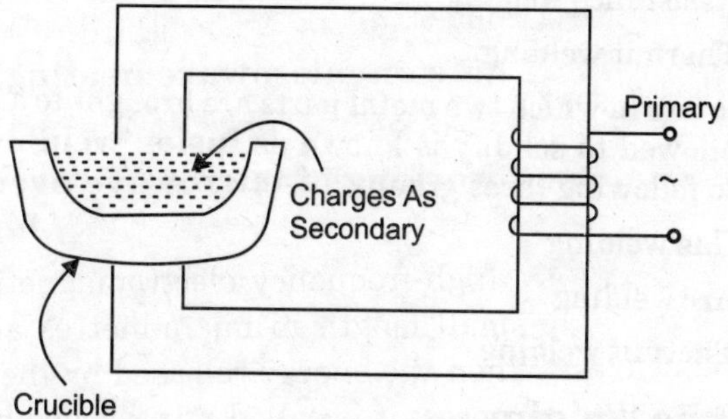

Fig. 3.50

4 Electric Welding

4.1. INTRODUCTION

Welding is a process in which two metal parts are joined by heating. The metal parts are heated to melting point which adhere on solidification. In some cases the pieces of metal to be joined are heated to plastic stage and are fused together. A considerable progress has been made in welding the metal pieces by heating them up to melting point. It may be defined as a process in which metals are joined by heating them to a suitable temperature with or without the application of pressure and addition of filler material. Plastic welding may be further subdivided into following three parts :

(*i*) Forge welding

(*ii*) Resistance welding

(*iii*) Thermit welding.

The process in which two metal parts are brought to a molten state and then allowed to solidify is known as **fusion welding.** It may be divided into following three groups :

(*i*) Gas welding

(*ii*) Arc welding

(*iii*) Thermit welding.

4.2. Electric Welding

Electric welding is defined as that branch of welding in which electric current is used to produce the large heat required for joining together into firm union, two pieces of metal.

Electrically welded joints are being used to replace rivetted bolted joints in the construction of ships, bridges steel frames, structures, tanks, pipe lines and for so many other purposes. In electric welding heat is produced either by striking an arc between an electrode and the metallic joint or by passing a heavy current through the joint. The former method is called **arc welding** and the later **resistance weld-**

ing. The electric welding can be carried out by any one or more of the following processes shown in chart.

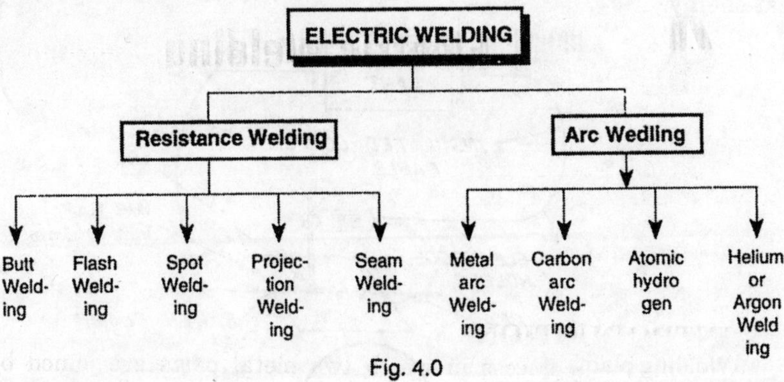

Fig. 4.0

4.3. Arc Welding

An electric arc is produced by bringing two conductors connected to a suitable source of electric current, momentarily in contact and then separating by small distance. The current continuous to flow across the small gap and gives intense heat. The heat developed is utilised to melt the part of work piece and the filler metal and thus form the joint. So the welded joint is a union of metal parts made by localised heat without any pressure.

The voltage necessary to strike the arc is higher than that required current to maintain it. The temperature is about 3000°C. According to B.S.S. 638, maximum voltage specified for welding is about 100 V for the safety of the operator. Current ranges from 30 to 500 A for manually operated welding and 75 to 600 A for automatic operated.

The electric welding is widely used for joining the metal parts to repair fractured castings and the fillings by the deposition of new metal on the worn parts. The heat developed by the arc is also used for cutting of metal. Arc welding is again sub-divided into four groups namely :

(*i*) Carbon arc welding

(*ii*) Metal arc welding

(*iii*) Atomic hydrogen arc welding

(*iv*) Helium or Argon arc welding.

4.3.1. Principle of Arc Welding. Current from a source (A.C. or D.C.) is obtained. One terminal is connected to the electrode and the other to work piece and the circuit is completed through air gap as shown in Fig. 4.1.

The gap is provided between the tip of the electrode and the surface of work piece by keeping the electrode at a distance of about 3 mm to 6 mm from the surface of the work piece. Due to the interruption by the air gap or gas, heat is produced and the temperature attained varies from 3700°C to 4000°C. Electrical energy is converted at the arc into heat energy.

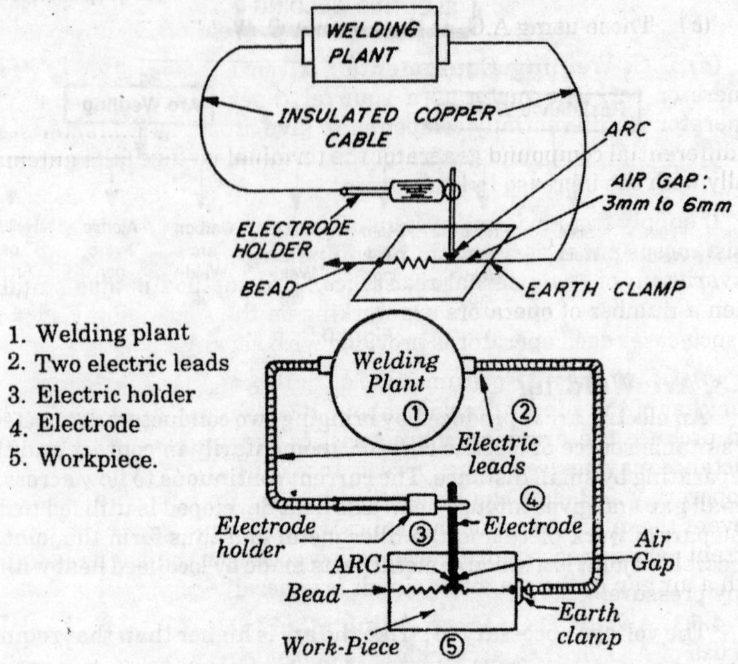

Fig. 4.1. Principle of arc welding.

4.4. Conditions for Successful Welding

The function of arc welding plant is to produce the required electrical conditions at the arc. Either D.C. or A.C. may be used, although aluminium and certain alloys can only be welded with D.C. The main requirements are a relatively high striking voltage to enable the arc to be struck and maintain stability, a relatively low voltage to enable the arc to be maintained without being violent, and a current of the correct value to melt the electrode and parent metal without burning.

The sparking voltage is usually in the region of 80 to 100 volts in the case of A.C. and 60 to 80 volts in the case of D.C. The arc voltage depends upon the type of metal and the flux of the electrode and varies between 20 to 35 volts. It is affected by the length of arc being maintained — a long arc requires a higher voltage than a short arc — but a good operator will maintain the correct length of arc and, therefore, the correct arc voltage. The current is also dependent upon the type of metal being welded and the gauge and type of electrode and is from 15 to 600 amps. for hand welding and upto 1200 amps. for automatic welding.

4.5. Electric arc Welding Equipments

Electric arc welding equipments falls into two natural sections

(*a*) Those using D.C. at the arc or D.C. Welding Equipments.

(*b*) Those using A.C. at the arc or A.C. Welding Equipments.

(*a*) **D.C. Welding Equipments :** It consists of mainly a motor generator set. The motor is a squirrel cage induction motor, and generator is differentially compound to give drooping characteristics. In differential compound generator the terminal voltage falls automatically with the increase in load current.

If supply from existing system is used for welding then a Ballast resistance is put in series with the equipments and control is obtained by variation of the external resistance. This method is also suitable when a number of operators are working on the same supply system. In such cases each operator is provided with separate ballast.

(*b*) **A.C. Welding Equipments :** A transformer reduces the voltage from that of the supply drawn to about 100 V. To regulate the current and produce the dropping characteristics required a resistance and reactance may be used. When a reactance is used it must be designed to operate well below the saturation-point of its magnetic circuit to prevent the introduction of harmonics which tend to lengthen the zero current period in each half cycle and allow cooling of the arc; a reactance with a air gap in the magnetic circuit is generally employed.

4.5.1. A.C. or D.C. Arc Welding Machines. On several factors, the use of A.C. or D.C. arc welding is dependent. By examining their relative merits and demerits, the use of A.C. or D.C. arc welding should be decided. The advantages of D.C. arc welding machine are as follows:

Advantages.

1. (*a*) It can select direct electrode positive also known as direct current reverse polarity (DCRP), which produces deeper penetration welds than direct current electrode negative (DCEN).

   (*b*) Electrodes made for welding copper, nickel and aluminium are generally of DCEP (direct current electrode positive) type.

2. It selects DCEN or DCSP where DCSP means direct current straight polarity.

   (*a*) DCEN or DCSP is used for electrodes having high metal deposition rates.

   (*b*) DCEN or DCEP can be used in positions other than flat.

The disadvantage of D.C. arc welding machine is that, it is very costly than an A.C. machine of the same quality current output and duty cycle.

Advantages of A.C. Arc Welding Machines

The advantages of A.C. arc welding machines are as mentioned below:

1. A.C. welds have usually moderate penetration.
2. Normally more diameter is required to have more A.C. current to get more filler metal deposit rates and faster welding speeds.
3 For equal quality, current output and duty rating. A.C. arc welding machines are less expensive than D.C. arc welding machines. Their disadvantage is that, not all SMAW electrodes can be used with alternating current. Finally, the choice of the welding machines depends upon the type of welds and economies of welding machine purchase and also depends upon personal preference.

4.6. Welding Accessories

In addition to the different types of welding plant described above, the following accessories should form a part of every operators equipment :

(*i*) One electrode holder, fitted with a length of flexible cable for current to plant.

(*ii*) One length of flexible cable, for connecting the job to the plant.

(*iii*) One face screen complete with coloured glasses.

(*iv*) Observer's face screen with coloured glasses.

(*v*) One pair of leather gloves.

(*vi*) One chipping hammer to remove slag from weld.

(*vii*) One wire brush to clean the weld after chipping.

A face screen or helmet should be made from fibre or other non-conducting materials and should have no screws or rivets that go right through the insulating material from which they are made. This is an important point, as otherwise there is a risk of the operator receiving an unpleasant, if not dangerous, shock in the face, should be accidentally touch such pieces of metal with his electrode holder.

4.7. Methods of Obtaining Drooping Characteristics

The high open circuit or stricking voltage and the low arc voltage is obtained by means of equipment designed to give a drooping characteristic. In this way the high voltage is available to enable the arc to be struck but as a result of the current flowing the voltage drops to the arc voltage.

It is impossible to keep the arc length constant in manual arc welding. If there is any change in the length of arc corresponding, there is a change in arc voltage. Since melting rate is directly proportional to the current and almost is independent of arc voltage, it is advisable to have as small current changes giving small changes in the arc voltage. This is possible only if slope of volt-ampere characteristic curve is steep in the welding range as shown in Fig. 4.2. From above consideration and also from arc stability point of view which will be discussed later on, voltage source should have drooping characteristics.

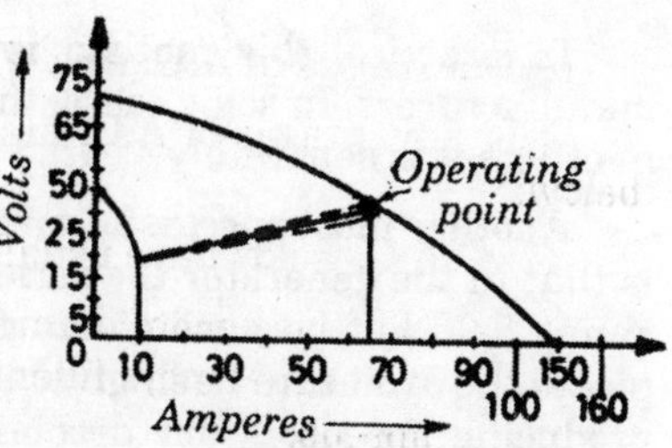

Fig. 4.2. Characteristics curve of welding generator.

This applies to both D.C. generators and to transformer equipment. The method of obtaining this characteristic, particularly in the case of generators, varying with different manufacturers.

4.8. Standard Method for D.C. Generators

The standard method of obtaining drooping characteristics from a D.C. generator is based on reversed series winding as shown in Fig. 4.3.

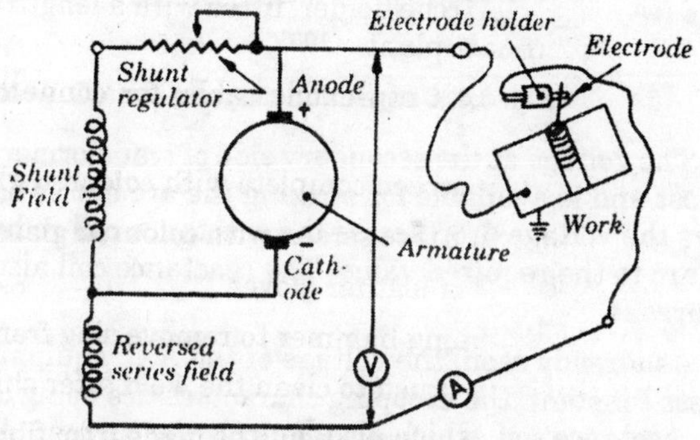

Fig. 4.3. Reversed series generator connection.

In this method the series coils are wound in a reverse direction to the shunt coils and, therefore, the field created by the series field is in opposition to that created by the shunt field. On open circuit, the shunt field only is operative and the maximum voltage is available to strike the arc. When the arc is struck however, current flows through the series winding, thus setting up opposition to the shunt field. The resultant field strength is, therefore, less than open circuit field strength and the voltage drops. By means of a shunt regulator inserted in series with the shunt field the open circuit voltage can be set to such a position that the voltage drops automatically to the appropriate arc voltage at the correct value.

In practice, this method is utilised with variations by several manufacturers. In some cases, the shunt field being self-excited whilst in others it is separately excited by an excitor fitted to the generator.

Another most successful method of obtaining ideal characteristics is that in the generator the series windings are in sympathy with the shunt field but because of the shape of the patented bifurcated pole pieces the armature itself gives the effect of a reversed series winding, producing the necessary drooping characteristics. This system combines rapid recovery and reaction with high efficiency.

4.9. Use of Reactance Coil for A.C. Equipment

The drooping characteristics is obtained in case of A.C. equipment by means of a reactance coil in series with the arc as shown in Fig. 4.4.

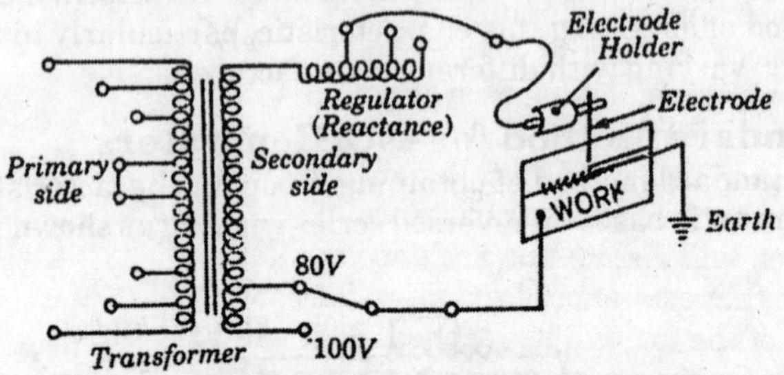

Fig. 4.4. Connection diagram for A.C. welding.

The voltage at the secondary side of transformer remains constant almost and is available for striking the arc but as soon as the current flows the voltage drop across the reactance coil reduces the voltage at the arc to the required value. The reactance coil also controls the flow of current.

As already seen, the voltage of the A.C. equipment remains more or less constant, the drooping characteristics being obtained by means of a reactance coil, which also controls the current. As the reactance in series with the arc, controlling the current is simply a matter of adjusting the amount of reactance in circuit. This is achieved by either having tapped coils, the tapping being brought out to a multi-point selector switch or by mechanical movement of the laminated iron core of the reactance coil. Again it is normal to provide a number of current ranges, this being done by tapping the reactance coil bringing the tapping out to plug points. In this way turns of the coil, can be cut out, thus changing the current range of the unit.

4.10. Electrical Properties of an Arc

As already discussed that an electric arc is an electric discharge in gases, accompanied by very bright glow and tremendously large amount of heat. Now in underneath paras we will discuss in details the electrical properties of an arc :

(i) **Air stability.** Electric arc has negative resistance characteristic. With the increase in arc current, the voltage across it decreases, *i.e.* resistance of the arc decreases. This can be seen from simple Ohm's law also, *i.e.*

$$I = \frac{V}{R} \text{ or } R = \frac{V}{I} \text{ ohms.}$$

Volt ampere characteristics of an electric arc is shown in Fig. 4.5.

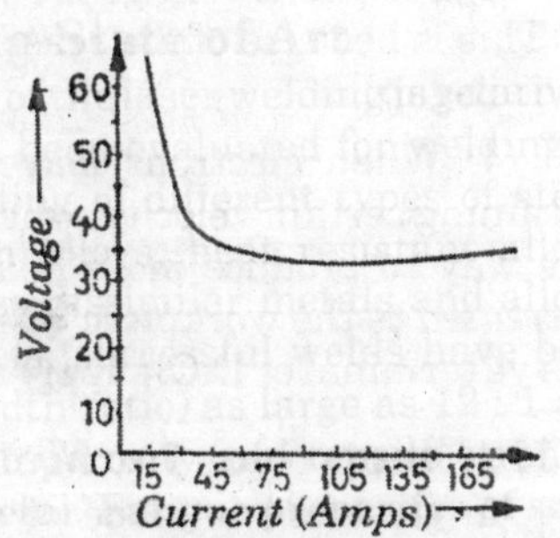

Fig 4.5. Volt Ampere Characteristics.

It is not possible in manual arc welding for welder to hold a constant arc weld length. The decrease in arc length will result in the decrease in arc resistance and hence building up of arc current. This will further reduce the arc resistance. In this way cause and effect will go on helping each other till the arc will be out of control. Similarly increase in the arc length will bring about increase in resistance of arc which will reduce the arc and its diameter resulting in further increase in the resistance of arc. Here the conditions again approaching extinction of the arc will be reached. So it will be noticed that it is very difficult job for the operator to maintain a stable arc.

Arc stability is achieved in A.C. by putting choke in the supply line which gives drooping characteristics. Choke also creates phase difference between current and arc voltage so that when arc current passes through its natural zero position, there exists sufficient re-striking voltage available at the electrode tip.

(2) **Arc blow.** An electric arc can be considered as very flexible conductor which can be deflected by various factors. The space round the arc and in the adjacent metal is always effected by magnetic fields which tend to deflect the arc. The phenomenon of arc deflection is known as **arc blow.**

Arc blow causes great hindrance especially when large currents are used because the intensity of magnetic field is proportional to the square of the currents. The magnetic field deflects the arc when it is not uniformly distributed relative to arc and the distribution of the field depends upon the position and shape of workpiece and our-gap etc. Arc blow happens more with direct current.

The effect of magnetic blow is that there is no proper penetration of the heat where it is required. This results in poor welds.

To avoid arc blow and get good quality of weld, an A.C. welding machine proves better. Bécause in A.C. current arc welding the flow of current changes direction rapidly and with that the magnetic field will also change directions very rapidly and this change will cancel the "arc blow" effect and stabilize the arc. If D.C. is to be used the arc blow can be avoided by welding away from the ground or changing the position of the work. In some cases, the magnetic condition is improved by giving one of welding leads a few turns round the work being welded.

(3) Sufficient voltage is required to strike an arc than to maintain it. This is because of the reason that some voltage is required to ionize the air-gap.

(4) When electrons leave the cathode they hit the anode after attaining certain velocity, where there momentum is destroyed. This is the way to produce more heat at anode than at cathode. As we are *increased* to the workpiece which is to be welded, it is always connected to +ve terminal or D.C. supply and welding electrode to the -ve terminal.

4.11. Electrodes for Metallic Arc Welding

An electrode is made of a metallic wire called core wire coated uniformly with flux. During fluxing the electrodes about 20 mm of its length is left bare at one end for inserting it into electrode holder. The coating of flux on the electrode is a bad conductor of electricity.

There are three types of electrodes which are used for arc welding:

(*a*) Bare wire rods.
(*b*) Dipped of lightly covered electrodes.
(*c*) Heavy coated electrodes.

Bare electrodes have no coating of flux on them whereas the coated electrodes are covered with flux. The coated electrodes are commonly used these days due to the following reasons :

Earlier the bare electrodes were used in arc welding but it was found that the weld made by the electrodes though good in tensile strength was considerably poor in ductility, resistance to impact and fatigue. The appearance of the weld was also not good. The poor performance of the bare electrode was due to vaporization of important elements of the weld metal during welding and also due to the presence of oxides and nitrides resulting from atmospheric contamination of weld metal.

In case of coated electrodes, the flux coated on the core wire is composed of special elements which, protect both the arc and the weld metal that passes through the range of liquid action and solidification during welding. The appearance and the physical properties of the weld are also improved considerably.

(*a*) **Bare Electrodes.** These are run usually at a lower welding voltage. Welding currents ranges from 60 to 310 amperes according to the size of electrode are used. With bare electrodes (Fig. 4.6) more difficulty is experienced in starting the arc.

(*b*) **Light Covered or Dipped Electrodes.** These provide better protection against oxidation of the weld. They are considered an improvement on bare electrodes and require a higher welding voltage and lower current than the bare rods.

Bare electrode deposit

ARC stream

Fig. 4.6. Bare electrode welding.

(*c*) **Heavy Coated Electrodes.** These are coated with a material of substantial thickness and solidity, calculated to provide protection to the electrode. Fig. 4.7 (*a*) and (*b*) illustrate this principle.

Electrode
Slag coating
Slag covering deposit
Core wire
Penetration

(a) Welding with coated electrode.

Fluxing compund containing alloys extruded between winding
Open spiral asbestos winding
Rod of special analysis

(b) Special type of coating for heavy coated electrodes.

Fig. 4.7.

Flux-covered electrodes have in general a spiral winding of asbestos yarn as the basis of the covering. Asbestos is a valuable flux in itself, as it fuses and combines with iron oxide into a compound silicate which, being very light and mobile, is easily removed after solidification, having a clean surface - an important advantage for work which requires more than one run of welding. Asbestos yarn also makes an excellent vehicle and support for the other constituents of the covering, and with a suitable binder makes a tough surfaced sheath to the metal core. These electrodes, therefore, travel and handle without loss and damage.

Bare or lightly dipped electrodes are operated on the negative pole. Heavily coated electrodes, especially those with asbestos covers, are generally used on the positive pole.

4.12. Advantages of Coated Electrodes

Following are the advantages of coated Electrodes :

1. **Maintaining the Arc.** It helps in maintaining the arc because the flux consists of arc stabilizing materials like calcium carbonate, titanium and potassium compounds.

2. **Preventing the weld from sudden cooling.** Basic salts from magnesium, silicon and calcium in the covering from slag which floats on the surface of metal thereby preventing sudden cooling of the weld. This in turn prevent weld from becoming brittle.

3. **Shielding the molten material from oxidisation.** The gases which are formed protect the well metal from oxidising action of atmospheric oxygen and nitrifying action of nitrogen of air. So we get good quality ductile welds.

4. **Used for higher currents.** The coating of flux makes the electrode insulated at its outer cover and thus the electrode may be used at higher currents without getting red hot.

5. **Prevention of spattering of metal.** Spattering of metal during welding also prevented by the use of flux.

6. **Used for A.C. supply also.** When the A.C. supply is used, arc cools at zero current position. This tends to de-ionize the path of arc. However, shielding gates keep the space ionized and, therefore, these electrodes can also be used with A.C. supply.

7. **Easy overhead and vertical welding.** Flux helps in making overhead and vertical welding easy. The slag made by the flux quick is freezing and by solidifying immediately prevents the liquid metal from falling down on the welder or earth.

4.12.1 Current Values for Arc Welding.

The following table shows recommended values of currents to be used with different sizes of welding electrodes :

Gauge	*Average recommended*	*Maximum*
4 SWG.	200 amps	270 amps
6 SWG.	170 amps	200 amps
8 SWG.	140 amps	170 amps
10 SWG.	115 amps	130 amps
12 SWG.	70 amps	90 amps
14 SWG.	45 amps	65 amps

4.13. Straight and Reverse Polarity in D.C. Welding

When an electrode is connected with the negative pole of D.C. welding plant, it is called straight polarity, whereas the electrode connected with positive pole of the plant is called reverse polarity connection. It should be noted that in A.C. welding there is no such polarity. It has been observed that during welding with D.C. current, about 65 to 75% of heat produced in the arc is collected on the positive pole and the rest on the negative pole of the welding plant. So, for welding of thick and thin jobs where variable amount of heat is required at the electrode, D.C. work is O.K because the polarity can be changed. For instance for welding thin jobs like sheets etc. the electrode is connected to the +ve pole so that it can melt easily even at low current used for thin sheet welding. The sheet is connected to -ve pole so that it has less heat to prevent burning through. For thick jobs where good penetration is needed, the electrode is kept negative and the job positive.

It should be seen that an electrode which is connected to the +ve pole (Reverse Polarity) will burn always much faster thān when connected to the –ve pole (Straight Polarity). For this reason, usually the bare and medium coated electrodes are connected to –ve pole and medium coated electrode to the positive pole. Thick electrodes are also connected to +ve pole because more heat is required to melt them.

4.14. Applications of Electric Arc Welding

The majority of arc welding is done by the hand process. The operator holding an electrode holder, into which the electrode is fitted in his hand and guiding the arc as required. When the electrode has been consumed it is replaced and the process is continued. The operator views the arc through a filter glass held in either helmet or hand-shield. The equipment already mentioned is primarily intended for this purpose but slight modification is used to supply current to the arc in other processes. These include D.C. and A.C. metallic arc welding, A.C. or D.C. arc welding. Atomic hydrogen welding and Hellium or Argon-arc welding. These are briefly described below :

1. **Metallic arc welding.** In the metallic arc system a metal rod is used as an electrode and the arc is struck between this electrode and work, which respectively form the two terminal. A.C. supply from the secondary of the welding transformer or D.C. supply from the D.C. welding generator depending upon whether A.C. or D.C. is used, is connected across the job and the electrode. The work is then suddenly touched by the electrode and then separated from it a little. This results in an arc between the job and the electrode. Due to the heat generated by the arc, a little portion of the work melts as also the tip of the electrode. The two pieces to be welded fuse together and when the electrode is removed the metal cools and solidifies giving a strongly welded joint. The arc drawn between the metal electrode and the work creates a temperature of over 3500°C in a small area concentrated at the point of weld and at the end of electrode.

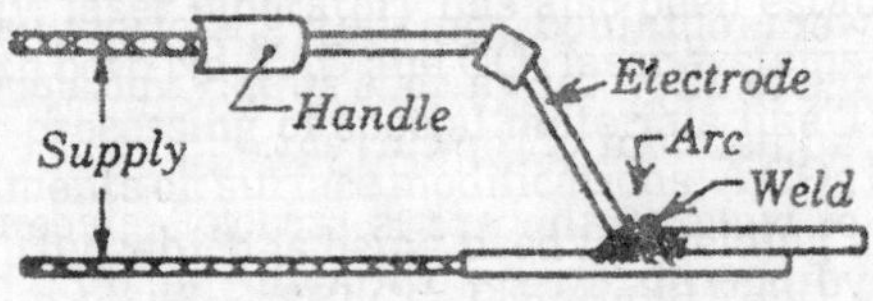

Fig. 4.8. Metal arc welding (Lap weld).

The soundest welds are produced with a short arc, the potential difference across the arc being 20 to 25 volt. In no case must it exceed 30 volt, as welds made under these conditions will not be sound.

The correct current to be used in any particular job cannot be stated as, owing to the cooling effect of the different masses of metal where the weld is made, it is necessary for different current values to be used on the same thickness of sheet, depending upon the type of joint made. A lap joint as shown in the figure will require more current than the butt joint in the same material owing to the greater dissipation of heat by the greater thickness of metal under the arc where the sheet metal over-laps.

2. **Carbon arc welding.** This method is normally used for welding copper and its alloys. The carbon electrode which is kept negative with respect to the work if D.C. is used. If the carbon electrode is made positive, the carbon particles have a tendency to go into the welded joint and cause brittleness. The electrode should, therefore, be kept negative and the work positive. For this type of welding only D.C. can be used. The heat from the arc forms a molten pool and the extra metal required to make the weld is supplied by a filler rod of the same composition as that of the molten metal. Two methods of carbon arc welding are used. In one method no flux is used and in the other method flux, either in the form of powder or paste is used to prevent the weld from oxidation. Former method is limited to non-ferrous metals and the later method is usually applied for ferrous metals.

3. **Atomic hydrogen arc welding.** The essentials of the atomic hydrogen arc welding process are (*i*) electrodes energy is supplied to an arc between two tungsten electrodes where it is transformed into heat. The source of power could be either D.C. or A.C. but A.C. supply should be chosen because it is more commonly available : (*ii*) molecular hydrogen is blown through this arc and transformed catalytically into the atomic form which acts as a vehicle for transfer of energy from the arc to the work, and (*iii*) in the direction away from the arc a sudden decrease of temperature (such as that obtained at the relatively cold surface of the weld area) causes the rapid decrease in the concentration of atomic hydrogen and a release of the heat of recombination.

To strike and maintain the arc, an open circuit voltage of 300 volt is necessary and for hand welding a current range upto 50 amp. is required. This power is obtained from the transformer having a number of primary tappings to allow for various supply voltages, and a tapped reactor to permit adjustment of the current.

In this type of welding the arc is struck between two tungsten electrodes and hydrogen is passed through the arc. Due to the high temperature (about 4000°C of the arc, hydrogen changes to its atomic form. When the atomic hydrogen travels to cooler regions in the vicinity of the arc it regains its molecular form. In doing so, it gives up the energy which it had received from the arc. Thus a very intense heat is generated which is used to melt the job to be welded. After cooling the two parts of the job are welded together. A filler rods may be used if additional metal is needed for making a joint.

This method is successfully used for welding stainless steel and moist non-ferrous metals etc.

4. **Helium or argon arc welding.** This method is used for welding aluminium alloys, magnesium and magnesium alloys. An arc is struck between electrode of tungsten and the work and helium or argon is used to give an inert atmosphere so that oxidation of the welded joint does not take place.

Standard D.C. or A.C. welding equipment may be used provided the open circuit voltage is around 100 volt for A.C. and 70 volt for D.C. For welding aluminium, and their alloys stainless and high alloy steels, nickel alloys and copper alloys upto 7/8" or 1/3 cm A.C. is suitable. D.C. may be used for all other common metals, and it is essential for the welding of copper and stainless steel and alloys over 1/2" thick.

4.15. Basic Features of an Arc Welding

Some of the basic features of A.C. arc welding are given below :

1. Good forceful arc.
2. Weld arc is easy to hold once it is obtained.
3. A good way to weld aluminium.
4. There is an absence of arc blow.
5. The most popular application is production welding on heavy gauge steel.

4.16. Arc Length

It is the distance between the tip of the electrode and the work surface upon which the molten globules are deposited. The length of the arc should be kept as short as possible so that the heat of arc results in simultaneous melting of the electrode and work. The advantages of short arc are given below :

(*a*) Maximum penetration.

(*b*) Slight over-lap.

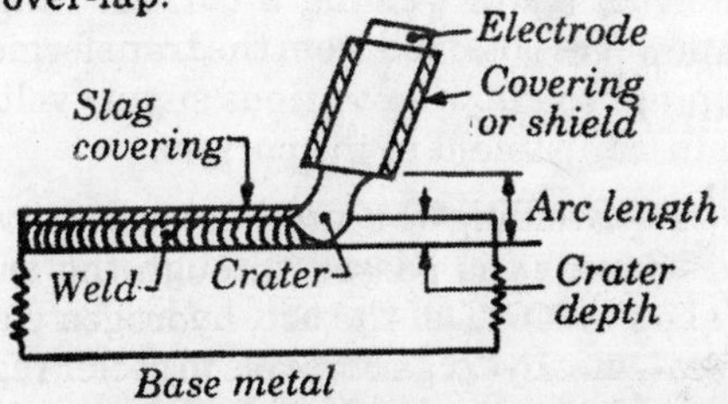

Fig. 4.9. A covered electrode arc weld in progress.

(c) Maximum strength of weld.
(d) Minimum porosity in the weld.
(e) Maximum ductility in the weld.
(f) Makes it possible to use alloy electrodes.

The length of the arc is an important variable in welding. The globules of molten metal should not be allowed to come in contact with air for a longer period because oxygen has bad effects upon the later developed mechanical properties of the weld. A short arc gives little chance for air to come into contact with molten metal. Whereas in a long arc the time of contact of molten metal with the air is longer. This way a long arc renders a wide shallow bead and enough heat is wasted in to air. The penetration is poor and more spattering of the metal is produced. Although electrode may be speedily consumed with long arc but the effective weld production is low and inefficient. The disadvantages of the long arc are :

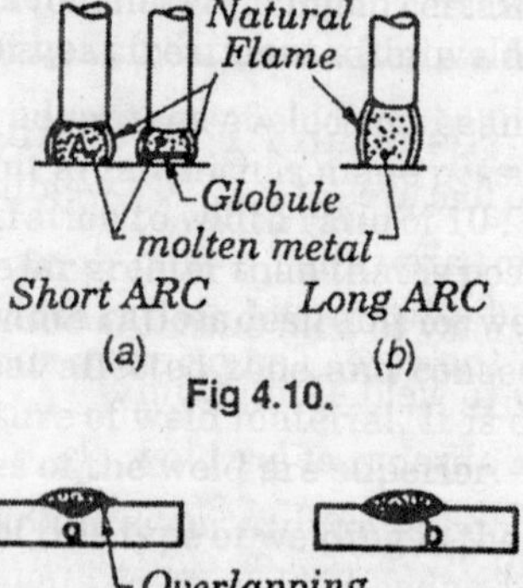

Fig 4.10.

Fig. 4.11. Weld beads. Long arc does not give good penetration of welding metal into the parent metal, thereby giving a bad overlapping as shown in at (a) Short arc gives good results shown at (b).

(i) Minimum penetration.
(ii) Excessive overlap.
(iii) Minimum strength.
(iv) Minimum ductility.
(v) Maximum porosity.
(vi) Uncontrolled deposit.

4.17. A.C. Welding Transformer

The use of alternating correct arc welding is increasing rapidly. The A.C. machines (Transformer) now produced are very efficient and easy to handle. Also electrodes have been developed for A.C. use which have ionizing compounds in the coatings. The ionized a.c. stream makes it easy to strike and maintain a stable arc.

Another factor which has increased the use of A.C. arc welding transformer is its initial cost. The cost of producing a transformer type A.C. arc welder is generally less than the cost of either a motor generator machine or a rectifier of equal welding capacity.

There are several types of A.C. transformer type arc welding machines as given below :

1. Adjustable reactor type.
2. Movable coil type.
3. Movable core type.

The electrically adjustable reactor type is becoming popular because the reactor current adjustment can be located near the operator by using an extension cord. The control may be either adjusted by the operators hand or foot.

The A.C. welding transformer consists of two windings, a primary and secondary. The primary welding is connected to power source. The secondary winding is connected to the electrode holder and ground and it is through the secondary winding that the welding arc current flows. The welding transformer is called a step down transformer. So the transformer consists of three main parts, the primary circuit, the secondary circuit, and the core. The primary circuit usually supplies single phase current. The secondary winding which is the arc circuit winding, is the step down type, which reduces the primary circuit voltage to an open circuit voltage of 80 volts and a closed arc circuit voltage between 20 and 40 volts. The core consists of laminated sheet steel of special magnetic properties. The welding arc current has a lower voltage and a higher amperage or current value than the primary circuit. Fig. 4.12 is the diagram of the electrical circuits in a typical transformer type arc welder. The ampere range of this arc welder is 25 to 295 amperes.

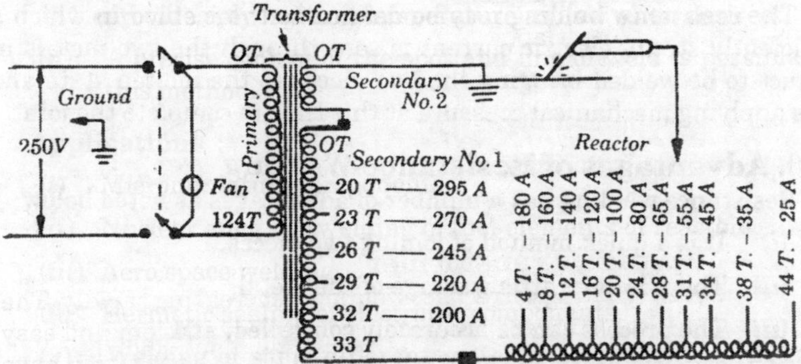

Fig. 4.12. Electrical circuit of typical A.C. transformer (Tap type) arc welder.

4.18. Resistance Welding

In 1885 Prof. Elihu Thompson in Philadelphia discovered the art of resistance welding. While experimenting with a speak coil operated from a bank of batteries the accidentally welded two steel rods of about 0.635 cm diameter.

The underlying principle of resistance welding is the generation of heat in the joint by passing a heavy current through the parts, this being followed by the application of mechanical pressure which welds the plastic metal and refines the grain structure. It will thus be seen that a resistance welding machine must be capable of supplying an electric current of the value required to produce the necessary heat, it must incorporate electrodes suitable designed to conduct the current to the parts at the right place, and it must be arranged to apply mechanical pressure to the weld at the correct position at time.

4.19. Generation of Heat

The heat generated, H by passage of an electric current, I through a resistance R is given by,

$$H = I^2 Rt$$

where R = The actual electrical resistance of the joint where the weld is to take place.

I = The current in amperes.

t = The time in seconds.

The temperature attained depends on this quantity and is affected by heat losses. The amount of current (generally A.C.) necessary is 4,400 to 5,000 amp. per sq. cm of area to be welded. The pressure varies from 280 to 565 kg. per sq. cm.

To realize the greatest heating effect, the current I should be large, because H varies as I^2. For this reason resistance welders utilize low voltage and high amperage. Summarising the above discussion :

The resistance welding may be defined as the method in which a sufficiently strong electric current is sent through the two metals in contact to be welded bringing the two pieces to the molten state and thus applying mechanical pressure at this time to complete the joint.

4.20. Advantages of Resistance Welding

Resistance welding has a number of advantages as noted below:

(*i*) It is a quick method of joining two pieces,

(*ii*) There is very little wastage of the metal,

(*iii*) The process can be accurately controlled, and

(*iv*) The welds are consistently uniform.

This type of welding is particularly well suited to all forms of automatic production. The following electric resistance welding processes are commonly used :

(*a*) Butt welding,

(*b*) Flash welding,

(*c*) Spot welding,

(*d*) Seam welding,

(*e*) Projection welding.

(*a*) **Butt Welding.** In this process, heat is generated by the contact resistance between two components. The faces of the components should be machined or edge prepared as shown in Fig. 4.13. The two

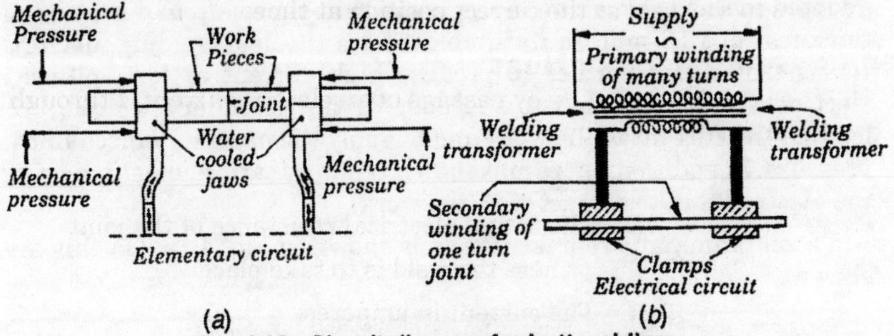

Fig. 4.13. Circuit diagram for butt welding.

parts or components are brought together and the pressure is applied along the axial direction by a spring. A heavy current is passed from the welding transformer (which is having comparatively larger number of turns on the primary and lesser number of turns on the secondary) which creates the necessary heat at the joint due to the comparatively high resistance of the contact area. The metal at the joint melts and the two parts fuse together producing a bulged joint.

Applications. The main applications of butt welding are :

(*i*) Where the parts are joined end to end or edge to edge.

(*ii*) For welding pipes, wires and rods.

(*b*) **Flash Welding.** This is similar to butt welding except for the difference that in this case current is applied to the parts before they are brought together so that when they meat arcing or flashing takes place. The two pieces to be welded are clamped strongly in a flash welding machine. The parts are brought together and the resistance to the current flow heats the contacting surfaces. As soon as the metal has been brought to its melting temperature, the current is shut off and the pieces are rapidly brought together under considerable pressure. When this action takes place the squeezed molten metal gives off sparks or flash, hence the name. As the metal is heated to its plastic state, the pieces are forced together under high pressure which forces fused metal and slag out of the joint making a good solid weld. Fig. 4.14 shows two pieces of metal before and after being flash welded together.

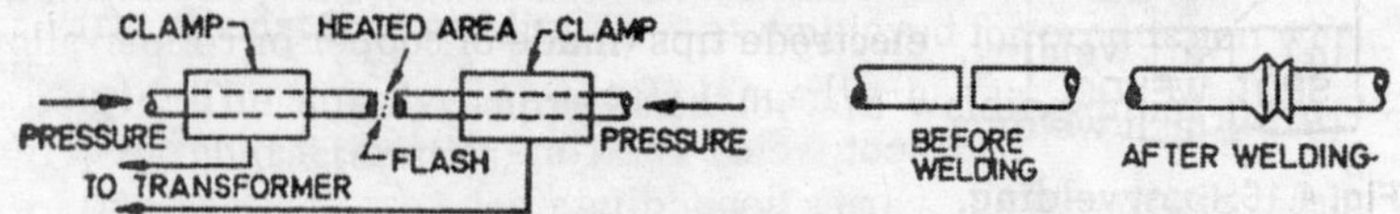

Fig. 4.14. Flash welding.

Applications. This method of welding is used extensively in production work, particularly in welding rods and pipes together. The flash method has almost entirely superseded the butt weld method.

(*c*) **Spot Welding.** This is the simplest and most universally adopted method of making lap welds in thin sheets up to a maximum thickness of 12.7 mm. In its simplest form the spot welding machine consists of a transformer to produce high current at low voltage : electrodes are connected to the ends of the secondary winding for leading the current to the work and to apply the necessary mechanical pressure. In addition, there must be a mechanical arrangement to bring the electrodes in to contact with the work.

A diagrammatic representation is shown in Fig. 4.15. On this are shown :

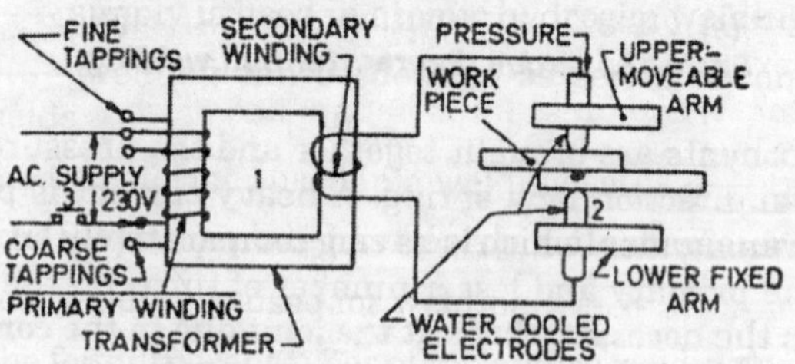

Fig. 4.15. Spot welding.

1. The transformer.
2. The electrodes.
3. The electrical control switch.

It is used to provide tappings on the primary winding of the transformer to regulate the secondary voltage, and thus the welding current. A common arrangement employs tapping at both ends of this winding, one end giving coarse adjustment and the other fine adjustment.

In the process the materials to be welded are overlapped and pressed between two water cooled electrodes and a impulse of current is passed through the assembly. The metals in the zone of pressure gets heated up to fusion and joint thus made gets cooled under pressure. For spot welding 5,000 Amp of current is required and the voltage between

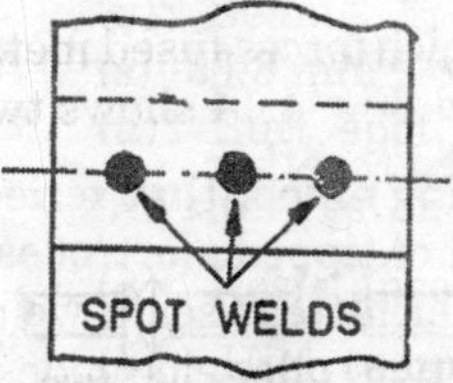

Fig. 4.16. Spot welding.

the electrodes is usually less than 3 volts. The open circuit voltage is less than 12 volts. The value of the welding current, the time for which the current flows and the pressure between the electrode tips (made of copper or copper alloy) are the main factors affecting the quality of the spot weld. The time period of flow of current varies depending upon the thickness of sheets and kinds of metal etc. for thin sheets this period is about 1/50 sec. for each 0.26 mm of total thickness of the two sheets to be joined.

Applications. (*i*) It is applied to welding of sheets.

(*ii*) It is used for fabricating all types of sheets metal structures where mechanical strength rather than water or air tightness is required.

(*iii*) It may be applied to all types of boxes, cores and enclosing cases etc.

Spot Welding Machine. A spot welding machine has a welding transformer and a timer. Timer controls the time for which the current flows. The primary winding of transformer has tappings and a particular tapping is selected by the selector switch '*S*' depending upon the current required for welding. For maximum welding current, the selector switch should be kept in position No. 1. Secondary of welding transformer is connected to two electrodes made of copper or bronze and cooled by water. When pressure is not applied on the pedal, electrodes are kept apart by the spring action. The sheets to be welded are placed overlapping each other between the two electrodes and pressure is applied by pressing the foot pedal. The spot weld is made between the

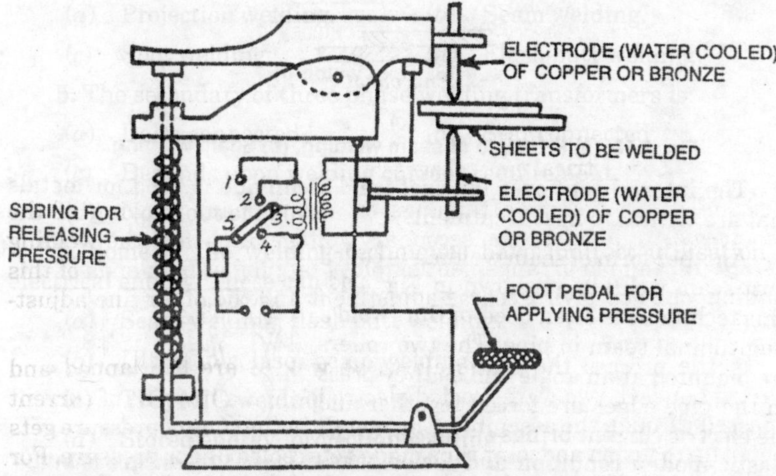

Fig. 4.17. Spot welding machine.

sheets. Thickness of the plates determine the duration magnitude of the current.

(*d*) **Seam welding.** It is similar to spot welding except that series of spots are produced by roller electrodes instead of tipped electrodes. Most seam welding produces a continuous or intermittent seam weld near the edge of two over-lapped metals, by using two roller electrodes. As these rollers travel over the metal the pieces are under pressure and the current passing between them heats the two pieces of the metal to the fusion point. This might be called a seam spot weld method. Fig. 4.18 (*a*) shows a schematic diagram of a typical seam welder in operation. The spaces can be timed to over lap or timed to have spaces between the spots. The main object of the overlapped spots is to produce gas and liquid leak-proof lap joints. Two types of seam welds are shows in Fig. 4.18 (*b*). If the current to the electrodes is turned off and on a series of overlapping spots form the seam. An uninterrupted flow of current to the electrodes will form a continuous seam as shown in Fig. 4.18 (*c*).

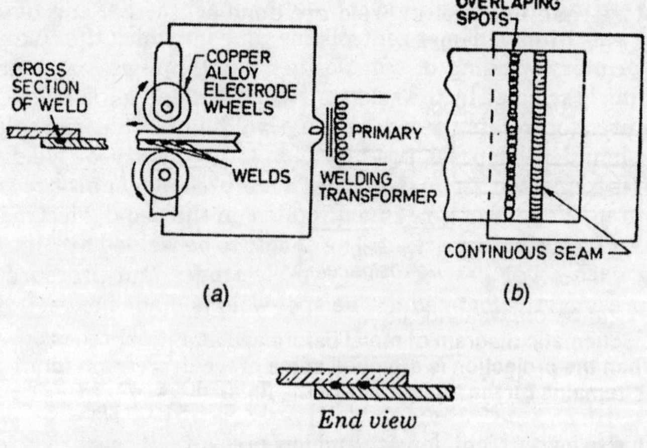

Fig. 4.18. (*a*) Principle of seam welding, (*b*) Seam welding.

The intermittent over lap spot weld technique is used for metals that are critical to heat treatment.

Another form of seam welding is called Butt-seam welding as shown in Fig. 4.19. This technique is often used to Butt weld a longitudinal seam in pipe. The two rollers are mounted at an angle and as they press on the pipe edges are forced together and the electric current brings the metal to the plastic molten condition and fusion takes place.

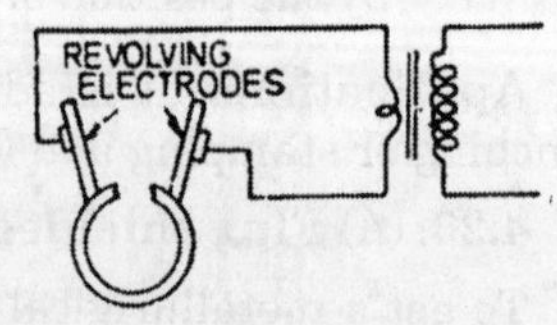

Fig. 4.19. Principle of butt. seam welding.

Applications. (*i*) It is used for making lap and butt welds.

(*ii*) It is quicker than spot welding operation.

(*iii*) It is used on many types of pressure tight or leak proof tanks for various purposes and a number of other products particularly of the relation type such as circular or rectangular containers, car body sections transformer radiator units.

(*e*) **Projection welding.** It is modified form of spot welding. Projection welding consists of forming slight projections on the sheet of metal. The projections are accurately formed in precise locations on the metal by a special set of dies. After the projections are formed, the raised portions on one piece are pressed into contact with another piece, while at the same time a heavy current is passed through the two pieces. When these raised portions touch the second sheet of steel, as they are clamped by the electrodes in a projection welder and the current is applied, current flows at the points, heats and fuses the two pieces together. Projection welding has following advantages over spot welding :

(*i*) More than one spot or weld are done at a time. Therefore, more output is obtained.

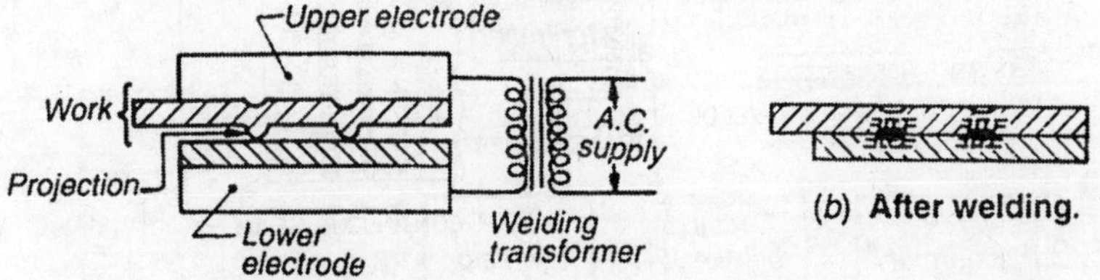

Fig. 4.20. Schematic diagram of metal before and after it is projection welded. When the projection is stamped some of the depression formed remains on the top surface of the finished weld.

(*ii*) Due to low current density and low pressure the electrode life is increased.

(*iii*) Good finished appearance is obtained because surface remains unindented by the electrodes.

(*iv*) It locates the welds automatically at certain desired points by the position of projections.

Applications. It is advantageous in assembling parts made by punching or stamping and for welding studs, nuts to plates.

4.20. (*A*) Flux shielded Arc Welding:

To get a metallurigical control, the electric arc provides the heat required for fusion, while a flux is used for shielding and cleaning. The electric arc has a-ve resistance characteristic and therefore, the current drawn in the arc is more. For prevenling the short circuit current occuring due to excedingly high currents, a choke is normally used to stabilise the arc, to facilitate various currents to be used. The choke or the ractor is provided with a number of tappings. The temperature of the arc is usually 4000°C, a temperature sufficient for melting most of the metals and for welding. Thus in electric arc welding process, heat is got by the arc, while a flux is used for shielding and cleaning purposes and for metallurigical control. In the metal arc welding (manual process), the electrodes are of short length and are covered by flux. The coating an the electrodes is of sufficient thickness.

4.20.1. Electronic Resistance Welding Unit

In Electronic controls in resistance welding we make use of SCR contactors or ignitron contractors, which allow the welding current to flow through the work pieces. Current flows in precisely controlled magnitude but also for a definite instant. This results in perfect welding of sheets. Sophisticated equipment viz. bodies of air planes, automobiles etc. need perfect welding which can't be controlled manually. Fig. 4.21 gives the basic requirements of the procedure for Electronic Resistance Welding.

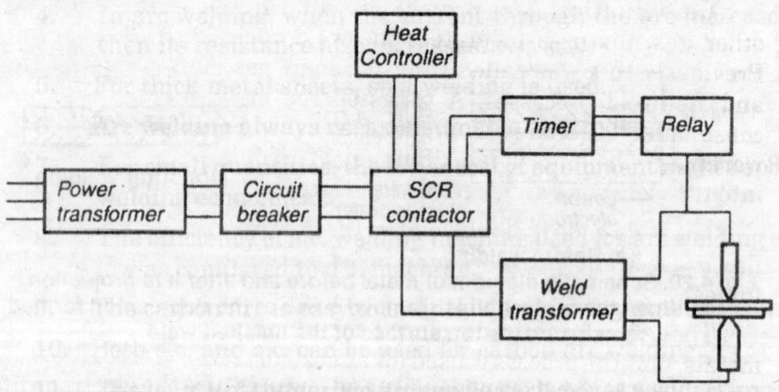

Fig. 4.21. (*a*) Block diagram.

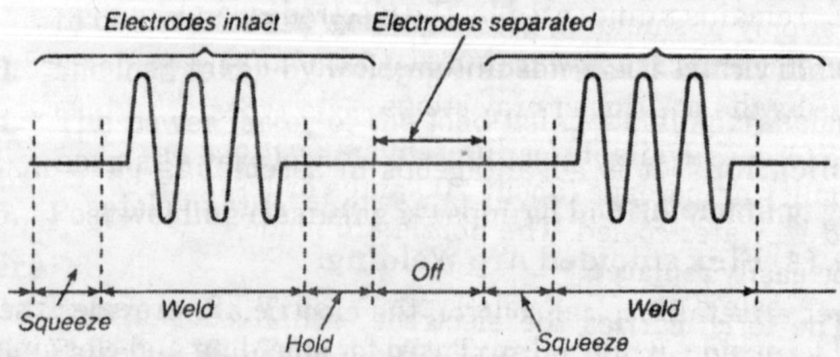

Fig. 4.21 (*b*) Flow of current.

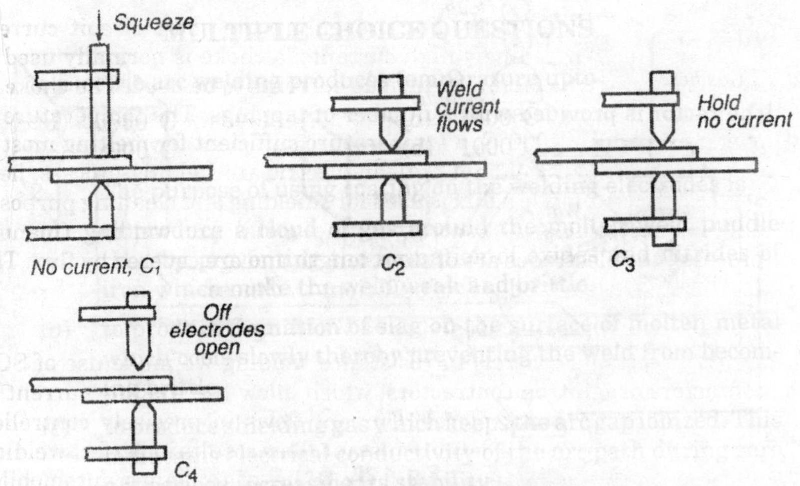

(*c*) **Off electrodes operation sequence.**

Fig. 4.21. Resistance welding controlled electronically.

GAS TUNGSTEN ARC WELDING

Introduction. This process uses an inert gas (helium or argon) or other gas mixtures to protect the weld zone from the atmosphere. Previously this was called tungsten inert gas (TIG) welding, Heliarc and Heliweld processes. A tungston rod or wire is used as a non-consumable electrode. An arc is established between the rod and the workpiece. On joints where filler metal is required, A welding rod is fed into the weld zone and melted with the base metal.

The weld zone is shielded from the atmosphere by the inert gas. No flax is used. The entire welding takes place without spatter, sparks or fumes.

Though originally developed for aluminium and other reactive metals, now the process is used for almost all metals. It is also employed for welding dissimilar metals and for applying hard facing materials to steel.

Description. The equipment consists of (1) a welding tourch or gun (2) a supply of shielding gas consisting of gas cylinder, regulator and flow meter and (3) a power source.

Welding tourch. There are various types of water cooled or air cooled torches available for different applications. Air cooled torches are of smaller ampereage (about 200 amps or less) while 300 Amp. torch would be water cooled. The nozzle are made of ceramic material and can be easily replaced.

The torch carries the electrode, usually of length 175 mm and diameter in the range 0.5 mm to 6 mm. The larger the diameter, the greater its current carrying capacity. The larger electrodes can be

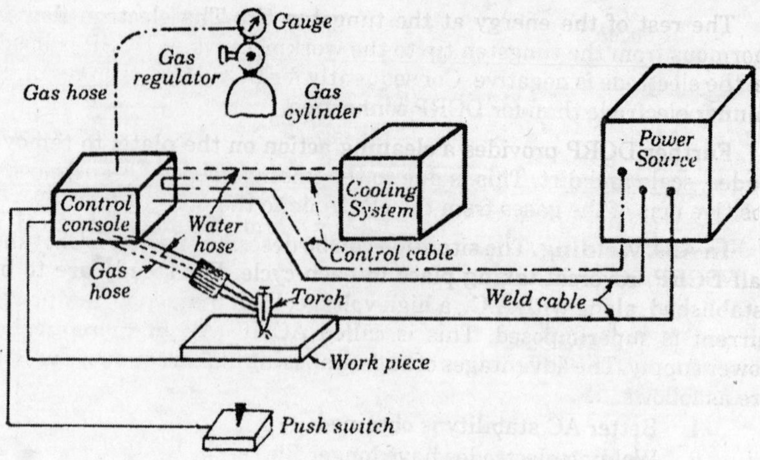

Fig. 4.22.

tapered at the end to increase its current density and improve arc stability.

The common electrode materials are pure tungsten, tungsten with 1 to 20% thorium and tungsten with 0.5% zirconium. Addition of thorium or zirconium improves electron emission and reduces contamination of tungsten.

Power source. The power supply for the process either A.C. or D.C. In D.C. welding the connection can be DCSP (straight polarity) or DCRP (Reverse polarity) in DCSP greater concentration of heat energy (about 2/3) occurs at the work piece while in the case of DCRP only about 1/3 of the energy is delivered to the workpiece.

Table 4.1. Recommended Electrodes for Various Welding Currents

Electrode Diameter (mm)	*Welding Currents, Amp.*			
	ACHF		*DCSP*	*DCRP*
	Using pure tungsten electrode	*Using thoriated electrodes*	*Using pure tungsten*	*or thoriated electrodes*
0.5	5–15	—	—	—
1.0	10–60	60–80	15–80	—
1.6	50–100	100–150	70–150	10–20
2.4	100–160	160–235	150–250	15–30
3.2	150–210	225–325	250–400	25–40
4.0	200–275	300–425	400–500	40–55
4.8	250–350	400–525	500–800	55–80
6.4	325–475	500–700	800–1100	80–125

The rest of the energy at the tungsten tip. The electron flow is enormous from the tungsten tip to the workpiece when DCSP is used, *i.e.* the electrode is negative. Consequently for DCSP the torch requires thinner electrode than for DCRP connection.

Further DCRP provides a cleaning action on the plate, to remove oxides, scale and dirt. This is due to the impingement of high velocity positive ions of the gases from the electrode to the plate.

In A.C. Welding. The situation can be described as half DCSP and half DCRP, reversal taking place in each cycle. For steady are to be established, along with A.C., a high voltage, high frequency, low power current is superimposed. This is called ACHF (AC high frequency) power supply. The advantages of superimposing high frequency current are as follows.

1. Better AC stability is obtained.
2. Welding electrodes have longer life
3. The arc may be started without touching the electrode to the workpiece.
4. A longer arc is possible. This is particularly useful in surfacing operation.
5. The use of wider current range for a specific diameter electrode is possible.

The shape of weld bead in A.C. welding is shown below for the relation between electrode diameter and current ranges.

Shielding Gases. Argon and helium shielding gases are used for the majority of applications. Argon is generally recommended because of its general suitability for a wide variety of metals and for the considerably lower flow rate required. Helium, however, is less expensive. For certain metals, mixtures of argon with helium are preferred.

The measured voltage across the arc is always higher with helium than with argon for a given arc gap.

Argon and helium are usually supplied in the purity level of 99.99%.

Welding technique. Filler metal, in the form of welding rod, need not to be used if proper reinforcement and complete fusion of the edges

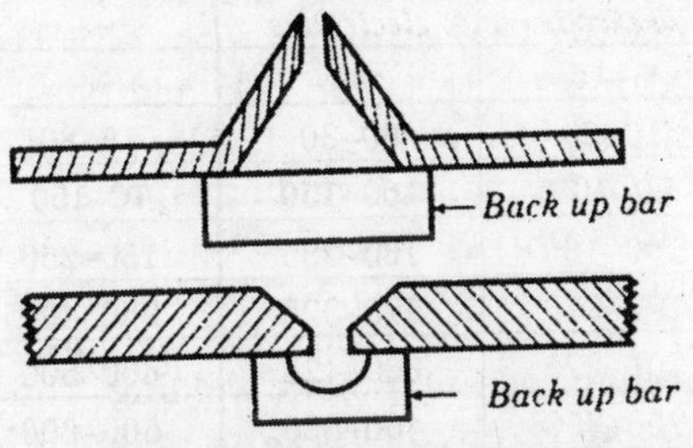

Fig. 4.23. Back Up Technique.

can be obtained without it. Proper cleaning of the surfaces prior to welding is essential.

While square edge butt joint can be used for thinner metal a single-vee grove joint (with a groove angle of 60°) is required on material thickness in the range 9 mm - 25 mm. A double vee groove joint is commonly used for thicknesses greater than 12 mm.

Weld Backup. On many GTAW applications a backup procedure is used. This is done for several reasons. On thin gauge material, backing is usually used to protect the underside of the weld from atmospheric contamination resulting in weld porosity and poor surface appearance. The weld back presents the weld puddle from dropping through by drawing away from the workpiece sense of the heat of the arc. A backup can also physically support the weld puddle.

The backup can be by (*i*) metal backup bars (*ii*) introducing on inert gas atmosphere on the weld underside (*iii*) a combination of (*i*) and (*ii*) or (*iv*) use of flux backing as a slurry or with glass tape. The backup gas, it used can be argon or helium.

A typical welding speed is 250 mm per minute for aluminium of 3 mm thickness, flow rate of argon : 9.4 litre/min.

Applications.

(*a*) GTAW is extensively used for welding, aluminum, stainless steel, titanium and other metals. (*b*) It is suited for thickness less than 4 mm, for large thickneses, the welding speed is very slow. (*c*) GTAW is also used for root runs in pipe welding for high pressure steam lines and for chemical plant.

4.20.3 MIG (Metal Inert Gas) Welding

Mig welding is the process of welding in which the electrode is in the form of a wire whcih is fed from a coil.

Principle of operation. The generic name metal inert gas (MIG) welding has been given to those welding processes which use a continuously fed small diameter, solid wire electrode, and a gas shroud. The principle operating features are shown schematically in figure below. The wire diameter is normally 0.8, 1.0, 1.2 or 1.6 mm and the shielding gas is usually CO_2 or argon with 2-5% O_2, or 5-25% CO_2. Argon-helium mixtures (With O_2 or CO_2 additions) can be used for special applications *e.g.* for welding stainless steel. MIG welding is slowly replacing MMA for manual welding especially for the welding of thin sections of ferrous and non ferrous materials, and when continuous operation or high deposition rates are required. Because of the continuous feeding of the wire electrode. MIG welding has found special applications in mechanised and robotic welding. The process can also be used at high current level. Almost twice that of MMA which can be exploited for welding thicker materials at higher speeds, but only in a mechanised operation.

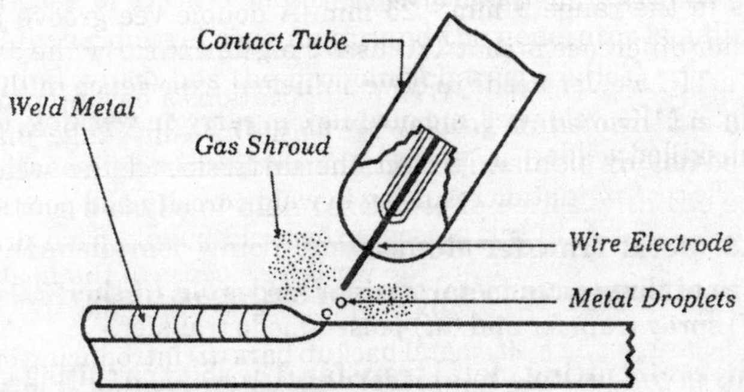

Fig. 4.24. Principle of operation for MIG welding.

Because the process is often used to increase the deposition rate/welding speed, care must be taken to avoid introducing weld defects such as lack of fusion. For this reason the process requires carefull tuning of the power source, and welder training, when high quality is required.

4.20.4. TIG Welding:

TIG welding is that process of welding in which the electrode is non-consumable. It is also known as "Tungsten inert-gas welding". The non-consumable electrode is that of tungsten.

Advantages of MIG welding

(*i*) Higher deposition rates
(*ii*) Less operator skill required
(*iii*) It has all position capability
(*iv*) Without starts and stops, long welds can be made.

Advantages of TIG welding

(*i*) Superior quality welds can be obtained
(*ii*) Welds can be made with or without using filler metal.
(*iii*) Free of spatter.
(*iv*) Low distortion.
(*v*) Precise control of welding variables.

Difference between MIG v/s TIG welding.

Both MIG and TIG welds are types of arc welding which utilises the concentrated heat of an electric arc to join together metals by fusion of the parent metal by a consumable electrode. Depending on the material to be welded and the electrode used, the process utilises either D.C. or A.C. for the welding arc.

Some of the Basic difference between MIG and TIG welds are that MIG welding is faster than TIG welding, as utilising TIG welding needs more skill than MIG welding. A solid wire is used in the MIG welding whereas TIG uses a flux cored electrode.

Another difference is that TIG uses tungsten to carry the arc, and a user of a TIG welder needs to have sufficient experience in the craft of welding. A MIG weld user, meanwhile can carry on working, despite being a unskilled welder.

4.20.6. Metal Transfer Modes.

There are three main metal transfer modes *viz.* (*i*) short circuiting or dip. (*ii*) spray transfer and (*iii*) pulsed mode transfer.

(*i*) Short circuiting metal transfer. The short circuiting mode is used at low current levels, typically less than 250 A with a 1.2 mm diameter wire at which levels it is impossible to operate with an open arc to give free plight in (spray) metal transfer. The transition from short-circuiting mode to spray mode is shown in Fig. 4.25. The mode is established by setting a low open circuit voltage so that the molten metal forming at the tip of the electrode wire transfers by short circuiting with the weld pool as shown in Fig. 4.25 (*b*).

If a free flight technique *i.e.* a spray type mode of metal transfer, were to be attempted below the threshold current level, the low arc forces would be insufficient to prevent large globules forming at the tip of the wire which would then transfer erractically across the arc under normal gravitational forces. Thus, in operating the short circuiting transfer mode, the initial selection of the welding parameters (open - circuit voltage, inductance, and wire feed speed) and the dynamic characteristics of the power source are crucial in stabilising the arc and metal transfer.

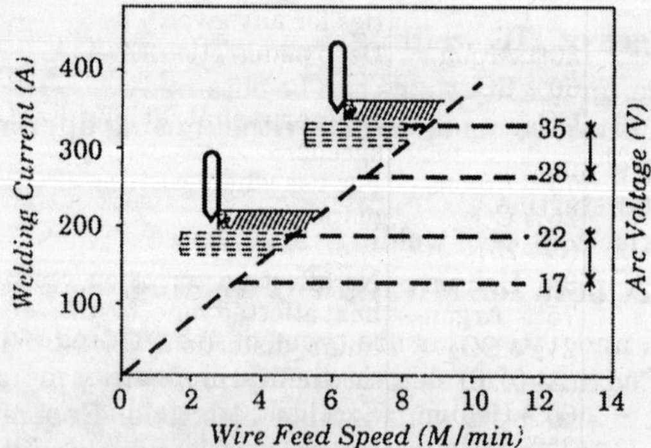

Fig. 4.25. (*a*) Normal operating speed in MIG welding. Note that the voltage must be reduced at low current levels to induce short circuiting metal transfer.

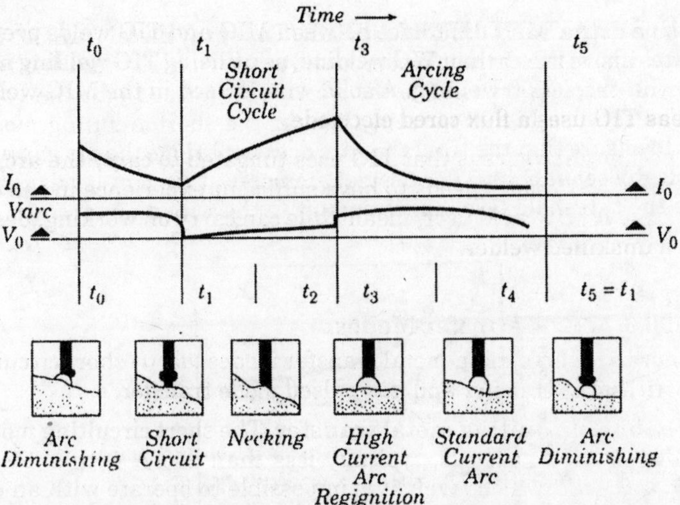

Fig. 4.25. (*b*) Short circuiting metal transfer in MIG welding. Note that a fluctuation in the welding current waveform occurs whenever the electrode short circuits to the workpiece.

Table 4.2. Recommended Shielding Gases for Short Circuit Transfer

Metal	*Thickness*	*Shielding Gas*	*Advantages*
Carbon Steel	Less than 3 mm	75% Argon 25% CO_2 (C—25)	High welding speeds without burn through. Minimum distortion and spatter. Best puddle control for out of position welding provides best mechanical properties for any given wire.
	Over 3 mm	75% Argon 25% CO_2 (C—25)	High welding speeds without burn through. Minimum distortion and spatter. Best puddle control for out of positions welding. Provides best mechanical properties for any given wire.
		50% Argon 50% CO_2	Deep penetration, low spatter.
		CO_2	Deeper penetration. Faster welding speeds.
Stainless Steel		75% Argon— 25% CO_2 (C—25)	For use where corrosion resistance is not mandatory.
		9% Helium— 75% Argon— 2½% CO_2 (A–1025)	No effect on corrosion resistance. Small heat affected zone. No undercutting: Minimum distortion. Good bead shape and mechanical properties.
High Yield Strength Steel		60% Helium 35% Argon— 5% CO_2 (A–415)	Excellent arc stability, wetting, characteristics, bead contour. Little spatter. High impacts.

***(ii)* SPRAY METAL TRANSFER.** The spray metal transfer mode operates above the threshold current level shown in Fig 4.26 and in this mode the droplets transfer to the workpiece in free flight the voltage must be set at a higher current than for the shortcircuiting mode in order to ensure that the tip of the wire does not bridge the arc gap which is typically set at 6 mm. The droplet diameter at current levels slightly above the threshold level approximates to the wire diameter.

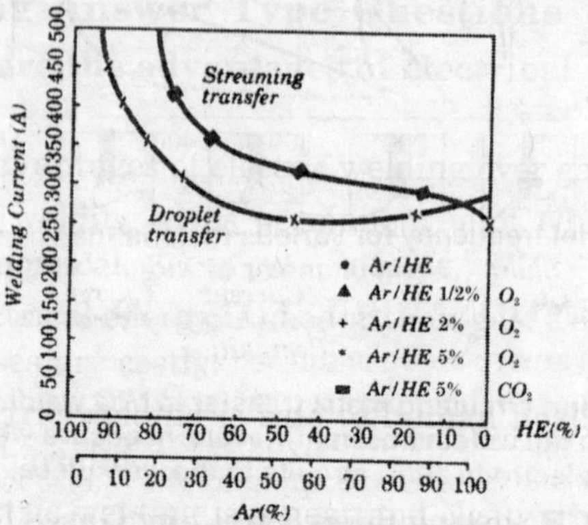

Fig. 4.26. Threshold values for spray metal transfer.

The threshold current level is determined by the wire diameter, low carbon steel wire in a range of argon helium gas mixtures are shown in fig below. It can be seen that the threshold value increases significantly in high helium gas mixtures, indicating that the spray transfer is more difficult to obtain.

The droplets are detached by a combination of an electromagnetic effect, a plasma jet and the formation of gas bubbles in the molten droplet. The droplets have velocity when they are detached from the tip of the wire but then accelerate across the arc gap.

The frequency at which the droplets transfers increases as the current level increases. The relationship between droplet frequency and the current for various materials combination is shown in Fig 4.27.

The droplet size is about the same as the wire diameter at the threshold level, but decreases significantly as the welding current increases. At current levels well above the threshold level the droplets transfer in a stream of fine droplets and the wire tip assumes a pencil point shape.

At very high current levels *e.g.* above 450A. for 1.0 mm diameter, low carbon steel wire, the end of the wire which is softened by resistive heating can rotate. Rotation of the wire is promoted by extending the electrode stick out *i.e.* the distance the wires extend from the contact tip to the arc. The principle benefits of extended stick out are increased deposition rate and an improved weld bead penetration profile (bowl shaped).

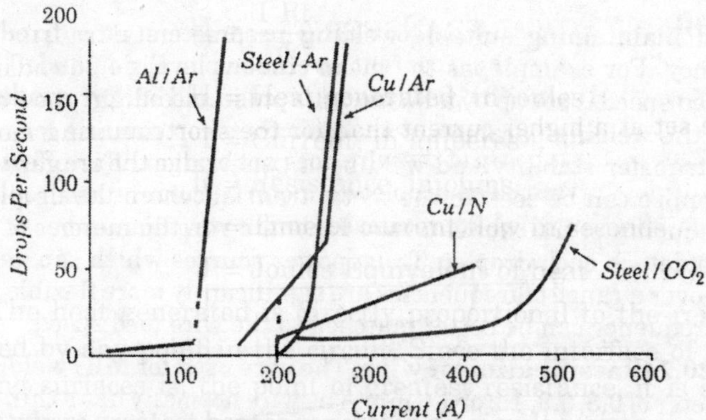

Fig. 4.27. Droplet frequency for various material-gas compositions of 1.6 mm diameter of wire.

(*iii*) CONVENTIONAL PULSED METAL TRANSFER

The pulsed transfer mode is a free flight technique *i.e.* non short - circuiting, which can be operated at welding currents below the threshold level. For this technique, high current pulses at a frequency of typically 50 Hz and 100 Hz are applied. The magnitude of the pulses must be sufficient to detach the droplets and project them across the arc. Droplet transfer to the weld poul is often completed during the back-ground period, the process of metal transfer is shown schematically in fig below. Where it can be seen that, although significant melting occurs during the background period, the high current pulse is required for droplet detachment.

The pulsed MIG technique is an attractive alternative to the short circuiting mode at low current levels because the open arc/free flight metal transfer greatly reduces spatter and provides greater tolerance to variations in the gun to work-piece distance. Until recently, the technique was not widely practised because of the difficulty in setting

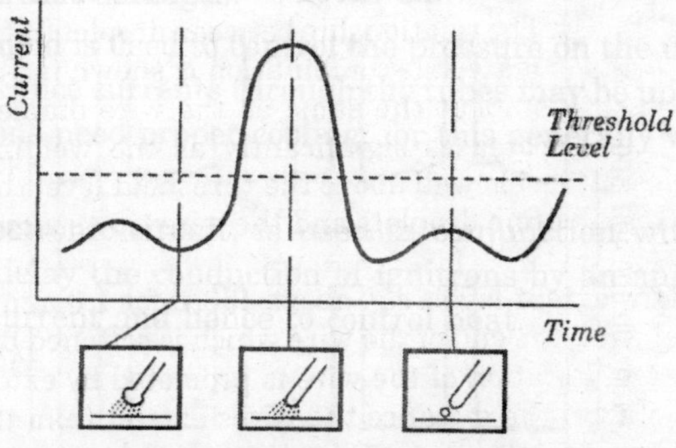

Fig. 4.28. Pulsed metal transfer in MIG welding.

up and maintaining suitable welding parameters at a fixed pulse frequency. For example, as the mean current level demanded by the wire feed speed is set up by adjusting the pulse and background current levels, the welding parameters are likewise a compromise for arc and metal transfer stability and weld bead penetration. The effect of the compromise can be seen in the metal transfer where the droplet size and frequency of transfer can vary depending on the number of pulses used to detach each droplet. Thus, power sources which can generate pulses over a range of frequencies are significantly more flexible, as the pulse frequency can be better matched to the wire feed speed.

4.20.7 Classification of wire. The wire used for MIG welding has a diameter of 0.8, 0.9, 1.2 or 1.6 mm. The wire normally has a thin copper coating, but this is not essential. The copper coating facilities feeding through the conduit and electrical pick-up in the contact tip.

The wire classifications are given in the following part of BS 2001 : 1983 Filler rods and wires for Gas shielding are welding.

Part 1 Ferritic steels
Part 2 Austenitic steels
Part 3 Copper and copper alloys
Part 4 Aluminum and aluminum alloys and magnesium alloys
Part 5 Nickle and Nickle alloys.

The corresponding AWS standards are :

A 5.7 copper and copper alloys
A 5.9 Stainless steel.
A 5.10 Aluminum and Aluminum alloys
A 5.14 Nickle & Nickle alloys
A 5.16 Titanium & titanium alloys
A 5.18 Carbon steel
A 5.28 Low Alloy steel.

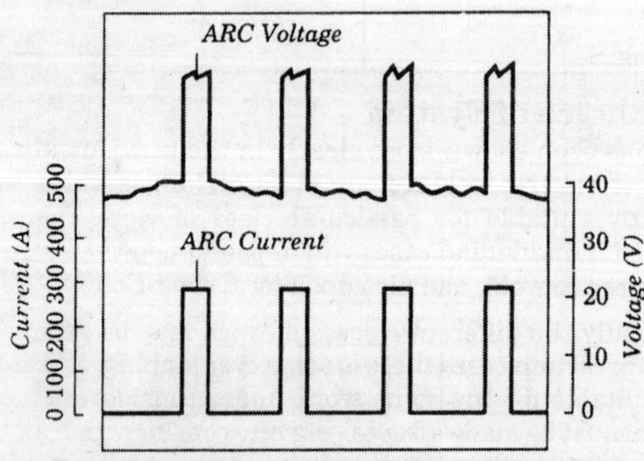

Fig. 4.29. Typical oscillogram of arc current & voltage for 2 m/min wire feed speed 50 Hz pulse frequency and 4 m pulse duration.

4.20.8 Power sources. Generally a DCRP connection can result in surface cleaning of metals such as aluminum. The oxide layer is removed. (DCSP is very rarely used. When employed it is used for aluminum and magnesium with a "burried-arc" or short circuiting metal transfer) AC supply is not convenient due to erratic arc behaviour.

4.20.9 SIGNIFICANCE OF GMAW. Weld Quality.

1. Since flux is not used, slag inclusions are eliminated.
2. Excellent protection for the weld from Hydrogen is achieved. Therefore no possibility of cracking in the weld or HAZ due to Hydrogen.
3. For the plain carbon steels, welding with CO_2 shielding lowers the cost.
4. The weld defects that arise are mostly due to faulty welding procedure.

Table 4.3. Systems of Electric Welding Applicable to Various Classes of Work

Nature of work		System applicable
Repair Foundary Heavy plate Constructional work	Seam welding Butt welding Spot welding	Arc
Seam welding light plates up to 9.5 mm. (3/8″) added thickness. Butt welding up to 40.64 sq. cm. (*i.e.* 16 sq. m) Spot welding		Resistance
Wire welding Heating Soldering		

4.21. Choice of System

It has been shown more clearly in the previous articles that each system of electric welding possesses certain features which make it eminently suitable for particular class of work. The system to be employed in individual cases will depend upon the nature of the work it is desired to weld, and also upon local conditions.

Broadly the different classes of work may be grouped as in Table 5.1, but in certain cases there is some over lapping, two or more systems being suitable for the same work under consideration, and the final decision must be made after taking into consideration all the conditions which determine the system to be employed.

4.22. Comparison Between Resistance and Arc Welding

S. No.	*Contrast Point*	*Resistance Welding*	*Arc Welding*
1.	Mode of heat development	Heat is developed due to flow of current through the contact resistance mainly.	Heat is developed due to an arc formed between the electrode and work piece.
2.	Source of Supply	Supply employed is usually a.c. only.	Supply can be a.c. or d.c.
3.	External Pressure	External pressure is required.	No external pressure is required and the equipment is simple and easy to control.
4.	Voltage level	Low voltage is required.	The striking voltage is high.
5.	Temperature	The temperature reached is not very high.	The temperature of the arc is very high.
6.	Nature of work	Cannot be used for repair work.	Most suitable for repair work.
7.	Power factor	Power factor is fairly good.	Power factor is very low.
8.	Production	Most suitable for mass production.	Not suitable for mass production.
9.	Electrodes	Roller, flat or bar type electrodes are used which are not consumable.	Flux coated metal electrodes are used which are consumable.

4.23. Requirements of a Good Weld

A good weld should have the following characteristics :

(*a*) Uniformly rippled surface of the weld.

(*b*) Even contour of the weld.

(*c*) Even width of the weld.

(*d*) Absence of surface defects like overlap, undercut, crack, crater and surface porosity.

(*e*) Absence of internal defects like blow holes, inclusion and incomplete penetration.

4.24. Comparison Between A.C. and D.C. Welding

S.No.	*Point of Comparison*	*A.C. Welding*	*D.C. Welding*
1.	Wear and tear	A.C. welding plant has no rotating plant so wear and tear is less.	D.C. welding plant has both generator and motor and has rotating parts so wear and tear is more.
2.	Efficiency	A.C. welding transformer has very high efficiency.	M.G. Set has low efficiency.

3.	Arc blow	The problem of arc blow is not much because it is controlled automatically.	Problem of arc blow is very swear and cannot controlled easily.
4.	Arc stability	Arc is not at all stable.	Arc is more stable.
5.	Cost of equipment	Equipment required is cheap.	Equipment required is costly.
6.	Electrodes	Only flux coated electrodes can be used.	Any type of electrode can be used.
7.	Energy consumption	Energy consumption is low as welding is always intermittent in nature.	Energy consumption is more because M. G. set is required to run continuously.
8.	Distribution of heat	In case of A.C. welding heat is equal in both poles.	Heat distributed is different at two poles 70% on +ve pole 30% on -ve pole
9.	Uniformity of heat production	The heat produced is not uniform because current passes twice through zero value.	Heat produced is uniform.
10.	Risk of shock	There is always danger of shock to operator due to high O.C. voltage [100 V required for striking the arc].	There is no danger of electric shock to operater due to low operating voltage.
11.	Effect of supply voltage.	In case A.C. weld the $1-\phi$ Transformer causes unbalance in $3-\phi$ supply system.	As M.G. set uses $3-\phi$ I.M. which is balanced load hence no unbalance in $3-\phi$ supply system is created.

4.25. Points Kept in View to Avoid the Weld Defects

A correct weld may be ensured if the following points are kept in view :

1. The joint preparation of the workpieces to be welded should be done properly.
2. The electrode of correct type and size should be selected.
3. According to the nature of the job the welding current should be selected. Job dimensions and material should be kept in view while selecting the current amperage and type of current (A.C. or D.C.).
4. Proper arc length should be maintained.
5. Correct welding technique should be used.

4.26. Laser Materials Processing

Laser has been hailed as one of ten engineering feats of this century by US National Academy of Engineering due to its diverse applications for the benefit of the mankind. Laser has many characteristics which can be utilized for different applications in engineering industries like pollution monitoring, non-destructive testing, holography, laser material processing etc. As most of these topics are being covered by other eminent experts, here we will mainly deal with laser material processing in engineering industries. Laser Material processing have revolutionized the manufacturing technology as it produces better quality products at very high production rates, thus meets as vital industrial requirement.

The main characteristics of the laser which are responsible for its high quality performance are its high power densities and focusability. The power density and interaction time can be varied to achieve the desirable conditions for different processes like welding, cutting, drilling heat treatment, surface modification etc. Moreover, a wide variety of materials ranging from wood, ceramics, plastic to high strength alloys can be processed with ease. In addition, laser can easily be transferred with the help of optics in difficult hazardous environment and inaccessible area which makes it more attractive.

Lasers have many applications in the nuclear industry due to easy transportability of the beam to long distances without much attention. It can be used for various applications in hazardous and inaccessible radioactive environments. The laser machine itself can be kept outside while only the beam is passed into the radioactive area. This avoids having a machine that will require regular and frequent maintenance in the hazardous area. Simple light weight, jigs and mechanical manipulators are all that is required within the hot zone. For example, laser can be and is being used for cutting of cladding tube (from radioactive spent fuel pin) for fuel reprocessing. Laser has been used for cutting of zincalloy-2 sheet for plate type fuel and welding nuclear fuel pin components at BARC.

Lasers have several applications in the plant industries. For example, they have been used for cutting the tube sheets, side plates and fins with higher speeds for the manufacture of regenerators for turbines causing net saving in the total cost. Lasers have also been used to drill holes in turbine power generator combustures and transition ducts made of Haste alloy-X (a Ni-Cr-Mo-Fe alloy) with speed at least 20 times higher than the other methods *e.g.* mechanical drilling, mechanical punching, electro-chemical machining and electro-discharge machining. Laser cladding of satellite on turbine disc alloy has been done for optimum corrosion/mechanical properties. Hard brittle ceramics such as SiN used in turbine engines have been cut ten times faster with a laser than with a diamond wheel. In addition, the other possible applications of lasers for power plant industries are cutting, drilling and welding of various materials such as orifice plates, valve plates, spring steels, stainless steels for plant construction etc.

In automobile industries small batches of parts are required from various sheet materials for which laser cutting has been found to be several times faster than the conventional methods of shear, nibble and grinding. In addition, greater flexibility in the design of parts have also been achieved. Lasers have also been used for drilling holes for manufacture of fuel filters, perforation of noise absorbing panels and manufacture of fuel jets. Laser cladding has been used for recovery of valves, crank shafts and cam shafts etc. The recovered parts have been found out to be of higher wear resistance than the new ones. Lasers have been used abundantly on production scale for hardening of components *e.g.* grey iron diesel engine cylinder liners, camshalts, gear teeth, internal bores of bearing housings etc.

In the aircraft industries lasers have been used for cutting of steels, titanium and its alloys and aluminium alloys. For example, titanium splice plate for horizontal stabiliser, stainless steel sheet for helicopter blade manufacture and aluminium alloys have been cut by laser with considerable economic advantages.

For the agriculture industries, lasers have been used for coating of ploughshears with an abrasion resistant layer composed of aluminium oxide, silicon oxide, tungsten carbide, iron, nickel and carbon. Other agricultural tools have also been cladded with molybdenum for higher abrasive wear resistance.

One of the important areas of laser applications for commercial advantages is obtaining products for tool industries for higher wear resistance. For example, laser hardening of cutting dies made of steels increased their wear resistance by 3-5 times. Laser hardening of cutting tools made of tungsten-cobalt alloys increased the wear resistance by 2 times. The wear resistance of thread joints for drilling equipment has been upgraded by 10 times by laser hardening. The use of selective hardening in tool production using lasers, has increased the durability of tools by a factor of 2-10 as compared with the other methods.

There are numerous other applications in which lasers have found widespread use on production scale *e.g.*

(*i*) production of high quality hermetic sealing welds in Ti and S.S. cases for the manufacture of heart pacemaker devices. Around 80% of total production is attributed to the laser.

(*ii*) laser cladding has been used to apply wear/corrosion resistant alloys on valve sealing surfaces for oil-fields, geothermal, nuclear power and chemical processing industries.

(*iii*) parts of textile machineries (hardening).

(*iv*) cutting of machine parts of low carbon steels in machine tool fabrication.

(*v*) Welding of parts (stainless steels and titanium alloys) for heat exchangers.

(*vi*) drilling of heat screens in aviation and rocket manufacture.

(*vii*) marking of measuring reals of instrument, press tools, cutters, saws, tubes and sheet articles.

In fact, the applications of LMP is ever increasing and it is difficult to define its limits in the field of material processing.

4.27. Laser Material Processing - Basic Concept

The thermal processes taking place are markedly different as the energy density of the heat sources is varied starting from radial conduction to vaporization conduction to vapourization loss to plasma formation over the sample. At relatively low energy density radial conduction mode is prevalent and is effective in heat treatment processes where no surface melting is required. Vaporization conduction mode is operative in thick welding or cutting which is popularly referred as "key-hole effect". In this process the rate of heating is very much higher than the rate of heat transfer by conduction mode and therefore temperature at a local spot rises above the melting point resulting in vapour formation. These vapours are held in the liquid metal surrounded by the base metal forming a "key-hole" and signifies the dynamic equilibrium of the heat and pressure conditions on moving object. Further increase in laser energy densities may lead to excess vapour loss and these conductions are created for cutting/drilling, and vapour deposition.

During these operations certain vapours are formed over the object which gets ionized by the incoming laser leading to its plasma formation. This plasma shields the job from incoming laser and therefore not preferred in some applications like welding, cutting etc. In fact there are techniques like pulsing or special gas flow arrangement or reduced pressure to offset the effect of plasma in welding. If plasma is near surface, its heat is utilized by the job. However, in many other applications the plasma formation can be utilized to the advantage. Certain chemical operation like reduction, oxidation etc. can be done to produce ultrafine powders or for surface modification. Plasma heat is also utilized in cladding operations where hard deposits are made. The quality of the plasma *i.e.* electron/ion/vapour density, temperature etc. are governed by the laser parameters, thermo-physical properties of the base material, environment and laser-metal plasma interaction. It is therefore very essential to study the laser-metal plasma interaction. Laser-metal plasma interaction can be studied by holography. Spectroscopy, photography and thermography which together can highlight the mechanism involved in any operation.

Laser parameters like energy, spot size, location of the focal point, pulse duration/repetition rate in pulsed welding, processing speed, pressure/environment are some of the important parameters which must be standardized for any operation. Changes in power density and interaction time can alter the conditions/process to a great extent. The standardization can either be done by experimentations or can also be visualized by involving theoretical models. These models are based on solution of heat and mass transfer equations with certain assumptions (for mathematical simplifications). Experiments are being conducted to validate these theoretical predictions.

A number of processes are possible with laser as heat source depending upon the energy density and time.

4.28. Laser Material Processing-State of Art

4.28.1. Laser Welding. The status of the laser welding is relatively advanced and a number of alloys have been evaluated for welding by lasers. Lasers are being used for welding of different types of steels (including stainless steels) aluminium alloys, heat resistant alloys, zirconium and titanium alloys, and also dissimilar metals and alloys. In the case of austenitic stainless steels, successful welds have been reported with aspect ratio (depth to width ratio) as large as 12 : 1 and welds have been found to be having no porosity and tensile strength equal to or more than that of parent metal. Fatigue properties of some of these welds have also been studied and it has been found to be either equal to or better than those of parent metal.

Aluminium and its alloys have been found to be difficult to weld due to their high reflectivity and thermal conductivity. These alloys have been found to be very sensitive to the intensity of the laser input energy and the welding variables. Though welding with acceptable micro-structures and bead profile have been achieved in certain alloys *e.g.* 2219 and 5456, porosity has been the biggest problem. Results of the mechanical testing of 5456 aluminium alloy show that the failure occurs in the weld zone with reduced ductility and high amount of porosity visible on the fracture surface. Laser, welding of zircalloy-2 has also been investigated. It has been shown that the laser weld zone has undergone much less loss of tin as compared to electron beam weld and Helium gas tungsten arc weld zones. Fusion zone in the laser weld consists of martensitic plates commonly encountered in quenched zirconium alloys. Mechanical testing results reveal that zircalloy-2 laser welds are stronger than base material. Work on the laser welding of zircalloy-2 appendages to fuel pin tubes has shown that lasers are very much suitable for this application and acceptable welds can be made having much higher load carrying capacity than those possible with resistance spot welding.

4.28.2. Laser Cutting. A large number of metals and alloys including steels, super alloys, Ti, Zr, Al and their alloys have been cut by lasers. Ti and its alloys have been laser cut very economically and successfully. The cut edges show a very good quality and in the case of smaller thickness thermally damaged zone is so small that it can be removed by a simple chemical etching. Lasers are specially suitable for cutting complicated shapes of both metallic and ceramic materials with the help of CNC. Thick materials can be cut with ease due to key-hole effect which increases the absorptivity of the beam. Lasers have been utilized for cutting the fuel rods inside the radio-active area for reprocessing.

4.28.3. Laser Hardening. Laser hardening is being applied industrially to increase the surface hardness and to obtain a better wear resistance. This technique has almost reached production status in automotive, road-building, agriculture machinery, mining and steel industries for treatment of various critical parts, hardening of tool dies, rolls of mills and bloomers etc.

4.28.4. Laser Surface Alloying. Laser surface alloying is being made use of for producing alloys for better corrosion, wear, high temperature and oxidation resistance properties on the surfaces of common or cheap metals and alloys. The direct implication of the development of laser surface alloying is the net saving in the consumption of precious metals *e.g.* Mn, Co, Cr, Ni, Mo etc. resulting into potential cost saving in the production of different alloys for different applications. For example, mild steel can be surface alloyed with Fe, -Cr, Fe-Ni, Fe-Cr-Ni systems to produce a new range of materials for different applications. An alloy of Fe-Mo can be produced on the surface of cast irons, high carbon steels or austenitic stainless steels for better wear and pitting resistance respectively. Likewise, Fe-Cr-Mn-C alloy can be developed for better wear resistance. Similarly, Pd can be alloyed with Ti to produce Ti-Pd alloy on the surface of Ti-metal for better corrosion resistance in very harsh environments *e.g.* HCl. H_2SO_4.

4.28.5. Laser Cladding. The laser cladding process is used to overlay a substrate with another material with a different chemistry by melting a thin inter facial layer to produce a metallurgical bond with minimum dilution. This provides the surface much better wear and corrosion resistance properties in response to the demand for developing special wear, friction and corrosion resistant materials for many industries. Cladding of Fe-Cr-Mn-C system on low carbon steels, satellite on high temperature alloys, TiC and WC on Fe, Ti, Al and Cu based alloys have been developed.

4.28.6. Laser Glazing. Laser glazing has been shown to be capable of producing cooling rates of 10^6 0/sec readily and 10^8 0/sec possibly in thin sections. Consequently, a variety of microstructures can be possible *e.g.* amorphous metallic solids, extended solid solutions, ultra fine eutectic structures etc. Work has been done to investigate the laser glazing process to improve the properties of metals and alloy surfaces. Investigations reveal that stainless steels 430 and 304, incoloy 800 and haste alloy have enhanced corrosion resistance due to preferrential formation of protective oxide scale such as SiO_2, CrO_2 and $MnCr_2O_4$. Laser glazing of A1 alloys *e.g.* (A1 - 1.2% Mn) also has improved the resistance to uniform corrosion in hydrochloric acid and sodium citrate solution. Laser glazed titanium alloys too have been found to have better resistance to stress corrosion cracking as compared to the base material.

4.28.7. Amorphization. Due to higher values of strength, ductility, corrosion resistance and radiation stability of the amorphous alloys as compared to the crystalline ones, they have many applications and have generated immense interest among the scientific community. Amorphous layers can be obtained on the surfaces of metals and alloys by laser surface treatment owing to very rapid cooling associated with it.

Although amorphous materials can be produced by a number of rapid quenching methods, the scope of their application is limited by the physical dimensions of the resulting glassy metals. Even the continuous ribbons and wires produced via commercial roller and expulsion fabrication methods are necessarily constrained in size. Lasers, however an overcome this limitation and can be used to produce amorphous layers on the surfaces of metals and alloys of almost any size and shape.

4.28.8. Laser Synthesis of Nitride and Carbide Layers on the Surfaces of Metals and Alloys. The use of lasers for the production of layers with special properties (high microhardness, corrosion resistance etc.) on different components is a highly promising method. For example the method of production of nitride layer by layer treatment in nitrogen atmosphere can be useful for increasing the wear resistance of components working in the dry friction units. This is also important for a number of components because the nitrides are characterized by high wear resistance, high chemical stability in various corrosive media, high resistance to fire, and a number of other properties significant for applications in practice. The advantages of this technique as compared to the traditional methods are :

(*i*) the possibility of strict localisation of nitride and carbide zones.

(*ii*) high rates of synthesis of nitride and carbide zones.

(*iii*) the possibility of producing nitride and carbide coatings with properties from those of the identical coatings produced by other methods.

4.28.9. Powder Production. Laser heated gas phase synthesis process can achieve process conditions to yield ideal powders having characteristics like small powder size, loosely agglomerated, narrow range of size distribution, spherical shape and highly controlled purity etc. Si, Si_3N_4, SiC powders have been synthesised using a 150 W CO_2 laser at a power density of 10^5 W/cm^2.

4.28.10. Laser Marking. Marking/numbering by laser is second most popular application, after laser cutting, in the industries, both metallic and ceramics material can easily be marked laser and a number of companies are marketing CNC based systems for commercial applications.

4.29. Hardening of Cam-Shaft Surfaces by Laser

A novel technology of surface melting, as a means of hardening selected surface, has been developed. The process is based on the use of high output carbon dioxide industrial lasers.

Because of their complex configuration, and because of the great wear to which camshafts are subjected, case hardening poses them serious problems. The new procedure is designed to overcome those difficulties, by means of laser powered surface melting. The system is very sensitive and flexible. Not only does it accommodate complex geometries, but can also meet the needs of different load profiles by varying penetration depth, within very close tolerances. The use of numerical control methods allows a single installation to process many different camshaft designs, without lengthy adjustment or tooling up procedures.

Throughput is very fast: up to two camshafts can be processed per minute. The new process already has been approved by at least one major automobile manufacturer. It makes it possible to maintain uniformly high product quality, while reducing scrap rates and keeping a tight rein on unit costs.

4.30. Present Status In India

In India, work on laser material processing was started in the eighties with *BARC* providing the initiative in different areas of laser material processing. Efforts in laser system development resulted in setting up a 100 W CO_2 laser for ceramic processing. *BARC* is also setting up a 1000 W CO_2 laser (axial flow type) suitable for laser material processing. Centre for Advanced Technology (CAT), Indore has now been identified as the centre for laser technology.

At BARC work on the various aspects of Laser Materials Processing have been carried out. R&D efforts have shown that Laser Material Processing is indeed an important technology needed for enhanced realization of nuclear power programme. The existing laser facility has been used for end-cap welding of thin fuel elements made of stainless steel & zircalloy-2. Acceptable quality welds have been produced with very little or almost no heat-affected-zone. Laser welds of these thin components were compared to both conventional arc and advanced electron beam welds and the former were observed to be better than the later ones in many aspects. Recently spacers have also been welded by laser on PHWR fuel rod to develop an alternative to the conventional method. Better quality welds were obtained when compared to the present resistance welding technique. Laser welds have been completely characterized by corrosion test, optical metallography, electron microprobe analyser, transmission electron microscopy, mechanical testing and useful data have been generated. Welding has been performed at low pressure to study the effect of metal-plasma on the penetration. It was observed that weld with higher penetration can be made at low pressure.

Laser cutting of stainless steel and zircalloy sheets has been carried out. Profile cutting has also been done using CNC. Drilling operation has been performed and recently fine holes were made on molybdenum sheets. Fine slits of stainless steel were also produced.

Laser glazing of zirconium alloy has been done to improve the surface properties of the alloy. Laser glazed surface is harder with martensite phase and corrosion rate of the surface depends on the loss of alloying elements due to the surface melting. Laser glazing has also been performed to desensitize the stainless steel. Laser hardening of cast irons has been done to improve the wear resistance & impact toughness of the material.

Work on laser surface alloying has been initiated. Fe-Cr-C system has been chosen for the study. The purpose of the study is to produce a stainless surface of Fe-Cr-C alloy on mild steel. Surface alloying is done by shining the laser on the Cr-coated mild steel to produce a surface alloy of Fe-Cr-C. It is observed that dilution depends mainly on the thickness of the chromium layer and the laser parameters. Corrosion tests has confirmed the stainless character of the surface alloyed mild steel. Similar experiments have been done with Fe-Mo system to produce higher Mo content in stainless for marine applications.

Experiments for nitriding of stainless steel by laser in presence of high pressure nitrogen gas has been done. It is observed that gas phase nitriding is not very effective. Recently stainless steel nitriding in liquid nitrogen media was done with success. Results are being investigated. However, gas phase nitriding has been done with Zr/Ti alloys.

Standardisation for laser deposition of super conducting thin film has been done and conditions to produce standard ($YBa_2\ Cu_3\ O_{7-x}$) oxide films have been established. However, work is to be completed to develop proper fast treatment to establish superconductivity of these films.

Other than the work mentioned above and being carried out at *BARC, Bombay,* laser laboratory has also been established in DMRL, Hyderabad with both Nd-YAG and CO_2 laser systems. The facilities are being used for processing of special materials like Ti-alloys, and conducting experiments on surface modifications. A CO_2 laser has been set up in WRI, Trichy, which is being used for welding and other applications for power industries. TELCO, Poona and Fine Lasercut Poona also have laser systems for cutting applications. Ordinance factory, Medak has recently commissioned a laser cutting systems.

Work has also been initiated to develop a suitable theoretical model for laser process using finite element method. Initially a two dimensional thermal analysis is being carried out to predict the temperature distribution during laser operation.

Thus, although, the work in LMP has been started in India (as outlined above), it has primarily been concentrated in the research laboratories. Very few industries have access to lasers partly due to their high cost and partly due to the lack of awareness & knowledge.

4.30.1. Future Trends

At present laser with 2-5 kW power are being used in the industries. High power laser upto 25 kW are being developed which will increase the speed of the processing. Processing of thick material shall also be attempted. Beam handling system and computer based job handling system with better accuracy will find its way to industries. Laser cutting and numbering will continue to be the attractive application for the industry. Laser surface modification will provide new combination of materials for many applications. New applications like laser assisted machining, laser deburring and production of ultra-fine powders shall find industrial acceptance.

4.30.2 Arc augmented laser welding (laser in conjunction with a tungsten inert gas system) to increase the welding speed without greatly affecting the quality of laser welds also needs attention. Laser application group at Culham has built a laser system with two operating stations to increase the productivity. They have also developed a flexible beam guide for use with 100-400 W laser called :Cobra: which can also be termed as 'Laser Torch'. Successful experiments have been reported by Peng Cheng Quiang of China to combine six CO_2 lasers to produce a laser beam of higher output power. Earlier, combination of two laser systems have been made at many laser application centres to augment the total power for material processing. Laser drilling with vibration by ultrasonic to increase the hole depth with good surface finish has also been tried.

In the industrial front, concept of small laser workshop has already become popular in United States. In U.K., department of trade and industry have set up a laser users club (UKAEA'S Culhum lab). USSR plans to set up laser centres for different industrial zones. In India, concept of providing laser services commercially has already become popular in diamond industries. In Poona, a small laser workshop is doing well and it appears that this approach may be successful in India. However, centralized efforts are also required to back up this approach so that talent of individual leading scientists and facilities of the different institutions can also be utilized in promoting this technology.

4.31. Line Contractor And Heat Control Ckts.

Line contractor and heat control Ckts.' is one of the types of electronic controls employed in resistance welding. The electronic controls are very accurate and precise in operation. So, this method can be expressed in detail as :-

In this, the welding transformer is connected to the line when a control signal is given and the heat is achieved by the accurate firing at ignitrons for a certain duration in each half cycle.

Working. When a thyratron conducts, the effect is to connect ignitor of the ignitron with its anode. Now, the thyratron can only conduct when the anode of the ignitron to which it is corrected is positive, if the thyratron is fired during a half cycle, an arc is established in the thyratron.

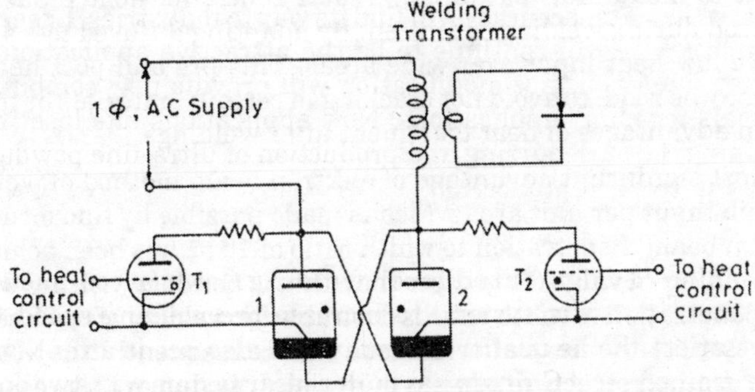

Fig. 4.30. Ignitron Contractor being fired by thyratron.

Referring to fig. 4.30 when the upper line is undergoing a positive half cycle of voltage, the anodes of ignitron 2 and thyratron T_2 are positive with respect to the cathodes. If thyratron T_2 is now fired current will pass through the ignitor of ignitron 2 and to the lower line. Ignitron 2 will therefore fire. Similar action takes places in ignitron 1 which is fired by thyratron 1.

The heat control Ckt. is such that the ignitrons conduct during a definite, predetermined angle in each half cycle only.

4.32. Electron Beam Welding

In electron beam welding, electron bambardment heating is used.

Certain materials such as tantalums, niobium and molybdenum are very much susceptible to impurities pick up in conventional inert-gas welding, welding in vacuum is the proper solution. It may however be noted that in inert gas tungsten electrode arc becomes unstable for pressure below 100 mm. of mercury. As such electron beam is the only method which can successfully be applied in vacuum Weld embrittlement specially in molybdenum is absent with this method of welding on account of two reasons, firstly the impurity pick up from weld atmosphere is nil, secondly volatile molybdenum oxide is released in vacuum.

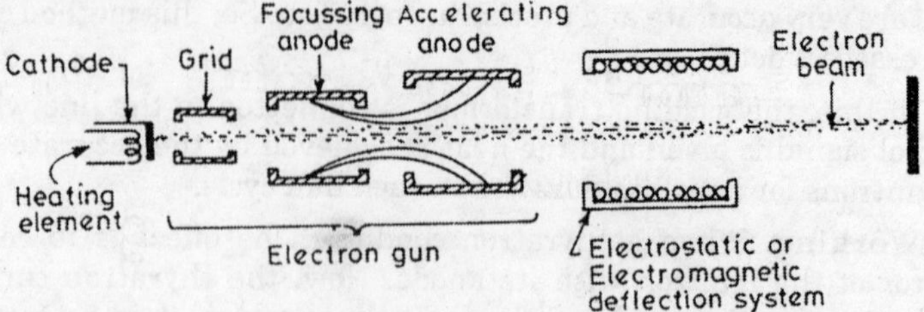

Fig. 4.31. Production of electron beam with the help of an electron gun.

By variation of beam current, voltage, position of filament with respect to modulator and focus, greater control of heat input can be exercised during welding operation. By adapting diffused focus we can achieve low heat input over wide areas. This pre and post heating of weld may be reqd. to avoid hot cracking in certain materials or to confer certain advantages of heat treatment after weldings.

Most significant advantage of electron beam method of welding is the high input per unit area which is made possible by fine focus of the electron beam. Penetration to width ratio of 10 : 1 has been achieved in butt welding - a value far greater than that is possible with arc welding. High penetration to width ratio is desirable for welding of thick sections. This restricts the heat affected zone and consequently the volume of coarse grained structure of weld material. It is due to this reason that mechanical properties of the weld are superior.

But a drawback of this type of welding is the restriction in size and movement of components to be welded owing to operation within an evacuated chamber.

Advantages of electron beam welding:. (*a*) Welding of certain non-ferrous materials results in contaminated weld if done in any gaseous atmosphere. Therefore evacuated space necessary for production of an electron beam provides an automatic solution to such problem. The example of materials requiring this type of welding is given by tantalum, niobium and molybdenum etc.

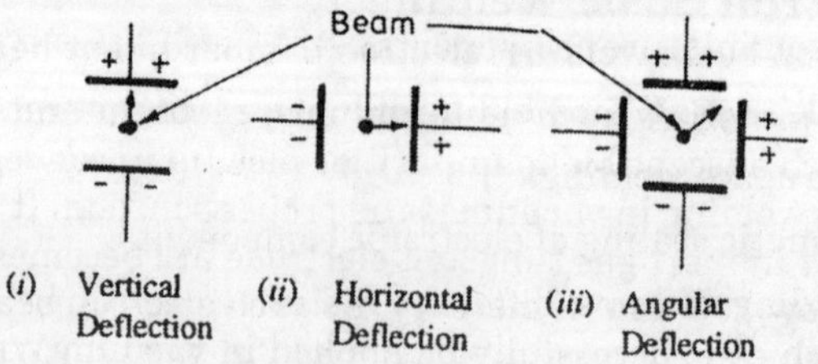

Fig. 4.32 (A). Electro Static Deflection of the beam.

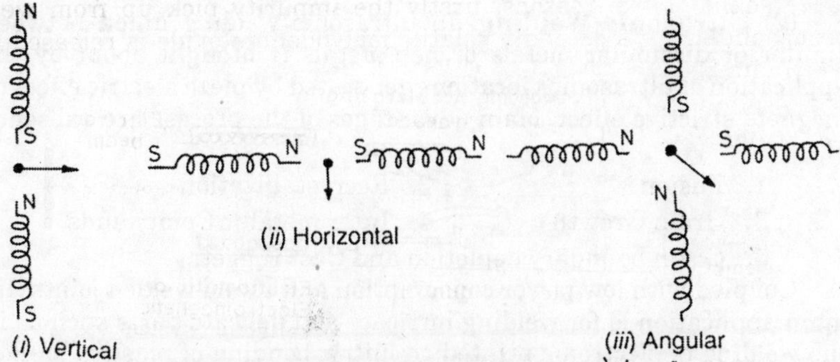

Fig. 4.32 (B). Electromagnetic deflection of the beam.

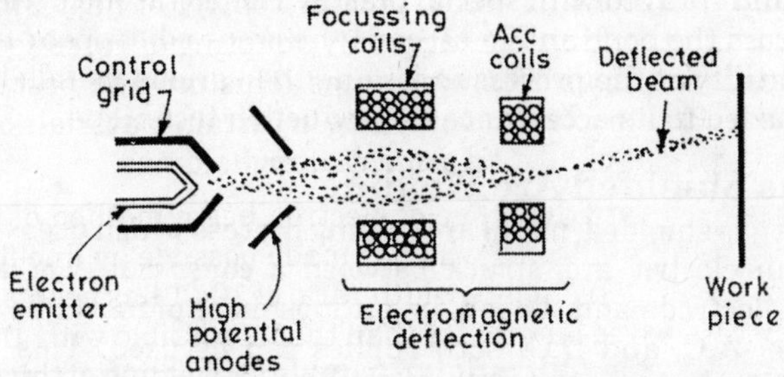

Fig. 4.32 (C). Modified version of Beam Welding unit.

(*b*) Better the focussing of electrons, deeper is the penetration and higher the heat input per unit area. It gives highly superior mechanical properties to the weld.

(*c*) Certain welds need heat treatment prior to welding and after the welding. This is at a temperature lesser than that of the welding temperature. It is quire convenient, by simply controlling the intensity and defocussing the electron beam to provide only heating temperature as and when desired.

(*d*) A precise control of the spot and line of weld is possible by this method.

Applications :

(*i*) Microminiature components.

(*ii*) Nuclear engg. for welding of fuel elements of reactors.

(*iii*) Aero space welding.

(*iv*) Hermetic sealing of electronic components.

(*v*) Welding of some different material components with each other.

(*vi*) Inner circuitry of microprocessor chips and IC's etc.

(2) **Ultrasonic Welding.** In ultrasonic welding union between similar or dissimilar metals or non-metals is brought about by the application of ultrasonic vibrations generated by piezo electric effect or megneto strictive effect. Main advantages of the process are : absence of,

1. Fusion
2. Recrystallization
3. Grain Growth
4. Inter metallic Compounds.
5. Grain boundary depletion and Cast rugget.

Coupled with low power consumption and sound welded joints. Its main application is for welding intricate and delicate parts such as in the welding of electronic printed circuitry, bonding of plastics, bi-metals, foils etc. It has been reported that in Soviet Union successful operations have been carried out in which bone fragments are put into position and sprayed with special plastic. The ultrasonic welder is then taken across the bone and in seconds the cracks disappear from sight. The versatility of the process in joining the various metals has shown that it has led to its acceptance as production technique.

4.33 Gas Shielded Arc Welding

It is a gas shielded, metal arc velding process which uses the intense heat of an electric arc, struck between a consumable or a non-consumable electrode and the job, in a atmosphere of inert gas like orgon or helium. A.C. and D.C. both suplies can be used. This process is employed for welding light alloys, non-ferrous metals such as copper, aluminium and their alloys.

SHORT QUESTIONS WITH ANSWERS.

Q.1. *What is welding ?*

Ans. Welding is a process in which metals are joined by heating them to a suitable temperature, with or without the application of pressure and addition of filler meterial.

Q.2. *What is the difference between plastic welding and fusion welding?*

Ans. In plastic welding, the pieces of metals to be joined are heated to a plastic state and then forced together by external pressure whereas in fusion welding, the material at the joint is heated to a molten state and allowed to solidify.

Q.3. *What is resistance welding?*

Ans. Resistance welding is that process in which heavy current is made to pass through the joint to be welded. This heavy current generates enough heat at the contact point. This heat which is due to I^2R loss melts the metal and causes fusion at the point of contact.

Q.4. *What is the difference between electric arc welding and resistance welding?*

Ans. The resistance welding process differs from arc welding, in that pressure is used but filler metal or fluxes are not used.

Q.5. *Why alternating current is found most suitable for resistance welding?*

Ans. Alternating current is found most suitable for resistance welding as it can provide any desired combination of current and voltage by means of a transformer.

Q.6. *What is meant by electrode?*

Ans. An electrode is a piece of wire or rod (*of a metal or alloy*), with or without flux covering, which carries current for welding.

Q.7. *What is the difference between carbon arc welding and metallic arc welding?*

Ans. In carbon arc welding, an arc is struck between a carbon electrode and the piece to be welded. A filler rod is used as an additional material, if required, whereas in metallic arc welding, a metal rod of the same material forms one of the electrodes and no filler rod is used because the electrode itself is used as a filler material.

Q.8. *What type of electric supply is suitable for electric arc welding?*

Ans. Both D.C. and A.C. supplies are used for electric arc welding.

In case of D.C., a motor generator set is used. The motor is generally a *Squirrel cage motor* and the generator is *Differential Compound* which has the drooping characteristics.

In case of A.C., a step down welding transformer is used. It is low voltage high current transformer. The stable arc can be obtained by putting a choke of high reactance in series with the secondary of the transformer which limits the current.

Q.9. *What is main draw back of electron beam process?*

Ans. Its high capital cost is the main draw back of electron beam process. Due to the requirement of vaccum pumps, the price of the equipment required is very high and it is expensive to operate.

Q.10. *In laser beam welding, the location of the focal point of the beam w.r.t. the surface of the work piece is very important. Why?*

Ans. When the beam is focused slightly below the surface area, the maximum penetration occurs. When the beam is focused on the surface or deep within the surface area the penetration is less.

OBJECTIVE TYPE QUESTIONS

1. Fill in the blanks in following sentences:

(*a*) In electric welding heat is produced either by --------or by --------

(*b*) There are two main methods of electric welding ----------and--

(*c*) In resistance welding current is passed through -------

(*d*) the pipes, rods and wires can be welded easily by -------

(*e*) In arc welding, heat is produced mainly by -------

(*f*) Resistance welding ---------- employed for production work.

(*g*) The supply source used for arc welding must have ------ characteristics.

(*h*) The power factor of the welding transformer is ------

(*i*) In the plastic or ------ welding, the pieces of meal to be joined are heated to plastic state and then forged together under external pressure.

(*j*) Project welding uses the same equipment as the ----- welding.

(*k*) The welding transformer must have a --------- leakage reactance.

(*l*) In atomic hydrogen welding, the electrodes used are made of ------

(*m*) TIG welding is ------- welding.

(*n*) MIG welding is ------- welding.

(*o*) The D.C. generator used in M.G. set employed in arc welding is of ----- type.

(*p*) ------- supply is used in atomic hydrogen welding.

(*q*) ------- welding can be defined as series of continuous spot welds.

(*r*) The transformer used in a welding set is ------ transformer.

(*s*) A 10 SWG electrode has approximate diameter of -----

(*t*) Butt welding is used for welding -----.

(*u*) The four different types of resistance welding process are ----.

Answers:

1. (*a*) By passing electric current, striking an arc

(*b*) Resistance welding, arc welding

(*c*) the metal pieces to be joined (*d*) butt welding

(*e*) electric arc (*f*) can be

(*g*) drooping (*h*) very low

(*i*) Pressure (*j*) spot

(*k*) High (*l*) Tungsten

(*m*) Tungsten inert gas (*n*) Metal inert gas

(*o*) Differential compound wound (*p*) A.C.

(*q*) Seam (*r*) Step down
(*s*) 3.3 mm (*t*) metal pieces.
(*u*) Butt, spot, projection and Seam.

2. Write 'Yes' or 'No' after the following sentences:-

(*a*) Resistance welding can be used for repair work.

(*b*) High temperatures can not be obtained through arc welding.

(*c*) Spot welding can be done without applying any pressure in electrodes.

(*d*) Projection welding can be employed for production work.

(*e*) In arc welding, when the current through the arc increases, its resistance also increases.

(*f*) Drooping characteristic is necessary for the supply source to be used for arc welding.

(*g*) Arc welding used consumable electrodes.

(*h*) The efficiency of a.c. welding set used for arc welding is low as compared to d.c. set.

CHOOSE THE CORRECT ANSWER

3. Arc blowing results in :

(*a*) Better coverage of weld surface
(*b*) Increased penetration of heat.
(*c*) Decline in current density, decreased heat penetration and poorer quality of welding.
(*d*) None out of above.

4. Rolling electrodes are specifically used for :

(*a*) Projection welding. (*b*) Seam welding.
(*c*) Spot welding. (*d*) Flash butt welding.

5. The secondary of three phase welding transformers is:

(*a*) Delta connected (*b*) Star connected
(*c*) Depends upon welding current requirement.
(*d*) None out of above.

6. Some of the welding techniques have indirect utilisation of electrical energy. These can be :

(*a*) Seam welding, flash butt welding.
(*b*) Ultrasonic, laser beam welding.
(*c*) TIG, MIG welding.
(*d*) Stored energy, projection welding.

7. Projection welding is a process :

(*a*) Identical to spot welding in characteristic
(*b*) More similar to steam welding
(*c*) Entirely different from above two.

8. It is through the :

(*a*) Cross magnetising effect of armature reaction
(*b*) Demagnetising effect of a armature reaction
(*c*) None of the two effects; that stabilisation of the arc is achieved in above mentioned generator.

9. The normal metal arc welding makes use of :

(*a*) some filler material in ferrous jobs.
(*b*) Non-ferrous electrodes for non-ferrous work
(*c*) Both as above.
(*d*) None as above.

10. Electronic beam welding is most suitable for :

(*a*) Massive structures
(*b*) Mass production of moderate jobs
(*c*) Minor but sophisticated works without any atmospheric contamination.
(*d*) None as above.

11. D.C. welding and A.C. welding have some different opposing characteristics. These include:

(*a*) Nature of sources (*b*) Effect on line regulation
(*c*) Reactors for arc stability (*d*) control panels necessity.
(*e*) magnitude of energy losses
(*f*) necessity of capacitors
(*g*) All as above.

12. MIG welding needs :

(*a*) Consumable electrodes
(*b*) Non-consumable electrodes without filler
(*c*) Non-consumable electrodes with filler
(*d*) Something other than above.

13. The suitable D.C. supply source for arc welding can be :

(*a*) Low voltage high current stacks of selenium rectifiers.
(*b*) A low voltage high current D.C. shunt generator.
(*c*) A cumulative compound D.C. generator with a strong series field.

(*d*) None as above. (*e*) All as above.

14. In electric arc produced for welding purposes, the hottest part is the;

(*a*) Cathode spot (*b*) Anode spot.

(*c*) Arc stream (*d*) work piece.

15. The electrodes used for TIG welding are :

(*a*) Uncoated rods which are non consumable

(*b*) Coated rods

(*c*) These are not rods but a continuous electrode rolled on a drum. The electrode wire is flux coated.

(*d*) Something other than above.

16. TIG welding needs :

(*a*) Consumable electrodes

(*b*) Non-consumable electrodes without filler

(*c*) Non consumable electrodes with filler

(*d*) Something other than above.

17. Arc blow is caused by :

(*a*) Magnetic effects arcing current

(*b*) Presence of ferrow-magnetic material within arc vicinity.

(*c*) Some other stronger magnetic field or evolutin of hot gases from the weld

(*d*) all as above

(*e*) None as above.

18. The electrodes used for MIG welding are :

(*a*) Uncoated rods which are non consumable

(*b*) Coated rods

(*c*) These are not rods but a continuous electrode rolled on a drum. The electrode wire is flux coated.

(*d*) Something other than above.

19. The source of power in MIG welding is :

(*a*) Rectified A.C.

(*b*) Rectified A.C. with filtering unit

(*c*) Pulsating D.C. with minimised ripple without incorporating filter circuit

(*d*) Can be A.C. or D.C.

20. The gas used in TIG welding is involved in :

(*a*) Heat generation (*b*) Production of flame.
(*c*) It is only for production of plasma enveloping the arc.
(*d*) All as above.

Ans.

2. (*a*) No. (*b*) No. (*c*) No. (*d*) Yes. (*e*) No. (*f*) Yes. (*g*) Yes. (*h*) No.

3. (*c*)	**4.** (*b*)	**5.** (*d*)	**6.** (*b*)	**7.** (*a*)
8. (*b*)	**9.** (*d*)	**10.** (*c*)	**11.** (*g*)	**12.** (*a*)
13. (*d*)	**14.** (*a*)	**15.** (*a*)	**16.** (*c*)	**17.** (*d*)
18. (*c*)	**19.** (*c*)	**20.** (*c*)		

TRUE OR FALSE STATEMENTS

1. Spot welding process can be done without applying any pressure on electrodes.
2. Arc can be produced either be a.c. or d.c.
3. For the electric arc welding the range of open ckt. Voltage is normally 100-250V.
4. In arc welding, when the current through the arc increases then its resistance also increases.
5. For thick metal sheets, spot welding is used.
6. Arc welding always uses consumable electrodes.
7. For small quantities, the lower cost of equipment makes spot welding economical.
8. The efficiency of a.c. welding machine used for arc welding is low as compared to d.c. machine.
9. The carbon arc is easy to maintain and is very stable.
10. Both d.c. and a.c. can be used for carbon arc welding.
11. The length of the carbon arc can be varied over a wider limits without causing the arc to go out.
12. Drooping characteristics are necessary for the supply source to be used for arc welding.
13. The load taken by welding transformer is highly inductive.
14. The power factor of the load using welding transformer is normally very low.
15. Post welding cleansing is required in MIG welding.

Answers

1. False	**2.** True	**3.** False	**4.** False
5. False	**6.** True	**7.** True	**8.** False
9. True	**10.** False	**11.** True	**12.** True
13. True	**14.** True	**15.** False.	

MULTIPLE CHOICE QUESTIONS

1. Electric arc welding produces temperature upto

(*a*) 5550°C (*b*) 3600°C
(*c*) 2000°C (*d*) 1000°C

2. The purpose of using coating on the welding electrodes is

(*a*) to produce a cloud of gas around the molten weld puddle thereby preventing the formation of oxides and nitrides of iron which make the weld weak and brittle.

(*b*) to produce formation of slag on the surface of molten metal which cools slowly thereby preventing the weld from becoming brittle

(*c*) to produce shielding gas which keeps the arc gap ionized. This maintains electrical conductivity of the arc path during zero arc periods increasing its stability

(*d*) to introduce special additives to alter the composition of the weld deposit

(*e*) to discharge all above functions

3. Increased heat due to shorter arc is harmful on account of

(*a*) burn through (*b*) excessive porosity

(*c*) under cutting of base material

(*d*). all of above

4. The problem with longer arc are

(*a*) less heat resulting in poor penetration

(*b*) exposure of deposited metal to oxidation increased, causing brittleness

(*c*) erratic and unstable arc

(*d*) all above

5. Magnetic arc blow is a problem with

(*a*) only d.c. arc welding (*b*) only a.c. arc welding

(*c*) with both a.c. and d.c. (*d*) neither d.c. nor a.c.

6. Major personal hazards in arc welding are

(*a*) harmful infra—red and ultra—violet rays from the arc

(*b*) flying sparks

(*c*) weld spatter (*d*) all of above

7. Problem of magnetic arc blow is severe in the case of welding at

(*a*) joint ends (*b*) corners

(*c*) complex or massive structures

(*d*) all above positions

8. Arc blow results in

(*a*) splashing out of metal from weld puddle

(*b*) non uniform weld beads

(*c*) shallow weld puddle giving rise to weak weld

(*d*) all above defects

9. Following form the resistance welding

(*a*) spot welding (*b*) seam welding

(*c*) projection welding (*d*) butt welding

(*e*) all above

10. The arc has

(*a*) linear resistance characteristics

(*b*) positive resistance characteristics

(*c*) negative resistance characteristics

(*d*) highly inductive characteristics

11. Arc can be produced by

(*a*) A.C. current only (*b*) D.C. current only

(*c*) either A.C. or D.C. current

(*d*) neither A.C. or D.C. current

12. A D.C. generator used for a.c. welding should have

(*a*) rising characteristics (*b*) drooping characteristics

(*c*) straight characteristics (*d*) all of above

11. The welding electric circuit is

(*a*) always earthed (*b*) never earthed

(*c*) through cables only (*d*) sometime earthed

14. The advantages of resistance welding is

(*a*) less skill is required (*b*) reduce distortion

(*c*) none of above (*d*) all of above

15. In dielectric heating current flows through

(*a*) metallic conductor (*b*) dielectric

(*c*) air (*d*) none of these

16. For arc heating, the electrodes are made of

(*a*) copper (*b*) aluminium

(*c*) graphite (*d*) ACSR conductor

Answers:

1. (*a*)	**2.** (*e*)	**3.** (*d*)	**4.** (*d*)	**5.** (*a*)
6. (*d*)	**7.** (*d*)	**8.** (*d*)	**9.** (*e*)	**10.** (*c*)
11. (*c*)	**12.** (*b*)	**13.** (*a*)	**14.** (*c*)	**15.** (*b*)
16. (*c*)				

EXERCISES

1. Define electric welding.
2. Name and explain the various welding methods.
3. Describe the source of welding current.
4. What is the difference between carbon arc welding and metallic arc welding ?
5. What do you understand by the negative resistance characteristics of an electric arc and in what way can an electric arc be stabilised ?
6. Mention the different types of welding electrodes.
7. What are the advantages of coated welding electrodes ?
8. Explain briefly the different welding processes under resistance welding. Why is it necessary to use 'welding transformer'.
9. Mention the three types of resistance welding. State their common applications.
10. What are the qualities of a good weld.
11. (*a*) What is resistance welding ?
    (*b*) What type of electric supply is suitable for electric arc welding ?
    (*c*) What properties would you seek for selection of electrode material for spot welding ?
    (*d*) How is the electrode in spot welding prevented from striking to the workpiece ?
12. Give the difference between the following :
    (*a*) Arc welding and resistance welding.
    (*b*) D.C. welding and A.C. welding.
13. Write a short note on arc stability.
14. Explain straight and reverse polarity in D.C. welding.
15. Define arc length and name the advantages of short arc length over long arc length.
16. What are the requirements of a good welding ?
17. What points should be kept in view to avoid the weld defects ?

Questions Bank with Answer

Short Answer Type Questions.

Q.1. What is welding?

Ans. Welding is the process in which metals are joined by heating them to a suitable temperature with or without application of pressure and addition of filler material.

Q.2. What is electrical welding?

Ans. The welding process in which electric current is used to produce a large amount of heat required for joining two pieces of metal is called electrical welding.

Q.3. Name the two forms of electrical welding.

Ans. The two forms of electric welding are

(i) Resistance welding (ii) Arc welding

Q.4. What is the resistance welding?

Ans. Resistance welding is that process in which heavy current is made to pass through the joint to be welded. This heavy current generates enough heat at the contact point. This heat which is due to I^2R loss melts the metal and causes fusion at the point of contact.

Q.5. What is arc welding.

Ans. An arc welding is a fusion welding process. The fusion of metal is brought about by heat, which is generated by establishing an arc between an electrode and the metal part to be welded. Both ac as well as dc can be used for the production of the arc, however dc gives a more steady arc and thus a uniform weld. But ac is economical and is usually preferred for arc welding.

Q.6. What is the difference between electric arc welding and resistance welding ?

Ans. In resistance welding process pressure is used but filler metal or flux is not used but in arc welding pressure is not applied but filler metal or flux is used.

Q.7. Why alternating current is found most suitable for resistance welding?

Ans. Alternating current is found most suitable for resistance welding as it can provide any desired combination of current and voltage by means of a transformer.

Q.8. What types of electric supply is suitable for electric arc welding?

Ans. Both D.C and A.C supplies are used for electric arc welding.

In case of D.C., a motor generator set is used. The motor is generally a Squirrel cage motor and the generator is a differential compound which has the drooping characteristics.

In case of A.C., a step down welding transformer is used. It is low voltage high current transformer. The stable arc can be obtained by putting a choke of high reactance in series with the secondary of the transformer which limits the current.

Q 9. Explain negative resistance characteristic of electric Arc.

Ans. The most important feature of electric arc is that it has –ve volt ampere characteristics, *i.e.,* its resistance decreases with increase in current. Due to this characteristic, it is not possible to obtain a steady arc. To over come this difficulty, the supply to the arc must be such that the voltage decreases rapidly when the current increases, *i.e.,* the source of supply has drooping characteristics as shown in fig. 4.1

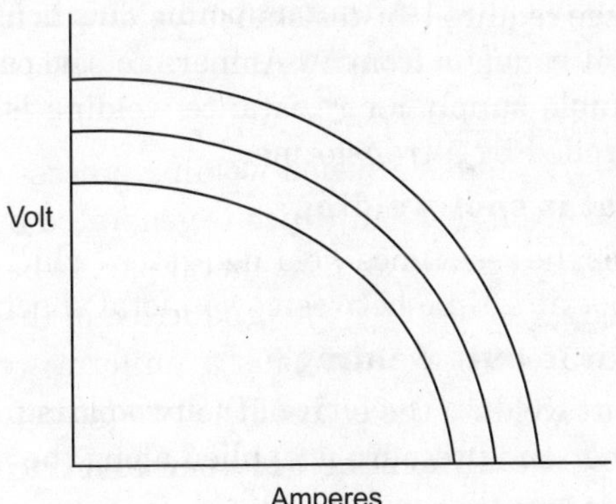

Fig. 4.33

Q.10. What is the principle of Arc production?

Ans. Electric arc is produced when contact is first made between the electrode and workpiece to complete the circuit and then separating the two by a small distance. Arc is a sort of an electric discharge through a path of ionised particles of air and produces an immense heat if continuous arc is maintained.

Q.11. What is a welding electrode ?

Ans. An electrode is a piece of wire or rod of a metal or alloy with or without flux covering, which carries current for welding.

Q.12. What are different types of resistance welding ?

Ans. (i) Butt welding (ii) Flash butt welding
(iii) Spot welding (iv) Projection welding
(v) Seam welding

Q.13. Name the two types of arc welding.

Ans. The two types of arc welding are
(i) Carbon arc welding (ii) Metal arc welding

Q.14. What is the difference between carbon arc welding and metallic arc welding ?

Ans. In carbon arc welding, an arc is struck between a carbon electrode and the piece to be welded. A filler rod is used as an additional material if required. In metallic arc welding, a metal rod of the same material forms one of the electrodes and no filler rod is used. The electrode itself is used as a filler material.

Q.15. What are the power supply requirement for resistance welding?

Ans. Voltage required for resistance welding is from 2 to 12 volt and the current range is from 50 Ampere to several 100 Ampere. The most suitable supply for resistance welding is AC supply as it can be controlled by a transformer.

Q.16. What is spot welding?

Ans. This is the resistance welding process which produces only one welded spot at a time between two metal sheets.

Q.17. What is butt welding?

Ans. In butt welding the ends of the two parts to be joined are brought together and pressure is applied along the axial direction.

Q.18. What is seam welding?

Ans. This is type of resistance welding process which makes continuous seam of weld on two overlapping metal sheets to be joined.

Q.19. What is percursion welding?

Ans. This is the type of electrical welding which utilizes electric arc for the production of heat and the pressure is utilized for producing joint.

Q.20. What is difference between spot and butt welding?

Ans. In spot welding the sheets to be welded are held overlapping between the two electrodes, one movable and the other fixed. Heavy current is passed through the electrodes and adequate pressure is applied at the spot where joint is to be produced.

In butt welding, the two parts to be welded are held in clamps one of which is fixed and other is movable. The ends of the two parts to be welded are brought together and pressure is applied along the axial direction. Heavy current is passed which generates required heat at the joint due to resistance of contact area.

Q.21. TIG stands for .... ***[June 2008]***

Ans. Tungustan inert gas arc welding.

Q.22. MIG stands for ....

Ans. Metal arc inert gas welding.

Q.23. What is TIG welding?

Ans. It is Tungsten Inert Gas arc welding. It is a type of arc welding in which an inert gas shield is provided around the arc. The arc is created between tungsten electode and work piece.

Q.24. In TIG welding electrodes are .... (Consumable/ Non consumable).

Ans. Non consumable.

Q.25. .... Electrodes are used in TIG welding.

Ans. Tungsten

Q.26. What is MIG welding?

Ans. It is gas metal Arc welding. In MIG welding a consumable electrode in an inert gas atmosphere is used to produce a weld. Therefore it is also called metal inert gas (MIG) welding.

Q.27. .... Electrodes are used in MIG welding.

Ans. Consumable.

Q.28. Which type of power supply is required for MIG welding?

Ans. DC with straight polarity.

Q.29. Which gas is commonly used in MIG welding?

Ans. Inert gas like Argon, Helium, Carbon-dioxide or a mixture of these gases. Now CO_2 is most widely used.

Q.30. Differentiate between TIG and MIG welding.

Ans. TIG welding is a type of arc welding in which an inert gas shield is provided around the arc. The arc is created between tungston electrode and work piece. In MIG welding a consumable electrode in an inert gas atmosphere is used to produce a weld.

Q.31. What is the welding control circuit.

Ans. Electrical circuit to control current, temperature and duration of current is called welding control circuit.

Medium/Long Answer Type Questions

Q.1. What are the advantages of electrical welding over gas welding?

Ans. Few advantages of electric welding over gas welding are:

(i) Electric welding gives permanent joint which is cheaper and economical.

The electrical energy is cheapr than gas. The H_2 and C_2H_2 both gases are costly.

(ii) The electric welding is safe to human health while gases may be harmful.

(iii) The electric welding is a neat and clean process.

(iv) The electrical energy can be controlled easily through a simple switch.

(v) Electrical energy gives more efficiency.

(vi) The electric welding takes lesser time as compared to the gas welding.

(vii) Welds made by resistance welding are cleaner than gas welding.

(viii) The welds made by resistance welding are localized.

(ix) The electric welding affords lighter weight through better utilisation of materials.

(x) It can join all commercial metals.

(xi) It can be done every where.

(xii) It provides design flexibility.

Q.2. Explain in brief the principal of electrical welding.

Ans. It is based on the well known principle that heat is generated by the resistance offered to the flow of electrical current. The amount of heat generated in the workpiece depends on the magnitude of the current, the resistance of the current conducting path and the time for which the current is allowed to flow. This is given by the relation:

$$H = \frac{I^2Rt}{J}$$

where H = Heat generated, in joules

I = Current, in amperes

R = Resistance, in ohms.

t = Time of current flow in seconds.

J = Joules Equivalent of heat

The heat generated is directly proportional to the resistance offered by any point in the circuit. Since the interface of the two joining surfaces is, the point of greatest resistance, it is also the point where greatest heat is generated. In simple resistance welding, a high-amperage, low voltage current is passed from one adjoining workpiece to the other, until metal becomes heated to a temperature which is high enough to cause localised fusion and the formation of a weld nugget. Then, pressure is applied, after the current is off, to sequeeze the two parts into a localized homogeneous mass.

Q.3. Draw and explain welding control circuit.

Ans. In resistance welding, many factors are to be controlled such as : Current, Pressure, Heat, Time. Manual control for the these may not be reliable. For precise control, electronic control circuits are used. The fig. shows two ignitron tubes used for controlling. The tubes conduct one by one for each half cycle of the supply. For the first half cycle, say ignitron II conducts. The voltage across this tube drops to low value and it will conduct current for a time enough to strike the arc. After this time, the tube will stop conducting. In next half cycle ignitron I will conduct and so on. Rectifiers are used to conduct current in proper direction and saivng ignitrons from reverse current damages.

A solenoid is used to control the pressure on the upper moving electrode. Since currents through the tubes may be upto 1000 amp, the ignitrons need proper cooling, for this generally water cooling is used.

An electronic circuit, is used in conjunction with the above circuit to delay the conduction of ignitrons by an angle to reduce the weld current and hence to control heat.

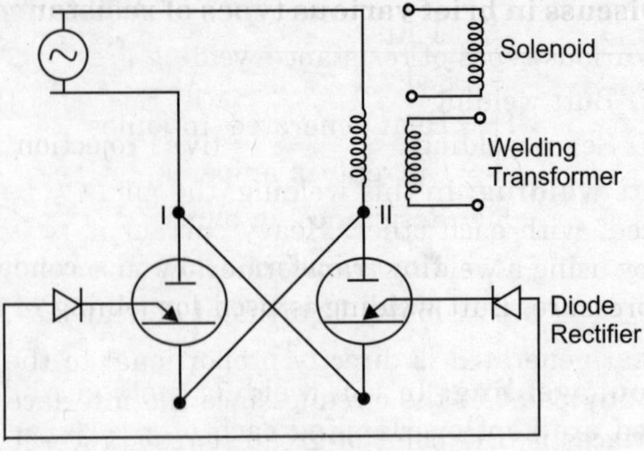

Fig. 4.34

Time control circuit is also an electronic circuit which works in conjunction with the above ignitron circuit.

Q.4. What is welding Transformer? Explain.

Ans. AC welding transformer is a step down transformer which reduces the voltage from that of supply mains to about 100 V. The stable arc is obtained by connecting a reactance in series with the secondary of a transformer. As soon as the current increases, the voltage drop in the reactor also increases. This reduces the arc current. The reactance reduces the power factor and increases the cost of the equipment. The resistance can also be used for current control but its use will reduce the efficiency of the system. Fig. 4.35 shows the arc welding transformer.

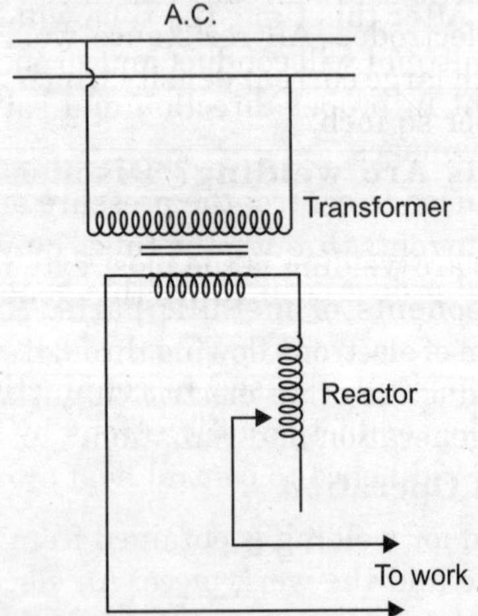

Fig. 4.35

Q.5. Discuss in brief various types of resistance welding.

Ans. Various types of resistance welding are

(i) Butt welding (ii) Spot welding

(iii) Seam welding (iv) Projection welding

(i) Butt welding: In this welding, the parts to be joined are kept "butted" with each other. Heavy current is passed through the parts by using a welding transformer. Later on they are joined by using pressure. Butt welding is used for joining of rods, pipes and wires.

(ii) Spot welding: In this welding method, the two parts to be joined are kept overlapping each other. Heavy current is passed through the parts by using a welding transformer. Later on pressure is applied by electrodes kept on the joint. After this, the spot is changed and a weld is obtained on the next spot. In this way complete weld is obtained spot by spot.

(iii) Seam welding: In this method, the parts to be joined are pressed between " Wheel electrodes". Current is passed and a seam (line) of spot welds is obtained. There may be 500 spots per meter. This method gives a leak proof joint. At the contact surface, weld ruggets formed continuously give a seam weld due to circular electrodes moving continuously under pressure at uniform velocity.

As it gives leak proof joint, this method is used for welding of Tanks, Transformers and Aircrafts.

(iv) Projection welding: It is similar to spot welding except that flat electrodes are used. This is suitable for punched or stamped parts where projections are already existing. The projections serve the purpose of electrodes. All resistance welding processes need 6-12 V supply with large current density which may be upto several thousand amp. per sq inch.

Q.6. What is Arc welding? Discuss its principle of operation.

Ans. Electric arc welding is the most extensively used method for joining components of metallic parts. An electric arc is a continuous stream of electrons flowing through some sort of medium between two conductors of an electric circuit and is accompanied by intense heat generation and radiation.

Principle of Operation

Heat required for welding is obtained from arc struck between a coated electrode and the work piece. An electric arc is obtained in following ways.

(i) Between a consumable electrode and the work piece.

(ii) Between a non-consumable electrode (carbon graphite or tungsten) and the work piece.

(iii) Between two non-consumable electrodes.

The arc melts the electrode end and the job and material droplets are transferred from the electrode to the job, and are deposited along the joint to be welded. The flux coating metals produce a gaseous shield and slag to prevent atmospheric contamination of the molten weld metal.

To strike an arc, the electrode is brought in contact with the work at the point where the welding is to be started. After a light contact the electrode is immediately withdrawn to a distance of (2 to 4 mm) from work. The following fig illustrates the principal of arc welding.

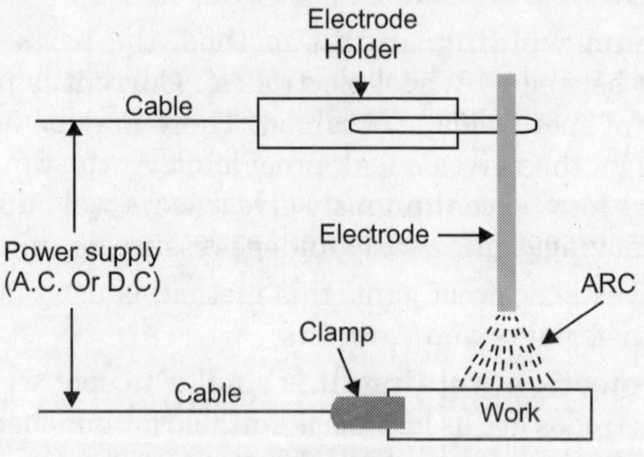

Fig. 4.36

Q.7. Give a list of equipment required for arc welding ?

Ans. The various equipment needed for arc welding includes the following items: **(1) Power equipment:**

(A) DC equipment: It may include.

(i) AC motor - DC Generator welding set

(ii) I.C. engine - DC Generator welding set

(iii) Transformer - Rectifier welding set

(B) AC Equipment

(C) Accessories :

Welding transformer set

(i) Cables,

(ii) Electrode holders or stringers

(iii) Ground clamps, Chipping hammer, wire burns

(ii) Safety equipment :

(1) Face shield/screen (2) Protective clothing

(3) Protective shoes (4) Exhaust fan etc.

Q.8. Distinguish between A.C. and D.C. arc welding.

Ans.

	A.C. arc welding	D.C. arc welding
1.	A welding transformer is cheaper than a d.c. generator.	A d.c. generator is costlier than a welding transformer
2.	A welding transformer is simpler in operation than a d.c. generator welding set.	A generator welding is more combersome in operation
3.	Maintenance of an a.c. transformer is easier and more economical since it has no moving parts.	A d.c generator carries moving parts and its maintenance cost is higher due to wear and frequent replacements needed.
4.	It is less suitable for use at low current with small diameter electrodes.	It is better suited for use at low current with small diameter electrodes..
5.	Maintenance of a short arc is difficult, except with iron power electrodes.	Maintenance of a short arc is easier.

Q.10. Write short note on

(i) TIG welding **(ii) MIG welding**

Ans. (i) TIG welding: In this process, electrode is held in an electrode holder. The electrode holder holds the electrode and ensures the supply of inert gas around the electrode. This creates an inert atmosphere aroung the electrode. Cooling during welding can be done by supplying water on air through the electrode holder. In this process no filler metal is used. This type of welding is applied to join aluminium-alloy, copper-alloys, magnesium-alloys, nickel-alloy etc. It protects the joint against oxidation. It also avoids the formatin of nitrides.

(ii) MIG welding: In this process, a continuous wire acts as an electrode. This wire is held by an electrode holder which feeds the wire into the arc space continuously. Feeding of the wire is controlled by an electric motor. The electrode holder also ensures the supply of inert gas to create an inert atmosphere around the arc. This type of welding process is applied for joining low temperature alloy sheets, aluminium alloys, magnesium alloys and copper alloys.

Q 9. What is difference between resistance welding and arc weldinc. Give specific applications of each?

Ans. Difference between resistance welding and arc welding.

	Resistance welding	Arc welding
1.	Heat is mainly developed due to flow of current through the contact resistance	Heat is developed due to an arc formed between the electrode and work piece.
2.	Supply employed is usually a.c	Supply can be a.c or d.c
3.	External pressure is required and the equipment is simple	No external pressure is required and easy to control
4.	Low voltage is required striking the arc	High voltage is required for
5.	Weld is obtained at low temperature.	Arc temperature is very high
6.	Cannot be used for repair work.	Most suitable for repair work.
7.	Power factor is fairly good	Power factor is very low
8.	Most suitable for mass production	Not suitable for mass production
9.	Roller, flat or bar type electrodes are used which are not consumable.	Flux coated metal electrodes are used which are consumable.

Applications of resistance welding : Resistance welding is used in welding pipes, wires, rods, manufacture of automobiles, refrigerators and sheet metal assemblies, welding studs, nuts etc. to metal parts and for continuous weld between metal parts.

Applications of Arc welding : Arc welding is perferably employed for welding of non-ferrous metals such as brass, copper and their alloys.

5 Illumination

5.1. Introduction

The major difference between illumination engineering and other forms of engineering is that whereas other forms of engineering largely deal with the exact sciences of mathematics, physics and chemistry with the economic factors always in mind, the illumination engineering should not only interest itself in the exact sciences and economics but also be thoroughly conversant with the physiology of eye, the peculiarities of our seeing process and psychological effects. An illumination engineer must also have appreciation of art and architecture in order to produce pleasing effects.

Just as brick, stone, steel etc. are used as structural material because of their typical characteristics, lighting should also be considered as a structural material to be used in a building to help it serve various functions. Lighting should be designed by the architect as definitely as when he uses other building materials. Unfortunately, it is a common practice, today, to treat lighting like furniture and decoration equipment that can be added after the building has been designed and completed. Even though it is difficult to visualise lighting as a structural element, this consideration must eventually exist as a conviction in the mind of the designer.

Poor lighting results in dangerous effects like eye strain, headaches, accidents due to insufficient lighting or to glare. Even then, it is astonishing to find that not much attention is paid to this aspect of engineering. If a large structure like a dam or a bridge or a transmission network is to be constructed, an engineer is always consulted but if a large public building is to be illuminated, the job is normally entrusted to a 'practical' man who uses rule-of-thumb methods rather than the scientific methods. Artificial sources of high luminosity are placed directly in the field of vision or so placed that reflected glare will be experienced from polished surfaces.

5.2. The Nature of Radiation

The question, what light is has puzzled mankind since time immemorial. The ancient Greeks thought of light as consisting of a stream of extremely minute corpuscles which were shot from a hot body which when impinged human eye produced some sensation known as light. Corpuscles theory was generally accepted till about 1850 even though Huygen had put forth his theory of waves in the year 1690. The corpuscles theory gave a satisfactory picture of reflections because if corpuscles could be considered as tiny perfectly elastic spheres, they would bounce off a polished surface with the angle of reflection equal to angle of incidence. This theory, however, was unable to explain the phenomenon of interference, where the addition of two lights may produce total darkness. But the wave theory also had its problems. To account for the motion of the light waves through free space, physicist thought of a medium - the ether. Later mathematicians proved that such a hypothetical medium must have quite contradictory properties and the idea of its existence was gradually given up. Also, the photoelectric effects could not be explained by wave theory.

It was found necessary to introduce a new fundamental principal which always applies when radiation is emitted or absorbed by matter. Energy is emitted and absorbed only in discrete quanta of magnitude hf, where h = Plank's constant ($h = 6.547 \times 10^{-27}$ erg-sec.) and f = frequency in Hz.

An extensive research in this field has revealed the existence of such discrete bundles or quanta of energy known as photons. Radiant energy is always emitted or absorbed in bundles and not continuously as envisaged in wave theory. It is to be noted that the valence electrons are responsible for the production of radiation in the visible region and these electrons in the outermost orbit are of utmost importance to the illuminating engineer. The energy of an atom is constant as long as the electron stays in a given stationary orbit and the atom is said to be in stationary state. Radiation is produced whenever an electron falls from an outer stationary orbit to one nearer the nucleus. Whenever, such a transition occurs, a single quantum (photon) of radiation is emitted. Thus the radiation from an atom is not continuous but is in the form of a quantum whenever an electron moves from one stationary orbit to another. But in practice as we deal with a large number of atoms, the radiation seems to be continuous in time though the individual atoms exit only in discrete quanta. Whenever an atom is excited *i.e.* energy is added to the atom is some way such as by collusion with an electron having a reasonably high velocity, the atom stays in the excited state (outermost electron goes to higher orbit) for a very short time of the order of 10 nano second and then reverts back to the normal state thereby emitting radiation energy.

Each photon may be considered to be associated with a wave which predicts how the photon will travel. By analogy to other waves we consider this fictitious wave also to have the three attributes frequency, wave length and velocity and the three quantities are related by the equation,

$$v = \lambda f \quad ...(5.1)$$

where λ is the wave length and f the frequency of the wave. The wave lengths used by illuminating engineers are very short and the units used are micron (10 metre) or Angstrom $\overset{o}{A}$ (10^{-10} metre). The complete spectrum of radiant energy extends over a tremendous range of frequency as shown in Fig. 5.0. The visual region extends from about 0.4 micron to 0.75 micron. The illumination engineer is interested in the conversion of electric energy into radiation energy of such frequencies as to be visible to the human eye and also in the conversion of radiant energy into electrical energy by photocells or thermocouples.

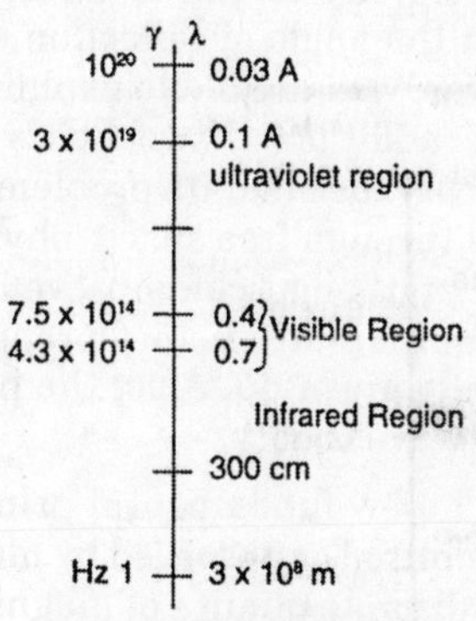

Fig. 5.0. Spectrum of Radiation Energy

5.2.1. Complete Spectrum :

The light spectrum consists of seven colours : Violet, Indigo, Blue, Green, Yellow, Orange and Red denoted by VIBGYOR. These radiations

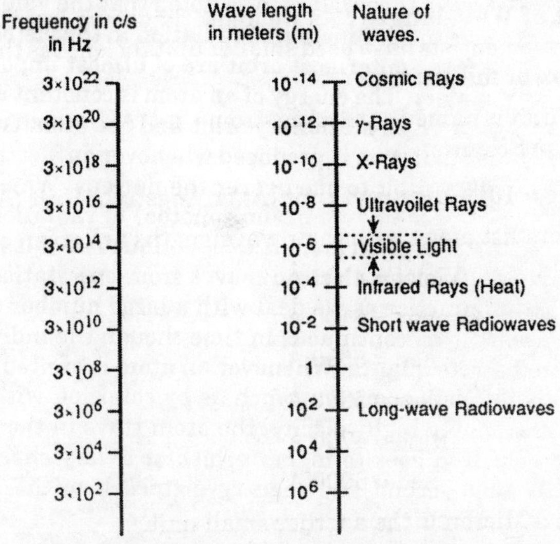

Fig. 5.01. Electromagnetic waves spectrum.

have different frequency and wavelength. Above and below are ultraviolet and Infra-red radiations.

The complete range of waves along with their frequency and wavelength is shown in fig 5.01.

From Fig 5.01 we noticed that the heat waves and radio waves are towards longer wavelength side, beyond the visible digit. X-rays and Gamma-rays are towards shorter wave length side. The light emitted together with its colour for the wave lengths within the visible spectrum is shown in fig 5.02.

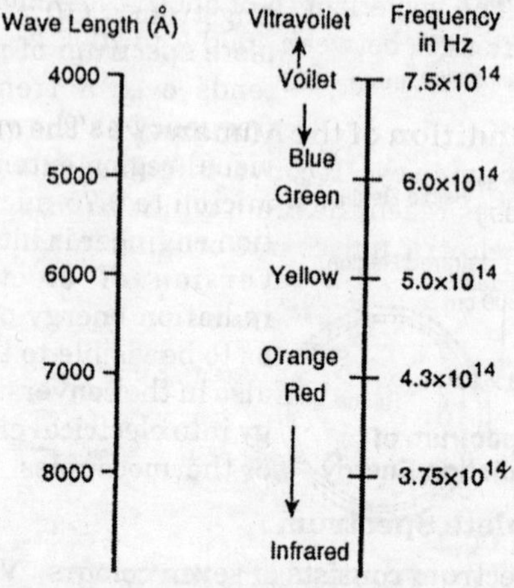

Fig. 5.02 Colour of light and wave length.

Unit of Wavelength : As we know that wave lengths of light are very short. Scientists have used smaller units of lengths than the metre, centimetre or millimetre.

This unit is named as the Angstrom Unit (AU) which is named after a Sweedish Scientist.

1 metre = 10^{10} Angstrom Unit (Å) or Angstrom Unit (Å) = 10^{-10} m

Colours list along with their wavelengths are given below :

Violet = 4100 Å

Orange = 6100 Å

Green = 5500 Å

Yellow = 5800 Å

Blue = 4700 Å

Red = 6000 Å

Note :- Micron is the another small unit.

1 micron = 10^{-6} metre

5.2.2. Human eye

The word sensivity is most important factor in Human Eye. Regarding sensivity we mean the sensations of light on human eye. All the wave lengths do not produce light sensation on eyes. The proper visibility of an object depends upon the following important factors :

1. **Distance of the object.** If distance is more, then visibility is decreased. So, the distance of the object is very important.

2. **Size of the object.** Visibility of the object depends on size of the object. If the size of the object is small in that case it can not be seen from a distance. The wave length of 5500 Å has most sensitive to the eye only the wavelength between 4000 Å to 7000 Å produce sensations to our eyes.

3. **Health condition of the human eye :** The greatest transducer of our body is the human eye. It converts falling light energy in to electric pulses. These pulses reach the brain which again acts as a transducer and converts the electric pulses into original image of the object.

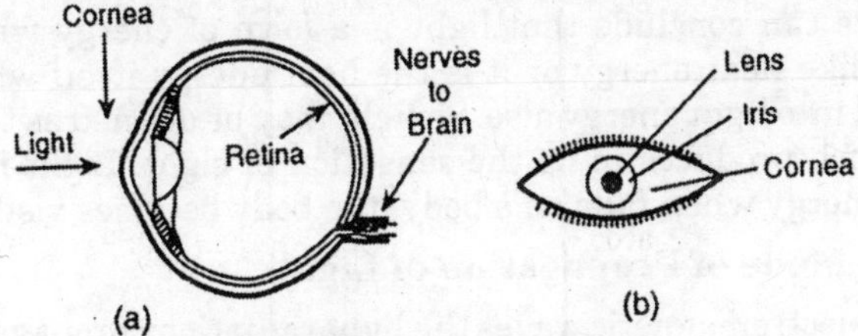

Fig. 5.03 (*a*, *b*) : Human Eye

5.2.3. Glare. The glare may be defined as the brightness within the field of vision of such a character as to cause discomfort, interference and annoyance with vision or eye-fatigue.

The size of the opening of the pupil in the numan eye is controlled by it *iris* [Figs. 5.03 (*a*) & (*b*)]. If the eye is exposed to a very bright source of light the large amount of light produces an intense image on the retina and the iris automatically protects the eye by contracting the pupil, thus reducing the intensity of the image. When the eye is towards some other object which is less bright is compared to the bright object already in the field of view, the iris will contract reducing the amount of light received on the retiña from every object in the field of view and making it difficult to see the object desired. At the same time, the portion of the retina which received image of the bright object remains fatigued. This phenomenon is called as **Glare** and a good lighting design should avoid it. Hence naked in candescent lamps and naked arc lamps should not be installed when they come in the field of view, but should be enclosed suitably in fittings of opaque glass or such material and design, that the brightness of the fitting is comparatively low. This effect of glare is familiar in connection with motor-car head lights.

5.3. Nature of Light

To understand illumination engineering it is essential to know the nature of light and fundamental principles underlying its production.

Light is a form of radiant energy. Various incandescent bodies are sources of light and light emitted by these incandescent bodies depend upon the temperature of the body. Heated bodies are of two types in colour (*i*) Red hot and white hot, while body is in red hot form, the energy radiated by it is of large wavelength and energy available is in the form of heat.

With increase in temperature red hot body becomes white hot and wave light of energy radiated by body becomes smaller and enters into the wavelength of light. At white hot state, body emits as well as heat energy. For example, if we take tungsten filament lamp, at low voltage, filament of lamp glows red emitting no light energy. But at rated voltage filament becomes white hot and emits light energy as well as some heat energy also. But radiation in form of heat is very small as compare to light energy radiated.

So, we can conclude that light is a form of energy which can be radiated like heat energy or it is the heat energy itself which can be converted into light energy also. So light may be defined as the external agent which produces in us the sensation of sight. Light is a form of radiant energy when falls on a body, the body becomes visible to eye.

5.3.1. Mode of Propagation of Light

Like electromagnetic waves the light radiations propagate through the medium in form of transverse waves. Transverse waves motion is one in which the medium particles vibrate up and down periodically at right angles to the direction of propagation of the disturbance.

According to Quantum Theory of Max Plank, light energy is emitted in form of small 'bundles' or 'packets' called quanta or photons.

According to modern theory light has dual nature, *i.e.* light energy is propagated in the form of waves while it is emitted and absorb in the form of packets or quantum of energy called photons.

Velocity of light is 3×10^8 m/sec.

When some obstacle comes in the way of radiations of light, either there is reflection or absorption. The reflected part of the energy results in illumination of the object. The absorbed part is converted into heat. But medium particles are neither illuminated nor heated.

5.3.2. Sensitivity of Eye.

The sensitivity of the eye to the lights of different wavelengths varies from person to person and according to age. The average relative sensitivity is shown in Fig. 5.1. The eye has greatest sensitivity for wavelengths of about 5,500 A, that is yellow-green can be seen under poor conditions of illumination when blue or red can not be seen under dim illumination the sensitive curve tends to shift as shown by the shaded region in Fig. 5.1. Therefore violet disappears first and red remains visible. Yellow disappears last as the illumination become very dim. As each colour disappears it becomes grey shade and finally black. The sensitivity of the eye to yellow green radiation is taken as unity 100% and the sensitivity to other wave lengths is expressed as a fraction or percentage of it. The relative sensitivity at a wavelength λ is written $k\lambda$ and is known as relative luminosity factor.

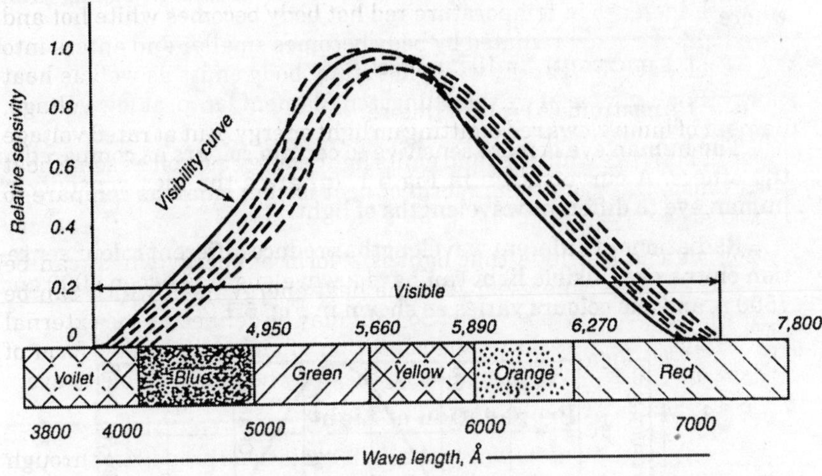

Fig. 5.1 Eye Sensivity Curve

5.4. Colour Sensation

Different light waves of same velocity have different frequencies. The colour sensation produced by different frequency light waves on the human eye is different. We know that velocity of light is approx. 3×10^8 m/sec.

If λ be the wave length in metres, the frequency in Hertz and v the velocity of propagation of the wave is in m/sec, then the velocity is equal to the product of frequency and wave length of any light wave.

i.e. $v = f\lambda$

or wave length, $\lambda = \dfrac{v}{f}$

For example a light wave having a frequency of 6×10^{14} Hz will have a wavelength,

$$= \frac{3 \times 10^8}{6 \times 10^{14}} = 0.5 \times 10^{-6} \text{ metres}$$

The practical unit of measuring wavelength is Angstrom units. So 1 metre is equal to 10^{10} Angstrom units (A.U.) so in the given example,

$$\text{The wavelength} = 0.5 \times 10^{-6} \times 10^{10} \text{ A.U.}$$
$$= 5000 \text{ A.U.}$$

Incidently this wavelength produces a sensation of green colour on the human eye.

It is convenient to measure the wavelength of high frequency waves in much smaller unit than centimeter, either in micron or angstrom where,

$$1 \text{ micron}(\mu) = 10^{-6} \text{ metre}$$
$$1 \text{ Angstrom (Å)} = 10^{-10} \text{ metre}$$

The human eye is more sensitive to certain colours as compared to the others. A curve shown in Fig 5.1. (*a*) gives the idea sensivity of human eye to different wavelengths of light.

Radiations of different wavelengths produce different colour sensation on the eye. Visible light can have a wavelength between 4000 and 7500 Å and the colours varies as shown in Fig. 5.1 (*a*).

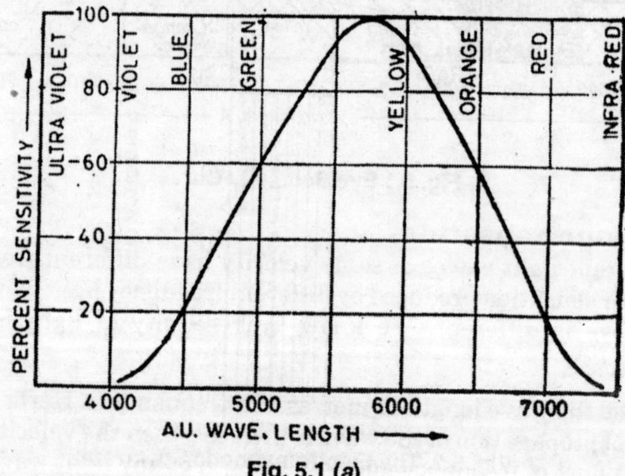

Fig. 5.1 (*a*)

Relative sensitivity of human eyes of standard observer varies very widely, being maximum at about 5550 A.U. which is wavelength of greenish-yellow colour.

That is why fluorescent tubes emitting greenish-yellow illumination are used in operation theaters of medical institutes of repute.

The bandwidth of radiations visible to human eye is small *i.e.* from 4.3×10^{14} Hz. Frequencies near 3.75×10^{14} Hz. come under infra-red rays and that near 8×10^{14}Hz come under ultraviolet rays. Both infrared and ultraviolet radiations are not visible to human eyes. Even then these radiations have various industrial, technical and medical importance. Part of ultraviolet radiations can however be converted into visible light by fluorescent substances. So we conclude that colour sensation produced by different frequency light waves on the human eye is different.

The band of frequencies producing sensation of vision can be clear from the study of electromagnetic spectrum showing different radiations having different frequencies as shown in Fig. 5.2.

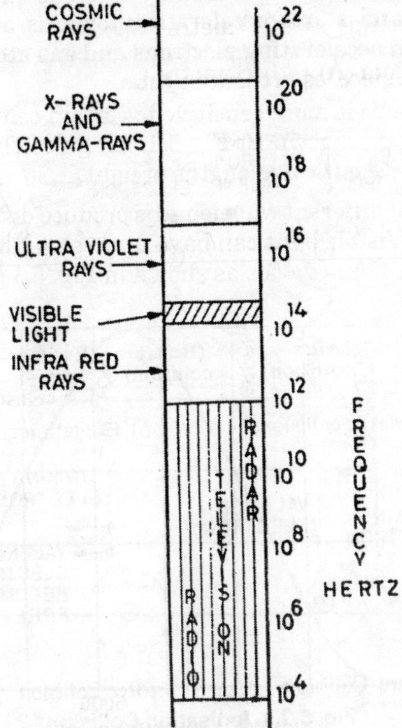

Fig. 5.2. The Electromagnetic Spectrum

5.5. Quantum Theory of Light

According to quantum theory, any excited atom is capable of emitting light when it departs with the extra energy, gained by it during collision with some high energy particles. The phenomenon occurs as below :

Consider a gas filled discharge tube filled with some mono-atomic gas like neon, organ or crypton at a low pressure. Connect its anode to positive of high voltage D. C. source and cathode to -ve of some source. Contact filament of heater to some anode source.

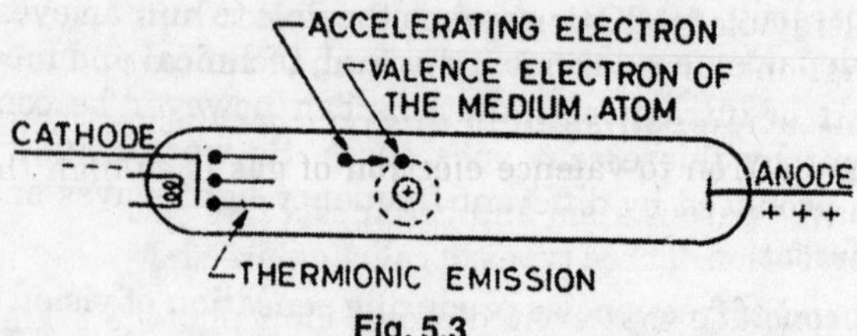

Fig. 5.3

When cathode is heated up with the help of hot filament, thermionic emission starts from the cathode. The electrons emitted are attracted towards anode and starts accelerating towards anode. But there is collision between accelerating electrons and gas atoms due to obstacle created by gas molecules present in tube.

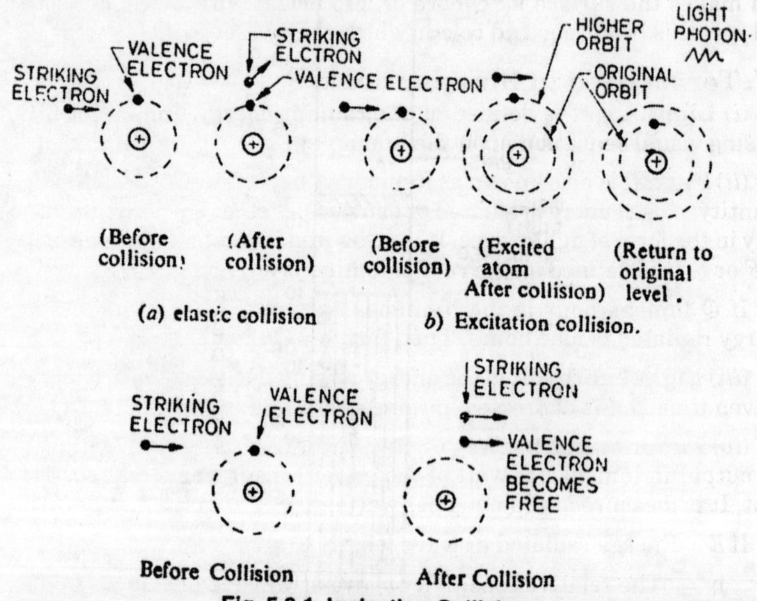

Fig. 5.3.1. Ionisation Collision.

There are three types of collisions :

(1) Elastic Collision. In this type the kinetic energy gained by the accelerating electron is so low that after collosion with some valence electron, of an atom of the gas medium it deviates from its original path. Energy supplied by it is not sufficient to cause any physical change in the atomic structure of the atom struck by it.

(2) Excitation Collision. In this type of collision quantum of energy given by striking electron is large enough to change the energy level of gas atom, electron coming in its path. The extra energy is insufficient to bring the valence electron to a conduction level. So electron remains attached to the parent atom, but shifts to a higher orbit temporarily. After few seconds these electrons return to original orbit and departs its extra energy in form of photons of lights.

(3) Ionisation collision. In this type of collision energy supplied by striking electron to valence electron of gas is so high that valence electron leaves parent atom and becomes a free electron. This process is called ionisation. Three types of collision are shown in Fig. 5.3.1.

So we conclude that excited atoms emit photons of light and the process continuous as long as emission in discharge tube occurs and anode potential is sufficiently higher than cathode.

5.6. Illumination

Illumination is different from light. Light is cause and illumination is the result of the light on surface on which it falls. Thus the illumination makes the surface look more or less bright with a certain colour and it is this brightens and colour which the eye sees.

5.7. Terms Used in Illuminations

(*i*) **Light.** Light is defined as the radiant energy from a hot body causing visual sensation upon the human eye.

(*ii*) **Flux.** It is also known as luminous flux. It is defined as the total quantity of light energy radiated or emitted per second from a luminous body in the form of light waves. It is measured in lumens and is denoted by F or ϕ. It is defined as the rate of luminous energy.

If Q lumens hour is the luminous output of source of light and energy radiated is for t hours. Then flux, $\phi = Q/t$ lumens.

(*iii*) **Light energy.** It is the energy obtained in visual radiations in a given time and is expressed in lumen hour and is denoted by Q.

(*iv*) **Luminous efficiency or Radiant efficiency.** It is defined as the output in lumens per watt of the power consumed by the source of light. It is measured in lumens per wattage.

If E = Energy radiated at wave length λ.

η = The relative sensitivity of eye at wave length λ.

K = Maximum possible efficiency if whole of the electrical input were transformed into radiating energy at 5550 A.U.

= 620 lumens/watt

Efficiency at wave length, $\lambda = \eta K$

$$\text{Total energy converted into visual effect} = K \int_{\lambda_1}^{\lambda_2} E \, \eta \, d\lambda$$

$$\text{Total energy radiated on all wave lengths} = \int_{0}^{\alpha} E \, d\lambda.$$

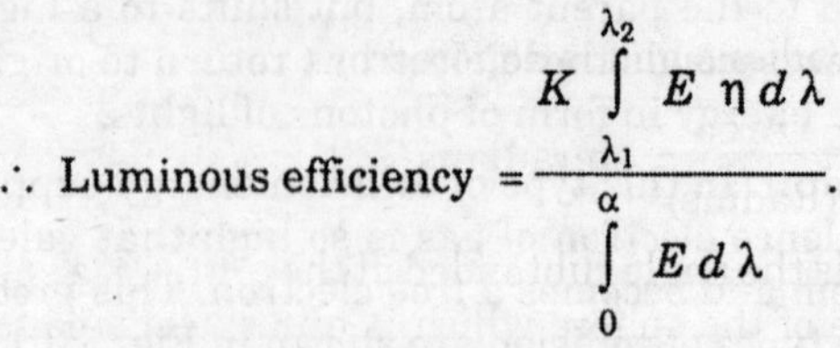

$$\therefore \text{ Luminous efficiency} = \frac{K \int_{\lambda_1}^{\lambda_2} E \, \eta \, d\lambda}{\int_0^{\alpha} E \, d\lambda}$$

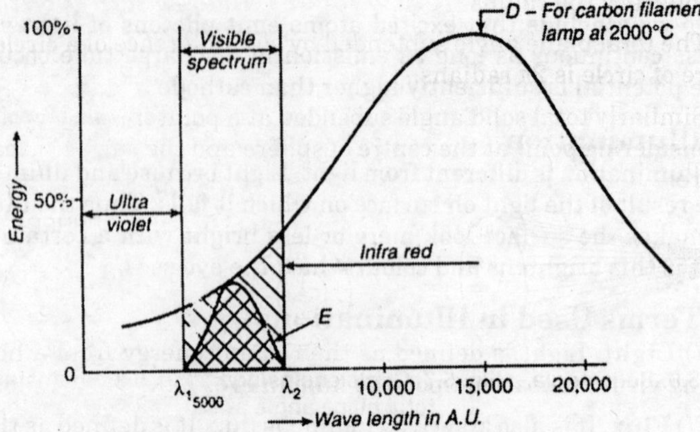

Fig. 5.4 (*a*) Special Distribution Curve

$$= \frac{\text{Area under curve E}}{\text{Area under curve D}}.$$

If a bulb is rated at 500 watts and 250 volts has an efficiency of 13 lumens/watt. Then total flux produced by bulb = $500 \times 13 = 6500$ lumens.

(*v*) **Plane Angle.** Plane angle is subtended at a point in the same plane by two converging lines.

This angle is measured in radians or degrees.

One radian is the angle subtended at the centre of a circle by an arc whose light is equal to the radius of the circle.

$$\theta = \frac{\text{Arc}}{\text{radius}}$$

$$= \frac{l}{r} \text{ radians}$$

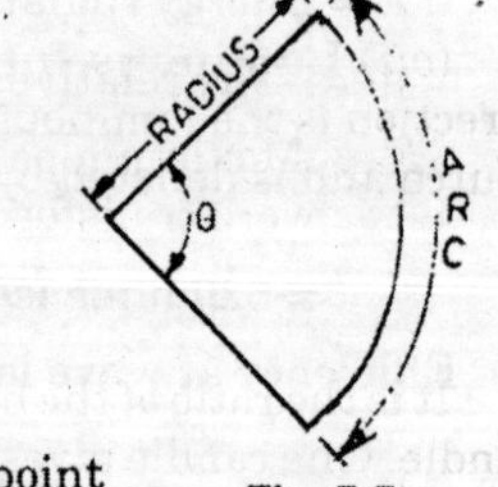

Fig. 5.5

and one Radians = $\left(\frac{180}{\pi}\right)$ degrees.

The largest plane angle subtended at a point may be 2π radians.

(*vi*) **Solid angle.** Solid angle is the angle generated by the line passing through the point in space and the periphery of the area or A solid angle enclose a volume by an infinite number of lines lying on a surface and meeting at a point.

It is measured in steradians and is denoted by ω and

$$\omega = \frac{\text{Area}}{(\text{Radius})^2} \text{ steradians}$$

Where one steradian is the angle subtended at the centre of a sphere by an area on the surface of the sphere which is numerically equal to the square of the radius.

The total plane angle subtended by circumference of a circle at the centre of circle is 2π radians.

Similarly total solid angle subended at a point in space is obtained by considering point at the centre of sphere and the surface area of the sphere.

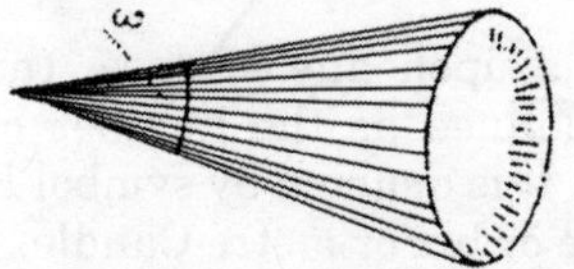

Fig. 5.6. Solid angle.

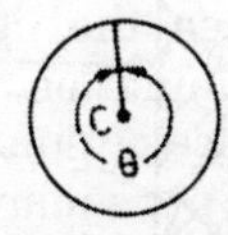

Fig. 5.7. Circle enclosing total plane angle at C.

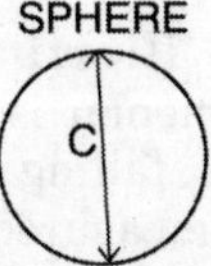

Fig. 5.8. Enclosing total solid angle at C

Plane angle $\theta = \frac{\text{Arc}}{\text{Radius}}$

$$= \frac{\text{Circumference}}{\text{Radius}} \text{ (Total angle)}$$

$$= \frac{2\pi r}{r} = 2\pi \text{ radians}$$

Total solid angle, $\omega = \frac{\text{Area of sphere surface}}{(\text{Radius of sphere})^2}$

$$= \frac{4\pi r^2}{r^2} = 4\pi \text{ steradians.}$$

(vii) **Luminous Intensity.** Luminous intensity in any particular direction is the luminous flux emitted by per unit solid angle by a point source and is denoted by, I.

$$I = \frac{F}{\omega} = \frac{\phi}{\omega} \text{ Lumens/steradian or Candla}$$

It is the ratio of the brightness of a source of light to that of standard candle. One candle gives out luminous flux of 4π lumens in space. Thus lumens emitted by one candle source of light is one lumen/steradian.

In scientific terms candla is defined as the luminous intensity in the perpendicular direction of a surface of 1/600,000 sq. metre of a full radiator at the temperature of freezing platinum under a pressure of 101,325 N/sq. metre.

(*viii*) **Lumen.** Lumen is defined as the amount of luminous flux given out in space represented by one unit of solid angle by a source having an intensity of one candle power in all directions.

Lumens = Candle power × Solid angle = C.P. × ω

Or

Total lumens given out by the source of one candla is 4 π lumens.

(*ix*) **Candle Power (C.P.).** Candle power is the light rendering capacity of a source in a given direction and is defined as number of lumens given out by the source in a unit solid angle in a given direction. It is denoted by symbol C.P.

i.e.
$$\text{C.P.} = \frac{\text{Lumens}}{\omega}$$

(*x*) **Illumination.** When the light falls upon any surface, the phenomenon is called illumination. It is defined as the number of lumens, falling on the surface per unit area. It is denoted by symbol E and is measured in lumens per square metre or lux or metre Candle.

If a flux of ϕ lumens falls on a surface of area A, then the illumination of that surface is $E = \phi/A$ lumens/m^2 or lux or metre-candle.

Bigger unit of illumination is phot. One phot = 10^4 lux

(*xi*) **Metre Candle.** It is also known as lux. It is the unit of illumination and is defined as the luminous flux falling per sq. metre on the surface which is every where perpendicular to the rays of light from a source of one Candle power and one metre away from it..**Or** it is the illumination produced by a source of one candle power on the inner surface of a sphere of one metre radius.

(*xii*) **Centimetre Candle.** It is also the unit of illumination and is defined as the luminous flux falling per square centimetre on the surface which is every where perpendicular to the rays of light from a source of one C.P. and one cm. away from it.

Or

It is illumination produced by a uniform source of one C.P. on the inner surface of a sphere of one cm. radius.

Or

It is measured in lumens/cm^2 or phot, 1 phot = 10^4 lumens/cm^2.

(*xiii*) **Foot Candle.** It is illumination produced by a uniform source of one C.P. on the inner surface of one foot radius.

Or

It is luminous flux falling per square foot on surface which is every where perpendicular to the rays of light from a source of one C.P. and one foot away from it.

$$1 \text{ foot candle} = 1 \text{ lumens/ft}^2$$

(*xiv*) **Mean spherical Candle power (M.S.C.P.).** It is defined as the average of candle powers in all directions and in all plans from the source of light.

$$\text{M.S.C.P.} = \frac{\text{Total flux in lumens}}{4\pi}$$

(*xv*) **Mean hemi-spherical Candle power (M.H.S.C.P.).** It is defined as the average of candle powers in all directions above or below the horizontal plane passing through the source of light.

(*xvi*) **Mean horizontal Candle power (M.H.C.P.).** It is average of all the candle powers in all directions in the horizontal plane containing the source of light.

(*xvii*) **Reduction factor.** Reduction factor of a source of light is defined as the ratio of its mean spherical candle power to its mean horizontal candle power.

$$\text{Reduction factor} = \frac{\text{M.S.C.P.}}{\text{M.H.C.P.}}$$

(*xviii*) **Reflection factor.** It is defined as the ratio of reflected light to incident light. It is always less than unity.

$$\text{R.F.} = \frac{\text{Reflected light}}{\text{Incident light}}$$

A polished surface like mirror reflects light incident over it. The part of light absorbed by such a surface is too small in comparison with the reflected part. Since all the incident light is not reflected, reflection factor is always less than one. Following are some typical values of R.F.

Source	*R.F.*
Mirror	0.80 to 0.85
Stainless steel	0.60 to 0.65
Chromium plating	0.55 to 0.60

(*xix*) **Lamp efficiency.** It is defined as the ratio of luminous flux to the power input. It is expressed as lumens/watt or watts/candle power.

$$\text{Lamp efficiency} = \frac{\text{Lumens emitted by source}}{\text{Wattage of source of light}}$$

(*xx*) **Space height ratio.** It is defined as the ratio of horizontal distance between two adjacent lamps and height of lamps from the working plane.

Space height ratio

$$= \frac{\text{Horizontal distance between two adjacent lamps}}{\text{Height of lamps above working plane}}$$

$$\text{S.H.R.} = \frac{d}{H}$$

The value of space height ratio vary normally between 0.8 to 1.2.

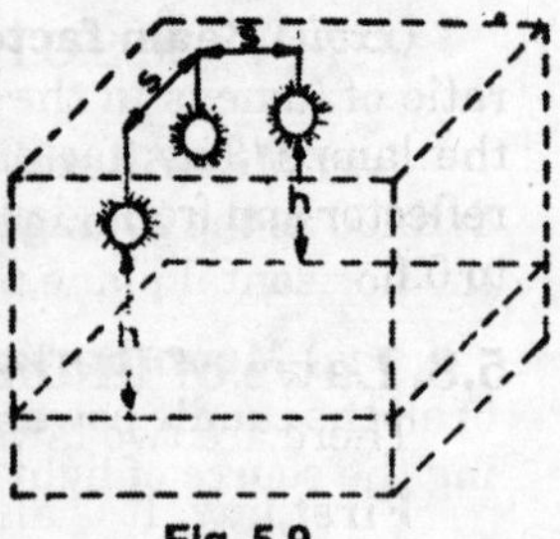

Fig. 5.9

(*xxi*) **Depreciation factor.** Due to accumulation of dirt, dust and smoke on lamps they emit less light as compare to that emit when they are new ones and similarly ceilings and walls etc. after being covered by dust etc. reflect less light as compare to that when they are new.

So the ratio of illumination under every thing is perfectly clean to the illumination under normally working condition is known as depreciation factor.

Depreciation factor

$$= \frac{\text{Illumination when everything is clean}}{\text{Illumination under normal working condition}}$$

It is more than unity.

(*xxii*) **Maintenance factor.** It is merely reverse of the depreciation factor. So it is always less than unity.

(*xxiii*) **Specific consumption.** It is the ratio of power input to the source of light to its luminous intensity. Its unit is watt per candela.

(*xxiv*) **Absorption factor.** In case of factories or places where atmosphere in full of smoke fumes, there is possibility of absorption of light energy.

Absorption factor

$$= \frac{\text{Net lumens available after absorption}}{\text{Total lumens emitted by source of light}}$$

(*xxv*) **Co-efficient of utilisation.** All the light emitted by the source does not reach the surface to be illuminated. Some portion of light directly falls upon surface and some light falls upon surface after reflection from walls and ceiling. All the light falling upon the walls and ceiling do not reach to surface because some light is absorbed.

∴ Coefficient of utilisation is the ratio of lumens reaching the working plane to the total lumens given out by the lamp.

Coefficient of utilisation

$$= \frac{\text{Lumens reaching on working plan}}{\text{Total lumens emitted by source of light}}$$

(*xxvi*) **Waste light factor.** In flood lighting scheme, a number of projector lamps are employed so that light of each projector not only overlaps that of its adjacent projector but at the edges of the area, some light falls outside. In order to compute total lumens required, theoretical lumens required are multiplied by waste light factor. Value of this factor is about 1.2 for rectangular areas and 1.5 for irregular objects such as statues and monuments etc.

(*xxvii*) **Beam factor.** It is used in turn of projector lamp. It is the ratio of lumens in the beam of a projector to the lumens given out by the lamps. This factor takes into account the absorption of light by reflector and front glass of the projector lamp. Its value lies between 0.3 to 0.6.

5.8. Laws of Illumination

There are two laws of Illumination :

First law. It is also known as Law of Inverse Square.

According to this law, the illumination of a surface is Inversely proportional to square of distance between the source and surface provided that the distance between the surface and the source is sufficiently large so that the source can be regarded as a point source.

(1) Inverse square law. Consider a point source S having an intensity I lumens/steradian. Let two surfaces having areas A_1 and A_2 be placed at distances r and $2r$ metres away respectively from the source. The two surfaces are enclosed in the same solid angle 'ω'.

Since the source gives I lumens per steradian, any surface enclosed by solid angle ω will receive a total flux = $I \times \omega$ lumens.

Thus, flux on area, $A_1 = I \cdot \omega$ lumens ...(*i*)

Also solid angle,

$$= \frac{\text{Area}}{(\text{Distance})^2} = \frac{A_1}{r^2} \text{ steradian} \quad ...(ii)$$

or Total flux on surface area,

$$= \frac{I \cdot A_1}{r^2} \quad \text{[Putting (ii) in (i)]}$$

Illumination, E_1 on surface A_1,

$$= \frac{\phi}{A_1} = \frac{I \cdot A_1}{r^2} \times \frac{1}{A_1}$$

$$= \frac{I}{r^2} \text{ Lux or Lumens/m}^2.$$

Similarly flux on any other surface, A_2 when placed $2r$ metres from a similar source enclosed in the same solid angle

$$= I \cdot \omega \text{ lumens.}$$

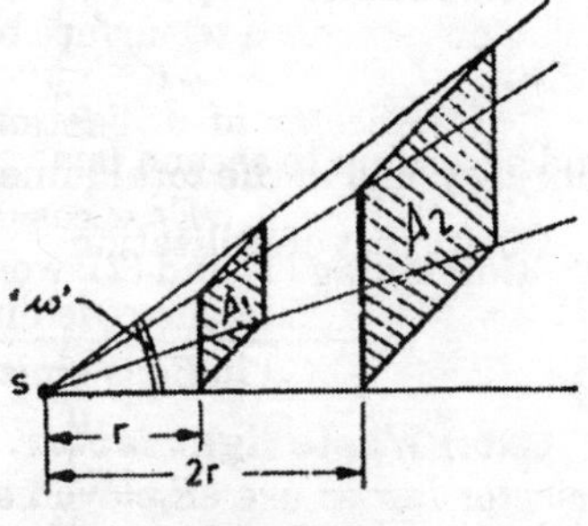

Fig. 5.10

But $$\omega = \frac{\text{Area}}{(\text{Distance})^2} = \frac{A_2}{(2r)^2} \text{ Steradians}$$

or $$\phi = \frac{I \cdot A_2}{(2r)^2}$$

and $$E_2 = \frac{I \cdot A_2}{(2r)^2} \times \frac{1}{A_2} = \frac{I}{4r^2} \text{ Lux}$$

or $$E_1 : E_2 : : \frac{I}{r^2} : \frac{I}{(2r)^2}$$

In other words, illumination at a surface is inversely proportional to the square of its distance from the source.

2. **Lambert's cosine law.** According to this law illumination of a surface varies directly as the cosine of the angle between the normal to the surface and direction of incident light.

i.e. $$E \propto \cos \theta$$

(*i*) In the normal surface case, $E = \dfrac{\phi}{\text{Area}}$

(*ii*) and in the inclined surface case, $E = \dfrac{\phi}{\text{Area}} \times \cos \theta$

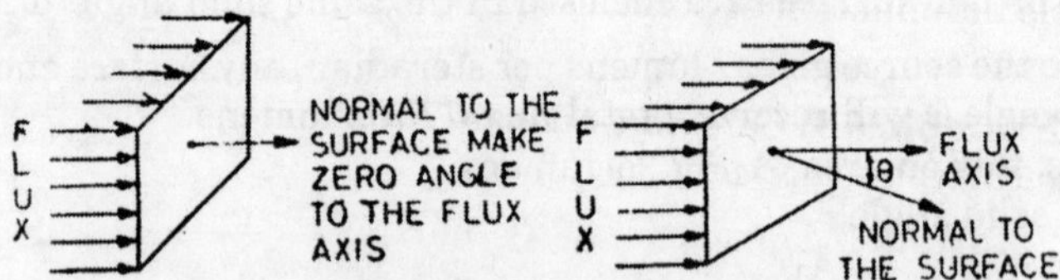

Fig. 5.11. (*a*) Normal surface. (*b*) Inclined surface.

5.9. Illumination at a Point

Let P be the point on a plane surface at a distance of d metre from a source of light, S. The source is at a height h metres from the surface. Source has a I, candle power (C.P.).

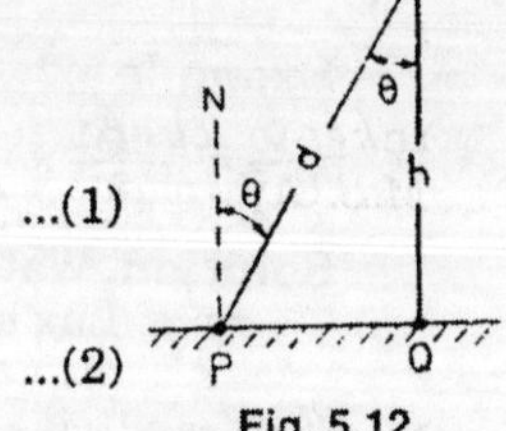

Fig. 5.12

Now by first law of illumination,

Illumination at point P,

$$E_P \propto \frac{1}{d^2} \quad \text{...(1)}$$

and according to second law

$$E_P \propto \cos \theta \quad \text{...(2)}$$

Combining (1) and (2), we get

$$E_P \propto \frac{1}{d^2} \cos \theta$$

or $$E_P = \frac{I}{d^2} \cos \theta \quad \text{...(3)}$$

where I is the luminous intensity of the source in candle power.

Now from Fig. 5.12,

$$\cos\theta = \frac{OS}{PS} = \frac{h}{d}$$

or
$$d = \frac{h}{\cos\theta}$$

Putting value of d in Eq. (3)

$$E_P = \frac{I}{\left(\frac{h}{\cos\theta}\right)^2}\cos\theta = \frac{I}{h^2}\cos^3\theta$$

Therefore, illumination at a point P,

$$= \frac{I}{d^2}\cos\theta = \frac{I}{h^2}\cos^3\theta$$

Similarly illumination at point O can also be calculated.

At point O, $\cos\theta = 1$ [as $\theta = 0°$]

and $d = h$

So illumination at point, $O = \frac{I}{d^2} = \frac{I}{h^2}$.

Example 5.1. *Calculate the total flux from the lamp having mean spherical candle power of 35.*

Solution. Mean spherical C.P. of lamp = 35

$$\therefore \quad \text{M.S.C.P.} = \frac{\text{Total flux}}{4\pi}$$

$$\therefore \text{ Total flux} = \text{M.S.C.P.} \times 4\pi$$

$$= 4\pi \times 35 = 4 \times \frac{22}{7} \times 35 = 440 \text{ lumens.}$$ **Ans.**

Example 5.2. *A 250 volt lamp has a total flux of 3000 lumens and takes a current of 0.8 amp. Calculate (i) Lumens/watt, (ii) M.S.C.P./watt.*

Solution. Wattage of lamp

$$= 250 \times 0.8 = 200 \text{ watts}$$

Flux emitted = 3000 lumens

Lumens/watt $= \frac{\phi}{\omega} = \frac{3000}{200} = 15$ **Ans.**

M.S.C.P. $= \frac{\phi}{4\pi} = \frac{3000}{4\pi} = 238.7$

M.S.C.P./watt $= \frac{240}{200} \times 1.2$ **Ans.**

Example 5.3. *A lamp of uniform intensity of 100 C.P. is enclosed inside glass globe. 25% of the light emitted by lamp is absorbed by the globe. Determine (i) Brightness of globe, (ii) C.P. of globe, dia. of globe is 20 cm.*

Solution. Dia. of globe = 20 cm

Surface area of globe

$$= 4\pi r^2 = 4 \times \pi \times (10)^2 = 400\,\pi \text{ sq. cm}$$

C.P. of the lamp = 100

Total lumens emitted by the lamp

$$= 4\,\pi \times 100 = 400\,\pi \text{ lumens}$$

Light absorbed by the globe = 25%

Therefore, light emitted from the globe

$$= 400\,\pi\,(1 - 0.25) = 400\,\pi \times 0.75 = 300\,\pi$$

$$\text{Brightness of globe} = \frac{\text{Light emitted by globe}}{\text{Surface area of globe}}$$

$$= \frac{300\,\pi}{400\,\pi} = 0.75 \quad \textbf{Ans.}$$

$$\text{C.P. of globe} = \frac{\text{Lumens emitted from globe}}{4\,\pi}$$

$$\frac{300\pi}{4\pi} = 75 \text{ C.P.} \quad \textbf{Ans.}$$

Example 5.4. *A source of light has an output of 500 C.P. in all directions below the lamp level. It is suspended at three metres above the ground. Calculate (i) Illumination at point A on ground directly below the lamp. (ii) Illumination at B, 5 metres away from A. Assuming uniform distribution of light. (iii) Total flux of light within a circular area of 1 metre dia. around the point A assuming uniform illumination.*

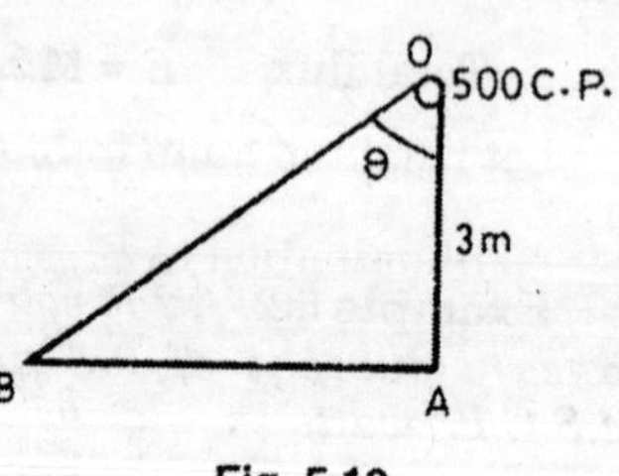

Fig. 5.13

Solution. C.P. of lamp

$$= 500$$

$$h = 3 \text{ m}$$

$$d = \sqrt{(5)^2 + (3)^2} = \sqrt{34} \text{ m.}$$

Illumination at A,

$$E_A = \frac{I}{h^2} = \frac{500}{9} = 55.55 \text{ lux} \quad \textbf{Ans.}$$

Illumination at B

$$E_B = \frac{I}{d^2} \cos\theta = \frac{500}{34} \cdot \cos\theta$$

$$\therefore \quad \cos\theta = \frac{3}{\sqrt{34}}$$

$$\therefore \quad E_B = \frac{500}{34} \times \frac{3}{\sqrt{34}} = 7.56 \text{ lux} \quad \textbf{Ans.}$$

Dia = 1 m

Surface area $= \frac{\pi}{4} D^2 = 0.7854 \text{ m}^2$

Total flux $= 55.55 \times 0.7854 =$ **43.64 lumens. Ans.**

Example 5.5. *The illumination at a point on a working plane directly below the lamp is to be 100 lumens / m*^{2}*. The lamp gives 256 C.P. uniformly below the horizontal plane. Determine the height at which the lamp is suspended and find illumination at a point on the working table 1.2 metres away from the vertical axis of the lamp.*

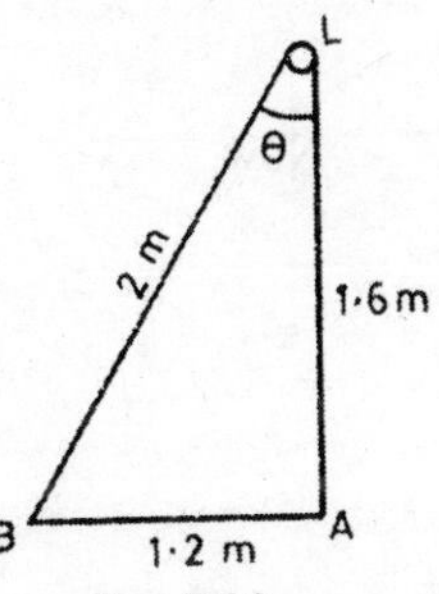

Fig. 5.14

Solution. Luminous intensity of lamp,

$$I = 256 \text{ C.P.}$$

Illumination at a point directly below the lamp

$$E = 100 \text{ lumens/m}^2$$

Let height of lamp from $A = h$ metre

$\therefore$ Illumination at E,

$$E_A = \frac{I}{h^2}$$

or $$100 = \frac{256}{h^2} \quad or \quad h^2 = \frac{256}{100}$$

or $$h = \sqrt{\frac{256}{100}} = 1.6 \text{ m} \quad \textbf{Ans.}$$

Illumination at B on the working table

$$E_B = \frac{I}{h^2} \cos^2\theta$$

Now ∵ $\cos\theta = \frac{1.6}{2} = 0.8$

∴ $$E_B = \frac{256}{(1.6)^2} \times (0.8)^3 = 51.2 \text{ lux} \quad \textbf{Ans.}$$

Example 5.6. *A lamp having an intensity of 1000 lumens per steradian, is hung 6 m over the centre of the floor of a hall. Find the illumination first below the lamp and corners of the hall, neglecting reflection from the walls and the floor.*

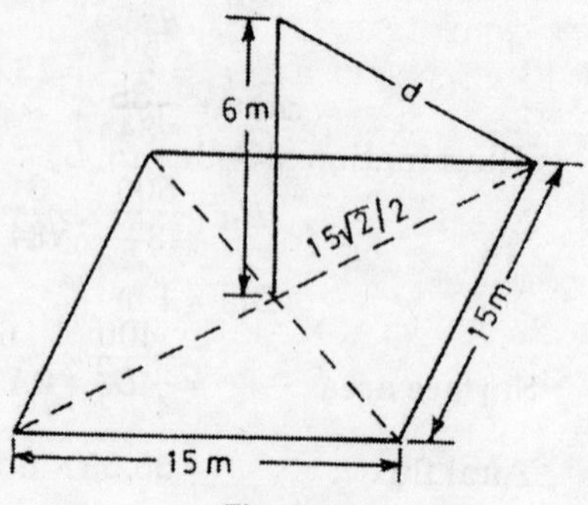

Fig. 5.15

Solution. C.P. = 1000

$h = 6$ m

E, below the lamp

$$= \frac{1000}{(6)^2}$$

$$= 27.77 \text{ lux} \quad \textbf{Ans.}$$

Now $$d = \sqrt{6^2 + \left(\frac{15}{\sqrt{2}}\right)^2}$$

$$= \sqrt{36 + \frac{225}{2}}$$

$$= \sqrt{36 + 112.5} = \sqrt{148.5} = 12.2 \text{ m}$$

$$\cos\theta = \frac{h}{d} = \frac{6}{12.2}$$

Illumination at the corners

$$= \frac{C.P.}{d^2} \cdot \cos\theta$$

$$= \frac{1000}{(12.2)^2} \cdot \frac{6}{12.2}$$

$$= \frac{6000}{1811.7} = 3.31 \text{ Lux.} \quad \textbf{Ans.}$$

Example 5.7. *A section of a road is to be illuminated by two lamps of 500 C.P. and 400 C.P. both horizontally 20 m apart and are suspended 6 m above the surface level. Calculate illumination at A directly below the lamp of 500 C.P. and at B directly below the lamp of 400 C.P. Also calculate illumination at C in the middle of points A and B.*

Solution. Illumination at point A,

= Illumination due to L_1 + Illumination due to L_2

Illumination at A due to, L_1

$$= \frac{I}{h^2} \cos^3 \theta = \frac{500}{36}$$

$$E_{AL_1} = \frac{500}{36} = 13.89 \text{ lumens/m}^2$$

Illumination at A due to, L_2

$$= \frac{I}{d^2} \cos \theta_1 = \frac{400}{\sqrt{[6^2 + (20)^2]^2}} \cdot \frac{6}{\sqrt{436}}$$

$$= \frac{400}{436} \times \frac{6}{\sqrt{436}} = 0.263 \text{ Lux}$$

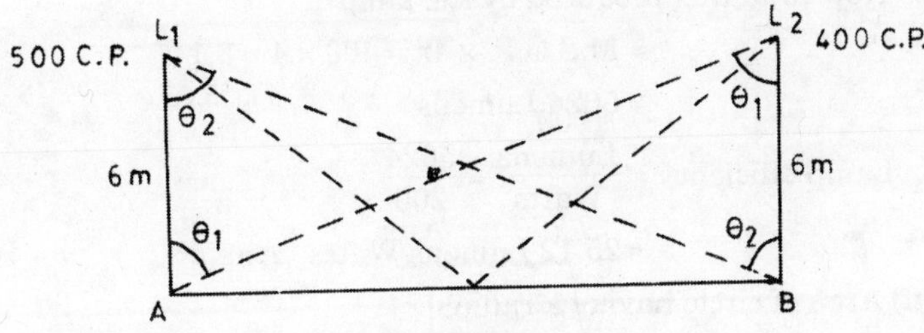

Fig. 5.16.

Total illumination at, A

$$= 13.89 + 0.263 = 14.153 \text{ Lux. } \textbf{Ans.}$$

Illumination at B = Illumination due to L_2

+ Illumination due to L_1

Illumination due to L_2,

$$E_{BL_2} = \frac{400}{36} = 11.11 \text{ Lux}$$

Illumination due to L_1,

$$E_{BL_1} = \frac{I}{d^2} \cos \theta_2$$

$$= \frac{500}{436} \times \frac{6}{\sqrt{436}} = 0.329 \text{ Lux}$$

Total Illumination at B

$$= 11.11 + 0.329 = 11.431 \text{ Lux. } \textbf{Ans.}$$

Illumination at C = Illumination due to L_1 + Illumination due to L_2

$$E_{CL_1} = \frac{I}{d^2} \cos\theta = \frac{500}{136} \times \frac{6}{\sqrt{136}} = 1.89$$

and $$E_{CL_2} = \frac{400}{136} \times \frac{6}{\sqrt{136}} = 1.51$$

Total Illumination at $C = E_{CL_1} + E_{CL_2}$

$= 1.89 + 1.51 = 3.40$ Lux **Ans.**

Example 5.8. *A lamp of 200 W having an M.S.C.P. of 400 is suspended 2 metres above a working surface. Calculate (i) Illumination directly below the lamp, (ii) Lamp efficiency, (iii) Total luminous flux in a radius of 0.20 m just below the lamp.*

Solution. (*i*) $$E = \frac{\text{C.P.}}{h^2} = \frac{400}{2 \times 2} = 100 \text{ Lux.}$$ **Ans.**

(*ii*) Now total flux produced by the lamp

$= \text{M.S.C.P.} \times 4\pi = 400 \times 4 \times 3.14$

$= 5024$ Lumens

$\therefore$ Lamp efficiency $= \frac{\text{Lumens}}{\text{Watts}} = \frac{5024}{200}$

$= 25.12$ Lumens/Watts **Ans.**

(*iii*) Area of circle having a radius

$= 0.20 \text{ m} = \pi r^2$

$= \pi (0.20)^2 = \frac{\pi}{25} \text{ m}^2$

Now luminous flux $= E \times \text{Area} = \frac{\pi}{25} \times 100$

$= 12.5$ Lumens. **Ans.**

Example 5.9. *Two lamps of 500 W with a lamp efficiency of 25 lumens per watt are mounted on two lamp posts 10 m apart. The posts have different heights of 3 m and 5 m respectively. Calculate the illumination at a point mid-way between the lamp posts.*

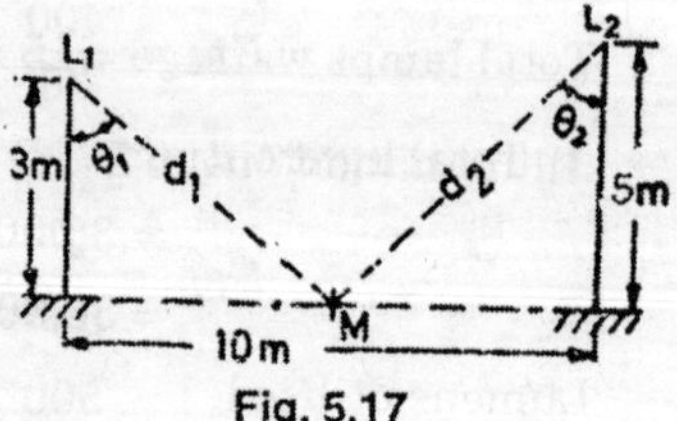

Fig. 5.17

Solution. Wattage of the lamps = 500 W/lamp

Output = Wattage × η = 500 × 25 = 12,500 Lumens

$\therefore$ C.P. of each lamp

$= \frac{\phi}{\omega}$

$$= \frac{12{,}500}{4\pi}$$

$$= 994.3 \text{ C.P.}$$

Illumination at point M due to L_1

$$= \frac{\text{C.P.}}{h_1^2} \cdot \cos^3 \theta_1 \qquad [d_1 = \sqrt{3^2 + 5^2} = \sqrt{34}]$$

$$= \frac{994.3}{3 \times 3} \cdot \frac{3}{\sqrt{34}} \cdot \frac{3}{\sqrt{34}} \cdot \frac{3}{\sqrt{34}} \qquad \left[\cos \theta_1 = \frac{3}{\sqrt{34}}\right]$$

$$= \frac{994.3}{34} \cdot \frac{3}{\sqrt{34}} = 15.1 \text{ Lux.}$$

Illumination at M due to L_2

$$= \frac{\text{C.P.}}{h_2^2} \cdot \cos^3 \theta_2 \qquad \left[d_2 = \sqrt{5^2 + 5^2} = \sqrt{25 + 25} = \sqrt{50}\right]$$

$$= \frac{994.3}{5 \times 5} \times \frac{5}{\sqrt{50}} \times \frac{5}{\sqrt{50}} \times \frac{5}{\sqrt{50}} \qquad \left[\cos \theta_2 = \frac{5}{\sqrt{50}}\right]$$

$$= \frac{994.3}{10} \times \frac{1}{\sqrt{50}} = 14.06 \text{ Lux.}$$

Total illumination $= E_1 + E_2$

$= 15.1 + 14.06 = 29.16$ Lux. **Ans.**

Example 5.10. *Determine the effective illumination of a room 12 m × 15 m, illuminated by 15 lamps of 200 watts each. The luminous efficiency of each lamp is given as 12 lumens / watt. Given coefficient of utilisation as 0.4. Let D.F. = 0.8 and waste light factor = 1.2.*

Solution. Output/watt = 12 lumens.

Total lamps wattage = 15 × 200 = 3,000 Watts.

∴ Total light output

= 3,000 × 12 [Output/Watt × Total Watts]

= 36,000 lumens.

Lumens utilised

$$= \frac{\text{Lumens emitted} \times \text{coeff. of utilisation} \times \text{D.F.}}{\text{W.F.}}$$

$$= \frac{36{,}000 \times 0.4 \times 0.8}{1.2} = \mathbf{9600 \text{ lumens.}} \quad \textbf{Ans.}$$

Room area $= 12 \times 15 = 180 \text{ m}^2$

Illumination $= \dfrac{\phi}{A} = \dfrac{9600}{180} = 53.3$ Lux. **Ans.**

5.9.1. Calculation of a lighting scheme.

Suppose it is desired to have E lumens/m² in a hall having area A m². Obviously total flux required on the working plane is E × A lumens. To send this much flux the output of the sources should be higher taking into consideration the depreciation factor, coefficient of utilisation etc. If the value of these coefficients is given < 1.

Solution. Then output of the source necessary

$$= \frac{\text{Lumens required}}{\text{Depreciation Factor} \times \text{Coeff. of utilisation}}$$

Note. (If any factor is per-chance given as more than one, then instead of dividing by, multiply the given factor).

Then wattage of lamps $= \phi/\eta$

The wattage/lamp is selected and accordingly the number of lamps found out. The spacing between the lamps is decided, keeping in view the space height ratio and the unformity of light distribution. The procedure will be clear from the numerical examples to follow.

5.9.2 Design of Lighting Scheme. For designing a lighting scheme the following factors should be taken into account :

(*i*) **Space-height ratio.** It is defined as the ratio of horizontal distance between lamps and the mounting height of the lamps, or space-height ratio

$$= \frac{\text{The horizontal distance between lamps}}{\text{Mounting height of lamps}} \quad ...(i)$$

In order to have a uniform illumination which can be only with reflectors, it is necessary that the value of this should be properly chosen. When reflectors are used, the value of ths ratio is given as between 1 and 2. The height of the source should be between 2.2 m to 2.45 m from the ground level.

(*ii*) Utilization factor. The total light flux radiated out by the source is not utilized on the working planes and its value is given as,

Utilization factor

$$= \frac{\text{Total lumens utilised on working planes}}{\text{Total lumens radiated by lamp}} \quad ...(ii)$$

The value of this co-efficient depends upon the following conditions:

(*a*) the area to be illuminated;

(*b*) height at which the lamps are fitted;

(*c*) the colour of surrounding walls, ceiling fittings etc.;

(*d*) the type of lighting-direct or indirect.

The value for direct light varies form 0.25 to 0.5 while its value for indirect light varies from 0.1 to 0.3.

(iii) **Depreciation factor.** When the lamps are covered with dust, dirt and smoke, they do not radiate out same amount of flux as when they do at the time of fitting new lamps.

Similarly, after some time the walls and surrounding in which lamps are fitted are covered with dirt and dust, so they do not reflect the same amount of light as compared with the initial conditions. The depreciation factor takes into account all such loss of flux.

Thus depreciation factor

$$= \frac{\text{Illumination under normal working conditions}}{\text{Illumination when everything is clean}} \quad ...(iii)$$

Its average value is 0.8.

The depreciation factor is also given as

$$= \frac{\text{Illumination when everything is clean}}{\text{Illumination under normal working conditions}} \quad ...(iv)$$

In this case, the value of depreciaiton factor is more than 1 and its value is 1.3 to 1.4. Table 5.1 indicates the light reflected from various coloured surroundings.

Table 5.1

Colour of surface	*Light reflected in percentage*
Light White	81%
Light Cream	69% to 75%
Light Green	65%
Light Grey	58%
Medium Grey	55%
Dark Tan	46%
Dark Grey	25%
Dark Olive Green	15%
Dark Red	12%
Natural	23%

Total lumens required. The total gross lumens output

$$= \frac{\text{Area (sq.ft.)} \times \text{illumination(ft. candles)}}{\text{Co-efficient of utilization} \times \text{Dep.factor (for values less than,1)}} \quad ...(v)$$

Also

$$= \frac{\text{Area (sq.ft.)} \times \text{illum(ft. candle)} \times \text{Dep. factor (for values more than, 1)}}{\text{Co–efficient of utilization}} \quad ...(vi)$$

5.9.3 Methods of Lighting Calculations

Some of the common methods of lighting calculations are given below :

(*a*) Watts per square metre method.

(*b*) Lumen or light flux method.

(*c*) Point to Point of Inverse square law method.

Description

(*a*) ***Watts per square metre method.*** Basically it is a "Thumb Rule' method. It is very handy for rough calculation or checking. While applying this method, an allowance of watts per square metre of area to be illuminated is taken according to the illumination desired on an average figure of overall efficiency of the system.

(*b*) ***Lumens or light flux method.*** This method is applied to those cases where the sources of light have to produce an approximate uniform illumination over the working plane or where an average value is required. Total lumens output is calculated from the efficiency of each lamp and the number of lamps used in the circuit. To calculate the lumens received on the working plane, the total lumens already calculated are multiplied by the co-efficient of utilisation. When the lamps and surroundings are not perfectly clean then while calculating the Lumens received on working plane, the depreciation factor or maintenance factor is taken into consideration.

Thus Lumens received on working plane

$$= \text{No. of lamps} \times \text{Wattage of each lamp} \times \text{efficiency of each lamp in terms of lumens/watt} \times \frac{\text{Coeffcient of utilisation}}{\text{depreciation factor}}$$

or

$$\text{No. of lamps} \times \text{Wattage of each lamp} \times \text{Efficiency of each lamp in terms of lumens/watt} \times \frac{\text{coefficient of utilisation} \times \text{maintenance factor}}{\text{Depreciation factor}}.$$

Coefficient of Utilisation or Utilisation factor

The light which is radiated by all lamps does not reach the woriking plane. When the installation is new, the ratio of lumens reaching the working plane to the total lumens given out by the lamp/lamps is known as utilisation factor or coefficient of utilisation. if the utilisation factor is on higher side, then more lumens will reach the working plane for the given lumens output of the lamps. The value of utilisation factor depend upon :

(*i*) The mounting height of lamps *i.e.* it decreases with the increase in mounting height of lamps.

(*ii*) Area to be illuminated *i.e.* utilisation factor increases with the increase in area to be illuminated.

(*iii*) Type of lighting *i.e.*, it is more for direct lighting and less for indirect lighting.

(*iv*) Colours of surroundings *i.e.* it is more for light colours and less for dark colours.

Its value varies from 0.25 to 0.5 and from 0.1 to 0.25 for direct and indirect lighting schemes respectively.

Maintenance factor. The illumination produced by a lighting installation decreases considerably after a year or two, partly due to the aging of lamps and partly due to the accumulation of dust on the lamps, on the transmitting and reflecting fixtures and on the ceilings and walls. So at the time of calculation of lumens received on the working plane, this fact is taken into account by including the maintenance factor which may be defined as the *ratio of the ultimate maintained metre candles on the working plane to the initial metre candles.* If the lamps and fixtures and surroundings are cleaned regularly, its value is more, say 0.8 and if there is more dust then its value decreases to 0.6.

Depreciation Factor. This is merely the reverse of maintenance factor. Its value in more than unity.

(*c*) ***Point to point or Inverse-squre law method.*** This method is applied where the illumination is required at a point due to one or more sources of light.

The illumination at any point within the range of the lamp can be calculated from the Inverse square law, if a polar curve of lamp and candle power of lamp reflected by its reflector in different directions is known. If two and more than two lamps are illuminating the same working plane, illumination due to each can be calculated and added. This method is not commonly used due to more complications involved in its calculations. However, it is used in some special problems such as flood lighting and yard lighting etc.

Example 5.11. *It is desired to illustrate a drawing, that with an average illumination of 200 lux. The hall is 30 × 20 m². The lamps are to be fitted 4 m from ground floor. Find the number of lamps and wattage/lamp for the lighting scheme. Given efficiency of the lamps available as 25 lumens/watt, depreciation factor 0.8 and co-efficient of utilisation 0.75, space height ratio between 0.8 and 1.2. Give satisfactory spacing arrangement.*

Solution. E, required

$$= 200 \text{ Lux or Lumens/m}^2$$

Floor area $= 30 \times 20 = 600 \text{ m}^2$

Total flux $= E \times A = 200 \times 600 = 120{,}000$ Lumens.

$$\text{Desired light output} = \frac{\phi}{\text{D.F.} \times \text{c.u.}}$$

$$= \frac{120{,}000}{0.8 \times 0.75} = 200{,}000 \text{ Lumens.}$$

$$\text{Wattage of lamps } = \frac{\phi}{\eta} = \frac{200{,}000}{25}$$

$$= 8000 \text{ Watts.} \quad \textbf{Ans.}$$

If we take 200 W lamps, then

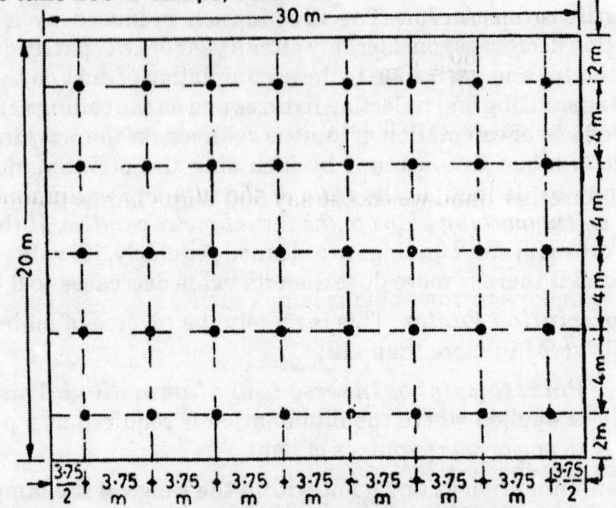

Fig. 5.18. Design of scheme for 40-lamps.

$$\text{Number of lamps} = \frac{8000}{200} = 40 \quad \textbf{Ans.}$$

Rating/lamp $= 200$ W **Ans.**

Taking five rows of eight lamps we have total distribution of 40 lamps.

Now lengthwise spacing comes to be, 30/8 = 0.75 m

Widthwise spacing comes to be, $\frac{20}{5} = 4$ m.

The lamps near the walls are kept at half the calculated spacing to keep unformity of light. If for example this distance is kept equal, then light between two adjacent lamps will be more as compared to illumination on the wall.

Now space height ratio lengthwise comes to be,

$$\frac{3.75}{\text{Height}} = \frac{3.75}{4} = 0.94 \text{ approx.}$$

Widthwise space height ratio

$$= \frac{4}{4}\left(\frac{\text{Space}}{\text{Height}}\right) = 1$$

Note. If in the given example we choose 80 lamps of 100 watts. Let us make 8 rows of 10 bulbs.

$$\text{Widthwise space } = \frac{20}{8} = 2.5$$

$$\text{Space height ratio } = \frac{\text{Space}}{\text{Height}} = \frac{2.5}{4} = 0.625$$

Lengthwise space comes

$$\frac{30}{10} = 3 \text{ m}$$

Spacing height ratio = 3/4 = 0.75 which does not satisfy the conditions.

If on the other hand we choose say 500 Watts lamps the number of bulbs comes $\frac{8000}{500} = 16$

Let us make four rows of four lamps each. The lengthwise spacing comes $\frac{30}{4}$ = 7.5 m.

$$\text{Space to height ratio } = \frac{7.5}{4} = 1.87$$

Widthwise space = 20/4 = 5 m.

Space to height ratio = 5/4 = 1.25

So it also fails to satisfy the given conditions of space height ratio lying between 0.8 to 1.2. Thus the scheme given in the problem is correct.

Example 5.12. *A room of 12 × 12 × 4 is to have direct lighting, giving illumination of 80 Lux, on a working plane 70 cm above the floor. Coefficient of utilisation is 0.5 and maintenance factor 0.8. If efficiency of lamps available is 14.75 lumens / watt, find the number of lamps and their rating.*

Solution. $E = 80$ Lux.

$$\text{Area} = 12 \times 12 = 144 \text{ m}^2$$

$$\text{Flux} = E \times A = 80 \times 144 = 11{,}520 \text{ Lumens.}$$

$$\text{Light required } = \frac{\phi}{\text{D.F.} \times \text{C.U.}} = \frac{11{,}520}{0.5 \times 0.8}$$

$$= 28{,}800 \text{ Lumens.}$$

$$\text{Wattage } = \frac{\text{Lumens}}{\eta} = \frac{28{,}800}{14.75} = 1955 \text{ Watts.}$$

We can use 100 Watts lamps, and number of lamps will be $\frac{1995}{100}$ say 20.

Note. Since lighting is required 70 cm above floor level, and if we use 30 cm suspension wire for each lamp effective height will be 4 m – 0.70 m –0.30 m = 3 m.

Now if we use 4 lamps/row and 5 such rows lengthwise, spacing can be worked out as 12/5 and 12/4, *i.e.* 2.4 m lengthwise and 3 m breadthwise.

The space height ratio comes as 2.4/3 and 3/3, *i.e.* between 0.8 and 1 which is normal.

The spacing arrangement can be as shown in Fig. 5.19.

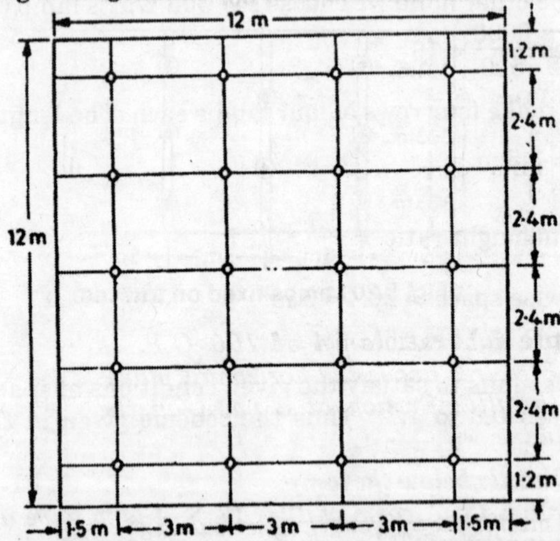

Fig. 5.19. 4-lamps in each row and 5 rows lengthwise make 20 lamps.

Example 5.13. *It is required to provide an illumination of 100 lumens/m² in a workshop hall 40 m × 10 m. Assume depreciation factor is 0.8 and coefficient of utilisation is 0.4 and efficiency of lamps is 14 lumens/Watt. Calculate number of lamps and their positions when seven trusses are provided at a mutual distance of 5 m.*

(P.U. Pb. Utilisation of Electrical Energy. Nov. 1985)

Solution. E, desired = 100 lumens/m²

$$\text{Area} = 40 \times 10 = 400 \text{ m}^2$$

Lumens required $= E \times A = 100 \times 400$

$= 40{,}000$ lumens.

Lamps output necessary

$$= \frac{\phi}{\text{C.U.} \times \text{D.F.}} = \frac{40{,}000}{0.4 \times 0.8}$$

$= 125{,}000$ lumens.

Efficiency/Watt $= 14$ lumens.

Wattage $= \dfrac{125{,}000}{14} = 8929$ Watts

If we take 500 W lamps, the number of lamps

$$= \frac{8929}{500} \approx 18 \text{ Bulbs.}$$

Since breadthwise trusses are provided, we have to use these trusses for fixing these bulbs.

The scheme of fixing these bulbs is shown in Fig. 5.20.

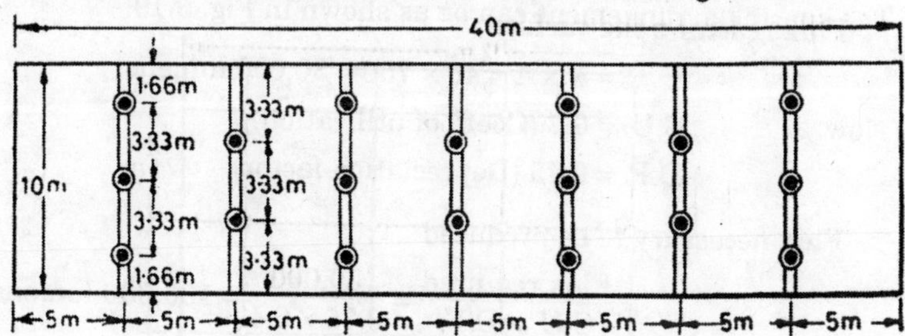

Fig. 5.20 Lamps fixed on trusses.

Example 5.14. *A lamp of 100 C.P. is suspended 3 metres above the horizontal plane. Calculate the illumination at a point on the horizontal plane.*

(*i*) *Directly below the lamp.*

(*ii*) *3 metres away from the vertical axis.*

(H.S.B.T.E. June 1986)

Fig. 5.21

Solution. (*i*) C.P. = 100

$h = 3$ m

Now illumination at a point Q directly below the lamp.

$$= \frac{\text{C.P.}}{h^2} = \frac{100}{(3)^2} = \frac{100}{9} = 11.11 \text{ lux} \quad \textbf{Ans.}$$

(*ii*) Illumination at a point P

$$= \frac{\text{C.P.}}{d^2} \cdot \cos\theta = \frac{100}{(3\sqrt{2})^2} \times \frac{3}{3\sqrt{2}}$$

where $d = \sqrt{3^2 + 3^2} = \sqrt{18} = 3\sqrt{2}]$

$$= \frac{100}{18} \times \frac{1}{\sqrt{2}} = 3.928 \text{ Lux.} \quad \textbf{Ans.}$$

Example 5.15. *A reading room 50 m × 15 m × 6 m, requires an illumination of 40 metre-candle on the reading table. Assuming a space height ratio of 1.2. Calculate (i) the number of lamps required (ii) C.P. of each. Assuming the utilisation factor as 0.4, depreciation factor as 0.75, efficiency of each lamp as 0.75 watts per candle power and the height of lamps above the reading table as 4 m. Draw a sketch of arrangement of lamps.*

Solution. Area of the room

$$= 50 \times 15 = 750 \text{ m}^2$$

E, desired $= 40$ m-candle $= 40$ Lux $= 40$ lumens/m^2

∴ Flux reaching the surface

$$= E \times A = 40 \times 750 = 30{,}000 \text{ lumens.}$$

Now C.U. = 0.7 (Coeff. of utilisation)

D.F. = 0.75 (Depreciation factor)

∴ Flux necessary to be produced

$$= \frac{\text{Flux required}}{\text{C.U.} \times \text{D.F.}} = \frac{30{,}000}{0.75 \times 0.4} = 100{,}000 \text{ lumens.}$$

$$\therefore \text{ C.P. desired } = \frac{\phi}{4\pi} = \frac{100{,}000}{4\pi}$$

Lamp, $\eta = 0.75$ Watts/C.P.

∴ Total wattage required

$$= \text{C.P.} \times \eta = \frac{100{,}000}{4\pi} \times 0.75 = 5965 \text{ Watts}$$

If we use 200 Watts lamps,

$$\text{Number of lamps } = \frac{5965}{200} \approx 30. \quad \textbf{Ans.}$$

(*ii*) C.P. of each lamp

$$= \frac{\text{Wattage}}{\eta} \quad (\text{when } \eta = \text{W/C.P.})$$

$$= \frac{200}{0.75} = 200 \times \frac{4}{3} = 266.67 \text{ C.P.} \quad \textbf{Ans.}$$

(*iii*) Height above working surface = 4 m

Space height ratio = 1.2

∴ Space between adjacent lamp = 1.2 × 4

= 4.8 m (say 5 m)

Since length and breadth of the hall is 50 m and 15 metres respectively, we have to take a space of 5 m for proper distribution of light instead of 4.8 m.

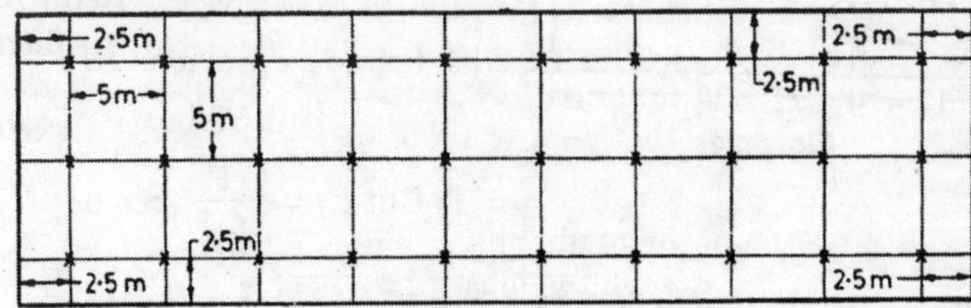

Fig. 5.22.

∴ Design of lighting scheme is as shown by keeping a distance of 5 metres between adjacent lamps.

Example 5.16. *Two arc lamps of 1000 C.P. and 500 C.P. respectively (assumed same in all directions) are suspended 15 metres above the ground level and are 30 metres apart. Find the intensity of illumination at a point on the ground in line with the two lamps and 15 metres from the base of the more powerful lamp.*

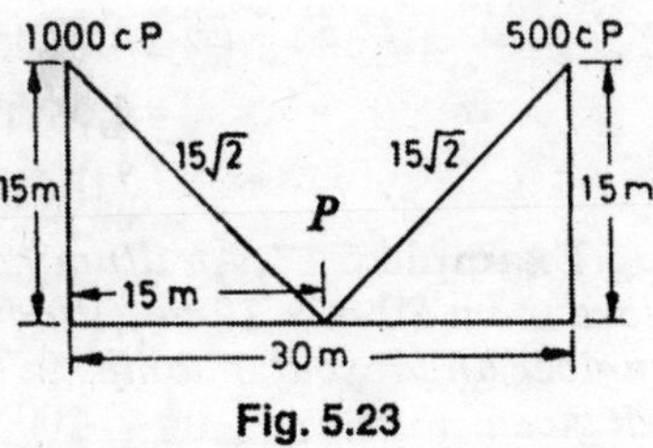

Fig. 5.23

Solution. Taking point P towards right to the more powerful lamp. We have

$$E = E_1 + E_2$$

where E_1 is illumination due to lamp of 1000 C.P. and E_2 is the illumination due to lamp of 500 C.P.

$$E_1 = \frac{1000}{(15\sqrt{2})^2} \cdot \frac{15}{15\sqrt{2}} \quad \because \left(\frac{\text{C.P.}}{d^2} \cdot \cos\theta\right)$$

$$= \frac{1000}{450} \times \frac{1}{\sqrt{2}} = 1.57 \text{ Lux.}$$

$$E_2 = \frac{500}{(15\sqrt{2})^2} \times \frac{15}{15\sqrt{2}} = \frac{500}{450} \times \frac{1}{\sqrt{2}}$$

$$= 0.785 \text{ Lux.}$$

Total illumination $= 1.57 + 0.785 = 2.355$ Lux.

(*ii*) Now let us take point P towards left of the more powerful source.

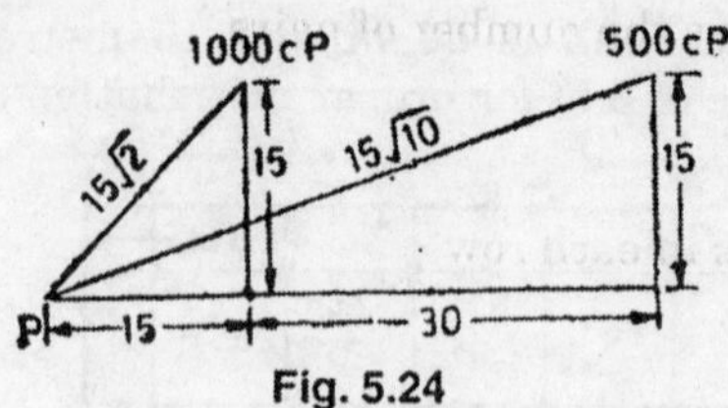

Fig. 5.24

Now E_1

$$= \frac{1000}{(15\sqrt{2})^2} \cdot \frac{15}{15\sqrt{2}}$$

(same as previous E_1)

$$= 1.57 \text{ Lux}$$

$$\text{But } E_1 = \frac{\text{C.P.}}{(d_2)^2} \cdot \cos\theta$$

where $d_2 = \sqrt{15^2 + 45^2} = 15\sqrt{1^2 + 3^2} = 15\sqrt{10}$

$$E_2 = \frac{500}{(15\sqrt{10})^2} \cdot \frac{15}{15\sqrt{10}} = \frac{500}{2250} \times \frac{1}{\sqrt{10}}$$

$$\approx 0.07 \text{ Lux}$$

$\because$ $E_1 + E_2 = 1.57 + 0.07$

$= 1.64$ Lux which is lesser than the previous case.

Example 5.17. *An illumination of 300 Lux is to be provided in a class room 20 m × 10 m with 40 W fluorescent lamps. Determine the number and layout of lamps in the lighting installation. Assume data not given.*

Solution. Given : Room area $= 20 \times 10 = 200 \text{ m}^2$

E, required $= 300$ Lux

Wattage of tubes $= 40$ W

Assumed :

Coefficient of utilisation $= 0.5$

Depreciation factor $= 0.75$

Efficiency of tubes $= 40$ lumens/Watt.

Now flux required $= E \times A = 300 \times 200$

$= 60{,}000$ lumens.

Flux necessary to be produced taking different factors into account,

$$\phi' = \frac{\phi}{\text{C.U.} \times \text{D.F.}} = \frac{60{,}000}{0.5 \times 0.75}$$

Wattage of lamp necessary

$$= \frac{\phi'}{\eta} = \frac{60{,}000}{0.5 \times 0.75 \times 40} = 4000 \text{ Watts.}$$

$\therefore$ Number of tubes

$$= \frac{4000}{40} \text{ (40 W/tube, given)} = 100.$$

Let us mount tubes in pairs. Thus the number of pairs

$$= \frac{100}{2} = 50.$$

Let us take 5 rows. So that pairs in each row

$$= \frac{50}{5} = 10.$$

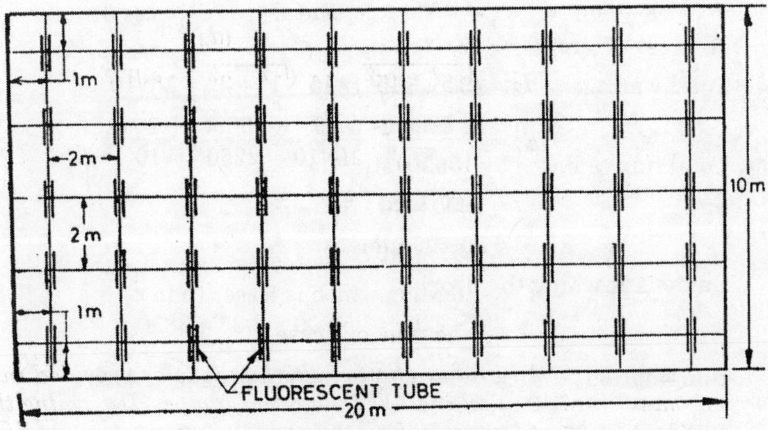

Fig. 5.25. Lighting Layout.

Example 5.18. *A source of 5000 lumens is suspended 6.1 m above ground. Find out the illumination (i) at a point just below the lamp and (ii) at a point 12.2 m away from the first, assuming uniform distribution of light from the source.*

(*A.M.I.E. Section B. Nov. 1982*)

6.1 m
$6.1\sqrt{5}$ m
P
12.2 m
Q

Fig. 5.25.0

Solution. C.P. $= \frac{\phi}{4\pi} = \frac{5000}{4 \times 3.14}$

(*i*) Illumination just below the lamp

$$= \frac{\text{C.P.}}{h^2} = \frac{5000}{4 \times 3.14 \times 6.1 \times 6.1}$$

$$= 10.7 \text{ Lux.}$$

(*ii*) Illumination at point Q

$$= \frac{\text{C.P.}}{d^2} \cdot \cos\theta$$

$$= \frac{5000}{5 \times 3.14 \times (6.1 \times \sqrt{5})^2} \times \frac{6.1}{6.1 \times \sqrt{5}} = 0.957 \text{ Lux.}$$

[where $d = \sqrt{(6.1)^2 + (12.2)^2} = 6.1\sqrt{5}$]

Example 5.19. *Determine average Illumination of a room measuring 9.15 m by 12.2 m illuminated by a dozen 150 watt lamps. The luminous efficiency of lamps be taken as 14 lumens/watt and the co-efficient of utilisation as 0.35.* *(A.M.I.E. Section B. Nov. ,1985)*

Solution.

Room area $= 9.15 \times 12.2 = 112 \text{ m}^2$

Wattage/lamp $= 150$ W

Number of lamps $= 12$

Total wattage $= 150 \times 12 = 1800$

$\eta = 14$ lumens/watt

Total flux radiated by 1800 watts

$= 1800 \times 14$

$= 25{,}200$ lumens.

Lumens reaching the floor

$= \phi \times \text{C.U.} = 25{,}200 \times 0.35 = 8820$

Illumination $= \dfrac{\phi}{\text{Area}} = \dfrac{8820}{112} = 79 \text{ lumens/m}^2$. **Ans.**

Example 5.20. *A lamp takes 3 amp. at 250 volts and emits 9420 lumens. Determine*

(i) Mean spherical candle power.

(ii) Efficiency of lamp.

Solution. Current = 3 amps.

Voltage = 250 volts

Total lumens emitted by lamp = 9420

$$\text{M.S.C.P.} = \frac{\text{Total lumens}}{4\pi} = 749.61 \quad \textbf{Ans.}$$

Power of lamp $= 250 \times 3 = 750$ watts

Efficiency of lamp $= \dfrac{9420}{750} = 12.56$ lumens/watts **Ans.**

Example 5.21. *A factory space of 33 m × 13 m is to be illuminated with an average illumination of 72 lumens per square metre by 200 W lamps. The co-efficient of utilisation is 0.4 and depreciation factor is 1.4. Calculate the No. of lamps required. The lumens output of a 200 W lamp is 2730 lumens.*

Solution. Area of factory

$= 33 \times 13 = 429$ sq. m

Illumination = 72 lumens/sq. m

U.F. = 0.4

D.F. = 1.4

Wattage of lamp = 200 watts

Lumens per lamp = 2730

Total illumination required from source

$$= \frac{E \times A \times \text{D.F.}}{\text{U.F.}}$$

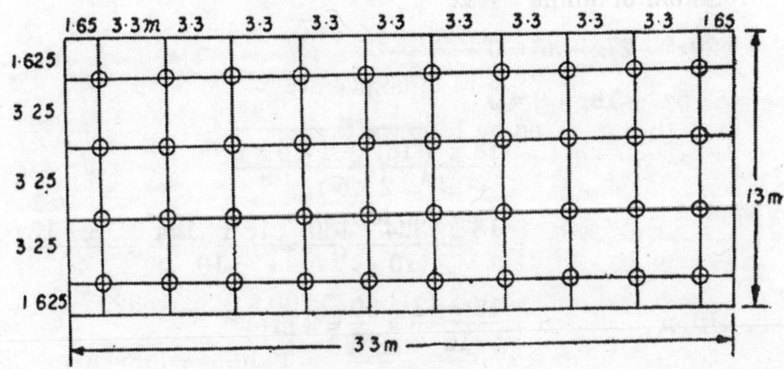

Fig. 5.25 (*a*)

$$= \frac{429 \times 72 \times 1.4}{0.4} = 108108 \text{ lux}$$

Total No. of lamps required

$$= \frac{\text{Total lumens required}}{\text{Lumens given by one lamp}}$$

$$= \frac{108108}{2730} = 39.6 \text{ (say) } 40. \quad \textbf{Ans.}$$

Example 5.22. *Two lamps of 36 and 16 C.P. respectively are one meter apart. Where should the screen be placed on a straight line passing through the lamps so as to have equal illumination on it.*

Solution. Say screen is placed at x metre from L_1

So distance from $L_2 = (1 - x)$ m

Illumination due to $L_1 = \frac{I}{d^2} \cos \theta = \frac{I}{d^2} = \frac{36}{x^2}$ lux

Illumination due to $L_2 = \frac{I}{d^2} = \frac{16}{(1-x)^2}$ lux

Since illumination of both lamps is same

$$\therefore \qquad \frac{36}{x^2} = \frac{16}{(1-x)^2}$$

Fig. 5.25 (*b*)

$$\frac{36}{x^2} = \frac{16}{1 + x^2 - 2x}$$

or $$36 + 36x^2 - 72x = 16x^2$$

or $$20x^2 - 27x + 36 = 0$$

$$5x^2 - 18x + 9 = 0$$

$$\therefore \quad x = \frac{18 \pm \sqrt{(18)^2 - 4 \times 5 \times 9}}{2 \times 5}$$

$$= \frac{18 \pm \sqrt{324 - 180}}{10} = \frac{18 \pm \sqrt{144}}{10} = \frac{18 \pm 12}{10}$$

$\therefore$ either $$x = \frac{18 + 12}{10} = \frac{30}{10} = 3 \text{ m}$$

or $$x = \frac{18 - 12}{10} = \frac{6}{10} = 0.6 \text{ m} \quad \textbf{Ans.}$$

Example 5.23. *A lamp of 500 C.P. is suspended 4 metres above a horizontal surface. Calculate the illumination.*

(*i*) *Directly below the lamp.*

(*ii*) *3 metres away from the vertical axis.*

Solution. Intensity of lamp = 500 C.P.

Height from surface = 24 m

Illumination at A = ? at 3 m away from *y axis.*

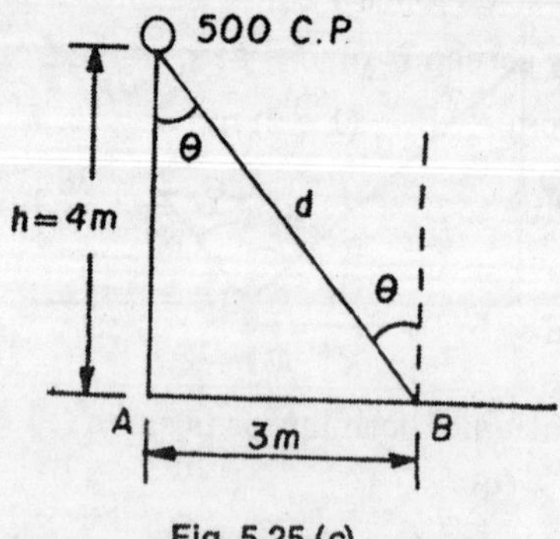

Fig. 5.25 (*c*)

$$d = \sqrt{(4)^2 + (3)^2} = 5 \text{ m}$$

$$\cos\theta = \frac{4}{5} = 0.8$$

Illumination at, $A = \frac{I}{d^2}\cos\theta$ [But θ = 0]

$$\frac{500}{25} = 20 \text{ lux}$$

Illumination at, $B = \frac{I}{d^2}\cos\theta = \frac{500}{25} \times 0.8 =$ **16 lux. Ans.**

Example 5.24. *Two lamps of 500 W, each with lamp efficiency of 25 lumens per watt are mounted on two lamp post 10 m apart. The posts have different heights of 3 m and 4 m respectively. Calculate the illumination at a point mid way between the lamp posts.*

Solution. Wattage of lamps = 500 watt

Efficiency of each = 25 lumens/watt

Total flux emitted = 500 × 25 = 12500 lumens

Intensity of each lamp

$$= \frac{\text{Flux}}{4\pi} = \frac{12500}{4\pi} = 994.7 \text{ C.P.}$$

$$d_1 = \sqrt{(3)^2 + (5)^2} = \sqrt{34} \text{ m}$$

$$d_2 = \sqrt{(4)^2 + (5)^2} = \sqrt{41} \text{ m}$$

Illumination at C = Illumination due to L_1

+ Illumination due to L_2

Illumination due to $L_1 = \frac{994.7}{34} \times \frac{3}{\sqrt{34}} = 15.1$ lux

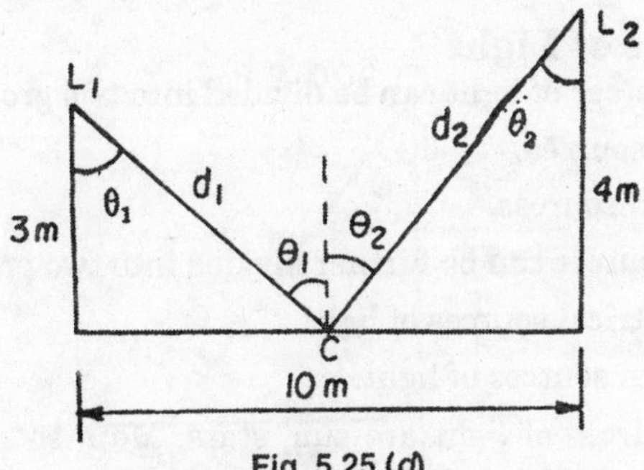

Fig. 5.25 (*d*)

Illumination due to $L_2 = \frac{I}{d_2^2}\cos\theta_2$

$$= \frac{994.7}{41} \times \frac{4}{\sqrt{41}} = 15.15 \text{ lux}$$

Illumination at $C = 15.1 + 15.15 = 30.25$ lux. **Ans.**

Example 5.25. *The illumination of drawing office 30 m × 10 m is to have a value of 250 lux and is to be provided by a No. of 300 W filament lamps. If the co-efficient of utilisation is 0.4 and depreciation factor is 0.9 and luminous efficiency of each lamp is 14 lumens per watt, determine the number of lamps required.*

Solution. Area of drawing office

$$= 30 \times 10 = 300 \text{ sq. m}$$

Illumination $= 250$ lux

Wattage of lamps $= 300$ watt

U.F. $= 0.4$

D.F. $= 0.9$

Luminous efficiency $= 14$ lumens/watt

Gross lumens output required from source

$$= \frac{E \times A}{\text{U.F.} \times \text{D.F.}} = \frac{300 \times 250}{0.4 \times 0.9}$$

$$= 208333.33 \text{ lumens}$$

Flux emitted by per lamp

$$= \text{Wattage} \times \text{efficiency}$$

$$= 300 \times 14 = 4200 \text{ lumens.}$$

$$\text{Total no. of lamps required} = \frac{\text{Gross lumens}}{\text{Lumens output of one lamp}}$$

$$= \frac{208333.33}{4200} = 49.6 = 50 \quad \textbf{Ans.}$$

5.10. Sources of Light

Various sources of light can be divided into two groups.

(*i*) Natural sources.

(*ii*) Artificial sources.

Artificial sources can be further divided into two groups

(*i*) Non-electrical sources of light.

(*ii*) Electrical sources of light.

Natural sources of light are sun, stars. Main sources of light in nature is sun.

The non-electrical sources of light are candles, kerosine lamps, petromax lamps etc. The electrical sources of light are incandescent lamps, mercury vapour lamps, fluorescent tubes, sodium vapour lamps etc. The quality, colour and quantity of light emitted by above various sources is different.

5.11. Electrical Sources of Light

Incandescent-Lamps

Filament materials

These are of carbon and tungsten, when the temperature increases wave-length decreases, till it reaches 7×10^{-5} cm and falls in the visible range. The light emitted is proportional to the 12th power of absolute temperature and this is the reason why lamps are run at a high temperature as far as possible.

Filament design

It has been already established that the current through the lamp is proportional to $d^{3/2}$. We can find the resistance of the filament by using the rated voltage. Thus $\frac{V}{I} = R = \frac{4\rho l}{\pi d^2}$.

Thus for a selected diameter of the filament, we can find the length of the filament.

Construction of lamp

If the space in the lamp is replaced with an inert gas, tungsten filament can reach a temperature of 2400 K without evaporation. Due to higher temperatures, loss of heat due to higher temperatures occurs. So the heat is reduced by gases like nitrogen or argon.

The convection loss is minimised by coiled coil filament. Spiral filaments are used now to prevent heat losses [Fig. 5.25.1 (*a*), (*b*)]. In

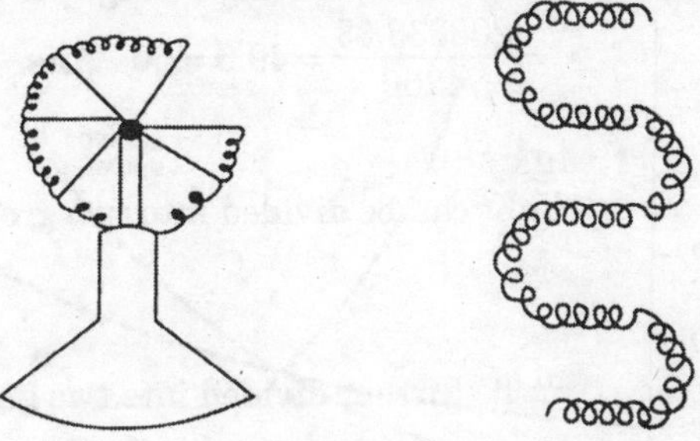

(*a*) (*b*)

Fig. 5.25.1 (*a*) Filament of an Incandescent Lamp, (*b*) Special Filament of an Incandescent Lamp.

the case of tungsten filament, light output $\propto E^n$ where $n = 4$ or 5 and power input $\propto E^n$ where $n = 1.8$ and E is the voltage.

During manufacture, all air is pumped out of glass envelope to prevent filament burning up when operating. Lamps larger than 40 watts are filled with inert gas (argon and nitrogen) to retard evaporation of tungsten. However, gradual evaporation causes a dark deposit on the inner side of the bulb blocking light. Now-a-days luminous efficiency of 10-30 lumen/watt is achieved [Fig. 5.25.1 (*c*)].

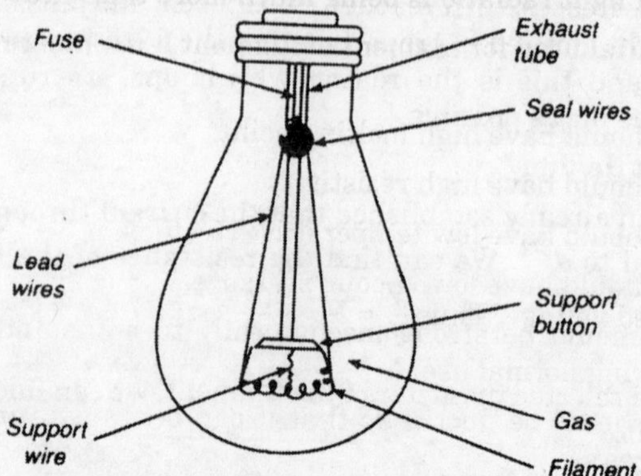

Fig. 5.25.1 (*c*) Incandescent Lamp with Filament.

Effect of voltage variation on the life of the lamp

The filament lamps are under operation under constant supply voltages. But a variation of ± 6% voltage at consumers terminals is permitted under the Indian Electricity Rules. Further drop of voltage in the electrical wiring may occur. Thus a voltage variation from + 6% to –8.5% may result. Hence the operating voltage varies from 212 to

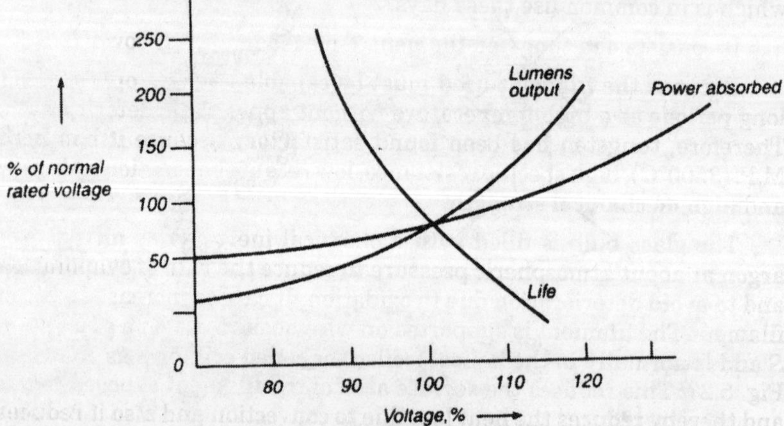

Fig. 5.26 Graph of Lumen Output, Power Absorbed and Life of a Filament Lamp with the Variation of Voltage.

224 V on a 230 V supply mains. A study of the behaviour of a 100 W filament gives the characteristic curves shown in Fig. 5.26 *d*] *for its life, lumen output and power absorbed.*

5.11.1 Incandescent Lamps. These are also known as filament lamps. These lamps work on the principal of Incandescence, hence known as Incandescent lamps. According to which, when current is passed through the wire, both heat and light are produced. When wire in red hot it emits more heat as compare to light. At white hot position, the amount of light radiations being much more than heat energy.

The material used for filament of filament lamp must have following properties :

(*i*) It should have high melting point.

(*ii*) It should have high resistivity.

(*iii*) It should have low temperature co-efficient.

(*iv*) It should have low vapour pressure.

(*v*) It should be strong mechanically to withstand vibrations during normal use.

(*vi*) It should be ductile so that it can be drawn into very thin wires.

The material used for filament is carbon, osmium, tantalum and tungsten.

Now-a-days tungsten in most commonly used metal filament because it fulfills above all properties. Moreover lamps with tungsten filament has higher efficiency than carbon, osmium and tantalum filament lamps.

Fig. 5.27 shows the construction of a gas filled incandescent lamp which is in common use these days.

It consists of a tungsten filament *T* placed in a glass cover *G*.

Because the filament used must be capable of being operated for long periods at a high temperature without appreciable deterioration. Therefore, tungsten has been found satisfactory because it has high M.P. (3500°C), high specific resistance, low rate of evaporation, ductility and high mechanical strength.

The glass bulb is filled with a chemical inert gas as nitrogen or argon at about atmospheric pressure to reduce the rate of evaporation and to avoid deterioration due to oxidation. It causes increase in life of filament.The filament is supported on wire hooks fixed on a glass stem *S* and is normally in the helical (coiled) or coiled coil form as shown in Fig. 5.27. This reduces the surface area of the filament exposed to gas and thereby reduces the heat loss due to convection and also it reduces space requirement for the large length of filament. The efficiency of a coiled coil lamp is high as compared to single coil lamps and it requires

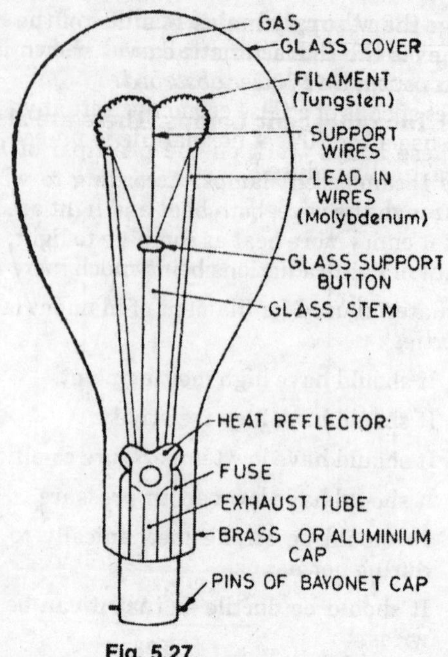

Fig. 5.27.

less number of supports for the filament. Lamps below 40 watts are not filled with gas. This is due to the fact that lot of heat will be wasted in these small sized bulbs due to conventional flow of gas which will transfer heat from the filament to the walls of bulb which are quite close to the filament. This will reduce the lumens emitted by the filament. In large bulbs the size of the filament is reduced by using coiled coils upto approximately 100 W and simple coiled filaments above 100 W. Heat reflector is used for bulbs of high power rating. The cap used in that case is also Edison screw cap.

Rating. The rating of the bulbs is specified in volts and wattage, *e.g.* 60 W, 230 V or 100 W, 250 V. Small bulbs used for indication in certain laboratory equipment have rating in milli-amps. If the bulbs of a given voltage rating are not operated at the rated voltage, there is direct influence on the :

(*i*) Lumens output,

(*ii*) Life,

(*iii*) Efficiency.

The above effect can be more appreciated by looking into Fig. 5.28. It is worth noting that a small increase of the order of 5%, in the operating voltage reduces the life by 50%. Similarly 1% change in the applied voltage changes the wattage by 1.5%, efficiency by 2% and output by 3.5%. Similarly we cannot ignore the fact that operation at lesser voltage than the rated value results in increased life period. The output on the other hand is drastically reduced so that in order to have the same flux output we have to use more number of bulbs. This as a matter of fact speaks of a false economy. Actually operation at reduced voltage and use after 70% of normal life is un-economical as the light output is drastically reduced.

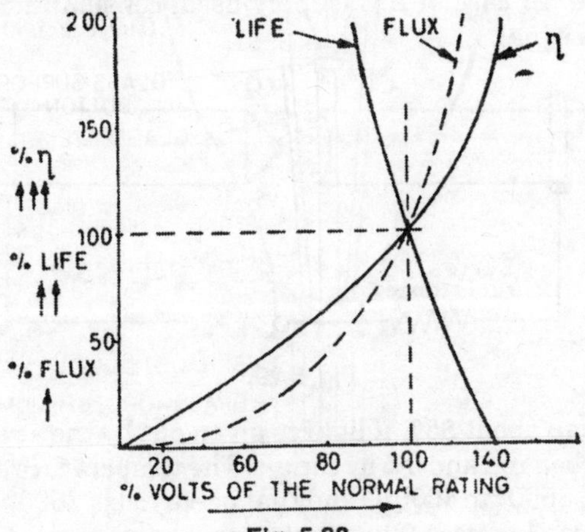

Fig. 5.28.

Effects of Glass Cover. The type and colour used for the glass cover produce a vital effect on the quality of light emitted from an incandescent bulb *e.g.* the milky glass cover gives a diffused light output which is almost *natural* in effect, like day light. The bluish glass cover gives the moon light effect which is pleasant only on hot summer nights. The green, yellow and red glass covers are useful for signalling purposes. The mercury reflector cover is useful for show room lights. Other miscellaneous coloured glass covers are used for decorative lighting.

5.11.2 Comparison with gas discharge lamps

The filament lamp is simple in construction and easy to install and maintain. The initial cost is less and it starts instantaneously. The lamp is free from stroboscopic effect. It has a low efficiency and a short life. Vibration very much shortens the life of the lamp and care is needed in mounting. It has a good colour equality with some tendency towards yellow. Frequency of stops and starts has no effect on the life of the lamp.

5.11.3 Arc Lamps : Carbon arc Lamp.

This is earliest type of lamp and is still used in cinema projectors and search lights when a very bright point source of light is required.

In this lamp two hard carbon rods are placed end to end and connected to D.C. supply. The voltage of D.C. supply should not be less than 45 volts. When current passed through the carbon rods they become incandescent due to high resistance. When the two carbon rods are separated apart by a small distance of two or three mm, an arc is struck between the two carbon rods and white light will be produced. The arc is maintained by transfer of carbon particles from +ve rod to -ve rod. So positive rod becomes hollow after some time and negative rod becomes pointed pencil. The rate of burning of +ve rod is twice than that of -ve rod. In case of A.C. supply used, consumption of both the carbon rods is equal.

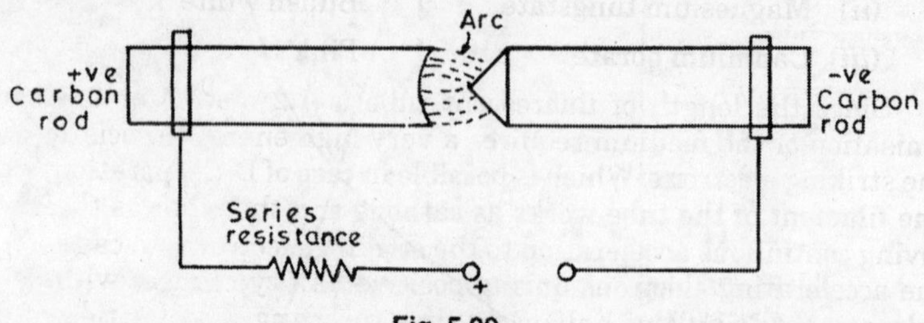

Fig. 5.29.

In this lamp about 85% if light is given out by the +ve carbon rod, 10% by -ve carbon rod and 5% by the air. The temperature of +ve carbon rod is between 3500° to 4000°C and that of -ve rod is 2500°C. Luminous efficiency of such lamps is 9 lumens/watt.

The electric arc has a -ve resistance characteristics so to avoid a short circuit, a series resistance is used for stabilising the arc.

5.11.4 Gaseous Discharge Lamps : Fluorescent Tube (Working Principle). The principle of working of a fluorescent tube will be clear from the following discussion :

We are familiar with the term excitation collision and excited atom from the study of article previously discussed. Consider the cathode filament emitting electrons after getting heated due to supply of current in Fig. 5.30. These electrons while accelerating collide with argon and mercury vapour atoms. The excited atoms of mercury give a radiation of 2537 A.U. wavelength in the given case due to low pressure inside the tube. Obviously this radiation is ultra violet and can produce no illumination if left as such. On the other hand, if radiation of this

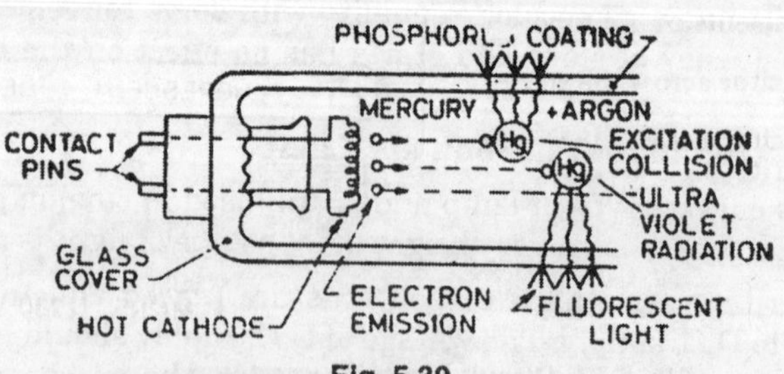

Fig. 5.30.

wavelength is made to strike with certain phosphor material, there is re-emission of light radiation of a different wavelength. This effect is called *fluorescence*, hence the name fluorescent tube.

Different colours of light are produced by different phosphors coated on the inner side of the tube glass cover. Some of the phosphors and colours given are mentioned below :

(*i*) Zinc silicate	Green
(*ii*) Magnesium tungstate	Bluish white
(*iii*) Cadmium borate	Pink

Since the length of fluorescent tube is 1.2 metres or 0.6 metre ionisation of the medium requires a very high energy associated with the striking electrons. Which is possible in case of D.C. operation, when one filament of the tube works as cathode and the other as the anode giving continuous acceleration to the striking electrons. In case of A.C. the accelerating electrons during positive half cycle meet with de-acceleration during -ve half cycle due to changed polarities of the electrodes. Thus a high voltage surge is essential to strike a discharge across the entire length of the tube during any half cycle of the supply, after which the discharge is maintained even at lesser than the supply voltage.

To study the effect let us consider the circuit shown in Fig. 5.31.

Working. When the circuit shown in Fig. 5.31 energised, nearly full voltage appears across the starter terminals, due to low resistance of filaments and negligible current through the choke. The starter is filled with some mono-atomic gas like argon. This gas ionises and a glow appears inside the starter. This warms up the bimetallic strip carrying moving contact. Soon after the strip bends and short circuits the starter terminals. This gives a circulation of strong current through the filaments F_1 and F_2, and the choke circuit. The filament is coated with oxides of barium and strontium, resulting in thermionic emission which can ionise the argon gas inside the tube. In the mean while the

F_1, F_2=filaments of the tube

C'=capacitor across the starter

C=capacitor across the tube circuit

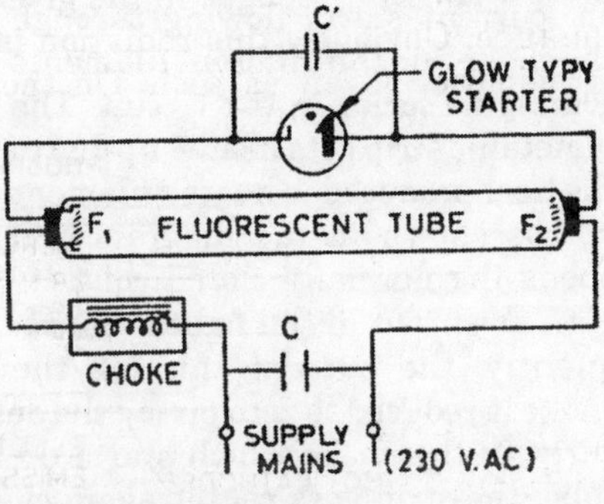

Fig. 5.31. Circuit for the fluorescent tube.

bimetallic strip of the starter cools down and breaks open the starter circuit. This sudden opening of the starter contacts results in an abrupt change of current/flux linking with the choke coil. A high voltage surge is induced in the choke which makes one the filament at a very high potential w.r.t. the other filament. This momentary high p.d. is enough to ionise the gas medium, *i.e.* mercury and argon present inside the tube. This results in passage of current between two electrodes inside the tube. As long as the ionisation persists collision between electrons and mercury atoms continue. Further emission from filaments is possible through the striking +ve ions with the filaments giving secondary emission. The current does not flow afterwards through the starter circuit due to the low p.d. across the tube since a part of the supply voltage drop across the choke circuit.

The choke also helps in keeping the current through the ionised medium within safe limits. The radiation from excited mercury atoms falling on the phosphor coating results in fluorescence as explained earlier. The average life of this fluorescent lamp is 4000 hours and its efficiency is about 40 lumens/watt.

Function of Capacitors. The capacitor across the starter terminals is to avoid introduction of noise signals into the reception of a radio/transistor receiver being operated near the tube. If the capacitor C' is missing across the starter, a radio receiver starts picking up a typical hum, which disturbs the quality of reception. The capacitor C across the tube main circuit is to improve the power factor which is otherwise low due to the presence of inductive chock in the tube circuit.

Tube circuit with thermal starter switch. The starter in the given case makes use of a bimetallic strip which keeps a pair of contacts closed when cold. A heating filament is also incorporated in the device, which comes in series with the main filaments of the tube as shown in Fig. 5.32. When the circuit is switched on, current starts flowing through the choke and the different filaments coming in series in the circuit. The bimetallic strip gets heated up due to the heat produced in heated filament HF as shown in the diagram and opens the normally closed contacts N.C. as shown in the figure. Consequently, the current through the choke is reduced to zero giving the desired change of flux to induce H.T. surge in the choke which starts the tube. When after starting of the tube, the strip goes cold it again closes the starter contacts. The p.d.

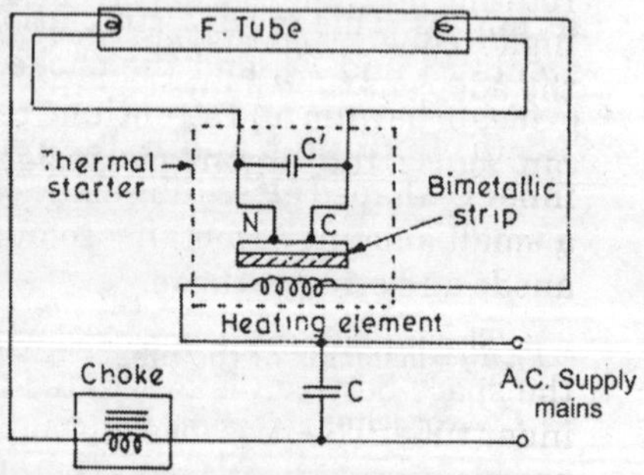

Fig. 5.32 Tube circuit with thermal starter switch.

across the tube does not send a strong current through the starter circuit to produce any remarkable action.

Note. In temporary lighting for decoration purposes, starters are not used by many electricians. Instead of starter they bring two leads from the filament, short the leads momentarily and open abruptly, giving the necessary start.

Tube circuit with D.C. mains. While operating across D.C. there is no necessity for choke and the capacitors. The current after ionisation has of course to be checked and kept within safe working limits. For this purpose an incandescent bulb is introduced in series with the filaments, which keeps glowing after the tube has started. The energy consumed by the series bulb is not so high as its rated value.

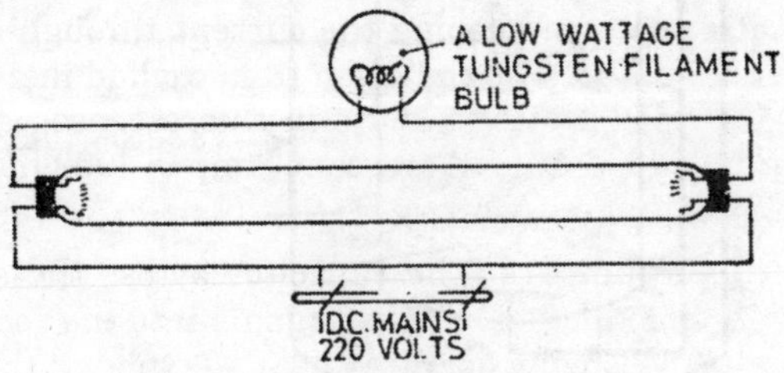

Fig. 5.33.

It is worth noting that while working on D.C. the tube light does not suffer from stroboscopic effect as on A.C. supply. This effect can be observed by looking into any thing rotating in the tube light. At certain r.p.m. an object rotating clockwise in actual practice will look like rotating anti-clockwise.

This can create a nuisance in the industrial activity involving rotating machine tool parts. The effect can be minimised either by using high frequency special supply circuits for tube lights or by distributing the adjacent tube lights on different phases of supply.

(*ii*) **Sodium Vapour Lamp.** Sodium vapour lamp consists of an inner U-shaped glass tube made of high resistance glass and containing a small amount of metallic sodium, Neon gas and two electrodes, one anode and other cathode.

The inner tube is enclosed with an outer glass tube of U-shape and the space between is evacuated to minimise the heat losses from the inner tube. The presence of neon gas serves to vaporise the sodium by producing sufficient heat. The glass tube is used of U-shape because long discharge paths are necessary. This lamp is used horizontally to help the sodium when spread along the tube.

When supply is given to the lamp the discharge take place in the neon gas first and red orange glow produces. The metallic sodium gradually vaporises and then ionises thereby producing a light of yellow colour.

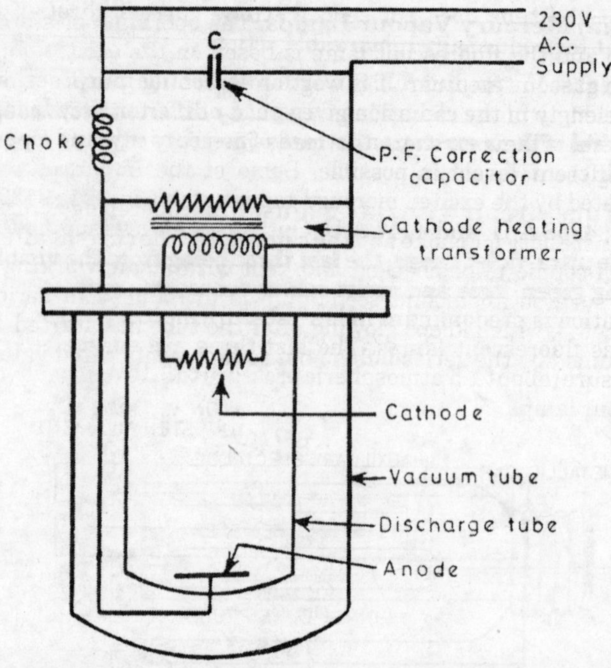

Fig. 5.34

This lamp is also called as a monochromatic lamp because it emits a light of yellow colour of wave length 5900 Å unit. The lamp takes about 5 to 6 minutes for starting and takes about 15 minutes for coming in full brilliancy. The luminous efficiency of the lamp is 50 lumens per watt and its average life is about 3000 working hours. Operating temperature for this lamp is about 300°C and lamps are manufactured in rating of 40, 60, 85 and 140 watts rating.

The cathode heating transformer is used to send a high voltage of about 480 volts at the time of starting the lamp so that luminous discharge inside the tube takes place. Under normal running conditions, normal rated supply is given to lamp.

Function of choke is to limit the current which tends to increase due to -ve temperature resistance characteristics of the discharge path. The condenser is used to improve the power factor, which becomes as low as 0.3 due to choke and transformer.

These lamps are suitable only for A.C. supply operation. These lamps are used for general outdoor lighting, and highways where colour distortion is not of much importance. These lamps are not suitable for indoor lighting.

(*iii*) **Mercury Vapour Lamps.** The operation of mercury vapour lamp and the fluorescent lamp is based on the excitation of atoms of some gaseous medium. It is worth mentioning that the frequency and wavelength of the radiation given out by different atoms depends upon the level of their excitation. In case of mercury vapours alone excitation to different levels is possible. Some of the important wavelengths radiated by the excited mercury vapour atoms include 2537 A.U., 1561 A.U., 4358 A.U. and 4047 A.U. Out of these wavelengths the first falls in the ultra-violet range, the last three being from the visible spectrum giving green, blue and violet colour sensations respectively. The first radiation is predominant in low pressure mercury vapour lamps working as fluorescent lamps. The last three are emitted in case of high pressure (about 1.5 atmospheric pressure to 5.10 Atmospheric) mercury vapour lamps.

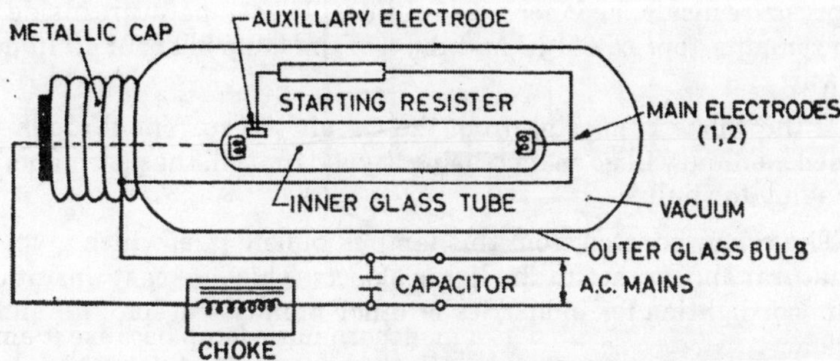

Fig. 5.35. Construction and Circuit for a high pressure M.V. Lamp.

Construction. The tube containing mercury vapour is made of hard glass. The outer glass cover protects the inner tube from coming into direct contact with atmospheric temperature variations. It also absorbs the ultra-violet radiations emitted from the lamp during working. Besides mercury the inner tube consists of a percentage of argon gas. There are two main electrodes made of tungsten wire and a starting electrode which is spaced quite close to one of the main electrodes. During connections to the supply mains the auxiliary electrodes through a high series resistance along with the main electrode (2) is connected to say neutral of the supply. The phase comes to the first main electrodes through the insulated part of the metallic cap.

Working. When the circuit is energised the supply voltage appears between the main electrode marked (1) and the starting or auxiliary electrode. The argon coming between these two electrodes is immediately ionised because distance between these two electrodes is very small and a glow appears between the said electrodes. A small current starts flowing through the starting resistor in series with the auxiliary electrode. This results in building up of pressure due to heating up of mercury which is originally in the condensed form. Ultimately the medium between the main electrodes is ionised and the current starts flowing between the two. It is no longer controlled by the high resistance in the starting circuit, but by the choke in the main circuit. (The resistance of the ionised medium is much smaller and the current chooses the low resistance path). The function of the capacitor is to improve the power factor of the circuit which will otherwise be quite low due to the high inductance choke present in the circuit. The lamp discussed above is called M.A. type and is rated at 250-400 Watts for A.C. mains voltage, *i.e.*, 230 V.

Caution. The lamp should always be mounted vertically, otherwise the inner tube can be damaged by the striking arc, thus rendering the device unserviceable.

The time taken during starting of the lamp is 5 to 6 minutes. The source will not start working again if the supply is once interrupted, till the pressure inside the inner tube subsides. The temperature inside the bulb remains approx. 650°C and the η of the lamp is about 40 lumens per watt.

If the choke is missing from the circuit the current through the ionised medium will go on increasing beyond control, thus permanently damaging the bulb.

The colour emitted from this lamp is bluish pale, which gives an un-natural appearance to the living objects. This makes it unsuitable for indoor lighting for industries or other premises despite its higher efficiency.

Lastly the lamp is suitable for A.C. mains only.

(*iv*) **Mercury vapour lamp without choke.** Another type of commercial purpose mercury vapour lamp is suitable for both A.C. as well as D.C. supply systems. The other major advantage with this type of lamp is the improvement of colour over that given out by an ordinary M.V. lamps discussed earlier. In this case the current is controlled by a series tungsten filament which glows besides the light emitted from mercury tube. Consequently the shortage of red colour in the light so emitted is compensated and quality of the light is improved.

Operation. The fundamentals of operation and construction of the M.A.T. type mercury vapour lamp can be given as follows :

A tungsten filament gets full supply voltage through a thermal switch with two N.O. and two N.C. contacts. It starts emitting its own light and simultaneous heating the inner mercury bulb. A stage comes when the mercury vapours can be ionised with a suitable supply voltage appearing across the main electrodes of the inner mercury tube. At this stage the thermal switch operates which (*i*) brings the tungsten filament in series with the main electrodes and (*ii*) short circuits a part of the tungsten filament. Thus a discharge occurs through the mercury bulb, besides a part of the tungsten filament gives its own share of light improving the quality of light emitted in total by the lamp. Absence of a choke permits use of both D.C. as well as A.C. supply mains. The bulbs are rated at 300/500 watts at supply mains voltage from 200-250 volts.

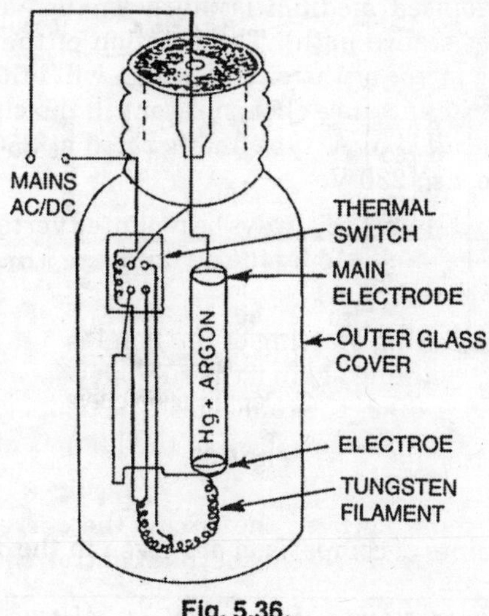

Fig. 5.36.

(*v*) **Neon lamps and tubes.** Neon lamp consists of a glass bulb filled with neon gas with a small percentage of helium. The two electrodes in the glass bulb are of pure iron spaced few mm apart. These lamps are operated on 110 volt **a.c.** or 150 volts **d.c.** supply.

Neon lamps give orange pink coloured light and its size is very small as compared to incandescent lamp. Its luminous efficiency is 15 to 40 lumens per watt and power consumption is about 5 watts. These lamps are used as indicator lamps and night lamps.

(*vi*) **Neon tubes.** It consist of a length of glass tubing having two electrodes made of iron or copper. Its operating voltage varies from 300 to 1000 volts per metre of tube length and high voltage is obtained from a step up transformer. The dia of glass tube varies from 10, 15, 20 and 30 mm.

Neon tubes are used in advertising and decorative treatment of buildings. The connection between letters in the tube are made through nickel wire and p.f. of neon tube is improved by using capacitor on the low voltage side of transformer.

5.12. Comparison of Different Sources

The below three figures can give an idea about the (1) output in lumens for each watt of power input to different sources (2) fall in total output with the passage time, *i.e.* depreciation effect (3) total life period of the three sources commonly used for lighting purposes in various situations.

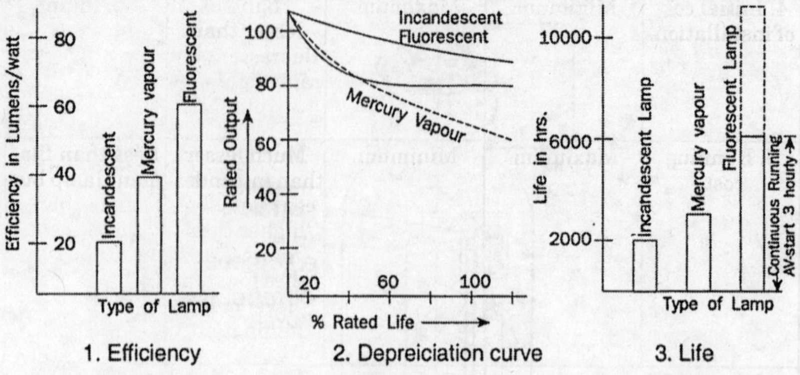

1. Efficiency 2. Depreiciation curve 3. Life

Fig. 5.37.

The other points of comparison are given in the Table :

Reference	*Incandescent lamp*	*Fluourscent lamp*	*Mercury vapour lamp*	*Sodium vapour lamp*
1. Starting time.	Instantaneous start.	Fraction of a second or little more.	Five to six minutes.	Five to six minutes.
2. Supply goes off and recovers next moment.	Lamp becomes momentarily off.	Tube goes off and restarts if the voltage remains sufficiently near rated value with a gap of fraction of second.	Lamp goes off and can not be re-started after the recovery of the voltage till the pressure falls to normal.	Lamp goes off and cannot be restarted after the recovery of voltage till the pressure falls to normal.
3. Colour of light.	Very near the natural light. No colour distribution is produced at proper illumination level.	Colours given, vary with different phosphor coatings. Each colour gives different impression for the same object, and differs from the natural look of the object.	Colour given by M.A. type M.V. lamp is very displeasing if a person has to constantly work in its light. Colour distortion is high.	Colour of light is yellowish and produces colour distortion.

Reference	Ineandescent lamp	*Fluourscent lamp*	*Mercury vapour lamp*	*Sodium vapour lamp*
4. Initial cost of installation.	Minimum.	Maximum.	High but lesser than fluorescent for same light output.	Maximum.
5. Running cost.	Maximum.	Minimum.	Much lesser than in-candescent lamp but higher than fluorescent tubes for same level of illumination.	Less than filament lamp but more than fluorescent tube.
6. Suitablility.	For direct lighting, Domestic, industrial, street lighting, flood lights.	For semi-direct lighting, domestic, industrial, commercial, roads and halls.	For open spaces like yards, parks, highway lighting when passing through towns.	For open spaces, highways and street lighting.
7. Stroboscopic effect.	Absent.	Present.	Present.	Present.
8. Change in voltage.	Effects efficiency, life and output to a great extent.	Effects starting. Light output changes not that remarkably as that of incandescent lamp. If too low, tube may not start at all.	Effects starting time and colour of radiation.	Affects starting and colour of light radiation.
9. Nature of supply.	Works equally well on both A.C. and D.C.	Different supplies require different accessories.	M.A. type works on A.C. only, M.A.T. type works on A.C. as well as D.C.	A.C. only.
10. Switching.	Frequent switching does not effect life of operation.	Switching effects life period.	Switching does not effect life period as no H.T. surges exist in A.C. starting, and no filament. Current changes produced during starting.	Effects life.

11. Local light.	Suitable for illuminating one object with proper reflector.	Not suitable for local lighting.	Not suitable for local lighting.	Not suitable for local lighting.
12. Change of colour.	Can be produced by changing the colour of the glass cover, at the cost of efficiency which is reduced.	Can be produced by using different phosphor coatings on the inner-side of the tube with reducing efficiency.	Colour improved in M.A.T. type by introducing a series tungsten filament, which improves the lumens output/efficiency to some extent.	No change in colour.
13. Utility	Maximum. Due to wide range of voltage ratings. Used on automobiles, trains, emergency lights, aeroplanes, signals at crossings and electric signals for raiways etc.	Quite high but voltage rating being limited. Use is confined to main's voltage only or complicated inverter circuits converting 12 volts D.C. into high volts A.C.	Very limited utility that too on mains voltage.	Used for street lighting system.
14. Position of working.	Any position.	Any position.	Vertical only.	Horizontal.
15. Luminous efficiency.	12 lumens/watt.	40 lumens/watt.	40 lumens/watt.	60 - 70 lumens/watt.
16. Average life.	1000 working hours.	4000 working hours.	3000 working hours.	3000 working hours.

5.12.1 Basic Principles of Light Control

When light falls on the surface, depending upon the nature of the surface, some portion of light energy is reflected. Some portion is transmitted through the medium of the surface and rest is absorbed. The ratio of reflected light energy to the incident light energy is known as reflection factor.

There are two basic types of reflection :

(*i*) Mirror or specular reflection and

(*ii*) Diffuse reflection.

In case of specular reflection a beam of light is reflected but not scattered and unless the eye is placed in the path of the reflected beam, the viewer is unaware of the existence of light. Moreover if his eye is placed in the path of reflected beam it sees, not the illuminated surface, but the light source. Surfaces causing specular reflections are silvered mirrors, highly polished metals etc. With diffuse réflection the reflected light is scattered in all directions and the viewers sees the illuminated surface, not the light source. Surfaces causing diffuse reflection are paper, frosted glass, chalk, dry earth, plaster etc. If the surface that is uniformly illuminated by a beam of light appears to be equally bright when viewed from all possible angles, the reflection is said to be perfectly diffuse. Perfect mirror surfaces and perfect diffusing surfaces are ideals that do not exist in nature. The reflection from any actual surface is partly specular and partly diffuse the proportional varying

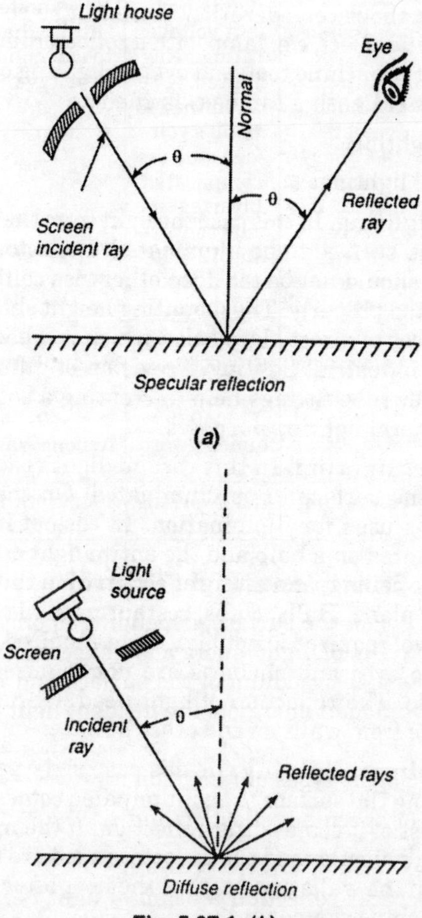

Fig. 5.37.1 (*b*)

widely. A surface that is almost free from mirror reflection is called a mat surface.

5.13. Types of Lighting Schemes

As far as designing a lighting scheme is concerned, before going into calculation of rating sources and spacing etc. we have to see whether the sources throw light directly over a working surface, or it reaches the surface, after reflection from some other surfaces. From that point of view the lighting schemes can be classified as :

(*i*) Direct lighting.

(*ii*) Indirect lighting.

(*iii*) Semi-direct lighting.

(*iv*) Semi-indirect lighting.

Besides the above consideration there can be sources mounted to illuminate specific object, *e.g.* lamp with a reflector illuminating the job and tools part of a machine tool, and overall lighting of a machine shop. These two types of lighting can be classified as :

(*v*) Local lighting.

(*vi*) General lighting.

(*i*) **Direct lighting.** In this case the light from the sources is thrown directly over the surface to be illuminated. Reflectors give additional help, but these should not be too deep otherwise ceiling and walls will not be illuminated properly. The mounting height should not be too low otherwise chances of direct glare exist, which is a hazard for accidents particularly in industrial lighting. Provision of diffusing sources like globes, covers for tube fittings help in creating a soothing effect, particularly for general lighting purposes.

(*ii*) **Indirect lighting.** In this case no light reaches directly from the source on the surface to be illuminated. On the other hand, the reflected light is used for illumination. In case of indirect lighting a reflector is mounted on a bulb and the entire light emitted by the bulb is thrown on the ceiling, and after reflection from the ceiling light falls on the working plane. Halls, clubs, restaurant halls and places where illumination level required is not high, light required, is free from glare and soothing to eyes and shadows are not required, use of indirect lighting is made. The reflection of light need not necessarily be from ceiling, it can be from walls or and other surface.

(*iii*) **Semi-direct lighting.** In this case sixty per cent or more of the light reaching the surface to be illuminated comes directly from the sources. The rest of it comes after reflection. If the mounting height of the lamps is high enough and the reflectors used are not deep, a part of the light falls on the walls and reaches the floor after reflection besides the directly reaching component.

(*iv*) **Semi-indirect lighting.** In this scheme more than sixty per cent light is thrown on the surface/surfaces reflecting the light. The rest comes on the surface to be illuminated directly from the sources. This will happen if the mounting height of the lighting source is low. The scheme is useful for indoor domestic lighting. It is free from glare and pleasing to eyes, with globes and reflectors.

(*v*) **Local lighting.** If the light is confined to illuminate a particular object, it is called local lighting *e.g.* a lamp mounted in a deep reflector and fitted on machine tool like lathe, planner etc. or an ordinary table lamp in a study room gives local lighting.

(*vi*) **General lighting.** If in a big hall number of lighting sources are fixed to give local lighting effect, there will be dark patches between bright spot scattered here and there. This will cause a lot of fatigue on the eyes of the workers, when they try to see something away from their job. This will happen due to time gap necessary for adjustment of pupils of the eyes to different levels of lighting. Therefore, in addition to local lighting some general lighting is compulsory. In a show room we find some local lighting sources focused on particular displayed items besides other sources of light which give general light to the entire show room. In industry, general lighting is also compulsory to minimise accidents and to avoid the feeling of loneliness/monotony on the workers working under local lights. The ratio of general to local lighting levels should not be more than one to three.

5.14. Requirements of Good Lighting

A good lighting is one, under the influence of which the eyes of persons, using the light do not feel any strain *e.g.* if the light is too dim we cannot observe the objects properly or cannot do reading or writing work. Whereas on the other hand, if the light is too bright our eyes start watering and are made so open forcibly. The good lighting does not have the above mentioned effects, but we find comfort while working under it.

The factors to be taken care of while planning a lighting scheme are:

(*a*) Illumination level.
(*b*) Absence of glare.
(*c*) Contrast.
(*d*) Shadows.
(*e*) Colour rendering.

(*a*) **Illumination Level.** The visibility of objects depends upon the magnitude of light falling over them and the light reflected by them. It further depends upon the following factors :

(*i*) **Size of the object.** Smaller the objects, greater is the level of illumination required for proper visibility.

(*ii*) **Distance between the sources and the object.** Greater is the distance between the source and the object higher is the degree of illumination necessary.

(*iii*) **Distance between the observer and the object.** Smaller distance requires lesser level of illumination and *vice versa*.

(*iv*) **State of object.** Stationary objects require lower level of illumination as compared to moving objects.

(*v*) **Period of observation.** Objects to be observed continuously for longer period require a higher level of illumination.

(*vi*) **Colour of object.** Darker the colour of the object greater is the level of illumination necessary.

(*vii*) **Contrast w.r.t. the background.** A black object in front of a white wall requires lesser level of illumination as compared to a white object placed at the same place.

(*viii*) **Purpose.** The requirement of illumination varies with the purpose, *e.g.* a badminton hall requires much higher level of illumination as compared to a cinema hall. Same consideration applies to a reading room versus a bed room.

Some idea can be framed about the level of illumination necessary for different applications as given below :

Place of application	*Lumens/m²*
Drawing room	80-200
Bed room	30-100
Study room	150
Bath room	80
Kitchen	80
Varandah	50
Toilet	100
Sewing	150
Library	150-200
Classroom	150
Drawing office	200
General office	80-110
Indoor sports	500
Battery charging room	100
Typing work	225
Coil and armature winding	200
Fitting shop	200-250
Machine shop	100-200
Compound	25-50
Very fine bench and machine work	800
Core making	100-150
Operation room (Hospital)	350
Operation table	3500

(*d*) **Absence of Glare.** Glare is the sensation experienced by human eye when some rays of light enter the eye directly from a source. *i.e.* Drivers use dipper at night when facing other vehicles to avoid glare. In case of a lighting scheme this glare can be of two types:

(*i*) Direct glare

(*ii*) Indirect glare.

In the first case, if the level of mounting the lamps is too low and there are no reflectors used, the effect will cause direct glare, *e.g.* a table lamp without reflector. Thus while using direct lighting, it should always be seen that no source is too low to throw light directly into the eyes of an observer.

In the second case the glare is caused by a shining object reflecting a major part of the incident light. This can be avoided by fixing the lamps at such a position that no direct light falls on shining surfaces *e.g.* chrome or nickel plated parts of machines or appliances.

Glare not only causes undue fatigue, but can be a cause of industrial accidents.

(*c*) **Contrast.** Whenever using sources of local illumination, care has to be taken for additional general light. As discussed earlier we can see that the pupil of human eye has to adjust its aperture depending upon the level of illumination. If we go from sun light into a dark room, we require sufficient time to locate the things properly inside the room. Therefore, if local lighting is not supported by general lighting one cannot see the objects away from the local source. This will cause a strain on the eyes of a person who has to move to and fro from the source of the local light. In no case the ratio of local to general illumination exceed more than 3 : 1.

(*d*) **Shadows.** The shadows play an important role in the design of a lighting scheme. In case of industrial lighting too deep and dark shadows cannot only be source of fatigue on the eyes of workers moving from brightly lit places to the shadows, but can be a source of an accident. On the other hand, dull shadows are a necessity to give a three dimensional look to any solid object. A third case is that of a place entirely free from shadows which can be the drawing office of any architect or a planning engineer, where uniform, bright and indirect lighting sources are used illuminating the drawlings perfectly without glare.

(*e*) **Colour rendering.** The looks of the same objects differ under different coloured lights. When there are other than the natural looks, we speak of colour direction. For most of the applications we prefer to have natural appearance of various objects. Some lamps, like sodium vapour/mercury vapour lamps produce such colour distortion that living objects give un-natural appearance. Thus the lamps are confined to outdoor illumination only. On the other hand , we can use lamps giving bluish light where we want a psychological feeling of coolness and calmness. In surgical operation theatres we can use greenish yellow tube lights producing maximum sensitivity for staff on duty. Using red colour lamps in a bed-room can badly effect the temperament of the inmates.

From above discussion we can guess the importance of colour of light emitted by different sources while selecting for a lighting scheme.

5.14.1. Lamp fittings and lighting systems

(*a*) Direct lighting

In this method, 90–100% of the light is directed downwards and 0–10% of the light upwards. This should be carefully used to avoid excessive glare, brightness or harsh shadows [Fig. 5.37.2 (*a*) and (*b*)].

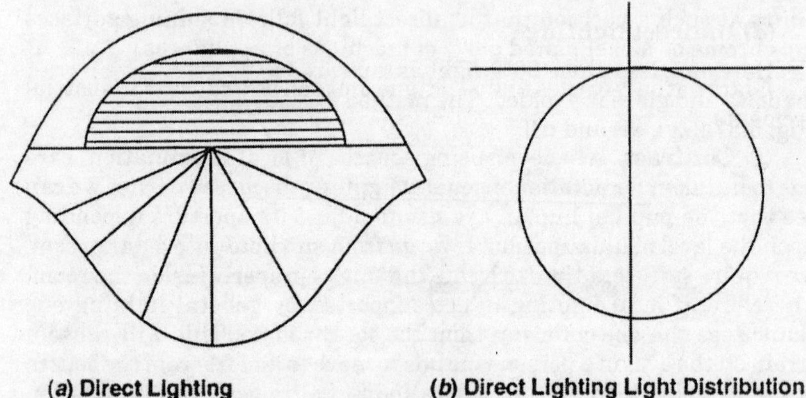

(*a*) Direct Lighting (*b*) Direct Lighting Light Distribution

Fig. 5.37.2

(*b*) Semi-direct

This method has 60–90% of light directed downwards and 10–40% upwards [Fig. 5.37.2 (*c*) and 5.37.2 (*d*)].

The distribution of light is shown in Figs. 5.37.2 (*c*) and (*d*).

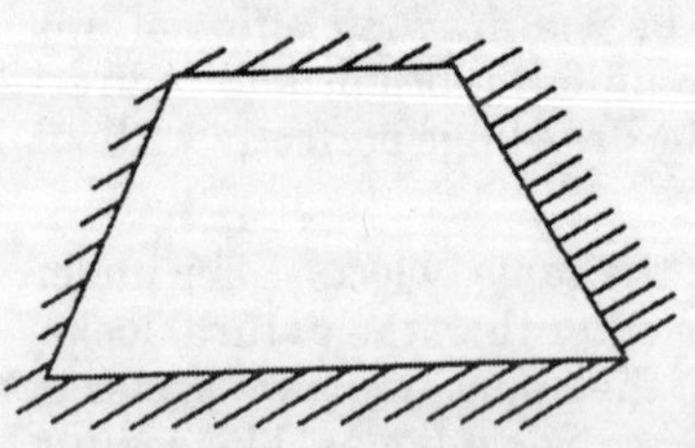

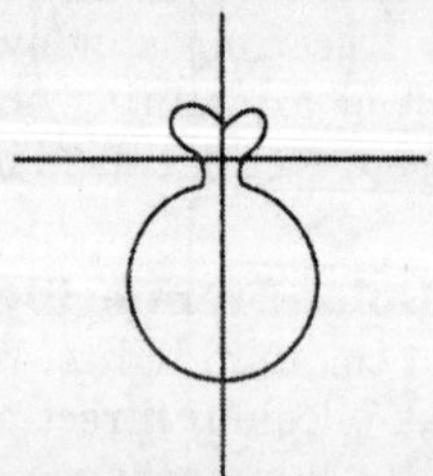

(*c*) Semi Direct Lighting (*d*) Semi Direct Light Distribution

Fig. 5.37.2

(*c*) **Diffusing**

This arrangement gives a uniform light distribution and is pleasing [Fig. 5.37.2 (*e*) and (*f*)].

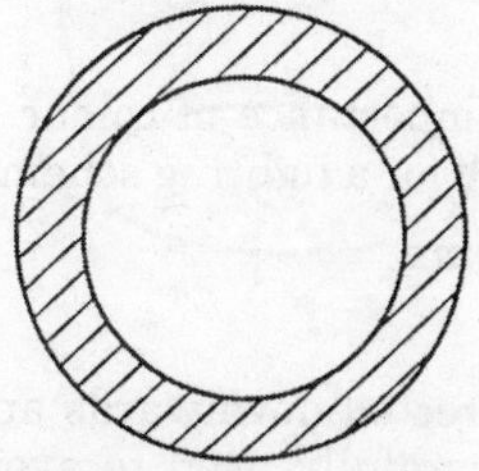

(*e*) Diffused Lighting

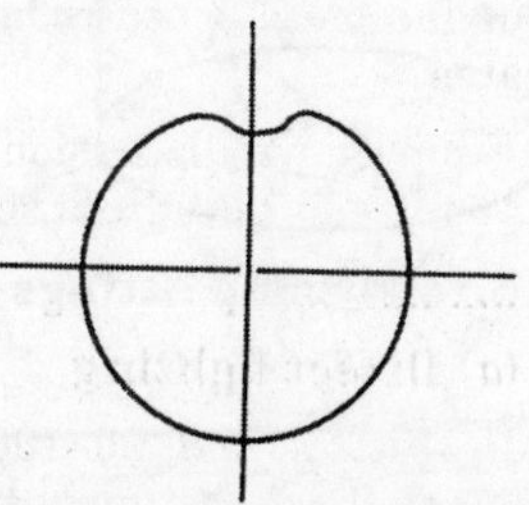

(*f*) Diffused Light Distribution.

Fig. 5.37.2

(*d*) **Indirect lighting**

Here not less than 90% light is upwards. The fittings have no shadows and glare is avoided. The method is costly as light is upwards [Fig. 5.37.2 (*g*), (*h*) and (*i*)].

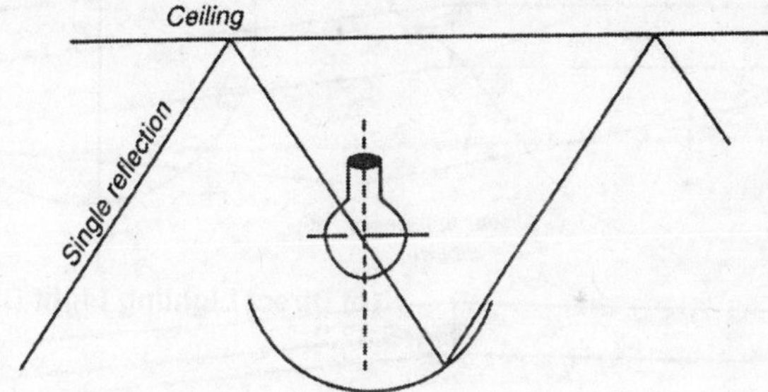

Fig. 5.37.2 (*g*) Indirect Lighting

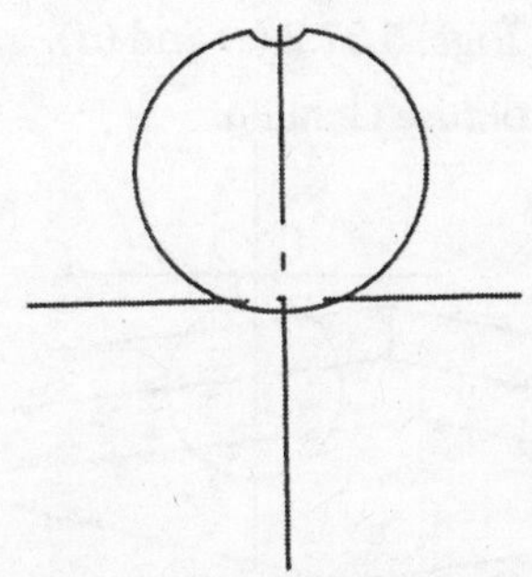

(*h*) Indirect Light Distribution

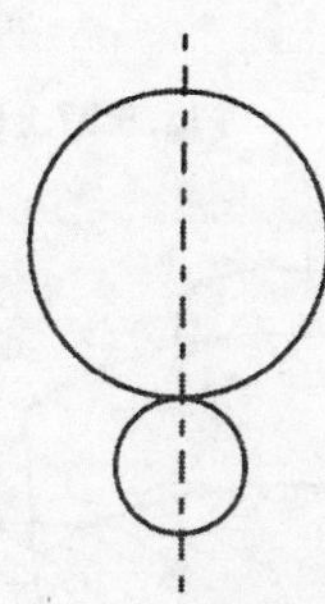

(*i*) Semi Direct Light Distribution

Fig. 5.37.2

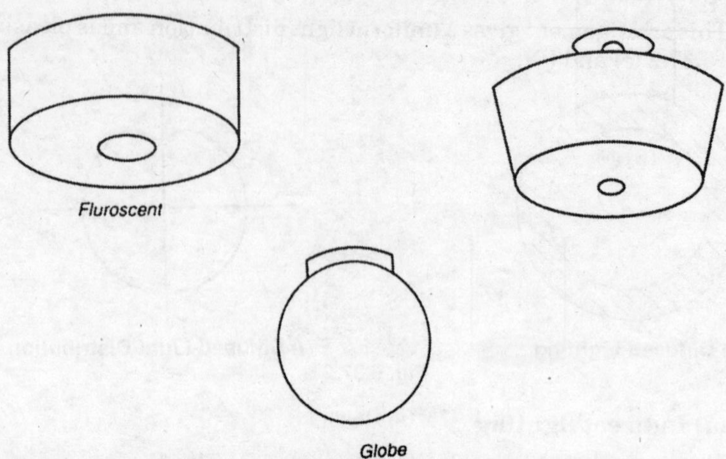

Fig. 5.37.2 (*j*) Diffusing Type Luminaires.

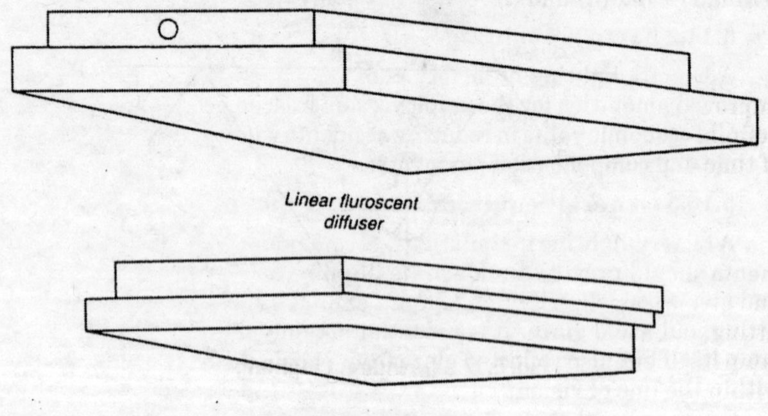

Fig. 5.37.2 (*k*) Large, General Diffuse Lighting.

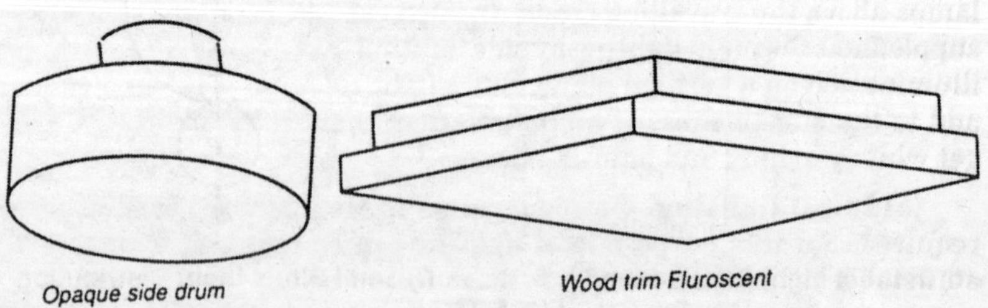

Fig. 5.37.2 (*l*) and (*m*) Diffuse Type Down Lighting.

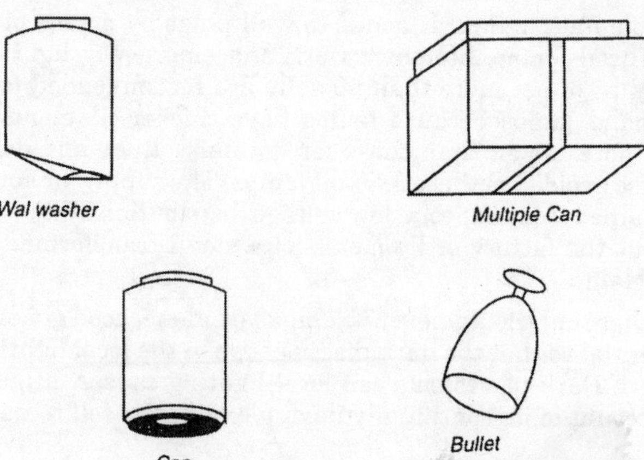

Fig. 5.37.2 (*n*) Down Lighting.

In semi-indirect lighting, 10–40% is downwards and the rest 60–90% light is upwards. Some of the luminaires used for mounting on the ceiling are shown in [Figs. 5.37.2 (*j*) to 5.37.2 (*n*)].

5.14.2 Factory Lighting.

Adequate lighting of factory is of very importance, as it provides improved amenities for the employees, increased production and has a definite economic value in reducing accidents with their consequent loss of time and compensation payments.

5.14.3 General requirements and Types of installations.

A factory-lighting installations, in common with other indoor equipments should provide an adequate illumination on the working plane and give a good distribution of light, employ simple and easily cleaned fitting and avoid glare. It is essential not only to avoid glare from the lamp itself but also reflected glare from polished surface which may be within the line of vision.

(*a*) **General lighting.** The usual scheme in factories and workshops is to mount a number of lamps at a sufficient height so that uniform distribution of light over the working plane is obtained. In large machine shops the height is governed by the necessity of keeping the lamps above the travelling crane. In such cases it is often desirable to supplement the main lighting by side lighting in order to give additional illumination on a vertical plane. Since light coloured walls and ceiling add to the effectiveness of an installation, therefore it is necessary to get white washing and painting done.

(*b*) **Local lighting.** On some points fairly intense illumination is required. For this purpose local lighting can be provided by means of adjustable fittings attached to the machine or bench in question or mounted or portable floor standards. Such lamps should be mounted in deep reflectors so that glare is avoided.

Portable hand lamps attached to wall plugs by means of trailing leads are used for maintenance work and emergency lighting. Low voltage lamps of not more than 50 volts are recommended for use as portable hand lamps because lamps have thicker filament and are therefore more robust than those for normal voltage and danger for shock is also avoided in these few volt lamps. The supply for such lamps can be obtained from a special low voltage distribution system running through out the factory or by means of a small transformer for each individual lamp.

Local lighting should never be employed alone, good general lighting is essential so that the dark places between the local lighting units are avoided. Dark places between local lighting causes fatigue to the eyes on account of its continually having to adjust itself to new conditions.

(*c*) **Emergency Lighting.** Some lights such as for (*i*) internal pilot lighting required for safe and speedy evacuation of personnel after main lighting circuit is off. (*ii*) External pilot lighting provided with careful shades leading to shelters required for evacuation of personnel. (*iii*) For control posts, first aid centres etc. (*iv*) Dials and gauges in important plants required to be watched regularly are required during an air raid when all the factory lights are off as a matter of air raid precaution. The circuit supplying the above emergency lights should be independentally controlled.

It is very desirable to provide auxiliary lighting from the source other than the main electric supply preferably from batteries or from small petrol driven generator set. If however, emergency light circuits are operated from main electric supply, these should be completely separated from main lighting circuit.

(*d*) **Industrial Lighting Fitting.** Reflectors from industrial purposes must be simple in design and easily cleaned. The requirements of most of the installations can be met by one of the following types of fittings :

(*i*) **Standard Reflectors.** These reflectors are made to accommodate lamps of rating from 40 to 1500 watts and designed. So that they give adequate and uniform illumination when they are mounted at a spacing equal to about 1.5 times their mounting height above the working plane.

(*ii*) **Diffusing Filling.** When more diffused light is required to that given by the standard reflector a diffusing glass screen may be fixed across a standard type of reflector. Such fittings are used where highly polished articles are dealt with.

(*iii*) **Concentrating Reflectors.** A reflector with a concentrated beam is employed in large machine shops and foundaries. Where the fittings are to be mounted on a considerable height above the working plane, in such places an ordinary reflector would have too wide angle of divergence and would waste a great deal of light on the walls.

(*iv*) **Enclosed Diffusing Fittings.** An opal globe completely enclosing the lamp giving a very even and well diffused light is used when light coloured walls and ceilings are there.

(*v*) **Angle Reflectors.** Angle reflectors are provided to illuminate vertical plane when concentrating type reflectors are used. These can be mounted on suitable stanchions or the walls.

(*vi*) **Maintenance.** In order to maintain the fittings in a condition of reasonable efficiency it is necessary to clean the light fittings periodically. The frequency of cleaning depends on the conditions in the particular factory under consideration and varies from once or twice a week, for very dirty surroundings to every four or six weeks under the best conditions.

(*vii*) **Types of Lamps.** The discharge lamps have been used in where colour rendering is not important. The fluorescent lamps are widely employed on account of its natural day light colour, its even illumination and absence of glare and in some cases, the fact that it gives rise to considerable less than filament lamps of the same light output.

5.14.4 Street Lighting.

The main objects of street lighting are :

(*i*) To make the traffic and obstruction on the road clearly visible in order to promote safety and convenience.

(*ii*) To make the street more attractive.

(*iii*) To increase the community value of the street.

The principle employed for street lighting is different from that of interior lighting. There are no walls and ceiling which reflect or diffuse light hence only direct lighting scheme can be employed and hard shadow and high contrast can not be avoided. Because the area to be illuminated being large, the value of illumination for economic reasons is very low compared to that for indoor lighting and the question of colour rendering is also of minor importance. High level of illumination is not necessary, because firstly one has seldom to look continuously at near by objects. Secondly it is the impression of whole scene that is required. In fact in case of interior lighting the objects are seen by light reflected by them but in case of street lighting the objects are seen in relatively bright background *i.e.* silhouetted against a relatively bright background. Further more, owing to the low illumination, the eye is in its most sensitive state and therefore glare must be avoided.

Two general principles are usually employed in the design of street lighting installation namely.

(*i*) The diffusion principle and

(*ii*) Specular reflection principle.

(*i*) **Diffusion Principle.** In this case the lamps fitted with suitable reflectors are used. The reflectors are so designed that they may direct the light downwards and spread as uniformly as possible over the road surface. In order to avoid glare the reflectors are made to have a cut-off of between 30° to 45° so that the filament is not visible except from underneath it. The diffusing nature of the road surface cause the reflection of a certain proportion of the incident light in the direction of the observer and therefore the road surface appears bright to the observer. The illumination at any point on the road surface is calculated by applying point to point or inverse square law method. Over certain proportion of the road the surface is illuminated from two lamps and the resultant illumination is the sum of the illuminations due to each lamp.

(*ii*) **Specular Reflection Principle.** In this case the reflections curved upwards so that the light is thrown on the road at a very large angle of incidence. It is observed that a motorist require to see objects about 30 metres aways. Thus in Fig. 5.37.3 the observer is shown about 30 metres from the object. Much of the light from the lamp L_3 is not reflected towards the observer. Where as most of the light from lamps L_1 and L_2 is reflected towards him. Thus the object will appear silhouetted against the bright road surface due to lamps at a long distance. The requirement of a pedestrian, who requires to see objects in his immediate neighbourhood, is also fulfilled in this method as some light from the lamps falls directly downwards. The method of street lighting is only suitable for straight section of the road. This method is more economical also as compared to the diffusion method of lighting but it suffers from the disadvantage that it produces glare for the motorists.

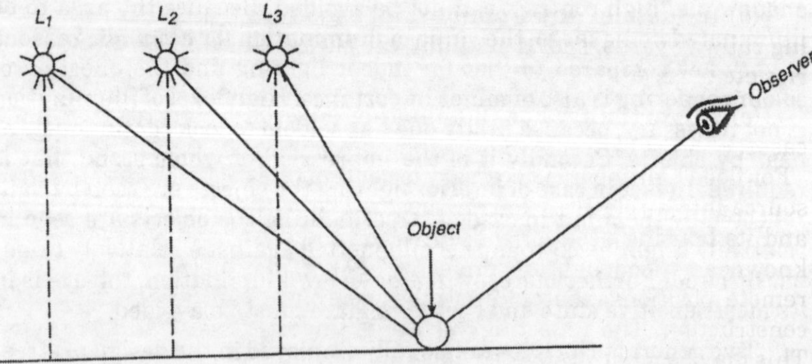

Fig. 5.37.3

5.14.5 Illumination Level for Street Lighting and Mounting-height of Lamps

The illumination required depends upon the class of street lighting installations. In class A installations *i.e.* in important shopping centres and road junctions, illumination level of 30 lumens/m^2 is required where as poorly lighted suburban streets, illumination level of 4 lumen/m^2 is sufficient. An average well-lighted street, will require illumination level between 8 to 15 lumens per square metre. Excellent illumination is considered when the distance apart is not more than 8 times the height of the luminaire ; that is, with the luminaire 8 metres above street level, they should be spaced not more than 64 metres (normal spacing for standard lamps is 50 metres with a mounting height of 8 metres) lamp posts should always be fixed at the junction of roads and as far as possible lamps near large trees should be avoided.

5.14.6 Types of Lamps for Street Lighting.

Mercury Vapour and Sodium discharge lamps have been found to have certain particular advantages for street lighting purposes the most important of these is the lower power consumption for a given amount of light, which inspite of the higher cost of the lamps makes the overall cost of an installation with discharge lamps less than that employing filament lamps. The colour and monochromatic nature of the light produced by discharge lamps does not matter much in street lighting installations.

5.14.7 Flood Lighting.

Flood lighting means flooding of large surfaces with light from powerful projectors. It is employed to serve one or more of the following purposes :

(*a*) **Aesthetic Flood Lighting.** For enhancing beauty of building at night such as public places, ancient buildings and monuments, religious buildings on important festive occasions etc.

(*b*) **Industrial and Commercial Flood Lighting.** For illuminating railway yards, sports stadium, car parks, construction sites, quarries etc.

(*c*) **Advertising.** For illuminating advertising boards and show cases.

For flood lighting it is necessary to concentrate the light from the light source into a relatively narrow beam. The particular type of reflector and its housing used for concentrating the light into narrow beam is known as "flood lighting projector". Since it is usually installed in remote positions, therefore it must be robust and weather proof in construction. The reflecting surface is important part in a projector. This may be made of silvered glass or chromium plate or stainless steel the efficiency of silvered glass is about 90% while that of polished metal is only about 70%. Metal reflectors being more robust are usually preferred.

The casing and its mounting are arranged so that the inclination of the beam can be varied in both a vertical and a horizontal direction on site. For permanent installations use of cast metal cases is made to achieve robustness and projection against weather and for temporary installations or those in sheltered situations, use of sheet metal casing is made. When 500 or 1000 watt lamps are used in projectors, ventilation may have to be provided for adequate cooling. This may be achieved by providing sufficient radiating surface. The front of the projector is usually of clear glass often bowed outwards to project it from the heat of lamp ; use of diffusing glass is made when a diffuse beam is required. As far as possible the projectors should not be visible to the passers-by. In some cases the projectors may be housed in ornamental stand standards.

Projectors are classified according to the beam spread :

(*i*) Narrow beam projectors with beam spread between 12 – 25°. These are used for distance beyond 70 metres.

(*ii*) *Medium angle projectors* : Projectors with beam spread between 25 – 40°. These are used for distance between 30 to 70 metres.

(*iii*) *Wide angle projectors* : Projectors with beam spread between 40 – 90°. These are used for distance below 30 metres.

For economic reason use of wide angle projector with high wattage lamp is encouraged over narrow beam projector with low wattage lamp. High wattage lamp is more efficient that low wattage lamp used in narrow beam projectors. Standard gas-filled tungsten filament lamps of 250, 500 or 1000 watts are used in medium and wide angle projectors which require accurate control of light, special lamps having bunched filaments and known as projector lamps are required.

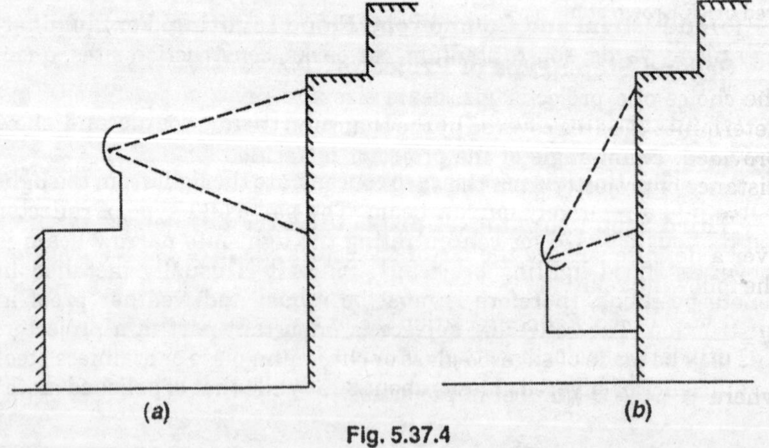

Fig. 5.37.4

5.14.8 Location and Mounting of Projectors. One of the most important factor which affects the choice of projector is the location of the projector. There are two possible locations of projector in practice. Fig. 5.37.4 (*a*) shows symmetric projector kept 20 to 35 metre away from the surface to be flooded and providing approximately parallel beam having beam spread of 25° to 30°. Fig. 5.37.4 (*b*) shows the case when the projector can not be located away from the building. In such a case an unsymmetric reflector mounted in a basement area or on a bracket attached to the building is used which directs more intense light towards the top of the building.

5.14.9 Flood Lighting Calculations.

The problem of flood lighting calculations may be roughly separated into 3 steps.

First Step. Illumination Level Required. The illumination level (lumens/m^2) required depends upon the type of building, the purpose of the flood lighting the amount of conflicting light in the vicinity etc. Typical figures are given below :

Building Surfaces	*Reflection Factors*	*Lux recommended for brightness of locality*		
		Low	*Medium*	*High*
White terra cotta cream terra cotta light marble	60–80	50	100	150
Light Grey Lime stone Bed Ford lime stone Buff lime stone Smooth buff ac Brick	40–60	70	120	200
Briar hill sand stone Smooth Grey brick medium grey lime Stone Common tan brick	20–40	100	150	250
Dark field Grey brick Common red brick Brown stone	10–20	120	180	300

Second Step : Type of Projector. Two consideration enter into the choice of a projector viz. beam size and light output. The former determines the area covered by the beam and the latter the illumination provided. Beam angle of the projector is decided keeping in view the distance of projector from the surface.

Third Step : Number of projectors. For any desired intensity over a definite surface the no. of projectors required is obtained from the following relation :

$$N = \frac{A \times E \times \text{depreciation factor} \times \text{waste light factor}}{\text{utilisation factor} \times \text{wattage of lamp} \times \text{luminous efficiency of lamp}}$$

where N = Number of projectors

A = Area of surface to be illuminated in m^2

E = Illumination level required in lumens/m^2.

The terms used in flood lighting calculations are described below :

(*i*) **Waste Light Factor.** Whenever a surface is illuminated by a number of light sources, there is always a certain amount of waste light on account of overlapping and falling of light beyond the edges of the area to be illuminated. This effect is taken into account by multiplying the theoretical value of lumens required by 1.2 for rectangular areas and 1.5 for irregular areas and objects such as statues and monuments etc.

(*ii*) **Depreciation Factor.** Dirst and Dust depositing on the reflector surface and on the surface of light reduces the effectiveness of the projector and 50 to 100% more light must be provided so that the illumination shall be adequate at the end of the interval between the cleaning periods. Depreciation factor is defined as the ratio of illumination under ideal conditions to the illumination under normal conditions.

(*iii*) **Coefficient of Utilisation.** It is also known as beam factor and is defined as the ratio of beam lumens to the lamp lumens. Its value lies between 0.3 & 0.5. This factor is due to the fact that all the light emitted by the projector is not along the direction of the beam but some of it is absorbed by the reflector and by the front glass. When the above losses are considerable the value of utilisation is low. The coefficient of utilisation factor goes up with deep reflectors having a good reflecting surface, such as silvered glass.

5.15. Unit of Measurement for Illumination

A foot candle (ft-cd) is a unit for the measurement of the intensity of light falling on a surface, and represents the number of lumens on one square foot of area.

If a 100 sq. ft. of a surface receive 400 lumens, then intensity of illumination

$$E = \frac{400}{100} = 4 \text{ foot candles}$$

It may also be referred to as 4 lumens per square foot $(lm/ft)^2$.

An other unit commonly used is lux, which represents an illumination of one lumen per square meter. If 1000 lumens fall on a 5 sq. m area, then intensity would be

$$\frac{1000}{500} = 200 \text{ lux.}$$

1 foot candle $(\text{lm/ft})^2$ = 10.76 lux

1 lux = 1 lumen per sq.–m

1 candle power = 12.57 lumen.

The units foot candle and the lux refer to the brightness of a surface on account of light flux incident on it. They should be distinguished from surface brightness of sources of light, for which different units are emloyed as mentioned below :

1 foot lambert = 1 lm/sq. ft.

1 lambert = 1 lm/sq. cm = 929 ft lambert

1 candela/sq. in = 452 ft lamberts = 0.155 cd/sq. cm.

1 stilb = 1 candela/sq. cm.

Candela (cd) should not be confused with candle power (C.P.).

5.16. Colour Temperature

In a popular sense, the colour temperature denotes the degree of whiteness of light. It is supposed to be that temperature at which a perfect black body glows when it is heated and emanates light of that particular colour. Generally speaking, when temperature is low the colour is red and if high the light is bluish, as may be seen from following:

Source of light	*Temperature*
Candle light	1900 ° K
Gas light	2360° K
Incandescent light	2848° K
Fluorescent warm white light	2750° K
Sky light	10,000° K

5.17. Stroboscopic Effect and Elimination

As the usual alternating current supply frequency is 50 Hz a discharge lamp will be extinguished 100 times per second. Although this effect is seldom noticeable in normals conditions. It is possible that this may in some instances give rise to a stroboscopic effect, that is moving object such as rotating parts of machinery illuminates by this light and may appear to be moving in disunity or jerks or rotating more slowly than their actual speed.

In actual applications, however where this effect might cause annoyance, it can be practically eliminated in a three lamp unit by connecting each lamp on a separate phase of a 3-phase system and it can be greatly reduced in a two lamp unit by the use of a two lamp control unit which employs a condenser in the ballast of one of the lamps. The current through the lamps is almost 90° out of phase and under these conditions light output of one of the lamps is at a maximum. This method has an additional of giving a overall power factor of nearly unity for the unit of two lamps.

It is the phenomenon which makes moving objects like fan blades to appear to be stand-still and a wave of the hand to appear as if it occured in a series of jumps. This effect is noticed when objects are lit by gas discharge lamps. The reason for this is that the light from these lamps is not continuously emitted but in spurts, with short intervals of no light, occuring 100 times a second when working on 50 Hz supply. The stroboscopic effect causes some unusual effects and is annoying. The blades of a revolving fan may appear stationary at some speed and to move in one direction or the other if the speed is raised or lowered or moving in anticlockwise direction to real direction. A filament lamp on the other hand has thermal capacity because of which the variations in the light output are generally evened out, and no stroboscopic effect is observed.

5.18. Ozonizer Lamp

This is also a discharge lamp, not intended for producing light. But ozone (O_3) which has the property of destroying smells as it is a strong oxidizing agent. The lamp consumes about 4 watt and operates on A.C. or D.C. on any voltages about 20 volt. It has a special glass bulb which transmits 1850 Å radiation which converts oxygen in the air to ozone. It also emits a small amount of germicidal radiation at 2537 Å. It is suitable for installation in hospitals, public urinals and air conditioned interiors.

5.19. Germicidal Lamp

It is a low pressure mercury vapour discharge lamp; similar to a fluorescent lamp. But without phosphor coating ; a special glass is used which freely transmits short wave ultra-violet light. Lamps rated at 15 W which are 18 in. and 36 in. long require a ballast, but the smallest lamp of 6W works on the glow discharge principle does not require any ballast and work off 230 V mains. It gives out a radiation 2537 Å which has germicidal property and is very useful destroying bacteria, germs, fungus and for purifying and sterilizing public bath, collieries, cold storage depots, medical laboratories, hospital dairy forms and food preservation factories. The number and size of lamps to be provided depends upon the size of the room. For example, for a small waiting room measuring 10 ft each way in a dispensary, four 15 W lamps, one mounted on each of the walls inside a trough at a height of 6 ft. above the floor, would effectively purify and de-contaminate the air in the upper portions of the room, which usually get infected. Care should be taken to ensure that no one is exposed to its direct radiations.

5.20. Is it Possible to Control, the Brightness of a Fluorescent Lamp ?

Ordinary fluorescent lamps are not well adopted for dimming. Instant start lamps are suited.

To control the brightness over a wide range, two conditions should be fulfilled *i.e.*, there should be some means of varying the discharge current, and it should be done without reducing the applied voltage.

One method of satisfying this difficult assignment is by the use of grid controlled gas tube called thyratrons. By varying the phase angle of the grid voltage, full control of anode current over a wide range of 100 : 1 is possible. At the same time, high re-starting voltage over each half cycle remains unimpaired.

5.21. Is Fluorescent Light Harmful to the Eyes ?

No, there is a false impression in some quarters that fluorescent lights are too bright and, therefore, injurious to the eyes. It should be remembered that sun light is a thousand times more bright than any artificial lighting. Yet no one can complain that day light is bad for the eyes.

5.22. Photometry

Except for the measurement of luminance of sources, measurements by visual comparisons have been replaced by photocell because these cells give more accurate and faster measurements. The more complex procedures and the apparatus involved in visual measurements are thus avoided. Also, if certain precautions are observed the measurements by photocells used for photometry measurements are the photo-voltaic cells and photo-emissive cells. The photo-voltaic cells is the most widely used one because of the simplicity of the associated circuits.

(*i*) **Photo-voltaic Cell :** The construction of a photovoltaic cell, also known as the barrier layer or rectified cell is shown in Fig. 5.38. It consists of a base plate made of either steel or aluminium and carries a layer of metallic selenium which is light sensitive. An electrically conducting layer of cadmium oxide is applied by sputtering over the selenium layer. The layer is sufficiently thin to allow light to reach the selenium and is electrically continuous as it acts as the –ve pole. The –ve contact is formed of a strip of wood's metal sprayed on the edge of the top surface. The base plate forms the +ve contact. The front surface of the cell is protected by a transparent varnish.

When light falls on the upper surface of the selenium, electrons are released from the surface which maintains a flow of current through the external circuit connected between the +ve and –ve contacts.

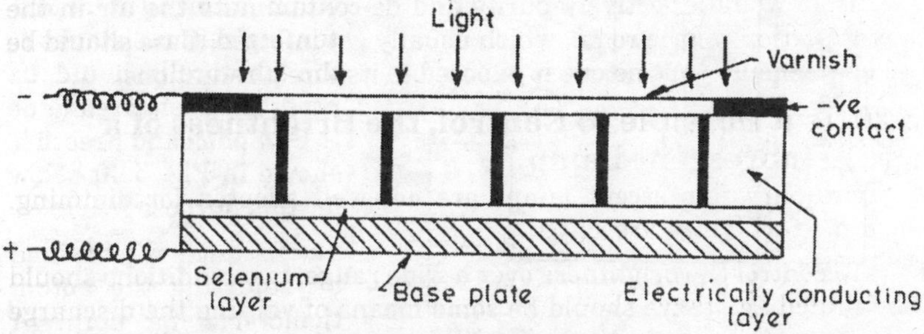

Fig. 5.38. Photo-voltaic cell.

In photometry the current output of a photocell should be proportional to the illumination which is achieved by keeping the external resistance at a low value. Also if greater accuracy is required then illumination should not be allowed to exceed 25 lumens/ft^2. The smaller the size of the cell, of course compatible with obtaining sufficient current to be properly measured, the better a linear relation is obtained between current and illumination. This is because, for such a cell, the resistance of the electrically conducting film is at a minimum. Also, since the current is small, the voltage drop due to the circuit resistance will be kept low.

Another limitation of the photocell is that if the light is incident at angles of 60° or above, the lacquer tends to reflect a significant amount of light which, therefore does not reach the selenium layer. Thus the reading is less than what it should be according to cosine law of illumination. Some compensation for this error can be made by using a matt lacquer. A better method is to omit the lacquer and cover the cell with a hemispherical dome of transparent plastic. Care should be taken to see that the whole of the cell is illuminated otherwise incorrect readings may result.

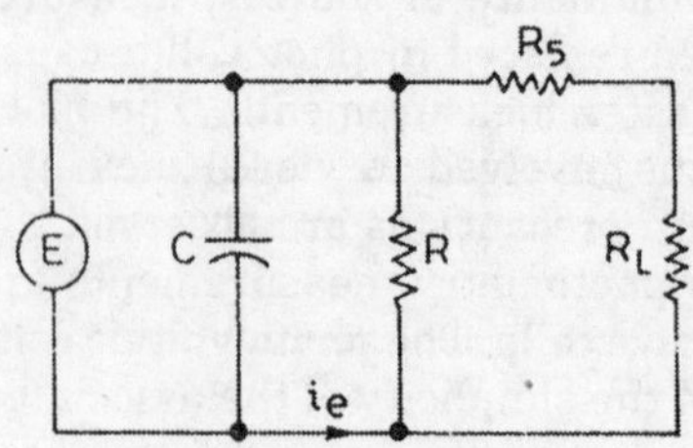

Fig. 5.39. Equivalent circuit of a Photo-voltaic cell.

The equivalent circuit of a photovoltaic cell is shown in Fig. 5.39 drawn above :

Here E is a perfect photoelectric generator which produces a current proportional to the illumination R_s the circuit resistance, C the effective capacitance, R the equivalent of the barrier layer resistance and R_L is the resistance of the external circuit. The temperature dependence and non-linearity errors are attributed to the variation of R with temperature and illumination. These effects can be reduced by making the effect of R_s and R_L as small as possible so that R is short circuited.

For more accurate work a circuit representing zero resistance to the photocell should be used. The basic circuit due to Campbell and Freeth is shown in Fig. 5.40 below for this purpose :

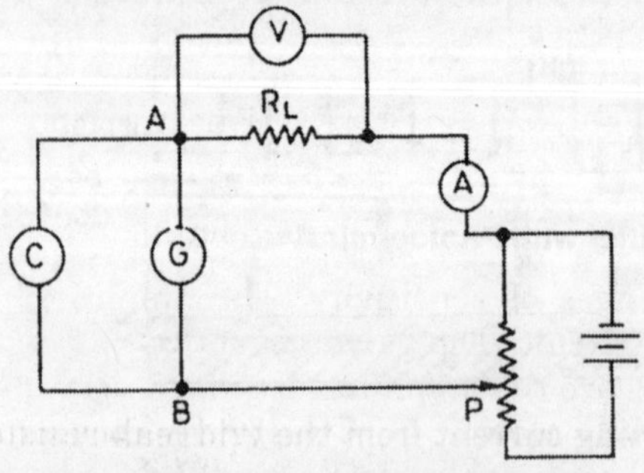

Fig. 5.40. Campbell-Freeth Circuit.

Balance is obtained by adjusting the potentiometer P *i.e.* point. A and B across which galvanometer is connected, are at the same potential which is equivalent to saying that A and B are short-circuited and hence the resistance R in Fig. 5.40 is short-circuited. Which is the requirement for obtaining linear relation between illumination and the current as no current is now diverted through R.

Photo-emissive cell : When greater precision in terms of linearity and stability is required, it is better to use photoemissive cell rather than photovoltaic cell. However, the circuit involved is more complex and requires the use of some sort of valve amplifiers.

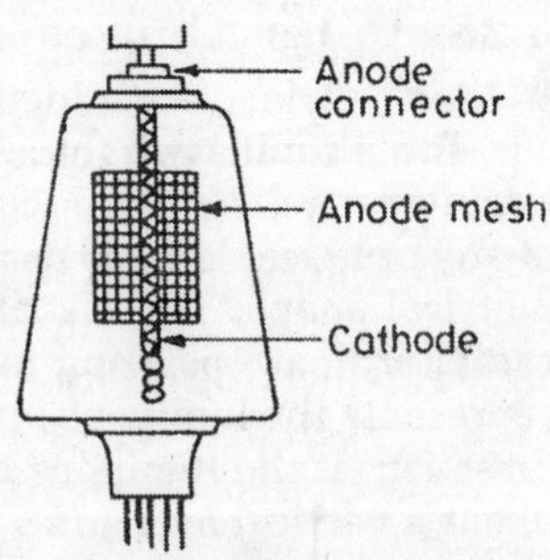

Fig. 5.41. A vacuum type photo-emissive cell.

The photoemissive cell has the anode made of a cylindrical wire mesh with cathode in its axis as shown in Fig. 5.41. The cathode is made of Bi-O-Ag-Cs cathode. The electrons are emitted from the cathode of the cell when light falls on it. The anode is normally 30 to 50 V at a higher potential as compared to the cathode. The spectral response of the cell is matched with the eye by using a suitable solution of cupric chloride and Potassium dichromate to correct respectively the blue and red ends of the response and cobalt ammonium sulphate to correct the region around 500 mH.

A typical d.c. amplifiers to be used along with the photocell is shown in Fig. 5.42.

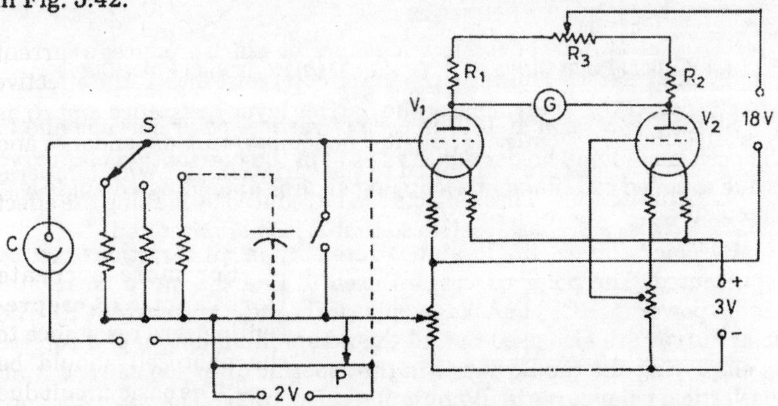

Fig. 5.42. D.C. Amplifier with Photoe-missive cell.

The switch S provides ranges of sensitivity. The amplifiers is basically a Wheatstone bridge circuit. The galvanometer G indicates the balance position. V_1 and V_2 are the amplifiers having a very high input resistance to avoid diverting current from the grid leak resistor R_1 and R_2 together with the other variable one R_3 from the other two arms of the bridge.

The out of balance current in the cell circuit changes the grid bias of V_1 thereby, its effective resistance is changed which is indicated by the galvanometer *G*. The dark current is balanced out by adjusting R_3 with *P* set to zero. When the light falls on the photocell the balance is restored by adjusting, *P*.

5.23. Polar Curves and Determination of MHCP & MSCP

The illumination flux emitted by a source can be determined from the intensity distribution curve. The luminous intensity or candle power of any practical lamp is not uniform in all directions due to its un-symmetrical shape. The distribution of light is given by polar curves. A radial ordinate pointing in any particular direction on a polar curve represents the luminous intensity of the source when viewed from that direction. If the luminous intensity is measured in a horizontal plane about a vertical axis and a curve is plotted between candle power and the angular position a horizontal polar curve is obtained. Similarly, vertical polar curve gives the relationship of candle power in vertical plane passing through the lamp at various angles. Typical vertical and horizontal polar curves for a filament lamp is shown in Fig. 5.43.

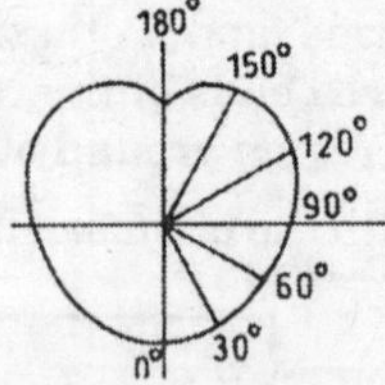

(a) Vertical Polar Curve

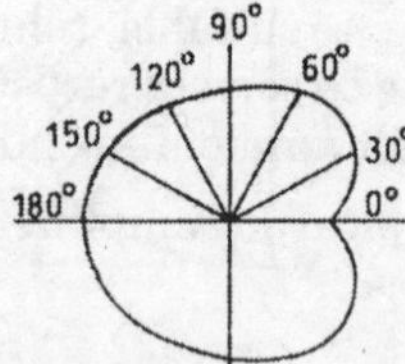

(b) Horizontal Polar Curve.

Fig. 5.43.

Slight Depression at 180° in case of vertical polar curve is due to the position of lamp holder whereas at 0° in the horizontal polar curve is due to coiled coil filament accupying an arc subtending an angle less than 360°.

By polar curves, the luminous intensity in all directions can be represented. The polar curves are used to find the mean horizontal candle power (MHCP) and mean spherical candle power (MSCP). The polar curves are also used to find the actual illumination of a surface by employing the candle power in that specific direction as read from the vertical polar curve in illumination calculations. *The mean horizontal candle power* of a lamp can be determined from the horizontal polar curve by taking the mean value of the candle power in a Horizontal direction.

Mean spherical candle power of the lamp can be determined from the vertical polar curve by Rousseau's construction.

Rousseau's Construction : The mean vertical distribution curve for the lamp is by the polar curve A. With O the pole of the polar curve as centre, a semicircle of any suitable radius OX is drawn as shown in fig 5.43.1. Parallel to XY is drawn X', Y' and from this line X', Y' ordinates are set equal to the corresponding radius on the polar curve as for example ordinate GH is equal to the radius OB, etc. These ordinates are set up along projection from the ends of radii of the polar curve. The graph drawn connecting these ordinates X' HZ is the Rousseau's curve. The mean heights of the graph which is equal to (Area X'HZY')/(Base X'Y') gives the MSCP of the given source of lamp. The scale will be the same as that of the polar curve or

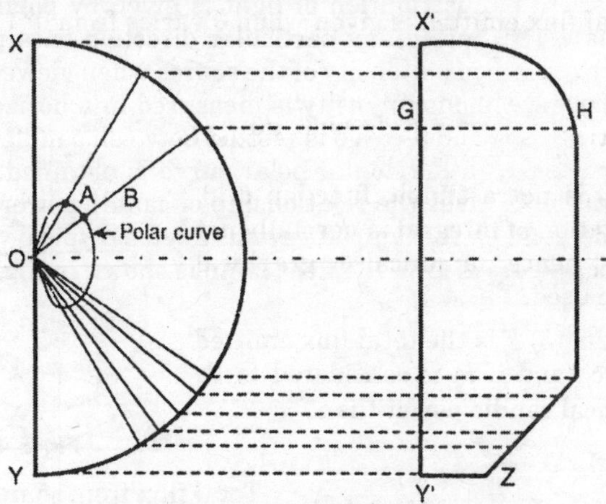

Fig. 5.43.1. Rousseau's Construction

$$\text{The mean ordinate of the curve} = \frac{\text{Area X'HZY'}}{\text{Length of X'Y'}}$$

The flux emitted by a source can be obtained by a polar solid which is related to the vertical polar curve. The polar solid is assumed to be concentric with a sphere and the flux emerging from the sphere through zones defined by the angular limits θ_1 and θ_2 is found as follows :

From the eq.

$$I = \underset{\delta w \to 0}{\text{Lt}} \frac{\delta F}{\delta w}, \qquad \text{...}(i)$$

the luminous intensity I_θ along θ is given as

$$I_\theta = \frac{df}{dw} \quad df = I_\theta \, dw$$

$$F\,\theta_1, \theta_2 = \int_{\theta_1}^{\theta_2} I_\theta - dw \qquad \text{...}(ii)$$

Now, from the Fig. 5.44 it is clear that the differential solid angle dw subtended by the differential strip area is given as :

$$dw = \frac{r\, d\theta\, 2\, \pi\, r \sin\theta}{r^2}$$

$$= 2\, \pi \sin\theta\, d\theta$$

Substituting in eqn. (2), we have :

$$F_{\theta_1, \theta_2} = \int_{\theta_1}^{\theta_2} I_\theta\, 2\pi \sin\theta\, d\theta$$

$$= 2\, \pi \int_{\theta_1}^{\theta_2} \sin\theta\, d\theta \qquad ...(iii)$$

The total flux emitted is given when θ varies from 0° to π radians *i.e.*

$$F = \int_0^{\pi} I_\theta \sin\theta\, d\theta \qquad ...(iv)$$

Since I_θ is not a simple function of θ, direct evaluation of integral is normally not possible and hence numerical or graphical methods are used.

From eqn. (3) F is the total flux emitted by the source and if I_θ is considered to be mean spherical candle power than

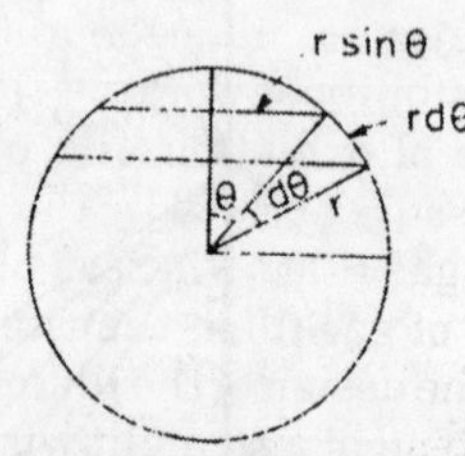

Fig. 5.44.

$$I_{\theta\text{ mean}} = \text{M. S. C. P.} = \frac{\text{Total flux from source}}{4\, \pi}$$

Also if $\theta_1 = \theta$ and $\theta_2 = \pi/2$, the integration covers the luminous flux emitted below the horizontal

$$F_{0,\, 90} = 2\, \pi\, I_\theta \int_0^{\pi/2} \sin\theta\, d\theta = 2\, \pi\, I_\theta$$

Therefore, I_θ = Lower mean hemispherical candle power

$$= \frac{\text{Flux of lower hemisphere}}{2\, \pi}$$

Similarly, the upper mean hemispherical candle power

$$= \frac{\text{Flux of upper hemisphere}}{2\, \pi}$$

Uses : The polar curve of a lamp helps in knowing the illumination of a lamp at a particular angle, with or without a reflector.

5.24. The Electric Lamp

Light sources are usually divided into two classes :

Incandescent sources and luminescent sources. An incandescent source is one which emits lights solely because of its high temperature equivalent a tungsten lamp whereas, a luminescent source emits light due to ionisation and excitation independent of temperature equivalent to mercury vapour lamp, neon lamp, sodium vapour lamp and firely. These sources usually radiate only at certain wave lengths.

(*i*) **Tungsten Lamp.** It has been observed that a black body is the most efficient radiator as it emits energy at every wave length. A black body when heated to 6250°C it emits its maximum energy within the visible spectrum. Even though it is not possible to realise a perfect black body, the materials used for lamp filaments approach it very closely. Tungsten with its melting point at 3500°C is now universally adopted as a material for incandescent lamps. The safe working temperature of such a filament is about 3000°C. The cold resistance of the filament is approximately one-sixteenth of its hot resistance. The filament resistance increases almost instantaneously. An inert gas, usually a combination of nitrogen and argon is filled in the bulbs which results in possibility of working the filament at higher temperature without any fear of oxidation and evaporation of the filament on account of the pressure exerted by the gas. Usually bulbs with rating 40 W and above are gas filled whereas below 40 watts they are vacuum type. Since the use of gas filled tungsten lamp, slight improvements have been made in the design of the filament such as the use of a coiled or helical filament instead of a straight wire and finally a coiled coil filament in which the single helix is coiled again on itself as shown. The coiled coil filament can be worked at a relatively higher temperature as a result the output per watt increase and the loss due to convection decreases. The efficiency in lumens per watt increases and the loss due to convection decreases.

The efficiency in lumens per watt increases as the size of wattage of the lamp increases. If the working voltage is marginally greater than the rated voltage, more light output is obtained but results in shorter life of the lamp operation of the filament at a voltage lower than the rated results in loss of light.

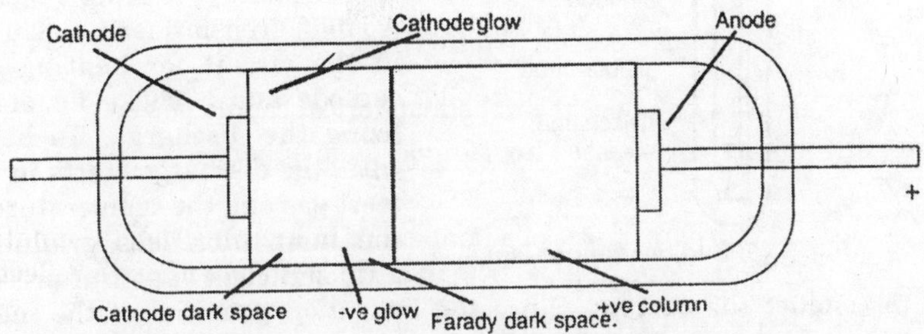

Fig. 5.45. Electrical Discharge in a gas at low pressure.

Electric discharge in a gas at low pressure

When an electric current flows between two electrodes in a gas at approximately 1 mm pressure, cathode glow is produced very close to the cathode, followed by the cathode dark space and a bright region to some extent called the -ve glow. Beyond this arc the Faraday dark space and the luminous +ve column. A decrease in pressure causes the cathode dark space to expand until it can be made to fill the whole tube. An increase in pressure causes the column to expand. In many of the commercial lamps the +ve column fills almost the whole tube.

If the cathode is heated so that it gives an adequate supply of electrons, the voltage drop at the cathode is low.

Thus hot cathode lamps such as the sodium lamp and the mercury vapour lamp using a mercury pool are essentially low voltage devices. In cold, cathode lamps the voltage drop at the cathode must be greatly increased to perhaps 10 times the ionization potential. This high drop accelerates the +ve ions towards the cathode they liberate electrons. In order to reduce power loss at the cathode, a long lamp is required which in turn necessitates a higher operating voltage for these lamps. The bombardment of the cathode also causes its gradual disintegration with consequent blacking of the tube, and there is a tendency for "clean up" of the gas which results in reduction in the life of the lamp.

(*ii*) **Sodium Vapour Lamp.** A sodium lamp consists of an inner bulb of special glass containing the sodium and the inert gas either neon or argon and is fitted with two filaments. This is enclosed in a larger bulb which is evacuated to prevent the escape of heat as much as possible. The efficiency of the bulb is high as the radiations from it are the yellow sodium lines which are near the maximum of the visibility curve. The normal operating temperature of the tube is around 300°C. A small transformer is included in the circuit for heating the cathode and a choke for stabilizing the discharge. To begin with, the discharge starts in the inert gas and the temperature of the lamp increases gradually until sufficient vapour is present to conduct the current. Since the ionization potential of the metal vapour is low, it will carry nearly all the current and the spectrum of the inert gas will practically disappear soon. Thus the bulb starts as a

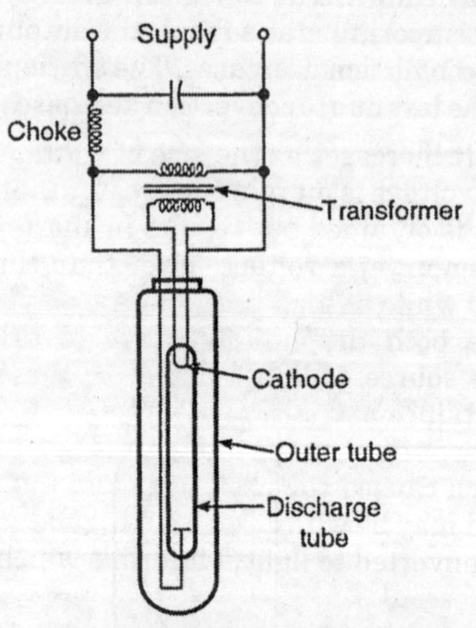

Fig. 5.46. Sodium vapour lamp.

neon lamp and then gradually changes from red to yellow as the temperature rises. Because of the monochr- omatic yellow light which makes the objects appear as grey, these lamps are used for street and high way lighting.

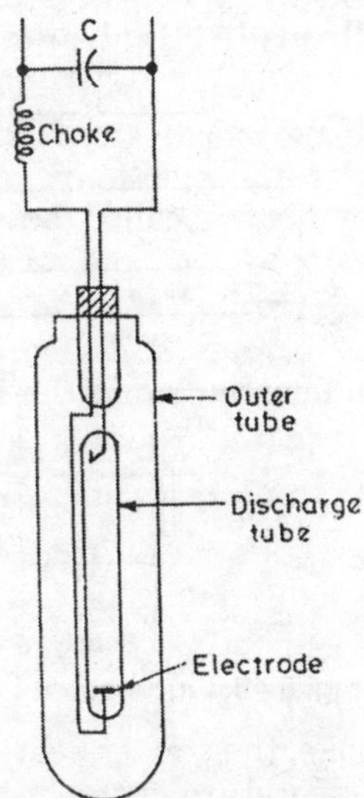

Fig. 5.47. Mercury vapour lamp.

(*iii*) **Mercury Vapour Lamp.** The mercury vapour lamp is similar in construction to the sodium vapour lamp and consists of a double glass bulb. Since the cathode is maintained in an incandescent state by ionic bombardment, no heating circuit is required. The operation of the mercury vapour lamp is at low mercury vapour pressure gives blue light and a high proportion of the ultra violet rays and hence are unsatisfactory. A more recent development is a high pressure mercury lamp in which a small but carefully measured drop of mercury is introduced in the bulb. The vapour pressure rises until all the mercury is vapourized when the light given out is with a bluish tinge. Here also an inert gas is introduced for initial discharge of the mercury. When the mercury vapour lamp is in use and say it goes out, it will not restart until it has cooled down and the vapour pressure fallen to a value sufficiently low to allow restricking of the discharge in the inert gas. The overall efficiency of the bulb is approximately 35 lumen per watt and the p.f. 0.65. A still more recent development in mercury lamps is an ultra pressure lamp operating at about 40 atmosphere and has an efficiency comparable with the sodium vapour lamps. **These lamps are used for outdoor and industrial lighting.**

(*iv*) **Fluorescent Lamp.** The mercury vapour lamps operating at low pressure emit radiation 2500 AU while at high pressure they emit radiation at about 3650 AU. Since both the emissions are in the ultraviolet region, are of no use as source of light. However, these radiations can be used to excite certain materials. When the excited molecules of these materials return to normal, they emit a radiation at a frequency different from that which caused the excitation and thus the radiation emitted by the material may be within the visible zone so that the ultra violet radiations are converted to light. Materials which possess this property are called fluorescent.

The fluorescent lamp is in the form of a tube 3 to 5 cms. in diameter and 5 m to 1.5 meter long with an electrode at each end which are in the form of coiled filament coated with an electron emitting material. The inside of the tube is coated with fluorescent powder and as the operating temperature of the tube is coated with fluorescent ap-

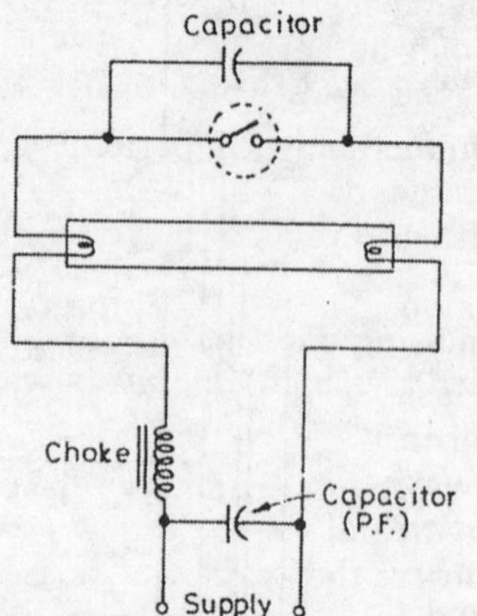

Fig. 5.48. Fluorescent-Lamp Circuit.

proximately 50°C, no outer tube is required. A series choke for stabilizing the discharge and a shunt capacitor across supply terminals for improvement of p.f. are used with the tube. The lagging p.f. is due to the choke only. Sometimes a small capacitor of 0.05 μ F is connected across the tube as shown in Fig. 5.48 to suppress radio interference as the tube draws non-sinusoidal current from the supply on account of the characteristics of the discharge.

When the supply is switched on, the current flows through the choke, the starter heater and the electrodes. The heater coil raises the temperature of the bi-metal contacts and they separate, thereby the current through the choke is interrupted and the consequent voltage pulse of appx. 1000 volts causes the tube to strike. Once the tube strikes the tube current flows through the starter heater and the bimetal contacts remain open.

Fluorescent discharge lamps may be operated on a d.c. supply also. This requires the use of a resistive ballast for arc stability in addition to the choke for producing the voltage pulse. The resistive ballast introduces very high power loss in the ballast. Also the migration of +vely charged mercury ions towards the cathode results in low light output from the anode end of the tube. For this reason whenever the tube is to be used on d.c., a reversing switch is normally used in the ckt. to change the direction of the current through the tube every few hrs.

It is usually preferred to use an inverter circuit whenever d.c. supply is available. The frequency is usually 10 KHz. With the advent of power transistors, this method of operation of tubes is extensively being used in trains, buses and aircrafts.

5.24.1 Energy saving considerations for fluorescent lighting system.

Due to the onset of energy crunch energy saving studies involving fluorescent lamps has attracted the attention of lighting engineers. The following are the guide-lines for energy saving in fluorescent lamp system :

1. Reduced wattage fluorescent lamps using Krypton-fill gas are used where ever possible but there is a limit to reduce wattage of the lamps without compromising reliability.
2. *Turning off lights* : Turning off lights for tea and lunch will save electricity but frequent starting affects the life of the

lamp; the shorter the burning/hr per start the shorter the life. Hence the decision depends on whether the money saved by reduced power bill will offset the cost of shorter life.

3. Luminaire cleaning in conjunction with periodic group relamping can maintain the same specified illumination level with fewer luminaires resulting in reduction of both initial and operating cost.
4. In commercial and merchandising applications three component phosphor lamps which give high efficacy as well as high colour rendering capability are to be used.
5. Lamps in which electronic components are added to increase the impedance of the lamp circuit and there by reduce current and wattage are to be used where the ambient temperature does not fall below 15°C.

5.24.2. Improved 35 watt low energy lamp-ballast system :

Energy conservation is the theme of the present day world. In accordance with this theme lamp manufacturers have developed a 35 watt low energy lamp with wave form modified ballast. This lamp could be physically and electrically inter-changeable with the existing 40 W lamp. The light out put is approximately the same as the 40 W lamp. The lamp uses an improved phosphor. This phosphor is a mixture of calcium fluorapatite and europium activated strontium chlorapatite. This phosphor yields 7% more lumens than a 35 W low energy lamp with usual standard ballast. It gives almost the same lumens as that of a 40 W lamp and more lumens per watt than the 40 W lamp with the standard ballast.

The ballast design is improved by using copper wire instead of aluminium wire and lower loss grade of steel for the magnetics. This resulted in a saving of 17 to 10 W of ballast losses besides improving lamp efficiency. The same materials and improved wave from modified

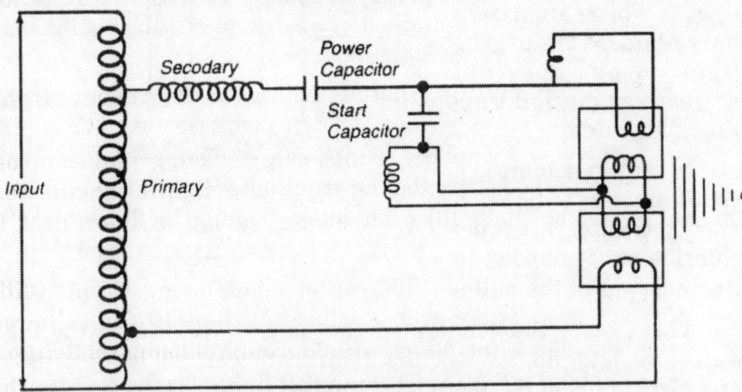

Fig. 5.48.1 Ballast Circuitry for 35 Watt Lamp.

ballast made the low energy lamp widely accepted. Normally a lamp requires a peak voltage at the starting and r.m.s voltage between the electrodes. The improved ballast design accommodates an internal starting aid which reduces the peak voltage necessary for break down. The r.m.s. breakdown voltage is reduced because of the Krypton added to the fill gas. Further the voltage gradient in the positive column is reduced there by resulting in a reduced lamp voltage. The Fig. 5.48.1 shows the ballast circuitry for a 35 watt wave from modified ballast with the lamp.

This lamp requires 3.6 volts for the cathode and 0.1 mfd for the starting capacitor. The starting capacitor is 0.1 mfd and is bigger than a 0.055 mfd capacitor used for 40 watt lamp. The secondary ballast impedance is tapped at different points to the primary coil. An increased number of turns has improved the wave form. Also the lumen output is more with modified lamp ballast as compared to its use used for 40 watt lamp. The secondary ballast impedance is tapped at different points to the primary coil. An increased number of turns has improved the wave form. Also the lumen output is more with modified lamp ballast as compared to its use on the standard type ballast.

5.25 Halogen Lamps

INTRODUCTION. Candle power of filament lamp falls off with regular use. This is due to the reason that the slow vaporization of the filament and partly due to black deposite formed on the inside of the bulb. This phen-omenon is more pronounced in carbon filament as compared to tungsten filament lamps.

It is special type of tungsten lamp as shown in fig. 5.49 and developed in 1959. Its bulb is comparatively smaller and it runs much hotter. Its light is whiter from ordinary filament lamps. The light output does not decrease as much during the life of lamp.

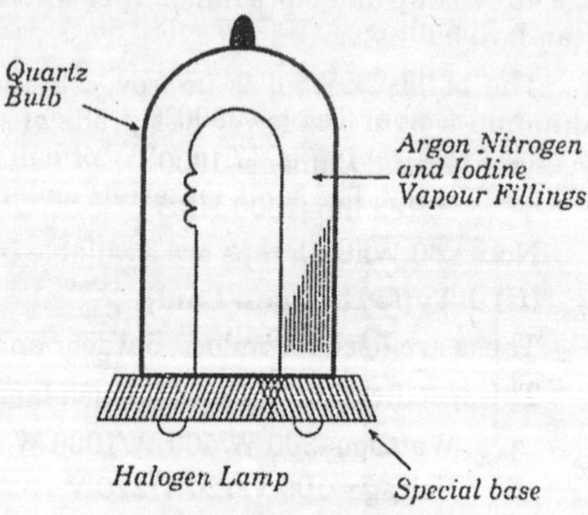

Fig. 5.49. Halogen Lamp.

In halogen lamps, a small amount of halogen vapour (iodine, bromine, chlorine etc.) is added to the inert gas of the bulb.

This causes the evaporated tungsten to resettle on the filament during cooling. So, the working temperature of the lamp is increased and blackening of the lamp is eliminated.

The efficiency of the halogen lamp becomes 25 to 23 lumen/watt, which is about 50% more than the ordinary incandescent lamps through out the life of lamp. As small envelope is used with resultant high gas filling pressure the life of the halogen lamps is upto double that of conventional tungsten filament lamps.

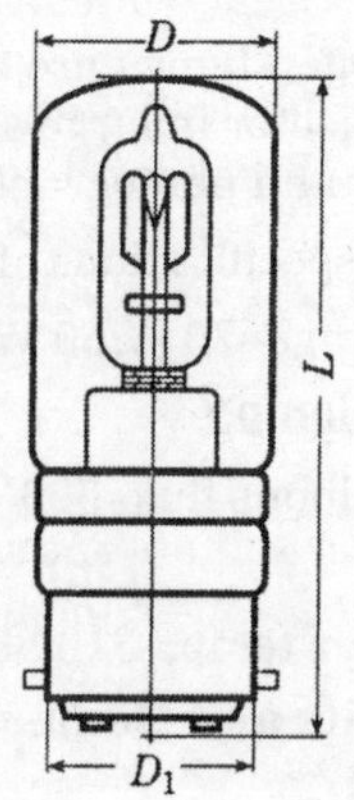

Fig. 5.49.1. Halogen lamp (JD Type).

In addition to this long tubular iodine lamps can also be made. These lamps can run only in horizontal position.

This is due to the fact that in vertical position of lamp, iodine will diffuse down due to high atomic weight. This will give rise to blankening in the upper part of the lamp.

Typical type of Halogen lamps used for different purposes :

(*i*) Halogen lamps for indoor, outdoor and flood lighting

These Halogen lamps are two types JD and J fig. 5.49.1 shows the JD type Halogen lamp.

(*i*) JD Type Halogen Lamp

The specifications of JD type Halogen are :

- Wattage–100 W
- Voltage–110/220 volts, 50/60 HZ
- Filament–Closed coil (C.C)
- Colour Temp.–2800°C
- Initial Lumens–1800
- Rated life–1000 HRS

Note : 60 Watts lamps are available in 110 volts.

(*ii*) J-Type Halogen Lamp

These are used for indoor, outdoor and flood lighting.

The specifications J type Halogen lamp are given below.

- Wattage–300 W/500 W/1000 W
- Voltage–100 V/120 V/220 V
- Luminous–6000/10500/21000 flux (Lm)
- Average life–2000/2000/2000 HRS
- Max. Colour–2900/2900/3000 Temp. (°K)

5.25.1 Halogen Lamps for photography and overhead projector

Halogen lamps are also frequently used in photography and overhead projector in large scale. The fig. shows the halogen lamp used for projector and photography.

The specifications of such type of lamps are

Wattage–20 W/50 W

Voltage–12 V

Luminous flux–500/1400 (Lm)

Colour temp.–3100/3300 (°K)

Merits and Demerits of Tungsten Halogen lamp

Merits :

1. It is miniature size and has high output 22-23 lumens/watts
2. No ballast
3. Good colour possible
4. Excellent optical control
5. Dimmable
6. Easy lumen maintenance
7. Longer life 2000 hours

Demerits :

1. Handling of lamp during maintenance is difficult
2. Low efficiency
3. Radiant-heat effects
4. Operating temperature effects lamp life

Uses of Halogen lamps

These lamps are used in public halls, sports halls, indoor, outdoor and flood lights, photo film and TV studios.

These are also used in overhead projector, photo copying machine and air port runways etc.

5.25.2 500 W Tungsten Halogen lamps

This is also another tungsten filament lamp. The bulb is filled with a halogen gas such as iodine, bromine, fluorine etc. The lamps have glass bulbs very close to the filament, the bulb is made of quartz to withstand high temperature. (Fig. 5.49.2).

The halogen gas helps to deposit back the evaporated molecules of the tungsten over the filament. To carry out this the evaporated molecules should be in the proximity of the filament. That is why the bulbs are made cylindrical in shape and set close to the filament.

As we know that the dazzling light thrown by head lights of modern automobiles or lamps used by video film makers. These small sized very bright sources of light are nothing but halogen lamps. Those used in

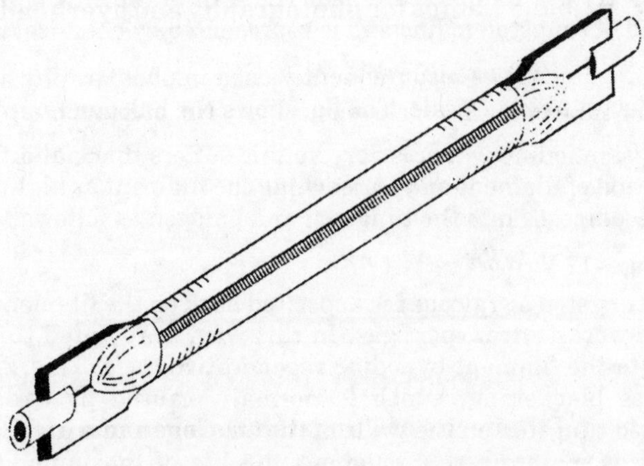

Fig. 5.49.2. 500 W Tungsten Halogen Lamp.

head lights or over head projectors are shown in Fig. 5.49(3) and those by video film makers in Fig. 5.49(4).

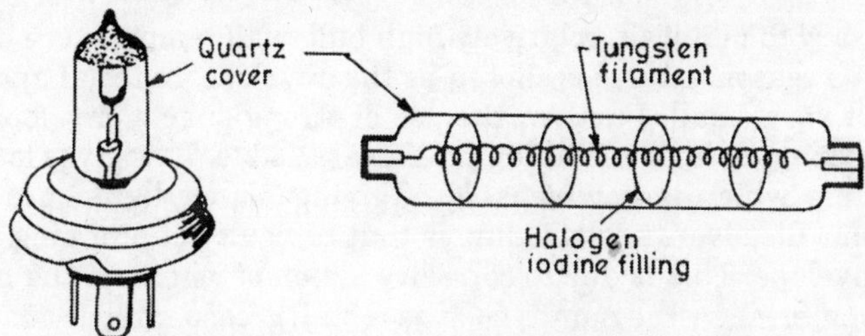

Fig. 5.49.3. Halogen bulb used in car head lights.

Fig. 5.49.4. Halogen doubler bulb used with suitable reflector mounting by video-graphers

These lamps are tungsten filament lamps with difference in shape, size, envelop/cover material used on one hand and gas filled within the outer cover, on the other.

The cover material as shown in the figure is transparent quartz. Due to its closeness to the filament, the cover temperature is as high as 250°C. Thus in some applications *viz.* over head projector it becomes essential to dissipate the generated heat with the help of a fan.

5.26. Types of Lamps

(*i*) **Halogen Lamp.** Candle power of filament lamp falls off with use. This is partly due to slow vaporization of the filament and partly due to black deposit formed on the inside of the bulb. This phenomenon is more pronounced in carbon filament lamps than in tungston filament lamps. In tungston halogen lamps, developed in 1959, iodine is added to the argon gas filling. The evaporated tungston of filament on reaching the relatively low temperature near the wall of the envelope forms a compound of tungston iodine and is represented by chemical reaction.

$$W + nx \rightleftharpoons nWX$$

where W = tungston, X = Halogen and n = No. of Atoms.

Tungston iodine which is very voltaic suffers thermal diffusion in the direction of filament and on reaching the filament at high temperature it decomposes into the tungston and halogen as follows :

$$WXn \rightarrow W + nX$$

The tungston so raleased is deposited back on the filament. Iodine, therefore works a transport gas. In this way, evaporated tungston is restored to the filament by iodine regenerative cycle. This, therefore, eliminates blackening, which is normally caused by a deposit of evaporated tungston on the walls of the envelope and a high luminous efficiency is maintained throughout the life of the lamp. As small envelope is used with resultant high gas filling pressure, the life of the halogen lamps is upto double that of conventional tungston filament lamps.

In order that regenerative cycle should function properly, it is necessary to maintain relatively high bulb wall temperature of 250°C. For this reason, quartz is chosen as the envelope material and its size is kept very small. Further, the use of strong quartz envelope makes employment of high gas filling pressure possible. This gives long life to the lamp which is now capable of giving higher light output. Conspicuous disadvantage of iodine is that difficulty of pumping it inside the envelope. This is due to corrosive action of iodine on the normally used material for the pumps such as metal, grease, rubber etc. and low vapour pressure. Above disadvantages are overcome by the use of halogen hydrocarbons such as monobromine methane (CH_3) (Br) or dibromine methane (CH_2) (Br_2). These materials besides being non-corrosive are easy to dose in correct quantity. At high temperature, bromine is liberated which starts its function as transport gas. It may further be noted that long tubular iodine lamps can only be run in horizontal position. This is due to the fact that in vertical position of lamp, iodine will diffuse down due to high atomic weight. This will give rise to blackening in the upper part of the lamp. Burning position of bromine lamp is less critical and that its output is slightly higher because absorption of light by the purple iodine vapour is absent.

Main characteristic of halogen lamps are miniature size, high output 22-23 lumens/watt, long life 2000 hrs. These lamps are used in public halls, factories, sport halls, photo film and T.V. studios, over head projectors, car lighting, signalling, illumination of buildings, playing fields, large gardens, fountains, car parks and airport runways etc.

(*ii*) **Neon Lamps.** These lamps are operated at a very low temperature of about 200° that is why, these are also called cold cathode discharge lamps. The electrodes are made from iron or nickle cylinders without any coating and practically these do not emit electrons. The electrodes are housed at the two ends of the tube which contains neon gas. Neon gas discharge gives a characteristic red light. Mixture of mercury vapour and argon gas gives bluish green colour whereas mercury and neon give blue colour. Voltage required for starting and maintaining the discharge may be as high as about 10,000 volts. The particular voltage is required for a particular length of the tube and it may be obtained from a transformer. Since the cold cathode lamps require about 1.5 to 2 times the operating voltage at the time of start, so these are provided with high reactance transformers. From the safety point of view, no part of the installation should have a voltage more than 5000 V above ground. Therefore, the transformers H.V. is limited to 10,000 volts and the centre point of the winding is earthed as shown in Fig. 5.50.

Neon lamps are generally used for advertising. Most of the letters have two ends at which electrodes are placed. In letters having more than two ends the tube path is repeated for some portion as shown in Fig. 5.51 for letter 'H' used in the above advertising. If the length of the advertising sign tube requires voltage more than 10,000 volts it is split up into parts.

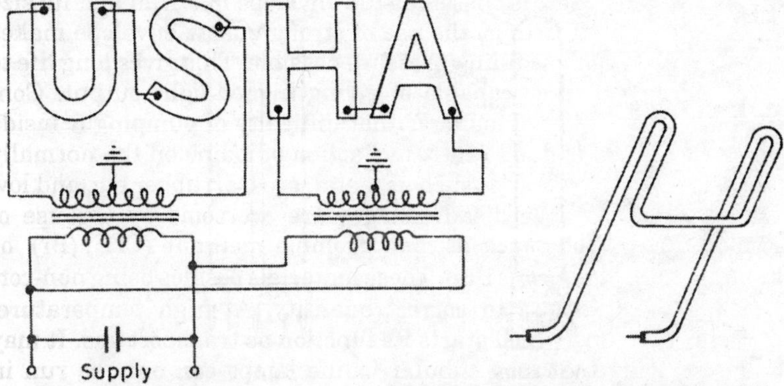

Fig. 5.50. Neon Sign. **Fig. 5.51. Tube Arrangement For 'H'.**

(*iii*) **Neon Indicator Lamps.** Most of the panels of control equipment, voltage reflectors etc. make use of indicating miniature bulbs. These are nothing but cold cathode Neon discharge lamps. The closeness of the electrodes doesn't demand a voltage higher than that of supply mains.

The ionisation of the gas within the cover takes place immediately after switching on the supply. A reddish pink glow appears across the electrodes under the influence of excited neon atoms. The retreat of electrons pushed to a higher level after collision results in a radiation of nearly 6536 A.U. The colour due to this radiation is in the visible spectrum *viz* red. In order to restrict the magnitude of circulating current a protective resistance is introduced in the supply circuit. To get a desired indication on the panel, the plastic cap of the lamp cover is given red, yellow or green colour. The body of the lamp cover is normally plastic fabricated. Metallic covers with plastic top (caps) are also available. The lamp size is approximately 1-25 cm in cylindrical formation with conical top (as shown in Fig. 5.52). The lamp diameter can be 3-4 mm.

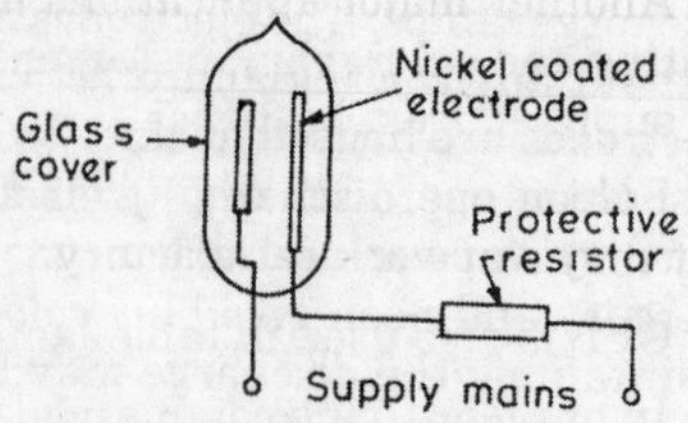

Fig. 5.52. Neon-indicator lamp.

5.27. CFL (Compact Fluorescent Lamps)

Compact Fluorescent Lamps (CFL)

The compact fluorescents lamps are now-a-days are frequently used for different purpose. The switch start, energy efficient and long lasting compact fluorescent lamps (CFL) are available in the market which gives some new and exciting possibilities of illumination. Compact fluorescent lamps are available in 5/7/9/11 watts, 220 V. fig. 5.52.1 shows the CFL available in single lamps *i e.* CFLS and double lamps CFLD.

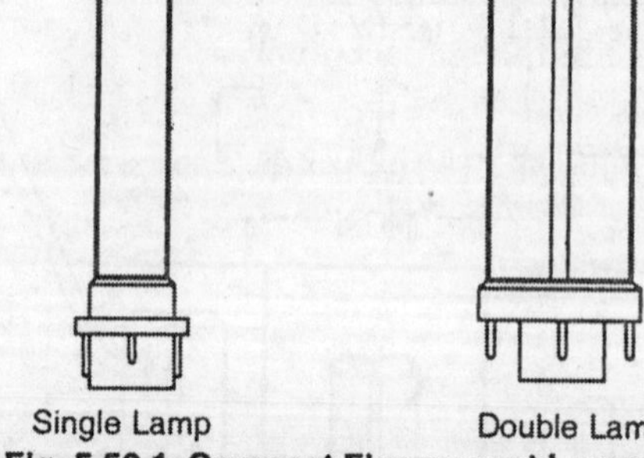

Fig. 5.52.1. Compact Fluorescent Lamps

The compact fluorescent is mounted on a shorter and more compact pin-base lamp. It may be single turn tube, double turn tubes and tripple trun tubes signle ended plug-in base with integral starter and R.F. suppression capacitor type C.F.L. is also available in market.

Applications of CFL

The compact size makes these lamps ideal for table lamps, recess/surface mounted down lighters and decorative fittings, which are widely used in hotel rooms, reception lobbies, restaurants, corridors, bedrooms, over dressing tables and also many other locations to enhance the surrounding ambience.

Another major applications include all areas requiring dimmable lighting and emergency system in departmental stores, office buildings etc. supplied from batteries.

For the operation of CFL electronic and magnetic ballast is used so that they can work satisfactory.

CFL is also used in table lamps.

In outdoor fixing commerical CFL luminair with acrylic cover and weather proof rubber gasket and side mounting pole fixing arrangement is used in street light.

5.27.1 Significant Advantages of CFL

The following are the main advantages of CFL over the ordinary incandescent lamp.

(1) Low energy consumption

(2) Low running costs

(3) Excellent colour rendering properties. Hence all colours look natural and desired ambience can be created in various interiors.

(4) Low operational costs : The power consumption of CFL is 80% less than incandescent lamps. Besides, these lamps lasts 8 times longer than the incandescent lamps.

(5) The compact size opened up possibilities for exciting and aesthetically appealing new luminaires.

(6) Average life of compact fluorescent lamp is 8000 burning hours.

The following figs. shows the comparison between CFL and GLS lamps irrespective of their power consumption.

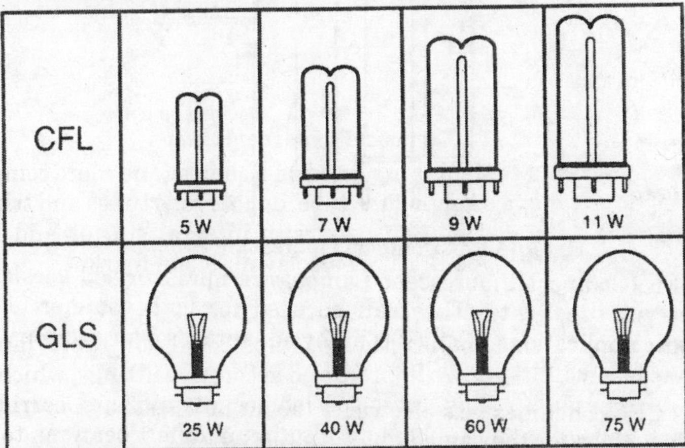

Fig. 5.52.2. CFL & GLS Lamps

Fig. 5.52.3 shows a small, economical and compact patti type fitting which is specially designed for use with CFL.

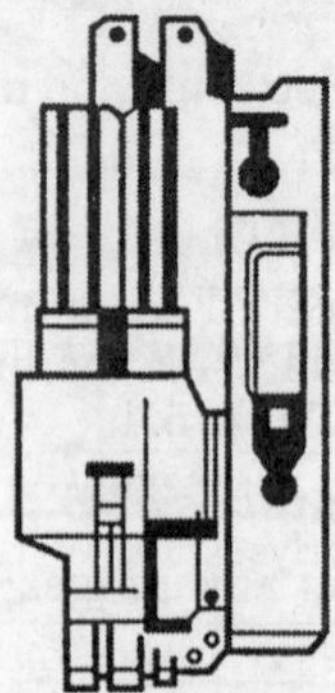

Fig. 5.52.3. Patti Type Fitting

This fitting can be a conveniently used any where.

Technical Data

Types of Lamps	*Wattage (W)*	*Luman (LM)*	*Lamp. Voltage (V)*	*Current (mA)*	*Dimention (m m)*		
					A	*B*	*C*
CFL 5W	5	250	34	180	69	85	108
CFL 7W	7	400	47	175	91	115	138
CFL 9W	9	600	60	170	129	145	168
CFL 11W	11	900	91	160	199	215	238

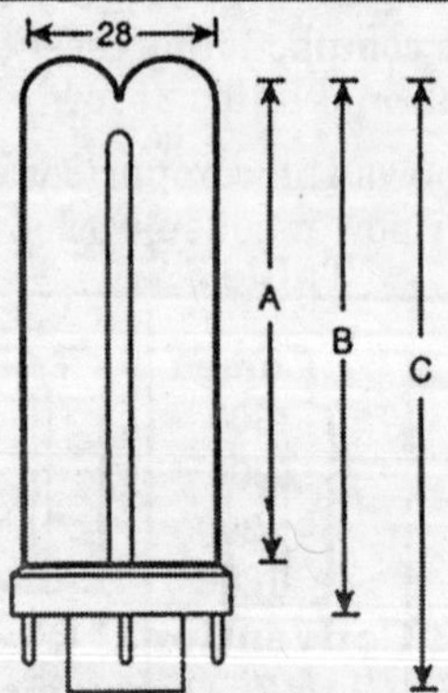

Fig. 5.52.4 Compact Fluorescent Lamps

CFL (Compact Fluorescent Lamps) are miniaturised versions of fluorescent tubelights. They can be used for both commercial and domestic applications. CFL offer many advantages over ordinary light bulbs as below :

(*i*) CFL consumes less electricity (about 80% less) as compared to ordinary bulbs for the same light output. The table 1 below shows the comparison between the electricity consumed by ordinary bulb and CFL:

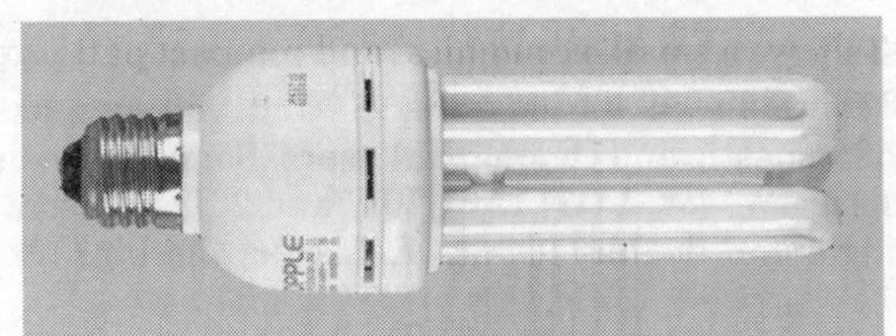

Fig. 5.53. Energy Saving Compact Fluorescent Lamps.

Table 5.1.

1.	*Rating*	20 W	15 W	11 W	7 W
2.	Replaces General Lighting System	100 W	75 W	60 W	40 W
3.	Life of lamp (Hrs)	10000	10000	10000	10000
4.	Units of Electrical Energy saved over the lamp life (kWh)	800	600	490	330
5.	Saving in Electricity Bill @ Rs. 2.50 per unit	2000	1500	1225	825
6.	Cost of 10 GLS lamps Rs. 130				
7.	Cost of 1 CFL lamp	775	775	760	760
8.	Net saving (5+6-7)	1355	855	595	195

(*ii*) **Low heat generation.** CFL generates 80% less heat than conventional bulbs. So the place where they are installed will need reduced air-conditioning load which will lead to further saving of electricity.

(*iii*) **Lighting Efficiency is higher.** CFL have a higher lumen per watt output than ordinary bulb. It means CFLs generate the same amount of light as conventional bulbs at lower wattages.

Table 2 shows comparison between ordinary (GLS) bulb and a Typical CFL :

Table 5.2.

CFL (Watts)	Lumen	GLS (Watts)
10	600	60
13	900	75
18	1200	100
26	1800	150

(*iv*) **Conservation of Conventional (Fossil) Fuel.** Considering the Advantages from S.No. (*i*) to (*iii*) one can say that by reducing the need for additional power required, CFL also helps in saving the fossil fuel reserves required for the power generation, cut short the rising cost and electricity, reduce the chances of power cuts/load shedding.

(*v*) **Type of voltage required.** CFL can be used on AC as well as on DC supply (176 to 340 V range). Therefore they are suitable for emergency lighting circuit. Some types of CFLs are also suitable for light Dimming systems.

5.27.1 Types of CFL available :

CFL are available in various shapes and voltages as per the requirement of domestic and commercial users. Some commonly available wattages are 5W, 7W, 11W, 15W, 20W and 23W and some shapes are Tubes (Single and double), Globe, Reflector, Circular, Concentric etc.

Conclusion:

In brief we can conclude that CFLs has the following advantages over conventional lamps :

- more compact format
- light weight
- higher luminous efficiency and greater energy saving
- instant flicker free starting
- flicker free operation
- can work on d.c. supply
- longer lamp life
- directly replacable with conventional lamp in the same lamp holder

The only disadvantage is the installation cost but this is overcome by the advantages as given above and the saving in electricity as shown in table 1.

5.28. Barriers of Efficient Lamps

Introduction. In our modern life electric lighting has become so common that it is difficult to conceive a situation without it. Such a situation, however, exists in remote villages, which are also reducing in number. Electricity consmuption including a share of lighting sector is an index of economic condition of a country. As a result with the development of the country, energy consumption for lighting will also grow. Shortage of energy calls for its conservation consequently, energy saving and development of efficient lamps are direct target for lamp manufacturers.

From the very beginning of innovation of electric lamps there was an effort for making efficient lamp. A close look into the limitations or barriers for making ideal lamps may be of interest. Some of barriers were historical, some are fundamental and some are technological.

5.28.1. A Long Way to Electric Lamps

The trends in technological innovations in lamp making are clear from the hisotry of lamp development. It is well known that a discovery in any technological field depends on other associated advancements and lamp-development is no exception to it.

Sir Humphrey Davy in 1810 produced a sustained discharge between two woood-charcoal pencils, which had no practical importance at that moment of time. Long before that, in 1800, Alessando Volta discovered the primary battery. Davy made a large battery on this basis to produce an arc radiating an intense light which was not known before. Practical arc light had to wait for solution of electrode material and control of arc length. The problem of arc control was first solved by Foucault and gradually automated by W.E. Staite in 1847. For the solution of electric power source we are to go back to the discovery of Michael Faraday, which again had no immediate application. But in April 1857, Federic Hale Homes installed a successful generator for the arc lamp installed at the Blackwall Light House. Several Practical generators became available in which Gramme's machine developed in 1870 made arc lighting a possibility.

Sir Joseph Wilson Swan and Thoms Alva Edison independently discovered carbon filament lamp in 1879. But there are evidences that in December 1878 Swan had demonstrated the carbon lamp in New-castle Chemical Society. Edison introduced carbon lamp, carbonised from bamboo filament. On the other hand, Swan in 1883 developed squirted cellulose filament, which lasted as the only suitable filament till replaced by tungsten filament in 1906. Search had began by 1884 to replace carbon filament by a better material to compete with gas-mantle lamp which was more efficient at that time.

The phenomenon of electric discharge in rarefied gas producing light was known in 18th century. Therefore, another direction of research was to develop gas discharge lamp using mercury. The first low pressure mercury discharge lamp was developed by Peter Cooper-Hewitt in 1901, which consisted of a long glass tube enlarged at one end used as mercury pool. The lamp had to be tilted to strike the lamp. As the light produced mainly in the UV region, it was not accepted. In 1910 a Frenchman Georges Claud introduced neon tube for advertisement purposes and in 1930 discovery of fluorescent coating laid the foundation of fluorescent lamp. In 1938 in America this low pressure mercury vapour discharge with fluorescent coating was introduced. In 1942 halophosphate phosphors were invented in UK.

Table 5.3. Ranges and Luminous Efficacies of Representative Types of Lamps

Lamp Type	*Rated Wattage (watt)*	*Efficiency (lm / W)*	*CRI (Ra)*
Tungsten Filament (GLS)	15-1000	12-18	100
Tungsten Halogen (Linear)	100-2000	15-22	100
Fluorescent (TL) (*a*) Agron filled (*b*) Krypton filled (*c*) High Frequency (argon filled)	20-80	65-96	56-85 and above
H.P. Mercury Coated (MBTF)	50-2000	38-58	40
H.P. Mercury Blended (MLL)		23 (approx)	70
Metal Halide (MBI)	250-2000	54-83	70-90

Metal Halide Coated (MBIF)	250-1000	64-85 (above)	
H.P. Sodium SON	50-1000	62-110	20
(*a*) SON/T	50-1000	79-130	
(*b*) SON/delux	150-400	78-88	
(*c*) SON DL/T	150-400	81-93	
L.P. Sodium SOX	35-250	120-200	45

5.28.2. Lamp Development

More than hundred years ago, various lamps were discovered and developed and found there applications in the lighting field. Due to limitations of efficacy, wattage, colour temperature and colour rendering qualities and light output none can be selected for universal applications. Depending upon the requirements, lamps are selected. Through more than one hundred years of development of light sources, their efficacy has improved as shown in Fig. 5.54 and Table 5.3 gives the technical data of various types of modern lamps.

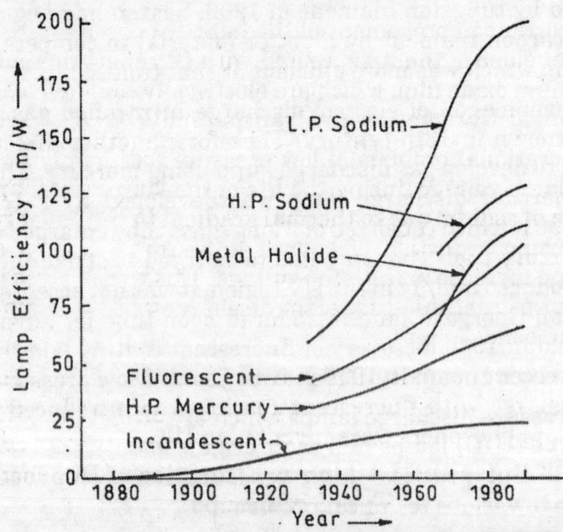

Fig. 5.54. Improvement in lamp efficacy over 100 years.

5.28.3. Tungsten-Halogen Lamp.

A knowledge of bulb-wall cleaning effect of chlorine was gained as early as 1881 by Edwin A. Scribner, along with the corrosive effect of halogen and loss of halogen in reaction with glass. The material and technology necessary to make a practical lamp were not known. In 1955, an incandescent lamp made from fused-quartz tube, with an axial filament of tungsten, filled with iodine and inert gas was successfully made. This discovery had created a renewed interest in the incandescent lamp and thorough study of regenerative cycle was made. At present wide range of halogne lamps, 4W- 10,000 W are available.

5.28.4. Low Pressure Solium Lamp.

Low pressure sodium lamps were developed in Holland and Germany nearly at the same time as High pressure mercury lamps, in 1930's. Unlike fluorescent lamps, the sodium resonance lines (589.0 nm and 589.6 nm) are in the visible region and therefore no phosphor is necessary. Thus, low pressure sodium radiation is nearly monochromatic yellow and the luminous efficacy is high, but CRI value is very low.

For low pressure sodium discharge it is necessary that the cold spot temperature should not be less than 260°C to get the required vapour pressure. Sodium reacts with ordinary glasses and during the life of lamp sodium is also absorbed. It was, therefore, necessary to develop a discharge tube which will not react upto 300°C. Initially borate glass was used, which was not suitable for glass processing and sealing with metal. oon ply glass tube consisting of sodalime tube coated inside with thin film (50 μm) sodium resistance borate glass was developed.

An important improvement of the lamp was brought by providing the outer envelope of the tube, coated with IR reflecting coating of tin oxide or indium oxide film, which are electrically conductive, have high reflectance near IR radiation and transparent to visible wavelengths.

One operational problem of low pressure sodium discharge is the rise in the lamp voltage during the life of the lamp. This happens due to migration of sodium due to thermal gradient to the cooler region and thereby causing reduction of vapour pressure of sodium and more depletion and increase in the lamp voltage. This was avoided by providing a cap clamped at the U-tube of the discharge tube to eleminate colder region in the bend.

5.28.5. High Intensity Discharge Lamp Development

High pressure discharge lamps which are often called High intensity discharge (HID) lamps had their origins along with low pressure counterpart. But practical high pressure lamps production were delayed due to various technical problems.

5.28.6. High Pressure Mercury Vapour Lamps

It was found that pressure-efficiency characteristics in a mercury vapour discharge improved when the pressure was increased at about one atmosphere and this led to high pressure mercury lamp development. First HP mercury lamp was developed in 1906 in Germany by R Kuch and T Retschinsky. But lamps for general lighting came in early 1930's. The practical problems were to find suitable glass for the discharge tube, glass to metal seal and starting at line voltage. Fused silica is affected by the barium oxide used for electrode emission and therefore, the temperature is restricted to 750°C. Molybdenum foil is used at the pinch and a combination of rare gas filling at a pressure of few mm of mercury is used. The starting problem is solved by an auxiliary electrode.

5.28.7. High Pressure Sodium Lamp

It was known for a long time that sodium discharge could be broadened by increasing the pressure and temperature away from the D-line resonance line by self absorption and re-emission. But no known discharge tube material could withstand highly corrosive hot sodium vapour for a long time. Initially synthetic saphire was used for lamps for laboratory studies only, due to high cost of material. Additional problems was alumina to metal seals that could resist sodium vapour above 700°C without thermal stresses. In USA during mid 1950's developed a new type of ceramic material called poly-crystalline alumina (pca) with optical properties as glass and in 1958 first laboratory lamp was made. Soon it was discovered that sodium, mercury and xenon mixture could produce a golden white light, characteristics of which depends on the ratios of these materials. By 1966 the 400 high pressure sodium lamp was in production.

5.28.8. Metal Halide Lamps

In recent time among all the HID lamps, metal halide lamps have drawn attention due to high efficacy and colour quality attained so far. This lamp, developed during 1970's and 1980's, is based on high pressure mercury discharge with added metallic iodides. The characteristics of the light output can be controlled by the choice of the metal halide and there are four basic types of these lamps vis.

(*a*) Scandium-sodium-iodide lamp

(*b*) Rare-earth metal (dysprosium)-iodide lamp

(*c*) Indium-thallium-sodium-iodide lamp.

(*d*) Tin-halide lamp.

Among these the first type is more popular and the last type is now preferred for warm colour temperature of 3000 K. Usually these lamps are elliptical shaped but tubular bulbs made from fused silica are also available.

5.28.9. Characteristics of Lamps

Suitability of any lamp depends on its characteristic parameters, to satisfy the lighting requirements. Lamps are developed to meet the increasing complexities of application areas. It should be noted that some parameters may be more important than the others, depending upon the requirements. The important parameters are as follows :

(*a*) Light output of the lamp is lumen.

(*b*) Efficacy of the lamp in lumen/watt.

(*c*) Lumen maintenance throughout the life of the lamp.

(*d*) Life of the lamp in hours of burning or percentage of lamp failures for burning hours.

(*e*) Colour Rendering Index (CRI) and Colour Appearance (Colour Temperature or Correlated Colour Temperature).

(*f*) Cost of the lamp and necessary starting devices.

5.28.9 (*A*) Comparison between CFL and GLS Lamps for Approximately Same Light Output

CFL Lamp (Ra = 80 - 85) Rating (Watt)	*GLS Lamp (Ra = 100) Rating (Watt)*
5	25
7	40
9	60
11	75
13	75
18	100
24	150
36	200

Average life of CFL is 5000 hrs. and that of GLS is 1000 hrs.

5.28.10. A Few Contemporary Lamps and their Problems

CFL Lamps : Miniature Fluorescent Lamps

There are two main reasons to reduce the size of fluorescent lamps as these lamps has excellent qualities except its size and temperature dependence. One reason being to find a replacement of GLS lamps and secondly to help a better optical design for more efficient luminaries. Table 2 shows comparative data of CFL and incandescent lamps.

In 1980's two well known manufacturers in Europe introduced the compact fluorescent lamp with designations PL and DULUX. This was possible mainly due to development of tri-phosphors allowing higher efficiency of UV conversion into visible wavelength leading to reduction of lamp power for the same light output and higher lamp wall loading leading to reduction in diameter of the lamp. The lamp lumen depreciation throughout life of the lamp became less with these new phosphors compared to old halophosphates.

The design parameters restricting CFL are the wattage which is proportional to the surface area or length for the same diameter and the control of vapour pressure at the given ambient temperature. The size is reduced by bending the tube in U-form or by twin or triple parallel tubes and insertion of glass wool. The gas filling of argon-neon mixture is replaced by krypton-neon or krypton-argon mixture.

A further reduction in size is being achieved by using double twin-tube (in 1986) and incorporation of electronic ballast integral with the lamp. Basic principle of these electronic ballasts is to increase the mains frequency of 50/60 Hz to a frequency of 20-50 KHz at which the efficacy of the lamp is highest and at the same time the ballast loss is reduced. A compact solid state electronic circuit generate this high frequency from the supply mains by first rectifying it to dc and then inverting it to high frequency by transistors or thyristors.

Discharge current in CFL lamp is high and its small volume increases the temperature. The tri-phosphors can withstand high wall loading and the mercury amulgum used for CFL allows higher vapour pressure.

5.28.11. Induction Lamps

Scientists were in constant search to eliminate some common drawbacks of discharge lamps and make a lamp which will not require any electrode and ballast or ignitors. Ultimately in 1991 they found a solution by lamp using electromagnetic induction at radio frequency with mercury vapour. This lamp has very long life. Electrode failure is the ultimate cause of failure of discharge lamp and limits the life of a lamp. The ballast losses are of also considerable amount reducing the circuit efficacy.

The electrodeless lamps are free some of the constraints like optimum UV conversion in low pressure mercury discharge and high electrode loss and reaction with electrode coating in HID lamps. In these lamps efficacy of 171 lm/RFW have been achieved and life is of 60,000 hrs.

The principle of operation of induction lamp is energy coupling of gas discharge with electromagnetic field by means of transformer principle. The induction coil acts as the primary and discharge from the secondary. The efficiency of field coupling can be achieved at high frequency in the microwave range. The practical problems facing the lamp was to make a very small RF generator which could be accommodated at the base of the lamp, and RFI restrictions. RF interference regulation Industrial, Scientific and Medical (ISM) use. Hence, it is restricted at 13.56, 27.12 and 40.68 MHz with very narrow bandwidth.

5.28.12. Incandescent Lamp Failure

The main reasons of incandescent lamp failure are :

(*a*) bulb wall blackening and (*b*) filament failure

The first effect is related with vapourization of the filament material at high operating temperature and deposition on the colder lamp wall. For effective visible radiation the operating temperature must be in the range of 2200°C – 3100°C and it must have selective emissivity. Tungsten is the best material so far found. For better efficacy, tungsten filament is maintained at temperature (approximately 2900°C) near the melting point resulting high evaporation. Poor efficacy of the incandescent lamp is due to large radiation in the infrared region.

Filament Failure. Microstructural property of the filament material determines the strength and dimensional rigidity throughout the life of the lamp. Pure tungsten is not suitable in this respect. After the lamp is first switched on recrystallization of the grains occur and the size and shape of the grains affect the stregth at high temperature. For pure tungsten, the grain boundary alignment helps to reduce the wire diameter which results in ultimate failure of the filament. But if the pure tungsten is blended with some additives then the mechanical properties are improved. Commonly used dopants are thoria or combination of potassium silicate and alumina (known as AKS dopant) which control the grain boundary formation according to requirements.

Tungsten is also used as emitter material for discharge lamps and to increase the life and electron emission capacity it is coated with one or more alkaline earth oxides, *e.g.* barium oxide or strontium oxide. For high wattage lamps, due to high sputtering off of oxide coatings, throriated tungsten emitter is used. For metal halide lamps, halogen would react with alkaline earth oxides to produce involatile oxides depositing on the bulb wall. But for thoriated tungsten emitter, both thorium and tungsten react with halogen to produce the regenerative transport cycle whereby the tungsten is redeposited on the electrode.

Failure Mechanism. Filament failure is a complex phenomenon. There are various studies made and theories postulated on the failure mechanism of incandescent lamp to find out ways and means to enhance the life of the lamp. These indicate that there is no particular reason and mode of failure, as it depends on various circumstances. It can, however, be summarised as:

1. Nonuniform evaporation from the filament due to mass loss resulting in non-uniformity of high temperature distribution leading to burn out.

2. Hot-spot growth due to wire defect and coil defect in case of coiled coil filament. Migration of voids in doped tungsten which is also necessary for dimensional stability. This happens due to collisions and coalescence of migrating bubbles, large void may be formed under the action of temperature and stress gradients. This bubble growth causes increase in porostiy as coalesced bubble has larger volume that sum of the smaller bubbles.

3. Coil defect due to variation of coil pitch, due to sag and other reasons causing temperature difference between the defect area and other parts of the filament.

4. Facet formation due to evaporation from a specific area of the filament causes change in shape of the wire, change in emissivity and electrical resistance of the filament. Electrical resistance of the filament will change both due to change in specific resistance ρ and filament geometry during operating, due to more evaporation from certain areas. The coils which develops facets during life causes reduction in tempera-

ture in that area due to increase in radiating areas. It has been found that there is no correlation between intitial hot spot and facet formation.

Experimental works have shown that hot spot failure mode is more common, but it can not be attributed as the sole reason and other failure modes are also valid and perhaps may act simultaneously. For example, in gas filled lamps it has been found that 66% final hot spot failure has coincided with first initial hot spot and 21% coincided with other initial hot spot. Failure due to minimum value of pitch at the first initial hot spot is only 30% and at the final hot spot is 44%.

5.28.13. Discharge Lamp Problems

It has been indicated earlier that initial problems of introduction efficient discharge lamps were material problems, *viz.*, glass for the lamp envelops, glass to metal seals. A common spot of early failure is due to electrode failure and sputtering of emission coatings. For sodium lamps sodium depletion is very common reason of failure. To improve upon the efficacy, colour and life, constant efforts are made which has also resulted in new types of lamps.

In case of low pressure mercury discharge lamps, the light production depends on UV generation and conversion to visible radiation. This naturally directed to development of efficient phosphors. It has been found that conversion efficiency can be improved by increasing the radiation in only few radiation bands which affect the CRI quality.

The colour rendering quality and luminous efficacy are two opposing characteristics demanded from a good phosphor. The tri-phosphors, mentioned earlier, developed in 1970's and known as Red, Green and Blue phosphors are rare earth compounds. The chemical compositions of the tri-phosphors are :

Red phosphor, emission peak at 611 nm : $(Y, Eu)_2O_3$

Green phosphor, Emission peak at 543 nm : CAT [(Ce, Tb), $MgAl_{11}O_{19}$

Blue phosphor, emission peak at 450 nm : BAM [(Ba, En) $MgAl^{10}O_{20}$]

The spectral power distribution due to three phosphors are linear sum of the individual phosphors. These phosphors are also capable of withstanding high wall loading.

5.28.14. Fundamental Limitations

The fundamental limitation of fluorescent lamps is due to quantum energy conversion. By Stoke's Law, at best one quantum of UV radiation can be converted into one quantum of visible radiation, *i.e.* conversion from shorter wavelength to longer wavelength. Now the quantum energy of a photon of visible wavelength band has half the energy of UV radiations. Therefore, the maximum theoretical efficiency of conversion is only 50%. The practical quantum ratio is less than one.

Another problem of fluorescent lamp is the strong sensitivity to temperature variations. The maximum luminous output occurs at ambient temperature of 25°C, corresponding to lamp wall temperature of 40°C, depending upon the lamp type.

An important factor for some of the discharge lamps is the colour shift during the life of the lamp. This information is not generally supplied by the manufacturers. If the quality aspect of light is important, the life of the lamp should be given in terms of colour shift. For example, the initial colour temperature of a low wattage metal halide lamp be 3000 K undergoing shift of co-related colour temperature (CCT) of 300 K in 6000 hrs. of burning, it will have CCT of 2700 K to 3300 K. This may not be acceptable for some application.

Considering the above factors, it appears the CFL lamp and Induction lamp have great prospect in contemporary lighting applications.

5.28.14 Recent Trends in Electronic Ballast for Fluorescent Tubes

Electronic ballast developed recently by lamp manufacturers helps fluorescent lamps to start at low voltages of 140V instantly. The ballast works at a P.f of 0.99 as compared to 0.67 in a conventional lamp. These lamps have longer life and avoids excessive current in the filament at the time of starting. There is a considerable saving in energy cost as the power factor is high. In addition 9W compact fluorescent lamps gives as much light as a 60 watt lamp. Such a lamp gives a work life of 8000 hrs. as against 1000 hrs. of a GLS lamp.

The circuit diagram for a 36 W fluorescent lamp with electronic ballast is shown in Fig. 5.54.1. The graphs in Fig. 5.54.2 shows the comparison between 36 Watt fluorescent lamp with electronic ballast with respect to the conventional lamps. The graphs are self explanatory.

Example 5.26. *Two lamps are hung at a height of 9 metres from the floor level. The distance between the lamps is one metre. Lamp one*

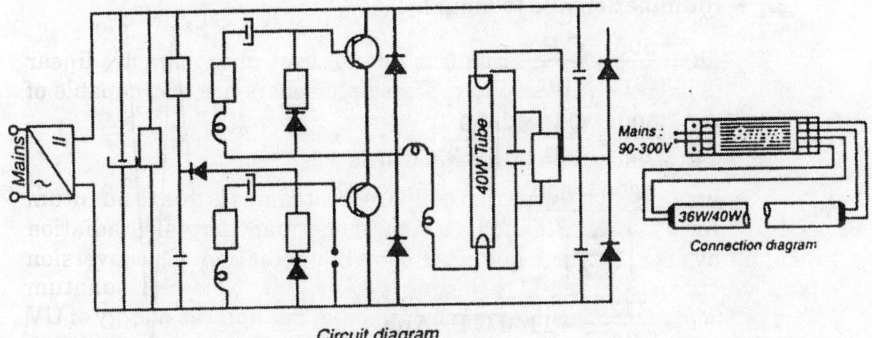

Fig. 5.54.1

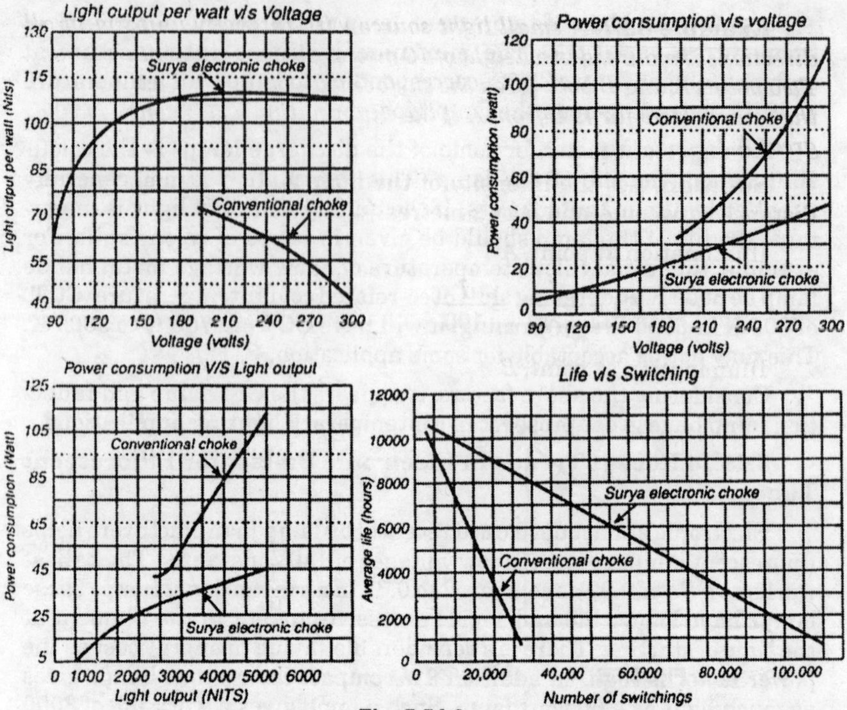

Fig. 5.54.2

is of 500 C.P. If the illumination on the floor vertically below this lamp is 20 lux, find the Candle Power of the lamp No. two.

Solution. Candle power of lamp, $L_1 = 500$ C.P.

Let candle power of lamp L_2 be C.P. x

Illumination at the point A on the floor level vertically below lamp, L_1 = Illumination due to lamp

L_1 + Illumination due to lamp L_2.

$$20 = \frac{500}{(9)^2} + \frac{\text{C.P. } x}{(9)^2} \cos^3 \theta$$

or
$$20 = \frac{500}{81} + \frac{\text{C.P. } x}{81} \cdot \left(\frac{9}{\sqrt{82}}\right)$$

or
$$\text{C.P. } x = \frac{1620 - 500}{\left(\frac{9}{\sqrt{82}}\right)^3}$$

$$= \frac{1120}{0.9817} = 1140 \text{ C.P. } \textbf{Ans.}$$

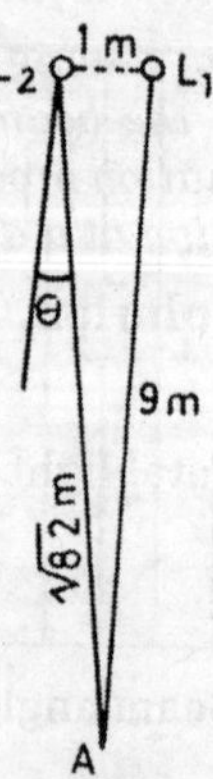

Fig. 5.54.3

Example 5.27. *A small light source with intensity uniform in all directions is mounted at a height of 10 metres above a horizontal surface. Two points A and B both lie on the surface with point A directly beneath the source. How far is B from A if the illumination at B is only 1/10 as great A ?*

Solution. Let the intensity of the lamp L, be I candela and the distance between A and B be x metres.

Illumination at point, A

$$E_A = \frac{I}{(10)^2} = \frac{I}{100}$$

Illumination at point, B

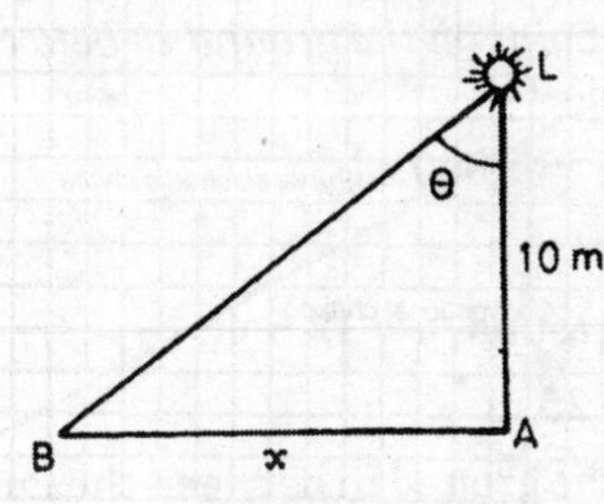

Fig. 5.55

$$E_B = \frac{I}{(10)^2} \times \left(\frac{10I}{\sqrt{(10)^2 + x^2}} \right)^3$$

$$= \frac{10\,I}{(100 + x^2)^{3/2}}$$

$$\because \quad E_B = \frac{1}{10} E_A$$

$\therefore$

$$\frac{10\,I}{(100 + x)^{3/2}} = \frac{1}{10} \times \frac{I}{100} \quad \text{or} \quad (100 + x^2)^{3/2} = 10{,}000$$

$$(100 + x^2) = (10{,}000)^{2/3}$$

or $$= 464$$

or $$x = \sqrt{464 - 100} = 19.1 \text{ metres.}$$ **Ans.**

Example 5.28. *A light source having an intensity of 500 C.P. in all directions is filled with a reflector so that it directs 80% of its light along a beam having a divergence of 15°. What is the total light flux emitted along the beam ? What will be the average illumination produced on a surface normal to beam direction at a distance of 10 metres ?*

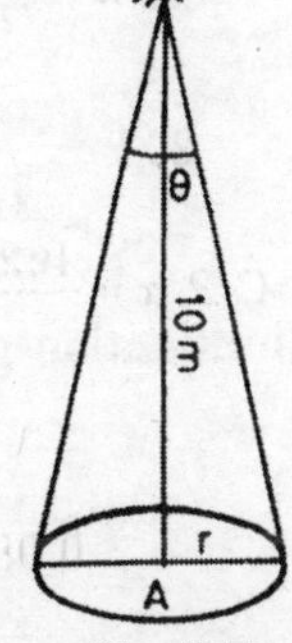

Fig. 5.56

Solution. Intensity of light source, I

$= 500$ *cd*.

Total light flux emitted along the beam

$= 0.8 \times 4\pi \times 500$

$= 5{,}227$ lumens **Ans.**

Beam angle, $\theta = 15°$

Distance of the surface from light source, l = 10 metres.

Radius of the circle to be illuminated, r

$$= l \tan \frac{\theta}{2} = 10 \tan \frac{15^\circ}{2} = 1.316524 \text{ m}$$

Area of the surface to be illuminated,

$$A = \pi r^2 = \pi \times (1.316524)^2 = 5.445 \text{ m}^2.$$

Average illumination, $= \frac{5.227}{5.445} = 960$ lux. **Ans.**

Example 5.29. *Two powerful street lamps of 1,000 candela and 800 candela (assumed uniform in all directions) are mounted 12.5 metres above the road level and are spaced 25 metres apart. Find the intensity of horizontal illumination produced at a point on the ground in between the lamp posts and just below the lamp posts.*

Solution. Candle power of lamp, L_1 = 1,000 cd.

Candle power of lamp, L_2 = 800 cd.

Height of lamp L_1 and L_2 from the ground, h = 12.5 m.

Distance between two lamp posts = 25 m

Illumination produced at a point on the ground in between the lamp posts.

E_C = Illumination due to lamp L_1 + Illumination due to lamp L_2

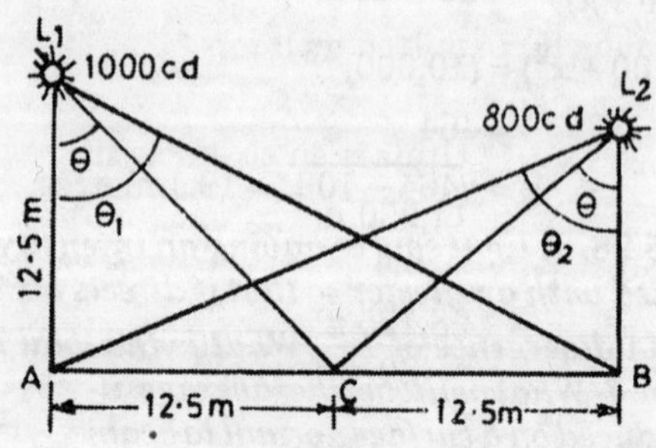

Fig. 5.57.

$$= \frac{1{,}000}{(12.5)^2} \times \left(\frac{1}{\sqrt{2}}\right)^3 + \frac{800}{(12.5)^2} \times \left(\frac{1}{\sqrt{2}}\right)^3 = 4.07 \text{ lux.} \quad \textbf{Ans.}$$

Illumination produced just below the lamp post having 1,000 cd lamp.

$$E_A = \frac{1000}{(12.5)^2} + \frac{800}{(12.5)^2} \times \left[\frac{12.5}{\sqrt{(25)^2 + (12.5)^2}}\right]^3$$

$$= 6.4 + 5.12 \times (1/\sqrt{5})^3$$

$$= 6.4 + 0.457 = 6.857 \text{ lux.} \quad \textbf{Ans.}$$

Illumination produced just below the lamp post having 800 cd lamp,

$$E_B = \frac{800}{(12.5)^2} + \frac{1000}{(12.5)^2} \times \left(\frac{1}{\sqrt{5}}\right)^3$$

$$= 5.12 + 6.4 \times (1/\sqrt{5})^3$$

$$= 5.12 + 0.57 = 5.69 \text{ lux} \quad \textbf{Ans.}$$

Example 5.30. *A small area 7.5 metre in diameter is to be illuminated by a lamp suspended at a height of 4.5 m over the centre of the area. The lamp having an efficiency of 20 lumens per watt is fitted with a reflector which directs the light output only over the surface to be illuminated, giving uniform candle power over this angle. Utilisation coefficient = 0.40. Find out the wattage of lamp. Assume 800 lux of illumination level from the lamp.*

Solution. Area of working plane,

$$A = \frac{\pi}{4} d^2 = \frac{\pi}{4}(7.5)^2 = \frac{225}{16} \pi \text{ m}^2.$$

Illumination on the working plane,

$$E = 800 \text{ lux}$$

Luminous flux reaching on the working plane,

$$F = E \times A = 800 \times \frac{225}{16} \pi = 11{,}250\,\pi \text{ lumens.}$$

Total luminous flux emitted by lamp

$$= \frac{F}{\text{Utilisation co efficient}}$$

$$= \frac{11{,}250\,\pi}{0.4} = 28{,}125\,\pi \text{ lumens.}$$

Wattage of lamp $= \dfrac{28{,}125\,\pi}{20} = 4{,}420$ watt. **Ans.**

Example 5.31. *A 60 c.p., 250 volt metal filament lamp has a measured candle power of 71.5 c.p. at 260 volts and 50 c.p. at 240 volts.*

(*a*) *Find the constant for the lamp in the expression $C = aV^b$, where C = candle power and V = voltage.*

(*b*) *Calculate the change of candle power per volt at 250 volts. Determine the percentage variation of candle power due to a voltage variation of ± 4% from the normal value.*

Solution. (*a*) $C = aV^b$ ...(*i*)

Substituting $C = 71.5$ c.p.

when $V = 260$ volts.

and $\quad C = 50$ c.p.

when $\quad V = 240$ volts.

We get $\quad 71.5 = a(260)^b \quad ...(ii)$

and $\quad 50 = a(240)^b \quad ...(iii)$

Dividing expression (*ii*) by expression (*iii*), we get

$$\left(\frac{260}{240}\right)^b = 1.43$$

$$b = 4.5 \quad \textbf{Ans.}$$

Substituting $b = 4.5$ in expression (*iii*), we get

$$a = \frac{50}{(240)^{4.5}} = 0.98 \times 10^{-9} \quad \textbf{Ans.}$$

Thus expression becomes,

$$C = 0.98 \times 10^{-9}\, V^{4.5} \quad ...(iv)$$

(*b*) Differentiating expression (*iv*) w.r.t. *V* for

$$V = 250 \text{ volts.}$$

We get $\quad \frac{dC}{dV} = 0.98 \times 10^{-9} \times 4.5\, V^{3.5}$

$$= 0.98 \times 10^{-9} \times 4.5 \times (250)^{3.5}$$

$$= 4.4\ V \quad \textbf{Ans.}$$

When voltage increases by 4%

$$\frac{C_2}{C_1} = (1.04)^{4.5}$$

and percentage change in candle power

$$= \frac{C_2 - C_1}{C_1} \times 100 = [(1.04)^{4.5} - 1] \times 100$$

$$= (1.193 - 1) \times 100 = 19.3\% \quad \textbf{Ans.}$$

When voltage falls by 4%

$$\frac{C_2}{C_1} = (0.96)^{4.5}$$

and percentage change in candle power

$$= \frac{C_2 - C_1}{C_1} \times 100$$

$$= [(0.96)^{4.5} - 1] \times 100 = (0.832 - 1) \times 100$$

$$= -16.8\% \quad \textbf{Ans.}$$

Example 5.32. *It is required to provide an illumination of 100 lux in a factory hall 30 m by 15 m. Assume that the depreciation factor is*

0.8, coefficient of utilisation is 0.4 and efficiency of lamp is 14 lumens / watt. Suggest the number of their ratings. The size of the lamps available are 100, 250, 4000 and 500 watts.

Solution. Area to be illuminated,

$$A = 30 \times 15 = 450 \text{ m}^3$$

Illumination required

$$E = 100 \text{ lux.}$$

Total lumens required

$$= A \times E = 450 \times 100 = 45{,}000$$

Total lumens to be given out by the lamps

$$= \frac{A \times E}{D.F. \times U.F.} = \frac{450 \times 100}{0.8 \times 0.4} = 140{,}625$$

Total wattage required

$$= \frac{140{,}625}{14} = 10{,}000 \text{ (say)}$$

Number of 100 watt lamps required

$$= \frac{10{,}000}{100} = 100$$

Number of 250 watt lamps required

$$= \frac{10{,}000}{250} = 40$$

Number of 400 watt lamps required

$$= \frac{10{,}000}{400} = 25$$

Number of 500 watt lamps required

$$= \frac{10{,}000}{500} = 20$$

40 lamps in 5 rows, each row having 8 lamps of 250 watts can be used giving spacing of 3.75 metres in length and 3 metres in width.

Ans.

Example 5.33. *It is required to provide an illumination of 100 lumens / m² in a workshop hall 40 m × 10 m and efficiency of lamp is 14 lumens / watt. Calculate the number and rating of lamps and their positions when trusses are provided at mutual distance of 5 metres. Take coefficient of utilisation as 0.4 and depreciation factor as 0.8.*

Solution. Area to be illuminated,

$$A = 40 \times 10 = 400 \text{ m}^2$$

Illumination required,

$$E = 100 \text{ lumens/m}^2$$

Total lumens required $= A \times E = 400 \times 100 = 40{,}000$

Coefficient of utilisation

$$U.F. = 6.4$$

Depreciation factor

$$D.F. = 0.8$$

Gross lumens required,

$$\phi = \frac{A \times E}{U.F. \times D.F.} = \frac{400 \times 100}{0.4 \times 0.8} = 125{,}000$$

Total wattage required

$$= \frac{125{,}000}{14} = 8{,}929$$

Since number of trusses are seven, therefore, each row will carry seven lamps having spacing of 5 metres. Breadthwise spacing may also be taken the same as length-wise, *i.e.* 5 metres giving 2 rows of lamps. Thus 14 lamps will be required.

Wattage of each lamp

$$= \frac{8{,}929}{14} = 638 = 750 \text{ Watts (say).} \quad \textbf{Ans.}$$

Example 5.34. *A drawing hall 30 metres by 15 metres with a ceiling height of 5 metre is to be provided with a general illumination of 120 lux. Taking a coefficient of utilisation of 0.5 and depreciation factor of 1.4. Determine the number of fluorescent tubes required, their spacing, mounting height and total wattage.*

Taking luminous efficiency of fluorescent tube as 40 lumens/watt for 80 watt tubes.

Solution. Floor are to be illuminated

$$A = 15 \times 30 = 450 \text{ m}^2$$

Illuminated required,

$$E = 120 \text{ lux}$$

Total lumens required at floor level

$$= A \times E = 450 \times 120 = 54000 \text{ lumen}$$

Utilisation factor $= 0.5$

Lumens required to be emitted by tubes or

Gross lumen, $\phi = \dfrac{A \times E \times D.F.}{U.F.}$

$$= \frac{54000 \times D.F.}{U.F.} = \frac{54000 \times 1.4}{0.5} = 151200 \text{ Lumens.}$$

Lumens output of each tube

$$= 80 \times 40 = 3200 \text{ lumens.}$$

Number of tubes required

$$= \frac{\text{Total lumens required}}{\text{lumens of each tube}} = \frac{151200}{3200} = 47.25 \text{ (48 say)}$$

Total wattage of tubes

$$= 48 \times 80 = 3840 \text{ watts.} \quad \textbf{Ans.}$$

24 twin tube units, each containing 2 tubes of 80 watts each can be provided in the drawing hall in six rows having spacing of 5 metres each row will have four units spaced 3.75 metres as shown in Fig. 5.58.

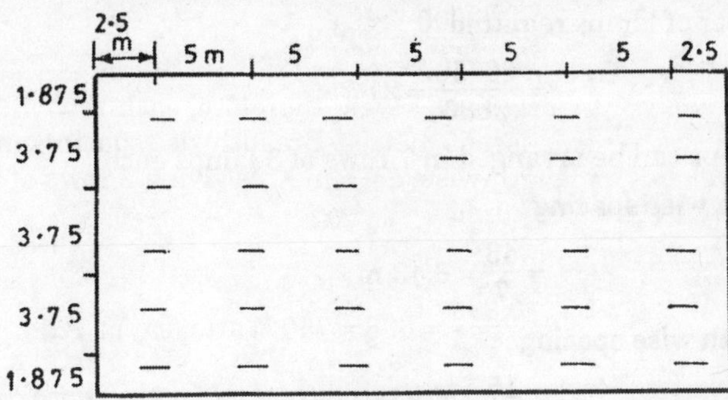

Fig. 5.58.

Assuming mounting height as 3.5 metres

(Length-wise) space height ratio

$$= \frac{5}{3.5} = 1.43$$

(Breadth wise) space height ratio

$$= \frac{3.75}{3.5} = 1.071$$

which is in between 1 and 1.5.

Example 5.35. *A hall measuring 15 m × 45 m is to be illuminated by suitable lamps to give an average illumination of 45 lux. Assuming mounting height of lamp to be 3 m above, the working plane, utilisation factor of 0.65 and depreciation factor of 1.3. Estimate the number and installation position of lamps. The lamps are to be selected from the following :*

Wattage of lamps	*75*	*100*	*150*	*200*
Lumens output	*880*	*1280*	*2120*	*2880*

Solution. The area to be illuminated

$$= 15 \times 45 = 675 \text{m}^2$$

Total lumens required

$$= 675 \times 45 = 30375 \text{ lumens}$$

Gross lumen required

$$= 30375 \times \frac{D.F.}{U.F.}$$

$$= 30375 \times \frac{1.3}{0.65} = 60750 \text{ lumens.}$$

Since the efficiency of higher wattage lamp is more and number of lamps required will be less so let us first try for 200 watt lamps.

Output of a 200 W lamp = 2880 lumens

Number of lamps required

$$= \frac{60750}{2880} = 21$$

21 lamps can be arranged in 7 rows of 3 lamps each.

Length wise spacing

$$= \frac{45}{7} = 6.43 \text{ m}$$

Breadth wise spacing

$$= \frac{15}{3} = 5 \text{ m}$$

Since height ratio in length and breadth-wise be 6.43/3 = 2.14 and 5/3 = 1.66, which is more than 1.5 and this arrangement can not provide uniform illumination.

Next trying 150 watt lamp size

Output of a 150 Watt. lamp = 2120 lumens

Number of lamps required

$$= \frac{60750}{2120} = 29.1 \text{ (say 30)}$$

Arranging 30 lamps in 10 rows of 3 lamps each

Length-wise spacing $= \frac{45}{10} = 4.5$ m

Breadth-wise spacing $= \frac{15}{3} = 5$ m

Space height ratio length-wise

$$= \frac{4.5}{3} = 1.5$$

Space height ratio breadth-wise

$$= \frac{5}{3} = 1.66$$

Instead of keeping 2.5 m distance between the lamp and wall on breadth side, let it be 3 metres and then the distance between two consecutive lamps will be 4.5 m., thus giving height ratio as 4.5/3 = 1.5. A suitable arrangement with 150 watt lamps can be as shown in Fig. 6.59.

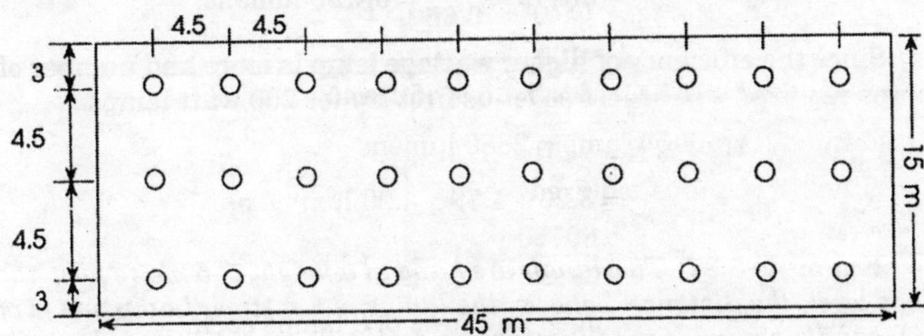

Fig. 5.59

Example 5.36. *A lamps of 100 C.P is placed 1 metre below a plane mirror which reflects 90% of light falling on it. The lamp is hung 4 metres above ground. Find the illumination at a point on the ground 3 metre away from the vertically below the lamp.*

Solution. Capacity of the source of light,

$$S = 100 \text{C.P.}$$

Height of source above ground,

$$SA = 4 \text{m}$$

Height of image of the source

$$S'A = 6 \text{m}$$

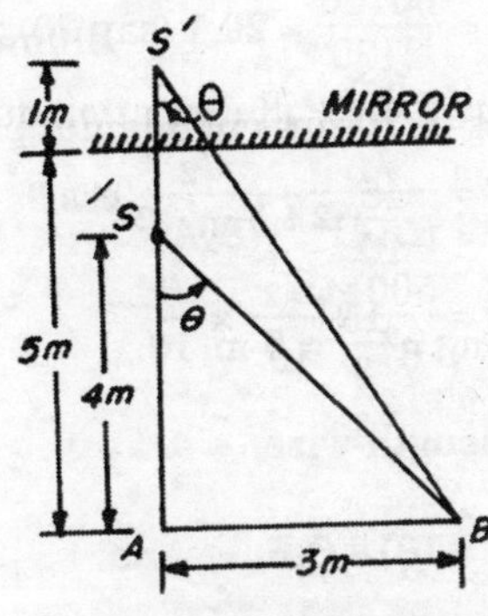

Fig. 5.60

The image of the source acts as the secondary source of light, So capacity of secondary source

$$= 0.9 \times 100 = 90 \text{C.P.}$$

Illumination at point B = illumination due to S + illumination due to S'

$$= \frac{100}{(SB)^2} \cos\theta + \frac{90}{(S'B)^2} \cos\theta_1$$

$$= \frac{100}{5 \times 5} \times \frac{4}{5} + \frac{90}{45} \times \frac{6}{\sqrt{45}}$$

$[\because \; S'B = \sqrt{45}]$

$$= 3.20 + 1.79 = 4.99 \text{ lux.} \quad \textbf{Ans.}$$

Example 5.37. *Two lamps are hung at a height of 6 metre form the floor level, the distance between the lamps is 8 metres, Lamp one is of 500 candle power. If the illumination on the floor vertically below this lamps is 20 lux. Find the candle power of second lamp.*

Solution. Intensity of lamp L_1,

$$I_1 = 500 C.P.$$

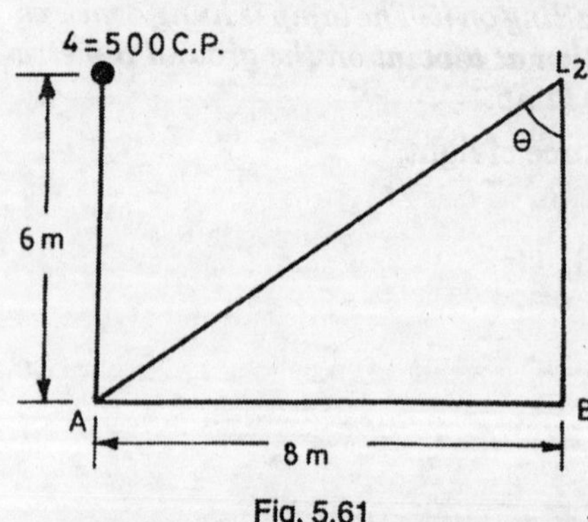

Fig. 5.61

Let the intensity of lamp, L_2 be I_2

Horizontal distance between L_1 and $L_2 = 8m$

Mounting height = 6 m

Distance

$$L_2A = \sqrt{8^2 + 6^2} = 10 \text{ m}$$

$$\cos\theta = \frac{6}{10}$$

Illumination at point, A

E = illumination due to L_1 + illumination due to L_2

$$E = \frac{I_1}{(L_1A)^2} + \frac{I_2}{(L_2A)^2} \cos\theta$$

$$20 = \frac{500}{6^2} + \frac{I_2}{10^2} \times \frac{6}{10}$$

or $$\frac{500}{36} + \frac{6I_2}{1000} = 20$$

or $$I_2 = 1018 \text{ C.P.} \quad \textbf{Ans.}$$

Example 5.38. *A hall 30 m long and 12 m wide is to be illuminated and the illumination required is 50 lumen per metre. Calculate the number, the wattage of each unit and the location and mounting height of the units, taking a D.F. of 1.3 and utilization factor of 0.5, outputs of the different types of lamps are given below :*

Watts	100	200	300	500	1000
Lumens	1615	3650	4700	9950	21500

Solution. Area to be illuminated

$$A = 30 \times 12 = 360 \text{ m}^2$$

Illumination required

$$E = 50m \text{ candles}$$

Depreciation factor = 1.3

Utilization - coefficient = 0.5

Gross lumens required,

$$\phi = A \times E \times \frac{D.F.}{U.F.} = 360 \times 50 \times \frac{1.3}{0.5} = 46{,}800$$

If 100 W lamps are used, number of lamps required

$$= \frac{46{,}800}{1{,}615} \cong 29. \quad \textbf{Ans.}$$

If 200 W lamps are used, number of lamps required

$$= \frac{46{,}800}{3{,}650} \cong 13. \quad \textbf{Ans.}$$

If 300 W lamps are used, number of lamps required

$$= \frac{46{,}800}{4{,}700} \cong 10 \quad \textbf{Ans.}$$

If 500 W lamps are used number of lamps required

$$= \frac{46{,}000}{9{,}950} \cong 5 \quad \textbf{Ans.}$$

If 1000 W lamps are used, number of lamps required

$$= \frac{46{,}800}{21{,}500} \cong 2 \quad \textbf{Ans.}$$

Let the mounting height be 5 metres.

Most suitable type of lamps will be 300 W, lamps required 10 lamps in two rows, each row having 5 lamps giving spacing of 6 metres in length as well as in width and space height ratio of 6/5 *i.e.* 1.2.

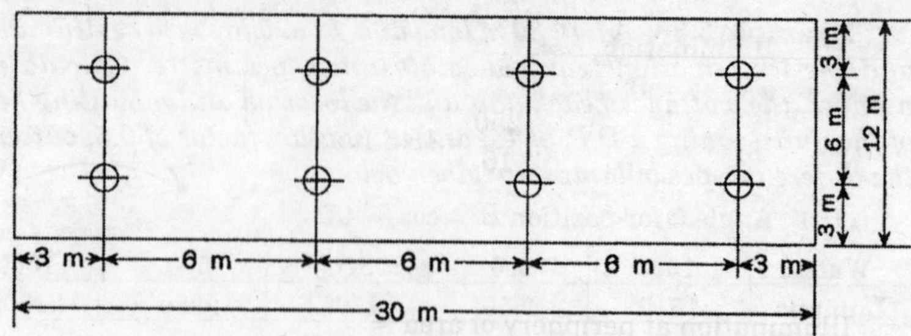

Fig. 5.62.

Lamps of lower wattage will be required more number off fittings and therefore, will increase the cost whereas lamps of higher wattage will be few in requirement giving space height ratio much more than desirable. Arrangement is shown in Fig. 5.62.

Example 5.39. *A light source having an intensity of 1000 C.P. in all directions is fitted with a reflector so that it directs 65% of its light along a beam having a divergence of 20°. What is the total light flux emitted along the beam and what will be the illumination produced on a surface 25 metres away from the source.*

Fig. 5.63

Solution. Intensity of light source, $I = 1000$ *C.P.*

Total light flux emitted along the beam

$$= 0.65 \times 4\pi \times 1000$$

$$= 8164 \text{ Lumens}$$

$$\theta = 20°$$

Distance of the surface from the light sourcre $l = 25m$

Radius of the circle to be illuminated

$$r = l \tan \frac{\theta}{2} = 25 \tan \frac{20°}{2} = 4.40817 \text{ m}$$

Area of the surface to be illuminated

$$A = \pi r^2 = \pi (4.40817)^2 = \mathbf{61.02\ m^2} \quad \textbf{Ans.}$$

$$\text{Average illumination} = \frac{8164}{61.02} = 133.79 \text{ lux} \quad \textbf{Ans.}$$

Example 5.40. *A 250 C.P. lamp is hung 4 m above the centre of a circular area of 6 m in diameter. Calculate the illumination at (i) Centre of area (ii) Periphery (iii) average illumination.*

Solution. Intensity of lamp, $I = 250C.P.$

$$\text{height, } h = 4 \text{ m}$$

(*i*) Illumination, $E = \frac{I}{h^2}$

$$= \frac{250}{16} = 15.6 \text{ lux}$$

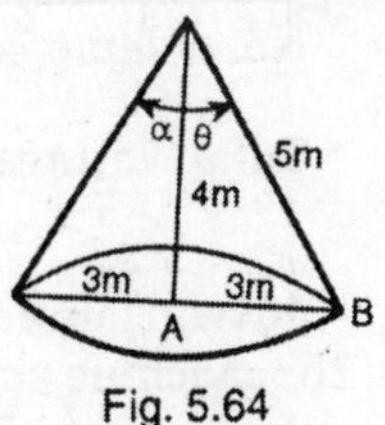

Fig. 5.64

(*ii*) Angle θ for position $B = \cos^{-1} 4/5$

$$= \cos^{-1} 0.8$$

Illumination at periphery of area =

$$E_B = \frac{C.P.}{h^2} \cos^3\theta = \frac{250}{(4)^2} \times (0.8)^3$$

$$= 8 \text{ lux}$$

Solid angle suspended at the area by the lamp

$$w = 2\pi\left(1 - \cos\frac{\alpha}{2}\right)$$

$$w = 2\pi(1 - \cos\theta) \qquad \because \theta = \frac{\alpha}{2}$$

$$= 2\pi(1 - 0.8) = 0.4\pi$$

Total Flux $= Iw = 250 \times 0.4\pi = 100\pi$

Area $A = \pi r^3 = \pi \times (3)^2 = 9\pi$

Average illumination $= \frac{F}{A}$

$$= \frac{100\pi}{9\pi} = 11.11 \text{ lux} \quad \textbf{Ans.}$$

Example 5.41. *A machine shop 50 m × 15 m × 6 m is desined to have an illumination of 150 lumens per sq-m on the working plane. The shop has trusses which are 3.3 m apart. Design a suitable installation using standard 80 watt flourscent lamp. Assume suitable values of utilization and depreciation factors and efficiency.*

Ans. Area to be illuminated $= 50 \times 15 = 750 \text{ m}^2$

Total lumens required $= 150 \times 750 = 112500$ lumen

Assume utilisation factor = 0.5

Depreciation factor = 1.4

Luminous efficiency = 50 lumen/watt

Gross lumens required $= \frac{112500 \times 1.4}{0.5} = 315000$ lumens

Lumens output of each tube $= 80 \times 50 = 4000$ lumens

No. of lamp reqd. $= \dfrac{315000}{4000} = 78.75$ say (80)

Total wattage of lamp $= 80 \times 80$

$= 6400$ W

40 twin lamp each containing two lamp of 80 watt each can provided in the machine shop is :

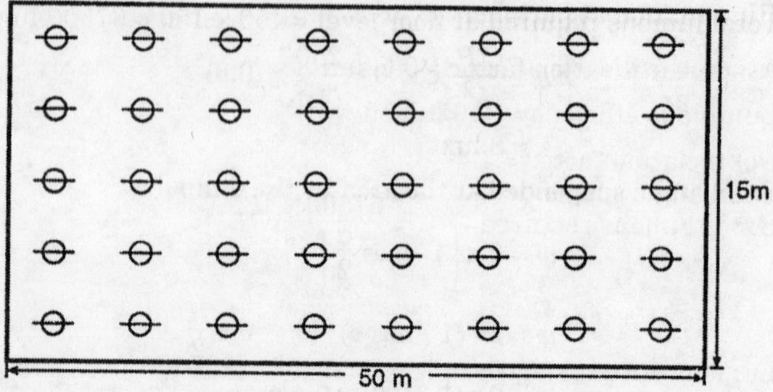

Fig. 5.65

Example 5.42. *A lamp has a uniform candle power of 300 in all direction and is provided with a reflector which directs 50% of the total emitted light uniformly on a flat circular disc 5 meter dia placed 5 meter vertically below the lamp. Calculate the illumination at the centre and at the edge of disc with and without the reflector.*

Ans. Candle power of lamp = 300

Height, $h = 5$ m

Diameter, $d = 5$ m

Illumination at the centre $= \dfrac{C.P.}{h^2}$

$= \dfrac{300}{25} = $ **12 lux Ans.**

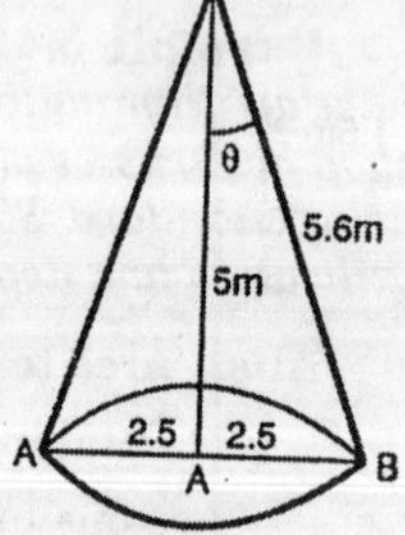

Fig. 5.66

Angle θ for position, *B*

$$= \frac{\cos^{-1} 5}{5.6} = \cos^{-1} 0.892$$

Illumination at the periphery of the area,

$$E_B = \frac{C.P.}{h^2} \cos^3 \theta = \frac{300}{(5)^2} \times (0.892)^3$$

= **8.52 lux Ans.**

Example 5.43. *A workshop 30 m × 15 m is to have an illumination of 150 lux on its working place, the lamps are lung 5 m above it. Give the lay out of a suitable installation using 80-watt flourscent lamp. Assume suitable data.*

Ans. Area to be illuminates = 30 × 15

$= 450 \text{ m}^2$

Average illumination = 150 lux

Total lumens required at floor level = 450 × 150 = 67500 lumens

Assume utilisation factor = 0.5 and

Luminous efficiency = 40 lumen/watt

Depreciation factor = 1.4

$$\text{Gross lumens required} = \frac{67500 \times D.F.}{U.F.}$$

$$= \frac{67500 \times 1.4}{0.5} = 189000 \text{ lumens}$$

Lumens output of each tube = 80 × 40

= 3200 lumens

$$\text{No. of tubes} = \frac{189000}{3200} = 59.06 (60 \text{ say})$$

Total wattage of lamp = 60 × 80 = 4800 watt

30 twin tube each containing two tubes of 80 watt, each can provided in the workshop in six rows having spacing 5 m. Each row will have five unit spaced 3 meter shown in Fig. 5.67.

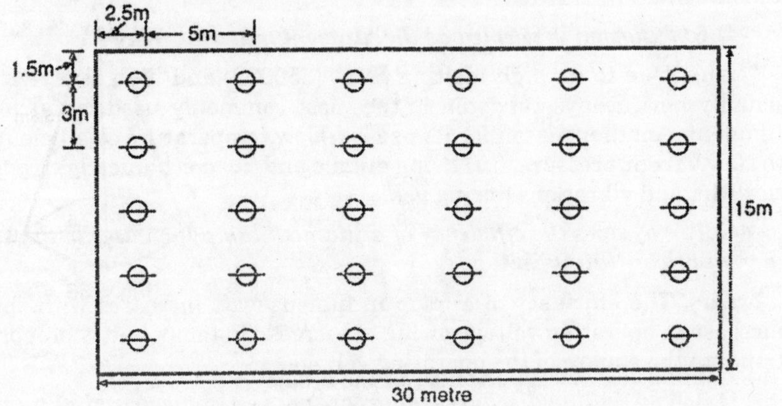

Fig. 5.67

SHORT QUESTIONS WITH ANSWERS

Q.1. *What is the difference between illumination and light.*

Ans. Light is the cause and illumination is the result of that light on surface on which it falls.

Q.2. *What are the advantages of artificial lighting produced electrically ?*

Ans. The following are the main advantages of artificial lighting which is produced electrically :

(*i*) Easy to Control

(*ii*) Cleaniness

(*iii*) Reliability

(*iv*) Low cost as well as steady output.

Q.3. *What are the two types of lighting used in commercial purpose lighting ?*

Ans. Direct and Indirect lighting

Q.4. *What is the difference between maintenance factor and depreciation factor ?*

Ans. Maintenance factor is defined as the ratio of illumination under normal working conditions to the illumination when every thing is properly clean. Depreciation factor is merely inverse of the maintenance factor.

Q.5. *What are the two laws of illumination ?*

Ans. There are two laws (*i*) Law of inverse squares (*ii*) Lambert's consine law.

Q.6. *Tungsten is selected as the filament material. Why ?*

Ans. Due to its high melting point (3500°C) and high resistivity usually now a days tungsten is the most commonly used metal for filament. Another reason for its use is (*i*) low temperature co-efficient, (*ii*) low vavour pressure, (*iii*) Being ductile and (*iv*) mechanically strong to withstand vibrations during use.

Q.7. *Why does the efficiency of a filament lamp increases with the increase in operating voltage ?*

Ans. The efficiency of a lamp in lumens/watt increases with the increase in operating voltage owing to increase in temp. and is proportional to the square of the operating voltage.

Q.8. *Why sodium discharge lamps are not used for general lighting?*

Ans. Sodium vapour discharge lamps are not used for general lighting because of having draw back of colour distortion or discrimination

Q.9. *What is flood lighting ?*

Ans. Flood lighting means flooding of large surfaces with light from powerful projectors.

Q.10. *Why thermal type starter for use in flourscent tubes have become obsolete these days.*

Ans. Because of more complicated construction, greater cost and greater power loss.

Q.11. *What are the two types of reflecting surfaces widely used for light fittings ?*

Ans. Specular and diffuse spread.

Q.12. *What is the amount of luminous per square metre should be provided for difficult or fine works.*

Ans. 1500.

Q.13. *What type of lamps are used at stages and video studios ?*

Ans. Halogen lamps.

Q.14. *What is the purpose of illuminating show case or window case?*

Ans. Illumination is done for equal visibility of window case objects varying in colour or fineness of detail or both.

Q.15. *What are the faults or accident which are likely to occur in show case or window case wiring ?*

Ans. (*i*) Damage to wiring insulation while placing merchandise or due to the weight of merchandise.

(*ii*) Damage to insulation or merchandise due to excessive heat in show case.

Q.16. *How illumination differs from product to another ?*

Ans. Consumer's goods of merchandise differ in colour, size, shape and fineness etc. and hence to make it attractive illumination must be different for different consumer's goods.

Q.17. *State the types of lamps used for illumination of show cases:-*

Ans. Following lamps are used for illuminating show cases:

(*i*) incandescent lamps

(*a*) Glow lamp

(*b*) Silvered bowl lamp

(*c*) Reflector or projector

(*ii*) Fluorescent tube lights and circline tube lights.

Q.18. *What are three kinds of reflector of light ?*

Ans. (*a*) specular reflector

(*b*) diffuse reflector

(*c*) spread reflector

Q.19. *Which is the most recent incandescent lamp which gives better illumination ?*

Ans. Halogen lamps.

Q.20. *How much lumens per square metre should be required for adequate visual representation ?*

Ans. 150 lumens per square metre should be provided for addequate visual.

OBJECTIVE TYPE QUESTIONS

1. Fill in the Blanks.

1. The illumination is measured in -----
2. There are ----- laws of illumination.
3. Sodium vapour lamp is a ----- lamp.
4. The other name of the lm per m^2 is -----
5. The ability of lamps to convert input watts into lumens is called -----
6. Expand CFL -----
7. The Amount of light entering the eye is controlled by -----
8. Neon lamps are basically ----- lamps.
9. ----- is commonly used for measuring mean spherical candle power.
10. Light waves travels with a velocity of -----
11. Power factor of a sodium vapour lamp is nearly ----- lagging.
12. Running cost of a flourescent lamp is ---------------- than incandescent lamp.
13. With the increase in voltage, life of the lamp -----
14. The resistance of a 200 W lamp is ----- than 100 W lamp.
15. Candle is the unit of -----
16. The function of a choke in a flourescent lamp is -----
17. Solid angle is measured in ----- which is equal to -----
18. The unit of luminous flux is ----
19. An electric lamp having max. running cost and minimum initial cost is known as -----
20. With 15 lumens per watt lumens efficiency is -----
21. Solid angle subtended by a sphere at its centre is -----

Answers

(1) lm per m^2 (2) Two (3) Discharge (4) Lux

(5) Luminous efficiency (6) Compact flourescent lamp.
(7) Pupil (8) Gas discharge
(9) Integrating sphere (10) 8.3×10^8 m/s
(11) 0.3 (12) less
(13) Decreases (14) Less
(15) Luminous intensity
(16) to produce high voltage surge
(17) steradian (18) Lumens
(19) incandescent lamp. (20) $\frac{100 \times 15}{4\pi}$
(21) 4π

TRUE/FALSE QUESTIONS WITH ANSWERS

1. Colour of light depends upon wavelength only.
2. For the same power output, low voltage lamps will be more sturdy.
3. Incandescent lamps are generally of copper filaments.
4. Halogen lamps are prefered for outdoor illumination of Buildings.
5. Sodium Vapour lamp is only suitable for operation on d.c. supply.
6. In flourescent tubes, low voltage is required for starting.
7. Power factor of flourescent lamps is high.
8. Interior lighting is designed by lumen method.
9. Greater the size of the room, lesser will be the co-efficient of utilization.
10. For the same lumens output, the operating cost of flourescent lamp is less than incandescent lamp.
11. In a lighting installation, selection of proper space-height ratio ensures uniformity of illumination.
12. Depreciation factor is merely the inverse of utilization factor.
13. In mercury vapour lamp, the harmful ultra-voilet rays are absorbed by outer bulb.

Answers

1. False 2. True 3. False
4. True 5. False 6. False
7. False 8. True 9. False
10. True 11. True 12. False
13. True.

CHOOSE AND MARK THE CORRECT ANSWER

1. For light radiation the temperature of a hot filament should be closer to :

(*i*) 500°C (*ii*) 1500°C

(*iii*) 2500°C (*iv*) 3500°C

2. The maximum sensitivity of human eye for light radiation occurs when the wave length of such radiation touches approximately:

(*i*) 5700 A.U.

(*ii*) 7500 A.U.

(*iii*) 4000 A.U.

(*iv*) It is same for all the wave length mentioned above.

3. The visible spectrum of light falls between the frequency range :

(*i*) 10^8 Hz to 10^{10} Hz (*ii*) 10^{10} Hz to 10^{12} Hz

(*iii*) 10^{12} Hz to 10^{14} Hz

(*iv*) 4.3×10^{14} Hz to 7.5×10^{14} Hz

4. Reflection factor is the ratio of :

(*i*) Reflected light to the incident light.

(*ii*) Output in lumens per watt of power consumed by a source.

(*iii*) Light usefully employed to the light emitted from the source.

(*iv*) Illumination produced by a source under normal working conditions to the illumination produced under ideal conditions.

5. Maintenance factor for a source is the ratio of :

(*i*) Reflected light to the incident light.

(*ii*) Output in lumens per watt of power consumed by a source.

(*iii*) Light usefully employed to the light emitted from the source.

(*iv*) Illumination produced by a source under normal working conditions to the illumination produced under ideal conditions.

6. Coefficient of utilisation is the ratio of :

(*i*) Reflected light to the incident light.

(*ii*) Output in lumens per watt of power consumed by a source.

(*iii*) Light usefully employed to the light emitted from the source.

(*iv*) Illumination produced by a source under normal working conditions to the illumination produced under ideal conditions.

7. Luminous efficiency is the ratio of :

(*i*) Reflected light to the incident light.

(*ii*) Output in lumens per watt of power consumed by a source.

(*iii*) Light usefully employed to the light emitted from the source.

(*iv*) Illumination produced by a source under normal working conditions to the illumination produced under ideal conditions.

8. Depreciation factor for a source is the ratio of :

(*i*) Reflected light to the incident light.

(*ii*) Output in lumens per watt of power consumed by a source.

(*iii*) Light usefully employed to the light emitted from the source.

(*iv*) Illumination produced by a source under normal working conditions to the illumination produced under ideal conditions.

9. Auxiliary electrode for initiating ionisation is observed in :

(*i*) High pressure mercury vapour lamp

(*ii*) Sodium vapour lamp

(*iii*) Halogen lamp

(*iv*) In some other discharge lamp.

10. Glare can be minimised by using :

(*i*) Suitable reflectors for direct sources.

(*ii*) Diffused lighting sources.

(*iii*) Indirect lighting.

(*iv*) All as above.

11. In industrial lighting schemes, it is desirable to have :

(*i*) Uniform lighting

(*ii*) Higher level of local and lower level of general lighting

(*iii*) Reverse of scheme (*ii*) is better

(*iv*) Depends upon types of machinery used.

12. Steradian is the unit of :

(*i*) Plane angle (*ii*) Solid angle

(*iii*) Luminous intensity (*iv*) Luminous flux.

13. Candle-power in other terms is same as :

(*i*) Lumens per steradian (*ii*) Lumens per metre square

(*iii*) Lux (*iv*) One Phot $\times 10^{-4}$

14. The human eye responds better to :

(*i*) Infra-red radiations (*ii*) Ultra-violet radiations

(*iii*) Both as above (*iv*) None as above.

15. The process of ionisation by collision involves :

(*i*) External source of radiation

(*ii*) It is an internal process where some accelerating electrons knock out valence electron from same mon-atomic gas medium.

(*iii*) It is a solid state process.

(*iv*) Something else.

16. According to quantum theory an atom is capable to release photon when :

(*i*) It gets ionised

(*ii*) When it is excited

(*iii*) When it regains normlacy after excitation.

(*iv*) None out of above.

17. An end to end filament exists in :

(*i*) Sodium vapour lamp

(*ii*) Halogen lamp

(*iii*) Choke-less mercury vapour lamp

(*iv*) None as above.

18. Write 'yes' or 'no' for the following sentences :

(*a*) Sun is the highest natural source of light.

(*b*) Velocity of light for every colour remains constant.

(*c*) Natural light consists of four colours.

(*d*) There is no difference between solid angle and plane angle.

(*e*) The unit of illumination is lumen.

(*f*) Brightness of fluorescent tube is less.

(*g*) Voltage fluctuation has no effect on the life of filament lamp.

(*h*) Indirect lighting is used in Cinemas, restaurants etc.

19. Fill in the blanks by appropriate words :

(*a*) Velocity of light, v = frequency × ....

(*b*) The unit of solid angle is ....

(*c*) The unit of illumination is .... or .... or ....

(*d*) In projectors .... lamp is used.

(*e*) The material of the filament in filament lamp is ....

(*f*) At high voltages, the life of filament lamp ....

(*g*) The sodium vapour lamp produces .... light.

(*h*) .... lamps are used for street lighting generally.

20. A sodium vapour lamp is a :

(*a*) Hot cathode device

(*b*) Cold cathode device

(*c*) The filamentary electrodes warm up during starting of the lamp like that of fluorescent tubes

21. The 'ignitor' used in sodium vapour lamp is identical to :

(*a*) Starter of fluorescent lamp

(*b*) Starting electrodes of MV lamp

(*c*) None of the two as above

22. The illumination level required for precision work is about

(*a*) 500-1000 lux (*b*) 250 lux

(*c*) 100 lux (*d*) 50 lux

23. Which of the following will need the lowest level of illumination ?

(*a*) Work shop (*b*) Displays

(*c*) Railway Platform (*d*) Garage

24. Which of the following will need the highest level of illumination ?

(*a*) Living room (*b*) Kitchen

(*c*) Proof reading (*d*) Hospital wards

25. The illumination level in houses is in the range of

(*a*) 20-50 lux (*b*) 100-200 lux

(*c*) 300-500 lux (*d*) 700 lux

26. The illumination level required for important traffic routes carrying fast traffic is about

(*a*) 30 lux (*b*) 100 lux

(*c*) 200 lux (*d*) 5 lux

27. A flourescent tube can be operated on

(*a*) both d.c. as well as a.c. (*b*) a.c. only

(*c*) d.c. only (*d*) satisfactorily only on d.c.

28. Flood lighting is not used for

(*a*) Reading (*b*) Aesthetic

(*c*) Advertising (*d*) Industrial

29. Direct lighting scheme is most efficient but is liable to cause

(*a*) Monotony (*b*) Glare

(*c*) Hard shadows (*d*) both (*b*) & (*c*)

30. 1 candle =

(*a*) 2π lumens (*b*) 4π lumens

(*c*) $2\pi r^2$ (*d*) none of these

31. Tungsten filament lamp contains

(*a*) Vaccum (*b*) nitrogen

(*c*) Oxygen (*d*) Hydrogen

32. A perfect diffuser surface is one that

(*a*) diffuses all the incident light

(*b*) absorbs all the incident light

(*c*) transmits all the incident light

(*d*) scatters light uniformly

33. Which of the following comparison between filament lamp and flourescent lamp is correct ?

(*a*) the flourescent lamp has higher dazzle

(*b*) the flourescent lamp produces sharper shadows

(*c*) the flourescent lamp produces greater brightness

(*d*) the average life of flourescent lamp is five to seven times higher.

ANSWERS

1. (*iii*)	2. (*i*)	3. (*iv*)
4. (*iv*)	5. (*iv*)	6. (*iv*)
7. (*ii*)	8. (*iv*)	9. (*i*)
10. (*iv*)	11. (*ii*)	12. (*ii*)
13. (*i*)	14. (*iv*)	15. (*ii*)
16. (*iii*)	17. (*iii*)	

18. (*a*) Yes (*b*) Yes (*c*) No (*d*) No (*e*) No (*f*) Yes (*g*) No (*h*) Yes

19. (*a*) Wavelength (*b*) Steradian (*c*) Lux, lumens/m^2, meter candle (*d*) carbon arc (*e*) tungston (*f*) decreases (*g*) yellowish (*h*) M.V. lamps.

20. (*b*)	21. (*c*)	22. (*a*)
23. (*c*)	24. (*c*)	25. (*b*)
26. (*a*)	27. (*a*)	28. (*a*)
29. (*d*)	30. (*b*)	31. (*b*)
32. (*a*)	33. (*d*)	

MULTIPLE CHOICE QUESTIONS

1. Which lamp cannot sustain much voltage fluctuations

(*a*) incandescent lamp (*b*) fluorescent lamp

(*c*) mercury vapour lamp (*d*) sodium vapour lamp

2. Filament lamp at starting will take current

(*a*) equal to its full running current

(*b*) less then its full running current

(*c*) more than its full running current

(*d*) none of these

3. A 100 W 230 V filament lamp, as its half rated voltage will take

(*a*) half rated current

(*b*) more than half rated current

(*c*) less than half rated current

(*d*) equal to one fourth rated current

4. Basic function of a reflector is to

(*a*) avoid glare

(*b*) provide better illumination

(*c*) protect the lamp (*d*) do all of above

5. If we seep the space height ratio more than the required value this will result in
   (*a*) non-uniform illumination on the working plane
   (*b*) cheap lighting installation due to reduced number of lamps required
   (*c*) less power requirement
   (*d*) all of above
6. The essential requirements of a good lighting system are
   (*a*) proper illumination level (*b*) avoidance of glare
   (*c*) low cost (*d*) all of above
7. Light
   (*a*) is a form of Elect. Energy (*b*) is a form of heat energy
   (*c*) consists of electro magnetic waves
   (*d*) consists of shooting particles.

Answers

1. (*a*)	2. (*c*)	3. (*b*)
4. (*d*)	5. (*d*)	6. (*d*)
7. (*c*)		

FILL IN THE BLANKS

1. Light of red colour has ----- wave length that of violet. (greater, less)
2. Unit of solid angle is ----- (degree, radian, steradian)
3. Unit of light flux is ----- (lumen, candle power)
4. Lumen/steradian is a unit of ----- (luminous intensity, intensity of illumination)
5. Solid angle subtended at a point in space is ----- (2π steradians. 4π steradians)
6. Unit of light flux is ----- (lumen, lumen hour, weber, candle power)
7. Rate of light energy radiations from a light source is called ----- (luminous flux, intensity of illumination)
8. One candle power source of light emits ----- lumens in space. (2π, 4π)
9. Lumen/steradian is a unit of ----- (luminous intensity, intensity of illumination)

10. One candle power source of light emits ----- lumen/staradian. (one, two)

11. While illumination refers to the light flux ----- per unit area, brightness refers to the light flux ----- per unit area. (emitted, received)

12. While unit of illumination is ----- that of brightness is ---- (nit, lux)

13. Ratio of reflected light to the incident light is called ----- (refraction factor, absorption factor)

14. Unit of luminous efficiency is ----- (watt/luman, lumen/watt)

15. Greater the rise of room ----- will be value of co-efficient of utillization. (less, more)

16. Maintenance factor and depreciation factor are -----
(two different concepts, two different names of same concept)

17. Greater the space height ratio ----- the number of lamps required by a given installation. (more, less)

18. Life of filament lamp is ---- burning hours. (1000, 2500, 5000)

19. Filament lamps upto ----- watts contains vacuum. (15, 25, 40)

20. Gas filling of filament lamp is done to -----
(Improve efficiency, to reduce evaporation of filament)

21. Use of inert gas and not air in bulb is done -----
(to reduce convection loss, to avoid filament combustion)

22. Inert gas mainly used in bulbs is ----- (neon, argon, CO_2)

23. Projection lamps contain 100% of ----- gas. (argon, N_2, Neon)

24. Fluorescent tube is a ----- pressure mercury vapour lamp. (low, high)

25. Filament of incandescent lamp is made of ----- (tungsten, thoriated nickle, copper)

26. Argon gas in fluorescent lamp helps in ----- (starting of tube, stabilizing the arc)

27. Coiled coil filament in incandescent lamps is primarily used to ----- (reduce the convection losses, concentrate the beam)

28. Sodium vapour are preferred in -----
(domestic applications, road way crossings)

29. Sodium vapour lamps are used where -----
(high level of illumination is required, glare is to be avoided)

30. Life span of ----- is affected by the number of switching on and off operations. (filament lamp, fluorescent tube)

31. Operation of ----- produces radio interference.

(filament lamp, discharge lamps)

32. Maintenance of ----- is more-cumbersome.

(filament lamp, discharge lamps)

33. In sodium vapour lamp, the colour of light at starting is ----- whereas the final colour attained is

(golden yellow, pink, white)

34. Colour of light depends upon -----

(wave length, frequency, both frequency and wave length, velocity, amplitude)

ANSWERS

1. greater	**2.** steradian
3. lumen	**4.** luminous intensity
5 4 π	**6.** lumen
7. luminous flux	**8.** 4 π
9. luminous intensity	**10.** one
11. received, emitted	**12.** lux, nit
13. refraction factor	**14.** lumen/watt
15. more	
16. two different names of same concept	
17. less	**18.** 1000
19. 25	
20. to reduce evaporation of filament	
21. to avoid filament combustion	**22.** argon
23. N_2	**24.** low
25. tungsten	**26.** starting of tube
27. reduce the convection losses	
28. road way crossings	
29. high level of illumination is required	
30. fluorescent tube	
31. discharge lamps	**32.** discharge lamps
33. pink, golden, yellow	**34.** frequency and wave length

TRUE/FALSE TYPE QUESTIONS

1. Candle power of street light lamp is uniform in all the directions.
2. Larger objects require more illumination
3. Distant objects require more illumination.
4. Stationary objects require more illumination than moving objects.

5. Objects which are to be seen for longer duration of time require less illumination.
6. Aged persons require more illumination than young persons.
7. Long shadows are produced by low mounting height of the lamps.
8. Glare is produced when light from the source directly goes into the eyes of the observer.
9. High mounting heights produce more glare.
10. Point sources of light produce primary glare.
11. Polished objects on floor produce primary glare.
12. Lighting installation making use of filament lamps is cheaper in running cost.
13. Lighting installation making use of filament lamps is cheaper.
14. Lower the pressure of the medium in discharge lamp, more will be the proportion of light rays of that characteristic colour.
15. Illumination of surface due to a point source will be more if angle of incidence becomes more.

ANSWERS

1. False	2. False	3. True	4. False
5. False	6. True	7. True	8. True
9. False	10. True	11. False	12. False
13. True	14. True	15. False	

EXERCISES

1. Define and explain the following terms in connection with illumination.
   (*i*) Illumination (*ii*) Luminous flux
   (*iii*) Coefficient of utilisation (*iv*) Depreciation factor
   (*v*) Reflection factor
2. State the laws of illumination.
3. With the help of a diagrammatic sketch, showing connections explain the principle of working of a fluorescent lamps with particular reference to starter and choke. Mention the special field of application of such lamps.
4. Define the following terms :
   (*i*) Mean horizontal candle power (*ii*) Luminous flux

(*iii*) Luminous intensity (*iv*) Illumination.

5. (*a*) Explain with the help of a circuit diagram the working of a fluorescent lamp.

(*b*) What are the advantages of fluorescent lighting over plain mercury discharge lighting ?

(*c*) Why condenser is used in the fluorescent tube circuit?

(*d*) Why are coiled coil gas filled incandescent lamps more efficient than evacuated lamps?

6. What factors determine the useful light flux reaching the working plane in indoor illumination ? Discuss each factor.

7. (*a*) Describe how glare is produced and suggest how it can be avoided?

(*b*) How can hard and long shadows be avoided ?

(*c*) Compare fluorescent and incandescent lamps on the basis of quality of light, capital and running cost.

7a. Why is tungsten selected as the filament material and upon what factors does its life depend ?

8. A drawing hall 30 m × 15 m with a ceiling height of 5 metres is to be provided general illumination of 120 lux. Taking a coefficient of utilisation of 0.5 and depreciation factor of 1.4, determine the number of fluorescent tubes required, their spacing, mounting height and total wattage.

Take luminous efficiency of fluorescent tubes as 40 lumens per watt for 80 watt tube.

[**Hint.** Factors > 1 have to be multiplied for calculation of gross or total lumens.] [**Ans.** 48 tubes]

9. Explain the different factors that are to be considered in the design of an area illumination. Explain, the terms 'utilisation coefficient', 'depreciation factors' and 'space height ratio' used in this connection.

A room 20 m × 5 m is illuminated by twenty, 200 watt lamps. Assuming the depreciation factor of 1.2 and utilisation coefficient of 0.6, find the average illumination on the floor. Assume the lamp efficiency of 15 lumens/watt. Show approximate location of the lamps. [**Ans.** 300 lux]

10. Give short account of inverse square law and cosine law as applied to illumination.

11. Describe the construction and working of high pressure mercury vapour lamp.

12. Explain the stroboscopic effect as referred to illumination. How can it be avoided while using fluorescent lamps?

13. A lamp of 500 c.p. is mounted at a height of 3 metres from the centre of a table 5 m × 3 m. Find the maximum and minimum illumination. [55.5 lux and 20.6 lux]

14. A room is illuminated with the help of twelve lamps of 200 W rating each giving an output of 15 lumens/Watt. The area of the room is

6 m × 12 m. If the depreciation factor is given as 0.6 and coefficient of utilisation as 0.8, find the averave illumination on the floor. [Ans. 240 lux]

15. A drawing hall 40 m × 10 m × 5 m has to be illuminated with 300 lux on the working plane. If the coefficient of utilisation is given as 0.75 and depreciation factor as 0.6, calculate the number of tubes required for the purpose. Take efficiency of the fluorescent tubes as 40 lumens/watt and rating of the tubes at 80 W, shows the arrangement of installation. [Ans. 83 tubes but use 84 in span]
16. Two lamps of 200 c.p. and 400 c.p. each are mounted 8 metres apart on lamp posts of 4 m height. Calculate illumination 3 m away from each post between the two lamps, on the line joining the bases of the two lamp posts. [Ans. 12.5 lux, 15.85 lux]
17. A factory space 33 m × 13 m is to be illuminated with an average illumination of 72 lumens/m^2, by 200 W lamps. The coefficient of utilisation is 0.4 and the depreciation factor is 1.4. Calculate the number of lamps required, the lumens output of 200 W lamps is 2730 lumens. *(Bombay Univ.)* [Ans. 40 bulbs]
18. State the laws of illumination. The illumination in a drawing office 30 m × 10 m is to have a value of 2.50 lux and is to be provided by a number of 300 W filament lamps. If the coefficient of utilisation is 0.4 and the depreciation factor 0.9. Determine the number of lamps required. The luminous efficiency of each lamp is 14 lumens/watt. [Ans. 50 bulbs]
20. Why the intensity of illumination produced by the sunlight is different in the (*i*) Morning (*ii*) Noon (*iii*) Evening hours ?

Questions Bank with Answer

Short Answer Type Questions

Q.1. Define illumination.

Ans. When light falls on a surface, it becomes visible, the phenomenon is called illumination. It is defined as the luminous flux falling on surface per unit area.

Q.2. Define luminous flux.

Ans. It is the amount of energy radiated per second by a body in the form of waves.

Q.3. What is the unit of luminous flux?

Ans. Lumen.

Q.4. The unit of illumination is ....

Ans. Lux.

Q.5. Define luminous intensity.

Ans. It is the luminous flux emitted by a source per unit solid angle in a particular direction.

Q.6. Define solid angle.

Ans. The angle subtended by a partial surface area of a sphere at its centre is called solid angle. It is given by the ratio of area enclosed to square of radius of sphere

$$i.e., \ \omega = \frac{\text{Area}}{(\text{radius})^2}$$

Solid angle is measured in steradian

Q.7. Define Lumen.

Ans. It is the unit of luminous flux. The luminous flux emitted per unit solid angle of a source of one candle power is called one lumen.

Q.8. Define candle power.

Ans. Light radiating capacity of a source is measured in candle power. It is the number of lumens given out by a source per unit angle in the given direction. One candle power = 4π lumens.

Q.9. Define steradian.

Ans. It is the unit of solid angle. One steradian is defined as the solid angle subtended at the centre of a sphere by its surface having area equal to square of its radius.

Q.10. Define glare.

Ans. If the eye is exposed to a very bright source of light, the pupil of eye contracts automatically in order to reduce the amount of light admitted and prevent damage to the retina. This reduces sensitivity of eye to see the surrounding objects. This effect is called glare.

Q.11. What is the wave length of visible light?

Ans. 0.75×10^{-6} m to 0.4×10^{-6} m.

Q.12. Human eye is most sensitive to .... colour.

Ans. Green.

Q.13. What is the nature of light?

Ans. Light is a form of electromagnetic waves. It travels in straight line with a velocity of 3×10^8 m/sec.

Q.14. Define Lux.

Ans. It is the unit of illumination and can be defined as luminous flux falling per unit (m^2) area of the surface which is perpendicular to the rays of light from a source of one candle power at a distance of one meter from it.

Q.15. Define brightness.

Ans. This is the luminous intensity per unit projected area of a source in the direction perpendicular to the source. It is given in candle power per m^2 or in lamberts or in foot candles.

Q.16. Define space height ratio.

Ans. This is the ratio of space (distance) between two nearest points and the height of the points from the ground. This is usually taken as unity.

Q.17. Which light source has highest efficiency incandescent/fluorescent/sodium vapour/merium vapour lamp?

Ans. Sodium vapour lamp.

Q.18. What is the colour of light emitted by sodium vapour lamp.

Ans. Yellow.

Q.19. Discharge lamps need a .... to maintain the discharge.

Ans. Ballast.

Q.20. Define depreciation factor.

Ans. All sources suffer a loss in their efficiency with the lapse of time *i.e.,* they suffer depreciation. Depreciation factor is the ratio of output of lamp at present to the light output at the time of installation.

$$DF = \frac{\text{Light output at present}}{\text{Initial output}}$$

Q.21. Define efficiency of a light source.

Ans. It is the output of a source in lumens per watt.

Q.22. What is the level of illumination required in drawing office?

Ans. 400 lux.

Q.23. What is flood lighting?

Ans. When large surfaces are illuminated with the help of powerful project or reflectors, the lighting scheme is known as flood lighting.

Q.24. What is light?

Ans. It is that part of radiant energy from a hot body which produces the visual sensation on human eye.

Q 25. Why choke is used in the circuit of a fluorescent lamp?

Ans. Choke in the circuit of a fluorescent lamp is connected in series with the tube which acts as a ballast. It provides high voltage surge for starting the glow in the tube

Q 26. Define wave length.

Ans. It is distance travelled in one cycle of a wave.

Medium/Long Answer Type Questions

Q.1. State and Explain laws of illumination. *V. Imp*

Ans. There are two laws of illumination

(i) Inverse square law: This law states that the illumination (E) of a surface is inversely proportional to square of distance (D) of the source from the surface. It is asumed that the surfce is sufficiently large so that the source can be regarded as a point source.

i.e. $$E \; a \frac{1}{d^2} \quad \text{........ (i)}$$

(ii) Lambert's cosine law: According to this law, the illumination on the surface is directly proportional to the cosine of the angle between the angle normal to the surface and the direction of incident light *i.e.,* d

$$E \, \alpha \cos \theta \quad \text{........ (ii)}$$

Combining (i) and (ii)

$$E \, \alpha \frac{1}{d^2} \cos \theta$$

or $$E = \frac{1}{d^2} \cos \theta$$

Where I is the illumination intensity of the source in candle power. Consider a source of illumination intensity I at distance d from the point of illumination making an angle θ with the direction of light and normal as shown in fig. 5.68.

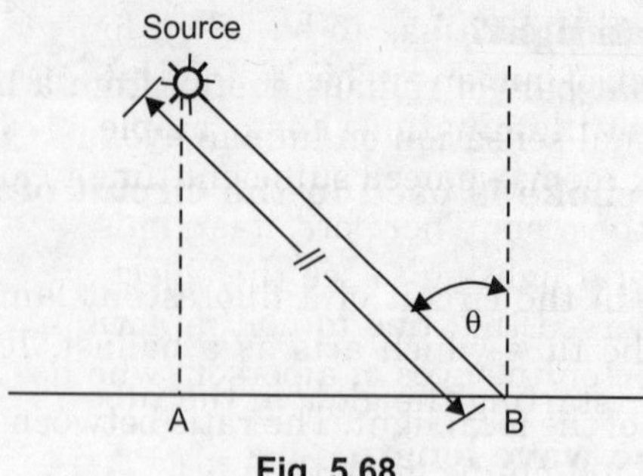

Fig. 5.68

If h is the height of source from the surface to be illuminated, from

$$h = d\cos\theta.$$

$$d = \frac{h}{\cos\theta}$$

$$E = \frac{1}{\left(\frac{h}{\cos\theta}\right)^2}\cos\theta = \frac{1}{h^2}\cos^3\theta$$

Q.2. Discuss in brief the main requirement of proper lighting.

Ans. The main requirements of proper lighting are

(i) It should provide adequate illumination.

(ii) It should provide uniformly distributed light all over the working plane.

(iii) It should avoid glare and hard shadows as far as possible.

(iv) It should provide light of suitable colour

(v) It should provide luminaries suspended at suitable height so that they may not come in the field of vision.

(i) Illumination level: This is the most vital factor in deciding the number and wattage of luminaries so that we are able to see and recognise the object properly. Degree of illumination, to give necessary brightness to the objects depends upon size of the object, its distance from the viewer, contrast between the object to be seen and its surroundings. It is also observed that moving objects require more illumination compared to the stationary objects. Depending upon these points, the illumination level for different purposes has been recommended by Bureau of Indian Standards.

(ii) Uniformity of illumination: Whenever using sources

of local illumination care has to be taken for additional general lighting. The pupil of human eye has to adjust its aperture depending upon the level of illumination. For example, if we go from sun light into a dark room, we need sufficient time to locate the things properly inside the room. Therefore, it is a must to support the local lighting by general lighting to see the object away from the local source. It is observed that due to non uniformity of illumination, it causes a strain on the eyes of a person, who has to move to and from the source of the local light. The ratio between local to general illumination should not be more than 3:1.

(iii) Colour of light: The appearance of the same object differ under different coloured lights. Therefore, the colour of the light is quite important for drawing offices or cloth show rooms and is a secondary factor in street lighting. For most of the applications, we prefer to have light colour of which appears to be natural. Day light flourescent tubes now-a-days make it possible to illuminate even the large spaces with artificial day light, giving good colour rendering and at sufficiently high level. On the other hand, high efficiently discharge lamps like sodium vapour lamps/mercury vapour lamp can be used for street lighting pruposes, where colour distortion is not of so importance.

(iv) Shadows: In lighting installations, formation of long and hard shadows are not desirable because it causes fatigue of eyes. There are some exceptional works (e.g., drawing offices), where shadowless light is essential, otherwise shadows will hinder the work. However, there are some works, where a certain amount of shadow is required (e.g., for decoration purpose).

(v) Glare: Glare is the sensation experienced by human eye, when some rays of light enter the eye. In case of lighting scheme, this glare can be of two types

(a) Direct glare (b) Indirect glare

In the first case, if the mounting height of the lamps is too low and there are no reflectors used, the effect will cause direct glare. For example a table lamp without reflector. Therefore while using direct lighting, it must be seen that no source of light should be too low to throw light directly into the eyes of an observer. In the second case, the glare is caused by a snining object, reflecting *a* major part of the incident light. This can be avoided by fixing the lamps at such a position that no direct light falls on the shining surfaces e.g chrome or mckle plated parts of machines or appliances.

Glare not only causes undue fatigue, but can also be a cause of accident.

Q.3. What is glare? Suggest methods to reduce it.

Ans. The light reaching directly or after reflection on our eyes produce a type of flash which is irritating and is a matter of concern. This is called glare and is not desirable while designing a lighting scheme.

To avoid glare, the lamps may be mounted at a height above the range of vision. The metal reflectors are provided with a skirt around the rim.

Q.4. Discuss the importance of glare, contrast and shadow in the lighting system design.

Ans. (i) Glare: Refer to Q 3

(ii) Contrast: The illumination should be contrast free. The pupil of eye adjusts its aperture dependiing upon level of illumination. If we go from sunlight into dark room, sufficient time is required to locate the things properly inside the room. Therefore while using source of local illumination, care has to be taken for additional generalised lighting to see the objects away from the local source. Ratio of local to general illumination should not be more than 3:1. The generalised lighting will reduce strain on the eyes of a person who has to move to and from the source of light.

(iii) Shadows: Refer to Q 2

Q 5. What do you understand by the following?

(i) Illumination level **(ii) Luminous efficiency**

(iii) Space height ratio **(iv) Co-efficient of utilization**

Discuss their importance in the design of lighting system.

Ans. (i) Illumination level: This is the most vital factor in deciding the number and wattage of luminaries so that we are able to see and recognise the object properly. Degree of illumination, to give necessary brightness to the objects depends upon size of the object, its distance from the viewer, constrast between the object to be seen and its surroundings. It is also observed that moving objects require more illumination as compared to the stationary objects. Depending upon these points the illumination level for different purpose has been recommended by Bureau of Indian Standards.

(ii) Luminous efficiency: It is the deciding factor for the selection of source of light from the available sources. Whole the electrical power supplied to lamp is not converted into luminous flux because most of it is lost in the form of heat radiations by conduction, convection and absorption. Only a small portion of radiant energy falls in visual range of wave length 4000° A to 7500° A. Efficiency of light source is expressed as luminous flux in lumens per watt. Efficiency of different light sources is as given below.

(a) Incandescent lamp = 13 lumen/watt
(b) Fluorescent lamp = 45 lumen / watt
(c) Sodium vapour lamp = 40 to 50 lumen / watt
(d) Mercury vapour lamp = 35 to 45 lumen / watt
(e) CFL = about 80 lumen / watt

Higher efficiency of light source means low operating cost in terms of power consumption. Luminous efficiency is not the only factor which decided the type of luminary to be used. Other factors such as colour of light, and its operating life are also to be considered.

(iii) Space height ratio: The ratio of space (horizontal distance) between the two adjacent lamps to the vertical height of the lamps above the working plane is called space height ratio.

In order to obtain almost uniform illumination over the working plane, the distance between the lamps should not be too much. An ideal scheme could be when there is a large number of small size lamps. But this would increase the installation cost. To have uniform illumination the ratio of spacing between the lamps and their height above the working plane (*i.e.,* space height ratio) should normally lie between 1 and 1.5.

(iv) Coefficient of Utilisation: A surface to be illuminated receives light either directly from the lamps or reflected from the ceiling and walls or both. In all cases, the total flux reaching the surface will never be equal to the flux emitted by the lamp, due to absorption by reflectors, ceilings and walls.

Hence, the ratio of lumens reaching at the working plane to the total lumens emitted by the light source is called co-efficient of utilisation or utilisation factor.

$$\text{Utilisation factor} = \frac{\text{Lumens reaching at the working plane}}{\text{Total lumens emitted by the source}}$$

Thus the flux reaching at the working plane can be calculated by multiplying the total flux emitted by the source by a factor called utilisation factor or co-eff. of utilisation. The value of utilisation factor depends upon the types of reflector, lighting schemes, condition of walls and ceiling etc. Usually, its value varies from 0.5 to 0.8.

Q.6. (a) Define luminous efficiency

(b) Compare luminous efficiency and operating life of mercury vapour lamp, sodium vapour lamp, metal hallide lamp and fluorescent lamp with incandescent lamp.

Ans. Luminous efficiency is amount of luminous flux emitted by a source of light in lumens per watt of power supplied to lamp.

Operating life of luminary is the no. of hours for which the source can burn during its life time. The following table gives the luminous efficiencies and operating life of various types of luminaries.

S.N.	Light Source	Efficiency	Operating life
1.	Incandescent lamp	12 lumen/watt	100 hours
2.	Sodium vapour lamp	40-50 lumen/watt	3000 hours
3.	Mercury vapour lamp	30-40 lumen/watt	3000 hours
4.	Fluorescent lamp	40-45 lumen/watt	3000 hours
5.	Compact Fluorescent	80 lumen/watt	8000 hours
6.	Halide lamps	30 lumen/watt	2000 hours

Q 7. What are different types of light sources?

Ans. Different types of light sources are

(i) Incandescent lamps

(ii) Discharge lamps

Incandescent lamps use a filament of tungston. When the filament is heated upto the stage of incandescence they start emitting light. Discharge lamps work on the principle of gaseous discharge. Discharge lamps are of the following types, (i) Mercury vapour lamps (ii) Sodium vapour lamps

(iii) Low pressure mercury vapour lamps (fluorescent lamps)

(iv) Compact fluorescent lamps.

Some other sources of light are:

(i) Halogen lamps

(ii) Neon lamps

Q.8. Explain the construction and working of incandescent lamp.

Ans. The different parts of gas filled lamp are shown in fig. The vital part of the lamp is a filament. The filament is made of specially treated tungsten. For getting higher efficiency, it is necessary to use working temperature more than 2000° C. The evaporation of tungsten can be prevented by filling the bulb with an inert gas-argon with a small percentage of nitrogen. The nitrogen is added to reduce arcing within filament lamps. Tungsten filament lamps have cold resistance almost one fifth of the hot resistance of the filament which is operated safely at temperature of about 2500°C to 2800°C in gas filled lamps.

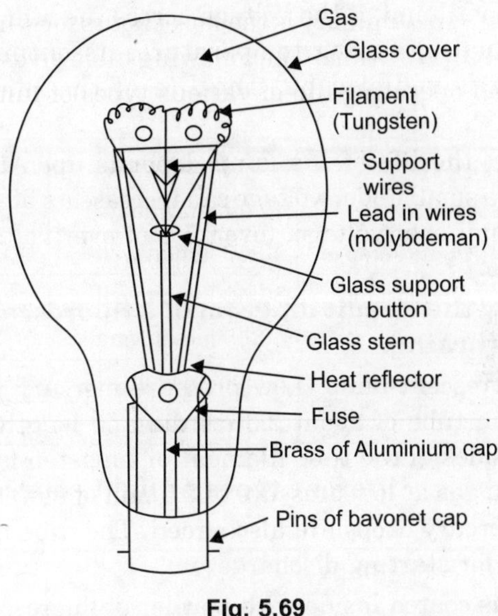

Fig. 5.69

The melting point of tungsten is 3400°C. Since there is presence of gas in the bulb, there is heat loss due to convection. This heat loss lowers the efficiency of the bulb. These heat losses can be reduced by reducing the surface area of the filament. This is achieved by using coiled filaments.

Lamps below 40 W are not filled with gas because a lot of heat will be wasted in these small sized bulbs due to convectional flow of gas. Coiled coil filaments are used upto 100 W lamps and simple coiled filaments above 100 W. The diameter of the filament increases with the increase in wattage, but as the diameter increases, the ratio of surface area to the cross-sectional area decreases. This reduces the cooling effect, resulting in higher operating temperature and increases the efficiency.

The efficiency of a 40 W lamp is 12 lumens/watt and that of 100 W lamps is about 14 lumens/watt.

The tungsten filament lamp is relatively inexpensive and has an average life of 1000 hours.

The filament lamps are very sensitive to the variation of voltage. If the voltage across filament lamp is increased, the current also increases thus causing the lamp or filament to operate at higher temperature. Thus the efficiency of the lamp is increased but its life is much reduced.

On the other hand, if the voltage across the lamp is decreased from its normal operating temperature, its output and power consumption is reduced but the life of the lamp increases considerably.

Therefore, the life of the lamp depends upon the operating voltage. Even a small under-voltage can increase its life considerably, whereas a small over-voltage, (even 5% of over voltage) shortens its life by 50%.

Q.9. Draw the circuit diagram of a fluorescent lamp and explain its working.

Ans. Fluorescent lamp (Low pressure mercury vapour lamp) : This is a thin tube of about 2.5 cm dia and length of 1,2 and 4 feet. On two sides of the tube filament of tungsten is provided. In the tube argon gas at low pressure of 2.5 mm of mercury column is filled. Few mercury drops are also added. The tube needs a choke and a starter for starting discharge.

The tube is coated inside with powder of fluorescent material, hence it is called as fluorescent lamp.

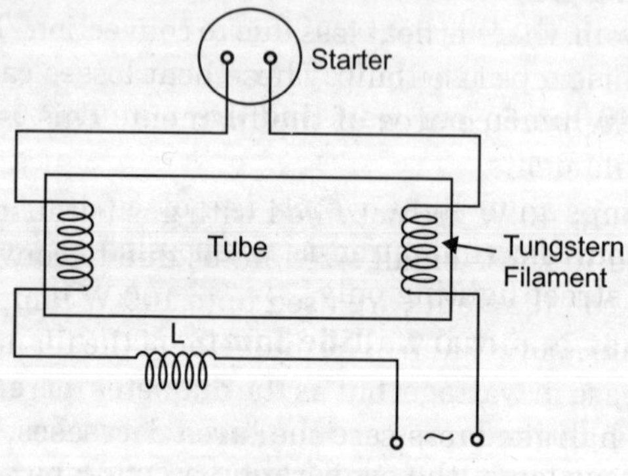

Fig. 5.70

When supply is switched ON, choke with the help of starter produces a very high voltage across filaments, which start emission. A capacitor can be connected across the supply to improve the power factor which is low due to choke.

Q 10. Draw the circuit diagram of a high pressure mercury vapour lamp and explain its working.

Ans. Mercury Vapour lamp (high pressure): It has two glass envelopes. The inner glass envelope has two main electrodes and one auxiliary electrode. Over this outer envelope is provided which acts as a cover. It has inside pressure of 1.5 atmospheres. The lamp has a screwed cap, and is connected with the supply through choke.

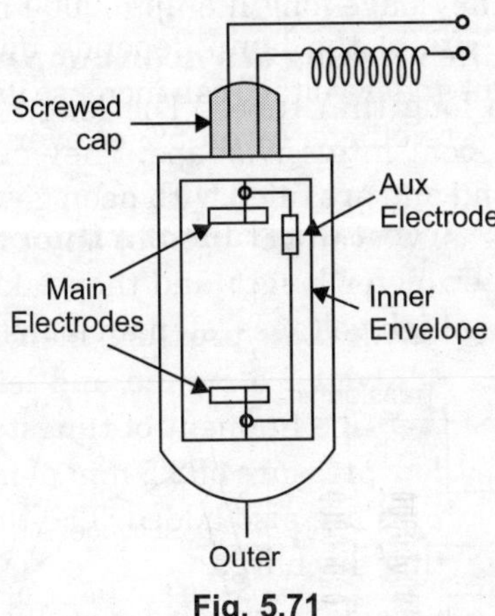

Fig. 5.71

When the lamp is switched ON, a discharge start between two electrodes. The lamp takes 5 minutes to start emission of light.

The lamp is to be used always in vertical position. It has an efficiency of 40 lumen per watt and are manufactured upto 500 watt rating.

These lamps give light which is not comfortable to eyes. Moreover, colour discrimination is not possible. These lamps are used only for street lighting only.

Q.11. What is metal hallide lamp? Explain its principle of working.

Ans. The halogen lamps are the latest incandescent lamps. In the incandescent lamp, the evaporation occurs which blackens its surface after some time. The Tungsten in the halogen lamp does not vaporise, as the lamp is filled with halogen (Iodine, Chlorine or Bromine) vapours. They have also tungsten filament but efficiency of Halogen lamps is about 30 lm/watt, which is about three times more than the incandescent lamps. Life of these lamps is also double of the incandescent lamps.

These lamps are small in size and do not need ballast for starting but their handling is difficult.

Halogen lamps are available upto 5 kW. They are used for decorating purposes, in T.V. studies in projectors and in flood lights.

In halogen lamps, the halogen vapours send the tungsten vapours back to filament by a chemical reaction *i.e.,* there is a sort of "regeneration cycle", so the lamp surface does not blacken and there is no depreciation of lumen output. The halogen lamps are small in size and they have long life upto 2000 hours.

Q.12. Explain the working of a neon lamp with neat sketch.

Ans. They are long thin tubes. The length may be upto 10 meters and can be bent into any shape. They have iron or nickel electrode at each end and are filled with neon gas. By mixing other gases in the neon gas, we can get light of any colour. The voltage required is 500 V per meter length and they take current in milli amperes. For getting high voltages, step up transformers are used.

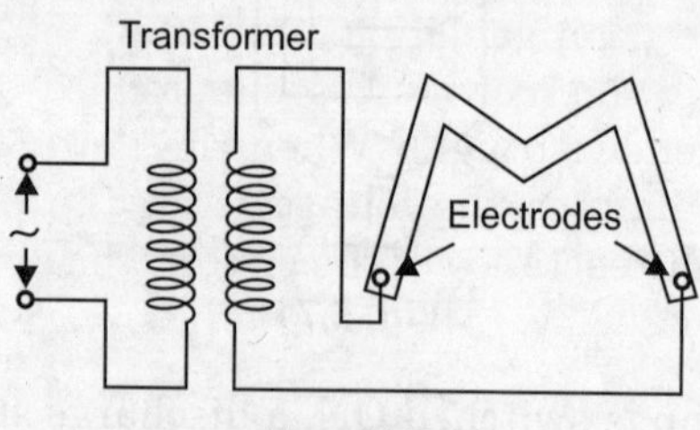

Fig. 5.72

These are used for advertisement purposes.

Q 13. Explain the construction and working of sodium vapour lamp. Ans. The sodium lamp consists of a dischrage tube made from a special heat resistance glass containing a small amount of metallic sodium, neon gas and two sets of electrodes. Neon gas is added to start the discharge and to develop sufficient heat to vaporise the sodium. Since long discharge paths are necessary, therefore the discharge tube is usually bent into U-shape to reduce the overall dimensions. The lamp operates at a temperature of about 300°C. To maintain this temperature and to reduce the net losses, the discharge tube is enclosed in a double walled flask as shown in fig. The lamp must be operated horizontally or nearly so, to keep the sodium well spread out along the tube. The double walled flask is interchangeable and can be fitted on to another U-tube. Care should be taken while replacing the inner-U-tube because if it is broken and sodium comes in contact with moisture, it may result in fire.

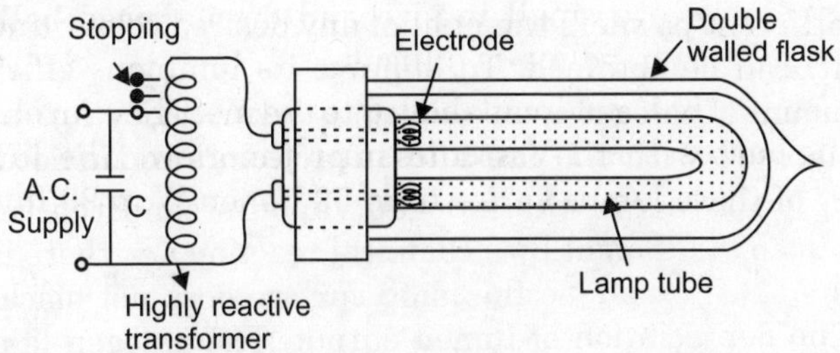

Fig. 5.73

Sodium vapour lamps require a higher voltage at the time of starting and low voltage during operation. The requirement is met by operating the lamp by a highly inductive reactance transformer or choke. At starting, a high voltage of about 450 V is applied across the lamp which is just sufficient to start the charge.

When the lamp is not in operation, the sodium is in the form of solid and when it is connected across the supply mains, discharge takes place in the neon gas, giving red-orange glow. The sodium vaporises gradually and then ionizes and produces the characteristic monochromatic yellowish light. When the lamp is fully operative after about 10-15 minutes, the voltage across the lamp falls to about 150 V. Due to high reactance of the circuit, the power factor is low and hence a capacitor must be used to improve the power factor.

The efficiency of a sodium vapour lamp is about 40-50 lumens/ watt. The average life is about 3000 working hours. These lamps are manufactured in 45, 60, 85 and 140, 400 watt ratings. These lamps are mainly used for street lighting, parks, railway yards, highways and airfield lighting where colour distortion is not so important.

Q.14. Describe the compact fluorescent lamp and mention its advantages over ordinary fluorescent lamp.

Ans. The compact fluorescent lamps are becoming more and more popular now-a-days because of their low power consumption, low running cost, longer life, attractive look, soothing light and low maintenance. These lamps are available in different sizes and designs as shown in fig. These may have single rod, double rod, tripple rod or spiral rod. These lamps are available in different power rating *i.e.,* 5/7/9/11/18/24 watt, 220 V.

Construction: Basically, it is low pressure mercury vapour lamp having two electrodes coated with electron emissive material placed at the two ends of a glass tube. In the tube one drop of mercury and argon gas are filled at low pressure. The tube is coated internally with some fluorescent material in the form of powder. By mixing different powders the light of any desired colour (including day light) can be obtained. To improve its luminous efficiency a small amount of halogen is also added to the inert gas. An electronic ballast is used with the lamp to start the glow. The luminous efficiency of these lamps varies from 50 lumen/W to 90 lumen/W.

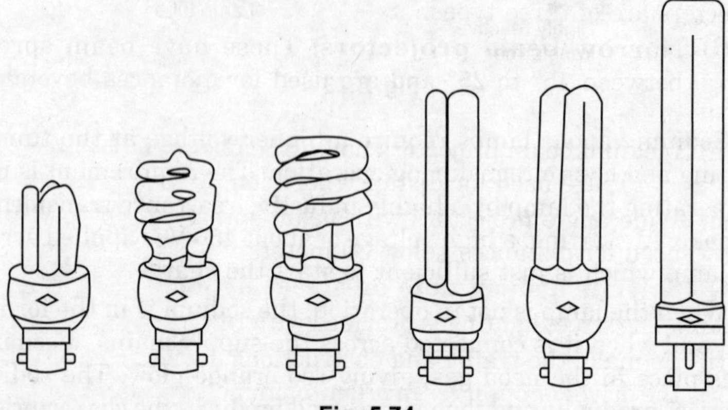

Fig. 5.74

Major advantages of CFL: The following are the main advantages of CFL over the ordinary incandescent lamp.

(i) Low energy consumption

(ii) Low maintenance costs.

(iii) It provides instant glow

(iv) It provides light without heating the surroundings

(v) Excellent colour rendering properties. Hence all colours look natural and desired ambience can be created in various interiors.

(vi) Low operational costs: The power consumption of CFL is 80% less than incandescent lamps.

(vii) The compact size opened up possibilities for exciting and aesthetically appealing new luminaries.

(viii) Average life of compact fluorescent lamp is 8000 working hours.

Q.15. What is Hood lighting. Discuss the luminaries used in flood lighting?

Ans. Flood lighting means flooding of large areas with light from power projectors. These lights serve the following purposes.

(i) Enhancing beauty or give decoration to a monument or important building

(ii) For lighting sports stadiums, parks etc.
(iii) For illuminating advertisement boards.
(iv) For illuminating places that need security.

Luminaries used for flood Lighting: Usually 250 W or 400 W metal hallide, mercury vapour or sodium vapour lamps are used. Light is concentrated through projectors with proper cooling. The projectors are of three types..

(i) Narrow beam projectors: These have beam spread between 15° to 25° and are used for distances beyond 70 m.

(ii) Medium beam projectors having beam spread between 25° to 40° are used for distance between 30 to 70 meter.

(iii) Wide angle projectors with beam spread of 40° to 90° are used for distances below 30 meters.

Q 16. Write a short note on street lighting.

Ans. The street lighting is provided with the following objectives:

(i) To make the traffic and obstructions on the road visible to the driver or observer in order to promote safety convenience
(ii) To make the street more attractive for the passangers
(iii) To increase the community value of the street.

While providing street lighting following factors are to be kept in mind.

(i) Level of illumination
(ii) Shadows and high contrast
(iii) Glare

Street lighting is either based on diffusion principle or specular reflection.

Level of illumination on various categories of roads varies from 1 lux to 20 lux. Luminaries used are sodium discharge lamps due to their high efficiency.

6 Electrolytic Processes

6.1. Introduction

The substances which decompose when an electric current is passed through them are called *electrolyte.* The process of decomposition of electrolyte by the passage of electric current through them is called *electrolytic process* or *electrolysis.* Electrolytes may be solids like caustic soda or Silver iodide but they are generally liquid. Pure liquids, like water do not suffer appreciable decomposition when electricity is passed through them and are called non ions which appears at anode are called anions and carry negative charge whereas the ions which appear at cathode are called cations and carry a positive charge.

6.2. Definition

The process of depositing a metal on the surface of some other metal by electrolysis is called electroplating.

6.3. Need of Electroplating

The electroplating is used for :

(*i*) To protect the metals against corrosion.

(*ii*) To give shining appearance to the articles.

(*iii*) To give reflecting properties to the reflectors.

(*iv*) To replace worn out material from a casting.

(*v*) To repair a damaged casting.

6.4. Basic Principles of Electrodeposition

If two electrodes are dipped in an electrolyte and potential is applied across them, the electrodes are named as +ve or anode and –ve or cathode. The molecules of the substance dissolved are "dissociated". They are broken up into two types of ions *i.e.* +ve, and –ve ions. Since the ions are electrically charged, therefore the chemical and general properties of an ion differ from those of the same atom when free of charge.

Let us consider a ring of iron to be given a copper coating. An arrangement is made as shown in Fig. 6.1.

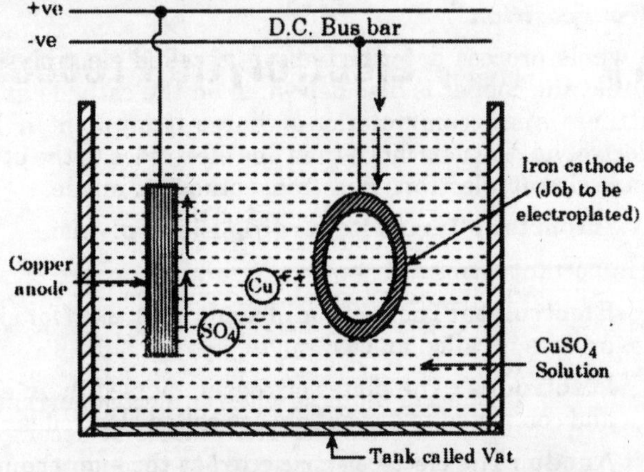

Fig. 6.1. Process of electro-depositic/electroplating.

The electrolyte used in the given example, is a solution of copper sulphate, ($CuSO_4$). The sulphate solution readily breaks into positive ions Cu^{++}, which are deficit in two electrons per ion and SO_4^{--} negative ions, having two surplus electrons per ion.

When the two electrodes shown in Fig. 6.1, get a d.c. supply from the bus bars, the negative ions move towards the anode made of copper. Each SO_4^{--} ion after transfer of two electrons to the anode becomes SO_4 radical. It attacks the copper anode to form $CuSO_4$ molecule, which dissolves once again in water to maintain the electrolyte density at original level.

The positive copper ions reach the negative electrode *i.e.* the cathode and receive two electrons each from the supply circuit to become Cu atoms. These get deposited on the cathode surface. The cathode as shown in the figure is the job or article to be copper plated. The copper deposited on the cathode surface is practically having the same mass as lost by the anode, in maintaining the electrolyte strength.

The sequence as above can also be expressed chemically as follows :

$CuSO_4$ solution in water splits into ions *viz.*

$$CuSO_4 \rightleftharpoons Cu^{++} + SO_4^{--} \quad ...(i)$$

At anode : SO_4^{--} less two electrons = SO_4 molecule in radical form.

$$SO_4 + Cu \rightarrow CuSO_4 \quad ...(ii)$$

At cathode : Cu^{++} plus two electron = Metallic Cu atom to be deposited on the matallic cathode surface. Conclusively we can say that the phenomenon of deposition of a metallic coating on the surface of some other metal, through the process of electrolysis is ***Electroplating or electrodeposition.***

The whole process described above is called electrolysis, but its effect is that the copper is also deposited on the cathode at the same time and there is no accummulation of charge at any point in the circuit. In the electrolyte, the ions move from one electrodes to the other but in the out side circuit, electrons flow from cathode to anode.

6.4.1. Important Terms Regarding Electrolysis:

The important terms are :

(1) **Electrolyte :** The solution of a salt when used for electrolytic process is called an electrolyte.

(2) **Electrodes :** The plates or rods immersed in an electrolyte and connected to d.c. supply are called electrodes.

(3) **Anode :** The electrode connected to the +ve terminal of the supply is called anode.

(4) **Cathode :** The electrode conected to the –ve terminal of the supply is called cathode.

(5) **Ions :** When a direct current is passed through an electrolyte, it gets chemically decomposed into two parts known as +ve and –ve ions.

(6) **Cations :** These are positively charged ions and they move towards the cathode.

(7) **Anions :** These are negatively charged ions and they move towards the anode.

(8) **Chemical equivalent weight :** Chemical eqivalent weight of a substance may be defined as the ratio of its atomic weight and valency.

i.e., Chemical Equivalent $= \dfrac{\text{Atomic weight}}{\text{Valency}}$

(9) **Electro-chemical Equivalent (E.C.E.) :** Electro-chemical equivalent of a substance is the amount deposited on passing a steady electric current of one ampere for one second through its solution.

(10) **Atomic weight :** The atomic weight of an element is a number which is the average of the masses of its various isotopes weight-relative to their abundance or the atomic weight is the ratio of the weight of an atom of the element to the weight of an atom of hydrogen.

(11) **Valency :** The valency of an atom or a group of atoms is the number of Hydrogen-atoms with which it will react chemically. Valency is always an integer (1, 2, 3, 4 etc.) but for a

given atoms or radical, it can have different value in different chemical reactions.

6.5. Application of Electrolysis

1. Electrolysis. Following are the applications of electrolysis :

1. Extraction of metal from their ores.
2. Extraction of zinc.
3. Extraction of Aluminium.
4. Refining of metals.
5. Production of chemicals.
6. Separating metal from their compounds.
6. Electrotyping.
8. Electro forming.
9. Electro deposition.
10. Electro cleaning.

1. Extraction of Metals from their Ores. There are two methods of extraction of metals from ores depending upon the physical state of the ore.

1. The ore is treated with a strong acid to obtain a salt and the solution of the salt is electrolysed to liberate the metal.
2. When the ore is in molten state it is electrolysed in the furnace.

2. Extraction of Zinc. The zinc ore which has mainly zinc oxide is treated with concentrated sulphuric acid, roasted and passed through various chemical processes to get rid of impurities like cadmium copper etc. by precipitation. The zinc sulphate solution thus obtained is then electrolysis. Electrolysis process is carried out in wooden box with inner lining of lead. The anodes are lining of lead and cathodes are of aluminium. Zinc is deposited on the cathode. The current density at the cathode is about 1000 Amp./m^2 and voltage drop per cell is about 3.5 volt.

3. Extraction of Aluminium. The ores of Aluminium are bauxite, cryoloite, baruite is treated chemically and reduced to aluminium oxide and then dissolved in fused cryolite and electrolysed. The furnace is lined with carbon Aluminium metal gets deposited at the cathode. The tempt. of furnace is about 1000° C and the furnace of area 15 m^2 will require a voltage of about 8 volt and a current of about 45000 Amp. Since electrolytic process requires a great amount of power. So such plants are located near hydro electric power station. At present whole of Aluminium required is produced in this way.

4. Refining of Metals. The metal extracted from its ore is not that much pure which would be used for electrical applications. The purity of copper obtained from its ore is about 98%. But copper to be used in electrical applications must have a purity of 99.95%. This purity of copper is obtained by electrolysis. The electrolyte solution used is copper sulphide. The anodes are made of impure copper extracted from its ore and pure copper is deposited on the cathode. At regular intervals the copper deposit is removed from the cathode and anode is replaced whenever required. The energy consumption is about 150 to 300 kwh per tonne of the refined copper.

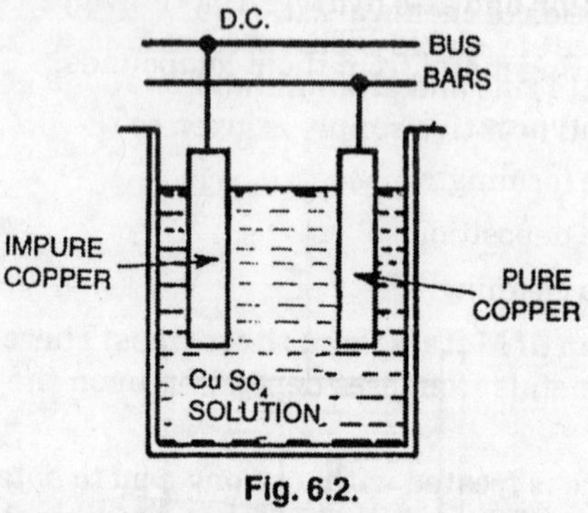

Fig. 6.2.

5. Production of Chemicals. Many chemicals such as caustic soda (KOH), chlorine gas, amonium sulphate, hydrogen, oxygen are produced by electrolysis on a large scale.

6. Separating Metals from their Compounds. Many metals are separated (from their compounds) by electrolysis *e.g.* an ore of aluminium contains about 70% aluminium oxide, silica and iron oxide. First pure aluminium oxide is obtained by suitable treatment and pure aluminium is obtained by electrolytic process.

7. Electrotyping. In this process wood cuts are reproduced in copper by the process of electroplating. First the mould is made of the type in wax, then it is coated with black lead to give it a metallic surface and then it is subjected to the process a film of copper is formed on the prepared surface.

8. Electroforming. The production or reproduction of an article by electrodeposition is known as electroforming. In the production of coins, models etc. mould is first made by having an impression of point in wax surface of the wax which has exact impression of the object is coated with powdered graphite in order to make it conducting. The mould is then dipped in an electroforming cell as a cathode. After obtaining coating of desired thickness the object is removed and the wax is melted out of the metal shell.

9. Electro-deposition. Electrodeposition is the process of depositing a metal over another metal or nonmetal by electrolysis. An electroplating is a very common example of such a process.

10. Electro-cleaning. The articles before electroplating should have a surface free from grease, oil etc. and they are cleaned by electrocleaning method. A solution of sodium phosphate is used as an electrolyte in the plating tank. The tank is connected to +ve terminal of D.C. supply and work piece to be cleaned is made the cathode and is suspended in the solution of sodium phosphate. Heavy current is passed through the solution and caustic soda is produced on the cathode which has a cleaning action and also hydrogen gas is evolved at cathode which removes the grease etc. quickly. This process is called cathodic cleaning and is applicable to zinc and aluminium. For anodic cleaning the work is made anode and negative supply is given to the plating tank.

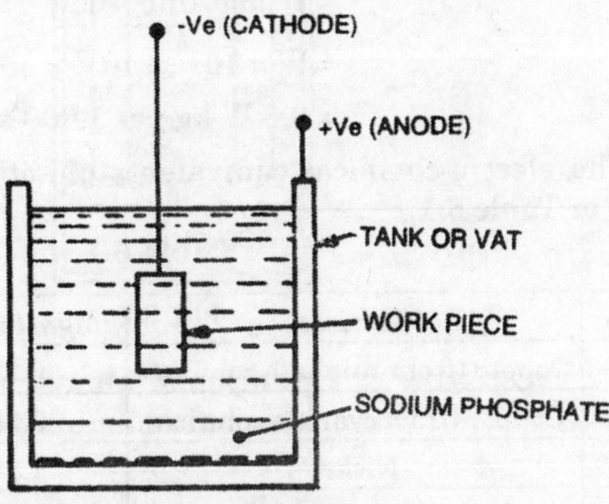

Fig. 6.3. Electro-cleaning.

6.6. Faraday's Law's of Electrolysis

Faraday was a scientist who deduced two laws on the phenomenon of electrolysis which are called Faradays laws of electrolysis.

First law states. The weight of a substance liberated from an electrolyte in a given time is proportional to the total quantity of electricity passed in that time.

If W is the weight of substance liberated in grams.

Then $\quad W \propto Q$

or $\quad W \propto I.T$

or $\quad W = ZIT$

where Q is the quantity of electricity passed,

= Strength of current (I) in amperes × Time in seconds during which the current flow (T).

i.e. $\quad Q = I.T$

where Z is a constant and is called electro-chemical equivalent of the substance (E.C.E.) and the value of this depends upon the nature of the substance.

If $I = 1$ amp.

and $T = 1$ sec.

Then $W = Z$.

Therefore, electro-chemical equivalent of a substance is the amount of that substance by weight liberated in a unit time by the passage of unit current. Its units will be milligramme per coulomb or gram per coulomb. Let us convert it into kilogram per 1000 ampere-hours as below :

$$Z = W \text{ mg/coulomb.}$$
$$= W \text{ mg/Amp. sec.}$$
$$= W \times 10^{-3} \times 10^{-3} \times 3600 \times 10^{2}$$
$$= 3.6 \text{ W kg per 1,000 ampere hours.}$$

The electro-chemical equivalents of some of the substances are given in Table 6.1.

Table 6.1.

S. No.	*Element*	*Mgm / coulomb*	*Kg / 100 Ah*
1.	Copper (from sulphate solution)	0.3290	1.1844
2.	Copper (from cyanide solution)	0.6580	2.3688
3.	Nickel	1.3043	1.0954
4.	Silver	1.1180	4.0248
5.	Zinc	0.3370	1.2132
6.	Tin	0.6140	2.2104
7.	Oxygen	0.0829	0.2984
8.	Hydrogen	0.0104	0.0374

Second law states. **If the same current flows for a given time through several electrolytes, the weights of substance liberated are proportional to their chemical equivalents.**

The chemical equivalent or equivalent weight of a substance is the weight of the substance which can displace or combine with unit weight of hydrogen or 8 grams of oxygen. Obviously the chemical equivalent of hydrogen is one by definition. The number of hydrogen atoms that can be displaced by an atom of another substance is called its valency. Therefore, the

Chemical Equivalent

$$= \frac{\text{Atomic weight of substance}}{\text{Valency}}$$

According to this law if we take the two electrolytes of copper sulphate and nickel sulphate in which same current flows for the same time, then

$$= \frac{\text{Weight of copper deposited by given quantity of electricity}}{\text{Weight of nickel deposited by same quantity of electricity}}$$

$$= \frac{\text{Chemical equivalent of copper}}{\text{Chemical equivalent of nickel}}$$

6.7. Definitions

Some of the important terms used in electrolytic processes are defined as under :

(*i*) **Current Efficiency.** Due to impurities which cause secondary reactions, the quantity of a substance liberated is less than that calculated from Faraday's laws,

i.e. $\text{Current efficiency} = \frac{\text{Actual quantity of substance liberated}}{\text{Theoretical quantity}}$

It should be noted that its value lies between 90 to 98%.

(*ii*) **Energy efficiency.** On account of secondary reactions the actual voltage required for the deposition or liberation of metal is higher than the theoretical value which increases the actual energy required. Energy efficiency is defined as the ratio of theoretical energy required to the actual energy required.

i.e. $\text{Energy efficiency} = \frac{\text{Theoretical energy}}{\text{Actual energy required}}.$

(*iii*) **Voltage.** The voltage required for the passage of the current through an electrolyte depends upon the potential drops at the electrodes and in the electrolyte. It is desirable that these drops should be reduced to minimum and for this purpose some agent is added to make the electrolyte a good conductor thus reducing the voltage drops.

For Example. In copper plating sulphuric acid is added to the copper sulphate bath. The normal voltage to pass current through most of electrolytes is about 1 to 2 volts.

Example 6.1. *Find the thickness of copper deposited on a plate of 2.25 cm^2 during electrolysis if a current of 1 ampere is passed for 100 minutes. (Density of copper is 8.9 gm per c.c. and E.C.E. of copper is 0.0003295 gm / coulomb).*

Solution. Electro-chemical equivalent (E.C.E.) of copper

= 0.0003295 gm/coulomb

Current strength, $I = 1$ amp.

Time for which the current is passed through the solution

$$T = 100 \text{ minutes} = 100 \times 60 = 6{,}000 \text{ sec.}$$

Wt. of copper deposited,

$$W = ZIT$$

$$= 0.0003295 \times 1 \times 6000 \text{ gm.} = 1.977 \text{ gm.}$$

Density of copper $= 8.9$ gm/c.c. (given)

Now density, (D) $= \dfrac{\text{Mass}}{\text{Volume}}$

or Volume of copper deposited,

$$V = \frac{\text{Weight}}{\text{Density}}\left(\frac{W}{D}\right) = \frac{1.977}{8.9} \text{ c.c.} = 0.222 \text{ c.c.}$$

Volume $=$ Area (A) $\times$ Thickness (t)

or $$t = \frac{V}{2.25} = \frac{0.222}{2.25} = 0.0986 \text{ cm. } \textbf{Ans.}$$

Example 6.2. *If a current of 10 amperes deposits 13.42 gm. of silver from a silver nitrate solution in 20 minutes, calculate the electro-chemical equivalent of silver.*

Solution. Strength of current,

$$I = 10 \text{ Amp.}$$

Time of flow of current,

$$T = 20 \times 60 = 1{,}200 \text{ sec.}$$

Weight of silver deposited,

$$W = 13.42 \text{ gm.}$$

Now E.C.E. of silver,

$$Z = \frac{W}{I.T} = \frac{13.42}{10 \times 1200}$$

$$= 0.001118 \text{ gm/coulomb. } \textbf{Ans.}$$

Example 6.3. *Calculate the quantity of electricity and the steady current required to deposit 5 gm. copper from copper sulphate solution, in one hour. Electro-chemical equivalent of copper is 0.3294 m. gm/coulomb.*

Solution. Weight of copper to be deposited

$$W = 5 \text{ gm.}$$

E.C.E. of copper, $Z = 0.329 \times 10^{-3}$ gm/coulomb

$$Q = I.T = \frac{W}{Z} \because W = ZIT \therefore IT = \frac{W}{Z}$$

$$= \frac{5}{0.3294 \times 10^{-3}} = 15{,}180 \text{ coulomb. } \textbf{Ans.}$$

Steady current required

$$I = \frac{Q \text{ in coulombs}}{\text{Time of deposite in sec.}} = \frac{15{,}180}{1 \times 60 \times 60}$$

$$= \frac{15{,}180}{3{,}600} = 4.216 \text{ Amp.} \quad \textbf{Ans.}$$

Example 6.4. *A rectangular metal plate having 5 cm × 4 cm × 1 cm as its dimensions is to be electroplated with nickel. How long will it take to deposit a layer of 0.1 mm thickness, when a current of 4.3 amperes flows through the circuit ?*

E.C.E. of nickel = 0.000304 gm/coulomb

and Density of nickel = 8.6 gm/c.c.

Solution. Volume of the metal plate before electroplating,

$$V = 5 \times 4 \times 1 = 20 \text{ c.c.}$$

Final volume of the metal plate after electroplating,

$$= 5.02 \times 4.02 \times 1.02 = 20.584 \text{ c.c.}$$

∴ Volume of nickel deposited

$$= 20.584 - 20 = 0.584 \text{ c.c.}$$

but as we know,

Weight or mass = Volume × Density.

So weight of the mass of the nickel deposited

$$= 0.584 \times 8.6 \text{ gm.}$$

According to law, if I be the current in the circuit, flowing for time T seconds. Then

$$W \text{ or } m = ZIT$$

i.e. $0.584 \times 8.6 = 0.000304 \times 4.3 \times T$ $[\because I = 4.3 \text{ amp}]$

$$\therefore \quad T = \frac{0.584 \times 8.6}{0.000304 \times 4.3} \text{ seconds}$$

$$= \frac{0.584 \times 8.6}{0.000304 \times 4.3 \times 60 \times 60} \text{ hours}$$

= 1 hours, 5 minutes, 2 seconds. **Ans.**

Example 6.5. *A rectangular plate 20 × 10 cm is to be coated with nickel with a layer of 0.2 mm thickness. Determine the quantity of electricity in ampere hour and time taken for the process. Current density is 190 amp/m^2 and current efficiency is 90% and specific gravity of nickel is 8.9.*

Solution. Area of the plate

$$= 20 \times 10 = 200 \text{ cm}^2.$$

Volume of material to be plated

$$= 200 \times 0.02 = 4 \text{ c.c.}$$

Weight of material to be deposited

$$= 4 \times 8.9 = 35.6 \text{ gm.}$$

Take Z for nickel $= 0.0003043$ gm/col.

$$W \text{ or } m = 0.0003043 \times 0.9 \times I.T. \quad [\because \; W = ZIT \times \eta]$$

or

$$I.T = \frac{35.6}{0.0003043 \times 0.9} \text{ amp. sec.}$$

$$= \frac{35.6}{0.0003043 \times 0.9 \times 3600} \text{ amp. hour.}$$

$$= 36.1 \text{ amp. hour.}$$ **Ans.**

Current density $= 190$ amp/m^2

So current = Current density × Area

$$= 190 \times \frac{200}{10^4} \times \frac{38}{10} = 3.8 \text{ amp.}$$

Time taken $$= \frac{\text{Amp. hours}}{\text{Amp.}} = \frac{36.1}{3.8} = 9.5 \text{ hours.}$$

or = 9 hours and 30 minutes. **Ans.**

Example 6.6. *If 96,500 coulombs of electricity liberate 1 gram equivalent of any substance, how long will it take for a current of 0.15 ampere to deposit 20 milligram of copper from a solution of copper sulphate ? Chemical equivalent of copper is to be taken as 32.*

Solution. Since as we know,

W or m $= ZQ$

and 32 grams of copper is liberated by 96,500 coulombs of electricity.

$$\therefore \quad Z = \frac{m}{Q} = \frac{32}{96,500} \text{ gm/coulombs.}$$

∴ The amount of electricity Q is required to deposit 20 mgm. of copper is

$$Q_1 = \frac{m_1}{Z} = \frac{20 \times 10^{-3}}{32/96,500} \text{ coulombs.}$$

$$= \frac{20 \times 10^{-3} \times 96,500}{32} \text{ coulombs.} = \frac{20 \times 96,500}{10^{-3} \times 32} \text{ coulombs.}$$

∴ Time required T to deposit 20 mgm. when the current,

$$I = 0.15 \text{ amp.}$$

$$T = \frac{Q_1}{I} = \frac{20 \times 10^{-3} \times 96{,}500}{32 \times 0.15} \text{ sec.}$$

$$= 402.1 \text{ sec.} = 6 \text{ minutes } 42.1 \text{ sec.} \qquad \textbf{Ans.}$$

Example 6.7. *A coating of nickel 1 mm thick is to be built on a cylinder 20 cm in diameter and 30 cm in length in 2 hours. Calculate the electrical energy used in the process if the voltage is 10 volts. E.C.E. of Ni is 0.000304. Specific gravity of nickel is 8.9.*

Solution. Total surface area of the cylinder

$$= \pi dl = \pi \times 0.2 \times 0.3 \text{ m}^2.$$

So volume of Ni deposited,

$$= \text{Surface area of cylinder} \times \text{Thickness of N}_i$$

$$= \pi \times 0.2 \times 0.3 \times 0.001 \text{ cu. metres.}$$

$$= 1.88 \times 10^{-4} \text{ cu. metres.}$$

$$\text{Mass} = \text{Specific Gravity} \times \text{Volume.}$$

So mass of Ni deposited

$$= 1.889 \times 10^{-4} \times 8.9 \text{ gm} = 1.678 \text{ gm.}$$

Now mass deposited,

$$m = ZIT$$

$$Z = 0.000304$$

$$T = 2 \times 60 \times 60 \text{ sec.}$$

So $\quad 1.678 = 0.000304 \times I \times 2 \times 60 \times 60$

$\therefore \quad I = \dfrac{1.678}{0.000304 \times 2 \times 60 \times 60} = 0.7663 \text{ amp.}$

$\therefore$ Energy consumed in two hours,

$$= \frac{VIT}{36 \times 10^5} \text{ kWh} = \frac{10 \times 0.7663 \times 2 \times 60 \times 60}{1000 \times 60 \times 60}$$

$$= 0.015326 \text{ kWh.} \qquad \textbf{Ans.}$$

Example 6.8. *An ammeter connected in series with a silver electrolytic cell reads 0.60 amp and it is found that 0.322 gm of metal deposit in a run lasting 10 minutes. What is the error of ammeter to this point of its range ?*

E.C.E. of silver = 1.118 milli. gm/coulomb.

Solution. Given mass,

$$m = 0.322 \text{ gm.}$$

Time taken, $T = 10 \text{ minutes} = 10 \times 60 \text{ sec.}$

$$Z = 1.118 \text{ milli. gm/coulomb}$$

$$= \frac{1.118}{1000} \text{ gm/coulomb}$$

Substituting these values in the formulae,

$$m = ZIT$$

$$0.322 = \frac{1.118 \times I \times 10 \times 60}{1000}$$

or $$I = \frac{0.322}{0.001118 \times 600} \text{ amp.} = 0.48 \text{ ampere.}$$

But the ammeter reads this value 0.60 ampere. So the error of the ammeter

$$= 0.60 - 0.48 = 0.12 \text{ amp.} \quad \textbf{Ans.}$$

Example 6.9. *Calculate the time taken to deposit a coating of nickel 0.05 cm thick on a metal surface by means a current of 20 A per sq. inch of surface. Nickel is a divalent metal of atomic weight 59 and density of 9 gm / cc. Silver has an atomic weight of 108 and E.C.E. of 1.118 mg / col.*

Solution. Surface area of the plate

$$= 2.54 \times 2.54 \text{ cm}^2.$$

Wt. of nickel deposited

$$= (2.54)^2 \times 0.05 \times 9 \text{ grams.}$$

Now $$\frac{\text{E.C.E. of Ni}}{\text{E.C.E. of Ag}} = \frac{\text{Chemical equivalent of Ni}}{\text{Chemical equivalent of Ag}}.$$

So E.C.E. of Ni $$= 0.001118 \times \frac{(59/2)}{108}$$

$$= 0.001118 \times \frac{29.5}{108} \text{ gm/coulomb}$$

$$= 0.0003053 \text{ gm/coulomb}$$

Now $$m = ZIT$$

or $$2.903 = 0.0003053 \times 20 \times T$$

or $$T = \frac{2.903}{0.0003053 \times 20} = 475 \text{ sec.}$$

$$= 7 \text{ min. } 55 \text{ seconds.} \quad \textbf{Ans.}$$

Example 6.10. *A steady current of 40 amp. is passed through a solution of copper sulphate for 1 hour. Find the weight of copper deposited. The chemical equivalent for copper is 31.8 and E.C.E. of hydrogen is 0.01044 m. gm / coulomb.*

Solution. $\dfrac{\text{E.C.E. of Cu}}{\text{E.C.E. of H}_2}$

$$= \frac{\text{Chemical equivalent of Cu}}{\text{Chemical equivalent of hydrogen}}$$

or E.C.E. of Cu,

$$= \text{E.C.E. of hydrogen} \times \frac{\text{Chemical equivalent of Cu}}{\text{Chemical equivalent of hydrogen}}$$

$$= 0.01044 \times 10^{-3} \times \frac{32.8}{1} = 0.000342 \text{ gm/coulomb}$$

Now $m = ZIT$

So $m = 0.000342 \times 40 \times (60 \times 60)$ grams

$= 49.31$ grams. **Ans.**

Example 6.11. *A group of 400 cells in a copper refining plant carries a current of 5,000 A. The voltage per cell is 0.25 V. If the plant working time per week is 100 hours, determine the annual output of refined copper from the cells and the energy consumption per tonne.*

E.C.E. weight of copper = 0.327 mg per coulomb.

Solution. E.C.E. of copper

$$Z = 0.327 \text{ mg/c} = 0.327 \times 10^{-6} \text{ kg/c.}$$

$$= 0.327 \times 10^{-6} \times 3600 \times 1{,}000 \text{ kg/1000Ah}$$

$$= 1.1772 \text{ kg per 1000 Ah}$$

Total number of ampere hours per annum

$$= 400 \times 5{,}000 \times 100 \times 52 = 104 \times 10^8 \text{ Ah}$$

Annual output of plant

$$= \frac{1.1772 \times 104 \times 10^8}{1000} \text{ kg} = \frac{1.1772 \times 104 \times 10^8}{1000 \times 1000} \text{ tonnes}$$

$$= 12{,}243 \text{ tonnes. } \textbf{Ans.}$$

Energy consumed per annum

$$= \text{Ah per annum} \times \text{voltage per cell}$$

$$= 104 \times 10^8 \times 0.25$$

$$= 26 \times 10^8 \text{ watt.hr.} = 26 \times 10^5 \text{ kWh.}$$

Energy consumption per tonne

$$= \frac{26 \times 10^5}{12{,}243} = \textbf{212.4 kWh/tonne.} \quad \textbf{Ans.}$$

Example 6.12. *18.258 gm of nickel is deposited by 100 A current flowing for 10 minutes, how much copper would be deposited by 50 A current in 6 minutes ? Atomic weight of nickel and copper are 58.6 and 63.18 respectively and valency of both is 2.*

Solution. E.C.E. of nickel

$$Z = \frac{m}{I \times T} = \frac{18.528 \times 10^{-3}}{100 \times 10 \times 60}$$

$$= 30.43 \times 10^{-8} \text{ kg C}^{-1}.$$

$$\text{E.C.E. of copper, } Z' = \frac{\text{Chemical equivalent of copper}}{\text{Chemical equivalent of nickel}} \times Z$$

$$= 30.43 \times 10^{-8} \times \frac{63.18/2}{58.6/2}$$

$$= 32.81 \times 10^{-8} \text{ kg C}^{-1}$$

Mass of copper deposited

$$= Z' \times I' \times T'$$

$$= 32.81 \times 10^{-8} \times 50 \times 6 \times 60$$

$$= 0.0059 \text{ kg} = 5.9 \text{ gm.}$$ **Ans.**

Example 6.13. *How much aluminium will be produced from aluminium oxide in 24 hours if the average current is 3000 A and current efficiency is 92%. Aluminium is trivalent and atomic weight is 27. The chemical equivalent weight of silver is 107.28 and 0.00111 gm of silver is deposited by one coulomb.*

Solution. E.C.E. of silver

$$= 0.00111 \text{ gm/c or } 111 \times 10^{-8} \text{ kg C}^{-1}$$

Chemical equivalent weight of silver

$$= 107.98$$

Chemical equivalent weight of aluminium

$$= \frac{27}{3} = 9$$

∴ E.C.E. of aluminium,

$$Z = \frac{\text{E.C.E. of silver} \times \text{Chemical equivalent weight of aluminium}}{\text{Chemical equivalent weight of silver}}$$

$$= 111 \times 10^{-8} \times \frac{9}{107.98}$$

$$= 9.252 \times 10^{-8} \text{ kg C}^{-1}$$

Current efficiency = 92% = 0.92

Average current, I = 3,000 A

$$t = 24 \times 60 \times 60 = 86,400 \text{ seconds.}$$

Mass of Aluminium produced,

$$m = ZIt \times \text{Current efficiency}$$

$$= 9.252 \times 10^{-8} \times 3000 \times 86,400 \times 0.92$$

$$= 22.062 \text{ kg} \quad \textbf{Ans.}$$

Example 6.14. *A copper refining plant, using 500 electrolytic cells, carries a current of 6000 A, voltage per cell being 0.25 volt. If the plant were to work 40 hours / week, calculate the energy consumption per tonne, assuming E.C.E. of copper as 0.3281 mg / coulomb of electricity.*

Solution. E.C.E. of copper, Z

$$= 0.3281 \text{ mg/c} = 32.81 \times 10^{-8} \text{ kg C}^{-1}$$

$$= 32.81 \times 10^{-8} \times 3600 \text{ kg/A.h}$$

$$= 0.00118116 \text{ kg/A.h}$$

Total number of ampere-hours per annum

$$= 500 \times 6000 \times 40 \times 52 = 624 \times 10^{-7} \text{ A h}$$

Annual output of plant

$$= 0.00118116 \times 624 \times 10^{-7} \text{ kg}$$

$$= 73,70,438 \text{ kg} = 7,370.438 \text{ tonnes}$$

Energy consumed per annum

$$= \text{A.h per annum} \times \text{voltage per cell}$$

$$= 624 \times 10^{7} \times 0.25 = 156 \times 10^{7} \text{ Watt.hours}$$

$$= 150 \times 10^{4} \text{ kWh.}$$

Energy consumption per tonne

$$= \frac{\text{Annual consumption in kWh}}{\text{Annual output in tonnes}}$$

$$= \frac{156 \times 10^{4}}{7,370.438}$$

$$= 211.66 \text{ kWh/tonne.} \quad \textbf{Ans.}$$

Example 6.15. *The worn out circular shaft 14 cm in diameter and 30 cm long is to be repaired by coating it with a layer of 1.5 mm of nickel. Determine the theoretical value of current required and the time taken if the current density used is 200 amps/m^2. Electrochemical equivalent of nickel is 0.000304 gm / coulomb electricity and density of nickel is 8.9 gms / cu-cm.*

Solution. Diameter of shaft

$= 14$ cm

Length of shaft $= 30$ cm

Therefore, surface area of shaft

$= \pi dl = \pi \times 14 \times 30$

$= 1320 \text{ cm}^2$

Mass of Nickel to be deposited

= Surface area × Thickness of coating × Density of metal

$= 1320 \times \frac{1.5}{10} \times 8.9 = 1762.2$ gms.

Theoretical value of quantity of electricity required

$$Q = \frac{m}{Z} = \frac{1762.2}{0.000304} = 5796.710 \text{ Amp./sec}$$

$$= \frac{5796.710}{3600} = 1610.2 \text{ Amp. hour.}$$

Now, current density used is 200 amperes/metre^2

Therefore, current = Current density × Surface area in m^2

$= 200 \times 1320 \times 10^{-4} = 26.4$ Amp.

Time required $= \frac{1610.2}{26.4} = 61$ hours. **Ans.**

6.8. Electro deposition

The process of depositing metal over another metal or non-metal by electrolysis is known as electro-deposition. Electroplating is a very common example of such a process. Table 6.2 gives the current density, solution and temperature of solution for various metals deposition.

Table 6.2

Metal Deposition	*Solution*	*Current density in Amp/in^2*	*Temperature of solution*
Nickel	Nickel sulphate	100 - 200	45° – 55°C
Chromium	Chromium acid	1800 - 2000	35°C
Silver	Double cynide of silver and sodium	30-50	Cold
Gold	Double cynide of potassium and gold	100-300 amp./ft^2	25°–30°C
Copper (from sulphate solution)	Copper sulphate	250 - 350	Warm or cold
Copper (from cynide solution)	Cuprous cynide and sodium cynide	30 - 40	50°C

6.9. Factors Effecting the Amount of Electro-Deposition

1. **Time.** Time is directly proportional to the quantity of electro-deposition. So, we can say that more mass will be deposited in more time and less mass will be deposited in less time if the other conditions remain same.

2. **Efficiency.** Greater is the efficiency, greater is the quantity of metal deposited for a given time.

3. **Current.** The value of current is also directly proportional to the mass of metal deposited. Greater is the current, greater is the quantity of metal deposited while other conditions remaining the same. But, if we increase the current beyond a certain limit which is fixed for different metals separately, the metal deposited will be of different colour such as blackish, which is known as burnt metal.

4. **Strength of solution.** If the strength of solution is more, then the mass of metal deposited will be more as compared to the dilute solution of electrolyte if the other conditions remain the same.

6.10. Factors Governing the Better Electro-Deposition

In order that the deposits have a fine grained and smooth appearances, suitable condition have to be provided. The factors which effect the appearance of the deposited metal are discussed in detail below :

(*i*) **Current density.** At low value of current density, the ions are released at slow rate, therefore the deposit will be coarse and crystalline in nature. At higher values of current density, the quality of deposit become more uniform and fine ground. If the current density is so high that it exceeds the limiting value for the electrolyte then a spongy and porous deposit is obtained. The recommended current densities per square metre for copper, gold, nickel and silver are 33 to 44, 110 to 330, 110 to 220 and 33 to 66 amperes respectively. **Current density means current per unit Area** *i.e.* **current density = I/A (Amp/m^2).**

(*ii*) **Electrolytic concentration.** These factors depend upon the current density because by increasing the concentration of electrolyte higher current density can be achieved. Increase of concentration of the electrolyte tends to give better deposit and it is generally recommended to use concentrated electrolyte.

(*iii*) **Temperature.** The temperature of the electrolyte is different for different metals to have better deposit, for example in chromium plating the temperature is maintained at 35°C. In copper plating at 50°C and in nickel plating it is between 50°C to 60°C.

(*iv*) **Addition of Agents.** The quality of a deposit is improved by the presence of an additional agent which may be an organic compound such as gums, rubber, alkalies and sugar etc. The additional agents are supposed to be absorbed by crystal nuclei and prevent their growth into large crystals.

(*v*) **Nature of the electrolyte.** The smoothness of the deposit largely depends upon the nature of the electrolyte, for example, silver from silver nitrate solution forms a rough deposit while that from cyanide solution forms a smooth deposit. The different electrolytes used for different metals are given below :

Types of electroplating	Electrolyte used
Copper plating	(*a*) $CuSO_4$ (Copper sulphate) + H_2SO_4 (*b*) Cuprous cyanide and sodium cyanide
Silver plating	Na $[Ag(CN)_2]$ Sodium argnto cyanide
Gold plating	Na$[AU(N)_2]$ Soudium Auro cyanide
Nickel plating	$(NH_4)_2$ SO_4. $NiSO_46H_2O$ (Nickel ammonium Sulphate)
Chrome plating	$Cr_2(SO_4)_3$ chromium sulphate.

(*vi*) **Nature of the metal upon which deposit is to be made.** These factors influence the growth of crystals since it is believed that the operation of crystals is in continuation of these in the base metals.

(*vii*) **Throwing power.** The throwing power of an electrolyte may be regardless as the quantity which produces a uniform deposit on a cathode having an irregular shape. Fig. 6.4 shows an article *ABCD* having an irregular shape. Although the potential applied between any point on the article and the anode is the same, yet the surface *AB* is far away from the anode as referred to surface *CD*. Therefore resistance to the current path from anode to *AB* is more than that in case of *CD*. Hence the thickness of the deposit on the surface *AB* will be smaller. The throwing power can however be improved in two ways :

(*a*) By increasing the distance between the object and the anode so that the relative variation in resistance between the anode and the different parts on the object is reduced.

(*b*) By using some colloidal matter which results in the increase of current density. For this reason solution of the cyanides of metals usually give a greater throwing power than solution of the sulphates.

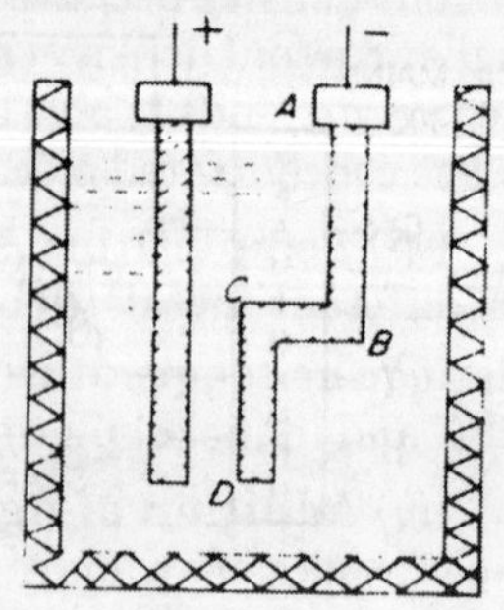

Fig. 6.4.

6.11. Electrical Equipment

Power required. Electric power required for plating processed has two important characteristics; it must be D.C. and the voltage required is generally very low

(1 – 16 volts) depending upon the nature of electrolyte and the rate at which the plating is to be done. The amperages are rather high say, upto 200 amperes or even more. The current density used varies with different metals. With high current density the deposit may be crystalline or powdery, and will not adhere well to the cathode. What is required is to regulate the current so that the deposited metal may be smooth and adherent. Much higher current densities can be employed if the solution be rapidly circulated by means of a pump or agitated by blowing in air.

Now-a-days, the generation and transmission of electric supply is done at high voltage A.C. Therefore, there must be some means to step down the high A.C. voltage to require low voltage and then converting this voltage to D.C. Two entirely different types of apparatus are used which are given below :

(*i*) The rectifier.

(*ii*) The motor generator set.

(*i*) **Rectifiers.** Rectifier sets, as used in electroplating actually consist of two separate apparatus (although generally housed in the

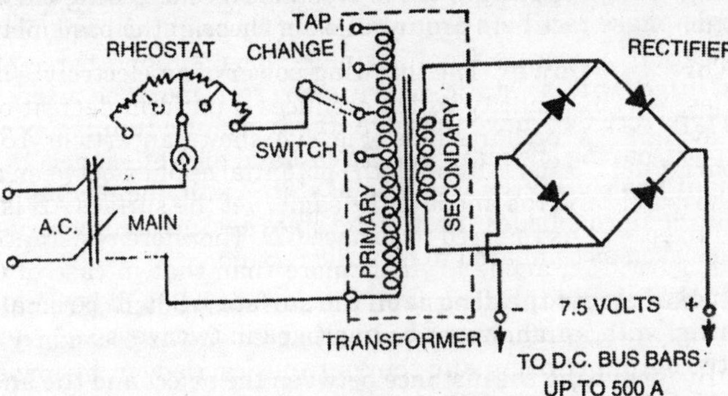

Fig. 6.5. (*a*) Method of controlling output from rectifier, for variable plating

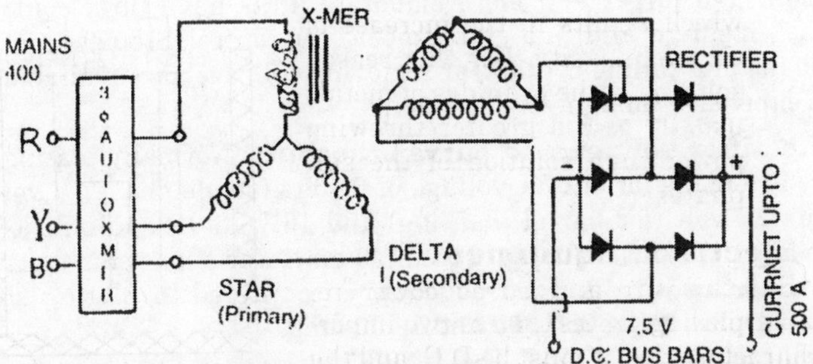

Fig. 6.5. (*b*) 3-ϕ supply system.

same case). These are (*a*) a transformer used to reduce the voltage of the A.C. supply to a suitable value for connecting to the rectifler (*b*) a plate rectifier to convert low voltage alternating current to low voltage direct current (*c*) the necessary control gear. The transformer takes alternating current from the mains, and delivers alternating current at a low voltage approximately to the low voltage required for the plating. This alternating current is then converted to low voltage direct current by means of the rectifier. A schematic diagram showing method of controlling output from rectifier, for variable plating requirements is shown in Fig. 6.5 (*a*) and (*b*).

The transformer consists of a laminated iron core with two sets of windings. A primary winding is connected to the high voltage A.C. supply (either single, two or three-phase), and a secondary coil of few turns is connected to the rectifier, and carries heavy current at low voltage.

The mercury arc rectifier cannot be used as it has low efficiency and low output, therefore, these days metal rectifiers of selenium type are used. They are totally immersed in oil, which protects them from the dust and fumes of the plating shop, thus greatly increasing their life time. Single-phase rectifiers are unsuitable for bright chrome plating. If it is desired to operate from a single-phase supply, three-phase rectifiers fitted with a static phase converter must be used. The selenium type rectifiers have the advantages of static running, occupying less space, having low maintenance cost and high efficiency. Recent development has shown the use of solid state rectifying devices using germanium or silicon diodes. These solid state devices occupy very-very small space even as compared to metallic rectifiers.

Control of electroplating rectifier. An electroplating rectifier is usually fitted with an ammeter and voltmeter on the secondary side showing the output current, and the control is carried out by means of tapping switches on the primary side or, on large units by means of an auto transformer. Smaller sets usually have two control switches "coarse" and "fine" giving a 11 control steps (not counting the "off" position) and larger and more elaborate units have three-controls - 'coarse', 'medium' and 'fine' giving a 63 - step control, Stepless controls, by means of a sliding contact, or a moving coil, are also available, but more expensive though very convenient.

Rectifier voltage and current outputs. Normally plating rectifiers are made for output voltage of 6 volts (single phase), 7 volts (3 phase), 12 volt (for barrel planting) and 16 volts (for sulphuric acid anodising). Output amperages from 20 amp. Jeweller's sets to 10,000 amp. But this size is not frequently encountered in plating shops, standard sizes being 250,500 and 1,000 amp.

(*ii*) **Motor generators.** The old fashioned method of converting mains current for plating use is the motor generator. This consists of an A.C. induction motor coupled with a dynamo (or "generator"), which produces a low voltage D.C. current. The arrangement for group supply to plating vats, using two dynamos and three bus-bars is shown in Fig. 6.6.

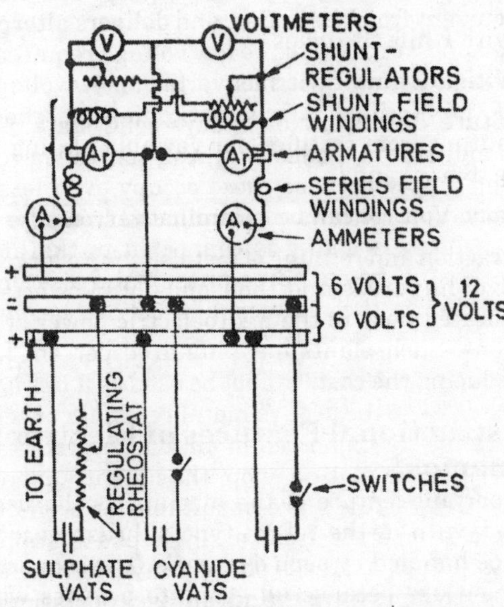

Fig. 6.6. Group supply to plating vats by dynamos and three bus-bars.

The magnetic field required by the dynamo can be generated either by a proportion of the generator's own output (self-excitation, compound wound types), or by D.C. current obtained direct from the mains by a separate small dynamo or by means of a small rectifier (separate excitation). Generally machines separately excited with a small rectifier are definitely preferable.

Resistance Boards (Control Panels). If a single plating vat only is operated, it is possible to connect it direct to the motor generator, or the rectifier. In fact, modern rectifiers, having suitable fine controls are often connected direct to a plating vat. Mostly, however, the source of plating current, whether motor generator or rectifier, is arranged to serve a number of vats and it is generally not possible to connect the vats in series with each other as they all require different plating voltages.

The usual arrangement in a standard plating shop is, therefore to have two heavy copper bars, known as 'bus bars' (originally probably "omnibus bars") connected to the source of current, and going round the shop along the wall, and the various plating vats draw current from these bus bars, in parallel to each other. In this arrangement a separate Control Panel, usually known in the plating trade as resistance board, is fitted near each vat, and the current is conducted from the bus-bars through the resistance board to the plating vat.

As far as their controls go, resistance boards are principally of two types :

(a) With Knife Switches.

(b) With a Rotary Switches.

Solid State devices. Now-a-days solid state devices employing germanium and silicon diodes or silicon controlled rectifiers are used for the power supply because these occupy even less space and have high efficiency. Voltage can be controlled very easily.

The extraction and refining of metals require large amount of power of the order of hundreds and thousands of kilowatts. Therefore such plants are located near to the hydroelectric power stations, or atomic power stations. Such plants have hundred percent load factor which assists in reducing the cost.

6.12. Constructional Features of Electroplating Dynamos

The important feature of the machine is the use of double commutator, and, with it, the robust type of brush gear, because of high currents to be handled by such dynamo's. The coppergraphite brushes are in common use because all graphite brushes will have too high resistance. Further, the mounting of the brush gear on the rocker necessitates careful insulation. Ample brush area must be provided to pick up the current without sparking.

Further, the speed of the machine should be as low as possible. High speed machines are cheaper in construction, but are much less efficient in operation. Owing to high voltage regulation, self-excited generators are not preferred as compared to separately excited generators.

6.12.1. Power supplies

For electroplating, DC generators are used over long period time. They are designed to give large currents between 6 and 20, V which is a low voltage. Generally, separately excited generators are used.

Now-a-days generators are no longer used, but rectifiers are used, as the supply available is AC. Hence, 0-35 V DC using transformer rectifiers and current ratings from few tens of amperes to 1000A or more are used. With the development of silicon controlled rectifiers or thyristors, which can control output voltages of power supplies, their use in power supplies for electro-chemical processes has increased. They are also compact and light in weight, even when cooling attachments are included. Voltage control is by using output transformers.

6.12.2. Types of rectifiers

There are three types of rectifiers in use; they are (1) silicon, (2) selenium, (3) germanium and (4) copper oxide. Silicon rectifiers are preferred. Germanium rectifiers, because of their tolerance for ambient temperatures, are less efficient due to a forward voltage drop of 1.1–1.2 V. They can be controlled easily, and due to this fact they are becoming more popular for electroplating works.

Germanium. They are very efficient, having a voltage drop per element at a maximum rated current of 0.5-0.6 only. They are however to be used for ambient temperatures below 45–50°C.

Selenium. This was mostly used till recently, but they have a limited life and therefore are used only for experimental work.

6.12.3. Power supply for extraction and refining metals

The power supply for extraction and refining of metals involves large amount of power to the extent of 100 MW and the current involved are 500-1000 A. Such plants are, therefore, located near hydroelectric plants, as they increase the load factor of hydroelectric stations as compared to steam power stations. Of course, this may require transport of raw materials, but this is not considered as a bottleneck.

6.13. Earthing

In case of rectifier unit fed from transformer, there is chance of low voltage side becoming accidentally connected to circuit of high potential. Whether it is safer to earth the vats or not is determined by the fact that it is difficult to insulate them and be certain that they will remain so. This suggests that the positive side of d.c. supply to the vats should be effectively connected to earth, the positive lead being selected to ensure that any leakage will be to the insulated pole. This is desirable, as any leakage will be such that it will deposit the plating material on to the work, which is the negative or cathode lead, also any tendency for a deposit to be carried to the frame of the vat is avoided. If negative of the supply were earthed, leakage current would be from anode to vat. This will deposit the metal on the lining of the vat.

By maintaining the vats at earth potential, the operator is protected from shock in the event of the low voltage supply becoming accidentally connected to a circuit of a potential which is of a value that would be dangerous.

6.14. Conditions for Electroplating

The following conditions must be satisfied for electroplating :

(*i*) The article to be plated must possess a chemically clean surface, *i.e.*, it must not have any rust, oil and dirt.

(*ii*) The article to be electroplated should form the cathode.

(*iii*) The metal to be deposited has to be in the electrolyte.

(*iv*) In order to keep the strength of the solution constant during electrolysis the anode must be of the metal to be deposited.

6.15. Plating Processes

Before any of the usual processes of electro-deposition upon metals can be restored to, it becomes necessary to impart to the surface of basic metal a uniformity which is essential for the production of a satisfactory finish upon the deposited metal. If the electroplating is done on an unclean and greased surface, the metal deposited will not be well adherent to the basic metal and is likely to peel off. Cleaning is necessary in the beginning to all the jobs even if dirt and grease etc. are not visible to the naked eye on the job's surface. There is a wide difference between the rough metal which emanates from the manufacturer and the finally finished product seen in the show room. A schematic diagram is shown in Fig. 6.7, giving the various plating processes.

These advanced processes are roughly, though not exactly, divided into two groups :

1. Mechanical groups which are usually accomplished with dry chemicals. The most important are :

   (*a*) Sand-blasting (*b*) Grinding
   (*c*) Polishing and (*d*) Buffing.

2. Chemical and electrolytic processes.

(*a*) **Sand Blasting.** This method of oxide or scale removal is dependent on the friction obtained on the surface of the work by articles of fine sand or steel balls projected upon it by means of a jet of compressed air. This method is particularly effective where there is no

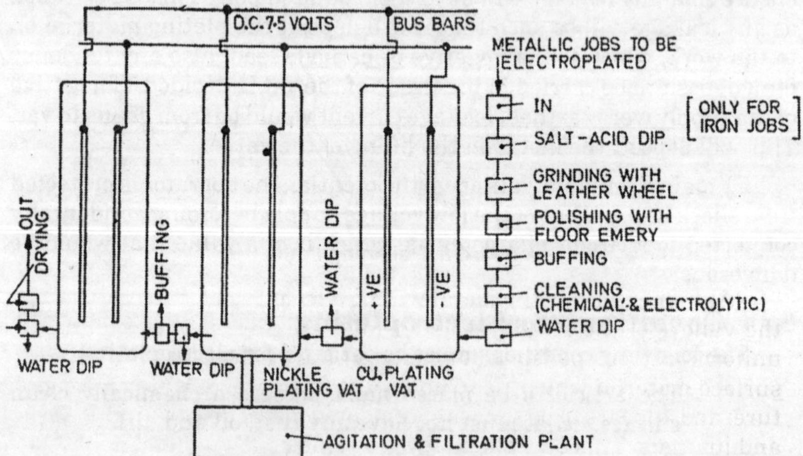

Fig. 6.7. Schematic diagram showing various processes of plating, *i.e.*, copper, nickel and chromium.

considerable amount of grease to interfere with the action of the sand. It is also often used to obtain a fine matt surface or otherwise smooth work.

Owing to the space occupied by the essential plant and its somewhat high cost, the sand blast method is not favoured, where small work has to be dealt with. However, in the cycle trade this method is found economical.

(*b*) **Grinding.** The removal of metal from a surface by the application of an abrasive is known as grinding. Grinding is used for rough cutting of metals.

The leather wheels of size 6", 8" and 10" dressed with glue and emery on its whole periphery are used. There are different grades or numbers of emery in common use. These vary from 6 to 200, these figures meaning the number of division per linear inch of the screen. For small work, No. 90 Emery is used. The speed of the grinding wheels is normally 2600 r.p.m.

(*c*) **Polishing.** Polishing is done by mops driven at peripheral speed of 2000 to 3000 metres/minute. Mops may be of leather, canvas or felt. The leather wheels are dressed with glue and No. 120 emery.

The mops shown in Fig. 6.8 should be run at high speeds. Lower speeds than mentioned above may cause the mop to drag and the article becomes hot and burnishing, and no polishing effect is obtained. Polishing is done for fine cutting for making a good ground for plating.

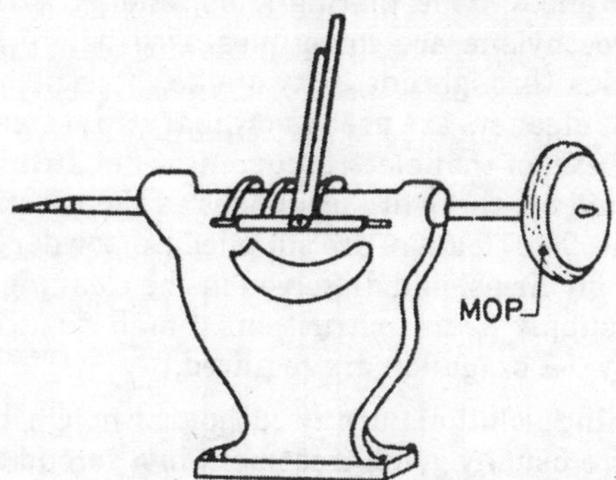

Fig. 6.8. Polishing Lathe.

At higher speeds the metal will flow from top of the scratches in to the valleys of the scratches under high pressure and temperature. A uniform lustrous polish thus results by the production of a thin film of surface material which, seen under the microscope, is free from structure, and appears to be a mass of cemented metal powder, hard, durable and lustrous. This superficial layer of broken down metal is called the Beilby layer.

(*d*) **Buffing.** This is a general type of polishing process, usually carried out with flexible wheels of cotton and a suitable abrasive or polishing composition. Buffing is done to make the surface of metal which is to be electroplated more fine. Buffing is also done on the same lathes in which polishing is done.

2. **Chemical and Electrolytic Processes.** The term chemical cleaning of metals applied to electroplating covers a number of operations, in which mechanical effects cannot be altogether excluded. For the most part, however, we are concerned with the removal, mainly by the processes of chemical action, of all those materials ordinarily present on the surface of metals subsequent to the processes of manufacture and mechanical cleaning.

A number of cleaning methods are used and these can be divided into (*a*) mechanical (scouring with lime etc.), (*b*) Physical cleaning, (*c*) Chemical cleaning, (*d*) Acid dips, (*e*) Electrolytic cleaning.

Physical Cleaning. In this process degreasing on the metal is done with solvents such as while spirit, solvent naptha, etc. at room temperature, or with trichloroethylene in specially constructed plants at higher temperatures or chemical, using alkaline cleaning solutions. The use of trichloroethylene has become wide spread with the progress of the industry. Trichloroethylene is a non-inflammable liquid, easy to handle and eminently suitable for preplating degreasing.

Chemical Cleaning. Chemical cleaning methods are of the greatest importance in the plating shop, and even though scouring and/or trichloroethylene are sometimes used before hand, the work invariably passes through one or two alkaline cleaners before being plated. Alkaline cleaners are used in ordinary mild steel tanks, similar to plating vats, except that electric current is not always required, and even if this is laid on, there are no anodes as the vat itself in that case is used as anode. The cleaners are supplied (in powder form or jelly) by the plating supply firms, and dissolved in the cleaning tank in accordance with the supply firms instructions. Small quantities only of the cleaner, usually 6-8 oz./gallon are required.

Most cleaning solutions are used hot, although brass and other copper alloys are usually given treatment in a second alkaline cleaner which contains cyanide, and is only used moderately warm, as the cyanide would decompose at higher temperatures. Special mild alkaline cleaners free from caustic soda must be used for zinc, tin and aluminium.

Acid dips. The chemically cleaned metals must be subjected to an acid dip in order to remove firms of oxides etc., which would interfere with the plating process. Standard acid dips usually required in a plating shop are 10% hydrochloric acid, 10% sulphuric and dipping acid or aquafortis (a mixture of nitric acid and sulphuric acid, with various additions).

The hydrochloric acid dip, also known as muriatic acid dip or simply 'muri' is mainly used for steel work and 10% sulphuric acid for non-ferrous metals. Dipping acid is required for the bright dipping of brass castings. Hydrofluoric acid dips are also sometime used, especially for zinc die castings.

Sulphuric acid and hydrofluoric dips may be kept in a lead lined container, but the hydrochloric acid and dipping acids require ceramic (earthenware) or other non-metallic materials or linings.

Electrolytic-cleaning. This method has of recent 'years come into extended use. It is a simple process. In a suitable liquid container in an iron tank which is made of anode, the work is suspended as a cathode and heavy current put upon it. Large volumes of hydrogen is given off and the grease is quickly removed.

6.16. Rinses

In all plating shops, special rinsing tanks with running water are extensively used. The ample supplies of water as well as drainage facilities are essential. The parts are rinsed or water dips between every process such as, cleaning, and dipping and plating, in order to prevent, carry over one processing solution to the other.

6.17. Copper Plating

All the processes before electroplating has already been fully explained for all types of plating. Now only brief description will be taken as follows. In copper plating the article to be elcctroplated is cleaned first by rough and then with fine emery wheels to get a polished surface. To remove grease and oil from the surface, it is further immersed in a boiling solution of caustic potash (KOH) and then washed in water, to put finally in copper plating baths which are two types.

(*i*) Acid type bath solution is made of

Copper sulphate ⟶ 150 to 200 gm

Sulphuric acid per

1000 c.c. of solution ⟶ 25 to 37 gm

Current density ⟶ 200 to 400 amp/sq-metre

Temperature ⟶ 25° to 50°C

Voltage ⟶ 4 to 9 volts

Time ⟶ 2 to 3 seconds.

The deposit obtained by this method is thick and rough and needs polishing.

(*ii*) Cyanide type bath solution is made of :

Copper cyanide ⟶ 25 gm

Sodium cyanide ⟶ 28 gm

Sodium carbonate ⟶ 6 — 12 gm

Sodium bi-sulphate per 1000 c.c. ⟶ 6 gm

Current density ⟶ 40 to 150 amp/sq.metre

Temperature ⟶ 35° to 40°C.

This gives thin and smooth deposit. In both of the above baths pure copper forms the anode.

Copper plating is mostly used for iron articles to prevent rusting. It is also employed as a preliminary coating for silver and nickel plating.

6.18. Nickel Plating

In nickel plating the article to be plated is cleaned and polished. Before plating it is put in a solution of potassium cyanide, then picked in dilute acid and finally water dipped.

Nickel does not adhere well to iron to steel articles, and furthermore, if after being plated upon steel, the article becomes scratched, the steel rusts, and the rust, getting beneath the nickel film, causes it to peel off. It is, therefore, recommended to first coat the iron with a film of copper, which, being a soft metal, is not readily removed by scratching. The nickel is then deposited upon the copper plating.

For brass and steel articles nickel bath solution is made of :

Nickel sulphate ⟶100 gm

Nickel chloride ⟶ 12 gm

Boric acid ⟶ 12 gm per 100 c.c.

Temperature ⟶ 20°C — 30°C

Current density ⟶ 10 to 20 amp/sq. metre.

For steel, brass, copper and zinc, nickel bath solution is made of:

Nickel sulphate ⟶ 180 to 240 gm (300 gm/litre)

Nickel chloride ⟶ 36 gm, 44 – 56 gm

Boric acid ⟶ 24 gm per 1000 c.c. 28 – 41 gm

Employed temperature ⟶ 40° to 65°C.

Current density ⟶ 250 — 500 amp/sq. metre

Voltage ⟶7 volt

Time ⟶ 30 minutes.

Anodes in both cases are used of pure nickel. For small electroplated articles, buffing is not necessary, but for large articles buffing should be done.

6.19. Chromium Plating

Chromium plating is employed where the surface of the article is to be protected from atmospheric corrosion. It gives a highly polished extremely hard coating, highly resistant to corrosion. The chromium plating bath solution is made of :

Chromic acid ⟶ 180 to 300 gm

Sulphuric acid ⟶ 2 to 3 gm per 100 c.c.

Temperature ⟶ 40° to 50°C.

Current density ⟶ 600 to 5000 amp/sq. metre

(Current density is higher for hard chromium plates than for decorative plating).

Time ⟶ 5 to 7 seconds.

Voltage ⟶ 7 to 9 volts.

Hard anti-monial lead electrodes are used as anodes if, necessary chromic acid is added to it. Steel with lead lining vats are employed for chromium plating.

It should be borne in mind that for chromium plating the electrolyte must be heated and the fumes are to be exhausted by exhaust fans.

Note. Nickel life depends upon the copper coating underneath the nickel coating and chromium life depends upon the nickel coating under the chromium plating. The filtration and agitation plant is only used in nickel plating.

6.20. Dry Out

After copper, nickel and chromium plating the article is washed in hot water and dried and buffed in case of nickel plating with fine polishing mops made of cotton rotated at high speeds. The drying out should be done in warm saw dust. The wood dust should be free from resinous matter.

Alternatively, the work may with advantage be dried in a current of warm air.

6.21. Combined Agitation and Filtration Plant

Modern workshop practice has thoroughly demonstrated the need for agitated solutions, with continuous filtration to ensure the complete freedom from dust, which otherwise so prominently encourages rough deposits system, some simple and other more elaborate now cater for this dual function. An example is shown in Fig. 6.9 which comprises a motor driven rotary air compressor, air reservoir with air filter, and the necessary lead filter coil and perforated agitating coils.

The solution is drawn from the far end of the plating tank and passed through the filter funnel containing a felt filter bag which is readily changed when necessary, and then returned to the opposite end of the tank.

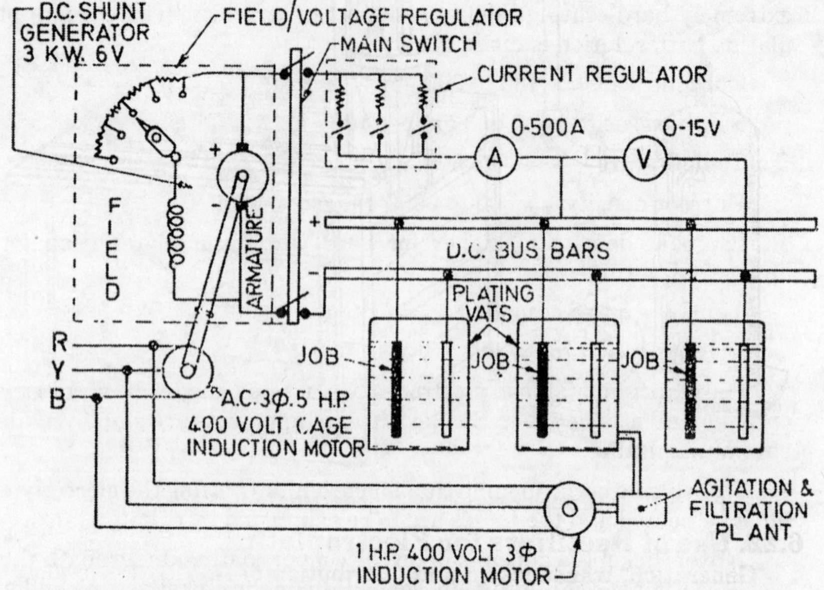

Fig. 6.9. Agitation and filtration plant.

The air reservoir contains some water through which the compressed air must pass, thus eliminating dust and oil before the air enters the solution. Regulation valves on the air supply provides control for both the filtration and agitation.

Acceleration of deposition with increased current density can also be obtained by applying a receprocating motion to the cathode bars, or alternatively and better, to both rods.

The overall arrangement of electroplating system is as shown in Fig. 6.10.

1. Normally M.S. jobs are copper plated.
2. Anodic cleaning means to put the job which is to be electroplated is made Anode and the tank is Cathode (-ve).
3. Cathodic cleaning job is to be made cathode (-ve) and the tank is made anode (+ve).
4. In plating, job is made cathode (-ve) and the material of which plating is done made anode.

Except in case of chrome, lead is used as anode. Because chrome has not his own metal whereas lead is insoluble.

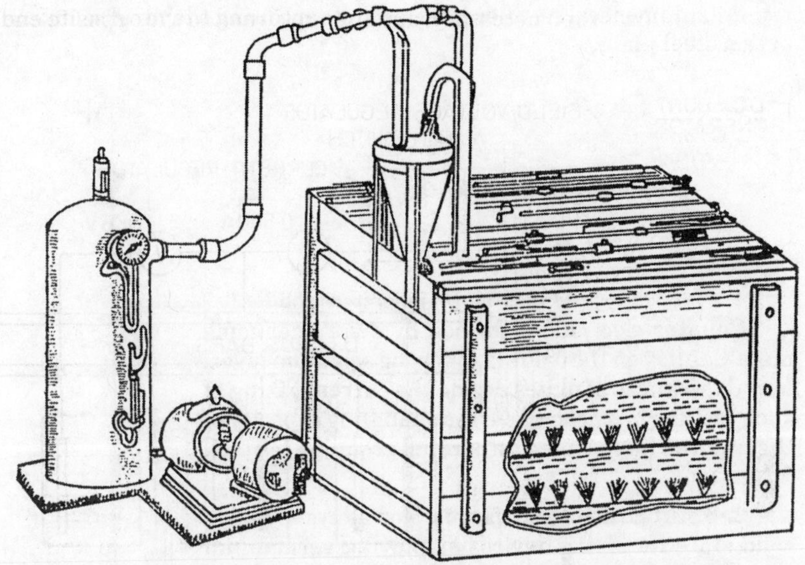

Fig. 6.10. D.C. power supply with generator for electroplating.

6.22. Use of Rectifiers for Electroplating

Generation, transmission and distribution of A.C. power is more economical than that of d.c. so instead of going into the purchase of D.C. electroplating generators, we use rectifiers which convert the A.C. power available by electricity boards into D.C. There are mainly 3 types of rectifiers used for this purpose :

1. Metal rectifier
2. Solid state rectifiers
3. Controlled rectifiers.

1. Metal Rectifiers : It is formed by coating a thin layer of copper oxide over a heavier metallic plate made of copper. It gives the property of unilateral conduction. It has low maintenance cost, occupies less space and has high operating efficiency.

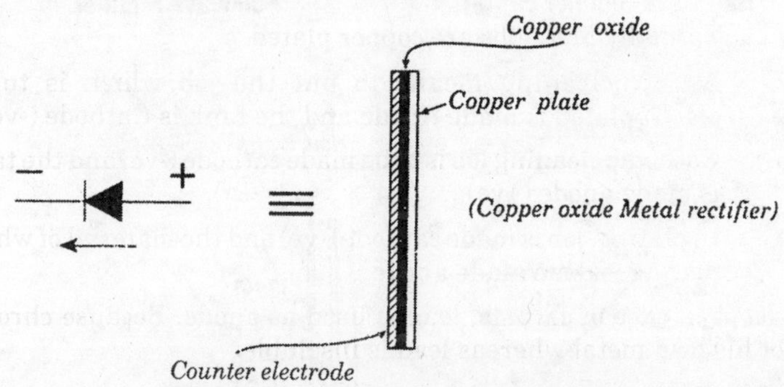

Fig. 6.11. (a) Copper oxide metal rectifier.

Selenium metal rectifier is formed by depositing a layer of selenium over a steel plate.

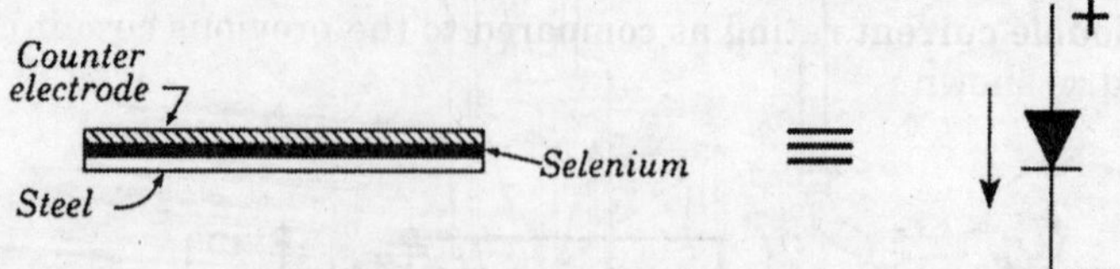

Fig. 6.11. (*b*) Selenium rectifier.

Counter electrode is formed by a spray of some metallic alloy on the oxide layer or the selenium layer. Metal thickness of plates decide the current rating of such rectifiers. A desired current handling capacity of the unit can be obtained by parallel grouping in stack formation.

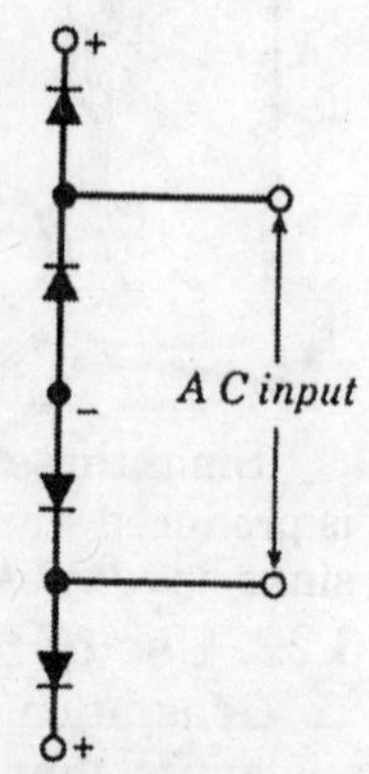

Fig. 6.11. (*c*) Stack circuit.

2. Solid State Rectifiers : Nowadays we have solid state rectifying devices employing germanium and silicon diodes. They occupy very small space as compared to metal rectifiers. They are also termed as silicon power diode rectifiers and can be used as half wave rectifier, full wave rectifier and full wave bridge rectifiers as shown in the fig.6.12

Step down t/f
A C mains
R_L
D.C. O/P

Half wave rectifier circuit.

Step down t/f
A.C. mains
D.C. O/P
R_L

Full wave rectifier.

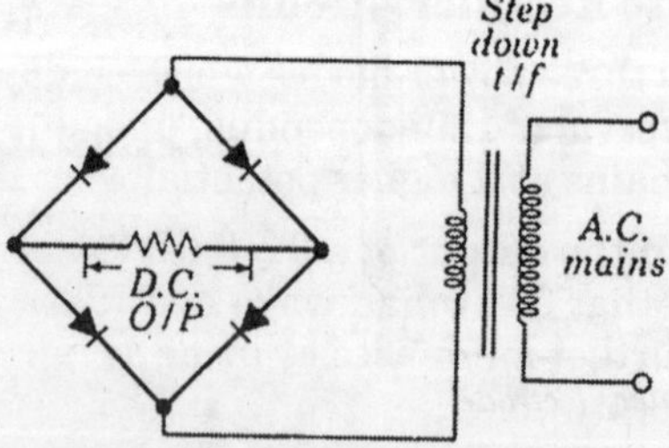

Fig. 6.12. Full wave bridge rectifier

Electroplating circuits require low and steady supply voltage but very high currents. We have silicon power diodes which can safely handle currents from 100 Amp to 400 Amp. Larger currents can be obtained by parallel grouping of such diodes. A full wave rectifier circuit with double current rating as compared to the previous circuit can be formed as shown :

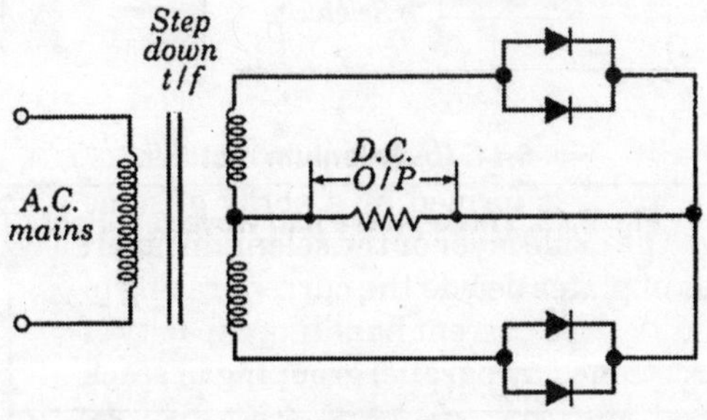

Fig. 6.13.

Since these power diodes handle very high current, so a lot of heat is produced which has to be dissipated. For this purpose metallic heat sinks have to be used.

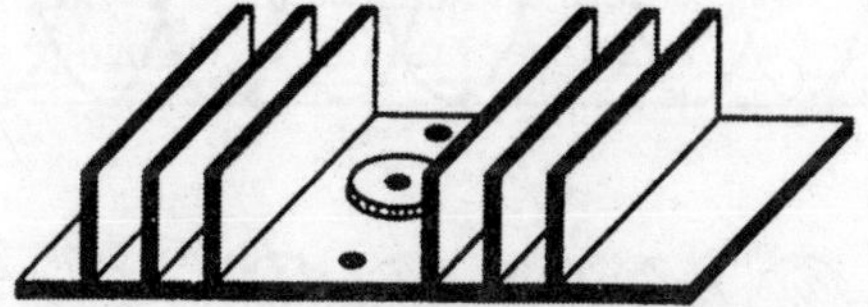

Fig. 6.14. Heat sink used with power diodes.

Sometimes rectifier unit along with the transformer is mounted in oil cooled metallic tanks to avoid excessive heating and derating.

For electroplating plants, the rectifier used should have low recovery period, low leakage current and less ripple factor. Ripple factor is very high in single phase half wave rectifiers so we use three phase rectifiers to minimise it. Use of single phase rectifiers result in un-balancing the distribution line. So we prefer to the polyphase rectifier circuit.

6.22.1 Polyphase Rectifier Circuits

This circuit uses three power diodes which conduct periodically in a sequence for an interval of 120° depending upon the time for which a particular phase remains at a higher potential w.r.t the other two.

Ripple content in the output of a 3 ϕ half wave rectifier ckt is far less as compared to that of 1 ϕ half wave and full wave circuits A.C. components can be further minimised by using 3ϕ full wave rectifier ckt as discussed below :

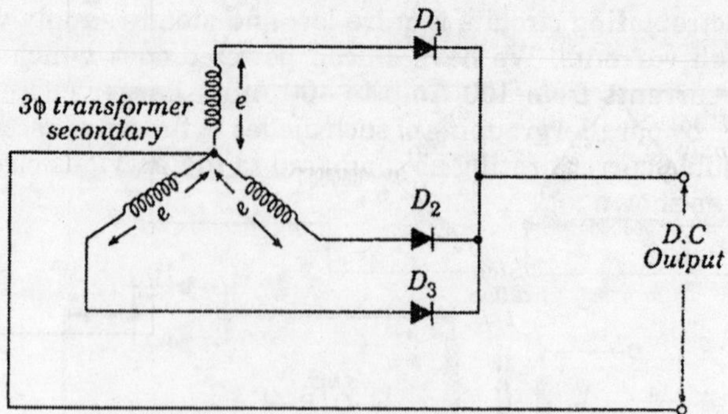

Fig. 6.15. Three phase half wave rectifier.

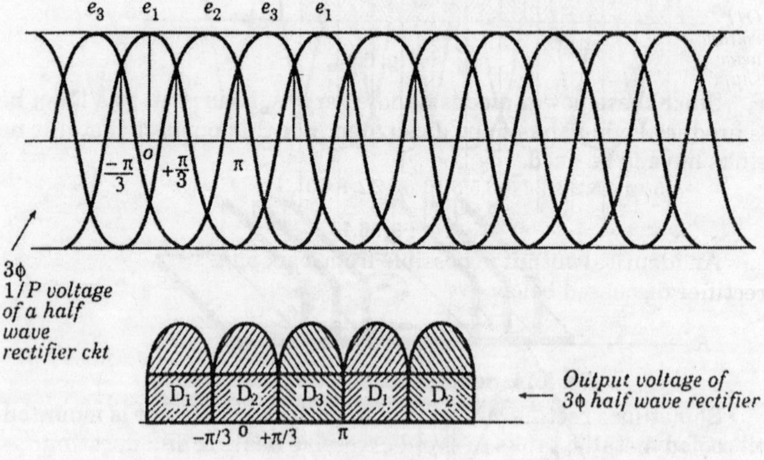

Fig. 6.16.

6.22.2 3 Φ Full wave Bridge rectifier circuit

In this circuit, primary is connected in delta and six diodes are connected in secondary as shown. When phase a is positive, D, conducts for 120°. During first 60°, it is in series with D_5 whose cathode is most -ve on phase band during next 60°, it is in series with D_6 whose cathode is most -ve in phase C. Similarly D_2 & D_3 conduct for 120° each thus producing a full wave rectified output which has minor ripple contents tappings may be provided on the 3 Φ transformer to control output voltage when input supply undergoes variations. Current control in the plating circuit can be accomplished through controlling resistance mounted on panel.

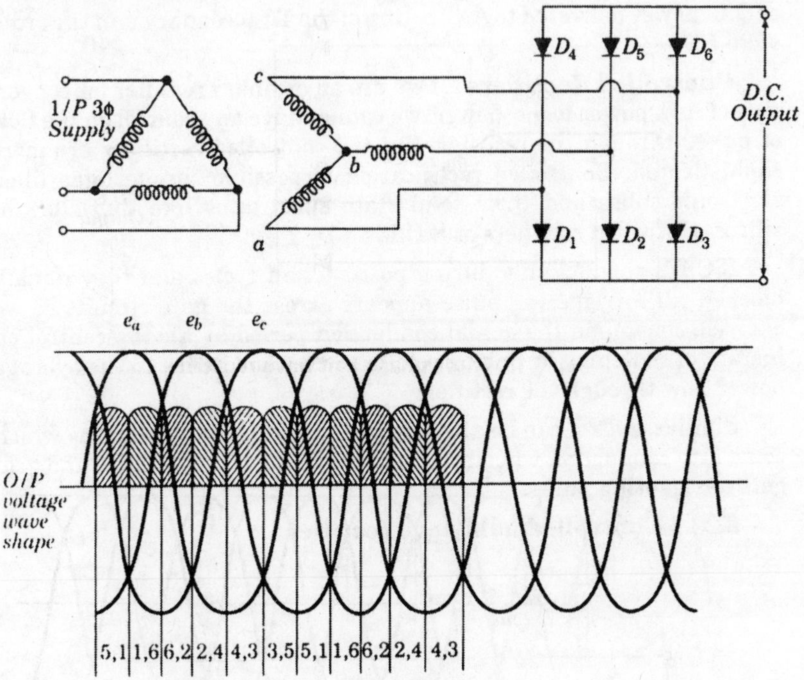

Fig. 6.17

An identical output is possible from a six phase six diode half wave rectifier discussed below.

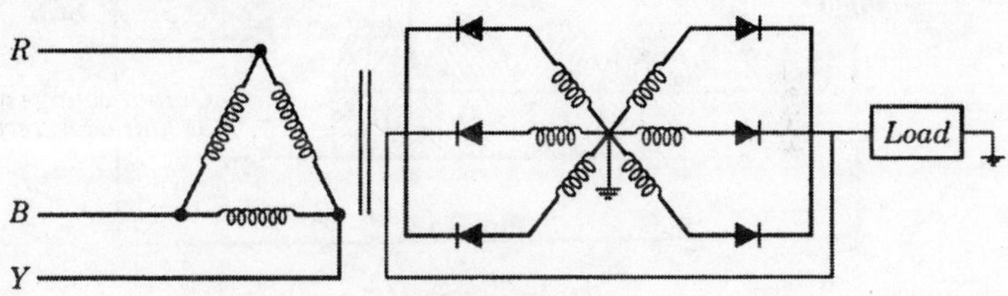

Fig. 6.18.

6.22.3 6 Φ half wave rectifier circuit

A six phase six diode half wave rectifier circuit shown above gives an output with minor ripple content.

It is clear that if the number of phases are increased, the output wave becomes smoother. Moreover current per diode in this case becomes half than that of a 3 Φ full wave rectifier six phase configuration does not require an independent/separate source of 3 Φ supply. It is possible to have the same directly from the supply mains. However secondary of the delta-star step down transformer should have centre tapped winding for each phase.

Ripple factor of a such a rectifier is as low as .035. Its ratio of rectification is 99.5% and transformer utilisation factor which is ratio of D.C. power delivered to A.C. rating of the T/f secondary is of the order of 95.5%.

Controlled Rectifiers : If we use an ordinary rectifier for conversion of A.C. power to d.c. power, we cannot have any control on the flow of power through the rectifier. But the controlled rectifiers are more sophisticated. Controlled rectification is possible through gas filled electronic tubes and their solid state substitutes like thyristors or silicon controlled rectifiers or S.C.R.s.

S.C.R.s conduct only during positive half cycles and they remain blocked till a triggering pulse appears across the gate circuit. So we exercise control on the actual conduction period by having control on instant of switching on during positive cycles and hence on the average power flow through the rectifier.

Similar control can be attained in thyratrons or phanstrons which come into conduction when its anode is +ve w.r.t. the cathode and grid gets a triggering pulse.

6.22.4 Controlled half wave rectifier

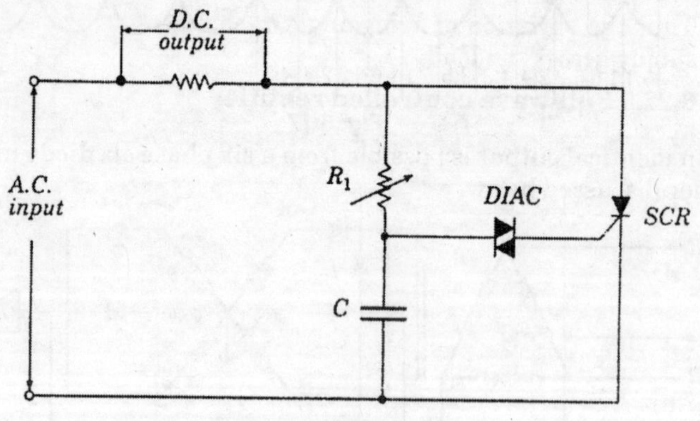

Fig. 6.19

As the A.C. power supply is switched on, S.C.R. can not conduct during negative half cycles. Even during positive half cycle, the SCR can conduct only when its gate is given proper potential so by varying R_1, the firing of SCR can be delayed as required.

At the start of +ve half cycle, current begin to flow through capacitor C and resistance R_1, so the capacitor starts charging. As soon as capacitor potential reaches breakover voltage of the diac, the diac becomes conducting providing enough current to the gate of SCR. This turns on the SCR. (At instant α) Charging time of capacitor depends on the value of R_1.

Average value

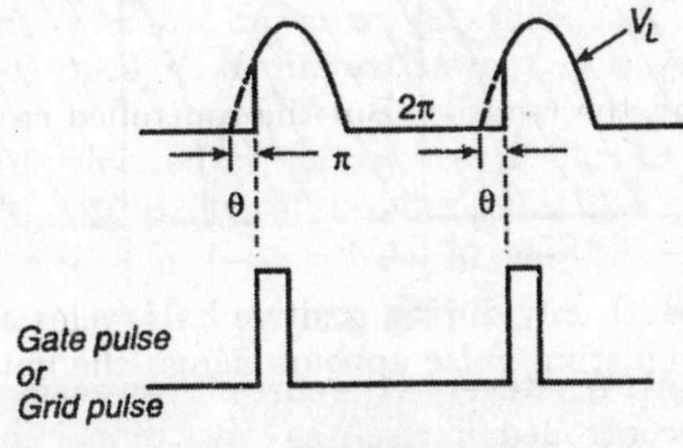

Fig. 6.20.

The average output voltage can be calculated as

$$V_{av} = \frac{1}{2\pi}\int_0^{\pi} v.\,d\alpha = -\frac{V_{max}}{2\pi}\,|\cos\alpha|_0^{\pi}$$

$$= \frac{V_{max}}{2\pi}(\text{Cos }\alpha \text{ Cos }\pi) = \frac{V_{max}}{2\pi}(1 + \text{Cos }\alpha)$$

Thus the variation of firing angle α, changes the average value of d.c. supply across the load.

6.22.5 Full wave controlled rectifier

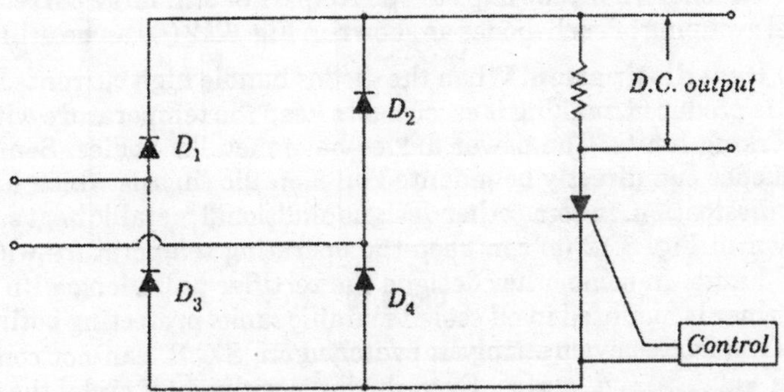

Fig. 6.21

A full wave controlled rectifier circuit shown above uses four diodes D_1, D_2, D_3 & D_4 which convert a.c. power to full wave rectified d.c. power. Then a S.C.R. is used to exercise control on this power. The conduction of S.C.R. is delayed from zero voltage of each half cycle by control circuit by an angle α. This delay angle α can be varied from the control circuit and we get an output as shown in the fig. 6.22.

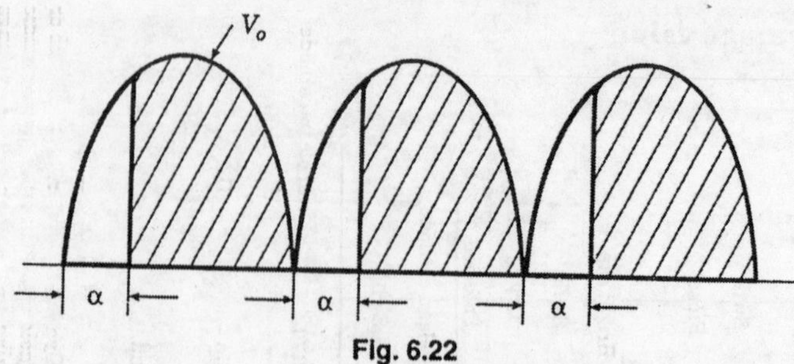

Fig. 6.22

Control panel : Whatever be the current/power source, the plating current has to be controlled precisely so every connection from the bus bar should get its current supply through a parallel combination of low resistors which can be connected in series with the plating electrodes through a set of Rotary switches or knife switches. In addition to this other accessories such as voltmeters, ammeters and main switch are also provided on the control panel.

6.23. Some Important Consideration while Selecting Rectifier Circuits for Electroplating Plants

(*i*) **Current Rating.** The electroplating circuits require low and steady supply voltage but remarkably high currents. As far as current rating is concerned silicon power diodes are available which can safely handle currents from 100 Amps to 400 Amps. For still large currents a parallel grouping of such diodes as shown in Fig. 6.12 (*f*) can be utilised.

(*ii*) **Heat dissipation.** When the diodes handle high currents a lot of heat is produced, making it essential to keep the temperature within safe working limits. The power diodes have metallic bodies. Some of these diodes can directly be mounted on metallic chassis which helps in heat dissipation. In some other designs additional metallic heat sinks as shown in Fig. 6.12 (*e*) can keep the operating temperature within desired limits. In some other designs the rectifier unit along with the transformer is mounted in oil cooled metallic tanks protecting both the devices from excessive heating and *derating*.

(*iii*) **Recovery period.** An other consideration in the *recovery period*. It becomes a problem if the operating temperature remains high. Lower this period better is the performance of the rectifier.

(*iv*) **Leakage current.** A large leakage current is a drawback for any rectifier circuit. The problem gets multiplied when a parallel grouping of the diodes (of low current/power rating) is adopted for the supply circuit.

(*v*) **Ripple factor.** The most serious consideration is the presence of A.C. component in the rectifier output. It is very high in single phase half wave rectifiers (R.F. stands at 1.21 for half wave 0.48 for full wave

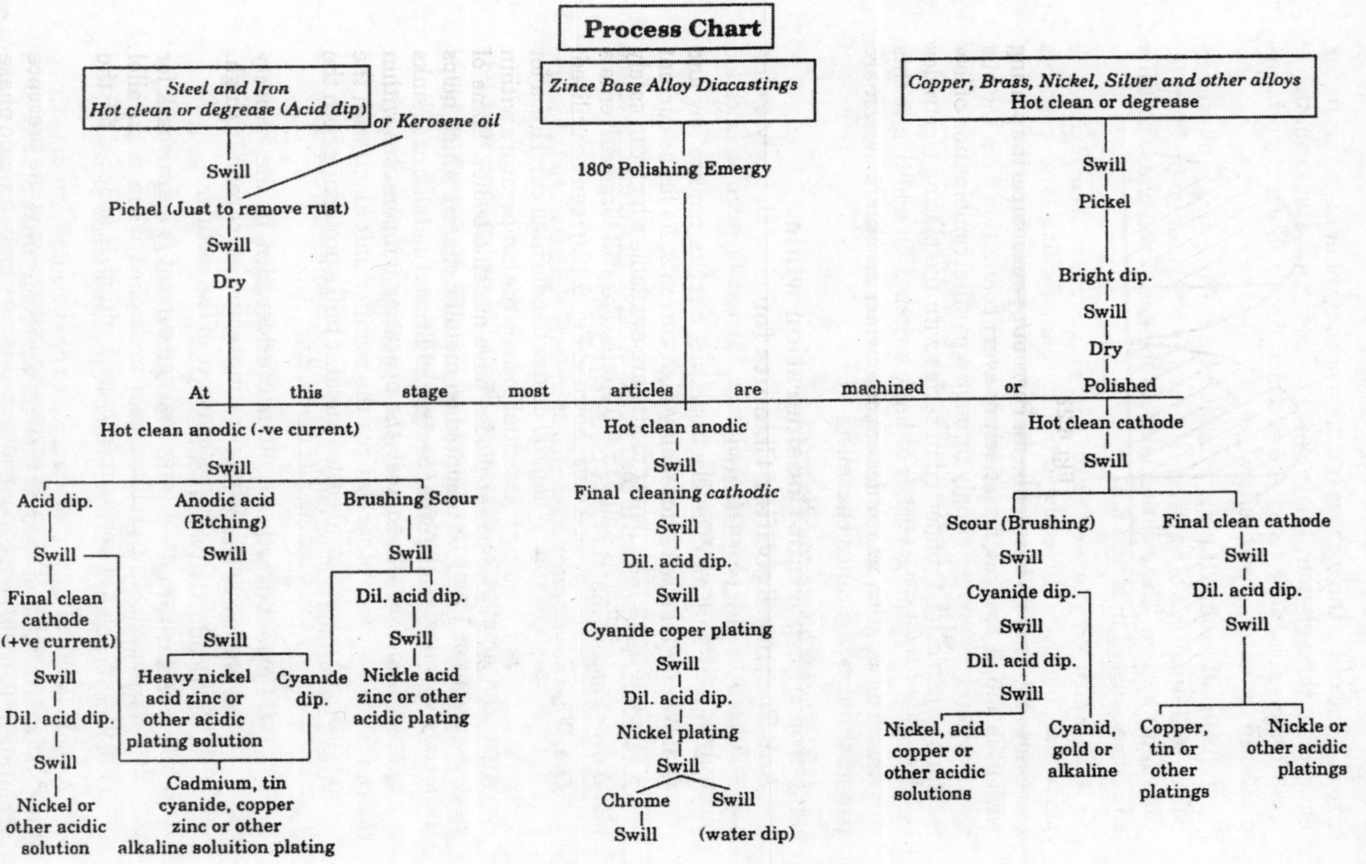
Process Chart
Steel and Iron
Hot clean or degrease (Acid dip)
or Kerosene oil
Swill
Pichel (Just to remove rust)
Swill
Dry
Zince Base Alloy Diacastings
180° Polishing Emergy
Copper, Brass, Nickel, Silver and other alloys
Hot clean or degrease
Swill
Pickel
Bright dip.
Swill
Dry
Polished
At this stage most articles are machined or
Hot clean anodic (-ve current)
Swill
Acid dip.
Swill
Final clean cathode (+ve current)
Swill
Dil. acid dip.
Swill
Nickel or other acidic solution
Anodic acid (Etching)
Swill
Swill
Heavy nickel acid zinc or other acidic plating solution
Cyanide dip.
Cadmium, tin cyanide, copper zinc or other alkaline soluition plating
Brushing Scour
Swill
Dil. acid dip.
Swill
Nickle acid zinc or other acidic plating
Hot clean anodic
Swill
Final cleaning cathodic
Swill
Dil. acid dip.
Swill
Cyanide coper plating
Swill
Dil. acid dip.
Nickel plating
Swill
Chrome
Swill
Swill
(water dip)
Hot clean cathode
Swill
Scour (Brushing)
Swill
Cyanide dip.
Swill
Dil. acid dip.
Swill
Nickel, acid copper or other acidic solutions
Cyanid, gold or alkaline
Final clean cathode
Swill
Dil. acid dip.
Swill
Copper, tin or other platings
Nickle or other acidic platings

circuit). The use of three phase rectifiers not only minimises this factor, but current through individual diode is reduced when star formation of transformer secondary feeds the rectifier circuit. Such circuits are discussed in the articles to follow.

(*vi*) **Unbalancing the distribution line.** When single phase rectifiers feed hundreds of amperes, it becomes a regulation problem for the supply system. The solution is again the use of polyphase rectifier circuits discussed in article 6.27.

(*vii*) **Reduced power factor.** The single phase transformers in particular, but even three phase transformers feeding large currents cause low p.f. This makes it necessary to use power capacitors along with such supply sources, for p.f. improvement.

SHORT ANSWERS—QUESTIONS

Q.1. *What is an electrolyte ?*

Ans. The solution which decomposes when an electric current is passed through it is called electrolyte.

Q.2. *What is an electrolytic process ?*

Ans. The process of decomposition (ionisation) of electrolyte (or a salt solution) by the passage of electric current through it is called electrolytic process or electrolysis.

Q.3. *What is an electro-deposition ?*

Ans. Electro-deposition is the process of depositing a coating of one metal over another metal or non-metal electrically.

Q.4. *What do you mean by electro-chemical equivalent of a substance ?*

Ans. The amount of substance deposited on passing a steady electric current of 1 amp. for one second through its solution is called the electro chemical equivalent of a substance.

Q.5. *Mention the factors on which the quality of electro-depositions depends :*

Ans. The various vectors on which the quality of deposition depends are :

(*i*) Current density (*ii*) Temperature

(*iii*) Nature of electrolyte

(*iv*) Conductivity and concentration

(*v*) Polarization (*vi*) Additional agents

(*vii*) Throwing power.

Q.6. *What is meant by throwing power and how it can be improved?*

Ans. It is the ability of electrolyte to produce uniform deposit on the article of irregular shape and is one of the important characteristics of plating or depositing bath.

Q.7. *What is an electroplating ?*

Ans. It is the art of depositing a superior metal over an inferior metal by means of electrolysis of an aqueous solution of suitable electrolyte.

Q.8. *What do you understand by 'Reverse current process' in electroplating ?*

Ans. Reverse current process in electroplating means that at regular intervals plating current is reversed for a second or so.

Q.9. *What for electroplating is done ?*

Ans. Electroplating is done to repair worn-out parts of the machinery. Sometimes it is also done for ornamentation and decoration purposes.

Q.10. *Why the article to be electroplated is thoroughly cleaned and degreased before putting it for electroplating ?*

Ans. If the article or object to be electroplated is not thoroughly cleaned and degreased, the deposit formed may not be well adherent to the base metal and is likely to peel off.

OBJECTIVE TYPES QUESTIONS WITH ANSWERS

A. Fill in the Blanks:

(*i*) In electrolysis, the salt solution is ionised into ----

(*ii*) In electrolytic processes, the electrical energy is converted into ---- energy.

(*iii*) The hydrozen and metal ions carrying positive charge are known as ----

(*iv*) The other name for negative ions is ----

(*v*) Power supply required for electrolytic processes is ---- current and at very ---- voltage.

(*vi*) Electro-deposition is obtained on ----

(*vii*) The voltage required to pass the necessary current through an electrolyte cell is of the order of ----

(*viii*) The name of the electrolyte and anode metal used for gold plating is ---- and ---- respectively.

(*xi*) Tho chemical deposition due to flow of current through an electrolyte is ---- proportional to the quantity of electricity passed through it.

(*x*) The current efficiency is defined as ----

(*xi*) The ratio of theoretical energy required to the actual energy required for depositing a given quantity of metal is known as ------------

(*xii*) The weight of metal deposited over the cathode is directly proportional to the quantity of ---- passed through the electrolyte.

(*xiii*) Electrolytic processes are conducted at ---- voltage and ---- current.

(*xiv*) Silver with valancy one and atomic weight 108 shall have equivalent weight ----

(*xv*) The solutions which decomposes when an electric current is passed through them are called ----

Answers

(*i*) ions (*ii*) chemical (*iii*) Cations (*iv*) Anions

(*v*) Direct, low (*vi*) cathode (*vii*) 1-2 V

(*viii*) double cyanide of potassium and gold, stainless steel

(*ix*) directly

(*x*) Actual quantity of substance liberated or deposited Theoretical quantity of substance

(*xi*) energy efficiency

(*xii*) electricity (*xiii*) low, High (*xiv*) 108

(*xv*) Electrolyte.

B. True or False Statements with Answers:

(*i*) In electrolytic processes, the electrical energy is converted in to heat energy.

(*ii*) Chemical plants load factor is about 50 percent.

(*iii*) In case of copper plating, material used is copper sulphate and nitric acid.

(*iv*) Addition of agents to electrolyte reduces the resistance of electrolyte.

(*v*) Power supply needed for electro-deposition of metal is A.C.

(*vi*) The quantity of electricity needed for electroplating is measured in watts.

(*vii*) In case the low current density is employed the deposit will be coarse.

(*viii*) For electroplating, low current, high voltage d.c. supply is required.

(*ix*) The voltage required to pass an electric current through the electrolyte depends upon the resistance of the electrolyte.

(*x*) When sulphate solution is used, the copper plating will be better.

(*xi*) Throwing, power can be improved by increasing the distance between anode and cathode.

(*xii*) The plant which is used for the manufacture of chemicals at large has a load factor 100%.

Ans.

(*i*) False	(*ii*) False	(*iii*) False	(*iv*) True
(*v*) False	(*vi*) False	(*vii*) True	(*viii*) False
(*ix*) True	(*x*) False	(*xi*) True	(*xii*) True.

C. Fill in the blanks of the following sentences :

(*a*) In electrolytic processes, the electrical energy is converted into .... energy.

(*b*) The solution of a salt used for electrolytic process is called ....

(*c*) The electrode connected to positive terminal is called ....

(*d*) The weight of a substance liberated during electrolysis is proportional to proudct of ....

(*e*) In the equation $W \propto I \times t$, the sign of proportionality can be replaced by a constant .... called ....

(*f*) The ratio of the theoretical value of energy to practical value is called ....

(*g*) The current required for electro-deposition is .... than that required for extraction of metals.

D. Write 'Yes' or 'No' after the following :

(*a*) The electrode connected to positive terminal of D.C. supply is called cathode.

(*b*) The quantity of electricity required for electroplating is measured in watts.

(*c*) The voltage required to pass an electric current through electrolyte depends upon the resistance of the electrolyte.

(*d*) Low voltage, current D.C. supply is required for electroplating.

(*e*) Addition of agents to electrolyte, reduce the resistance of electrolyte.

3. The six phase rectifier circuit meant for electroplating needs.

(*a*) Special A.C. generator

(*b*) Normal three phase mains

(*c*) the system as (*a*) and (*b*) is just a theoretical possibility.

4. In the process of electroplating the circuitry involved is

(*a*) Polarised (*b*) Non-polarised

(*c*) Depends upon nature of plating (*d*) None out of above.

5. The existence of a counter electrode is observed some where is the

(*a*) Plating vats

(*b*) Electro-chemical cleaning baths

(*c*) D.C. supply sources

(*d*) Nothing as above is connected with the plating system.

6. The capacitor bank installed in the rectifier system of any electroplating plant is meant for

(*a*) smoothing the effects of loads variation.

(*b*) minimising the ripple content of the D.C. supply.

(*c*) to improve power factor and line regulation of the mains feeding the rectifier system.

(*d*) All as above.

7. The object undergoing surface plating work as :

(*a*) Cathode (*b*) Anode

(*c*) Depends upon nature of supply source :

8. The compound gensets used for the purpose are

(*a*) Differentially excited (*b*) Cumulatively excited

(*c*) Depends upon plating load.

9. The perferred vat polarity is

(*a*) Positive (*b*) Negative

(*c*) Zero potential without any polarity (*d*) An arbitrary choice.

10. The current efficiency in some electrolytic process is the ratio of :

(*a*) Actual current density to the calculated current density for a given mass of coating

(*b*) Mass of metal actually liberated to the calculated mass for a given current density.

(*c*) None out of above.

11. Mopping is an other name of

(*a*) Grinding (*b*) Polishing

(*c*) Abrasion (*d*) None out of above.

12. The filtration of electrolyte is necessary

(*a*) To remove impurities going into the electrolyte along with the main salt.

(*b*) To remove suspended salt particles and other suspended impurities from the electrolyte.

(*c*) To make the agitation process more effective

(*d*) Only some of the plating salts need filtration.

13. The process as above is

(*a*) Continuous (*b*) Intermittent

(*c*) Done only once before plating commencement

(*d*) Varies from electrolyte to electrolyte.

14. Spongy coating of electroplating speaks of :

(*a*) Under current density (*b*) Over current density

(*c*) Excessive electrolyte density (*d*) Poorer electrolyte density.

15. A process known as hall bloating is done :

(*a*) Prior to subjecting a surface to electroplating

(*b*) After plating a surface

(*c*) Done in between primary and secondary plating layers.

16. Ripple factor is being minimised

(*a*) By incorporating filter circuits along with the rectifier plant

(*b*) By using single phase bridge rectifiers

(*c*) With the help of multiphase rectifier unit without using additional filter network.

(*d*) Power capacitors are doing the need fuel.

17. The metal being deposited is available in form of

(*a*) Constituent of electrolyte (*b*) One of the electrodes

(*c*) Both as above (*d*) None out of above.

18. The shunt fields in such arrangement are :

(*a*) Connected in parallel to each other

(*b*) Connected in anti parallel to each other

(*c*) Connected in series across the outers

(*d*) Field of generator (1) excited by armature (2) output and vice-versa.

19. Chrome plating is done as

(*a*) Primary layer (*b*) Secondary layer

(*c*) Tertiary layer.

20. Polarisation on cathode surface can be checked through :

(*a*) Limiting current magnitude (*b*) Agitation of electrolyte

(*c*) Periodical reverse plating (*d*) All as above.

Answers

1. (*a*) chemical (*b*) electrolyte (c) anode (*d*) current and time (*e*) Z, electro-chemical equivalent (*f*) energy efficiency (*g*) less.

2. (*a*) No (*b*) No (*c*) Yes (*d*) Yes (*e*) Yes

3. (*b*) 4. (*a*) 5. (*c*) 6. (*c*) 7. (*a*) 8. (*b*) 9. (*a*) 10. (*b*) 11. (*b*) 12. (*b*) 13. (*a*) 14. (*b*) 15. (*a*) 16. (*c*) 17. (*c*) 18. (*d*) 19. (*c*) 20. (*d*).

EXERCISES

1. What is electroplating ? Describe it in details.
2. State and explain Faraday's laws of electrolysis.

3. What is meant by the term electro-deposition ?
4. What are the different applications of electrolysis ?
5. What is the necessity of electroplating ?
6. What factors govern the rate of electro-deposition process ?
7. Does electroplating take place with low voltage high current or high voltage low current ? Give reasons.
8. Why is it necessary to clean a job before electro-plating and how is it done ?
9. What is the effect of increasing the current density on the quality of electroplating ?
10. Which type of supply A.C. or D.C. is used in electro plating ? Give reasons.
11. What is the special constructional features of electroplating dynamo ?
12. Why buffing is necessary before electroplating ?
13. How electro-cleaning is done ?
14. What are the various factors on which quality of electrodeposition depend ?
15. A rectangular plate 20 cm × 10 cm is to be coated with nickel with a layer of 0.2 mm thick. Determine the quantity of electricity in ampere hours and the time taken for the process. Current density is $190 A/m^2$ and current efficiency is 60%. Specific gravity of nickel is 8.9. **[Ans. 35.1 A/hr ; 9 hours 13 minutes]**
16. Why positive side of D.C. supply meant for electroplating is earthed?
17. What do you understand by the term electro-deposition and what factors govern the deposition process ?
18. Determine the amount of silver deposited in 20 minutes when a current of 15 A is passed through the solution of silver nitrate. Assume E.C.E. of silver as 0.001118 gm per coulomb of electricity. **(Ans. 20.124 gm)**
19. Why polishing is done at higher speeds from 2000 to 3000 metres per minute ?

Questions Bank with Answer

Short Answer Type Questions.

Q.1. What is electrolysis?

Ans. The process of decomposition of electrolytes by the passage of electric current through them is called electrolytic process or electrolysis.

Q.2. What is an electrolyte?

Ans. The solution which decomposes when an electric current is passed through it is called an electrolyte.

Q.3. What is an electro-deposition?

Ans. The process of depositing a coating of one metal over another metal or non-metal by electrolytic process is callled the electro-deposition.

Q.4. What is electroplating? *[Dec 2008/07, June 2006]*

Ans. Electroplating is an art of depositing a superior metal over an inferior metal by means of electrolysis of an aqueous solution of a suitable electrolyte.

Q.5. What is meant by anodising ?

Ans. Anodising is a process of anodic oxidation in which a thin uniform passive film is produced artificially by the passage of electric current.

Q.6. What is galvanising?

Ans. It is the process of depositing layer of zinc on Iron.

Q.7. The deposit will be coarse when .... current density is used. Low/high. *[June 2009]*

Ans. Low.

Q.8. Low voltage high current AC supply is required for electroplating. State T/F. *[June 2009]*

Ans. False. For electroplating low voltage high current DC supply is required.

Q.9. Define throwing power.

Ans. Throwing power in an electrodeposition solution is defined as the ability of the solution to produce uniform deposits on irregular surfaces.

Q.10. What is meant by electro-chemical equivalent of a substance ?

Ans. Electro-chemical equivalent (ECE) of a substance is the amount of substance deposited on passing a steady electric current of 1A for one second through its solution.

Q.11. What is electroforming?

Ans. The reproduction of object by electro-deposition on some sort of mould form is called electroforming.

Q.12. In electrolytic refining of metals impure metal is made... Anode/ cathode.

Ans. Anode

Q.13. What is an electrode.

Ans. The plates or rods immersed in an electrolyte and connected to DC supply are called electrodes.

Q.14. What are cations?

Ans. These are + vely charged ions and move towards cathode.

Q.15. What are Anions?

Ans. These are negatively charged ions and move towards Anode.

Q.16. What is an Anode?

Ans. The electrode which is connected to positive terminal of supply is called Anode.

Q.17. What is a Cathode?

Ans. The electrode which is connected to negative terminal of supply is called Cathode.

Q.18. What are the power supply requirements of an electrolytic process?

Ans. 10 to 12 V DC and current from 100 to 200 amperes.

Q 19. Why the article to be electro-plated is thoroughly cleaned and degreased before putting it for electro-plating?

Ans. In case the object to be electroplated is not cleaned and degreased, the deposit formed may not be well adherent to the base metal and is likely to peel off.

Q.20. For what purpose electro-plating is done.

Ans. Electro-plating is done for ornamentation and decoration purpose. Sometimes it is done, with a view to repair worn out parts of the machinery.

Medium/Long Answer Type Questions

Q.1. State and Explain Faraday's laws of electo-deposition or electrolysis.

Ans. Michal Faraday gave two laws of electrolysis.

First law: The deposition due to the flow of current through an electrolyte is directly proportional to the the quantity of electricity passed.

i.e., $m \alpha Q$

α I. t

= Z.I.t

Where m = mass of deposition

I = current through electrolyte
t = time
Z = a constant called electro chemical equivalent (ECE) of the substance.

IF I = 1 amp, and t = 1 sec
then Z = m

The ECE of a substance can be defined as the amount of substance deposited on passing a current of 1 amp for 1 sec through the electrolyte. It is given in "mg per coulamb" or "kg per coulamb".

Second Law: "When same quantity of electricity is passed through several electrolytes, the masses of the substances deposited are proportional to their "Chemical equivalents" or chemical weights.

If m_1 and m_2 are masses of the substances deposited by same amount of current, and E_1 and E_2 are their chemical equivalents or chemical weight then,

$$\frac{m_1}{m_2} \text{ a } \frac{E_1}{E_2}$$

Q.2. What is electro deposition or electrolysis? Explain the process.

Ans. When a chemical compound is dissolved in a liquid, the molecules of the electrolyte are split up into "Cations" (+ ve ions) and Anions {-ve ions) and these ions move freely in the solution. Now if electricity is passed through two electrodes in the electrolyte the cations are attracted to the negative electrode (cathode) and anions attracted towards the positive electrode (anode).

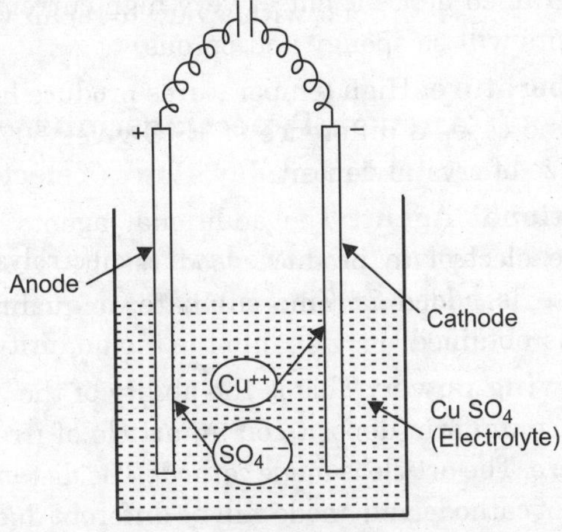

Fig. 6.23

For example, if copper sulphate is dissolved in water immediately it is broken into Cu^{++} and SO_4 ions. When electricity is passed through electrodes, the copper ions are attracted to cathode and sulphate ions to anode.

The each copper ion (Cu^{++}) on reaching cathode (-ve plate) takes two electrons and becomes metallic copper and is deposited here. Similarly each sulphate ions(SO_4^{--}) on reaching anode (+ve plate) gives up its two electrons and becomes "radical", and attacks water to make sulphuric acid ($H_2 SO_4$) and oxygen is liberated.

$$CuSO_4 \rightarrow Cu^{++} + SO_4$$

$$2H_2O + 2SO_4 \rightarrow 2H_2SO_4 + O_2$$

In this process, anode is consumed. The mass of Cu deposited on the cathode is equal to that removed from the anode.

The whole of the above process is called electrolysis and is the basis of all electrolytic processes.

Q.3. Discuss in brief the various factors governing electrodeposition.

Ans. The process of depositing a coating of a metal over another metal (or non metal) is called electrodeposition. It includes processes like electroplating, electrotyping, repair of worn parts etc.

The factors affecting quality of deposition are as under.

(i) **Nature of Electrolyte:** The electrolytes which can give complex ions such as cylindes, provide high quality deposits.

(ii) **Current Density:** High current density gives uniform and fine grained deposit, but at very high current density the deposits will be spongy and porous.

(iii) **Temperature:** High temperatures produce large crystals. In some cases a difference of 15°C gives 50% increase in the size of crystal deposit.

(iv) **Additional Agent:** The additional agents when mixed in the electrolyte produce good results *e.g.* If sugar or glucose is added in Zinc sulphate, a quality deposit of Zinc is obtained.

(v) **Throwing power:** This is the ability of the electrolyte to produce uniform deposits on an article of irregular shape and size. The article is made cathode The distance of various parts of cathode and anode will be different due to irregular shape of the article. Due to unequal distance, the resistance of current path through the electrolyte for various parts of the article will be different. But as the potential difference between cathode and anode will be same, the current density will be more at the nearer portion of the article. This will result in uneven deposits.

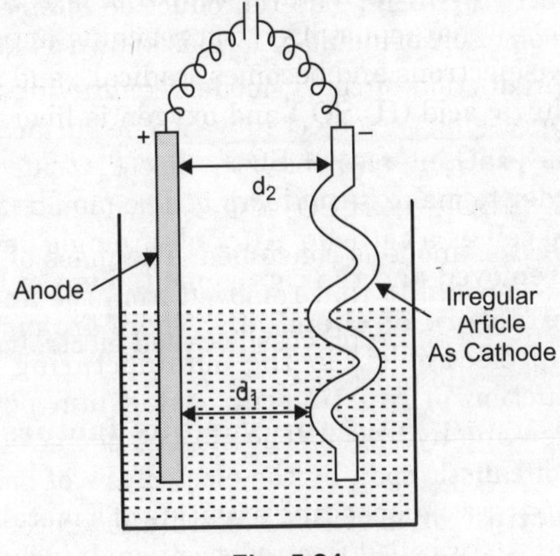

Fig. 5.24

The throwing power can be improved by :

(i) Increasing distance between anode and cathode.

(ii) Reducing voltage drop at cathode

Note that cyanide solution of metals have better throwing power than sulphate solution.

Q.4. Discuss in brief various applications of electrode position.

Ans. Various applications of electrodeposition are

(i) Electroplating: Electroplating is the process in which a superior metal is deposited on an inferior metal through electrolysis *i.e.,* nickel or chrome is deposited on articles of Iron. Gold and silver is deposited on copper. The purpose of electroplating may be repair of worn out articles, decoration or ornamentation. The article is made cathode, the metal to be deposited is made anode and electrolyte is the salt solution of metal to be deposited.

(ii) Anodising: Anodising is a process in which a film of oxide is produced artifically by the passage of current. This is very hard coating and is formed to protect the base metal from corrosion. It also has a decorative appearance. This is also used to produce decorative articles by giving separate colours. The article to be anodized is thoroughly cleaned and made anode. Carbon rod is made cathode. An electric current of 10-30 amperes per square meter is passed through an electrolyte of sulphuric acid and chromic acid. The required voltage is about 10-50 volts.

(iii) Electroforming: The reproduction of objects by electro deposition on some sort of mould or form is known as Electroforming.

In the reproduction of coins, models, engravings etc. a mould is made by impressing the object in wax. The surface of the mould which bears the exact impressions of the object is coated by powdered graphite in order to make it conducting. The mould is dipped in an electroforming cell as a cathode. After obtaining a desired coating, the article is removed and the core is melted out of the metal.

(iv) Manufacture of chemicals: There are many industrial applications of electrolysis in the manufacturing of chemicals such as production of caustic soda, potassium permanganate, chlorine, oxygen and hydrogen. The most important out of them is manufacture of caustic soda by the electrolysis of brine.

(v) Extraction of metals: Electrolytic processes are used in the extraction of metals like aluminium, copper, magnesium, sodium and zinc.

(vi) Refining of metals: In the process of refining of metals, the impure metal obtained from its ore from metallurgical reduction is made anode. The salt of the metal to be refined is made eletrolyte. Whenever potential difference is applied across the electrodes, the impure metal gets dissociated into positively and negatively charged ions. The positively charged ions consist of pure metal. Since positively charged ions are attracted by the cathode, these ions give their positive charge to the cathode. The metal deposited on the cathode has purity as high as 99.95 %.

Q.5. Write a short note on electroplating of non conducting materials. ***[June 2006]***

Ans. A layer of none conducting materials such as rubber can also be deposited on metals by the process of electrodeposition. The rubber obtained from trees in the form of latex consists of a very fine colloidal particles suspended in water Rubber solution is made electrolyte. The rubber particles are negatively charged. On electrolysis, the negative rubber particles dissociate from solution and get deposited on anode.

Q.6. Write a short note on anodising and its applications. ***[Dec 2008]***

Ans. Anodising is the process in which a film of oxide is produced artificially by the passage of current. The article to be anodised is thoroughly cleaned and made anode and a carbon rod is made cathode. An electric current of about 10 to 30 amperes per square meter is passed through electrolyte which is solution of sulphuric acid and chromic acid.

Application of Anodising :

(i) Protection of base metal from corrosion

(ii) Production of decorative articles by giving separate colours.

Q.7. Define electroplating. Explain the process. What are the equipment and accessories required in electroplating?

Ans. Electroplating is defined as the process of covering the articles made of inferior metals by a thin coating of superior metal such as iron with nickel, chromium, silver or gold. In this process, the article to be plated is made the cathode, electrolyte is the salt solution of metal of the same metal to be deposited or of some inert material. Anode is made of the metal to be deposited. Substances, other than metals can be electroplated by first coating their surfaces with powdered graphite. Material to be electroplated is first cleaned and polished before electro deposition is done.

Electroplating is done with one or more of the following objectives.

(i) Protection of metallic surfaces against corrosion.

(ii) Giving shining appearance to articles

(iii) Replacing the worn out material

(iv) Providing reflecting properties to reflectors.

Equipment and Accessories required in electroplating: The equipment and accessories used in electroplating are shown in Fig. 5.25.

DC power supply is used for which a dynamo, battery or rectified AC Current density employed is 120 to 150 A/m^2 at 10 to 12 volt of potential difference. Heating coil is required to maintain the suitable temperature. Temperature should be aroung 40-50° C which is measured by a thermometer. The tank or vat is made of steel with lining of lead.

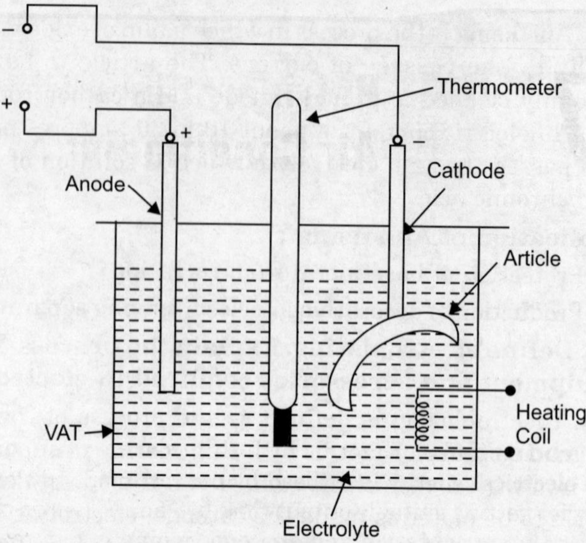

−
+
+
−
Thermometer
Cathode
Anode
Article
Heating
Coil
VAT
Electrolyte

Fig. 5.25

7 Refrigeration and Air Conditioning

7.1. Introduction

The electrical energy is utilised by converting it into one form or the other for getting suitable output from it. The utilisation of electrical energy for the refrigeration and air conditioning is fast replacing the old methods. Actually, the basic principles of refrigeration and air conditioning are concerned with mechanical engineering but since electrical energy is utilised for driving the machines used there, so it becomes necessary for the electrical engineers to have basic knowledge about this topic. In this chapter, we shall focus our attention mainly upon the working of refrigeration and air conditioning machines from the electrical point of view.

7.2. Refrigeration

The process of reducing the temperature of a body from the general level of temperature of the surroundings is called ***refrigeration.*** It is a process in which the heat is removed from the objects in a controlled way in order to cool them. It is achieved by the evaporation of a liquid refrigerant, thereby extracting heat from the medium to be cooled.

7.3. Air-Conditioning

Air conditioning means control of temperature, humidity, purity and movement of air. The working conditions and comfort for human beings is improved by airconditioning. The temperature of the building to be air conditioned is brought down during summer and increased during winter without causing any discomfort. The air after necessary control and adjustment is circulated throughout the building. This circulation of air is either effected naturally or forced by a fan. The humidity is maintained at about 60% being most suitable for human beings.

The use of air-conditioning is being extended fastly to multiroom buildings, restaurants and night clubs, trains and cars etc.

7.4. Refrigerator

This is a machine by which the temperature of substances is lowered by storing them in this machine. Generally, this machine has refrigerant (gas or fluid) which is circulated through its different stages to extract the heat from the substances. It is thermally insulated from the atmosphere.

7.4.1. Refrigeration System

The system used for lowering or producing low temperature in a body or a space, whose temperature is already below the temperature of its surroundings, is called the refrigeration system. In this case heat is being usually pumped from the low level to the higher one and it is rejected at high temperature.

7.4.2. Tonne of Refrigeration (TR)

In refrigeration and air-conditioning, the rate of heat transfer is measured in Tonnes. It is a practical unit of refrigeration and is defined as the amount of refrigeration effect produced by the uniform melting of one tonne (1,000 kg) of ice from and at 0°C in 24 hours. It is equivalent to heat extraction rate of 50 kcal./minute. While large commercial plants are designed in tonnes and domestic refrigerators are designed by the capacity (Volume) of the Cabinet. One tonne capacity of refrigeration plant needs approximately 3/4 kW output motor.

Illustration

A refrigerator is of 1 tonne capacity, it means its ice making capacity is 1 tonne in 24 hours. The medium house-hold refrigerator is of approximately 1/10 tonne capacity *i.e.* they can make 2000 lb. ice in 24 hours. (1 tonne = 2000 lbs.)

7.5. Principle of a Refrigerator

A refrigerator works on the principle of "*cooling caused by evaporation.*" To illustrate this, take a small amount of eather or spirit in a small test tube. Dip this test tube in a small beaker containing water as shown in Fig. 7.1. Then blow air in the test tube by a pipe, so that eather or spirit volatizes. It will be observed that ice is deposited near the base of test tube.

This happens because when the liquid in the test tube is evaporated it draws heat from its surroundings which lowers the surroundings temperature. Similarly, any volatile liquid can be circulated around the substances to be cooled and the same result can be obtained. One can also observe this by dipping finger in petrol or spirit. When the finger is taken out cooling is felt. This is because when petrol or spirit evaporates it draws heat from the finger thus cooling it.

7.5.1. Refrigerant

Refrigerant : It is a heat carrying medium which during their cycle (i.e., compression, condensation, expansion and evaporation) in the refrigeration system absorbs heat from a low temperature system (Cold body) and delivers it to a higher temperature system (Hot body).

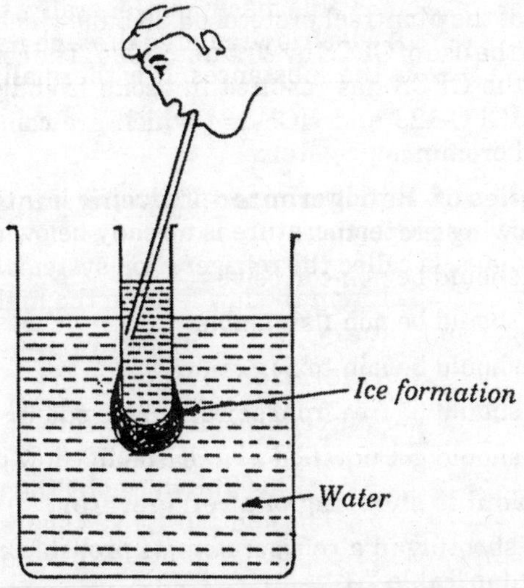

Fig. 7.1. Cooling Caused by Evaporation.

Any liquid that can be alternately liquified and vaporised can be used as a refrigerant. Thus CO_2 is used as a refrigerant in ships. But piping problems and machinery required for handling required other alternatives. Now due to its special characteristic Freon is used as a refrigerant. Its formula is CCl_2F_2 or $CHClF_2$. Other common refrigerant are :

1. Ammonia (toxic),
2. SO_2 (again toxic)

Due to the destructive nature of chemicals known as chlorofluorocarbons or CFCS trade name Freons which are extensively used in refrigerators and airconditioners a depletion of the Ozone layer in the stratosphere occurs. How these freons are harmful ? It has been found that CFCS are harmful because when they escape into the atmosphere and reach the stratosphere they are exposed to ultraviolet rays and are broken apart with the release of chlorine atoms. These chlorine atoms attack the ozone molecules (0_3) and release oxygen atoms forming chloromonoxide molecules. These chloromonoxide molecules are attacked by the free oxygen atoms detaching the chlorine atoms. Which in turn attack ozone molecules and the process continues. Thus the CFCS are harmful to the 'OZONE' in the stratosphere because a depletion of the ozone layer allows the 'Ultraviolet rays' reach the Earth's atmosphere and these rays attack the skin of human beings causing burns and cataract of the eyes or skin cancer. It is estimated the release of CFCS from use in refrigerators etc. causes a depletion of the ozone layer in the stratosphere in the course of next 100 years.

In turn the result is the exposure to ultraviolet rays with consequent disturbance of the Ecological system. The realisation has led to the signing of the Montreal protocol on 20.5.89 which will result in the reduction in the use of CFCS by 20% upto 1994. The continuous research in limiting the CFCS has resulted in ozone friendly substitutes like HCFC-22, HCFC-123 and HCF-134 which are compounds free from chlorine and bromine.

Properties of Refrigerants :- The refrigerants used should possess the following properties :

(*i*) It should be non-corrosive

(*ii*) It should be non-flammable

(*iii*) It should be non-toxic

(*iv*) It should be free from objectionable odour

(*v*) It should get liquified at a reasonable low pressure.

7.5.2. Trouble shooting of a refrigerator

Trouble shooting in a refrigerator, its probable cause and remedy are presented in Table 7.1.

Table 7.1. Trouble, cause, remedy of refrigerator

Trouble	*Probable cause*	*Remedy*
Unit not working	Fuse blown out or Low voltage supply Relay not working Service cord not good A lead might have broken to compressor trimmer	Replace fuse Change relay if necessary Change the cord Repair or replace broken leads
Refrigeration section hot	Door seal not good Inside light remains More food stuffs being packed inside Door opening operated more frequently Cold control knob at too warm	Adjust the door seal Test the light switch for fault Reduce the load stuff Advise the customer Keep the knob at cold
Refrigeration section is very cold	Refrigeration section and control knobs at the coldest position Defective air flow Air flow heater not good	Keep the control knob at the warmer position Change the control Change the heater
Freezer and refrigerator are too hot	1. Fan motor not in working condition and cold control set too hot or inoperative 2. Evaporator fins blocked with ice 3. Less refrigerant 4. Space behind cabinet is less 5. Door opened repeatedly	Check and replace the motor Defrost heater Detect any leak and refill Shift the cabinet for making more space behind the cabinet Instruct the customer to open the door only when required

Unit runs continuously	6. Door seal is bad less air around the cabinet Door seal in bad state Room temperature Door opened excessively	Repair the door Make space behind the cabinet Repair the door seal Provide good ventilation Instruct the customer to restrict opening
Noisy operation	Flooring is bad Cabinet is hot in level condition Compressor grounded mechanically	Repair the floor Level the cabinet Change compressor mounts.

7.6. Vapour Compression Cycle

The process of refrigeration is most commonly accomplished by the evaporation of a liquid refrigerant which extracts heat from the medium to be cooled. The refrigerant is generally a gas such as carbon dioxide (CO_2), Sulphur dioxide (SO_2), Ammonia (NH_3), Freon (F-12) [CCl_2F_2] and methyl chloride etc. The refrigeration cycle is chiefly composed of four stages *i.e.* compression, condensing, expansion and evaporation as shown in line diagram in Fig. 7.2. The purpose of refrigeration cycle is to remove heat from the refrigerant by again putting it into liquid form in order that it may be again used for extracting heat from the substances.

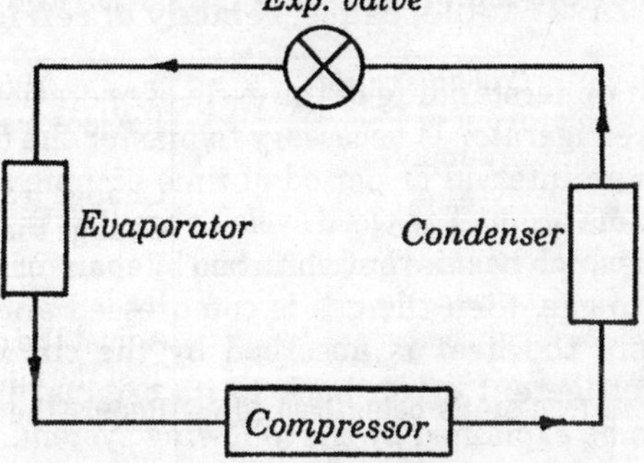

Fig. 7.2. Vapour Compression Cycle.

The refrigeration cycle is shown in Fig 7.2 and Fig. 7.3. Compressor and motor are assembled in a single unit in case of refrigerators and is known as sealed unit, which is shown at A in the Fig. 7.3. The refrigerant used is Freon-12 which has the property that it boils at a temperature of $-20°C$ when its pressure is 1.54 kg/cm^2. When this liquid is passed over any surface/surrounding whose temperature is more than $-20°C$, it absorbs heat from that surface/surrounding, thus cooling the surface and in turn getting vaporised. This is the principle used. The working of the cycle given above is as shown in Fig. 7.2.

When the supply to compressor motor unit *A* as shown in Fig. 7.3 is switched on, the compressor extracts refrigerant from the evaporator E in vapour form. The compressor after extracting the refrigerant in the form of vapour at low pressure (1.54 kg/cm^2) from the evaporator, compresses it to a pressure of 12 kg/cm^2. At this high pressure during compression, the temperature of the refrigerant vapour rises to 90°C. Whereas the boiling point of liquid at 12 kg/cm^2 pressure is 45°C.

This high pressure and high temperature liquid passes through the tubes of condenser B when the surrounding temperature might be something like 30°C. Thus heat is radiated by this liquid to the surroundings by the process of radiation and on doing so the temperature of the liquid falls to 40°C, pressure still remaining the same, *i.e.* 12 kg/cm^2.

This high pressure refrigerant liquid, now goes to the receiver *C* from where it is supplied to the evaporator E through capillary tube D which is about 1.5 metre long and has a diameter of 0.04″. As soon as this liquid refrigerant comes out from the capillary tube to evaporation, the pressure of liquid falls to 1. 54 kg/cm^2. At this pressure, the boiling point of liquid as already mentioned is – 20°C whereas the surroundings of the evaporator have quite a high temperature say + 20°C. Then the liquid freon refrigerant absorbs latent heat from the surroundings, thus cooling the surroundings and itself in turn getting evaporated to be again taken by the compressor for liquification and repetition of the cycle.

A thorough understanding of the cycle of operation *i.e.* what takes place inside a refrigerator is necessary to master the field of refrigeration. A cycle is an interval or period of time occupied by one round of events in the same order. The word cycle as applied here means a series of operations in which heat is first absorbed by the refrigerant, changing it from liquid to gas, then the gas is compressed and forced into the condenser where the heat is absorbed by the circulating air, thus bringing the refrigerant again back to its original liquid state. The whole cycle can be explained by the following system.

7.7. Vapour Compression System

Fig. 7.3 shows the different stages of vapour compression system of a typical refrigerating machine. The four main stages of the vapour compression cycle are discussed below :

(*i*) **The Vapourizing Process :** From the expansion valve, the low pressure refrigerant at low temperature enters the evaporator which is placed near or around the material to be refrigerated. Before entering the evaporator, the refrigerant is in the liquid state and when it absorbs heat from the material to be refrigerant, it is transformed from a liquid to a vapour state. This process is called **evaporation.**

(*ii*) **The Compressing Process :** The compressor draws the vapours from the evaporator through the return piping and then compresses these vapours until their temperature is raised above that of the condensing medium. The compressor is driven by external mechanical energy which may be supplied by an engine or electric motor. Generally, electric motors are employed for this purpose. This process is called **compression**.

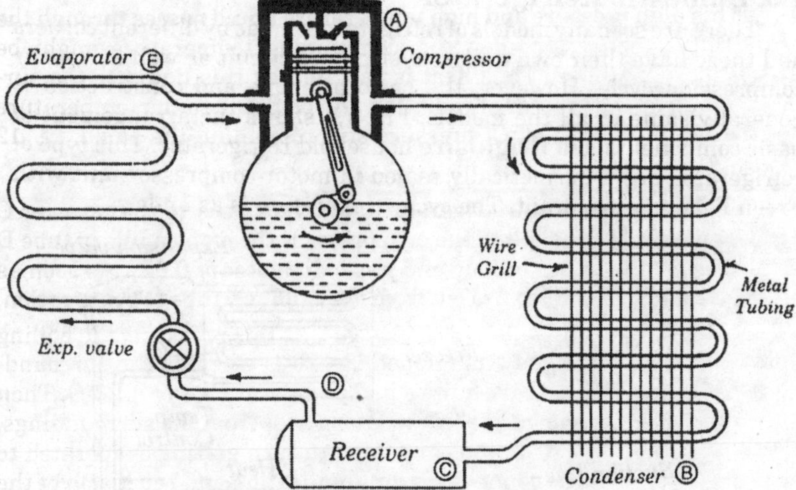

Fig. 7.3. Vapour Compression Process.

(*iii*) **The Condensing Process :** When the temperature of the vapour is raised above that of the condensing medium by the compressor, then the heat of vaporization will flow from the vapour to the condensing medium and so condense the refrigerant to a high pressure liquid. This high pressure liquid then flows to the receiver where it is stored until it is supplied to the cooling unit through the expansion valve. This process is completed in **condenser**.

(*iv*) **The Pressure Reducing Process :** As the compressor withdraws the refrigerant vapour from the evaporator, the cooling unit-must be supplied with more low temperature, low pressure refrigerant capable of absorbing heat. This is done by a liquid control valve known as an expansion valve. This valve reduces the pressure of the high pressure liquid from the receiver to a low pressure liquid capable of absorbing heat. This also maintains a constant supply of liquid in the evaporator and acts as a dividing point between high and low pressure sides of the system. The process is called **expansion**.

7.8. Practical Applications of Vapour Compression Cycle

The vapour compression refrigeration cycle is employed for the following purposes :

(*i*) Domestic refrigerator.

(*ii*) Water cooler.

(*iii*) Air conditioner.

(iv) Ice plant.

(v) Cold storage etc.

7.9. Domestic Refrigerator

There are so many models of refrigerators made by different corners and these have their own design of electrical circuit as well as vapour compression cycle. However, the basic principle and construction is generally same in all the models. Fig. 7.4 shows the arrangement of basic components in a **Frigidaire** household refrigerator. This type of refrigerator has a hermetically sealed in motor-compressor unit with Freon F-12 as refrigerant. The cycle of operation is as under :

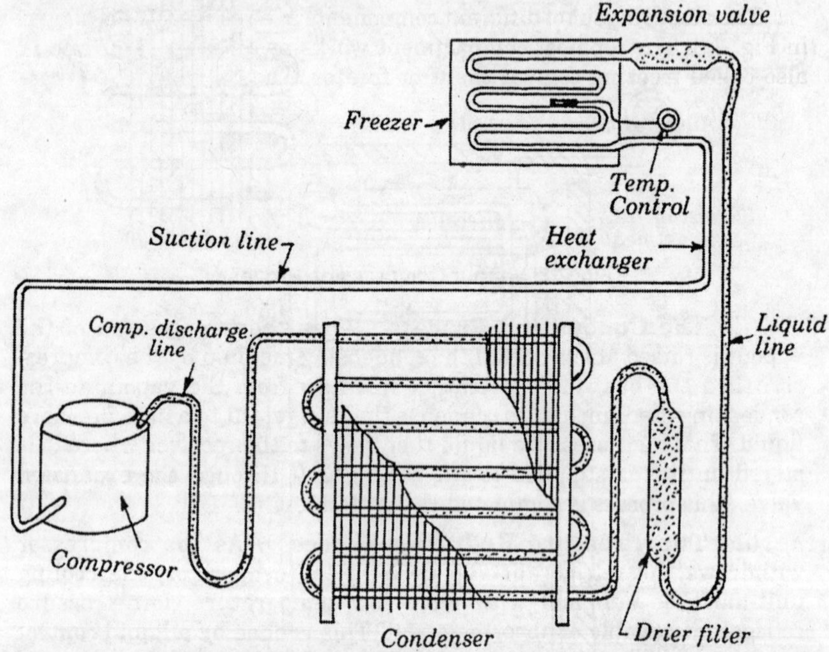

Fig. 7.4. Mechanical Circuit of Refrigerator.

Heat enters in cabinet through the insulation and is also carried in by the food, water and the opening and closing of the door of the refrigerator. This heat laiden air in the cabinet while passing over the freezer gives up its heat to it because the freezer is colder than the air passing around it.

While absorbing it, the refrigerant in the freezer vapourizes or changes in to a gas. This low pressure gas is then drawn through the suction line, pass the heat exchanger, and then into the compressor. In passing through the compressor the heat laiden gas is raised from the low pressure in the suction line to a higher pressure thereby raising its temperature.

It is then forced from the compressor through the discharge tube into the condenser. This heat laiden gas is distributed through the condenser where the heat is removed from the vapours within the condenser by means of natural air circulation over the outside of the condenser.

This removal of heat from the vapours causes them to get liquified and then flow to the bottom of the condenser. The liquid refrigerant is delivered to the freezer through a drier, liquid line and restrictor. The drier absorbs any moisture if present in the refrigerant. The supply of the refrigerant to the evaporator is controlled by the refrigeration demand.

The actual layout of different components in a refrigerator is shown in Fig. 7.5. The topmost compartment works as an evaporator and is also called freezing compartment or freezer. The control valve of the

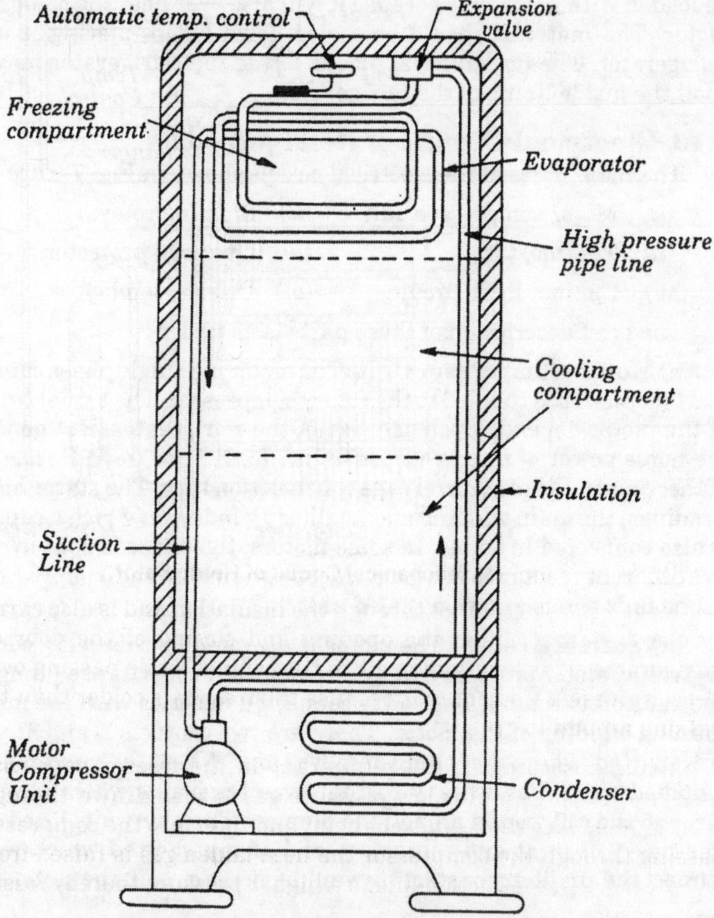

Fig. 7.5. Layout of Components in a referigerator.

temperature controlling device (thermostat) is placed here. The space below the freezer is generally divided into two compartments by placing shelves. Here the eatables, fruits, vegetables, cold drinks etc. are stored. A sealed motor-compressor unit is also installed at the bottom. The condenser is shown in the bottom compartment in the above figure but in domestic refrigerators the condenser is fitted at its back. In modern domestic refrigerators, a capilliary tube is used in place of expansion valve. The size of refrigerators is given in litres such as 165 litres, 200 litres and 300 litres.

In most of the refrigerators, automatic door closing mechanism is provided which helps in keeping the refrigerator cool at lesser running of the compressor. If there is no such mechanism, the door may remain open slightly and the outside air will circulate in the refrigerator and will prolong its cooling period. The compressor will have to run for longer periods thus over heating the motor. The refrigerator should not be loaded with hot things because it will also over heat the compressor motor. The material should be cooled down before placing it in the refrigerator. The hot material placed inside the refrigerator may also spoil the inside lining of the refrigerator.

7.10. Electrical Circuit of Refrigerator

The main parts of the electrical circuit shown in Fig. 7.6 are :

(*i*) Motor compressor unit. (*ii*) Starting relay.

(*iii*) Thermostat. (*iv*) Overload protector.

(*v*) Cabinet light circuit. (*vi*) Drier assembly.

The brief description of these parts is as under :

(*i*) **Motor-Compressor Unit :** The motor is a single phase capacitor start motor and is coupled to the rotory compressor. The actual capacity of the motor depends upon the size of the refrigerator. But generally the horse power of motors varies from 7/32 H.P. to 1/8 H.P. The rotor of these motors is of squirrel cage construction type. The stator has two windings, the main winding and auxiliary winding in which a capacitor is also connected in series. In some motors, the phase is split by using two different reactance windings. Mostly the motor and compressor are in one unit and is known a sealed unit.

(*ii*) **Starting relay :** The motor is operated by this relay which is also called motor protective relay. It essentially consists of a plunger on which a coil is wound. The coil is connected in series with the main or running winding of the motor. There are two contacts L and M which are shorted when the plunger moves up. When the motor is just switched on, it draws heavy current from the mains which also flows through the coil wound around the plunger. The plunger is lifted up by this heavy current and contacts L and M are shorted which in turn connect the auxilliary or starting winding in the circuit, the motor starts

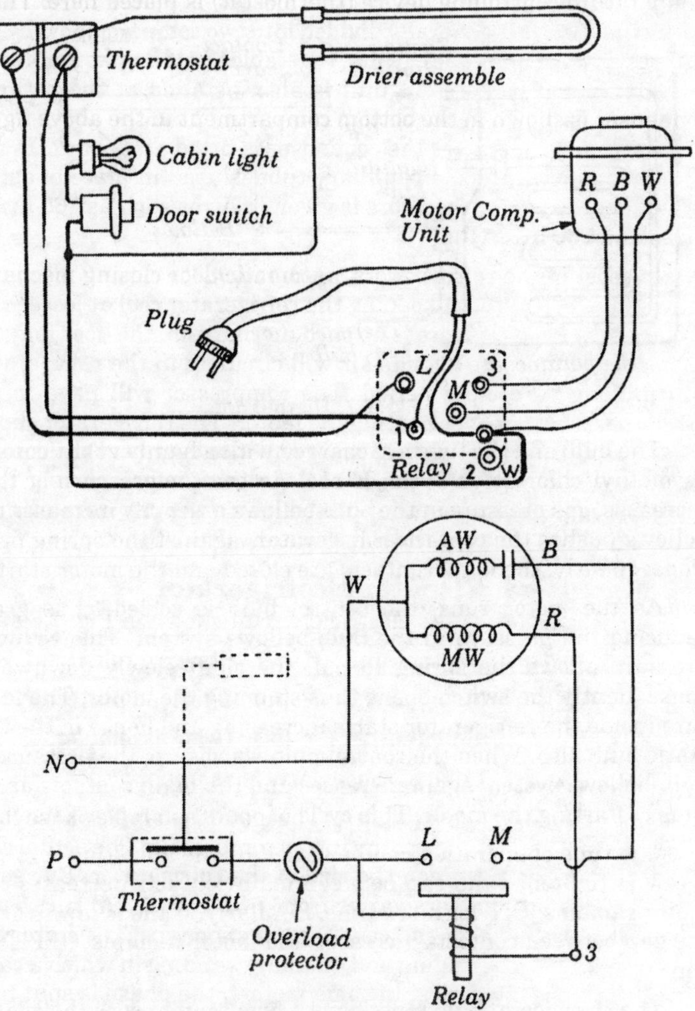

Fig. 7.6. Electrical Circuit of Refrigerator

rotating and when it picks up speed, the current drawn by it decreases. The decrease in current allows the plunger to come down thus opening contacts *L* and *M* and disconnecting starting winding. Now the motor continues to run on main winding.

(*ii*) **Thermostat :** A thermostatic switch is used on all the household refrigerators by which the temperature is controlled, and it is called thermostat.

It consists principally of a thermostatic bulb which is placed near the evaporator or freezer. This bulb is connected to the bellows by means of a capillary tube as shown in Fig. 7.7.

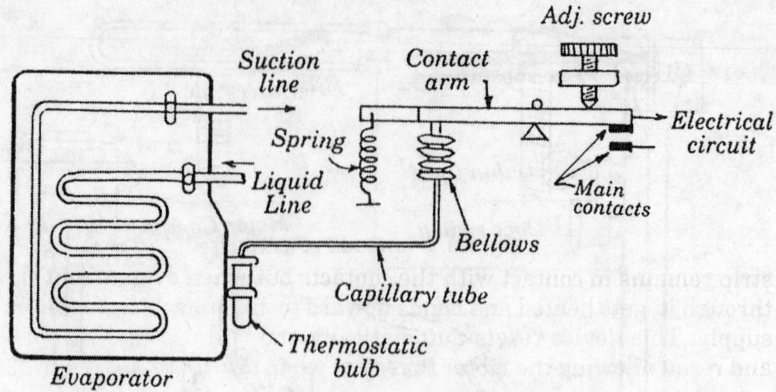

Fig. 7.7. Thermostat.

The bulb and the tube are charged with a highly volatile fluid such as methyl chloride or Freon F-114. As the temperature of the bulb increases, gas pressure in the bulb-bellows assembly increases and the bellows pushes the contact arm upwards against the spring pressure. Consequently, the main contacts are closed and the motor starts.

As the motor runs, the control bulb is cooled, thus gradually reducing the pressure in the bulb-bellows system. This reduction in pressure allows the spring to pull the shaft slowly downward and consequently the switch opens thus stopping the motor. The temperature inside the refrigerator starts increasing and hence in the thermostatic bulb also. When the control bulb warms up, the pressure in the bulb-bellows system again increases and the main contacts are closed thus restarting the motor. This cycle of operation repeats itself.

The time of operation can be varied by the adjusting screw. If the screw is tightened the gap between main contacts decreases and the motor remains 'ON' for more time. Similarly, if the screw is loosened, the gap between contacts increases and motor remains 'OFF' for more time.

The thermostat bulb must have a firm contact with the evaporator otherwise the temperature will not be effectively controlled. If the thermostat bulb is taken out of the refrigerator and placed in the room, the gas pressure in the bulb will not decrease and the motor will remain 'ON' continuously. This will cause overloading, extra energy consumption and the motor may get damaged.

(iv) **Overload Protector :** The relay also contains a thermal device which opens the circuit of the motor when overloaded. This device is called overload protector. The enlarged view of this is shown in Fig. 7.8. It is a circular bimetallic strip which is placed over two contacts and fixed with the help of a screw passing through its centre. Normally the

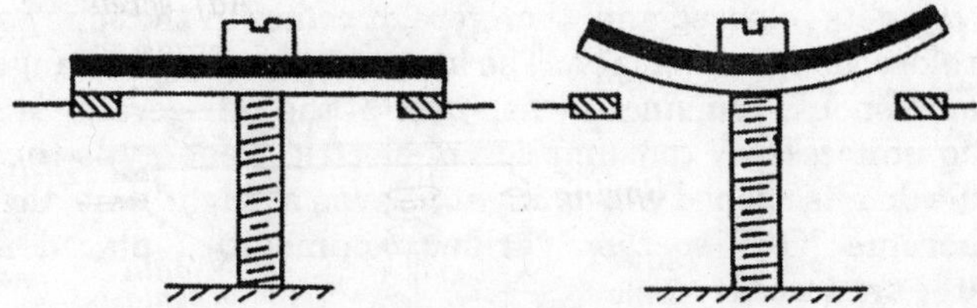

Fig. 7.8. Overload Protector.

strip remains in contact with the contacts but when overcurrent flows through it, gets heated and bends upward to disconnect the motor from supply. This device resets automatically and will continue to cut out and reset, allowing the motor to restart when overload conditions have passed.

(*v*) **Cabinet light circuit :** This circuit contains the lamp of small wattage fitted inside the cabinet and a switch which is installed at the hinged portion of refrigerator door. This switch is called door switch and is so installed that when the door is closed, the lamp remains off. When the door is opened, the knob of door switch moves outward thus switching on the lamp and illuminating the cabinet.

(*vi*) **Drier assembly :** This assembly is not employed on all the domestic refrigerators. This is a high resistance wire passing around the drier to remove the moisture from the refrigerator. This is connected across the supply and is switched on intermittently by the thermostat.

To facilitate the practical connections of refrigerators, another electrical circuit with colour code of wires is also shown in Fig. 7.9. In most of the Indian domestic refrigerators, these circuits are used.

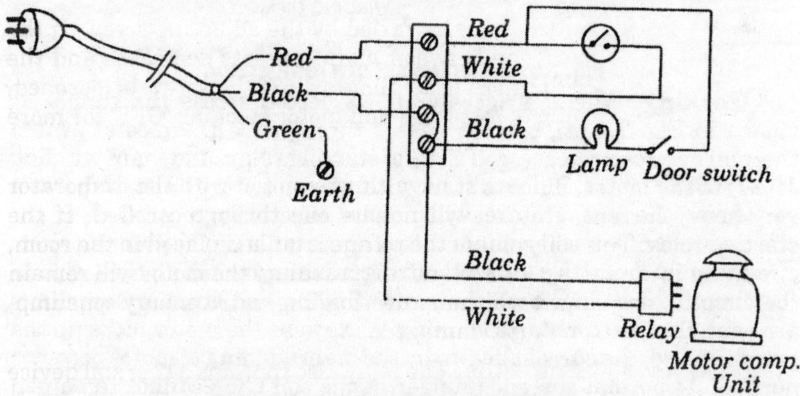

Fig. 7.9. Practical Electrical Circuit.

Lamp and switch. As shwon in Fig. 7.9 (*a*) when the door of the refrigerator remains closed, switch in series with the lamp (push button type remains opened under the pressure of door and as a result, the lamp remains off but as soon as the door is opened, the pressure on the push button gets released and, therefore, it comes in the "on" position and, therefore, lamp becomes on. The idea of making this arrangement is that lamp should remain off when door of the refrigerator is closed so that no unnecessary consumption of electrical energy is there. As soon as the door is opened which might be even at night time, the lamp should become "ON" so that various commodities placed in the refrigerator are traced.

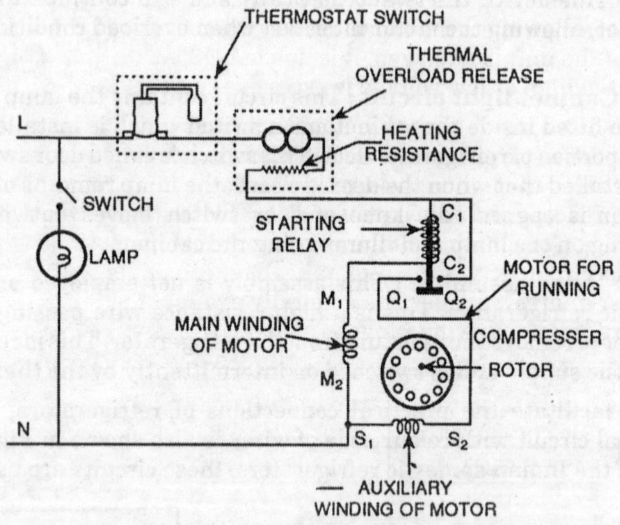

Fig. 7.9 (*a*) Electric ckt of a refrigerator.

Working. When refrigerator is connected across the supply as shown in Fig. 7.9 (*a*), current passes through the thermostat switch thermal overload release, coil C_1C_2 of starting relay, and main winding $M_1 M_2$ of the motor. Since to start with, the motor is at rest, it draws a very heavy current. This heavy current flows through coil C_1C_2 of the starting relay. This coil gets energised and it pulls up the plunger, short circuiting in contacts a_1a_2 and putting auxiliary winding S_1S_2 also in the circuit. Now since both the main winding and auxiliary windings are energised, motor starts running. As soon as the motor picks up the normal speed, the current drawn by the main winding of motor becomes normal. At normal current plunger in the coil C_1C_2 cannot remained pulled and it is released down, opening the contacts a_1a_2, thus auxiliary windings get out of circuit. Therefore, the purpose of the starting relay is to put auxiliary winding in circuit at the time of starting the motor and to disconnect it as soon as the motor starts working at normal speed.

In case starting relay fails to close, motor will not start as explained above. But once it closes, and it fails to open, then auxiliary winding shall also keep on drawing the current, resulting is increased current drawn by the motor. In that case, either the fuse shall be blown off or thermal over load shall trip out.

7.11. Need of Voltage Regulator

In our country, however the specified voltage for domestic use is 230 V but more commonly this voltage fluctuates. It is sometimes very low or very high. Since the motor, relay and other accessories are designed to operate very efficiently at one particular voltage, so it is necessary to maintain the voltage applied to refrigerator, as constant as possible. Whereas, the low voltages cause overloading of the motor, high voltage may effect the capacitor. Most of the times when the motor or relay of the unit is damaged, it is due to high or low voltage because the result of high or low voltage is excessive current flow.

To stabilize the voltage, a voltage stabilizer or voltage regulator is used with domestic refrigerators. The voltage regulator may be manual or automatic. In manual voltage regulator, one has to check the voltmeter reading and adjust it off and on. This is simply a tap-changing transformer. Automatic voltage regulator employs relays and other electronic circuitry which automatically keeps the output voltage within very narrow range. The generally available automatic voltage stabilizers are of following ranges :

S.No.	Input voltage	*Output voltage*	*Capacity*
1.	165 to 280 volts	220 to 240 volts	0.5 K VA
2.	135 to 280 volts	220 to 240 volts	0.5 K VA

The automatic regulator is costly as compared to manual but is more advantageous because it automatically takes care of the refrigerator during night time and other periods of neglect. The voltmeter of such regulators, need not to be switched on all the time which enhances its life.

The refrigerator should not be switched 'ON' or 'OFF' frequently because it also causes overload on the motor.

7.12. Water Cooler

The operation of a water cooler can be studied under three subheadings :

(*i*) Water circulation.

(*ii*) Vapour compression cycle.

(*iii*) Electrical circuit.

(*i*) **Water circulation :** The water cooler is generally used to provide cold drinking water at public places such as big offices, railway

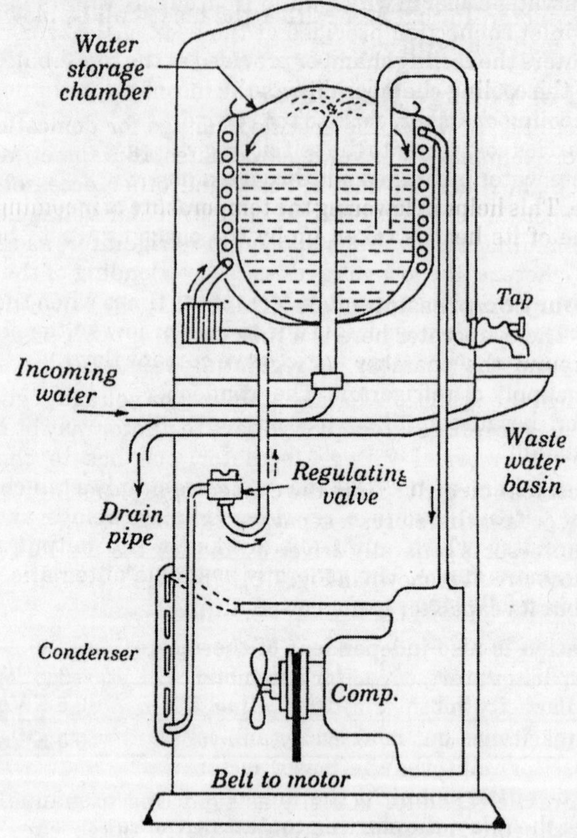

Fig. 7.10. Water Cooler.

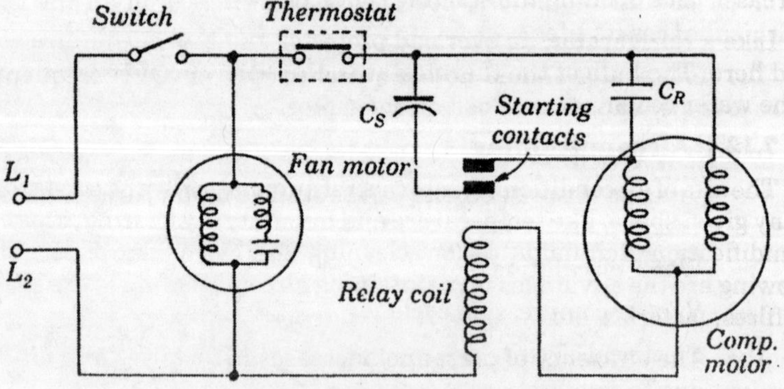

Fig. 7.11. Electrical Circuit.

stations, colleges and schools. Large quantity of water is circulated through the water cooler during its operating period. Fig. 7.10 shows a pressure type water cooler in which water from outside enters the cooler through the inlet connection provided at the back of the cooler. From here water enters the cooling chamber provided at the top. A baffle plate on the top of the cooling chamber directs the incoming water down the sides of the cooling chamber, close to the refrigerated walls. The outlet water tube carries cold water to the self closing valve pressure regulator tap. The waste water is drained out through a pipe which passes along the inlet pipe. This helps in lowering the temperature of incoming water because some of its heat is taken up by the outgoing waste but cold water.

(*ii*) **Vapour compression cycle :** This is just similar to that of a refrigerator. The evaporator here is a tube carrying refrigerant which is wound around the chamber. A regulating valve is also used to regulate the supply of refrigerant. The condenser is cooled by forced air circulation because unlike a refrigerator, here the operation is generally constant.

(*iii*) **Electrical circuit :** Since the condenser is cooled by forced air circulation by a fan, therefore, a separate 1- phase motor is used for this. The compressor is generally driven by another motor fitted outside it. The complete circuit is shown in Fig. 7.11. The fan motor is a capacitor start motor but its capacitor is not cut-off during running.

Its operation is also independent of thermostat. The compressor motor is a capacitor start, capacitor run motor. The starting capacitor C_S is of large capacity but intermittent rating. The running capacitor is of smaller capacitance but heavy duty and constant rating type. The running capacitor improves the power factor, overload capacity and starting torque of the motor. At the time of start when heavy current flows through the relay, the starting contacts are closed and (C_S) comes in the circuit. When motor picks up speed, the current through the relay decreases thus opening the starting contacts.

Like a refrigerator, an overload protector and thermostat are also used here. The bulb of the thermostat is dipped in the water chamber of the water cooler.

7.12.1. Airconditioning

The aim of airconditioning is to control automatically the conditions in any given space, like temperature and humidity, by heating, cooling, humidification, dehumidification, cleaning and movement of air. The following are the advatages in maintaining airconditioning of the space in offices, factories, etc.

1. The efficiency of personnel increases in factories and offices due to comfort.

2. Due to the comfort of airconditioning, commercial establishments will have more sales.
3. As the premises remain clean, there is more economy all over.
4. Food preservation plants, electronic laboratories and computer rooms, etc., require airconditioning for their working.
5. Static electricity due to dry air is avoided.

7.12.2. Types of airconditioning system

The particular type of airconditioning system to be adopted depends on the type of building to be airconditioned. Further, the air temperature and relative humidity are to be varied. Space is required for duct-work and pip-work. There are three main types of airconditioning systems :

(*a*) In a central plant, conditioned air is treated and led into various rooms by duct-work.

(*b*) **Air and water system :** In a central plant, air is treated and most of the heating and cooling is got by passing air over hot or cold water flowing through coils from a boiler and refrigeration plant. Although the central plant is smaller, space is taken by the room heating and cooling units.

(*c*) Self-contained units in which the airconditioning is carried out are placed inside the individual rooms against the outside walls, so that air from outside can be drawn into the unit through the short duct in the wall. Here space is occupied by the units in the rooms.

7.12.3. Room airconditioner

This is nothing but a unit airconditioner, suitable for placing in any room required to be airconditioned. Thus it can be used in any room of a house, office or shop.

They can be of portable type or window type. Portable airconditioners are used where it is needed and thus can be moved from room to room. They are very convenient for installation.

On the other hand, the window airconditioner has to be installed in a window properly, so that fresh air can enter and leave through the airconditioner, after it gets airconditioned. Further the location should be such that the air circulated inside the room properly. Also there should be no openings left when the unit is installed, as otherwise warm air from outside could enter the unit.

The airconditioner has three main systems :

1. Electrical, 2. Refrigerant and 3. Air path. The electrical system consists of, motor compressor, unit control switch, fan and fan motor, starting and running capacitors, starting relay and thermostat and wiring.

The refrigerant system has compressor, condenser, drier strainer, capillary tube, evaporating tubing and accumulator. The air path has an air discharge and intake grills, air filter, vent door and control, etc.

7.12.4. Operation of room airconditioner

The refrigeration cycle of a room airconditioner is the same as that of the domestic refrigerator, the fan in the evaporator draws the recirculated air into the unit through the side. The air after passing through the air filter evaporator is discharged through the grill in the front of the unit into room. The condenser and the compressor compartment are outside the room, while the evaporator compartment is separated by an insulated partition. Air which flows through the compressor compartment goes to the central section of the condenser, which is covered by the fan housing. The condensed water from evaporator collects into the base and goes to the condenser end of the unit.

7.12.5. Cooling capacity of an airconditioner

This refers to the ability of the unit to extract heat from a given room and is measured in BTU/hour. Larger the size of the room, larger should be the capacity of the airconditioner. For describing the heat capacity of larger units, we use the term ton of refrigeration. One ton of refrigeration is equal to 12000 BTU/hr *i.e.* 3000 kcal/hr. Thus, an airconditioner giving (60,000 BTU/hr *i.e.* 3000 kcal/hr, 15,000 kcal/hr) is referred to as a 5 ton air conditioner.

7.12.6. Working of an electrical system

The following are the electrical components of a room airconditioner.

1. motor compressor with thermal overload protection, 2. starting relay, 3. starting capacitor, 4. running capacitor, 5. fan motor with capacitor, 6. thermostat and 7. on-off switch combined with speed control.

The electrical circuit diagram of the airconditoner is shown below in Fig. 7.11 (*a*). A starting relay is in series with the motor running winding. When the starting current which is high flows through the motor starting winding, it starts. Also the starting relay contacts close and connect and starting capacitor to the starting winding. As the motor gradually reaches the required speed, the current in the running winding decreases, making the relay open, thus cutting the supply to the starting winding of the motor and the starting capacitor.

The supply is connected to thermostat switch through point A as shown in fig. 7.11 (*b*). As a result of this the fan motor gets connected across the supply and fan motor starts working even if the compressor motor is in off position due to operation of thermostat switch. The speed of fan motor is low if "speed change switch ON" is in 'O' position as the inductance gets connected in series with the motor. The speed of this fan motor becomes high, if "speed change switch" is in 'Hi' position as no inductance is connected in series with motor. As against this the speed of the fan motor of water cooler is only one (connections as shown by the chain line) as no inductance gets connected.

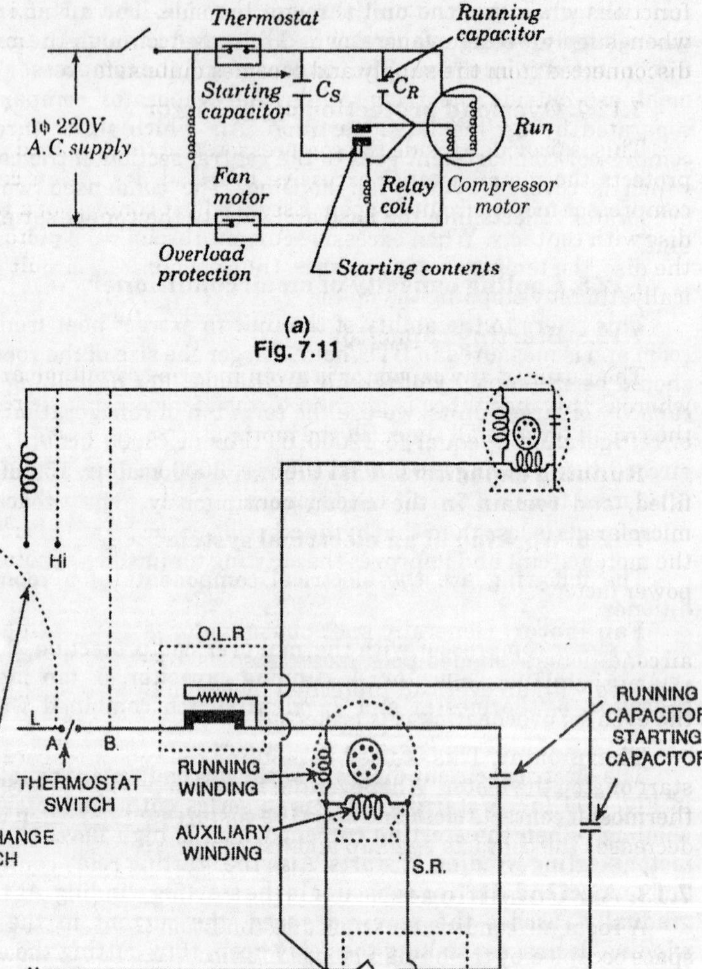

(a)
Fig. 7.11

(b)
Fig. 7.11 Electric Circuit Diagram of an Air Conditioner.

If the thermostat switch gets in 'ON' position, the compressor motor starts running as supply is being fed to both running winding and auxiliary. The compressor motor used as capacitor start, capacitor run. After the motor has picked up its normal speed, the voltage relay trips and auxiliary winding goes out of circuit whereas main winding still continues getting supply voltage. Over load release (O.L.R.) which functions when the load on the motor becomes more. This is possible when supply voltage falls down. Thus the compressor motor gets disconnected from the supply and becomes quite safe.

7.12.7. Overload protection of the motor

This is provided inside the compressor, ar.d the overload protector, protects the motor against excessive overload. Its action causes the compressor motor circuit to open. Essentially it consists of a bimetallic disc with contacts. When excessive current through the motor flows in the disc, the temperature increases, thus opening the circuit automatically, thereby stopping the motor.

7.12.8. Starting capacitor

The rating of any capacitor is given in terms of voltage and kVAR, whereas its capacitance is given in terms of microfarad. It remains in the circuit for a small time, till the motor starts.

Running capacitors. They are heavy duty capacitors and are oil filled, and remain in the circuit continuously. The capacitance in microfarads is less than that of the starting capacitors. This remains in the motor circuit and improves the starting torque of the motor and the power factor.

Fan motor. Generally split-phase motor is used, while in some airconditioners shaded-pole motoı also is used. The fan motor is provided with an overload protection placed in the winding, so that in the event of overheating it is protected.

Thermostat. This is used to monitor the room temperature and start or stop the motor. Whenever the room temperature increases, the thermostat contacts close and start the compressor and when the temp. decreases the contacts open up and the compressor stops.

7.13. Air Conditioner

A room air conditioner is used to condition the air of a particular space occupied by the human being *i.e.* office, room etc. is shown in Fig. 7.12. It has automatic operation to cool and maintain the humidity of air.

It consists of a two parts case. The two parts are divided by a partition with a small opening at the top as the outdoor part and indoor part. The outdoor partition consists of a hermetically sealed motor-compressor unit, condenser, motor driven fan and a tray. This portion is divided into two parts providea with a left hand side opening.

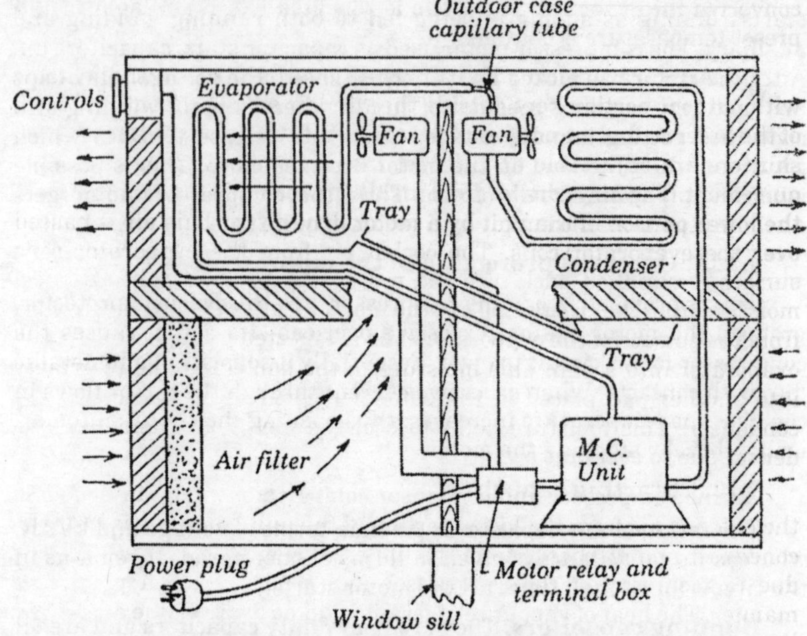

Fig. 7.12. Room Air Conditioner.

The indoor partition consists of an evaporator, motor driven fan, remote bulb refrigerant control, a control panel, an air filter, power connector and a tray. This partition is further sub-divided into two parts by an opening on the right hand side. A pipe line connects the two trays in the inner and outer parts.

Like a water cooler, the complete operation of an air conditioner can also be studied under three sub-headings :

(*i*) Vapour compression cycle.

(*ii*) Air circulation.

(*iii*) Electrical circuit.

(*i*) **Vapour compression cycle :** The refrigerant from the condenser is controlled and supplied to the evaporator through a capillary tube. Evaporator is further connected to the compressor by a suction pipe line. As the unit is switched on, the low pressure vapour through the suction pipe line is drawn from the evaporator and sent to the compressor. Here it is compressed to high pressure and delivered to the condenser. In the condenser, the vapours are condensed and heat is removed from them. The condensed vapours collect at the bottom coils of the condenser and are passed through the filter and capilliary tube to the evaporator at low pressure. The low pressure liquid refrigerant in the evaporator quickly picks up heat from its surface and gets converted into vapours. This cycle is repeated again and again till a preset temperature is obtained.

(*ii*) **Air circulation :** The air conditioner is fitted in the window with outdoor portion kept outside the window sill. The front and back of the inner and outer portion of the cooler is fitted with shutters. These shutters are adjustable at different inclinations according to the requirement. The air is drawn from inside the room through a filter from the lower portion of the unit by a motor driven fan. This air is passed over the evaporator coils. The cooled air from the evaporator coils surface is supplied back into the room. Due to dehumidification of moisture from the circulating air, the moisture after flowing down ward from the surface of the evaporator coils is collected at the bottom of the evaporator into a pan. This moisture in the pan flows due to gravity into another pan placed in the outdoor portion at the bottom of the condenser. This moisture helps to cool the compressor and the condenser due to evaporation.

In the compressor and condenser compartment, the fan draws in the out side air from the lower portion of the unit, circulate it over the condenser and discharge from the upper portion of the unit to outside due to mounting of the compressor and condenser in a particular manner. The heat of vapour refrigerant is taken away by the air during its passage through the condenser and it is cooled into liquid form. The unit automatically stops after the desired temperature is reached inside the room. The automatic operation of the unit is accomplished by a valve at the control panel.

(*iii*) **Electrical Circuit :** The electrical circuit is shown by a single line in the above figure. It is just similar to that of a water-cooler with only difference that in this case two fans are used for circulating the air. The thermostat has the temperature sensitive element which is located in the return air circuit.

The size of room air conditioner is given in tons such as $\frac{1}{2}$ ton, 1 ton and 2 ton etc. The horse power of motors varies from $\frac{1}{4}$ H.P. to $\frac{3}{4}$ H.P. Fan motor is two-speed motor (High and low).

7.14 Thermo-Electric Refrigeration

This system of refrigeration works on the principle of Peltier effect according to which heat energy is evolved at one junction and absorbed at the other one when direct current is passed through a junction of two dissimilar metals like antimony and bismuth. It consists of a number of thermoelectric module assemblies in series joined by copper strips, as shown in the fig. below. Each thermo-electric module is built up of a large number of thermo-electric module couples. When the direct current is passed through the thremo - electric module assemblies in the direction shown in fig. junctions at the top of the assembly get heated up and those at the bottom are cooled.

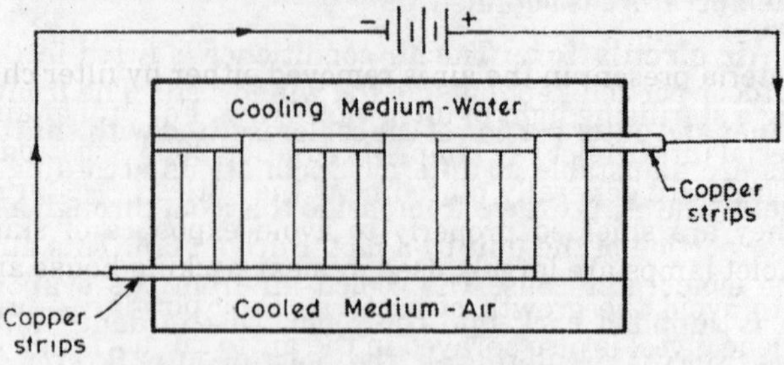

Fig. 7.13. Thermo-Electric Refrigeration.

Thus bottom junctions abstract heat from the surroundings and produce refrigerating effect and top junctions require cooling by water. Module assembly, therefore, abstracts heat from the medium at bottom and rejects the same to the medium at the top. If the direction of flow of current is reversed, heat production at the junctions is reversed, *i.e.* now top of the modules will be cooled and bottom ones get heated up. If it is an air-conditioner, air will get heated up instead of getting cooled. Thus summer and winter operating of air-conditioner can easily be obtained just by reversing the direction of flow of current. The main advantages of this system of refrigeration are absence of moving parts and case of automatic control by controlling the magnitude of current. The only draw back of this system of refrigeration is its very high initial cost.

7.15 Air—Purification

Proper filtration, cleaning and purification of air is necessary to keep it free from dust, dirt and other impurities which are very injurious for health. The dust particles are liable to clog the nose, throat and lungs of the workers seriously affecting their health. The dirt present in air acts as a carrier of germs and is responsible for disease spreading. In case of certain industries like chemicals the dirt produced as a by-product is very harmful for health if inhaled. The presence of dirt in the air used for air cooling of electric machines and diesel sets in power plants is liable to clog their air passage resulting in reduced efficiency and life. Various types of air impurities other than gas or vapours, called the Aerosols, are also present. These are dust, fumes, smoke (due to combustion of organic materials like tobacco, wood, oil, coal etc.); pollen (from wood, tree, flower and other vegetation--may be harmless or harmful); bacteria (living micro-organism).

There are various methods of cleaning, filtering and purifying the air from the above mentioned impurities. Mists or fogs in the air are usually removed by raising the temperature of air. This results in the formation of small suspended water particles into vapours making them invisible. The impurities like gases and vapours are removed by the process of condensation Pollen grains will adhere to these wetted surfaces.

Bacteria present in the air is removed either by filter chemicals or by using a sterilising light of ultra violet rays. For proper air sterilisation special ultra violet ray lamps of radiation 2600 angstrom units are used. These are generally placed near the outlet grille on its outside end. They are shielded properly to avoid exposure of skin or eyes. Ultraviolet lamps are largely used in meat packing house and storage rooms to avoid the growth of bacteria. Fine powder of propylene of triethylene glycol is also sprayed in the air to kill bacteria.

To remove dust particles from air, methods using equipment like centrifugal devices, water filters or air washers, electrostatic precipitators are used.

7.16 Conditions of Comfort

This is the type of air-conditioning whose objective is to produce comfortable and healthy conditions for the occupants. In other words, it is the process of treating air so as to control simultaneously the prevailing conditions of temperature, humidity, cleanliness and distribution in the conditioned space to meet the comfort and health requirements of the occupants. It includes the air conditioning of spaces, places and surroundings which are occupied by human beings.

Some of important places intended for human occupancy and requiring comfort air conditioning are the residences, departmental stores, public buildings including theatres, cinema halls, dancing halls, skating rings, restaurants and night clubs, hospitals etc. In order to understand the comfort air-conditioning the human requirements should be known.

7.17 Human Requirements

Our body is thermostat with inner temperature of 98.4° F and outer temperature of about 26.5°C. Under normal conditions our body emits 2500 K-Cal per 24 hrs.; 45% of which is dissipated by radiation, 30% by convection and evaporation is governed by surrounding temperature, air temperature and humidity of air. It is necessary for our body to dissipate the above proportion of heat in order that we may feel convenient and comfortable. For producing comfort it is necessary to maintain surroundings atmosphere at a temperature between 22°C to 25°C during summer and 17°C to 20°C during winter and humidity between 40 to 60%.

In Indian Railway air conditioned coaches the temperature is maintained at 24°C during summer and 20°C during winter, as those are most accepted temperature by passengers.

7.18 Effective Temperature

Comfort requirements for human being except radiation, air odour and cleanliness generally depend upon dry and wet bubble temperature of the air and its movement rate. An arbitrary index which combines into a single value the degree of sensation of warmth or cold felt by the human body due to the effect of air, temperature, humidity and air movement is called the effective temperature. Its numerical value can not be measured directly but is taken as that temperature of the still saturated air which induces the same sensation of warmth or coolness as those created by the air surrounding a person. An atmospheric condition will have an effective temperature of 20°C if it induces the same sensation of warmth or coolness as created by still saturated air at 20°C.

From the investigations and results of various studies it has now been proved that the effects of humidity shown by effective temperature scale are not great if the air temperature and the human activities are such that very little heat is lost by evaporation of the perspiration. For an individual doing an activity in sitting position with air temperature between 23°C and 25°C a relative humidity in the range of 30 to 65% will have no effect on his comfort. At the same temperature a higher rate of activity may result in a sensation of warmth but a relative humidity of 40% will be more comfortable than 60% relative humidity under similar conditions. In case of a large no. of persons doing different activities like sitting, playing, dancing etc. in the same space a 23.5°C temperature with 40% or less relative humidity generally provides most comfortable condition for all.

7.19 Central Air Conditioning Systems

The central air conditioning systems serving the needs of large buildings or spaces are generally called the year-round air conditioning systems. They have a cooling capacity of 25 tons and circulate about 300 m^3/minute of conditioned air. Depending upon its type such a system may contain some of the following equipments :

(*a*) Heating coils supplied with steam or hot water.

(*b*) Cooling and dehumidifying coils–both for chilled water and for direct expansion of the refrigerants.

(*c*) Blower and driving motor.

(*d*) Sprays for cooling and dehumidifying or for washing the air.

(*e*) Air-cleaning equipment containing filters, electrostatic precipitators, odour removing equipment and germicidal lamps.

(*f*) Various types of control devices.

The devices assembled as an air-conditioning unit are housed and enclosed in sheet-metal housings or plenums. The plenum is usually covered with insulating board over its entire outer surface and is stoutly braced. The refrigeration condensing units and steam or hot water boilers for air conditioning in this system are either located in the same room or in the basement or in an adjoining room.

The various factors affecting the choice of the equipment for design and installation of the central air-conditioning systems are capacity of the plant, filters and mixing plenums, refrigeration and heating equipment, insulation, noise and vibration.

(*i*) ***Capacity of the plant.*** For each 30 m^3/minute of air handled, heating and cooling coils handling air must provide 0.185 m^2 of face area (four row coil) at face velocities of about 150 m/minute. One of the limiting factors influencing the design of this system is the physical size of the coil and filter banks.

The horizontal space between flows of a building is usually kept between 3.65 and 4.87 m because the operating expenses of the system depends upon the floor area utilised. Thus for a single air-conditioning unit a practical limit of 1200 m^3/minute or 100 tons of cooling capacity is set. For the spaces or buildings requiring more than this capacity two or more zones may be planned with an air-conditioning system installed to handle each unit.

The extent of the zoning to be specified is generally determined by the nature of the buildings. A theatre or departmental store installation requiring spaces of large cubical structure are most economically conditioned by only a few (preferably one) central station air conditioning units. Office buildings, schools, colleges and similar buildings requiring many rooms necessitate greater zoning. Air conditioning units of 300–900 m^3/minute are usually employed. Considering all the factors discussed above, it seems more economical to supply chilled water, hot water, steam or refrigerant through pipes to various smaller air-conditioning units than erect massive supply and return ducts for serving an extremely large unit.

(*ii*) ***Filters and mixing plenums.*** Before filtering, the return air and fresh outside air should be brought into a plenum for proper mixing. The amount of outside air to be introduced is usually controlled by motorised dampers. Depending upon the nature of occupancy, the dampers are controlled by a thermostat to admit less outside air during extreme outdoor temperature or by smoke and odour sensing devices to increase the amount of ventilation air according to requirement in the conditioned space.

Vertically assembled and cleanable types of filter banks are generally used. For central systems of less than 300 m^3/minute capacity disposable filters are used. The various types of filters and air cleaning devices commercially available accomplish all the following cleaning operations :

(*a*) Removal of solid particles of visible size (dust, dirt, lint, pollen, cinders etc.)

(*b*) Removal of particles of microscopic size (smoke, haze, etc). Electrostatic precipitators are used for this purpose.

(*c*) Proper control of odour—for this purpose activated charcoal is usually used.

(*d*) Killing of bacteria—for this purpose germicidal lamps and certain chemicals are used.

(*iii*) ***Refrigeration and heating equipment.*** The design of central station system is greatly affected by these equipments. The direct-expansion (DX) refrigerant coils and hot water coils are usually employed in systems requiring only one or two zones. The air conditioning units in these systems are generally located close to the cooling and heating equipments. The chilled water is used and circulated as a cooling medium in the air-conditioning coils for many zones at remote locations which are served by a single cooling and heating plant. The heating medium is generally the hot water or steam. For serving the various zones water is generally pumped to the mechanical rooms.

(*iv*) ***Insulation.*** If the supply ducts are not running through the conditioned spaces, they should be properly insulated. Generally 25 mm thick insulating board is used to insulate the outer surface of the entire housing for the central station air-conditioning unit. In order to prevent the heat losses and gains, all the steam, hot water, chilled water and suction refrigerant piping should be properly insulated.

(*v*) ***Noise and vibration.*** The noise and vibrations produced in the air conditioning units are very important factors for considerations while designing and selecting equipments. The undue noise and vibrations are considered undesirable.

7.20 Heating of Buildings

The different types of equipment used for heating of buildings are given below.

1. ***High Temperature Radiators*** : The high temperature radiator essentially consists of a high resistance element wound on porcelain or refractory former and mounted into decorative frames. Resistance elements get heated to high temperature (1400°C to 1600°C) due to flow of electric current through them and 50 to 70% of heat so produced is dissipated by radiation and remaining by convection. Advantages of this type of heater are that it is portable, cheap in initial cost, simple in construction and attains its highest temperature quickly. The main draw back of this type of heater is that heat dissipated is localised. These are used for domestic purposes or where intermittent heating is required. The usual sizes are from 0.5 kw to 3 kw.

2. ***Tubular Heaters*** : The most common type of low temperature heater is 5 cm diameter steel tubes of length ranging from 0.6 to 5 meters and containing resistance elements arranged on mica, fire clay, or porcelain former. Such tubes are mounted around the skirting board of room to be heated. The normal loading is 200 watts per metre length. The temperature of the external surface is maintained at about 200°F. About 90 percent of the heat is transferred by convection. These are spaced to produce convection currents in the room so as to heat whole of it. These are specially suitable for thermostatic control, by which means they are switched on and off automatically as the room temperature falls below or rises above the desired value. When fixed at the base of a cold wall or underneath a window they prevent cold down-draughts, which cause cold feet to the occupants of the room.

3. ***Convector Heaters*** : In these heaters, heating elements are mounted with in sheet-metal cases which admits cool air from bottom. Air after getting heated up, comed out from grills at the top. The front and top may be of moulded plastic material for the sake of pleasing appearance. These heaters may be either portable or fixed, and may be handled without fear of burns. In these type of heaters, heating element attains low temperature of 66° to 93°C and proportion of heat transmitted by convection is large as compared to heat transmitted by radiations.

A very satisfactory method of heating a room is to use convector heaters to warm the room to about 13°C, together with a small radiant heater for local heating.

4. ***Panel Heaters*** : In this system of heating large panels of fire clay or some other suitable heat resisting materials are used along the walls or ceiling of a room. The resistance elements are embedded in these panels. The heat produced is utilised in the form of convection or radiations. This system of heating of buildings is advantaneous from the decoration point of view and gives a uniform distribution of heat. Loading are of the order of 400 watt/m^2 for the low temperature panels and 6,000 watt/m^2 for the high temperature panels Low temperature panels operating at temperature from 27°C to 65°C may be fixed in the walls or ceiling of a room and plastered over. High temperature panels operating at temperature up to 288°C are used on walls and ceiling but are not fixed in the wall surface. They are some times be suspended from the ceiling.

5. ***Thermal Storage System*** : Roughly this sytem is similar to the coal, coke, or oil-fired central heating system, except that the heat is supplied to the storage tank electrically. Heated water is circulated through the pipes and radiators situated through out the building. There are two types of heaters used in this system ; the immersion heater for a load up to 200 kw and the electrode type for loads exceeding 200 kw. In immersion heaters, resistance wire is embedded in an

insulating medium like magnesium oxide inside a copper or steel tube, which is bent to the desired shape. Heat loading is 7-8 W/cm^2. Single tubes in ratings of 4-5 kw are available. For higher ratings, groups of tubes are used. The electrode type is only suitable for use on a.c., usually 3 phase, since d.c. may cause electrolysis. In electrode boilers water is heated by placing cast iron electrodes directly in the water. For 3 ϕ supply three electrodes are placed in the water and the tank is earthed so as to form a star-connected system. Sodium carbonate is added, if necessary, for increasing the conductivity. Control of power input may be carried out by varying the height of the electrodes in the water, by means of a movable shield surrounding the electrodes, by moving one electrode relative to the other, or, in large units, by varying the water level by means of a pump. A low-voltage boiler operating at 400-V is usually employed for a requirement up to 500-700 kW while for larger requirement a high-voltage boiler operating at voltages up to 11000 V or even higher is used.

The advantage of this system is that the water can be heated during night (off peak periods) and is supplied throghout the day, thus use of cheap power available during off peak periods can be made. This system of heating also provides hot water service for the building.

6. ***Floor Warming System*** **:** In this system a number of heating cables are embedded in the floor of a room or otherwise contained in the floor space. This heat is given to the concrete or other floor material slowly and heat from the floor is used in warming the air in the room, and in warming the walls and the furniture. The electrical power intake is controlled by thermostat fixed on the room wall in a suitable position. The cables used for heating are made from alloys of nickel, chromium, iron or other resistance materials and are insulated with asbestos mineral insulation, butyl rubber, or other heat-resisting insulating materials. For mechanical protection lead, copper, aluminium, or PVC sheating can be used. Loading of from 100 to 150 watts per metre^2 of floor are generally satisfactory.

In this system also the advantage of cheaper power available during night-hours can be taken but in such a case it will be desirable that one or more convectors or radiant heaters for use towards the end of the floor-warming output cycle are installed.

7.21 Calculation of Rating of Electrical Equipment

The amount of heat required for personal comfort in an enclosed space such as a room or office depends upon the following factors ;

1. Number of changes of air per hour.
2. Area of windows.
3. Situation of walls, external or internal (most rooms have at least one external wall).

4. The exposure of the ceiling (open to the roof or with heated bed room over).
5. Material of which walls, floor and ceilings are composed.
6. The temperature of outside air.
7. Whether the building is to be heated continuously or intermittently.

The method of calculations will be clear from the following examples.

7.21.1 Desert Cooler

As the name suggests, desert cooler is suitable for places where the humidity content of air is very low and temperature is very high. Such conditions are more in desert areas, that is why these coolers are called desert coolers. Desert cooler operates on the principle of "evaporative cooling". It is a process in which sensible heat is removed and moisture added to the air.

Construction and Working

A desert cooler consists of the following main parts:

(*i*) Blower/Fan (*ii*) Water circulating pump
(*iii*) Water wetted pads (*iv*) Water tank
(*v*) Float valve (*vi*) Water taps.

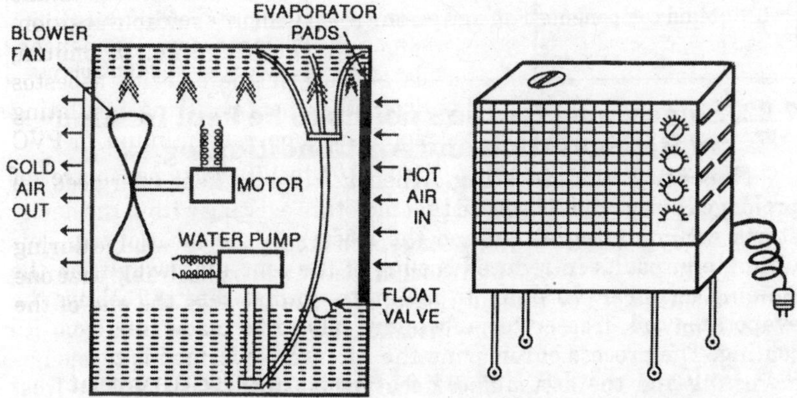

(*a*) Main parts of desent cooler **(*b*) External view of desent cooler**

Fig. 7.14

The water is filled in the water tank of the cooler from mains of water supply and the level of the water tank is controlled by a float valve. The circulating water pump lifts the water and supplies it at the top of the cooler to the water distribution system which consists of channels with small holes through which the water-drops fall continuously on the wetted pads. The water which drops back from the pads is re-circulated. The pump can be made of stainless steel, brass or plastic.

The Blower (centrifugal fan) sucks the air over the wetted pad and delivers it to the space to be cooled through an opening provided on the fourth side of desert cooler cabinet.

To have longer life of the desert cooler and better performance, the pads should be changed every year and the holes provided on the channel (Water distribution system) must be cleaned regularly. The water tank of the desert cooler should be cleaned immediately after the season and must be coated with the corrosion resisting paint of red oxide and with other types of paints.

Points of Comparision Between Desert Cooler and Window type Air-Conditioner

Sr. No.	*Description*	*Desert cooler*	*Window Type Airconditioner*
1.	Initial cost	Less	Very high
2.	Location	Placed inside or in the window of the room.	Fitted either in the wall or window.
3.	Utility power	In the dry season can be used to cool the air.	It can be used for both cooling and heating purposes.
4.	Power reqd. per unit cooling	More	Less
5.	Humidity control	Not possible	Possible
6.	Main components	Fan and Motor	Complete refrigeration unit : Compressor, condenser, evaporator and controls etc.

7.22. Some Consideration points to be kept in mind in Refrigeration and Air Conditioning

Necessity of defrosting. When a fridge is kept energised for prolonged period, it is observed that moisture existing within the device starts accumulating all around the freezer in the form of ice. This ultimately results in reduced cooling of the contents, away from the freezing chamber. In order to maintain normal rate of cooling, the evaporator coil, freezer cover boundry has to be made free from ice coating. The process of removing the ice layer so formed is termed as *defrosting* and the accumulated coating is called frosting. The frost works of an insulation, which retards the coefficient of performance (C.O.P.) of the cooling system [C.O.P. = Heat extracted/Work done].

Methods of defrosting. There are several possible methods used for defrosting.

(*i*) **Manual Defrosting.** In this case the contents are taken out from the fridge as far as possible or covered adequately to avoid ice melt water drops, entering into the cooked food stuff etc. The supply is then switched off so that warming up of the freezer and the evaporating coils starts taking place, the ice layer melts gradually. The water so formed is drained out, the lining dried properly and the fridge is re-energised.

(*ii*) **Semi-Auto defrosting.** We know that as long as the fixed and the sliding contacts shown in the thermostat (Fig. 7.15) remain closed, the compressor motor continues running. This keeps the fridge 'ON' till the contents of the sensing bulb decrease in volume to the extent that the bellows shrinks and the main spring pushes the lower hinged plate downwards, which results in opening of the sliding and fixed contacts.

If the motor has to be switched off without disconnecting the fridge from the supply, a defrost push is provided by some manufacturers in the centre of cooling adjustment knob shown in Fig. 7.15. When this 'push' is depressed the upper as well as the lower hinged plates help in opening the motor contacts. The motor remains off till the sensing bulb contents warm up, expand the bellows to an extent that the spring is compressed to an extent that the defrost 'push' is pushed back to original position and the motor contacts close once again. The restart of the motor takes place automatically after defrosting.

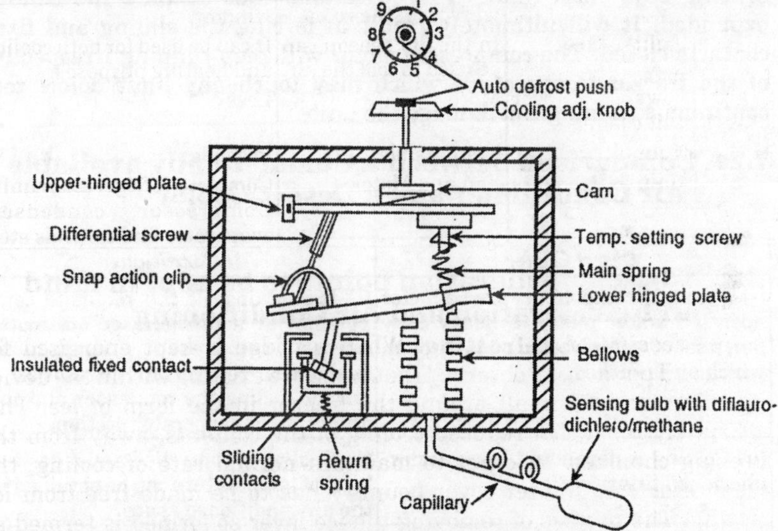

Fig. 7.15. Thermostat with defrost arrangement.

(*iii*) **Auto defrost.** There are several methods possible for auto-defrost. Some possible methods are discussed below :

(*a*) **Through electronic timers.** These timers are set to de-energise the compressor motor once in every twenty four hours and re-energise the same after time necessary for defrosting.

(*b*) **The pressure sensing transducer.** Since the frost accumulation results in operation of the compressor at a low suction pressure, a pressure sensing transducer operates a relay which opens the motor contacts. The contacts close once again automatically after defrosting of the evaporator coils with the help of thermostat as explained earlier.

(*c*) **Through temperature sensing transducer.** Since frost, accumulation does affect the thermal state of the cooling system, a deviation from the temperature setting of the device operates a temperature sensing transducer which de-energises the motor circuit. The circuit is brought to closing state through the thermostat after defrosting.

Note. There are some other methods like 'Revese cycle defrost', hot gas defrost method' 'Heating of finned coils electrically' etc. which are beyond the scope of this book.

7.23. Effect of Keeping the Sensing Bulb of the Thermostat or the Thermostat itself at Room Temperature

From the functioning of the thermostat it is evident that the bellows size and pressure neutralise the main spring tension. If the temperature sensing bulb is maintained at room temperature the contents of the sensing bulb shall retain a large volume thus keeping the bellows expanded. It will ultimately result in forcing the sliding and fixed contacts closed. The compressor motor will keep running irrespective of the freezer temperature which may touch any limit below zero continuous running can damage the unit.

7.24. Comparison between a Commercially available Air Conditioner and a Desert Cooler

Desert Cooler	*Air Conditioner*
(*i*) The main components are : a motor operated water pump, a heavy duty fan, a set of water wetted pads through which air is sucked.	(*i*) It contains all the requisite components of a refrigerator *viz* motor driven compressor, condenser, evaporator, refrigertant circulation, a fan for cooling the condenser and the other fan for evaporator section.
(*ii*) Cooling is dependent upon atmospheric conditions.	(*ii*) Cooling is controlled and a desired level of temperature can be obtained in the air conditioned space.
(*iii*) It adds to the humidity of the atmosphere within the enclosure cooled by it.	(*iii*) The air going through it is filtered, cooled and de-humiditised by the coldest component i.e. the evaporator.
(*iv*) No heating arrangement is provided in commonly used coolers.	(*iv*) Both heating and cooling arrangements are available.
(*v*) No cooling is produced if the atmosphere is over humid.	(*v*) Cooling is produced irrespective of the atmospheric humidity.
(*vi*) Initial cost is quite low.	(*vi*) Initial cost is more than ten times of a desert cooler.

Desert Cooler	*Air Conditioner*
(vii) The pump stops functioning if supply voltage falls below certain level. The fan still runs.	*(vii)* The device is over loaded and relays disconnect it from the supply till voltage level is sustained.
(viii) Cooling is affected by water evaporation and the increased humidity in the hot air sucked in by the device.	*(viii)* Cooling is affected by refrigerant evaporation. The water added to the device is used to increase heat dissipation of the condenser section.

Example 7.1. *It is proposed to condition 3,000 m^3 of air per hour from a temperature of 5°C to 20°C. It is further necessary to evaporate 5 kg of moisture per 1,000 m^3 of air per hour to control humidity. Estimate the power required if heat required to raise the temperature of 1 m^3 of air through 1°C is 1,220 J and latent heat of evaporation is 2450 × 10^3 J/kg.*

Solution. Heat required to raise the temperature of air

H_1 = Volume of air to be conditioned × heat required to raise the temperature of 1 m^3 of air through 1°C × difference of temperatures

$$= 3,000 \times 1,220 \times (20 - 5) \text{ Joules}$$

$$= 549 \times 10^5 \text{ J}$$

$$\text{The moisture present in the air } = \frac{5 \times 3,000}{1,000} = 15 \text{ kg}$$

Latent heat required to evaporate the moisture

$$H_2 = 15 \times 2,450 \times 10^3 = 36.75 \times 10^6 \text{ J}$$

Total heat required,

$$H = H_1 + H_2 = (54.9 + 36.75) \times 10^6 = 91.65 \times 10^6 \text{ J/hour}$$

$$\text{Power required} = \frac{91.65 \times 10^6}{3,600 \times 1,000} = 25 \text{ kw} \quad \textbf{Ans.}$$

Example 7.2. *A drawing office 6m × 5m × 4m high is to be heated. The temperature of the room is to be kept at 25°C when the outside temperature is 5°C. The air is to be renewed every 15 minutes. Find the necessary rating of the electric heater. The heat loss from the walls is 836 × 10^3 J/°C/hour.*

Solution. Volume of air to be conditioned per hour

$$= 6 \times 5 \times 4 \times \frac{60}{15} = 480 \text{ m}^3\text{/hour}$$

Heat required to raise the temperature of 1 m^3 of air through 1°C = 1,220 J

Heat required to raise the temperature of 480 m^3 air through (25 – 5) *i.e.* 20°C

$$H_1 = 480 \times 1220 \times 20 = 11.712 \times 10^6 \text{ J/hour}$$

Heat losses from the walls

$$= 836 \times 10^3 \times 20 \times 1 = 16.72 \times 10^6 \text{ J/hour}$$

Total heat required,

$$H = H_1 + H_2 = 11.712 \times 10^6 + 16.72 \times 10^6 = 28.432 \times 10^6 \text{ J/hour}$$

$$\text{kw rating of heater} = \frac{28.432 \times 10^6}{3{,}600 \times 1{,}000} = 8 \text{ kw} \quad \textbf{Ans.}$$

7.25. Fault Finding in Sealed Type Refrigeration Systems

	Trouble	*Possible cause*	*Remedy*
1.	Compressor fails to start. No humming after switch on.	1. Fuse blown off. 2. Compressor motor burnt. 3. Control contacts open 4. Wiring defective or damaged.	Check and replace fuse. Check motor continuity and get the motor rewired. Check contact points. Locate wiring fault.
2.	Compressor hums but fails to start	1. Supply voltage too low. 2. Relay too hot 3. Leaky motor capacitors 4. Relay contacts open 5. Starting capacitor gone weak 6. Bearing friction too high 7. Piston gone too tight 8. Defective relay	Check stabiliser input and output voltages. See cause of heating and current drawn by the unit. Check capacitors and replace. Check relay. Check starting capacitor. Check bearings. Check compressor. Check and replace.
3.	Compressor starts but trips after a short period	1. Starting winding remains in the circuit. 2. Stickly relay contacts. 3. Voltage gone low after starting. 4. Starting capacitor gone weak. 5. Grounded compressor. 6. Welded relay contacts under high operating conditions. 7. During servicing motor connections made improper.	Check centrifugal switch. Check relay points. See voltage fluctuation. Check and replace capacitor. See for earth fault. Check relay contacts. Check motor circuit.

	Trouble	*Possible cause*	*Remedy*
4.	**Compressor over heats**	**1. Running at low voltage** **2. Condenser choked** **3. Poor ventillation** **4. Bearing friction too high.** **5. Piston too light.**	**Check supply/stabiliser output.** **Check and clean the condenser.** **Unit mounted, if close to the walls should be shifted away, better near a window.** **Check bearings.** **Check piston.**
5.	**Compressor starts after repetitive attempts**	**1. Low supply voltage.** **2. Weak starting capacitor.** **3. Air in the system.** **4. Wrong thermo-setting.** **5. Pitted relay contacts.**	**Check for voltage.** **Check and replace.** **Let the sealed unit recharged.** **Check and reset the thermostat.** **Check the contacts for replacement.**
6.	**Unit runs out cooling poor**	**1. Leaky door Gasket.** **2. Frequent door opening.** **3. Too hot contents.** **4. Under charged sealed unit.** **5. Defective thermal insulation.** **6. Defrosting not done at required time.** **7. Oil entry into cooling coil.**	**Check and replace.** **Avoid frequent and prolonged door opening.** **Check and remove hot contents.** **Get refrigerant recharging.** **Check and replace insulation.** **Defrostand remove thick ice coating from the cooling coils.** **Put steaming hot water in the freezer.**
7.	**Unit runs but no cooling.**	**1. Refrigerant leaked out.** **2. Over charged refrigerant during repairs.** **3. Excessive ambient temperature.** **4. System choked.** **5. No ventillation.** **6. Improper compression.**	**Get the unit recharged.** **Purge and overcharge.** **Provide desired temperature gradient.** **Get the choking removed.** **Provide requisite ventillation.** **Check compressor unit.**
8.	**Freezer works but food spoils.**	**1. Over loaded food shelves.** **2. Hot food insertion.** **3. Fridge door left open.** **4. Ice coating thick on cooling tubes.** **5. Condenser tubes exposed to heat/sun rays.**	**Remove extra load.** **Remove hot stuff.** **Close the door firmly.** **Arrange defrosting.** **Provide requisite ambient temperature.**

	Trouble	*Possible cause*	*Remedy*
9.	Prolonged cycling.	1. Temperature control switch maladjusted. 2. Capillary choked. 3. Bellows hardened. 4. Door Gasket damaged. 5. Over loading.	Readjust thermosetting. Check capillary tube. Check thermostat. Replace gasket. Don't exceed capacity of the unit.
10.	Short cycling	1. Thermosetting misadjusted. 2. Refrigerant over charged. 3. Refrigerant under charged. 4. Choked condenser. 5. Compressor defective.	Readjust temperature controller. Get extra charged purged. Get adequate charge. Get condenser unit restriction free. Get compressor checked.
11.	Rattling of the device.	1. Uneven floor mounting. 2. Knocking due to oil in the compression cycle. 3. Fluctuating voltage touching a lower level repeatedly. 4. Presence of air in the system. 5. Defective bearings. 6. Repetitive failure of supply.	Balance the unit. Get the oil purged. Switch off the unit for the time being or use a constant voltage transformer (CVT) as a supply source. Get the air removed. Get motor/compressor bearings replaced. Switch off supply till regularity.

SHORT QUESTIONS WITH ANSWERS

Q.1. ***What do you mean by refrigeration ?***

Ans. The refrigeration means or defined as the process of removing heat from a substance to be cooled and passing it on to some other substance under controlled conditions. It also includes the process of reducing and maintaining the temperature of cooled body below the general temperature of its surroundings.

Q.2. ***What do you mean by an air-conditioning ?***

Ans. The meaning of air-conditioning can be defined as the process to simultaneous control of heating/cooling, cleaning, humidifying air circulations either for comfort for human beings or animals or for the proper performance of some industrial or scientific processes.

Q.3. What is the unit of Refrigeration ?

Ans. Unit of refrigeration is expressed in ton of refrigeration.

Q.4. ***Mention the types of Refrigeration system :***

Ans. The types of refrigeration system are :

(*i*) Vapour compression refrigeration system

(*ii*) Vapour absorption refrigeration system

(*iii*) Thermo-electric refrigeration system.

Q.5. *Why is it necessary to maintain constant voltage for the refrigerator ?*

Ans. Because of the reason, as high voltage or low voltage causes the excessive current to flow which may damage the motor, relay and capacitor.

Q.6. *How the voltage be maintained constant ?*

Ans. By means of voltage stabilizer or voltage regulator, the applied voltage is kept constant.

Q.7. *Reason for not putting the very hot things into the refrigerator.*

Ans. Very hot things should not be placed in the refigerator because it will over heat the compressor motor.

Q.8. *What is a thermostat switch ?*

Ans. A thermostat switch is provided on all the house hold refrigerators by which the temperature is controlled or maintained within safe value.

Q.9. *What is the function of a relay in a refrigerator ?*

Ans. The function of the relay is to start the motor by putting main winding and auxilary winding of split phase motor across the supply.

Q.10. *In what manner the sizes of refrigerators & air conditioners expressed ?*

Ans. The sizes of the refrigerators are expressed in the range of 65, 135, 165 and 285 litres and the sizes of the air conditioners are mentioned as ½ ton, 1 ton, 1.5 ton and 2 ton etc.

Q.11. What are the main qualities of a good refrigerant used in refrigeration cycle ?

Ans. The main important qualtities of a refrigerant are :

(*i*) Low boiling point

(*ii*) Freezing point should be lower than any temperature to be maintained in the refrigerating system.

(*iii*) Easy to liquefy at moderate temp.

(*iv*) Non poisonous

(*v*) Safe to handle

(*vi*) Non-corrosive to metals

(*vii*) Non-inflammable

(*viii*) Cost should be low

(*ix*) Easily available

(*x*) Easy to detect leakages

(*xi*) Thermal conductivity should be more.

Q.12. *What are the applications of vapour compression cycle ?*

Ans. The vapour compression refrigeration cycle is used in most of the modern refrigerator system in large industrial plants requiring rapid removal of large amount of heat and in domestic refrigerators.

Q.13. *How many stages are of vapour compression cycle ?*

Ans. It consists of four stages :

(*i*) Compression

(*ii*) Condensation

(*iii*) Expansion and

(*iv*) Vaporization

Q.14. *Why is ammonia the only common refrigerant used in the absorption system ?*

Ans. Because ammonia is cheap, readily available and has the highest efficiency.

Q.15. *Give the concept of Desert coolers.*

Ans. Desert coolar is a machine used for places where the humidity content of air is very low and temperature is high. In summer season it is used for **cooling the air.**

1. Fill in the blanks :

(*i*) The most comfortable temperature for human being is

(*ii*) The mostly used refrigerant is ___________

(*iii*) The A.C's are rated in _________

(*iv*) One Ton is equal to _________ kilo calories per-hour.

(*v*) Name a eco friendly refrigerant _________

(*vi*) Out of the refrigeration system, the ___________ system is the most important system from the point of commercial and domestic utility.

(*vii*) In a domestic vapour compression refrigerator, the refrigerant commonly used is _____________

(*viii*) The vapour abortion refrigerator uses ___________ as a refrigerant.

(*ix*) Aqua ammonia is a mixture of __________ and __________

(*x*) The sizes of the refrigerators are ------- rated in ___________ (*xi*) Thermo-electric refrigeration system works on the principle of ____________

(*xii*) For a one Ton air-conditioner, the electric motor used is of ____________ H.P.

(*xiii*) The motor of a water cooler is ____________ duty type.

(*xiv*) A desert cooler works on the principle of ________

Answers

(*i*) 26 °C (*ii*) Freon (*iii*) Tons (*iv*) 3000 (v) R-134 (*vi*) Vapour Compression (*vii*) F-12 (*viii*) Aqua-ammonia (*ix*) Ammonia, water (*x*) litres (*xi*) peltier effect. (xii) 1 (one) (*xiii*) Heavy. (*xiv*) Evaporative cooling.

2. True or False Type Questions

(*i*) The unit of refrigeration is tonne.

(*ii*) A room air-conditioner should preferably, be fitted to a window which is not exposed to the direct sun rays.

(*iii*) The use of air-conditioner at peak load periods should be the minimum.

(*iv*) The main function of a condenser in a refrigeration system is to remove the heat absorbed by the evaporator.

(*v*) One ton of refrigerating machine means that one ton of refrigeration is used.

(*vi*) Defrosting of refrigerator may be done by stopping the compressor for a short period.

(*vii*) Nitrogen is used as a refrigerant.

(*viii*) Ammonia is normally used in ice plants.

(*ix*) Compressor is the heart of the refrigeration system.

(*x*) An evaporator is used is the low pressure side of a refrigerating system.

(*xi*) During refrigeration cycle, heat is rejected by the refrigerant in a condensor.

(*xii*) Room air conditioners have capacity from 3 to 5 tons.

(*xiii*) Grills are used for distribution of air in an air-conditioned system.

(*xiv*) The thermostat switch used for refrigerator can be used for water coolers.

(*xv*) There is no difference between the motors of water cooler and a domestic refrigerator.

Answers

(*i*) True	(*ii*) True	(*iii*) True	(*iv*) True
(*v*) False	(*vi*) True	(*vii*) False	(*viii*) True

(*ix*) True (*x*) True (*xi*) True (*xii*) False
(*xiii*) True (*xiv*) False (*xv*) False.

3. Fill in the blanks by suitable words to complete the following sentences :

(*i*) Refrigeration is a process of .... the temperature.

(*ii*) The suitable humidity for human beings is .... %.

(*iii*) Working conditions in a room are .... by air conditioning.

(*iv*) The principle of refrigeration is "...........".

(*v*) In almost all the refrigerators, vapour .... cycle is employed.

(*vi*) The gas or fluid used in the refrigerator is called ....

(*vii*) Common refrigerants are .... and ....

(*viii*) Vapour compression cycle has .... stages.

(*ix*) The temperature inside a refrigerator is automatically controlled by a ....

(*x*) The topmost compartment of a refrigerator where evaporator coil is wound is known as ....

(*xi*) The size of refrigerators is given in ....

(*xii*) The size of motors used in refrigerators varies from .... H.P. to .... H.P.

(*xiii*) The volatile fluid used in thermostat bulb is .... or ....

(*xiv*) The cabinet light is operated by a .... switch.

(*xv*) The fan motor of water cooler is .... duty type.

(*xvi*) The fan motor of a water cooler is a .... speed motor.

(*xvii*) The motor of a air-conditioner is a .... speed motor.

(*xviii*) The size of room air conditioner is given in ....

(*xiv*) The cycle of operation of room air conditioner can be ....

(*xx*) Voltage regulators are used to .... the voltage.

4. Write 'Yes' or 'No' for the following sentences :

(*a*) By air-conditioning, the temperature of a room is only raised above that of surroundings.

(*b*) The suitable humidity for a human being is about 40%.

(*c*) Refrigerator is thermally insulated from the atmosphere.

(*d*) Cooling caused by evaporation is the basic principle on which the operation of refrigeration and air conditioning is based.

(*e*) Nitrogen is used as a refrigerant.

(*f*) Vapour absorption cycle is used in refrigerator.

(*g*) Ammonia is generally used in ice plants.

(*h*) There is no change of state of refrigerant during operation.

(*i*) The capacity of refrigerator is given in h.p.

(*j*) Thermostat works as a link between electrical circuit and vapour compression cycle or mechanical circuit.

(*k*) Voltage stabilizer steps up the voltage.

(*l*) There is no difference between the motors of a domestic refrigerator and water cooler.

(*m*) The automatic closing tap is always used on water cooler.

(*n*) Air conditioner only controls the temperature of the building.

(*o*) There is no difference between air conditioning and air cooling.

Answers

1. (*a*) reducing (*b*) 60 (*c*) improved (*d*) cooling caused by evaporation (*e*) compression (*f*) refrigerant (*g*) carbon dioxide, Ammonia, Freon, Methyl Chloride (*h*) four (*i*) thermostat (*j*) freezer (*k*) litres (*l*) 7/32, 1/8 (*m*) four F-114, methyl 1 chloride (*n*) door (*o*) heavy (*p*) single (*q*) two (*r*) tons (*s*) reversed (*t*) stabilize.

2. (*a*) No. (*b*) No (*c*) Yes (*d*) Yes (*e*) No (*f*) No (*g*) Yes (*h*) No (*i*) No (*j*) Yes (*k*) No (*l*) No (*m*) Yes (*n*) No (*o*) No.

OBJECTIVE TYPE QUESTIONS

1. A cooling system that does warm up, the atmosphere in the event of low ambient conditions.

(*a*) Room cooler (*b*) Desert cooler
(*c*) Air conditioner (*d*) None out of the these.

2. A coolest part in the refrigeration cycle is

(*a*) compressor (*b*) condenser
(*c*) evaporator (*d*) Receiver.

3. A system in which water is added to cool some part of the system and not the transformed to the space being cooled.

(*a*) Room cooler (*b*) Resent cooler
(*c*) Air cooo1er (*d*) Home out of these.

4. The refrigerant while being pumped into a cooling system by the manufacturer is in the form of.

(*a*) Vapour/Gas (*b*) Liquid
(*c*) Mixed as in (*a*) and (*b*)

5. A cooling system that de-humititises the atmosphere is :

(*a*) Vapour/Gas (*b*) Liquid
(*c*) Mixed as in (*a*) and (*b*)

6. The refrigerant while passing the condenser is in the form of.

(*a*) Vapour/Gas (*b*) Liquid
(*c*) Mixed as in (*a*) and (*b*)

7. The process of defrosting can be

(*a*) Manual (*b*) Semi-auto

(*c*) Automatic (*d*) All as above.

8. Frosting is observed around

(*a*) Manual (*b*) Semi-auto

(*c*) Automatic (*d*) All as above.

9. The cooling system that stops cooling when the humidity is excessive in the space being cooled as :

(*a*) Manual (*b*) Semi-auto

(*c*) Automatic (*d*) All as above.

10. The sealed unit is one system as above encloses :

(*a*) Compressor (*b*) Motor

(*c*) Evaporator (*d*) Compressor and motor both.

11. In a common refrigerator of today, during the period of defrosting the electric supply.

(*a*) Is disconnected (*b*) Remains connected

12. The heat extracted by the refrigerant is dissipated into the surrounding medium through

(*a*) compressor (*b*) condenser

(*c*) evaporator (*d*) Receiver.

13. Temperature setting inside the deep freezer is accomplished through ;

(*a*) Thermal relay

(*b*) Thermostate adjustment

(*c*) Voltage stabiliser setting

(*d*) Some other device in the electrical circuit.

14. In a refrigeration cycle, the flow of refrigerant is controlled by

(*a*) compressor (*b*) Expansion value

(*c*) condensor (*d*) Evaporator

15. Which of the following motor is used in household refrigerators ?

(*a*) D.C. shunt motor

(*b*) Reluctance motor

(*c*) Single phase induction motor

(*d*) Synchronous motor

16. Types of water coolers are :

(*a*) Instantaneous or pressure type

(*b*) Storage type

(*c*) Exhaust fan type

(*d*) both (*a*) and (*b*)

17. Air-conditioning is the simultaneous control of ----------- in a confined space.

18. The conditioned air supplied to the room must have the capacity to take up

(*a*) Room sensible heat only (*b*) Room latent heat only

(*c*) Room sensible heat and room latent heat

(*d*) None of these.

Answers

1. (*c*) 2. (*c*) 3. (*c*) 4. (*a*) 5. (*c*) 6 .(*b*) 7. (*d*) 8. (*c*) 9.(*a*) 10. (*d*) 11. (*b*) 12. (*b*) 13. (*b*) 14. (*b*), 15. (*c*), 16. (*d*), 17. (*d*), 18 (*c*)

EXERCISE

1. Explain by means of a sketch refrigeration cycle. What is the essential difference in the cooling system of refrigerator, air conditioner and a water cooler.
2. (*a*) With the aid of a sketch of a thermostat, explain how the temperature inside a refrigerator is adjusted.
   (*b*) Why it is not advisable to put very hot things into the refrigerator?
   (*c*) Why is an automatic door closing mechanism provided in a modern refrigerator ?
3. What is the fault in refrigerator due to when motor continues the work even though temperature inside the refrigerator is very low ?
4. Draw electric circuit diagram of an air-condition and explain its working. What is the main difference between the fan motor of an air-conditioner and a water cooler.
5. Draw electric circuit diagram of a refrigerator and explain the function of starting relay and overload release. How can the temperature inside the refrigerator be adjusted.
6. (*a*) Describ e vapour compression refrigeration cycle and show how cooling effect is produced.
   (*b*) Why is it necessary to maintain constant voltage for the refrigerator ? How will you maintain the constant voltage.
7. (*a*) Describe the construction and working of a domestic refrigerator. Also draw its electric circuit diagram.
   (*b*) How are the sizes of refrigerators and air-conditioners expressed ? Give a brief note on the characteristics of an air conditioners.

Questions Bank with Answer

Short Answer Type Questions.

Q.1. Define Refrigeration ?

Ans. The Refrigeration can be defined as the process of removing heat from a substance to be cooled and passing it on to some another substance under controlled conditions. It also includes the process of lowering and maintaining the temperature of a cooled body below the general temperature of its surroundings.

Q.2. Define Air conditioning.

Ans. Air conditioning means the simultaneous control of heating/cooling, cleaning, humidifying and air circulation either for comfort of human beings or of animals or for the proper performance of some industrial or scientific process.

Q.3. Which type of motor is used in compressor of window air conditioner?

Ans. Single phase capacitor start and capacitor run induction motor.

Q.4. Which type of motor is used to drive compressor of domestic refrigerator?

Ans. Split phase single phase motor.

Q.5. What is a thermostat switch?

Ans. Thermostat switch is used in refrigerators to control its temperature.

Q.6. Unit of Refrigeration is ....

Ans. Tons of refrigeration.

Q.7. What do you mean by "Tons of refrigeration"?

Ans. It is the practical unit of refrigeration. It is defined as the amount of refrigeration effect produced by melting of 1 Ton of ice at 0°C in 24 hours. It is equal to heat extraction at the rate of 50 K cal per minute. Large commercial refrigeration and air conditioning plants are rated in Tons.

Q.8. What is a refrigerant?

Ans. Refrigerant is a substance which is used in refrigeration and Air conditioning system. It absorbs heat from a cold body (or space) and delivers to a hot body (or space).

Q.9. Name the commonly used materials as refrigerants.

Ans. (i) Air (ii) NH_3 (iii) CO_2 (iv) Chloro-fluoro carbon

Q.10. Mention at least two desirable properties of an ideal refrigerant.

Ans. (i) It should be colourless, odourless and non-poisonous.

(ii) It should have low specific heat.

(iii) It should have high value of latent heat of vaporisation.

Q.11. Which system of refrigeration is commonly used in modern refrigerators and air conditioners?

Ans. Vapour compression refrigeration system.

Medium/Long Answer Type Questions.

Q 1. Explain vapour compression refrigeration cycle.

Ans. This is the most popular system of refrigeration employed in domestic and industrial refrigeration and air conditioning plants. It works on the principle that when liquid evaporates, it absorbs heat. Once the refrigerant is charged in the system, the same refrigerant is used again and again. The main components of vapour compression refrigeration system are shown in fig. 6.1

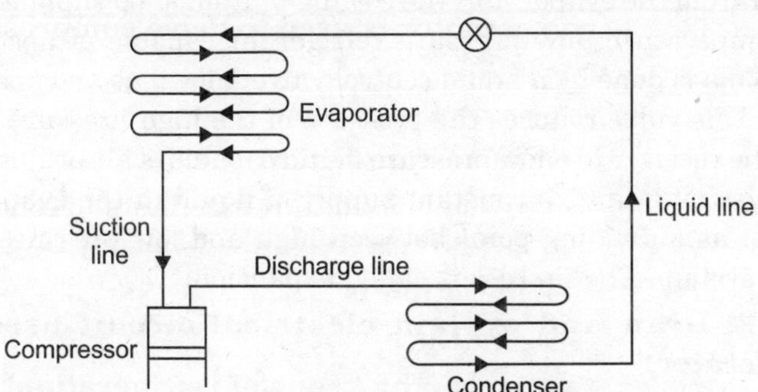

Fig. 7.16

The vapour compression cycle is chiefly composed of four stages ie compression, condensation, expansion and vaporisation. The purpose of refrigeration cycle is to remove heat from refrigerant by again putting it into liquid form so that it may be used again and again.

The four main stages of vapour compression cycle are discussed as below.

(i) Vaporisation: From the expansion valve, the low pressure refrigerant at low temperature enters the evaporator which is placed near or around the material to be cooled down. Before entering the evaporator, the refrigerant is in the liquid state and when it absorbs heat from the material to be refrigerated it is transformed from a liquid to a vapour state. The process is called evaporation.

(ii) Compression: The compressor draws the vapours from the evaporator through the return piping and then compresses these vapours until their temperature is raised above that of the condensing medium. The compressor is driven by external mechanical energy which may be supplied by an engine or electric motor. Generally, electric motors are employed for this purpose. This process is called compression.

(iii) Condensation: When the temperature of the vapour is raised above that of the condensing medium by the compressor, then the heat of vaporization will flow from the vapour to the condensing medium to condense the refrigerant to a high pressure liquid. This high pressure liquid then flows to the receiver where it is stored until it is supplied to the cooling unit through the expansion valve. This process is called condensation.

(iv) Expansion: As the compressor withdraws the refrigerant vapour from the evaporator, the cooling unit must be supplied with low temperature, low pressure refrigerant capable of absorbing heat. This is done by a liquid control valve known as an expansion valve. This valve reduces the pressure of the high pressure liquid from the receiver to a low pressure liquid capable of absorbing heat. This also maintains a constant supply of liquid in the evaporator and act as a dividing point between high and low pressure sides of the system. The process is called expansion.

Q.2. Draw and explain electrical circuit used for refrigerator.

Ans. Schematic diagram of domestic refrigerator is shown in fig. 6.2 with various components.

Working: When the electric supply is connected to the refrigerator, the current passes through the thermostat switch to the thermal over load release coil C_1C_2 of the starting and main winding M_1M_2 of the motor. Since the motor is at rest, it draws heavy current. The heavy current passes through the coil C_1C_2 of the starting relay and the coil gets energised, pulling the plunger upward connecting points a_2 to a_1 *i.e.,* putting auxiliary winding S_1S_2 also into circuit. The moment both the windings are energised, the motor starts running. When the motor gains normal speed, the current drawn by the main winding also becomes normal. At normal current, the plunger in the coil of starting relay cannot remain pulled up and is released down opening the points a_1, a_2 and auxiliary winding S_1S_2 gets out of circuit. The main function of the starting relay is to put auxiliary winding in the circuit at the time of starting the motor and to disconnect it when the motor gains normal speed. When starting relay does not close, the motor will not start and in case after closing it does not open, the thermal overload will trip or the fuse will blow off.

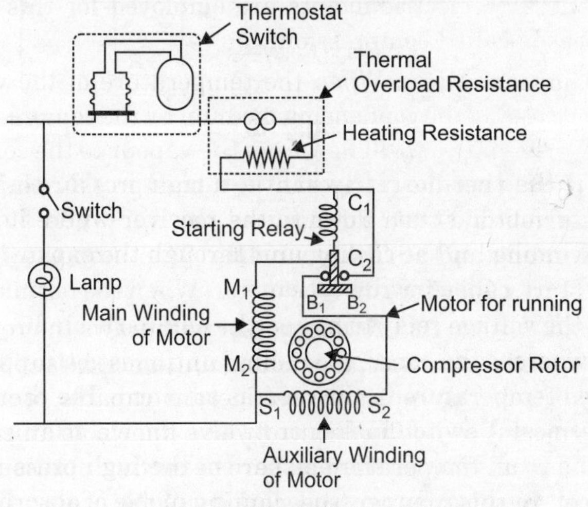

Fig. 7.17

Q 3. Draw and explain electrical circuit of air conditioner.

Ans. Electrical circuit of an air conditioner is shown in fig.7.18 When the supply is connected to the thermostat switch at point a which in turn connects to the selector switch. The fan motor and compressor motor are in off position. When the selector switch knob is turned clockwise, the fan motor gets connected across the supply and the fan motor starts working. Normally, two speed selector switch is provided *i.e.* low and high.

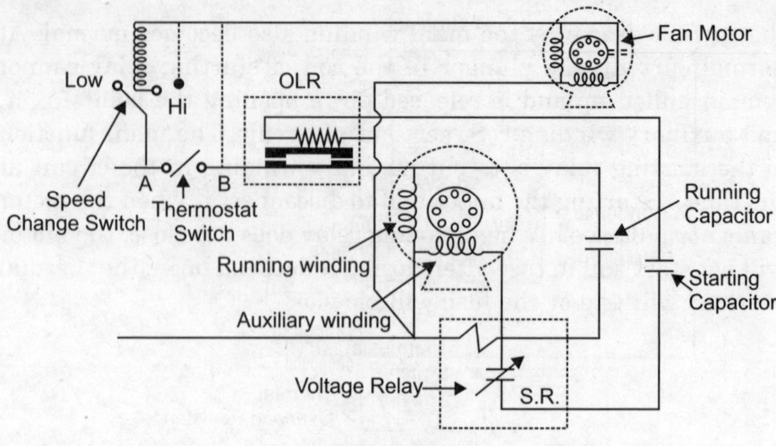

Fig. 7.18

Now if the thermostat switch is turned in ON position, the compressor motor starts running as the supply is fed to both running winding and auxiliary winding. The compressor motor is capacitor start, capacitor run type motor. When the normal speed is attained, the voltage relay trips and the auxiliary winding goes out of circuit whereas the running winding continues getting supply till the desired temperature in the room is attained. The thermal bulb of the thermostat switch makes or breaks the electric circuit after sensing the room temperature as per the setting position low cool or high cool. In this manner, the working of the compressor motor *i.e.* starting and stopping becomes automatic. In case the supply voltage falls down, the load on the motor becomes more, the over load release comes into function. Thus the compressor motor gets disconnected from the supply and becomes quite safe.

Q.4. Draw the electric circuit of a water cooler and explain its working.

Ans. The electric circuit is shown in the fig. 7.19.

There are two motors. One for running compressor, is a capacitor start capacitor run induction motor. When the motor picks up the speed, the relay coil cuts off the starting winding (SW) alongwith the starting capacitor (Cs) and the motor continues to run on the running winding (RW) only.

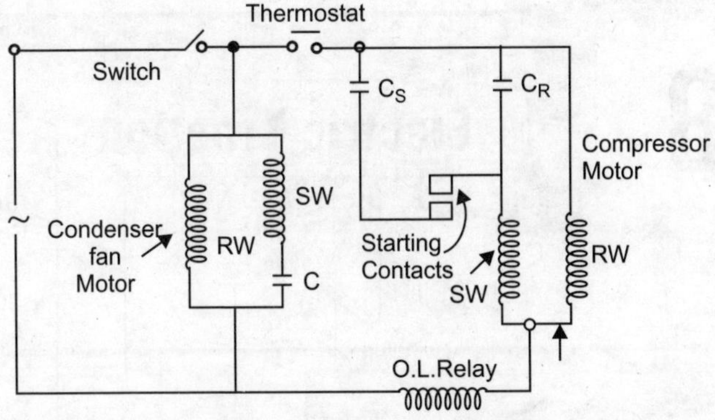

Fig. 7.19

There is a separate motor for the fan, which is used to cool the condenser. This is *a* capacitor start motor but capacitor is not removed during running, also its operation is independent of the thermostat.

The thermostat disconnects the motor from the supply when the water is cooled to the desired temperature. An O.L relay is also provided, which protects the motor from overloads.

Q 5. Differentiate between water cooler and refrigerator. Draw electrical circuit for water cooler and refrigerator.

Ans. Refrigerator and water cooler are both based on vapour compression refrigeration cycle. Refrigerator is used to preserve food stuff, medicines etc whereas water cooler is used to cool water. Refrigerator consists of single motor for driving the compressor. The water cooler consists of two motors, one for driving the fan for cooling the condensor and other for driving the compressor. Overload protector and thermostat switch are used in both water cooler and refrigerator but in water cooler the bulb of thermostat is dipped in the water chamber.

Electric circuit of water cooler : Refer to Q.4

Electric circuit of refrigerator : Refer to Q. 2

8 Electric Traction

8.1. Introduction

Electric traction means a locomotion in which the driving force is obtained from electric motors. One of the practical applications of electricity, which enters into the everyday life of many of us, is its use in the service of transport - the electric propulsions of vehicles - electric trains, trolley buses, tram cars and in the latest developments such as diesel-electric and gas turbine electric vehicles. There are many advantages of electric traction over other forms of locomotion. If the electrical energy could be generated cheaply and initial cost, which is enormous, could be born, then this system of locomotion would replace steam locomotion entirely for electrification.

8.2. Ideal Traction System

The most important requirements of the driving equipment for traction service are as follows :

(*i*) Maximum tractive effort should be exerted at starting in order that a rapid acceleration may be attained.

(*ii*) The equipment should be capable of overloads for short periods.

(*iii*) The wear caused on the track should be minimum.

(*iv*) The locomotive or train unit should be self-contained and able to run on any route.

(*v*) Braking should be possible without excessive wear on brake shoes and if possible the braking energy should be regenerated and returned to the supply.

Electric traction is one of the major utilities of electricity at present. Day by day, the transportation is changing over to electric traction, for obvious advantages.

8.3. Advantages of Electric Traction

Following are the advantages of electric traction system :

(*i*) **Cheapness.** It is cheapest method of all the other methods of traction.

(*ii*) **Acceleration and Braking.** These are smooth and rapid.

(*iii*) **Cleanliness.** It is free from smoke and flue gases and hence it is most suitable type of traction system to be used for underground and tubular railways.

(*iv*) **Maintenance Cost.** Maintenance and repair cost of electric traction is about 50% than that of steam traction system.

(*v*) **Starting time.** It can be started without any loss of time whereas steam traction requires minimum 2 hours before a steam locomotive can be put into operation.

(*vi*) **High starting torque.** Electric traction system makes use of d.c. or a.c. series motor which has a very high starting torque ; high starting torque makes acceleration rate very high.

(*vii*) **Braking.** In electric traction, regenerative braking is used which feeds back about 40% of the energy.

(*viii*) **Saving in high grade coal.** Steam locomotive requires high grade coal which is deficient in our country. Electric energy required for running electric locomotive is taken either from hydropower station or from a thermal power station which is run from low grade coal.

(*ix*) **Better co-efficient of adhesion.** In electric traction co-efficient of adhesion is better due to absence of unbalanced forces produce by reciprocating masses. This reduces the weight to H.P. ratio of locomotive and causes less wear and tear of the track.

(*x*) It has great passenger carrying capacity at higher speed as compared to steam locomotive.

(*xi*) The fans and lights in train can be connected directly to the supply lines and there is no need for providing extra generators and batteries.

(*xii*) **Centre of gravity.** The height of the electric locomotive is quite less so the centre of gravity is lower as compared to steam locomotive whose height is more. The lower height of locomotive enables it to take curves even at higher speed.

8.4. Disadvantages of Electric Traction

Following are the disadvantages of electric traction :

(*i*) Higher initial expenditure is involved in electric traction.

(*ii*) Failure of supply is a problem to be faced in electric traction.

(*iii*) The electrically operated vehicles have to move only on electrified track.

(*iv*) For the achievement of electric braking and control, additional equipment is required. In case of D.C. series motor the regenerative braking can not be easily achieved.

(*v*) When a.c. energy is utilised for traction then precautions are to be taken to prevent the distribution network to interfere with the adjacent telegraph and telephone lines.

8.5. Systems of Track Electrification

In India electrification of the track has been confined to the urban and suburban lines in the vicinity of large cities *i.e.* Bombay, Calcutta, Madras and now from Delhi on-ward also. Two types of vehicles are used for electric traction. The first type of vehicles receive power from a distribution net work while the second type of vehicles generate their own power. In the second category comes, diesel engine electric drives. The former type of vehicles are used on a.c. or d.c. power from the overhead line. Accordingly there are two main systems of electric traction.

1. **D.C. Systems.** The d.c. traction in India exists only in Bombay region (Western Railway) and some parts of Madras. In this system the electric motors used are d.c. series motors. The operating voltage is generally about 600 volts for suburban railways and tram cars. For main line railways the operating voltage is from 1500 volts to 3000 volts. The motors receive power from an over-head line with the help of a pantograph and the railway steel track is the return conductor. The overhead wire is fed from the various substations. These sub-stations receive power from three phase 11kV or 3–ϕ, 33 kV or 3ϕ, 66 kV transmission lines. Then this a.c. power is converted into d.c. power by using mercury arc rectifiers or rotary converters. The modern trend is to use semiconductor rectifiers. The distance between sub-stations depend upon the factors like voltage drop and density of traffic. For sub-urban services the distance between the sub-stations is 3 to 5 km whereas for main line service it is about 40 to 50 km.

2. **A.C. Systems.** The a.c. system in India is being employed from Howrah to Tundla [Recently extended upto New Delhi] and from Madras to Tambram. The modern development in electric traction is to use 1–ϕ a.c. supply. The a.c. system used in practice is further of four types :

(*a*) **Three Phase A.C. System.** This system employs 3–ϕ slip ring induction motors, speed control being obtained by a combination of pole changing and rotor resistance method. The main advantage of this system is that regenerative braking is obtained immediately as the speed exceeds the synchronous speed, without any change is conditions. The voltage and frequency at which the motor iN made to operate are about 3600 V and $16\frac{2}{3}$ cycles/second.

The major disadvantage is the use of two overhead conductors. (Third being the rail itself) and hence the system is almost out of use.

(*b*) **Single phase standard frequency system.** This system is also known as composite system of traction. This system is employed in India on South-Eastern and Eastern railway. The system has a single overhead wire supplied at 25 kV 50 c/s which is the standard industrial frequency. A transformer is mounted on the locomotive and it steps down the voltage which is further rectified and supplied to the traction motors.

The sub-stations are supplied at a high voltage of upto 132 KV which is stepped down to 25 kV by transformer installed at each substation.

With the development in semiconductor rectifiers, the system has become very popular and is being increasingly adopted throughout the world. All modern electric traction work is being done by this system. As compared to mercury arc rectifiers, semi-conductor rectifiers are more reliable, more efficient and simple to maintain. The driving force is obtained from D.C. series motors.

(*c*) **Single phase low frequency system.** Single phase 15 KV, $16\frac{2}{3}$ cycles/sec system is used in West Germany, Sweden and Australia for the main line service. In some urban areas of U.S.A. 11 kV at 25 cycles/sec. is also employed. A step down transformer is carried in the traction unit which steps down the voltage to about 400 V for the use of traction motors.

Each substation is supplied at a high voltage at the standard frequency. The voltage at the substation is stepped down and frequency is also converted by a motor-alternator set. Series motor is employed for traction. Due to the commutating difficulties at normal frequency on such motors, a low frequency supply is essential. The main disadvantages of this system is that a special low frequency power distribution network is required.

(*d*) **Single phase to three phase system** *[Kondo System]*. In this system 1–ϕ high voltage a.c. system is employed for distribution network. The locomotives carry a phase converter which converts 1–ϕ a.c. into 3–ϕ. The 3ϕ supply is connected to three phase induction motor for getting the necessary driving force. The system has the advantage of low cost distribution and cheap and robust construction of induction motor (I.M.). The voltage used for distribution network, is 16,000 V at 50 cycles/second. This system is adopted in Hungary.

8.5.1. Method of Supply Power to railway trains

There are two systems for supplying power to railway trains (*a*) overhead system (*b*) conductor rail system.

(*a*) **Overhead System.** This system is adopted usually when distribution network is fed at high voltage. The current is collected from the overhead network with the help of collector. The rail track is used as return path of the conductor.

(*b*) **Conductor Rail System.** It is adopted for heavy electric traction. High conductivity steel rails are used for supplying power. These conductor rails are properly insulated. The current is collected with the help of collector shoe. The use of this system is restricted upto 0.8 kV only.

8.6. Advantages of 25 kV A.C. System over D.C. System

Advantages of 25 kV A.C. system over other types of electrification systems particularly D.C. system are given below :

(1) **Light overhead catenary.** Use of high voltage in the overhead system reduces the current in the line which makes use of small sized conductors. Small sized conductors being light, it is very easy to support them. Thus supporting structures required are quite light.

Although the high voltage needs more insulation which becomes quite expensive but reduction in size of conductor has overwhelming effect.

(2) **Saving in sub-stations.** In high voltage A.C. system, since the current drawn is less, so voltage drops which are mainly due to reactance of the line are also quite less. This provides longer distances between two sub-stations.

(3) Since the distance between the sub-stations is large, so more flexibility can be used in setting up sub-stations as compared to D.C. system in which the distance between the sub-stations is quite small. Proper location of sub-station near to national high voltage grid which incidentally in this country runs closer to the railway tracks, reduces the capital expenditure on the transmission line which is to feed the sub-station.

In case of D.C. system, since the sub-stations cannot be appropriately located near the grid, so separate transmission lines have to be run for feeding the sub-stations which becomes quite expensive.

(4) A.C. sub-stations have lesser equipment and it is easy to maintain them.

(5) Starting efficiency of A.C. locomotive is higher than that of D.C. locomotive.

(6) 25 kV A.C. system is much cheaper to D.C. system as far as cost of fixed installation is concerned.

8.7. Disadvantages of 25 kV A.C. System

(*i*) Single phase A.C. system imposes unbalancing effect on the supply system.

(*ii*) It produces interference with telecommunication lines.

8.8. Block Diagram of an A.C. Locomotive

The Block Diagram of a A.C. Locomotive is shown in Fig. 8.1. It collects the current from an overhead line through pantograph and supplies to the motors. The main function of various components of an a.c. locomotive systems are as under :

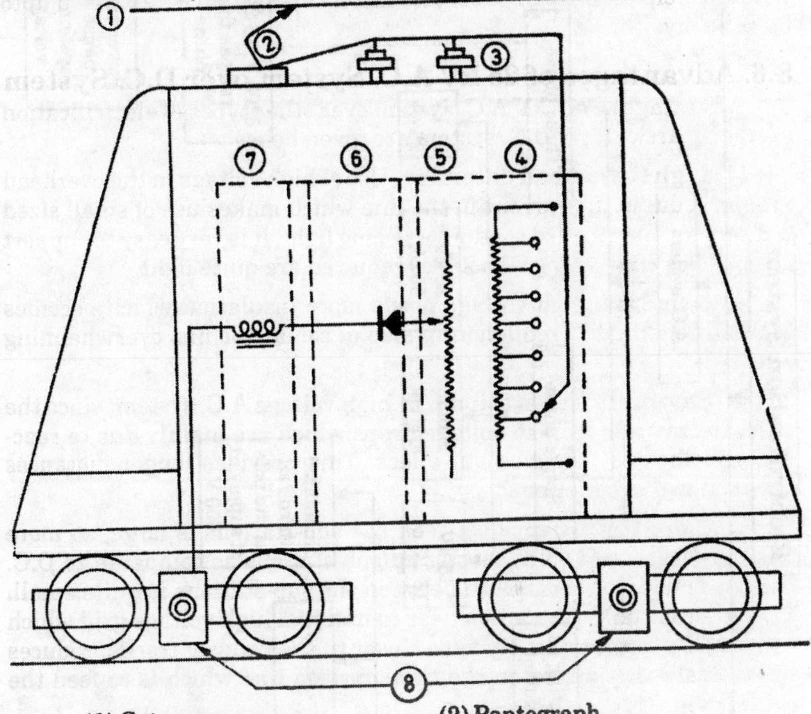

(1) Catenary. (2) Pantograph.
(3) Circuit-breakers. (4) On load tap changer.
(5) Transformer. (6) Rectifier.
(7) Smoothing reactor. (8) D.C. motor used for traction.

Fig. 8.1. A.C. locomotive.

1. **Catenary.** This is an overhead wire of special cross-section. The catenary is supplied at 25 kV 50 c/sec. supply A.C. High voltage is tapped from catenary by pantograph.

2. **Pantograph.** This is a device used for collecting current from the catenary. Pantograph is mounted on the top of roof of the vehicle and carries a sliding shoe for contact with overhead catenary. The pantograph is raised or lowered as and when required, from the driver's cabin by air pressure.

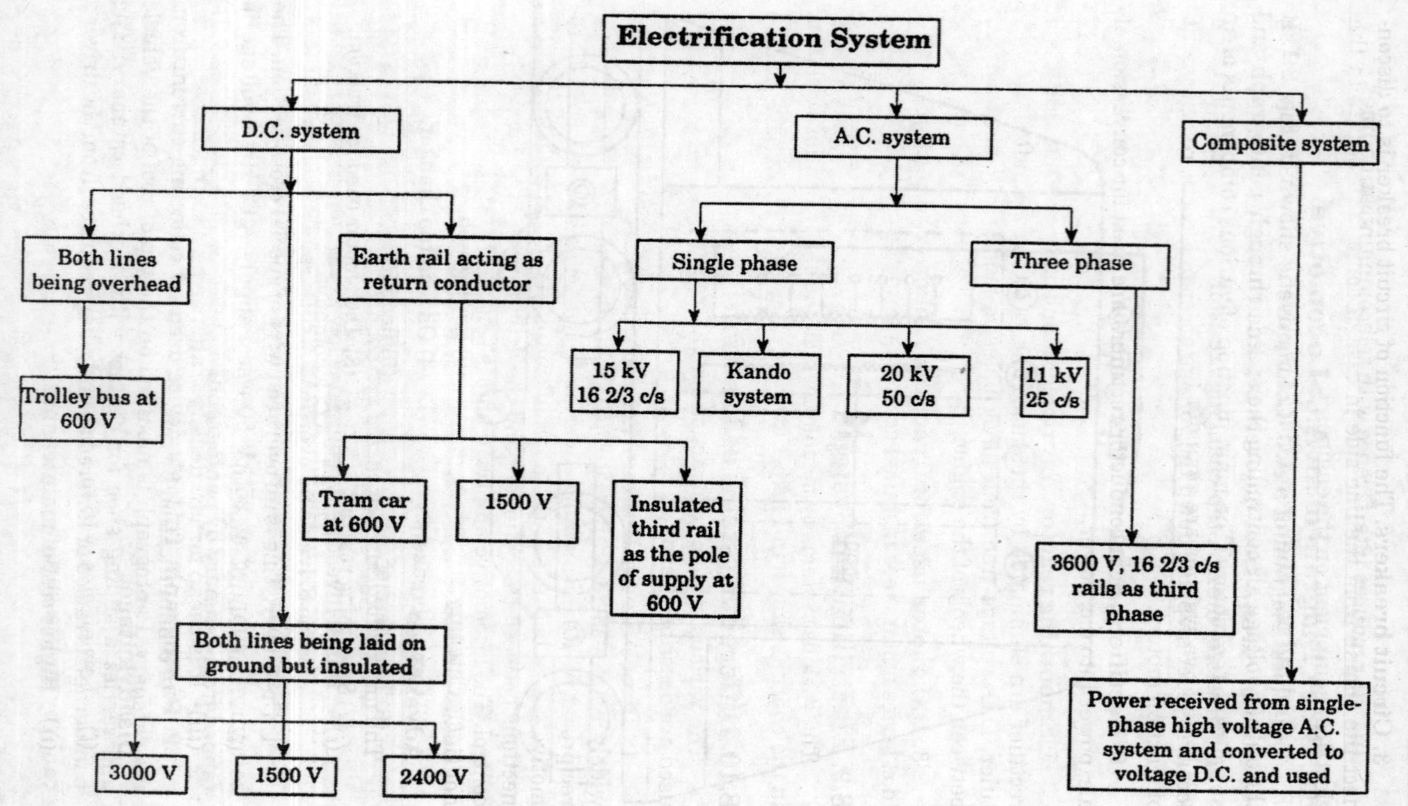
Electrification System
D.C. system
A.C. system
Composite system
Both lines being overhead
Earth rail acting as return conductor
Single phase
Three phase
15 kV 16 2/3 c/s
Kando system
20 kV 50 c/s
11 kV 25 c/s
Trolley bus at 600 V
Tram car at 600 V
1500 V
Insulated third rail as the pole of supply at 600 V
3600 V, 16 2/3 c/s rails as third phase
Both lines being laid on ground but insulated
Power received from single-phase high voltage A.C. system and converted to voltage D.C. and used
3000 V
1500 V
2400 V

Fig. 8.1 (*a*)

3. **Circuit breakers.** The function of circuit breaker is to disconnect the engine from the line side in case of some fault occurs in the electrical equipment on the engine.

4. **On load tap changer.** On load tap changer is used for varying the output voltage for controlling the speed of motor.

5. **Transformer.** A tap changing transformer is installed in the locomotive for stepping down the voltage according to the requirement of traction motors.

6. **Rectifiers.** Semiconductor rectifiers are used for conversion of a.c. power into d.c. power.

7. **Smoothing reactor.** Any ripples found on the output side of the rectifier are eliminated by the smoothing reactor fitted after the rectifier. D.C. voltage free from any ripples is fed to D.C. motor fitted between the wheels of the engine.

8. **D.C. motor used for Traction.** A number of traction motors (d.c. series motors) are installed according to the power of locomotive.

8.9. Electrification Systems

The various systems by means of which the electrification is shown in the chart as shown in Fig. 8.1 (*a*).

8.10. Different Systems of Traction

The various systems of traction for rail road operation commonly used are described below :

1. **Direct Steam Engine.** Reciprocating steam engine is most widely employed for rail road operation. In this type of drive, the reciprocating steam engine is invariably used for getting the necessary motive power because of its inherent simplicity, the simplicity of connections between the cylinders and the driving wheels and easy speed control. The locomotive and train unit is self-contained, therefore, it is not tied to a route.

Advantages

The main *advantages of such a system are :*

(*i*) Initial investment required is low in comparison with that of electric drive as no overhead structure, distribution system and power generation is required.

(*ii*) It has simple controls.

(*iii*) It is cheap for low density traffic areas and in initial stages of communication by rail.

Disadvantages

(*i*) Low efficiency (6 to 8 per cent).

(*ii*) Higher cost of maintenance.

(*iii*) Limited over load capacity.

(*iv*) Greater wear of track because of the oscillating parts, atmospheric contamination and corrosion of steel structures and buildings due to smoke.

2. **Direct Internal Combustion Engine.** The characteristic of an I.C. engine is that it produces almost constant torque at all speeds. To increase the starting torque and also for speed control a gear box has to be provided. Such an arrangement is only applied to small trolleys, wherein a petrol engine is used. The efficiency of internal combustion engine at its normal speed is about 25 per cent.

Advantages

(*i*) Initial investment required, being only the cost of vehicle and garage is very low.

(*ii*) Speed control, and breaking system employed is very simple.

(*iii*) It is cheap drive for the outer suburbs and country districts.

Disadvantages

(*i*) Its overload capacity is limited on account of the speed torque characteristics of the engine.

(*ii*) Speed control is possible only by employing a gear box.

(*iii*) The maintenance and running costs are higher.

(*iv*) As the fuel oil is to be imported in our country, therefore, there is drain on foreign exchange which is vital for expanding economy of the country. In war time this system may be completely paralysed.

(*v*) It is not suitable for heavy railway work as the arrangement possible for speed control and utilizing the torque of engine to full capacity when running in full speed is not altogether satisfactory for it.

3. **Internal Combustion Engine Electric Drive.** In this system the reduction gear and gear box are eliminated as a diesel engine drives an D.C. generator coupled to it at a constant speed, which supplies power to electric motors fitted with the wheels. The *locomotives* of this type of drive are becoming widely used because these have been found considerable favour for railway work.

Advantages

(*i*) No modification of existing track is required for conversion from steam traction to diesel engine electric traction.

(*ii*) Low capital cost, as it does not involve track electrification.

(*iii*) Greater tractive effort as compared to steam engine and hence greater rate of acceleration at starting.

(*iv*) Better riding qualities due to having smooth acceleration and breaking.

(*v*) Time for which the locomotive is out of service for maintenance is less and hence higher availability.

(*vi*) Absence of smoke and dirt and hence convenience of passengers.

(*vii*) Overall efficiency is greater than that of steam locomotives (about 25% in comparison of 8% or so in case of steam locomotives).

Note. For the above mentioned reasons this system is widely used for high speed main line operation in countries where oil is plenty. It is also ideally suited on long routes passing through deserts, where coal and water for a steam engine would have to be transported.

Disadvantages

(*i*) Maintenance and operating cost is high.

(*ii*) Over load capacity is limited as diesel engine cannot be over loaded.

(*iii*) Life of diesel engine is comparatively shorter.

(*iv*) Special cooling system is required for cooling the diesel engine in addition to motor-generator set.

(*v*) For the same power diesel electric locomotive is costlier than steam or electric locomotive because it carries diesel engine, main generator and traction motors *oil capable of handling full locomotive power.*

Note. A comparison could be effective between a steam locomotive and a diesel electric locomotive, by way of their comparative fuel consumption. While a steam locomotive uses 28 kg of coal per km, a diesel electric locomotive uses only about 7.5 kg of diesel oil per km. Of course, the initial cost is much higher, about twice as much.

4. Battery Electric Drive. In this drive the *locomotive* carries the secondary batteries which supply power to D.C. motors employed for driving the vehicle. The capacity being small, it is used for shunting in railway yards, for traction in mines, for local delivery of goods in large towns and large industrial plants.

Other advantages are low maintenance cost and absence of fumes.

5. Electric Drive. This is the most extensively developed and widely used system of traction. In this system the vehicle draws electrical energy from the distribution system (From a contact wire suspended above the track or from an additional conductor rail laid along side it) fed at suitable points from either a centration power station or sub-stations. The advantages and disadvantages of electric drive are given below :

Advantages

(*i*) Due to the absence of smoke and dust it is very clean.

(*ii*) Maintenance and repair cost is about 50 per cent of that of steam locomotive.

(*iii*) Rapid acceleration and braking increases the schedule speed. Thus more trains can be run during peak hours and greater traffic handled.

An electric locomotive has an acceleration of 1.5 to 3.5 km. p.h.p.s. for urban service. Trolley buses have acceleration of 4 to 6 km p.h.p.s. The steam locomotive has an acceleration of only 0.8 km. p.h.p.s. Hence the schedule speed may be from 50 to 100 per cent higher than that corresponding to steam operation over short runs and with frequent stops.

(*iv*) It effects fuel saving where hydro-electric power is cheap.

(*v*) Traction motors can withstand high over loads for short time, whereas steam and diesel locomotives cannot. This is helpful on mountain grades.

(*vi*) For city and suburban service the flexibility of operation saves a lot of time at the terminus. An electric train with multiple unit control can be operated from both ends and is ready for the return trip without much loss of time.

(*vii*) Electric locomotive can be put into service immediately whereas steam locomotive takes about two hours to get up steam.

Disadvantages

(*i*) In vicinity it causes interference to the telephone lines.

(*ii*) Whole system becomes standstill in case of power failure.

(*iii*) A very high capital cost is involved for power generation and the associated transmission and distribution.

Conclusion

Note. The main consideration in the electrification of railways or other traction services is one of the economics. If operation by electric traction proves to be cheaper than any other method (*i.e.* electrical energy is available at cheap rates and the density of the traffic is high) and necessary capital can be raised then no other factor or disadvantages would prevent the conversion of steam railway to electric railway as there is no technical difficulty in adoption of electric traction.

8.11. Electric Traction Systems

Electric traction systems may be divided into two categories or groups namely :

(1) The group consisting of self-contained locomotives. This is further sub-divided into :

(a) Battery driven road vehicles.

(b) Diesel electric trains and ships.

(c) Petrol electric trucks and lorries.

(2) The group consisting of vehicles which receive electric power from a distributing network fed at suitable points from either a central power stations or sub-stations suitably speed. This group is further sub-divided into :

(a) Systems operating with A.C. such as railways.

(b) Systems operating with D.C. such as trolley buses, tramways and railways.

8.12. Railway Service Categories

Railway service may be classified into following categories.

(1) **Main Line Service.** This involves operation over a long distance within frequent stops. The operating speeds are high and the accelerating and braking periods are relatively unimportant.

(2) **City Service.** This involves operation over sections where the distance between the stops in the order of a kilometer. It requires a high rate of acceleration and braking in order to maintain reasonable schedule speed.

(3) **Suburban Service.** In this case the distance between the stops is larger, up to about 6 kilometers. Here, also high values of acceleration and braking are required.

8.13. Review of Existing Electric Traction Systems in India

Before independence in our country, electric trains were working in western region near Bombay and in southern region Madras. The tram cars, however were also working in some cities. All these systems operated, and are still operating on D.C. supply of 1500 V and 600 V for trains and tramcars respectively. After independence traction system working on single-phase A.C. and diesel electric locomotive were developed. In the Eastern Zone of our country sufficient progress has duly been made in laying 25 kV at 50 c/s single-phase system and work is still in progress.

Single-phase 15 kV, $16\frac{2}{3}$ c/s system is used in Germany. Sweden and Austria for the main line service. Single-phase 11 kV at 25 c/s is employed in some countries 25 kV at 50 c/s used in our country on south Eastern and Eastern Railways. In Kando single phase 16 KV, 50 c/s supply is used. A phase converter which is installed inside the locomotive changes this single-phase supply to 415V, 3-phase supply at the same frequency and this supply runs a 3ϕ induction motor. Three-phase system is used in some mountain regions of some countries such as Italy, France, and Switzerland.

Ultimately the whole of the country will have a net work of electric trains with the development of power projects.

Railways made their first appearance in India in April 1853 when a section from Bombay to Kalyan (32 miles) was opened to traffic. This was followed by Calcutta-Raniganj (120 miles) and Madras-Arkanam line (39 miles). The management of railways was, left to private companies who used to run the system in different territories on 99 year lease basis.

Electric traction was first introduced on British Railways in the year 1890 at 600 V D.C. using third rail. In India first electrification was undertaken during the period 1925 to 1932. This included 304 route km or 692 track km. On Central Railway between Bombay V.T. to Igatpuri and Poona, 62 route km or 210 track km on Western Railway between Church Gate and Virar, 32 route km. or 77 track km of meter gauge between Madras Beach and Tambaram. This gives total of 398 route km or 979 track km. of electrification which remained static for a period of about two decades. As regards electrification around Calcutta is concerned, project studies had started in 1920, but it was only in 1954 when traffic had become three times within period of seven years after partition of India that it was decided to electrify main line sections between Howrah to Burdhawan and Sheoraphuli to Trakeswar branch line on 3000 V D.C. system. This work of 142 route km or 443 track km was completed in August 1958. After 1954 the developments were so rapid that Government of India obtained a report from French National Railways on comparative merits and demerits of A.C. and D.C. system. From these reports it was found that A.C. system using industrial frequency was advantageous both financially and operationally. Railways Board in 1957 after making sure that single phase traction load will not produce adverse effects of unbalance and harmonics in three-phase power supply, decided to adopt 25 kV, 50 c/s single-phase system of electrification for all future schemes.

So far up to the end of March 1972, 3982 route km or about 10,200 track km have been electrified. On this basis India now occupies second position in Asia, first being Japan. Among the world, India occupies eleventh position. It is expected to electrify about 1702 route km during the fourth five year plan.

As proportion of the broad gauge route km the electrified track of 3928 route km now amount to approximately 13% of total B.G. route km. For the further electrification it is contemplated that the major track routes interlinking the metropolitan cities of Calcutta, Delhi, Bombay and Madras should be electrified because the traffic density on these routes is very heavy while the actual length of these track routes, constitutes only about 14% of the total B.G. track metreage, they carry almost 50% of the goods traffic.

8.13.1. Comparison Between Pure A.C. and D.C. Systems

1. DC series motors are cheaper, lighter for equivalent h.p., higher efficiency and needs less maintenance as compared with AC series motors.
2. DC series motors are capable of giving high acceleration than a.c. series motors because former develops more starting as well as running torque for same size than later.
3. Many speeds are possible with a.c. series motor by using tap changer method whereas in case of d.c. series motor except chopper method number of speeds are limited.
4. Number of substations needed for a given track kilometerage will be less with a.c. than with d.c. system.
5. Regenerative braking is more efficient and has lesser complications in case of d.c. than with a.c. system.
6. Overhead distribution system is lighter and less costly in a.c. than in d.c. system.
7. Rail conductor system of track electrification, which is less costly is possible with d.c. system than with a.c. system.
8. A.C. system has high interference with communication lines than d.c. system.
9. Due to high impedance of track, in case of a.c. system more elaborate negative boosting is required than in d.c. system.

8.14. D.C. and A.C. Motors for Traction Service

Series and compound motors are employed for D.C. traction systems whereas single-phase series and three-phase induction motors are used on A.C. systems. The service conditions being extremely severe, the motors have to conform to a very robust design and must be necessarily of totally enclosed type to protect them from water and dust etc.

8.15. General Requirements/Features

1. The motors used for traction purpose must have high torque during starting and acceleration in order to accelerate the heavy mass.
2. Motor should have suitable speed-torque characteristic.
3. In traction work usually more than one motors are required, so it is necessary that these motors should be capable of being operated in parallel. For small speed variations of motors,

there should not occur vide differences in the torques developed and currents drawn by various motors *i.e.* speed-torque and speed-current curves should not be flat.

4. During start heavy current rushes which causes considerable voltage fluctuation of supply line, so the traction motors should be so designed to withstand voltage fluctuations without undue effect on their performance.
5. Motors should be in a position to withstand temporary in-truptions in power supply occurred while locomotive crosses from one supply section to another.
6. Motors used should be amenable to simple speed control methods.
7. Motor should be amenable to easy and simple methods of rheostatic and regenerative braking mechanisms.
8. When regenerative braking is used, the distribution system is subjected to higher voltages during regeneration. The distribution system and the drive equipment must be designed to account for this.
9. Motors are subjected to very arduous duty. Large current rush may produce large armature reaction magnetic flux along the brush axis, resulting in the formation of arc be-tween brush and the commutator segments. This arc or flash over must be avoided. Motor selection should be capable of taking heavy loads without flash over.
10. Motor should be robust in construction, have high power/ weight ratio which is achieved by employing high speed motors. Maximum surface speed of armature is limited to 55 m/sec, and that of commutator to 45 m/sec. Length of the armature ranges between 22 cm to 24 cm for meter gauge and 40 to 45 cm for broad gauge.
11. Motor should be mechanically protected so as to prevent ingress of dirt, water, mud etc. into it.
12. Motor should have high mechanical and electrical efficiency.

8.16. D.C. Motors

Characteristics of Series Wound Motors. In the series wound motor, the field winding is connected in series with the armature as shown in Fig. 8.2.

The series field winding consists of a few turns of thick wire having low resistance. It may be seen that current passing through the field winding is the same as that in the armature. If the mechanical load on the motor increases, the armature current also increases. Hence, the flux in series motor increases with the increase in armature current and *vice-versa*.

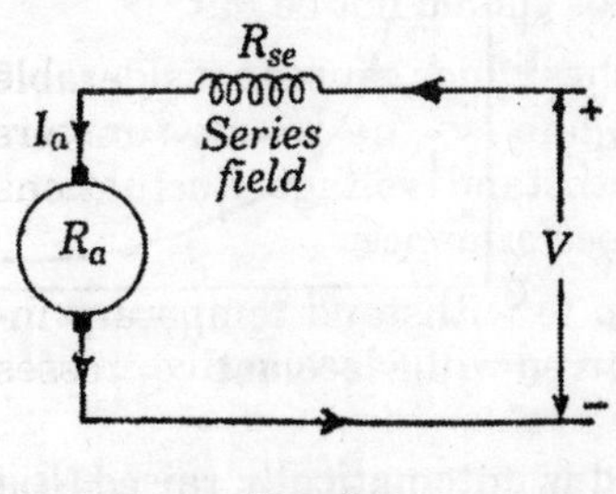

Fig. 8.2.

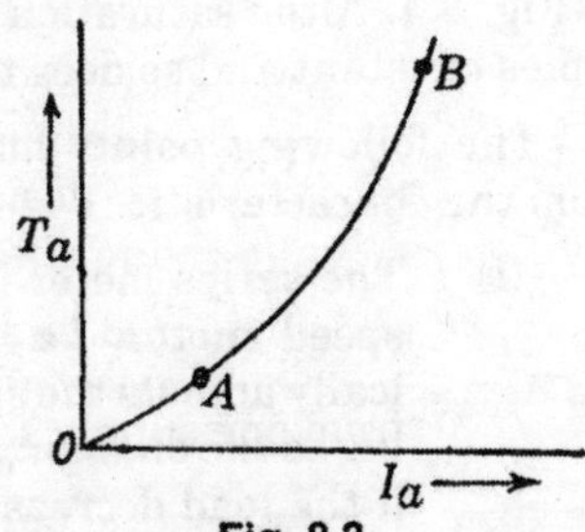

Fig. 8.3.

(a) T_a/I_a Characteristics

We know that $T_a \propto \phi I_a$. Since the field winding of the motor carries the armature current I_a therefore, up to saturation, flux ϕ and I_a have linear relationship, *i.e.*

Upto magnetic saturation, $\phi \propto I_a$

$$\therefore \quad T_a \propto I_a^2$$

Hence, upto magnetic saturation, the armature torque is proportional to the square of the armature current. If I_a is doubled, T_a is almost quadrupled. Therefore, T_a/I_a curve is a parabola (upto magnetic saturation curve *OA* as shown in Fig. 8.3).

However, after magnetic saturation, the torque is directly proportional to the armature current, following a straight line path (curve *AB* in Fig. 8.3).

It may be seen from the characteristic that prior to magnetic saturation $T_a \propto I_a^2$, *i.e.* the series motor exerts a torque proportional to the square of armature current. Hence, series motors are used where high starting torque is required *e.g.*, in hoists, electric traction etc.

(b) N/I_a Characteristics

The speed *N* of series motor given by $N \propto \dfrac{E_b}{\phi}$

where $\qquad E_b = V - I_a(R_a + R_{se})$

When the armature current increases the back emf E_b decreases due to $I_a(R_a + R_{se})$ drop whereas flux ϕ increase. However, $I_a(R_a + R_{se})$ drop is quite small under normal conditions and may be neglected.

$$\therefore \quad N \propto \frac{1}{\phi} \propto \frac{1}{I_a} \text{ up to magnetic saturation.}$$

Thus, upto magnetic saturation, the N/I_a follows the hyperbolic path as shown in Fig. 8.4. After saturation the flux becomes constant and so does the speed.

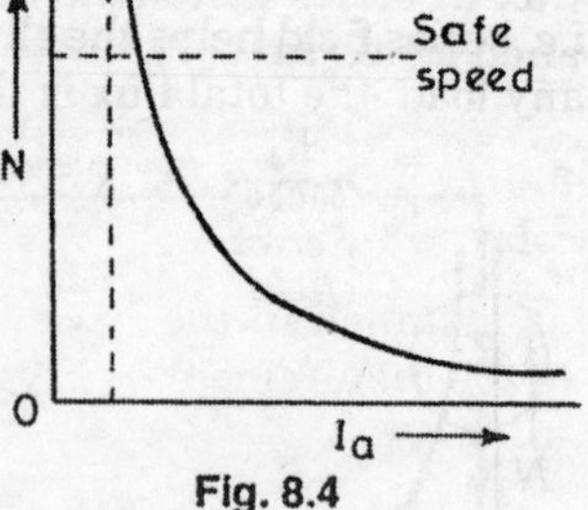

Fig. 8.4

The following points may be noted from the characteristic :

(*i*) The series motor is a variable speed motor, *i.e.* it automatically adjusts the speed according to the changing load. Thus, if the load decreases, its speed is automatically raised (Fig. 8.4). This is a variable feature in a drive of which the speed is required to adjust itself automatically to compensate for changes on the load *e.g.* for traction work.

(*ii*) As the armature current reaches the no load value, the flux becomes very small. Hence the speed rises to an excessive high value ($\because N \propto 1/\phi$) at no load. This is dangerous for the machine, may be destroyed due to centrifugal forces set up in the rotating parts. Therefore, a series motor should never be started on no load. However, to start a series motor, mechanical load is put first and then motor is started.

Important Note. The minimum load on a D.C. motor should be great enough to keep the speed within limits. If the speed becomes dangerously high the motor must be disconnected from supply.

3. T_a/N Characteristics. The T_a/N characteristic of a motor is shown Fig. 8.5. It is clear from the curve that series motor develops a high torque at low speed and *vice-versa.* It is because an increase in torque requires an increase in armature current, which is also the field current. The result is that flux is strengthened and hence the speed drops, therefore, $N \propto (1/\phi)$. Similarly, at low torque, the motor speed is high.

Compound Wound Motors. A compound motor is a compromise between the shunt and series motor and combines the features of both. In a compound motor, there are two field windings; one of which is connected in series with the armature and the other in parallel with it. Depending on the direction of the magnetising effects of series and shunt windings; compound motors are classified into the following two types :

(*i*) Cumulative compound motor.

(*ii*) Differential compound motor.

Characteristics of Cumulative Compound Motor. In a cumulative compound motor, series field winding is so connected that flux produced by it is in the same direction as that of shunt field winding, *i.e.* series field helps the shunt field as shown in Fig. 8.6. Therefore, for any load, the total flux is the sum of shunt and series fluxes, *i.e.*

$$\phi = \phi_{se} + \phi_{sh}$$

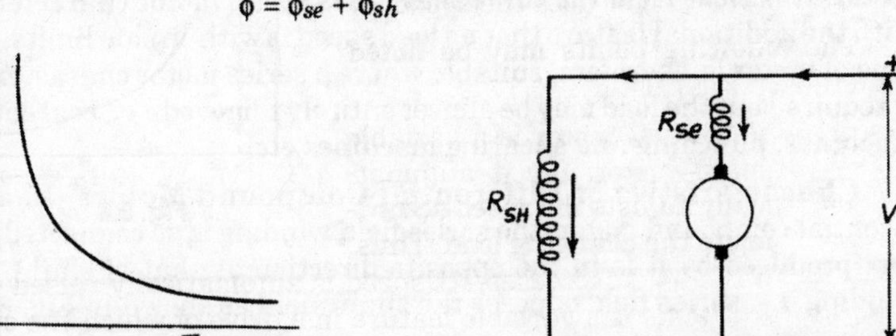

Fig. 8.5. Fig. 8.6

Hence, the resultant flux of the machine would depend upon current, in the series and shunt field windings.

1. T_a/I_a Characteristic. We know that $T_a \propto \phi \cdot I_a$. In this case, $\phi = \phi_{se} + \phi_{sh}$ so that total flux is strengthened. Consequently, torque is increased in fact, the increase in, I_a.

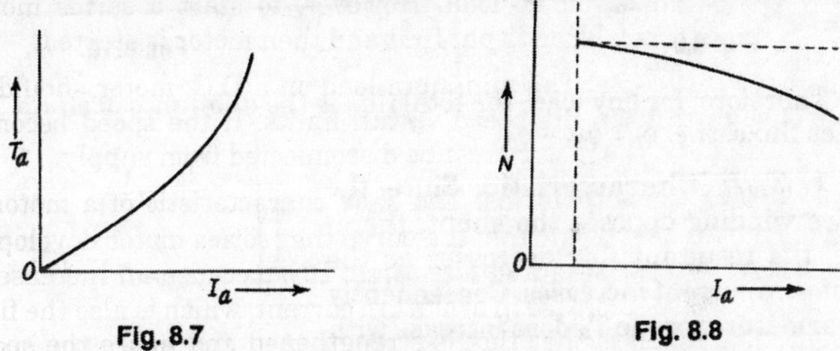

Fig. 8.7 Fig. 8.8

T_a with I_a in this case is greater than that of the shunt motor (see Fig. 8.7).

2. N/I_a Characteristic $N \propto \dfrac{E_\phi}{\phi}$, **where** $\phi = \phi = \phi_{se} + \phi_{sh}$

When load increases, the series field strengthens the shunt field so that total flux increases. Hence, the speed of the motor falls as in a series motor (Fig. 8.8). However, the speed will would be less than it,

if the machine were running on either field alone. From the N/I_a characteristic, the following two points are worth noting :

(*i*) The speed of cumulative compound motor is reduced on heavy loads as in the case of an series motor.

(*ii*) The speed of the motor is within safe limits at no load. It is because shunt field winding produces flux even on no load.

3. N/T_a Characteristic. The N/T_a or mechanical characteristic is shown in Fig. 8.9. It lies between that of shunt motor and of a series motor. It is clear from the curve that it has series motor characteristic with the additional feature that no load speed is with-in safe limits. This type of motor is, therefore, suitable where a series motor characteristic is required but the load may be almost entirely removed *e.g.*, coal cutting machines, punching and shearing machines etc.

Characteristics of Differential Compound Motors. In a differential compound motor, the series field winding is so connected that flux produced by it is in the opposite direction as that of shunt field winding, *i.e.* series field opposes the shunt field as shown in Fig. 8.10.

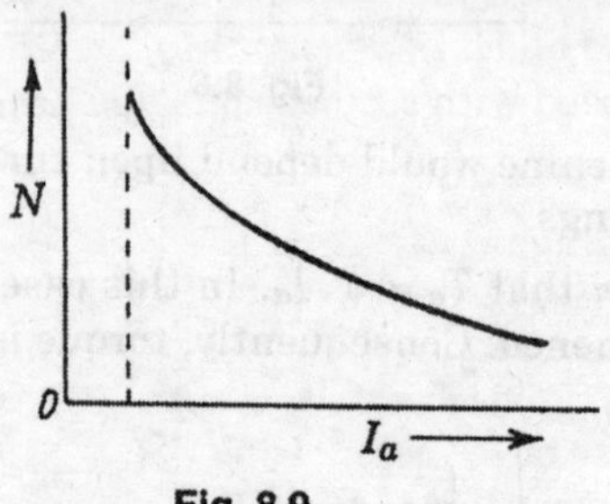

Fig. 8.9

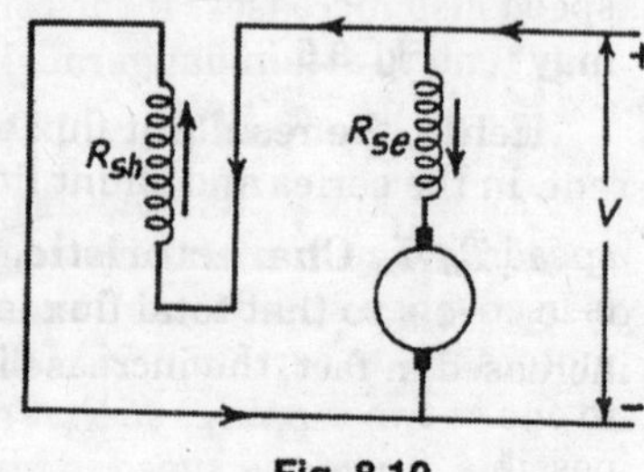

Fig. 8.10.

Therefore, for any load, the total flux is the difference of shunt and series fluxes, *i.e.* $\phi_t = \phi_{sh} - \phi_{se}$.

1. T_a/I_a Characteristic. Since the series winding opposes the shunt, therefore, the resultant flux decreases as the armature current increases. Consequently the armature torque T_a does increase with I_a but does not as rapidly as in a series motors. The T_a/I_a characteristic is shown in Fig. 8.11.

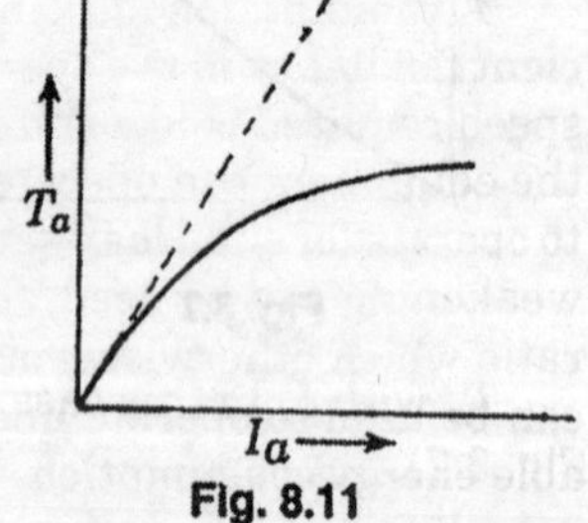

Fig. 8.11

2. N/I_a Characteristic. With the increase in load, the resultant flux decreases and hence the motor speed increases ($\because N \propto E_b/\phi$). The greater the load, the lesser is the flux and hence faster would the motor tend to run (Fig. 8.12). This property of the motor is quite dangerous since a heavy load would result in such a high speed that motor might burst under the centrifugal forces.

Therefore, in connecting compound motors, care should be taken to avoid differential connection.

3. N/T_a Characteristic. The N/T_a characteristic or mechanical characteristic of a differential compound motor is shown in Fig. 8.13.

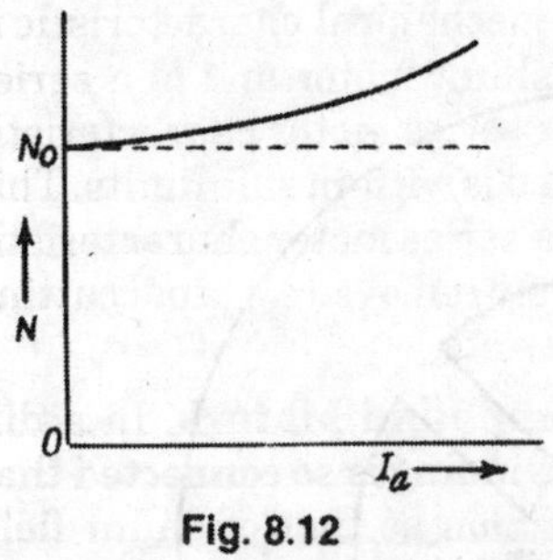

Fig. 8.12

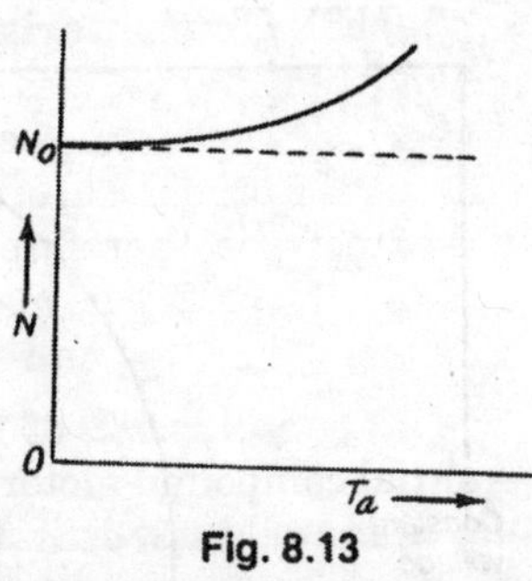

Fig. 8.13

It is clear from the curve that as the torque increases, the motor speed also increases. If the torque is too much increased, the machine may tend to attain dangerously high speed with serious consequences.

8.16.1. Speed Control by field weakening

As we know that the armature voltage control is used whenever speeds lower than normal speeds are required, and field control method is used when speeds more than normal are required. In case of traction motors due to certain design difficulties, field control is usually limited to one or two tappings so that an increase of 15 to 30% in speed is made possible. Since the speed is inversely proportional to the flux, thus by weakening the field, the speed is increased. For weakening the field either a diverter is provided or the field is tapped.

A combination of both the field and voltage control provides sufficient flexibility in the operation of traction motors. For city service, the speed required is low and frequent starting and stopping are required, the equipment can operate with full field and when the same vehicle is to operate for suburban services where higher speeds are required, field weakening can be used, thus eliminating the need for change in gear ratio which otherwise would be required. Thus one type of equipment can be used to operate under different service conditions with reasonable energy consumption.

8.16.2. Metadyne Speed Control

This speed control system is based on constant current system. In this system metadyne converter is used which takes power at constant voltage and variable current and delivers the same at constant current and variable voltage. In series parallel control or resistance control systems there is waste of electrical energy in the starting resistance and the jerks are also experienced when the controller of the starter moves on notches. In metadyne speed control since current throughout the starting period remains constant, uniform tractive effort is developed. This gives very smooth drive and high coefficient of adhesion.

The converter has two pole d.c. armature and four pole field magnet as shown in Fig. 8.13 (*a*). There are two sets of brushes, one set connected to a constant voltage supply main and the other is connected

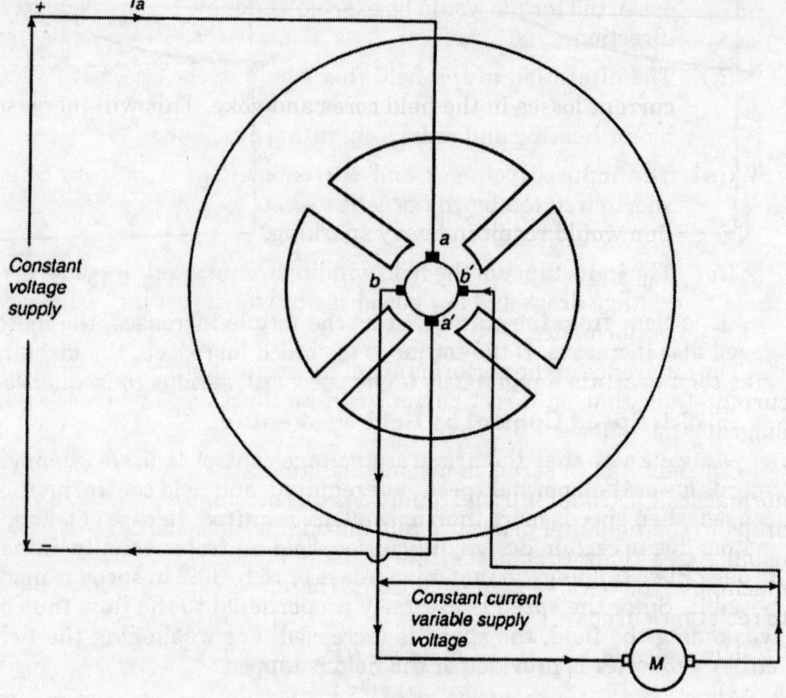

Fig. 8.13 (*a*) Metadyne Converter.

to the load or traction motor. During operation the metadyne converter draws current I_a from the mains. The other set of brush will feed constant current at varying voltage to the traction motor.

8.17 A.C. Series Motor

D.C. shunt or series motors rotate in the same direction regardless of the polarity of the supply. This, it might be expected that either motor would operate on A.C. It has been found, however, that the shunt motor develops but little torque when it is connected to an A.C. supply. The reason of it is that the field winding, owing to its high reactance, causes the field current to lag the armature current by such a large angle that a very low net torque results. However in series motors, the field and armature currents being the same, the main field and armature currents are in phase, therefore, theoretically same torque is developed with a given alternating current as with a like amount of direct current in series motor. If an ordinary D.C. series motor were connected to an A.C. supply, it would operate, but not very satisfactorily. The reasons are :

(i) Since the field and armature currents, both reverse every half cycle, the torque would be exerted at double frequency in one direction.

(ii) The alteration in the field flux would cause excessive eddy current losses in the field cores and yoke. This will increase motor heating and reduce operating efficiency.

(iii) The induced voltages and currents in the armature coils, short circuited by the brushes when undergoing commutation would result in heavy sparking.

(iv) The inductance of the field winding would result in abnormal voltage drops and low power factor which results in the poor performance.

The difference in the performance of the series motor on alternating current from that on direct current can be illustrated by the vector diagrams, as follows :

Consider a D.C. series motor having laminated stator field and of normal design. Under full load conditions this machine will have voltage drops in the field and armature but the sum of these voltage drops is usually less than 10% of the line voltage V except in case of small machines. The back *e.m.f.* developed is, therefore, large in comparison to resistance drops.

Numerically $$N = E_b + I(R_a + R_{se})$$

Also $$E_\phi = K \phi N$$

Where K is any constant, ϕ is the flux per pole and N is the speed of the machine.

Thus with direct current, throughout the load range E_b is nearly constant and slightly less than applied voltage. Now consider the same machine connected to an a.c. voltage of the same magnitude. The flux created will be alternating and in time phase with the current (refer Fig. 8.14). The back e.m.f. E_b developed in the armature, $K \phi N$ will be in time phase with flux, ϕ.

The current causes resistance drops as in the case of D.C. operation, and it also causes large resistance drops in the armature and in the field. Hence with a.c. operation the back e.m.f. developed must be less than with D.C. operation. This means that for a given torque the speed for the A.C. operation must be less than that for D.C. operation (refer Fig. 8.15). The power factor of the motor is also poor.

Hence some modifications are necessary in a d.c. series motor that is to operate satisfactorily on a.c.

Improvements

1. In order to reduce the reactance of the series field, a.c. series motors are built with as few fields turns as possible.

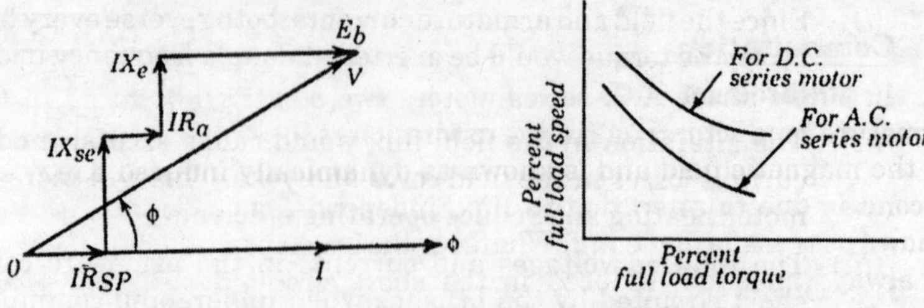

Fig. 8.14 Fig. 8.15

2. Reduction in number of turns on the field winding would also reduce the flux per pole, ϕ for a given current would tend to increase the speed and reduce the torque available for a given current. Hence to develop the required load torque the number of armature conductors would have to be increased proportionately.
3. The increase in armature conductors would increase the inductive reactance of the armature so that the net inductive reactance of the total machine circuit would not be improved. In order to overcome this difficulty the machine is provided with a compensating winding, which will neutralize as effectively as possible the m.m.f. of the armature. If the compensating winding could completely neutralize the armature m.m.f., the armature winding would have no inductive reactance.
4. In order to lower the transformer e.m.f., in the short-circuited element of the armature winding the motor is provided with a large number of poles with lesser flux per pole.
5. The armature coils are single turn coils and brush of less width are used in order not to short circuit more than two coils at a time.
6. The air gap is made very small because of very weak field, which is necessary to obtain high power factor.
7. All parts of magnetic circuit of an A.C. series motor are laminated in order to reduce the eddy current losses.
8. The operating voltage is kept low (300 to 400 V) in order to reduce the inductance, because high voltage motor with proportionally low current would require a large number of turns to produce the given flux.
9. Since inductive reactance is directly proportional to the frequency so A.C. series motor are built in large size for traction service and are designed to operate at low frequency of 25 Hz or 16 ⅔ Hz. However, fractional horse power motors may be designed to operate satisfactorily at commercial frequency of 50 Hz.

Commutation

In single-phase A.C. series motors two e.m.f. are induced in the armature conductors : one is due to rotation of the armature conductors in the magnetic field and is known as dynamically induced e.m.f. and second is due to alteration in flux, independently of rotation and is known as transformer e.m.f. Whatever the brush position may be, there is always an e.m.f. induced in the short circuited elements of the armature winding. Thus when the brushes are in normal position, *i.e.* perpendicular to the main field axis, transformer induced e.m.f. is in the short-circuited element and when the brushes are along field axis, dynamically induced e.m.f. is in it. Hence a.c. series motor gives more sparking than an ordinary d.c. series motor. Excessive sparking and poor power factor are prevented by providing compensating winding, which carries the armature current and neutralize the armature reactance. In Fig. 8.16 the field flux is produced along the axis *AB* and armature flux is produced along axis *CD* and in a direction opposite to that of the armature flux at every instant. In this case the compensating windings carries current equal to armature current but opposite in direction due to conductive effect, so it is known as conductively compensated series motor.

The compensating winding, instead of being connected in series with the main field winding, can be short-circuited, as shown in Fig. 8.17. It then acts like a short-circuited secondary of the transformer, of which the primary is the armature winding. Since in this coil the compensating winding is excited inductively so motor with short-circuited compensating winding is known as inductively compensated

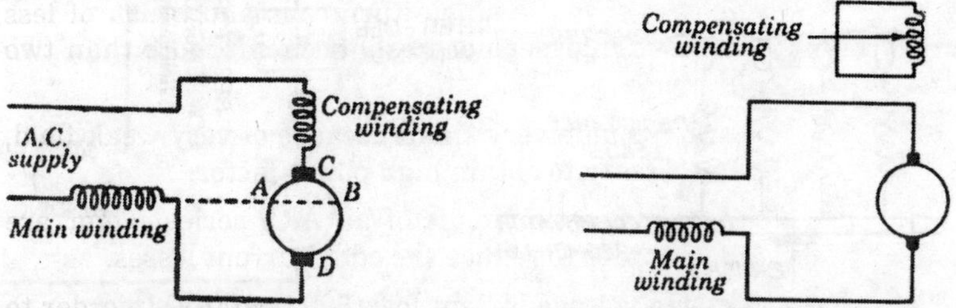

Fig. 8.16. Conductively compensated a.c. series motor.

Fig. 8.17. Inductively compensated a.c. series motor.

series motor. The current in the compensating winding will be almost directly proportional to the armature current and practically 180° out of phase of it. Therefore, the action of the compensating winding will be almost identical with that of a conductively compensating winding. The resistance and inductive reactance characteristics of the armature circuit of the inductively compensated series motor will be increased over their values for the armature alone by a certain amount through the coupling of the compensating winding to the armature in accordance with transfer action. Thus, the total resistance and reactance characteristics of the inductively compensated series motor are practically the same as those for an equivalent conductively compensated series motor. In fact, the performance characteristics are practically identical for the two types of motors.

The single-phase A.C. series motors are provided with commutating or interpoles. Although these poles do not help in reducing the sparking at start and practically effective at lower speeds, their use enables the circulating currents to be reduced to lower values over a range of speeds and loads, so that good commutation is achieved for normal running conditions. Commutating poles increase the efficiency of the machine. Output of the machine with a given size of armature core is also increased by employing commutating poles.

The earliest method of reducing the sparking due to transformer e.m.f. was by use of high reactance leads from the winding tappings points to the commutator segments, as shown in Fig. 8.18. Such high resistance leads form path for main current as well as circulating current, so copper loss due to main current is increased, but that due to circulating current is reduced. Therefore, the value of this resistance used to be so decided that it may result in minimum copper loss.

Now this method has been replaced by shunting the interpole winding with non-inductive resistor, as shown in Fig. 8.19. In this method the rotational e.m.f. in a short-circuited coil has a component in phase opposition to the transformer e.m.f.

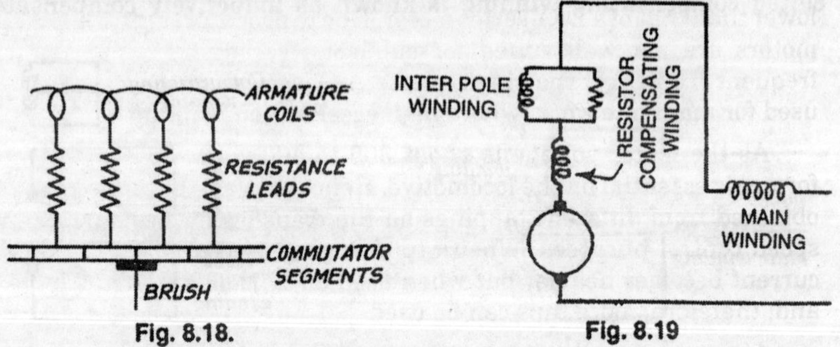

Fig. 8.18. Fig. 8.19

Vector diagram

Vector diagram of single-phase A.C. series motor is shown in Fig. 8.20. The resistance drops IR_{se}, IR_p, IR_c and IR_a due to resistances of series field, interpole winding, compensating winding and of armature respectively are in phase with the current. The reactance drops IX_{se}, IX_p, IX_c and IX_a due to reactances of series field, interpole circuit, compensating winding and of armature respectively are in quadrature with the current and leading. Counter e.m.f. E is essentially 180° out of phase of the current. The terminal voltage V is vector sum of reversed counter e.m.f. and voltage drops in armature, compensating winding, interpole and main field winding circuit.

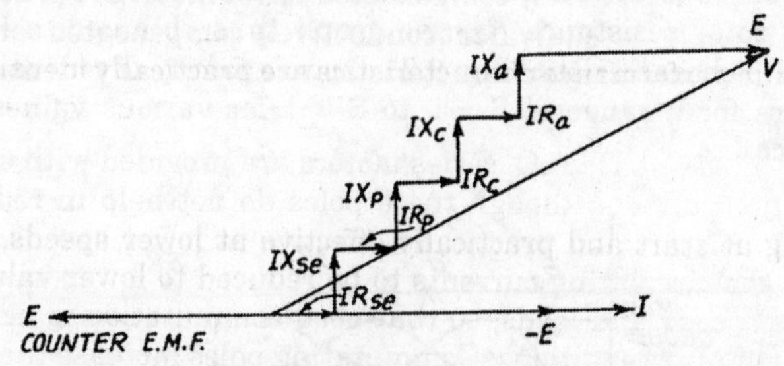

Fig. 8.20

Characteristics

The single-phase series motor has practically the same operating characteristics as d.c. series motors. The torque or tractive efforts varies nearly as square of the current and the speed varies inversely as the current nearly. The inductively compensated A.C. series motor also operate satisfactory on D.C. system and has increased output and efficiency.

Single-phase compensating series motor, which can be built in size upto several hundred kW are usually employed for electric traction work. For given kW output the single phase A.C. series motors have 1.5 to 2 times weight than corresponding D.C. motors. At start the power factor is low, therefore, for a given current the starting torque, is much lower than that of a D.C. series motor. Due to poor starting torque, these motors are not well suited for suburban services where stops are frequent. However, the single-phase A.C. series motor is extensively used for main line work, where high acceleration is not important.

As the motor voltage is about 300 to 400 volts step-down transformer is essential on the locomotive. Hence the variable voltage can be obtained from different tappings on the transformer for starting and speed control purposes. When tap changing is used on L.T. side the current becomes heavier but when used on H.T. side it is much lower and, therefore, more taps can be used.

At start low voltage is applied cross the motor. With the increase in speed the applied voltage is gradually increased. At start higher torque is required, and as the torque reduces the current reduces and speed increases automatically because of series characteristics of motor.

8.18. Three-phase Induction motor

Torque-slip characteristics. We have seen that rotor torque under running condition is given by,

$$T = \frac{K_2 SR_2}{R_2^2 + (SX_2)^2}$$

If a curve is drawn between the torque and slip for a particular value of rotor resistance R_2, the graph thus obtained is known as torque-slip characteristic. Fig. 8.21 shows a family of torque slip characteristics for a range of S = 0 to S = 1 for various values of rotor resistance.

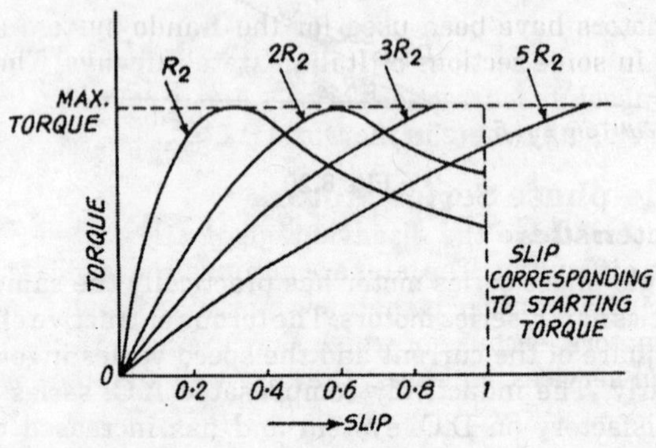

Fig. 8.21

The following points can be noted :

(i) As $S = 0$, $T = 0$ so that the torque/slip characteristic starts from the origin.

(ii) At normal speed, slip is small so that SX_2 is negligible as compared to R_2.

$$\therefore \quad \text{Torque, } T \propto \frac{S}{R_2}$$

$$\propto S \text{ as } R_2 \text{ is constant.}$$

Hence, the torque/slip curve is a straight line from zero slip to a slip that approximately corresponds to full load.

(iii) As slip is increased beyond full load slip, the torque increases and becomes maximum at $S = R_2/X_2$. This maximum torque in an induction motor is known as pull-out torque or break down torque. Its value is at least twice the full load torque when the motor is operated at rated voltage and frequency.

(iv) When the slip is increased beyond that corresponding to maximum torque, the term $S^2 X_2^2$ increases very rapidly so that R_2^2 may be neglected as compared to $S^2 X_2^2$.

$$\therefore \quad \text{Torque, } T \alpha \frac{S}{S^2} X_2^2$$

$$\propto \frac{1}{S} \text{ as } X_2 \text{ is constant.}$$

Thus, the torque is now inversely proportional to slip and hence, the torque-slip curve is a regular hyperbola (See Fig. 8.21).

(*v*) It may be seen from the torque-slip characteristics that addition of resistance to the rotor circuit does not change the maximum value of torque but it only changes the value of slip at which maximum torque occurs.

These motors have been used for the Kando system adopted in Hungry and in some sections of Italian state railways. These motors are however absolute for traction services and is not being used for any further extension anywhere in the world.

8.19. Single-phase Series Motors

These motors have the disadvantage of a low power factor and commutation difficulties. The later are minimized by using low frequencies. Another disadvantage is their low starting torque as compared to D.C. series motors, because of which they are not well suited for city and suburban services. However, they are well adapted for main line services.

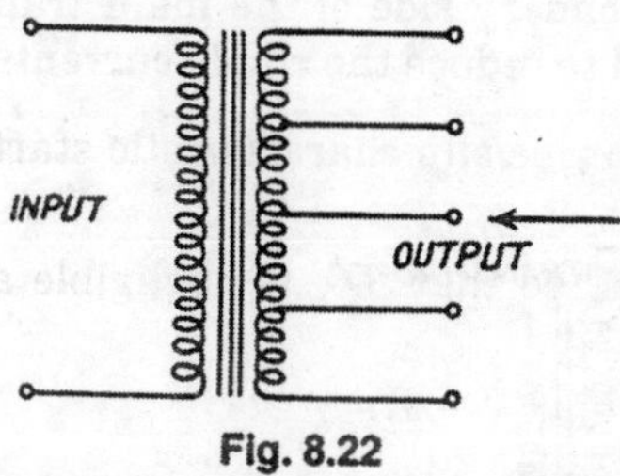

Fig. 8.22

Speed control in single-phase series motor is done by variation of applied voltage. Usually, the supply is obtained through a transformer, provided with 'on load tap changing device' so that a variable voltage is easily available. However, the difficulty with on load tap changing is that when changing over from one tap to the other, the number of turns on the winding in between are short momentarily, producing a heavy current rush. Hence a tapping choke is provided as shown in Fig. 8.23 along a short circuiting switch S.

At the time of changing from tap 1 to 2, the switch S is opened, allowing the turns to be shorted by the tapping choke, which prevents short circuit current between them. Switch S is now opened, and the change over to tap 2 is effected.

8.19.1. Tap changing on H.V. traction system.

Tap changing on high voltage A.C. traction system at power frequency is generally through an auto transformer connected to the traction supply as shown in Fig. 8.24.

The odd number taps (1, 3, 5) are connected to one side of the tapping choke, while the even numbers to the other side.

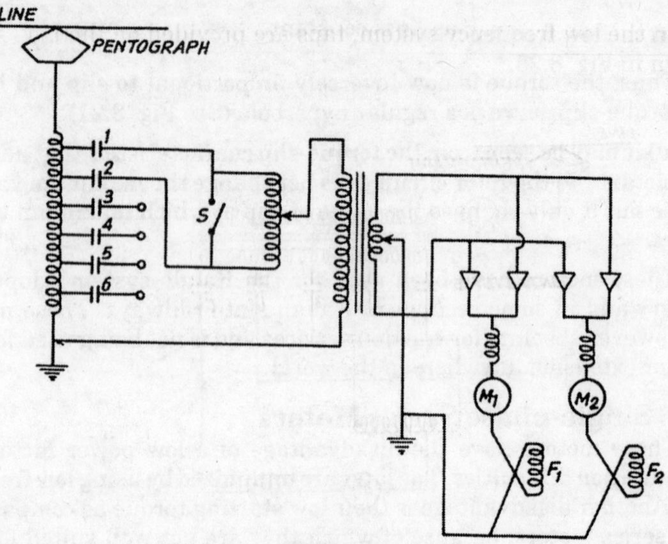

Fig. 8.23

Full wave rectifiers are provided for D.C. supply to the traction motors. They are connected to the secondary side of the main transformer. Smoothing chokes are provided to reduce the ripple currents.

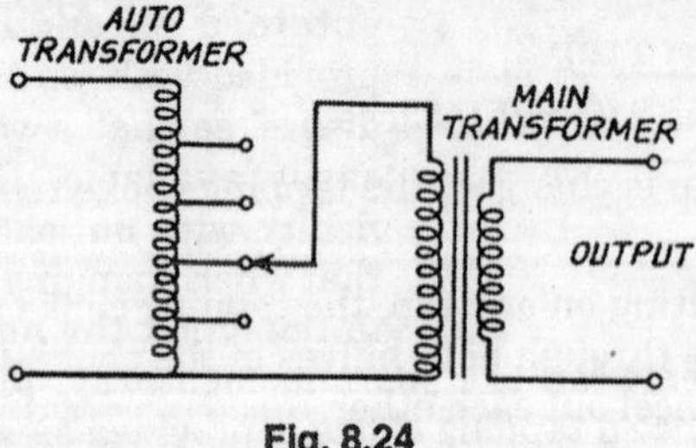

Fig. 8.24

8.19.2. Voltage Variation in A.C. Systems

In A.C. system, the voltage is varied to control the starting, current, as well as for speed control. This is true for either system, whether at power frequency or at a lower frequency. A transformer is provided in the locomotive for stepping down the supply voltage. Tappings are provided on the secondary side for varying the voltage.

For high voltage tap changing, an additional auto-transformer is generally-employed. As the current in the primary winding is lower, more tappings can be provided on the H.V. side of the main transformer; because in that case, to get a small voltage at start, the number of turns on the primary winding will have to be large.

In the low frequency system, taps are provided on the L.V. side as shown in Fig. 8.25.

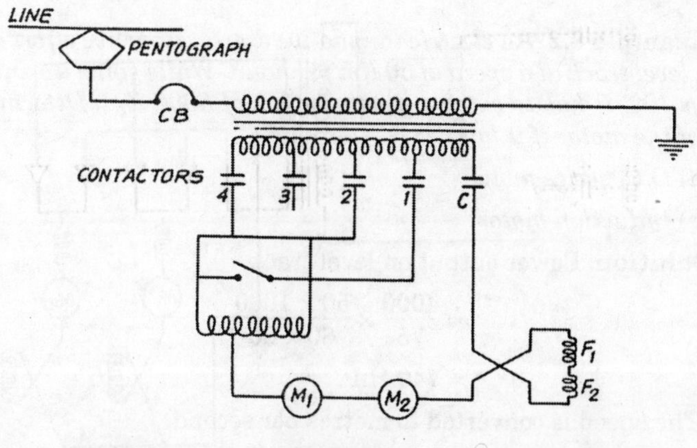

Fig. 8.25

Tapping choke is provided for a smooth change over from one tap to another. Switch *S* is short circuiting it. The operation is as follows.

On tap 1, the contactor 1 and switch *S* is closed.

2, 3, 4 are open.

For change to tap 2.

Open *S*, Close 2, Open 1, Close *S*.

While the tap is changed, the tapping choke tends to stabilize the current.

(While operating on any tap, the tapping choke is shorted, so that the current flows through both halves of the choke ; and the resulting ampere-turns cancel out each other.)

Example 8.1. *An electric locomotive driven by two D.C. series motors, exerts a tractive effort of 2000 Nw while running at a speed of 50 Km per hour. Determine the current taken from the supply, operating at 500 V, assuming an efficiency of 80%.*

Solution.

Power requirement = Tractive effort × Speed (It is given in watts, if the tractive effort is in newtons and the speed in metres/second)

Power required $= 20{,}000 \times \dfrac{50 \times 1000}{60 \times 60} = 278$ kW.

Input to the motors

$$= \frac{278}{0.8} = 347.5 \text{ kW.}$$

Current drawn from the supply

$$= \frac{347.5 \times 1000}{500} = \mathbf{695\ Amp.}$$

Example 8.2. *An electric locomotive exerts a tractive effort of 4000 Kg on level track at a speed of 50 Km per hour. While going up an incline of 1 in 100, it has to exert a tractive effort of 5000 Kg. Determine the H.P. of the motor if it has*

(a) D.C. series motor

(b) Induction motor.

Solution. Power output on level track

$$= \frac{4000}{75} \times \frac{50 \times 1000}{60 \times 60}$$

$$= 740 \text{ H.P.}$$

(The speed is converted to metres per second)

However, the motor has to exert greater torque while going up the incline. But it is not necessary that the motor power would also have to be increased in the same ratio. In case of a series motor

$$\text{Output } \alpha \sqrt{\text{Torque}}$$

Hence for a series motor, the H.P. required

$$= 740 \times \sqrt{\frac{5000}{4000}} = 827 \text{ H.P.}$$

In case of induction motor (or any shunt characteristic motor)

$$\text{Output } \alpha \text{ Torque}$$

$$\therefore \text{ H.P. required} \quad = 740 \times \frac{5000}{4000} = \mathbf{925\ H.P.}$$

(This clearly indicate the advantage of series motors.)

8.20. Auxiliary Equipment

A traction system comprises of the following equipments, apart from the driving motors :

1. Overhead conductors for track electrification. The overhead system is used with D.C. tramways, trolley-buses and locomotives and also with A.C. locomotives. The track is used as the return conductor. The voltage in the overhead system is 600 V and above.

For tramways and trolley buses, the trolley wire is of the section shown in Fig. 8.26 and a typical trolley collector-wheel is shown in Fig. 8.27.

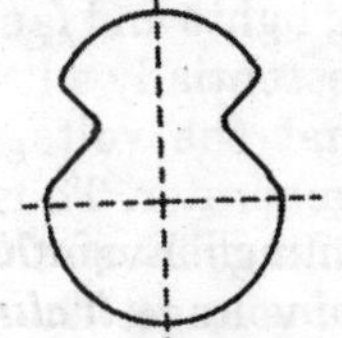

Fig. 8.26. Trolley-wire section.

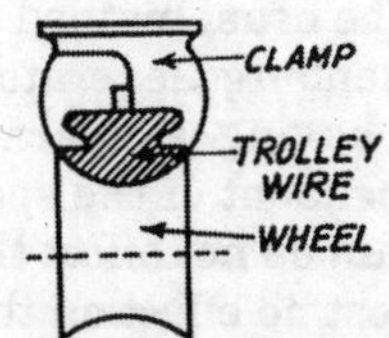

Fig. 8.27. Collector wheel and trolley wire.

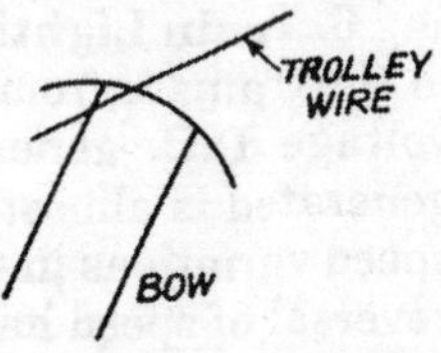

Fig. 8.28. The bow collector.

The trolley collector is used up to speeds of about 32 kmph as beyond this speed there is every possibility of the wheel jumping off the trolley wire.

The bow collector consists of a light metal strip pressing against the trolley wire and the framework is mounted on the roof of the car. There is no possibility of its jumping off the trolley wire.

The overhead conductors designed for the current carrying capacity as per load requirements on the system, are mounted on poles usually 100 metres apart. The conductors are supported on a cross-arm arrangement, to a point midway above the track.

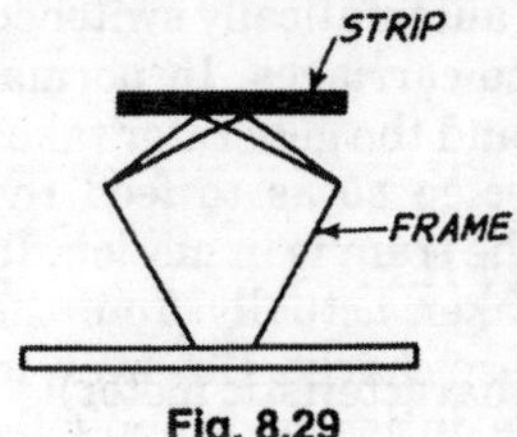

Fig. 8.29

2. Pantograph. The pantograph collector is used, where the vehicles run at high speeds, *i.e.* in railways and where currents to be collected are large (2000 to 3000 A). The pantograph is mounted on a framework which can be raised or lowered by compressed air or springs. Compressed air for raising is normally used (Fig. 8.29).

A flexible arrangement of the shape of a pantograph is provided above the locomotive, which makes contact with the overhead conductor. It has a revolving pulley at its top making contact with the conductor, thus, carrying the current to the motor.

It is flexible in operation, so that it makes a firm contact with the conductor, and at the same time, does not have much wear and tear, and avoids arcing between the conductor and the receiving point.

3. Transformer or Auto-transformer. Depending on the system of electrification used as described earlier. They are generally provided with tap changing device as well.

4. Rectifier Unit. For A.C. systems using D.C. motors for traction metal rectifiers are generally used. They are available in larger sizes as well these days. They require little maintenance and are generally preferred for this reason.

5. Driving Axles and Gear Arrangements. The driving motors are connected to the driving axle through a gear arrangement ; the usual ratios being 4 : 1 or 6 : 1.

6. Train Lighting. The usual method of providing lights and fans in the trains is from a Rosenburg Generator which is essentially a low voltage D.C. generator having a characteristic that the voltage generated is almost independent of the speed of the generator. Thus, speed variations in the train do not effect the output voltage. Even the reversal of speed has almost no effect on the generated voltage. Thus, it is essentially a constant voltage generator.

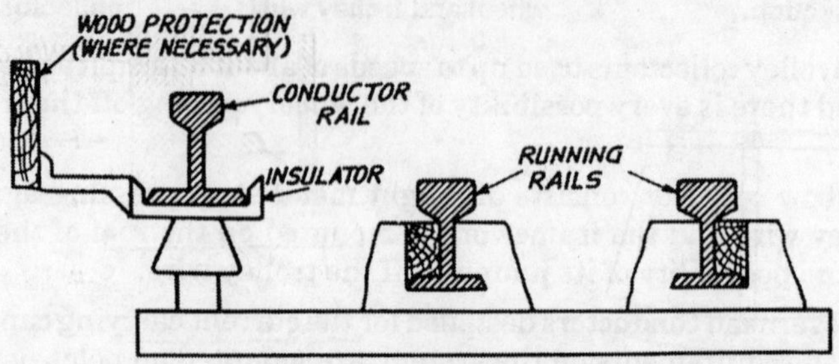

Fig. 8.30 Conductor-rail system.

The generator is coupled to the driving wheels of the locomotive. When the train is at rest, the supply circuit get automatically switched on to set of batteries provided underneath the carriages. In normal running, the battery supply gets disconnected and the generator takes up the load. Usually, the generator is connected so as to feed the batteries and continues to charge them when the train is in motion. In some cases, the supply for train lighting is taken actually from the batteries and the generator merely serves to keep charge. However, in most cases, the generators accomplish both the duties, *viz.* supply for train lighting while running, as well as, charging the batteries, so that the supply can be drawn from them while the train is at rest.

7. Multiple Control. An electric train comprises of a number of units, each units comprises of a motor coach, passenger coaches or goods wagons. The number of units depend on the intensity of traffic on the line. The motor coach comprises of the traction motors, the driving axles, control gear and protective equipment. Many trains have more than one motor coach, instead of an electric locomotive. There are two or four motors in each coach, which increases the adhesive weight of the train consequently developing greater tractive effort. However, their control gets more complicated. Generally, a multiple unit control is incorporated, whereby it is possible to control all the motors from one motor coach.

8. Conductor Rails. Instead of the trolley wire, a conductor rail for feeding is laid out. Some suburban railways at 600 V employ this method. The rail is mounted on insulators. For collecting the current a shoe as shown in the Fig. 8.31 is used.

The shoe presses on to the rail with a force of about 15 kg with a current collection rating from 300 to 500 amperes per shoe.

Since the conductor rail is not always laid on the same side of the track, shoes are attached to the motor coach or locomotive on both sides. Conductor rail with a side running contact is also used (as shown in Fig. 8.32).

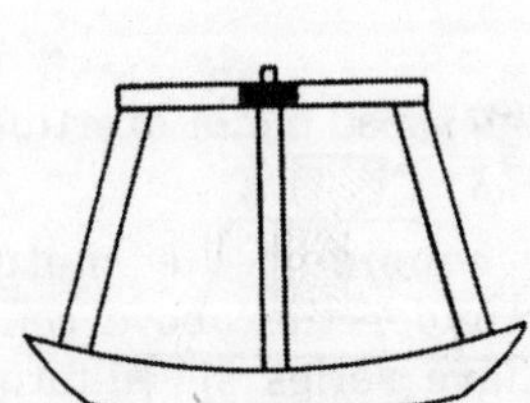

Fig. 8.31. Current collecting shoe.

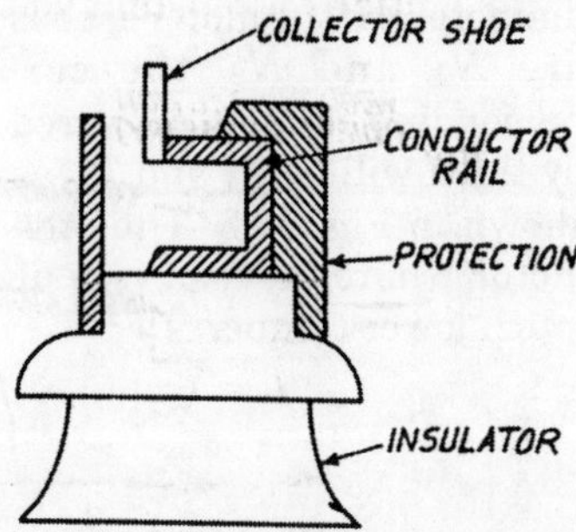

Fig. 8.32. Conductor rail with side running contact.

Although most of electric traction services use the overhead line for supply, yet in congested areas and underground cables, the supply is through conductor rails. However, their use is limited only to L.T. (voltages not exceeding 600 volt in general). In this system, high conductivity steel rails are used, insulated from each other. The current is collected from the rails by collector brushes, known as current shoes.

8.21. General Features

The essential electric and mechanical features of traction motors are :

(1) **High starting torque.** It has been seen that during the accelerating period, tractive effort is needed to provide acceleration and also to overcome train resistance, whereas during the free running period through the velocity is high only sufficient tractive effort is needed to overcome train resistance. (In case of gradients extra tractive effort will be needed in both cases). In order to produce a high tractive effort at starting the motor should develop a high tractive torque.

(2) **Series characteristics.** It is desirable that a traction motor has a series speed torque characteristic because of the following considerations :

(*a*) **Parallel running.** A traction unit may have a number of identical motors fitted to different wheels. As long as the wheel diameters are same the operating speed of motors is same and hence load on each motor is same. However, due to different rates of wear the wheel diameters may differ in course of time and this effects load sharing.

First consider two motors with shunt speed-torque characteristics running in parallel fitted to wheels with unequal diameters. If the motor speeds are N_1 and N_2, the corresponding torques delivered by them will be T_1 and T_2, as shown in Fig. 8.33. Thus the motors share the load very unequally, the high speed motor exerting a much lower torque.

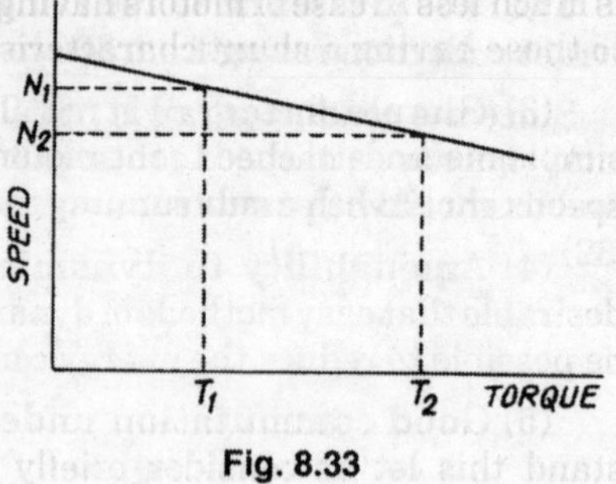

Fig. 8.33

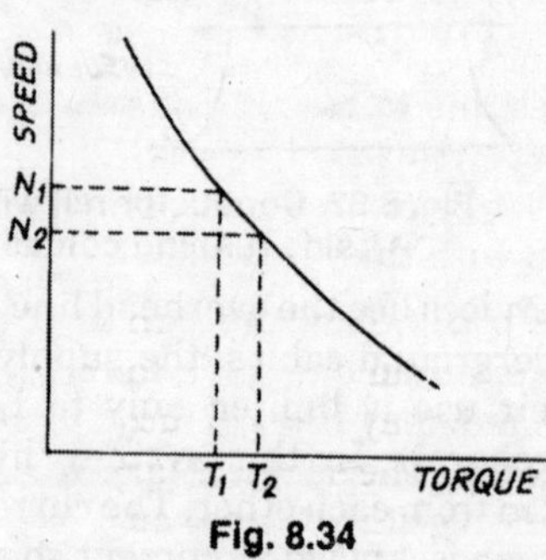

Fig. 8.34

Next consider the motors operating under the above conditions to have series speed-torque characteristic. It will be seen in Fig. 8.34 that the torques T_1 and T_2 are nearly equal.

This explain the desirability of series characteristic for traction motors.

(*b*) **Self-protective property.** Because of the series speed-torque characteristic, the speed of the motor reduces as load torque increases. The exact relationship neglecting saturation is, $n \propto 1/\sqrt{T}$ more where T is torque and n is the speed.

The h.p. output of the motor is proportional to, torque $\times n$. (From the formula, output in watts = torque in newton-metres $\times 2\pi n$

i.e. $$h.p. \propto T \times n \propto \sqrt{T}.$$

Hence by increase of load torque to 3 times the full load torque, the h.p. increases to about 1.73 times the full load h.p. Thus the motor protects itself against excessive over loading.

(In case of motors with a shunt speed-torque characteristic the speed is almost constant and h.p. $\propto$ T).

(*c*) **High free running speed.** Motors with a series characteristic have high speeds at low loads. This is of advantages when the train is running unload or when the locomotive itself has to be run.

(*d*) **Reduces KVA demand during overloads.** In case of overloads (like ascending a deep gradient) the KVA demand from the lines is much less in case of motors having a series characteristic as compared to those having a shunt characteristic (for reasons already explained).

(3) **Simple speed control.** The motor should be amenable to simple methods of speed control and also a large number of economical speeds should be possible.

(4) **Amenability to dynamic or regenerative braking.** It is desirable that easy methods of dynamic or regenerative breaking should be possible to reduce the energy consumption.

(5) **Good commutation under sudden voltage rise.** To understand this let us consider briefly the phenomenon of commutation. When a coil on the armature passes from the region of a north pole to that of the adjacent south pole, the current in it undergoes reversal. This reversal is however delayed because of the inductance of the coil. To facilitate this reversal an external e.m.f. is suitably injected into the coil by providing interpoles.

There may be sudden voltage rises due to opening of sub-station circuit breakers in order to clear a short circuit in the track. This gives rise to a sudden current rush which has to be commutated. The traction motor should be able to effectively deal with this.

(6) **Temporary interruption of supply.** Temporary interruption of supply are likely to occur at cross overs from one track to the other. When the supply is restored the back e.m.f. may have reduced and this is likely to cause a heavy current surges. The motor characteristics should be such that it is possible to minimise the current surge.

(7) **Robustness.** The traction motor should be of robust construction to withstand the continuous vibration to which it is subjected.

(8) **Minimum weight and overall size.** The weight of the motor should be the minimum possible so that the arrangement and hardware for fixing the motor are simple. Also reduced weight effect a saving in the specific energy consumption. The space available for fixing the motor is limited and so the overall dimensions should be the minimum possible.

8.21.1. Motors used in Traction System

Previously, d.c. series motors were usually used in Traction system because of their high starting torque and capability for high torque overloads. With an increase in torque, the flux also increases; therefore, for the same increase in torque, the increase in motor current is less compared to other motors. Thus during heavy torque overloads, power overload on the source and thermal overloading of the motor are kept limited to reasonable values. The motor speed-torque characteristic is also suitable for better sharing of loads between motors. Further due to a large inductance in the field, sharp fluctuations in supply voltage do not produce sharp peaks in armature current. Thus the motor commutation remains satisfactory, which does not happen in a separately excited motor, unless an additional inductance is connected in the armature circuit.

With the availability of semi-conductor converters, separately excited motor is now preferred over series motor. With independent control of armature and field, the speed-torque characteristic of separately excited motor can be shaped to satisfy the traction requirements in the optimum manner. Further, because of low regulation of its speed-torque characteristics, the coefficient of adhesion has higher value. On the other hand, series motor has a number of limitations. The field of a series motor cannot be easily controlled by semiconductor switches. If field control is not employed, the series motor must be designed with its speed equal to the highest desired speed of the drive. The higher base speeds are obtained by using fewer turns in the field winding This, however, reduces the torque per ampere at start and therefore, acceleration. Further, there are a number of problems with regenerative and dynamic brakings of a series motor. On the other hand, regenerative and dynamic brakings of a separately excited motor are fairly simple and efficient, and can be carried out down to very low speeds.

Currently compound motor is being preferred for traction applications as it incorporates the advantages of both series and separately excited motor.

Due to the availability of reliable variable frequency semiconductor inverters, squirrel-cage induction motor and synchronous motor are now finding applications in traction. Because of a number of advantages associated with these motors, they are likely to replace dc motors for traction applications.

Some of the important advantages of squirrel-cage induction motors over dc are : ruggedness; lower maintenance; better reliability; lower cost, weight, volume and inertia; higher efficiency; and ability to operate satisfactorily with sharp supply voltage fluctuations and in dirty environment. The major drawback of dc motor is the presence of commutator and brushes, which require frequent maintenance, particularly when the flashovers at the commutator occur due to sharp voltage fluctuations. In terms of advantages mentioned for squirrel-cage motor in comparison with dc motors, the synchronous motor lies in-between the two and has one important advantage over squirrel-cage induction motor, that it can be operated at leading power factor. Thus permitting the use of load-commutated thyristor inverter which is cheaper and occupies less volume and weight compared to forced commutated thyristor inverter required by induction motors. The weight and volume of an induction motor drive can also be kept low by using GTO (gate turn-off thyristor) inverter, but is more expensive than a load commutated thyristor inverter.

8.21.2. Traction Motor Control

Operation of a dc separately excited motor for traction applications can be divided into three regions. First two are identical to those shown in Fig. 8.34 (A), *i.e.* constant torque and power regions. In constant

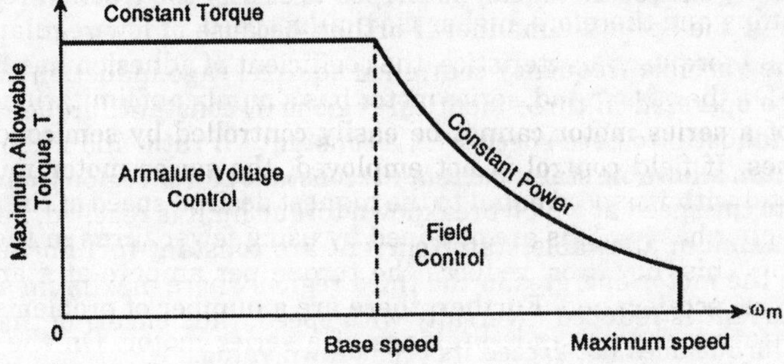

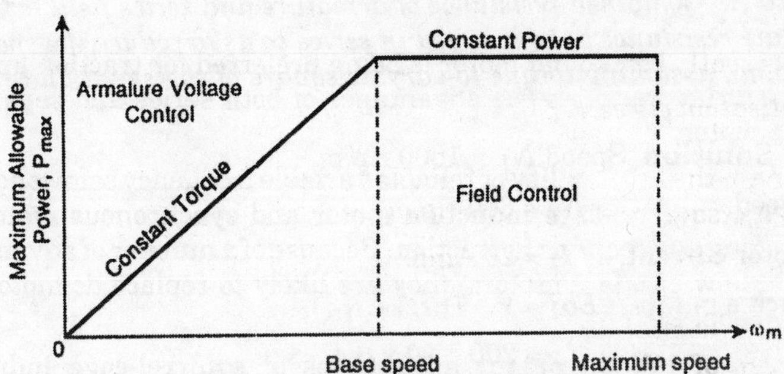

Fig. 8.34 (A) Torque and power limitations in combined armature voltage and field control.

torque region, from zero to base speed, the field current is maintained constant at the rated value and the armature voltage is controlled. In constant power region, which is carried out above base speed, the armature voltage is maintained constant at the rated value and field current is controlled. In both these regions, the armature current is allowed to reach rated value on continuous basis. The limit of constant power operation is reached when a decrease in field current to increase motor speed leads to sparking at the brushes at the rated armature current. The motor is said to reach the commutation limit. Operation at higher speeds (and lower field currents) can now be carried out by progressively decreasing maximum allowable armature current. This is the third region of operation in which available output power of the motor progressively decreases with the increase in speed. Traction motor can be operated in third region because the torque required at high speeds is much less compared to the accelerating torque. The form of third region is determined by whether or not the motor is compensated and the type of power modulator. For a non-compensated motor, the ratio of maximum allowable armature current to field current is maintained constant. In a compensated motor, the maximum allowable aramture current is varied inversely with speed. A compensated machine is always preferred because it allows greater degree of field weakening and therefore, higher maximum speed.

The variable frequency controlled squirrel cage induction motors are also operated in three identical regions (*i*) constant torque region from standstill to base speed with a constant V/f ratio and a constant maximum allowable stator current (*ii*) constant power region from base speed to the speed at which breakdown torque limit is reached, here 'V' and maximum allowable stator current are constant (*iii*) For higher speeds the motor operates in the third region where maximum allowable current is reduced inversally with speed, thus ensuring that the motor torque does not exceed its breakdown value.

Example 8.3. *A 200 V D.C. series motor runs at 1000 r.p.m. and takes 20 A. Combined resistance of armature and series field is 0.4 Ω. Calculate resistance to be inserted in series so as to reduce the speed to 800 r.p.m., assuming torque to vary as square of the speed and linear magnetisation curve.*

Solution. Speed $N_1 = 1000$ r.p.m.

Speed, $N_2 = 800$ r.p.m.

Motor current, $I_1 = 20$ Amp.

Back e.m.f., $Eb_1 = V - I_1(R_a + R_{se})$

$= 200 - 20 \times 0.4$

$= 192$ volts

As $T \propto N^2$

So, $$\frac{T_2}{T_1} = \left(\frac{N_2}{N_1}\right)^2$$

or $$\frac{I_2\,\varphi_2}{I_1\,\varphi_1} = \left(\frac{N_2}{N_1}\right)^2 \qquad [\because T \propto \phi\, I_a]$$

or $$\frac{I_2}{I_1} \times \frac{I_2}{I_1} = \left(\frac{800}{1000}\right)^2$$

$[\phi \propto I, N_2 = 800 \text{ and } N_1 = 1000]$

or $I_2 = \sqrt{0.64} \times I_1 = 0.8 \times 20 = 16$ Amp.

Let the resistance of R ohms be connected in series with the motor circuit.

Back e.m.f., $Eb_2 = V - I_2(R + R_a + R_{se})$

$$= 200 - 16\ (R + 0.4) = (193.6 - 16R) \text{ volts}$$

Since $\dfrac{Eb_2}{Eb_1} = \dfrac{N_2}{N_1} \times \dfrac{\phi_2}{\phi_1}$

$\therefore \quad 193.6 - \dfrac{16\,R}{192} = \dfrac{800}{1000} \times \dfrac{16}{20}$

$[\because \phi_2 = I_2 \text{ and } \phi_1 = I_1]$

or $\quad R = \dfrac{193.6 - 0.64 \times 192}{16} = 4.42\ \Omega.$ **Ans.**

Example 8.4. *Two d.c. traction motors run at speed 700 r.p.m. and 750 r.p.m. respectively when each takes a current of 50 A from 500 V mains. Each motor has an effective resistance of 0.3 Ω. Calculate the speed and voltage cross each machine when mechanically coupled and electrically connected in series and taking a current of 50 A from 500 volt mains, the resistance of each motor being unchanged.*

Solution. Consider the two motors be A and B of

Speed, $\quad N_A = 700$ r.p.m. and $N_B = 750$ r.p.m.

Resistance of each motor, $R_m = 0.3\ \Omega$

Applied voltage across each motor, $V = 500$ volts

Back e.m.f. of motor. A when taking a current of 50 A,

$$E_{bA} = V - IR_m = 500 - 50 \times 0.3 = 485 \text{ Volts.}$$

Back e.m.f. of motor, B when taking a current of 50 A,

$$E_{bB} = V - IR_m = 500 - 50 \times 0.3 = 485 \text{ Volts.}$$

When the machines are connected in series and mechanically coupled, the speed of each motor will be same, say N r.p.m., current will be same and equal to 50 Amp. (given) and the sum of the voltage across the two motors will be equal to 500 Volts.

Let the voltage across motor A and B be V_A and V_B respectively.

Now $\quad V_A + V_B = 500 \quad$ ...(*i*)

Back e.m.f. of motor A, E'_{Ab}

$$= E_{bA} \times \frac{N}{N_A} = 485 \times \frac{N}{700}$$

Voltage across motor A,

$$V_A = E_{bA} + IR_m$$

$$= \frac{485}{700} \times N + 50 \times 0.3 = \frac{485}{700} \times N + 15$$

Back e.m.f. of motor B,

$$E_{bB'} = E_{bB} + \frac{N}{N_B} = 485 \times \frac{N}{750}$$

Voltage across motor B,

$$V_B = E_{bB'} + IR_m = \frac{485}{750} N + 15$$

Putting the values of V_A and V_B in expression (i), we get

i.e. $$\frac{485}{700} N + 15 + \frac{485}{750} N + 15 = 500$$

or $$N\left(\frac{485}{700} + \frac{485}{750}\right) = 500 - 30$$

or $$N\left(\frac{1}{700} + \frac{1}{750}\right) = \frac{470}{485}$$

or $$N = \frac{470}{485} \times \frac{700 \times 750}{700 + 750}$$

$$= \frac{470}{485} \times \frac{525{,}000}{1450}$$

$$= \frac{470}{485} \times \frac{10{,}500}{29}$$

$$= 351 \text{ r.p.m.}$$ **Ans.**

P.D. across machine A,

$$V_A = \frac{485}{700} N + 15$$

$$= \frac{485}{700} \times 351 + 15$$

$$= 258 \text{ volts.}$$ **Ans.**

P.D. across machine B,

$$V_b = \frac{485}{750} N + 15$$

$$= \frac{485}{750} \times 351 + 15$$

$$= 242 \text{ volts.}$$ **Ans.**

Example 8.5. *A motor coach is being driven by two identical d.c. series motors. First motor is geared driving to wheel having diameter of 90 cm and other motor to driving wheel having dia. of 86 cm.*

The speed of the first motor is 500 r.p.m. when connected in parallel with the other across 600 V supply. Find the motor speeds when connected series across the same supply. Assume armature current to remain same and armature voltage drop of 10% at this current.

Solution. Speed of 1st motor,

$$N_1 = 500 \text{ r.p.m.}$$

Back e.m.f., $E_{b_1} = 600 - \frac{10}{100} \times 600 = 540$ volts.

When the motors are connected in series across 600 volt supply as shown in Fig. 8.35.

Let the supply voltage across motors (1) and (2) be V_1 and V_2 volts and speed N_1' and N_2' respectively.

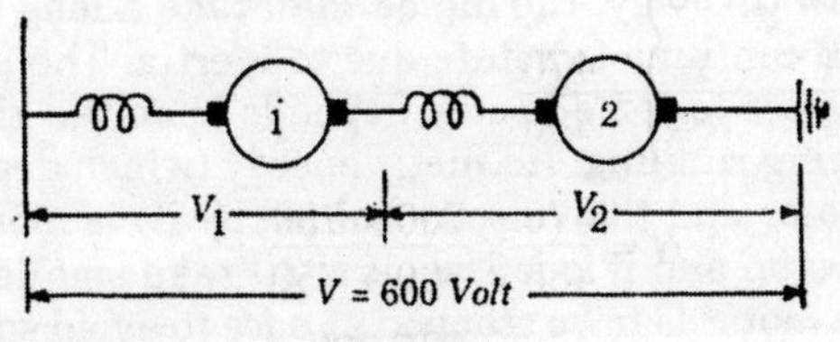

Fig. 8.35.

Since speed, $N \propto V - \frac{IR}{\phi}$

Current through the motor remains the same, therefore, flux produced by it also remains the same and $N \propto (V - IR)$

$$\frac{N_1'}{N_2'} = \frac{V_1 - IR}{V_2 - IR} = \frac{V_1 - \frac{10}{100} \times 600}{V_2 - \frac{100}{100} \times 60}$$

$$= \frac{V_1 - 60}{V_2 - 60} \quad \text{...(i)}$$

Also $\quad N_1' D_1 = N_2' D_2$

$$\therefore \quad \frac{N_1'}{N_2} = \frac{D_2}{D_1} = \frac{86}{90} \quad \text{...(ii)}$$

$$\therefore \quad \frac{V_1 - 60}{V_2 - 60} = \frac{86}{90}$$

or $\quad 90V_1 - 5400 = 86V_2 - 5160 \quad$ ...(iii)

or $\quad 90V_1 - 86V_2 = 5400 - 5160 = 240$

also $\quad V_1 + V_2 = V = 600V \quad$ ...(iv)

Solving equation (*iii*) and (*iv*), we have,

$$V_1 = 294.55 \text{ Volts.}$$

and $\quad V_2 = 305.45$ volts.

Now the speeds of the motors can be calculated as follows :

$$\frac{N_1'}{N_1} = \frac{E_{b1}'}{E_b}$$

or $\quad N_1' = N_1 \times \frac{E_{b1}'}{E_b} = 500 \times \frac{294.55 - 60}{600 - 60}$

$$= 217 \text{ r.p.m.} \quad \textbf{Ans.}$$

and $$N_2' = N_1' \times \frac{D_1}{D_2} = 217 \times \frac{90}{86}$$

$$= 277 \text{ r.p.m.} \quad \textbf{Ans.}$$

8.22. Braking characteristics

If any electric motor is to be stopped, electric supply to it is disconnected. In this way, the motor shall take a long time to stop as the motor shall continue to rotate due to inertia. The amount of time taken by the motor to come to stop depends upon the kinetic energy it has gained during running. If kinetic energy (which depends upon the mass of the motor and the velocity gained by it) is more it shall take longer time to stop and if it is less, it shall take smaller time to stop. Sometimes the motor is to be stopped at once to avoid some accident or in other cases the motor is to be stopped quickly as in case of a planer at the end of its stroke. Thus the motor cannot be left to come to stop of its own after disconnecting the supply. Some braking system must be used so that it stops within the pre-determined short time.

A good braking system must possess the following important features :

(*i*) It should be fast and reliable.

(*ii*) The braking force must be capable of being controlled.

(*iii*) Kinetic energy of the rotating parts of the motor and its driven machine must be suitably dissipated and suitable means must be provided.

Failure of any part of the braking system must result in the application of the brakes as a protection against some accident.

Following are the two types of the braking system normally employed:

(*i*) **Mechanical braking.** In mechanical braking, the stored energy of the rotating parts is dissipated in the form of heat by a brake shoe of brake lining which rubs on a wheel or brake drum.

(*ii*) **Electric Braking or electrodynamic braking.** In this type of braking system, the kinetic energy of the moving part of the system is converted into electrical energy which in turn is dissipated as heat in a resistance or in certain cases may be returned to the supply.

8.23. Advantages of Electric Braking Over Mechanical Braking

(*i*) Electric braking is quite fast.

(*ii*) It is quite cheap as far as maintenance part is concerned due to the fact that no replacement of brake shoes or lining is needed as in the case of mechanical braking system.

(*iii*) Since electric braking is quite fast, higher speeds can be maintained. This results in higher capacity of the system.

(*iv*) In some cases of electric braking such as when the electric train is moving down gradient or the heavy weight is lowered by the crane, a part of the electric energy produced during the braking period, can be fed back to the supply system. This results in saving in the operating cost. This is not feasible in the mechanical braking system.

(*v*) Heat produced during the period of electric braking is in no way harmful but heat produced on the lining etc. during the application of mechanical brakes may result in the failure of brakes.

(*vi*) Electric braking is free from fires and is more smooth than mechanical braking.

8.24. Disadvantages of Electric Braking

(*i*) Electric braking can stop the motor but it cannot hold it stationary. During the period of braking, electric motor works as a generator but as soon as the motor comes to stop, the generating action stop and thus electric brakes do not remain applied when the motor comes to stop. Thus it becomes essential to provide mechanical brakes in addition to electric braking.

(*ii*) Electric braking cannot be applied to all types of electric motors. Thus the choice of electric motors becomes limited.

(*iii*) Its initial cost is very high.

The following types of electric braking are employed :

(*i*) **Plugging or Reverse current Braking.** This method involves, reconnection of motor to supply in such a way that motor develops a torque in opposite direction to the movement of the rotor. The system speed decelerates till zero speed and then it will accelerate in opposite direction. It becomes necessary to disconnect the supply of the system as soon as it comes to rest.

This method is inefficient as not only the kinetic energy of the moving masses is wasted but it requires additional energy from the supply to develop counter torque. This method is applicable to d.c., induction and synchronous motors.

In this method of electric braking system, the connections of the supply to the armature are reversed so that the motor tends to revolve in the opposite direction and in doing so, it comes to rest. As soon as it comes to rest, the supply gets automatically cut-off so that the motor does not start revolving in the opposite direction. At that time mechanical brakes may also be applied, if need be. This method of braking can be applied to both D.C. as well as A.C. motors.

(*a*) **Plugging as applied to D.C. shunt motor.** The direction of the torque developed in D.C. shunt motor can be reversed either by interchanging the field connection or the armature (*i.e.* either the direction of magnetic field is reversed or that of armature current). Since field winding has usually large time constant due to heavy inductance in case of shunt motors, it is the armature current that is normally reversed.

Fig. 8.36 (*a*) shows the normal working of the shunt motor. E_b opposes the applied voltage. The armature current flows from *A* to *B* in the armature. Fig. 8.36 (*b*) shows when plugging is applied. Since the direction of the armature still remains the same, therefore, the direction of back e.m.f. E_b also remains the same. Armature connections having been reversed, applied voltage *V* and back e.m.f. E_b now help each other and are in the same direction. These conditions are worse than those of

(*a*) Normal running (*b*) Braking.

Fig. 8.36.

starting motor direct on line, as now nearly double the supply voltage is impressed directly on the armature having low resistance. A very heavy current is likely to flow which is given by

$$I = \frac{V + E_b}{R_a}$$

where R_a = Resistance of the armature.

In order to limit the current during the braking period, additional resistance *R* is, therefore, inserted in the armature circuit.

∴ The current at the instant of braking

$$= \frac{V + E_b}{R + R_a}$$

Braking torque is proportional to product of flux and armature current, armature current can be varied by changing the external resistance in the armature circuit during braking period. Braking torque can, therefore, be regulated by varying the magnitude of the resistance inserted in the armature circuit.

(*b*) **Plugging as applied to D.C. series motor.** While braking D.C. series motor by the method of plugging, as soon as the current in the armature is reversed, the current in the field also stands reversed. This result in the production of the torque in the same direction as it was, when the motor was working normally. In order to produce torque in the reverse direction so that motor tends to stop, direction of field current should remain unchanged.

For that the connections have been shown in Fig. 8.37 (*b*). Fig. 8.37 (*a*) shows the connections of D.C. series motor before plugging is applied. As shown in Fig. 8.37 (*b*), the direction of field current remains unchanged whereas direction of armature current I_a gets reversed and given by as in the previous case,

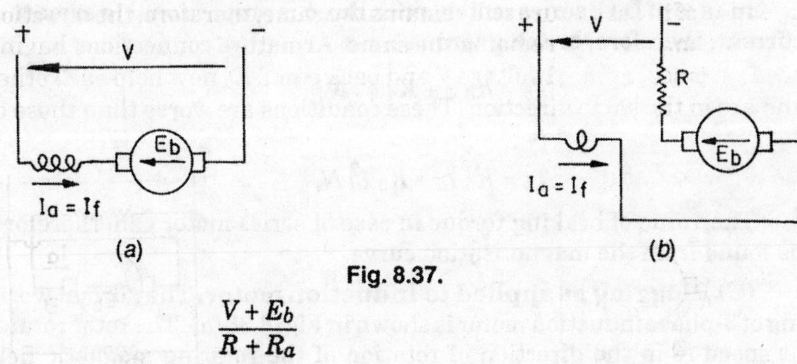

Fig. 8.37.

$$\frac{V + E_b}{R + R_a}$$

where R is the additional resistance connected in series to limit the current.

Braking torque $T_b \alpha \propto I_a$

Where ϕ is the flux and I_a is the armature current during braking period.

$$T_b = K_1 \phi I_a$$

$$I_a = \frac{V + E_b}{R + R_a}$$

$$T_b = K_1 \phi \left(\frac{V + E_b}{R + R_a}\right)$$

$$E_b \propto \phi N$$

where N = speed in r.p.m. of the motor.

$$E_b = K_2 \phi N$$

$$T_b = K_1 \phi \frac{V + K_2 \phi N}{R + R_a}$$

$$= \frac{K_1 \phi V + K_1 K_2 \phi^2 N}{R + R_a} = \frac{K_1 \phi V}{R + R_a} + \frac{K_1 K_2 \phi^2 N}{R + R_a}$$

$$\frac{K_1 V}{R + R_a} = K_3 = \text{Constant}$$

$$\frac{K_1 K_2}{R + R_a} = K_4 = \text{Constant}$$

$$\therefore \qquad T_b = K_3 \phi + K_4 \phi^2 N.$$

Now in case of shunt motors flux ϕ is constant

$$K_3 \phi = K_5$$

$$K_4 \phi^2 = K_6$$

Braking torque in shunt motor $= (K_5 + K_6 N)$ for any given value of resistance R.

In case of D.C. series motor, since flux ϕ is proportional to armature current, therefore, braking torque

$$T_b = K_3 \phi + K_4 \phi^2 N$$

$$\phi \propto I_a$$

$$T_b = K_5 I_a + K_5 I_a^2 N.$$

The value of braking torque in case of series motor can, therefore, be found from the magnetisation curve.

(C) Plugging as applied to Induction motor. The normal working of 3-phase induction motor is shown in Fig. 8.38 (*a*). The rotor rotates at speed N in the direction of rotation of the rotating magnetic field (rotating at speed N_s).

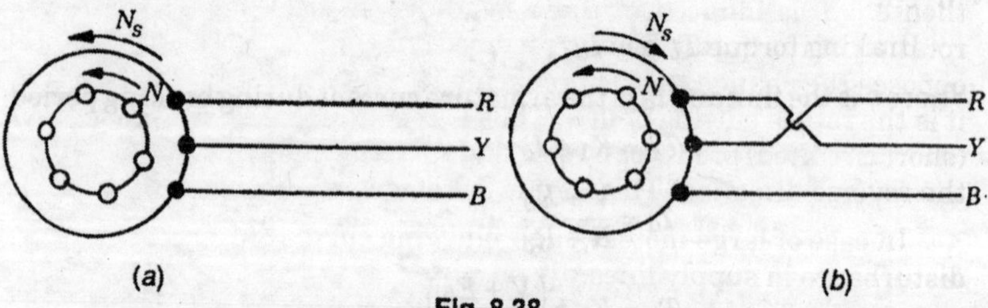

Fig. 8.38.

When braking by the method of plugging is applied to three phase induction motor, the two phases to the stator are interchanged. This is shown in Fig. 8.38 (*b*). This interchanging of phases to the stator, results in the rotation, of rotating magnetic field at synchronous speed in the reverse direction, *i.e.* clockwise. Now the rate of cutting of the flux with the rotor conductors gets almost doubled, inducing almost two times the e.m.f. in the rotor at standstill. This high value of e.m.f. induced in the rotor windings requires additional insulation to be provided so that this high voltage is with-stood by the winding.

Rotation of magnetic field in the reverse direction produce torque in the reverse direction, thus applying brakes to the motor. Care should be taken that as soon as the motor comes to stop the supply to the stator should be cut off, failing which the motor would start revolving in the opposite direction.

(*D*) Plugging as applied to Synchronous Motor. Normally, three phase voltage is applied to the stator of synchronous motor producing rotating magnetic field and D.C. voltage is fed to the rotor field windings producing permanent poles. If rotor is made to revolve at synchronous speed, the two opposite poles of the stator and rotor get locked and appear to move in the clockwise direction as shown in Fig. 8.39. If the direction of rotation of rotating magnetic field is reversed by interchanging the two phases to the stator, the braking torque is not produced.

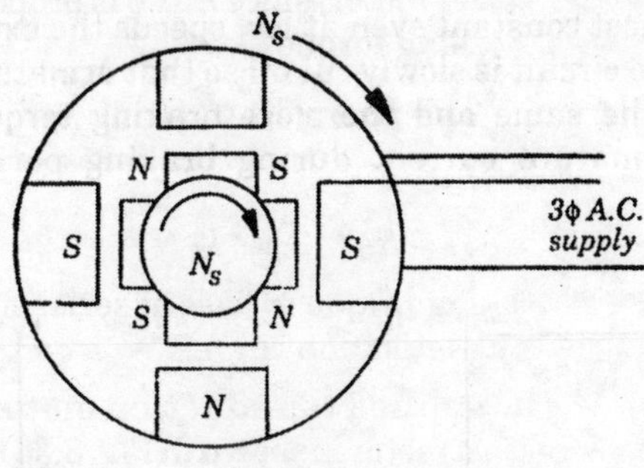

Fig. 8.39.

But if damper windings are placed in the pole shoes of the rotor, then e.m.f. is induced in these windings as soon as the direction of rotation of the stator field is reversed. As per Lenze's law, this e.m.f. opposes the change that is producing it. The change that is producing it is the rate of cutting of flux. Thus e.m.f. induced in damper windings (short circuited) produces circulating current, which produces torque in the reverse direction. This torque is known as braking torque.

In case of large motors, high plugging current can create a severe disturbance in supply lines.

(*ii*) Rheostatic Braking or Dynamic Braking. In this method of braking, the supply to the motor is disconnected during the braking period. The field continues to get energised in the same direction. Thus the motor starts working as a generator and all the kinetic energy of the moving parts in converted into electrical energy which is further dissipated in external resistance connected across the motor during the braking period.

This method of braking has the advantage over the plugging, that during braking, the energy is not drawn from the supply.

(*a*) Rheostatic Braking as applied to the D.C. shunt motor. Fig. 8.40 (*a*) shows the normal working of D.C. shunt motor. When rheostatic braking is to be applied, the armature is disconnected from the supply and is connected across suitable resistance, so that the flow of current at any time is not more than full load current. The field continues to draw current from the supply but this current is very small. This is shown in Fig. 8.40 (*b*). The machine starts working as a generator. The amount of e.m.f. induced *i.e.* E_b shall depend upon the speed. At high values of speeds, E_b is more and therefore armature current is more but at lower values of the speeds, the armature current is less. Since braking torque is proportional to the product of flux and armature current and flux remaining constant in case of shunt motor, braking torque is directly proportional to armature current. Thus at high values of speeds, the braking torque shall be more and at low values of speeds, the braking torque is less. In order that, braking torque remains almost constant even at low speeds the external resistance in the armature circuit is slowly cut off so that armature current remains practically the same and therefore braking torque remains almost uniform. Armature current during braking period is equal to *i.e.*

$I_a = \dfrac{E_b}{R + R_a}$.

(*a*)

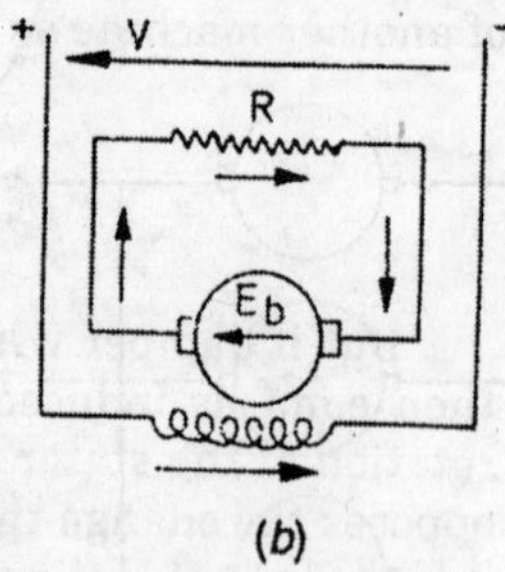

(*b*)

Fig. 8.40.

(*b*) Rheostatic braking as applied to D.C. series motor

Fig. 8.41 (*a*) shows the normal working of D.C. series motor. When rheostatic braking is to be applied, the armature is disconnected from the supply and is connected across the resistance R.

If after disconnecting the supply, the armature is connected as shown in Fig. 8.41 (*b*), the direction of both armature and field currents are reversed. This results in production of torque in the same direction as in Fig. 8.41 (*a*) and therefore no braking torque is produced. For braking torque to be produced, only the direction of current in the armature should be reversed and in the field, it should remain the same. For that, connections as shown in Fig. 8.41 (*c*) are made.

The armature current during the period of braking is given by,

$$I_a = \frac{E_b}{R + R_a}$$

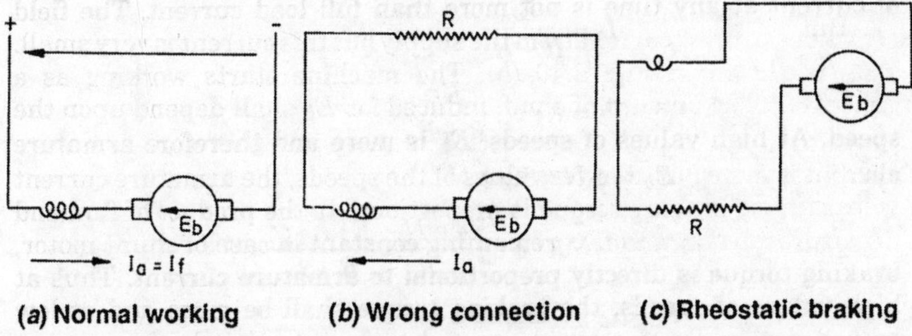

(*a*) Normal working (*b*) Wrong connection (*c*) Rheostatic braking

Fig. 8.41

The resistance inserted in the circuit should be less than the critical resistance, otherwise the generator shall not be self exciting.

In case there are more than one machines as in electric traction, we can connect the motors through equalising connection as shown in Fig. 8.41.1 (*a*) below or by exciting one machine by the armature current of another machine as shown in Fig. 8.41.1 (*b*).

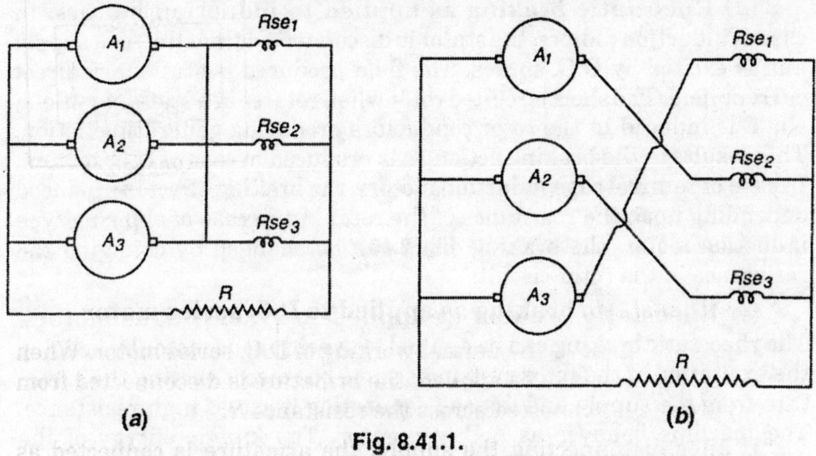

Fig. 8.41.1.

(*c*) Rheostatic Braking as Applied to D.C. Compound Motors

Rheostatic braking to D.C. compound motors is applied in a similar way as to D.C. shunt motors.

Relation between the braking torque and speed for D.C. motor during rheostatic braking is determined as below :

$$T_b = \text{Braking Torque} \propto \phi\, I_a$$

$$= K_1\, \phi\, I_a.$$

But $$I_a = \frac{E_b}{R + R_a}.$$

$$\therefore \quad T_b = K_1 \phi . \frac{E_b}{R + R_a}$$

also $$E_b \propto \phi N = K_2 \phi N$$

$$\therefore \quad T_b = K_1 \phi . \frac{K_2 \phi N}{R + R_a} = \frac{K_1 K_2 \phi^2}{R + R_a} N$$

$$R + R_a = \text{Constant}$$

$$\therefore \quad \frac{K_1 K_2}{R + R_a} = \text{Constant} = K_3$$

$$\therefore \quad T_b = K_3 \phi^2 N.$$

In case of shunt motors since flux ϕ is constant, therefore, torque is proportional to the speed but in case of D.C. series motor, since flux is proportional to the armature current, therefore torque can be determined graphically from the magnetisation curve.

(*d*) Rheostatic braking as applied to induction motors. In case of induction motors, the stator is disconnected from the A.C. supply and is excited by D.C. source. The field produced is stationary direct current field. The short circuited rotor when rotates in a stationary field, e.m.f. is induced in the rotor conductors producing generating action. This results in the braking action as is produced in case of D.C. motors. In case of squirrel cage induction motors, the braking effect is produced depending upon the resistance of the rotor but in case of slip ring type induction motor, the braking effect can be changed by changing the resistance in the rotor circuit.

(*e*) Rheostatic braking as applied to synchronous motors. The rheostatic braking can be applied to synchronous motor by keeping the excitation of the rotor system as same but disconnecting the armature from the supply and instead connecting it across high resistance. The machine behaves as A.C. generator. The kinetic energy of the rotating parts gets converted into electrical energy which gets dissipated in the resistance.

(*iii*) Regenerative braking. In the first two methods, the mechanical energy of the revolving parts was converted into electrical energy which was further dissipated in the external resistance. In case of plugging, even energy was drawn during the braking period also and was wasted. In case of regenerative braking, the machine is made to work in a way that it generates energy and feeds it back to the supply system.

As shown in Fig. 8.41.2 the shunt machine draws power from supply if applied voltage 'V' is greater than induced voltages E_b.

If induced voltage E_b becomes greater than the applied voltage, the machine instead of taking power from the supply (motoring action) sends back power to supply (generating action). This can be achieved if,

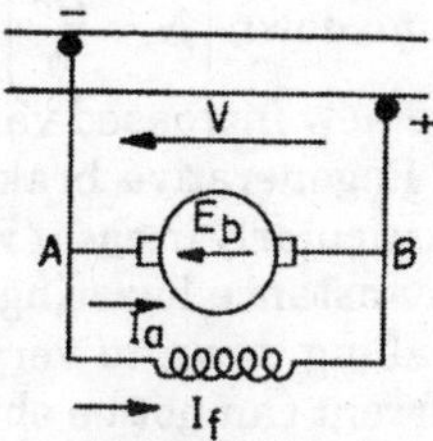

Fig. 8.41.2.

(*i*) Supply voltage V is reduced.

(*ii*) The field excitation is increased so that E_b becomes greater than applied voltage V.

(*iii*) The speed of the motor becomes greater than the no-load speed of the motor, in that case again the value of E_b becomes greater than the applied voltage V.

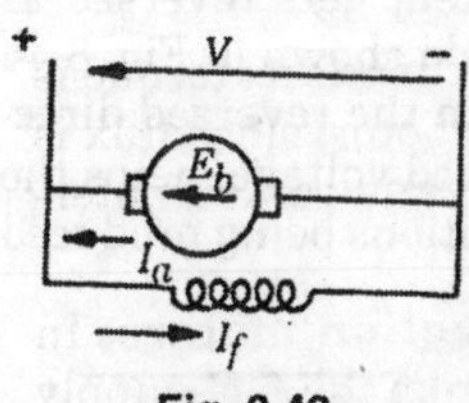

Fig. 8.42

Thus whenever the value of E_b becomes greater than V the direction of current in the armature gets reversed, the direction of field remaining same, the torque produced is in the opposite direction. Thus braking effect is produced. When the torque is reversed, the motor tends to stop thus reducing the speed and thus reducing E_b. In order that braking torque is still applied, the field excitation is increased so that E_b still remains greater than applied voltage. Thus it is not possible to obtain regenerative braking down to very low speeds as E_b falls tremendously down and it cannot be increased by the increase in excitation because too much increase in field cannot be obtained.

Thus regenerative braking cannot stop the motor. It can only be applied to overhauling loads which tend to drive the motor at speeds greater than the no-load speed of the motor. This can take place in a crane motor due to action of lowering load or in a motor of an electric locomotive moving down-gradient.

(*a*) Regenerative braking applied to D.C. shunt motors. Regenerative braking can be applied easily to D.C. shunt motors without any change of connections. Fig. 8.43 shows the normal working of D.C. shunt motor in which case applied voltage V is greater than the induced e.m.f. E_b and the motor is drawing current from the supply. If due to overhauling loads, the speed of the motor becomes greater than the normal, E_b becomes greater than V and as a result direction of armature current, gets reversed. This feeds back the power to supply and produces retarding torque due to which the speed of the motor comes down.

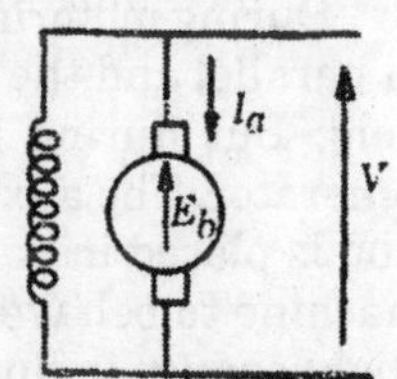

Fig. 8.43

Another method by which E_b could be increased was by increasing the excitation. Thus by increasing the excitation, the speed of the motor shall also come down $\left(N \propto \frac{E_b}{\phi}\right)$. The motor will thus take up a reduced speed at the new increased value of excitation.

Regenerative braking can be easily applied to D.C. shunt motors, particularly in cases where it is required to hold a load at a certain speed for instance lowering a hoist. It is not possible to obtain regenerative braking down to very low speeds, since a sufficient increase in field current can not be obtained.

(b) Regenerative braking applied to D.C. series motors. Regenerative braking cannot be applied to D.C. series motors as such because as the direction of current is reversed in the armature, for regenerative purpose, the direction of field current also gets reversed. Thus series field connections must be reversed. If series field connections are not reversed, the direction of field current gets reversed as shown in Fig. 8.44 (*b*). The original direction of field shown in Fig. 8.44 (*a*). Reversal of field current generates e.m.f. E_b in the reversed direction. Thus this e.m.f. instead of opposing the applied voltage, helps the applied voltage which results in short circuit conditions being produced although for a short duration.

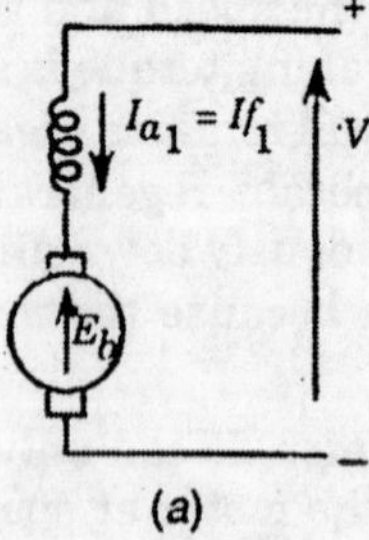

(a)

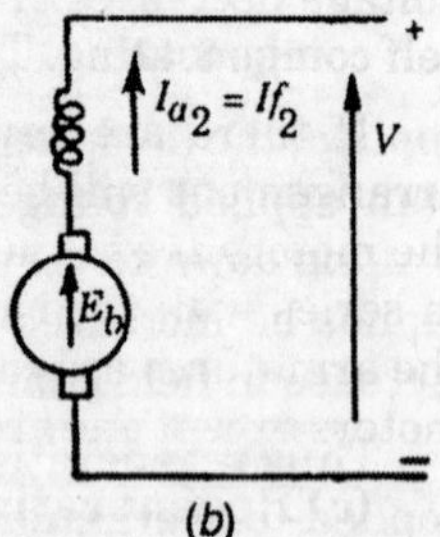

(b)

Fig. 8.44

In case of D.C. series motor, the direction of the current in the field as well as in the armature is reversed during the period of regenerative braking. Thus the back emf E_b and the supply voltage V will aid each other causing short circuit. In order to avoid short circuiting, it is therefore necessary to reverse the field connections.

Fig. 8.44.1 represents method of regenerative braking as employed in case of trolley buses, tramways etc. where usually one D.C. series motor is used.

During motoring the auxiliary field windings are bunched together in parallel and the whole bunch is connected in parallel to main series field. But during regenerative braking, the motor starts acting as generator. The auxiliary fields are arranged in series and the combination is placed in parallel across the main series field, thus making the machine to behave as a differentially compound generator. If there is slight change in line voltage the shunt field will immediately cause, the

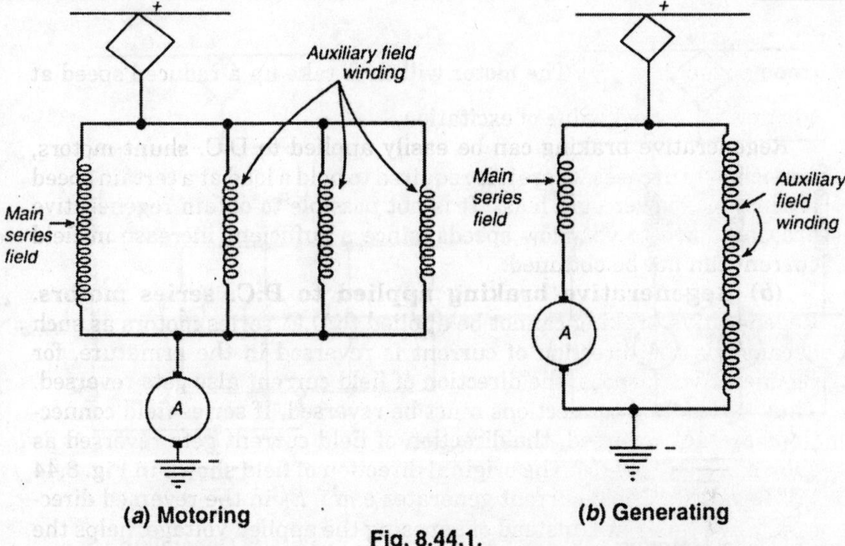

(*a*) Motoring (*b*) Generating

Fig. 8.44.1.

increase in the generated e.m.f. The reverse will be the case if the supply voltage decreases or tends to decrease. Thus the arrangement will be self compensating. This method is known as French Method.

If there are several motors as in case of electric railways the arrangement will be as shown in Fig. 8.44.2. During normal running the motors are connected in parallel with the field windings connected in series with their respective armatures, but during regeneration all the armatures are connected in parallel and series field windings of all motors except one are connected in series and placed across the supply.

(*c*) Regenerative Braking in Composite Traction System

To apply regenerative braking in composite traction system, the d.c. generated by the series machines is firstly converted into a.c. of proper magnitude and frequency. This conversion of d.c. to a.c. is achieved by the use of SCR as shown in Fig. 8.44.3.

The fields of the two motors F_1 and F_2 are excited and controlled separately by separate group of SCR's connected in symmetrical bridge arrangement.

8.25. Main group of SCR's work as inverters

The regenerative braking is firstly controlled by the excitation control of motor fields and secondly by controlling the counter e.m.f. of inverters by means of phase angle control of the inverter operated thyristors. With full electronic control of SCR's, it has been possible for driver to simply select a reference voltage corresponding to desired speed, the SCR gate control

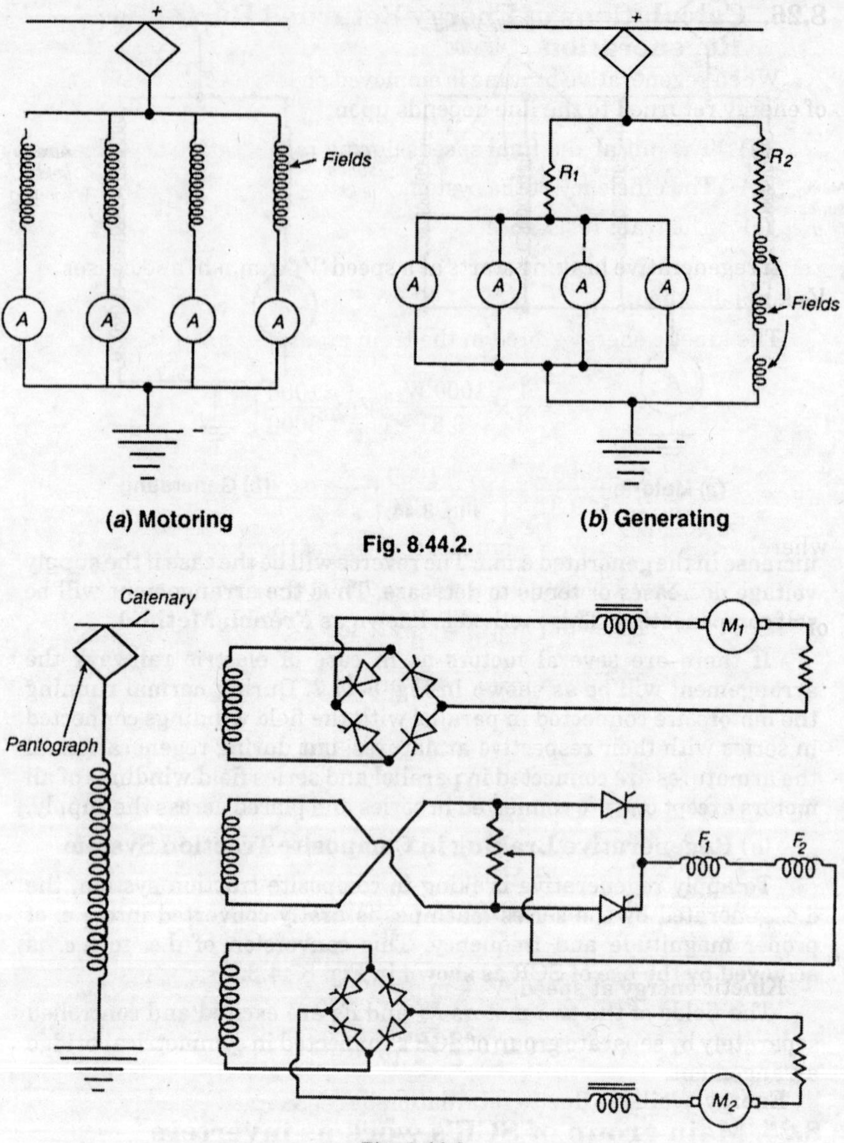

Fig. 8.44.2.

Fig. 8.44.3.

circuit will bring the locomotive to that speed and maintain it by the regulation of motor voltage.

The main disadvantage of adopting this arrangement is of low power factor at starting which can be overcome by providing low loss capacitor of suitable capacity permanently across the circuit.

8.26. Calculations of Energy Returned During Regeneration

When regenerative braking is employed on level track, the amount of energy returned to the line depends upon,

(a) The initial and final speeds during regenerative braking.

(b) The efficiency of the system.

(c) The train resistance.

If regenerative braking starts at a speed 'V_1' k.m.p.h. and ceases at V_2 k.m.p.h., then,

The kinetic energy stored in the train at a speed of V_1 k.m.p.h.

$$= \frac{1}{2} \times \frac{1000\, W_e}{9.81} \left(V_1 \times \frac{1000}{3600} \right)^2 \text{ kg metres}$$

$$= \frac{1}{2} \times \frac{1000\, W_e}{9.81} \left(V_1 \times \frac{1000}{3600} \right)^2 \times 9.81 \text{ Newton metres.}$$

where W_e = equivalent weight of the train in tonnes.

[Now 1 Newton metre/sec = 1 Watt

or 1 Newton metre = 1 watt sec

$$= \frac{1}{3600} \text{ watt hrs.]}$$

$$= \frac{1}{2} \times \frac{1000 W_e}{9.81} \left(V_1 \times \frac{1000}{3600} \right)^2 \times 9.81 \times \frac{9.81}{3600} \text{ watt hr.}$$

$$= 0.01072\, V_1^2 \frac{W_e}{W} \text{ Watt hrs per tonne}$$

(W = Dead weight of train)

Kinetic energy at speed, V_2 k.m.p.h.

$$= 0.01072\, V_2^2 \frac{W_e}{W} \text{ Watt hrs/tonne.}$$

Energy available due to retardation be regenerative braking,

$$= 0.01072 \frac{W_e}{W} (V_1^2 - V_2^2) \text{ Watt-hrs/tonne.}$$

If S is the distance in km covered during regenerative braking, the retardation, β in k.m.p.hs. during braking is,

$$= \frac{V_1^2 - V_2^2}{2S \times 3600}$$

$$\therefore \qquad S = \frac{V_1^2 - V_2^2}{2\beta \times 3600} \text{ km.}$$

If r = Specific resistance of the train in newtons per tonne.

Work done in overcoming this resistance, r for a distance, S

$$= 1000\, r \times S \text{ Newton metre/tonne}$$

$$= 1000\, r \times S \text{ watt sec./tonne}$$

$$= \frac{1000}{3600} r \times S \text{ watts hr/tonne}$$

$$= 0.278\, r \times S \text{ watt hr/tonne}$$

$$= 0.278 \times r \times \frac{(V_1^2 - V_2^2)}{2\,\beta \times 3600} \text{ Watt hr/tonne}$$

∴ Energy returned to the line,

$$\therefore \qquad = 0.01072\, V_1^2\, \frac{W_e}{W}(V_1^2 - V_2^2)$$

$$- 0.278 \times r \times \left[\frac{V_1^2 - V_2^2}{2\,\beta \times 3600}\right] \text{ Watt hr/tonne}$$

$$= V_1^2 - V_2^2 \left[0.01072 \frac{W_e}{W} - \frac{0.0139\, r}{3600\,\beta}\right] \text{ Watt hr/tonne}$$

If η is the efficiency of the system, then energy returned

$$= \eta\,(V_1^2 - V_2^2) \left[0.01072 \frac{W_e}{W} - \frac{0.01039\, r}{3600\,\beta}\right] \text{ Watt hr/tonne}$$

If the track is inclined at a per cent gradient 'G', all through the distance 'S' and the train is moving down this gradient, r in the alone expression should be replaced by (r – 98.1 G). If the train is moving up this gradient then it should be replaced by (r + 98.1 G).

Example 8.6. ***A 400-tonne train travels down a gradient 1 in 70 for 120 secs during which period its speed is reduced from 80 Km / hr to 50 Km / hr by regenerative braking. Find the energy returned to the lines if tractive resistance is 5 Kg / tonne and allowance for rotational inertia is 7.5%. Overall efficiency of motors is 75%.***

Solution.

Weight of the train,

$$W = 400 \text{ tonnes}$$

Rotational inertia $= 7.5\%$

So, effective weight,

$$W_e = 400 \times 1.075 = 430 \text{ tonnes.}$$

Track resistance, $r = 5 \times 9.81$ Newton metre/tonne

Gradient, $G = \frac{1}{70} \times 100 = \frac{10}{7}\%$

Overall efficiency, $\eta = 75\%$

Period for which regeneration takes place = 120 seconds.

Speed reduces from 80 k.m.p.h. to 50 k.m.p.h.

$\therefore$ Energy available for recuperation = 0.1072 $W_e (V_1^2 - V_2^2)$ Watt hrs.

$$= 0.01072 \times 430(80^2 - 50^2)$$

$[V_1 = 80$ k.m.p.h. $V_2 = 50$ k.m.p.h.$]$

$$= 0.01072 \times 430 \; [6400 - 2500]$$

$$= 0.01072 \times 430 \times 3600$$

$$= 18{,}150 \text{ watt hr} = 18.150 \text{ kw hr}$$

Tractive effort required during retardation,

$$= W(r - 98.1\, G)$$

$$= 400\left(49 - 98.1 - \frac{10}{7}\right)$$

$$= 400(49 - 140.14)$$

$$= 400 \times (-91.14) = -36{,}451 \text{ Newtons}$$

Also the distance travelled by train during retardation period,

$$S = \frac{V_1^2 - V_2^2}{2\,\beta \times 3600} \times 1000 = \left[\frac{V_1 + V_2}{2} \times \frac{1000}{3600}\right] \times t$$

$$\left(\because t = \frac{V_1 - V_2}{\beta}\right)$$

$$= \left(\frac{80 + 50}{2} \times \frac{100}{3600}\right) \times t$$

$$= \frac{130}{2} \times \frac{1000}{3600} \times 120 = \frac{13}{6} \times 1000$$

$$= 2.17 \times 1000 = 2170 \text{ metres.}$$

Energy available during train movement,

$$= 36{,}451 \times 2170 \text{ Newton metre}$$

$$= \frac{36{,}451 \times 2170}{3600 \times 1000} = 22 \text{ kW hr.}$$

Net energy returned to the supply system

$$= \eta \times (18.15 + 22)$$

$$= 0.75 \times 40.15 = \mathbf{30.12\ kWh.}$$

Example 8.7. *Due to regenerative braking, a 400 tonne train has its speed reduced from 80 to 50 km per hour in a distance of 1.5 km down a uniform gradient of 1 in 100. Find the energy returned to the line. Specific train resistance is 50 Newtons per tonne, and efficiency of equipment is 80%. Allow 10% for rotational inertia.*

Solution. The forces helping the motion of the train are :

(*a*) Tractive effort produced by the motors during retardation.

(*b*) Tractive effort due to gradient.

The forces opposing the motion of the train are :

(*c*) Tractive effort responsible for regeneration.

(*d*) Tractive effort absorbed in friction.

For dynamic equilibrium of the train, we have,

$$a + b = c + d \quad \text{...(i)}$$

Now braking retardation,

$$\beta = \frac{V_1^2 - V_2^2}{2 \times S \times 3600}$$

$$= \frac{80^2 - 40^2}{2 \times 1.5 \times 3600} = \frac{6400 - 1600}{2 \times 1.5 \times 3600}$$

$$= \frac{4800}{2 \times 1.5 \times 3600} = 0.445 \text{ k.m.p.h.p.s.}$$

Tractive effort produced by the motors during retardation

$$= 277.8 \times W_e \times \beta$$

Also tractive effort due to gradient,

$$= 98.1 \times W \times r$$

and tractive effort absorbed in friction

$$= W \,.\, r.$$

∴ Equation (*i*) becomes

$$277.8\, W_e\, \beta + 98.1\, WG$$

= Tractive effort responsible for regeneration + Wr.

∴ Tractive effort responsible for regeneration,

$$= 277.8\, W_e\, \beta + 98.1\, WG - W \,.\, r$$

$$= 277.8 \times 1.1 \times 400 \times 0.445 + 98.1 \times 400 \times 1$$
$$- 400 \times 50$$

$$= 54{,}400 + 39{,}240 - 20{,}000$$

$$= 93{,}640 - 20{,}000 = 73{,}640 \text{ Newtons}$$

Work done by this force or effort,

$$= 73{,}640 \times 1500 = 110.5 \times 10^6 \text{ Newton metre}$$

$$= 110.5 \times 10^6 \text{ watt sec}$$

$$= \frac{110.5 \times 10^6}{10^3 \times 3600} \text{ kWh}$$

∴ Energy returned to the line,

$$\eta = 80\% = 110.5 \times \frac{0.8}{3.6}$$

$$= \mathbf{24.6 \text{ kWh.}} \quad \textbf{Ans.}$$

Example 8.8. *A train weighing 500 tonnes is going down a gradient of 20 in 1000. It is desired to maintain train speed at 40 km.p.h. by regenerative braking. Calculate the power fed into the line. Tractive resistance is 40 Newton / tonne and allow rotational inertia of 10% and efficiency of conversion of 75%.*

Solution. Dead weight of train,

$$W = 500 \text{ tonnes.}$$

Train speed, $V = 40$ km.p.h.

Tractive resistance, $r = 40$ N/tonne

Gradient, $G = \frac{20}{1000} = 2$ per cent

Efficiency of conversion,

$$\eta = 75\% = 0.75$$

Tractive effort required while going down the gradient,

$$= W \,.\, r - 98.1 \text{ WG}$$

$$= 500 \times 40 - 98.1 \times 500 \times 2$$

$$= -78{,}100 \text{ Newtons.}$$

i.e. tractive effort of 78,100 newtons is available.

Power available, $P = \frac{F_t \times V}{3600}$ kW

$$= 78{,}100 \times \frac{40}{3600} = 867.8 \text{ kW}$$

Power fed into the line,

= Power available × Efficiency of conversion

$$= 867.8 \times 0.75 = \mathbf{650 \text{ kW.}} \quad \textbf{Ans.}$$

8.27. Traction Mechanics

Under this article of traction mechanics we will mainly concern ourselves with the applied mechanics side of the traction.

8.28. Units Used in Traction Mechanics

Units of speed and distance etc. in the mechanics of train are altogether different from those used in solving the problems of ordinary applied mechanics. Various such units used and their notations are given as follows :

S = Distance in kilometres (km).

V = Speed in kilometres per hour (kmph).

α or β = Acceleration or retardation in kilometres per hour per sec. (km.p.h.p.s.).

W = Weight in tonnes.

T = Time in seconds.

F = Tractive effort or Force in Newtons.

T = Torque in Newton metres.

8.29. Speed Time Curves for Different Services

It is the curve drawn between instantaneous speed of train in kilometres per hour along *y-axis* and time in seconds (or minutes) along *x-axis.* The speed-time curve gives complete information of the motion of the train. This curve gives the speed at various times after the start of run directly.

Further, the distance travelled by the train during a given interval of time can be obtained by determining the area between the curve and the time axis corresponding to this interval.

Such speed time curves are of fundamental importance in electric traction, as when the speed time curve is given and the resistances to motion of the train or vehicle are known, the energy required for propulsion can be calculated. Moreover, when the characteristic curves of the motors are known the energy input to the train or vehicle can also be calculated.

Typical speed-time curve for main line service is shown in Fig. 8.46. It mainly consists of (*i*) initial acceleration (*ii*) constant speed run or free run (*iii*) coasting and (*iv*) Retardation or braking.

(*i*) **Acceleration.** It consists of two parts such as :

(*a*) **Acceleration while notching up or constant acceleration.** During this period of run (0 to t_1), starting resistance is gradually cut so that the motor current is limited to a certain value and the voltage across the motor is gradually increased. Thus tractive effort is constant and therefore acceleration remains constant during this period. *Tractive effort is defined as the force in Newtons exerted by driving wheel on the rail to produce movement.* Sum of the tractive efforts exerted by all the driving wheels of a locomotive is called *Gross tractive effort* which is proportional to the combined torque exerted by the motors. Motor

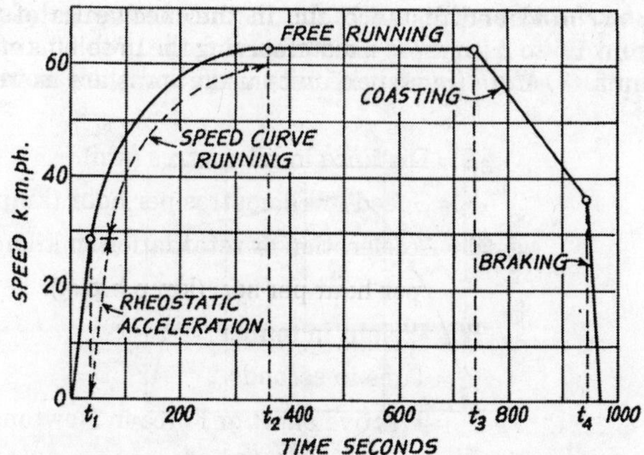

Fig. 8.46. Speed time curve for main line service.

current, during notching up period, fluctuates between certain maximum and minimum limits.

(*b*) **Acceleration on speed curve or speed curve running.** During speed curve running (t_1 to t_2), the current starts decreasing with the increase in speed according to the characteristics of the series motor and finally the current taken by the motor becomes constant. The voltage acting across the motor remains constant. During this period, though the train accelerates but acceleration decreases with the increase in speed and finally becomes zero at the speed at which the tractive effort developed by the motor becomes exactly equal to the resistance to motion of the train.

(*ii*) **Free Running or Constant Speed Run** (t_2 to t_3). This period occurs, on level track, when the power output from the driving axles balances the rate at which energy is expended against the resistance to motion. At the end of speed curve running *i.e.* at t_2 the train attains the maximum speed.

(*iii*) **Coasting period** (t_3 to t_4). At the end of free running period, supply to motors is cut off and train is allowed to run under its own momentum. Due to train resistance, speed of the train gradually decreases.

(*iv*) **Retardation or Braking Period** (t_4 to t_5). At the end of coasting period brakes are applied to bring the train to stop.

8.30. Typical Speed Time Curves

Such type of curves for electric trains operating on passenger services are given in Figs. 8.46, to 8.48. Each curve corresponds to a particular class of traffic, (1) ***Urban or city service,*** where the distance between stop is of the order of say 1 km. or so ;

(2) ***Suburban Service,*** where the distance between the stops may average from 1.5 to 2 km over a distance of from 15 to 20 km from the city terminus. (3) *Main line service,* where the stops are in frequent.

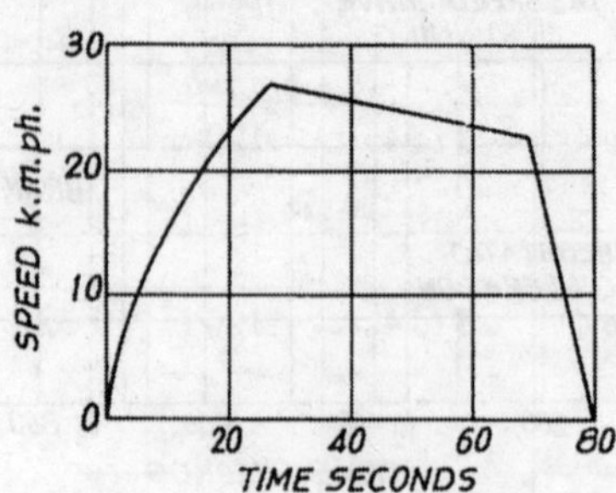

Fig. 8.47. City service.

In Fig. 8.47, which is a representative of city service, relatively high values must be adopted for the acceleration and retardation (*i.e.* between 1.5 and 4 km.p.h.s.) in order to obtain a moderately high average speed between the stations. The short distance between the stations does not permit of a free running period since it is desirable to include a short coasting period in order to obtain a reasonable energy consumption. This class of traffic requires a frequent service of trains.

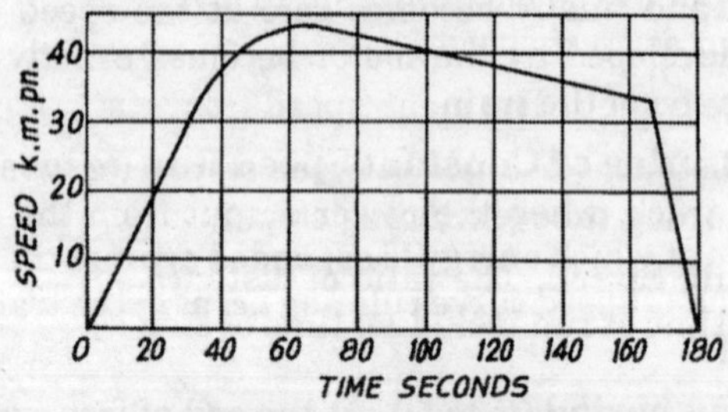

Fig. 8.48. Suburban service.

In suburban services (Fig. 8.48) the lower distance between the stations permits of a longer coasting period. In this case also, relatively high values for the accelerations and braking retardation are required in order to render the service as attractive as possible. Main line service (Fig. 8.46) is characterized by the long periods of free running at high speeds the accelerating period being relatively unimportant.

8.31. Characteristics of each type of service are given below:

No.	*Type of Service*	*Accelera-tion in km.p.h.p.s.*	*Retarda-tion in km.p.h.p.s.*	*Mini-mum Speed in km.p.h.*	*Dis-tance be-tween station in km.*	*Special Remarks*
1	Urban	1.5 to 4	3 to 4	120	1	Free running period is absent and coasting period is small
2	Subur-ban	1.5 to 4	3 to 4	120	1 to 6	Free running period is absent and coasting period is long
3	Main Line	0.6 to 0.8	1.5	160	More than 10 km	Long free run-ning and coast-ing periods. Acceleration and braking periods are small com-paratively.

8.32. Various Limiting Factors

When considering train movement from one station to another the various limiting factors are :

(1) **Crest Speed.** The maximum speed attained by the train during the run is known as crest speed.

(2) **Average Speed.** The mean speed from start to stop :

i.e. $$\frac{\text{Total distance (distance between stops)}}{\text{Running time (excluding stops)}}$$

(3) **Schedule Speed.** The ratio of distance covered between two stops and total time of run including time of stop is known as schedule speed.

Mathematically schedule speed

$$= \frac{\text{Total distance}}{\text{Total time including stops}}$$

8.33. Factors Affecting Schedule Speed

Following are the factors which affect the schedule speed of a train engaged in a given service (*i.e.* with a given distance between stations).

(*i*) Acceleration and breaking retardation (*i.e.* value of α and β)

(*ii*) Maximum or crest speed, V_m.

(*iii*) Duration of stop.

For a given distance between the stations, greater the value of acceleration and retardation, less will be the actual time of run and more the schedule speed. Similarly for given distance between stops and fixed value of acceleration and retardation, greater the value of maximum speed reached, less will be the time of run and more the schedule speed. Average speed remaining same, increase in duration of stop decreases the schedule speed. Station stoppage time has very much effect on the schedule speed in case of urban or suburban services where running time itself is very small. It is due to this reason that station stops in the case of urban and suburban services are kept very small. When the distance between stations is much and duration of actual run long, station stop has less conspicuous effect on the schedule speed.

8.34. Simplified Speed-Time Curves

When comparative performances for a given service at various schedule speeds are required (for example, in preliminary calculations for time tables etc), the actual speed-time curves of Figs. 8.46, 8.47 and 8.48 are replaced by simplified speed-time curves, which do not involve a knowledge of the motor characteristic. These simplified speed-time curves have simple geometric shapes, and in consequence, the relationship between the acceleration, retardation, average speed and distance can be deduced by simple mathematics.

For example in Fig. 8.49, *OABC* is the actual speed-time curve. In case of simplified trapezoidal speed time curve *OA′B′C*, speed curve running and coasting periods are replaced by constant speed period. On the other hand, in case of simplified quadrilateral speed-time curve *OA"B"C*, speed curve running and coasting periods are extended. Trapezoidal speed-time curve gives closer approximation of the conditions of main line service, where long distances are involved and quadrilateral speed time curve for urban and suburban services.

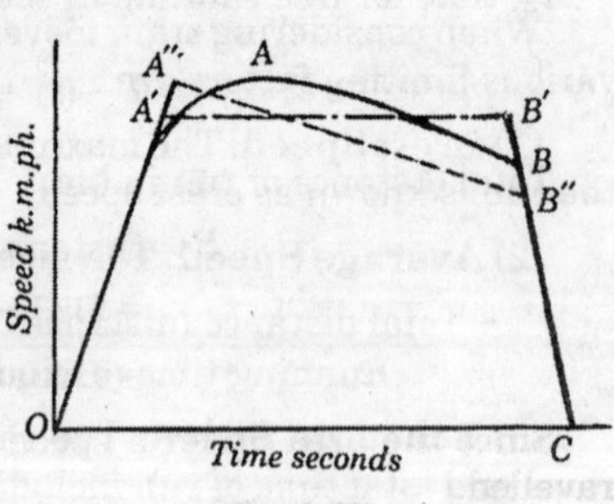

Fig. 8.49

8.35. Relationship Between Principal Quantities in Speed-Time Curves

Considering (*i*) Trapezoidal speed-time curve :

In Fig. 8.50

Let, S = Distance between stops in km.

α = Rate of acceleration in km/hr/sec.

β = Rate of retardation in km/hr/sec.

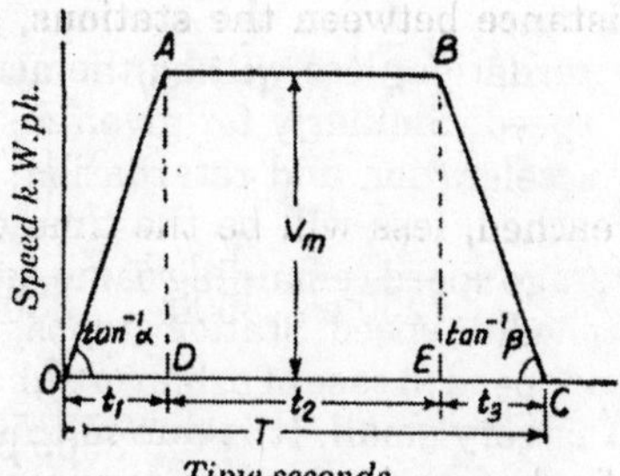

Fig. 8.50. Trapezoidal curve.

V_m = Maximum speed in km/hr.

T = Total time of run in seconds.

t_1, acceleration period or time in second,

$$= \frac{V_m}{\alpha}$$

t_3, retardation period or time in second,

$$= \frac{V_m}{\beta}$$

t_2, time for free running in second,

$$= T - (t_1 + t_3) = T - \left(\frac{V_m}{\alpha} + \frac{V_m}{\beta}\right)$$

Total distance of run in km,

S = Distance travelled during acceleration
+ Distance travelled during free run
+ Distance travelled during braking.

Since the area under a speed-time curve represents the distance travelled.

Hence, the total distance in km,

$$S = \text{Area } OABC$$

$$= \text{Area of } \Delta OAD + \text{Area of rect. } ABED + \text{Area of } \Delta BEC$$

$$= \frac{1}{2} V_m . \frac{t_1}{3600} + V_m . \frac{t_2}{3600} + \frac{1}{2} V_m . \frac{t_3}{3600}$$

Substituting the values of t_1, t_2 and t_3 we have,

$$S = \frac{1}{2} V_m . \frac{V_m}{\alpha . 3600} + \frac{V_m}{3600} . t_2$$

$$+\frac{1}{2}V_m \cdot \frac{V_m}{\beta\, 3600}$$

$$\left[\because T-\left(\frac{V_m}{\alpha}+\frac{V_m}{\beta}\right)=t_2\right]$$

or
$$=\frac{V_m^2}{7200\,\alpha}+\frac{V_m}{3600}T-\frac{V_m^2}{3600\,\alpha}-\frac{V_m^2}{3600\,\beta}+\frac{V_m^2}{7200\,\beta}$$

or
$$=\frac{V_m\,T}{3600}-\frac{V_m^2}{7200\,\alpha}-\frac{V_m^2}{7200\,\beta} \quad \text{...(A)}$$

or
$$\frac{V_m^2}{3600}\left(\frac{1}{2\alpha}+\frac{1}{2\beta}\right)-\frac{V_m\,T}{3600}+S=0$$

or
$$V_m^2\left(\frac{1}{2\alpha}+\frac{1}{2\beta}\right)-V_m\,T+3600\,S=0$$

this is a quadratic equation for V_m.

Taking $\frac{1}{2\alpha}+\frac{1}{2\beta}=K$, we get, (Pl. Note)

$$KV_m^2-V_m\,T+3600\,S=0$$

or
$$V_m=T\pm\sqrt{\frac{T^2-4K\times 3600\,S}{2K}}$$

or
$$V_m=\frac{T}{2K}\pm\sqrt{\frac{T^2}{4K^2}-\frac{3600\,S}{K}}$$

The +ve sign gives very much high value of V_m which is not possible in practice.

Hence negative (–ve) sign is adopted.

$$\therefore \quad V_m=\frac{T}{2\,K}-\sqrt{\frac{T^2}{4K^2}-\frac{3600\,S}{K}} \quad \text{[C.T.M.]}$$

or
$$S=\frac{V_m\,T}{3600}-\frac{V_m^2}{7200}\left(\frac{1}{\alpha}+\frac{1}{\beta}\right)$$

[Considering equation (A)]

Taking R.H.S. to the L.H.S., we get

$$\frac{V_m^2}{7200}\left(\frac{1}{\alpha}+\frac{1}{\beta}\right)-\frac{V_m\,T}{3600}+S=0$$

or
$$V_m^2\left(\frac{1}{\alpha}+\frac{1}{\beta}\right)-2\,V_m\,T+7200\,S=0$$

If $\left(\frac{1}{\alpha}+\frac{1}{\beta}\right)=K$ (Pl. Note)

Then,

$$K V_m^2 - 2\,T\,V_m + 7200\,S = 0 \quad \text{...(1)}$$

This is quadratic equation for V_m

$$\therefore \qquad V_m = \frac{2\,T \pm \sqrt{(2\,T)^2 - 4\,K \times 7200\,S}}{2\,K}$$

$\therefore$ In case of an equation $ax^2 + bx + c = 0$ the solution is given by,

$$x = \frac{-b \pm \sqrt{b^2 - 4ac}}{2\,a}$$

So, $$V_m = \frac{T}{K} \pm \sqrt{\frac{T^2}{K^2} - \frac{7200\,S}{K}}$$

Positive (+ve) sign gives much high value of V_m which is not possible in practice.

Hence negative (–ve) sign is adopted.

$$\therefore \qquad V_m = \frac{T}{K} - \sqrt{\frac{T^2}{K^2} - \frac{7200\,S}{K}}$$

Equation (1) can be put as,

$$KV_m^2 - 2\left(\frac{3600\,S}{V_{av}}\right) V_m + 7200\,S = 0$$

$$\therefore \qquad V_{av} = \frac{3600\,S}{T}$$

$$\therefore \qquad T = \frac{3600\,S}{V_{av}}$$

or $$KV_m^2 = \frac{7200\,S}{V_{av}} . V_m - 7200\,.\,S = 0$$

or $$K = \frac{7200\,S}{V_{av}} \cdot \frac{V_m}{V_m^2} - \frac{7200\,S}{V_m^2}$$

or $$= \frac{7200\,S}{V_{av}} . \frac{1}{V_m} - \frac{7200\,S}{V_m^2}$$

or $$K = \frac{7200\,S}{V_m^2}\left(\frac{V_m}{V_{av}} - 1\right)$$

$$\frac{1}{\alpha} + \frac{1}{\beta} = \frac{7200\,S}{V_m^3}\left[\frac{V_m}{V_{av}} - 1\right] \quad \text{...(2)}$$

$$\left[\text{where } V_{av} = \text{Average velocity} = \frac{S}{T/3600} = \frac{3600\,S}{T}\right]$$

If V_{av}, V_m and S are given in a particular case, α and β can be adjusted by using equation (2), to suit the traction motor characteristics and convenience of passengers respectively.

Equation (2) can be put in the following formulae,

i.e.
$$V_m = \sqrt{\frac{7200\,S}{K}\left(\frac{V_m}{V_{av}} - 1\right)}$$

If S and K are known, by choosing a suitable value for V_m/V_{av}, we can predict the value of V_m which can be used for elementary calculations. The value V_m/V_{av} depends on the shape of the trapezoid (simplified speed-time curve and may be taken as 1.25 for the common shape of the curve).

(*ii*) **Quadrilateral Speed Time Curve.** In Fig. 8.51,

Let T = Total time of run in seconds.

V_1 = Max. speed at the end of acceleration in km/hr.

V_2 = Speed at the end of coasting in km/hr.

α = Acceleration in km/hr/sec.

β_c = Coasting retardation in km/hr/sec.

β = Braking retardation in km/hr/sec.

Acceleration time in seconds,

$$t_1 = \frac{V_1}{\alpha}$$

Coasting time in seconds,

$$t_1 = \frac{V_1 - V_2}{\beta_c}$$

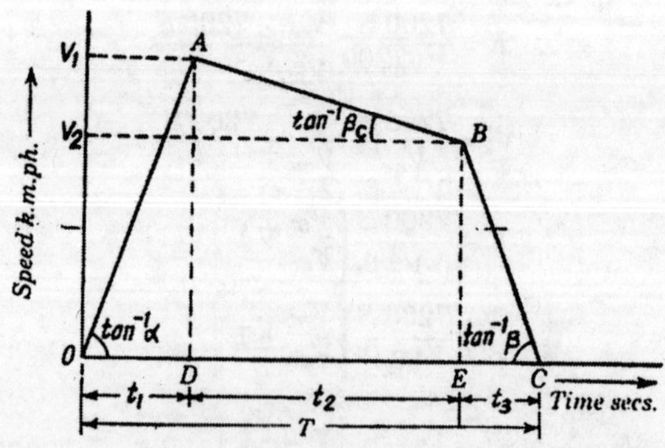

Fig. 8.51. Quadrilateral speed-time curve.

Braking time in seconds,

$$t_3 = \frac{V_2}{\beta}$$

Total distance, travelled in km,

$$S = \text{Distance travelled during acceleration} + \text{Distance travelled during coasting} + \text{Distance travelled during retardation.}$$

$$= \text{Area } OABC$$

$$= \text{Area of } \Delta\, OAD + \text{Area of } ABED + \text{Area of } \Delta\, BEC$$

$$= \frac{1}{2} V_1 \times \frac{t_1}{3600} + \frac{V_1 + V_2}{2} \times \frac{t_2}{3600} + \frac{1}{2} V_2 \times \frac{t_3}{3600}$$

$$= \frac{V_1\, t_1}{7200} + \frac{V_1\, t_2}{7200} + \frac{V_2\, t_2}{7200} + \frac{V_2\, t_3}{7200}$$

$$= \frac{V_1}{7200}(t_1 + t_2) + \frac{V_2}{7200}(t_2 + t_3)$$

or
$$S = \frac{V_1}{7200}(T - t_3) + \frac{V_2}{7200}(T - t_1) \qquad [\because\ t_1 + t_2 + t_3 = T]$$

or
$$S = \frac{T}{7200}(V_1 + V_2) - \frac{V_1\, t_3}{7200} - \frac{V_2\, t_1}{7200}$$

$$= \frac{1}{7200}\left[T\,(V_1 + V_2) - \left(V_1 \times \frac{V_2}{\beta} + V_2 \times \frac{V_1}{\alpha}\right)\right]$$

$$= \frac{1}{7200}\left[T(V_1 + V_2) - V_1 V_2 \left(\frac{1}{\alpha} + \frac{1}{\beta}\right)\right] \qquad ...(3)$$

(C.T.M.)

Also,
$$V_2 = V_1 - \beta_c\, t_2$$

$$= V_1 - \beta_c\,(T - t_1 - t_3)$$

$$= V_1 - \beta_c\left(T - \frac{V_1}{\alpha} - \frac{V_2}{\beta}\right)$$

$$= V_1 - \beta_c\left(T - \frac{V_1}{\alpha}\right) + \beta_c \times \frac{V_2}{\beta}$$

$$V_2 - \frac{\beta_c}{\beta} \times V_2 = V_1 - \beta_c\left(T - \frac{V_1}{\alpha}\right)$$

or
$$V_2\left(1-\frac{\beta_c}{\beta}\right)=V_1-\beta_c\left(T-\frac{V_1}{\alpha}\right)$$

or
$$V_1=\frac{V_1-\beta_c\,T+\frac{\beta_c}{\alpha}V_4}{1-\frac{\beta_c}{\beta}} \quad ...(4)$$

By solving equations (3) and (4) we can get the value of D, V_1 and V_2.

Example 8.9. *A train runs with an average speed of 40 kmph. Distance between stations is 2 km. Values of acceleration and retardation are 1.5 km.p.h.s. and 2.5 km.p.h.p.s. respectively. Find the maximum speed of train assuming trapezoidal speed-time curve.*

Solution. Acceleration,

$\alpha = 1.5$ km.p.h.p.s.

Retardation, $\beta = 2.5$ km.p.h.p.s.

Distance of run, $S = 2$ km.

Average speed, $V_a = 40$ km.p.h.

$$\text{Time of run}, T=\frac{S}{V_a}\times 3600=\frac{2}{40}\times 3600 = 180 \text{ seconds}$$

$$K=\frac{1}{\alpha}+\frac{1}{\beta}=\frac{\alpha+\beta}{\alpha\,\beta}=\frac{1.5+2.5}{1.5\times 2.5}=\frac{4}{3.75}=1.06.$$

Substituting various values in equation,

$$V=\frac{T}{K}-\sqrt{\frac{T^2}{K^2}-\frac{7200\,S}{K}}$$

$$=\frac{180}{16/15}-\sqrt{\frac{(180)^2}{(16/15)^2}-\frac{7200\times 2}{16/15}}$$

$$=\frac{180\times 15}{16}-\sqrt{\frac{180\times 180\times 15\times 15}{16\times 16}-\frac{144{,}400\times 15}{16}}$$

$$=\frac{2700}{16}-\sqrt{\frac{180\times 180\times 225}{(16)^2}-\frac{216{,}000}{16}}$$

$$=\frac{2700}{16}-\frac{1}{4}\times\sqrt{\frac{7{,}290{,}000}{16}-216{,}000}$$

$$=\frac{675}{4}-\frac{1}{4}\times\sqrt{455{,}625-216{,}000}$$

$$=\frac{675}{4}-\frac{1}{4}\times\sqrt{239{,}625}$$

$$=\frac{675}{4}-\frac{489}{4}=\frac{186}{4}=\mathbf{46.5\ kmph.}\quad \textbf{Ans.}$$

Example 8.10. *A train has a schedule speed of 30 kmph over a level track, distance between stations being 1 km. Station stopping time is 20 sec. Assuming braking retardation of 3 kmphps. and maximum speed 25 per cent greater than average speed, calculate acceleration to run the service.*

Solution. Schedule speed,

30 kmph, $\beta = 3$ km.p.h.p.s.

Distance of run, $S = 1$ km

Schedule time of run

$$= \frac{S}{\text{Schedule speed}} \times 3600$$

$$= \frac{1}{30} \times 3600 = 120 \text{ secs.}$$

Actual time of run = 120 – 20 = 100 sec.

Now $\quad V_a = \frac{1}{100} \times 3600 = 36$ kmph.

$\left|\because V_a = \frac{S}{T} \times 3600\right|$

$$V_m = 1.25 \times 36 = 45 \text{ kmph.}$$

Substituting these various values in the following equation, we get

$$\frac{1}{\alpha} + \frac{1}{\beta} = \frac{7200\, S}{V_m^2}\left(\frac{V_m}{V_a} - 1\right)$$

or
$$\frac{1}{\alpha} + \frac{1}{3} = \frac{7200 \times 1}{45 \times 45}(1.25 - 1)$$

or
$$\frac{1}{\alpha} + \frac{1}{3} = \frac{8}{9}$$

or
$$\frac{1}{\alpha} = \frac{8}{9} - \frac{1}{3} = \frac{5}{9}$$

or
$$\alpha = \frac{9}{5} = \mathbf{1.8\ kmphps.} \quad \textbf{Ans.}$$

Example 8.11. *Assuming a simplified speed-time curve for a train which has got a schedule speed of 50 km/hr between two stops 2 km apart. Find the crest speed of the train over the run, if the stop is of 30 seconds duration and the values of acceleration and retardation are 2.5 km/hr./sec. and 3.4 km/hr./sec.*

Solution. Acceleration,

$$\alpha = 2.5 \text{ kmphps}$$

Retardation, $\quad \beta = 3.4$ kmphps

Distance of run, $\quad S = 2$ km

$$\text{Now schedule speed} = \frac{\text{Distance between stop}}{\text{Time of run + Time of stop}}$$

$$\text{Time} = \frac{\text{Distance between stops}}{\text{Schedule speed}}$$

$$= \frac{2 \times 3600}{50} = 144 \text{ seconds.}$$

Time of run, $T = 144 - 30 = 114$ sec.

By putting the above given various values in the equation of V_m we get

$$V_m = \frac{T}{K} - \sqrt{\frac{T^2}{K^2} - \frac{7200\,S}{K}}$$

where $$K = \frac{1}{\alpha} + \frac{1}{\beta} = \frac{1}{2.5} + \frac{1}{3.4}$$

$$= \frac{2.5 + 3.4}{2.5 \times 3.4} = \frac{5.9}{2.5 \times 3.4} = 0.695$$

So $$V_m = \frac{114}{0.695} - \sqrt{\left(\frac{114}{0.695}\right)^2 - \frac{7200 \times 2}{0.695}}$$

$$= 164 - \sqrt{(164)^2 - 20{,}700}$$

$$= 164 - \sqrt{26{,}900 - 20{,}700} = 164 - \sqrt{6200}$$

$$= 164 - 10 \times \sqrt{62} = 164 - 10 \times 7.88$$

$$= 164 - 78.8 = \mathbf{85.2 \text{ kmph.}} \quad \textbf{Ans.}$$

Example 8.12. *An electric train is to have a braking retardation of 3.2 kmphs. If the ratio of maximum speed to average speed is 1.3, the time for stop is 26 seconds and acceleration is 0.8 km / hr / sec. Find its schedule speed for a run of 1.5 km. Assume trapezoidal speed time curve.*

Solution. Acceleration,

$\alpha = 0.8$ km/hr/sec

Retardation, $\beta = 3.2$ km/hr/sec

Distance of run, $S = 1.5$ km

Let the actual time of run be T seconds.

$$\text{Average speed, } V_a = \frac{3600\,S}{T} = \frac{3600 \times 1.5}{T} = \frac{5400}{T} \text{ km/hr.}$$

$$\text{Maximum speed, } V_m = 1.3\,V_a = 1.3 \times \frac{5400}{T} = \frac{7020}{T} \text{ kmph.}$$

$$\text{Since } V_m^2 \left(\frac{1}{\alpha} + \frac{1}{\beta}\right) - 2\,V_m\,T + 7200\,S = 0$$

or
$$V_m^2\left(\frac{1}{2\,\alpha}+\frac{1}{2\,\beta}\right)-V_m\,T+3600\,S=0$$

$\therefore$
$$V_m^2=\frac{V_m\,T-3600\,S}{\frac{1}{2\,\alpha}+\frac{1}{2\,\beta}}=\frac{\frac{7{,}020}{T}\times T-3600\times 1.5}{\frac{1}{2\times 0.8}\times\frac{1}{2\times 3.2}}.$$

$$=\frac{7020-5400}{5/6.4}$$

or
$$V_m^2=\frac{1620\times 6.4}{5}$$

or
$$V_m=\sqrt{\frac{1620\times 6.4}{5}}=45.53$$

and
$$V_a=\frac{V_m}{1.3}=\frac{45.53}{1.3}=35.028\text{ km.p.h.}$$

Actual time of run, $T=\frac{3600\,S}{V_a}=\frac{3600\times 1.5}{35.028}=154$ seconds

Schedule time, T_s = Actual time of run + Time of stop

$= 154 + 26 = 180$ seconds.

Schedule speed, $V_s=\frac{S\times 3600}{T_s}=\frac{1.5\times 3600}{180}$

$=$ **30 km.p.h. Ans.**

Example 8.13. *A Suburban electric train has a maximum speed of 65 kmph. The schedule speed including a station stop of 30 sec. is 43.5 km.p.h. If the acceleration is 1.3 km.p.h.p.s. Find the value of retardation when the average distance between stops is 3 km.*

Solution. $V_m = 65$ km.p.h.

Schedule speed, $T_s = 43.5$ km.p.h.

Acceleration, $\alpha = 1.3$ km.p.h.s.

Distance of run, $S = 3$ km.

Total time of run including stop,

$$T_s=\frac{S}{V_s}\times 3600$$

$$=\frac{3}{43.5}\times 3600=248\text{ sec.}$$

Time for actual run,

$$T = T_s - \text{Duration of stop}$$

$$= 248 - 30 = 218\text{ seconds.}$$

Average speed,

$$V_a = \frac{S}{T} \times 3600$$

$$= \frac{3}{218} \times 3600 = 49.5 \text{ km.p.h.}$$

Using the relation,

$$\frac{1}{\alpha} + \frac{1}{\beta} = \frac{7200\,S}{V_m^2}\left[\frac{V_m}{V_a} - 1\right]$$

or $$\frac{1}{\beta} + \frac{1}{1.3} = \frac{7200\,S}{(65)^2}\left[\frac{65}{49.5} - 1\right]$$

or $$\frac{1}{\beta} + \frac{1}{1.3} = \frac{21{,}600}{4225}\,[1.3 - 1]$$

or $$\frac{1}{\beta} + \frac{1}{1.3} = \frac{6480}{4225}$$

or $$\frac{1}{\beta} = \frac{6480}{4225} - \frac{1}{1.3}$$

$$= \frac{15}{10} - \frac{10}{13}$$

$$= \frac{195 - 100}{130} = \frac{95}{130}$$

or $$\beta = \frac{130}{95} = \textbf{1.3 km.p.h.p.s.} \quad \textbf{Ans.}$$

Example 8.14. *An electric train has a schedule speed of 25 km.p.h. between stations 800 metres apart. The duration of station stop is 20 seconds, the maximum speed is 20 per cent higher than the average running speed and the braking retardation is 3 km.p.h.p.s. Calculate the rate of acceleration required to operate this service.*

Solution. Distance of run,

$$S = 0.8 \text{ km}$$

Schedule speed, $V_m = 25$ km.p.h.

Duration of stop = 20 sec.

$$\frac{V_m}{V_a} = 1.2$$

Retardation, $\beta = 3$ km.p.h.p.s.

Schedule time of run,

$$T_s = \frac{3600\,S}{V_a} = \frac{300 \times 0.8}{25} = 115.2 \text{ sec.}$$

Actual time of run,

$$T = T_s - \text{ Duration of stop}$$

$$= 115.2 - 20 = 95.2 \text{ sec.}$$

$$V_a = \frac{S}{T} \times 3600$$

$$= \frac{0.8 \times 3600}{95.2} = 30.3 \text{ km.p.h.}$$

and $$V_m = 1.2 \times V_a = 1.2 \times 30.3 = 36.36 \text{ km.p.h.}$$

Now, we have

$$\frac{1}{\alpha} + \frac{1}{\beta} = \frac{7200\, S}{X_m^2}\left[\frac{V_m}{V_a} - 1\right]$$

or $$\frac{1}{\alpha} + \frac{1}{\beta} = \frac{7200 \times 0.8}{(36.36)^2}\,[1.2 - 1]$$

or $$\frac{1}{\alpha} + \frac{1}{3} = 0.8730$$

or $$\frac{1}{\alpha} = 0.8730 - 0.333$$

$$= 0.54$$

or $$\alpha = \frac{1}{0.54}$$

$$= \mathbf{1.85 \text{ km.p.h.s.}} \quad \textbf{Ans.}$$

Example 8.15. *A train is required to run between two stations 1.6 km. apart at an average speed of 40 km.p.h. The run is to be made to a simplified quadrilateral speed-time curve. If the maximum speed is to be limited to 64 km / hr, acceleration to 2.0 k.m.p.h.p.s. and coasting and braking retardations to 0.16 km.p.h.p.s. and 3.2 km.p.h.p.s. respectively, determine the duration of accelerating, coasting and braking periods.*

Solution. Average speed,

$$V_a = 40 \text{ km.p.h.}$$

Distance of run, $S = 1.6$ km.

Maximum speed, $V_1 = 64$ km.p.h.

Acceleration, $\alpha = 2.0$ km.p.h.p.s.

Coasting retardation,

$$\beta_c = 0.6 \text{ km.p.h.p.s.}$$

Braking retardation,

$$\beta = 3.2 \text{ km.p.h.p.s.}$$

Duration of acceleration,

$$t_1 = \frac{V_1}{\alpha} = \frac{64}{2.0} = 32 \text{ sec.} \quad \textbf{Ans.}$$

Actual time of run,

$$T = \frac{3600\,S}{V_a} = \frac{3600 \times 1.6}{40} = 144 \text{ sec.} \quad \textbf{Ans.}$$

Consider the speed before applying brakes be, V_2

Then duration of coasting,

$$t_2 = \frac{V_1 - V_2}{\beta_c} = \frac{64 - V_2}{0.16} \text{ sec.}$$

Duration of braking, $t_3 = \dfrac{V_2}{\beta} = \dfrac{V_2}{3.2}$ sec.

Since actual time of run,

$$T = t_1 + t_2 + t_3$$

$$\therefore \quad 144 = 32 + \frac{64 - V_2}{0.16} + \frac{V_2}{3.2}$$

or $$V_2\left(\frac{1}{0.16} - \frac{1}{3.2}\right) = 32 + 400 - 144$$

or $$V_2\,(6.25 - 0.3125) = 288$$

or $$V_2 = \frac{288}{6.25 - 0.3125}$$

$$= 48.5 \text{ km.p.h.}$$

Duration of coasting,

$$t_2 = \frac{V_1 - V_2}{\beta_c} = \frac{64 - 48.5}{0.16} = \textbf{96.85 sec.}$$

Duration of braking,

$$t_3 = \frac{V_2}{\beta} = \frac{48.5}{3.2} = \textbf{15.15 sec.}$$

8.36. Mechanics of Train Movement

Consider the essential driving mechanism of an electric vehicle. The armature of the driving motor has a pinion of diameter d_1, meshed with the gear wheel on the driving wheel.

Let the driving motor exert a torque, T (in N-m) which will be transferred to the road wheel

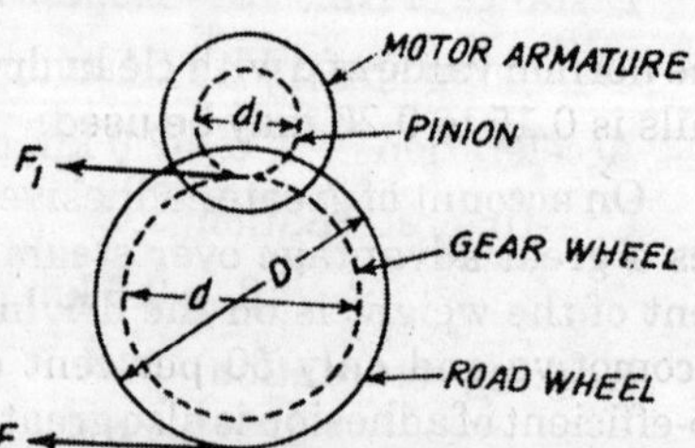

Fig 8.52. Transmission of tractive effort.

through the gear. F_1 is the tractive effort at the pinion.

Tractive effort at the edge of the pinion is given by the expression,

$$T = F_1 \times \frac{d_1}{2}$$

or
$$F_1 = \frac{2T}{d_1}$$

Tractive effort transferred to the driving wheel,

$$F = \eta F_1 \left(\frac{d}{D}\right) = \eta \cdot \frac{2T}{d_1}\left(\frac{d}{D}\right)$$

$$= \eta T \left(\frac{2}{D}\right)\left(\frac{d}{d_1}\right) = \eta T \frac{2\gamma}{D}$$

where η = efficiency of transmission

γ = gear ratio $= \frac{d}{d_1}$

d = dia. of gear wheel

D = dia. of driving wheel.

The maximum frictional force between the wheel and the track $= \mu W$, where μ is the co-efficient of adhesion between the wheel and the track and W is the weight of the train on the driving axles (called adhesive weight). Slipping will not take place unless tractive effort, $F > \mu W$. For motion of trains without slipping, tractive effort, F should be less than or at the most equal to μW but in no case greater than μW.

8.37. Adhesive Weight (W)

The total weight to be carried on the driving wheels is called the adhesive weight.

8.38. Co-efficient of Adhesion, (μ)

The magnitude of the tractive effort which can be usefully employed for propulsion, depends upon the weight coming over the driving wheels and their adhesion with the track. The co-efficient of adhesion is the ratio,

i.e.
$$\mu = \frac{\text{Tractive effort to slip the wheels}}{\text{Adhesive weight}}$$

the normal value of μ with clean dry rails is 0.25 and with wet or greasy rails is 0.15 to 0.20 may be used.

On account of greater adhesive weight that electric traction posses-ses a great advantage over steam traction. In a motor coach 100 per cent of the weight is on the driving wheels, 70 per cent in an electric locomotive and only 50 per cent or less in a steam locomotive. The co-efficient of adhesion is also greater in electric traction than in a steam traction for the following two reasons :

(*a*) In electric traction the torque exerted is continuous whereas in steam traction the torque is pulsating which causes jolting and skidding.

(*b*) The driving wheels are distributed over much greater length in the case of electric traction whereas in steam traction they are very close to each other. The tractive effort in electric traction is much greater, therefore, electric train can be made to accelerate at a faster rate. This results in saving of time,.especially when the distance between stops is small *e.g.*, in urban and suburban areas.

In order to increase the tractive effort of a locomotive, it is not enough to increase the h.p. of the traction motors alone, but at the same time weight on the driving wheels has to be increased. Railway standards provide certain maximum axle load. As such adhesive weight can only be increased by increasing the number of driving axles. In steam locomotive this is done by using coupling rod to connect driving wheels.

As coefficient of friction is more at starting than at braking so also coefficient of adhesion at any speed will be less than that at lower speed or at starting. This is due to the cyclic variation of axle loads caused due to the acceleration and retardation to which unsprung masses are subjected as they have to follow track irregularities. Relationship between the coefficient of adhesion is given by eq. (*i*) used by French National Railways.

$$\mu_{ar} = \mu_{as} \frac{8 + 0.1V}{8 + 0.2V} \qquad ...(i)$$

or by equation (*ii*) given by Curtius and Kniffler.

$$\mu_{ar} = \frac{7.5}{V + 44} + 0.16 \qquad ...(ii)$$

where μ_{ar} = coefficient of adhesion at running.

μ_{as} = coefficient of adhesion at starting

V = speed of locomotive in km.p.h.

8.39. Tractive Effort

The effective force, necessary to propel the train, at the wheels of vehicle is called the tractive effort. It is tangential to the driving wheels and measured in Newton. It is a vector quantity.

Tractive effort is the force developed by the traction unit at the wheel rims for moving the traction unit and its train.

Draw bar pull is the force exerted by the traction unit through the draw bar for moving the train.

Thus draw bar pull is less than the tractive effort by the force required to move the traction unit.

Tractive effort exerted by the traction unit has to perform the following functions :

(a) To give necessary linear and angular acceleration to the train mass.

(b) To overcome the gravity component of the weight of the train.

(c) To overcome the wind and frictional resistance of the train.

(d) To overcome curve resistance.

8.40. Tractive Effort for Propelling a Train

The total tractive effort required to run a train on track

= Tractive effort during acceleration + Tractive effort to overcome the effect of gravity + Tractive effort used in overcoming frictional resistance.

i.e. $$F_t = F_a \pm F_g + F_r.$$

(a) **Tractive effort for Acceleration, (F_a).** Let the weight of the train be W tonnes. If it is to be accelerated at a rate of α km/hr/sec., then the accelerating force or tractive effort required for linear acceleration, F_a will be given by,

$$F_a = m\,\alpha \text{ (Mass} \times \text{Acceleration)}$$

Where weight of train = 1000 W kgf.

Mass of train, m = 1000 W kg.

Acceleration, α in km.p.h.p.s. $= \alpha \times \frac{1000}{3600}$ m/sec^2

Putting the values, we get

i.e. $$F_a = m\,\alpha = 1000\, W \times 0.2778\,\alpha$$
$$= 277.8\, W\,\alpha \text{ Newtons.}$$

When a train is accelerated in a linear direction, its rotating parts like the wheels and the armatures of motors have to be accelerated in an angular direction. Therefore, the accelerating mass of the train is greater than the head mass of the train by 8 to 15 per cent. For most of the trains the effective or accelerating weight W_e is 1.1 times their dead weight W.

Hence $F_d = 277.8\, W_e\,\alpha$ Newtons.

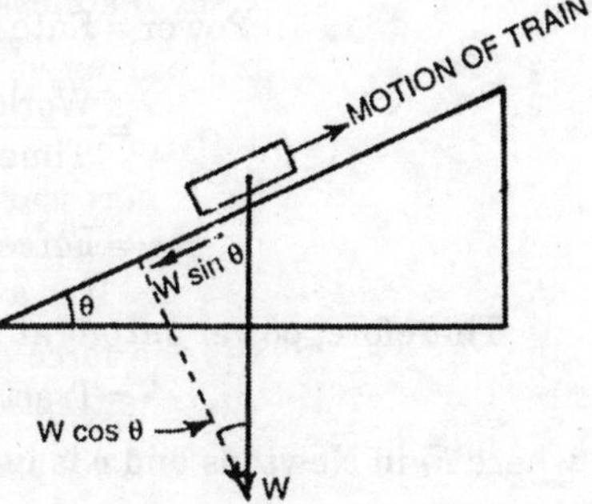

Fig. 8.53

(b) **Tractive effort, to overcome the effect of gravity or to move the train up a gradient, (F_g):** The dead weight of the train W tonnes can be assumed to be made up of two parts *viz.*. W cos θ and W sin θ, where θ is the angle of inclination of the track to the horizontal (Fig. 8.53). Since the component W cos θ is perpendicular to the direction of motion of the train it has no effect on its motion but the other component $W \sin \theta$ tries to oppose the motion of the train. Hence an additional tractive effort, $F_g = W \sin \theta$ tonnes must be provided for running the train on the inclined track.

If G = Per cent gradient of the track.

then $G = 100 \sin \theta$.

The tractive effort, F_g (in Newtons) becomes,

$$F_g = \pm 1000\, W \sin \theta \times 8.81 \text{ Newtons}$$
$$= \pm (100 \sin \theta) \times 98.1\, W \text{ Newtons}$$
$$= \pm G \times 98.1\, W \text{ Newtons} = \pm 98.1\, WG \text{ Newtons.}$$

If the train is moving down the gradient then the component of dead weight along the track, *i.e.* $W \sin \theta$ will help the motion of the train.

(*c*) **Tractive effort to overcome the frictional resistance (F_r).** If r is the specific train resistance in Newtons per tonne of the dead weight, then the tractive effort required to overcome this resistance

i.e. $F_r = W \times r$ Newtons

where r is a function of velocity for a given train.

For a normal train the value of specific resistance has been 40 to 70 Newtons/tonne.

Hence the total tractive effort required,

$$F_t = F_a \pm F_f + F_r = 277.8\, W_e\, \alpha + 98.1\, WG + Wr$$

The +ve sign is used when the train is going up the gradient and the –ve sign is used when it is going down the gradient.

8.41. Power Output from the Driving Axles

Power = Rate of doing work

$$= \frac{\text{Work}}{\text{Time}} = \frac{\text{Force} \times \text{Distance}}{\text{Time}}$$

$$= \text{Force} \times \frac{\text{Distance}}{\text{Time}} = \text{Force} \times \text{Speed.}$$

Therefore, power output at any instant from the driving axles

$$= \text{Tractive effort} \times \text{Speed} = F_t \times v$$

where F_t in Newtons and v is in m/sec.

$$= F_t \times v \times \frac{1000}{3600} \text{ Watts.}$$

$\therefore$ 1 N m/sec. = 1 Watt.

$$= \frac{F_t \times v}{3600} \text{ kW} \qquad [v \text{ is in kmph.}]$$

Example 8.16. *The peripheral speed of a railway motor cannot be allowed to exceed 45 metres per second. If the gear ratio is 16/73, the armature diameter 40 cm, and the wheel diameter 90 cm, what is the limiting value of the speed of the train ?*

Solution. Circumference of armature,

$$= \frac{\pi D}{100} \text{ metre} = \frac{\pi \times 40}{100} \text{ metre}$$

Peripheral speed of armature = 45 metres/sec.

Max. revolutions made by armature in one second,

$$= \frac{\frac{45}{\pi \times 40}}{100} = \frac{45 \times 100}{\pi \times 40}$$

Max. revolutions made by the wheel in one second,

$$= \frac{4500}{\pi \times 40} \times \frac{16}{73}$$

Circumference of the wheel $= \dfrac{90 \times \pi}{100} = 0.9\,\pi$ metres

Maximum distance travelled by the train in one second,

$$= 0.9 \times \frac{4500 \times 16}{\pi \times 40 \times 73} \text{ metres}$$

Maximum speed of the train $= \dfrac{0.9\,\pi \times 4500 \times 16}{40 \times 73} \times \dfrac{3600}{1000}$

= 80 kmph. **Ans.**

8.42. Energy Output from Driving Axles

Energy output from driving axles for a given time, can be found by finding the value of average power during the time and multiplying it by the time. Considering the trapezoidal speed-time curve.

Total energy output for the run, = Energy required during acceleration + Energy required during free run.

= Average power during acceleration × Acceleration period

+ Average power during free run × Duration of free run.

$$= \frac{1}{2} \cdot \frac{F_t \times V_m}{3600} \times \frac{t_1}{3600} + \frac{F'_t V_m}{3600} \times \frac{t_2}{3600} \text{ kWh}$$

where V_m = Maximum speed in km.p.h.

t_1 = Acceleration period in seconds

t_2 = Time of free run in seconds

F_t, F'_t = Tractive efforts during accelerating and free running period in Newtons.

8.43. Specific Energy Output

The specific energy output of the driving wheels is the energy output in watt-hr. per tonnes-km, *i.e.*

$$\frac{\text{Energy output in watt hr.}}{\text{(Wt. of trains in tonnes)}} \times \frac{1}{\text{(Distance of run in km.)}}$$

8.44. Specific Energy Consumption

The total energy input to all the traction motors in a train is known as the "energy consumption" of the train. The specific energy consumption is the energy consumption in watt-hr, per tonne-km.

Hence, specific energy consumption

$$= \frac{\text{Specific energy output}}{\text{Efficiency of gear} \times \text{Efficiency of motor}}$$

$$= \frac{\text{Energy output in watt-hours}}{\eta_{gear} \times \eta_{motor}\ \text{(Wt. of train in tonnes} \times \text{Distance of run in km)}}$$

8.45. Determination of Specific Energy Output

Consider the simplified speed-time curves.

Let the track have a gradient of G per cent through out its run.

(*a*) Energy output for accelerating the train from rest to V_m

$$= \frac{1}{2}\frac{F_t\,V_m}{3600} \times \frac{t_1}{3600}\ \text{kWh.}$$

$$= \frac{1}{2}\frac{F_t\,V_m}{3600} \times \frac{V_m}{3600\,\alpha} \qquad \left[\because\ t_1 = \frac{V_m}{\alpha}\right]$$

$$= \frac{1}{2}\frac{V_m^2}{(3600)^2\,\alpha} \times (277.8\,W_e\,\alpha + 98.1\,WG + W.r)\ \text{kWh.}$$

$[\because\ F_t = 277.8\,W_e\,\alpha + 98.1\,WG + W.r]$

(*b*) Energy output to run the train at the speed V_m against the gradient and resistance to motion

$$= \frac{F'_t \times V_m}{3600} \times \frac{t_2}{3600}\ \text{kWh}$$

$$= \frac{F'_t \times D'}{3600}\ \text{kWh}$$

$\left[\text{Since } \frac{V_m\,t_2}{3600} = D', \text{ the distance travelled during free run}\right]$

$$= (W.r + 98.1\,WG) \times \frac{D'}{3600}\ \text{kWh.}$$

Total energy output for the run,

$$= \frac{V_m^2}{2\,\alpha(3600)^2}[277.8\,W_e\,\alpha + 98.1\,WG + Wr]$$

$$+ [Wr + 98.1\,WG] \times \frac{D'}{3600}\text{ kWh.}$$

$$= \frac{V_m^2 \times 1000}{2\,\alpha \times (3600)^2}[277.8\,W_e\,\alpha + 98.1\,WG + Wr]$$

$$+ [W\,r + 98.1\,WG] \times \frac{D' \times 1000}{3600}\text{ Wh.}$$

$$= \frac{V_m^2}{2\,\alpha} \times \frac{1000}{(3600)^2} \times 277.8\,W_e\,\alpha + \frac{V_m^2}{2\,\alpha}$$

$$\times \frac{1000}{(3600)^2}[98.1\,WG + Wr]$$

$$+ \frac{1000\,D'}{3600}[W\,r + 98.1\,WG]$$

$$= 0.01072\,V_m^2\,W_e + (98.1\,WG + Wr)$$

$$\left[\frac{V_m^2}{2\,\alpha \times 3600} + D'\right] \times \frac{1000}{3600}$$

$$= 0.01072\,V_m^2\,W_e + \frac{1000}{3600}[98.1\,WG + Wr]\,(D'' + D')$$

where $\frac{V_m^2}{2\,\alpha \times 3600} = D''$, distance moved during accelerating period.

$$= 0.01072\,V_m^2\,W_e + 0.2778(98.1\,WG + Wr] \times D_1$$

($D_1 = D'' + D'$ = Distance travelled during acceleration and free run in km)

Specific energy output,

$$= \frac{\text{Energy output for the run in watt-hr}}{\text{Wt. of train in tonnes} \times \text{Distance of run in km.}}$$

$$= \frac{0.01072\,V_m^2\,W_e + 0.2778\,(98.1\,WG + W.\,r) \times D_1}{W\ \ D}$$

$$= \frac{0.01072\,V_m^2}{D} \times \frac{W_e}{W} + 0.2778 \times \frac{D_1}{D}(98.1\,G + R)$$

watt-hr/tonne-km.

If the track is level one,

$G = 0$, then,

Specific Energy output,

$$= \frac{0.01072\, V_m^2}{D} \times \frac{W_e}{W} + 0.2778 \times \frac{D_1}{D} \times \frac{\gamma}{\eta}$$

Watt-hr/Tonne-km

Specific energy consumption

$$= \frac{\text{Specific energy output}}{\text{Efficiency of motors}}$$

$$= \frac{0.01072\, V_m^2}{D\, \eta} \times \frac{W_e}{W} + 0.2778 \times \frac{D_1}{D} \times \frac{r}{\eta}$$

Watt-hr per tonne-km.

8.46. Factors which Effects the Specific Energy Consumption

As can be seen from the expressions for specific energy output for acceleration, train resistance and gradients, the various factors affecting specific energy consumptions (see Fig. 8.54). Typical values of specific energy consumption are :

(*a*) **Distance between stops.** Greater the distance between stops, less will be the specific energy consumption are :

Main line service 18 to 31 watt-hr. per tonne km. Suburban service 50 to 75 watt-hr per tonne-km.

(*b*) **Acceleration and retardation.** D_1 which is the distance for coasting for which the power remains on increasing with higher values of acceleration and retardation giving an increase in specific energy consumption.

(*c*) **Gradient.** Steep gradients will naturally involve more energy consumption even though regenerative braking is employed.

(*d*) **Train resistance.** This depends upon the nature of route, speed of the train and shape of front and rear of the train. More the train resistance, greater will be the specific energy consumption.

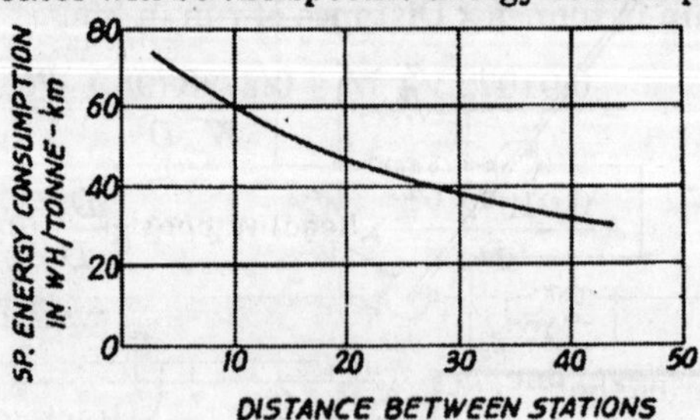

Fig. 8.54

(e) **Train equipment.** More efficient train equipment will reduce the specific energy consumption.

(f) The maximum speed.

(g) Weight of the train.

(h) Moment of inertia of the rotating parts taken into account by effective weight, W_e.

8.47. Dead Weight, Accelerating Weight and Adhesion Weight

(i) **Dead weight.** The total weight of locomotive and train to be pulled by the locomotive is known as *dead weight*.

(ii) **Accelerating weight**. The dead weight of the train, *i.e.* the weight of locomotive and train can be considered to be divided into two parts *viz.*,

(a) The weight which requires angular acceleration such as weight of wheels, axles, gears etc. and

(b) The weight which requires linear acceleration.

Hence the effective weight, which is greater than dead weight is called as accelerating weight. Accelerating weight is taken 5 to 10 per cent more than dead weight.

Adhesive weight. The total weight to be carried on the driving wheels is known as *adhesive weight.*

8.47(*a*). Negative Booster

As the train car goes away from the substation the potential drop increases. The effect of this increased drop is to reduce the voltage of the trolley wire and to increase the voltage between any two points in the rail track being employed as return path.

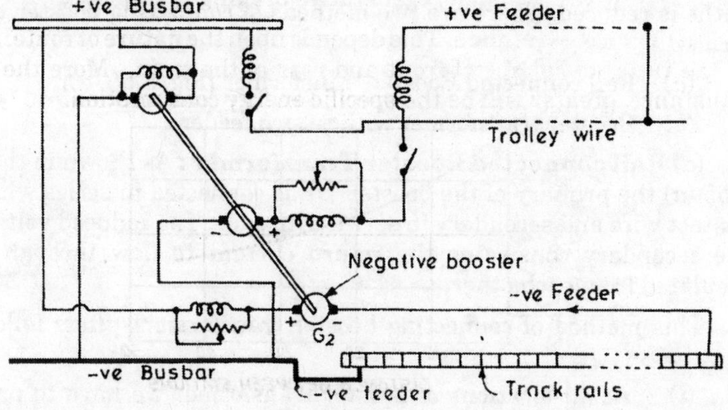

Fig. 8.55.

As already said the potential of track rail is kept within the limit of 7 watts by employing a negative booster.

The negative booster is a separately excited generator whose armature is connected in series with the negative booster feeder and field is connected in series with the feeder supplying the corresponding sections of the trolley wire. It is driven by a d.c. Motor-M. The "boost" on the feeder is adjusted by means of a divertor rheostat connected in parallel with the field winding.

Sometimes, a positive booster (a series exited generator) is also connected on the positive side in addition to negative booster on negative side in order to compensate the voltage drop in trolley wire.

8.47(*b*). Train Resistance

When the train moves, there is a force opposing the motion on account of friction and wind etc. this force is known as friction force and depends upon various factors such as shape, size and condition of track. This is expressed in newtons/tonne of the dead wt. For a normal train the value of specific resistance has been 40 to 70 newtons per tonne.

8.47(*c*). Booster Transformer

In a.c. traction system, return current which flows from the locomotive to the track soon leaks to the ground within short distance and returns to the substation through its earth. These ground currents cause heavy interference with the communication lines in the vicinity. Logical step to minimise the interference due to ground currents is to constrain these currents to flow through some restricted path. This is what **booster transformer** is supposed to do.

It consist of two windings of 1 : 1 ratio. The primary winding is connected in series with contact wire. Any amount of current flowing through primary requires to be balanced by equal current in the secondary and therefore tendency of currents flowing through stray paths is reduced. There are two methods of connecting booster transformer :

(*a*) Rail connected Booster transformer [Fig. 8.55 (*a*)]

(*b*) Booster transformer with return feeder.

(*a*) **Rail connected Booster Transformer :** As shown in the Fig. 8.55 (*a*) the primary of the booster T/F is connected in series with the contact wire and secondary in series with rails. The induced voltage in the secondary constrains the return current to flow through rails insulated from each other.

This method of connecting booster transformer suffers following drawbacks :

(*i*) At all locations of booster transformer we have to provide insulated rail joint with small neutral section in the OHE

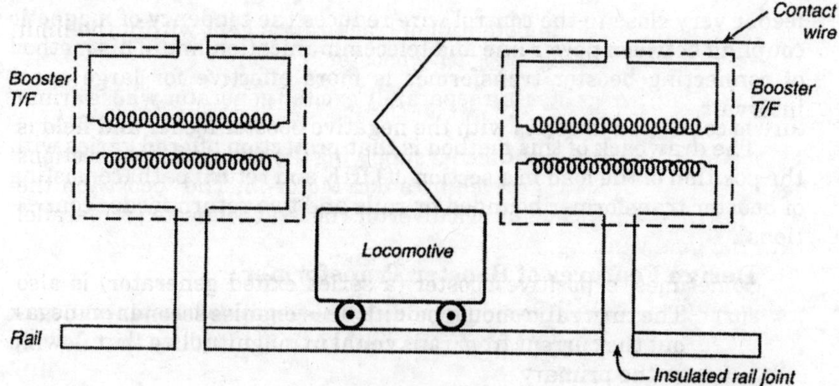

Fig. 8.55 (*a*)

(*ii*) Insulation puncture between rails of insulated joint may cause short circuiting the secondary and make booster transformer ineffective.

(*iii*) Voltage rise of 200 V between rails and 100 V between rail and earth are not uncommon. This may endanger life.

(*iv*) Protection offered being not more, it requires closer spacing of booster transformer.

(*b*) **Booster Transformer with Return feeder :**

The Figure 8.55 (*b*) shows midway between booster transformers, rails are connected to return feeder.

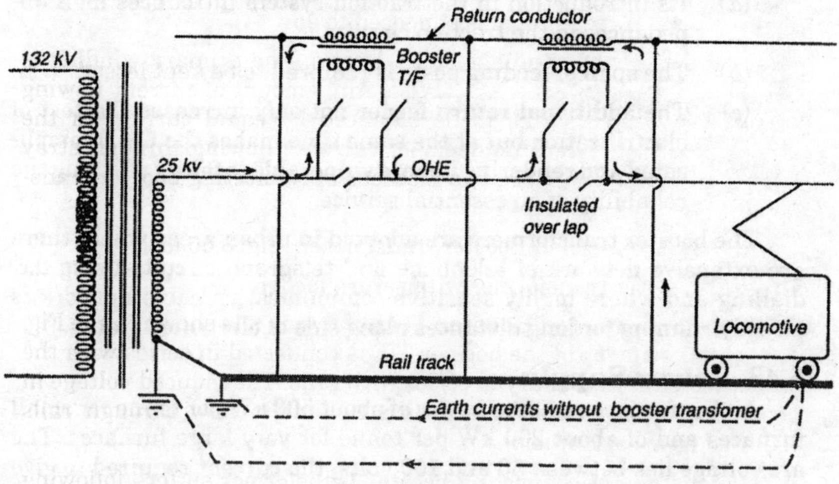

Fig. 8.55 (*b*)

By this arrangement rails are relieved of return current which now flows through the return feeder back to substation. Running the return feeder very close to the control wire reduces the tendency of magnetic coupling between power line and telecommunication line. This method of connecting booster transformer is more effective for large space intervals.

The drawback of this method is that protection offered varies with the position of the load in a section of OHE and return path (consisting of booster transformer bounded by rails and two return feeder connections).

Design Features of Booster Transformer :

(*a*) The turn ratio should be 1 : 1. This enables secondary to suck out the current from rails equal in magnitude to that flowing in the primary.

(*b*) Since booster transformer is connected in series with OHE on primary side and in series with return conductor on secondary side, it should be capable to withstand severe mechanical stresses occurring on account of fault currents which may rise upto 4000 to 5000 amps.

(*c*) Insulation of primary of the booster transformer is kept same as that of 25 kV OHE.

(*d*) Secondary attains high potential in case of short circuiting, therefore insulation level of secondary is kept at 3 kV which is considered quite within safe limits.

Drawbacks of Booster Transformer :

(*a*) Its introduction in the traction system introduces high impedances in the traction circuits.

(*b*) The span of feeding posts is required to be kept lesser.

(*c*) The additional return feeder not only increases the cost of electrification but at the same time makes the OHE complicated and rendering it more vulnerable to faults affecting the reliability of an essential service.

The booster transformers are adopted in urban areas where there are extensive network of telephone and telegraph circuits using d.c. dialling and where highly sensitive communication cable network is already running for long distances along side of electrified tracks.

8.48. Power-Supply

An arc furnace requires power of about 500 kW per tonne for small furnaces and of about 200 kW per tonne for very large furnaces. The arc voltage lies between 50 and 150 volts, the current required to give the above mentioned power, is therefore, of the order of several hundred amperes. This requires a special furnace transformer with output of low voltage and large current.

Since the leads from the transformer to the furnace are to carry heavy currents, so leads, consisting of rectangular copper strip spaced a few mm apart as current carrying conductors are mostly employed.

Control

The equivalent circuit of the arc along with the T/F is shown in Fig. 8.55 (*c*) where R_T and X_T are the equivalent resistance and reactance of the furnace transformer referred to the secondary side R_L and X_L are the resistance and reactance of leads, R_E is the resistance of electrodes and $E_A = IR_A$, the voltage drop across the arc.

The power input can be controlled by raising or lowering the electrodes, which results in variation of the arc resistance R_A. The power input can be also controlled by changing the tapping of the transformer which results in variation of voltage across the furnace. For complete control of furnace temperature and to achieve best operating conditions both voltage and electrode controls are employed.

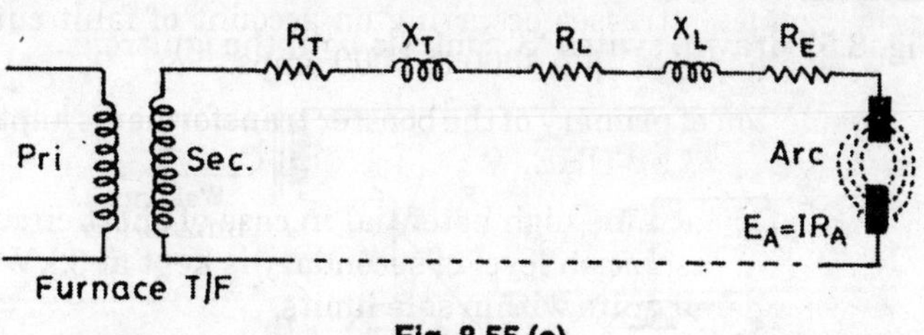

Fig. 8.55 (c)

8.49. Rectification of Power

Ignitrons

Ignitron is a gaseous tube having mercury pool cathode, carbon anode and ignitor made of boron carbide. It does not have any heating element or grid control as in thyaratron. Since this is a cold cathode tube, electrons are to be made available before tube can start conducting. This function is performed by ignitor which is kept dipped in mercury pool. Symbol of ignitron is as shown in the Fig. 8.56.

Ignitrons can handle current from 40 to 10,000 amps. It may also be noted that **electronic current** flow from ignitor to cathode spoils the ignitor and, therefore, should not be allowed.

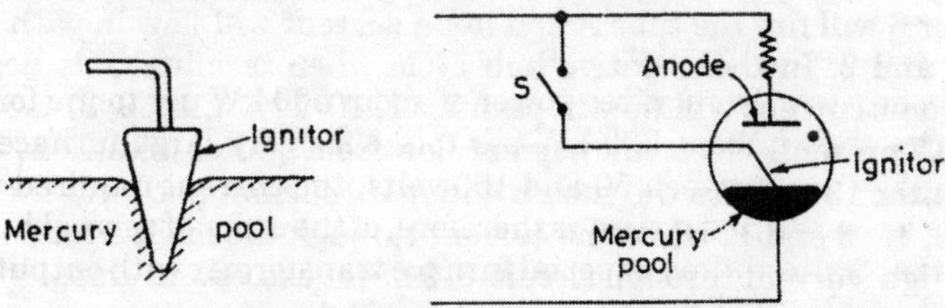

Fig. 8.56. Symbol of Ignitron.

Need of Ignitron Contractor

When we have to handle large currents, mechanical switches, if employed to control them, have following drawbacks :

(*a*) Due to large mass and inertia of mechanical switches, switching operation is very inconvenient. It is impossible to achieve switching operation required many times a second.

(*b*) Switching operation by mechanical switches can not have as much precise control as required in certain applications.

To overcome above limitations of mechanical switches, we resort to switching contactors, employing two thyratrons or ignitrons connected back to back.

Ignitron Contactor

Now, we can discuss the use of ignitron as a contactor for controlling heavy currents.

In Fig. 8.57 drawn, switch 'S' controls both the ignitrons.

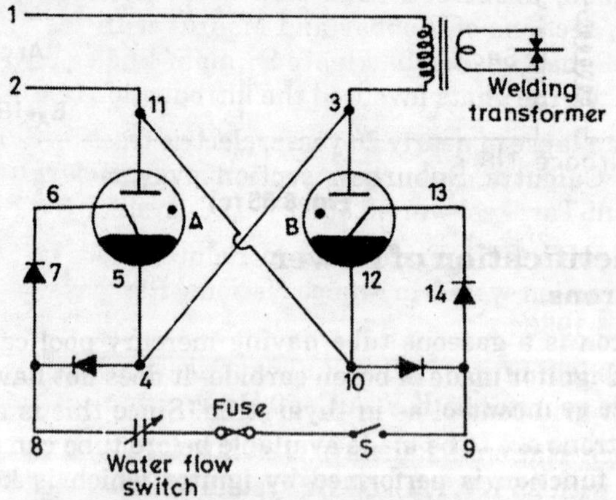

Fig. 8.57. Ignitron Contactor.

When terminal 1 is negative and switch is closed, electronic current flows in the path indicated by 1, 3, 4, 5, 6, 7, 8, 9, 10, 11 and 2. Diode 14, prevents electronic current flow from ignitor to cathode in tube B. Ignitor 6 will fire the tube A and main current will flow in path 1, 3, 4, 5, 11 and 2. In the negative half cycle when terminal 2 is negative, electronic current will flow in path 2, 11, 10, 12, 13, 14, 9, 8, 4, 3 and 1. Diode 7 presents electronic current flow from ignitor to cathode in tube A. Ignitor 13 will fire the tube B and main current path will then be 2, 11, 10, 12, 3 and 1. While the current through the main circuit may be very high, current through circuit. 8, 9 never exceeds 40 amps. It switch 'S' is opened, no ignitor will be working and therefore there is no current flow in the main circuit. Since very heavy current flows through the ignitron say 1000 amps. and arc drop be taken constant at 10 V, there will be losses to the tune of 10 kW. As such ignitrons are always water-cooled. In case of very heavy load, temperature of water becomes too much, thermostat contacts, which are normally closed as shown, will open and ignitors will stop conducting.

8.49.1 Introduction to OHE System

The Indian Railways are Asia's largest and World's second largest Railway system under a single management extending over 61836 Route Kilometres consisting of 33,669 Route Kilometres of B.G. (1676 mm), 23921 RKM of Metre Gauge (1000 mm) and 4246 RKM of Narrow Gauge (0.762 mm and 0.610 mm) Railway lines.

D.C. Electric traction on 1500 V, D.C. made its beginning in India in 1925, when the first section of 16 KM on Central Railway from Bombay-VT to Kurla (Via Harbour Branch) was electrified. By 1930, the D.C. traction system was extended upto PUNE and Igatpuri. The Southern Railway Metre Gauge line from Madras Beach to Tambaram was electrified in 1931.

By 1936, about 388 RKM had been electrified serving mainly suburban sections of Bombay and Madras with the only exception of Bombay-Pune and Bombay-Igatpuri main line sections where heavy gradients on the ghats favoured the introduction of Electrification.

After a lapse of nearly 20 years, electric traction was introduced in 1958 in Calcutta Suburban section over Howrah-Burdwan and Sheorapuli-Tarakeshwar on 3000 V, D.C. system.

In the year 1951, France first introduced 25 K.V. A.C. single phase 50 cycle traction system in French National Railways. This was followed by U.K. in 1956.

In the late fifties, Indian Railways also have, as a matter of policy decided to go in for 25 K.V. A.C. 50 Hz Traction system.

French National Railways, *i.e.* S.N.C.F. had the maximum experience with 25 KV, A.C. 50 Hz system of Electrification and so, they were appointed as Technical Consultants for the Electrification Projects by Indian Railways.

The S&T System design was based on the data already available with the Indian Railways or on the experience of M/S. SNCF, France.

The first section of 25 K.V. A.C. Electrification was carried out for 75 Route Kilometres (RKM) between Raj-Kharsawan - Dongaposi on South Eastern Railway in August 1960.

Since then, tracks to the extent of more than 7000 RKMs have been put to 25 KV A.C. traction systems on Indian Railways with nearly 405 RKM in Bombay area still continuing to be operated on 1500 V D.C. traction system.

Broad categories of traction drives

Type of Traction	*Traction Motor*	*Type of conversion and device used*	*Controls*
ac single phase 15 kV 16 2/3 Hz or 25 kV 50 Hz or 60 Hz	dc	(*i*) ac to dc by Ignitrons or Excitrons	Voltage control by onload tap changes
		(*ii*) ac to dc by full wave bridge using semiconductor diodes	Do
		(*iii*) ac to dc by Thyristor bridge rectifiers using phase control with line commutation	Electronic control starting with analog circuits to present day micro-computers controls.
	ac (3-phase cage motors)	(*i*) Single-phase ac to 3-phase ac using Thyristor Convertor-Inverter with forced commutation techniques	Earlier analog and then digital electronic controls using PWM technique.
		(*ii*) Single phase to 3-phase using GTO thyristors	PWM technique with Direct Self Control scheme using micro-computers.
dc (750 V, 1500 V, 3000 V)	dc	(*i*) dc to dc using electromechanical control	High-power resistance switching (*i.e.* Rheostatic control)
		(*ii*) dc to dc using forced commutated thyristors (chopper control)	Starting with analog electrons to present day digital microcomputer controls
	ac (3-phase cage motors)	(*ii*) dc to 3-phase ac using GTO/IGBT based inverters	Digital micro-computer control with PWM technique.

Various attempts to introduce new technologies for railway traction drives in Indian Railways

Sl. No.	*Type of Technology*	*Zonal Railway of application/type of traction vehicle*	*Quantity*	*Year of commissioning*	*Executing Agency*
1	dc drive, fully controlled single step thyristor bridge, analog controls	South Eastern Railways for 3000 HP, WAG 1 type locomotives	2 Nos	late 60s	AEG, Germany
2.	dc drive, fully controlled single step thyristor bridge, regeneration facility, analog controls	do	1 No	1978-79	Developed by BARC supplied by ECIL
3.	dc drive, fully controlled two step thyristor bridge, regeneration facility, analog controls	do	9 Nos.	1982-85	Supplied by ECIL combining technologies of BARC and BBC, Switzerland
4.	dc chopper with regeneration facility, analog controls	Central Railways for EMU suburban service in Mumbai, Motor coach rating 1000 kW	10 Nos	Prototype developed by 1986 Commercial supplies made during 1992-94	Developed by BARC supplied by ECIL and BHEL
5.	dc drive, half controlled two step thyristor bridge, microprocessor control, without regeneration	South Eastern Railways 6000 HP WAG 6 type of locomotives	18 Nos (complete import of locomotives)	1988-89	ASEA, Sweden and Hitachi, Japan
6.	3-phase ac drive, GTO based inverter, microprocessor control, regeneration facility	6000 HP prototype locomotives being tested in Northern Railways	30 Nos (complete import of locomotives) and CKD/SKD parts)	1995-96	ABB, Switzerland
7.	dc drive, hald controlled two step thyristor bridge, microcomputer control, without regeneration	Prototype being commissioned in Central Railways for 5000 HP WAG 5 type of locomotive	1 No	1996	Developed by BARC/RDSO, engineered by ECIL
8.	3-phase ac drive, GTO based inverter and microcomputer control	1000 kW rated EMU motor coaches	2 Nos.	–	Currently under design stage at BHEL

8.49.2 Power System

Before we proceed further, it is necessary that we study the Power Supply arrangements of 25 K.V. A.C. traction system, as certain items have bearings on signalling system.

25 K.V. A.C. 50 Hz single phase power supply for electric traction is derived from the **Grid System** of State Electricity Boards through **Traction Sub-stations** located along the route of the electrified sections at distances of 45 to 80 KMs. Traction sub-stations receive 3-phase supply at 132 KV.

To ensure continuity of supply under all conditions, the high voltage feed to the traction sub stations is invariably arranged either from two sources of power or by a double circuit transmission line, so that even if one source fails, the other remains in service.

At each traction sub-station, normally two single phase transformers of 12.5 MVA are installed, one of which is in service and the other is 100% standby.

These transformers step down the voltage from 132 KV to 25 KV for feeding the traction overhead Equipment (OHE) 25 KV feeders carry the power from the sub stations to **Feeding Posts** located near the tracks.

One end of the 25 KV winding of each 132/25 KV transformer is solidly earthed at the sub-station.

8.49.3 Feeding and Sectioning Arrangements

Since the Generation and Transmission systems are 3 phase, the single phase traction load causes unbalance in the supply system, which has undesirable effects on the Generators, if certain specified limits are exceeded.

To keep the unbalance on the 3 phase grid system within specified limits, power for traction is tapped off the grid system across the different phases at adjacent sub-station in cyclic order.

Thus it becomes necessary to separate electrically the overhead equipment systems fed by adjacent sub-stations. This is done by providing a "Neutral Section" which ensures that the 2 phases are not bridged by the pantographs of passing electric locomotives.

The sketch (Fig. 8.57 (*a*)) shows the General Power Supply arrangements, Neutral Sections, Sectioning and Sub sectioning arrangements.

8.49.4 Sectioning and Paralleling Post (SP)

These posts are situated midway between feeding posts. At these posts, a neutral section is provided to make it impossible for the pantograph of an electric loco or EMU to bridge the different phases of

25 KV supply while passing from the zone fed from one sub station to the next one.

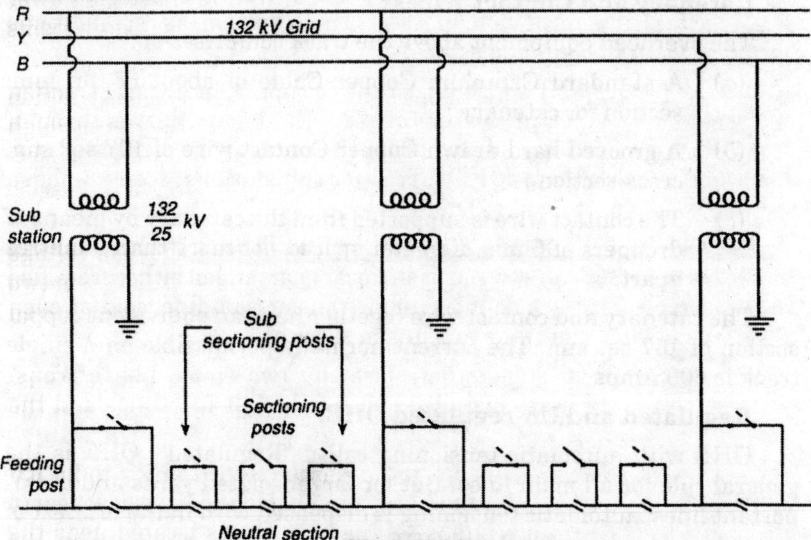

Fig. 8.57 (*a*)

Sketch showing the power supply, neutral section, sectioning and subsectioning arrangement.

Since the neutral section remains dead, caution notices are provided well in advance to enable the Elec. loco driver to COAST through it.

Special care is taken in fixing the location of neutral sections, on level gradients, *far away from signals* level crossing gates etc. to ensure that the train coasts through the neutral section which is a *dead section* of about 41 metres, at a sufficiently high speed, to obviate all possibility of its stopping and getting stuck within the neutral section.

In automatic signalling sections, if an automatic signal is at ON, the train has to go at a slow speed upto next stop signal, as per GR & SR. With a Normal dead section of about 41 metres, if a train passes at slow speed, it will get stuck up. To avoid such situations, the dead section is reduced to about 5 metres by an arrangement using PTFE Insulators (POLYTETRA FLOURIDE ETHYLENE INSULATORS are imported), in automatic signalling sections.

8.49.5 Sub Sectioning and Paralleling Post (SSP)

One or more SSPs are provided between each FP & SP depending upon the distance between them.

On double track sections, facilities are available to parallel the OHE of the UP & DOWN tracks, at SPs and SSPs. Where paralleling facilities are not available the Sectioning Posts and Sub section facilities are indicated as 'S' and 'SS' respectively.

8.49.6 Overhead Equipment

Catenary and Contact Wires :

The overhead equipment above the track comprises of :

(*a*) A standard Cadmium Copper Cable of about 65 sq. mm. section for catenary ;

(*b*) A grooved hard drawn Copper Contact wire of 107 sq. mm. cross-section ;

(*c*) The contact wire is supported from the catenary by means of droppers of 5 mm diameter spaced not more than 9 metres apart.

The catenary and contact wire together have an equivalent copper section of 157 sq. mm. The current normally permissible on a single track is 600 Amps.

Regulated and Un-regulated OHE.

OHE with automatic tensioning called "Regulated". OHE is the general rule for all main lines. But for large isolated yards and unimportant lines, automatic tensioning is dispensed with in the interest of economy and only unregulated OHE is provided.

Height of Contact Wire

The normal height of contact wire for regulated OHE is 5.55 m above Rail level. For un-regulated OHE in areas with a temperature range of 4°C to 65°C, the height is 5.75 m and in areas with a temperature range of 15°C to 65°C, the height is 5.65 m. In certain cases, such as underbridges, the height may be as low as 4.65 m on BG and 4.02 m on MG.

Span of Supporting Wires

On straight tracks, the catenary system is normally supported at maximum intervals of 72 m for B.G. and 63 m on MG.

Stagger

The contact wire is "Staggered" so that as the pantograph glides along, the contact wire sweeps to and fro across the bearing surface of the panto upto a distance of 200 mm on *either side* of centre line of straight track. In case of curved track, the stagger is 300 mm on *one side*. This ensures a uniform wear of the wearing strips of the pantographs.

Anchor

The OHE conductors are terminated at intervals of about 1.5 KM to 2.0 KM and suitably anchored. An overlap span is provided generally as shown in the sketch 8.57 (*b*), the conductor heights being so adjusted that the pantograph glides from one conductor to the other smoothly without shock.

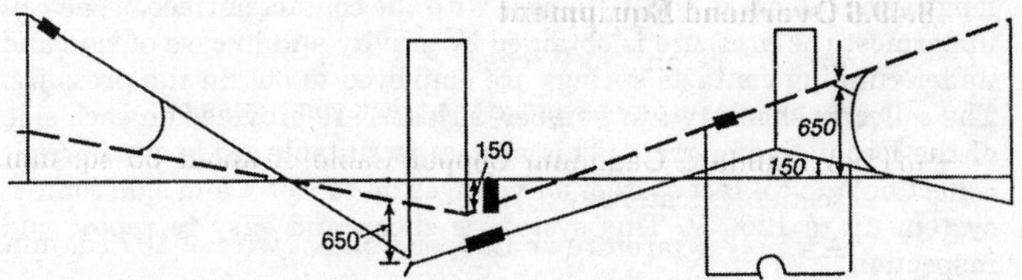

Fig. 8.57 (*b*) Over lap span.

8.50. Systems of Supplying Power in Electric Traction

Economics of Supply Systems :

There are two methods of supplying power to locomotives, tramways or trolleys. These are : (*i*) Conductor rail system (*ii*) Overhead system. Current collection from overhead system is far superior to that from the rail system. This is because both theoretically and experimentally current collection is more difficult from a rigid body than from an elastic one. Further the insulation of the 3'rd rail at high voltages used on single phase a.c. traction would also be impracticable proposition and endanger the safety of the personnel.

1. Conductor-rail-system : This system of supply has been used in many countries mainly for heavy electric traction services, since, in these cases the currents required by the trains may be large. The current is supplied to the electrical operated vehicle either through one rail conductor or through two rail conductors. In case of one rail conductor the track rail is employed as the return conductor. The advantage of using insulated return rail is elimination of electrolytic action due to return currents on other public services buried in the vicinity of the railway tunnels. The rails are mounted on insulators parallel with the track rails at a distance of 0.3 to 0.4 m from the running rail with their upper most surface acting as contact surface and are fed at suitable points from substations. Conductor rail with a side running contact, as shown in the fig. below is also employed. Some railways employ an under running contact. Latter systems are supposed to be more protective against accidental contact.

In view of less cost, the steel is employed for rail conductor. Since for conductor rails low resistivity is more important and not the hardness as in case of track rails, so these are made from special steel of low carbon and manganese content. To reduce the voltage drop at joints, conductor rails are bounded together by copper conductors rivetted or welded to the rails.

The current is conveyed from the conductor rail to the train equipment by means of collector shoe which is flat steel shoe about 20 cm in length and 7.6 cm in width. In case with the contact surface of the rail uppermost the pressure is obtained by gravity and in case of side and under running contacts springs are employed to obtain the pressure. The collector shoes, two in number, at least, are provided on each side of the locomotive or train. This system is suitable for heavy current collection, top contact system for voltages up to 750 V and side contact system up to 1200 V. This system is cheap and easy to repair and inspection.

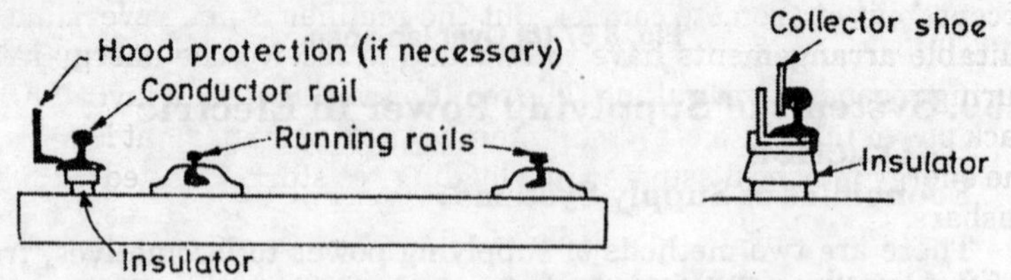

Fig. 8.58. Conductor Rail System.

2. Over Head System : This system is adopted when the trains are to be supplied at high voltage. In this system with high operating voltage the trains requiring high power may be supplied through conductors of relatively small cross-section and the collection of the current required by a train can be done by a collector with a sliding contact. Overhead construction is universal for all a.c. railways and is also used with d.c. tramways, trolley buses and locomotives operating at voltages of 1500 volts and above. In all these cases the running rails are utilized as the return conductor, therefore, with d.c. and single phase system only one overhead wire is required for each track.

8.51. The Location of and Distance between Sub-Stations: (From : System of Power Supply)

The type of sub-stations depends upon the nature of the primary supply and the system of electrification.

Sub-stations for 1-systems : If the primary supply is only for the purpose of feeding a traction system on single phase as in the case with most European railways, the generators may be single-phase, $16\frac{2}{3}$ c/s machines. The sub-stations may be supplied at 66-132 kV. If the supply is available from a national grid, it has to be converted to suit traction requirements and transformers and frequency changers have to be employed.

On a-c traction system at standard 50 or 60 c/s frequency, the grid sub-stations are tapped from the 66-132 kV busbars to obtain 25 kV single-phase supply either through scot-connected T/F or straight T/F's. Transformers of capacity 5, 7.5 and 10 MVA are used.

Spacing : The sub-station spacing is based on the permissible voltage drop expressed as a percentage of the normal voltage. It should not exceed 20 percent under normal conditions and 40 percent under abnormal conditions. The spacing of sub-station on 25 kV is about 1½ times the spacing on 3000-volt d.c. system.

D.C. sub-stations : These sub-stations are now equipped with mercury-arc rectifiers which have superseded the rotary converter on account of their much higher all-day efficiency, reliable operation acceptability to remote control. But the rectifier is non-reversible and suitable arrangements have to be made to receive the energy return during regenerative braking. The rectifier working as an inverter feeds back power into the a.c. system where such an arrangement is not used, the energy may be dissipated in a loading resistor connected to the d.c. busbars.

8.52. Overhead Equipment

General Considerations :

The overhead contact-wire equipment as employment for tramways and trolley buses and for railways of 1500 V and over is more difficult to design than the conductor-rail equipment on account of the greater difficulty of ensuring that the collector and contact wire remain in contact at a fairly even pressure.

The materials used for the contact wire are hard-drawn copper cadmium copper, or silico-bronze, particulars of which are given in table below. In spite of its higher electrical resistance, silico bronze is in fairly common use on account of its higher tensile strength and also its greater wear-resisting properties.

Table - 1

	Ultimate strength (N/m^2)	*Sp. resistance (ohm-m)*	*Temperature coefficient* (*per* °C)	*Modulus of elasticity : E* ($(N/m)^2$)
H.D. copper	37×10^7	1.76×10^{-8}	16.6×10^{-6}	124×10^9
Cadmium copper	41.7×10^7	2.035×10^{-8}	16.6×10^{-6}	124×10^9
Silico-bronze	49.4×10^7	3.86×10^{-8}	17.0×10^{-6}	124×10^9

The actual quality of the wire is laid in ISS which also specifies eight standard sizes ranging from the smallest, having a cross-sectional area equal to No. O.S.W.G. (53.57 mm^2) upto a size of 193.5 mm^2. To facilitate connection to the supports the wire is grooved. The dimensions of the groove are also laid down in the B.S.S. or I.S.S.

8.52.1 Current Collectors

The following types of current collecting systems are employed in electric traction system :

1. **Conductor rail-system.** (Refer Article 8.50)

2. **Cable collector.** It is used specially at places where range of operation is small and where it is hazardous to have current collection from overhead wire by other means, such as in gathering service of gassy mines.

It consist of a long cable which pays out or winds upon a power driven wheel/reel carried on the locomotive when it travels into or out of working place.

3. **Pole collector.** For tramways grooved gun metal wheel trolley collector or grooved slider shoe with carbon insert, attached to the end of long pole provided on the top of the car is used.

Main drawback of trolley collector is that it has to be rotated by 180°C before tram car can have motion in reverse direction. It is suitable upto a speed of 22 to 30 km.p.h. The other drawback is the poor contact between wheel and trolley wire, giving rise to high current density, which in turn results in heavy arching. Its use is prohibited in gassy mines.

4. **Bow collectors.** Bow collector consist of two trolley collector poles at the end of which is placed a light metal strip upto one metre long for current collection. Main advantage of bow collector is that it can be used for higher speeds. The disadvantage of the type of collector is that it has to run always in trailing. To overcome this disadvantage provision of duplicate bows or an arrangement of reversing the bows for motion in the reverse direction is made.

5. **Pantograph Collector.** The shortcomings of irreversible operation and low current collection capacity are overcome by employing pantograph collector. Main function of pantograph collector is to maintain a link between the overhead system and power circuit of the locomotive at varying speeds in different climate and wind conditions on varying stiffness of OHE.

There are basically two forms of pantograph namely **"diamond"** type and single ended **"faiveley"** type. Collector steps are forced against the contact wire by upward action of pantograph springs. These strips slide along the bare metal of the contact wire as the train moves. Steel strips are used in India and France as collector strips and metallized carbon strips are used on the EMU and locomotives in Britain.

Pantograph may be raised or lowered from cabin by any of these methods :

(*a*) Air raised gravity lowered

(*b*) Air raised spring lowered

(*c*) Spring raised air lowered.

For 25 kV a.c. system, normal working pressure of pantograph is between 6.5 to 9 kg.

8.52.1 Single and Compound Catenary Construction in Railways

Single Catenary Construction : Single Catenary construction in polygonal system as employed on 25 kV single phase line is shown in the Fig 8.59. It consists of a solid grooved copper or cadmium copper wire 107 mm^2 in cross sectional area, supported at intervals by droppers from standard cadmium copper cable 65 mm^2 in cross-sectional area, suspended in a catenary between supporting structures spaced upto 72 m. The dropper spacing is 9 m except the first and second droppers which are 2.25 m and 6.75 m from support respectively. This type of construction provides sufficient flexibility for speeds upto 120 km.p.h. without any difficulty at current collection. Besides this it is relatively less costly, is easy to erect and does not call for heavy maintenance. It is considered suitable for most light weight a.c. system and for branch and loop lines and secondary routes where operating speeds are low.

Difficulty of current collection arises when operating speed goes up. This is because the elasticity of ONE being greatest at mid span, the amount of wire push up is small near the supporting points and large at mid span. As a result of this, slider of the pantograph will proceed along the contact wire with up and down motion. Therefore, when the train speed increases and as a result, the slider motion approaches the resonance frequency of the vibration system of the pantograph and the overhead wire, the vibrating amplitudes of the two will grow large until the slider fails to follow the movement of the contact wire, thus giving rise to a contact break phenomenon. Therefore, for high speeds specially where heavier conductors are used, improved operation by the use of more complicated supporting arrangement between the contact and main catenary wire such a

(*i*) Compound catenary construction

(*ii*) Stitched catenary construction or modified simple catenary construction.

(*iii*) Modified Y-compound catenary construction.

(*iv*) Continuous mesh catenary construction.

Compound Catenary Construction : It consists of 3 wires, the main catenary, auxillary catenary and contact wire, all in the same vertical plane as shown in Figure 8.59.

The contact wire is suspended from stranded cadmium copper auxillary catenary, besides increasing the current carrying capacity, is to provide for more uniformity in elasticity. The hard spots at the point of suspension are avoided because registration is done from the auxillary catenary and the contact wire is merely suspended from it. This construction enables automatic tensioning to be applied to the contact wire without causing the displacement of the supporting bracket assembly. Weight tensioned or regulated compound catenary is used on lines where speed range is 190 to 224 kmph.

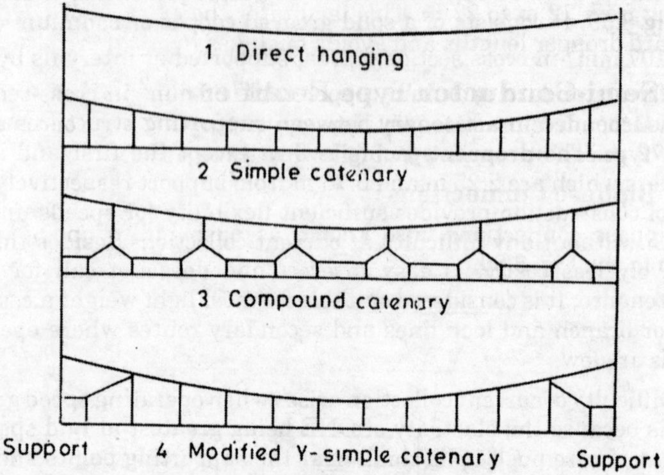

Fig. 8.59.

8.52.2 Position of contact wire in relation to track

The minimum height of contact wire is decided on the basis of rolling stock gauge and the permissible clearances depending upon the system voltage. Normally mid span height of contact wire above rail level is 5.50 m under worst conditions. The height is reduced under bridges and tunnels to the extent required. In case of Broad Gauge railways, the permissible minimum height of contact wire is 4.80 m, in order to allow diesel electric locomotive of 4.42 m height to ply on the electrified routes. The minimum height in case of meter gauge railways is 4.02 m, which permits movement of electric locomotive of 3.68 m high.

At electric loco sheds and over locomotive inspection pits, minimum height of contact wire of 5.80 m is maintained, which provides adequate/sufficient head room space for a workman working on the top of the electric locomotive. At level crossing a minimum height of 5.50 m is maintained for both broad gauge and meter gauge railways.

8.52.3 Span Length

The maximum span length between traction supports is governed by the consideration of contact wire being blown off the pantograph pan under maximum wind conditions (wind pressure of 98 kg/sq. m for coastal areas and 74 kg/sq. m for interior areas is taken).

Maximum span length of 72 m is taken on straight tracks, wind pressure of 88 kg/sq.m and 67.5 m for wind pressure of 98 kg/sq.m for coastal areas. On curves the span length is reduced to maintain the contact wire within working width of the pantograph under maximum wind conditions.

Span lengths have been standardised in the multiple of 4.5 m varying from 72 m to 27 m. This facilitates the erection of OHE with standard dropper lengths and avoids mistakes in erection.

8.53. Semi-Conductor Type Rectifier

In locomotive practice, rectifiers of mercury arc type or semi-conductor type can be connected in following three ways :

1. Biphase Connections :

Biphase connections, also known as push pull connections are shown in the Fig. 8.60.

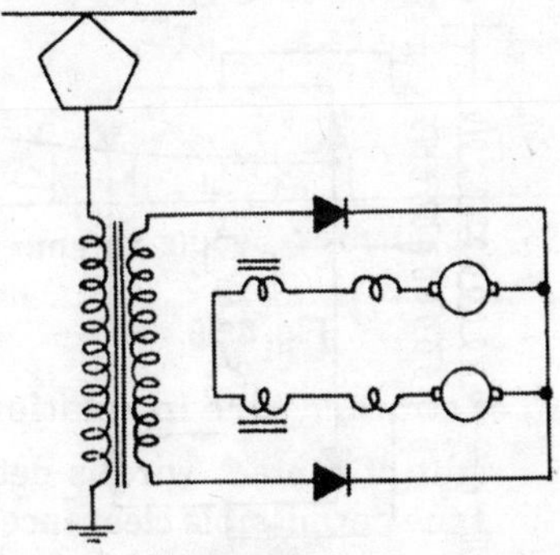

Fig. 8.60.

These connections have the advantage of requiring only two rectifier units. Their main disadvantages are that each rectifier unit is subjected to twice the reverse peak voltage and only half of the transformer winding is utilized a time. This is more significant in large equipment than in motor coach sizes.

2. Bridge Connections :

In bridge connections shown in Fig. 8.61, both of the drawbacks of biphase connections are removed but it requires double number of rectifier units. Biphase method is suited to supply two motors in parallel and bridge method is suited for two motors in series as shown in the fig. drawn above and below respectively. Also biphase circuit is generally used with mercury arc rectifiers.

3. Three Wire Bridge Circuit :

This is similar to bridge circuit except that mid-point of T/F is connected to the mid-point of the load as in fig. drawn below. Two motors even though connected in series have their voltage fixed. This method is, therefore, superior from wheel slip point of view. Normally, mid wire does not carry any current except when one wheel axle slips.

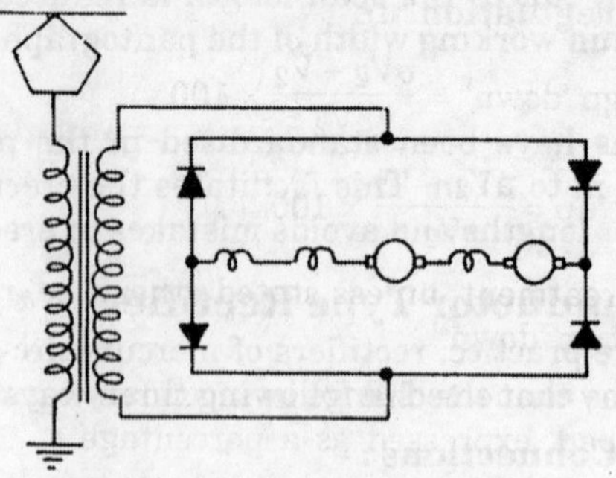

Fig. 8.61. Bridge Connections.

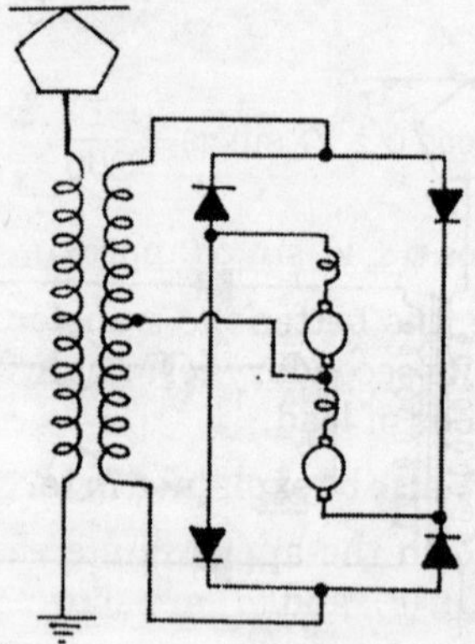

Fig. 8.62. Three Wire Bridge Circuit.

If two or more group of rectifiers operate in parallel, even load share between them is ensured by providing 1 : 1 ratio balancing transformers with their two windings connected in the a.c. connections of the two groups.

8.54. T/F and Regulation of Voltage

1. When a transformer is loaded with a constant primary voltage, the secondary voltage decreases because of its internal resistance and leakage reactance.

Let oV_2 = secondary terminal voltage at no-load

$= E_2 = KE_1 = KV_1$ because at no-load

the impedence drop is negligible.

V_2 = secondary terminal voltage on full-load

The change in secondary terminal voltage from no-load to full-load is $= oV_2 - V_2$.

This change divided by oV_2 is known as the regulation 'down.' If this change is divided by V_2 *i.e.* full-load secondary terminal voltage, then it is called **regulation 'up'**

$$\therefore \ \% \text{ age regn 'down'} = \frac{oV_2 - V_2}{oV_2} \times 100$$

$$\text{and } \% \text{ age regn. 'up'} = \frac{oV_2 - V_2}{V_2} \times 100$$

In further treatment, unless stated otherwise, regulation is to be taken as regulation 'down.'

As we can say that the change in secondary terminal voltage from no-load to full-load, expressed as a percentage of no-load Secondary voltage, is :

$$= V_r \cos \phi \pm V_x \sin \phi \text{ (approximately)}$$

Or, more accurately,

$$= (V_r \cos \phi \pm V_x \sin \phi) + \frac{1}{200} (V_x \cos \phi + V_r \sin \phi)^2$$

$$\therefore \ \% \text{ age regn } = V_e \cos \phi \pm V_x \sin \phi \text{ (approximately)}$$

The lesser this value, the better the transformer, because a good transformer should keep its secondary terminal voltage as constant as possible under all conditions of load.

(2) The regulation may also be explained in terms of primary values.

In Fig. 8.63(a), is shown the approximate equivalent circuit of a transformer and in Fig. (b), (c), and (d) are shown the vector diagrams corresponding to different power factors.

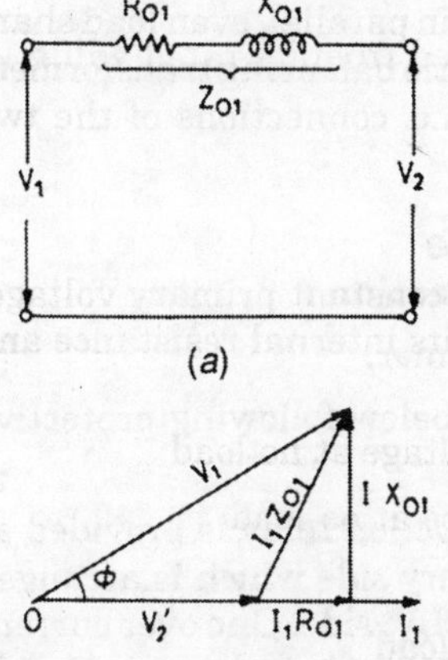

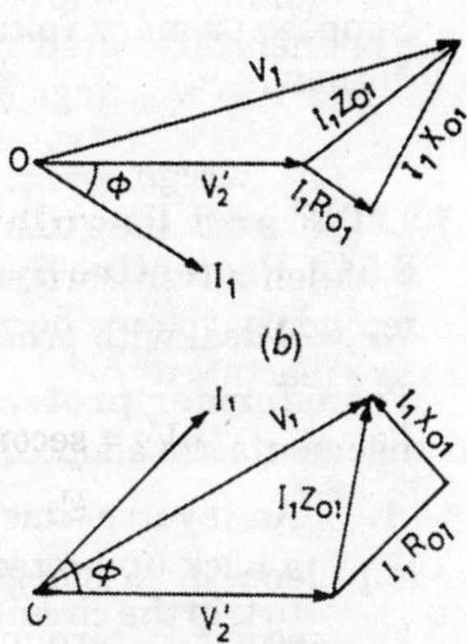

Fig. 8.63.

The secondary no-load terminal voltage as referred to primary is $E'_2 = E_2/K = E_1 = V_1$ and if the secondary full load voltage as referred to primary is $V'_2 = (V_2/K)$ then,

$$\therefore \quad \% \text{ regn} = \frac{V_1 - V'_2}{V_1} \times 100.$$

From the vector diagram, it is clear that if angle between V_1 and V'_2 is neglected, then the value of numerical difference $V_1 - V'_2$ is given by $(I_1 R_{01} \cos\phi + I_1 X_{01} \sin\phi_1)$ for lagging P.F.

$$\therefore \% \text{ regn} = \frac{I_1 R_{01} \cos\phi + I_1 X_{01} \sin\phi}{V_1} \times 100$$

$$= V_r \cos\phi + V_x \sin\phi$$

where $\frac{I_1 R_{01} \times 100}{V_1} = V_r$ and $\frac{I_1 X_o \times 100}{V_1} = V_x$

As before, if angle between V_1 and V'_2 is not neglected, then the value of % age regulation will be :

$$= (V_r \cos\phi \pm V_x \sin\phi) + \frac{1}{200}(V_x \cos\phi \mp V_r \sin\phi)^2$$

(3) In the above definitions of regulation, primary voltage was supposed to be kept constant and the change in secondary terminal voltage were considered.

As the transformer is loaded, the secondary terminal voltage falls (for a lagging P.F.). Hence, to keep the output voltage constant, the primary voltage must be increased. The rise in primary voltage reqd. to maintain rated output voltage from no-load to full-load at a given power factor expressed as %age of rated primary voltage gives the regulation of the transformer.

Suppose primary voltage has to be raised from its rated value V_1 to V'_1 then,

$$\%\text{regn} = \frac{V_1' - V_1}{V_1} \times 100.$$

8.54.1 Protective System for A.C. Traction :

We will deal with protection on transformer.

Transformer protection : As shown below following protective arrangements are adopted for transformer.

1. An inverse time over current induction relay is provided as a back up protection on the primary side which is arranged to trip the circuit breaker CB_1 on H.V. side. One over current protection induction relay is also provided on the L.V. side which will trip 25 kV transformer *C.B.*

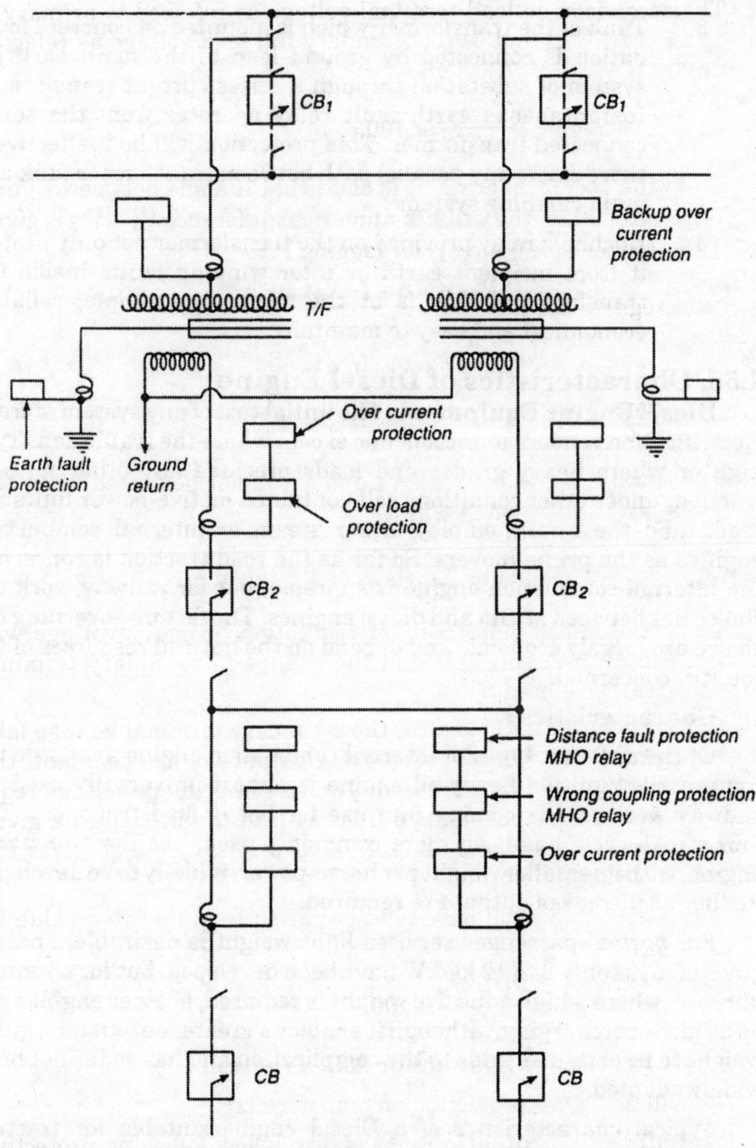

Fig. 8.63 (*a*)

2. **One overload thermal relay is also provided on the secondary side for protection against sustained overloads. With the use of this relay full benefit of allowable overload for specified time can easily be obtained as it is made to operate corresponding to thermal characteristic of the transformer.**

3. Tank of the transformer which is mounted on concrete foundation is connected by ground lead to the main earthing system of substation through a series current transformer. Instantaneous earth fault relay operates from the series connected transformer. This protection will be ineffective if there exists any parallel path between transformer tank and main earthing system.
4. Buchholz relay provided on the transformer not only protect it from incipient earth or inter winding faults inside the transformer but it is at the same time simple, reliable, economical and easy to maintain.

8.55. Characteristics of Diesel Engine

Diesel Engine Equipment : The initial cost of any system of track electrification renders it uneconomic except where the traffic density is high or where heavy grades and loads preclude any other form of traction, under other conditions self-contained motive-power units are reqd., and these may employ either steam or internal combustion engines as the prime-movers. So far as the road traction is concerned the internal-combustion engine is supreme, but for railway work the choice lies between steam and diesel engines. The factors governing the choice are largely economic and depend on the natural resources of the country concerned.

Characteristics :

Of the different types of internal combustion engine available the compression-ignition heavy-oil engine is almost universally used for railway work and is coming into use for heavy load transport. The four-stroke cycle has been more commonly used, but the two-stroke engine, with its smaller weight per horse-power, is likely to be developed further as increased output are required.

For normal passenger services light weight is desirable, and engines of mass only 9 to 12 kg/kW have been developed, but for shunting purpose, where a high adhesive weight is required, heavier engines can be used. Supercharging, although it enables a greater output for a given weight to be obtained, adds to the complication and has so far not been widely adapted.

Typical characteristics of a Diesel engine suitable for traction service are shown in Fig. 8.64. It can be seen that the torque obtainable is appx. constant and independent of the speed. In common with all internal-combustion engines, it has no starting torque and must be started by some auxillary means such as compressed air or electrically, as in the ordinary motor car, furthermore, it has only a very small overload capacity, and any attempt to obtain more than about 10% overload results in stalling of the engine.

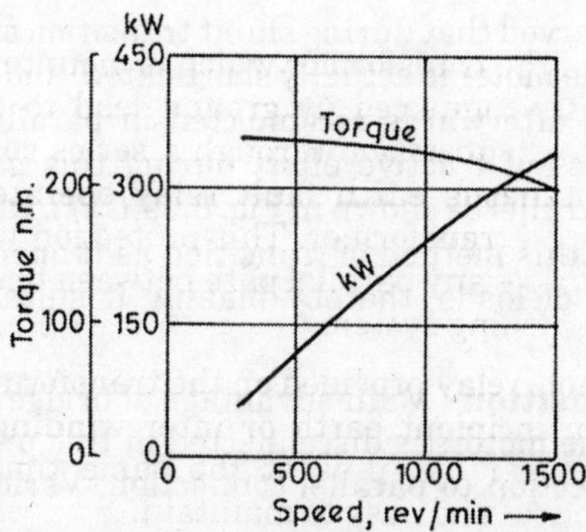

Fig. 8.64. Diesel-engine Characteristics.

8.56. Effective weight of Train

The total weight of locomotive and train to be pulled by the locomotive is known as dead weight. The dead weight of the train *i.e.* the weight of locomotive and train can be considered to be divided into two parts.

(*i*) The weight, which requires angular acceleration such as weight of wheels, axles, gears etc. (*ii*) and, the weight which requires linear acceleration.

Hence, the effective weight, which is greater than dead weight is called as accelerating weight. Accelerating weight is taken 5 to 10 percent more than dead weight.

The total weight to be carried on the driving wheels is known as the adhesive weight.

8.57. Sequence Diagram of Shunt and Bridge Transition

Shunt Transition : According to this method, transition from connections shown in Fig. 8.65 (*b*) to 8.65 (*d*) is carried out in four steps as shown in Fig. 8.65 (CS_1) to (CS_5).

1. After whole of the external resistance has been cut out as shown in Fig. 8.65 (b) or (CS_1) when motors are in series, first-step is to re-insert some resistance as shown in Fig. 8.65 (CS_2).
2. Second step is to short-ckt. one motor as shown in Fig. 8.65 (CS_3).
3. Third step is to open one end of the short ckted. motor as shown in Fig. 8.65 (CS_4).
4. Fourth step is to connect this open end of the motor with corresponding terminal of the other motor. This is shown in Fig. 8.65 (CS_5) which is same as Fig. 8.65 (*d*).

It will be observed that during shunt transition steps from Fig. 8.65 (CS_3) to (CS_5), one motor is actually short-ckted. and then disconnected from supply and afterwards reconnected in parallel. This, therefore results in the loss of tractive effort during this period till motor is reconnected in parallel as shown in Fig. 8.65(CS_5). Inspite of this, many engineers prefer this method of transition as it operates satisfactorily over wide range of loads, thereby making it suitable for locomotive applications.

Bridge Transition : Main advantage of bridge transition method is that none of the motors is disconnected in this process of transition from series connection to parallel connection. Various steps invalued are :

1. In the first step, individual resistances are shorted out as shown in Fig. 8.65(CB_1) which is equivalent to Fig. 8.65 (*b*). Motors are now in full series.
2. In the second step, bridge link is to moved that motors are put in series by passing the resistances which are at the same time unshorted as shown in Fig. 8.65 (CB_2).

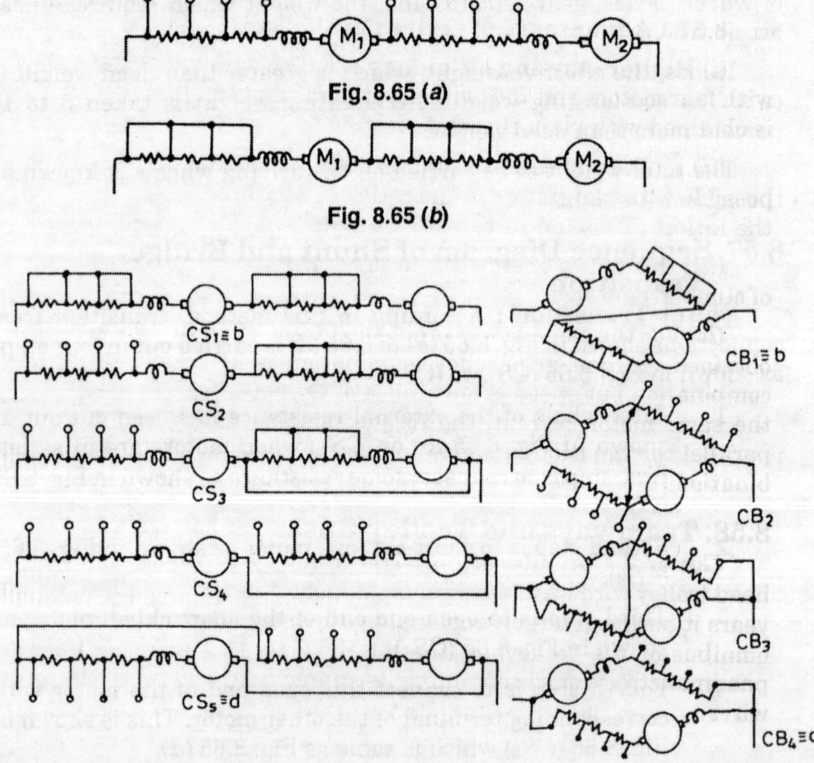

Fig. 8.65 (*a*)

Fig. 8.65 (*b*)

Fig. 8.65 (*c*)

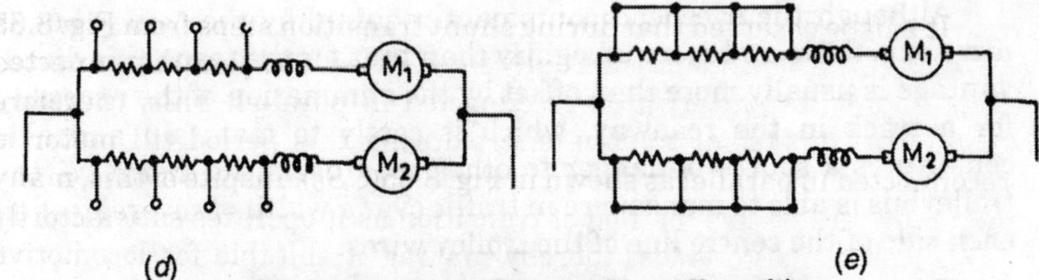

Fig. 8.65. Bridge Transition and Shunt Transition.

3. In the third step, a portion of starting resistance is connected in parallel with each motor as shown in Fig. 8.65(CB_3).
4. In the fourth step, bridge link is removed. Now motors are parallel with individual resistance inserted in each motor as shown in Fig. 8.65(CB_4). This is equivalent to Fig. 8.65 (*d*).

Bridge transition method is very much suitable for operation, under predetermined conditions of current and load such as those obtained in multiple unit trains.

Now, the diagrams of these transitions are as drawn in Fig. 8.65 (*a* to *e*) :

8.57.1 Advantages of series parallel starting :

(*i*) **Higher starting efficiency :** 66.67% with two motors, 72.72% with four motors and 75% with six motors where as only 50% efficiency is obtained with plain rheostat starting.

(*ii*) **Different economical speeds :** Two economical speeds are possible with two motors in the ratio of 1 : 2; three with four motors in the ratio 1 : 2 : 4 and three with six motors in the ratio of 1 : 2 : 6.

(*iii*) Due to low energy loss in the starting resistances they are not of cumbersome size.

(*iv*) For the same power input the torque of different magnitude is obtained. Torque in parallel combination is 1/4th of that in series combination but speed is four times of that of series combination. For the same motor current, the torque remains the same but speed in parallel combination becomes double the speed of that of series combination.

8.58. Tramways and Trolley Buses

General Consideration : The tramway supplied from an overhead trolley wire was a familiar sight in most large towns, but in recent years it has been largely replaced by the internal-combustion engined omnibus or the trolleybus, the latter being an electrically operated pneumatictred vehicle obtaining its supply from two overhead contact wires.

Although the overhead equipment required by the trolley bus is more expensive and more unsightly than that for the tram, this disadvantage is usually more than offset by the elimination of the necessity for a track in the roadway, which is costly to maintain and also constitutes a source of danger to other road users, furthermore, the trolleybus is able to manoeuvre in traffic over a width of several feet on each side of the centre line of the trolley wires.

Experience shows, however, that for the very dense traffic of large cities the ordinary tramway is the most economical means of transport available. The trolleybus is most suited for medium traffic densities, and many outlying tramway routes can be advantageously converted to trolleybus operation when the time comes for capital to be expended on track renewal.

Although the i.c.-engined omnibus is more expensive to maintain and operate than a trolleybus, urban and rural services operating with a service interval of more than 15 or 20 minutes are most economically dealt with by this type of vehicle, as the cost of the overhead equipment for a trolleybus service would be too great relative to the number of passengers carried and the revenue obtained.

Tramways

The chief features in connection with the electrical equipment of a tram are the provision of at least two driving axles in order to secure the necessary adhesion and the fact that control must be possible from either end. Apart from this the equipment is generally similar to that required for railway work, although of course the output is much smaller, not usually exceeding 45 kW or 55 kW. Direct current at about 600 V is almost universal and is conducted from overhead contact wires by a trolley or bow collector. The running rails form the return path. The use of two or more driving axles lends itself to the use of two motors with series-parallel control, with, in more modern equipments, a field tapping to give higher speeds without sacrificing any starting torque.

Two drum-type controllers, one at each end, are usually employed for control of the vehicle. These controllers are connected in parallel, but have suitable interlocking arrangements to prevent their being used simultaneously. With the increasing outputs required by present-day tramcars the controller is tending to become so large that its operation is fatiguing to the driver, and master controller equipment similar to that already described for use on railways is employed in recent designs.

The trolleybus

The chief features wherein the trolleybus differs from the tramcar are :

1. As the total weight is limited, that of the electrical equipment must be cut down to the minimum possible value.

2. The adhesion between a rubber-tyred wheel and the roadway is sufficiently high to enable only a single driving axle, and therefore a single motor, to be used.
3. Only one driving position is required, but starting stopping, and speed control must be carried out by means of a foot operated pedal in order that the driver shall have his hands free for steering and hand-brake operation.
4. As the vehicle is insulated from earth by the rubbertyred wheels precautions must be taken to prevent the setting up of any potential difference between the vehicle framework and the road-way.

8.58.1 Protective Devices on Tramways and Trolley Buses :

(*a*) **Circuit Breakers :** It is to protect electrical equipments against excessive overloads. Circuit breaker is installed between the current collector and the main wiring.

(*b*) **Radio Interference Suppressor :** The trolley bus is insulated from the earth. Due to sparking and arcing at the collector while in operation on controller or on motors, an e.m.f. at radio frequency can be induced. This radio frequency emf will be radiated through trolley wire. To suppress the radio frequency emfs, Ratio interference suppressors are installed.

(*c*) **Lightning Arrestor or Surge Divertor :** It is used when vehicle is operating in areas susceptible to lightening. The lightning arrestor installed for this purpose consist of a block of carborandum held between positive and negative electrodes. The positive electrode has a small air gap in series and is connected to positive trolley wire and the negative electrode to that of negative trolley wire or rail track depending upon the type of system adopted.

8.58.2 Transmission :

As mentioned above, only a single driving axle is necessary, invariably the rear axle, but, as with the ordinary motor car, some form of differential action must be incorporated between the two wheels in order to prevent wheel slip at curves.

Either one or two motors may be employed ; a single motor is preferable from the point of view of weight and cost is the usual practice. Two motors, however, one coupled to each driving wheel, have been used in some cases, and this arrangement eliminates the need for a differential gear and also enables low economical running speeds to be obtained by series parallel control.

The motor may be installed under the bonnet of the vehicle, where it is easily accessible, or under the centre of the car body, which eliminates a long shaft between the motors and the differential gear. Illustrations of typical layouts are shown in Fig. 8.66.

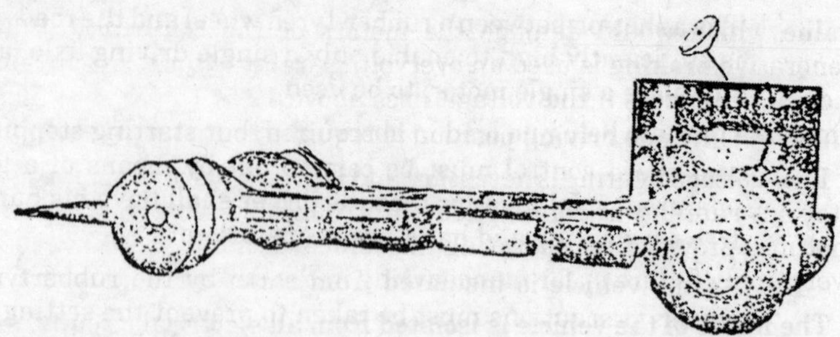

Fig. 8.66. General Arrangement of Sunbeam-B.T.H. Trolleybus Chassis.

Motors. In addition to the general requirements for a traction motor, a motor for a trolleybus is more severely limited with regard to its dimensions and weight since the overall weight and size of the vehicle is governed by Ministry of Transport Regulations, and any saving in the motor results in greater passenger-carrying capacity. Noise must also be considered since vehicle noise is much less than with rail-traction equipment. A further wear, of feature is the practical necessity in order to save break lining wear of employing rheostatic or regenerative braking.

The motor output required is generally between about 45 kW and 75 kW, and a compound wound motor is usually employed, the shunt winding ampere-turns assisting those of the series winding when motoring speed control is effected by resistance in the shunt field and tappings or a divertor on the series field. During rheostatic braking the series field opposes the shunt field and can be disconnected entirely if required ; braking down to about 6.5 km/h. can be obtained. If arrangements are made for regenerative braking it occurs automatically due to the shunt field at speeds above about 20 or 25 km/h.

Mass as low as 5 kg. per kW is obtainable by the use of high-peripheral speeds of 61 m/s for the armature and 48.25 m/s for the commutator, these being obtained as a result of improvements in material and, more particularly, in methods of manufacture.

Control Equipment : Foot operated master controllers are employed, with two pedals, one controlling starting, speed variation and regenerative braking, and the other controlling the rheostatic and mechanical brake or compressed-air breakes. The contactors are operated from the 600-V trolley supply, although in some cases the braking contactors are supplied from a low voltage battery source to ensure that they shall still be operative if the supply fails.

Automatic circuit-breakers are fitted which trip and disconnect the supply if a fault occurs. An over-current relay is also provided which trips the circuit-breakers; this relay is set for a lower current value than the circuit breakers and is intended to operate when the current reaches a value which would damage the motors or the contactors. When regenerative braking is used an overvoltage relay is necessary to open the circuit-breakers if the voltage rises above 12 to 15 per cent above normal due to there being no load to absorb the regenerated energy.

Low-voltage lighting is necessary and is provided from a motor-generator set running in conjunction with a battery. Batteries are also sometimes provided for operating the vehicle at a low speed in case of power supply failure or for manoeuvring in the depot.

The frame of the vehicle is isolated from all electrical circuits, and it is necessary to take precautions against the possibility of its becoming alive, due to leakage, and causing electric shocks to passengers when boarding or alighting. If one pole of the supply is earthed and a leakage to the frame occurs near to the other pole, the frame will become alive to earth at the full system voltage; if both poles of the supply are isolated from earth (as is usually the case) any current flowing through a person alighting will also have to flow through the insulation resistance of one pole of the supply; this may be between 10,000 and 100,000Ω, so that this condition is less dangerous although unpleasant. To guard against such dangers every trolleybus must have its insulation resistance tested once per day, preferably when coming off service as the circuits are then hot and damp and, therefore in their worst condition; all metal work on the platform must also be covered with insulating material.

8.59. Control Equipment for Locomotives

Although a small industrial locomotives employ plain drum controllers similar to those used for industrial work, the currents dealt with in railway work are too heavy for this type of equipment and the master controller system is used. This system lends itself easily to the multiple unit method of control, in which any number of locomotives or motor coaches can be completely controlled from a master controller situated at either end of any one of them. In this system, all three motors can be controlled by any one of the controllers. It is seen that control circuits must be carried from vehicle to vehicle and this is done by means of flexible jumpers plugged into sockets at the ends of the coaches.

Three types of master controller equipment are in common use:

1. All electric equipment.
2. Electro-pneumatic equipment.
3. Camshaft equipment (with electric or pneumatic drive).

In any of these the operation may be entirely by hand or it may be semi-automatic; in the former the closing of the accelerating contactors is governed by the position of the controller handle, so that the rate of acceleration is completely under the control of driver. With the semi-automatic arrangement the position of the handle governs the number of contactors which will close but the sequence in which they operate and the time-interval between their individual operations are governed by time-lag devices as in fully automatic equipment. The latter method is sometimes used for suburban services, as a consistently uniform acceleration is ensured with a minimum energy consumption, but the former is preferable in locomotives, where there is a probability of wide divergences in the loads hauled and the possibility of wheel slip.

With all types a Dead Man's Handle device is required in order to stop the train automatically should the driver faint or become incapacitated in any way. This usually takes the form of a contact attached to the knob of the controller handle ; if the driver releases his hold on the handle the knob rises, causing the contact to close and operate the main circuit breaker and apply the brakes.

1. All electric Equipment

This is similar in principle to that used for industrial work, but there are certain difficulties associated with its application to traction service quite apart from the severe vibration which such equipment has to withstand. The chief of these are the impracticability of building contactors to operate at full voltage if above 600 V, and also the wide fluctuations in supply voltage which are liable to occur and which make the design of suitable contactors somewhat troublesome. The difficulties can be overcome by using a special low-voltage generator to give a supply at 100 V. or lower for the operating circuits, and although such a scheme is sometimes used, the modern tendency is to use one of the alternative systems. The multiple unit equipment operates on the all electric principle.

2. Electric-pneumatic System

The contactors are closed and opened by compressed air, and the valves admitting air to the operating cylinders are worked electrically by solenoids energised from the master controller. This system leads to simplicity in the electrical connections and enables the control system to be quite independent of the supply. Furthermore, the use of compressed air enables a larger mechanical force to be obtained with smaller and simpler equipment than by any other method. As in many cases compressed air is necessary for other purposes (brakes or pantograph operation) its supply does not involve any additional complication. A low-voltage supply is, of course, required for operation of the air valves, and this is obtained from an auxillary motor-generator set, which is usually also in necessary for other purposes.

3. Camshaft Control

The contactors are operated directly by means of cams mounted on a camshaft drived by an electric or a pneumatic motor. This type of control has the advantage of extremely simple electrical connections, and is being widely adopted. The position of the master controller handle governs the angle through which the camshaft is turned and therefore fixes the number of contractors which are closed.

Auxillary Equipment

In addition to the main traction motors, a considerable amount of auxillary equipment has to be installed on a locomotive or motor coach, the total output of which may amount of to 5 or 6 per cent of that of the main motors. Any or all the following may have to be provided :

1. **Motor-generator Set.** For supplying a low voltage for lighting, control system, and other purposed. Between 30 and 100 V is usually employed. The set must be of special design so that fluctuations of the voltage do not have any effect on the low voltage supply. Using this for lighting avoids light flicker, which would occur if the lighting were supplied from the $16\frac{2}{8}$ or 25 Hz supply.

2. **Battery.** To supply auxillary power in cases of emergency and to control gear in case of a supply failure, a battery is usually connected in parallel with the motor-generator set.

3. **Air Compressor.** For supplying compressed air for operation of electro-pneumatic control gear, brakes, and pantograph-operating gear. At 1,500 V or under, the motor of the compressor can be supplied direct from the line, otherwise it is usually driven from the auxillary generator.

4. **Heating Equipment.** Electric heating for the locomotives and possible for the train may be required. This may be supplied directly for the line up to 3,000 V. Where an electric locomotive has to haul steam-heated rolling stock an electrically heated boiler is often provided.

8.59.1 Support & Insulation of Trolley Wires in Tramways

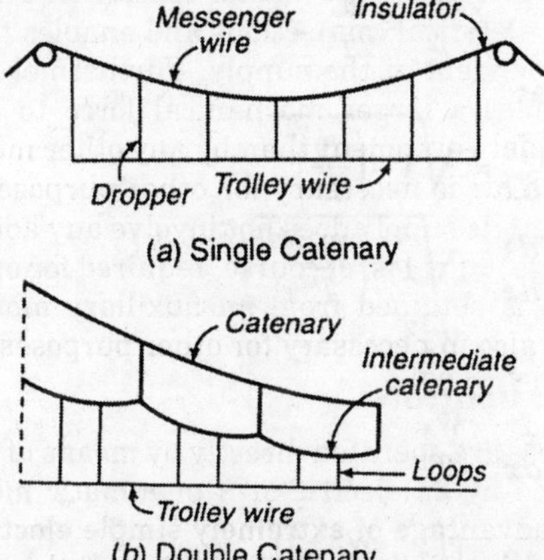

Fig. 8.67 Suspension Arrangements for trolley wire.

Trolley Wires

The trolley wire must be suspended with the minimum of sag so that it remains practically horizontal. Small spans have therefore to be used and if they are not possible, some method is employed.

8.60 Sag and Tension Calculation for a Trolley Wire.

A flexible wire when suspended horizontally between two supports it would assume the form of a catenary. But in case of tramways the sag in the trolley wire rarely exceeds 1 or 1½ percent of the span length and therefore the wire may be considered to hang in the form of parabola.

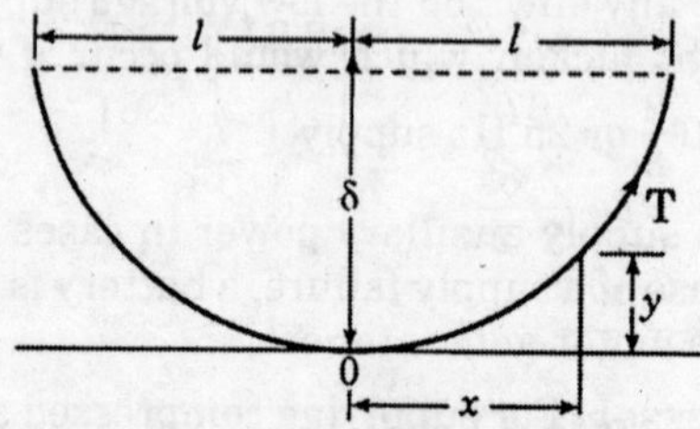

Fig. 8.67 (c)

If $2l$ is the span length in meters, w is the weight in kg per meter length of wire and T is the tension of wire in kg.

Then by taking moments about one of the supports we get,

$$T.\delta = Wl \times \frac{l}{2}$$

or $$\delta = \frac{Wl^2}{2T} \quad ...(i)$$

The equation of the parabola is given by the expression

$$y = \frac{\delta x^2}{l^2} = \frac{w}{2T} . x^2 \quad ...(ii)$$

We know that

$$\frac{ds}{dx} = \sqrt{1 - \left(\frac{dy}{dx}\right)^2}$$

or $$\frac{ds}{dx} = \sqrt{1 - \left(\frac{Wx}{T}\right)^2}$$

Since $$y = \frac{Wx^2}{2T}$$

$$\therefore \quad \frac{dy}{dx} = \frac{Wx}{T}$$

or $$\int dS = \int \left(1 + \frac{W^2 x^2}{T^2}\right)^{1/2} . dx$$

or $$S = \int \left(1 + \frac{W^2 x^2}{T^2}\right)^{1/2} . dx$$

$$= \int \left(1 + \frac{W^2 x^2}{2T^2} + \frac{3 W^4 x^4}{8T^4} + ...\right) dx$$

$$= x + \frac{W^2 x^3}{6T^2} + ...$$

when $x = l$ *i.e.* half of the span, the length of conductor in half of the span,

$$S = l + \frac{w^2 l^3}{6T^2} = l + \left(\frac{2T\,\delta}{l^2}\right)^2 \times \frac{l^3}{6T^2}$$

$$= l + \frac{2}{3}\frac{\delta^2}{l} \qquad ...(iii)$$

Since $$\delta = \frac{Wl^2}{2T}$$

$\therefore$ $$W = \frac{2T\,\delta}{l^2}$$

The change in temperature causes a change in length which will cause the change in sag and tension in span. It should be noted that decrease in temp. reduces the sag and increases the tension. The reduction in the sag causes straching which on the contrary tends to decrease the sag. In view of this combined effect the elasticity of the conductor material play an important role in determining the exact values of the sag at different temperatures.

Example. 8.16 (*a*) *A trolley wire of a tramway is suspended from two poles 40 metres apart. If the diameter of the wire is 1 cm and weight per metre is 0.5 kg, find its sag of the tension applied is 500 kg. Find also the total length of wire required.*

Solution. Span length, $2l = 40$ metres

or $l = 20$ metres

Weight per metre, $W = 0.5$ kg

Tension applied, $T = 500$ kg

Sag, $$\delta = \frac{Wl^2}{2T} = \frac{0.5 \times 20^2}{2 \times 500}$$

$= 0.2$ metre **Ans.**

Total length of wire required, $2S$,

$$= 2\left[l + \frac{2}{3}\frac{\delta^2}{l}\right] = 2\left[20 + \frac{2}{3} \times \frac{(0.2)^2}{20}\right] = 10.00267 \text{ metre} \quad \textbf{Ans.}$$

8.60.1 Neutral Section.

In d.c. traction system, supply of OHE is maintain from both ends by parallelling the substations to reduce the voltage drop. To sectionalize the boundary of two substations for convenient maintenance purpose, a simple insulated overlap is provided. When a pantograph bridges the OHE, it doesn't cause any damage to the system.

In case of single phase a.c. traction system the above arrangement is not permitted, because adjacent substations are tapped different phases of three phase supply in order to maintain even loading of the three phase system. The momentarily passing of pantograph under insulated overlap will cause short circuiting between two phases of the supply system thereby damaging the pantograph and the OHE system.

In order to prevent bridging of two different substation sections fed by two different phases, a small length of OHE called "Neutral Section" is interposed, which insulates both the phases from each other. The neutral section is insulated at both ends and is not connected to any source of supply. Its main function is to permit physically smooth and electrically sparkless passage of pantograph from one section to another.

Important precautions are taken to locate the neutral section at such a place so that the locomotive or EMU stock should not be brought to stop at this section as there will be no electric supply available. It becomes necessary to locate them far away from stopping signals so that train can pick up enough momentum to enable it to coast through the neutral section without any difficulty.

8.60.2 Electrolysis by Earth Current

In case where track is used as return conductor or the rail conductor system is adopted for electrification, the current has the chance to leak out at certain places. This current when leaked out follows a path of low resistance. Thus it flows towards the water pipes, cable sheaths and gas mains.

This current when leaves the water will cause the corrosion at the leaving places, because of the electrolysis. The water pipe acts as anode and the surrounding moisture and salts in the earth acts as electrolyte. The current while leaving the pipe carries away some metal.Following methods are adopted for reducing the effect of electrolysis:

(a) The resistance of the return path should be very low. To achieve this, good bonding is to be done at the track joints.

(*b*) Insulating joints over the water pipes should be provided to prevent the current for entering into the pipes.

8.61. Trolley-Bus-Routes

Serious drawback of tramway is the lack of monoeuvrability in congested areas and noise. These are overcome by trolley bus system of traction. In this system the vehicle is similar to omnibus having traction. In this system the vehicle is motor. Since there is not track as in of tramways, there are two overhead trolley wires fed at 600 V d.c. Since coefficient of adhesion between rubber tyred wheel and road is higher than that in the case of tram car, it is possible in this case to employ single axle drive without running the risk of slip. Motor employed is generally 80 to 150 h.p. compound wound where speed control can be carried out by inserting resistance in the shunt field or using driverter across series field. In order that driver should have his using driverter steer the vehicle and apply hand brake, master controllers are foot operated. One pedal controls the starting, speed control and regenerative braking if any, and second pedal controls rheostatic and compressed air brakes. Employment of regenerative braking for trolley bus or tram car, though used in past is being discouraged because of difficulty of ensuring that supply system is always in position to absorb the energy generated by the motors. In trolley bus system due to high adhesion between road and rubber tyres, power transmission gear is liable to be damaged if plain rheostatic braking is used. To avoid this situation, stabilised rheostatic braking is employed. In ordinary rheostatic braking applied to shunt motors, field remains connected to the supply and armature after being disconnected from supply is connected across loading rheostat. In stabilized braking a portion of loading resistance is included in the shunt field circuit. Portion R of loading rheostat Fig. 8.68 carries field as well as armature currents.

It will be observed that the voltage applied across the field is equal to supply line voltage less the voltage drop across resistance R. Any tendency for braking current to increase will reduce the voltage applied across shunt field thereby reducing field flux. In this way, any tendency for braking current to increase, and thereby overstressing the transmission gear is automatically countered by the fall in field flux. Reverse is also true. Net result fall this arrangement is to maintain the braking current substantially constant over a wide range of speed. On first notch, contactors (C_1), (C_2), (C_3) and C_4 are connected. This put full resistance R in field circuit. Another rate braking can be obtained if

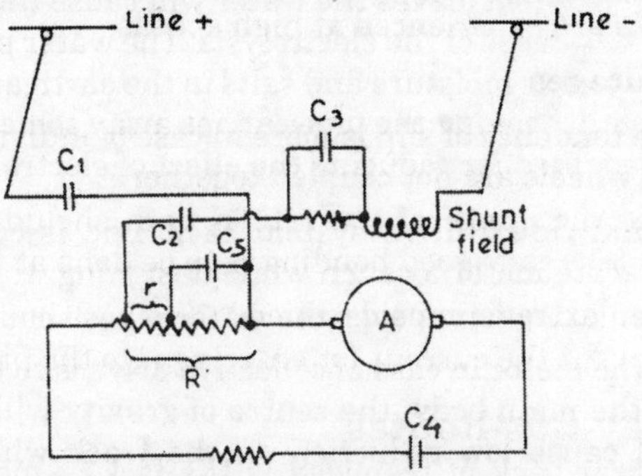

Fig. 8.68 For Trolley-Bus Routes.

contactor C_5 is also closed. In that case value of resistance in field circuit is r. If battery is provided on the vehicle, latter can be manoeuvred from the overhead line in case of emergency. If vehicle is halted in the dead section, it can be changed over to battery operation to clear the dead section. Trolley bus is insulated from ground by rubber tyres. It is, therefore, very important to see that no leakage develops which may cause shock of to passengers while boarding or alighting if supply systems has one pole earthed. Conditions are less dangerous if we use insulated supply system. In the event of leakage, insulation resistance of one supply pole will be in the circuit and current will be very small. To guard against these dangers, trolley buses are once a day insulation tested preferably when coming off from service when they are hot and damp and therefore, they are in their worst condition. Trolley buses have more passenger carrying capacity, higher acceleration and braking retardation than oil engined buses. These are, therefore, used for medium traffic density as obtained in inner suburbs. Oil engined buses on the other hand are used for outer suburbs and country side where traffic density is low.

8.61.1. Types of Drive

There are basically two methods of driving electric locomotive :

(*a*) **Individual Drive :** In this type of drive each axle of the locomotive has a separate motor usually smaller in size.

(*b*) **Collective drive :** In this case one motor is provided on the locomotive (larger in size), which provides driving power collectively to all the axles of the locomotive.

Following are the advantages and disadvantages of the individual drive when compared with the collective drive.

Advantages

1. Connecting rods are to be provided in the collective drive system whereas in case of individual drive it is not necessary.
2. Collective drive is not suitable for high speed locomotive, on account of stresses and vibrations in the connecting rods and jerks are experienced at high speeds.

Disadvantages

1. The tendency of slip is more in case of individual drive as all the wheels are not coupled together.
2. In individual drive system each axle is connected with a separate motor as such while designing, the space factor is given extreme consideration.
3. As the motor in case of collective drive is to be mounted well up the main body, the centre of gravity will be raised. This will cause low reduction on the track when compared to individual drive.

Conclusion : The individual drive is advantageous for high speed locomotives, whereas for low speed locomotives which are generally employed for passenger and goods trains, collective drive is preferred.

8.61.2 Transmission of Drive

The electric locomotives are specifically designed to have springs between the driving axle and the main body. These spring bears the whole weight of the locomotive. This arrangement eliminates maximum percentage of vibrations and reduces the damage to the track.

The method of transmission of drive from electric motor to the axle must be flexible. There are two types of transmission :

(*a*) Gearless transmission (*b*) Geared transmission.

Gearless Transmission

1. **Direct Drive.** In this type bipolar motor having flat poles is employed as shown in Fig. 8.69 (*a*). The armature at the bipolar is directly mounted on the driving axle. The poles and body of this motor are rigidly fixed with the frame.

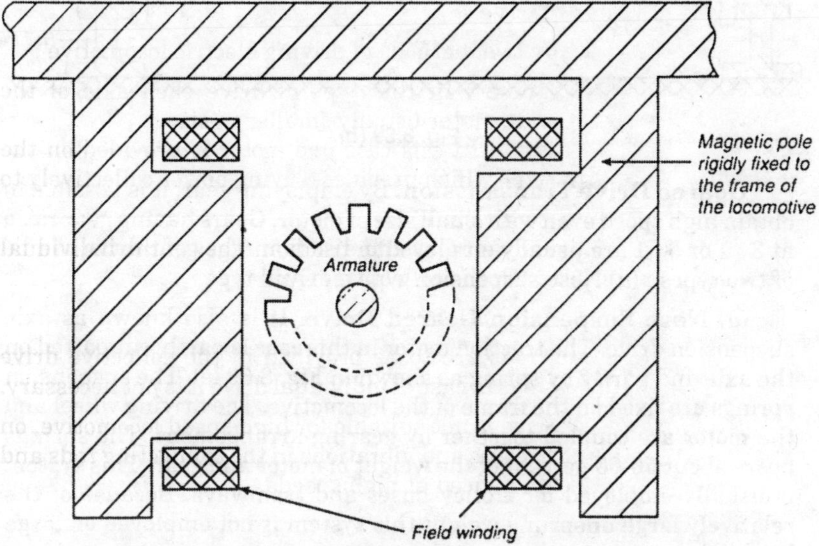

Fig. 8.69 (*a*)

The size of the motor armature is limited by the diameter at the driving wheel. The use of this method is limited mainly due to two reasons. Firstly, if the diameter of the driving wheel is higher than the desired value then the motor will run it low speed and secondary due to large unspring weight of the motor armature and low centre of gravity of the locomotive.

2. **Direct Quill Drive.** In this type of gearless transmission a hollow shaft (Quill) surrounds the driving axle as shown in Fig. 8.69

(*b*). The driving axle and quill are connected together with the help of springs. The motor armature is mounted on the quill. This system eliminates the chances of any relative motion between the motor armature and the driving axle. In this system the unsprung weight of the armature is zero. The size and speed of the motor are limited by the driving wheel as in the case of direct drive method.

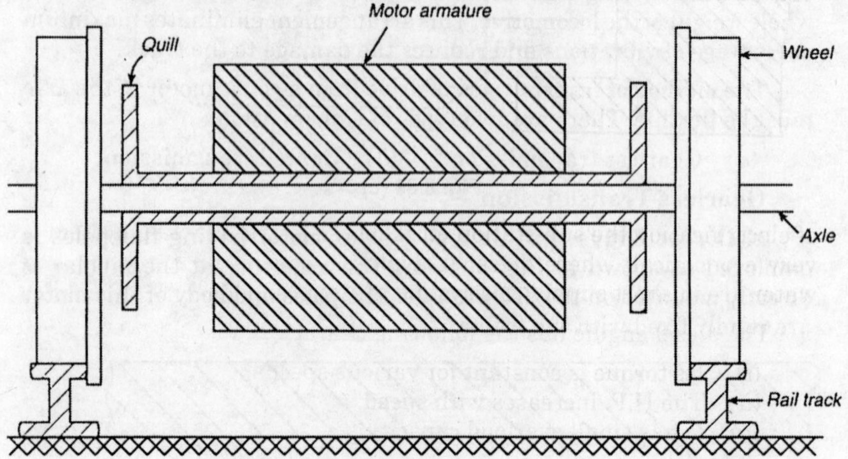

Fig. 8.69 (*b*)

Geared Drive Transmission. By employing gear, it is possible to obtain high speed even with small size of motor. Gears having gear ratio at 3 : 1 or 5 : 1 are usually employed in traction. The gear drive can be of two types : (*a*) Nose suspension type (*b*) Quill type.

(*a*) **Nose Suspension Geared Drive.** It is also known as axle suspension drive. The traction motor in this case is partly supported on the axle and partly by spring as shown in Fig. 8.69 (*c*). The suspension springs are fixed on the frame of the locomotive. The driving wheel and the motor are coupled together by gearing arrangement. The springs bears about 50-60 percent of the weight of motor and gear. This system is usually employed for trolley buses and tramways. Because of the relatively large unsprung weight this system is not employed on large locomotives.

(*b*) **Quill Type Geared Drive.** In order to raise the centre of gravity, to reduced unsprung weight and to accomodate the motors of larger size and dimensions this Quill type method is employed. Similar to direct quill method the driving axle is surrounded by a quill. However in this case the gear is placed on the quill.

8.61.1. Diesel Electric Traction

In this system a D.C. Generator is run by Diesel Engine. The generator supplies power to D.C. Traction motor. It eliminates the cost

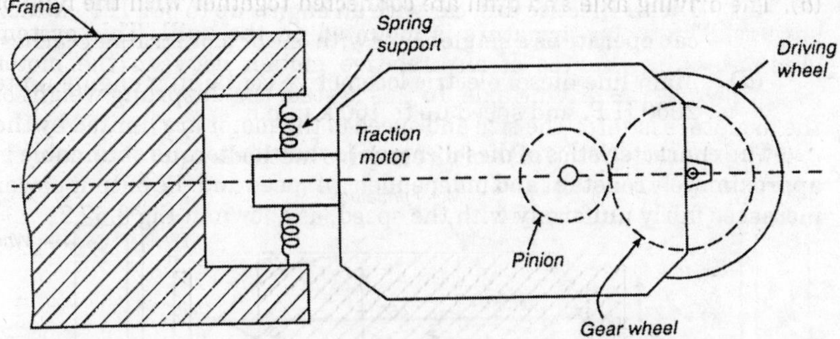

Fig. 8.69 (c)

of electrification the sub-station etc. Diesel Electric Traction system is very economical, where diesel is available cheaply. Where coal and water are in short supply this system is also economical.

The diesel engine has the following characteristics:

(*i*) Its torque is constant for various speeds.
(*ii*) The H.P. increases with speed.
(*iii*) It has small overload capacity.

For traction purposes, the torque should be inversally proportional to the speed, so the constant torque of the diesel engine should be modified by a **Torque converter** to be suitable for traction system. This can be achieved by :

(*i*) **Electrical Transmission :** For supplying power to D.C. traction motors, the system employs a diesel engine coupled to a d.c. generator. The Fig 8.69.1 shows the arrangement in Diesel Electric Traction.

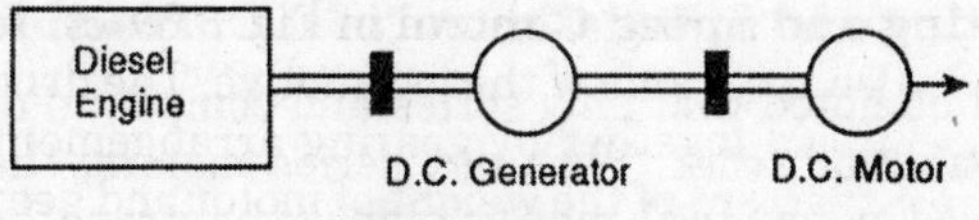

Fig. 8.69.1

(*ii*) **Mechanical Transmission :** In this system a variable gear is employed between engine shaft and the driving wheel. Due to lack of control facility and flexibility, this system is not very popular.

The most popular diesel electrical units are :

(*a*) Diesel electric shunting locomotive containing diesel engine up to 500 H.P. and speed up to 60 k.m.p.h.

(*b*) Diesel electric multiple unit (DMU) where each motor coach has an engine up to 200 H.P. and speed up to 110 k.m.p.h.

(c) Diesel Electric rail, car with an engine up to 600 H.P., which can operate as a single car or with one or more trailer coaches.

(d) Main line diesel electric locomotive containing engine up to 2000 H.P. and speed up to 160 k.m.p.h.

The characteristics of diesel engine is that the torque obtainable is approximately constant and independent of speed and the output power increases fairly uniformly with the speed, as shown in Fig 8.69.2.

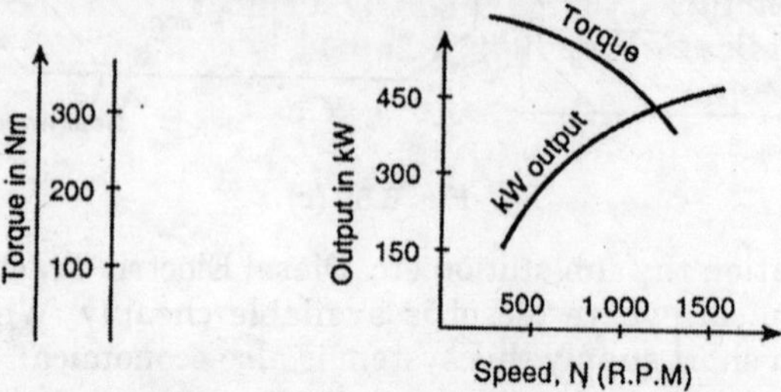

Fig. 8.69.2. Diesel-Engine-Characteristics.

In diesel electric traction system, the series motors are supplied power directly from the generator instead of through dropping resistance in series as employed in straight electric traction system. The starting & speed control is obtained by varying the excitation of the generator.

Regenerative braking is not possible in diesel electric locomotives but if required the rheostatic braking can be employed.

8.61.2. Starting and Speed Control of Traction Motors

The Traction motors may be D.C. or A.C. motors. First we shall discuss starting & speed control of D.C. motors and after words the starting & speed control of A.C. motors will be discussed.

(a) Starting and Speed Control of D.C. Traction Motors :

It should be noted that only series and compound d.c. motors are suitable for traction work. With d.c. series motors, the current and torque produced at standstill may be reduced by strengthening the field or lowering the terminal voltage.

With the d.c. compound motors the start may be made with full armature current in the series field coils and maximum current in the shunt field coils. A starting resistance inserted in the armature ckt. is reduced in steps untill the armature and series field are connected across the line.

For starting and speed control of D.C. series motors the different methods used are as follows :-

1. **Rheostatic Control :** Series motor can be started by connecting an external resistance in series with the main ckt. of the motor. At the

time of start of the d.c. series motor, the back e.m.f. developed by the motor is zero, therefore, in this condition, full or maximum resistance is connected in series with the armature which is of such a value that the voltage drop across it with full load rated current is equal to the line voltage. At the time when the motor picks up the speed, the back e.m.f. developed by the motor increases and with that the external resistance is gradually cut off in order to maintain the constant current throughout the accelerating period. Fig 8.69.3 shows a traction motor circuit with rheostatic starting. In this method, a lot of power is wasted in the external circuit.

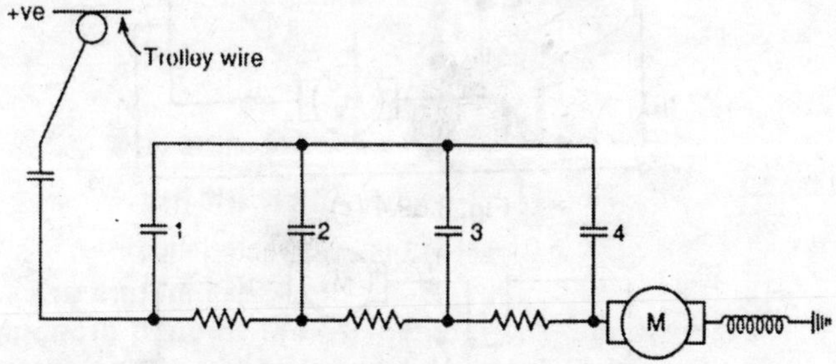

Fig. 8.69.3. Basic Traction Motor circuit.

2. **Series-Parallel Control :** The main drawback of wastage of electrical power in rheostatic control is partly over-come in this method by using two or more motors.

When used Two motors then these are connected in series through a starting resistance '*R*'. The starting resistance '*R*' is gradually cut off to zero value, as the motor speeds up. As the external resistance is totally removed, the village across each motor will be equal to half the supply voltage as shown in Fig 8.69.4 (*b*). This is the first running position.

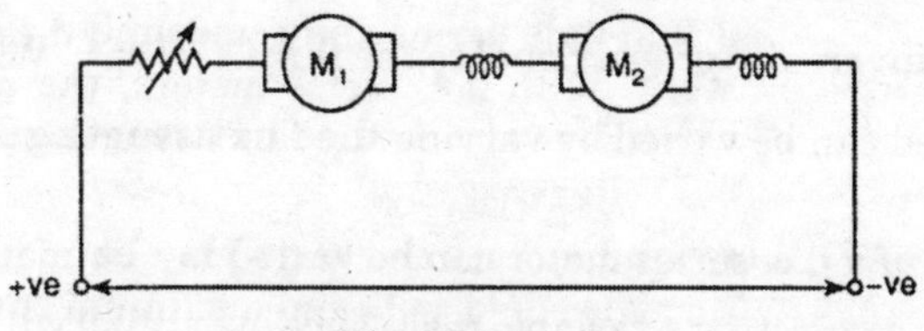

Fig. 8.69.4. (*a*)

In this position, each motor will run at 1/2 of its normal speed. Since, no external resistance lies in the ckt., therefore, there is no loss of energy and the motor will run at their maximum efficiency.

In the second running position, the two motors are connected in parallel and in series with an external resistance '*R*' as shown in Fig 8.69.4 (*c*).

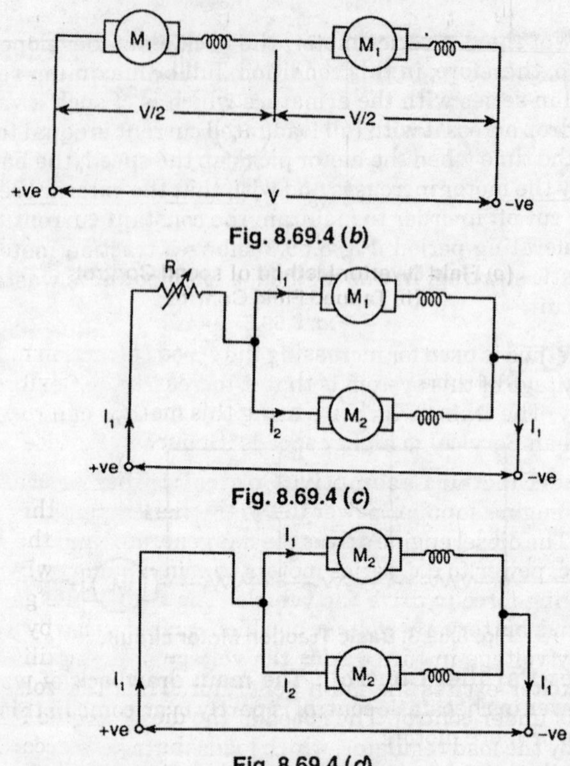

Fig. 8.69.4 (*b*)

Fig. 8.69.4 (*c*)

Fig. 8.69.4 (*d*)

The external resistance is gradually cut off as the motors speed up. When the external resistance is totally removed from the ckt. as shown in Fig 8.69.4 (*d*) the second running position is obtained. In this position, the voltage across each motor is equal to the full line voltage.

3. **Field Control Method :** As we already know that the speed of d.c. motor is inversally proportional to the flux *i.e.* $N \propto \frac{1}{\phi}$. Which means that the speed can be varied by varying the flux assuming line voltage constant.

The flux of a d.c. series motor can be varied :

(*i*) By connecting a variable resistance in parallel with the series field winding which is known as **field divertor.**

(*ii*) By cutting out some of the series field turns.

In both of these cases, the flux can only be reduced. Therefore, this method is also known as **Field Weakening method** and speeds above normal can be obtained. By using this method, speed of the motor can be raised by 20 to 30% of the normal speed due to design difficulty with the traction motors.

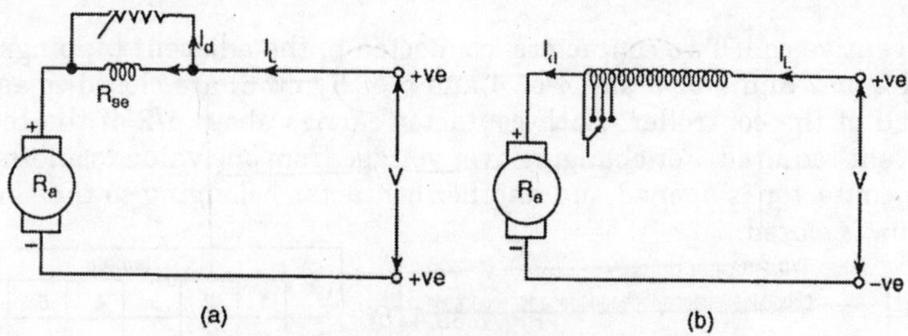

(*a*) Field Divertor Method of speed Control
(*b*) Tapped Field Control
Fig. 8.69.5

This method is used for increasing the speed of traction motors. The main advantage of this system is that it increases the flexibility of the train utility. The trains which are using this method can run from low speeds (Urban Service) to higher speeds (Suburban Service.)

4. **Diesel Electric Locomotive Control :** In diesel electric locomotive, diesel engine supplies power to the traction motors through a d.c. generator. The diesel engine drives the d.c. generator and the generator supplies d.c. power to d.c. series motors (*traction motor*) which in turn supply driving force to drive the vehicle. The field of the generator is excited from a battery, the voltage of which can be varied by a rheostat. The battery voltage in turn varies the voltage of an auxiliary exciter and this exciter excites the main generator. Thus the voltage of the generator is under control. The speed of the diesel engine is normally controlled by the load regulator, which loads the engine according to its controller setting. The function of the regulator is to maintain its output constant. If for any reason, the engine demands more fuel than called for by the setting, then the regulator reduces field excitation and thus prevents the stalling of engine.

(*b*) Speed Control and Starting of Single Phase Series Motors

In this method of speed control and starting, variable voltage is applied to a single phase series motor. The operating voltage required for the motor is 320 volts and this voltage is obtained from a transformer which form as important part of the locomotive. The variable voltage both for starting and speed control can be had from the tappings provided on the L.T. side of the transformer. The main advantage of this method is for starting purpose, so there is no loss of power and each control point becomes a running point.

A protective cool is used for reducing the circulating current flowing on Account of short circuiting of a section of the winding of the transformer between the tappings. This coil has high resistance and is connected between the terminals of two groups consisting of tappings 1, 3, 5 and 2, 4, 6. The motor is connected to the centre point of the preventive coil. Two contactors, connected to the adjacent tappings (1 and 2 or 2 and 3 or 3 and 4 or 4 and 5 or 5 and 6) are closed to each notch of the controller. Each contactor carries about 1/2 of the total current required. For changing over voltage from one value to another, one contactor is opened and another contactor belonging to the same group is closed.

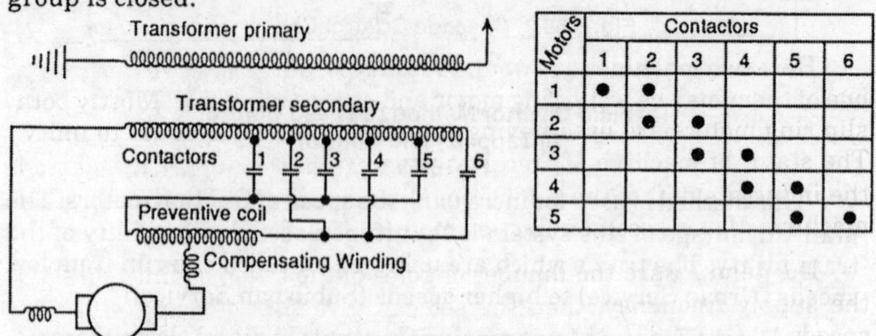

Motors	Contactors 1	2	3	4	5	6
1	●	●				
2		●	●			
3			●	●		
4				●	●	
5					●	●

Fig. 8.69.6. Contactor Method of Tap Changing.

(*c*) Starting and Speed Control of 3 – ϕ Induction Motors

Starting and speed control of 3 –ϕ induction motors can be done by the following four methods :

(*i*) Rheostatic Control (*ii*) Pole Changing Control (*iii*) Cascade Control and (*iv*) Combination of Cascading and Pole Changing.

(*i*) **Rheostatic Control :** This is the non-efficient and simplest method of speed control. In this method metallic rheostats are inserted in the rotor circuit of the motor. Sometimes liquid rheostats are preferred to be used because of non-inductance and the resistance can be gradually cut out resulting in a uniform torque. This method gives only one economical running speed with starting resistance completely cut-out. The main disadvantage is that 50% of energy supplied during starting is wasted in the rheostats. This method of speed control is applied in light locomotives and motor coaches where a single economical speed is sufficient and wastage of energy is not of so importance.

(*ii*) **Pole-Changing :** This is the simplest form of multi-speed control method. In this method, number of poles of the stator winding can be changed by changing the inter connections of the stator winding with the help of multiple control switches. The choice of number of poles on a pole changing winding are in the ratio 2 : 1, 3 : 2, 4 : 3. This method has the advantages of simplicity, good speed regulation for each setting, high operation efficiency and moderate initial cost and maintenance.

(*iii*) **Cascade Control :**

Different speeds are possible if two induction motors are mechanically coupled together and also connected electrically as shown in Fig 8.69.7.

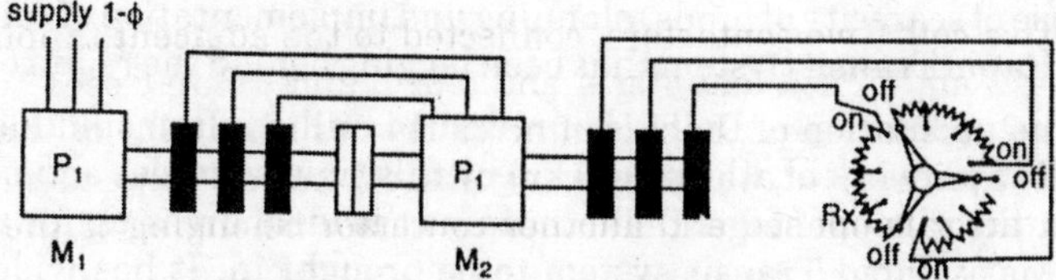

Fig. 8.69.7. Cascade Control Ckt.

Here two machines are wound rotor type, out of the two machines, one of them may be a slip ring motor and other cage motor. Mostly both slip ring motors are used having transformation ratio equal to unity. The stator of machine M_1 receives power from the supply mains and the induced e.m.f. of the rotor from slip rings is given to the stator of machine, M_2 *i.e.* second induction motor.

If P_1 and P_2 are the number of poles of the two machines and f is the supply frequency, then the set can give the following different speeds :

$$N = \frac{120 f}{P_1 + P_2}$$

Further speed control can be obtained by putting external resistance in the rotor circuit of the second machine. The power factor and efficiency of this arrangement is very low.

(*iv*) **Combination of Cascading and Pole Changing Control.**

In this method, again two slip ring induction motors of the same number of poles are required and their windings are so arranged that the number of poles can be changed. With the same number of poles of the two machines, one speed is obtainable and for more speeds, number of poles are changed.

Cascade control is used for two speed goods locomotive (25 kmph and 50 kmph) and a combination of cascade and pole changing is employed for the four-speed passenger locomotive (37, 50, 74 and 102 kmph) in some types of locomotives.

8.61.3. Introduction to Metro Railways

With the growth in size of cities, the number of vehicles on the roads for making trips have also gone up. This necessitates a firm policy shift to discourage private modes and encourage public transport, once the level of traffic along any travel corridor in one direction exceeds 25,000 persons per hour. This necessitates the introduction of a rail based system. The Mass Rapid Transit Systems (MRTS) are capital intensive and have long gestation period. It has been observed that in most of the developed countries, planning for mass transit system starts when city population size is more than 1 million; the system is in position by the time the city population is 3 to 4 million and once the population exceeds 4-million or more planned extensions to the Mass Rapid Transit systems is vigorously taken up. In developing countries including India, because of scarecity of funds, planning and implementation of rail based Mass Rapid Transit Systems has been lagging behind the requirements.

The population of Delhi is of round 14 million. It should have had an MRTs network of atleast 150 km by this time, whereas actually it is still at the take-off stage. Delhi contains the ideal dress-up for an very good Mass Rapid Transit system to be brought in. It has wide roads (Roads Cover 25% of the city area) where road possession for construction is not difficult except old city areas. Implementation will also not involve demolition of large scale private properties. Most of the land required is under Govt. control and hence can be easily acquired.

For want of efficient mass transport system, the number of motor vehicles has increased from 6 Lakhs in 1982 to 32 lakhs in 1999 and further increased to 40 Lakhs up to 2001 and is projected to go up to 50 Lakhs by 2008. The number of motor vehicles in Delhi is now more than that of Mumbai, Calcutta & Chennai put together. The result is highly congestion on Delhi roads, ever slowing speeds, increase in road accidents, fuel wastage and environmental pollution with motorized vehicles alone contributing to about $\frac{2}{3}$*rds* of the atmospheric pollution.

Today on roads of Delhi, the traffic is hetrogenous mix of Cycles, Scooters, Buses, Cars; Auto-Rickshaws and rickshaws jostling with each other. This has resulted in a chaotic situation so much so that due to road accidents, the average number of deaths by accidents per day has increased to 4 to 5 and of those injured to 15. The position is expected to deteriorate in future.

To rectify this situation the Govt. of India and the Govt. of Delhi in equal partnership have set up a Company named as **Delhi Metro Rail Corporation Ltd.** Under the Companies act, 1956 which has been given a mandade to construct 65 kms of Metro Rail tracks in Delhi by 2005 which has now completed by 2006.

The citizens are enlightened and have welcomed introduction of people friendly MRTS though they have initially faced some difficulties during the implementation phase. Added to this Delhi has an advantage in its excellent railway network comprising two rings and six spurs totaling about 130 km within the Urban area. Few points to be noted as follows:

(*i*) METRO is an electric train

(*ii*) There are 7 to 10 coaches in a train

(*iii*) Upto now the coaches were imported but soon it be built at Rail Coach Factory at Kapurthala, Punjab.

(*iv*) It has luxury coaches with ultra modern facilities and fully air-conditioned. When trains stops at a station the doors open automatically. When train starts again the doors are closed automatically.

(v) At present the train runs on 9 to 10 routes in Delhi. But by 2020, the target is to cover whole of Delhi.

(vi) The train stops after every 4 to 5 minutes after covering 7 to 10 kilometers of distance.

(vii) The train covers Urban and sub-urban areas in Delhi.

The Metro in Delhi has been built at a cost of Approximately $ 2 billions.

New Delhi's Metro Rail Chief named Elattuvalapil Sreedharan, has received France's top civilian award, **Knight of the Legion of Honour**, for his outstanding contribution in the field of transport infra-structure in the country.

Costing $ 510 million, a 6.5 km long tunnel connects the heart of the city with the old city Qrs. About 150,000 people presently use the metro.

8.61.4 Introduction to Electric Multiple Unit (E.M.U.)

The Electric Multiple Unit (E.M.U.) refers to car used for mostly suburban train services which have multiple prime movers (either diesel engines or electric motors) for each car *i.e,* same car that carries passengers also has the motive power, as opposed to the normal situation where the passengers are in coaches that are not self-propelling and a locomotive hauls the train.

The Indian Railway has electric multiple units in operation in several sub-urban sections (Bandel-Katwa, Calcutta, Chennai and Mumbai etc.)

The Mumbai region with 1.5 V, D.C. traction has several models of EMUs. Most models have D.C. traction motors with Rheostatic Control (Resistance banks to vary the input power supply) D.C. EMU's are also used on the PUNE-Lonovala section.

The BHEL has shortly developed A.C., D.C. EMU's for use in the Mumbai are in both the 25 KV A.C. and 1.5 KV D.C. traction regions. The new A.C.-D.C. EMU's have also 3- ϕ induction motors and thyristor control recently.

Chopper EMU's. In year 1981, Indian Railway contracted with the Bhabha Atomic Research Centre (BARC) to develop energy - efficient control systems for the city Mumbai EMU'S. The BARC design included Chopper (Thyristor) Control of the motor power supply instead of rheostatic control, thereby eliminating the waste of power in the resistance Grids. The EMU's were also provided with the capability for regenerative braking to convert the kinetic energy of the rake back to electrical energy fed to the catenary when braking.

The Chopper motor coaches make a characteristic humming or whining noise when beginning to accelerate, different from the standard Mumbai EMU's.

8.62. Power Electronics in Indian Railways an Analysis

Continuous development in semiconductor technology for the last three decades or more have led to the availability of a host of high power semiconductor devices and fast, versatile VLSI (Very Large Scale Integration) chips enabling proper control of such power devices. These developments have enabled a phenomenal growth of power electronics which basically deals with processing of electric power to make it suitable for various applications. A major application of power electronics in this period has been in the field of railway traction all over the world. This has led to benefits like smoother drives for passenger comfort, energy conservation, higher operational speeds, cost effective maintenance besides helping the growth of electronics industry in general. Indian Railways is one of the largest railway system in the world and offers a big scope for application of power electronics. From the late '60s few efforts have been made to introduce such systems with mixed results. Greater challenge lies, before the Indian industry to make use of this particular application area for its growth.

8.63. Introduction

Since the advent of Silicon Controlled Rectifier (SCR) in 1957 continuous developmental efforts have resulted in power semiconductor devices with ever increasing voltage/current ratings, switching speeds and other improved characteristics. Parallely, availability of sophisticated micro electronic circuits with high speed and versatility aided the process of evolution of a host of practical control schemes for processing of bulk electric power using these power devices. Present day high power thyristors, Gate Turn Off (GTO) thyristors, Insulated Gate Bipolar Transistors (IGBT), MOSFETS etc. have enriched the area of power electronics so much that it finds application from almost every field of industry and service sector down to cozy computer rooms and home appliances.

Transportation has remained one of the major application areas of power electronics. The modern electric locomotives, train sets, tram cars, trolley buses etc. all use today's dc or ac drive systems for control of the traction motors. The important advantages of power electronics are :

(*a*) it enables energy efficiency.

(*b*) it provides stepless control thereby increasing passenger comfort.

(*c*) better adhesion between wheel and rail due to smooth control.

(*d*) its easy adaptability to digital control provides flexible operations, wide range of diagnostic features and very compact size of equipment

(e) significant reduction in maintenance requirements in the long run.

(f) due to all these, it accelerates growth of electric traction—an environmentally clean transport technology.

However, one important disadvantage of such systems is the electrical noise and harmonics introduced in the power system. Special techniques are needed to minimise the levels of such disturbances.

Indian Railways is one of the largest railway system in the world scattered over 62400 route kilometer (Rkm) and currently deploys over 2000 electric locomotives and 1200 electric motor coaches over its more than 12,000 km of electrified track. Therefore, it has already become one of the most potential application field of power electronic drives and systems. In this paper an effort has been made to highlight the attempts made to introduce such systems and their results, the current scenario in this regard and possible scope for the future.

8.64. Electric Traction : A Brief Background

Low-voltage dc traction was first introduced in Europe and USA around 1880. Subsequently low frequency high voltage ac traction (16 2/3 Hz in Central Europe and 25 Hz in USA) was introduced. It was only in the 1950s when rectifiers, first mercury arc then semiconductor, were developed with required current and power ratings for traction. That the high voltage, industrial frequency traction lines were established ; 25 kV, single-phase 50 Hz or 60 Hz has become the standard for long distance mainline railway electrification ; 750 V or 1500 V dc power supply, however, is still usually the most economic for urban and street railways.

Electric Traction in Indian Railways was introduced around Mumbai in 1925 with 1500 V dc (which is still existing). In 1957 electric traction at 3000 V dc was introduced in Howrah-Bandel-Burdwan main line and Tarakeswar branch. This was subsequently changed to 25 kv, 50 Hz followed by gradual electrification at this voltage and frequency of most of the track and some of the branch routes of Indian Railways. As of March 31 1995 a total of 12,266 route km stands electrified which is about 19% of the total route km on Indian Railways.

With regard to on-board traction drives many schemes have been used during the last 50 years or so. In Table 1, a categorised format is given.

Out of the schemes mentioned above the dc drives using phase control technique in ac traction or chopper controllers in dc traction have been by far the most widely used schemes till the mid '80s. The majority of traction drives in use at present in the developed Railways fall in these categories. For example, the famous 'Shinkansen' or 'Bullet' trains in Japan or the earlier generation of TGV trains in Europe use dc drives. However, efforts were continuing for a long time to develop

Broad categories of traction drives

Type of Traction	*Traction Motor*	*Type of conversion and device used*	*Controls*
ac single phase 15 kV 16 2/3 Hz or 25 kV 50 Hz or 60 Hz	dc	(*i*) ac to dc by Ignitrons or Excitrons	Voltage control by onload tap changes
		(*ii*) ac to dc by full wave bridge using semiconductor diodes	Do
		(*iii*) ac to dc by Thyristor bridge rectifiers using phase control with line commutation	Electronic control starting with analog circuits to present day micro-computers controls.
	ac (3-phase cage motors)	(*i*) Single-phase ac to 3-phase ac using Thyristor Convertor-Inverter with forced commutation techniques	Earlier analog and then digital electronic controls using PWM technique.
		(*ii*) Single phase to 3-phase using GTO thyristors	PWM technique with Direct Self Control scheme using micro-computers.
dc (750 V, 1500 V, 3000 V)	dc	(*i*) dc to dc using electromechanical control	High-power resistance switching (*i.e.* Rheostatic control)
		(*ii*) dc to dc using forced commutated thyristors (chopper control)	Starting with analog electronics to present day digital microcomputer controls
	ac (3-phase cage motors)	(*ii*) dc to 3-phase ac using GTO/IGBT based inverters	Digital micro-computer control with PWM technique.

Various attempts to introduce new technologies for railway traction drives in Indian Railways

Sl. No.	*Type of Technology*	*Zonal Railway of application/type of traction vehicle*	*Qunatity*	*Year of commissioning*	*Executing Agency*
1.	dc drive, fully controlled single step thyristor bridge, analog controls	South Eastern Railways for 3000 HP, WAG 1 type locomotives	2 Nos	late 60s	*AEG, Germany*
2.	dc drive, fully controlled single step thyristor bridge, regeneration facility, analog controls	do	1 No	1978-79	Developed by BARC supplied by ECIL.
3.	dc drive, fully controlled two step thyristor bridge, regeneration facility, analog controls	do	9 Nos.	1982-85	Supplied by ECIL combining technologies of BARC and BBC, Switzerland
4.	dc chopper with regeneration facility, analog controls	Central Railways for EMU suburban service in Mumbai, Motor coach rating 1000 kW	10 Nos	Prototype developed by 1986 Commercial supplies made during 1992-94	Developed by BARC supplied by ECIL and BHEL
5.	dc drive, half controlled two step thyristor bridge, microprocessor control, without regeneration	South Eastern Railways 6000 HP WAG 6 type of locomotives	18 Nos (complete import of locomotives)	1988-89	ASEA, Sweden and Hitachi, Japan
6.	3-phase ac drive, GTO based inverter, microprocessor control, regeneration facility	6000 HP prototype locomotives being tested in Northern Railways	30 Nos (complete import of locomotives) and CKD/SKD parts)	1995-96	ABB, Switzerland
7.	dc drive, hald controlled two step thyristor bridge, microcomputer control, without regeneration	Prototype being commissioned in Central Railways for 5000 HP WAG 5 type of locomotive	1 No	1996	Developed by BARC/RDSO, engineered by ECIL
8.	3-phase ac drive, GTO based inverter and microcomputer control	1000 kW rated EMU motor coaches	2 Nos.	–	Currently under design stage at BHEL

cost-effective drive systems for ac traction motors (3-phase cage motors) for following advantages of such motors compared with dc series or separately excited machines.

(*a*) High maximum speed

(*b*) Robustness are reliability with a low maintenance requirement

(*c*) Simple cooling arrangement with enclosed frames

(*d*) High power/weight ratio

(*e*) Low cost/power ratio

(*f*) High Voltage operation

(*g*) Inherent regenerative braking capability.

Primarily due to the advent of high power GTO thyristors together with the availability of a host of fast and versatile VLSI circuits, the 3-phase ac drive system as mentioned in the above table has come to the forefront from the mid '80s since this type of drive offers advantages like very high speed operation, unity power factor control, energy conservation and less maintenance. In the developed Railways of the world any new procurement of traction vehicles today is most likely to be with this state-of-art technology. Famous trains in Europe like new generation TGVs, Intercity Expresses etc. all use this technology.

Even though electric traction in Indian Railways is as old as 1925, its accelerated growth came only after 1980. Before that diesel locomotives provided the main thrust alongwith steam locomotives. All our electric locomotives and Electric Multiple Unit (EMU) trains in a.c. traction area use electro mechanical controls with on-load tap-changers followed by diode rectifiers. In dc traction area, *i.e.*, in Mumbai area use of rheostatic control is in practice for the dc locomotives or EMUs. Obviously such technologies which are more than 50 years old have disadvantages like low speed operation, costly and elaborate maintenance requirements, energy wastage etc.

Even our prestigious trains like Rajdhani and Shatabdi Expresses are equipped with such technology.

8.64.1. Feeding and Distributing System of Power Supply to Traction System

For electrification, the network is divided into number of circuits. Each circuit is fed through distributors. These distributors are fed at suitable points known as feeding points from cables coming out of the sub-station and these cables are termed as feeders.

The function of the feeder is to maintain constant voltage at the feeding points whereas the function of the distributor is to maintain practically the constant voltage throughout the circuit.

The feeders are designed to keep the annual operating cost, *i.e.* cost of losses, interest and depreciation of feeding cable as minimum. The distributors are designed to keep the variation of the voltage in the distributors within permissible value.

8.64.2 Traction Power Supply System

25 kV, 50 Hz single phase traction supply to overhead equipment (OHE) is obtained through a step down transformer which in turn gets supply from state electricity Boards (SEB's) at 220/132/110 or 66 kV. The High voltage winding of the traction T/F located at traction substation (TSS) is connected across two phases of the supply received from SEB. On low voltage side of the traction transformer, one terminal of the 25 kV winding is connected to the overhead catenary system and other terminal to the rails of the track. The TSS's are normally located 40 to 60 km apart.

The supply to the OHE is fed through feeder circuit breakers and interrupters provided at Feeding Post (FP).

In order to maintain the balance on the phases of grid supply, two different phases are tapped at adjoining sub-stations. As the phases on OHE will also be different in the feeding zones of two adjoining TSS's therefore, a neutral section on OHE is provided between two TSS's. A switching station at neutral section is provided with a facility to bridge the electric supply on two sides of the neutral section, in case of a TSS failing and paralleling the supply of two tracks (up and down) for load sharing by both lines and better voltage regulation. This switching station is called "Sectioning and Paralleling Post" (SP).

Between the TSS and adjoining neutral section, the OHE is further divided into sub-sectors for minimising the section in case of fault and ease of maintenance of OHE. The switching station provided at such a location is known as "Sub-sectioning and Paralleling Post" (SSP).

A general power supply arrangement for double line section is shown in Fig. 8.69.8.

8.64.3 Computer Based Electric Traction Power Management

As more and more routes are being electrified, control of the power supply with least possible interruptions has been one of the serious concern. The quantity of the detailed information which has to be handled by the supervisory personnel for managing the system has rapidly increased in the recent years, making it difficult manually to supervise all the information and to take appropriate and timely action. Computer based supervisory control and Data Acquisition (SCADA) system makes the task easier.

A short interruption of few minutes brings the whole operation stand still. Unregulated supply is liable to effect the punctuality of the

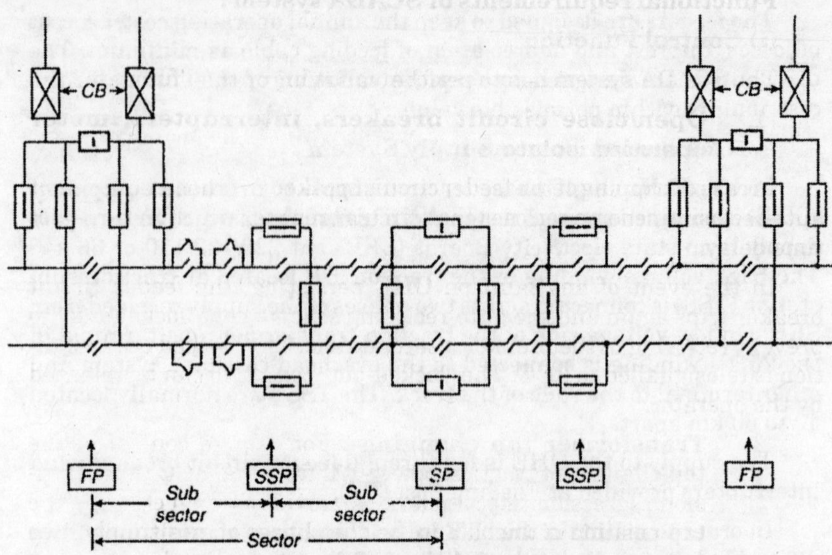

Fig. 8.69.8 General Power Supply Arrangement of 25 kV AC Traction.

train. Data acquisition, its analysis followed by timely corrective action helps railways in checking the tariff penalties payable to electricity authorities. Besides it, electric supply control system has emerged as a very important tool for the safety of train movement/operation. As soon as any abnormality in the rolling stock is noticed by the train passing staff, the electric supply can be got disconnected through the advise to the traction power controller, the train is checked and thus a mishap can be avoided. On double line section, when an accident takes place on one line causing permanent earth fault in overhead equipment, the movement of trains on the other line is automatically checked which otherwise could have resulted into another disaster.

Therefore Electric Traction Distribution has to have a reliable, effective SCADA system.

8.64.4 SCADA System for 25 kV Single Phase 50 Hz, AC Traction Power Control

The SCADA system enables the traction power controller (TPC) to control the power supply from a master station called Remote Control Centre (RCC) by monitoring and operating various equipments at remote locations called controlled stations. The system also enables collecting, storing and processing of various data relating to Traction Power Supply.

Functional requirements of SCADA system :

***(i)* Control Function**

The SCADA system has to perform following control functions :

1. **Open/close circuit breakers, interrupters, motor operated isolators :**

In case of tripping of the feeder circuit breaker on fault, a single-shot auto-reclosing scheme recloses that circuit breaker once after a pre-set time delay

In the event of any fault on OHE persisting, the feeder circuit breaker trips again and the auto-reclosing scheme gets "locked-out" to prevent reclosing of the breaker a second time. The "locked out" condition is telesignalled to RCC. The "locked-out" condition can be released by the operator.

2. **Transformer tap changing :** For remote control of the tap-changer of traction power T/F from the RCC, a progression/regression telecommand is provided which changes the tap position of the T/F in field and the tap position will be telesignalled to the RCC in confirmation of the tele command.

3. Interlock release/request facility for CB/interrupters control at boundary post :

When a controlled station separates the zone controlled by two adjoining RCC's, control breakers/interrupters at this controlled station is so arranged that the breaker/interrupters can be operated from one RCC, only when an interlock is released from the other RCC.

***(ii)* Data Acquisition Functions**

1. **Monitoring of the state of catenary :** State of catenary in each sub-section is monitored and the field conditions are displayed on the VDU. The catenary state is also utilised for under voltage relay operation at SPs in case the voltage is less than the selected value.

2. **Monitoring of the status of devices.** The status of devices are monitored and displayed on VDU (Visual display units) by using different colours.

3. **Monitoring of telesignals.** Various telesignals are monitored and displayed on the VDU.

4. The position on the tap changer as monitored and displayed on the screen.

Measurement Function

1. Two current parameters are measured at each TSS through CT's.
2. One voltage parameter is measured at each traction sub-station (TSS) taking reference from either of feeder PTs through

a change over device. In the event of failure of any of the PT, the change over device automatically connect the healthy PT for measurement.

Other Functions

1. **Auto fault localisation :** All the subsectors are energised one after another in a particular sequence till the faulty sector is identified. This whole operation is carried out by opening minimum number of interrupters.

2. **Under voltage tripping at SP (Sectioning and Paralleling Post) :**

Instantaneous type under voltage circuits are provided to trip the bridging interrupters of the catenary voltage drops below a preset limit. The operating range for the circuit is adjustable between 15 kV and 22 kV continuously in steps of 1000 volts.

3. **Maximum demand (MD) calculation and indication of MVA :**

SCADA software at master station multiplies the sum of feeder currents at individuals TSS and feeder voltage. This value is called MVA (instantaneous) and is displayed on the VDU against each TSS. MD is also computed here. In case MD value exceeds the set value an alarm is raised by the system and the value gets stored in the system.

4. **Limit setting for the telemetered parameters :**

Provisions for setting the "low" and "high" limit of the voltages and currents are available on the system. The settings are operator selectable. Any set limit violations operates the alarms.

8.65. Indian Efforts

However, right from the late '60s attempts have been made to introduce newer technologies in this field including the ongoing introduction of the latest 6000 HP 3-phase ac drive locomotives. The Table 2 shows at a glance the various efforts made upto the present day. This is followed by some analysis of the nature of such efforts in the next section.

8.66. The AEG Experiment

It was very creditable on the part of Indian Railways to plan introduction of thyristor locomotives in the late '60s when this technology was just making its debut in the developed world. The plan was to replace tap-changer-excitron combination in existing 3000 HP WAG-1 type of locomotives. A contract was awarded to AEG, Germany for two systems. These were commissioned in South Eastern Railways. The system employed fully-controlled, single-step thyristor bridge rectifiers for armature supply of the 1500 HP dc traction motors and a thyristor dual amplifier for automatic reversal of the field

The experiment was not very fruitful. Since individual SCR ratings were not very high in those days, AEG had to use a large number of devices in series and parallel which created problems. Automatic reversal of field by the thyristor dual amplifier could not be commissioned and use of mechanical reverser had to be resorted to. There were reports of frequent traction motor failures also.

8.67. The First Indigenous Thyristor Locomotive

This experience led to Railways looking for alternatives. Commercial procurement actions had to be abandoned due to high costs. This in turn led the Railways to build indigenously and thus came ECIL-BARC effort. In 1973-74, Research Design and Standards Organisation (RDSO) initiated first attempts to get the technology developed in the country. Power Electronics group of Reactor Control Division. Bhabha Atomic Research Centre (BARC), Mumbai was identified for the development and production was to be carried out by Electronics Corporation of India Ltd. (ECIL), Hyderabad for application in 3000 HP WAGI locomotives in SE Railways.

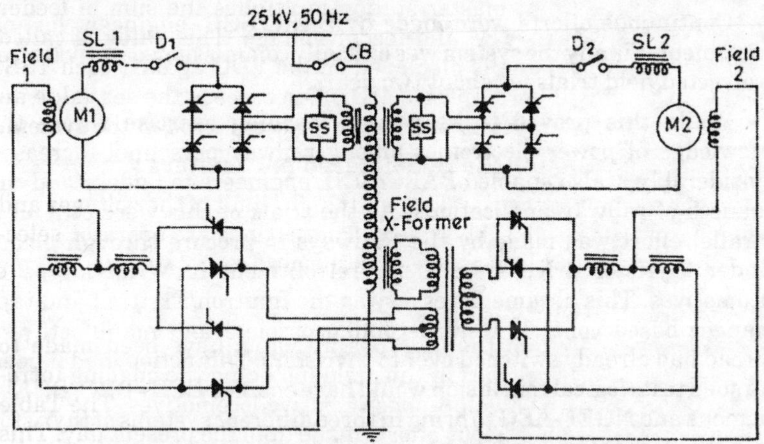

M1, M2--1250V, 1500 HP Motors. * CB--Circuit Breakers D1, D2--DC Breakers * SS--Surge Suppressor * SL--Smoothing Reactor

Fig. 8.70. Basic power circuit for BARC/ECIL design.

By 1978, BARC was ready with the prototype system. The power circuit diagram of this scheme is shown in Fig. 8.70. The main features of this very important indigenous development were:

(a) Totally in-house development at BARC with assistance from ECIL and RDSO.

(b) Fully controlled, single-step, thyristor bridge rectifier for armature supply of the 1500 HP dc traction motors.

(c) Forced air-cooling for converters.

(*d*) Regeneration capability.

(*e*) Analog controls with current and speed loop operation.

(*f*) Automatic reversal of field current (maximum 1100 A) by thyristor dual amplifier.

(*g*) Series-parallel operation (3 S × 3P) of 2000 V, 1250 A rated thyristors.

(*h*) Design and fabrication of 1100 A rated smoothing reactor.

However, the system was not completely trouble-free. Frequent tripngs with failure of armature thyristors were the main problems. In-depth analysis indicated few major deficiencies in the first design.

(*a*) Weak trigger pulses causing hot spots in thyristors.

(*b*) Inadequate input surge suppressor design.

(*c*) Use of some analog components that needed to be avoided in such hostile environments inside a locomotive.

(*d*) Inadequate EMI/EMC protection schemes.

Continuous efforts were made by BARC/ECIL engineers for improvement. Finally the system was officially commissioned in 1980 after protracted field trials for about two years.

While this proved to be a good learning exercise where the knowledge of power electronics among railway personnel increased considerably it also enabled BARC/ECIL engineers to understand the nuances of railway applications. As the trials of this were still on a parallel effort was made by the Railways to procure through global tender thyristor drive systems for retrofitment in WAM-4 type of locomotives. This became necessary as the Ignitron/Excitron and tap-changer based controls were becoming obsolete and manufacturers abroad had already switched over to thyristors. This action, had it been pursued to its logical conclusion would have enabled ECIL-BBC, BHEL-Siemens and NGEF-AEG to bring in three different systems of thyristor dc drive technology and three manufacturing organisations in the country. However, this was not pursued.

The interaction with Brown Boveri Company (BBC), Switzerland which ECIL established enabled ECIL to expedite the work on WAG-1 project on which certain improvements of BBC design were incorporated. This led to the following design :

8.68. Upgradation of the First Indigenous System

During 1980-81 a joint decision was taken by Railways and ECIL to modify the armature power circuit from single-step to two-step sequential bridge to derive the advantages of power factor improvement and reduction in harmonics. Also the technology for critical portion of the control electronics was imported from BBC, Switzerland. Thus with the combination of BARC-designed power converters (armature now changed to 2-step scheme) and imported control electronics from BBC. ECIL completed supply of 9 more systems. By 1983-84 this new system was successfully established in the SE Railways. The modified armature power circuit of this scheme is shown in Fig. 8.71. Other features of this design remained atmost same as in the previous case. However, for some reasons Railways did not pursue further applications of this design though it was considered to be a significant achievement.

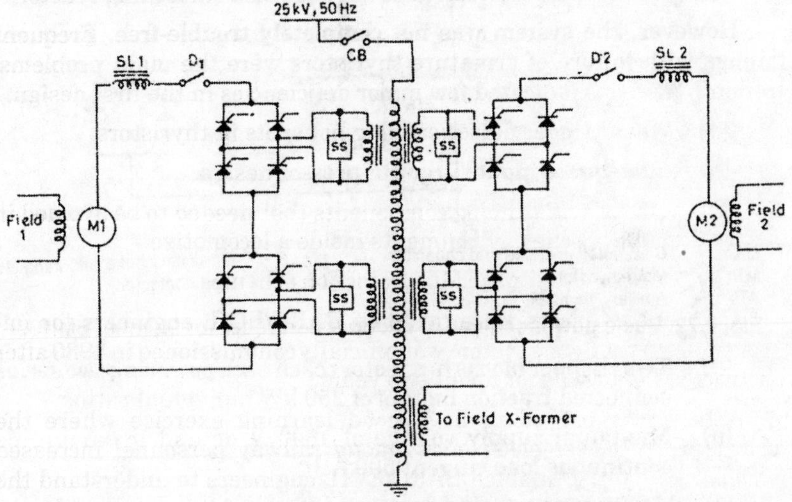

Fig. 8.71. Modified BARC/ECII design.

8.69. Chopper Controller for Mumbai EMUs

Suburban rail transport which employs EMU trains provides another major application area of power electronics. In the absence of coasting and with only acceleration and braking the suburban system is a potential area of energy saving on a continuous basis both during acceleration and braking phases. As mentioned earlier, Mumbai suburban train service depends on 1500 V dc traction. The EMU trains there normally have 9 coaches out of which 3 are motor coaches each having 4 × 250 kW rated dc traction motors. The speed control technique in practice is simple rheostatic control. Due to frequent start stop operations in a suburban service energy wastage is significant apart from maintenance problems in an electro-mechanical control. Thyristor chopper was an obvious choice and development work on this was initiated again at BARC sponsored by Department of Electronics, Government of India and the Railways by the late '70s.

With active technical support by RDSO engineers, prototype chopper system of BARC was ready by end of 1985 and was put on extensive field trial in an EMU motor coach. The trials were successfully completed by middle of 1986 and energy saving of more than 25% was demonstrated compared to the conventional coaches.

Fig. 8.72 shows the basic power circuit of this chopper. Controls were analog. The main features of this development are :

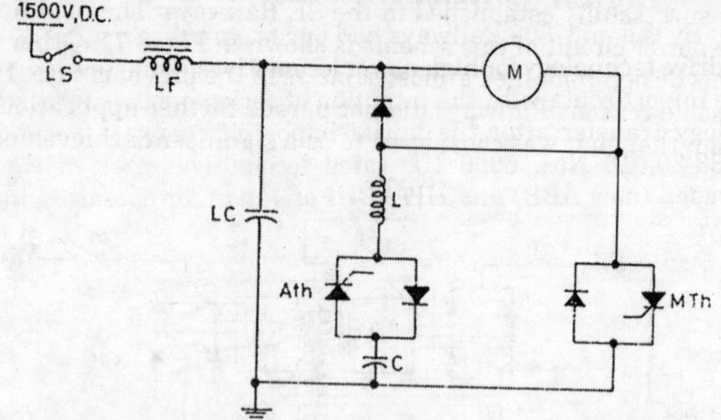

L, C - Commutating choke and capacitor
MTh - Main thyristor
ATh - Auxilary thyristor
LF, LC - Line filter choke and capacitor
LS - Line contactor

Fig. 8.72. Basic power scheme of chopper controller for Mumbai EMUS.

(a) Two chopper blocks in a motor coach each powering two series connected traction motors of 250 kW individual rating.

(b) Maximum supply voltage of 1800 V dc and maximum dc continuous load current 500 A.

(c) Maximum operating frequency 330 Hz.

(d) Maximum duty cycle 0.9.

(e) Minimum 'ON' time per cycle 0.3 m/sec.

(f) Use of 2000 V, 1500 A rated devices for the main and auxiliary thyristors in suitable series-parallel combination.

(g) Design and fabrication of the critical line filter, motor filter and commutation chokes at BARC.

(h) Regeneration capability.

(i) Forced air-cooling.

With the energy-saving potential of choppers demonstrated in actual operation, Central Railway authorities saw the cost-effectiveness of this design. BARC technology was transferred to ECIL and BHEL for carrying out proper engineering. Railways placed purchase order of 10 Nos of chopper controllers on each of these firms in 1988.

These commercial systems have been supplied and commissioned. Three EMU trains fitted with these chopper controllers are on regular commercial service now and are reported to be running trouble-free. Therefore, this was another important and successful indigenous development on traction drives.

8.70. ASEA/HITACHI Technology Introduction of Digital dc Drives

By the mid '80s Railways had once again taken steps to introduce dc drive technology for high power locomotives. This time it was through the import route with the intention of series production through technology transfer after the initial import of complete locomotives. By 1988-89, 18 Nos, 6000 HP rated locomotives arrived from ASEA, Sweden (now ABB) and HITACHI of Japan for operation in SE Railways.

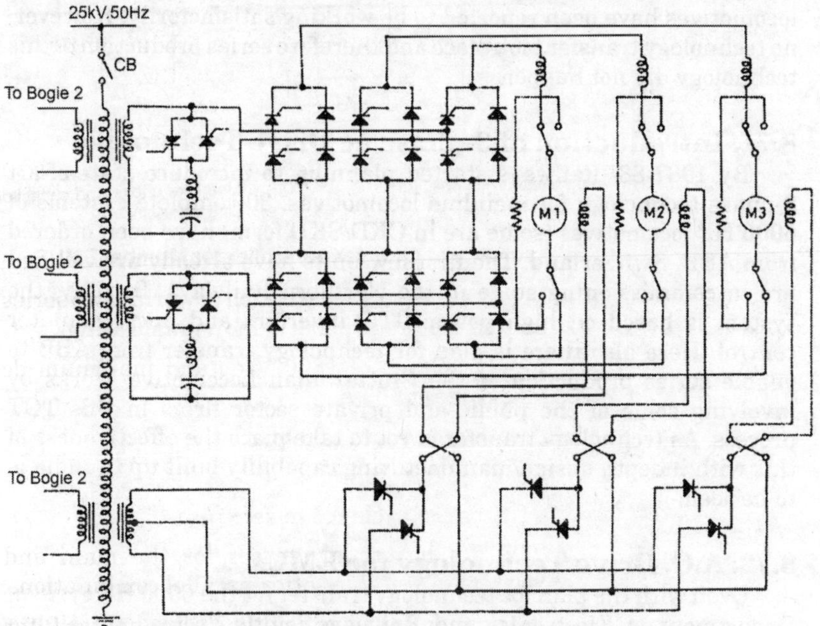

Fig. 8.73. Power schematic for ASEA 6000 HP locomotive.

The basic power circuit scheme of the ASEA System is shown in Fig. 8.73. The main features of this design are :

(*a*) Half-controlled, two-step thyristor bridge for the armature converters each powering 3 dc traction motors with individual rating of 785 kW.

(*b*) Simple mid-point converter for separately excited field with mechanical reversing arrangement.

(*c*) Dynamic Braking.

(*d*) Forced air-cooling of converters.

(*e*) Motorola 68000 processor based control electronics.

(*f*) Use of specially designed block diagram based language that produces very compact and fast codes for ease of application programming.

(*g*) Provision of a thyristor-switched Power Factor Correction mechanism to improve input pf and to reduce line side harmonics.

(*h*) The design also included a separate static single-phase to 3-phase converter for supply to the auxiliary machines in the locomotive. (All locomotives of the Indian Railways otherwise use 180 kW rated ARNO converter--a rotating machine--for this application).

The HITACHI design is also similar to the ASEA design. The locomotives have been reported to be working satisfactorily. However, no technology transfer took place and therefore series production of this technology did not happen.

8.71. Introduction of 3-phase ac Drive Technology

By 1987-88, Railways started planning to introduce state-of-art 3-phase technology for mainline locomotives. 30 complete systems of 6000 HP locomotives (some are in CKD/SKD form) have been ordered from ABB, Switzerland. The first few units have already arrived and are in commissioning stage in the Northern Railways. Broadly, the system is based on high power GTO inverters and microcomputer control. Here also there is plan for technology transfer from ABB to enable series production at the Chittaranjan Locomotive Works by involving some of the public and private sector firms in this TOT process. As technology transfer is yet to take place the effectiveness of this with indepth design/manufacturing capability built up in India is to be seen.

8.72. A.C. Drive Technology for EMUs

Even with the plan for technology transfer for the 6000 HP system, Department of Electronics and Railways jointly decided to initiate indigenous development of this technology to create design strength within the country and to enable meaningful assimilation of the imported technology. A project on this have been entrusted to BHEL for development of GTO based ac drive systems for the ac EMUs operating in Calcutta, Delhi and Madras regions. This 1000 kW rated design may later be upgraded for high power locomotives. The project is currently in the design and fabrication stage.

8.73. Digital dc Drive for Retrofit

There exist over 2000 electric locomotives in the Indian Railways system that are fitted with on-load tap-changers and associated electromechanical controls as mentioned earlier. These have their own problems as tap changer has long become an obsolete item. Department of Electronics and Railways once again thought to come out with an indigenous digital dc drive technology that could be cost-effective for retrofit option in these locomotives whereas the ac drive systems, cost of which is ought to be much higher than dc drives, could be deployed in new series production of locomotives.

Based on this planning a development activity was sponsored by DoE/Railways in 1992. BARC and RDSO of Railways have been in the lead for the design of converter and electronics. ECIL has fabricated the power converters for the prototype system meant for 5000 HP WAG 5 type locomotives. The basic power schematic is almost the same as in Fig. 8.73 with some minor difference in field thyristor connections. The main features of this design are :

(*a*) Half-controlled, two step thyristor bridge for armature converters each powering 3 dc traction motors with individual rating of 630 kW.

(*b*) Dynamic Braking.

(*c*) 16-bit 80196 processor based control electronics.

(*d*) Specially designed block diagram based language for ease in application programming.

(*e*) Provision of thyristor switched Power Factor Correction Scheme.

The prototype system is currently under commissioning in the Central Railways.

8.74. D.C. Drive Systems by BHEL and NGEF

During the '80s both BHEL and NGEF also undertook to develop their own technologies of Thyristor dc drive systems for mainline locomotives of Indian Railways. While BHEL depended on the expertise gained through their earlier collaboration with Siemens, Germany, NGEF relied on the support from their collaborator AEG, Germany. Both companies supplied prototype systems to the Railways. The efforts of these two also did not make much headway in getting converted to commercial orders.

These various efforts have underlined the importance of cooperation between Railways and manufacturers with a policy support to nurture local development as a necessary condition to successfully implement modernisation projects.

8.75. Other Power Electronics Applications in Railways

There are other power electronic applications in Railways. The two major areas are :

(*a*) Single-phase to three-phase static converters in each electric locomotive for supply to auxiliary machines.

(*b*) Static inverters for air-conditioning coaches.

While the rating of the first is about 200 kVA, the second application requires about 60 kVA rating per coach. Considering the vast number of locomotives and ac coaches in the Indian Railways there exists a potentially large business in this area.

The first development effort for single-phase to 3-phase static converter to replace the existing rotating ARNO machines was carried out in the early '80s at Central Electronics Engineering Research Institute (CEERI), Pilani. A 150 kVA thyristor based system was developed and field tried. However, it was very bulky in size and could not be productionised. Recently BHEL has developed a 220 kVA system based on GTOs and initial round of field testing has been completed. Development efforts based on IGBT devices have also been initiated by a few companies for the static inverters for ac coaches.

8.76. Retrospection

It is evident there have been a number of efforts for the last 25 years to modernise the traction vehicles in the Indian Railways and majority of these are based on indigenous strengths. This is only to be expected since a large railway system like ours can never depend upon total and continuous imports. Obvious economical advantages have always led to the thrust on local development and manufacture. But unfortunately these efforts have not produced the desired results so far. An objective analysis in this regard shows that there are a number of problem areas which need to be addressed to for greater successes in future. The points that emerge are as follows :

(*a*) Dedicated thrust from the user is essential. It is most important to have managers in the user organisation who are accountable for the result from the beginning and actually take the leadership role in such projects. In actual practice the developing agencies have to deal with a number of railway engineers taking up this role for short durations during the entire project period. Such a situation definitely hampers smooth progress and delays creep in.

(*b*) Another important requirement is the technical leadership by the user organisation. A dedicated and technically up-to-date core group is a must in the Railways who should actually drive the developing agencies in all the phases of development, manufacture and field trials. Most of the developed Railways of the world can boast of such groups in various fields of railway technology. It is heartening to note that the RDSO of the Indian Railways in recent years have slowly started building up a few such groups.

(*c*) On the part of the developing agencies and manufacturers also a lot more professional approach is required. In almost all the development projects considered here there were inordinate delays most of which could have been avoided by the development agency/manufacturer itself by better project management. Such delays actually made the final outcome, even if successful, almost an obsolete one in this fast changing technology scenario. Evaluating one's real strength and proper exposure of the engineers to the intricacies of designing such critical systems are essential before taking on such challenging projects. Industry needs to take on development based commercial orders rather than always providing 'Proven Technology' based on CKD/SKD imports.

(*d*) Lack of integrated infrastructure for testing such systems is another major area of concern. The systems considered here require considerable trial runs for proving reliability and also elaborate measurements of interference currents through the rails to ascertain whether signalling circuits are affected or not. In absence of a dedicated test track, designers have to test their systems on normal railway tracks used for regular commercial service. With the lowest priority given to such test vehicles by the traffic controllers an inordinate delay takes place to complete trials. Moreover, there is lack of proper testing arrangement in the Railways for quick results of interference tests. This results in further delay to clear the system from signalling angle. Such problems can be reduced greatly if an integrated infrastructure for testing is built up, say, adjacent to the Chittaranjan Locomotive Works (CLW).

(*e*) Lack of a sense of pride in our own capability.

However, these efforts made over the years have yielded definite positive results. Considerable knowledge base has been built up in the Railways in the field of power electronics. The organisations participating in these exercises, particularly the Bhabha Atomic Research Centre and the associated companies like ECIL and BHEL have developed skill and infrastructure to design, manufacture and test such systems to a level that would enable us now to take up greater challenges more confidently. The Centre for Electric Transportation (CET), Bhopal established by BHEL, who are also producing 5000 HP electric locomotives at their Jhansi operations, is one of the finest test facilities in this part of the world for inhouse testing of complete electric locomotives, motor coaches and other traction vehicles.

It may, however, be mentioned here that there are few other organisations also in the country engaged in development and manufacture of high power electronics who have so far not got any opportunity to participate in these activities.

8.77. Future Scope : Some Suggestions

Electric traction will continue to play a dominant role in the Railways since it is an environmentally clean technology compared to diesel traction and also its running cost is low. Indian Railways are now electrifying about 500 km of its track annually with nearly 13000 km of track already having been electrified. Relying more and more on fast and efficient electric traction should enable us transferring greater part of goods movement to rail from road helping the nation to reduce consumption of scarce petroleum products. Introduction of rail-based suburban transport system in more than a dozen cities and expansion of existing suburban rail systems in the four metros are also urgently needed. More number of trains require to be urgently introduced on the trunk routes to cater to the evergrowing passenger traffic.

All these are possible today only by adopting high-speed energy-efficient traction vehicles (coupled with introduction of sophisticated signalling systems, of course.) And power electronic equipment will be the mainstay of such new generation of rolling stocks.

There is also ample scope of modernisation of the existing fleet of electric locomotives and motor coaches in the Railways. As mentioned earlier tapchanger technology, which is employed in these, is thoroughly obsolete with all its attendant problems. As a retrofit measure these can be equipped with digital dc drives procured locally. This may be a very cost-effective and appropriate way to modernise the existing fleet. An indication of the possible market for power electronic equipment in this regard is given below:

Number of electric locomotives existing (as of 31.3.95) : 2302

Number of existing motor coaches for EMU operation (as of 31.3.95) : Approximately 1200

Average number of electric locomotives produced annually : Approximately 160

Average number of electric motor coaches produced annually : Approximately 100.

Presently the cost of a 5000 HP locomotive with tapchanger-diode rectifier technology is approximately Rs. 5 crores. The cost of a digital dc drive system for such a locomotive is approximately Rs. 1 crore. A conservative estimate of 100 locomotives annually for retrofit as well as new ones (till the 3-phase ac drive technology establishes in India) can be made to consider the above power electronic system. This opens a market of Rs. 100 crore annually to meet the needs of locomotives alone. Further, the Railways have plans to manufacture 6000 HP 3-phase ac-drive locomotives under technology transfer from ABB. The power electronic packages are proposed to be procured from Indian companies involved in this TOT process. The cost of the GTO inverters and associated controls will be substantially higher than that of the digital dc drive system mentioned above.

Other requirements for power electronic products for EMU trains, static converters for auxiliary machines in locomotives, inverters for ac coaches are also there.

While there is a vast scope of application in our Railways we have to demonstrate our capabilities in a big way. Why can we not build our own state-of-art high speed trains drawing on the strengths we have built in infrastructure and manpower and with the advantages of a single administrative set up of Railways ? A centrally coordinated demonstration project led by the Railways and pooling in all the resources in the industry, academic/R&D institutions and the Government can go a long way in establishing ourselves as one of the leaders in the field. After all the same approach followed by our defence and space departments for developing missile and satellite technologies has enabled us to join the select club of countries who own such technologies. In the field of Railways also there are a number of examples of successful collaborative efforts in the developed countries. As an example, the project on Intercity Express in Germany during the mid '80 can be cited. A 7500 HP, 350 km/hour train set with 3-phase drive and microcomputer controls was developed for the German Railways by a consortium of AEG, BBC and Siemens AG for the electrics and Kruss-Maffei, Krupp Industrietechnik and Thyssen Henschel for the mechanicals. The project was sponsored by the West German Ministry of Research and Technology.

8.78. Conclusion

India with its large railway network and the growth of large metropolitan centres is building up a large market for railway transport equipment. Clean technologies due to environmental considerations, energy conservation to minimise wastage of power, faster/reliable railway transportation which today's power electronics provides have to be resorted to in a major way.

The design, manufacturing, testing and commissioning, and operational infrastructure have been built up as a result of a number of approaches followed in the last 25 years. A few inter-disciplinary projects centrally coordinated by Railways by pooling in the talent in the country and utilising the local infrastructure would enable introduction of state-of-the-art technologies in the Indian Railway system in the most economical way. The success of space, defence and atomic energy can be replicated by the Railways also. It is necessary to work on functional goals and use the local technology base available as an economic option and suitably upgrade technology base as one goes along.

8.78.1. Performance Analysis of d.c. Motor Drive in Electric Traction with Wheel Slip Control

INTRODUCTION

A majority of railway traction drives in use today still utilize dc motors. The most appreciated dc motor for traction purpose is dc series motor because of its inherent high starting torque capacity. In early days most of the dc motor traction drives used inefficient resistance control to control the speed of the motor. However, with time, these drives were replaced with much efficient thyristor based drives which enabled the use of high power ac supply in electric traction. With closed loop dc drive, it became possible to control acceleration of the motor by controlling the motor torque. This was done by controlling the voltage from a low value at start to the high value at high speed. Thus the motor torque can be controlled such that, it accelerates with high value acceleration.

However, a high acceleration of the dc motor alone does not guarantee a high acceleration of the train which also depends upon the wheel-rail interaction.

Thus, the rotation of the wheel (a consequence of motor speed) may not completely result in translational motion of the train. When this happens, the wheel starts slipping on the track. This results in wearing of the wheel and also unnecessary wastage of precious electrical energy without yielding any acceleration to the train. Senini, *etal* offer a valuable insight into wheel-rail interaction and provide a mathematical formulation for this complex interaction by dividing entire mechanics of traction load into wheel dynamics and train dynamics. Hori, *et al* suggest design of model based controller and optimal slip ratio controller for slip controller and experimental implementation of the scheme. They also introduce the concept of gradient control for effective slip control.

In this chapter, an attempt has been made to simulate the traction load and apply it to a dc drive with effective slip control strategies. The motor used for analysis is dc series motor which has been conventionally used as a traction motor in India. The converter configuration used is two stage sequence control of two half controlled converters connected in series. This converter configuration has advantages of better power factor and lower current ripples.

MATHEMATICAL MODELLING OF TRACTION LOAD

The mechanics of traction deals with the nature of traction load and the dynamics of wheel-rail interaction. This includes slip-spin effects of the wheel on the track. These effects are governed by the complex interaction between motor, its controller and wheel-rail adhesion characteristics. This interaction is crucial to the motion of the locomotive but it can also introduce certain characteristics which are undesirable, such as wheel slip/spin.

Motion of Locomotive

The motion of the locomotive is induced by the force created when the wheel pushes against stationary track. The torque applied by the motor is transferred as a force at the rim of the wheel. This force at the rim of the wheel is transferred to a force which propels the train through a mechanism known as adhesion. Adhesion may be defined as the ability of the wheel to exert the maximum tractive effort without sliding on the rail. This adhesion force is the virtue of friction between the two surfaces (the wheel and the rail). Under the condition of acceleration there is a dynamic unbalance in the adhesion force and force exerted by the motor. This causes slipping motion of the train on the track. Friction causes the wheel to catch up and prevents the slipping motion. However, the surface could not possibly have such an effect on the wheel once the wheel has achieved pure rolling motion and constant angular and linear velocity. Thus friction becomes zero when the train moves on a flat track at constant velocity. Under pure rolling the train moves under the effect of rolling friction.

Rolling friction is caused primarily by the interference of small indentations formed as one surface rolls over another. This is because the wheel and the surface will undergo deformations due to their particular elastic characteristics. At the contact points, the wheel flattens out while a small trench is formed in the surface of the rail. The normal force is now distributed over the actual contact area rather than the point just below the center of the wheel. The load seen by the motor is also the result of adhesion as the force transferred by the adhesion is also seen as the load on the motor.

Coefficient of Adhesion

As defined earlier, adhesion is the ability of the wheel to exert the maximum tractive effort on the rail and still maintain the persistence of the contact without sliding. Thus the coefficient of adhesion can be approximately defined as the ratio of maximum tractive effort that can be developed without slipping and the weight on the driving axle. Thus, the force of adhesion is given by

$$F_a = \mu N \tag{8.1}$$

where μ is the coefficient of adhesion and N is the normal reaction force at the point of contact.

The coefficient of adhesion is affected by a number of factors. Contaminants like oil, grease, water, snow and mud reduces its value. Its value also depends upon the torque speed characteristic of the motor. The motor with lower speed regulation has higher value of coefficient of adhesion. The value of coefficient of adhesion decreases slowly with wheel slip after initial sharp rise. The measure of wheel slip is provided by a factor called wheel slip ratio which is given by

$$\lambda = \frac{V_w - V_t}{V_w} \tag{8.2}$$

where V_w is the peripheral velocity of wheel (referred as wheel velocity) and V_t is the velocity of train.

A change in slip ratio causes change in coefficient of adhesion. his is shown in Figure 8.74. Besides, the coefficient of adhesion decreases with wheel speed.

Nature of Traction Load

The traction load comprises of the various frictional forces that act on the system and the gravitational pull that act if the train is moving on an inclined track. The various frictional forces that act on the traction system can be classified as:

Internal Friction

This comprises of the friction at the bearings, guides etc.

External Friction

It comprises of the rolling friction between the wheels and the rails, and friction between wheel flanges and rail. This depends upon the nature of the track.

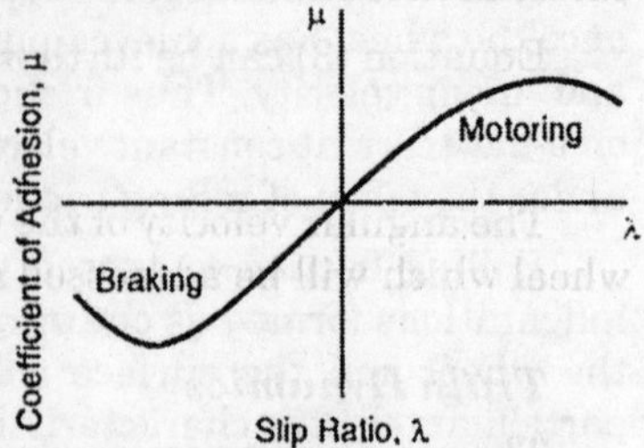

Fig. 8.74 Variation of coefficient of adhesion with wheel slip ratio for various track conditions.

Air Friction

This is independent of the weight of the train but depends upon the shape and the size of the train, velocity and direction of wind and speed of the train.

All these components collectively form the train resistance. The train resistance can also be identified in terms of common classification of friction such as windage, viscous friction, coulomb friction and stiction which has a very large value. However, owing to large value of the inertia, particularly due to the weight of the vehicle, accelerating torque forms the major component of total torque in the accelerating range.

The overall dynamics of the train can be broadly classified as

- Wheel Dynamics; and
- Train Dynamics

Wheel Dynamics

The wheel dynamics is nothing but dynamics of a rotating mass. This can be generally represented by a first order system comprising of the load and the inertia. The load torque can be split into two components, the conventional fricitonal elements and windage, and the component unique to the wheel (torque due to adhesion force).

Thus, the differential equation defining the wheel dynamics is given as

$$T_e = J\frac{d\omega}{dt} + T_l + F_a r \tag{8.3}$$

where $F_a = \mu N$; T_e, motor torque referred to the wheel; J, the combined moment of inertia of the wheel and motor referred to the wheel; r, the radius of the wheel; T_l, load torque contributed by friction (external friction), windage and grvitational component and is given by

$F_{ext} + Mg \sin(\theta)\, k$

where k is the minimum distance between the radius of the wheel and the horizontal line passing through the centre of gravity; F_{ext}, the external friction.

Equation (3) can be further solved as

$$\omega = \int \frac{T_e - T_l - F_a r}{J} dt \tag{8.4}$$

The angular velocity of the wheel yields the peripheral speed of the wheel which will be addressed as wheel speed.

$$V_w = \omega r \tag{8.5}$$

Train Dynamics

The train dynamics governs the translational motion of the train (vehicle). This can be given by a first order equation in terms of mass, train resistance and tractive effort. This is given as

$$M\frac{dv}{dt} = TE - R - Mg \sin\theta \tag{8.6}$$

where M is the mass of the train; TE, the total tractive effort and R is the train resistance.

This equation can be solved as

$$V_t = \int \frac{TE - R - Mg \sin\theta}{M} dt \tag{8.7}$$

Thus, the mechanical load model is represented by Figure 8.75.

ELECTRICAL TRACTION DRIVE WITH WHEEL SLIP CONTROL

Wheel Slip Control

It may be observed from Figure 8.76 that the entire slip range is divided into two regions, namely, stable and unstable region. In the stable region of operation the value of coefficient of adhesion increases with slip. In the stable region, as the motor torque increases, the wheel speed increases, resulting in immediate increase in the value of slip, which causes increase in force of adhesion. This results in increase of the mechanical load on the system causing wheel-slip to improve. Thus there is a self regulating action.

However, in the unstable region, coefficient of adhesion decreases with wheel-slip. In the unstable region, as the motor torque increases, the wheel-slip increases causing reduction in force of adhesion. This results in decrease of mechanical load on the system causing further acceleration of wheel. The slip ratio increases further. This is a cumulative action resulting in slipping of the wheel on the track.

The goal of all slip control methods is to control the slip in order to prevent wear of the wheels and the rail and to use the present adhesion effectively. Optimizing methods adds a search of the maximum adhesive force. This is achieved when the slip is controlled towards the

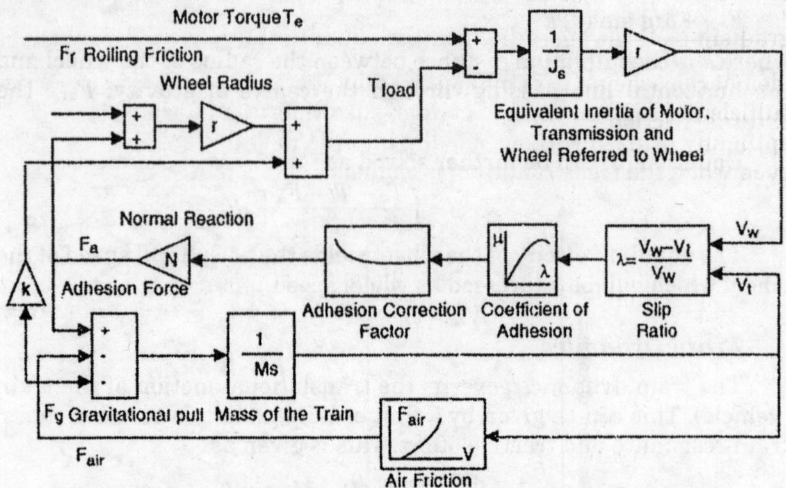

Fig. 8.75 Block diagram for entire mechanical load system

peak of the slip curve within the stable region as shown in Figure 8.76. The two slip control strategies suggested towards this end are

- Slip control to the reference value; and
- Gradient control method.

In slip control to the reference value, the slip error is processed by a slip ratio controller (a PI controller) which produces a torque command for the motor. Thus, the motor accelerates at a reference slip under the stable region such that the adhesion force is optimally utilized. Once the train reaches the desired speed, the speed controller takes control and maintains the speed at desired value. To obtain an optimal reference slip, it is sufficient to know the μ - λ characteristic of the track. Once, this characteristic is known, the optimal slip can be estimated by a choosing a value of slip which is near the peak of the μ - λ curve.

However, this method has a drawback that when the track condition changes the reference slip may no longer be optimal and it may happen that for the same operating slip, the train operates in the unstable zone of μ - λ characteristic. To overcome this disadvantage, gradient control scheme is adopted.

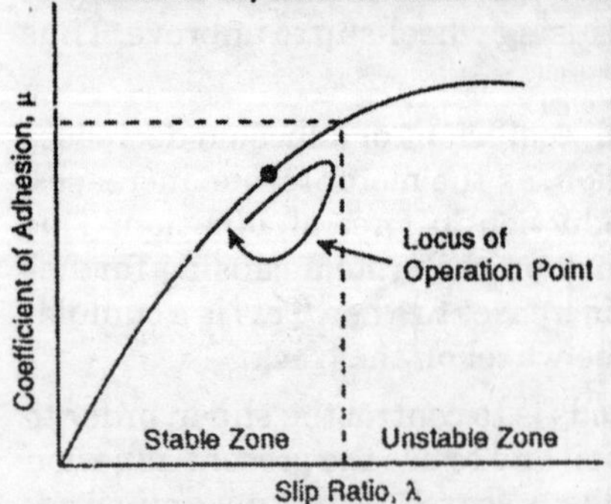

Fig. 8.76 Image of desired slip control.

Gradient Control Scheme

In this scheme, the gradient of μ - λ curve for the track is calculated and the operating point is controlled by a PI controller, such that, the gradient $\left(\frac{d\mu}{d\lambda}\right)$ at operating point is equal to the reference value. The idea of the scheme is to make the train operate at an operating point sufficiently high on the μ – λ curve, such that, the force of adhesion is optimally utilized without moving into the unstable zone of operation even when the track condition is changed.

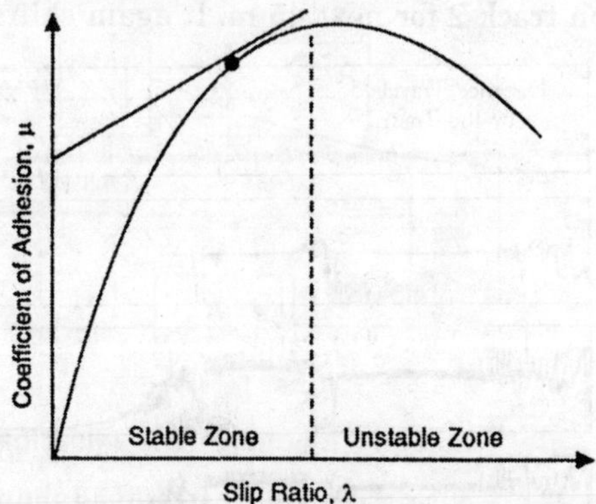

Fig. 8.77. The μ – λ curve showing stable and unstable regions and gradient at operating point.

It may be noted in Figure 8.77 that the gradient of μ – λ curve is positive in the stable region and unstable in the negative region. Further, as one moves towards the peak of the curve, the slope continuously decreases and at the peak of the curve it becomes zero. Thus one can ensure an optimal utilization of adhesion force by selecting a small positive value of gradient as a reference value. A block diagram representation of a gradient control scheme is given in Figure 8.78.

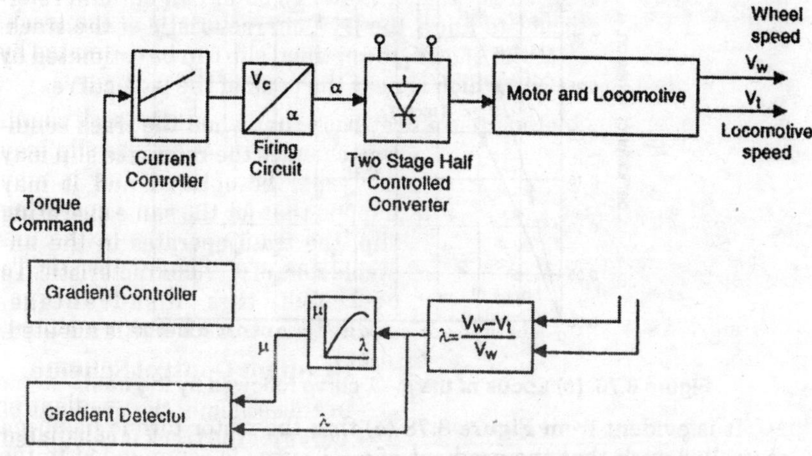

Fig. 8.78. Block diagram of gradient control scheme.

The gradient detector detects the slope of the $\mu - \lambda$ curve at the point of operation. This slope is then made to follow up the reference gradient with the help of gradient controller.

Figures 8.78 (*a*)-8.78 (*c*) shows the response of the drive when subjected to different track conditions. The train moves first 35 m on track 1. The track condition changes immediately at this point and the train moves on track 2 for next 45 m. It again shifts to track, 1 after this.

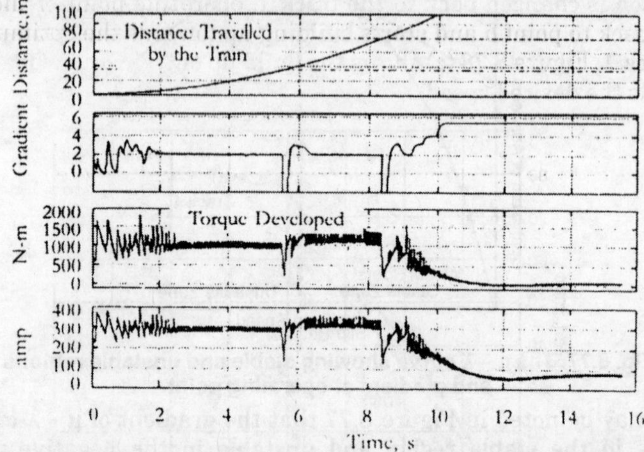

Figure 8.78. (*a*) The response of the system showing corresponding armature current and torque developed by the motor as the slip ratio is being controlled through gradient control to the commanded value while the train builds up to the desired value of speed on a variable track.

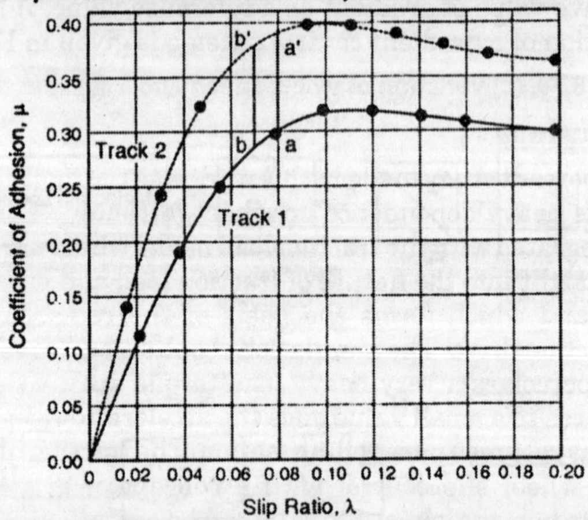

Figure 8.76. (*b*) Locus of the $\mu - \lambda$ curve followed by the train.

It is evident from Figure 8.78 (*a*) that the motor torque has been controlled such that the gradient of $\mu - \lambda$ curve is commanded to the reference value even when the track condition is changed. Figure 8.78

(*b*) shows the locus of the $\mu - \lambda$ curved followed by the train. It reveals that the train repeatedly moved from stable to unstable zone. This was the result of the oscillation due to the tuning of controller. However, soon after, the train settled at an optimum slip governed by the reference value of gradient. This is given by point *a* in Figure 7 (*b*). When the track condition changed the point of operation immediately shifted from *a* to *a′*. Then, after a small transient, the train settled at optimum slip for track 2. This is shown by point *b′*. Now, when the track condition is changed back to the track 1, operating point of the train drops back to point *b* and starts building up towards the optimum slip for track 1. Figure 8.78 (*c*) shows the build up of train and wheel speed towards the desired speed.

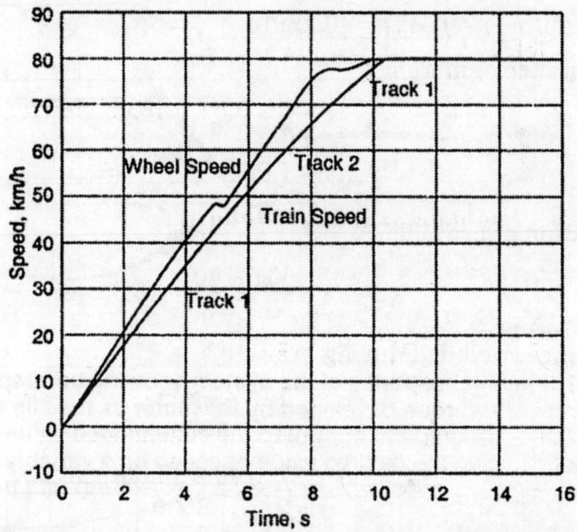

Figure 8.78. (*c*) Variation of wheel speed and train speed with time.

Conclusions

An important study made in this work was the modelling of traction load and its heavy dependence on wheel-rail interaction. The electric drive is integrated with the traction load model which provides extremely useful insight into the nature of traction load and pattern of tractive effort demand which forms the basis of the requirements from an electric traction drive. The simulation results clearly indicate the traction drive perceives a very heavy load torque during the acceleration which reduces to a small value once the acceleration period is over and the train has resumed pure rolling motion. The traction drive is modeled to provide wheel slip control with a concept of gradient control to demonstrate how the wheel slip ratio can be controlled to make optimal use of force of adhesion between train wheel and the track so as to allow maximum acceleration. The model prepared here can be used to provide a simulation study of electric traction drive performance employing wheel slip control under various conditions such as motion of the train on a flat track on a changing track.

Example 8.17. *A 300 tonnes E.M.U. train is started with a uniform acceleration and reaches a speed of 40 km.p.h. in 24 sec. on level section. Find the specific energy consumption, assuming a simplified trapezoidal curve, with rotational inertia as 8% retardation as 3 km.p.h.p.s., the distance between two stations as 3 km, efficiency of motor as 0.9 and train resistance 5 kg / tonne.*

Solution.

Distance between two stations, $D = 3$ kms

Maximum speed, $V_m = 40$ km/hr.

Resistance of motion,

$$r = 5 \times 9.81 = 49.05 \text{ Newtons/tonne}$$

$$\frac{W_e}{W} = 1.08$$

Duration of braking,

$$t_4 = \frac{V_m}{\alpha} = \frac{40}{3} = 13.33 \text{ sec.}$$

Distance travelled during braking,

$$= \frac{1}{2} \times \frac{V_m \times t_3}{3600}$$

$$= \frac{1}{2} \times \frac{40 \times 13.33}{3600} = 0.0741 \text{ km}$$

$$D' = 3 - 0.0741 = 2.9259 \text{ km}$$

Specific energy output

$$= \frac{0.01072\, V_m^2}{D} \times \frac{W_e}{W} + 0.2778\, r \times \frac{D'}{D} \text{ wh/t km.}$$

$$= \frac{0.01072 \times (40)^2}{3} \times 1.08 + 0.2778\, r \times 49.05 \times \frac{2.9259}{3}$$

$$= 6.17 + 13.29 = 19.46 \text{ wh/tonne km.}$$

Specific energy consumption

$$= \frac{19.46}{0.9} = \mathbf{21.6 \text{ wh/tonne-km.}} \quad \textbf{Ans.}$$

Example 8.18. *Calculate the specific energy consumption if a maximum speed of 12 20 metres / sec and for a given run of 1525 metres an acceleration of 0.366 m/sec² are desired. Train resistance during acceleration is 52.6 Newtons / 1000 kg and during costing is 6.12 New-*

tons/1000 kg, 10% being allowable for rotational inertia. The efficiency of the equipment during the acceleration period is 50%. Assume a quadrilateral speed time curve.

Solution. Let W is the dead weight of train in kg

Effective weight of train,

$$W_e = 1.1\,W$$

Acceleration, $\alpha = 0.366 \text{ m/sec}^2$

Max. speed, $V_m = 12.2$ m/sec.

Distance of run, $D = 1525$ metres.

Acceleration period,

$$t_1 = \frac{V_m}{\alpha} = \frac{12.2}{0.366} = 33.3 \text{ sec.}$$

Tractive effort required during acceleration,

$$F_t = W_e\,\alpha + W \times r$$

$$= 1.1\,W \times 0.366 + 0.0526\,W$$

$$= 0.4552\,W \text{ newtons}$$

Total energy required for the run

= Energy required during acceleration as there is no free run

= Average power during acceleration × Acceleration period

$$= \frac{1}{2} F_t \,.\, V_m \times t_1 = \frac{1}{2} \times 0.4552\,W \times 12.2 \times 33.3 \text{ joules}$$

$$= 92.57\,W \text{ joules or watt sec.}$$

$$= \frac{92.57\,W}{3600} = 0.0258\,W \text{ watt hr.}$$

Specific energy output

$$= \frac{\text{Energy required for the run}}{\text{W} \times \text{D}}$$

$$= \frac{0.0258W}{W \times 1525}$$

$$= 1.685 \times 10^{-5} \text{ Wh per kg m.}$$

Specific energy consumption

$$= \frac{\text{Specific energy out put}}{\eta} = \frac{1.685 \times 10^{-5}}{0.5}$$

$$= \mathbf{3.37 \times 10^{-5} \text{ Wh/kg m.}} \quad \textbf{Ans.}$$

Example 8.19. *An electric locomotive of 100 tonnes can just accelerate a train of 500 tonnes (trailing weight) with an acceleration of 1* *km.p.h.p.s. on an up gradient of 1/1000. Tractive resistance of the track is 45 Newtons per tonne and the rotational inertia is 10%. If this locomotive is helped by another locomotive of 120 tonnes. Find, (i) the trailing weight that can be hauled up the same gradient under the same conditions and (ii) the maximum gradient the trailing hauled remaining unchanged.*

Assume adhesive weight expressed as percentage of total dead weight to be the same for both the locomotive.

Solution. Dead weight of train and locomotive,

$$W = 100 + 500 = 600 \text{ tonnes}$$

Effective weight, $W_e = 1.1\, W = 1.1 \times 600 = 660$ tonnes

Gradient, $G = \dfrac{1}{1000 \times 100} = 0.1\%$

Acceleration, $\alpha = 1$ km.p.h.p.s.

Tractive resistance, $r = 45$ Newtons/tonne.

Required tractive effort,

$$F_t = 277.8\, W_e\, \alpha \times 98.1\, WG + W.r$$

$$= 277.8 \times 660 \times 1 + 98.1 \times 600 \times 0.1 + 600 \times 45$$

$$= 216{,}234 \text{ Newton}$$

$$= \frac{216{,}234}{1000 \times 9.81} = 22 \text{ tonnes} \quad ...(i)$$

Max. tractive effort of 1st locomotive

$$= \mu W_L \times \frac{\text{Adhesive Wt}}{\text{Dead Wt}}.$$

$$= \mu \times 100 \times 0.8 = 80\, \mu \quad ...(ii)$$

Consider adhesive weight of locomotive be 80% of total dead weight.

Equating equations (*i*) and (*ii*), we have,

$$80\, \mu = 22$$

or $$\mu = \frac{22}{80} = 0.275$$

where μ = Co efficient of adhesion.

With two locomotive total adhesive weight

$$= (100 + 120) \times 0.8 = 176 \text{ tonnes}$$

Tractive effort that can be developed

$$= \mu \times 176 = 0.275 \times 176 \text{ tonnes} \quad ...(iii)$$

Let the trailing weight which can be hauled, up the same gradient under the same conditions be W_1 tonnes.

Now the total dead weight,

$$W' = (100 + 120 + W_1) \text{ tonnes.}$$

Tractive effort required,

$$F_t' = 277.8 \times W'_e\, \alpha + 98.1 W'G + W'.r$$
$$= W'(277.8 \times 1.1 + 98.1 \times 0.1 + 45)$$
$$= 360.39\, W' \text{ Newtons}$$
$$= \frac{360.39}{9.81 \times 1000} W' \text{ tonnes}$$
$$= 0.036738\, W' \text{ tonnes} \quad ...(iv)$$

Equating expression (*iii*) and (*iv*), we get

$$0.036738\, W' = 0.275 \times 176$$

or
$$W' = \frac{0.275 \times 176}{0.036738} = 1320 \text{ tonnes.}$$

(*i*) Trailing weight $= W' - 100 - 120$

$$= 1320 - 220 = \mathbf{1120\ tonnes.} \quad \textbf{Ans.}$$

(*ii*) Total hauled load = 500 + 100 + 120 = 720 tonnes

Tractive effort required,

$$F_t = \left(277.8 \times \frac{W_e}{W''} \times \alpha + 98.1\, G + r\right) W''$$
$$= (277.8 \times 1.1 \times 1 + 98.1\, G + 45) \times 720\, N.$$
$$= \frac{(350.58 + 98.1\, G) \times 720}{1000 \times 9.81} \text{ tonnes} \quad ...(v)$$

Equating expressions (*iii*) and (*iv*), we get

$$0.275 \times 176 = \frac{(350.58 + 98.1\, G) \times 720}{1000 \times 9.81}$$

or
$$350.58 + 98.1G = \frac{0.275 \times 176 \times 9810}{720}$$

or
$$350.58 + 98.1\, G = 659.45$$

or
$$G = \frac{659.45 - 350.58}{98.1} = \mathbf{3.15\%.} \quad \textbf{Ans.}$$

Example 8.20. *An electric train has quadrilateral speed time curve as follows :*

(*i*) *Uniform acceleration from rest at 2 km.p.h.p.s. for 30 sec,*

(*ii*) *Coasting for 50 sec.*

(*iii*) *Uniform braking to rest for 20 seconds.*

If the train is moving a uniform up gradient of 10/1000, train resistance is 40 Newtons/tonne, rotational inertia effect 10% of dead weight and duration of stop 30 seconds. Find the schedule speed.

Solution.

Acceleration, $\alpha = 2$ km.p.h.p.s.

Acceleration period,

$$t_1 = 30 \text{ seconds.}$$

Coasting period, $t_2 = 50$ seconds.

Maximum speed, $V_1 = \alpha\, t_1 = 2 \times 30 = 60$ km.p.h.

Let the coasting retardation be B_c

$\therefore$
$$F_t = 277.8\, W_e\,(-B_c) + 98.1\, WG + W \times r.$$

$$O = -277.8 \times 1.1\, W\beta_c + 98.1 \times 1 \times W + 40\, W$$

or
$$\beta_c = \frac{98.1 + 40}{277.8 \times 1.1} = \frac{138.1}{277.8 \times 1.1}$$

$$= 0.453 \text{ km.p.h.p.s.}$$

$$V_2 = V_1 - \beta_c t_2 = 60 - 0.453 \times 50 = 37.4 \text{ km.p.h.}$$

Braking retardation,

$$\beta = \frac{V_2}{t_3} = \frac{37.4}{20} = 1.87 \text{ km.p.h.p.s.}$$

Distance travelled,
$$S = \frac{V_1\, t_1}{7200} + \frac{(V_1 + V_2)\, t_2}{7200} + \frac{V_2\, t_3}{7200} \text{ km.}$$

$$= \frac{60 \times 30}{7200} + \frac{90 + 37.4}{7200} \times 50 + \frac{37.4 \times 20}{7200}$$

$$= 1.03 \text{ km.}$$

Time of schedule,
$$T_s = t_1 + t_2 + t_3 + \text{Duration of stop}$$

$$= 30 + 50 + 20 + 30 = 130 \text{ sec.}$$

Schedule speed,
$$V_s = \frac{3600 \times S}{T_s}$$

$$= \frac{3600 \times 1.03}{130} = \textbf{28.4 km.p.h.} \quad \textbf{Ans.}$$

Example 8.21. *(a) Describe the power circuit of the driving motor for a 25 kV. single-phase, 50 Hz, a.c. traction.*

(b) An arc carrying train weighing 5000 tonne is to be hauled down a gradient 1 in 50 at a maximum speed of 30 km.p.h. and started on a level track at an acceleration of 0.08 m/sec^2. How many locomotive, each weighing 75 tonne, will have to be employed ?

Train resistance during starting = 3 kg/tonne

Train resistance at 30 km.p.h. = 5 kg/tonne

Co-efficient of adhesion = 0.3

Rotational inertia = 10 per cent.

Solution. (*a*) For part (*a*) please refer the previous discussed article.

(*b*) Enough number of locomotives are needed which not only produce requisite tractive effort to start the train at level but they should also apply sufficient braking effort on down gradients. Down gradient force due to gravity $= W \sin \theta$

$$= 5000 \times \frac{1}{50} = 100 \text{ tonnes}$$

Track resistance at 30 km.p.hr.

$$= \frac{5 \times 5000}{1000} = 25 \text{ tonnes}$$

Braking force to be applied by brakes

$$= 100 - 25 = 75 \text{ tonnes}$$

One locomotive can produce maximum braking force of

$$= 75 \times 0.3 = 22.5 \text{ tonnes}$$

Number of locomotive required for braking

$$= \frac{75}{22.5} = 3.33 \textit{ i.e. } 4 \text{ locomotives.} \quad \textbf{Ans.}$$

Tractive effort required to haul the train on level

$$= \left(\frac{1000 \times 1.1 \times 5000 \times 0.1}{9.81} + 5000 \times 3\right) \text{kg.}$$

$$= \frac{1}{1000}\left(\frac{1000 \times 1.1 \times 5000 \times 0.1}{9.81} + 5000 \times 3\right)$$

$$= \textbf{71 tonnes}$$

This also requires 4 locomotive.

Hence we will select 3 locomotives which can perform both functions of starting and braking.

Example 8.22. (*a*) *State the factors that effect specific energy consumption and their influence on it.*

(*b*) *An electric train has a quadrilateral speed time curve as follows:*

(*i*) *Uniform acceleration from rest at 2.5 km.p.h.p.s. for 25 sec.*

(*ii*) *Coasting for 50 sec.*

(*iii*) *Duration of braking 25 sec.*

If the train is moving along a uniform up gradient of 1 in 100 with a tractive resistance of 45 N/tonne, rotational inertia 10% of dead weight, duration of stops at stations 20 sec, and overall efficiency of transmission gear and motor 80%. Calculate the schedule speed and specific energy consumption of run.

Solution. (*a*) Refer the Article previously discussed.

(*b*) Time of acceleration,

$$t_1 = 25 \text{ sec.}$$

Time of coasting, $t_2 = 50$ sec.

Time of braking, $t_3 = 25$ sec.

Acceleration, $\alpha = 2.5$ km.p.h.p.s.

Maximum speed, $V_1 = \alpha . t_1 = 2.5 \times 25 = 62.5$ km.p.hr.

Let the coasting retardation be β_c,

$$F_t = 277.8\, W_e(-\beta_c) + 98.1\, WG + W.r$$

$$O = -277.8 \times 1.1 \times W \beta_c + 98.1W + 45 \times W$$

$$\left[\because G = \frac{1}{100} = 1\%\right]$$

or

$$\beta_c = \frac{98.1 + 45}{277.8 \times 1.1} = \frac{143.1}{277.8 \times 1.1}$$

$$= 0.458 \text{ km.p.h.p.s.}$$

$$V_2 = V_1 - \beta_c\, t_2 = 62.5 - 0.458 \times 50$$

$$= 39.6 \text{ km./hr.}$$

Braking retardation,

$$\beta = \frac{V_2}{t_3} = \frac{39.6}{25} = 1.58 \text{ km.p.h.p.s.}$$

Distance travelled, S or D

$$= \frac{V_1 \times t_1}{7200} + \left(\frac{V_1 + V_2}{7200}\right) t_2 + \frac{V_2\, T_3}{7200} \text{ km}$$

$$= \frac{62.5 \times 25}{7200} + \left(\frac{62.5 + 39.6}{7200}\right) \times 50$$

$$+ \frac{39.6 \times 25}{7200}$$

$$= \frac{1562.5}{7200} + \frac{102.1 \times 50}{7200} + \frac{9900}{7200}$$

$$= 0.21 + 0.70 + 1.39 = 2.30 \text{ km.}$$

Schedule time, $T_s = t_1 + t_2 + t_3 +$ Duration of stop

$$= 25 + 50 + 25 + 20 = 120 \text{ sec.}$$

Schedule speed, $V_s = \dfrac{3600\,S}{T_s}$

$$= \frac{3600 \times 2.30}{120} = \textbf{69 km.p.h.} \quad \textbf{Ans.}$$

Now applying equation for specific energy output, *i.e.*

$$= \frac{0.01072\, V_m^2}{D} \times \frac{W_c}{W} + 0.2778$$

$$\times \frac{D_1}{D}(98.1\,G + r) \text{ watt, hr/tonne km.}$$

$$= \frac{0.01072 \times (62.5)^2}{2.30} \times 1.1 + 0.2778$$

$$\times \frac{0.21}{2.30}(98.1 \times 1 + 40)$$

$$\left[\because D_1 = \frac{V_2\,t_1}{7200} = 0.21,\ G = 1\%\right]$$

$$= \frac{0.01072 \times 62.5 \times 62.5 \times 1.1}{2.30}$$

$$+ \frac{0.2778 \times 0.28 \times 9.81}{2.30}$$

$$+ \frac{0.2778 \times 0.21 \times 40}{2.30}$$

$$= 20.027 + 0.249 + 1.014 = 21.29$$

Specific energy consumption

$$= \frac{\text{Specific energy output}}{\eta} = \frac{21.29}{0.80}$$

$$= \textbf{26.61 watt-hr./tonne-km.} \quad \textbf{Ans.}$$

Example 8.23. *The torque-armature current characteristic of a series traction motor are given as :*

Armature current (Amp.)	*5*	*10*	*15*	*20*	*25*	*30*	*35*	*40*
Torque (Nw-m),	*20*	*50*	*100*	*155*	*215*	*290*	*360*	*430*

The motor resistance is 0.3 ohm. If the motor is connected across 130 V, deduce the speed armature current characteristics.

Solution. $V = 230$ volts

Resistance of motor

$$R_m = R_a + R_{sc} = 0.5 \text{ ohm}$$

Armature current I_a in amperes	5	10	15	20	25	30	35	40
Torque T in $Nw\text{-}m$	20	50	100	155	215	290	360	330
Back e.m.f. $E_b = (V - I_a R_m) V$	227.5	225	222.5	220	217.5	215	212.5	210
Speed $N = \dfrac{9.55 (V - I_a R_m) I_a}{T}$ in r.p.m.	533.16	429.75	318.73	271.1	241.5	212.4	197.3	186.56

Example 8.24. *A train with an electric locomotive weighing 300 tonnes is to be accelerated up a gradient of 1 in 33 at an acceleration of 1 km.p.h.p.s. If the train resistance, co-efficient of adhesion and effect of rotational inertia are 80 Nw. per tonne, 0.25 and 12.5 per cent of the dead weight respectively. Determine the minimum adhesive weight of the locomotive.*

Solution. Train with electric locomotive weight,

$$W = 300 \text{ tonnes}$$

Gradient, $G = \dfrac{1}{30} = \dfrac{10}{3} = 3.3\%$

Acceleration, $\alpha = 1$ km.p.h.p.s.

Train resistance, $r = 80$ Nw per tonne

Co-efficient of adhesion

$$\mu = 0.25$$

Acceleration weight of the train will be 12.5% more than the dead weight, *i.e.*

$$W_e = 300 \times 1.125 = 337.5 \text{ tonnes}$$

In view of equation of tractive effort, *i.e.*

$$F_t = 277.8\, W_e\, \alpha + 98.1\, WG + W \times r$$

$$= \left[277.8 \frac{W_e}{W} \alpha + 98.1\, G + r\right] W$$

$$= \left[277.8 \times 1.125 \times 1 + 98.1 \times \frac{10}{3} + 80\right] 300$$

$$= (312.5250 + 407) \times 300$$

$$= 719.5250 \times 300$$

$$= 215857.5000 \cong 215857.5 \text{ Newton}$$

$$= \frac{215857.5}{9.81 \times 1000} \qquad [\because \ 1 \text{ kgf} = 9.81 \text{ Nw}]$$

$\cong$ 22 tonnes

Adhesive weight $= \dfrac{\text{Tractive effort}}{\mu} = \dfrac{22}{0.25}$

$=$ **88 tonnes. Ans.**

Example 8.25. *The schedule speed with a 200 tonnes train on an electric railway with stations 777 metres apart is 27.3 km per hour and the maximum speed is 20 per cent higher than the average running speed. The braking rate is 3.22 km.p.h.p.s. and the duration of stop is 20 seconds. Find the acceleration required. Assume a simplified speed-time curve with the free running at the maximum speed.*

Solution. Distance of run, S = 777 metres = 0.777 km

Schedule speed, V_s = 27.3 km.p.h.

Retardation, β = 3.22 km.p.h.p.s.

Schedule time of run,

$$T_s = \frac{3600 \times S}{V_s} = \frac{3600 \times 0.777}{27.3} = 102.5 \text{ sec.}$$

Actual time of run,

$$T = T_s - \text{Duration of stop} = 102.5 - 20 = 82.5 \text{ sec.}$$

Average speed, $V_a = \dfrac{3600 \times S}{T} = \dfrac{3600 \times 0.777}{82.5} = 33.9$ km/hr.

Maximum speed, $V_m = 1.2\, V_a$

$= 1.2 \times 33.9 = 40.7$ km/hr.

Since, $V_m^2 \left(\dfrac{1}{2\alpha} + \dfrac{1}{2\beta} \right) - V_m \,.\, T + 3600\, S = 0$

$$\therefore \quad \frac{1}{2\,\alpha} + \frac{1}{2\,\beta} = \frac{V_m \,.\, T - 3600\, S}{V_m^2}$$

or $$\frac{1}{2\,\alpha} + \frac{1}{2 \times 3.22} = \frac{40.7 \times 89.5 - 3600 \times 0.777}{0.7^2} = 0.338.$$

or $$\frac{1}{2\,\alpha} = 0.338 - \frac{1}{6.44} = 0.338 - 0.155 = 0.183$$

α = **2.73 km.p.h.p.s. Ans.**

Example 8.26. *A train weighing 400 tonnes has speed reduced by regenerative braking from 40 to 20 km.p.h. over a distance of 2 km as a down gradient of 20%. Calculate the electrical energy and average power returned to the line. Tractive resistance is 40 Newtons / Tonne and allow rotational inertia of 10% and efficiency of conversion 75%.*

Solution. Dead weight of train,

$$W = 400 \text{ tonnes}$$

Accelerating weight of train,

$$W_e = 1.1\ W = 1.1 \times 400 = 440 \text{ Tonnes}$$

Initial speed $V_1 = 40$ km.p.h.

Final speed $V_2 = 20$ km.p.h.

Track resistance $r = 40$ newtons/tonne

Distance travelled during braking period = 27 km

Gradient, $G = 20$ per cent

Efficiency of conversion,

$$\eta = 75\% = 0.75$$

Energy available due to reduction in speed

$$= 0.0107\ W_e(V_1^2 - V_2^2)$$

$$= 0.0107 \times 440\ (40^2 - 20^2)$$

$$= 5{,}649.6 \text{ watt-hr} = 5.65 \text{ kWh}$$

Tractive effort required while going down the gradient

$$F_t = W.r - 98.1\ WG$$

$$= 400 \times 40 - 98.1 \times 400 \times 20$$

$$= -7{,}68{,}800 \text{ Newtons}$$

i.e. tractive effort of 7,68,800 newtons is available.

Hence energy available on account of moving down the gradient a distance of 2 km.

$$= \frac{F_t \times D \times 1000}{1000 \times 36000} \text{ kWh} = \frac{768{,}800 \times 2 \times 1000}{1000 \times 3600}$$

$$= 427.11 \text{ kWh}$$

Total energy available

$$= 5.65 + 427.11 = 432.76 \text{ kWh}$$

Energy returned to the line

$$= 0.75 \times 432.76 = \mathbf{324\ kWh.} \quad \textbf{Ans.}$$

Average speed $= \dfrac{40 + 20}{2} = 30$ km./hr.

Average time taken to cover 2 km.

$$= \frac{2}{30} = \frac{1}{15} \text{ hour}$$

Average power,

$$P = \frac{\text{Energy returned the line}}{\text{Time}}$$

$$= \frac{324}{1/15} = \mathbf{4{,}860\ kW.}\quad \textbf{Ans.}$$

Example 8.27. *A motor coach is being driven by two similar motors taking supply at 650 V. If first motor is geared to driving wheel having dia of 96 cm and if speed of first motor is 400 r.p.m. When connected in parallel with other motor across 650 V supply. Find out the speed of motors when connected in series. Assume armature current to remain same and armature voltage drop of 10% at this current.*

Solution. We know that $N \propto \dfrac{V - IR}{\phi}$.

Since current through out during series parallel running remains constant, flux produced by it also remains constant.

$$\therefore \quad N \propto (V - IR)$$

Let V_1 and V_2 be the voltage across each motor when they are in series (Fig. 8.55), then,

$$\frac{N_1}{N_2} = \frac{V_1 - IR}{V_2 - IR}$$

If
$$\frac{N_1}{N_2} = K$$

$$\therefore \quad K = \frac{V_1 - IR}{(V - V_1) - IR}$$

$$\therefore \quad KV - KV_1 - KIR = V_1 - IR$$

$$\therefore \quad V_1 (1 + K) = KV - KIR + IR$$

$$\therefore \quad V_1 = \frac{K(V - IR) + IR}{1 + K} = \frac{V - IR}{1 + \frac{1}{K}} + \frac{IR}{1 + K}$$

$$= \frac{V - IR}{1 + \frac{N_1}{N_2}} + \frac{IR}{1 + \frac{N_1}{N_2}}$$

Similarly,
$$V_2 = \frac{V - IR}{1 + \frac{N_1}{N_2}} + \frac{IR}{1 + \frac{N_2}{N_1}}$$

$$IR = \frac{10}{100} \times 650 = 65 \text{ Volts}$$

$$\frac{N_1}{N_2} = \frac{D_2}{D_1} = \frac{86}{90}$$

$$V_1 = \frac{650 - 65}{1 + \frac{90}{86}} + \frac{65}{1 + \frac{86}{90}}$$

$$= \frac{586 \times 86}{176} + \frac{65 \times 60}{176}$$

$$= \frac{50,310 + 5850}{176} = 319 \text{ Volts}$$

$$V_2 = 650 - 319 = 331 \text{ Volts}$$

Speed of first motor with 650 V supply is 400 r.p.m., it's speed with 319 volt supply can be found out as follows :

If N_1' be the speed of first motor with 650 V supply and N'_1 be the speed, with 319 volt supply, then

$$\frac{N'_1}{N_1} = \frac{V - IR}{V_1 - IR} = \frac{650 - 65}{319 - 65}$$

$$\therefore \quad \frac{400}{N_1} = \frac{585}{254}$$

$$\therefore \quad N_1 = 400 \times \frac{254}{585} = \mathbf{174 \text{ r.p.m.}} \quad \textbf{Ans.}$$

$$\therefore \quad N_2 = N_1 \times \frac{90}{86} = 174 \times \frac{90}{86}$$

$$= \mathbf{182 \text{ r.p.m.}} \quad \textbf{Ans.}$$

Example 8.28. *An electric train has an average speed of 42 kmph on a level track between stops 1,400 m apart. It is accelerated at 1.7 kmphps and is braked at 3.3 kmphps. Draw the speed time curve for run. Estimate the sp. energy consumption. Assume tractive resistance as 50 N / tonne and allow 10% for rotations inertia.*

Solution. Acceleration,

$$\alpha = 1.7 \text{ km. p.h. p.s.}$$

Retardation, $\beta = 3.3$ km. p.h. p.s.

Distance of run, $S = 1,400$ m $= 1.4$ km

Average speed, $V_a = 42$ km. p.h.

Time of run,

$$T = \frac{S}{V_a} \times 3,600 = \frac{1.4}{42} \times 3600 = 120 \text{ seconds.}$$

Using equation, $V_m = \dfrac{T}{2K} - \sqrt{\dfrac{T^2}{4K^2} - \dfrac{3600\,S}{K}}$

where $$K = \frac{1}{2\alpha} + \frac{1}{2\beta} = \frac{1}{3.4} + \frac{1}{6.6} = 0.4456$$

$\therefore$ Maximum speed,

$$V_m = \frac{120}{2 \times 0.4456} - \sqrt{\frac{(120)^2}{(2 \times 0.4456)^2} - \frac{3600 \times 1.4}{0.4456}}$$

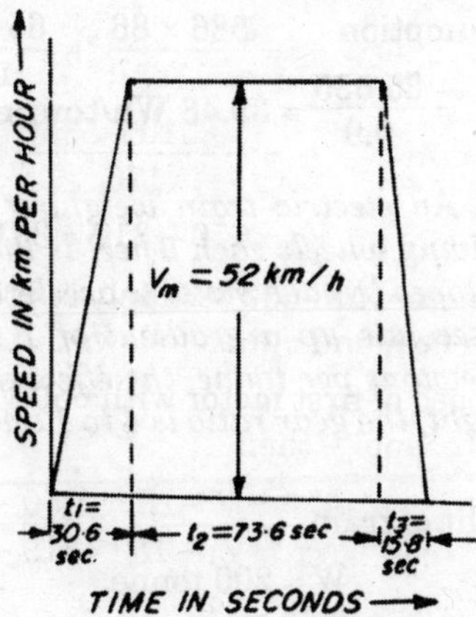

Fig. 8.79 Speed-Time Curve

= 52 kmph.

resistance to motion, $r = 50$ N/tonne

$$\frac{W_e}{W} = 1.1.$$

Acceleration period

$$t_1 = \frac{V_m}{\alpha} = \frac{52}{1.7} = 30.6 \text{ seconds.}$$

Braking period, $t_2 = \dfrac{V_m}{\beta} = \dfrac{52}{3.3} = 15.8$ sec.

Free running period

$$t_2 = T - (t_1 + t_3)$$

$$= 120 - (30.6 + 15.8) = 73.6 \text{ sec.}$$

Distance travelled during braking

$$= \frac{1}{2} \times V_m \times \frac{t_3}{3600} = \frac{1}{2} \times 52 \times \frac{15.8}{3600} = 0.114 \text{ km.}$$

$$D_1 = 1.4 - 0.114 = 1.286 \text{ km.}$$

Sp. energy output,

$$= \frac{0.01072 \times V_m^2}{D} \times \frac{W_e}{W} + 0.2778\, r\, \frac{D_1}{D}$$

$$= \frac{0.01072 \times 52^2}{1.4} \times 1.1 + 0.2778 \times 50 \times \frac{1.286}{1.4}$$

$$= 35.535 \text{ Wh/tonne-km.}$$

assuming efficiency of motors as 90%

Sp. energy consumption

$$= \frac{35.535}{0.9} = \textbf{39.48 Wh/tonne-km.} \qquad \textbf{Ans.}$$

Example 8.29. *An electric train weighing 200 tonne has eight motors geared to driving wheels each wheel is 90 cm diameter. Determine the torque developed by each motor to accelerate the train to a speed of 48 kmph in 30 seconds up a gradient of 1 in 200. The tractive resistance is of 50 newtons per tonne, the effect of rotational inertia is 10% of the train weight, the gear ratio is 4 to 1 and gearing efficiency is 80 per cent.*

Solution. Weight of train

$$W = 200 \text{ tonne.}$$

Diameter of driving wheels

$$D = 90 \text{ cm} = 0.9 \text{ m.}$$

Percentage gradient

$$G = \frac{1}{200} \times 100 = 0.5\%$$

Tractive resistance, $r = 50$ newtons/tonne

Gear ratio, $\gamma = 4$

Gearing efficiency, $\eta = 0.80$

Equivalent accelerating weight of the train.

$$W_e = 1.10W = 1.1 \times 200 = 220 \text{ tonne}$$

Maximum speed, $V_m = 48$ kmph.

Time taken in accelerating the train to a speed of 48 kmph.

$$t_1 = 30 \text{ seconds.}$$

Acceleration, $\alpha = \frac{V_m}{t_1} = \frac{48}{30} = 1.6$ kmphps.

Tractive effort required,

$$F_t = 277.8\, W_e\, \alpha + 98.1\, WG + W \,.\, r$$

$$= 277.8 \times 220 \times 1.6 + 98.1 \times 200 \times 0.5 + 200 \times 50$$

$$= 1,17,596\, N.$$

Total torque developed

$$T = \frac{F_t \times D}{\eta \times 2\gamma} = \frac{1,17,596 \times 0.9}{0.8 \times 2 \times 4} = 16,537 \text{ N-m.}$$

Torque developed by each motor

$$= \frac{16537}{8} = \textbf{2,067 N-m.} \qquad \textbf{Ans.}$$

Example 8.30. *A 250 tonne motor coach having 4 motors, each developing 5,000 Nm torque during acceleration, starts from rest. If up gradient is 25 in 1000, gear ratio 5, gear transmission efficiency 88%, wheel radius 44 cm, train resistance 50 N/tonne addition of rotational inertia 10%, calculate the time taken to reach a speed of 5 kmph.*

If the supply voltage were 1500 V d.c. and efficiency of motors 83.4%, determine the current drawn per motor during notching period.

Solution. Weight of train

$$W = 250 \text{ tonnes.}$$

Diameter of driving wheels

$$D = 2 \times 44 = 88 \text{ cm} = 0.88 \text{ m.}$$

Percentage gradient

$$G = \frac{25}{1000} \times 100 = 2.5\%$$

Tractive resistance, $r = 50$ N/tonne

Gear ratio, $\gamma = 5$

Gearing transmission efficiency,

$$\eta = 0.88$$

Equivalent accelerating weight of the train.

$$W_e = 1.1 \times 250 = 275 \text{ tonne.}$$

Total torque developed

$$T = 4 \times 5000 = 20{,}000 \text{ N-m.}$$

Tractive effort, $F_t = \eta \, T \dfrac{2\,\gamma}{D} = \dfrac{0.88 \times 20000 \times 2 \times 5}{0.88}$

$= 200000$ N ...(*i*)

Let the acceleration of the train be α, kmphps then

$$F_t = 227.8 \, W_e \, \alpha + 98.1 \, WG + W \, . \, r$$

$$= 227.8 \times 275 \, \alpha + 98.1 \times 250 \times 2.5 + 250 \times 50$$

$$= 6{,}395 \, \alpha + 73812.5 \quad \text{...(ii)}$$

Comparing expression (*i*) and (*ii*) we get,

$$76{,}395 \, \alpha + 73{,}812.5 = 2000000$$

or $$\alpha = \frac{126187.5}{76395} = 1.65 \text{ kmphps}$$

Time taken for the train to attain a speed of 45 kmph

$$t = \frac{V_m}{\alpha} = \frac{45}{1.65}$$

$$= \textbf{27.25 seconds.} \quad \textbf{Ans.}$$

Power output from driving axles

$$= \frac{F_t \times V}{3600} = \frac{200000 \times 45}{3600} = 2500 \text{ kW.}$$

Power input $= \frac{\text{Power output}}{\eta_{max}} = \frac{2500}{0.834} = 3000 \text{ kW.}$

Total current drawn $= \frac{3000 \times 1000}{1500} = 2000\ A.$

Current drawn per motor

$$= \frac{2000}{4} = 500 \text{ A.} \quad \textbf{Ans.}$$

Example 8.31. *An electric train accelerates uniformly from rest to a speed of 48 km/hour in 24 sec. It then coasts for 69 secs. against a constant resistance of 58 N/tonne and is braked to rest at 3.3 km/hour/second in 11 seconds. Calculate (i) the acceleration (ii) coasting retardation and (iii) the schedule speed, if the station stops are of 20 seconds duration. What would be the effect on schedule speed of reducing the station stops to 15 second duration, other conditions remaining the same? Allow 10% for the rotational inertia.*

Solution. Time of acceleration

$$t_1 = 24 \text{ seconds.}$$

Time of coasting, $t_2 = 69$ seconds.

Time of braking, $t_3 = 11$ seconds.

Maximum speed, $V_1 = 48$ seconds.

(*i*) Acceleration $\alpha = \frac{V_1}{t_1} = \frac{48}{24} = 2$ **kmphps. Ans.**

Let the coasting retardation β_c kmphps.

$$F_t = 277.8\ W_e(-\beta_c) + 98.1\ WG + W \,.\, r$$

$$= -277.8 \times 1.1\ W \beta_c + 98.1\ W + 58\ W$$

(*ii*) or
$$\beta_c = \frac{58}{277.8 \times 1.1}$$

$$= 0.19 \text{ kmphps.} \quad \textbf{Ans.}$$

Speed at the end of coasting period

$$V_2 = V_1 - \beta_c \times t_2 = 48 - 0.19 \times 69$$

$$= 34.9 \text{ kmph.}$$

Distance travelled, $S = \frac{V_1 t_1}{7200} + \frac{(V_1 + V_2)\, t_2}{7200} + \frac{V_2 t_3}{7200}$ km

$$= \frac{48 \times 24}{7200} + \frac{(48 + 34.9) \times 69}{7200} + \frac{34.9 \times 11}{7200}$$

$$= 1.00 \text{ km.}$$

Schedule Time, $T_s = t_1 + t_2 + t_3 +$ duration of stop

$= 24 + 69 + 11 + 20 = 124$ seconds.

(*iii*) Schedule speed

$$V_s = \frac{S \times 3600}{T_s} = \frac{1.00 \times 3600}{124}$$

$= 29$ kmph. **Ans.**

When the duration of stop is 15 seconds

Schedule time $T'_s = 24 + 69 + 11 + 15 = 119$ seconds.

and schedule speed,

$$V'_s = \frac{1.00 \times 3600}{119} = 30.25 \text{ kmph.} \quad \textbf{Ans.}$$

Example 8.32. *An electric train accelerates uniformly from rest to a speed of 50 km-ph in 25 seconds. If then coasts for 1 minute 10 seconds against a constant resistance of 70 N per tonne and is braked to rest at 4 kmphps in 10 seconds. Calculate the schedule speed, if the station stops are of 15 second duration.*

Solution. Time of acceleration

$t_1 = 25$ seconds.

Time of coasting, $t_2 = 70$ seconds.

Time of braking, $t_3 = 10$ seconds.

Maximum speed, $V_1 = 50$ kmph.

Let the coasting retardation be β_c km. ph. ps.

Assume $W_e = 1.1$ W

$$F_t = 277.8\, W_e(-\beta_c) + 98.1\, WG + W \,.\, r$$

$$= -277.8 \times 1.1\, W(-\beta_c) + 98.1\, W \times 0 + 70\, W$$

or

$$\beta_v = \frac{70}{277.8 \times 1.1} = 0.229 \text{ kmphps.}$$

Speed at the end of coasting period

$$V_2 = V_1 - \beta_c \times t_2$$

$= 50 - 0.229 \times 70 = 34$ kmph.

Distance travelled,

$$S = \frac{V_1\, t_1}{7200} + \frac{(V_1 + V_2)\, t_2}{7200} + \frac{V_2\, t_3}{7200}$$

$$= \frac{50 \times 25}{7200} + \frac{(50 + 34) \times 70}{7200} + \frac{34 \times 10}{7200}$$

$= 1.037$ km.

Schedule Time, $T_s = t_1 + t_2 + t_3 + \text{duration of stop}$

$= 25 + 70 + 10 + 15 = 120$ seconds.

Schedule speed $V_s = \dfrac{3600 \times S}{T_s}$

$= \dfrac{3600 \times 1.0375}{120} = \mathbf{31.125}$ **kmph. Ans.**

Example 8.33. *An electric train weighing 100 tonnes has a rotational inertia of 10%. This train while running between two stations which are 2.5 km apart has an average speed of 50 km/hour. The acceleration and retardation during braking are respectively 1 km/hour/second and 2 km per hour per second. The percentage gradient between these two stations is 1% and the train is to move up the incline. The track resistance is 40 N/tonne. If the combined efficiency of the electric train is 60%, determine (i) maximum power at driving axle (ii) total energy consumption and (iii) specific energy consumption. Assume that journey estimation is being made in simplified trapezoidal speed time curve.*

Solution. Weight of train,

$$W = 100 \text{ tonnes}$$

Effective weight of train,

$$W_e = \left(1 + \frac{10}{100}\right) W$$

$$= 1.1\, W = 1.1 \times 100 = 110 \text{ tonnes.}$$

Train resistance, $r = 40$ N/tonne

Percentage gradient

$$G_1 = 1\%$$

Acceleration $\alpha = 1$ kmphps.

Retardation, $\beta = 2$ kmphps.

Distance of run, $D = 2.5$ km.

Average speed, $V_a = 50$ kmph.

Duration of run $= \dfrac{3600 \times D}{V_a} = \dfrac{3600 \times 2.5}{50} = 180$ sec.

$$K = \frac{1}{2\,\alpha} + \frac{1}{2\,\beta} = \frac{1}{2} + \frac{1}{2 \times 2} = 0.75$$

Maximum speed

$$V_m = \frac{T}{2\,K} - \sqrt{\left(\frac{T}{2\,K}\right)^2 - \frac{3600\, D}{K}}$$

$$= \frac{180}{2 \times 0.75} - \sqrt{\left(\frac{180}{2 \times 0.75}\right)^2 - \frac{3600 \times 2.5}{0.75}}$$

$= 71$ **kmph.**

Acceleration period

$$t_1 = \frac{V_m}{\alpha} = \frac{71}{1} = 71 \text{ seconds.}$$

Braking period, $t_3 = \frac{V_m}{\beta} = \frac{71}{2} = 35.5$ **seconds.**

Time of free run $t_2 = 180 - (71 + 35.5) = 73.5$ **seconds.**

Tractive effort required

$$F_t = 277.8\, W_e\, \alpha + 98.1\, W \,.\, G + W \,.\, r$$

$$= 277.8 \times 110 \times 1 + 98.1 \times 100 \times 1 + 100 \times 40$$

$$= 44{,}368 \text{ N}$$

(*i*) **Maximum power at the driving axle**

$$= \frac{F_t \times V_m}{3600} = \frac{44{,}368 \times 71}{3600} = 875 \text{ kW.} \quad \textbf{Ans.}$$

$$\text{Total energy output} = \frac{1}{2} F_t \times \frac{V_m}{3600} \times \frac{t_1}{3600} + \frac{F_t' V_m}{3600} \times \frac{t_2}{3600}$$

where F_t' **is effort required during free run and is equal to (98.1 WG + Wr).**

i.e. $98.1 \times 100 \times 1 + 100 \times 40 = 13{,}810$ **N.**

$$\text{Total energy output} = \frac{1}{2} \times 44{,}368 \times \frac{71}{3600} \times \frac{71}{3600} + 13{,}810 \times \frac{71}{3600} \times \frac{73.5}{3600} = 14.19 \text{ kWh.}$$

(*ii*) **Total energy consumption**

$$= \frac{14.19}{0.6} = 23.65 \text{ kWh.} \quad \textbf{Ans.}$$

(*iii*) **Specific energy consumption**

$$= \frac{\text{Total energy consumption in watt hours}}{\text{Weight of train in tonnes} \times \text{Distance of run in km}}$$

$$= \frac{23.65 \times 1000}{100 \times 2.5} = \textbf{94.6 Wh/tonnes/km.} \quad \textbf{Ans.}$$

Example 8.34. *A 500 tonne goods train is to be hauled by a locomotive up a gradient of 1 in 40 with an acceleration of 1.5 kmphps. Determine the weight of the locomotive and number of axles, if axial load* *is not to exceed 24 tonnes. Co-efficient of adhesion is 0.3, track resistance 45 N / tonne and effective rotating masses 10% of dead weight.*

Solution. Let the weight of the locomotive be W tonnes.

Dead weight of train and locomotive

$$W = (500 + W_L) \text{ tonnes.}$$

Acceleration, $\alpha = 1.5$ kmphps.

Track resistance, $r = 45$ N/tonne.

Gradient, $G = \frac{1}{40} \times 100 = 2.5\%$

Co-efficient of adhesion,

$$\mu = 0.31$$

Effective weight of train and locomotive,

$$W_e = 1.1(500 + W_L)$$

Tractive effort, $F_t = 277.8\ W_e\ \alpha + 98.1\ WG + Wr$

$$= \left[277.8 \frac{W_e}{W} \alpha + 98.1 \times G + r\right] \times W$$

$$= [277.8 \times 1.1 \times 1.5 + 98.1 \times 2.5 + 45] \times W$$

$$= 748.62\ W \text{ Newtons}$$

$$= \frac{748.62\ W}{9.81 \times 1000} = 0.07631\ W \text{ tonnes}$$

$$= 0.07631(500 + W_L) \text{ tonnes} \qquad ...(i)$$

Maximum value of tractive effort that can be possible with W_L adhesive weight of locomotive

$$F_t = \mu\ W_L \qquad ...(ii)$$

$$= 0.31\ W_L$$

Comparing expression (*i*) and (*ii*) we get,

$$0.31\ W_L = 0.07631(500 + W_L)$$

or $$W_L = \frac{0.07631 \times 500}{0.31 - 0.07631} = 163.3 \text{ tonnes}$$

Number of axles required

$$= \frac{163.3}{24} = 7 \textbf{ Ans.}$$

Example 8.35. *Two d.c. series motors of a motor coach have resistance of 0.1 Ω each. These motors draw a current of 500 A from 600 V mains during series parallel starting period of 20 seconds. If the acceleration during starting period remain uniform, determine :*

(*i*) *Time during which motor operates in*

(a) *Series*

(b) *Parallel*

(ii) ***The speed at which the series connections are to be changed if the speed just after starting period is 70 kmph.***

Solution. No. of motors operating = 2

Line voltage, $V = 600\ V$

Current per motor, $I = 500\ A$

Motor resistance, $R = 0.1\ \Omega$

Maximum speed, $V_m = 70$ kmph

Back e.m.f. of each motor in full series position.

$$E_{b_{se}} = \frac{V}{2} - IR$$

$$= \frac{600}{2} - 500 \times 0.1$$

$$= 250\ V$$

Back e.m.f. of each motor in full parallel operation,

$$E_{b_p} = V - IR$$

$$= 600 - 500 \times 0.1$$

$$= 550\ V$$

Since motors takes 20 seconds to build up 550 V, therefore time taken to build up 250 V, assuming smooth acceleration and building up of e.m.f. at constant rate.

(i) Period of series operation

$$T_{se} = 20 \times \frac{250}{550}$$

$$= \mathbf{9.09\ sec.}\quad \textbf{Ans.}$$

Period of parallel operation,

$$T_{parallel} = T - T_{se}$$

$$= 20 - 9.09$$

$$= \mathbf{10.91\ sec.}\quad \textbf{Ans.}$$

(ii) Speed at which the series connections are to be changed

$$= \alpha\, T_{series} = \frac{V_m}{T} \cdot T_{series}$$

$$= \frac{70}{20} \times 9.09$$

$$= \mathbf{31.82\ kmph.}\quad \textbf{Ans.}$$

SHORT ANSWERS WITH QUESTIONS

Q.1. *What do you mean by Electric Traction ?*

Ans. The movement of trains, locomotives and vehicles by electrical energy is known as Electric Traction or Electric Traction means a locomotive, which is equipped with traction motors to get the driving force.

Q.2. *What is the value of voltage used as overhead supply for trains in India ?*

Ans. 25 KV.

Q.3. *Mention the most important requirements of the main line railway service.*

Ans. The most important requirements of the main line railway service are :

(*i*) High maximum speed

(*ii*) Minimum cost of overhead structure.

Q.4. *Name the system (D.C or A.C.) which is preferred for sub-urban services and road transport where stops are frequent and distances of coverage are short.*

Ans. D.C. System.

Q.5. *Name the system (A.C., or D.C.) which is preferred for main line railway service.*

Ans. Single phase (1– ϕ) A.C. System.

Q.6. *What do you mean by Schedule speed of a train ?*

Ans. Schedule speed of a train is the ratio of distance covered between two stops and total time of run including times of stop.

Q.7. *What are the factors which affect the schedule speed of a train?*

Ans. The following are the factors on which the schedule speed of train when running on a given service is affected :

(*a*) Maximum or Crest speed

(*b*) Time duration of stop

(*c*) Acceleration and retardation

Q.8. *Under what circumstances series-parallel method of speed-control of d.c. series traction motors adopted ?*

Ans. When motors employed are two, four six or more in numbers, in that case the series-parallel method of speed-control is adopted.

Q.9. *Name the different categories of Railway service.*

Ans. There are three types of railway services namely :

(*i*) **Urban or City Service :** In this type, the distance between two stops is very less (about 1 km.). It requires high rate of acceleration and retardation.

(*ii*) **Sub-Urban Service :** In this type, the distance between two stops is about 6-7 km. In this service also, high value of acceleration and retardation is required.

(*iii*) **Main Line Service :** In this type, the distance between two stops varies from 25 km to 45 km. The operating speeds are high. Acceleration and braking is not of so importance.

Q.10. *What are the different systems of Electric Traction ?*

Ans. There are two systems of Electric traction as follows :

(*i*) **Electric Drive :** In this system, traction motors receive electrical power from overhead lines.

(*ii*) **Diesel-electric Drive :** In this system of drive, the traction motors (D.C. Series motors) receive electrical power from D.C. Generator which is driven by diesel-engine

Q.11. *Why the accelerating weight of an electric train is greater than its dead weight ?*

Ans. The rotating parts of the electric train such as wheels and motors also accelerate in an angular direction with the linear acceleration of the train, and therefore, the tractive effort required is equal to the arithmetic sum of tractive effort required to have the angular acceleration of rotating parts and tractive effort required to have the linear acceleration. Hence the equivalent or accelerating weight of the train is taken higher than the dead weight.

Q.12. *What do you mean by the term adhesive weight ?*

Ans. Adhesive weight means the total weight to be carried on the driving wheels.

Q.13. *What do you mean by the term co-efficient of adhesion ?*

Ans. Co-efficient of adhesion means, as the ratio of tractive effort to slip the wheels and adhesive-weight. The co-efficient of adhesion vary with the increase in speed.

Q.14. *Why 3 - ϕ induction motors are not used in electric traction ?*

Ans. Because of the following reasons, the 3– ϕ induction motors are not used for electric traction :

(*i*) Drawing higher current at start

(*ii*) Developing low starting torque

(*iii*) Flat speed-torque characteristics

(*iv*) Constant speed operation

(*v*) Complicated speed control and

(*vi*) Complicated over head feeding system.

Q.15. *Why the single phase a.c. series motors are not suitable for urban and suburban services ?*

Ans. For urban and suburban services large starting torque is an essential feature to provide high acceleration and frequent starting is involved where as the single phase a.c. series motors develop low starting torque and poor power factor at start. Because of this reason single phase series motors are not used for urban & suburban services.

Q.16. *Name the current collection systems employed for electric locomotives, trolleys or tram ways.*

Ans. There are two systems of current collection :

(*i*) Overhead system and (*ii*) Conductor-rail system

Q.17. *What stands for TPC ?*

Ans. TPC stands for Traction Power Controller.

Q.18. *Why mechanical braking is not totally replaced by electrical braking?*

Ans. Due to the reason that electric braking can not hold the vehicle stationary.

Q.19. *Why the master controller is normally equipped with 'dead mans handle' ?*

Ans. In case the train driver faints due to some reason in that case the motor may be stopped.

Q.20. *Where Metro is operating and give the meaning of E.M.U. ?*

Ans. Metro is an overhead and underground electric rail service started in Delhi and E.M.U. stands for electric multiple unit.

Q.21. *How the electric supply is fed to d.c. series motors ?*

Ans. An electric locomotive takes electric supply of 25 kV, 50 Hz through a pentograph. The supply is stepped down and rectified in the locomotive and fed to D.C. series motors. The railway track acts as a return conductor.

1. Fill in the Blanks:

(*i*) For main line railways single phase ----- series motor is used.

(*ii*) The modern electrification of track is done for ----- system.

(*iii*) The other name for composite traction is ----- system.

(*iv*) The most used traction motor is ----- motor.

(*v*) The distance between two consecutive such stations using D.C. electrifications is ----- than those using single phase a.c. electrification.

(*vi*) ----- Drive is used for electric traction.

(*vii*) The system of traction involving the use of electricity is called the -----

(*viii*) For regeneration braking it is necessary for supply voltage to -----

(*ix*) The D.C. motor used for electric traction is -----

(*x*) In Kando system of track electrification, single phase A.C. is converted in to -----

(*xi*) D.C. system used for tramways has a voltage of about ----- volts.

(*a*) 750 (*b*) 1500

(*c*) 3000 (*d*) 2400

Answers

(*i*) compensated (*ii*) Composite

(*iii*) kundo (*iv*) D.C. series

(*v*) electric (*vi*) electric traction

(*vii*) Drop (*viii*) series motor

(x) Three phase (*x*) 750 V, D.C.

2. True or False Questions:

(*i*) Distance between stops divided by the actual time of run is called the schedule speed.

(*ii*) In case of main line service, free running and coasting periods are longer but acceleration and braking periods are smaller.

(*iii*) Gears are used in electric traction to increase the tractive effort at the driving wheel.

(*iv*) In a buck and boost method of control, the number of machines used are four.

(*v*) The rectifiers used in electric traction are based on the principle of silicon controlled rectifiers.

(*vi*) The sub-stations for Track electrification are only arranged for remote control operation.

(*vii*) Metadyne converter output is, constant current and constant voltage.

(*viii*) Series parallel control is widely used in electric traction.

(*ix*) Battery-electric drive is usually used for urban and sub-urban service.

(*x*) Cost of A.C. electrification of track is higher than that of d.c. electrification.

(*xi*) During starting, the train resistance increases linearly with speed.

(*xii*) The acceleration is constant during notching up.

Answers

(*i*) False	(*ii*) True	(*iii*) True
(*iv*) True	(*v*) True	(*vi*) True
(*vii*) False	(*viii*) True	(*ix*) False
(*x*) False	(*xi*) True	(*xii*) True.

3. Objective Type Questions:

(*i*) The current collector that can be used with various different speeds under all the wind conditions and stiffness of OHE is called the ....

(*a*) Messenger collector (*b*) Trolley collector

(*c*) Bow collector (*d*) Pantograph collector

(*ii*) In suburban trains, the traction motors are installed on :

(*a*) Locomotive only (*b*) Locomotive and coaches

(*c*) Coaches only

(*iii*) The type of d.c. motor used in electric traction is:

(*a*) series (*b*) shunt

(*c*) separately excited (*d*) None of the above

(*iv*) Two d.c. shunt motors having identical characteristics are used to drive a train car with unequal wheel diameters. The two motors share load equally if they are connected in :

(*a*) Series (*b*) Parallel

(*c*) The loading will always be unequal.

(*v*) A Locomotive with a mass of 50,000 kg on a track whose co-efficient of adhesion is 20 percent will produce a tractive effort of (approximately)

(*a*) 1 kN (*b*) 100 kN

(*c*) 25 kN (*d*) 250 kN

(*vi*) In underground traction, the supply system is :

(*a*) 500 V to 1,000 V D.C. (*b*) 25 kV, 50 Hz

(*c*) 50 kV, 50 Hz (*d*) 25 kV, 25 Hz

(*vii*) In suburban services as compared to urban services

(*a*) the coasting period is longer

(*b*) the coating period is smaller but free running period is longer.

(*c*) the coasting and free running periods are smaller

(*d*) None of the above is true

(*viii*) In India, Diesel locomotives for traction systems are manufactured at ....

(*a*) Calcutta (*b*) Bangalore

(*c*) Ajmer (*d*) Varanasi

(*ix*) Electric traction as compared to other different traction systems has the advantages of

(*a*) Better speed control

(*b*) Higher acceleration and braking retardation

(*c*) Very clean system so, ideally suitable for the underground railways

(*d*) All of the above

(*x*) Track electrification in Kando system

(*a*) single phase a.c. is converted into 3–ϕ, a.c.

(*b*) 1–ϕ a.c. is converted to d.c.

(*c*) 3–ϕ a.c. is converted into single phase a.c.

(*d*) 3–ϕ a.c. is converted into d.c.

(*xi*) In case of underground traction, the supply system is

(*a*) 25 kV, 25 Hz (*b*) 25 kV, 50 Hz

(*c*) 500 V to 1000 V D.C. (*d*) 50 kV, 50 Hz

(*xii*) Long distance railways operates on

(*a*) 25 kV, 3–ϕ, A.C. (*b*) 15 kV, 3 – ϕ, A.C.

(*c*) 25 kV, 1–ϕ, A.C. (*d*) 600 V D.C.

(*xiii*) The ultimate speed at which trains run on main line railway service is:

(*a*) 120 kmph (*b*) 200 kmph

(*c*) 110 kmph (*d*) 160 kmph

(*xiv*) Tractive effort desired for an electric locomotive can be increased by

(*a*) Using kW motors

(*b*) Increasing the supply voltage

(*c*) Increasing the dead weight on the driving axles.

(*d*) both (*a*) and (*c*)

(*xv*) Main traction systems used in India are of those using ---- locomotives.

(*a*) Electric (*b*) Steam engine

(*c*) Diesel engine (*d*) All of the above

Answers :

(*i*) (*d*)	(*ii*) (*a*)	(*iii*) (*a*)	(*iv*) (*a*)
(*v*) (*b*)	(*vi*) (*a*)	(*vii*) (*a*)	(*viii*) (*d*)
(*ix*) (*d*)	(*x*) (*a*)	(*xi*) (*c*)	(*xii*) (*c*)
(*xiii*) (*d*)	(*xiv*) (*d*)	(*xv*) (*d*)	

4. Fill in the blanks with appropriate words from the words given in the brackets.

(*i*) On main line railways more importance is given to ______.
(acceleration, breaking, speed)

(*ii*) The modern electrification of track is being done ________ system.
(D.C., A.C., composite)

(*iii*) Three phase a.c. system is employed for traction on ______ railway.
(Southern, Eastern, No where)

(*iv*) The system of traction employed for urban railways is _______ .
(d.c., a.c., composite)

(*v*) Textile industry requires _____ motor.
(d.c., shunt, schrage, compound)

(*vi*) For punches and shears ______ is used.
(d.c., compound, motor, series, shunt)

(*vii*) For small domestic appliances ______ motors are generally used.
(Universal, shunt, split phase induction)

(*viii*) The reversal of rotation of motors for electric braking is known as ______ (regenerative braking, plugging, dynamic braking)

5. Write 'Yes' or 'No' for the following sentences :

(*i*) Power factor of the system is improved by the group drive.

(*ii*) Initial cost in group drive is more.

(*iii*) In case of lifting loads, the torque required is independent of speed.

(*iv*) Torque varies as the square of the speed in case of blowers and centrifugal pumps.

(*v*) In crushing and deformation loads, torque is inversely proportional to speed.

(*vi*) Torque in a shunt motor is proportional to the armature current.

(*vii*) The starting torque of capacitor start induction motor is very high.

(*viii*) The starting torque of shaded pole motor is very poor.

3. A d.c. motor in which the r.p.m. are inversely proportional to the armature current, is a :

(*a*) D.C. series motor.

(*b*) D.C. shunt motor.

(*c*) Differentially Compounded D.C. motor.

(*d*) Cumulatively Compounded D.C. motor.

4. A drive having mixed properties of a d.c. shunt and D.C. series is :

(*a*) D.C. series motor.

(*b*) D.C. shunt motor.

(*c*) Differentially Compounded D.C. motor.

(*d*) Cumulatively Compounded D.C. motor.

5. A d.c. motor showing linear load torque characterisation is a :

(*a*) D.C. series motor.

(*b*) D.C. shunt motor.

(*c*) Differentially Compounded D.C. motor.

(*d*) Cumulatively Compounded D.C. motor.

6. A single large motor when used to drive various machines with individual control on either of the machines, is classified as :

(*a*) Group drive

(*b*) Individual drive

(*c*) Some other factors specify the system.

7. A drive that has a tendency to pick-up r.p.m. when load increases is :

(*a*) D.C. series motor.

(*b*) D.C. shunt motor.

(*c*) Differentially Compounded D.C. motor.

(*d*) Cumulatively Compounded D.C. motor.

8. A group of motor when used to drive various machines come under the so called :

(*a*) Group drive (*b*) Individual drive

(*c*) Some other factors specify the system.

9. A drive in which speed is inversely proportional to torque is :

(*a*) D.C. series motor.

(*b*) D.C. shunt motor.

(*c*) Differentially Compounded D.C. motor.

(*d*) Cumulatively Compounded D.C. motor.

10. During rheostatic braking of a D.C. series motor :

(*a*) It is connected across a resistor and disconnected from the mains
(*b*) Its field connections are reversed along with the step mentioned in (*a*)
(*c*) Only step (*b*) is taken but the supply remains connected.

11. With the provision of external control circuitry the torque is improved to a greater extent is the case of :
(*a*) Single cage induction motor (*b*) Double cage induction motor
(*c*) Slip ring induction motor
(*d*) All induction motors have poor starting torque.

12. Braking through plugging without any additional corrective measures is possible for :
(*a*) Three phase induction motor (*b*) D.C. series motor
(*c*) Synchronous motor (*d*) All as above.

13. The starting torque without involving external control, is higher with a :
(*a*) Single cage induction motor (*b*) Double cage induction motor
(*c*) Slip ring induction motor
(*d*) All induction motors have poor starting torque.

14. A D.C. motor in which variation of torque between Zero and T_{max} (permissible) fails to remarkably affect the drive r.p.m. is :
(*a*) D.C. series motor. (*b*) D.C. shunt motor.
(*c*) Differentially Compounded D.C. motor.
(*d*) Cumulatively Compounded D.C. motor.

15. A d.c. motor having armature current proportional to under root its torque within certain limits of the field flux is a :
(*a*) D.C. series motor. (*b*) D.C. shunt motor.
(*c*) Differentially Compounded D.C. motor.
(*d*) Cumulatively Compounded D.C. motor.

16. Provision of damper winding in case of a synchronous motor is done so that it :
(*a*) Needs lesser exciting current
(*b*) Can be used as an alternator, which is self excited
(*c*) Needs no extra drive to bring it to initial synchronous r.p.m.
(*d*) Can be brought to a half through plugging.

17. In the composite system of traction supply :
(*a*) The single phase supply is converted to three phase within the locomotive.
(*b*) The single phase supply converted to D.C. within the locomotive.
(*c*) The two methods as above are adopted simultaneously.

(*d*) The two methods are individually adopted depending upon the type of drive.

18. Out of the given set of characteristics the more efficient electrical characteristics for a traction motor are :

(*a*) Running characteristics. (*b*) Starting characteristics.

(*c*) Speed control. (*d*) Braking characteristics.

Answers

1. (*i*) Speed (*ii*) Composite (*iii*) No. where (*iv*) d.c. (*v*) Schrage (*vi*) d.c. Compound (*vii*) Universal (*viii*) plugging.

2. (*i*) No (*ii*) No (*iii*) Yes (*iv*) Yes (*v*) Yes (*vi*) Yes (*vii*) Yes (*viii*) Yes.

3. (*a*) **4.** (*d*) **5.** (*b*) **6.** (*a*) **7.** (*c*) **8.** (*b*)

9. (*a*) **10.** (*b*) and (*c*) **11.** (*b*) **12.** (*a*) **13.** (*b*)

14. (*b*) **15.** (*a*) **16.** (*d*), (*c*) **17.** (*d*) **18.** (*b*).

EXERCISES

1. State the main features for an ideal traction system.
2. State the advantages of electric drive in case of traction system.
3. What are the advantages of electric traction over other forms of traction with special reference to the suburban traffic ? What are technical aspects of railway electrification ?
4. State the merits and/or demerits of electric traction with reference to Electric Power Crisis in India.

   Do you advise completely independent power supply for electric traction ? Justify.
5. Discuss the various systems of track electrification :
6. Write short notes on the following :

(*i*) Traction system in India.

(*ii*) Diesel electric traction.

(*iii*) Battery locomotive.

(*iv*) D.C. *versus* A.C. traction.

(*v*) A.C. and D.C. systems of electric traction.

(*vi*) Single-phase traction at 50 Hz.

(*vii*) Single-phase low frequency A.C. system of electrification.

7. What do you understand by speed-time curves ? What is its use in practice ?
8. Define crest speed, average speed and schedule speed and discuss the factors which effect the schedule speed of a train.
9. Explain clearly "Free running", 'Coasting' and 'Braking' with reference to electric traction system.

10. Draw the speed time curve for an electric train accelerating uniformly to a speed V_m. The power is cut off and after a time t when it attains V_m, coasting is allowed to continue up to the end of the run, the braking being neglected. Drive a suitable equation for V_m.

11. Explain how an actual speed-time curve for an electric train service can be replaced by a curve having simple geometric shape deduce, from first principles, the relationship between acceleration, maximum speed, running time and distance between stops assuming a simplified speed-time curve.

12. Prove that the crest speed of main line service with trapezoidal speed-time curve can be obtained by the formula.

$$V_m = \frac{T}{K} - \sqrt{\frac{T^2}{K^2} - \frac{7200\,S}{K}}$$

where $$K = \frac{1}{\alpha} + \frac{1}{\beta}$$

13. Derive an expression for the tractive effort.

14. Explain the terms, "Dead weight", "Effective weight" and "Adhesive weight" in a locomotive.

15. For a typical trapezoidal speed-time curve, show how you would calculate the energy consumed during two consecutive stops. Use either graphical or semi-graphical method.

16. Explain the terms "Specific energy output" and specific energy consumption.

17. Explain the terms Adhesive weight and co-efficient of adhesion.

18. Comment on the following statements :

(*i*) The adhesion co-efficient of a locomotive employing D.C. shunt motor will be higher than the using D.C. series motor.

(*ii*) A higher ratio of adhesive weigh to total weight is desirable for suburban trains, where the distance between the consecutive stops is very small.

19. Write short notes on the following :

(*i*) General feature of traction motors.

(*ii*) Suitability of D.C. series motor for traction service.

(*iii*) A.C. series motor.

(*iv*) Three-phase induction motor.

(*v*) Selection of Traction motors.

20. Explain briefly how regenerative braking differs from other methods of braking traction motors. With neat and explanatory notes, demonstrate the usefulness of this method of braking with (*a*) D.C. series motors (*b*) A.C. motors in traction.

21. Write short notes on the following :

(*i*) Electric braking.

(*ii*) Dynamic braking.

(*iii*) Regenerative braking used in traction.

22. Describe the main difficulties encountered in regenerative braking of D.C. series motor and describe any one scheme for over coming these difficulties.

23. Describe the power circuit of the driving motor for 25 kW, 1-phase, 50 Hz a.c. traction.

24. Write brief notes on the following :

(*a*) Different methods of electric braking in a traction system.

(*b*) Suitability of D.C. and A.C. motors for traction and discuss their performance characteristics.

(*c*) Discuss the relative advantages of steam, electric and diesel electric traction now existing in India.

25. What do you understand by tractive effort ? Derive an expression for the same.

26. Discuss in details why series motors are ideal for D.C. or A.C. traction.

27. Draw the speed-time curves of electric trains for (*i*) Main line service, and (*ii*) Suburban service and explain their characteristics.

28. State the factors that effect specified energy consumption and their influence on it.

29. State the advantages and disadvantages of electric braking.

30. Discuss regenerative braking as applied to

(*i*) Separately excited D.C. motors

(*ii*) D.C. shunt motors and

(*iii*) D.C. series motors.

PROBLEMS

1. A train runs at an average speed 45 km per hour between stations 2.5 km apart. The train accelerates at 2 k.m.p.h.p.s. and retards at 3 k.m.p.h.p.s. Find its maximum speed assuming a trapezoidal speed time curve. Calculate also the distance travelled by it before the brakes are applied.

[**Ans.** 50.268 km.p.h., 2.383 km]

2. The schedule speed with a 200 tonne train on an electric railway with stations 777 metres about is 27.3 km. per hour and the maximum speed is 20 per cent higher than the average running speed. The braking rate is 3.22 km.p.h.p.s. and the duration of stop is 20 seconds. Find the acceleration required. Assume a simplified speed-time curve with free running at the maximum speed.

[**Ans.** $\alpha = 2.73$ km.p.h.p.s.]

3. A suburban electric train has a maximum speed of 65 km.p.h. The schedule speed including a station stop of 30 seconds is 43.5 km.p.h.

If the acceleration is 1.3 km.p.h.p.s. Find the value of retardation when the average distance between stops is 3 km.

[**Ans.** β = 1.21 km.p.h.p.s.]

4. An electric train is accelerated uniformly from rest to a speed of 40 km.p.h. The period of acceleration being 20 seconds. If it coasts for 60 seconds against a constant resistance of 50 N-Tonne and is brought to rest in a further period of 10 seconds by braking, determine :

(*i*) The acceleration.

(*ii*) The coasting retardation.

(*iii*) The braking retardation.

(*iv*) Distance travelled.

(*v*) Schedule speed with station stops of 10 seconds duration. Allow 10 per cent for rotational inertia.

[**Ans.** α = 2 km.p.h.p.s., β_c = 0.1636 km.p.h.p.s. ;
β = 3 km.p.h.p.s., S = 0.736 km. V_s = 27.5 km.p.h]

5. The speed-time curve of an electric train on a uniform rising gradient of 1 in 100 comprises :

(*i*) Uniform acceleration from rest at 2 km.p.h.p.s. for 30 seconds.

(*ii*) Coasting with power off for 70 seconds.

(*iii*) Braking at 3 km.p.h.p.s. to a standstill.

The weight of the train is 250 tonnes, the train resistance on level track being 5 km/tonne, and allowance for rotary inertia 1%.
Calculate the maximum power developed by traction motors and total distance travelled by the train. Assume transmission efficiency as 97%.

[**Ans.** 3,258 kW, 1.12 km]

6. 400 tonnes goods train is to be hauled by a locomotive up a gradient of 2% with acceleration of 1 km.p.h.p.s. Co-efficient of adhesion is 20%, track resistance 40 N/tonne and effective rotating masses 10% of the dead weight. Find the weight of the locomotive and number of axles if the axle load is not increased beyond 22 tonnes.

[**Ans.** 152.6 Tonnes, 7]

7. A 500 tonnes goods train is to be hauled by a locomotive up a gradient of 20% with an acceleration of 1.2 km.p.h.p.s. Co-efficient of adhesion is 25%, track resistance 40 N/tonne and effective rotating masses 10% of dead weight. Find the weight of the locomotive and number of axles if load is not to exceed 20 tonnes.

[**Ans.** 160 tonnes, 8]

8. Determine the maximum adhesive weight of a loco required to start a 2340 tonne (inclusive of loco) on 1 : 150 gradient and accelerate it at 0.1 km.p.h.p.s. Assume co-efficient of adhesion as 0.25, train resistance 4 kg/tonne and rotary inertia as 8%.

[**Ans.** 128.5 tonnes]

9. A train weighing 400 tonnes has speed reduced by regenerative braking from 40 to 20 km.p.h. over a distance of 2 km, on a down gradient of 20%. Calculate the electrical energy and over-power returned to the line. Tractive resistance is 40 Newtons/tonne and allowed rotational inertia of 10% and efficiency of conversion 75%.

[Ans. 324 kWh., 4860 kW]

10. Find the time taken by a 200 tonnes motor coach train having 4 motors each developing a shaft torque of 5000 Newton-metres during the accelerating period to attain a speed of 50 km/hr. Starting from rest on gradient of 1 in 300. The gear ratio of the motor is 3.5 and the gear efficiency 95%. Wheel diameter is 1 metre. The specific resistance of the train is 47 Newtons per tonne and the accelerating weight of the train is 10% higher than its dead weight.

[Ans. t_1 = 26.7 seconds]

11. An electric train accelerates uniformly from rest to a maximum speed of 50 km/hr in 25 sec. After this it coasts for 70 seconds against a constant resistance of 50 Newtons/tonne and is stopped in 10 seconds by producing a retardation of 3.5 km.p.h.p.s. with the help of brake. Find

(*i*) Acceleration.

(*ii*) The coasting retardation.

(*iii*) The schedule speed if the time of station stops is 15 sec. Accelerating weight of the train is 10% higher than its dead weight.

[Ans. 2 km.p.h.p.s., 0.1635 km.p.h.p.s., 39.228 km.p.h.]

12. Ore carrying trains weighing 5000 tonne each are to be hauled down a gradient of 1 in 66 at a maximum speed of 40 km/hr and started on a level track at an acceleration of 0.1 m/sec. How many locomotives, each weighing 75 tonnes, will have to be employed ?

Train resistance during starting = 3 kg/tonne.

Train resistance at 40 km.p.h = 4.7 kg/tonne.

Co-efficient of adhesion = 1/3

Rotational inertia = 1/10

[Ans. 3 Loco]

13. An electric train has quadrilateral speed time curve as follows.

(*i*) Uniform acceleration from rest at 2 km.p.h.p.s. for 30 seconds.

(*ii*) Coasting for 50 seconds.

(*iii*) Duration of braking 20 seconds.

If train is moving a uniform up gradient of 1%. Tractive resistance is 40 N/tonne, rotational inertia effect 10% of dead weight, duration of station stop 15 seconds and overall efficiency of transmission gear and motor as 75%. Find the schedule speed and specific energy consumption of run.

[Ans. V_s = 32.4 km.p.h., specific energy consumption = 67.4 Watt-hr/tonne-km]

14. A train runs with average speed of 40 km.p.h. Distance between stations is 2 km. Values of acceleration and retardation are 1.5 km.p.h.p.s. and 2.5 km.p.h.p.s. respectively. Find the maximum speed of train assuming trapezoidal speed line curve. **[Ans.** 46.5 km.p.h.]

15. A train with an electric locomotive weighing 300 tonnes is to be accelerated up a gradient of 1 in 100 at an acceleration of 1 km.p.h.p.s. If the train resistance, co-efficient of adhesion and effect of the rotational inertia are 75 Nw per tonne, 0.3 and 15% of the dead weight respectively, determine the minimum adhesive weight of the locomotive. **[Ans.** 67.73 tonnes]

16. The relationship between the current and torque of a series motor determined by a static test, is as follows :

Current (amperes)	10	20	30	40	50	60	70	80
Torque (N-m)	45	130	231	350	470	612	768	911

Deduce the speed-current curve of the motor when supplied at a constant voltage of 500 volts, having given the resistance of main circuit of motor 0.5 ohm.

17. A 230 V d.c. shunt motor has an armature resistance of 0.06 ohm and with full field has an e.m.f. of 225 V at a speed of 1500 r.p.m. It is coupled to an over hauled load with a torque of 225 N-m. Determine the lowest speed at which the motor can hold the load by regenerative braking.

18. An electric train weighing 200 tonnes accelerates uniformly on the level to a speed of 43 km/hr. in 20 sec. Power is then cut-off and the train coasts until the speed is 34 km.p.h. When the brakes are applied and the train is brought to rest in 10 sec. Allowing a train resistance of 5 kg per tonne during acceleration and 6.4 kg per tonne during coasting, and 12% for rotational inertia, calculate :

(*i*) The maximum horse power developed at the driving axles.

(*ii*) The specific energy output.

Questions Bank with Answer

Short Answer Type Questions.

Q.1. What is Traction?

Ans. The act of drawing or state of being drawn or propulsion of a vehicle is called traction.

Q.2. What is electric traction?

Ans. A system of running a vehicle by means of electrical energy is called electric traction

Q.3. Define scheduled speed.

Ans. The ratio of distance covered between two stops and the total time of run including time of stop is called scheduled speed.

$$\text{Scheduled speed} = \frac{\text{Distance between stops}}{\text{Actual time of run + time of stop}}$$

Q.4. Which type of motor is commonly used in electric traction? ***[June 2008]***

Ans. (i) DC series motor

(ii) Slip ring induction motor

Q.5. In railways more importance is given to .... acceleration/speed/braking.

Ans. Braking.

Q.6. Which type of braking is commonly used in electric traction?

Ans. Regenerative braking.

Q.7. Motor recommended for locomotive drive is .... DC series/shunt/compound.

Ans. DC series.

Q.8. Why DC series motors are preferred for electric traction?

Ans. Main requirements of electric traction motor are high starting torque and good speed control. Both these features are satisfied by DC series motors. Therefore DC series motors are preferred for electric traction.

Q.9. What are the specifications of electrical supply used for electric traction?

Ans. (i) DC system - 300V to 300V.

(ii) Single phase to direct current system - 25 KV, 50 Hz.

Q.10. What is a pentagraph?

Ans. Pentagraph is employed in electrical locomotives for collection of current from over head lines. It is mounted on the top of the vehicle and carries a sliding shoe for contact with the over head wire.

Q.11. What is EMU?

Ans. It stands for electric multiple unit. It is an electric train in which each coach has its own prime mover and passagers.

Q.12. What is metro.

Ans. It is railways running in metropolitan cities.

Q.13. In underground traction system, the supply is ....

Ans. 500V to 1000 VDC.

Q.14. In kando system of track electrification single phase supply is converted into ....

Ans. Three phase.

Q.15. What is the value of voltage used in overhead supply for trains in India?

Ans. 25 KV, single Phase 50 Hz.

Medium/Long Answer Type Questions

Q.1. What are the main advantages and disadvantages of electrical traction? Explain braking used in traction.

Ans. Following are the advantages of electrical traction.

(i) It is clean and pollution free system.

(ii) No separate arrangement is required for fans and lights inside the train.

(iii) Better speed control and braking.

(iv) Needs less maintenance.

(v) It is economical and efficient system. Its efficiency is 40% as compared to 8% in steam locomotive.

(vi) Less weight to horse power rating.

(vii) Less fire hazards

(viii) Requirement of less terminal space.

(ix) Electrical locomotive can be started immediately whereas steam locomotive requires at least two hours starting time.

(x) Less wear and tear of track

Disadvantages:

(i) The system can be used only for electrified routes.

(ii) It involves heavy initiail expenditure

(iii) Power break down for a short period can paralyse the whole system.

(iv) Interferes with communication lines.

(v) Additional equipment required for regeneration.

Braking used in traction is regenerative braking.

Q.2. Draw the block diagram of electrical locomotive and describe various equipment and accessories used.

Ans. The block diagram of an A.C Locomotive is shown in the fig. 8.80. It collects current from an over head line through a sliding contact by means of a pantagraph and supplies to the traction motors. An air circuit breaker is also provided on the roof of the locomotive. The main functions of various coomponents of an a.c locomotive system are as under.

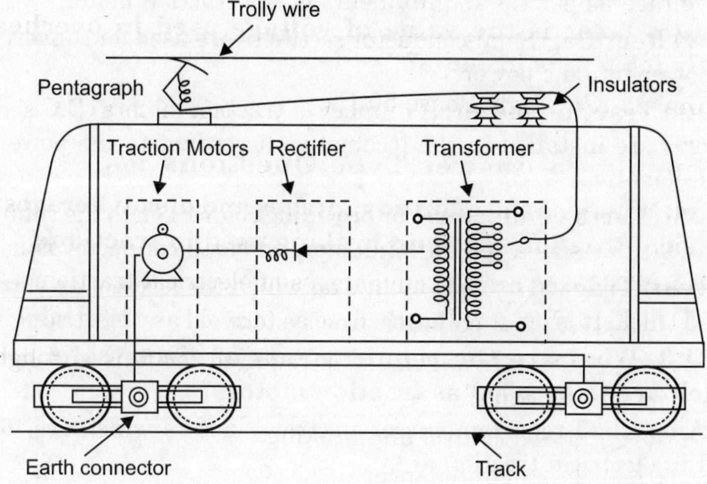

Fig. 8.80

(i) Trolly wire: This is an over head wire materials used for trolly wire made of hard drawn copper or alloys of copper. This wire is suspended with another over head wire called catenary or messenger wire as shown in fig. 8.81 (a) A high voltage of about 25 k V is supplied to the catenary The cross-section of the trolly wire is circular grooved as shown in fig. 8.81(b).

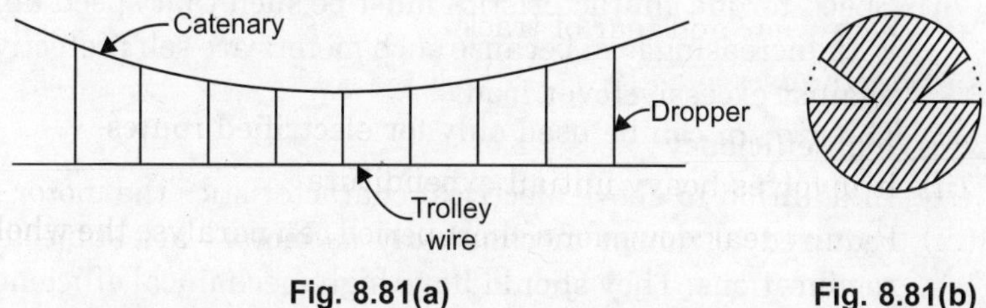

Fig. 8.81(a) **Fig. 8.81(b)**

(ii) Pentagraph: This is employed in railways for collection of current from trolly wires. Pentagraph is mounted on the top of the vehicle and carries a sliding shoe for contact with the overhead trolly wire. The pentagraph is raised or lowered from the driver's cabin by air pressure.

(iii) Insulators: Pin type porcelain insulators are also installed on the roof of the locomotive for carrying the incoming supply mains up to the transformer

(iv) Transformer: A tap changing transformer is also installed in the locomotive. The function of the transformer is to step down the voltage as per the requirements of the traction motor.

(v) Rectifier: Semi-conductor rectifiers are used for converting a.c power into d.c power.

(vi) Traction motor: A number of traction motors (D.C series motors) are installed in the locomotive according to the power of the locomotive.

(vii) Earth connector: For providing connection to the running rail, which acts as a return conductor, an earth connector is used.

(viii) Track: The railway line on which the locomotive runs is called track. It is properly earthed, so as to avoid any electric shock.

Q 3. What are the requirements of electrical motors which are to be used as traction motors?

Ans. Chief requirements of electrical motors which are to be used in electrical traction are,

(i) High starting torque
(ii) Simple speed control
(iii) Possibility of regenerative braking
(iv) Capability of withstanding voltage fluctuations
(v) Overload capacity
(vi) Possibility of parallel running as electrical locomotive normally requires 4 to 6 motors.
(vii) Speed torque characteristics must be such that speed falls with increasing load because such motors are self protective against excessive over load.
(viii) high efficiency
(ix) In addition to above electrical characteristics the motor is required to robust mechanically as motors are subjected to vibrations. They should have high mechanical efficiency and low torque to weight ratio.

Q 4. Describe the function of a pentagraph with neat sketch.

Ans. The pentagraph is employed in electric traction for collection of current from trolly wires. It is mounted at the top of a vehicle and carries sliding shoes for contact with the over head trolly wire as shown in fig. 8.82.

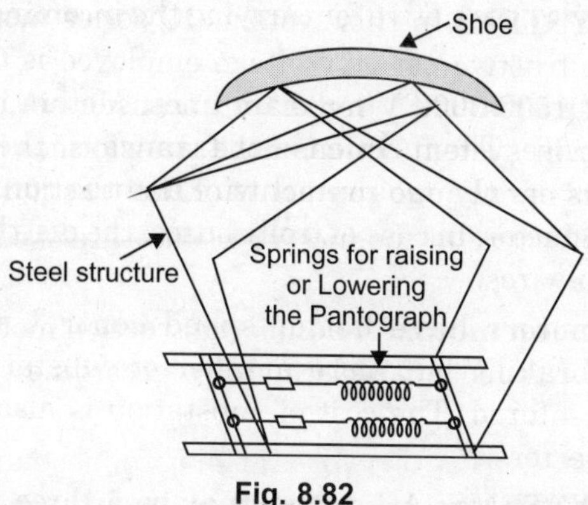

Fig. 8.82

It is made of steel. It can be raised or lowered with the help of springs from driver's cabin. Pentograph is usually used on main lines at speeds upto 150 km. It can operate in either direction of train and there is no risks of leaving contact of the over head conductor.

Q.5. Define scheduled speed and state the factor affecting scheduled speed of train.

Ans. Scheduled speed : The ratio of distance covered between two stops and the total time of run (including time of stop) is called schedule speed of the vehicle.

$$\text{Scheduled speed} = \frac{\text{Distance between two stops}}{\text{Actual time of run} + \text{time of stop}}$$

Factor affecting scheduled speed : Scheduled speed of a vehicle depends upon the following factors.

(i) Acceleration and Retardation: The increase of acceleration will result in decrease of actual time of run and therefore increase in schedule speed. Similarly increase in the retardat in during braking will affect the schedule speed.

(ii) Maximum speed: For a constant distance run and with fix acceleration and retardation actual time of run will decrease and therefore schedule speed will increase with increase in maximum speed.

(iii) Duraton of stop: For a given average speed the schedule speed will increase by reducing the duration of stops.

Q 6. What are different systems of track electrification? Which system of track electrification is adopted in India?

Ans. Various systems of track electrification are

(i) DC system (ii) AC system (iii) Composite system.

(I) DC SYSTEM: In this system DC series motors are used for getting the tractive power. Voltage employed is 600-750 V for tramways and 1500-3000 V for main lines. Motors receive power from distribution system. In case of tramways, the distribution system consists of only one contact wire and the running rail acts as a return conductor. In case of trolly buses, the distrbution system consists of two wires.

The d.c motors have better speed control and efficient regenerative braking but more number of substations at short distance are required. The cost of substation is also more due to a.c to d.c converters.

(ii) AC SYSTEMS: AC system may be a three phase system or single phase system.

(a) Single phase a.c system: In this system, single phase ac series motors are used. The voltage employed is 15000 to 25000V at $20\frac{1}{3}$ to 25 Hz. This voltage is stepped down to 400 V by a step down transformer in the vehicle itself.

This system is much used at main lines where stops are not so frequent.

(b) Three phase AC system: In this system, three phase induction motors at 3000V and $20\frac{1}{3}$ Hz are used. The induction motors are simple robust and have high efficiency of operation. They have a property of automatic regenerative braking without any additional equipment. The distribution system consists of two overhead wires and railway track as the third phase.

This system is not much used except in some hilly areas, where output power required is high and regeneration at large scale is possible.

(iii) Composite system: This utilises the merits of all the above three systems. There are two composite systems.

(a) Single to three phase or Kando system: In this system, single phase supply at 15000 V- 50 Hz is converted into three phase 400 V-10 Hz by using silicon rectifiers and given to three phase induction motors The induction motor develops high starting torque at low frequency with an easy speed control.

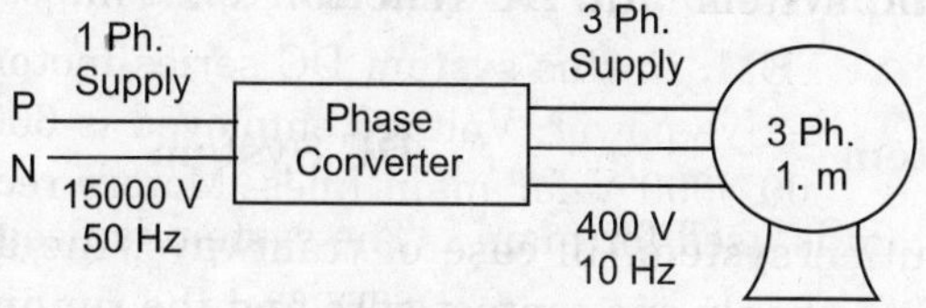

Fig. 8.83

(b) Single phase to Direct current system : In this system 25 KV-50Hz. 1 ph supply is used for overhead distribution. This is stepped down in the locomotive by a transformer and then converted into dc supply which runs dc series motor. In this way, high efficiency of ac distribution and desirable characteristics of dc series motors are combined together.

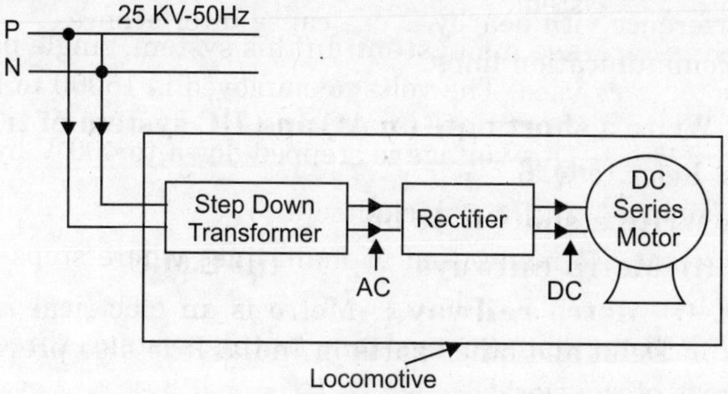

Fig. 8.84

Single phase dc system is adopted in India and in many countries. This system has the following advantages .

(i) Line current is reduced due to High ac voltage, this results in the reduction of cross section of the conductors and size of poles. The steel required in the track is half of that required in the dc system,

(ii) On account of high voltage, no. of substations are reduced. Moreover the ac stations are simple and cheaper than the dc stations,

(iii) The ac locomotive has greater co-efficeint of adhesion as compared to dc locomotive. This increases the hauling (pulling capacity) of ac locomotives.

(iv) The dc series motors have ideal traction characteristics.

(v) The ac locomotive has higher starting efficiency than dc locomotives.

Q 7. Compare AC and DC traction system.

Ans.

AC System	**DC System**
1. The system is used for main line service.	The system is used for urban service.
2. Capital cost is low	Capital cost is high
3. The cost of distribution system is low.	The distribution system is costly
4. The maintenance cost is high.	The maintenance cost is low.
5. The ac vehicle is heavy	The dc vehicle is lighter
6. The ac system causes interference vith nearby telecommunication lines.	The dc system does not cause interference.

Q.8. Write a short note on AC and DC system of traction.

Ans. Refer to Q. 6

Q 9. Write a short note on

(i) Metro railways **(ii) E.M.U**

Ans. (i) Metro railway : Metro is an electrical railways running in Delhi and other parts in India. It is also prevalent in many parts of the world.

This is a solution to the tremendous road traffic in Delhi and other metropolitan cities. To face this problem, Govt. of India and govt. of Delhi set up a company which was named as "Delhi Metro Rail Corporation Ltd."

This is to cover the city in three level, ground level, elevated track and underground track, covering urban and sub urban area of Delhi.

Metro coaches are luxury coached with all modern facilities. They have "self opening and closing" doors. Railway coach factory, Kapurthala (Pb) is manufacturing these coaches.

(ii) E.M.U: This is also an electric train and stands for "Electric Multiple unit". It is called "Multiple" as each of its coach contains its own prime mover along with its passengers. Traditionally coaches do not carry prime mover but only passengers and a separate locomotive carrying prime mover pulls them.

Usually dc series motors are used to provide the necessary tractive force. The supply used is 25 AV ac, which is stepped down, rectified in the coach and is given to the motors. Regenerative braking is employed.

The E.M.U s are running successfully in Mumbai, Kolkata and other cities in India.

Q.10. Write a short note on electric traction.

Ans. Electric traction means the locomotive which is equipped with d.c series motor or single phase a.c series motor or a three phase induction motor. The electrical energy required for driving electric motor is drawn from over head dsitribution system which is fed at suitable places either from a central power station or a substation.

AC and DC system are both available for electric traction but single phase to direct current system is mostly used in india and many countries of the world due to its number of advantages over other systems. Electric current is collected by means of Pantograph installed at the top of locomotive. As the supply voltage is 25 KV, it is first stepped down by means of a transformer and then rectified to run DC series motors for running the locomotive. Usually four or six motors in parallel are employed to provide the required tractive force.

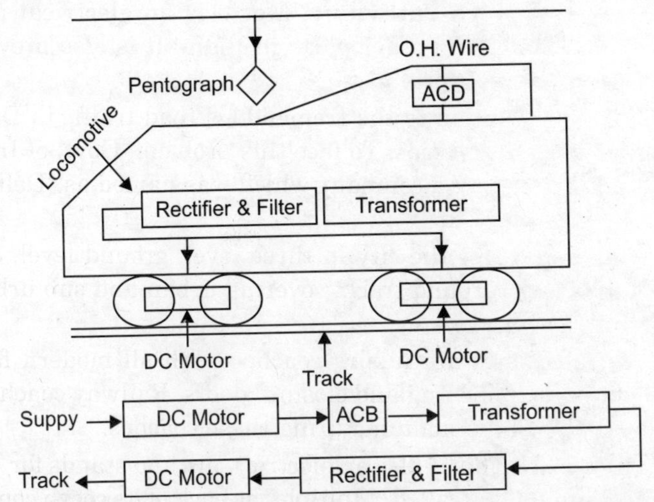

Fig. 8.85 Shows the block diagram of an electrical locomotive.

Electrical traction has a number of advantages over other system of traction. Some of them are of high efficiency, pollution free, better speed control, less maintance required, less vibrations, superior braking, instant operation.

Few disadvantages of electric traction are

(i) High initial cost

(ii) Interference with communication lines and possibility of supply failure which may paralyse the whole system.

9 Economic Aspects

9.1 Economic Aspects of Utilising Electric Energy

Financial planning and economics plays an important role in all the engineering works. In most of the cases the cost decides whether a certain project will be viable *or* not, although political and other considerations may sometimes intervene. The maximum utilization of advantageous tariff rates should be made. The design and construction work of electrical power generating plants and other systems is undertaken for the purpose of generating electrical energy to be sold at profit, so every effort is made to produce electrical energy as cheap as possible. The problem of determining the cost of any scheme is often difficult because the cost varies considerably with time and tariff conditions. The schemes involving the use of aluminium, which used to be considered very expensive a century ago, are now considered to be cheap due to lower cost. The electrical equipment should be grouped, installed and utilized in such a way so that the maximum demand may not rise haphazardly thereby levy of penal tariff by the utility company *or* Electricity Board. Also the sizes of the cables carrying current should be adequate. The capacitors for improvement of power factor should be installed so that economic utilization of the equipment and cables be made. The electric energy may be used in off peak load periods thereby availing concessional rates of tariff levied by the Electricity Board. Economic problems occur in the fields of generation, transmission, distribution and utilization of electrical energy. Thus the electrical engineer has to adopt the cheapest and most convenient scheme having study of the relative costs of the possible schemes.

9.2 Costing of Electric Energy

In general the cost of generating electrical energy can be divided into the following:

(*i*) **Fixed charges.** We know the rating of plant in a generating *or* receiving station must be equal to the simultaneous maximum demand which it has to meet (assuming no spares). This means that a fixed sum proportional to the demand to be met has to be invested for the plant required. Thus fixed charges are calculated to include interest on cost of plant, buildings, transmission and distribution expenses, part of staff salaries etc. These charges are independent of maximum demand and energy output.

(*ii*) **Semi-fixed charges.** These charges are dependent upon the installed capacity of the plant and are independent of the electrical energy output of the plant. These charges include :

(*a*) Interest and depreciation of the capital cost of the generating plant, transmission and distribution network, building and other civil engg. works etc. Capital cost of the plant also include the interest paid during the construction of the plant. Salaries of engineers employed on the design, development and construction of the power station. It also includes cost incurred on account of transportation labour octroi etc. to bring the equipment on site and commissioning the same. In case of nuclear power station capital cost of station also included the cost of initial charge of nuclear fuel minus the salvage value of fuel at the end of its useful life.

(*b*) All type of taxes, insurance premium paid on policies to cover risks of accidental break down, fire in the factory and yearly compensation given to the workers is also included.

(*c*) Salaries of field staff which depend upon the area occupied by generating plant or installed capacity of plant or the area of power distribution.

(*d*) Rent paid for land actually utilised for construction of plant.

(*iii*) **Operating costs or running charges.** The cost depends upon the energy output. The running cost of the plant is due to the annual cost of fuel, lubricating oil, water, maintenance and repair charges, and wages and salaries of operating staff.

In case of thermal plant, cost of feed water for the boiler including the cost of water softening and water conditioning and cooling water required by the steam condenser.

The running *or* operating cost is approximately proportional to the units generated.

While deciding any scheme for any given service, the choice must be such that the total operating cost, *i.e.*, the sum of fixed, semi-fixed and running charges should be minimum.

Total annual cost incurred in the power station is represented by the expression:

$$E = a + b \text{ kW} + c \text{ } k\text{Wh}$$

where a, b and c are constants. Fixed and semi-fixed cost being independent of the amount of energy generated is also called **"Standing Cost"**. Typical values of constants b and c in the above expression for various types of power plants are given below :

Types of Power Plant	Value of b	Value of c
Diesel Power Plant	₹ 200-280/kW	₹ 1.50 paise per unit or kWh
Steam Power Plant	₹ 200-300/kW	10-12 paise per unit or kWh
Nuclear Power Plant	₹ 500-1000/kW	6-8 paise per unit or kWh
Hydro-Power Plant	₹ 425-900/kW	3-5 paise per unit or kWh

In deciding any scheme for any given service, the choice must be such that the total operating cost (sum of annual fixed cost, semi-fixed cost and operating cost) must be minimum.

9.3 Interest and Depreciation

Usually money is borrowed from banks of insurance companies or other financial institutions for big projects such as generating plant and at the end of a year, the undertaking is required to pay the interest on the capital cost. Even if equipment is purchased out of reserve cash, the normal rate of interest must still be allowed for, since the money might otherwise be earning this interest. If a capital outlay required for a certain installation is P and the rate of interest per unit is r per annum then an amount of rP per annum must be provided as interest, and this is added to the annual operating cost. Rate of interest to be paid depends upon the money market position and the credit of the borrower and varies between 10 to 12 per cent (0.10 to 0.12 per unit) per annum.

Every plant after use has wears and tears and progressively becomes less fit for the use. In other words, the values of plant decrease with use. In order that it should give proper service it frequently requires periodical maintenance and repairs. The useful life of the plant is that after which repair becomes so frequent and expensive that it is found economical to replace the plant by a new one. The useful life of the plant depends upon the conditions of service under which plant is working and how it is protected from injurious elements and effectiveness and promptness of repairs and maintenance carried out.

Besides physical depreciation due to wear and tear, decay or corrosion as described above, equipment may suffer functional depreciation due to its becoming obsolete on account of new technical improvements or being inadequate due to increase in load. The value of the plant or equipment at the end of useful life is called *salvage value.* Salvage value some time becomes negative when its junk value is nil and it costs to remove the old equipment in order to make room for the installation of the new equipment.

Total amount of depreciation of the sinking fund, as is usually called, is the difference between the initial value of the plant minus the salvage value of the plant at the end of useful life. Sound economic policy demands that certain amount of money should be set aside every year from the earnings of the plant so that we should be in possession of sufficient amount of money at the end of useful life so as to enable us to purchase new plant to carry on business as usual or repay the loan. The amount so set aside annually towards sinking fund is called **'depreciation'.**

The rate of depreciation is found by the following three methods:

(i) Straight line depreciation.

(ii) Diminishing or reducing balances.

(iii) Sinking fund depreciation.

(i) **Straight Line Method.** In this method provision is made for setting aside each year an equal proportion part of the depreciable cost depending upon the useful life of the plant

i.e. Annual depreciation charge

$$= \frac{\text{Initial cost} - \text{Salvage value}}{\text{Number of years of useful life}}$$

$$= \frac{P - S}{n}$$

Though this method is very popular because of its simplicity but does not take into account the amount of interest earned by the amount set aside yearly. Fig. 1.1 illustrates this method graphically.

(ii) **Diminishing Value Method.** In this method provision is made for setting each year a fixed rate, first applied to the original cost and then to the diminishing value: such rate depending upon the useful life of the plant. This method results in distribution of the total

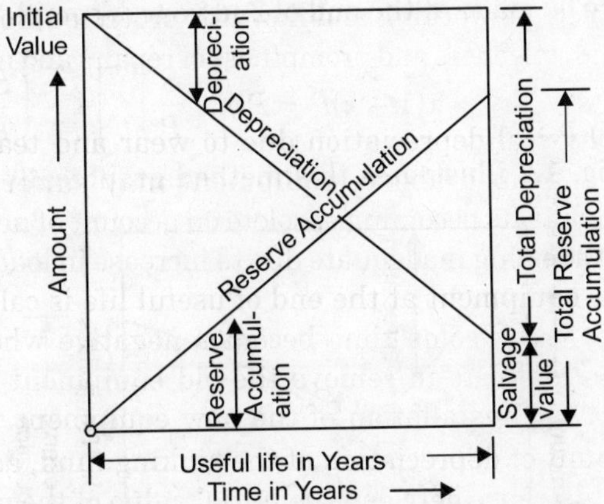

Fig. 9.1. Straight line method of depreciation.

expansion of the plant (depreciation plus maintenance cost) over its total useful life because in this method depreciation charges are heavy in early years when maintenance charges are low and depreciation charges are low in later years when maintenance charges are heavy. The serious disadvantage of this method is that a heavy burden towards depreciation charges is imposed in the early years when the plant is to develop and build up its income. In this method also the amount of interest earned by the reserve accumulation is ignored.

Let capital cost of the plant be, P

Salvage value after n years be, S

Annual unit depreciation be, x

The value of plant after one year $= P\,(1 - x)$

The value of plant after two years.

$$= P(1 - x)(1 - x) = P(1 - x)^2$$

The value of plant after n years,

$$S = P(1 - x)^n$$

or $$(1 - x) = \left(\frac{S}{P}\right)^{1/n}$$

or $$x = 1 - \left(\frac{S}{P}\right)^{1/n}$$

Deposit to be made at the end of 1st year of completion of plant

$$= xP = P\left[1 - \left(\frac{S}{P}\right)^{1/n}\right]$$

Deposit to be made at the end of 2nd year of completion of plant

$$= x(1 - x)P = P \times \left(\frac{S}{P}\right)^{1/n}\left[1 - \left(\frac{S}{P}\right)^{1/n}\right]$$

and so on. Fig. 1.2 illustrates this method graphically.

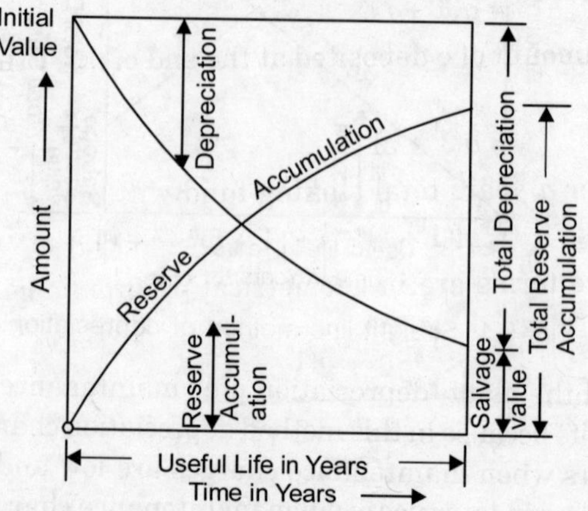

Fig. 9.2. Diminishing value method of depreciation.

(iii) **Sinking Fund Method.** In this method provision is made for setting aside each year such a sum as, invested at certain interest rate compounded annually, will give the amount equal to replacement cost of the installation at the end of its useful life. As compared to straight line method, it requires smaller annual amounts and also the amounts for annually are uniform.

Let capital cost of the installation be, P

Salvage value after n years be, S

Annual rate of interest be, r

Cost of replacement after n years,

$$Q = P - S.$$

If the annual deposit be q, it will earn interest $r\,q$ in one year, so that it will worth $q + rq = q\,(1 + r)$ at the end of one year. Thus its value will be multiplied by the ratio $(1 + r)$ every year and it will worth $q\,(1 + r)^2$ at the end of two years, $q\,(1 + r)^3$ at the end of three years and $q\,(1 + r)^n$ at the end of n years.

Thus an amount of q deposited at the end of first year, in $n - 1$ years will be

$$= q(1 + r)^{n-1}$$

An amount of q deposited at the end of second year, in $n - 2$ years will be

$$= q(1 + r)^{n-2}$$

Similarly the amount of q deposited at the end of third year, in $n - 3$ years will be

$$= q(1 + r)^{n-3}$$

and the amount of q deposited at the end of $(n - 1)$th year, in one year will be

$$= q(1 + r)$$

Thus after n years total sinking fund will be

$$= q[(1 + r)^{n-1} + (1 + r)^{n-2} + (1 + r)^{n-3} + ...(1 + r)]$$

The above terms are in geometrical progression and

$$\text{Total sum} \quad = \frac{\frac{1}{1+r}q - q(1+r)^{n-1}}{\frac{1}{1+r} - 1}$$

$$= q\frac{1 - (1+r)(1+r)^{n-1}}{1 - (1+r)} = q\frac{(1+r)^n - 1}{r}$$

This must be equal to the cost of replacement Q, so that

$$Q = q\frac{(1+r)^n - 1}{r}$$

or annual deposit, $q = \dfrac{Qr}{(1+r)^n - 1}$

This method is illustrated in Fig. 1.3.

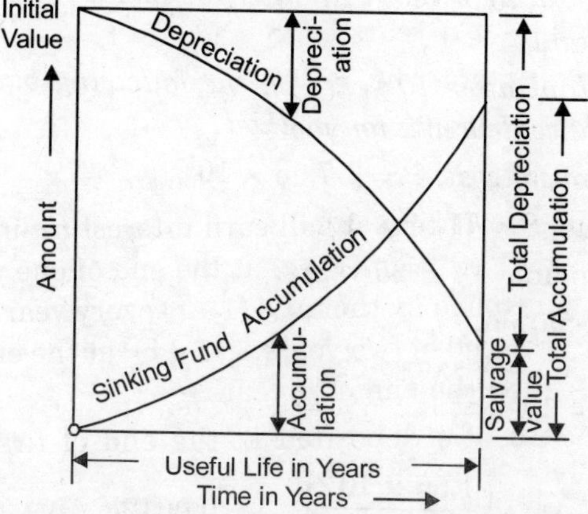

Fig. 9.3. Sinking fund method of depreciation.

Example 9.1. *A plant costing ₹ 65,000 has a useful life of 15 years. Find the amount which should be saved annually to replace the equipment at the end of that time.*

(i) By the straight line method and

(ii) By the sinking fund method if the annual rate of compound interest is 15%.

Assume that the salvage value of equipment is ₹ 5000.

Solution. Initial cost of the plant, P = ₹ 65,000

Salvage value of the plant, S = ₹ 5,000

Useful life of the plant, n = 15 years.

(i) **By straight line method**

Annual deposit to be made,

$$q = ₹\ \frac{P - S}{n} = \frac{65,000 - 5,000}{15}$$

$$= ₹\ 4,000.00 \quad \textbf{Ans.}$$

(ii) **By sinking fund method**

Annual deposit,

$$q = ₹\ \frac{Qr}{(1 + r)^n - 1} = \frac{(65,000 - 5,000) \times 0.15}{(1 - 0.15)^{15} - 1}$$

$$= ₹\ \frac{60,000 \times 0.15}{(1.15)^{15} - 1} = ₹\ 1,261 \quad \textbf{Ans.}$$

Example 9.2. *A plant cost ₹ 7.56 × 10⁵ and it is estimated that after 25 years it will have to be replaced by a new one. At that instant its salvage value will be ₹ 1.56 × I0⁵. Calculate,*

(i) the annual deposit to be made in order to replace the plant after 25 years and

(ii) value of the plant after 10 years on the following basis : 'Reducing balance depreciation method'.

Solution. Plant initial cost, P = ₹ 7.56×10^5

Plant salvage value, S = ₹1.56×10^5

Useful life of the plant, n = 25 years.

Annual unit depreciation,

$$x = 1 - \left(\frac{S}{P}\right)^{1/n}$$

$$= 1 - \left(\frac{1.56 \times 10^5}{7.56 \times 10^5}\right)^{1/25} = 0.0612 \quad \text{Ans.}$$

Value of plant after 10 years,

$$= ₹\ P(1 - x)^{10}$$

$$= ₹\ 7.56 \times 10^5 (1 - 0.0612)^{10}$$

$$= ₹\ 4.02 \times 10^5$$ **Ans.**

9.4 Formulation of Electrical Tariffs

Tariff is the rate for payment *or* schedule of rates on which the charges to be recovered from the consumer of electric energy are computed.

The main objective of the tariff is to distribute equitably the cost of supplying electric energy among the various classifications of use such as lighting, heating, industrial and bulk supply etc. All types of tariffs must cover the following costs :

(*a*) Recovery of cost of capital investment in generating plant, in transmission and distribution system equipment and in other forms of the operating equipment.

(*b*) Recovery of cost of operation and maintenance of equipment and plant.

(*c*) Recovery of cost of auxiliary equipment, metering equipment, billing and collection expenses and miscellaneous services.

(*d*) A reasonable return on the total capital investment.

(*e*) Recovery of cost of generating electricity at the generating station.

The **principle factors** involving the formulation of tariff are enumerated below :

(*i*) That the proper return is secured from each consumer of electric energy.

(*ii*) That the consumers are made to make more extended use of electric energy. Tariff should be attractive.

(*iii*) The consumers of electric energy are charged according to what the energy costs. Tariff must be fair.

(*iv*) The tariff should be simple and cheap and should be easy to understand by the consumers.

(*v*) The tariff should ensure the recovery of total cost of generation, transmission and distribution.

9.5 Various kinds of Tariffs in Application that forms commercial policy in sale of Electrical Energy

The different kinds of tariffs in common use are as follows :

(a) Simple tariff. (*b*) Flat rate tariff.
(c) Block rate tariff. (*d*) Two-part tariff.
(*e*) Maximum demand tariff. (*f*) Power factor tariff.
(*g*) Three part tariff. (*h*) Off peak tariff.
(*i*) Step rate tariff or step meter rate.
(*j*) Modified step rate tariff or modified step meter rate.

(*a*) **Simple tariff.** This is the simplest kind of tariff. Here, the cost of energy is charged on the basis of units consumed. The rate per unit can be calculated as :

Charges per kWh

$$= \frac{\text{Annual fixed charges} + \text{Annual running charges}}{\text{Total number of units supplied to the consumers}}$$

This method has following drawbacks although the procedure for fixing the rates is very simple for the supplier and calculation verification by the consumer.

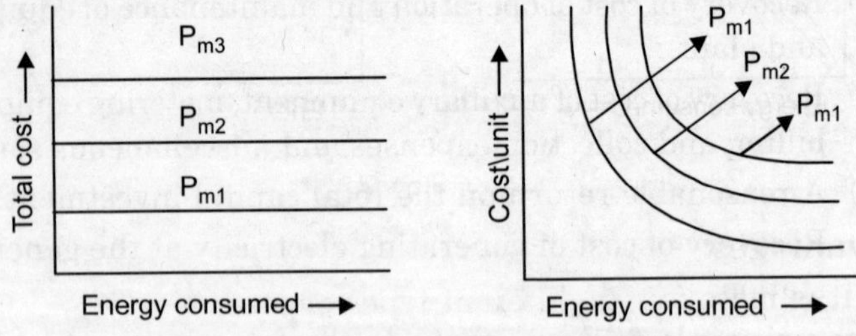

Fig. 9.4.

(*i*) The cost per kWh delivered is higher.
(*ii*) There is no discrimination among the different categories (such as domestic, commercial and industrial) of consumers.
(*iii*) The simple tariff can be made suitable by having some below mentioned modifications:
  (*a*) By allowing a discount to the consumer who consumes more electricity than an average consumer.
  (*b*) Special tariff be framed for different categories of consumers such that the domestic consumers may be charged higher rates than in case of industrial consumers and commercial consumers.
  (c) By encouraging the consumers of electric energy to use electricity during off peak load period thereby giving them a special discount. Its use is restricted to industrial loads only.

(*b*) **Flat rate tariff.** This kind of tariff differs from the former (simple tariff) one in the sense that the different categories of consumers are charged at different rates. In this case the flat rate is slightly higher for domestic (light and fan) loads than for industrial (power) loads. The rate for each category of consumers is derived by taking into account its load factor and diversity factor.

This method is very popular with the consumers since it can be easily understood by the consumers and the calculations at the supplier's end are very simple. If the energy consumed by an industrial consumer is x units and the flat rate per unit is Rs. a per unit then the charges of the bill are Rs. ax.

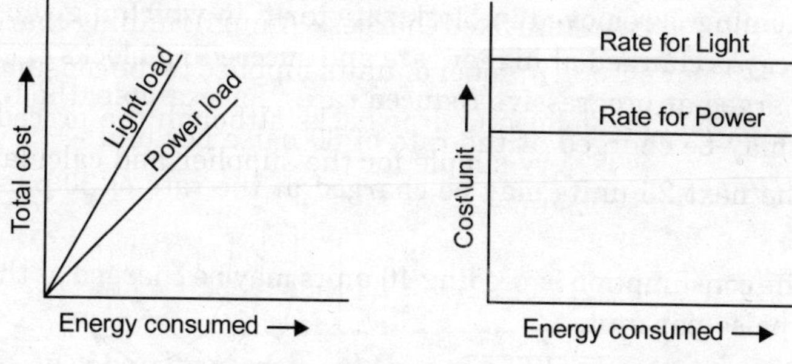

Fig. 9.5

Advantages:

(*i*) It is more fair to different types of consumers.

(*ii*) It is quite simple in calculation.

Disadvantages:

(*i*) Consumers are not encouraged to consume more energy because same rate per unit of energy consumed is charged irrespective of the quantity consumed.

(*ii*) Separate meters are required to measure energy consumed for light load and power loads.

(*iii*) Suppliers does not get any return for the connection given to the consumer if he does not consume any energy in a particular period.

In some public utilizes for certain categories of supplies *e.g.* agricultural loads, flat demand rate based on per kW or BHP connected load is charged. Here the main advantage of this tariff is the elimination of the metering equipment, meter reading, billing and accounting costs. This tariff is known as **flat demand rate**

tariff. The main disadvantage of this tariff is that there is tendency on the part of the consumers to keep their appliances 'on' even when not required. But this tariff can be used where number of hours of use are known and energy consumption is readily predictable such as street lighting, sign lighting and signal lighting etc.

(*c*) **Block rate tariff.**

The disadvantage of the flat rate tariff is that the charges are made for the total quantity of energy consumed at the same rate irrespective of the magnitude or energy consumed by the different consumers. The fixed charges are spread over a greater number of units generated *or* consumed thus lowering the rate per unit. This shortcoming is removed in block rate tariff in which a given block of energy is charged at higher rate and succeeding blocks of energy are charged at progressive reduced rates, for example, the first 15 units may be charged at the rate of 60 paise per unit.

The next 25 units may be charged at the rate of 30 paise per unit.

The consumption exceeding 40 units may be charged at the rate of 30 paise per unit.

This tariff is a modified form of two part tariff and is used in all the electricity boards for domestic, commercial and small industrial consumers.

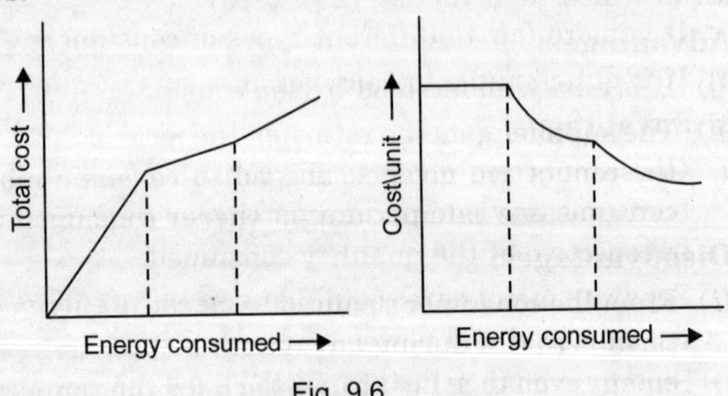

Fig. 9.6.

Advantages:

(*i*) By giving an incentive, the consumers are encouraged to consume more energy. This increases the load factor of power system and hence reduces per unit cost of generation.

(*ii*) Only one energy meter is required to measure consumption.

Disadvantages:

(*i*) The supplier does not get any return for the connection given to consumer if consumer does not consume energy in a particular period.

This type of tariff is mostly applied to domestic and small commercial consumers.

(*d*) **Two-part tariff.** The charges include fixed charges independent of energy consumed and proportional to per kW of maximum demand and the running (operating) charges per kWH of the total energy consumed. The tariff can be expressed in the expression:

$$C = Rs.\ (\mathbf{a}.KW + \mathbf{b}.KWh)$$

where Rs. a is the charge per kW of maximum demand assessed and Rs. b is the charge per kWh of energy consumed.

This tariffis mostly applicable to the medium industrial consumers.

Here, the charge made on the maximum demand covers the fixed charges such as interest and depreciation on the capital cost of building and equipment, taxes and insurance charges and operating cost which is independent of energy applied by it. The charge varies with the variation in energy supplied.

There is a disadvantage of this tariff to the consumers because they have to pay also the fixed charges unnecessarily during the month in which their industry is closed.

Advantages:

(*i*) It is easily understood by the consumers.

(*ii*) The supplier gets the return in the form of fixed charges for the connection given to the consumer even if he does not consume any energy in a particular period.

Disadvantages:

(*i*) From the consumer's point of view the disadvantages of the tariff is that if during any month he does not consume any energy even then he is to pay the fixed charges unnecessarily.

(*ii*) Since the maximum demand of the consumers is not measured therefore there is always error in assuming maximum demand.

(*e*) **Maximum demand tariff.** This tariff is very similar to the two-part tariff except that in this case the maximum demand is actually measured by a maximum demand indicator instead of merely assessing it on the basis of predetermined value. Thus the drawback of two-part tariff is removed. This tariff is applicable to large industrial consumers and bulk consumers.

(*f*) **Power factor tariff.** The equipment and plant efficiency depends on the power factor of the load. In order to increase the utility of the plant and equipment to the maximum, the plant must be operated at the most economical power factor. Because of this reason, the following types of power factor tariffs are applied. By applying power factor tariffs, the consumers are penalised for keeping poor power factor of their loads.

(*i*) **kWh and kVARH Tariff.** In this type of tariff the consumer is charged separately on the basis of kWh and kVARH consumption. With the improvement of power factor the kVARH decreases and thus the consumer tries to improve power factors of his load in order to decrease the charges on account of less kVARH registered by the kVARH meter.

(*ii*) **kVA Maximum Demand Tariff.** This is a modified form of two part tariff. The consumption is measured in kWh while the maximum demand is measured in kVA instead of kW. Because with low power factor of the installation the kVA part of load will be larger, the consumer tries to improve power factor of his installation so as to avoid more demand charges.

(*iii*) **Average power factor tariff or sliding, scale tariff.** In this type of tariff, some reference power factor, say 0.8 lagging, may be taken and for every decrease of 0.01 over the average power factor (0.8), the consumer is levied surcharge. Similarly a discount may be allowed for each 0.01 by which the average power factor rises above 0.8. Such types of tariff are rarely used.

(*g*) **Three-part tariff.** In this type of tariff, the cost consists of the fixed charges, semifixed charges and the running *or* operating charges. Thus three part tariff is composed of three elements and is mathematically represented by the equation :

Total cost = $a + b$ (kW *or* kVA) + c kWh.

Constant charges a, are also called consumer charges and are on account of giving service connection and expenses incurred in connection with metering and office establishment. Semi-fixed charges are proportional to kW *or* kVA maximum demand. These are variable charges and include the initial cost of plant as well as the operating cost which is independent of total energy supplied by it. The charges vary with the maximum demand.

Part '*c*' is the variable charge based on energy consumed and recovers the cost of fuels and other operating charges etc.

This type of tariff is generally applicable to bulk supplies.

(h) **Off peak tariff.** The load on the power station is very low during night hours (10 P.M. to 6 A.M.) while in the evening peak loads occur. During the off peak period, the plant, the transmission and distribution system equipment will be lying idle. In order to encourage the consumers to use electric energy during off peak load period, the energy can be supplied without incurring any extra capital investment and thus is profitable.

This tariff is advantageous for process as water heating by thermal storage, refrigeration and pumps etc.

Step Rate Tariff. According to this tariff, the cost of 1 kWH is not fixed but goes down in steps as the energy consumption goes up. The variation of total cost and the cost per unit with the energy consumed is shown in Fig. 9.7. The main advantage of this type of tariff is that it recognises the fact of lower generation cost as a result of higher energy consumption consequent to improvement of P.F. This tariff therefore promotes the use of electricity. The drawback of this tariff is that the total cost of energy decreases after crossing the step. Thus there is tendency with the consumer just approaching the limit of the step, to any close the step so that his whole of energy may be charged at lower rate.

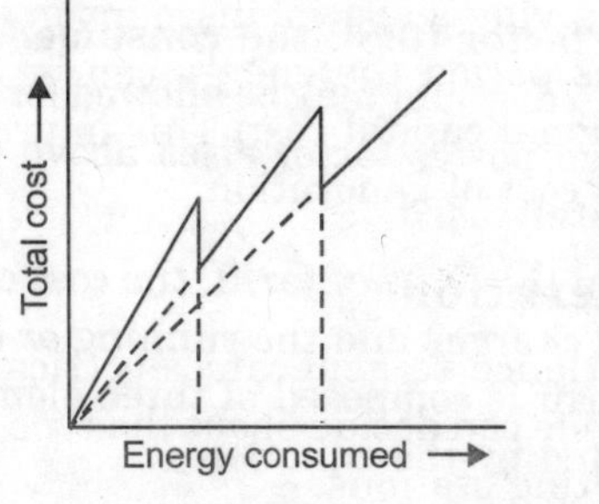

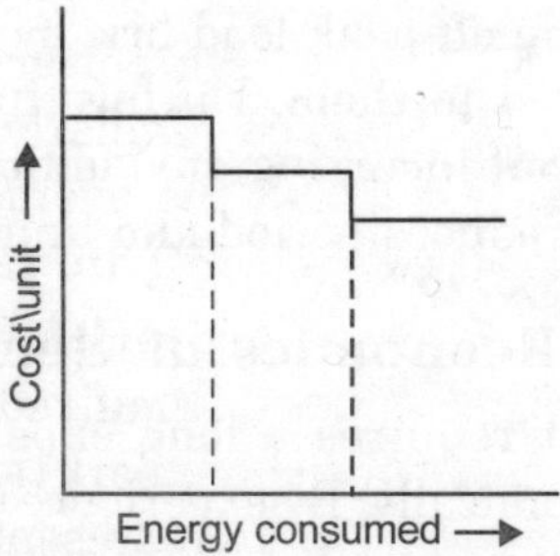

Fig. 9.7

Modified Step Rate Tariff. This tariff removes the snag of lowering of total cost at the cross over from one step to the lower step usually attendant with "step rate tariff" by the modification of the rate. As a result of this total cost instead of decreasing now remains constant as shown in Fig. 9.8 and the cost per unit does not fall abruptly to lower value but it drops very steadily as shown in Fig. 9.8 *(b)*. The defect still persists as any consumption of energy between a and b is not charged.

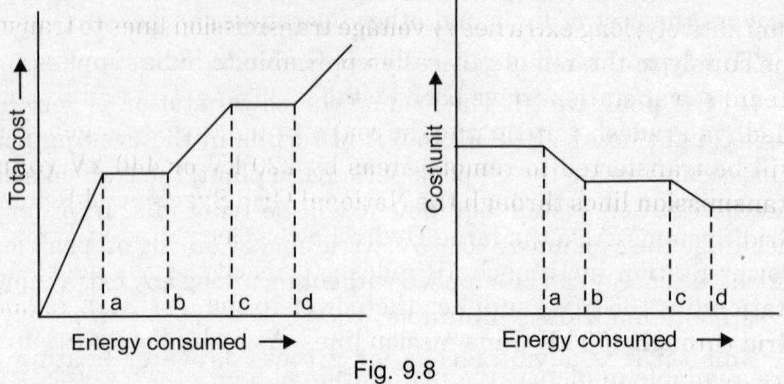

Fig. 9.8

Off Peak Load Hours. The hours during which the load on the generating station is very low as compared to its rated capacity are called off peak load hours.

Usually the load on the generating station is maximum in the morning and early morning, whereas during night *i.e.* from 10 P.M. to 6 A.M. the load on the generating station is very low and considered as off peak load period. During this period, a large proportion of the generation and distribution equipment lie idle. This reduces the Load Factors of generating station which in turn increase per unit cost of generation.

Thus, the consumers are encouraged to use electrical energy during off peak load hrs. by giving a special discount in the tariff applied to them. During this period the energy can be supplied without incurring any additional capital cost. This improves P.F. and reduces period and unit cost of generation.

9.6 Economics of Generation

It requires a long experience to generate electrical energy economically. By experience one can decide about the type, location and the rating of the generating stations.

The generating stations may be hydro, nuclear, diesel *or* steam etc. The availability of natural resources available in the area decides mainly the type of generating stations to be installed in that area. Hydro-stations do not require running (operating) charges except the salaries and wages of the operating staff while steam stations require coal and involve heavy running (operating) charges and transportation of coal to the distant places is a problem due to transportation limitations.

So steam power stations are best suited near coal pits alternatively, long extra heavy voltage transmission lines to transmit the supply to the remote places are economical. That is why large steam power stations have been installed at Singrauli and Korba in Madhya Pradesh very near to the coal pits and the power generated will be transferred to remote areas by 220 kV or 440 kV voltage transmission lines through the National Grid System and Regional Grid Systems. Also the large Hydro Power Generated at Bhakhra Power Station in Himachal Pradesh (Capacity of the generating station 6 × 165 MW) supplies the power to the Northern regional Grid through 400 kV transmission lines. As hydro power resources are available in plenty, the installation of hydro power stations in the region are economical though the steam power stations will work as base load in the region. Nuclear power stations are best suited where hydro as well as steam power stations are not feasible because of non-availability of hydro resources and coal. Diesel power stations are installed for small quantity of power requirements in order to maintain continuity of supply in the areas such as hospitals, radio-stations, cinema houses and telephone exchanges etc.

In case of D.C. supply system, the power station should be as near as possible to the centre of the load so that the transmission cost and losses in transmission are minimum.

To other aspects of design of power station are reliability, minimum capital and operating costs. For this purpose the layout should be such that the maintenance and the repairs can be carried out easily ; the layout must be well planned and compact ; the equipment must be standardized one so that the capital cost may be less and replacement of worn-out parts becomes easy and the equipment used should be simple to understand so that less operating staff is required. The equipment must be automatic so as to reduce the operating costs provided the reliability of the station is not affected. The automatic equipment shall be such that it could be operated manually if desired. Also there should be provision for extension of the power station for coping with the future demand without incurring further heavy expenditure.

When a power project is taken into hand the project engineer should have with him the following information :

(*i*) Probable load

(*ii*) Future load development

(*iii*) Location of the loads, because as in case of hydro-electric generating stations, the transmission cost is also required to be considered.

For deciding the rating and type of the generating plant, the design engineer must be familiar with the following important terms. However, with the concept of National and Regional grid system ; decisions of the Union Government of India for installing large steam power stations near coal pits; and advent load despatching system, the flow of electric power can be controlled now-a-days and the power can be fed to the area as per its requirements of load at any time.

(*i*) **The load curves.** The load on the power station *or* the grid system is seldom constant ; it varies from time to time. The daily variation in load on the power station *or* grid from time to time hourly *or* half hourly, can be plotted on a graph with load on *y*-axis and the time on *x*-axis. The curve so obtained is known as **daily load curve.** From this daily load curves of a particular month, the monthly load curve can be plotted by calculating the average value of power at particular time of the day. Similarly if such monthly load curves of a particular year are considered, the actual load curve can be obtained. These curves bear a great importance in generation. From the load curves, the following informations can be obtained:

(a) The area under the curve represents the total number of units generated in a day.

(*b*) The variations in load during different hours of the day.

(*c*) The maximum demand on the station on the particular day is represented by the peak of the curve.

(*d*) The average load on the station is given by the area under the load curve divided by the number of hours.

(*e*) The load factor is then, the area under the load curve divided by the total area of the rectangle in which it is contained.

Thus load curves help in deciding the size of the units to be installed and also in preparing the operation schedule for the generating units.

Choice of size and No. of Generating Units.

The size and no. of generating units depend upon the load curve of generating station, the initial cost of power plant depends upon the choice of size and no of units to be installed and the running cost depends upon their working. While deciding the size of units no. and operation schedule the following important points must be kept in view :

(*i*) A single unit of max. capacity should not be chosen because it will give very poor efficiency at light load and in case of break down whole the supply will fail.

(*ii*) Size and no. of units chosen should be such that at all loads the units which are to be operated should give maximum efficiency. This is only possible when all of the units are used at their rated capacity.

(*iii*) If large no. of units are chosen, more space will be required which will increase the per unit cost of energy.

(*iv*) All the machines of same size should be preferred due to the fact that they require less space, give better appearance and require less spare parts.

(*v*) One spare set of capacity of that of largest units is installed as stand by unit, so that maintenance and repair work of the working unit may be carried out without interruption in the supply.

(*vi*) The capacity of the power station should be 20% more than the expected maximum demand and growth of the demand expected in near future should also be considered.

(*a*) **Advantages of Plotting Load Curves**

(*i*) Load curves are helpful in deciding the total installed capacity of the plant required and most economical size of various generating units.

(*ii*) It helps to estimate generating cost.

(*iii*) It also helps us to decide the operation schedule of power station *i.e.* how, when and in what sequence various sets should be started, run and shut down.

(*iv*) In thermal power plants load curves give full information to the boiler operators regarding the incoming load peaks who keep the boiler in ready condition to cope up with the impeding demand. Because in these plants with the increase in load the demand for steam for turbine driven generator also increases.

(*v*) By knowing the probable position of any valley in the load curve proper planning can be done regarding maintenance, general cleaning of equipment and boiler de-slagging etc. which may require shutting down of some of the apparatus.

(*vi*) Lesser the load peaks and smaller the voltages in load curve greater will be the amount of energy generated by the given plant. Thus fixed charges will be now distributed over greater number of units which reduces the overall cost of energy per unit.

Load duration curve :

This curve gives the duration in hours for which the load has either remained equal to or exceeding the given value. For example, consider the load curve in Fig. 9.9. *(a)*.

For instance for load P_3, duration in hours is found to be equal to [(1 + 1) = 2] hrs. In the same way duration for load P_1 is 4 hrs. In this way load duration curve can be drawn which is shown in Fig. 9.9 *(b)*.

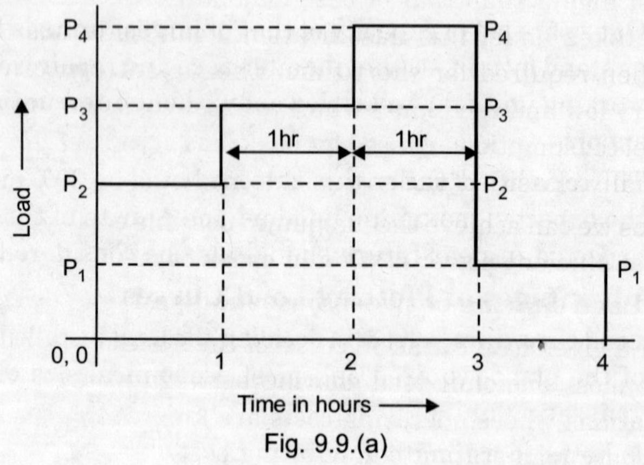

Fig. 9.9.(a)

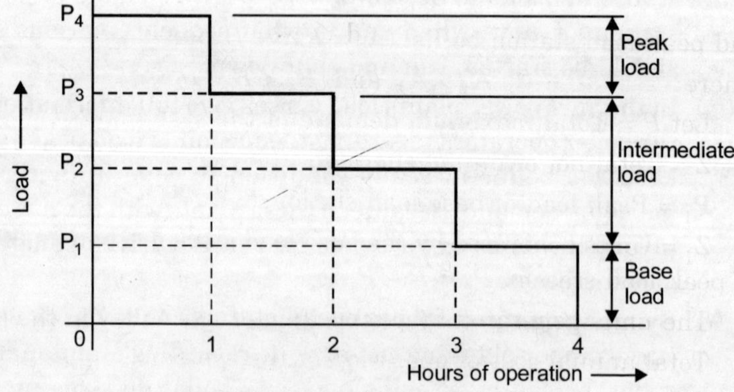

Fig. 9.9. *(b)* Load Duration Curve.

Load duration curve enables clear analysis of generating economic problems possible. What the area under the load curve is divided into three parts *viz* base load, intermediate load and peak load, so accordingly there will be three types of generating stations namely *base load stations, Intermediate load stations and peak load stations.* Base load station take up the load on the lower region of the load duration curve of whole grid system. These stations are highly efficient and operate on three shift basis through-out the year. Fixed and semi-fixed cost of these plants is usually high. Continuous operation of base load stations at high load factor improves the capacity factor of these stations and this makes the operation of costly plant an economic preposition. Hydro and nuclear power plants belong to this category. Intermediate loads operate on two or single shift basis. Capital cost of these stations is lower and fuel cost higher than that of base load stations. Thermal stations may be taken under this classification. Peak load stations operate only when required for short time. Their plant capacity factor is very very low and the plants which are less costly and even having high fuel consumption are used for peak load stations. Diesel stations are usually classified under this category.

Thus we can achieve the maximum possible economy by loading the most efficient power station in order of merit of low fuel cost.

(*b*) **Load sharing between base load and peak load station.**

For a given load duration curve of the system we will now find out the most economic load sharing between base load and peak load stations whose operating costs are known. Let the operating cost of base load station be :

$$\text{Rs. } (A_1 \times KW + B_1 \times KWh)$$

and peak load station be Rs. $(A_2 \times KW + B_2 \times KWh)$

where $$A_1 > A_2 \text{ and } B_1 < B_2$$

Let P = Total maximum demand of the system.

Z = Total no. of units generated.

P_1 = Peak load of base load station.

Z_1 = Units generated by the base load station. Then peak load of peak load station, $P_2 = P - P_1$.

The units generated by peak load station, $Za = Z - Z_1$

Total annual cost of operation of the system,

$$C = A_1 P_1 + A_2 (P - P_1) + B_1 Z_1 + B_2 (Z - Z_1)$$

$$\frac{dC}{dP_1} = A_1 - A_2 + B_1 \frac{dZ_1}{dP_1} - B_2 \frac{dZ_1}{dP_1}$$

For minimum cost $\frac{dc}{dP_1} = 0$

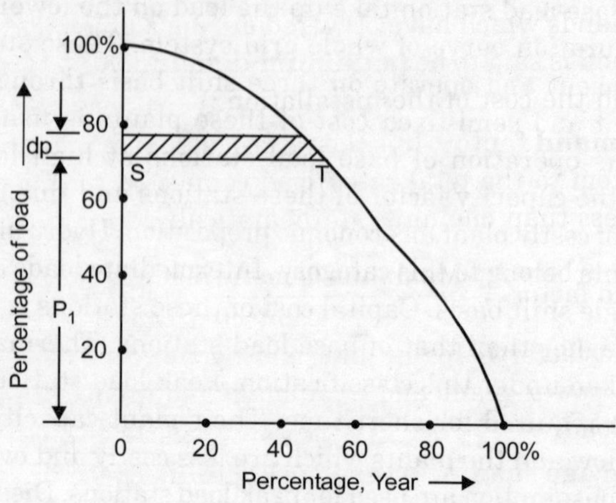

Fig. 9.10

$$\therefore \quad A_1 - A_2 + B_1 \frac{dZ_1}{dP_1} - B_2 \frac{dZ_1}{dP_1} = 0$$

or $$\frac{dZ_1}{dP_1}(B_2 - B_1) = A_1 - A_2$$

$$\frac{dZ_1}{dP_1} = \frac{A_1 - A_2}{B_2 - B_1} \text{ hours}$$

Since Z_1 = Units generated by base load station *i.e.* Area under straight line ST.

dZ_1 = shaded area of the strip = $dP_1 \times$ ST or $\frac{dZ_1}{dP_1}$ = ST.

Thus, for economic load sharing, the area under the load duration curve is so divided by Horizontal straight line ST that its magnitude is given by

$$ST = \frac{A_1 - A_2}{B_2 - B_1} \text{ hours}$$

In other words, for economic load sharing, peak load station will work for $\left(\frac{A_1 - A_2}{B_2 - B_1}\right)$ hours per year.

(*ii*) **Connected load.** The sum of the continuous ratings of all the equipment connected to the supply system is known as **connected load.**

(*iii*) **Maximum demand.** It is not necessary that all the connected load be put on the system at one time. The greatest of all the demands which have occurred during a given period on the power station is said to be **maximum demand.** The term decides the size and the cost of the installation.

(*iv*) **Demand factor.** It is the ratio of actual maximum demand on the system to the total rated load connected to the system. It is always less than one unit. Mathematically,

$$\text{Demand factor} = \frac{\text{Maximum demand made by the load}}{\text{Connected load}}$$

Demand factor helps in determining the capacity and hence cost of the power plant and equipment required to cater to the connected load.

(*v*) **Average load or average demand.** The average load *or* average demand of a power station is the average of all the loads occurring at the various moments of time. It can also be expressed as the energy delivered in a given period of time.

i.e. Daily average load

$$= \frac{\text{Units (kWh) supplied in a day}}{\text{Period in hours in a day}}$$

$$= \frac{\text{kWh supplied in a day}}{24}$$

Similarly, Monthly average load

$$= \frac{\text{Units supplied in a day}}{720}$$

and Yearly average load

$$= \frac{\text{Units supplied in a year}}{24 \times 365}$$

(*vi*) **Load factor.** Load factor can be expressed in two ways :

(*a*) Operating load factor. (*b*) Connected load factor.

Operating load factor is the ratio of the average load on the station to the maximum demand on the station during that period. For example, if the daily average load of a station is 60 MW and the maximum demand on the station during the day is 100 MW, then the operating load factor of the station will be 60% for that day.

Since average load is always less than the maximum demand, load factor is always less than unity.

Connected load factor is the ratio of average power input to the connected load and relates only to the receiving equipment.

Connected load factor

$$= \frac{\text{Average power input}}{\text{Connected load}}$$

$$= \frac{\text{Average power}}{\text{Max. demand}} \times \frac{\text{Max. demand}}{\text{Connected load}}$$

$$= \text{Load factor} \times \text{Demand factor.}$$

(vii) **Diversity factor.** Since maximum demands of different elements of load connected to a station do not occur simultaneously but are spread over a period of time thus, the maximum demand of the whole group of loading devices (or simply loads) is less than the sum of maximum demands of each loading device (loads). *Diversity factor is defined as the ratio of sum of the maximum demands of individual consuming units in a group during a specified period to the maximum demand of the whole group during the same period. Mathematically, the Diversity factor is written as*

Diversity factor

$$= \frac{\text{Sum of individual maximum demands of different units in a group}}{\text{Maximum demand of the entire group}}$$

The value of diversity factor is always greater than unity.

(viii) **Capacity factor or plant factor.** The extent, to which the generating plant will be used, is measured by this factor. It is the ratio of actual energy produced to the maximum possible energy that the plant could produce during the same period or alternatively it can be expressed as the ratio of average load to the rated capacity of the plant. This factor is exclusively applied to the generating plant. But the load factor may be applicable to the generating plant as well as the receiving equipment. Mathematically,

Plant capacity factor

$$= \frac{\text{Average demand of the plant}}{\text{Rated capacity of plant}}$$

$$= \frac{\text{Average demand}}{\text{Max. demand}} \times \frac{\text{Max. demand}}{\text{Rated capacity}}$$

$$= \text{Load factor} \times \text{Utilization factor}$$

where utilization factor is expressed below :

For taking care of the future expansion and increase in load, every generating plant has a reserve capacity and, therefore, the total installed capacity of the plant is usually greater than that actually required.

(ix) **Utilization factor.** It is the ratio of the maximum demand on the plant to the rated capacity of the plant. Utilization factor is always less than unity. Mathematically,

$$\text{Utilization factor} = \frac{\text{Max. demand on the plant}}{\text{Rated capacity of the plant}}$$

Types of Loads

(1) Residential or domestic load. This consists mainly of lights, fans, domestic appliances such as heaters, refrigerators, air conditioners for room, radio receivers, television sets, electric cookers, electric water heaters, small motors for pumping and small electrical appliances for use in the home.

(2) Commercial Load. This type of load mainly consists of lighting for shops, advertisement etc. Fans and electrical appliances used in commercial establishments—shops, restaurants, market places etc. It includes small motors and load similar to residential loads but larger. This class of load is spread over more hours of the day compared to residential load and load factor may be between 25% and 30%. Commercial loads are not as large as industrial loads.

(3) Industrial Loads. This load depends upon the power range repaired by the industry. For example, cottage industries will not require more than 5 kW each. Small industries take up to about 25 kW. Medium industries require between 25 kW and 100 kW. Large industries will take up between 100 to 500 kW and heavy industries will require more than 500 kW. The last two types of load need power over a longer period and fairly distributed throughout the day.

(4) Government Load. It may be classified as a separate type of load when it has separate feeders and special working conditions *e.g.* defence factories.

(5) Municipal Load. Street lighting is one of the loads of municipalities. This type of load is practically constant throughout the hours of darkness, so that demand factor is 100%. This load can be supplied by a distributor run completely separately and independent of other distributors for lighting loads of other types.

This is necessary because all street lights have to be put on and off approximately together. This is done by means of time switches so that lights are automatically controlled at the required time. Load factor for street lightning may be taken as 20% to 25% on a 24 hour basis.

Another type of municipal load is for water supply and drainage. Water is pumped to overhead tanks from the source of water supply. This needs pumps driven by the motors. This type of load can be easily fitted at a suitable time on the load curve so as to improve the load factor of whole supply system of the town, if it has its own power station. The normal load-on the power station will be much less at 11 p.m. or midnight. If the pumping load is supplied during this off peak period, it will improve the load factor.

(6) Irrigation. This type of load is required for supplying water for fields as required for crops in different seasons, by means of suitable pumps driven by electric motors.

(7) In some areas, there is a mining load.

(8) Traction Load. Tram cars, trolley buses, suburban railways and main railway systems come under this type of load.

(9)Bulk Supplies to Different Distributing Licensees. For some towns, it is more economical to buy power from the electric grid nearby than to have their own power stations. The load on the main system which supplies several towns, bulk supply for distribution purposes then consists of all the load requirement of these towns.

(10) In some location, there may be special industries such as chemical plants, textile industries, paper industries and extraction plants etc.

Prediction of Future Loads. Steam stations take 2 to 3 years for planning and building, nuclear stations take 3 to 6 years and hydro-stations take 4 to 8 years for the same purpose. Life expectancy of thermal plant is 20 to 25 years, that of nuclear plant is 15 to 20 years and that of hydro-station is very long. It is, therefore, essential that the installed capacity of power station should not only be suitable for the load conditions at the time of commissioning of the station but should have provision for future expansion also. Therefore for casting the future load requirement is essential part of power project designs. The forecast of load is done by two methods *namely statistical method* and *field survey method :*

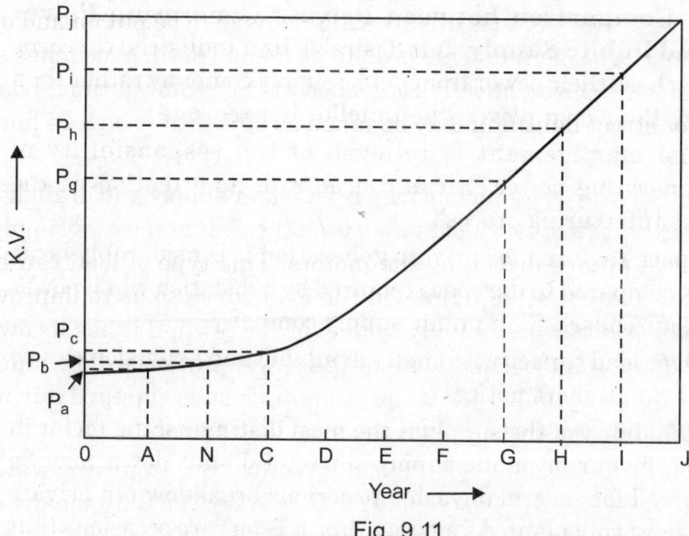

Fig. 9.11

(*a*) **Statistical Method.** Fundamental law of statistics is that every phenomenon in nature obeys same mathematical law which can be represented by a smooth curve. Load development in any area except due to unforeseen developments such as the estabishment of large and major industries, can not be exception to this law. According to this method data of annual maximum demand pertaining to the area is collected for past several years.

A graph is then drawn between maximum demand P_b, P_b, ... Pg for the years A, B... G respectively. Now, in order to predict the maximum loads during future *H, I,* and *J* years we will extrapolate the curve and get this P_h, P_i, and P_j respectively as shown in Fig. 9.11 (c).

(*b*) **Field Survey Method.** In this case, the existing requirement of the area is found out under different classes of load *i.e.* residential, commercial, industrial, municipal and agricultural. Future load requirements are then forecast taking into account various factors such as population growth, natural resources of the region, standard of living of the people, topography and climate of the region, the cost of alternate form of energy easily available in the region etc. After taking into account the energy losses in distribution network which is usually between 10 to 15% and diversity factor between various class of loads, the annual demand and annual energy requirement can be computed.

9.6.1. Comparison between Private Generating Power Plants and Public Supply. It is desirable that industrial concerns should purchase their power from public supply company rather than generating their own power due to following reasons :

1. The management is relieved of the responsibility of generating power. It can concentrate its attention on the manufacturing process.
2. Space required for private generation is considerably large as compared to the space required by substation when power is purchased from public supply company.
3. Over load capacity is readily available and extension can be made at short notices.
4. Reliability of the supply is the most distinguishing factor in the favour of public supply since total shut down may be forced for several days due to serious breakdown in private generating plant. As against this, it is on rare occasions that public supply failure occurs and that too for a maximum duration of an hour or so.
5. Power supply from public supply company will be found cheap.

Even if private generating plant is used due to various reasons given hereafter it is found desirable to have a connection from the public supply. This will not only serve as reserve but it will also supply power at the times of light load such as at night or at weekend when it becomes uneconomical to run one generating unit of private generating station.

Inspite of the following advantages some time private generation is also chosen:

(*a*) **Use of Exhaust Steam** : Certain industries require steam in their process work such as in paper mills, textile mills, sugar mills etc.

Steam is generated at high pressure and throttled down in pressure in steam turbine for driving alternators. Thus, we get electrical energy as by-product and it becomes cheap to generate in above cases. The only difficulty, however, is balance power load and steam load. In case power load is more, pass out type turbines with condensing plant are installed so as to condense steam not required by the process work.

In case the steam load is more than power load then steam from high pressure stage is throttled down to low pressure required.

(*b*) **Waste Heat Stations** : In certain industries wastes produced are used as fuel. Since no fuel cost is incurred for power generation it becomes economical to have private generating station. In steel industry coke oven gas and blast furnace gas having calorific value of 4450 and 890 kcal/m^3 respectively are by-products.

These gases may be used in gas engines driving generators or burnt under boiler for generating steam used in steam turbine driving alternators.

In sugar industry, the left over of sugarcane after crushing usually called as begasse is burnt under boilers for generating steam which may be used for power generation and process work.

(*c*) **Pit Head Generation** : Low grade coals are uneconomical to transport to far off places for power generation. These can be best utilized to generate power at the pit head where it is put to use in various types of jobs such as for coal cutting, coal haulage and coal transport.

(*d*) **Diesel Plant** : Specially where power requirements are small say of the order of 500 kW, the diesel plants become more economical due to higher overall efficiency than public supply in countries where diesel oil is available at cheap cost.

(*e*) **Emergency Service** : Private generating plant is a must in addition to public supply in industries where stoppage of process would involve huge losses or in essential public service such as in hospitals where failure of power, say in operation, can't simply be allowed.

9.7. Effect of Diversity Factor on Tariff

Diversity factor play an important part in the fixation of tariff and hence in the cost of the supply of electrical energy. Higher the value of this factor, the lower will be the overall cost per unit generated.

The diversity factor is defined as,

$$\frac{\text{Sum of the individual maximum demand of each consumer}}{\text{Actually maximum demand of the group of consumers connected to the station}}$$

its effect can be studied by considering two cases :

First case. Let us consider a small power station of capacity 2000 kW, delivering supply to two consumers *X* and *Y* having their daily load curves as shown in Fig. 9.12. Also let us consider the fixed element of the generating cost be ₹ *a* per kW per day. Then the fixed charges from the consumers *X* and *Y* would be ₹ 2000 a.

Since sum of peak loads of X and Y is 2400 kW, hence the tariff that would be set shall have ₹ $\frac{2000a}{2400}$ = ₹ $\frac{5}{6}$ a, as the demand charge per kW per day. In this case then, the diversity factor would be $\frac{2400\,\text{kW}}{2000\,\text{kW}}$, *i.e.* 1.2.

Second Case. Now let the same power station supply consumers, X, Y and Z as per load curves shown in Fig. 9.13. The load curves of consumers X and Y are the same as before, whereas the consumer Z is adjusted such that his peak load occurs as shown in Fig. 9.13.

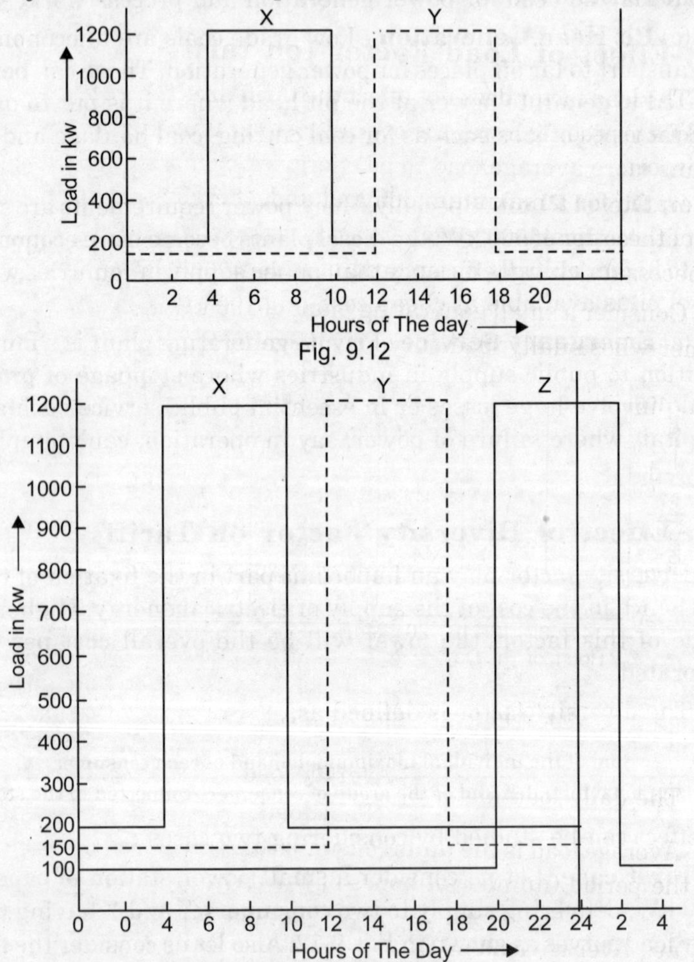

Fig. 9.13 Daily load curves of consumer *X, Y and Z.*

The sum of peak loads is 1200 + 1200 + 1200 = 3600 kW. Since the installed capacity of the station is 2000 kW, the fixed charges to be recovered is again ₹ 2000 a. Hence the tariff that would be set up shall have ₹ $\frac{2000a}{3600}$ *i.e.* ₹ $\frac{4}{9}$ a, as the demand charge per kW per day.

In this case the diversity factor is $\frac{3600}{2000} = 1.8$.

Thus the above two cases serve to show how the *demand (fixed) charge* of tariff decreases as the diversity factor is increased where by the consumer gets electric energy at a cheaper rate.

9.8. Effect of Load Factor (on tariff)

The load factor does not affect the tariff directly but an increased load factor goes to lower the overall cost per unit. Higher load factor means more average load to be energised or more number of units generated for a maximum demand and, therefore, overall cost per unit of electrical energy generated is reduced due to distribution of unnecessary charges on account of idle capacity of the plant.

Consider a small generating plant of 100 kW supplying to a consumer whose daily load curve is shown in Fig. 9.14.

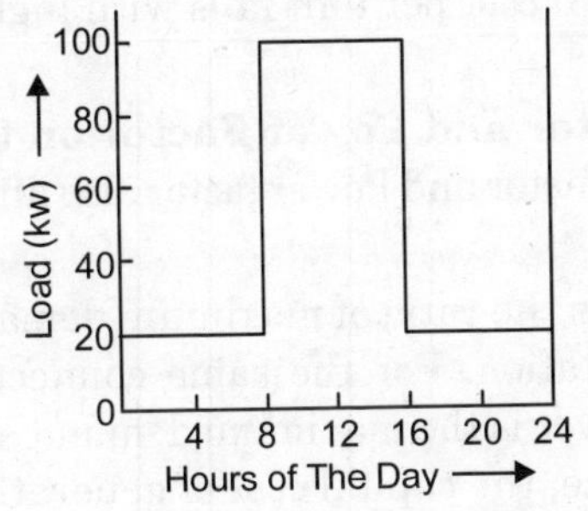

Fig. 9.14.

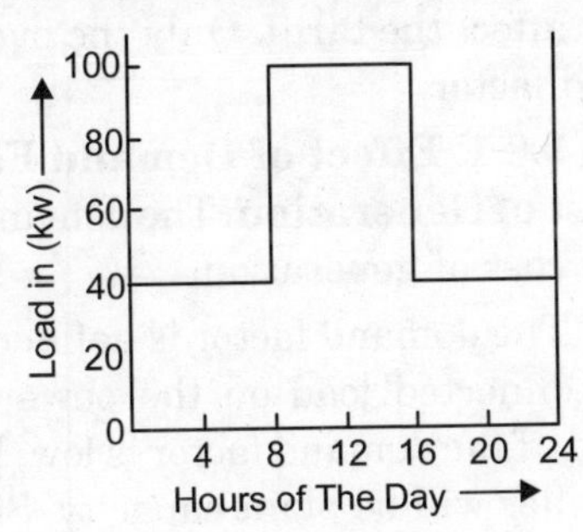

Fig. 9.15.

$$\text{The daily load factor} = \frac{\text{Average load}}{\text{Peak load}}$$

Average load is the units consumed or area of load curve divided by the period (number of hours of the day).

i.e. Average load $= \frac{8320}{100} = 83.2$ and Daily load factor

$$= \frac{46.67}{100} = 0.4667.$$

It will be seen from the load curve and the calculations given above, that the energy cost will be proportional to 1120 units produced in the two part tariff where the fixed (demand) charges will correspond to 100 kW.

Now consider the same plant supplying load to a consumer whose load curve is as shown in Fig. 9.15.

Here daily load factor

$$= \frac{640 + 800}{100} + 24 = 38.4$$

and the number of units generated

$$= 40 \times 16 + 100 \times 8 = 1440 \text{ kWh.}$$

Now the energy charges will be distributed over more number of units (i.e. 1440 kWh instead of 1120 units in the previous case) and also the idle capacity of the plant is utilized. However, the fixed charges will again be calculated on the basis of maximum demand of load, *i.e.* 100 kW.

As more units are produced, more expenses are incurred and hence greater is the energy cost in the tariff. Hence load factor does not affect the tariff. Only the overall cost per unit falls with higher load factor.

9.8.1. Effect of Demand Factor and Power Factor on the Cost of Generation. The demand factor and Power factor also affect the cost of generation.

The demand factor is defined as the ratio of maximum demand to connected load on the power station. For the same connected load, if the demand factor is low, lower is the maximum demand and smaller will be plant capacity. Since, the capital cost of generating station depends upon the capacity of the plant, therefore, the fixed charges due to capital investment will be much reduced with lower demand factor. Thus the cost of energy generated will be quite less.

Hence per unit cost of generation will be less at lower demand factor.

The power factor is defined as the ratio of true power to apparent power *i.e.*

$$\text{Power factor} = \frac{\text{True Power}}{\text{Apparent Power}}$$

or $$\text{P.F.} = \frac{\text{kW}}{\text{kVA}} \quad \text{So kVA} = \frac{\text{kW}}{\text{P.F.}}$$

The alternators installed at the generating stations are rated in kVA. Therefore, for the same power to be delivered, if the P.F. at the generating station is high the alternators of lower rating are required. This reduces the capital cost which in turn reduces the fixed charges and the cost of generation.

Thus at higher P.F. the cost of generation per unit will be less.

9.9. Commercial Policy in Sale of Electrical Energy Such as off Peak Tariffs and Bulk Supply Tariffs

Use of energy at offpeak periods improves the diversity factor. This in turn improves the load factor of the generating station resulting in the decrease of overall cost. The tariff should be such as to give the advantage of low cost of generation during off peak period to the consumer creating demand during those period. There may be installed special meters (which have two sets of dials selected automatically by means of time switches) for the measurement of energy consumption during offpeak period.

Similarly, the supply companies allow lower rate for energy consumed by the bulk or industrial consumers. Bulk consumers consume large amount of energy and thus the supply company is served from providing distribution network thereby avoiding unnecessary losses in distribution. Also the average load which the ratio of units generated to the time-period is increased and hence the load factor is also increased whereby reducing the overall cost of the electric energy. That is why the bulk consumers are charged at less the rate per unit of electric energy.

Factors Affecting the Framing of Tariff. Determination of various costs incurred is not so difficult task but passing on these costs among various classes of consumers to arrive at equitable distribution is very complex. It requires considerable engineering judgement and more often the tariff produced is found to be as is common with many situations in life, a huge compromise as it is neither possible to satisfy all consumers nor any attempt can justify which meets all the consumers hereafter enumerated:

1. Almost in all the industries except in case of power generation, production is very less affected by seasonal variation in demand. Because in the lean demand period the products can store and which can release at the time of high

demand. Thus our production activity remains more or less constant and that the installed capacity of the plant and machinery has to match the annual average demand. In the case of electrical generation, the case is quite different. Energy can not be stored on large scale, it has to be produced as and when required. As such the capacity of generating plant has to match with the maximum demand and the generating equipment has to be held in readiness to meet the consumer's full requirement at any time. It will, therefore, be observed that every consumer has a certain share in the generation plant, Transmission and distribution network. Since nature justice requires that whatever expenses are incurred in generation, the supplying authority should pass it on to the consumer, it is in the fairness of the game that fixed charges on that portion of the plant which is exclusively allocated to the consumer should be borne by him in addition to running charges for the amount of energy used. A tariff therefore should involve max. demand charges and total energy charges.

2. For a given kW demand, a consumer having low P.F. will be drawing more current. He will therefore be not only utilizing the greater part of the plant but there will be more copper losses taking place in transmission and distribution network. Tariff should give due consideration to this aspect.
3. With the improvement in load factor a given plant will be generating more amount of electrical energy. The fixed charges of plant will now be spread over more no. of units. The over all cost per unit thereby decreases. The tariff should therefore provide low rates for high consumption. This is the reason why domestic or commercial power is sometimes charged less than light and fan load because power load involve high consumption which improves power factor.
4. Use of energy at off peak periods improves diversity factor. This in turn improves the load factor of the generating station resulting in the decrease of over all cost. The tariff should be such as to pass the advantage of low cost of generation during off peak periods to the consumers creating demand during those periods. For the measurement of energy consumption during the off peak periods there are special meters which have two sets of dials selected automatically by means of time switches.

5. Tariff should take into account the service rendered by the electricity to the consumer. Since the requirement of energy of heating and cooling is more than that for lighting, it will not pinch the consumer to pay little more for lighting than for heating.
6. The tariff should also take into account the ability of the consumer to foot the electricity bill. For instance, even though electricity may be used for the same purposes of lighting and power in residential and commercial consumers, the rate of energy for commercial consumers is sometimes more than those of domestic consumers. In case of industrial consumer on the other hand since the amount of energy is very much, any increase in energy bill will increase the cost of production. In case of domestic consumer, the amount of energy involved is very small and he can afford to pay higher rate than industrial consumer in tough competition. Therefore many a time rates for industrial consumers are lower than those for domestic and commercial consumers.
7. Tariff should take into account whether the supply to consumers is given at High Tension or Low Tension. In case of supply at H.T. consumer is expected to have his own substation. Supply authority, is therefore, relieved of the investment in the substation equipment and also of responsibility of maintaining the same. Therefore tariff for supply at H.T. should be lower than that for supply at L.T.
8. Tariff should be simple so that not only the billing should be an easy process but the consumer should also be in a position to understand it.
9. Tariff should be such that it should not only promote the sale of energy but it should also compel the consumers to improve their P.F. and load factor.
10. Tariff should be such that it should reduce the complication of separate wiring and metering. For instance, if there is separate tariff for lighting and power, there will be separate connection meter and wiring for both. Similarly if there is a separate tariff for domestic and commercial consumers, it will require separate meter, connection and wiring. More often in majority of Indian cities almost all the premises are used for both residential and commercial purpose-lower portion on ground floor being used as shops and upper portion for residential purpose. Sometimes it becomes difficult to decide the nature of premises.

Example 9.3. *Complete the total bill and average unit cost for consumption of 250 kWh at block rate tariff quoted as below :*

(*i*) *First 15 kWh at 50 paise per kWh.*

(*ii*) *Next 25 kWh at 40 paise per kWh.*

(*iii*) *Next 60 kWh at 35 paise per kWh.*

(*iv*) *Exceeding 100 kWh at 30 paise per kWh.*

Solution. The total bill

$$= 15 \times 0.50 + 25 \times 0.40 + 60 \times 0.35 + 150 \times 0.30$$

$$= 7.50 + 10.00 + 21.00 + 45.00 = ₹\ 83.50.$$

Now, the unit cost $= \dfrac{83.50}{250} = 0.334$ paise per kWh. **Ans.**

Example 9.4. *A two part tariff is given as follows :*

(*a*) **Demand charges :**

(*i*) *First 1 KW of maximum demand at ₹ 25 per KW per month,*

(*ii*) *Next 4 KW of maximum demand at ₹ 20 per KWper month.*

(*iii*) *Exceeding 5 KW of maximum demand at ₹ 16 per KW per month.*

(*b*) **Energy charges:**

(*i*) *First 100 units at 40 paise per unit,*

(*ii*) *Next 200 units at 30 paise per unit.*

(*iii*) *Exceeding 300 units at 25 paise per unit.*

Calculate the monthly bill for a consumption of 2000 units at a maximum demand of 20 KW.

Solution, (*a*) The demand charges

$$= 1 \times 25 + 4 \times 20 + 15 \times 16 = ₹\ 345.$$

(*b*) The energy charges

$$= 100 \times 0.40 + 200 \times 0.30 + 1700 \times 0.25$$

$$= 40 + 60 + 425 = ₹\ 525.$$

Hence total monthly bill = ₹ 345 + 525 = ₹ 870. **Ans.**

Example 9.5. *Explain the economic basis for tariff charges in which the consumer's power factor and maximum demand are involved. Assuming a 400 V, 3-phase supply, show how these quantities are measured or estimated.*

A load having a maximum value of 1000 KW at a power factor of 0.8 has an annual load factor of 40%. Find the cost per unit to the consumer if the tariff is ₹150 per year per KW demand plus 4 paise per unit plus 2 paise per reactive KVAH.

Solution. The annual consumption charges will consist of the demand charges, the energy charges and the reactive kVAH cost. We shall calculate them individually and then find the overall cost per unit.

Demand Charges. Maximum demand (Given)

$$= 1000 \text{ kW}$$

Therefore, annual demand charges

$$= 1000 \times ₹\ 150 = ₹\ 150{,}000$$

Energy Charges. Here the maximum load (peak load) of 1000 kW is at a power factor of 0.8 (*i.e.* $\cos\phi = 0.8$) and an annual load factor of 0.4.

Thus energy consumption would be = $1000 \times 0.4 \times 365 \times 24$,

$$= 3504 \times 10^3 \text{ units}$$

The annual energy charges would be = $3504 \times 10^3 \times 0.04$

$$= ₹140{,}160$$

Reactive kVAH cost

Given maximum load = 1000 kW at 0.8 power factor

∴ Reactive power

$$= \frac{\text{kW}}{\text{Power factor}} \times \sin\phi = \frac{1000}{0.8}\sin\phi$$

$$= \frac{1000}{0.8} \times 0.6 \text{ kVAR} = 750 \text{ kVAR}.$$

Again, this reactive power of 750 kVAR is at an annual load factor of 0.4.

Thus the number of reactive units (kVAH) consumed shall be

$$= 750 \times 365 \times 24 \times 0.4 = 2628 \times 10^3$$

Hence the kVAH cost works out to be

$$= 2628 \times 10^3 \times 0.02$$

$$= ₹\ 52{,}560$$

And the total annual cost of supply

$$= 150{,}000 + 140{,}160 + 52{,}560$$

$$= ₹\ 342{,}720.$$ **Ans.**

and the cost/unit

$$= \frac{342{,}720}{3504 \times 10^3} = 9.8 \text{ paise per kWh.}$$

Example 9.6. *A generating station has got maximum demand of 80 KW. Calculate the cost per KWh delivered from the following data :*

Capital cost $= ₹\ 16 \times 10^6$

Annual cost of fuel and oil

$= ₹\ 14 \times 10^5$

Taxes, wages and salaries

$= ₹\ 9 \times 10^5$

The rate of interest and depreciation is 12%.

Annual load factor is 60%.

Solution. First of all, we shall calculate the interest and depreciation charges on capital cost which will be

$$= ₹\ 16 \times 10^6 \times \frac{12}{100} = ₹\ 19.2 \times 10^5$$

Now the annual cost of generation would be

$$= ₹\ 19.2 \times 10^5 + 14 \times 10^5 + 9 \times 10^5$$

$$= ₹\ 42.2 \times 10^5.$$

The number of units delivered from the generating station having maximum demand of 80 kW and annual load factor of 60%

$$= 80 \times 365 \times 24 \times \frac{60}{100}$$

$$= 420{,}480 \text{ kWh, } i.e. \ 4.2048 \times 10^5 \text{ kWh}$$

Thus the cost of kWh delivered

$$= \frac{42.2 \times 10^5}{4.2048 \times 10^5} = 10.036 \text{ paise.} \quad \textbf{Ans.}$$

Example 9.7. *A factory to be set up is to have a fixed load of 760 KW at 0.8 p.f. The electricity board offers to supply energy at the following alternate rates :*

(*i*) *Low voltage supply at ₹ 32 per KVA maximum demand per annum plus 10 paise per KWh;*

(*ii*) *High voltage supply at ₹ 20 per KVA maximum demand per annum plus 10 paise per KWh. The high voltage switchgear costs ₹ 60 KVA and switchgear losses at full load amount to 5%. Interest depreciation etc. charges for the switchgear are 12% of capital cost.*

If the factory is to work for 48 hours per week, determine the economical tariff.

Solution. The given is a factory having fixed load of 760 kW at 0.8 p.f.

The kVA demand $= \dfrac{760}{0.8} = 950.$

The factory is to work for 48 hours per week. Therefore, the units consumed per week = 760 × 48 = 36,480 kWh and the units consumed per annum

$$= 36{,}480 \times 52 = 18.9696 \times 10^5 \text{ kWh}$$

(i) If low voltage supply is taken,

Demand charges per annum = ₹ 32 × 950 = ₹ 30,400

Annual energy charges $= 18.9696 \times 10^5 \times \dfrac{10}{100} =$ ₹ 189,696.

Total annual energy bill = 189,696 + 30,400 = ₹ 220,096.

(ii) If high voltage supply is taken,

Annual energy charges $= 18.9696 \times 10^5 \times \dfrac{10}{100} =$ ₹ 189,696

Now since the switchgear losses at full load are 5%, *i.e.* the switchgear rating would be,

$$= \frac{950}{0.95} = 1000 \text{ kVA.}$$

Thus the high voltage switchgear costs,

= 60 × 1000 = ₹ 60,000

Therefore, Interest, depreciation charges,

$$= 60{,}000 \times \frac{12}{100} = ₹\ 7200$$

and demand charges per annum

= 1000 × 30 = ₹ 30,000.

Now the total annual energy bill will be

= ₹ 189,696 + ₹ 30,000 + ₹ 7,200

= ₹ 226,896

From the above two cases it is seen that the low voltage supply tariff is economical one.

Example 9.8. A consumer *has an annual consumption of 176,400 KWh.*

The charge is ₹ 120 per KW of maximum demand plus 4 paise per KWh.

(*i*) *Find the annual bill and the overall cost per kWh if the load factor is 36%.*

(*ii*) *What is the overall cost per KWh, if the consumption were reduced 25% with the same load factor ?*

(*iii*) *What is the overall cost per KWh, if the load factor 27% with the same consumption as in (i).*

Solution.

(*i*) The load factor = 0.36

The annual consumption = 176,400 kWh.

∴ The maximum demand

$$= \frac{176{,}400}{0.36 \times 365 \times 24} = 55.936 \text{ kW}$$

The annual bill will be demand charges per annum plus the annual consumption (energy) charges, *i.e.*

= ₹ 55.936 × 120 + ₹ 176,400 × 0.04

= ₹ 6712.32 + ₹ 7056 = ₹ 13768.32 **Ans.**

and the overall cost per kWh

$$= \frac{13768.32}{176{,}400} = 7.8 \text{ paise.} \quad \textbf{Ans.}$$

(*ii*) Now the annual consumption is reduced to

176,400 × 0.75 = 132,300 units.

The load factor = 0.36

∴ The maximum demand

$$= \frac{132{,}300}{0.36 \times 365 \times 24} = 41.952 \text{ kW}$$

Therefore, the annual bill

= (41.952 × 120 + 132,300 × 0.04)

= (5034.24 + 5292) = ₹ 10326.24

and the overall cost per kWh

$$= ₹\ \frac{10{,}326.24}{132{,}300} = 7.8 \text{ paise.} \quad \textbf{Ans.}$$

(*iii*) The annual consumption

= 176,400 kWh

The load factor = 0.27

∴ The maximum demand

$$= \frac{176{,}400}{0.27 \times 365 \times 24} = 74.58 \text{ kW}$$

Now, the annual bill

$$= (74.58 \times 120 + 176{,}400 \times 0.04)$$

$$= 8949.60 + 7056 = ₹\ 16005.60$$

Thus overall cost per kWh

$$= ₹\frac{16005.60}{176{,}400} = 9.07 \text{ paise.}$$ **Ans.**

Example 9.9. *A consumer taken a steady load of 200 KW at a power factor of 0.8 lagging for 10 hours per day and 300 days per annum. Estimate his annual payment under each of the following tariffs :*

(i) 10 paise per KWh plus ₹ 100 per annum per KVA;

(ii) 10 paise per KWh plus ₹ 100 per annum per KW plus 2 paise per KVARH.

Solution. Annual consumption at a steady load of 200 kW

$$= 200 \times 300 \times 10 = 6 \times 10^5 \text{ kWh}$$

kVARH consumption

$$= \text{kVA} \sin \phi \times 300 \times 100$$

$$= \frac{200}{0.8} \times 0.6 \times 300 \times 100 = 4.5 \times 10^6 \text{ kVARH}$$

$$\text{kVA rating of load} = \frac{200}{0.8} = 250 \text{ kVA}$$

(*i*) The annual charges

$$= (6 \times 10^5 \times 0.10 + 250 \times 100)$$

$$= 60{,}000 + 25{,}000 = ₹\ 85{,}000. \text{ Ans.}$$

(*ii*) The annual charges

$$= (6 \times 10^5 \times 0.10 + 200 \times 100 + 4.5 \times 10^6 \times 0.02)$$

$$= (60{,}000 + 20{,}000 + 90{,}000)$$

$$= ₹\ 170{,}000.$$ **Ans.**

Example 9.10. *A consumer requires one million units per year and his annual load factor is 50%. The tariff in force is ₹ 120 per KW per annum plus 5 paise per unit. Estimate the saving in his energy charges which would result if he improved his load factor to 100%.*

Solution. Annual consumption

$$= 1 \times 10^6 \text{ kWh.}$$

(*i*) When load factor is 50% ;

The maximum demand

$$= \frac{1 \times 10^6}{0.50 \times 365 \times 24} = 228.31 \text{ kW.}$$

Therefore, annual energy bill

$$= (228.31 \times 120 + 1 \times 10^6 \times 0.05)$$

$$= (27,397.20 + 50,000) = ₹\ 77,397.20$$

(ii) When load factor is 100%,

The maximum demand

$$= \frac{1 \times 10^6}{100 \times 365 \times 24} = 114.155 \text{ kW.}$$

The annual energy bill

$$= (114.155 \times 120 + 1 \times 10^6 \times 0.05)$$

$$= (13,698.60 + 50,000) = ₹\ 63,698.60$$

Thus saving in energy charges,

$$= 77,397.20 - 63,698.60 = ₹\ 13,698.60 \textbf{ Ans.}$$

9.10. Effects of Power Factor

First of all, we shall illustrate what is power factor. Consider an inductive circuit drawing a current *I* amps. from the supply lagging behind the supply voltage by an angle ϕ. The vector diagram is given in Fig. 9.16.

The current *I* can be resolved into two components

$$OY = I \sin \phi$$

and $$OX = I \cos \phi.$$

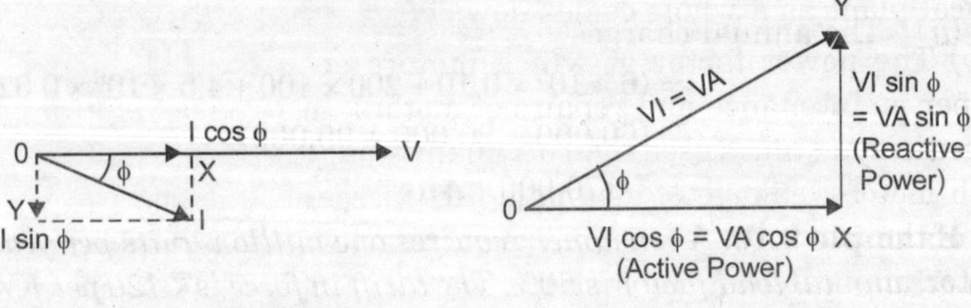

Fig. 9.16 Fig. 9.17

The component $I \cos \phi$ is known as the in-phase or **active component,** whereas $I \sin \phi$ is known as out of phase or **reactive component** of current.

If all the components of the Δ *OXY* are multiplied by voltage *V*, then $VI \cos \phi$ will represent the real power component and is expressed in watts or kW while the quadrature component $VI. \sin \phi$ will represent reactive power and is expressed in *VA* or *KVA*. We get a power triangle as shown in Fig. 9.17.

From triangle *OXY*,

$$\text{power factor, } \cos \phi = \frac{OX}{OY}$$

$$= \frac{VI \cos \phi}{VI}$$

= Ratio between the true or active power and the apparent power

Reactive power,

$$\text{kVAR} = \text{kVA} \sin \phi = \frac{KW}{\cos \phi} .\sin \phi = \text{kW} \tan \phi.$$

9.11. Causes of Low Power Factor

(*i*) All a.c. motors (except over excited synchronous motors and certain type of commutator motors) and transformers operate at lagging power factor. The p.f. falls with the decrease in load. For example an induction motor has a reasonable higher p.f. of 0.85 at full load, 0.8 at 75% of full load, 0.7 at half full load, 0.5 at 25% of full load and as low as 0.1 on no load.

(*ii*) Arc lamps and electric discharge lamps operate at low lagging power factor.

(*iii*) Due to increase in supply mains voltage, which usually occurs during low load periods such as lunch hours, night hours etc. The magnetizing current of inductive reactances increase and p.f. of the electrical plant as a whole comes down.

(*iv*) The power factor at which motors operate falls due to improper maintenance and repairs of motors. In required motors, less wire is some times used than originally wound motors, therefore in such motors leakage of magnetic flux increases and the power factor of the motor decreases. In case of heavily worn out bearings the rotor may catch at the stator.

Some metal is some times removed from the rotor by turning instead of replacing the defective bearings. In doing so, the length of air gap between stator and rotor increases due to which greater magnetising current is required and therefore, power factor drops.

(*v*) Industrial heating furnaces such as arc and induction furnaces operate on very lagging power factor.

(*vi*) A transformer is an inductive device and draws a magnetising current which causes the total current to lag behind the voltage *i.e.* it operates at lagging P.F. At normal loads, magnetising current does not affect P.F. much, but at light loads magnetising current decreases the P.F. effectively.

The average power factors of some of the common appliances are given below.

Type of load	**Power Factor**
Incandescent lamps	0.98 -1.0
Inductrom heaters	0.85
Fluorescent lamps	0.6-0.8
Neon lamps used for advertisement	0.4-0.5
Arc lamps used in cinemas	0.3-0.7
Fans	0.5 - 0.8
Induction motors	0.5-0.85
Fractional H.P. Motors	0.4-0.75
Resistance furnaces	0.6-0.9
Arc Welders	0.85
Induction furnaces	0.6
Arc furnace	0.3-0.4
Resistance welders	0.4-0.75

9.12. Disadvantages of Low Power Factor

The load current is given by the expression,

$$I_L = \frac{P}{\sqrt{3}V \cos \phi}$$

Where P is the power in watts. For a given load, if the power factor is low, the load current will be higher. The higher current due to low power factor affects the systems and have the following disadvantages:

(*i*) The rating of the alternators and transformers being proportional to their output current, the cross-sectional area of the busbar, and the contact surface of the switchgear shall be of higher capacity as the output current would contain the reactive component of the current too. Thus more material for busbars and contact surface of switchgear is required.

(ii) With low power factor and for the same power transmitted, the transmission line or cable will have to carry more current and hence the size of the conductor of transmission line or cable will be high. More copper or aluminium is required to deliver the same load at lower power factor.

(iii) As copper losses are proportional to the square of the current hence are inversely proportional to the square of the power factor. This means more copper losses occur at low power factor and hence result in poor efficiency of the system.

(iv) Low power factor (lagging) causes poor regulation because of large voltage drop in alternators, transformers, transmission lines and cables. Hence to maintain the voltage within permissible limits, extra voltage regulating equipment is required, thus resulting in the extra cost.

Thus we can conclude that the low power factor results in a higher capital cost for the alternators, transformers, switchgears, transmission lines, cables and distribution lines etc.

Low power factor is caused by the reactive component of the alternating current being taken by the transformers, induction motors, arc lamps, arc furnaces and induction furnaces etc. due to their lagging power factors. The lagging power factor is because of presence of inductance (reactance) in the circuit and also because of reactance of the equipment. The magnetizing current drawn by the equipment from the supply causes low lagging power factor. More the magnetizing current more low the power factor would be.

9.13. Power Factor Improvement

From the foregoing paragraph, it is worthwhile to improve the power factor of the system or the equipment. The following methods are usually adopted for power improvement:

(a) **By the use of special type of motors having high power factors.** The motors which operate at power factor of approximately unity can be used *e.g.* compensated induction motors, schrage motors and synchronous motors. The use of such motors is limited because of their high cost and expensive maintenance cost than ordinary induction motors. The synchronous motors are used to operate at

leading power factor and hence compensate the lagging power factor of other equipment too.

(*b*) **By the use of induction motors along with phase advancers.** An a.c. exciter or phase advancer is mounted on the shaft of the induction motor. This exciter supplies the exciting current to the rotor circuit of the motor at slip frequency. The use of phase advancers is usually limited to motors above 200 h.p. Shunt and series type phase advancers are available according to whether the exciting winding of the advancer is connected in parallel or in series with the rotor winding of the induction motor.

Types of Phase Advancers. Phase advancers are the A.C. commutator machines, whose function is to improve the P.F. of induction motors. The use of these machines becomes economical for induction motors only above 100 HP. Various types of phase advancers are :

(*i*) Expedore, Leblanc or Scherbius phase advancer.

(*ii*) Susceptor or frequency converter type phase advancer.

(*iii*) Kapps vibrator.

(*i*) **Expedore.** It is a d.c. machine with dummy stator. The armature of this m/c is fed from the slip rings of main induction motor.

Slip frequency currents fed to the commutator machine will set up rotating magnetic field rotating at slip speed in space. So long the speed of armature in space is less than slip speed, d.c. armature winding behaves like a choke in the rotor circuit of the induction motor.

At slip speed it behaves like a pure resistance. If, however, armature is made to run at a speed greater than slip speed and in same direction as the rotating magnetic field, it behaves like a

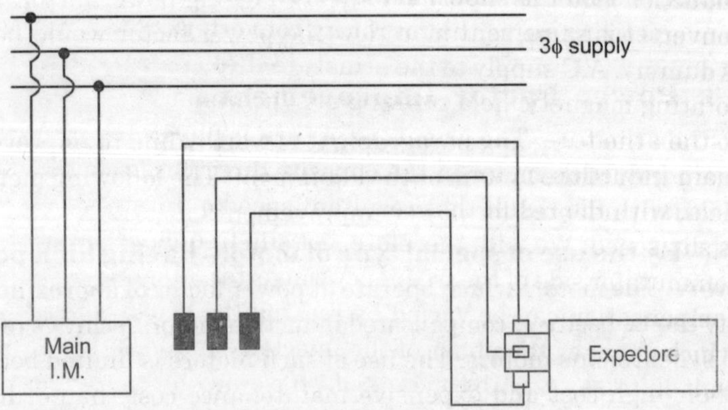

Fig. 9.18 (a)

condenser. Expedore machine is connected as shown in figure 9.18 (a) where main induction motor drives the armature of expedore. In this case since the speed of armature of expedore is **advantage** Fixed., we don't have any control on the phase compensation.

If the armature of expedore is driven by a separate auxiliary motor whose speed can be varied as shown in Fig. 9.18 *(b)* we can control the amount of phase compensation.

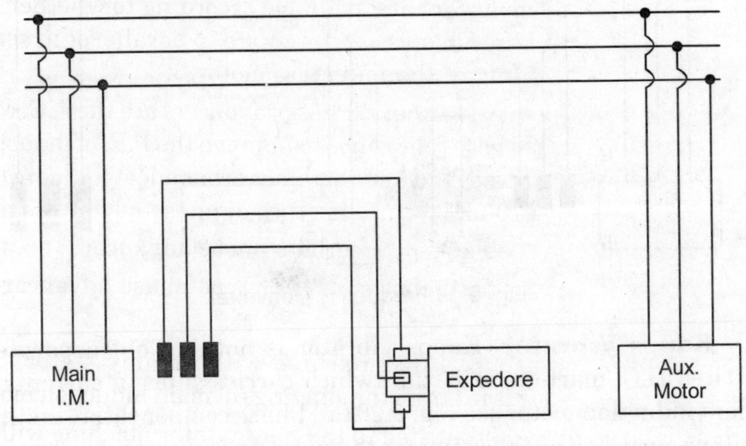

Fig. 9.18 *(b)*

(ii) Susceptor or frequency converter type phase advancer. P.F of slip ring induction motor can be improved by injecting into the rotor an e.m.f having the same frequency of supply as that induced in the rotor circuit. This is achieved by the use of frequency converter which is electrically as well as mechanically coupled to the main Induction motor as shown in Fig. 9.19. Construction of the frequency converter is same as that of rotary converter except that its stator is dummy. A.C supply to the armature through slip rings will set up rotating magnetic field rotating at synchronous speed with respect to the armature. The armature in turn is physically rotated by the main induction motor in the opposite direction to that of magnetic field, with the result that resultant speed of rotating field in space is slip speed. We will, therefore, get slip frequency supply from the commutator side of the frequency converter. Magnitude of the phase compensation can be varied by rocking the brush gear as a whole, which changes the phase of the injected e.m.f. Another way to look at it is to say that by means of phase advancer the magnetising current of Induction motor is supplied through rotor instead of stator.

Since the frequency in the rotor is 2-3% of supply frequency, motor excited at this frequency requires only a fraction of KVA necessary when they are magnetised at full supply frequency.

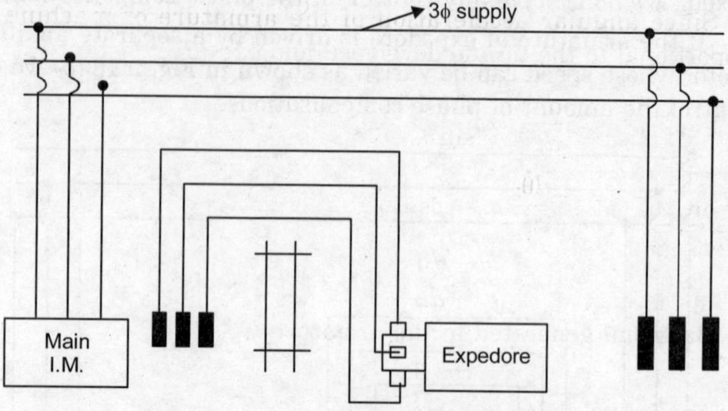

Fig. 9.19 Frequency Converter.

Kapps Vibrator. Kapps vibrator is nothing but separately excited D.C. machine armature which carries slipring currents of main induction motor see Fig. 9.20 (a). Phase compensation method will be clear by the following mathematical treatment. Mechanical torque developed by the armature of the machine is proportional to the product of magnetic flux and armature current.

$$T \propto \phi . i \quad \text{...}(i)$$

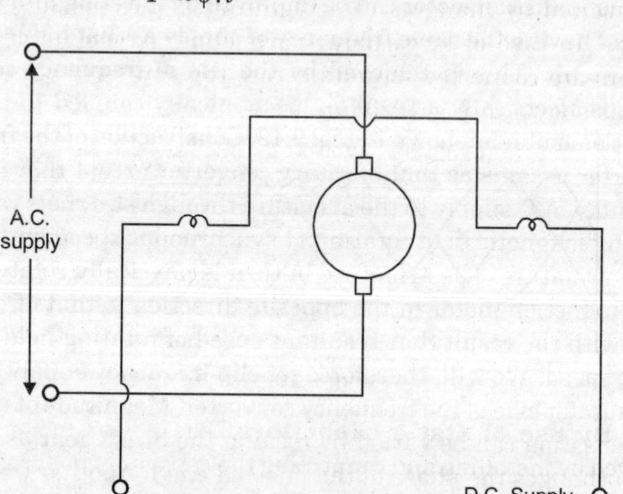

Fig. 9.20 (*a*)

The flux of the machine is constant and the current in the machine is alternating and given by, $i = i_m \sin\theta$ ...(ii)

$$\therefore \quad T \propto \phi i_m \sin\theta$$

Since angular acceleration of the armature of machine is proportional to the torque developed and ϕI_m are constant

$$\frac{d^2\theta}{dt^2} \propto \frac{T}{\sin\theta}$$

or

$$\frac{d\theta}{dt} \propto -\cos\theta$$

$$e \propto -\frac{d\theta}{dt}$$

Back emf generated in the armature

$$\propto \cos\theta$$

Applied voltage $v = -e$

$$\propto \sqrt{-\cos\theta}$$

Power taken by the machine, $W = i.v.$ ...(iii)

$$\propto \sin\theta\,(-\cos\theta)$$

Graphical representation of equation *(i)* to *(iv)* is given in Fig. 9.20. *(b)* from which it will be observed that current taken by the machine leads the applied voltage by 90° which explains the phase compensation action. Magnitude of the phase compensation can be varied by changing the magnitude of d.c. excitation of the machine. In case of large induction motors, Armature of Kapp's vibrators are connected in delta across the slip rings of induction motor as shown in Fig. 9.20 (*c*).

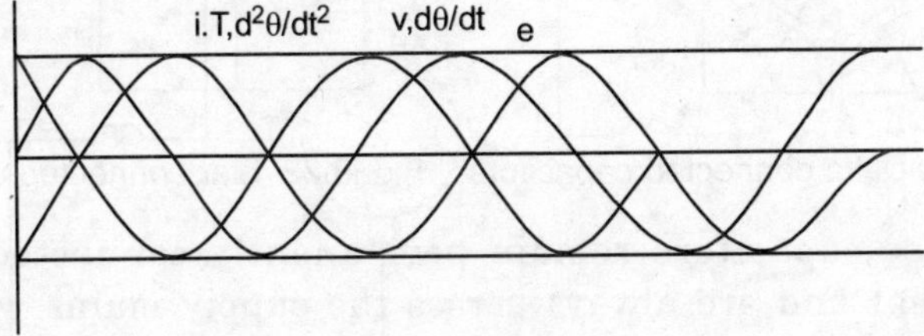

Fig, 9.20(b)

(*c*) **By use of static capacitors.** The power factor can be improved by installing and connecting the capacitors in parallel with the equipment operating at lagging power factor such as fluorescent tubes, induction motors etc. The static capacitors have small power losses, require little maintenance and also are very convenient in

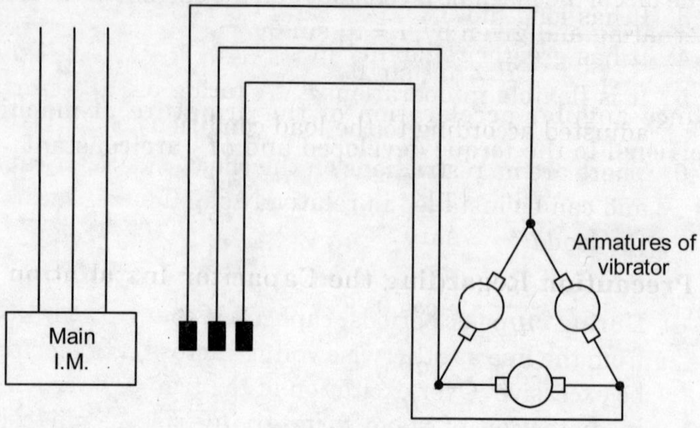

Fig. 9.20 (c)

small sizes. Also the Electricity Boards install static capacitors on 11 kV busbars in the sub-station for improving the power factor of their system. The leading current drawn by the static capacitors neutralizes the lagging current drawn by the equipment hence the wattless component of power is reduced thus resulting in the improvement of the power factor. The shunt capacitors may be connected in delta or in star as shown in Figs. 9.21 and 9.22 respectively.

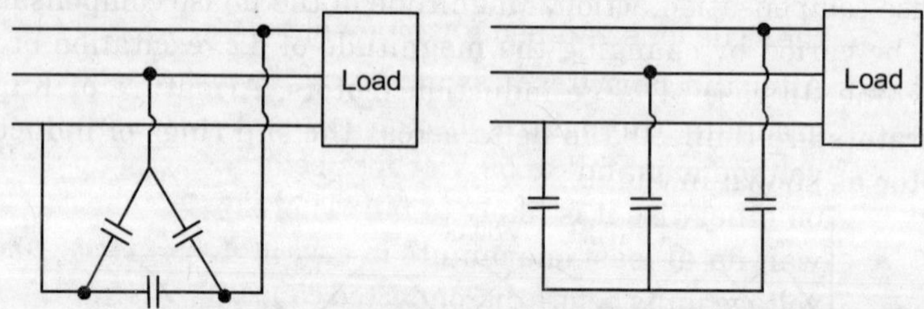

Fig. 9.21. Delta connected capacitors. Fig. 9.22. Star connected capacitors

These capacitors remain permanently connected to the equipment and are always across the supply mains when the equipment is switched on to the supply mains.

Advantages of Capacitor:

1. Capacitor losses are less than 0.5% of the rating as against 1.5%, for large units of synchronous compensators of rating 50%, 100 MVA.
2. Capacitors are easy to install, operate and maintain.

3. It has long life.
4. It has greater reliability in service.
5. It is flexible in operation, KVA rating of capacitor can be adjusted according to the load conditions.
6. There are no restrictions on the choice of site for capacitors and can be installed in relatively small banks located near the load.

Precaution Regarding the Capacitor Installation :

(*i*) During off peak periods, capacitors should be switched off from the line as otherwise voltage across Transformers will be excessive. Over excitation of the transformer will result in distortion of wave form giving rise to objectionable harmonic currents. For the same reason capacitor units should be switched in only after full load of the ckt. whose p.f. is to be improved, is switched on.

(*ii*) Switching currents of capacitors are many times the full rated current. Therefore wiring leads should be of twice the size chosen for normal current carrying capacity. For the same reasons ammeters and current transformers when used should have twice the range of capacitors installed load. In that case the pointer of ammeter will not overshoot. The scale range on initial surge and will be near the middle of the scale for sustained reading where scale is least crowded.

(*iii*) After the capacitor is switched off its residual charge falls gradually through its discharge resistance. The capacitor voltage must fall to 50 V or less before it is again switched on otherwise it is likely to fail. Therefore it is necessary to wait for at least one minute in case of low capacity and low voltage units before disconnected capacitor is again switched on.

Camparison between static capacitors and synchronous compensators

Topic of comparison	Static capacitor	Synchronous compensator
1. First cost	1	1 to 1.2 times
2. Loss	Less than 1, 1/2%	1, 1/2 to 4%
3. Variation of Reactive Power	In steps only give rise to surges of current & voltage	Continuous & line control

4. Supply of lagging kVAR	Not Possible	Possible
5. Over load	Not possible without damage	Possible for short time
6. Flexibilities of installation	Simple-several small units can be installed along line	More difficult-large units are preferred for economy & efficiency
7. Effect of Harmonic voltages	Possibility of large harmonic currents resonance	No effect.

(d) **By Capacitance Boosters.** The power factor can be improved by connecting capacitors in series with the line as shown in Fig. 9.23. The inductive reactance of the line is neutralized by the capacitive reactance of the capacitors. The capacitance of the capacitor (condenser) to neutralize the line inductive reactance is given by,

$$C = \frac{1}{\omega^2 L} = \frac{1}{(2\pi f)^2 L}$$

where f is the supply frequency and L *is* the line inductance per phase.

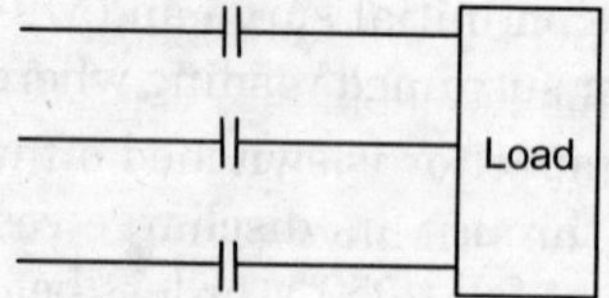

Fig. 9.23. Capacitors in series with load.

(e) **Synchronous condensers.** The synchronous motor takes a leading current when it is over-excited. An over-excited synchronous motor when running on no load is known as **synchronous condenser** and behaves like a condenser or capacitor, the capacitive reactance of which depends upon the motor excitation. The power factor can be improved by using synchronous condensers just like shunt capacitors connected across the supply as shown in Figs. 9.21 and 9.22.

The advantage of synchronous condenser over static capacitor is that power factor can very easily be controlled by variation of the field excitation.

9.14. Location of Power Factor Improvement Apparatus

The best location for the power factor improvement equipment to be installed is where the apparatus or equipment responsible for low power factor operating. In distribution systems if static capacitors are to be installed for improving the power factor, the correct location would be the receiving end and most suitably the receiving end sub-station and the static capacitors should be connected to the 11 kV busbars. The static capacitors installed in this way would relieve not only the generators but also the lines carrying the excessive currents due to low power factor. The small static capacitors are installed very near to the induction motors such as pump motors for tubewells, arc lamps and arc furnaces etc. If on the other hand, synchronous condensers or static capacitors were installed in a generating station, only the generators would be relieved of excessive currents, the distribution and transmission system would still have to carry currents more than the normal. The synchronous condensers are installed at large sub-stations or generating stations for regulation of power factor of the power system.

9.15. Economic Limits of Power Factor Improvement

It may appear on first thought that a power factor of unity or very near to unity will be the most desirable one for a consumer because for the same kW demand, the kW demand will diminish and accordingly the consumer have to bear less demand charges. On the other hand, he will have to incur expenditure on the power factor improvement apparatus. The most economical limit for the improvement of power factor thus can be found as follows :

Case: I

kW demand constant. Let the kW demand (assumed constant) of the consumer be represented by *OA*. If, in the beginning, the power factor (lagging) is $\cos \phi_1$, the demand in kVA is given by *OB*. Let $\cos \phi_2$ be the power factor after improvement such that the new kVA demand is *OC* as shown in Fig. 9.24.

Reduction in kVA demand

$$= OB - OC = \frac{KW}{\cos \phi_1} - \frac{KW}{\cos \phi_2}$$

If Rs. *X* is the demand charge per kVA then the saving in demand charge

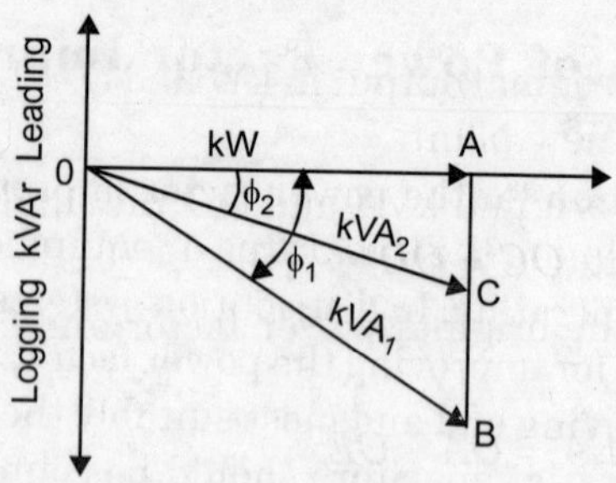

Fig. 9.24 Vector diagram - Power factor improvement.

$$= \text{Rs.}\, X\left(\frac{KW}{\cos\phi_1} - \frac{KW}{\cos\phi_2}\right)$$

Against this saving, the kVAR (leading) to be provided by power factor improvement apparatus

$$= AB - AC = CB$$

i.e. $(KW \tan\phi_1) - (KW \tan\phi_2)$ or $KW(\tan\phi_1 - \tan\phi_2)$.

If Rs. Y is the annual interest and depreciation charges on capital investment per kVA of improvement apparatus, then total annual charge on this account will be,

$$= \text{Rs.}\, Y\, KW(\tan\phi_1 - \tan\phi_2)$$

Net saving, $S = X.KW\left(\dfrac{1}{\cos\phi_1} - \dfrac{1}{\cos\phi_2}\right) - YKW(\tan\phi_1 - \tan\phi_2)$

For maximum saving: $\dfrac{dS}{d\phi_2}$ should be equal to zero.

i.e. $$-X.\, \text{kW} \sec\phi_2 \tan\phi_2 + Y.\text{kW} \sec^2\phi_2 = 0$$

or $$\sin\phi_2 = \frac{Y}{X}$$

or $$\cos\phi_2 = \sqrt{1-\sin^2\phi_2} = \sqrt{1-\left(\frac{Y}{X}\right)^2}$$

From the above expression it can be seen that the power factor $\cos\phi_2$ to be determined is independent of the original power factor $\cos\phi_1$ and is governed by the relative costs of supply and the power factor improvement apparatus (equipment).

Case : II

When kVA demand is constant. We consider the case when the additional kW capacity is met by the installation of new plant.

Let OE be the original output in kW and OA the output after the installation of the new plant.

Let OC be the original kVA and OD the final kVA. (since kVA demand is constant, $OC = OD$).

Let $\cos \phi_1$ be the original power factor and $\cos \phi_2$ be the final power factor.

Now $\quad EA = OA - OE$

$$= OD \cos \phi_2 - OC \cos \phi_1 = KVA(\cos \phi_2 - \cos \phi_1)$$

From triangle FBC, CF represents the additional power (kW) supplied by the additional generating plant and CB represents the additional kVA generated by this plant.

Therefore, additional generated kVA.

$$= \text{CB} = \frac{CF}{\cos \phi_1} = \frac{EA}{\cos \phi_1}$$

$$= \frac{kVA(\cos \phi_2 - \cos \phi_1)}{\cos \phi_1}$$

Now consider the case when the generator kVA can be kept down the original value by installation of a phase advancing plant having a kVAR rating equal to DB.

$$\therefore \quad DB = AB - AD = OB \sin \phi_1 - OD \sin \phi_2$$

$$= \frac{OA}{\cos \phi_1} \cdot \sin \phi_1 - OD \sin \phi_2$$

$$= \frac{OD \cos \phi_2}{\cos \phi_1} . \sin \phi_1 - OD \sin \phi_2$$

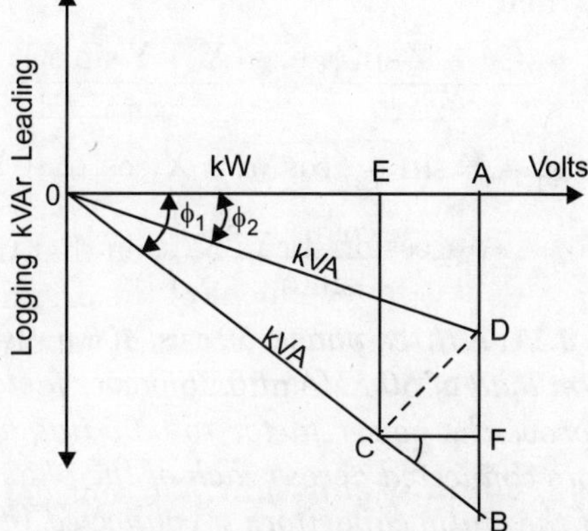

Fig. 9.25. Vector diagram – Power factor improvement.

$$= OD\left(\frac{\cos\phi_2.\sin\phi_1}{\cos\phi_1} - \sin\phi_2\right)$$

$$= KVA\ (\tan\ \phi_1 \cos\ \phi_2 - \sin\ \phi_2)$$

If the first alternative costs Rs. X per kVA and the second alternative costs Rs. Y per kVA, the two costs are respectively,

$$\text{Rs. } X \cdot KVA \left(\frac{\cos\phi_2 - \cos\phi_1}{\cos\phi_1}\right) \text{ and}$$

$$\text{Rs. } Y \cdot KVA\ (\tan\ \phi_1 \cos\ \phi_2 - \sin\ \phi_2)$$

Thus the phase advancing plant will be cheaper if

$$Y\ KVA\ (\tan\ \phi_1 \cos\ \phi_2 - \sin\ \phi_2) < X.\text{kVA}\left(\frac{\cos\phi_2 - \cos\phi_1}{\cos\phi_1}\right)$$

Or $$Y(\tan\ \phi_1 \cos\ \phi_2 - \sin\ \phi_2) < X\left(\frac{\cos\phi_2 - \cos\phi_1}{\cos\phi_1}\right)$$

Or $$(Y\tan\ \phi_1 \cos\ \phi_2 - Y\sin\ \phi_2) < \left(\frac{X\cos\phi_2}{\cos\phi_1} - X\right)$$

Or $$(X - Y\sin\ \phi_2) < \left(\frac{X\cos\phi_2}{\cos\phi_1} - Y\tan\phi_1\ \cos\phi_2\right)$$

Or $$\frac{X - Y\sin\phi_2}{\cos\phi_2} < \frac{X - Y\sin\phi_1}{\cos\phi_1}$$

In the limiting case, the maximum cost of the phase advancing plant which would justify its installation is when its working cost is equal to the working cost of generating plant to give the same increase in power, *i.e.* when

$$\frac{(X - Y\sin\phi_2)}{\cos\phi_2} = \frac{(X - Y\sin\phi_1)}{\cos\phi_1}$$

or $$X\cos\ \phi_1 - Y\sin\ \phi_2.\ \cos\ \phi_2 = X\cos\ \phi_2 - Y\sin\ \phi_1.\ \cos\ \phi_2$$

or $$Y = X\ \frac{\cos\phi_2 - \cos\phi_1}{\sin(\phi_1 - \phi_2)}$$

Example 9.11. A *three-phase, 50 c/s, 400 volts induction motor has a maximum load of 50 KVA at 0.75 power factor (lagging). It is desired to improve the power factor to 0.95 (lagging) by means of static capacitors connected across each of the phases. Calculate the capacitance of the static capacitors if connected in delta across the load.*

Solution. Original power factor of the load, $\cos \phi_1 = 0.75$ (lag)

Improved power factor of the load, $\cos \phi_2 = 0.95$ (lag)

Load of the motor = 50 × 0.75 = 37.5 *KW*

Now,

$\tan \phi_1 = \tan \cos^{-1} 0.75 = 0.8816$, $\tan \phi_2 = \tan \cos^{-1} 0.95 = 0.3288$

Leading kVAR supplied by the static capacitors,

$$= \text{kW} (\tan \phi_1 - \tan \phi_2)$$

$$= 37.5(0.8816 - 0.3288) = 37.5(0.5528) = 20.75$$

Leading kVAR supplied by each of the three capacitors

$$= \frac{20.75}{3} = 6.92 \quad \text{... (i)}$$

Phase current of the capacitor, $I_{Cph} = \dfrac{V}{X_C}$

where *Xc is* the capacitive reactance

$$= \frac{V}{1/\omega C} = \omega CV = 2\pi fCV$$

$$= 2 \times \frac{22}{7} \times 50 \times C \times 400 = 125{,}600C$$

where *C* is the capacitance of each capacitor to be found out.

kVAR of each phase

$$= \frac{V_{ph.} I_{cph}}{1000} = \frac{400 \times 125{,}600C}{1000} \quad \text{...(ii)}$$

Equating equation (*i*) and *(ii)*

$$\frac{400 \times 125{,}600C}{1000} = 6.92 \text{ or } C = \frac{6.92 \times 1000}{400 \times 125{,}600}$$

$$= 137.7 \times 10^{-6} \textbf{ Farads} \text{ or } 137.7\mu F. \quad \textbf{Ans.}$$

Example 9.12. *A synchronous motor having a power rating of 50 KW is connected in parallel with a load of 200 KW at a power factor of 0.8 lagging. If the combined load has a power factor of 0.9, what is the value of leading KVAR supplied by the motor and at what power factor is it working ?*

(P.U. II Year Elect, and Mech. Engg. April, 1984).

Solution. Motor load, $P_1 = 50$ kW, Other load, $P_2 = 200$ kW

Power factor of load, P_2

$$= \cos \phi_1 = 0.8 \text{ (lagging)}$$

$$\therefore \quad \phi_1 = \cos^{-1} 0.8$$

and $\tan \phi_1 = \tan \cos^{-1} 0.8 = 0.75$

Total load $= P_1 + P_2 = 50 + 200 = 250$ kW

Power factor of combined load,

$$\cos \phi_2 = 0.9 \text{ (lagging)}$$

$$\therefore \quad \phi = \cos^{-1} 0.9$$

and $\tan \phi_2 = \tan \cos^{-1} 0.9 = 0.484.$

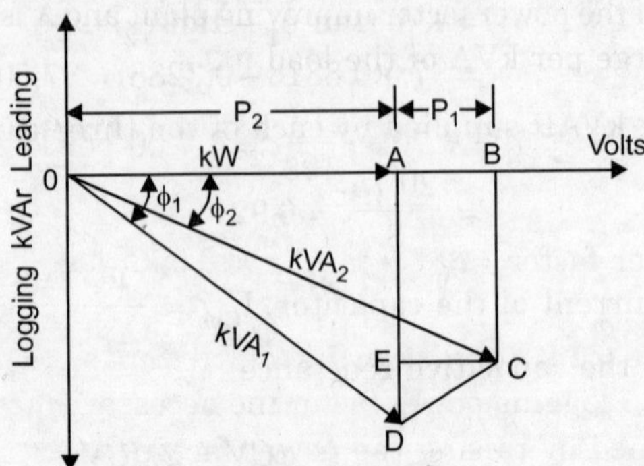

Fig. 9.26.

As shown in Fig. 9.26, from the vector diagram, reactive kVAR (lagging) supplied by the motor,

$$ED = AD - AE = AD - BC = P_2 \tan \phi_1 - (P_1 + P_2) \tan \phi_2$$

$$= 200 \times 0.75 - 250 \times 0.484$$

$$= 150 - 121 = 29 \text{ kVAR} \textbf{ Ans.}$$

kVA rating of the motor,

$$= \sqrt{(KW)^2 + (KVAR)^2} = \sqrt{(50)^2 + (29)^2} = 57.8$$

Hence power factor of motor,

$$\cos \phi = \frac{P_1}{57.8} = \frac{50}{57.8} = 0.865 \text{ (leading)} \quad \textbf{Ans.}$$

Example 9.13. *(a) Deduce an expression for the most economical angle of lag at which a consumer should take power when the charge is based on a fixed sum per KVA of maximum demand plus a flat rate per KWh.*

(b) Calculate the value of the new power factor when the tariff is ₹90 per KVA of maximum demand plus a flat rate per KWh. Assume additional cost of condensers etc. is ₹ 70 per KVA of such plant. Rate of interest and depreciation is together to be taken as 10%.

Solution, (*a*) See paragraph 9.15, case 1.

(*b*) From the foregoing paragraph 1.15, case 1 the value of new power factor,

$$\cos \phi_2 = \sqrt{1 - \left(\frac{Y}{X}\right)^2}$$

where Y is the interest and depreciation charges on the capital investment of the power factor improving plant and X is the maximum demand charge per kVA of the load.

Hence, $$Y = ₹\ 70 \times \frac{10}{100} = ₹\ 7.00 \text{ per kVA}$$

$$\therefore \quad \text{Power factor} = \sqrt{1 - \left(\frac{7}{90}\right)^2} = 0.996 \text{ (lagging)} \quad \textbf{Ans.}$$

Example 9.14. *For increasing the kW* capacity of a power plant working at 0.7 lagging power factor the necessary increase in power can be obtained by raising the power factor to 0.9 or by installing additional plant. What is the maximum cost per kVA of power factor correction apparatus to make its use more economical than additional plant at ₹ 800 per kVA ?

Solution. From foregoing paragraph 9.15 case II, it is seen that the kVA demand of the power plant is fixed and by installation of additional plant, the additional generated kVA

$$= \frac{\text{KVA demand} (\cos \phi_2 - \cos \phi_1)}{\cos \phi_1}$$

The cost of the additional plant,

$$= ₹\ 800 \times \frac{0.9 - 0.7}{0.7} \text{ per kVA} = ₹\ 800 \times \frac{0.2}{0.7}$$

$$= ₹\ \frac{1600}{7} = ₹\ 228.57$$

kVA rating of the phase advancing plant (power factor correction apparatus) if installed,

$$= \text{kVA demand} (\tan \phi_1 . \cos \phi_2 - \sin \phi_2)$$

kVA rating of the plant per kVA demand

$$= (\tan \phi_1 . \cos \phi_2 - \sin \phi_2)$$

Now, $$\tan \phi_1 = \tan \cos^{-1} (0.7) = 1.0203$$

$$\sin \phi_2 = \sin \cos^{-1} (0.9) = 0.483$$

and $\cos \phi_2 = 0.9.$

$\therefore$ kVA rating of the plant per kVA demand

$$= (1.0203 \times 0.9 - 0.483) = 0.4353.$$

Let ₹ Y be the cost per kVA rating of the power factor correction plant. Then for the maximum cost per kVA of the p.f. correction plant to make its use more economical than additional plant,

$$Y(0.4353) = ₹\ 228.57 \text{ or } Y = ₹\ \frac{228.57}{0.4353} = ₹\ \mathbf{525.} \quad \textbf{Ans.}$$

Example 9.15. *A system is working at its maximum KVA capacity with a lagging power factor of 0.71. An anticipated increase of load could be met by:*

(i) raising the power factor of the system to 0.87 by the installation of phase advancers.

(ii) installing extra generating plant, cables etc. to meet the increased power demand.

Estimate the limiting KVA of phase advancing plant which would justify its use if the cost for generating plant is ₹ 60 per KVA. Interest and depreciation charges may be assumed to be 1% in each case.

Solution. Let the original kVA be 1000 and the vector diagram be as shown in Fig. 9.27.

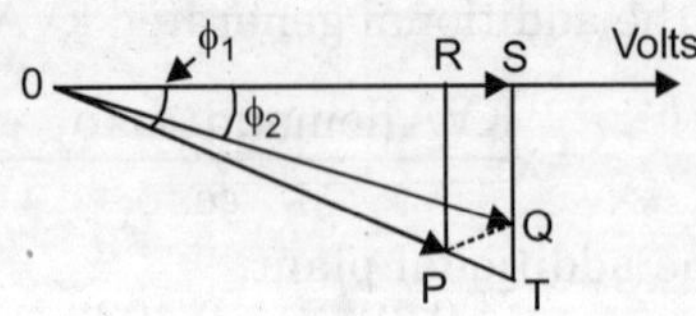

Fig. 9.27

Increase in true power

$$= 1000 (\cos \phi_2 - \cos \phi_1)$$
$$= 1000 (0.87 - 0.71) = 160 \text{ kW}$$

Total true power (load)

$= (OS \times OP) +$ Increase in true power $= (1000 \times 0.71 + 160)$

$= 870$ kW which operates at 0.87 power factor lagging.

kVAR of necessary phase advancing plant $= QT$

$$= OS (\tan \phi_1 - \tan \phi_2) = 870 \times (0.9917 - 0.5669)$$
$$= 870 \times 0.4248 = 369.6 \text{ kVAR.}$$

If extra generating plant is installed, rating of plant required, = 160 kW which operates at a p.f. of 0.71.

Therefore, kVA of new plant required

$$= \frac{160}{0.71} = 225.28 \text{ kVA.}$$

Let cost of generating plant be ₹ X per kVA, then

$$X = \frac{10}{100} \times 60 = ₹\ 6.00$$

Let Y be the annual working cost per kVAR of phase advancing plant. Then, we have for maximum cost of phase advancing plant, ₹ 6 × 225.28 = ₹ Y × 369.6

Thus

$$Y = 6 \times \frac{225.28}{369.6} = ₹\ 3.65 \quad \textbf{Ans.}$$

It means that ₹ 3.65 × $\frac{100}{10}$ = **₹ 36.5 per kVAR** is the maximum capital cost of phase advancing plant to justify its installation. Above this price, additional generating plant will prove to be more economical.

Check. By use of the formula,

$$\frac{X - Y \sin \phi_2}{\cos \phi_2} = \frac{X - Y \sin \phi_1}{\cos \phi_1}$$

$$\frac{6 - Y \times 0.4932}{0.87} = \frac{6 - Y \times 0.7042}{0.71}$$

or
$$Y = 6\left[\frac{0.7042}{0.71} - \frac{0.4932}{0.87}\right] = 6\left[\frac{1}{0.71} - \frac{1}{0.87}\right]$$

Solving, Y = ₹ 3.65 (as before).

Example 9.16. *The load on the mains of a supply system is 1000 KW at a power factor of 0.707 lagging. What must be the KVA rating of phase advancing plant which takes leading current at a power factor of 0.1 in order to raise the power factor of the whole system to 0.95 ?*

Solution. From phasor diagram (Fig. 9.28), the original load at a power factor $\cos \phi_1$ (= 0.707 lagging) = $\frac{1000}{0.707}$ kVA

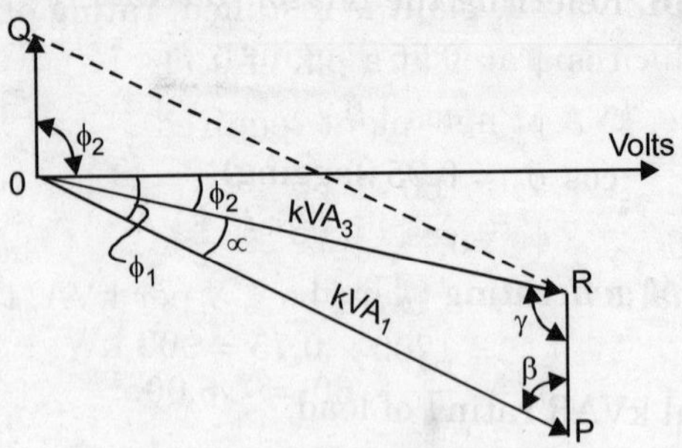

Fig. 9.28. Phasor diagram.

OQ, is the load due to phase advancing plant leading the voltage by an angle ϕ_2, where cos ϕ_2 = 0.1 and *OR* which is the resultant *of OP* and *OQ* lags the voltage by an angle ϕ_3, where cos ϕ_3 = 0.95.

We thus have,

$$\cos\phi_1 = 0.707, \quad \phi_1 = 45°$$
$$\cos\phi_2 = 0.1, \quad \phi_2 = 84°\ 16'$$
$$\cos\phi_3 = 0.95, \quad \phi_3 = 18°\ 12'$$

Considering triangle *OPR* in which,

$$\alpha = \phi_1 - \phi_3 = 45° - 18°\ 12' = 26°\ 48'$$
$$\gamma = \phi_2 + \phi_3 = 84°\ 16' + 18°\ 12' = 102°\ 28'$$
$$\beta = 108° - (26°\ 28' + 102°\ 28') = 50°\ 44'$$

Then $$\frac{kVA_2}{kVA_1} = \frac{\sin 26°28'}{\sin 102°28'} = \frac{0.4509}{0.9764} = 0.4618$$

So that $$kVA_2 = kVA_1 \times 0.4618 = \frac{1000}{0.707} \times 0.4618 = 653.2$$

Thus kVA rating of the phase advancing plant = **653.2 Ans.**

Example 9.17. *The load on a certain installation may be considered constant at 1200 KVA, 0.75 lagging power factor for 3000 hours per annum. The tariff is ₹65 per KVA of maximum demand plus 2 paise perKWh.*

(a) Determine the annual charge for electrical energy.

(b) Power factor improving apparatus is installed to improve the power factor to 0.95 lagging. Determine the KVAR required and the new annual charge if the power factor improving apparatus costs ₹60 per KVAR, annual interest and depreciation charges are 10% of the capital cost and the losses in the apparatus are 5 per cent of the kVAR rating.

Solution. Referring the phasor diagram (Fig. 9.29).

$\cos \phi_1 = 0.75$ (lagging)

Thus, $\phi_1 = \cos^{-1} 0.75$

and $\cos \phi_2 = 0.95$ (lagging)

and $\phi_2 = \cos^{-1} 0.95$

Now, original kW rating of load,

$= 1200 \times 0.75 = 900$ kW

Original kVAR rating of load,

$= 1200 \times \sin(\cos^{-1} 0.75)$

$= 1200 \times 0.6613 = 793.56.$

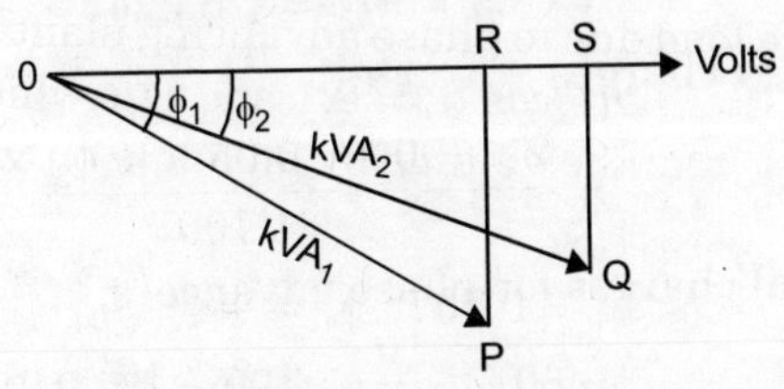

Fig. 9.29. Power diagram.

Let x be the kVAR rating of phase advancer required for power factor improvement.

Then, Final kVAR $= 793.56 - x$ ...(*a*)

Also there are losses of 5% in the phase advancing apparatus.

Thus loss due to x kVAR rating,

$= 0.05x$ kW.

Hence new kW demand

$= 900 + 0.05x$ ...(*b*)

Now from (*a*) and (*b*)

$$\tan \phi_2 = \frac{793.56 - x}{900 + 0.05x} = 0.3288$$

[$\because \tan \phi_2 = 0.3288$ as $\cos \phi_2 = 0.95$]

or $x = 489.6$ kVAR

$= 490$ kVAR (say) **Ans.**

(*a*) **Annual charges.** If phase advancing apparatus is not installed:

(*i*) Maximum demand charges,

$= 65 \times 1200 =$ ₹ 78,000

(*ii*) Energy charges

$$= \frac{2 \times 900 \times 3000}{100} = ₹\ 54{,}000.$$

∴ Total-annual charges,

$$= 78{,}000 + 54{,}000 = ₹\ 132{,}000 \quad \textbf{Ans.}$$

(b) **Annual charges.** When phase advancing apparatus is installed.

Maximum demand in kVA

$$= \frac{900 + 0.05x}{0.95} = \frac{900 + 0.05 \times 490}{0.95} = 973.$$

(i) Maximum demand charges,

$$= ₹\ 65 \times 973 = ₹\ 63{,}245$$

(ii) Energy charges

$$= \frac{2 \times (900 + 0.05 \times 490) \times 3000}{100} = ₹\ 55{,}470$$

(iii) Annual charges for phase advancers,

$$= 60 \times \frac{10}{100} \times 490 = ₹\ 2{,}940$$

Thus total annual charges

$$= ₹\ (63{,}245 + 55{,}470 + 2{,}940) = ₹\ 121{,}655$$

We thus find that with the installation of p.f. improvement plant, there is a net saving of ₹ (132,000-121,655) = ₹ 10,345 in the total annual charges.

9.16. Improvement of Load Factor

As already explained earlier above the higher load factor means more average load or more number of units generated for a given maximum demand and, therefore, overall cost per unit of electrical energy generated is reduced due to the distribution of the fixed charges over the more number of units generated.

With the improvement of load factor the un-utilized capacity of the plant is utilized. Alternately it can be argued that for given average load, any increase in the load factor would result in corresponding proportional reduction in the maximum demand on the station. This results in the reduction of installed capacity and consequently semi-fixed charges are also reduced.

The effect of the load factor on the overall cost per kWh will be visualized from the following table :

Load Factor	0.2	0.4	0.6	0.8	1.0
Fixed charges per kWh	18	9	6	4.5	3.5
Running charges per kWh	8	8	8	8	8
Total charges per kWh	26	17	14	12.5	11.5

9.17. Economic Choice of Equipment – Initial Cost and Efficiency

There are a variety of equipment available in the market to perform the same functions but each type of equipment having some special features over the other. It is upto the engineer who is to weigh the effects of various features quantitatively and decide which of the equipment will be economical in the long run. In some situations it may be desirable to install more costly equipment which have less losses or *vice versa.* Total annual expenditure on the equipment consists of two parts, *first part being composed of annual interest and depreciation charges on capital cost of the equipment and second part represents the monetary value of annual losses taking place.* Annual losses in turn depend upon the daily duty cycle of the equipment. If the number of working hours of the equipment during the year are more, the cost of annual energy losses with more efficient equipment will be less and it is desirable to choose more efficient but costly equipment. On the other hand, if the working hours per year of the equipment are less, then the cost of annual energy losses will be less as compared to the fixed charges and it may be economical to choose a less efficient and less costly equipment. It can be seen that the same equipment may prove to be economical under one set of conditions while in the other set of conditions its choice may be uneconomical. It can thus be summarised that the various factors which influence the choice of the equipment are as follows

(*a*) rate of interest and depreciation ;
(*b*) cost of electric energy per unit;
(*c*) demand charges of two part tariff if applicable ;
(*d*) duty cycle of the equipment.

9.18. Capitalization of Losses

The economic choice of the equipment does not take into account the capital cost of the equipment but also the capitalized cost of the losses. If C_a and C_b be the costs of the equipment A and B respectively and L_a and L_b be their annual losses and P be the percentage rate of annual interest and depreciation, then the choice of that equipment will be economical for which the sum of capital cost plus the capitalized cost of the losses is less. If C_a is greater than C_b and consequently L_a is less than L_b, then equipment A will be economical as compared to equipment B.

Mathematically, when $C_a + \frac{L_a}{P} \times 100 < C_b + \frac{L_b}{P} \times 100$

Then $\quad C_a - C_b < \frac{L_b - L_a}{P} \times 100.$

9.19. Reduction of Energy Costs

The energy cost depends upon the following charges :

(i) **Fixed cost.** Tho cost independent of maximum demand and energy output. It is due to the annual cost of central organisation, interest on the capital cost land (especially if some land is held for future development) and salaries of high officials.

(ii) **Semi-fixed cost.** The cost depending upon the maximum demand but independent of energy output. The semi-fixed cost is due to annual interest and depreciation on the capital cost of building and equipment, taxes and insurance charges and salaries of management and clerical staff, cost of transmitting and distributing the electrical energy. The semi-fixed cost is approximately proportional to the maximum demand.

(iii) **Operating cost.** This cost depends upon the energy output. The running or operating cost is due to annual cost of fuel, lubricating oil, water, maintenance and repairs, wages and salaries of operating staff etc. The operating cost is approximately proportional to the units generated.

By installing the equipment chosen economically and having load factors very near to unity, the energy costs can be reduced. Unutilized capacity of the plant will be utilized. As the running or operating charges in case of hydroelectric plants are very nominal, the installation of more hydroelectric plants, the energy charges are reduced. If the diversity factor of the load is higher, then with the same number of consumers the capacity of the plant required will be smaller and consequently the fixed charges due to capital investment will be reduced very much, thereby reducing the energy costs. Thus it can be summarized that the energy costs can be reduced by :

(a) Installation of hydroelectric plants
(b) Improvement of load factor.
(c) Improvement of diversity factor

9.20. Cost of Renewals

In some instances, the life of the equipment and, therefore, the cost of renewals can be widely influenced by the mode of operation. An example of this nature is afforded by the metal-filament lamp, curves of life and output which are given in Fig. 9.30. The life of a lamp at its rated voltage is quoted by the manufacturers be about 1,000 hours, and in most cases it is desirable to operate lamps at their rated voltage.

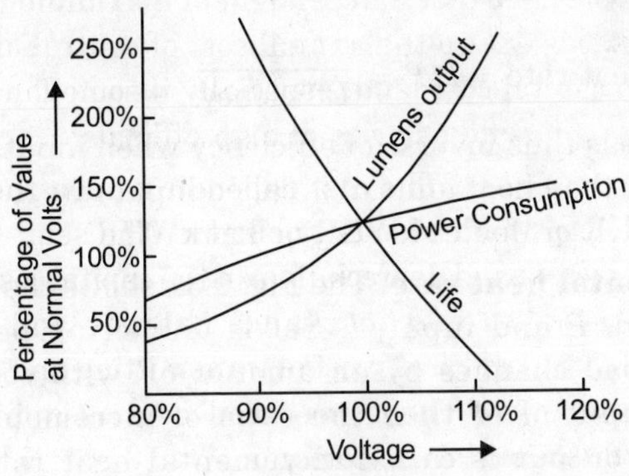

Fig. 9.30. Effect of voltage on output, power consumption and life of filament lamps.

It should be kept in mind that a lower energy cost would result in a greater saving, while if the energy cost is very high it may even pay to operate lamps at a voltage higher than their rated value.

As a rule, all items which recur every ten years or oftenr, and which are not large enough to embarrass the year's accounts unduly, can conveniently be regarded as current repairs and paid for out of hand, whereas additions, betterments, and renewals of all major items of the plant whose choice is a matter of individual selection, should be paid for out of the capital account (depreciation reserve), a new fund being then instituted for the new part. It, therefore, depends in part upon the magnitude as well as the frequency of the renewal; thus in a storage battery, new acid or alkali even though required only once or twice in a life time, would usually be regarded as maintenance, whereas new plates would be virtually a new battery and would be purchased out of the depreciation fund established for the purpose.

Naturally the repairs and renewals policy must be taken into consideration when estimating the total useful life and salvage value of each item, since a generous policy of replacements out of revenue may keep the plant in such a condition that the depreciation fund virtually has only to provide against obsolescence. In any case, however, the estimates should assume that the plant is maintained in working order out of revenue.

9.20.1. Heat Rate.

It is the rate of input expressed in heat units per unit of output. If input is '*I*' and output '*P*'.

Then $\text{Heat rate} = I/P = \dfrac{1}{\text{Efficiency}}$

Heat rate is thus inverse of efficiency when input is expressed in units other than heat units, it is called input rate. It is expressed in B.T.O/kWHR or Joules/kWHR or kal/kWHR.

Incremental heat rate. The Fig. 9.31 shows the relationship between load, *P* and input, *I*. Set is delivering power *P* with input *I*. If load changes by an amount δP with corresponding change in input of δI then the ratio of incremental input to incremental output is called incremental heat rate or simply incremental rate.

$\therefore$ Incremental heat rate = $\delta I/\delta P$

In the limit at the load point = *P*

$$\text{Incremental heat rate} = \frac{dI}{dP}$$

Condition for max. efficiency. Since efficiency is inverse of heat rate therefore when efficiency is maximum, heat rate will be minimum.

$$\text{Heat Rate (HR)} = I/P$$

For minimum value of heat rate $\dfrac{d(HR)}{dP} = 0.$

$$\therefore \quad \frac{d(HR)}{dP} = \frac{I\dfrac{dP}{dP} - P\dfrac{dI}{dP}}{P^2} = 0. \quad \text{or } I - P\frac{dI}{dP} = 0.$$

or $$\frac{dI}{dP} = I/P.$$

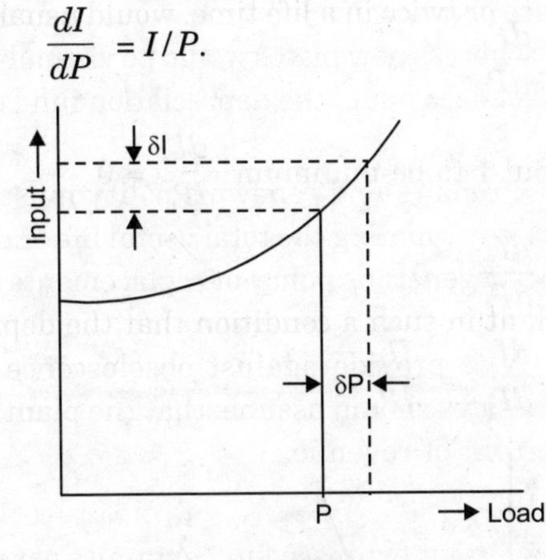

Fig. 9.31

i.e. condition for max. efficiency is that it will take place at load at which incremental heat rate is equal to heat rate.

9.20.2. Economic load sharing by alternators.

If a no. of generators are working in parallel, it is imperative from economy point of view that load shared by different generators should be such as to get max. operating n of the whole system. Considering two generators 1 and 2 total load on them is fixed *i.e.* P and if P_1 and P_2 are the load shared by generators then

$$P = P_1 + P_2 \qquad ...(i)$$

differentiate *(i)* we get $\dfrac{dP}{dP_1} = \dfrac{dP_1}{dP_1} + \dfrac{dP_2}{dP_1} = 1 + \dfrac{dP_2}{dP_1}$ $\because$ P is constant.

$$\therefore \quad \frac{dP}{dP_1} = 0$$

$$\therefore \quad \frac{dP_2}{dP_1} = -1 \qquad ...(ii)$$

But $$I = I_1 + I_2 \qquad ...(iii)$$

Differentiating *(iii)* with respect to P_1, we get

$$\frac{dI}{dP_1} = \frac{dI_1}{dP_1} + \frac{dI_2}{dP_1}$$

$$= \frac{dI_1}{dP_1} + \frac{dI_2}{dP_2} \times \frac{dP_2}{dP_1}$$

$$= \frac{dI_1}{dP_1} - \frac{dI_2}{dP_2}$$

$$\because \quad \frac{dP_2}{dP_1} = -1$$

For total input I to be minimum $\frac{dI}{dP_1} = 0$

$$\therefore \quad \frac{dI_1}{dP_1} - \frac{dI_2}{dP_2} = 0$$

or $$\frac{dI_1}{dP_1} = \frac{dI_2}{dP_2}$$

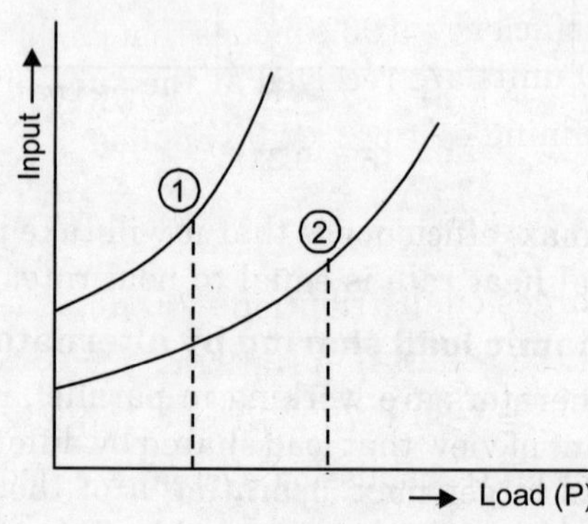

Fig. 9.32

SOME MORE CLASSIFIED EXAMPLES

Example 9.18. *A generating station has a maximum demand of 75 MW and a yearly load factor of 40 per cent. Generating costs inclusive of station capital costs are ₹60 per annum per KW demand plus 10 paise per KWh transmitted. The annual capital charges for the transmission system is ₹20 lacs and for the distribution system ₹ 15 lacs, the respective diversity factors being 1.2 and 1.25. The efficiency of the transmission system is 90% and that of the distribution inclusive of sub-station losses is 85%. Find the yearly cost per KW demand and the cost per KWh supplied (a) at the sub-stations, (b) on the consumer's premises.*

Solution, (*a*) Capital charges on transmission,

$$= ₹ \ 20 \text{ lacs} = ₹ \ 200 \times 4^4.$$

Capital cost of the generating station including fixed cost of the generation

$$= ₹ \ 60 \times 75 \times 10^3 = ₹ \ 450 \times 10^4.$$

Therefore, total fixed charges for supply to sub-stations,

$$= ₹\ (450 + 200) \times 10^4 = ₹\ 650 \times 10^4.$$

Aggregate of all maximum demand of sub-stations,

= Simultaneous maximum demand on the station × Diversity factor of sub-stations

$$= 57 \times 10^3 \times 1.2 = 90 \times 10^3 \text{ kW}.$$

Thus yearly cost per kW demand of sub-stations,

$$= \frac{₹650 \times 10^4}{90 \times 10^3} = ₹\ 72.22 \quad \textbf{Ans.}$$

Transmission efficiency of 90% means that out of every 100 units generated only 90 units are received at the sub-stations.

Therefore, Running cost per kWh supplied at the sub-stations

$$= \frac{1.00}{0.90} \times 10 = 11.11 \text{ paise. } \textbf{Ans.}$$

(*b*) Capital charges on distribution system = ₹ 150 × 10^4

Capital cost of the generating station

$$= ₹\ 60 \times 75 \times 10^3 = ₹\ 450 \times 10^4.$$

Capital charges on transmission system = ₹ 200 × 10^4.

∴ Total fixed charges for supply to the consumers

$$= ₹\ (150 + 450 + 200) \times 10^4 = ₹\ 800 \times 10^4.$$

Aggregate of all maximum demand of consumers

= Maximum demand on stations × Diversity factor of transmission × Diversity factor of distribution

$$= 75 \times 10^3 \times 1.2 \times 1.25 = 11.25 \times 10^4 \text{ kW}.$$

Thus yearly cost per kW demand on consumer's premises,

$$= \frac{₹800 \times 10^4}{11.25 \times 10^4} = ₹\ 71.11 \qquad \textbf{Ans.}$$

Similarly, running cost per kWh supplied to consumers,

$$= \frac{1.00}{0.90 \times 0.85} \times 10 = 13.07 \text{ paise.} \quad \textbf{Ans.}$$

Example 9.19. An *industrial load can be supplied on the following alternative tariffs :*

(a) H.V. supply at ₹72 per KW per annum plus 4 paise per KWh,

(b) L.V. supply at ₹ 75 per KW per annum plus 4.5 paise *per KWh.*

Transformers and switchgear suitable for the H.V. supply cost ₹50 per KW, the full load transformation losses being 2%. The fixed charges are 25% per annum on the capital cost of the H.V. plant and the installation works at full load. If there are 48 working weeks in a year, find the number of working hours per week above which the H.V. supply is cheaper.

Solution. Let x be the number of working hours per week above which the H.V. supply is cheaper. In other words, the two tariffs give equal annual costs at x.

Assume a load of 1 MW = 10^3 kW.

Transformation losses being 2%, the rating of transformers and

$$\text{switchgear} = \frac{10^3}{0.98}\text{ kW}$$

Cost of transformers and switchgear,

$$= \frac{50 \times 100}{0.98} = ₹\ 51.02 \times 10^3$$

Fixed charges per annum on H.V. Plant,

$$= 0.25 \times 51.02 \times 10^3 = ₹\ 12{,}755$$

Units consumed by the load,

$$= (48 \text{ weeks} \times x \text{ hours}) \times (1000 \text{ kW})$$
$$= 48{,}000\, x\text{kWh}.$$

On the H.V. supply the total units to be paid for,

$$= \frac{48{,}000x}{0.98} = 48{,}980\ x$$

(*a*) Annual cost on H.V. supply

= Fixed cost + Running cost + Fixed charges on H.V. plant

$$= ₹\left(72 \times \frac{1000}{0.98} \times \frac{4}{100} \times 48{,}000x + 12{,}755\right)$$

$$= ₹\ (73{,}470 + 1{,}920x + 12{,}755) = ₹\ (86{,}225 + 1920x).$$

(*b*) Annual cost on L.V. supply = Fixed cost + Running cost

$$= ₹\left(75 \times 1{,}000 + \frac{4.5}{100} \times 48{,}000x\right) = ₹\ (75{,}000 + 2{,}160x).$$

Equating the two annual costs,

$$86{,}225 + 1{,}920x = 75{,}000 + 2{,}160x$$

or $$240\,x = 11{,}225$$

$$\therefore \quad x = \frac{11{,}225}{240} = \mathbf{46.77} \text{ say } \mathbf{47 \text{ hours.}}$$

i.e. Above 47 working hours per week the H.V. supply is cheaper.

Example 9.20. *There are two offers from the suppliers in connection with the purchase of 1 MVA transformer by an electric supply company.*

Transformer I : Full-load efficiency 98%, core loss 8 KW

Transformer II : Full-load efficiency 98.5%, core loss 6 KW.

The service needed per year is 2000 hours at 1 MVA, 1000 hours at 500 KVA, and 1000 hours at 200 KVA. If the transformer I is costing ₹50,000 estimate the maximum price that could be paid for transformer B. Take interest and depreciation at 10% per annum and cost of energy 5 paise per KWh. Assume the system working at unity power factor.

Solution. In each transformer the losses will be capitalized and will be considered separately.

Transformer I

Full load output = 100 kVA × 1 = 100 kW (∵ *p.f.* = 1)

Full load efficiency = 0.98

$$\text{Full load input} = \frac{1000}{0.98} = 1020.4 \text{ kW.}$$

Full load loss = 1020.4 – 1000 = 20.4 kW

Core loss, *Pi* = 8 kW (Given)

Full load copper loss, P_c = 20.4 – 8 = 12.4 kW.

Thus the transformer losses at 1 MVA

$$= P_i + P_c = (8 + 12.4)\text{kW} = 20.4 \text{ kW}$$

The transformer losses at 500 kVA

$$= P_i + \left(\frac{1}{2}\right)^2 P_C = 8 + \frac{12.4}{4} = 11.1 \text{ kW}$$

The transformer losses at 200 kVA

$$= P_i + \left(\frac{1}{5}\right)^2 P_C = 8 + \frac{12.4}{25} = 8.5 \text{ kW}$$

Total energy losses per year

$$= 20.4 \times 2000 + 11.1 \times 1000 + 8.5 \times 1000$$

$$= 40{,}800 + 11{,}100 + 8{,}500 = 60{,}400 \text{ kWh.}$$

Cost of energy wasted per annum

$$= ₹\ \frac{5}{100} \times 60{,}400 = ₹\ 3020$$

Transformer II

Full load efficiency = 0.85

Full load output = 1000 kVA

$$\text{Full load loss} = \frac{1000}{0.985} = 1015.2 \text{ kW}$$

Full load loss = 1015.2 – 1000 = 15.2 kW

Core loss, *Pi* = *6* kW (Given)

Full load copper loss

= 15.2 – 6 = 9.2 KW

The transformer losses at 1000 kVA

$$= P_i + P_c = 6 + 9.2 = 15.2 \text{ kW}$$

The transformer losses at 500 kVA

$$= P_i + \left(\frac{1}{2}\right)^2 P_C = 6 + \frac{9.2}{4} = 8.3 \text{ kW}$$

The transformer losses at 200 kVA

$$= P_i + \left(\frac{1}{5}\right)^2 P_C = 6 + \frac{9.2}{25} = 6.37 \text{ kW}$$

Total energy losses per year

= 15.2 × 2000 + 8.3 × 1000 + 6.37 × 1000

= 30,400 + 8300 + 6370 = 45,070 kWh

Cost of energy wasted per annum,

$$= \frac{5}{100} \times 45{,}070 = ₹\ 2253.50$$

Let the cost of transformer, II be ₹ X.

Extra cost of transformer, II ₹ (X – 50,000)

Annual interest and depreciation,

$$= ₹\ (X - 50{,}000) \times \frac{10}{100} = ₹\ (0.1X - 5{,}000)$$

Hence transformer II costs annually ₹ (0.1X – *5000) more than transformer I* and annual saving due to less wastage of energy in transformer *II*

= ₹ (3020 – 2253.50) = ₹ 766.50

In limiting case annual extra cost of transformer II will be equal to saving in energy cost,

i.e. 0.1 *X* – 5000 = 76,650

or *X* = ₹ 57,665

Thus, the maximum price that could be paid for transformer II is ₹ **57,665. Ans.**

Example 9.21. *A proposed power station has to supply load as follows :*

Time in hours:	*1-8*	*8-12*	*12-17*	*17-20*	*20-23*	*23-01*
Load in KW:	*10,000*	*20,000*	*25,000*	*18,000*	*35,000*	*20,000*

After drawing the load curve, find out the load factor. Also choose suitable generator units to supply this load maintaining reliability of supply. Prepare operational schedule for the machines and calculate plant use factor.

[*Util. of Elect. Power Winter 1977 (New Scheme)*]

Solution. The load curve from the data given is as shown in Fig. 9.33.

The number of units supplied,

$$= \text{Area under the load curve}$$
$$= 20,000 \times 1 + 10,000 \times 7 + 20,000 \times 4 + 25,000 \times 5 + 18,000 \times 3 + 35,000 \times 3 + 20,000 \times 1$$
$$= 20,000 + 70,000 + 80,000 + 125,000 + 54,000 + 105,000 + 20,000 = 474,000 \text{ kWh}$$

$$\text{Average load} = \frac{474,000}{24} = 19,750 \text{ kW.}$$

Maximum demand = 35,000 kW (from load curve)

$$\therefore \quad \text{Load factor} = \frac{19,750}{35,000} = 0.56. \quad \textbf{Ans.}$$

Suitable generator units to cater this load will be,

10 MW, 10 MW, 5 MW, 5 MW, 5 MW, 3 MW, 3 MW, 2 MW, 2 MW.

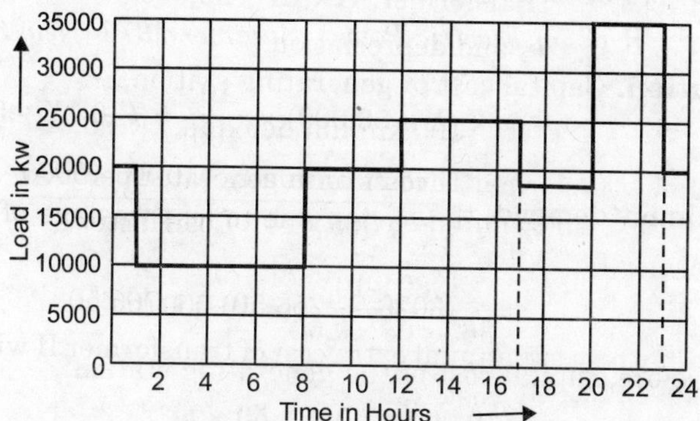

Fig. 9.33. Load curve.

(**Note** : One 5 MW, one 2 MW and one 2 MW units are as spare).

The operational schedule for the above chosen units will be:

1 — 8 hours	One, 10 MW unit.
8 — 12 hours	Two, 10 MW units.
12 — 17 hours	Two, 10 MW units and one 5 MW unit.
17 — 20 hours	One 10 MW unit, one 5 MW unit and one 3 MW unit.
20 — 23 hours	Two, 10 MW units, two 5 MW units, one 3 MW unit and one 2 MW unit.
23 — 1 hours	Two, 10 MW units.

$$\text{Plant use factor} = \frac{\text{Maximum demand on the plant}}{\text{Rated capacity of the plant}}$$

$$= \frac{35,000}{45,000} = 0.778.$$ **Ans.**

Example 9.22. *Calculate the minimum two part tariff to be charged from the consumers of a supply system from the following data :*

Generating cost per unit = 4 paise

Generating cost per KW of maximum demand = ₹ 50

Total energy generated per year = 4380 × 10^4 units

Load factor of the generating station = 50%

Annual charges for the distribution = ₹ 125,000

Diversity factor for the distribution = 1.25

Total loss between stations and consumer = 10%.

[*Util. of Electric Power, Summer 1978 (New Scheme)*]

Solution. Capital cost of generating station,

= Maximum demand × ₹ 50.

$$\text{Maximum demand} = \frac{\text{No. of units generated per year}}{365 \times 24 \times \text{Load factor}}$$

$$= \frac{4380 \times 10^4}{365 \times 24 \times 0.50} = 10,000 \text{ kW}$$

Therefore, capital cost of the generating station

$= 10,000 \times 50 = ₹\ 50 \times 10^4$

Annual charges for the distribution

$= ₹ 12.5 \times 10^4$

Therefore, total fixed charges for supply to consumers

$= ₹ (50 \times 10^4 + 12.5 \times 10^4) = ₹ 62.5 \times 10^4$

Aggregate of all maximum demands of consumers

= Maximum demand on station × Diversity factor of distribution

$= 10{,}000 \times 1.25 = 12{,}500$ kW.

Thus, yearly cost per kW demand on consumer's premises

$$= \frac{₹ 62.5 \times 10^4}{12.500} ₹ \mathbf{50}$$

Similarly, running cost per kWh supplied to consumers

$$= \frac{1.0}{0.9} \times 4 = 4.44 \text{ paise.}$$

Hence two part tariff for supply of 10,000 kW load to consumers is:

(i) Annual charges per kW of maximum demand = ₹ 50

(ii) Annual energy cost per kWh supplied = 4.4 paise **Ans.**

Example 9.23. *Annual load during curve of a system of loads is a straight line with maximum of 12 MW at the beginning and a value of 2 MW at the end of the year. Annual costs of both the base load and peak load stations are given below :*

₹80,000 + ₹75 per KW + 3 paise per KWh for base load station

₹60,000 + ₹55 per KW + 4 paise per KWh for peak load station.

Determine the duration of time when peak load station will work in order to obtain the maximum annual cost. Also find the lowest over all cost per unit.

Solution. Let the operating cost of base load station be,

$= ₹ (A_1 \times \text{kW} + B_1 \times \text{kWh})$

and the peak load station be

$= ₹ (A_2 \times \text{kW} + B_2 \times \text{kWh})$

where $A_1 > A_2$ and $B_1 < B_2$

Then for economic load sharing the area under the load duration curve is so divided by such horizontal straight line *(XY)* that its magnitude is given by,

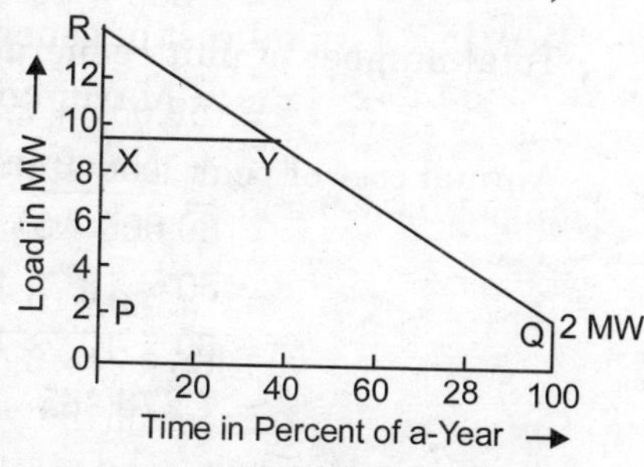

Fig. 9.34.

$\frac{A_1 - A_2}{B_2 - B_1}$ hours.

In other words, for economic load sharing, peak load station will work for $\frac{A_1 - A_2}{B_2 - B_1}$ hours per year.

For maximum economy, peak load station should work for

$$\frac{A_1 - A_2}{B_2 - B_1} = \frac{75 - 55}{4 - 3/100} = 20 \times 100 = 2000 \text{ hours.}$$

The load duration curve is as shown in Fig. 9.34.

The load shared by the peak load station *(RX)*

$$= \text{RP} \times \frac{XY}{PQ} = 10 \times \frac{2000}{(365 \times 24)} = 2.283 \text{ MW}$$

Load on base load station, *OR*

$$= 12 - 2.283 = 9.717 \text{ MW.}$$

Units generated by base load station

$$= \frac{XY + PQ}{2} \times XP + OP.PQ$$

$$= \left[\frac{2000 + 365 \times 24}{2} \times 7.717 + 2 \times 365 \times 20\right] \times 10^3$$

$$= [5380 \times 7.717 + 2 \times 8760] \times 10^3$$

$$= (41{,}517.46 + 17{,}520) \times 10^3$$

$$= 59.037 \times 10^3 = 59.037 \times 10^6 \text{ kWh.}$$

Units generated by peak load station,

$$= \frac{1}{2} RX \,.\, XY = \frac{1}{2} \times 2.283 \times 10^3 \times 2000$$

$$= 2.283 \times 10^6 \text{ kWh.}$$

Total number of units generated

$$= (59.037 + 2.283) \times 10^6 = 61.32 \times 10^6 \text{ kWh.}$$

Annual cost of peak load station

$$= 60{,}000 + 55 \times 2.283 \times 10^3 \times 0.04 \times 2.283 \times 10^6$$

$$= 60 \times 10^3 \times 125.565 \times 10^3 + 91.32 \times 10^3$$

$$= 60 \times 10^3 \times 125.565 \times 10^3 + 91.32 \times 10^3$$

$$= ₹\ 276.885 \times 10^3$$

Annual cost of base load station

$= 80{,}000 + 75 \times 9.717 \times 10^3 + 0.03 \times 59.037 \times 10^6$

$= 80 \times 10^3 + 728.775 \times 10^3 + 1771.11 \times 10^3$

$= ₹\ 2579.885 \times 10^3$

Total annual cost

$= ₹\ (2579.885 + 276.885) \times 10^3$

$= ₹\ 2856.77 \times 10^3$

Overall unit cost

$$= \frac{₹2856.77 \times 10^3}{61.32 \times 10^6}\ ₹\ 0.046$$

= 4.6 paise per kWh. **Ans.**

Example 9.24. (*a*) *Show that the economical limit to which the power factor of lagging load can be raised is independent of the original value of the power factor when the tariff consists of a fixed charge per KVA plus a flat rate per unit.*

(b) A consumer is charged ₹. 30 per annum per kVA maximum demand, plus a flat rate per KVA. Phase advancing plant can be purchased for ₹30 per KVA, the cost for depreciation, housing and interest on capital being 12 per cent. Find the most economical angle of lag to which the load can be improved under these conditions.

Solution (*a*) For solution of (*a*), please refer to previously discussed relevant paragraph.

(*b*) From the given data, we have,

(*i*) Cost per annum per kVA maximum demand,

$X = ₹\ 30$

(*ii*) Cost per annum per kVAR *of p.f.* improvement plant,

$Y = ₹\ 30 \times 0.12 = ₹\ 3.6$

Hence the most economical angle of lag ϕ_2, is given by,

$$\sin \phi_2 = \frac{Y}{X} = \frac{3.6}{30} = 0.12$$

Therefore, $\phi_2 = \sin^{-1}(0.12) = 6.9°$. **Ans.**

[**Note.** When the phase advancing plant does not consume power, *i.e.* it is wattless then its kVA rating is same as kVAR rating. If, however, it consumes power say *w* kW, then the kVA rating would be $[\sqrt{(KVAR)^2 + (w)^2}]$

Example 9.25. *The maximum demand of a factory is 3000 KVA at a power factor of 0.7 (lagging) and the tariff is ₹60 per annum per KVA of maximum demand. Determine the KVAR rating of capacitors* *which would result in maximum economy if the capital cost is ₹40 per kVAR and interest charges are at the rate of 10% per annum. What would be the annual saving in electricity charges when the capacitors are installed ?*

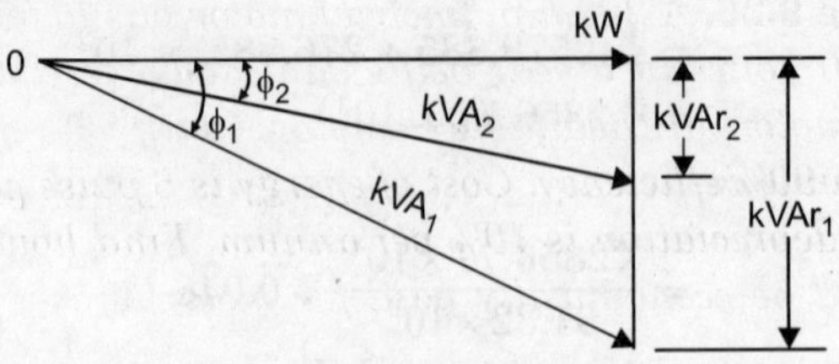

Fig. 9.35.

Solution. First of all find out the most economical angle of lag as in the previous example. Thus,

$$\phi_2 = \sin^{-1}\left(\frac{40 \times 0.1}{60}\right) = 3.8°$$

$$\phi_1 = \cos^{-1}(0.7) = 45.5°$$

Now we have to compare the cost of supply without and with capacitors. When no capacitors were installed the load of the factory was 3000 kVA at an angle of lag, $\phi_1 = 45.5°$ and power drawn by it was $3000 \times 0.7 = 2100$ kW.

After the installation of capacitors the same amount of power that is, 2100 kW is drawn at angle of lag $\phi_2 = 3.8°$. Thus there is reduction in kVA demand and this goes to affect the reduction in bill. However, the energy consumption is the same in two cases and so we do not calculate the energy cost for computing the difference.

Cost of supply without capacitors

$$= ₹\ (3000 \times 60) = ₹\ 180 \times 10^3$$

Cost of supply with capacitors

$$= ₹\ [\text{kVA } 2\ (\tan\phi_1 - \tan\phi_2) \times 40 \times 0.1 + \text{kVA}_2 \times 60]$$

$$\text{kVA}_2 = \frac{3000 \times 0.7}{\cos 3.8°} = 2110 \text{ kVA.}$$

The kVAR rating of the capacitor

$$= 2110\ (\tan\phi_1 - \tan\phi_2)$$

$$= 2110\ (1.0176 - 0.0664) = 2000.$$

Hence the annual cost of supply

$= ₹\ (2000 \times 40 \times 0.1 + 2110 \times 60)$

$= ₹\ (8000 + 126{,}600) = ₹\ 134{,}600.$

Hence the annual saving = ₹ (180,000 - 134,600)

= ₹ 45,400. **Ans.**

Example 9.26. *A 200 h.p. (metric) motor has to operate at full load for 2000 hours in a year being shut down for the remaining period. Two motors are available — motor A with 89% efficiency and motor B with 90% efficiency. Cost of energy is 6 paise per KWh and interest and depreciation is 10% per annum. Find how much more amount could be economically paid for motor B over the price of motor A ?*

Solution. Power drawn by motor,

$$A = \frac{\text{H.P.} \times 735.5}{\text{Efficiency} \times 1000} = \frac{200 \times 735.5}{0.89 \times 1000} \text{ kW}$$

Cost of energy consumed by A per year

$$= ₹ \left(\frac{200 \times 735.5}{0.89 \times 1000} \times 2000 \times 0.06 \right)$$

Cost of energy consumed by *B* per year

$$= ₹ \left(\frac{200 \times 735.5}{0.9 \times 1000} \times 2000 \times 0.06 \right)$$

Saving per year on energy bill, if motor *B* is used,

$$= 200 \times 0.7355 \times 2000 \times 0.60 \left(\frac{1}{0.89} - \frac{1}{0.9} \right)$$

$$= 735.5 \times 24 \left(\frac{0.01}{0.89 \times 0.9} \right) = ₹\ 220.40.$$

Capital cost corresponding to annual charges of ₹ 220.40 at 10% rate of interest and depreciation

$$= \frac{22.40}{0.1}\ ₹\ 22.04$$

We can, therefore, spend ₹ 2204 more on motor *B* than on motor A.

Example 9.27. A *generating station has a maximum demand of 80,000 KW and yearly load factor of 50%. Generating costs inclusive of station capital cost are ₹ 70 per annum per KW demand plus l. 5 paise per KWh transmitted. The annual capital charge for the transmission system is ₹ 30 lakhs and for distribution system ₹ 25 lakhs respective diversity factors being 1.2 and 1.3. The efficiency of* *the transmission system is 90% and that of distribution system inclusive of sub-station losses is 83%. Find the yearly cost per kW demand and the cost per KWh supplied on the consumer premises.*

Solution. Maximum demand on the power station

= 80,000 kW

Generating cost inclusive of station capital cost per annum

= ₹ 70 × 80,000 = ₹ 56,00,000

Capital charges for the transmission system per annum

= ₹ 30,00,000

Hence total fixed charges associated with supply of energy on the sub-station ;

= ₹ 56,00,000 + ₹ 30,00,000

= ₹ 86,00,000

Aggregate of all maximum demand by the sub-stations = Max demand of power station × Diversity factor

= 80,000 × 1.2 = 96,000 kW

Capital charges for the distribution system per annum

= ₹ 25,00,000

Total fixed charges associated with the supply of energy on the consumer premises

= ₹ 86,00,000 + ₹ 25,00,000

= ₹ 1,11,00,000

Aggregate of maximum demands of all consumers = Aggregate of all maximum demands on the substation × Diversity factor

= 96,000 × 1.3 = 124,800 kW.

Annual cost per kW demand on the consumer premises

$$= \frac{1,11,00,000}{124,800} = ₹\ \mathbf{88.94\ Ans.}$$

Since transmission and distribution efficiencies *are* 90% and 83% respectively so for every *KWh* transmitted only 0.9 kWh reaches the sub-station and 0.9 × 0.83, *i.e.* 0.0747 kWh reaches the consumer premises and costs 1.5 paise.

Cost per kWh delivered at the consumer premises

$$= \frac{1.5}{0.747} = \text{Paise } 2. \quad \textbf{Ans.}$$

Example 9.28. *A 250 V, 10 H.P. motor is to be selected for a workshop from two motors A and B. The cost of each motor is same* *but the losses at full load are different as given below :*

	Motor A	*Motor B*
Stray loss	*1000 watts*	*900 watts*
Shunt field loss	*250 watts*	*200 watts*
Armature copper loss	*300 watts*	*450 watts*

The motor has to work on full load for 8 hours, half load for 4 hours and quarter load for 4 hours each day. Which motor should be selected?

[*Utilisation of Electrical Power July 1975*]

Solution. Each motor will be considered on the basis of total losses and will be treated separately.

Motor A

Stray loss = 1,000 watts

Shunt field loss = 250 watts

Constant loss = Stray loss + Shunt field loss

= 1000 + 250 = 1250 watts

Full load armature copper loss = 300 watts

Half load armature copper loss,

$$= \left(\frac{1}{2}\right)^2 \times 300 = 75 \text{ watts}$$

Quarter load armature copper loss,

$$= \left(\frac{1}{4}\right)^2 \times 300 = 18.75 \text{ watts}$$

Copper loss per day,

$$= (300 \times 8 + 75 \times 4 + 18.75 \times 4) \times 10^{-3} \text{ kWh}$$

$$= 2.775 \text{ kWh}$$

Constant loss per day

$$= 1250 \times 16 \times 10^{-3} = 20 \text{ kWh}$$

Total losses per day = 2.775 + 20 = 22.775 kWh

Motor B

Stray loss = 900 watts

Shunt field loss = 200 watts

Constant losses = 900 + 200 = 1100 watts

Full load armature copper loss = 450 watts

Half load armature copper loss

$$= \left(\frac{1}{2}\right)^2 \times 450 = 112.5 \text{ watts}$$

Quarter load armature copper loss

$$= \left(\frac{1}{4}\right)^2 \times 450 = 28.\ 125 \text{ watts}$$

Copper loss per day

$$= (450 \times 8 + 112.5 \times 4 + 28.125 \times 4) \times 10^{-3} \text{ kWh}$$

$$= 4.1625 \text{ kWh}$$

Constant losses per day

$$= 1100 \times 16 \times 10^{-3} = 17.6 \text{ kWh}$$

Total losses per day = 4.1625 + 17.6 = 21.7625 kWh

Since total energy loss per day in motor *B* is less than that in *A*, hence **motor B** will be selected.

Example 9.29. *A sub-station supplies power by four feeders to its consumers. Feeder No. 1 supplies six consumers whose individual daily maximum demands are 70 KW, 90 KW, 20 KW, 50 KW, 10 KW, and 20 KW, while maximum demand on the feeder is 200 KW. Feeder No. 2 supplies four consumers whose daily maximum demands are 60 KW, 40 KW, 70 KW and 30 KW while the maximum demands on feeder No. 2 is 160 KW. Feeder No. 3 and 4 have a daily maximum demands of 150 KW and 200 KW respectively while the maximum demand on the station is 600 KW.*

Determine the diversity factors for the consumers No. 1, feeder No. 2 and for four feeders.

(Power System, July 1980)

Solution. Diversity factor for feeder No. 1, =

$$\frac{\text{Sum of individual max demands of consumers suppplied by feeder No.1}}{\text{Maximum demand in feeder No.1}}$$

$$= \frac{70 + 90 + 20 + 50 + 10 + 20}{200} = 1.3 \quad \textbf{Ans.}$$

Similarly Diversity factor for feeder 2,

$$= \frac{60 + 40 + 70 + 30}{160} = 1.25 \quad \textbf{Ans.}$$

Diversity factor for four feeders,

$$= \frac{\text{Sum of individual max. demands of four feeders}}{\text{Max. demand on the power station}}$$

$$= \frac{200 + 160 + 150 + 200}{600} = 1.183 \quad \textbf{Ans.}$$

Example 9.30. *The annual fixed cost of a certain thermal power station amount to ₹400 per KW of installed capacity. The fuel end operating costs are estimated at 5.0 paise per KWh generated, determine the total cost per unit of electrical energy generated for station, having load factors of 1.0, 0.8, 0.6, 0.4. Draw a graph showing the valuation of cost / KWh versus the load factor.*

[*Power System Nov. 1982*]

Solution. Let us consider 1 kW installed capacity.

Maximum demand of Power station,

= Installed capacity of 1 kW (assuming utilisation factor to be unity)

Units generated per annum for load factor of unity,

$$= 1 \times 8760 \times 1.0 = 8{,}700 \text{ kWh}$$

Units generated per annum for load factor of 0.8,

$$= 1 \times 8 + 60 \times 0.8 = 7{,}008 \text{ kWh}$$

Similarly units generated per annum for load factor of 0.6,

$$= 1 \times 8760 \times 0.6 = 5{,}256 \text{ kWh}$$

Units generated per annum for load factor of 0.4,

$$= 1 \times 8760 \times 0.4 = 3{,}504 \text{ kWh}$$

Total cost per unit of Electrical energy generated for station load factor of unity,

= Fixed cost per kWh + fuel and operating cost per kWh

$$= \frac{400 \times 100}{8{,}760} + 5.0 = 9.566 \text{ paise/unit. } \textbf{Ans.}$$

Similarly total cost per unit of Electrical energy generated for station load factor of 0.8

$$= \frac{400 \times 100}{7{,}008} + 5.0 = 10.7 \text{ paise/unit. } \textbf{Ans.}$$

Total cost per unit of Electrical energy generated for station load factor of 0.6,

or $$= \frac{400 \times 100}{5{,}250} + 5.0 = 12.61 \text{ paise/unit. } \textbf{Ans.}$$

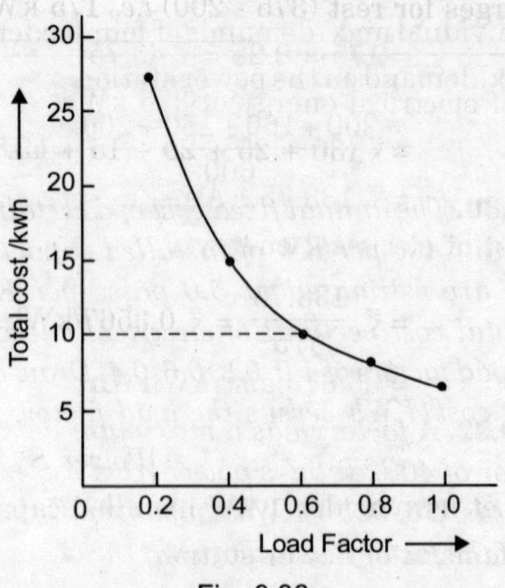

Fig. 9.36

Total cost per unit of Electrical energy generated for station load factor of 0.4,

$$= \frac{400 \times 100}{3,504} + 5.0 = 16.41 \text{ paise/unit. } \textbf{Ans.}$$

Graph showing the variation of cost per kWh versus load factor is shown in Fig. 9.36.

Example 9.31. *Compute the cost of electrical energy and average cost for consuming 375 KWh under 'block rate tariff' as under :*

First 50 KWh at 60 paise per KWh

Next 50 KWh at 50 paise per KWh

Next 50 KWh at 40 paise per KWh

Next 50 KWh at 30 paise per KWh

Excess over 200 kWh at 25 paise per kWh.

[*Utilisation of Electrical Power, July 1983*]

Solution. Energy charges for first 50 kWh

= 0.60 × 50 = ₹ 30.00

Energy charges for next 50 kWh at 50 paise/kWh

= 50 × 0.50 = ₹ 25.00

Energy charges for next 50 kWh at 40 paise/kWh

= 50 × 0.40 = ₹ 20.00

Energy charges for next 50 kWh at 30 paise/kWh

= 50 × 0.30 = ₹ 15.00

Energy charges for rest (375 - 200) *i.e.* 175 kWh

$= 175 \times 0.25 = ₹\ 43.75$

Total cost of electrical energy of 375 kWh

$= ₹\ (30 + 25 + 20 + 15 + 43.75)$

$= ₹\ 133.75$ **Ans.**

Average cost of electrical energy

$$= ₹\ \frac{133.75}{375} = ₹\ 0.3567/\text{kWh}$$

$= 35.67$ paise/kWh **Ans.**

Example 9.32. *A factory has a maximum demand of 1000 KW and a load factor of 40%. It buys power from a company at the rate of ₹45 per KWplus ₹. 0. 025 / KWh consumed. Calculate the annual electricity bill of the factory, also work out the overall cost of one unit of electricity.*

Solution. Maximum demand = 1000 kW

Yearly consumption = Max. demand in kW × load factor × 8760

$= 1000 \times 0.4 \times 8{,}760 = 35{,}04{,}000$ units

Annual demand charges

$= 45 \times 1000 = ₹\ 45000$

Annual energy charges

$= 0.025 \times 3504000 = ₹\ 87600$

Annual Electricity charges

$= 45{,}000 + 87{,}600 = ₹\ 132600$

Overall cost of one unit of electricity

$$= ₹\ \frac{132600}{3504000} = ₹\ 0.0378$$

= 3.78 paise. Ans.

Example 9.33. *A 3 ϕ synchronous motor is connected in parallel with a load of 700 KW at 0.7 power factor lagging and its excitation is adjusted until it raises the total power factor to 0.9 lagging. Mechanical load on the motor including losses is 150 KW. Find out the power factor of the synchronous motor.*

[*A.M.I.E. (Sec. B) Utilisation of Electrical Power Winter 1983*]

Solution. Load, $P_1 = 700$ kW

Load power factor, $\cos \phi_1 = 0.7$ (lagging)

Motor load, $P_2 = 150$ kW

Total load $= 700 + 150 = 850$ kW

Power factor of combined load, $\cos \phi = 0.9$ (lagging)

Reactive kVA supplied by motor,

= Reactive kVA drawn by the load of 700 kW - Reactive kVA drawn by the combined load of 850 kW

$$= P_1 \tan \phi_1 - P \tan \phi$$

$$= 700 \tan (\cos^{-1} 0.7) - 850 \tan (\cos^{-1} 0.9)$$

$$= 714.14 - 411.67 = 302.47$$

Power factor of synchronous motor,

$$\cos \phi_2 = \left(\frac{\tan^{-1} \text{reactive kVA supplied by the motor}}{\text{load supplied by the motor including losses}} \right)$$

$$= \cos \left(\frac{\tan^{-1} 302.47}{150} \right)$$

$$= \cos 63.62° = 0.444 \text{ (Lead)} \textbf{ Ans.}$$

Example 9.34. *A 340 KW, 3300 V, 50 Hz, 3ϕ star-connected induction motor has full load efficiency of 85% and power factor of 0.8 lagging. It is desired to improve the power factor to 0.96 lagging by using a bank of three capacitors. Calculate*

(i) The KVA rating of the condenser bank :

(ii) The capacitance of each limb of the condenser bank connected in delta.

(iii) The capacitance of each capacitor. If each one of the limbs of the delta connected condenser bank is formed by using 6 similar 3300 V capacitors.

Solution. Motor output = 340 kW

$$\text{Motor input,} \quad P = \frac{\text{Output}}{\eta} = \frac{340}{0.85} = 400 \text{ kW}$$

Power factor, $\cos \phi_1 = 0.8$ (lagging)

Power factor after improvement, $\cos \phi_2 = 0.96$ (lagging)

(i) kVA rating of condenser bank

= Leading kVAR supplied by the static capacitors

$$= P(\tan \phi_1 - \tan \phi_2)$$

$$= 400 \tan (\cos^{-1} 0.8) - \tan (\cos^{-1} 0.96)$$

$$\cos \phi_1 = 0.8; \tan \phi_1 = \tan (\cos^{-1} 0.8) = 0.75$$

$$\cos \phi_2 = 0.96; \phi_2 = \cos^{-1} (0.96) = 18° \ 12'$$

$$\tan \phi = \tan 18° 12' = 0.3288$$
$$= 400\,(0.75 - 0.3288) = (0.4212 \times 400)$$
$$= 168.4800 \text{ KVAR} \quad \textbf{Ans.}$$

(ii) Since reactive kVA output of the capacitors when connected in delta (Fig. 9.21) is given by $6\pi CFV^2 \times 10^{-9}$ where C is the capacitance in micro farads.

$$\therefore 6\pi\, FCV^2 \times 10^{-9} = 168.480$$

or
$$C = \frac{168.480 \times 10^9}{6\pi \times 50 \times (3300)^2} = 17.86\ \mu F \quad \textbf{Ans.}$$

(iii) Since 6 similar capacitors are connected in parallel (voltage rating of each capacitor being same as that of supply voltage) to form each limb of delta connected condenser bank.

So capacitance of each capacitor

$$= \frac{17.86}{6} = 2.977\ \mu F \quad \textbf{Ans.}$$

Example 9.35. A *50 h.p. induction motor has power factor 0.9 and efficiency 0. 9 at full load, power factor 0.6 and efficiency 0.7 at half load. At no load the current is 25% of the full load current and power factor 0.1. Capacitors are supplied to make the line power factor 0.8 at half load with these capacitors in circuit, find the line power factor at*

(i) full load and, (ii) no load.

Solution. Input to motor at full load,

$$P_F = \frac{50 \times 735.5}{0.9 \times 1{,}000} = 40.86 \text{ kW}$$

kVAR (lagging) drawn by motor at full load,

$$= \text{kVAR } F_1 = P_F \tan \phi F_1$$
$$= 40.86 \tan(\cos^{-1} 0.9) = 19.79$$

Input to motor at half full load

$$P_H = \frac{\frac{1}{2} \times 50 \times 735.5}{0.7 \times 1000} = 26.27 \text{ kW}$$

kVAR (lagging) drawn by motor at half full load,

$$\text{kVARH}_1 = P_H \tan \phi H_1$$
$$\text{kVARH}_1 = 26.27 \tan \cos^{-1} 0.6 = 35.02$$

Full load current,

$$I_F = \frac{\text{Output in h.p.} \times 735.5}{\sqrt{3} \times V_2 \times \cos\phi F_1 \times \eta}$$

$$= \frac{50 \times 735.5}{\sqrt{3} \times V_2 \times 0.9 \times 0.9} = \frac{26.212}{V_2} \text{ Amp.}$$

Current at no load,

$$I_0 = 0.25\, I_F$$

$$I_0 = \frac{26.212}{V_2} \times 0.25 = \frac{6.553}{V_2} \text{ Amp.}$$

Input to motor at no load,

$$P_0 = \frac{\sqrt{3} V_L I_O \cos\phi_O 1}{1000} \text{kW}$$

$$= \sqrt{3}\,\frac{6.553}{V_2} \times \frac{V_L \times 0.1}{1000} = 1.135 \text{kW}$$

kVAR (lagging) drawn by motor at no-load

$$\text{kVAR}_{01} = 1.135 \tan(\cos^{-1} 0.1) = 11.293$$

kVAR (lagging) drawn from mains at half load with capacitors.

$$\text{kVAR}_{H2} = P_H \tan \phi H_2 = 26.17 \tan(\cos^{-1} 0.8) = 19.7$$

kVAR supplied by capacitors,

$$\text{kVAR}_C = \text{kVARH}_1 - \text{kVARH}_2 = 35.02 - 19.7 = 15.32$$

kVAR drawn from mains at full load with capacitors,

$$\text{kVARF}_2 = \text{kVARF}_1 - \text{kVAR}_C = 19.79 - 15.32 = 4.47$$

(i) Line power factor at full load

$$= \cos\left(\tan^{-1}\frac{\text{kVARF}_2}{\text{P}_F}\right) = \cos\left(\tan^{-1}\frac{4.47}{40.86}\right)$$

$$= \cos 6.24° = 0.994 \text{ (lagging)}. \textbf{ Ans.}$$

kVAR drawn from mains at no load with capacitors,

$$\text{kVAR}_{02} = \text{kVAR}_{o1} - \text{kVAR}_C = 11.293 - 15.32 = -4.027$$

(ii) Line power factor at no load

$$= \cos\left(\tan^{-1}\frac{-4.027}{1.135}\right)$$

$$= \cos(-74.26°) = 0.271 \text{ (leading)} \textbf{ Ans.}$$

Example 9.36. *It is necessary to choose a transformer to supply a load which varies over 24 hours period in the manner given below: 500 KVA for 4 hours, 1000 KVA for 6 hours, 1500 KVA for 12 hours* *and 2000 KVA for the rest of the period. Two transformers each rated at 1500 KVA have been quoted. Transformer I has iron losses of 2.7 KW and full load copper loss of 8.1 KW, while transformer II has an iron loss and full load copper loss of 5.4 KW each.*

(i) *Calculate the annual cost of supplying losses for each transformer, if electrical energy costs 10 paise per KWh.*

(ii) *Determine which transformer should be chosen if the capital cost of the transformer I is ₹ 1000 more than that of transformer II and annual charges of interest and depreciation are 10%.*

(iii) *What difference in capital cost will reverse the decision made in (ii)* above ?

Solution. Transformer I

Energy losses per day due to iron loss = 24 × 2.7 = 64.8 kWh

Energy losses per day due to coper losses,

$$= \left(\frac{500}{1500}\right)^2 \times 8.1 \times 4 + \left(\frac{1000}{1500}\right)^2 \times 8.1 \times 6$$

$$+ \left(\frac{1500}{1500}\right)^2 \times 8.1 \times 12 + \left(\frac{2000}{1500}\right)^2$$

$$\times 8.1 \times (24 - 4 - 6 - 12)$$

$$= 3.6 \times 21.8 + 97.2 + 28.8 = 151.2 \text{ kWh}$$

Total energy losses per annum

$$= (64.8 + 151.2) \times 365 = 78840 \text{ kWh}$$

Annual cost of supplying losses

= 0.10 × 788.40 = ₹ 7884.0 Ans.

Transformer II

Energy loss per day due to iron loss = 24 × 5.4 = 129.6 kWh

Energy loss per day due to copper loss,

$$= \left(\frac{500}{1500}\right)^2 \times 5.4 \times 4 + \left(\frac{1000}{1500}\right)^2 \times 5.4 \times 6$$

$$+ \left(\frac{1500}{1500}\right)^2 \times 5.4 \times 12 + \left(\frac{2000}{1500}\right)^2 \times 5.4 \times 2$$

$$= 2.4 + 14.4 + 64.8 + 19.2 = 100.8 \text{ kWh.}$$

Total energy loss per annum

$$= (129.6 + 100.8) \times 365 = 84096 \text{ kWh.}$$

Annual cost of supplying losses

$$= 0.10 \times 84096 = ₹\ 8409.60. \quad \text{Ans.}$$

(ii) The capital cost of transformer I = ₹ 1000 more than that of transformer II

Extra annual charges in case transformer I is chosen,

$$= ₹\ \frac{1000 \times 10}{100} ₹\ 100.00$$

Annual Saving in energy cost due to losses

$$= ₹\ (8409.60 - 7884.00) = ₹\ 525.60.$$

Since annual saving energy cost of supplying losses is much more than extra annual charges due to extra cost of transformer II, therefore transformer I will be chosen. **Ans.**

(iii) Decision taken in *(ii)* above will be reversed *i.e.* transformer II will be chosen in case charges due to extra capital cost of transformer I exceeds the annual saving on the energy cost due to losses.

i.e $$x \times \frac{10}{100} > ₹\ 525.60$$

where x is capital cost of transformer I is excess

or $$x > ₹\ 525.60 \times \frac{100}{10} \text{ over that of transformer II}$$

or $$x \text{ exceeds } = ₹\ 5256.00. \quad \textbf{Ans.}$$

Example 9.37. *The motor of a 30 h.p. condensate pump has been burnt beyond economical repairs. Two alternatives have been proposed to replace it by,*

Motor A. Cost

= *₹6000 :η at full load = 90% : η at half load = 86%.*

Motor B. Cost

= *₹4000 : η at full load = 85% : η at half load = 82%.*

The life of each motor is 20 years and its salvage value is 10% of the initial cost. The rate ofinterest is 5% annually. The motor operates at full load for 25% of the time and at half load for the remaining period. The annual maintenance cost of motor A is ₹. 420 and that of motor B is ₹240. The energy rate is 10 paise per KWh.

Which motor will you recommend ?

Solution. Motor A

Capital cost = ₹ 6000.00

Annual interest on capital cost $= ₹\ \frac{6000 \times 5}{100} = ₹\ 300.00$

Annual depreciation charges,

$$= \frac{\text{Original cost} - \text{Salvage value}}{\text{Life of motor in years}}$$

$$= ₹\ \frac{6000 - \frac{10}{100} \times 6000}{20} = ₹\ 270.00$$

Annual maintenance cost = ₹ 420.00

Energy input per annum,

$$= \frac{30 \times 0.7355}{0.9} \times 0.25 \times 24 \times 365$$

$$+ \frac{\frac{1}{2} \times 30 \times 0.7355}{0.86} \times 0.75 \times 24 \times 365$$

$$= 53691 + 84283 = 137974 \text{ kWh}$$

Annual energy cost,

$$= ₹\ 0.10 \times 137974 = ₹\ 13797.40$$

Total annual cost

$$= ₹\ (300 + 270 + 420 + 13794.4)$$

$$= ₹\ 14784.40.$$

Motor A

Capital cost = ₹ 4000

Annual interest on capital cost

$$= ₹\ \frac{4000 \times 5}{100} = ₹\ 200.00$$

Annual depreciation charges

$$= ₹\ \frac{4000 - \frac{10}{100} \times 4000}{20} = ₹\ 180.00$$

Annual maintenance cost = ₹ 240.00

Energy input per annum

$$= \frac{30 \times 0.7355}{0.85} \times 8760 + \frac{\frac{1}{2} \times 30 \times 0.7355}{0.82}$$

$$\times 0.75 \times 8760$$

$$= 56850 + 88394 = 1,45,244 \text{ kWh}$$

Annual energy cost

= ₹ 0.10 × 1,45,244 = ₹ 14525.40

Total annual cost

= ₹ (200 + 180 + 240 + 14254.40)

= ₹ 15144.40.

Since total annual cost of motor B is (₹ 15,144.40 - ₹ 14784.40) *i.e.* ₹ 354 more than that of motor A. So motor A will be recommended. **Ans.**

Example 9.38. *A consumer requires an induction motor of 50 h.p. (metric). He is offered two motors of the following specifications:*

Motor (A) efficiency 88% and p.f. =0.9;

Motor (B) η = 90% and p.f. = 0.81.

The consumer is being charged on a two part tariff of ₹ 70 per KVA of the maximum demand plus 5 P. per unit. The motor (B) power factor is raised to 0.89 by installing condensers. The motor B costs ₹ 150.00 less than A.

The cost of condenser is ₹ 60 per KVAR. Determine which motor is more economical and by how much. Assume rate of interest and depreciation as 10% and working hours of motors as 2400 hours in a year.

Solution. Motor A

Output = 50 × 0.7355 = 36.775 kW

Motor input in kW

$$= \frac{36.775}{0.88} = 41.8\text{kW}$$

Motor input in kVA,

$$= \frac{41.8}{\cos\phi} = \frac{41.8}{0.9} = 46.43$$

Energy consumed per annum

= Motor input in kW x Working hours in year

= 41.8 × 2400 = 100296 kWh

Annual energy charges

= ₹ 70 × 46.33 + ₹ 0.05 × 1,00,296

= 3250.20 + 5014.8 = ₹ 8265.00

Fixed annual charges of motor A in excess over those of motor B,

$$= ₹\ 150 \times \frac{10}{100} = ₹\ 15.00$$

Total annual charges = ₹ 8265 + ₹ 15 = ₹ 8280

Motor B. Motor input in kW,

$$= \frac{36.775}{0.9} = 40.86$$

Motor input in kVA with condensers,

$$= \frac{40.86}{0.89} = 45.91$$

Energy consumed per annum,

$$= 40.86 \times 2400 = 98067 \text{ kWh}$$

Annual energy charges

$$= ₹\ 70 \times 45.91 + 0.05 \times 98067$$

$$= ₹\ 3213.70 + ₹\ 4903.30 = ₹\ 8117.00$$

Leading kVAR supplied by condensers,

$$= 40.86[\tan(\cos^{-1} 0.81) - \tan(\cos^{-1} 0.89)]$$

$$= 40.86(0.724 - 0.512) = 8.65$$

Capital cost of condensers,

$$= ₹\ 60 \times 8.65 = ₹\ 519.00$$

Annual interest and deperciation charges,

$$= \frac{519 \times 10}{100} = ₹\ 51.90$$

Total annual charges,

$$= ₹\ (8117 + 51.90) = ₹\ 8168.90$$

Motor (B) will be economical and annual savings

$$= 8280 - 8168.90 = ₹\ 111.10 \quad \textbf{Ans.}$$

Example 9.39. *A 20000 KW generating station has the generating costs :*

Capital cost ₹ 65 × 10^5

Annual Taxation ₹ 1.5 × 10^5

Annual Salaries and wages ₹ 6 × 10^5

The calorific value of fuel burnt is 54.5 × 10^5 cal / Kg. The cost of fuel is ₹ 55.00 per tonne. Assume plant heat rate 29.0 × 10^5 cal / KWh at 100 per cent capacity factor and 38.0 × 10^5 cal. / KWh at 60 per cent capacity factor. Calculate the generating cost at 60 per cent and 100% capacity factor. Interest and depreciation is 10%.

Solution. Capital cost = ₹ 65×10^5

Annual interest and depreciation

$$= ₹\ 65 \times 10^5 \times \frac{10}{100} = ₹\ 6.5 \times 10^5$$

Annual taxation = ₹ 1.5×10^5

Annual salaries and wages = ₹ 6×10^5

Units generated at 100% capacity factor

$$= 20000 \times 8760 = 1752 \times 10^5$$

Coal required per annum

$$= \frac{\text{Units generated} \times \text{Plant heat rate}}{\text{Calorific value of coal}}$$

$$= \frac{1752 \times 10^5 \times 29.0 \times 10^5}{54.5 \times 10^5} = 932.26 \times 10^5 \text{ kg}$$

$$\text{Annual fuel cost} = ₹\ \frac{55 \times 932.26 \times 10^5}{1000} = ₹\ 51.274 \times 10$$

Total annual costs,

$$= ₹\,(6.5 \times 10^5 + 1.5 \times 10^5 + 6 \times 10^5 + 51.274 \times 10^5)$$

$$= ₹\ 65.274 \times 10^5$$

Generating cost per unit,

$$= ₹\ \frac{65.274 \times 10^5}{1752 \times 10^5} = ₹\ 0.0372 \text{ per kWh}$$

$$= 3.75 \text{ paise/kWh} \quad \textbf{Ans.}$$

Units generated at 60% capacity factor,

$$= 20000 \times 0.6 \times 8760 = 1{,}051.2 \times 10^5$$

Coal required per annum,

$$= \frac{1051.2 \times 10^5 \times 38.0 \times 10^5}{54.5 \times 10^5} = 732.95 \times 10^5 \text{ kg}$$

Annual fuel cost,

$$= ₹\ \frac{55}{1000} \times 732.95 \times 10^5 = ₹\ 40.31 \times 10^5$$

Total annual costs,

$$= (6.5 \times 10^5 \times 1.5 \times 10^5 + 6 \times 10^5 + 40.31 \times 10^5)$$

$$= ₹\ 54.31 \times 10^5$$

Generating cost per unit,

$$= ₹\ \frac{54.31 \times 10^5}{1051.2 \times 10^5} = ₹\ 0.0517 \text{ per kWh}$$

= 5.17 paise per kWh **Ans.**

Example 9.40. *A generating station has the following daily loads:*

0 to 6 hrs.	4000 kW
6 to 8 hrs.	3000 kW
8 to 12 hrs.	8000 kW
12 to 14 hrs.	2000 kW
14 to 18 hrs.	7500 kW
18 to 20 hrs.	3000 kW
20 to 24 hrs.	4500 kW

Draw load duration curve and determine the (i) Load factor (ii) Plant capacity factor. Assuming the capacity of the plant 11,000 kW.

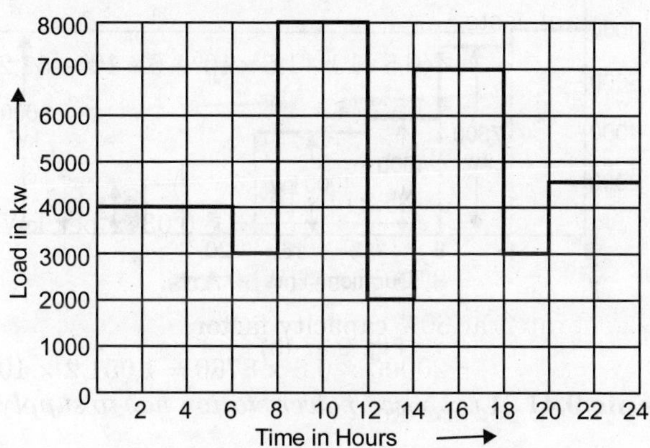

Fig. 9.37 (a)

Solution. Load curve and duration curve are drawn as shown in Fig. 9.37 (a) and 9.37 (b).

Capacity of the plant = 11,000 kW (given)

Max. demand of the plant = 8,000 kW

In 24 hrs. units generated,

$$= 4{,}000 \times 6 + 3{,}000 \times 4 + 8{,}000 \times 4$$
$$+ 2{,}000 \times 2 + 75{,}00 \times 4 + 4{,}500 \times 4$$

$= 24{,}000 + 12{,}000 + 32{,}000 + 4{,}000 + 30{,}000 + 18{,}000$

$= 1{,}20{,}000$ kWh.

$$\text{Average Load} = \frac{\text{Unit generated}}{\text{Time in hrs.}} = \frac{12{,}000}{24} = 5.000 \text{ kW}$$

$$\text{Load factor} = \frac{\text{Average load}}{\text{Maximum demand}} \times 100 = \frac{5{,}000}{8{,}000} \times 100$$

$= 62.5\%$ **Ans.**

Plant capacity factor

$$= \frac{\text{Average load}}{\text{Rated capacity of Power Plant}} \times 100$$

$$= \frac{5{,}000}{11{,}000} \times 100 = 45.45\% \textbf{ Ans.}$$

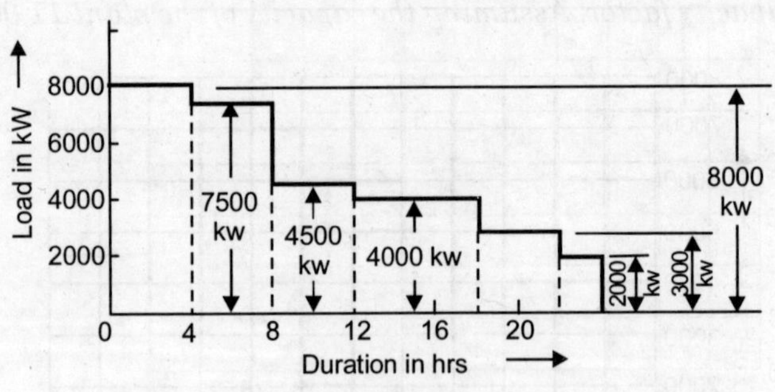

Fig. 9.37 (b)

Example 9.41. *A proposed power station has to supply load as follows:*

Time in Hrs.	01-08	08-12	12-17	17-20	20-23	23-01
Load in kW	10,000	20,000	25,000	18,000	35,000	20,000

After drawing the load curve, find out the load factor. Also choose suitable generator units to supply this load, maintaining reliability of supply. Prepare opration schedule for the machines and calculate plant use factor.

Solution. Load curve and load duration curves are drawn as shown in Fig. 9.38 (a) and (b).

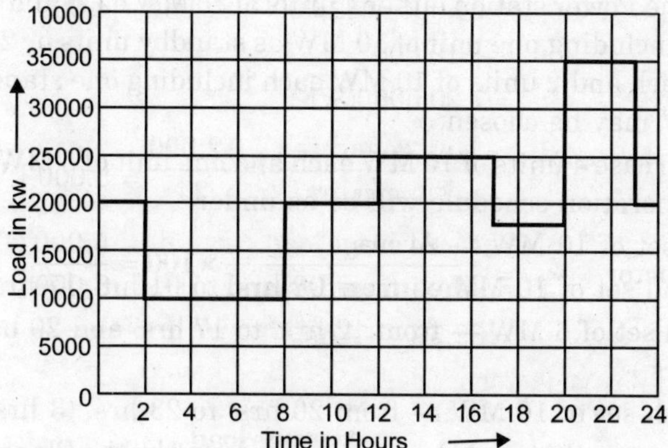

Fig. 9.38(*a*)

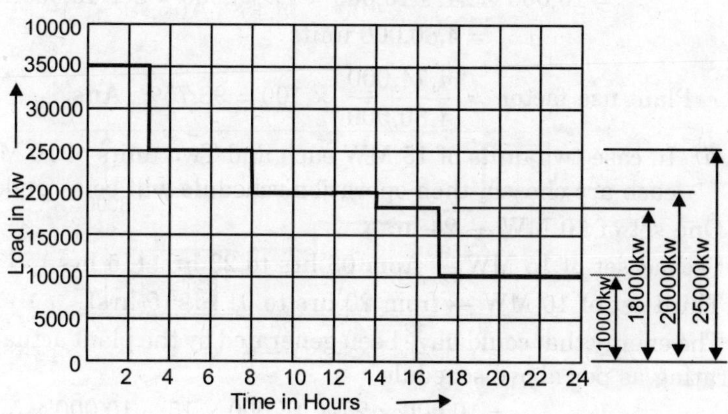

Fig. 9.38(*b*)

Maximum demand on the power station = 35,000 KW Units generated in 24 hrs.

$$= 35{,}000 \times 3 + 25{,}000 \times 5 + 20{,}000 \times 6 + 18{,}000 \times 3 + 10{,}000 \times 7$$

$$= 4{,}74{,}000 \text{ kwh.}$$

$$\text{Average Load} = \frac{\text{Units generated}}{\text{Duration in hrs.}} = \frac{474000}{24} = 19{,}750 \text{ KW}$$

$$\text{Load factor} = \frac{\text{Average load}}{\text{Maximum demand}} \times 100$$

$$= \frac{19{,}750}{35{,}000} \times 100 = 56.43\% \quad \textbf{Ans.}$$

For the Power station either 4 units of 10 MW each and one unit of 5 MW including one unit of 10 MW as standby units or 2 units of 15 MW each and 2 units of 10 MW each including one standby unit of 15 MW may be chosen.

(*i*) In case 4 units of 10 MW each and one unit of 5 MW chosen, operation schedule will be as under :

One set of 10 MW — 24 hrs.

Second set of 10 MW — from 08 hrs. to 01 hr. (17 hrs.)

Third set of 5 MW — from 12 hrs. to 17 hrs. and 20 hrs. to 23 hrs. (8 hrs.)

Fourth set of 10 MW — from 20 hrs. to 23 hrs. (3 hrs.)

The energy that could have been generated by the Plant actually operating as per above schedule

$$= 10,000 \times 24 + 10,000 \times 17 + 5000 \times 8 + 10,000 \times 3$$

$$= 4,80,000 \text{ units}$$

$$\therefore \text{ Plant use factor } = \frac{4,74,000}{4,80,000} \times 100 = 98.75\% \textbf{ Ans.}$$

(*ii*) In case two units of 15 MW each and Two units of 10 MW each are chosen, then operation schedule will be as under:

One set of 10 MW — 24 hrs.

Second set of 15 MW — from 08 hrs to 23 hrs. (15 hrs.)

Third set of 10 MW — from 20 hrs to 01 hrs (5 hrs)

The energy that could have been generated by the plant actually operating as per above schedule

$$= 10,000 \times 24 + 15,000 \times 15 + 10,000 \times 5$$

$$= 5,15,000 \text{ units.}$$

$$\text{Plant use factor } = \frac{4,74,000}{5,15,000} \times 100 = 92\% \textbf{ Ans.}$$

Highlights

(1) For any project to be constructed, whether it is engineering or non-engineering, it is essential to first find out its cost. From cost, the cost calculated, it is determined as to whether the project is worth being undertaken or not. Determination of cost itself poses a big problem as the cost keeps on varying tremendously with time and tariff condition :

Economic problem arises in the field of generation, transmission, distribution and utilisation of Electric Power and an Electrical Engineer' is supposed to propose the cheapest and the most convenient scheme after studying the relative advantages and disadvantages of all other possible schemes.

The cost of generating Electrical Energy can be classified into the following three elements.

1. Fixed cost. This cost has got nothing to do with the maximum demand and energy output of the station and is quite independent of it. This is due to the interest on the capital cost of the land used for the station and land kept for future extension and also salaries of very high officials.

2. Semi-fixed cost. This cost is due to the annual interest on the capital cost of the building and equipment. It also includes taxes and insurance charges to cover the risk of fire to the building, any accidental break down of the Machine and Worker's compen sation. It also includes the salaries of management staff including that of clerical staff.

This semi-fixed cost is independent of energy output of the station but depends upon the maximum demand and can be taken approximately proportional to the maximum demand.

3. Running cost. Running cost is also sometimes called operating cost and is dependent upon the energy output of the station that is to say the total number of units generated annually and is, therefore, due to annual cost of fuel, lubricating oil, water, maintenance and repairs, wages etc. of operating staff.

More is the number of units generated, more shall be fuel consumed and more shall be its cost and similarly the other costs as mentioned above also depend upon total number of units generated.

Before any scheme is decided to be executed, it should be seen that total cost of that scheme which includes fixed cost, semi-fixed cost and running cost is the least of all the schemes studied.

2. Interest and Depreciation

Normally the money for undertaking the project is taken as loan from the Banks and its interest is to be paid annually and while calculating the total annual operating cost, this interest to be paid to the Bank should also be taken into account. Even if the money is not taken from the Bank, the fact should be taken into consideration that had this money remained in the Bank it should have earned some interest. Thus whether the money is taken from the Bank or used from one's own's side, the interest on it should be taken into account for calculating the total overall annual cost.

Similarly every building and equipment has some life and after that life is finished, the building/equipment is required to be replaced. Thus it becomes necessary to set-aside certain amount annually to produce sufficient sum at the end of its probable life for replacing the plant and machinery by new ones. This amount is called depreciation and depends upon the type of Machine. This should also be taken into consideration while calculating the total annual operating cost of the station.

3. Electrical energy must be ready to meet the needs and requirements of all users, from the largest to the smallest upon demand. To do this, the growing requirements of electricity users must be anticipated well in advance and provision made to meet them upon their earliest need. The size of a unit to be installed depends upon the characteristics of these loads and how they combine.

4. Loads Curves

The various consumers indirectly connected to a generating station have their power load demands varying at different times in day and as a result the load on a power station cannot remain constant. The daily variation in load with respect to time can be found out if the graph is plotted between the load on the station and the time, load in kW being taken on '*Y*' axis and time in hours being taken on *X*-axis. Such type of graph or curve is kwown as daily load curve.

If daily load curves are plotted for all the days of a month and average values of the load are calculated at one particular hour on each day and these daily average values are plotted so that these come on the Y-axis and corresponding days of the month come on X-axis, then the curve so obtained is known as monthly load curve.

Similarly if the monthly load curves are plotted for all the 12 months of the year and average load, calculated for any particular day for each month and if this value is plotted on Y-axis and corresponding months of the year on X-axis, then the curve so obtained is known as yearly load curve.

A typical 24 hour load curve for any generating station is shown as below. The load varies between a maximum at about 7-30 P.M. to minimum at 3.30 A.M. Between these the load is constantly changing.

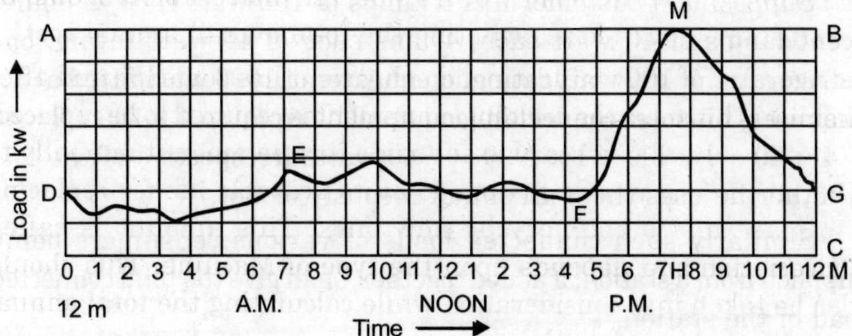

Fig. 9.39 Dally load curve.

The load curves are quite important as they give the following useful information.

1. At any particular hour load can be determined.

2. If the area under the whole curve is found out it shall give the product of k.W. (on y axis) and hours (on x-axis), *i.e.*, kilowatt hours generated in the whole day.

3. The greatest value of load, as shown by point M in Fig. 9.39, is called the peak or maximum demand on the station. Thus maximum demand can also be found out from the load curve.

4. The ratio of the area under the load curve ODEFMGC to the total area of the rectangle OAMBC in which the area is contained gives the load factor.

5. The average load from the load curve can be found out by finding out the area under the load curve ODEFMGC and dividing it by 24 hours.

5. Load Duration Curve

It is also a curve drawn between load in kW and time in hours but it differs from ordinary load curve in a way that ordinates are rearranged such that the greatest load is on the left and least load on extreme right and in between loads reducing from left to right.

Corresponding to any kW ordinate of this curve abscissa will give the number of hours during which that number of kW of load have been on the plant.

The area of both load curve and load duration curve gives the total number of units generated for the period considered.

6. Connected Load

It is the sum of the rated maximum values of all the loads in kW on consumer's premises:

Supposing a customer has 4 lamps of 100 watt each, 2 flourescent lamps of 40 watt each, 4 fans each of 40 watt rating, one refrigerator of 100 watt rating, one heater of 1000 watt fitted at his premises, his total connected load shall be equal to 4 × 100 + 2 × 40 + 4 × 40 + 1 × 100 + 1 × 1000 = 1740 watts irrespective of the fact whether he uses them all simultaneously or not.

Similarly such connected loads of various consumers being supplied from a station, if added together, shall give the total connected load of the station.

7. Maximum Demand

If the number of consumers are being supplied from a particular power station, it is not necessary that all the consumers be switching on all the connected load at a time. Some may be using heaters, at one time, others may not, some may be switching on all the lamps, the others may be switching on some as a result the maximum demand is not the sum of the connected loads of all the consumers, but a time shall come in a day when maximum load shall be switched on by consumers though it may not be the total connected load.

The greatest of all the demands which have occurred at a time during a given period on the power station is known as maximum demand. Thus maximum demand on a given station for which the load curve in Fig. 9.39 has been drawn is MH kilowatts. The size and cost of the station depends upon the maximum demand.

8. Average Load

If the number of kilowatt hours supplied by a station in one day is divided by 24 hours, then value so obtained shall be a daily average load.

Thus daily average load

$$= \frac{\text{kilowatt hours delivered in one day}}{24}$$

Monthly average load

$$= \frac{\text{kilowatt hours delivered in one month}}{30 \times 24}$$

Yearly average load

$$= \frac{\text{kilowatt hours delivered in one year}}{365 \times 30 \times 24}$$

9. Load Factor

It has already been seen that area under the load curve gives the total number of units generated. Thus area of the load curve *OAB*, as shown in Fig. 9.40, shall give the total number of units generated.

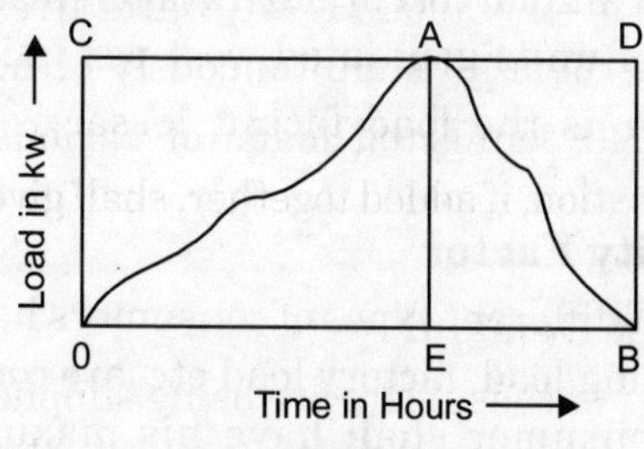

Fig. 9.40

Area of curve *OCADB* gives the total number of units that could have been generated had the maximum demand *AE* remained constant during the whole period *OB*.

The ratio

$$\frac{\text{No. of units actually generated in a given period}}{\text{Number of units that could have been generated with the same maximum demand}}$$

is called load factor of a station.

Thus load factor,

$$= \frac{\text{Area of curve } OAB}{\text{Area of curve } OC\,ADB} = \frac{\text{Average load} \times \text{Time } OB}{AE \times OB}$$

$$= \frac{\text{Average load}}{AE} = \frac{\text{Average load}}{\text{Maximum demand}}$$

Thus load factor is also the ratio between average load in *kW* to maximum demand in *kW*.

If the period considered is a day, the load factor is daily load factor and if period considered is a month, it is monthly load factor and if period considered is a year, it is called yearly load factor.

Since average load on the station shall always be lesser than the maximum demand on the station, therefore, load factor shall always be lesser than one.

10. Effect of load factor on the Cost of Power Generation

$$\text{Load Factor} = \frac{\text{Average load in } kW}{\text{Max. load in } kW}$$

$$= \frac{\text{Average load in } k\text{Whr.}}{\text{Max. load in } k\text{Whr.}}$$

Thus higher the load factor, greater is the average load which means for a given maximum demand, more is the number of units generated. Total annual cost of generation remaining the same more is the number of units generated, cost per unit shall get reduced.

Thus higher is the load factor, lesser is the overall cost of generation.

11. Diversity Factor

If number of different types of consumers having loads like tube well load, Lighting load, factory load etc. are connected to a station, each type of consumer shall have his maximum of demand at different time say a consumer with tube well load may have a maximum demand at one time in a day whereas a consumer with lighting load will have his maximum demand at any time in the night & consumer with factory load may have his maximum demand at some other time in a day.

Thus maximum demands of all the consumers supplied from a station do not occur simultaneously and, therefore, maximum demand on a station is always lesser than the sum of maximum demands of all consumers connected to a station.

This ratio between the sum of maximum demands of all the consumers connected to a station to the maximum demand of the power station is known as diversity factor.

$$\text{Diversity factor} = \frac{\text{Sum of Individual consumers maximum demand}}{\text{Maximum demand of a station}}$$

Since sum of the individual consumers maximum demand is always greater than the maximum demand of a station, diversity factor is always greater than one.

As an example consider the case of a consumer which has to meet the following loads.

Group A. 120 kW motor load working between 8 A.M. to 6 P.M.

Group B. 80 kW of lighting load between 6 P.M. to 10 P.M.

Group C. 60 kW of pumping load between misprint.

Here the sum of maximum demands of all the consumers of group A, B and C is 120 + 80 + 60 = 260 kW but on careful examination it shall be revealed that the power station to which only these loads are connected need not be designed for 260 kW as all the loads do not remain connected to station for all the time, if the load figures are analysed.

Time	*Load*	*Total load*
8 A.M. to 4 P.M.	Motor load of 120 kW	120 kW
4 P.M. to 6 P.M.	(1) Motor load of 120 kW.	120 + 60
	(2) Pumping load of 60 kW.	= 180 kW.
6 P.M. to 10 P.M.	(1) Lighting load of 80 kW.	80 kW. + 60 kW.
	(2) Pumping load of 60 kW.	= 140 *kW*.
10 P.M. to 8 A M.	(1) Pumping load of 60 kW.	60 kW.

Thus during a period of 24 hours, at no time on the station, the load is more than 180 kW. Thus max. demand on the station is 180 kW. whereas the sum of the max. demand of the consumers is 260 kW.

Diversity factor of the station

$$= \frac{260}{180} = \frac{13}{9} = 1.45$$

12. Effect of Diversity factor on Cost of Generation

Had diversification of the loads not been taken into consideration, the station designed would have been of 260 kW whereas now it shall be designed for 180 kW.

Thus greater is the diversification of load, the greater is the diversity factor and, therefore, the lesser is the maximum demand on the station. Thus the size of the station gets reduced for a given number of consumers. Therefore the capital cost which is dependent upon the size of the station also gets reduced resulting in lower overall rate of generation of electricity as the fixed charges due to capital investment will be much reduced.

Thus higher is the value or diversity factor, lower is the overall rate of generation of electricity.

Thus the supply companies should always make efforts to improve the load factor and diversity factor by inducing ihe consumers to use the electricity during off peak time even if they may have to charge at reduced rates.

13. Capacity Factor or Plant factor

While designing a power station, care is taken of future expansion and increased loads and, therefore, the capacity of the power station is made always larger than the actually reguired at that time by keeping certain units as reserve units.

Thus average load on the station is always lesser than the capacity of the power plant. Ratio between the average load on the station plant to its rated capacity is known as capacity factor or plant factor and its value is lesser than unity.

The capacity/plant factor

$$= \frac{\text{Average load on the plant}}{\text{Rated capacity of the plant}}$$

14. Demand Factor

Actual maximum demand on the station is lesser than the total connected load to the station. The ratio between the actual maximum demand on the station to the total rated load connected to the station is known as demand factor.

Therefore demand factor

$$= \frac{\text{Maximum demand on the station}}{\text{Total connected load to the station}}$$

Its value is always lesser than one

15. Base Load and Peak Head Stations

Since the head on generating station keeps on varying, so every station has to cope up with the varying demand. Thus depending upon the increasing or decreasing of the load the individual generators are either pressed into service or shut down. Still there is some minimum load on the station which remains constant through out the day. This unvarying load is known as "base load" and the maximum load occuring in the load curve excluding the base load is called "Peak Load". Base loads and Peak loads are shown in Fig. 9.41.

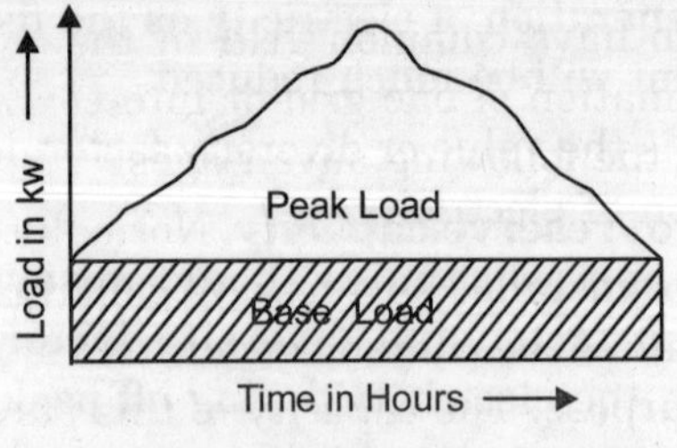

Fig. 9.41.

The generating station cannot be designed on the basis of peak load otherwise for most of the time, many generators shall remain idle, thus reducing the load factor and increasing the cost of generation.

The stations which are designed for base loads and required throughout the day are known as base load stations

The stations installed to share the load of base load stations during peak hours are known as peak load stations.

Normally thermal power stations and hydro power stations are used as Base load stations because of their low running cost whereas diesel power stations which do not take much time in starting and whose running cost is quite high are used as peak load stations.

16. Interconnection of Power Stations and its Advantages

There are certain states in india like Himachal Pradesh which have large resources for generating Power but its power consumption at present and in the near future is going to be very less where as other states like Punjab and Haryana etc., have large requirement of power but have very feable resources for generating power.

In the light of this the best course is to interconnect two systems i.e., power stations in Himachal Pradesh and the ones in Punjab and Haryana with each other through Transmission lines known as tie lines so that extra power generated in Himachal Pradesh may be transmitted to Punjab and Haryana when these states need it. The two systems connected like this are known as interconnected systems.

When the number of system power stations interconnected is more than two, the "Grid" is said to have been formed.

Because the demand of power in different states of India is different, from power potential they have, the necessity of interconnecting the various stations has arisen as a result, and efforts are being made to have common grid in the country known as national grid. Formation of one grid or interconnection of various stations possesses the following advantages.

1. Reduction of reserve capacity. Normally every station has one extra generator installed as a reserve generator which is made to run when any one of the running generators are shut for repairs or maintenance purpose. The capacity of this reserve generator is normally equal to that of the largest generator in that station.

If number of power stations are there in a country, every station has its own reserve capacity which is being used for a very short period in a year and if reserve capacities of all the generating stations are added, it amounts to a very large value which remains idle for most part of the year, if the power stations are inter-connected, then reserve capacity is reduced as only one reserve-generator can be had at any station which can feed power to any other station during emergency or during maintenance or during faulty period. Reduction in reserve capacity shall increase the load factor and efficiency of operation.

2. Reliability of supply is increased. If stations are interconnected, then during fault at any station, the power can be fed from other stations and thus reliability of supply gets increased.

3. Increased size of generators. With interconnected stations, larger generator can be used resulting in reduced Capital cost per kW. Larger generators can be used so that in case of failure of any one, power could be transferred from other station. If the stations are not interconnected and generator units are also very large, in that case failure of one generator results in large reduction of power to be generated at the faulty station.

4. If stations are interconnected, the sharing of load between two can be arranged so that the more efficient station can work at high load factor and inefficient one works only for peak load hours.

5. Interconnection of hydro power station and thermal power station makes the operation very economical.

During raining season, when water is available in sufficient quantity, the hydro power station can be made to run as a base load station as running cost of it is quite low and thermal power station is run as peak load station where as during dry season it is otherwise, i.e. Thermal power station is run as base load station and hydro-power station as peak load station.

6. As the load curves of different stations to be interconnected usually differ, the effective capacity of the whole system get increased because of the improvement in diversity factor.

Example 1. *The following is the load cycle of a house per day.*

	Load	*Time*
1.	*1 kW (Lamp load)*	*6 A.M. to 8 P.M.*
2.	*500 W (Lamp load)*	*7 P.M. to 10 P M*
3.	*Fan load, 500 W*	*12 Noon to 8 P.M.*
4.	*Heater load, 2 kW*	*6 A.M. to 7 A.M. and*
		5.00 P.M. to 6.00 P.M.

Determine the Diversity factor, Maximum demand and load factor.

Solution

Time	*Load*
6 A.M. to 7 A.M.	Heater load of 2 kW + 1 kW of lamp load = 3 kW.
7 A.M. to 12 Noon	1 kW lamp load
12 Noon to 5 P.M.	1 kW lamp load+fan load of 0.5 kW = 1.5 kW.
5 P.M. to 6 P.M.	Lamp load of 1 kW + Fan load of 0.5 kW + Heater load of 2.00 kW. = 3.5 kW.
6 P.M. to 7 P.M.	Lamp load of 1 kW + Fan load of 0.5 kW. = 1.5 kW.
7 P.M. to 8 P.M.	Lamp load of 1 kW + lamp load of 0.5 kW + Fan load of 0.5 kW = 2.00 kW.
8 P.M. to 10 P.M.	Lamp load of 0.5 kW.

Thus maximum demand is between 5.00 P.M. to 6.00 P.M. and is equal to 3.5 kW.

Average demand

$$= \frac{3\times1+1\times5+1.5\times5+3.5\times1+1.5\times1+2\times1+0.5\times2}{24\text{ hours}}$$

$$= \frac{23.5}{24}\text{ kW}$$

$$\text{Load factor} = \frac{\text{Average demand}}{\text{Maximum demand}}$$

$$= \frac{23.5}{24}\times\frac{1}{3.5} = 0.279. \quad \textbf{Ans.}$$

Diversity factor

$$= \frac{\text{sum of consumers' maximum demand in kW}}{\text{Maximum load on the station in kW}}$$

$$= \frac{1+0.5+0.5+2}{3.5} = \frac{4}{3.5} = \mathbf{1.14} \quad \textbf{Ans.}$$

Example 2. The maximum *demand of a hydro electric Plant* is 200 *M.W.The annual load factor being 60% calculate the total electrical enery generated per year in B.O.T. units.*

(AMIE Sec B. Elect.power)

Solution. Maximum demand

$$= 200 \text{ M.W.}$$

$$\text{Load factor} = \frac{\text{Average demand}}{\text{Maximum demand}}$$

Load factor = 0.6

Maximum demand = 200 M.W.

$\therefore$ Average demand = Maximum demand x load factor

$= 200 \times 0.6 = 120$ M.W.

Number of units of electrical energy generated

= Average demand in kW × no. of hours in one year

$= 120 \times 10^3 \times 8,760$

$= 1.051 \times 10^6$ kWh **Ans.**

Example 3. *A generating station has a maximum demand* of 40 M.W. and has a connected load of 75 MW. The number of units *generated annually are 200 × 10*^{6}*. Calculate (a) load factor (b) the demand factor.*

Solution. No of units generated annually = 200×10^6

$$\text{Average demand} = \frac{200 \times 10^6}{8,760} = 2.28 \times 10^4 \text{ kW}$$

Maximum demand = 40 MW = 40×10^3 kW

$$\text{Load factor} = \frac{\text{Average demand}}{\text{Maximum demand}} = \frac{2.28 \times 10^4}{40 \times 10^3} = \frac{2.28}{4}$$

= 0.57 **Ans.**

$$\text{Demand factor} = \frac{\text{Maximum demand on the station}}{\text{total connected load on the station}}$$

$$= \frac{40}{75} = 0.533. \quad \textbf{Ans.}$$

Example 4. *The cost of generation at a station is ₹ 500 per kW per annum plus 5 paisa per unit. Calculate the overall cost per unit if the load factor is (i) 100% (ii) 50% (iii) 20%.*

Solution. Let the maximum demand on the station be P, kW

Annual fixed cost = ₹ 500 × P

If load factor = ₹ 100% = 1

Average load = Max. load × load factor = P × 1 = P kW

No. of units generated annually = P × 8,760

Annual cost of generation of energy *i.e.* Annual running charges

$$= P \times 8{,}760 \times \frac{5}{100} = ₹\ 438\ P$$

Total annual cost of generation

= ₹ 500 P + ₹ 438 P = ₹ 938 P

∴ Cost of generation per unit

$$= \frac{938 \times 10^2 \times P}{8760 \times P} = 10.7 \text{ paise}$$

(ii) If load factor = 50%

Average load = P × 0.5 = 0.5 P

No. of units generated annually

= 0.5 × P × 8,760 = 4,380 P

Cost of energy generation per unit

$$= \frac{500P + 4380P \times \frac{5}{100}}{4380P}$$

$$= ₹ \left(\frac{500}{4{,}380} + \frac{5}{100}\right)$$

$$= \left(\frac{50000}{4{,}380} + 5\right) \text{ Paise} = 16.4 \text{ Paise}$$

(iii) If load factor = 20%

Average load = P × 0.2 kW

No. of units generated annually

= P × 0.2 × 8,760 = 1,752 P

Cost of energy generated per units

$$= \frac{500 \times P + 1752P \times \frac{5}{100}}{1752 \times P}$$

$$= ₹ \left(\frac{500}{1{,}752} + \frac{5}{100}\right)$$

$$= \left(\frac{50,000}{1752} + 5\right) \text{ paise} = (28.5 + 5) = 33.5 \text{ paise } \textbf{Ans.}$$

Example 5. *A power system having demand of 100 MW has a load factor of 60%. It can be supplied by one of the following schemes.*

(i) Steam station capable of supplying the whole load

(ii) A nuclear station capable of supplying the whole load

(iii) A hydro station capable of supplying the whole load

(iv) A steam power station in conjunction with a hydro-station, the latter supplying 100 × 10^6 units per annum with a maximum out put of 40 MW.

	Steam	*Hydro*	*Nuclear*
Capital cost per kW of installed capacity	*₹ 600*	*₹ 1500*	*₹ 2000*
Interest and depreciation of capital	*12%*	*10%*	*10%*
Operating cost per unit	*5P*	*1P*	
Transmission cost per unit	*Negligible*	*0.5P*	*negligible*

Work out ihe total cost of the scheme per annum, in each.

[UPSC.IEC. Generation, Transmission and Distribution of Power Station

Solution

(i) Steam station capable of supplying the whole load

Max. demand = 1,00 M W

Load factor = 30%

Average load = 100 × 0.3

= 30 MW

Total units generated per annum

$= 30 \times 10^3 \times 8760.$

Capital cost shall be calculated on the basis of maximum demands

Capital cost of steam station,

$= 600 \times 100 \times 10^3$

$= ₹\ 6 \times 10^7$

interest and depreciation per annum,

$$= \frac{12}{100} \times 6 \times 10^7$$

$= ₹\ 72 \times 10^5$

Total operating cost per annum,

$$= ₹\ \frac{15}{100} \times 30 \times 10^8 \times 8760$$

$$= ₹\ 13140 \times 10^3$$

$$= ₹\ 131.4 \times 10^5$$

Transmission cost = nil.

Total cost of steam station per annum

$$= 72 \times 10^5 + 131.4 \times 10^5$$

$$= ₹\ 203.4 \times 10^5. \quad \textbf{Ans.}$$

(iii) Nuclear Power station capable of supplying the whole load.

Capital cost of nuclear power station

$$= ₹\ 100 \times 10^3 \times 2000$$

$$= ₹\ 2000 \times 10^5$$

Interest and depreciation per annum

$$= \frac{10}{100} \times 2000 \times 10^5 = ₹\ 200 \times 10^5$$

Total units generated as above per annum

$$= 30 \times 10^3 \times 8760$$

Total operating cost per annum

$$= \frac{2}{100} \times 30 \times 10^3 \times 8760$$

$$= ₹\ 5256 \times 10^3 = ₹\ 52.56 \times 10^5$$

Transmission cost per annum

= nil

Total cost of nuclear power station per annum

$$= ₹\ (200 \times 10^5 + 52.56 \times 10^5)$$

$$= ₹\ 252.56 \times 10^5 \quad \textbf{Ans.}$$

(iii) Hydro. Station capable of supplying the whole load

Capital cost of hydro-power station

$$= ₹\ 100 \times 10^3 \times 1500$$

$$= ₹\ 1500 \times 10^5$$

Interest and depreciation per annum

$$= ₹\ \frac{10}{100} \times 1500 \times 10^5 = ₹\ 150 \times 10^5$$

Total units generated as above per annum

$$= 30 \times 10^3 \times 8760$$

Total operating cost per annum

$= \frac{1}{100} \times 30 \times 10^3 \times 8760$

$= ₹\ 26.28 \times 10^5$

Transmission cost per unit

$= ₹\ 0.005$

Total transmission cost

$= 30 \times 10^3 \times 8760 \times 0.005$

$= ₹\ 13.14 \times 10^5$

Total cost of Hydro-power station per annum

$= ₹\ 150 \times 10^5 + ₹\ 26.28 \times 10^5 + ₹\ 13.14 \times 10^5)$

$= ₹\ 189.42 \times 10^5$. **Ans.**

(iv) A steam station in conjunction with hydro-station

Max. output of Hydro-station

$= 40$ MW

Max. output of steam station

$= 100 - 40 = 60$ MW

Capital cost of steam station with Max. output of 100 MW

$= ₹\ 600 \times 60 \times 10^3 = ₹\ 360 \times 10^5$

Capital cost of Hydro-station with max. output of 40 MW

$= ₹\ 40 \times 10^3 \times 1500$

$= ₹\ 600 \times 10^5$

Interest and depreciation per annum on capital cost of steam power station

$= \frac{12}{100} \times 360 \times 10^5 = ₹\ 43.2 \times 10^5$

Interest and depreciation per annum on capital cost of Hydro-power station

$= \frac{10}{100} \times 600 \times 10^5$

$= ₹\ 60 \times 10^5$

Total interest and depreciation per annum of combination

$= (₹\ 43.2 \times 10^5 + 60 \times 10^5)$

$= ₹\ 103.2 \times 10^5$

Units supplied by hydro-station

$= 100 \times 10^6$

Units supplied by steam station

$$= 30 \times 10^3 \times 8760 - 100 \times 10^6$$

$$= (262.8 \times 10^6 - 100 \times 10^6)$$

$$= 162.8 \times 10^6.$$

Operating cost of Hydro-station

$$= \frac{1}{100} \times 100 \times 10^6$$

$$= ₹\, 10^6$$

Operating cost of steam station

$$= \frac{5}{100} \times 162.8 \times 10^6 = ₹\, 8.14 \times 10^6$$

Total operating cost of combination

$$= ₹\, 1 \times 10^6 + ₹\, 8.14 \times 10^6$$

$$= ₹\, 9.14 \times 10^6$$

Transmission cost of steam station

= Nil

Transmission cost of Hydro-station

$$= ₹\, 0.005 \times 100 \times 10^6$$

$$= ₹\, 0.5 \times 10^6$$

Total operating cost of combination per annum

$$= ₹\, 0.5 \times 10^6$$

Total cost of combination per annum

$$= ₹\, (103.2 \times 10^5 + 91.4 \times 10^5 + 5 \times 10^5)$$

$$= ₹\, 199.6 \times 10^5$$ **Ans.**

Example 6. *A generating station supplies the following loads : 15000 kW, 12000 kW, 8000 kW and 5000 kW. The station has maximum demand of 20000 kW and the annual load factor is 50 per cent. Find out :*

(i) The number of units supplied annually.

(ii) The Diversity factor.

Solution.

(i) Maximum load = 20000 kW

Annual load factor = 0.5

$$\text{Annual load lactor} = \frac{\text{Average load}}{\text{Max. load}}$$

or

$$0.5 = \frac{\text{Average load}}{20{,}000}$$

$\therefore$ Average load = 20,000 × 0.5 = 10000 kW

Number of units supplied annually

= Average load × No. of hours in one year

= 10000 × 8760

= 87600,000 = 876 × 10^5 kWh. **Ans.**

(ii) Diversity factor

$$= \frac{\text{Sum of Individual Maximum demands}}{\text{Maximum demand of power station}}$$

$$= \frac{15000 + 12000 + 8000 + 5000}{20,000}$$

$$= \frac{40,000}{20,000} = 2 \text{ } \textbf{Ans.}$$

Example 7. *A diesel generating station supplies, 500 kW, 480 kW, 300 kW and 180 kW loads and has maximum demand of 750 kW.*

Calculate :

(i) Annual load factor

(ii) The Diversity factor

(iii) The Demand factor. The number of units supplied annually are 16 × 10^5.

Solution :

(i) Number of units supplied annually = 16 × 10^5

Yearly Average demand on the station

$$= \frac{16 \times 10^5}{8760} = 182.65 \text{ kW}$$

Maximum demand on the station

= 750 kW

Annual load facteor

$$= \frac{\text{Yearly Average demand}}{\text{Maximum demand}}$$

$$= \frac{182.65}{750} = 0.243$$

= 24.3%. **Ans.**

(ii) Sum of the Maximum demand of consumers

$$= 500 + 480 + 300 + 180 = 1460 \text{ kW}$$

Maximum demand of the station

$$= 750 \text{ kW}$$

Diversity factor

$$= \frac{\text{Sum of Maximum demand of the consumers}}{\text{Maximum demand on the station}}$$

$$= \frac{1460}{750} = 1.947 \quad \textbf{Ans.}$$

(*iii*) Demand factor

$$= \frac{\text{Maximum demand on the station}}{\text{Total connected load on station}}$$

$$= \frac{750}{500 + 480 + 300 + 180}$$

$$= \frac{750}{1460} = 0.5137. \quad \textbf{Ans.}$$

Example 8. (*a*) *Why tube-wells normally work at night ?*

(b) Why do supply authorities encourage night shift in factories?

(c) What is expected to the load factor of a nuclear power station?

Solution. (a) Tube wells are worked at night because :

1. The supply voltage is high due to system load being less. This increases the efficiency of pump sets.

2. To obtain great diversity among the various types of loads.

3. To obtain better load factor so that cost per unit will be less.

4. To obtain better system power factor. Since the induction motor draw power at lagging p.f. they will compensate the leading currents due to capacitance to transmission lines.

(*b*) Night shifts are encouraged in factories to get a good diversity and better load factor. Both these will reduce the cost of generation.

(*c*) The load factor of a nuclear power station expected to be as near to unity as possible.

Example 9. *A generating station has a maximum demand 35,500 kW and has a connected load of 65,000 kW. The number of units generated annually are 25.6 ×10⁷. Calculate.*

(*a*) *The load factor.*

(*b*) *The demand factor.*

Solution. Given, maximum demand

$= 35{,}500$ kW

Connected load $= 65{,}000$ kW

Number of units generated

$= 25.6 \times 10^7$/annum.

$$\therefore \text{ Average load} = \frac{\text{Number of units generated / annum}}{\text{Number of hours in a year}}$$

$$= \frac{25.6 \times 10^7}{8760} \text{ kW}$$

$$(a) \text{ Load factor} = \frac{\text{Average load}}{\text{maximum demand}}$$

$$= \frac{25.6 \times 10^7}{8760} \times \frac{1}{35{,}500} = 0.822. \textbf{ Ans.}$$

$$(b) \text{ Demand factor} = \frac{\text{Maximum demand}}{\text{Connected load}}$$

$$= \frac{35{,}500}{65{,}000} = 0.546. \quad \textbf{Ans.}$$

Example 10. *A power station is to supply four regions of load whose peak loads are 10,000 kW, 5,000 kW, 8,000 kW and 7,000 kW. The diversity factor of the load at the station is 1.5 and the average annual load factor is 60%. Calculate the maximum demand on the station and the annual energy supplied from the station.*

Solution. Given peak loads

$= 10{,}000$; $5{,}000$; $8{,}000$; $7{,}000$ kW.

Diversity factor $= 1.5$

Load factor $= 60\%$

$$\text{Diversity factor} = \frac{\Sigma \text{ individual maximum demand}}{\text{System maximum demand}}$$

$\therefore$ System maximum demand

$$= \frac{10{,}000 + 5{,}000 + 8{,}000 + 7{,}000}{1.5}$$

$$= \frac{30{,}000}{1.5} = 20{,}000 \text{ kW.} \quad \textbf{Ans.}$$

Average load $=$ Maximum demand $\times$ load factor

$= 20,000 \times 0.6$

$= 12,000$ kWh.

Energy supplied

$=$ Average load × Number of hrs/Annum

$= 12,000 \times 8,760$ kW-hrs/Annum.

$= 105.12 \times 10^6$ kW-hrs/Annum. **Ans.**

Objective Type Questions

1. The cost which depends upon the maximum demand but it is independent of energy output is called........
2. The decrease in value caused by wear due to constant use of equipment is called.......
3. The total electrical energy delivered in a given period divided by the time in that period is called.......
4. The cost of the money borrowed from banks for purchase of equipment is called.......
5. For same no. of consumers, higher the diversity factor of the station lower will be the maximum demand of station. (False/True).
6. At high diversity factor cost of generation per unit will be.......
7. At high demand factor cost of generation per unit will be.......
8. What is the power factor ?
9. The value of the power factor may be (a) more than one. *(b)* Less than one (c) more or less than one *(d)* more than 100.
10. In inductive circuit power factor is.......(lagging/leading).
11. The efficiency of transmission lines is independent of system P.F. (a) True or False.
12. To improve the P.F. of an induction motor a capacitor is connected in.......(series or parallel).
13. An under excited synchronous motor operated at........P.F.
14. Can we change the P.F. of synchronous motor by changing excitation ?
15. What is tariff?
16. The tariff in which rate per unit of energy fixed is called......
17. The tariff in which different types of consumers charge differently is called......

18. The tariff in which consumers charged on the basis of max. demand and units consumed by them is called.......

19. The fixed tariff consists of the interest on the capital cost, depreciation, insurance and other taxes. (Yes or No).

20. Maximum demand tariff is basically a two part tariff with the only difference that the max. demand is actually measure. (True or False).

Answers

1. Semifixed cost **2.** Depreciation
3. Average load **4.** Interest
5. True **6.** Low
7. Low
8. The cosine of the angle between voltage and current in an a.c. system is called P.F.
9. Less than one **10.** Lagging
11. False **12.** In parallel with the motor
13. Lagging P.F. **14.** Yes
15. The rate at which electrical energy is sold to a consumer is called tariff
16. Simple tariff **17.** Flat rate tariff
18. Two part tariff **19.** True
20. True

Objective Type Questions

1. Diversity factor of value 2.5 gives a saving of .... percent in the generating installation.
   (a) 50 (b) 60 (c) 25 (d) 40
2. The load factor of residential load is usually
   (a) 10 to 15% (b) 50 to 60%
   (c) 60 to 70% (d) 30 to 40%
3. For heavy industry load factor may be taken as:
   (a) 25 to 40% (b) 10 to 20%
   (c) 50 to 70% (d) 70 to 80%
4. Area under a load curve provides
   (a) average demand (b) maximum demand
   (c) energy consumed (d) none of the above.

5. Load curve is normally useful in deciding the
   (a) size of the generating units
   (b) operating schedule of generating units
   (c) total plant installed capacity
   (d) all of the above
6. Power plant load curve has always
   (a) positive slope
   (b) zero slope
   (c) negative slope
   (d) any combination of (a), (b) and (c)
7. Power plant annual operating expenditure consists of
   (a) fixed charges (b) a running charges
   (c) semi-fixed charges (d) all of the above
8. Diversity factor is always
   (a) less than unity (b) equal to unity
   (c) more than twenty (d) more than unity
9. Load factor for heavy industries may be considered as
   (a) 25 to 40% (b) 10 to 25%
   (c) 50 to 70% (d) 70 to 80%
10. Maximum demand of an installation is shown by its
   (a) greatest average power demand
   (b) instantaneous maximum demand
   (c) average maximum demand over a definite interval of time during a particular period.
   (d) average power demand during an interval of 1 minute
11. Annual depreciation cost is find by
   (a) straight line method (b) sinking fund method
   (c) both (a) and (b) (d) none of the above
12. A consumer has to pay lesser fixed charges in
   (a) two part tariff (b) flat rate tariff
   (c) maximum demand tariff (d) any of the above
13. Load factor (L.F.) of a power plant is termed / defined as
   (a) maximum demand / average load.
   (b) average load / maximum demand
   (c) average load × maximum demand
   (d) average load × maximum demand

14. Power station load factor (L.F.) is generally
    (a) equal to unity (b) more than unity
    (c) less than unity (d) equal to zero
15. Load curve indicates
    (a) During the period average power used
    (b) average kwh energy cosumption during the period
    (c) either of the above two
    (d) none of the above.
16. For high value of diversity factor (D.F.) a power plant of given installed Capacity will be in a position to supply :
    (a) more number of consumers.
    (b) less number of consumers.
    (c) neither of the above two (a) and (b)
    (d) either (a) or (b)
17. Which generating plant can never have 100 precent load factor (L.F.)?
    (a) base load plant (b) peak load plant
    (c) nuclear power plant (d) Hydro electric plant
18. Rough estimation of power demand can be made by
    (a) statistical methods (b) load survey method
    (c) mathematical method (d) Economic perameters
    (e) all the above

Answers

1. (b)	2. (a)	3. (d)	4. (c)	5. (d)
6. (d)	7. (d)	8. (d)	9. (d)	10. (c)
11. (c)	12. (c)	13. (c)	14. (c)	15. (b)
16. (a)	17. (b)	18. (c)		

Q.19. What considerations/factors determine the economical limit of power factor correction?

Ans. It is governed by the relative costs of the supply and power factor correcting equipment.

Q.20. What are the different types of loads on a power system?

Ans. The main types of loads on a power system are commercial, agriculture, industrial, domestic, municipal corporations & traction etc.

Q.21. Why power improvement is necessary?

Ans. Low power factor causes the ralings of transfromers and generators, cross sectional area of the bus bars and the contact surface of the switchgear, in size of the feeders and distributors, energy losses, voltage drops in transformers, generators, transmission lines and distributors to increase.

Q.22. What type of tariff is normally applied in case of bulk consumers.

Ans. Three-part tariff.

Q.23. What type of information is given by load curves?

Ans. The various information that can be taken from load curves are:

(i) Variations of load during defferent hours of the day.

(ii) Total number of units generated in a day

(iii) Maximum demand on the power station

(iv) Load factor and

(v) Average load

Q.24. For domestic consumers what type of tariff is employed?

Ans. Block rate tariff.

Q.25. Which type of tariff encourages the consumers to keep the load factor and power factor high.

Ans. KVA maximum demand tariff encourages the consumers to keep the power factor and the load factor high.

Q.26. Mention some useful life-span of various types of equipment.

(i) Buildings 30 to 50 years

(ii) Transmission lines 20 to 25 years

(iii) Motors and generators 15 to 20 years

(iv) Lamps 1000-3000 burning hours.

EXERCISES

1. Four Possible answers for each of the following questions have been given, out of them, only one answer is correct, write the correct answer:

   (*a*) Diversity factor of any power station is (less than unity/ equal to unity/more than unity/none of the above) (*b*) The load on a power station is (always constant/never constant/seldom constant/none of the above)

(*c*) The area under the load curve divided by the number of hrs. represents the (average load/peak load/connected load/none of the above)

(*d*) Load factor of any power station is (less than unity/equal to unity/more than unity/none of the above)

2. Fill up the following by suitable words selected from those given in the brackets.

(*a*) .... cost is independent of maximum demand and energy output (fixed/semi fixed/operating)

(*b*) .... cost is approximately proportional to units generated (fixed/semi fixed/operating)

(*c*) Demand factor of any power system is always .... (less than unity/equal to unity/greater than unity)

(*d*) Load factor is the ratio of......... and .... (average load, maximum load, connected load)

(*e*) Diversity factor of any power station is .... (less than unity/equal to unity/greater than unity).

(*f*) Power factor is the ratio of .... (kW and kVA/kW and kVAR/kVAR and kVA)

(*g*) .... Motor can be used for power factor improvement (squirrel cage induction/slipring induction/synchronous)

(*h*) The best location for the power factor correction equipment to be installed at......... end of transmission line (receiving/sending)

3. Explain the important considerations to devise a 'two-part' tariff for bulk power supply to a medium size industrial concern. How is it different from that of a small consumer?
4. State the influence of load factor, diversity factor and power factor on the cost of supply.
5. Define load factor and maximum demand. Explain how the load factor affects the cost of energy generated.
6. Define the load factor and diversity factor. How does the load factor and diversity factory influence the cost of generation ?
7. What are the principal factors involved in fixing of a tariff?
8. Explain what is meant by 'two part tariff and give the economic basis for adopting such a tariff.
9. What is the effect of load factor on the charges for electricity consumed?

10. Can a power factor clause be included in a tariff?
11. What are disadvantages of poor power factor in an electrical system and what is the economical limit to which power factor can be improved?
12. What are different methods of improving power factor of an installation?
13. Explain how power factor improvement plant can increase the KW capacity of a generating station of given kVA capacity.
14. Under what conditions would you recommend installation of power factor improvement plant for augmenting the kW loading capacity of generating station in preference to installing additional generating equipment?
15. What is the advantage of power factor improvement? Outline the methods of improving power factor.
16. Derive an expression for the most economical value of power factor which may be attained by a consumer, taking into consideration the cost of plant for power factor improvement, interest and depreciation charges thereon.
17. Deduce an expression for the most economical angle of lag at which a consumer should take power when the charge is based on a fixed sum per kVA of maximum demand plus a flat rate per kWh.
18. Write short notes on the following :
    (a) Capitalization of losses.
    (b) Economic choice of equipment
    (c) Electric supply tariffs
    (d) Economics of power factor improvement.
19. Explain that in the case of a generating plant the actual earning capacity is governed by the kW output whereas the possible earning capacity is governed by the kVA output.
20. Define the following terms for generating station :
    (a) Load factor (b) Utilization factor
    (c) Load curves
21. What are the factors which influence the tariff design in an electric supply system?
22. (a) Derive the condition for most economical power factor for a plant in terms of annual tariff per kVA of maximum demand and annual cost of phase advancing plant per kVAR.
    (b) What are the principal factors involved in fixing a tariff?
    (c) Distinguish between (i) two part tariff, and (ii) Three part tariff.

OBJECTIVE TYPE QUESTIONS OF INDUSTRIAL UTILISATION

10

Industrial Utilization

10.1 Objective Type Questions

1. Direct drive is used for power transmission only when:
   (*a*) Large amount of power is involved
   (*b*) Negligible slip is required
   (*c*) Speed of the driven machine equals the motor speed
   (*d*) High speed motor is to drive a low speed machine
2. A group of motors when used to drive various machines come under which:
   (*a*) Group drive
   (*b*) Individual drive
   (*c*) Some other factors specifying to the system
3. A single large motor when used to drive various machines with individual control on either of the machines, is classified as:
   (*a*) Group drive
   (*b*) Individual drive
   (*c*) Some other factors specifying the system.
4. The basic elements of an electric drive are:
   (*a*) electric motor and the transmission system.
   (*b*) electric motor, and the transmission and control system.
   (*c*) The transmission and control system.
   (*d*) electric motor and conversion equipment.
5. These days electric drive is becoming more and more popular because of the reason:
   (*a*) It provides easy and smooth control.
   (*b*) It is simple, clean, compact and reliable.

(*c*) It is cheaper in initial as well as in maintenance cost.

(*d*) All of the above.

6. In case of individual drive

(*a*) each operator has complete control of their machine.

(*b*) load-losses are not eliminated.

(*c*) the initial cost is less than that of group drive.

(*d*) both (a) & (b)

7. For paper mill the type of drive used requiring constant speed operation and flexibility of control is:

(*a*) multi motor drive

(*b*) group drive

(*c*) individual drive

(*d*) individual or multi-motor drive.

8. Out of the given set of characteristics the more significant electrical characteristics for a traction motor are:

(*a*) running characteristics

(*b*) starting characteristics

(*c*) speed control

(*d*) braking characteristics

9. A D.C motor showing linear load torque characteristics is a:

(*a*) D.C series motor

(*b*) D.C shunt motor

(*c*) Differentially compounded D.C motor

(*d*) Commutatively compound D.C motor

10. The starting torque, without involving external control is higher with a:

(*a*) Single case induction motor

(*b*) Double case induction motor

(*c*) Slip ring induction motor

(*d*) All induction motors having poor starting torque

11. In case of three phase induction motor the supply is connected across the stator. In the case of wound rotor induction motor, the supply is connected across:

(*a*) Slip rings

(*b*) Stator alone

(*c*) Slipring and the Stator both

(*d*) after starting it is withdrawn from the rotor.

12. Provision of damper winding in the case of a synchronous motor is done so that it:
    (*a*) needs lesser exciting current
    (*b*) can be used as an alternator, which is self excited
    (*c*) needs no extra drive to bring it to initial synchronous R.P.M.
    (*d*) can be brought to a half through plugging.
13. Braking through plugging, without any additional corrective measures is possible for:
    (*a*) Three phase induction motor
    (*b*) D.C series motor
    (*c*) Synchronous motor
    (*d*) All of above
14. In the composite system of Traction supply:
    (*a*) the single phase supply is converted to three phase with in the locomotive
    (*b*) the single phase supply is converted to D.C. with is the locomotive.
    (*c*) the two methods as above are adopted simultaneously
    (*d*) the two methods are individually adopted depending upon the type of drive.
15. A drive in which speed in inversely proprtional to torque is:
    (*a*) D.C series motor
    (*b*) D.C shunt motor
    (*c*) Differentially compounded D.C motor
    (*d*) Commutatively Compounded D.C motor

Answers

1. (*c*)	2. (*b*)	3. (*a*)	4. (*b*)	5. (*d*)
6. (*d*)	7. (*d*)	8. (*b*)	9. (*b*)	10. (*b*)
11. (*b*)	12. (*d*)	13. (*a*)	14. (*d*)	15. (*a*)

10.2 Fill in the blanks with appropriate word in the following :

1. The starting torque of a 3-phase induction motor can be increased by ....
2. Textile industry requires .... motor.
3. The starting torque of dc series motor is ....
4. W = ... Ixt.
5. For small domestic appliances .... motors are generally used.

6. If the field of a synchronous motor is under excited the power factor will be ....
7. The torque of a shunt motor .... with the increase in load.
8. The motor of a water cooler is .... duty type.
9. The voltage regulators are used to .... the voltage.
10. DC shunt motor is a .... motor.
11. The motor used generally for electric traction is ....
12. The various methods used for electric braking are ....
13. Regenerative braking .... (can/cannot) bring the motor to rest.
14. Current in the armature of a shunt motor during plugging is given by .... where Vis the applied voltage. E_b is the back emf and R_a is the armature resistance.
15. The motor used for cranes is .... motor, if DC supply is used and .... motor if AC supply is used.
16. AC series motors are used for ....
17. Rheostatic braking .... (can/cannot) bring the motor to rest.
18. .... drive is used for electric traction.
19. For regeneration braking, it is necessary for supply voltage to ....
20. The dc motor used for electric traction is ....
21. The ac motor used for running the lathe is ....
22. Continuous rating of motor is defined as ....
23. In the individual drive the operator has .... control of his machine.
24. Where flexibility of control and consistency of speed is required, ... of drive is most suitable.
25. While changing over from engine drive to motor drive, ... system is more economical.

Answers

10.3 Fill in the Blanks.

1. adding resistance in the rotor circuit
2. schrage motor
3. very high
4. V
5. single phasse a.c

6. lagging
7. increases
8. intermittent
9. regulate
10. constant speed.
11. d.c series motor
12. plugging, regenerative, rheostatic
13. cannot
14. (V – Eb) Ra
15. series; 3-phase slipring
16. High speed domestic appliances
17. cannot
18. Electric (d.c series motor)
19. be less than induced emf
20. series motor
21. 1–Phase capacitor motor
22. a motor which can run continuously for 24 hours without over heating at its rated load. It can also be overloaded up to 25% more than its rated capacity for 2 hours.
23. complete
24. Individual drive
25. group drive.

10.4 Multiple Choice Questions

Select the correct answer in the following:

1. The direction of rotation of a dc series motor can be changed by

   (a) interchanging supply terminals
   (b) interchanging field terminals
   (c) interchanging armature terminals
   (b) both b and c.

2. The motor normally used for crane is

   (a) slip ring induction motor
   (b) synchronous motor
   (c) dc shunt motor
   (d) dc differentially compound motor

3. The resistance of earth wire should be

   (a) infinite *(b)* high *(c)* reasonable *(d)* very low

4. Electric braking is preferred because it is

   (a) smooth
   (b) maintenance cost is less
   (c) energy is saved in regenerating braking
   (d) *all* of these.

5. For rheostatic braking of two series motors connected in parallel
   (*a*) equalizar connection is better
   (*b*) cross connection is better
   (*c*) both are equally good
   (*d*) none of these.
6. The maximum horse power upto which 440V electric motor are used, is
   (*a*) 200 HP (*b*) 50 HP (*c*) 20 HP (*d*) 10 HP
7. Which of the following motor is preferred for blowers ?
   (*a*) wound rotor induction motor
   (*b*) dc shunt motor
   (*c*) squirrel cage induction motor
   (*d*) dc series motor.
8. A steel mill requires a motor having high starting torque, wide speed range and precise speed control. which of the motors will you choose:
   (*a*) dc shunt motor (*b*) synchronous motor
   (*c*) dc series motor (*d*) slipring induction motor
9. During regenerative braking of elctric motors, they are.
   (*a*) disconnected from the supply
   (*b*) reverse connected to the supply
   (*c*) made to run as generator
   (*d*) made to stop.
10. While plugging dc motors — connections are reversed
    (*a*) supply (*b*) armature
    (*c*) field (*d*) both armature & field
11. Rheostatic braking may be applied to an induction motor provided.
    (*a*) separate dc source for field excitation is available
    (*b*) it is squirrel cage type
    (*c*) it is a slipring type
    (*d*) variable external resistance is available.
12. In a centrifugal pump, the mechanical load is ....
    (*a*) lifting (*b*) friction (*c*) deformation (*d*) linear
13. The starting torque of a d.c. series motor is
    (*a*) high (*b*) very high (*c*) low (*d*) poor

14. The torque of a shunt motor .... with the increase in load.
    (*a*) increases (*b*) decreases
    (*c*) remains same (*d*) none of these
15. The characteristics of d.c. compound motor are .... the series and shunt motor.
    (*a*) like (*b*) unlike (*c*) in between (*d*) similar to
16. The starting torque of a three phase induction motor can be increased by ....
    (*a*) increasing the rotor resistance
    (*b*) decreasing the rotor resistance
    (*c*) increasing the stator resistance
    (*d*) decreasing the stator resisiance
17. The maximum starting torque in the induction motor is developed when ....
    (*a*) $X_{2s} = R_2$ (*b*) $X_{2s} = \frac{R_2}{2}$ (*c*) $X_{2s} = 2R_2$ (*d*) $X_{2s} = 3R_2$
18. The direction of split phase motor can be reversed by ....
    (*a*) reversing leads of main winding
    (*b*) reversing leads of auxiliary winding
    (*c*) both (*a*) and (*b*)
    (*d*) reversing the supply terminals
19. If the field of a synchronous motor is over excited the power factor will be ....
    (*a*) lagging (*b*) leading (*c*) unity (*d*) zero
20. The speed of a schrage motor can be controlled by ....
    (*a*) shifting the brushes
    (*b*) changing the supply voltage
    (*c*) adding resistance in stator winding
    (*d*) decreasing the current
21. The type of drive preferred for d.c. series motor is ....
    (*a*) belt drive (*b*) direct coupling
    (*c*) chain drive (*d*) rope drive
22. For traction work, the best suited motor is
    (*a*) slip ring induction motor (*b*) synchronous motor
    (*c*) d.c. series motor (*d*) d.c. shunt motor

23. For small domestic appliances .... motors are generally used.
    (*a*) universal (*b*) shunt
    (*c*) split phase induction (*d*) 3-phase induction
24. For punches and shears .... motor is used
    (*a*) dc compound (*b*) dc series
    (*c*) dc shunt (*d*) ac I-phase capacitor
25. Textile industry requires .... motor
    (*a*) dc shunt (*b*) schrage (*c*) dc compound (*d*) dc series

Answers

1. (*d*)	2. (*a*)	3. (*d*)	4. (*d*)	5. (*b*)
6. (*a*)	7. (*d*)	8. (*c*)	9. (*c*)	10. (*b*)
11. (*c*)	12. (*a*)	13. (*b*)	14. (*a*)	15. (*c*)
16. (*a*)	17. (*a*)	18. (*c*)	19. (*b*)	20. (*a*)
21. (*b*)	22. (*c*)	23. (*a*)	24. (*a*)	25. (*b*)

10.5 True or False Questions

State whether the following statement *is* true or false:

1. Power factor of the system is improved by the group drive.
2. Initial cost of group drive is more.
3. In case of lifting loads, the torque required is independent of speed.
4. Torque varies as the square of the speed in case of blowers and centrifugal pumps.
5. In crushing and deformation loads, torque is inversely proportional to speed.
6. Torque in a shunt motor is inversely proportional to the armature current.
7. In a series motor, the starting torque is proportional to the square of armature current.
8. A d.c. series motor can be started safely without load.
9. The characteristics of a compound motor lie in between those of series and shunt motors.
10. A 3-phase induction motor exerts maximum torque when the rotor resistance is equal to its reactance.
11. The starting torque of capacitor start induction motor is very poor.

12. The starting torque of shaded pole motor is very high.
13. A synchronous motor is a constant speed motor.
14. A d.c. shunt motor has a fairly constant speed.
15. The power factor of a 3-phase induction motor is very low at light loads.
16. The synchronous motor works on leading power factor when over excited.
17. The speed torque characteristics of a schrage motor are similar to shunt motor.
18. The starting torque of repulsion motor is low.
19. Plugging cannot be applied to induction motor.
20. Types of enclosure does not affect the selection of motor for a particular service.

Answer

1. false	2. false	3. true	4. true	5. true
6. false	7. true	8. false	9. true	10. true
11. false	12. false	13. true	14. true	15. true
16. true	17. false	18. true	19. true	20. false

10.6 Control of Industrial Motors

Q.2.1. What do you mean by a chopper? How chopper's are classified.

OR

What are different types of choppers? Discuss their types and applications.

Ans. Chopper is a device which converts fixed DC into variable DC. In a chopper circuit, a thyristor is used to switch ON supply to the load for certain intervals of time and switch OFF supply for another certain intervals of time. The cycle of operation is repeated. The thyristor acts as a high speed switch. For turning ON of thyristor, gate signal is applied and to turn it off, some method of forced commutation is applied. By varying the ON and Off period, the voltage across load can be controlled.

1. Types of Choppers:

***(i)* Step down chopper:** In a step down chopper, the load voltage can be less than or equal to the source voltage and power flows from D.C source to load.

A basic step down chopper circuit is shown in fig. 2.1.

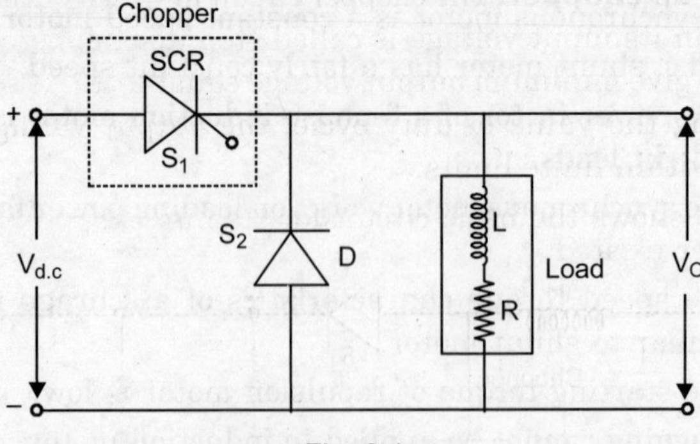

Fig. 2.1

It is basically a series switch which can be turned ON and OFF at high frequency to get train of square pulses across the load. The SCR used in the circuit acts like a switch. Diode connected across the load provides path for the inductive load current when SCR is open.

When SCR is not conducting, the voltage across load is zero. When SCR is turned ON the load voltage is $V_0 = V_{dc}$ and load current rises exponentially as shown in fig. 2.2

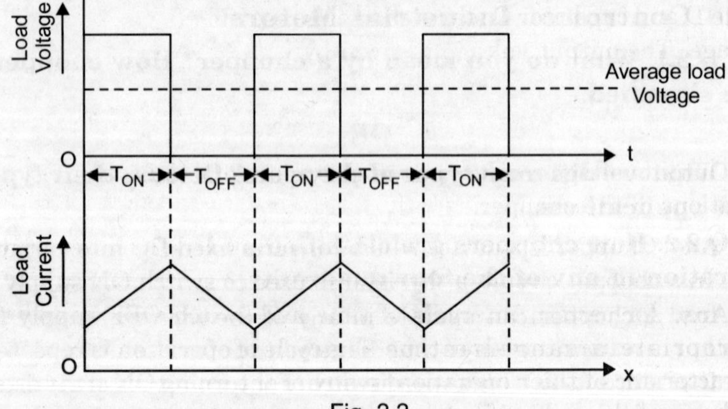

Fig. 2.2

The average voltage across load depends upon T_{ON} and T_{OFF} durations and is given by

$$V_O = \alpha . V_{dc}$$

where $$\alpha = \frac{T_{ON}}{T_{ON} + T_{OFF}} = \text{Duty cycle of chopper.}$$

2. Step up chopper: The chopper circuit in which output voltage is more than its input voltage is called step up chopper. They are designed to give minimum output voltage equal to its input voltage. By changing the value of duty cycle, the output voltage can be increased within finite limits.

Fig. 2.3 shows the basic circuit for step up chopper

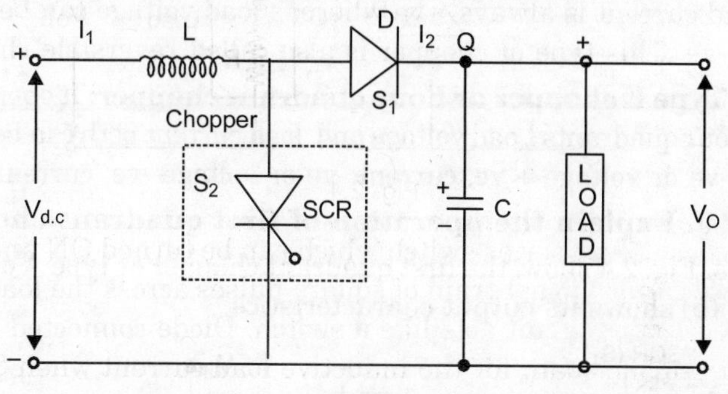

Fig. 2.3

When SCR is turned ON, the inductor current begins to rise linearly and energy is stored in the inductor. When SCR is turned off, the stored energy in inductor L is transferred to capacitor through diode D and capacitor C starts charging resulting in the rise of load voltage. The output voltage is given by

$$V_O = \frac{T_{ON} + T_{OFF}}{T_{ON}} \quad Vdc$$

Output voltage can be controlled by changing the T_{ON} and T_{OFF} durations of the chopper.

Q.2.2. How choppers are classified on the basis of their operation in any of the four quadrants.

Ans. A chopper can operate in any of the four quadrants by aporopriate arrangement of semiconductor devices. This characteristic of thier operation in any of the four quadrants form the basis of thier classification into.

***(i)* First quadrant or Type - A chopper:** In this type of chopper both the average output voltage and current are positive. The circuit works only in the first quadrant.

***(ii)* Class B chopper or Second quadrant chopper:** In this type of chopper the output current is restricted to -ve values and output voltage is + ve and higher than the input voltage. It is also called step up chopper.

***(iii)* Class C chopper or Two quadrant type A chopper:** It is the combination of two single quadrant choppers class A and class B.The load voltage is always + ve whereas the current can be either + ve or -ve.

***(iv)* Class D chopper or two quadrant chopper type B:** It is a two quadrant chopper operating in first ard fourth quadrant. The load current is always + ve whereas load voltage can be either +ve or -ve. This type of chopper is also called reversible chopper.

***(v)* Type E chopper or Four quadrant chopper:** It operates in all the four quadrants Load voltage and load current both can be either +ve or -ve or voltage + ve, current -ve or voltage -ve, current +ve.

Q.2.3. Explain the operation of first quadrant chopper.

Ans. Fig. 2.4 show the first quadrant chopper or type A chopper and fig (b) shows its output characteristics.

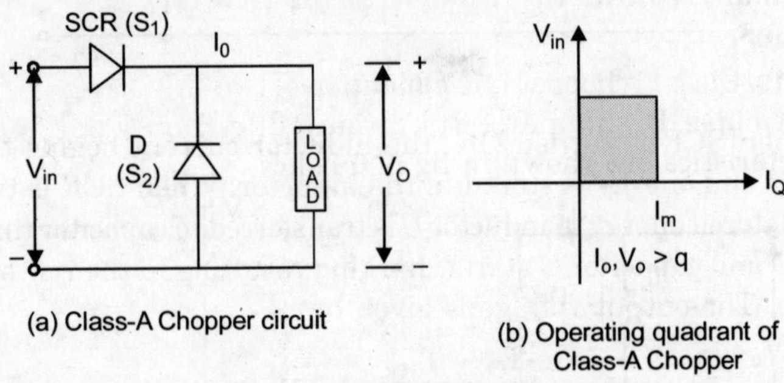

(a) Class-A Chopper circuit

(b) Operating quadrant of Class-A Chopper

Fig. 2.4

The chopper operates by periodic opening and closing of switch S_1 (SCR). When S_1 is ON the current I_0 flows from source to load. When Switch S_1 is OFF, the source is disconnected, but the load current (I_0) flows in the same direction through diode D. Both load voltage and current are + ve and the power flow is only from source to load. It is also called step down chopper as the average value of output voltage is always less than the input voltage.

Q.2.4. Explain the operation of class B chopper or second quadrant chopper

Ans. Fig.2.5 fshows the circuit for class B chopper, and its operating characteristics.

It is also called step up chopper as the power is transferred from lower voltage to higher voltage. This is possible only when load contains a d.c source like a battery E or d.c motor. Above certain speed d.c motor behaves as generator. When the switch S is turned

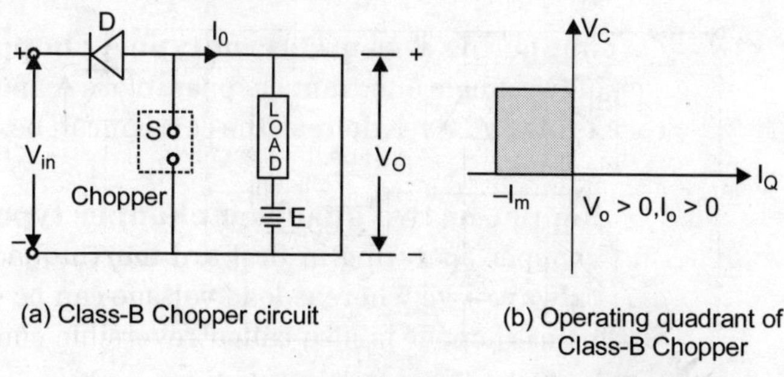

(a) Class-B Chopper circuit

(b) Operating quadrant of Class-B Chopper

Fig. 2.5

ON the load current begins to flow and an energy is stored in the inductive load. When switch S is turned off, the energy stored in the load circuit inductor is returned to the input supply via diode D.

This configuration is used for regenerative breaking of d.c motor.

Q.2.5. Explain the operation of two quadrant class C chopper.

Ans. Class C chopper is combination of two single quadrant class A and class B choppers. Its circuit diagram and operating characteristics are shown in fig. 2.6 below.

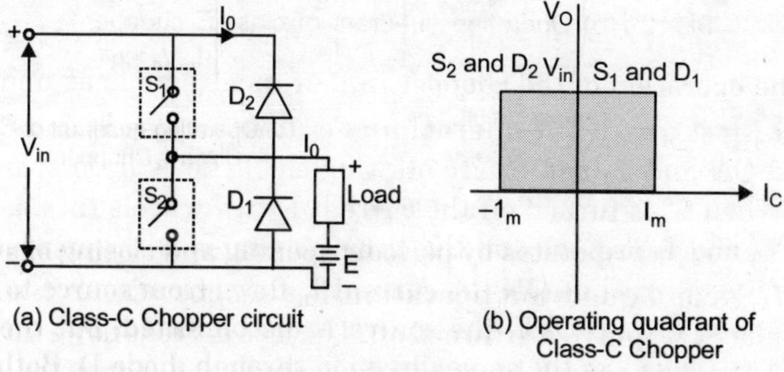

(a) Class-C Chopper circuit

(b) Operating quadrant of Class-C Chopper

Fig. 2.6

It uses two thyristors shown by switches S, and S_2 and two diodes D_1, and D_2. The load voltage is always +ve whereas the load current can be either +ve or -ve.

If I_0 is + ve, either S, and D_1 conduct but not simultaneously. If I_0 is -ve, then either D_2 or S_2 conduct. Power can flow from source to load or load to source.

Q.2.6. Explain the operation class E four quadrant chopper

Ans. Fig. 2.7 shows the circuit diagram of class E four quadrant chopper and its operating characteristics.

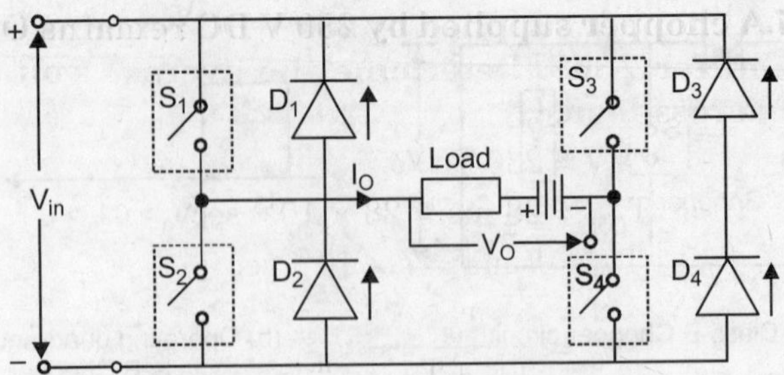

Fig. 2.7(a) : Class-E Chopper circuit

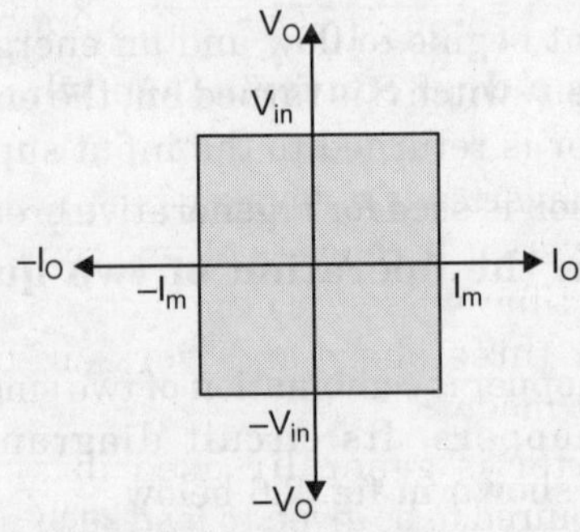

Fig. 2.7(b) Operating quadrant of class-E chopper

The operation of the chopper in four quadrants is as under.

***(i)* First quadrant operation:** For this purpose S_4 and S_1 are turned ON and S_2 and S_3 are off. Current I_0 flows from source to load. When S_1 is turned off the current I_0 free wheels through D_2. Both V_0 and I_0 are +ve and the circuit operates is 1st quadrant.

***(ii)* Second quadrant operation:** For this operation S_2 is turned ON while S_1 S_3 and S_4 are turned off. The load current flows through S_2 and D_4. This operation gives + ve voltage and -ve current.

***(iii)* Third quadrant operation:** In this operation S_1 and S_4 are turned off while S_2 and S_3 are turned ON simultaneously. The load is connected to source through S_2 and S_3 and both V_0 and I_0 are -ve. When S_2 is off, the load current free wheels through S_2 and D_4.

***(iv)* Fourth quadrant operation:** In this operation voltage is + ve but current is -ve. For this operation S_1, S_2 and S_3 are turned off while S_3 is turned ON. The polarity of EMF E is to be reversed for this operation. When S_4 is ON, +ve current flows through S_4 and D_2. Energy is stored in the inductor when S_4 is turned off, the current flows to source through D_3 and D_2.

Q.2.7. A chopper supplied by 230 V DC remains ON for 28 mS and off for 7 mS. Determine the average voltage that appears across load.

Ans.
$$V_{in} = 230 \text{ V}$$
$$T_{ON} = 28 \text{ ms} = 28 \times 10^{-3} \text{ sec}$$
$$T_{off} = 7 \text{ ms} = 7 \times 10^{-3} \text{ sec}$$
$$V_0 = \frac{T_{ON}}{T_{ON} + T_{OFF}} V_{in}$$
$$= \frac{28 \times 10^{-3}}{28 \times 10^{-3} + 7 \times 10^{-3}} \times 230 = 184 \text{ Volt}$$

Q.2.8. What is a dual converter? In what applications it is used?

Ans. A dual converter is a device which is used to convert AC into DC and Vice versa. It consists of two converter bridges. One bridge acts as a rectifier while the other acts as an inverter. It may be single phase or three phase and works in circulating or non circulating current modes.

A dual converter is generally used in applications where reversible DC is required to be given to load such as in speed reversal of DC motors and regenerative breaking.

Q.2.9. Draw the circuit diagram of a single phase Dual converter and explain its working.

Ans. Fig. 2.8 shows the circuit diagram of single phase dual converter

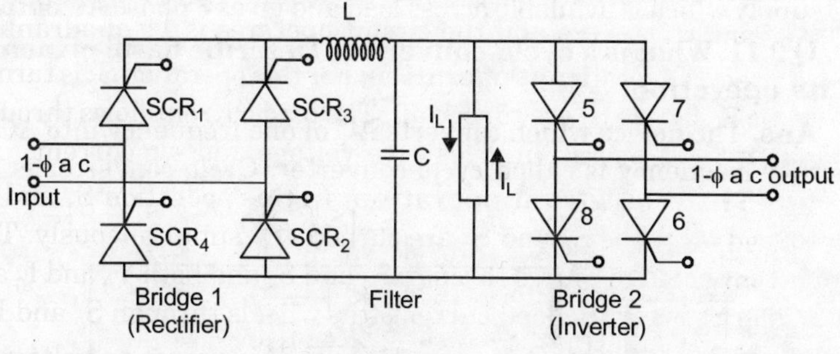

Fig. 2.8

It consists of two converter bridges. Both bridges consist of 4 SCRs. The first bridge works full wave rectifier and the second bridge acts as an inverter. L-C filter network is used to minimise the ripples in the rectified d.c. The d.c voltage across load is controlled by controlling the gate current or firing angle of SCRs in the first bridge.

The second bridge acts as an inverter circuit as long as bridge 1 is gated. In case bridge 1 is not gated, the bridge 2 can be used as another rectifier by feeding it with 1 ϕ AC supply.

Q.2.10. Draw and explain the circuit diagram for three phase dual converter.

Ans. Fig.2.9. Shows a three phase dual converter.

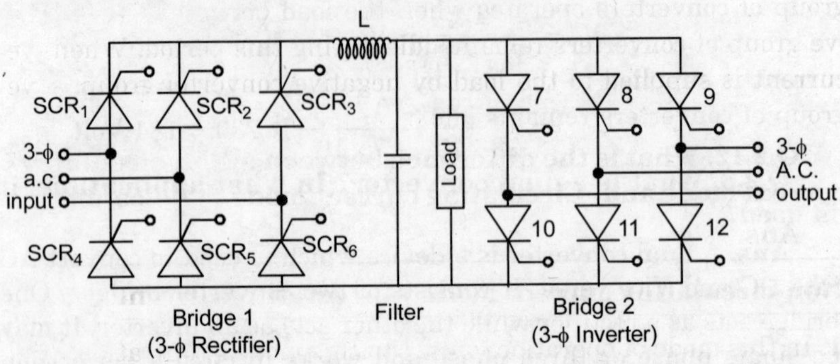

Fig.2.9.

In three phase dual converter, two bridges are formed. The first bridge consists of six SCRs and performs the function of a 3 ϕ controlled full wave rectifier. The second bridge having six SCRs acts like a 3 ϕ inverter. LC network acts a filter to remove ripples from rectified output.

Three phase AC supply is fed to the input of a.c bridge which rectifies it. The second bridge which acts as an inverter is fed from DC supply which is available across load and gives 3 phase AC output.

Q.2.11. What is a cyclo-converter? Describe basic principle of its operation.

Ans. The device which converts AC of one frequency into AC of another frequency is called cyclo-converter. Cyclo-converter is

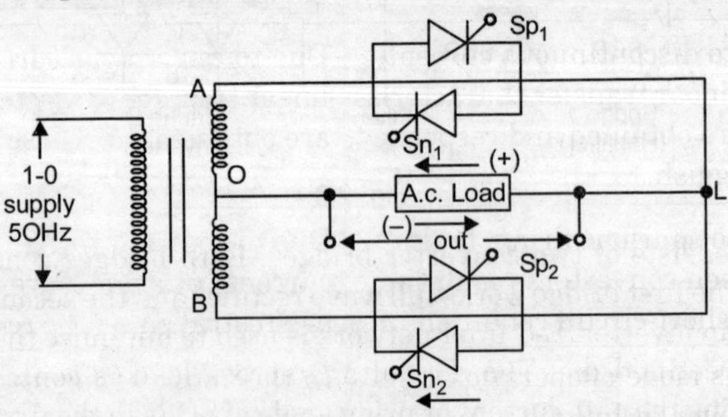

Fig. 2.10

basically a.c to a.c power converter where the alternating voltage at supply frequency is converted directly to a different frequency without any dc link as shown is fig. 2.10.

It consists of two groups of dual converters, one forming the + ve converter group and the other forming the -ve converter group feeding the common load Input to both the groups is same. The + ve group of converters operates when the load current is + ve. The -ve group of converters remains idle during this period. When -ve current is supplied to the load by negative converter group, + ve group of converters remains idle

Q.2.12. What is the difference between a Non-circulating current mode and Circulating current mode cyclo-converter

Ans.

Non-circulating current mode	Circulating current mode
1. In this mode of operation, only one converter operates at a time and the second converter remains in a blocking state.	In this mode of operation, one converter operates as a rectifier and the other converter operates as an inverter.
2. Converters may operate in discontinuous current mode.	Converters operate in continuous current mode.
3. Reactors may be needed to make load-current continuous.	Reactors are needed to limit circulating current.
4. Since no cirulating current flows through the converters, efficiency is higher.	Circulating current flows through the converters and hence increases the losses. Hence efficiency is low.
5. Due to discontinuous current, non-linear transfer characteristics are obtained and response is sluggish.	Due to continuous current, linear transfer characteristics are obtained. Response is fast.
6. Due to spurious firing, faults between converters result in dead short-circuit conditions.	Due to spurious firing, fault current between converters are restricted by the reactor.
7. In this mode of operation, the crossover technique is complex.	In this mode of operation, the cross over technique is simple.

8. Loss of control for 10 to 20 mS is observed in this mode of operation.	Since converters do not have to pass through blocking unlocking and safety intervals of 10 to 20 mS, hence control is never lost in this mode of operation.
9. The complete scheme is cheaper compared to circulating current mode.	The complete scheme is expensive.

Q.2.13. Draw the circuit diagram of single phase cyclo converter and explain its working.

Ans. Fig. 2.11. shows the circuit diagram of single phase bridge converter and its out put wave form

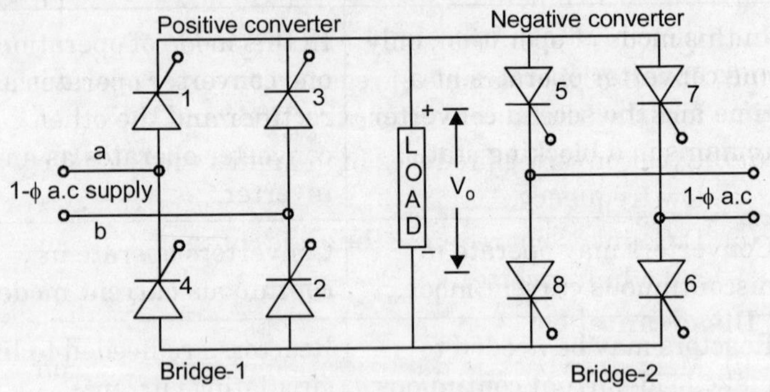

Fig. 2.11 (a)

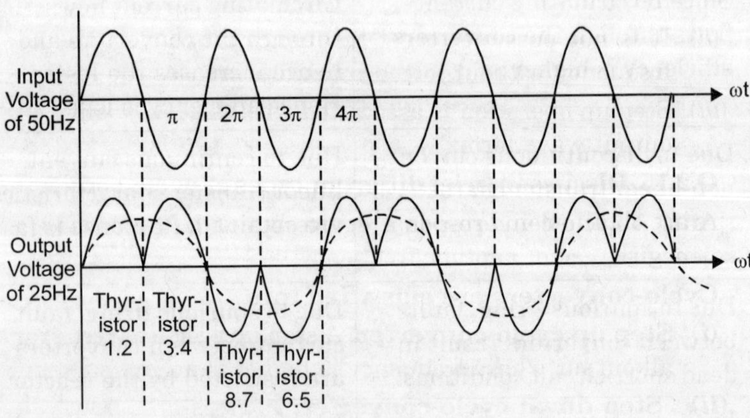

Fig. 2.11(b)

During +ve half cycle of AC input supply, the thyristor 1 and 2 are triggered during -ve half cycles, the thyristors 1 and 2 are automatically commutated and thyristors 3 and 4 are triggered simultaneously. In both the half cycles the direction of current flow through load is same and two positive halfs are produced across load. The next +ve and -ve half cycles can be utilized to produce two negative half across load by triggering the thyristor pairs in the second bridge. The output wave forms are shown in fig. 2.10 (b)

Q.2.14. What are the advantages and disadvantages of cyclo-converters.

Ans. Advantages:

(*i*) Operation of the cyclo-converter is based on line commutation, therefore, absence of forced commutation makes the circuit simple and cheap.

(*ii*) Due to an inherent capacity of power flow in either direction, the circuit can operate at any power factor.

(*iii*) The circuit is capable of regenerative operation with active loads, without any complexity of control circuit.

(*iv*) Cycloconverter produces high quality sinusoidal output at low frequency,

(*v*) D.C. link is not required thus it is cheaper and simpler than a d.c link inverter.

Disadvantages:

(*i*) Main drawback is the limitation of frequency control, the frequency can not be more than 1/3 or 1/2 of the input frequency.

(*ii*) It is not possible to get frequency more than the input frequency without forced commutation.

(*iii*) Step up operation is associated with the low quality of the output wave form.

Q.2.15. Discuss in brief different types of cyclo-converters.

Ans. A cyclo-converter is a device which converts ac of one frequency into ac of another frequency.

Cyclo-converters are classified into

(*i*) **Step up cyclo-converter:** A step up cyclo-converter gives all output whose frequency is higher than that of input.

(*ii*) **Step down cyclo-converter:** It gives an output whose frequency is higher than the frequency of input.

The cyclo-converters are also classified as

(*i*) Single phase to single phase cyclo-converter.

(*ii*) Three phase to single phase cyclo-converter

(*iii*) Three phase to three phase cyclo-converter

(*iv*) Six pulse cyclo-converter.

The cyclo-converter may also be classified into

(*i*) Non circulating current cyclo-converter

(*ii*) Circulating current cyclo-converter.

Short Answer Type Questions:

Q.2.16. What is an electric drive?

Ans. The economic combination of electric motor, control equipment and energy transmission mechanism is called electric drive.

Q.2.17. What are conventional methods of speed control of D.C. motor?

Ans. (*i*) Armature voltage control

(*ii*) Armature resistance control

(*iii*) Field flux control

Q.2.18. What is single phase drive?

Ans. In single phase drive single phase AC is converted into DC by controlled rectifiers and d.c is fed to motor. By varying the firing angle the voltage applied across motor can be varied and thus speed can be controlled.

Q.2.19. What is a chopper drive?

Ans. The drive in which variable d.c is obtained by means of chopper to control the speed of d.c motor is called chopper drive.

Q.2.20. What is a power modulator?

Ans. It controls the flow of power from source to motor.

Q.2.21. What is phase locked loop (PLL) control of DC drives?

Ans. It is closed loop control which is required for precise control of servo systems.

Q.2.22. What is static scherbius drive?

Ans. It is used to control the speed of phase wound motor below synchronous speed.

Q.2.23. The slip power recovery scheme of speed control of slip ring induction motor .... the efficiency but ....

Ans. Increases, reduces the power factor.

Q.2.24. Voltage source inverter, allows a variable frequency supply to be obtained from a .... source (AC/DC).

Ans. DC.

Q.2.25. Cyclo-converter is used for .... drives (AC/DC).

Ans. AC

Q.2.26. What is a half wave drive?

Ans. The drive in which a single thyristor is used to control input power is called half wave drive.

Q.2.27. What do you mean by slip control?

Ans. Slip control means the speed control of Induction motor by varying the slip.

Medium/Long Answer Type Questions

Q.2.28. Draw and explain the block diagram of electric drive.

Ans. Fig. 2.12 shows the block diagram of electric drive.

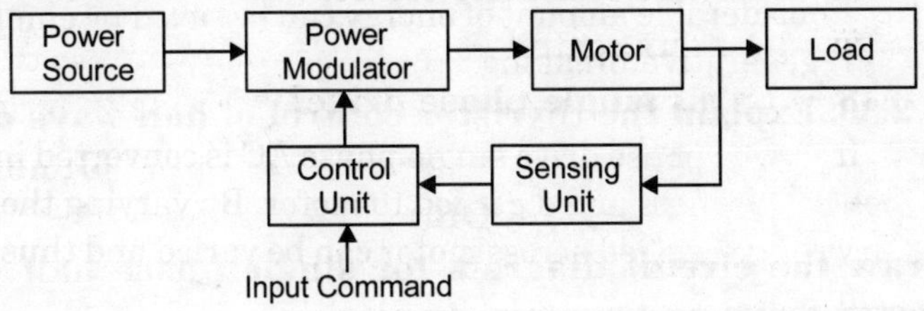

Fig. 2.12

It consists of a power source, power modulator, motor, load, sensing unit and control unit.

(*i*) **Power source:** Power source may be single phase or 3 phase AC supply or DC supply.

(*ii*) **Power modulator:** It controls flow of power from source to the motor and converts electrical energy of source to the form which suits the motor, *i.e.*, if source is d.c and induction motor is employed, then modulator is required to convert d.c into variable frequency AC.

(*iii*) **Motor:** A motor having speed torque characteristics compatible with load is selected.

(*iv*) **Load:** Load may be fan, pumps, machine tools, drills, trains etc.

(*v*) **Control unit:** It generates commands for the protection of power modulator and motor. It operates at much lower voltage and power levels. It consists of firing circuits which may employ linear ICs and Transistors,

(*vi*) **Sensing unit:** Sensing of certian drive parameters such as motor current and speed which may be required for protection or closed loop operation of system is done in the sensing unit. The counter electromotive force is detected and compared with reference voltage and the error so obtained is amplified and fed to control unit.

Q.2.29. What are the advantages of electric drives.

Ans. Following are advantages of electric drives

(*i*) Flexibility (*ii*) Economical

(*iii*) They are available in wide range of torque, speed and power.

(*iv*) High efficiency (*v*) Easy control

(*vi*) They are non polluting (*vii*) Requires less maintenance

(*viii*) Have long life (*ix*) Instant restart

(*x*) Can be operated in all the four quadrants of operating plane.

(*xi*) Considerable amount of energy can be saved by employing regenerative breaking

Q.2.30. Explain the thyristor control of half wave drive. [June 2009]

OR

Draw the circuit diagram for single phase half wave converter drive and explain its working.

Ans. Fig.2.13. shows the circuit arrangement for single phase half wave converter drive.

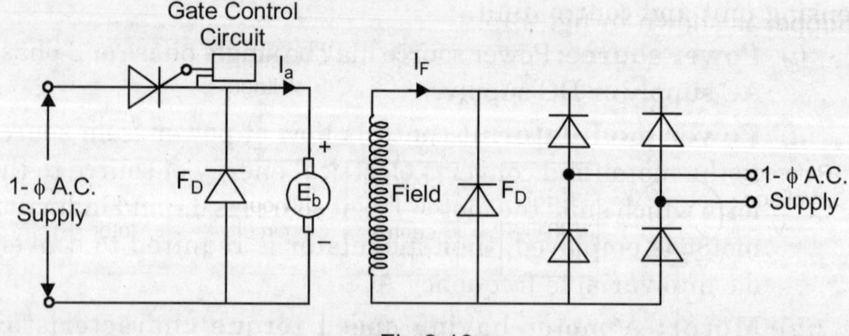

Fig.2.13

Separate converters are used for armature and field circuits. Field circuit is supplied from an independent full wave rectifier source.

The armature is connected through a single thyristor and a free wheeling diode. The armature current I_a flows during +ve half cycle. There is no current in the armature and motor deaccelerates till the triggering pulse is applied in the next cycle. The speed is controlled by controlling the firing angle $\propto$ of thyristor.

Q.2.31. Describe a single phase full wave converter drive

Ans. A full wave converter drive consists of four thyristors connected in the form of a bridge as shown in fig. 2.14.

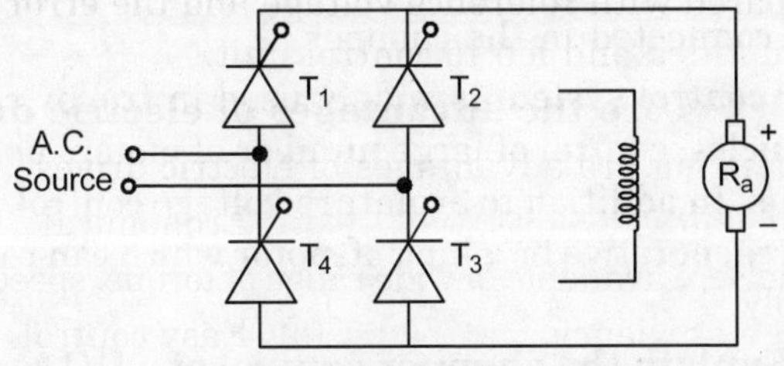

Fig. 2.14.

When field control is required, the field is fed from a separate source. During +ve half cycle, thyristor T_1 and T_3 are gated from firing angle α to π. In the -ve half cycle the thyristor T_2 and T_4 are gated from $(\pi + \alpha)$ to 2π. The current starts flowing with the turn ON of thyristor T_1 and T_3 at $\omega t = \alpha$ The motor gets connected to source. During -ve half cycle T_1 and T_3 are turned off and T_2 and T_4 are turned ON when triggered at $(\pi + \alpha)$. Armature voltage can be varied by changing the firing angle α.

Q.2.32. Draw and explain the block diagram for speed control of d.c motor using chopper drive.

Ans. Block diagram for the spread control of d.c motor using chopper is shown in fig. 2.15.

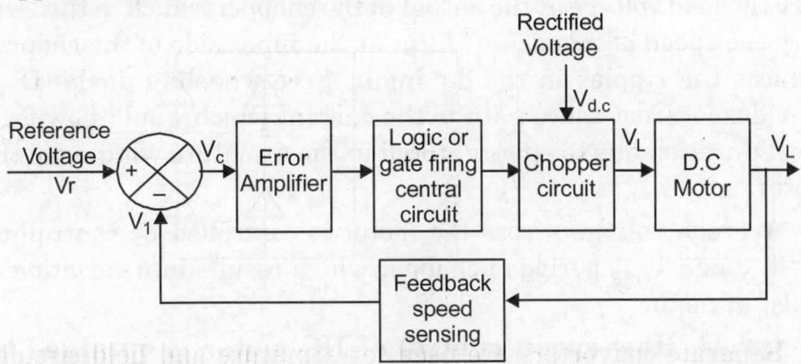

Fig. 2.15.

Rectified and filtered DC is applied to chopper. Logic circuit decides the firing of SCRs. The ON-OFF duration of SCRs is decided by this unit depending upon the signal obtained from comparator through an error amplifier. Speed feed back from d.c motor is converted into equivalent voltage signal by means of a technogenerator, which is given to the comparator. Comparator compares this voltage signal with the reference signal. If there is a difference between the two, an error signal is generated which is amplified and sent to logic circuit which decides ON-OFF duration of the SCRs connected in the chopper.

Chopper control system is widely used in trolley cars, battery operated vehicles, control of large number of motors operated from a common bus. In addition to armature voltage control, d.c chopper can provide regenerative breaking of motor which can return energy back to supply.

Q.2.33. Explain the chopper control of a DC series motor.

Ans. The circuit arrangement for controlling the speed of DC series motor using chopper is shown in fig. 2.16.

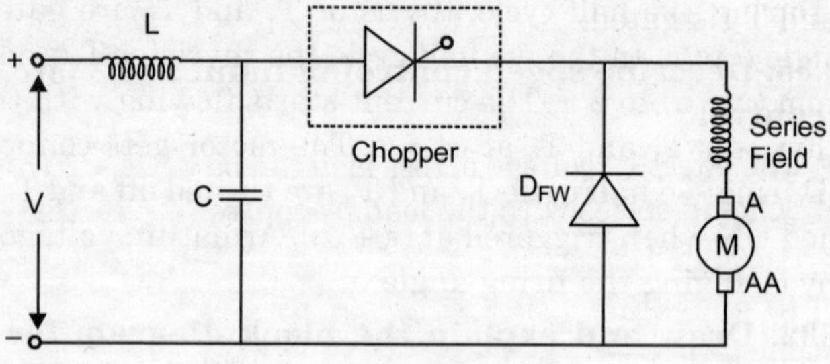

Fig. 2.16

The variation in ON-time and OFF - time of the thyristor will vary the load voltage at the output of the chopper, which in turn will vary the speed of motor. L-C filter at the input side of the chopper reduces the ripples in the d.c input. Free wheeling diode (D_{FW}) provides low resistance path to the current which would flow for a short duration due to energy stored in the armature winding of the motor.

Average voltage across the motor is controlled by controlling the T_{ON} and T_{OFF} period of chopper which results into variation of speed of motor.

Q.2.34. How speed control of DC motor is obtained by using dual converter.

Ans. Fig 2.17. shows a single phase dual converter for speed control of d.c motor

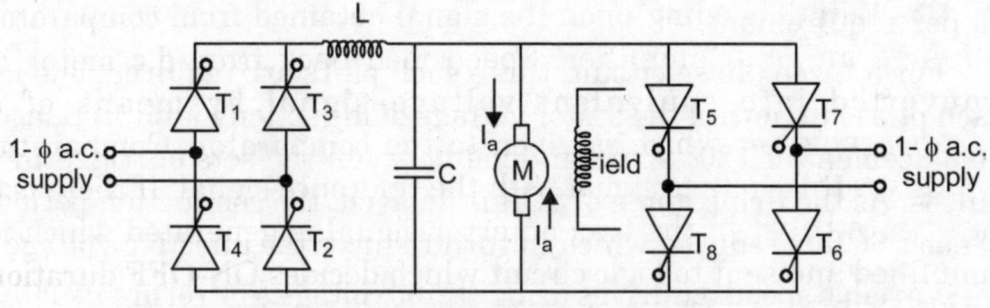

Fig 2.17

Four thyristors T_1, T_2, T_3, T_4 form the first bridge. It serves as full wave fully controlled bridge, which rectifies AC into DC. L and C are included in the circuit to remove ripples. During +ve half cycle of AC supply, thyristors T_1 and T_2 conduct simultaneously and in the -ve half cycle. Thyristors T_3 and T_4 conduct. Speed of motor is controlled by controlling the firing angles of thyristors.

To reverse the direction of rotation of motor, first bridge is commutated and the second bridge consisting of thyristors T_5, T_6, T_3, T_7 and T_8 is gated.

Q.2.35. Describe speed control of induction motor using phase control.

Ans. The effective voltage across load can be varied by connecting a pair of SCRs in series with the load by controlling the firing angle of SCRs.

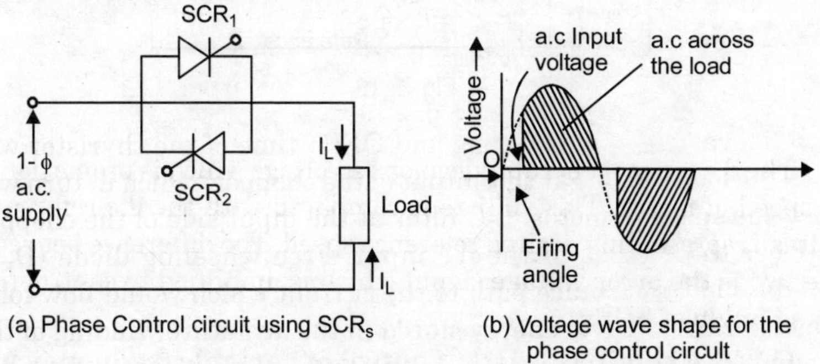

(a) Phase Control circuit using SCR_s (b) Voltage wave shape for the phase control circuit

Fig 2.18

In positive, half cycle, SCR_1 is triggered and the direction of current through load is from top to bottom. In the -ve half cycle of AC supply, SCR_1, is turned off and SCR_2 is triggered and the direction of current is now reversed. By varying the firing angles of SCR_1 and

SCR_2, the magnitude of AC voltage across the load can be changed. If the load is single phase motor the speed of motor can be changed as per requirement.

For a three phase circuit, three such pairs are required, one in each phase to control the stator voltage of induction motor. A phase displacement of 120° is maintained between the sets of triggering pulses. As the firing angle of SCR is delayed, the conduction period of each SCR is reduced which in turn reduces the effective voltage.

Variable speed a.c drives using stator voltage control are usually closed loop drive as shown in fig. 2.19

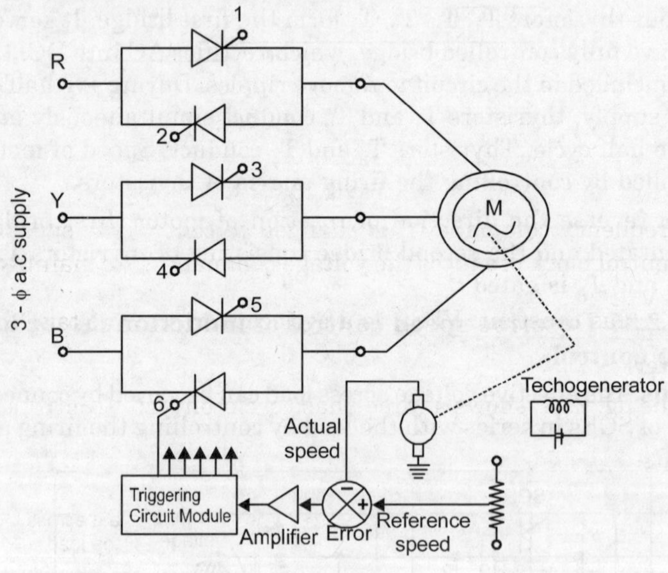

Fig. 2.19

The d.c technogenerator develops a voltage which is proportional to speed of motor. This voltage is compared with the d.c reference voltage representing desired reference speed. The difference between the two is the error voltage signal which is amplified to control the firing angle of SCRs.

Q.236. Explain thyristor control of variable frequency AC drives.

OR

Describe cyclo-converter controlled AC drives.

Ans. The speed of 3 phase induction motor is determined by frequency and number of poles. The synchronous speed is given by

$$Ns = \frac{120f}{P}$$

Where f = frequency of AC supply and P is the number of poles. The synchronous speed and hence the speed of motor can be controlled by varying the supply frequency.

Motor is fed from a variable freqency variable voltage source (VFS) as shown in fig. 2.20

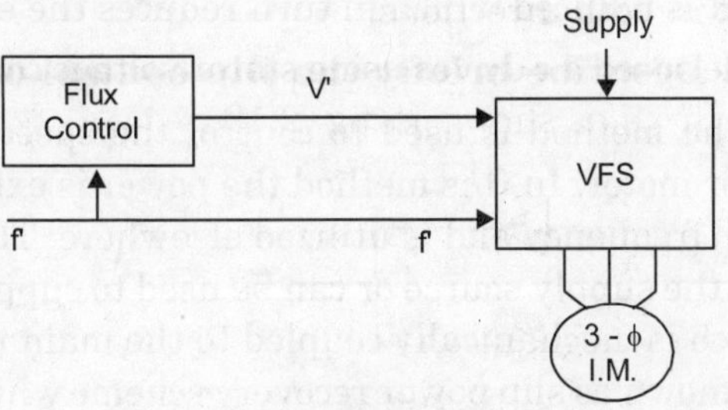

Fig. 2.20

Frequency command (f) is changed to control the speed. The flux control block produces the voltage command V′ to maintain the ratio $\frac{V'}{f'}$ as constant. VFS is a cycloconverter or voltage source inverter.

The fig. 2.21. shows a three phase cycloconverter controlled AC drive.

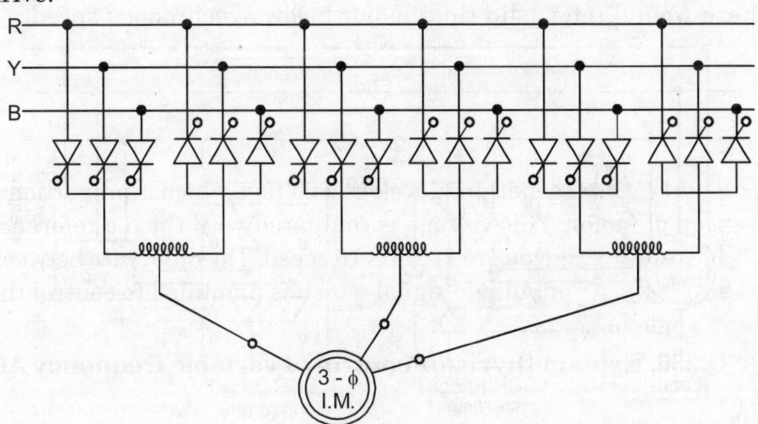

Fig. 2.21

A cyclo-converter allows variable frequency and voltage supply from a fixed voltage and frequency AC supply. It requires total of 18 thyristors. The + ve and -ve group rectifiers must supply one half cycle of low frequency output current. As the load is inductive, the low frequency output current lags the output voltage, the each rectifier group must continue conduction even after output voltage has changed sign. During this period, the group is in inverter mode and the power is returned to the a.c supply. Such operation continues till the load current is zero and starts in the reverse direction. Then the other group of thyristors start conduction and energy is transferred in both directions.

Q.2.37. Describe briefly the slip control of AC drives.

Ans. The method is used to control the speed of three phase wound rotor motor. In this method the power is extracted from the rotor at slip frequency and is utilized elsewhere. This power can be fed back to the supply source or can be used to supply an additional motor, which is mechanically coupled to the main motor. This type of drive is known as slip power recovery scheme which improves the over all efficiency of the system. The speed of motor can be controlled in subsynchronous and super-synchronous regions both by taking out and feeding power to rotor respectively. If the power fed to rotor has phase opposition with respect to rotor emf, the rotor current and hence the torque decreases. Therefore the motor speed decreases. If the power is fed to the rotor, the machine operates in super synchronous region.

Fig. 2.22 shows an arrangement for the speed control of three phase wound rotor (slip ring) motor below synchronous speed.

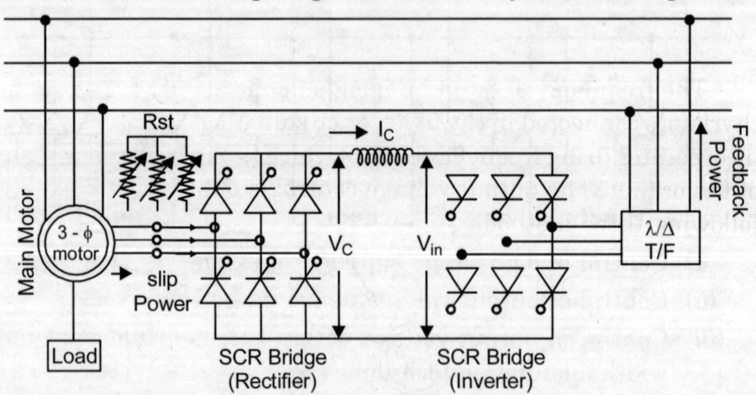

Fig. 2.22

The slip frequency power is rectified by a 3 phase diode bridge and is fed to a line commutated inverter bridge. The inverter returns the slip power to a.c supply through a suitable transformer. The

SCR bridge operates in inverter mode by restricting the firing angle between 90° and 180°. The machine operates in subsynchronous region. The power feed back can be cont controlled by controlling the inverter firing angles.

To operate such a drive at super synchronous speed, the operation of converter and inverter is reversed *i.e.*, inverter acts as rectifier and converter as inverter. This will inject slip frequency power in the rotor, thereby reversing the volume of slip and the speed above synchronous is obtained.

Q.2.38. Describe voltage controlled inverter drives.

Ans. Inverter provides variables frequency and voltage at the output, run variables speed drives, the control of circuit voltage is required with the frequency in order to avoid saturation of magnetic circuit. Fig 2.23 shows a three phase bridge inverter.

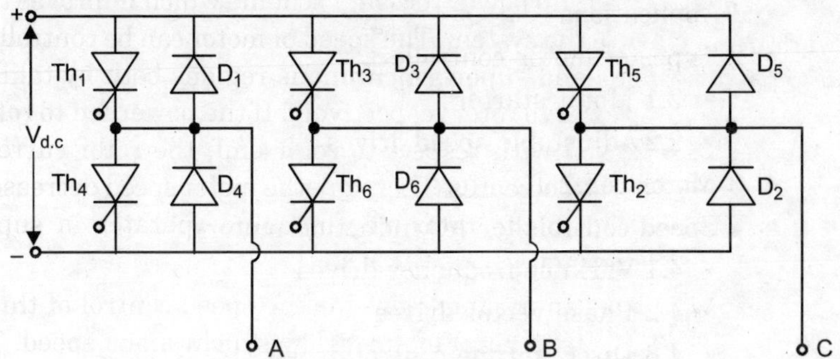

Fig. 2.23

The frequency of output depends upon the switching speed of thyristors connected in the inverter circuit. Most triggering circuits incorporates firing frequency control which is used to vary frequency of the output. The output voltage control is obtained by any of the following three methods.

(*i*) Control of d.c voltage supplied to inverter.

(*ii*) Control of output voltage delivered by the inverter,

(*iii*) Control of output voltage within the inverter using pulse width modulation techniques.

The d.c voltage supplied to the inverter can be controlled by a chopper when the source is DC. If the input voltage to inverter is obtained by rectifier then a phase control rectifier can give the variable d.c voltage.

The inverter output voltage can be controlled by introducing an a.c voltage regulator between the inverter and load.

The most suitable means of controlling the output voltage is to incorporate pulse width modulation technique.

2.39. Motor controller

From Wikipedia, the free encyclopedia

A motor controller is a device or group of devices that serves to govern in some predetermined manner the performance of an electric motor.[1] A motor controller might include a manual or automatic means for starting and stopping the motor, selecting forward or reverse rotation, selecting and regulating the speed, regulating or limiting the torque, and protecting against overloads and faults.[2]

2.40. Contents

2.40.1. Applications

Every electric motor has to have some sort of controller. The motor controller will have differing features and complexity depending on the task that the motor will be performing.

The simplest case is a switch to connect a motor to a power source, such as in small appliances or power tools. The switch may be manually operated or may be a relay or contactor connected to some form of sensor to automatically start and stop the motor. The switch may have several positions to select different connections of the motor. This may allow reduced-voltage starting of the motor, reversing control or selection of multiple speeds. Overload and overcurrent protection may be omitted in very small motor controllers, which rely on the supplying circuit to have overcurrent protection. Small motors may have built-in overload devices to automatically open the circuit on overload. Larger motors have a protective overload relay or temperature sensing relay included in the controller and fuses or circuit breakers for overcurrent protection. An automatic motor controller may also include limit switches or other devices to protect the driven machinery.

More complex motor controllers may be used to accurately control the speed and torque of the connected motor (or motors) and may be part of closed loop control systems for precise positioning of a driven machine. For example, a numerically controlled lathe will accurately position the cutting tool according to a preprogrammed profile and compensate for varying load conditions and perturbing forces to maintain tool position.

2.40.2 Types of motor controllers

Motor controllers can be manually, remotely or automatically operated. They may include only the means for starting and stopping the motor or they may include other functions.[2,3,4]

An electric motor controller can be classified by the type of motor it is to *drive* such as permanent magnet, servo, series, separately excited and alternating current.

A motor controller is connected to a power source such as a battery pack or power supply and control circuitry in the form of analog or digital input signals.

(*a*) Motor starters

Main article: Direct on line starter

Main article: Motor soft starter

A small motor can be started by simply plugging it into an electrical receptacle or by using a switch or circuit breaker. *A* larger motor requires a specialized switching unit called a motor starter or motor contactor. When energized, a direct on line (DOL) starter immediately connects the motor terminals directly to the power supply. A motor soft starter connects the motor to the power supply through a voltage reduction device and increases the applied voltage gradually or in steps.[2,3,4]

(*b*) Adjustable-speed drives

Main article: Adjustable-speed drive

An adjustable-speed drive (ASD) or variable-speed drive (VSD) is an interconnected combination of equipment that provides a means of driving and adjusting the operating speed of a mechanical load. An electrical adjustable-speed drive consists of an electric motor and a speed controller or power converter plus auxiliary devices and equipment. In common usage, the term "drive" is often applied to just the controller.[3,4]

2.40.3 Motor control centers

Fig. 2.24: right A small, early 1960's-vintage motor control center for 480 volt motors.

A motor control center (MCC) is an assembly of one or more enclosed sections having a common power bus and principally containing motor control units.[1] Motor control centers are in modern practice a factory assembly of several motor starters. A motor control center can include variable frequency drives, programmable controllers, and metering and may also be the electrical service entrance for the building. Motor control centers are usually used for low voltage three-phase alternating current motors from 230 volts to 600 volts. Medium-voltage motor control centers are made for large motors running at 2300 V to around 15000 V, using vacuum contactors for switching and with separate compartments for power switching and control.[5]

Motor control centers have been used since 1950 by the automobile manufacturing industry which used large number of electric motors. Today they are used in many industrial and commercial applications. Where very dusty or corrosive processes are used, the motor control center may be installed in a separate air-conditioned room, but often an MCC will be on the factory floor adjacent to the machinery controlled.

A motor control center consists of one or more vertical metal cabinet sections with power bus and provision for plug-in mounting of individual motor controllers. Very large controllers may be bolted in place but smaller controllers can be unplugged from the cabinet for testing or maintenance. Each motor controller contains a contactor or a solid-state motor controller, overload relays to protect the motor, fuses or a circuit breaker to provide short-circuit protection and a disconnecting switch to isolate the motor circuit. Three-phase power enters each controller through separable connectors. The motor is wired to terminals in the controller. Motor control centers provide wire ways for field control and power cables.

Each motor controller in an MCC can be specified with a range of options such as separate control transformers, pilot lamps, control switches, extra control terminal blocks, various types of bi-metal and solid-state overload protection relays, or various classes of power fuses or types of circuit breakers. A motor control center can either be supplied ready for the customer to connect all field wiring, or can be an engineered assembly with internal control and interlocking wiring to a central control terminal panel board or programmable controller.

Motor control centers (MCC) usually sit on floors, which are often required to have a fire-resistance rating. Firestops may be required for cables that penetrate fire-rated floors and walls.

2.40.4. Speed controls for AC induction motors

Recent developments in drive electronics have allowed efficient and convenient speed control of these motors, where this has not traditionally been the case. The newest advancements allow for torque generation down to zero speed. This allows the polyphase AC induction motor to compete in areas where DC motors have long dominated and presents an advantage in robustness of design, cost, and reduced maintenance.

(*a*) Variable frequency drives

Main article: Variable-frequency drive

(*b*) **Phase vector drives**

Main article: Vector control (motor)

Phase vector drives (or simply **vector drives)** are an improvement over variable frequency drives (VFDs) in that they separate the calculations of magnetizing current and torque generating current. These quantities are represented by phase vectors, and are combined to produce the driving phase vector which in turn is decomposed into the driving components of the output stage. These calculations need a fast microprocessor, typically a DSP device.

Unlike a VFD, a vector drive is a closed loop system. It takes feedback on rotor position and phase currents. Rotor position can be obtained through an encoder, but is often sensed by the reverse EMF generated on the motor leads.

In some configurations, a vector drive may be able to generate full rated motor torque at zero speed.

(*c*) Direct torque control drives

Direct torque control has better torque control dynamics than the Pi-current controller based vector control. Thus it suits better to servo control applications. However, it has some advantage over other control methods in other applications as well because due to the faster control it has better capabilities to damp mechanical resonances and thus extend the life of the mechanical system.

Main article: Direct Torque Control

2.40.5. Brushed DC motor speed or torque controls

Main article: Torque and speed of a DC motor

These controls are applicable to brushed DC motors with either a wound or permanent magnet stator. A valuable characteristic of these motors is that they are easily controlled in torque, the torque being fairly directly proportional to the driving current. Speed control is derived by simply modulating the motor torque.

Fig. 2.25: An industrial grade first quadrant PWM DC-motor controller

(*a*) SCR or thyristor drive

Main article: Thyristor drive

SCR controls for DC motors convert AC power to direct current, with adjustable voltage. Small DC drives are common in industry, running from line voltages, with motors rated at 90V for 120V line, and 180V for a 240V line. Larger drives, up to thousands of horsepower, are powered by three phase supplies and are used in such applications as rolling mills, paper machines, excavators and ship propulsion.

DC drivers are available in reversing and non-reversing models. The waveform of the current through the motor by a single-phase drive will have strong ripple components due to the switching at line frequency. This can be reduced by use of a polyphase supply or smoothing inductors in the motor circuit; otherwise the ripple currents produce motor heating, excess noise and loss of motor torque.

(*b*) PWM or chopper drives

Main article: Pulse-width modulation

PWM controls use pulse width modulation to regulate the current sent to the motor. Unlike SCR controls which switch at line frequency, PWM controls produce smoother current at higher switching frequencies, typically between 1 and 20 kHz. At 20 kHz, the switching frequency is inaudible to humans, thereby eliminating the hum which switching at lower frequency produces. However, some motor controllers for radio controlled models make use of the motor to produce audible sound, most commonly simple beeps.

A PWM controller typically contains a large reservoir capacitor and an H-bridge arrangement of switching elements (thyristors, Mosfets, solid state relays, or transistors).

2.40.6. Servo controllers

Main article: Servo drive

Main article: Servomechanism

Servo controllers is a wide category of motor control. Common features are:

- precise closed loop position control
- fast acceleration rates
- precise speed control

Servo motors may be made from several motor types, the most common being

- brushed DC motor
- brushless DC motors
- AC servo motors

Servo controllers use position feedback to close the control loop. This is commonly implemented with encoders, resolvers and Hall effect sensors to directly measure the rotor's position. Others measure the back EMF in the undriven coils to infer the rotor position and therefore are often called "sensorless" controllers.

A servo may be controlled using pulse-width modulation (PWM). How long the pulse remains high (typically between 1 and 2 milliseconds) determines where the motor will try to position itself. Another control method is pulse and direction.

2.40.7 Stepper motor controllers

Main article: stepping motor

A stepper, or stepping, motor is a synchronous, brushless, high pole count, polyphase motor. Control is usually, but not exclusively, done open loop, *i.e.,* the rotor position is assumed to follow a controlled rotating field. Because of this, precise positioning with steppers is simpler and cheaper than closed loop controls.

Modern stepper controllers drive the motor with much higher voltages than the motor nameplate rated voltage and limit current through chopping. The usual setup is to have a positioning controller, known as an **indexer,** sending step and direction pulses to a separate higher voltage drive circuit which is responsible for commutation and current limiting.

2.40.8 Relevant circuits to motor control

H-bridge

Main article: H-bridge

DC motors are typically controlled by using a transistor configuration called an "H-bridge". This consists of a minimum of four mechanical or solid-state switches, such as two NPN and two PNP transistors. One NPN and one PNP transistor are activated at a time. Both NPN or PNP transistors can be activated to cause a short across the motor terminals, which can be useful for slowing down the motor from the back EMF it creates.

References

1. National Fire Protection Association (2008). "Article 100 Definitions" (http://www.nfpa.org /aboutthecodes/About The

Codes.asp?DocNum=70&cookie%5Ftest=l). *NFPA 70 National Electrical. Code.* 1 Batterymarch Park, Quincy, MA 02169: NFPA. pp. 24. http://www.nfpa.org/aboutthecodes/About the Codes.asp?DocNum=70 &cookie%5Ftest=1. Retrieved January 2008.

2. abc Siskind, Charles S.(1963). *Electrical Control Systems in Industry*. New York: McGraw-Hill, Inc... ISBN 0070577463.

3. abc National Fire Protection Association (2008). "Article 430 Motors, Motor Circuits and Controllers" (http://www.nfpa.org/aboutthecodes/AboutTheCodes.asp?DocNum=70&cookie%5Ftest=1). *NFPA 70 National Electrical Code.* 1 Batterymarch Park, Quincy, MA 02169: NFPA.pp. 298 http://www.nfpa.org/aboutthecodes/AboutThe Codes.asp?DocNum=70&cookie%5Ftest=l. Retrieved January 2008.

4. abc Campbel, Sylvester J. (1987). *Solid-State AC Motor Controls.* New York: Marcel Dekker, Inc...ISBN 0-8247-7728-X.

5. abc Robert W. Smeaton (ed) Switchgear and Control Handbook 3rd Ed., Mc Graw Hill, New York 1997.

2.3.Intelligent Motion control of Electrical Drives

All India seminar on Intelligent motion control of Electrical Drive was conducted in the Udaipur local centre of the Institution of Engineers (India) where different professors of Electrical engineering of various universities and Engineering Colleges attended All India seminar under the aegis of electrical engineering Division on the above topic in year Dec 2010. In that, seminar mainly emphasized on optimization of the motion control performance by intelligent, self-learning or self-organizing control utilizing artificial intelligent techniques like expert systems, fuzzy logic and neutral network techniques with embedded protection and smart devices. One of the professors, Queensland university , Australia, in his address said that motion control solutions and systems were in continual development requiring the sustained efforts of control, electrical and mechanical engineers. He further emphasized that an increase in precision and reduction in cycle time was required in packaging machines, plastic injection modelling glass, wood and ceramic industries, metal forming machine and a number of other tasks in processing machines. He further added that conflicting motion-control requirement, for decreased cycle time, increased operating speed and increased accuracy are made more challenging by mechanical resonance, finite resolution, sensor imperfection and noise. Papers on various sub-themes were presented. During discussion recommendations the emerged included: Engineers should adopt DSP and FPGA in design of motion control systems liberally for performance enhancement. Engineers should also enhance and optimize the motion control performance further by intelligent, self-learning or self-organising control utilizing artificial intelligent techniques like expert systems, fuzzy logic and neutral network with embedded protection and smart devices. A sustained effort of control electrical and mechanical engineers in finding motion control solution is the need of the hour. Efforts should be made for direct converter drive for linking two different power systems having different frequencies and voltages.

10.7 Short Questions with Answers

1. ILLUMINATION

Q.1.0 Explain the Principle of Lighting.

Ans. Principle of Lighting.

When light falls on a surface, depending upon the nature of surface, some portion of light energy is reflected, some portion is transmitted through the medium of the surface and rest is absorbed. The ratio of reflected light energy to the incident light energy is known as reflection factor.

There are two types of reflection :

(i) mirror or specular reflection and

(ii) diffuse reflection.

In case of specular reflection a beam of light is reflected but not scattered, and unless the eye is placed in the path of reflected beam, the viewer is unaware of the existence of light. Moreover if his eye is placed in the path of the reflected beam, he sees, not the illuminated surface, but the light source. Surfaces causing specular reflection are silvered mirrors, highly polished metal etc. With diffuse reflection the reflected light is scattered in all directions and the viewer sees the illuminated surface, not the light source.

Q.1.1. Explain the nature of light and of its propagation.

Ans. Nature of Light.

To understand illumination engineering it is essential to know the nature of light and fundamental principles underlying its production.

Light is a form of radiant energy. Various incandescent bodies are sources of light and light emitted by these incandescent bodies depend upon the temperature of the body. Heated bodies are of two types in colour *(i)* Red hot and white hot, white body is in red hot form. The energy radiated by it is of large wavelength and energy available in form of heat

With increase in temperature red hot body becomes white hot and wave light of energy radiated by body becomes smaller and enters into the wavelength of light. At white hot state, body emit as well as heat energy. For *e.g.*, if we take tungsten filament lamp at low voltage, filament of lamps glows red emitting no light energy. But at rated voltage filament becomes white hot and emits light energy as well as some heat energy also. But radiation in form of heat is very small as compared to light energy radiated.

So, we can conclude that light is a form of energy which can be radiated like heat energy or it is the heat energy itself which can be converted into light energy also. So light may be defined as the external agent which produces in us the sensation of light. Light is a form of radiant energy when falls on a body, the body becomes visible to eye.

Mode of propagation of light: Like electromagnetic waves the light radiations propagate through the medium in form of transverse waves. Transverse waves motion is one in which the medium particles vibrate up and down periodically at right angles to the direction of propagation of the disturbance.

According to Quantum Theory of Max Plank, light energy is emitted in form of small 'bundles' or 'packets' called quanta or photons.

According to modern theory light has dual nature, *i.e.* light energy is propagated in the form of waves while it is emitted and absorbed in the form of packets or quantum of energy called photons. Velocity of light is 3×10^8 m/sec.

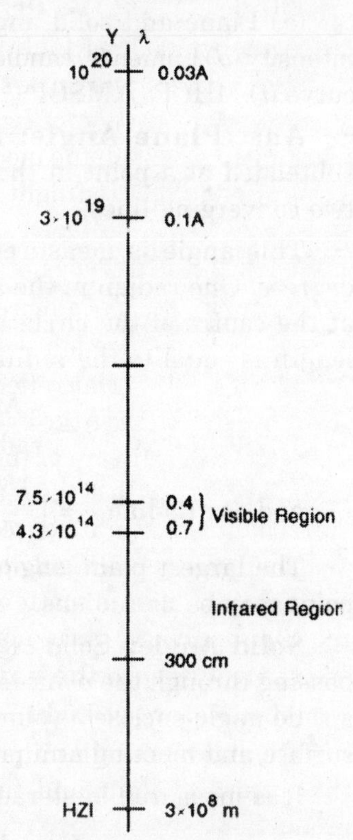

Fig. 1.1. Spectrum of Radiation Energy.

When some obstacle comes in the wave of radiation of light, either there is reflection or absorption. The reflected part of energy results in illumination of the object. The absorbed part is converted into heat. But medium particles are neither illuminated nor heated.

Q.1.2. State the units of wave length.

Ans. The wave lengths used by illuminating engineers are very short and the units used are micron (10 meter) or Angstrom A (10^{-10} meter). The complete spectrum of radiant energy extends over a tremendous range of frequency as shown in fig 1.1. The visual region extends from about 0.4 micron to 0.75 micron. The illumination engineer is interested into conversion of electrical energy into radiant energy of such frequencies as to be visible to the human eye and also in the conversion of radiant energy into electrical energy by photo cells or thermo couples.

Q.1. 3. Define.

(*a*) Plane and solid angles (*b*) Luminous Flux (*c*) Luminous intensity (*d*) Lumen (*e*) candle (*f*) Illumination (*g*) Brightness (*h*) Polar curve (*i*) MHCP (*j*) MSCP (*k*) MHSCP.

Ans. Plane Angle: Plane angle is subtended at a point in the same plane by two converging lines.

This angle is measured in radians or degrees. One radian in the angle subtended at the centre of the circle by an arc whose length is equal to the radius of the circle

$$\theta = \frac{\text{Arc}}{\text{radius}} = \frac{l}{r} \text{ radians}$$

and one Radian $= \left[\frac{180}{\pi}\right]$ degrees

Radius
θ
ARC

Fig. 1.2

The largest plane angle subtended at a point may be 2π radians.

Solid Angle:- Solid angle is the angle generated by the line passing through the point in space and the periphery of the area or a solid angle enclose a volume by an infinite no. of lines lying on a surface and meeting at a point.

It is measured in steradians and is denoted by ω and

$$\omega = \frac{\text{Area}}{(\text{Radius})^2} \text{ steradians}$$

Where one steradian is the angle subtended at the centre of a sphere by an area on the surface of the sphere which is numerically equal to the square of the radius.

The total plane angle subtended by circumference of a circle at the centre of circle is 2π radians.

Similarly, total solid angle subtended at a point in space is obtained by considering point at the centre of sphere and the surface area of the sphere.

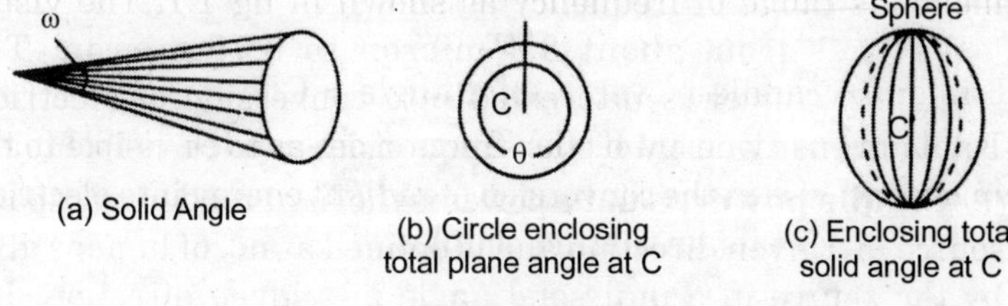

(a) Solid Angle (b) Circle enclosing total plane angle at C (c) Enclosing total solid angle at C

Fig. 1.3

$$\text{Plane angle,} \quad \theta = \frac{\text{Arc}}{\text{Radius}} = \frac{\text{Circumference}}{\text{Radius}} \text{ (Total angle)}$$

$$= \frac{2\pi r}{r} = 2\,\pi \text{ radians}$$

$$\text{Total solid angle, } \omega = \frac{\text{Area of sphere surface}}{(\text{Radius of sphere})^2}$$

$$= \frac{4\pi r^2}{r^2} = 4\,\pi \text{ steradians}$$

***(b)* Luminous Flux :** It is defined as the total quantity of light energy radiated or emitted per second from a luminous body in the form of light waves. It is measured in lumens and is denoted by *F* or ϕ. It is defined as the rate of luminous energy.

If Q lumens hour is the luminous output of source of light and energy radiated is for *t* hours, then flux $\phi = \frac{Q}{t}$ lumens.

***(c)* Luminous Intensity:** Luminous Intensity in any particular direction is the luminous flux emitted by per unit solid angle by a point source and is denoted by, *I.*

$$I = \frac{F}{\omega} = \frac{\phi}{\omega} \text{ Lumens/steradian or candela}$$

It is the ratio of the brightness of a source of light to that of standard candela. One candle gives out luminious flux of 4π lumens in space. Thus lumens emitted by one candela source of light is one lumen/steradian.

In scientific terms candela is defined as the luminous intensity in the perpendicular direction of a surface of 1/600,000 sq. meter of a full radiator at the temp. of freezing platinum under a pressure of 101,325 N/sq. meter.

***(d)* Lumen :-** Lumen is defined as the amount of luminous flux given out in space represented by one unit of solid angle by a source having an intensity of one candle power in all directions.

Or

Lumens = candle power × solid angle = C.P. × ω

Total lumens given out by the source of one candela is 4π lumens.

***(e)* Candle power.** Candle power is the light rendering capacity of a source in a given direction and is defined as no. of lumens given out by the source in a unit solid angle in a given direction. It is denoted by symbol, C.P.

$$\text{C.P.} = \frac{\text{Lumens}}{\omega}$$

***(f)* Illumination.** When light falls upon any surface, the phenomenon is called illumination. It is defined as the no. of lumens, falling on the surface per unit area. It is denoted by symbol, E & is measured in lumens per square meter or lux or meter candle.

If a flux of ϕ lumens falls on a surface of Area, A *then* the illumination of that surface is $E = \phi/A$ lumens/m^2 or lux or metre candle.

Bigger unit of illumination is phot. One phot = 10^4 lux.

***(g)* Brightness:** When the eye receives a great deal of light from an object we say it is bright and brightness *is* an important quantity in illumination, it is all the same whether the light is produced by the object or merely reflected from it. The object send out light as though each small piece of its surface were of a certain luminous intensity. Generally the brightness of an object is not the same from all points of view. When brightness is considered in terms of measurable quantity, the term luminance is employed.

Brightness or Luminance is defined as the luminous intensity per unit projected area of either a surface source of light or a reflecting surface and is denoted by L.

If a surface of Area A has an effective luminous intensity of I candelas in a direction θ to the normal then the luminance of that surface is

$$L = \frac{1}{A\cos\theta} \text{ Candela/m}^2 \text{ or nits}$$

i.e. nit is defined as candela/m^2. Bigger unit of luminance is stilb which is defined as candela/cm^2. Lambert is also the unit of brightness which is lumens/cm^2. Foot lambert is lumens/ft^2.

(h) **Polar Curves:** The illumination flux emitted by a source can be determined from the intensity distribution curve. The luminous intensity or candle power of any practical lamp is not uniform in all directions due to its un-symmetrical shape. The distribution of light is given by polar curves. A radial ordinate pointing in any particular direction on a polar curve represents the luminous intensity of the source when viewed from that direction. If the luminous intensity is measured in a horizontal plane about a vertical axis and a curve is plotted between candle power and the angular position, a horizontal polar curve is obtained. Similarly, vertical polar curve gives the relationship of candle power in vertical plane passing through the lamp at various angles. Typical vertical and horizontal polar curves for a filament lamp is shown in Fig 1.4.

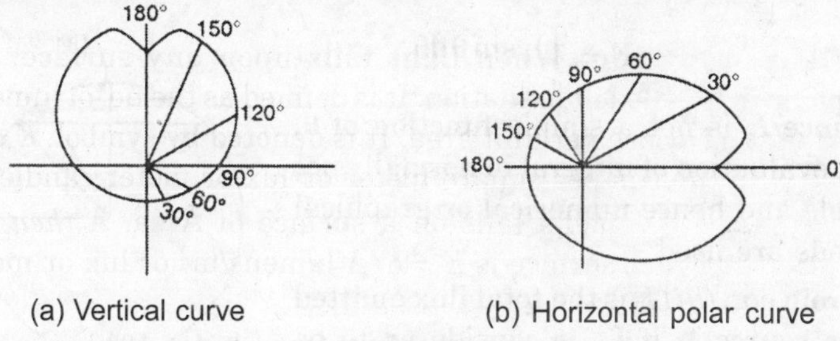

Fig. 1.4

Slight depression at 180° in case of vertical polar curve *is* due to the position of lamp holder whereas at 0° in the horizontal polar curve is due to coiled coil filament occupying an arc subtending an angle less than 360°.

The flux emitted by a source can be obtained by a polar solid is assumed to be concentric with a sphere and the flux emerging from the sphere through zones defined by the angular limits θ_1 and θ_2 is found as follows :

From the eq. : $I = \underset{\delta\omega 0}{Lt} \frac{\delta F}{\delta\omega'}$ ....(*i*)

the luminous intensity I_θ along θ is given as:

$$I_\theta = \frac{df}{d\omega}$$

$$dF = I_\theta \, d\omega$$

$$F\theta_1, \theta_2 = \int_{\theta_1}^{\theta_2} I\theta.d\omega \qquad ....(ii)$$

Now from the Fig. 1.5 drawn below it is clear that the differential solid angle dω subtended by the differential strip area is given as :

$$d\omega = \frac{rd\theta\ {}_2\pi r \sin\theta}{r^2}$$

$$= 2\ \pi\ \sin\ d\theta$$

Substituting in eqn (2), we have:

$$F\theta_1, F\theta_2 = \int_{\theta_1}^{\theta_2} I\theta.2\pi \sin\theta\, d\theta$$

$$= 2\pi \int_{\theta_1}^{\theta_2} \sin\theta\, d\theta \qquad ...(iii)$$

The total flux emitted is given when θ varies from θ° to π radians *i.e.*

$$F = \int_0^{\pi} I_\theta \sin\theta\, d\theta \qquad ...(iv)$$

Since I_θ is not a simple function of θ, direct evaluation of integral in normally not possible and hence numerical or graphical methods are used

From eqn *(iii) F* is the total flux emitted by the source & if I_θ is considered to be mean spherical candle power then

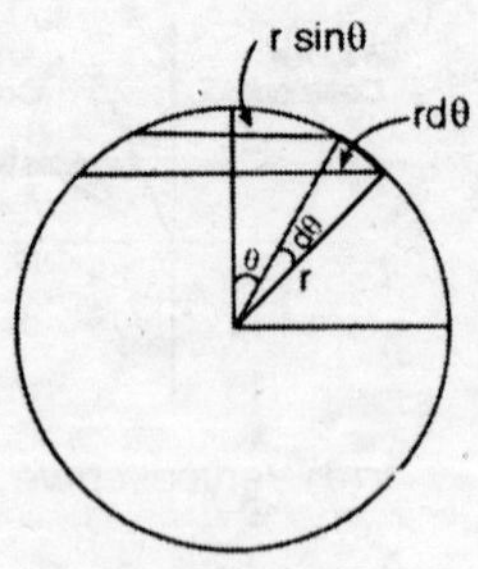

Fig. 1.5

$$I_\theta \text{ mean} = \text{M.S.C.P.} = \frac{\text{Total flux from source}}{4\pi} \qquad ...(iv)$$

Also if $\theta_1 = \theta$ and $\theta_2 = \pi/2$ the integration covers the luminous flux emitted below the horizontal

Fo,

$$90 = 2\pi\ I\theta \int_0^{\pi/2} \sin\theta\, d\theta = 2\pi I_\theta$$

Therefore,

$$I_\theta = \text{lower mean hemispherical candle power}$$

$$= \frac{\text{Flux of lower hemisphere}}{2\pi}$$

Similarly, the upper mean hemispherical candle power

$$= \frac{\text{Flux of upper hemisphere}}{2\pi}$$

Q.1.4 Explain the production of light by:
(*a*) Excitation (*b*) Ionisation (*c*) Fluorescence and phosphorescence.

Ans. (*a*) Excitation: In this type of collision quantum of energy given by striking electron is large enough to change the energy level of gas atom, electron coming in its path. The extra energy is insufficient to bring the valence electron to a conduction level. So electron remain attached to the parent atom, but shifts to a higher orbit temporarily. After few secs. thes electrons (e$^-$)return to original orbit and departs its extra energy in the form of photons of lights.

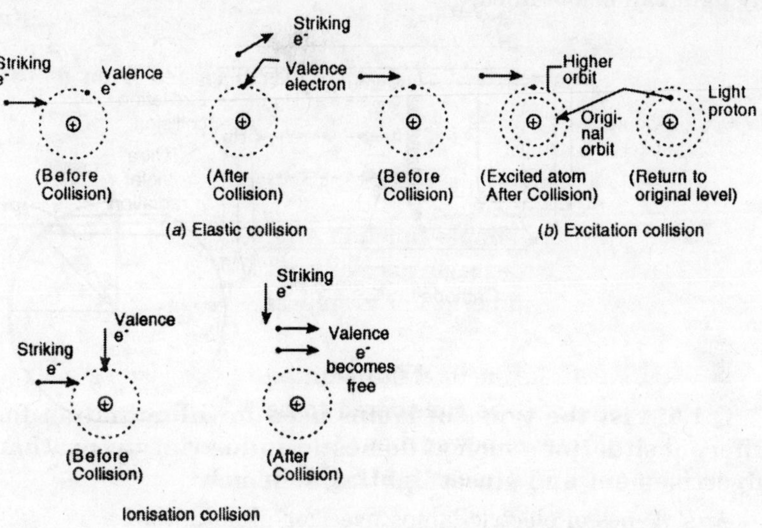

Fig. 1.6

(*b*) Ionisation: In this type of collision energy supplied by striking e^- to valence e^- of gas is so high that valence e^{-ves} leaves parent atom and becomes a free e^-. This process is called ionisation.

So we conclude that excited atoms emit photons of light and the process continues as long as emission in discharge tube occurs and anode potential is sufficiently higher than cathode.

(*c*) Fluorescence and phosphorescence: Consider the cathode filament emitting $e-^{ves}$ after getting heated due to supply of current.

These e^{-ves} while accelerating collide with argon and mercury vapour atoms. The excited atoms of mercury give a radiation of 2537 A.U. wavelength in the given case due to low pressure inside the tube. Obviously this radiation is ultra violet and can produce no illumination if left as such. On the other hand, if radiation of this wave length is made to strike with phosphorous material, there is re-emission of light radiation of a different wavelength. The effect is called fluorescence hence the name fluorescence tube.

Phosphorescence. The coating materials used depend upon the colour effect desired and may consist of zinc silicate, cadmium silicate or calcium tungstate. These organic chemicals are known as phosphorus and their powder transforms short wave invisible radiation into visible light. This effect is called the phosphorescence. By mixing the various powders light of any desired colour including day light can be obtained.

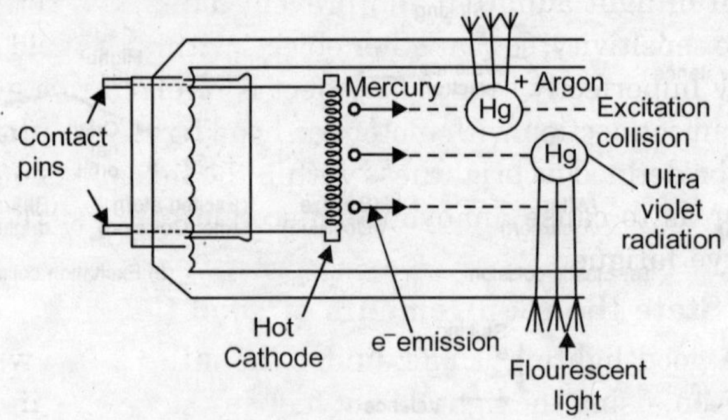

Fig. 1.7

Q.1.5. List the types of lamps used for illumination for different situations such as domestic, industrial, decoration, advertisement and street lighting schemes.

Ans. Types of electric lamps used for illumination:

Different methods to produce light are:

(i) By passing the current through a wire (also called filament), raising its temperature to incandescence (As in incandescent Lamps).

(ii) By estabilishing an arc between two electrodes (As in Arc lamps)

(iii) By maintaining an electric discharge through a vapour or gas (Discharge lamps)

The different types of lamps used for domestic, industrial, decoration & street lighting are given below:

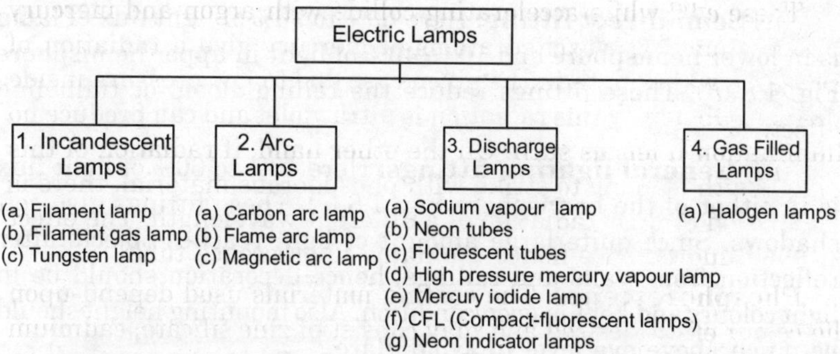

Q.1.6 Define glare.

Ans. Glare. The size of the opening of the pupil in the human eye is controlled by its iris. If the eye is exposed to a very bright source of light the iris automatically contracts in order to reduce the amount of light admitted and prevent damage of retina. This reduces the sensitivity, so that other objects within the field of vision can be only imperfectly seen. This effect is referred to a glare and is familiar in connection with motor car head lights. In other words glare may be defined as brightness within the field of vision of such a character as to cause annoyance, discomfort, interference with vision or eye fatigue.

Q.1.7. State the requirements of good lighting.

Ans. A good lighting is one, under the influence of which the eyes of person, using the light do not feel any strain *e.g.* if the light is too dim we cannot observe the objects properly or cannot do reading or writing work. Whereas on the other hand, if the light is too bright our eyes start watering and are made to open forcibly. The good lighting does not have the above mentioned effects, but we find comfort while working under it. The factors to be taken care of while planning a lighting scheme are:

(*a*) Illumination level (*b*) Absence of glare

(*c*) Contrast (*d*) Shadows (*e*) Colour rendering

Q.1.8. List the lamp fittings used in domestic and industrial applications.

Ans. Lamp fittings are classified according to the way light reaches the object such as:

(*i*) **Direct fittings:** Here 90% to 100% light flux of lamp and fitting is in lower Hemisphere and 0—10% of light flux in upper hemisphere (in fig. 1.8). These fittings are efficient, cheap and give hard light and have tunneling effect *i.e.* ceiling remains dark.

(*ii*) **Semi-direct fitting:** Here 60% to 90% of light flux of lamp is in lower hemisphere and 10 to 40% of light in upper hemisphere Fig. 1.8 (*b*). These fittings reduce the ceiling gloom or tunneling effect.

(*iii*) **General lighting fittings:** Here 40% to 60% of light flux is in either of the hemisphere Fig. 1.8 (*c*). These fittings give soft shadows. Since quite large amount of light reaches objects after reflection from walls and ceilings, hence decoration should be in light colours and kept in good condition. Also mounting height should be much above eye level to avoid glare.

(*iv*) **Semi-direct fittings:** Here 10 to 40% of light is in lower hemisphere and 60 to 90% of light flux is in upper hemisphere Fig. 1.8 (*d*). These fittings give soft shadows with low efficiency.

(*v*) **Indirect fittings:** Here 0—10% of light is in lower hemisphere and 90% to 100% of light flux is in upper hemisphere in Fig. 1.8 (*e*). These fittings give practically no shadows or no glare even with high intensity sources.

This system of indirect lighting is most expensive as large proportion of light is lost in ceiling which is required to have matt surface of high reflection factor.

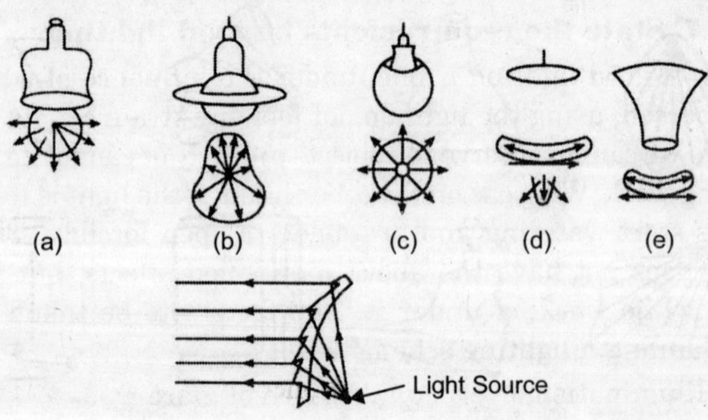

Fig. 1.8

Q.1.9. Sketch different lamp fittings. State use and advantages of each type.

Ans. Having made the classification of light fittings we will now consider these fittings individually giving their scope of applications. Before that we should however understand the importance of providing ventilation which not only reduces heat but also keeps reflectors clean due to chimney effect. Therefore vented reflectors are specially well suited for dusty areas such as foundries where it is tough to keep the reflectors clean. These reflectors also allow some light to pass on the ceiling cutting contrast.

(*i*) **Dispersive reflector:** In Fig 1.9 (a), it is a widely used fitting in industry. Its angle of cut off is 70 and maximum C.P. is obtained at 45° from the vertical. It gives uniform illumination with space height ratio of $1\frac{1}{2}$. It requires about *1* watt/m^2 to get illumination of 6 Lux. These are used for moderate ceiling height.

(*ii*) **Concentrating reflector:** In Fig. 1.9 (*b*) this reflector has shape of deep parabola. Maximum C.P is obtained upto 25° to the vertical. Its angle of cut off is 60° to the vertical. Uniform illumination with these fittings is obtained with space height ratio of one. It is suitable for high mounting height as required in workshop bays having electric over head travelling cranes.

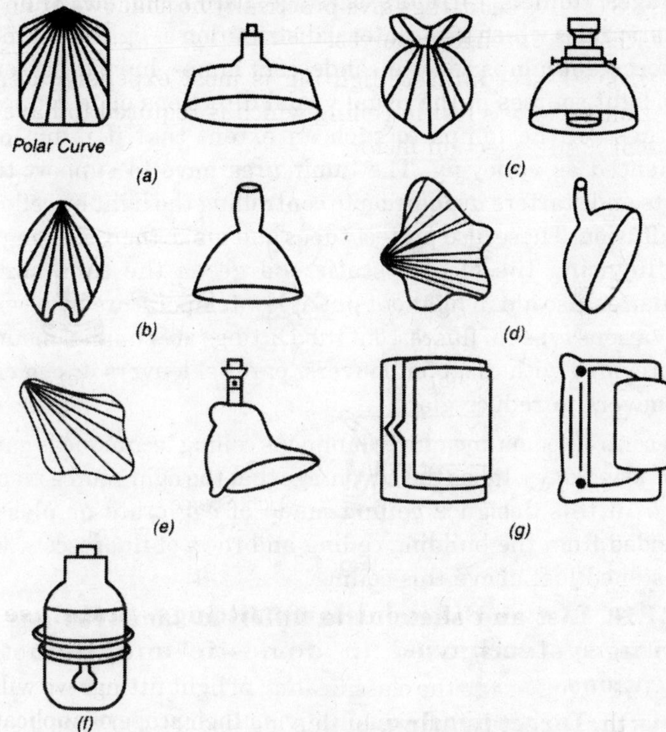

Fig. 1.9

(*iii*) **Glass steel diffusing fittings:** In Fig. 1.9 (c), these fittings give high quality of light. It minimizes glare from polished surfaces and give less shadows. Maximum C.P. is from 0 to 45° to the vertical. Uniform illumination requires space height ratio of $1\frac{1}{4}$. Some light goes to ceiling which reduces the ceiling gloom.

(*iv*) **Angle reflector:** These are used where suspended lamps would cause obstruction or to illuminate vertical surfaces which may not receive light from overhead system of lights. There are various types of shapes which determine the light distribution. Parabolic and elliptical type angle reflectors are shown in Fig. 1.9 (*d*) and (*e*) respectively. Angle of tilt of principle axis is *45°* angle of cut off 17° behind and 84° in front. Max. C.P. is from 15° to 75° in front.

(*v*) **Well glass fitting:** These are used in places where acid fumes or moisture prevails.

(*vi*) **Bulk head fittings:** In Fig. 1.9 (*g*) these are used in places where sufficient place is not available to fit in normal fittings such as garages, tunnels, corridors, carriage, bath room, etc. They have prismatic glass which gives lateral distribution of light. Luminaires for fluorescent lamps as for incandescent lamps, have to screen the bright light sources in the usual visual directions or to reduce the brightness of the lamps to such an extent that it is no longer experienced as annoying. The luminaires have to support tubes, ballasts and starters in addition to controlling the light by reflection and diffusion. These also protect tubes and make their starting easy by influencing the electro static charge on the tube surface. Luminaires also aid in light out put at low temperature by retaining heat. Various types of fluorescent tube fittings are channel mounting type provided with diagonal louvers, parallel louvers or egg crates type louvers to reduce glare.

In control rooms and offices luminous ceiling technique is gaining favour now-a-days. It has the advantage that the light source is spread evenly. In this design a ceiling made of egg crate or plastic is suspended from the building ceiling and rows of fluorescent tubes are fastened just above this ceiling.

Q.1.10. List and sketch the different lamp fittings and lighting systems used in domestic and Industrial applications.

Ans. (*a*) **Direct lighting.** In this method 90-100%, of the light is directed downwards and 0-10% of the light upwards. This should be carefully used to avoid excessive glare, brigthness or harsh shadows as shown in fig 1.10 (*a*) & (*b*).

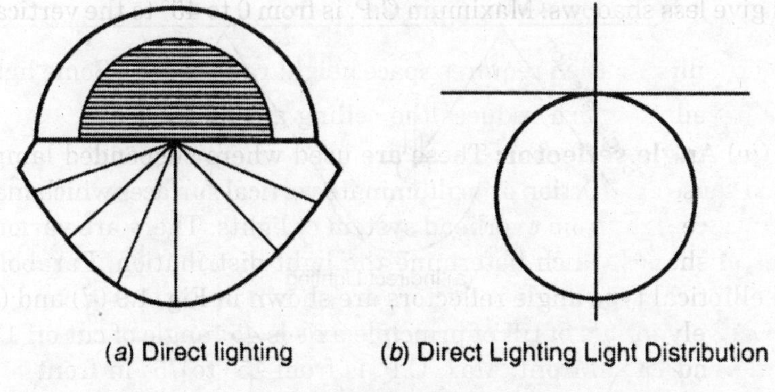

(*a*) Direct lighting (*b*) Direct Lighting Light Distribution

Fig. 1.10.

(*b*) **Semi-direct.** This method has 60-90% of light directed downwards and 10-40% upwards [Fig. 1.11 (*a*) and (*b*)].

The distribution of light *is* shown in Figs. 1.2 (a) and *(b).*

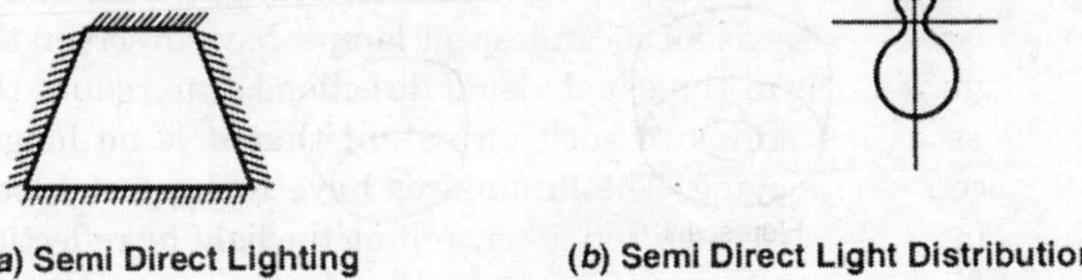

(*a*) Semi Direct Lighting **(*b*) Semi Direct Light Distribution.**

Fig. 1.11.

(*c*) **Diffusing.** This arrangement gives a uniform light distribution and is pleasing [Fig. 1.12 (*a*) and (*b*)].

(*a*) Diffused Lighting **(*b*) Diffused Light Distribution.**

Fig. 1.12.

(*d*) Indirect lighting

Here not less than 90% light is upward. The fittings have no shadows and glare is avoided. The method is costly as light is upwards [Fig. 9.13 (*a*), (*b*) and (*c*)].

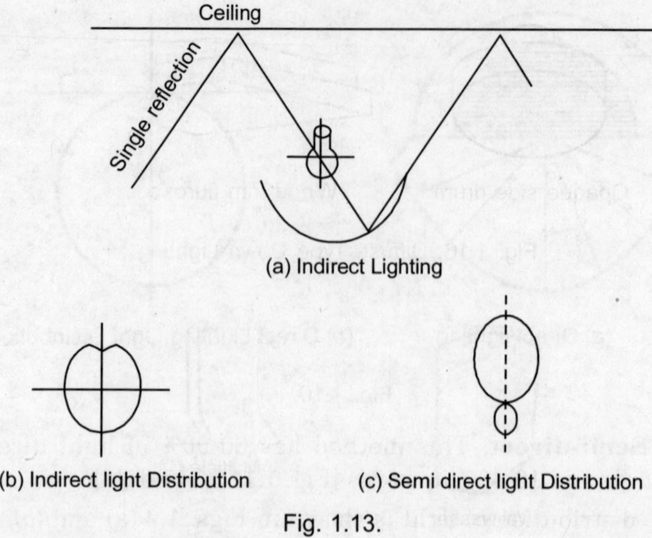

Fig. 1.13.

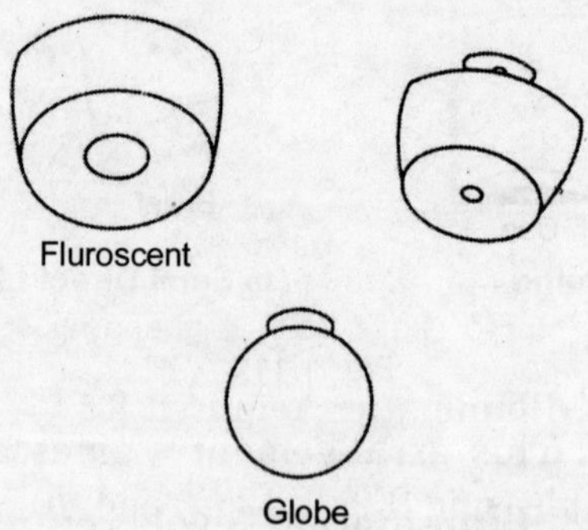

Fig. 1.14. Diffusing Type Luminaires.

In semi-indirect lighting, 10-40% is downwards and the rest 60-90% light is upwards. Some of the luminaires used for mounting on the ceiling are shown in [Figs. 1.14 to Fig. 1.17]

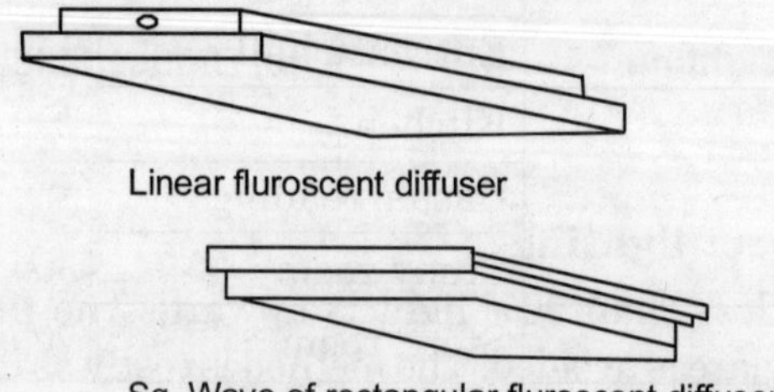

Fig. 1.15. Large. General Diffuse Lighting.

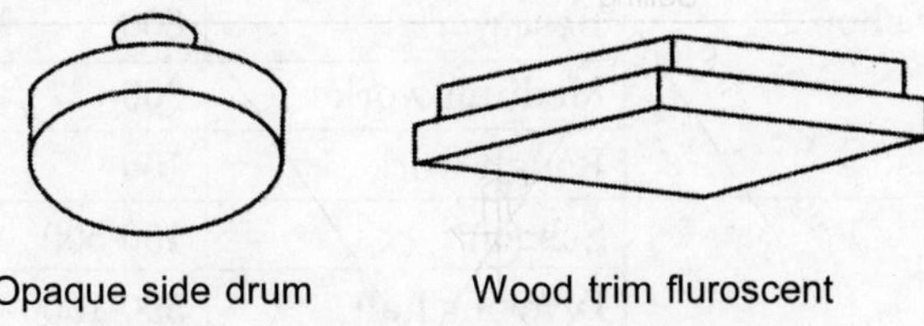

Fig. 1.16. Diffuse Type Down Lighting.

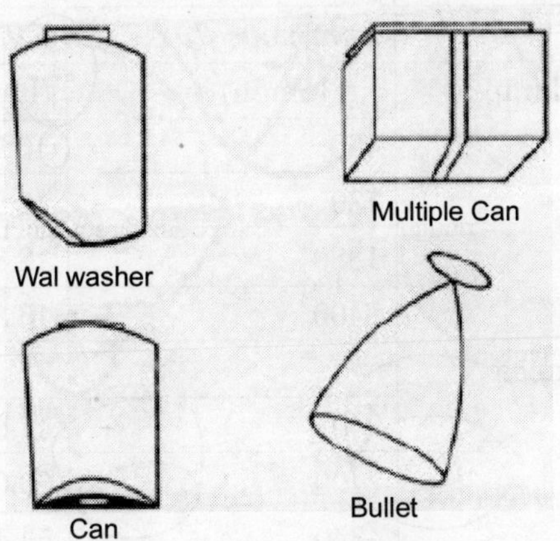

Fig. 1.17. Down Lighting.

Applications. Illumination required for various applications are given in Table 9.1

Table 9.1: Illumination levels required for various applications

Place	*Description*	*Illumination (lux)*
Offices	General office	300
	Conference hall	300
	Entrance hall	150
Houses	Kitchen	200
	Casual reading	150
	Study room	300
Shops	Store room	200
	Other areas	150-300

Assembly shop	Best works	800
	Medium work	300
	Rough work	150
School	Schools	150-500
	Drawing hall	500-1000
Sports ground	Foot ball	100-200
	Tennis court	250-500

Wattage	*Luminous flux*	*Efficiency*
Incandescent lamps	(in lumen)	(lumen/watt)
40	430	10.7
60	730	12.2
100	1380	13.8
500	8400	16.8
Fluorescent tubes		
20	940	47.0
40	2400	60.0
80	4350	54.4
High pressure mercury vapour lamp		
80	3400	42.5
250	12500	50.0
1000	55000	55.8

Q.1.11. Design a simple lighting scheme for (*i*) Indoor (*ii*) Factory (*iii*) Outdoor (*iv*) street lighting by light flux method.

Ans. Indoor:

(*a*) **Illumination level**: This is the most vital factor which affects the total wattage of the lamps in an installation. Level of illumination depends upon many considerations which dealt in table given below:

Location	*Illumination Level (Lux)*
Entrance and Hall ways	100
Living rooms	300
Dining rooms	150

Bed rooms	300
Dressing table bed heads	200
Table games	300
Kitchen	300
Bathroom	100
Stairs	100
Garage	70
Study	300

(*b*) **Quality of Light:** By quality of light we mean the absence of glare, contrast, deep shadows, presence of polished and glossy material will cause indirect glare unless diffused light is used. If all the people in a room are not doing same job, some will require more light than others. This may cause contrast glare and as such to avoid this general and local illumination have to be co-ordinated. For certain applications we require completely shadow free light such as in drawing office. This will need indirect light fittings employing high power sources. In practice however for many applications some shadows should be formed to give the protection of solid objects. This is very important as shadow free light is some times potential source of accidents.

(*c*) **Physical conditions of premises:** Most differentially feature of indoor lighting is that light reaches the object not only directly from the source but some portion of it does reach the object after suffering reflection from the ceiling and walls, colour and decoration of the same deciding very much the portion of the reflected or diffused light. Mounting height of the lamp plays another important role. To avoid glare and long shadows mounting height should not be less than 8 ft. above floor. Even then filament lamps require diffused lightings. Mounting height above 12 ft. brings its own problems. Cleaning and replacing become more difficult as ordinary step ladder can not be used above 12 ft. height. Special arrangement has to be done for access to the lamp. Therefore mounting height should be between 8 to 12 ft. or lamps should be mounted very high to avoid shadows and glare. With the increase of mounting height we can use high power sources which will not only reduce the running cost but at the same time installation and wiring cost would be reduced. In order to avoid dark areas of illumination and get only uniform distribution of light, lamps can not be spaced beyond certain limit, for most of the reflectors space height ratio lie between 1 to 2. In this way height affects the spacing of lamps which in turn decides the number of lamp fittings. In many cases position of beams and trusses is fixed. This decides to some extent the spacing of lamps and we have to work backwards to decide the mountings height of lamps. In some cases, placement and orientation of the lamps may be decided by the position of machines, desks, benches and light obstructions such as columns, girders and tall m/c. For instance if illumination on desk along the wall is required we may have to use wall bracket fittings as any pendent fitting would cause objectionable shadows. As our work places are usually oriented to suit the day time lighting, positioning the lamp fitting near window makes the direction of light falling on working surfaces. Similar to that of day light thus taking account of our working habits which have been adjusted to sideway light from windows. This will therefore do away with the psychological dimming effects due to twilight.

(*d*) **Colour of light:** For certain applications such as packing department etc. colour of the light does not matter much if different components have not to be distinguished from each other by their colours, we can use highly efficient discharge lamps which are not free from colour distortion effect. In packing department of biscuits it is essential to distinguish between under baked, properly baked and charred biscuits by way of colour only. As such we can not afford to have coloured light for such applications.

(*e*) **Financial economy:** Economy is a factor which can not be ignored in any engineering application. Installation cost depends upon the type of lamp and fittings used. Discharge lamps may be very efficient but are more costly as compared to filament lamps. Recurring costs of maintenance, cleaning and relamping have also to be taken into account. For high mounting heights, cost on this accounts will be more. Some times good appearance is the main consideration with certain customers. We have therefore to think of employing costly decorative type fittings.

(*ii*) **Factory lighting:** Adequate lighting of factories is of very importance, as it provides improved amenities for the employees, increased production and has a definite economic value in reducing accidents with their consequent loss of time and compensation payments.

General requirement and Types of installations: A factory lighting installations in common with other indoor equipment should provide an adequate illumination on the working plane and give a good distribution of light employing simple and easily cleaned fittings and avoid glare. It is essential not only to avoid glare from the lamp itself but also reflected glare from any polished surface which may be within the line of vision.

(*ii*) **General lighting:** The usual scheme in factories and workshops is to mount a number of lamps at a sufficient height so that uniform distribution of light over the working plane is obtained. In large machine shop the height is governed by the necessity of keeping the lamps above the travelling cranes. In such cases it is often desirable to supplement the main lighting by side lighting in order to give additional illumination on a vertical plane. Since light coloured walls and ceilings add to the effectiveness of an installation, therefore it is necessary to get white washing or painting done.

Local lighting: On some points fairly intense illumination is required. For this purpose local lighting can be provided by means of adjustable fitting attached to the machine or bench in question or mounted on portable floor standards. Such lamps should be mounted in deep reflectors so that glare is avoided. Portable hand lamps attached to wall plugs by means of trailing leads are used for maintenance work and emergency lighting. Low voltage lamps of not more than 50 volts are recommended for use as portable hand lamps because lamps have thicker filament and are therefore more robust than those for normal voltage and danger for shock is also avoided in these few volt lamps. The supply for such lamps can be obtained from a special low voltage distribution system running through out the factory or by means of a small transformer for each individual lamp.

Local lighting should never be employed alone, good general lighting is essential so that the dark places between the local lighting units are avoided. Dark places between the local lighting units cause fatigue to the eyes on account of its continually having to adjust itself to new conditions.

Emergency lighting: Some lights such as for (*i*) internal pilot lighting required for safe and speedy evacuation of personnel after main lighting circuit is off. (*ii*) External pilot lighting provided with careful shades leading to shelters required for evacuation of personnel. (*iii*) For control posts, first aid centres etc. (*iv*) Dials and gauges in important plants required to be watched regularly are required during an air raid when all the factory lights are off as a matter of air raid precaution. The circuit supplying the above emergency lights should be independently controlled. It is very desirable to provide auxilliary lighting from the source other than the main electric supply from batteries, small petrol driven generator set.

(*iii*) **Outdoor lighting system:** Following are two main application of outdoor lighting system:

(*i*) Street lighting employing dispersive fittings.

(*ii*) Flood lighting employing projector fittings.

Dispersive fittings reflect light comparatively to small area in the ground around them. As such these have to be fitted on poles which are to be created within the area to be illuminated. Projector lamps on the other hand have accurate control of light and are used to illuminate large areas from long distance. As such poles for projector lamps are to be erected around the area to be illuminated. This gives obstruction free area which is a great advantage. In order to reduce direct glare, these lamps are mounted very high.

(*iv*) **Street lighting:** Problems of street lighting are quite different from those of indoor lighting in following accounts:

(*i*) There are no walls and ceilings which reflect or diffuse light. The light reaching the objects is direct from the lamp it is therefore not possible to avoid hard shadows and high control which form essential qualities of good lighting.

(*ii*) Since areas illuminated are usually large, it is neither practicable for economic reasons nor necessary to have same level of illumination as used for indoor lighting. High level of illumination is not necessary because firstly one has seldom to look continuously at nearby objects. Secondly it is the impression of whole scene that is required. In fact objects on roads are made visible by not being themselves illuminated but by being silhouetted against a relatively bright back-ground.

Road lighting serves two types of users. Pedestrians who look at short distances make use of diffusion principle in the perception of objects; vehicle drivers who look at long distances make use of specular reflection principle. Same lighting installation can serve both type of users simultaneously because of particular behaviour of road surface. When the angle of incidence is small road behaves as a diffusing surface and some portion of incident light is diffused in the direction of the observer. This gives perception at small distances. It is essential that road appear bright and uniform which is only possible by using special reflectors which give polar curve such that C.P. is least at an angle of cut off. Illumination at any point can be calculated by square law equation.

On the other hand when a ray of light falls on the road at very high incident angle specular reflection takes place from the matt surface of the road. It is this light which gives the road necessary brightness. In figure rays of light from lamps B and C from very high angle of incidence at obstruction and therefore suffer specular reflection which help observer to see the obstruction. Ray from lamp A falls at a very small angle of incidence and therefore does not reach the observer who is a vehicle driver. On the other hand reflected rays from A are useful to pedestrian who is very near to obstruction. It is therefore observed that pedestrian sees the objects by principle of diffused reflection of light given by lamps nearby and vehicle drivers see the objects by principle of specular reflection of light given by the lamps far off. All lamp fittings should be mounted solid so that they should not swing due to breeze. Lamps should be placed on the outside of the curve at road since this improves Silhouette effect. Lamp post should always be fixed at the junction of two roads as far as possible we should aid lamps near large trees.

(*v*) **Point to Point Method of Calculation:** This method of calculation is used for outdoor installation specially in street lighting where light comes to the object directly from the source and there is no reflected light either from walls or ceiling. Illumination at any point due to point source is given by either of the equation (1) or (2)

$E_{AB} \cos\theta$ or $E_{CD} = \frac{1}{h^2} \cos^3\theta$. Total illumination at any point will be sum of illumination due to various light sources. Equation (1) and (2) holds good only for point sources. Any practical source of light employing diffused fitting of usual design can be considered point source provided its distance from the point where illumination required is not less than 10 times the linear dimension of the fitting or five times the greatest dimension in case of fluorescent lamp. In case of street light the C.P. of lamp is diff. at diff. angles. It is therefore necessary to use proper value at C.P. as read from polar curve.

Q.1. 12. State the laws of illumination, or Explain the laws of illumination.

Ans. There are two laws of illumination:

First Law. It is also known as law of inverse square. According to this law, the illumination of a surface is inversely propotional to square of distance between the source and surface provided that the distance between the source and surface is sufficiently large so that the source can be regarded as the point source.

(1) **Inverse Square law.** Consider a point source S having an intensity I lumens/steradians. Let two surfaces having area A_1 and A_2 be placed at a distance *r* and 2*r* meters away resp. from the source. The two surfaces are enclosed in the same solid angle 'ω'.

Since the source gives I lumens per steradian, any surface enclosed by solid angle ω will receive a total flux = I × ω lumens

Thus flux of area, A_1 = I.ω radians ...(*i*)

Also solid Angle ω.

$$= \frac{\text{Area}}{(\text{Dis tan ce})^2} = \frac{A_1}{r^2} \text{ steradian} \quad (ii)$$

or Total flux on surface area. $= \frac{I.A_1}{r^2}$ ...*Putting (ii) in (i)*

Illumination, E_1 on surface, A_1

$$\frac{\phi}{A_1} = \frac{I.A_1}{r^2} \times \frac{1}{A_1} = \frac{I}{r^2} \text{ Lux or Lumens/m}^2$$

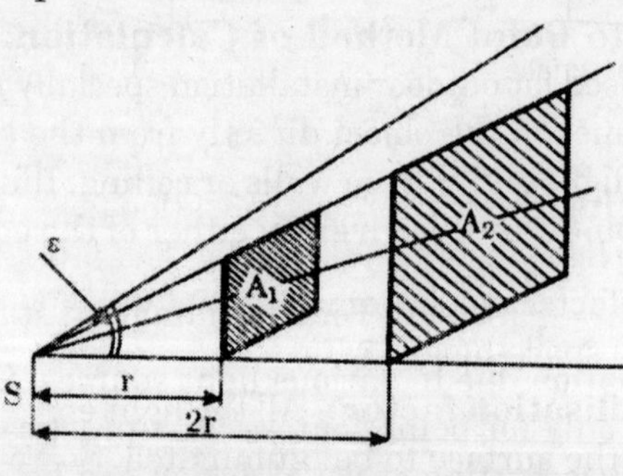

Fig, 1.18

Similarly flux on any surface, A_2 when placed 2*r* meters from a similar source enclosed in the same solid angle = *I*.ω lumens

But $$\omega = \frac{\text{Area}}{(\text{Dis tan ce})^2} = \frac{A_2}{(2r)^2} \text{ Steradians}$$

$$\phi = \frac{I.A_2}{(2r)^2}$$

$$E_2 = \frac{I.A_2}{(2r)^2} \times \frac{1}{A_2} = \frac{I}{4r^2} \text{ Lux}$$

$$E_1 : E_2 : \frac{1}{r^2} : \frac{1}{(2r)^2}$$

In other words, illumination at *a* surface is inversely proportional to the square of its distance from the source.

2. Lambert's Cosine law. According to this law illumination of a surface varies directly as the cosine of the angle between the normal to the surface and direction of incident light.

$E \propto \cos\theta$

(*i*) In the normal surface area, $E = \dfrac{\phi}{\text{Area}}$

(*ii*) In the inclined surface case, $E = \dfrac{\phi}{\text{Area}} \times \cos\theta$

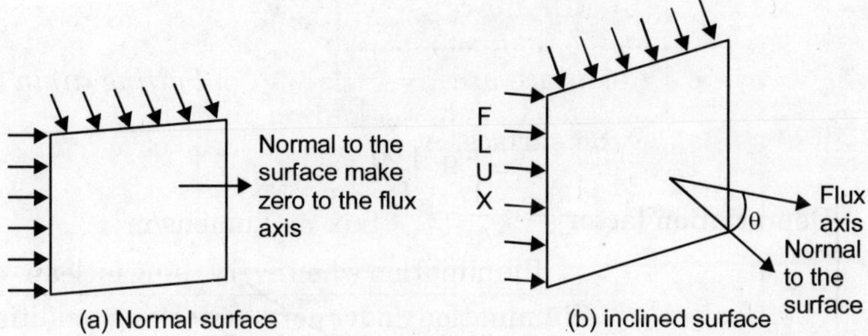

Fig. 1.19.

Q.1.13. Define the terms: (*a*) Utilisation factor (*b*) Depreciation factor (*c*) Waste light factor (*d*) Reflection factor (*e*) Reduction factor (*f*) Absorption factor (*g*) Luminous efficiency (*h*) Specific energy consumption (i) space-height ratio.

Ans. (*a*) **Utilisation factor :-** All the light emitted by the source does not reach the surface to be illuminated. Some portion of light directly falls upon surface and some light falls upon surface after reflection from walls and ceiling. All the light falling upon the walls and ceiling do not reach the surface because some light is absorbed.

Coefficient of utilisation is the ratio of lumens reaching the working plane to the total lumens given out by the lamp.

$$\text{Coefficient of utilisation} = \frac{\text{Lumens reaching on working plane}}{\text{Total lumens emitted by source of light}}$$

(*b*) **Depreciation factor :-** Due to accumulation of dirt, dust and smoke on lamps they emit less light as compared to that emit when they are new ones and similarly ceilings and walls etc. after being covered by dust etc. reflect less light as compared to that when they are new.

So the ratio of illumination under every thing perfectly clean to the illumination under normal working condition is known as depreciation factor.

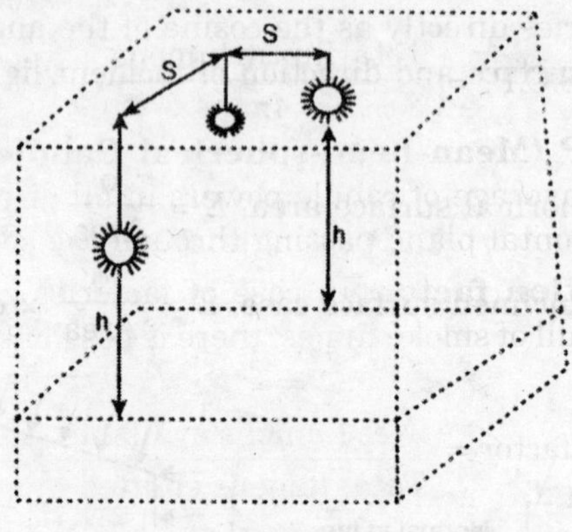

Fig. 1.20

Depreciation factor

$$= \frac{\text{Illumination when every thing is clean}}{\text{Illumination under normal working condition}}$$

It is more than unity.

(*c*) **Waste light factor:** In flood lighting scheme, a no. of projector lamps are employed so that light of each projector not only overlaps that of its adjacent projector but at the edges of the area, some light falls outside. In order to compute total lumens reqd., theoretical lumens reqd. are multiplied by waste light factor. Value of this factor is about 1.2 for rectangular areas and 1.5 for irregular objects such as statues and monuments etc.

(*d*) **Reflection factor:** It is defined as the ratio of reflected light to incident light. It is always less than unity.

$$\text{Reflection factor} = \frac{\text{Reflected light}}{\text{Incident light}}$$

(*e*) **Reduction factor :-** Reduction factor of a source of light is defined as the ratio of its mean spherical candle power to its mean horizontal candle power.

$$\text{Reduction factor} = \frac{\text{M.S.C.P.}}{\text{M.H.C.P.}}$$

(*i*) **MHCP (Mean horizontal candle power):** It is the average of all the candle powers in all directions in the horizontal plane containing the source of light.

(*j*) **MSCP (Mean spherical candle power):** It is defined as the average of candle powers in all directions and in all planes from the source of light.

$$\text{M.S.C.P.} = \frac{\text{Total flux in lumens}}{4\pi}$$

(*k*) **MHSCP (Mean hemi-spherical Candle power):** It is defined as the average of candle powers in all directions above or below the horizontal plane passing through the source of light.

(*f*) **Absorption factor:** In case of factories or places where atmosphere is full of smoke fumes, there is possibility of absorption of light energy.

$$\text{Absorption factor} = \frac{\text{Net Lumens available after absorption}}{\text{Total lumens emitted by source of light}}$$

(*g*) **Luminous efficiency:** It is defined as the output in lumens per watt of the power consumed by the source of light. It is measured in lumens per wattage. If a bulb is rated at 500 watts and 250 volts has an *n* of 13 lumens/watt Then,

Total flux produced by bulb = 500 × 13 = 6500 Lumens

(*h*) **Specific energy consumption:** It is the ratio of power input to the source of light to its luminous intensity. Its unit is watt per candela.

(*i*) Space-height ratio: It is defined as the ratio of horizontal distance between two adjacent lamps and height of lamp from the working plane.

$$= \frac{\text{Horizontal distance between two adjacent lamps}}{\text{Height of lamps above working plane}}$$

$$\text{S.H.R.} = \frac{\text{d}}{\text{H}}$$

The value of space-height ratio varies normally between 0.8 - 1.2.

Explain Questions: The Types of Lamps

Types of Lamps

It will be found that two kinds of lamps are used in showcase and window-box lighting. Lighting control or effective light is achieved by dimmers and in a majority of cases by various lamp shades.

The bulbs used are:

- incandescent lamps
- tube lights
- silvered bowl lamps of higher capacity varying from 200 watts to 500 watts.

The No.of lamps depend upon the size and requirements, series of lamps can be fixed. It will also be found that light is thrown on the merchandise by fixing the lamp outside a showcase.

The lamps are basically divided into 3 categories. ☐ Combustion ☐ Incandescent ☐ Electric discharge

2. Types of bulbs/incandescent lamps

Glow lamps
Moonlight lamps
Luminous lamps
Daylight lamps
Tree light lamps
Photo flood lamps
Movie flood lamps
Photo flash lamps
Silvered bowl lamps
Projector lamps
Reflector lamps
Fluorescent lamps
Halogen lamps.
C.F.L. Lamps (Compact Fluorescent Lamps)
L.E.D. Lamps (Light emitting diode Lamps)

Definition: A light emitting diode (L.E.D.) is a semi conductor device that emils visible light when an electric current passes through it.

3. Points to remember while Designing Illumination:

Lighting for commercial purposes is divided into many parts such as built in direct lighting, indirect lighting, core, lighting, spot lighting etc.

To achieve the above lighting there are ceiling fixtures, side wall fixtures, portable fixtures and other luminaries available.

Many elements employed in lighting involve special dimensional or planning consideration other than proper spacing and capacity of outlet that save them.

Illumination is done by the architect, illuminating engineer or a layman for producing the desired results with various degrees of excellence.

There are unlimited decorative possibilities and they can provide almost any desired illumination level without shade under complete control.

The appearance, character and efficiency are determined by the glassware or translucent medium forming the surface of the panel. The choice of material for the panel, its colour and decoration are determined by the necessity for harmony with the rest of the structure. The desired luminous elements have either covering or lightly diffusing translucent material or an exposed uniformly illuminated diffusing background.

- Though efficient planning of a lighting installation is; the job of the lighting engineers the electrician is sometimes called in to advise on lighting requirements of small premises.
- The number of lumens required on the working plane in square sect and lumens/ft^2 desired. The lumens provided by the lamps must however be greater than this figure to allow for depreciation of the installation owing to dust and dirt on the lamps and their fittings.
- Atleast 150 lumens per square metre should be provided for adequate visual performance on rough or unskilled work. Up to 1500 lumens per square meters should be provided for difficult or fine works.

Most sources, radiate light in all directions and are too bright to be viewed comfortably. The light must therefore be controlled to direct it where it is required and to soften its brilliance.

Reflection of light may be of three kinds.

- Specular reflection
- Diffuse reflection
- Spread reflection

4. Specular reflection

When light strikes a minor like surface it is reflected at the same angle and in the same plane as it strikes, for example a car lamp.

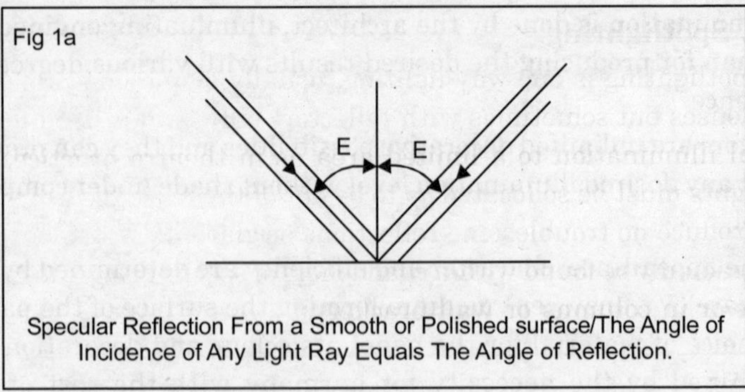

Specular Reflection From a Smooth or Polished surface/The Angle of Incidence of Any Light Ray Equals The Angle of Reflection.

4.1 Diffuse reflection

Diffuse reflection is useless for the precise control of light, but it can be used to reflect light in a general direction.

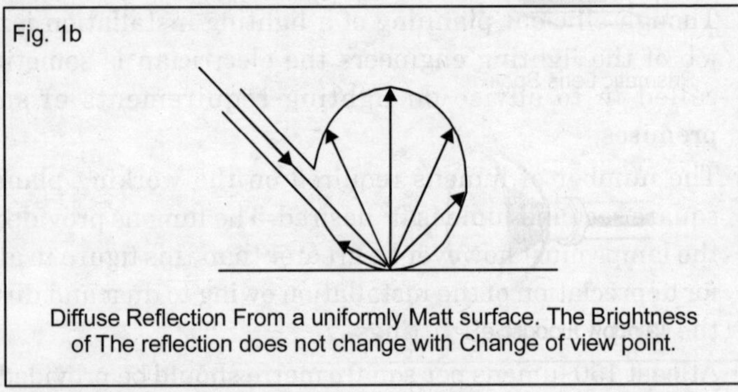

Diffuse Reflection From a uniformly Matt surface. The Brightness of The reflection does not change with Change of view point.

4.2 Spread reflection

Unpolished metals and satin-finish mirrored surfaces have reflection characteristics between specular and diffuse. Vitreous and synthetic enamels are widely used for reflecting surfaces of lighting fittings. Vitreous enamels is the more hard working.

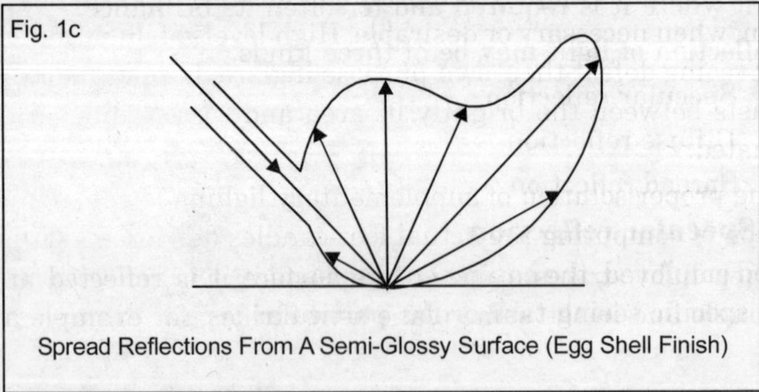

Spread Reflections From A Semi-Glossy Surface (Egg Shell Finish)

5. Spotlighting

Spotlighting is one way lighting, usually employing projectors with lenses but sometimes with reflectors only, and is used to give special illumination to a limited area as in theatre practice. The spotlights must be so located *as* to be out of the direct line of vision and produce no troublesome reflections or glare. Possible locations may gradually be found within chandeliers of special design, in ceiling beams or in columns or wall brackets.

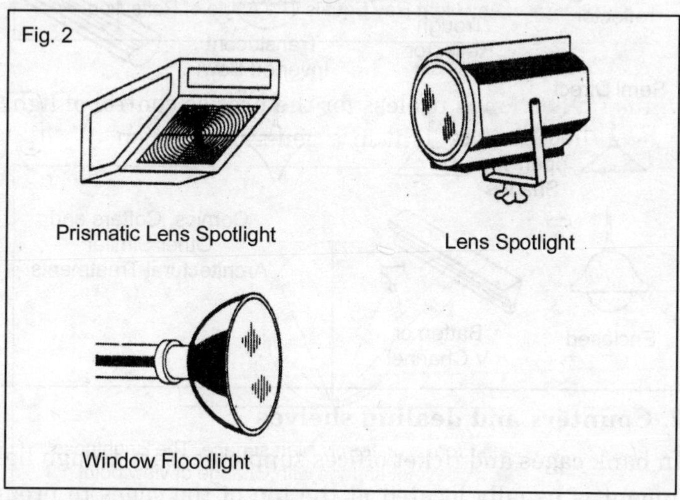

Fig. 2

6. Supplementary lighting

Severe visual tasks and modern merchandising methods require higher levels of illumination to eliminate eye strain or to highlight special displays than are practical or economical to obtain from a system of general lighting alone. Supplementary lighting as the name implies, should be employed in conjunction with a general lighting system, when necessary or desirable. High levels of illumination for better seeing defeat their own purpose if glare, harsh shadows and contrasts between the brightly lit area and sorroundings are not eliminated.

The proper solution of supplementary lighing is not so much a question of computing the actual foot-candles delivered as it is the method employed, the equipment used and its location with respect to the specific seeing task or the particular type of display.

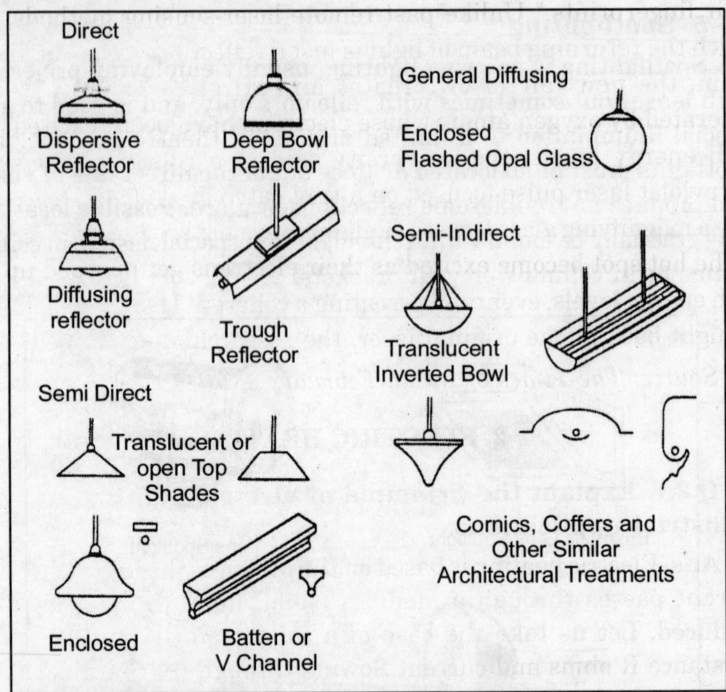

7. Counters and dealing shelves

In bank cages and ticket offices supplementary trough lighting equipment is usually located at the top of the cages to produce a band of light lengthwise on the counter. Troughs may be covered with diffusing glass or fitted with longitudinal louvers to shield the lamps.

Short Note on Air Laser Technology:

Air Laser Technology to Detect Bombs from a Distance

American scientists claim to have developed a new laser sensing technology that will allow soldiers to detect hidden explosives from a distance. Researchers at Princeton University who were behind the invention said their technique involves generating a beam of laser light out of nothing but air. It will also help researchers measure airborne pollutants. "We are able to send a laser pulse out and get another pulse back from the air itself," Richard Miles, professor of mechanical and aerospace engineering at Princeton, said. "The returning beam interacts with the molecules in the air and carries their fingerprints." Unlike past remote laser-sensing methods, in which the returning beam of light is just a reflection of the outgoing beam, the new "air laser" creates an entirely new laser beam generated by oxygen atoms whose electrons have been "excited" to high energy levels, the university said in a release. Using an ultraviolet laser pulse focused on a tiny patch of air, similar to the way a magnifying glass focuses sunlight into a hot spot, oxygen atoms in the hot spot become excited as their electrons get pumped up to high energy levels, eventually creating a coherent laser beam aimed straight back at the original laser, the researchers explained.

Source: The Times of India, February 1, 2011.

2. ELECTRIC HEATING

Q.2.1. Explain the principle of electric heating and its industrial applications.

Ans. Electric heating is based on the principle that when electric current passes through a medium (soild, liquid or gas) heat is produced. Let us take the case of a soild material which has a resistance R ohms and current flowing through it is I Ampere for t seconds the heat produced in the material will be

$$H = I^2Rt \text{ Joules}$$

$$= \frac{I^2Rt}{4.2} \text{ Calories} = 0.24\, I^2\, Rt \text{ Calories}$$

This is by virtue of "joules law".

Industrial Applications of electric Heating:

(*i*) Melting of metals

(*ii*) Heat treatment of metals like annealing, tempering, soldering, brazing and case hardening etc.

(*iii*) Electric welding

(*iv*) Moulding of glass for making glass appliances

(*v*) Baking of Insulators

(*vi*) Moulding of plastic components

(*vii*) Enamelling of copper conductors

(*viii*) Heat treatment of pointed surfaces

(*ix*) Making of plywood

(*x*) Vitreous enamelling of wire wound resistors, danger plates and other metallic components.

Q.2.2. State the advantages of electric heating.

Ans. (*i*) **Clean and neat Atmosphere:** There is no coal, dust or smoke and operators hand do not go black while operating an electrical heating appliance.

(*ii*) **No pollution:** Absence of flue gases does not result in pollution of atmosphere and there are no heat losses involved through smoke or flue gases.

(*iii*) **Controlled temperature:** The temperature can be controlled within ± 5°C, which is not possible in non electrical heating processes.

(*iv*) **Ease of control:** The heating can be started instantaneously or stopped at a required time keeping a time, gap between switching off and cooling of the heating circuit. Automatic switching controls are possible.

(*v*) **Localised application:** A workpiece can be heated upto particular depth for heat treatment, whereas the piece as a whole receives heat in non electrical heating process.

(*vi*) **Low Ambient Temperature:** The temperature around an electric furnace is much lower as compared to that around non-electrical furnace which is troublesome for the workers.

(*vii*) **Uniform Heating:** The heating can be generated from within workpiece resulting in uniform heating through induction heating.

(*viii*) **Heating of Bad Conductor of Heat and Electricity:** Wood, plastic and bakery items can be uniformly and suitably heated with dielectric heating process.

(*ix*) **Highest efficiency of Utilisation:** Heat produced electrically does not go waste through chimney and other by-products. Most of the heat produced is utilised by the material being heated electrically. This results in the highest efficiency as compared to other processes.

(*x*) **Cheap furnaces:** The electrical furnaces do not require big space for installation, no storage of coal and fire wood is necessary, no chimney to be constructed, no extra heat insulation necessary. This brings the cost involved to a lower value as compared to non-electrical furnaces.

(*xi*) **Mobility of job:** Pieces undergoing heat treatment can be mounted on a conveyer passing through the heating cabinets making use of electric heaters.

Q.2.3 List the requirements of good heating material.

Ans. Materials requirements for Element Making: The indirect resistance furnaces make use of a wide range of temperature and different shapes, sizes and materials come to fulfil different requirements. A good element material should have following important properties *viz:*

(*i*) **Low temperature coefficients of Resistance:** Since Resistance of a conducting element varies with the temperature, this variation should be small in case of an element. Otherwise when switched ON from room temp, to go upto say 1200° C, the low resistance at initial stage will draw excessively high currents at the same operating voltage.

(*ii*) **Resistance Coefficient positive:** If tempertaure coefficient is negative the element will draw more current when hot. A higher current means more wattage, a higher temperature or a still lower resistance, which can cause instability of operation.

(*iii*) **High melting point:** Its melting point should be sufficiently higher than its operating temperature. Otherwise a small rise in the operating voltage will destroy the element.

(*iv*) **High Specific Resistance:** The resistivity of the material used for making element should be high. This will require small lengths and shall give convenient size.

(*v*) **High oxidizing temperature:** Its oxidizing temperature should be higher than its operating temperature otherwise oxidised layer from the surface will flake off changing the resistance of filament and giving it a smaller life.

(*vi*) **Ductile:** To have convenient shapes and sizes, the material used should have high ductility and flexibility. It should not be brittle and fragile.

(*vii*) **Should withstand vibrations:** In most industrial processes, quite strong vibrations are produced. Some furnaces have to open or rock while hot. The element material should withstand the vibrations while hot and should not break open.

(*viii*) **Anti-corrosive:** In many heat treatment processes the heating atmosphere is corrosive. When exposed to this atmosphere or any other chemical fumes the filament should not corrode otherwise its resistance will change and life will shorten.

(*ix*) **Mechanical Strength:** The material used should have sufficient mechanical strength of its own.

Q.2.4 (*a*) State the material employed for heating.

Ans. No single element metallic or non-metallic is used for making heating elements. On the other hand, different alloys have been tried to give different operating temperature. The following list gives the names, composition and safe operating temperature for different alloys used as heating elements.

S.No.	*Name of alloy*	*Composition*	*Safe operating temperature*
1.	Constantan	Nickel 45% Copper 55%	400° C
2.	Nickrome	Nickel 80% Chromium 20%	1500° C
3.	Kanthal	Chromium Aluminium Cobalt, Iron	1300° C
4.	Silicon Carbide	1450	—

Q.2.5 (*b*) Design of heater element and problem thereon.

Ans. Design of Heating element: Assume that the whole of the heat energy is dissipated solely by radiation.

Heat dissipated according to Stefan's Law,

$$H = 5.72K_c\ [(T_1/100)^4 - (T_2/100)^4]\ \text{watts/m}^2 \qquad ...(1)$$

Electrical Input, $P = \dfrac{V^2}{R}$

Where V is the supply voltage and R is the resistance of heating element and is given by the expression,

$$R = \rho\frac{l}{a} = \frac{\rho l}{\pi d^2/4} = \frac{4\rho l}{\pi d^2}$$

For a circular wire of diameter d, Length l and resistivity, ρ Electrical Input,

$$P = \frac{V^2}{R} = \frac{V^2}{4\rho l/\pi d^2} = \frac{\pi d^2 V^2}{4\rho l}$$

$$\frac{1}{d^2} = \frac{\pi V^2}{4P\rho l} \quad ...(2)$$

Surface Area, $S = \pi d$

Heat dissipated $= \pi dl.H$

At steady temparture we know, Power Input = Heat dissipated

$$P = \pi.dl \times H$$

$$\frac{\pi d^2 v^2}{4\rho l} = \pi l \times H$$

$$\frac{d}{l^2} = \frac{4\rho H}{V^2} \quad ...(3)$$

Solving expression (2) and (3), length and diameter of wire can be determined.

For ribbon type of conductor, let ω be the width and t the thickness, Electrical Input,

$$P = \frac{V^2}{R} = \frac{V^2}{\rho l/A} = \frac{V^2}{\rho l/\omega\times t} \qquad (\because a = \omega\times t)$$

$$p = \frac{V^2.\omega t}{\rho t}$$

$$\frac{\rho t}{\omega} = \frac{V^2}{P\rho} \quad ...(4)$$

Since at steady temperature,

Power Input = Heat dissipated

$$P = 2\omega l.H$$

(neglecting the side area as the thickness is negligible)

$$\frac{V^2 t}{\rho l} = 2l.H$$

$$\frac{t}{l^2} = \frac{2H\rho}{V^2} \quad ...(5)$$

By solving eqn (4) & (5) length l and width (ω) for a ribbon of thickness, t will be evaluated.

Q.2.6 Explain with sketches:

(*a*) Direct Resistance heating and

(*b*) Indirect Resistance heating.

Ans. Resistance heating: This method is based upon the l^2R loss. Whenever current is passed through a resistive material heat is produced because of $l_2 R$ loss. There are two methods of resistance heating as mentioned below:

(*a*) **Direct Resistance heating:** In this method of heating, the material or charge to be heated is taken as a resistance and current is passed through it. The charge may be in the form of powder, pieces or liquid. The two electrodes are immersed in the charge and connected to the supply. In case of D.C. or single phase A.C., two electrodes are required but there will be three electrodes in case of three phase supply. When material pieces are to be heated a powder of high resistivity material is sprinkled over the surface of charge to avoid direct short circuit. The current flows through the charge and heat is produced. The method has high n since heat is produced in the charge itself. Though automatic temperature control is not

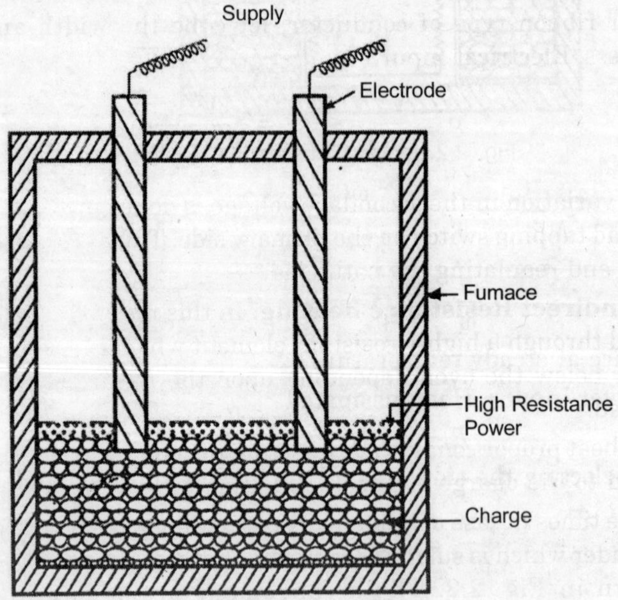

Fig. 2.1. Direct Resistance Heating.

possible in this method but it gives uniform heat and high temperature. One of the major application of the process is salt bath furnaces having an operating temparture between 500°C to 1400°C.

An immersed electrode type medium temperature salt bath furnace is shown in Fig. 2.1. The bath makes use of a supply voltage across two electrodes varying between 5 to 20 volts. For this purpose a special double wound transformer is reqd. which makes use of 30 primary and a l-i|) secondary. This speaks of an unbalanced load.

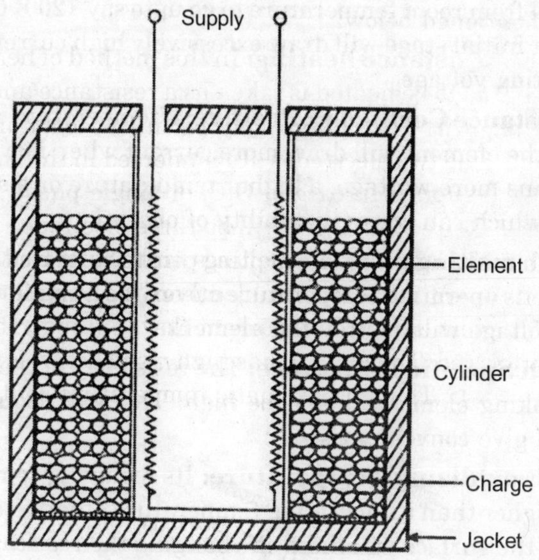

Fig. 2.2. Indirect Resistance Heating.

The variation in the secondary voltage is done with the help of an off load tapping switch on the primary side. This is necessary for starting and regulating the bath load.

(*b*) **Indirect Resistance heating:** In this method the current is passed through a highly resistant element which is either placed above or below the oven depending upon the nature of job to be performed.

The heat proportional to I^2R losses produced in heating element delivered to the charge either by radiation or by convection.

Some times in case of industrial heating the resistance is placed in a cylinder which is surrounded by the charge placed in the jacket as shown in Fig. 2.2. The arrangement provides a uniform temperature.

Automatic control temperature can be provided in this case. Both A.C. and D.C. supplies can be used for this purpose. At full mains voltage depending upon the design of heating element, the common *e.g.* of this type of heating is resistance oven.

Q.2.7 State the industrial application of the above heating,

Ans. (*i*) **Low temperature co-efficients of Resistance:** Since resistance of a conducting element varies with the temperature, this variation should be small in case of an element otherwise, when switched ON from room temperature to go upto say 1200° C, the low resistance at initial stage will draw excessively high currents at the same operating voltage.

(*ii*) **Resistance Coefficient Positive:** If temperature coefficient is negative the element will draw more current when hot. A higher current means more wattage, a higher temperature or a still lower resistance, which can cause instability of operation.

(*iii*) **High melting point:** Its melting point should be sufficiently higher than its operating temperature otherwise a small rise in the operating voltage will destroy the element.

(*iv*) **High Specific Resistance:** The resistivity of the material used for making element should be high. This will require small lengths and give convenient size.

(*v*) **High oxidizing Temperature:** Its oxidizing temperature should be higher than its operating temperature otherwise oxidised layers from the surface will flake off changing the resistance of the filament and giving it a smaller life.

(*vi*) **Ductile:** To have convenient shapes and sizes, the material used should have high ductility and flexibility. It should not be brittle and fragile.

(*vii*) **Should withstand vibrations:** In most industrial processes quite strong vibrations are produced. Some furnaces have to open or rock while hot. The element material should withstand the vibrations while hot and should not break open.

(*viii*) **Anti-Corrosive:** In many heat treatment processes the heating atmosphere is corrosive. When exposed to this atmosphere or any other chemical fumes the filament should not corrode, otherwise its resistance will change and life will shorten.

(*ix*) **Mechanical Strength:** The material used should have sufficient mechanical strength of its own.

Q.2.8 Explain the different method of temp. control with Diagram.

Ans. If a constant voltage *is* continuously energising an electric furnace, the temperature will gradually be increasing with the passage of time. It will cross the desired level of heating at one stage. It is at this stage when some controlling mechanism should come into action and should stop further rise of temperature. In doing so the heat generated by the heating elements has either to be cut off momentarily or reduced. In either case the temperature will try to fall from a predetermined level. At this stage the procedure has to be reversed, *i.e.* the supply if it has been reduced or stopped has to be fully restored. In either case we can easily guess that the temperature will vary about a mean value but it will not be exactly constant. How closely it remains near a mean value depends upon the *response* of the furnace to the controlling action and the *mode* of control.

The control exercised on larger furnaces is automatic which can be electromagnetic or electronic involving servo-mechanism operations. The modes of control can be :

(*i*) Two step (*ii*) Three step (*iii*) Continuous.

The two steps can be (a) ON - OFF, (b) Low-High.

As indicated by the name ON - OFF mode, when the temperature reaches the desired level the supply to the heating elements should be switched off. In case of low and high mode the circuit has not to be switched off but the heat generated should be reduced. In three step control the levels of temperature involved are three *viz.* low - medium - high. In case of continuous control, the control mechanism continuously varies the energy input depending upon the temperature inside the furnace so that the input is reduced when temperature rise and *vice versa.*

The basic control involves following adjustments which can vary the temperature inside a resistance furnace :

(*i*) Intermittent switching, *i.e.* ON - OFF control.

(*ii*) Variation of number of heating elements.

(*iii*) Variation in circuit configuration.

(*iv*) Change of applied voltage.

(*i*) **Intermittent Switching.** As mentioned earlier by switching the heating supply OFF for some time the temperature rise can be stopped beyond a desired limit. Of course the switching has to be done before the temperature actually reaches the desired limit. Even after switching off the supply, the temperature rises to some extent in totally enclosed furnaces. After some time the temperature starts falling off. When the fall reaches again a predetermined value the supply is again switched ON. Since entire furnace is switched ON and OFF in this process there is lot of arcing at the switching contacts. As such the method is suitable for small furnaces only. Air circuit breakers are ideal if switching action has to be frequent to keep the temperature within close working limits.

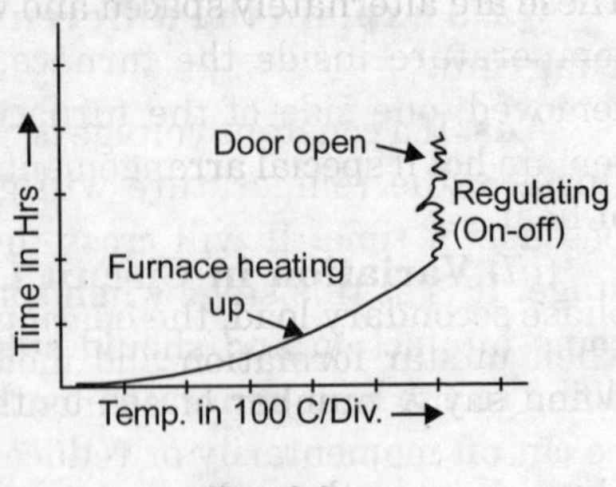

Fig. 2.3

The figure 2.3 shows the results of ON-OFF control on temperature as a function of time.

(*ii*) **Variation of Number of Heating Elements.** In this case instead of cutting off the supply to the entire furnace, the number of heating elements are changed in the supply circuit. Obviously with smaller number of heating elements, supply of heat is reduced and *vice versa.*Lesser the supply of heat smaller should be the temperature and so on. This is applicable for continuous mode and multistep modes of temperature control. High/low, two step mode can also work on this practice.

Note. In case of a three-phase circuit, care should be taken while removing from heating circuit, equal number of elements are switched off from each phase. This is necessary to keep balanced load conditions. Secondly from a row of heating elements the switched off elements should be equally spaced *e.g.* If six elements are connected per phase and while regulating, three have to be removed out of circuit & if either of the switch is opened three elements will be switched off. If S_1 is opened, elements A, C and *E* will come out.

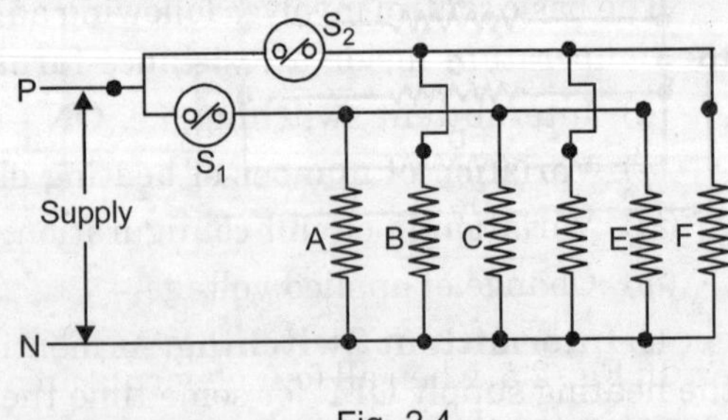

Fig. 2.4

These are alternately spaced and will keep a uniformity of operating temperature inside the furnace. If on the other hand, *ABC* are removed, one side of the furnace will cool down and other shall remain hot if special arrangements do not exist for forced circulation of heat.

(*iii*) **Variation in Circuit Configuration:** In case of three-phase secondary load, the heating elements will give different heat when in star formation and more heat when in delta formation. When say *X* number of elements connected in parallel come per phase in star the voltage across these elements is $\frac{V_L}{\sqrt{3}}$ volts. If suppose line voltage is 400 volts the voltage in star formation per phase will be 231 volts. Since power is $\frac{V^2}{R}$ watts it will be *one-third of* the power when the same elements get full 400 volts. This will be possible when the elements appear in delta formation (See Fig. 2.5.)

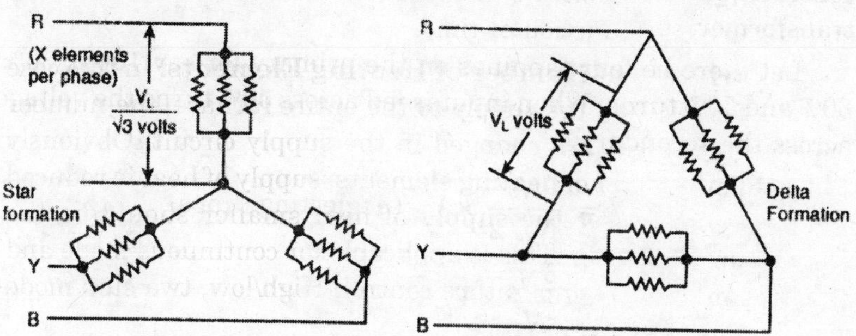

Fig. 2.5. Star/delta configuration of heating elements.

In single phase circuits series parallel grouping can cause a change in power dissipation resulting in change of temperature.

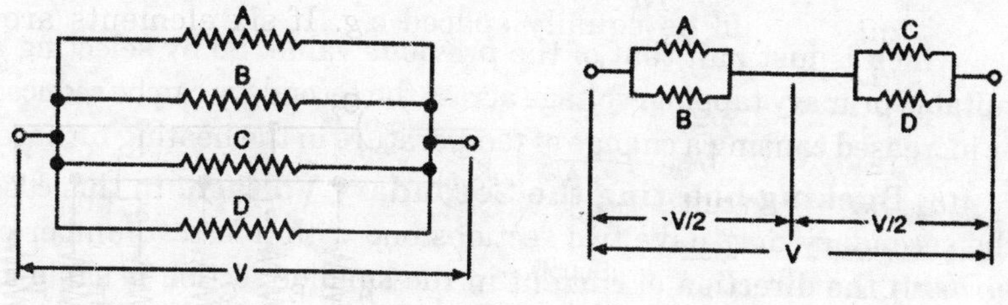

Fig. 2.6

In Fig. 2.5, when all four elements are in parallel, voltage across each element of the same resistance is *V* volts. In the series parallel formation the voltage across each element is reduced to V/2 volts. Obviously power dissipated in the second case will be *one-fourth* of the previous value. (Neglecting temperature coefficient and assuming constant resistance of the elements in either connections.)

(*iv*) **Change of Applied Voltage.** (*a*) This is of course very simple to guess that lesser the magnitude of applied voltage smaller will be the power dissipated and lesser will be the temperature produced. In case of furnace transformer having high voltage primary; it is advisable to have a tapping control on the high voltage side. It is because *of lesser* magnitude *of current* on this side. For simplicity let us take into consideration a single phase circuit involving a multi-tap step down transformer.

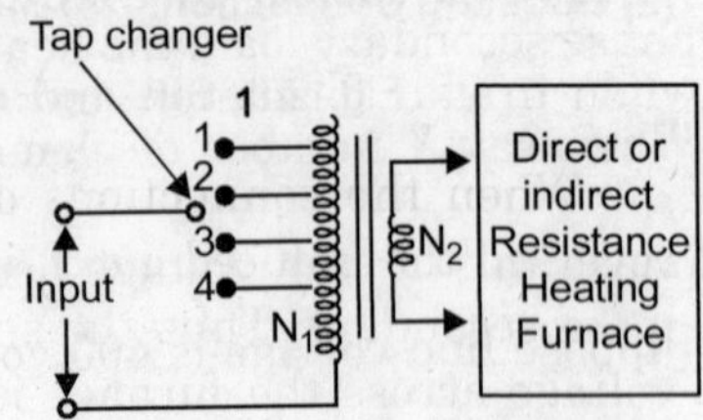

Fig. 2.7 Tapping the input.

Let there be four tappings on the primary for say 100%, 80%, 60% and 50% turns. When input *is* fed across 50% turns the voltage across the secondary,

$$= \frac{N_2}{N_1/2} \times V_{in} \text{ (neglecting drops)}$$

$$= \frac{2N_2}{N_1} \times V_{in} \quad \text{...}(i)$$

whereas when hundred per cent primary is in the circuit the voltage across the secondary

$$= \frac{N_2}{N_1} \times V_{in} \quad \text{...}(ii)$$

which is just half that of the previous value. So by selecting a suitable primary tapping, voltage across the secondary can be reduced or increased causing a change of temperature in the heating furnace.

(*b*) **Bucking-boosting the Secondary Voltage.** In this case the secondary may have two sections one with a *lesser* number of *turns.* If the direction of current in the smaller section is same as that of remaining secondary, its induced e.m.f. helps or adds to the voltage induced in the rest of the secondary. This action is boosting action, and results in increased secondary voltage. The effect is shown in Fig. 2.8.

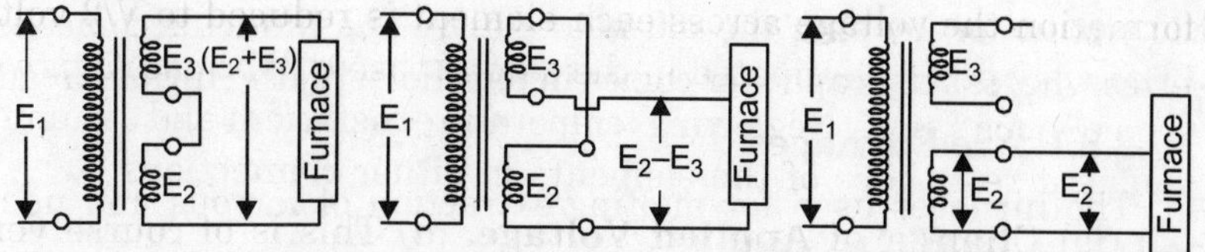

(a) Boosting Connections ***(b)*** **Bucking Connections** **(c) Without boosting**

Fig. 2.8. Bucking/Boosting of secondary voltage.

When the connections of this extra secondary winding are reversed the induced e.m.f. in this winding is in opposition to the main secondary. Thus the second effect is *Bucking* effect and the voltage across the furnace is reduced. The third possibility is to neglect the voltage developed across the extra secondary thus the three different voltages possible against the furnace can be,

(*i*) $E_2 + E_3$ boosting

(*ii*) $E_2 - E_3$ Bucking

(*iii*) E_2 no bucking-no boosting.

(*c*) **Auto-transformer Control.** For very small resistance furnaces, the auto transformer control can work for changing the applied voltage and as such the temperature. The circuit used is shown in Fig. 2.8. Required voltage can be applied with the help of voltage selector.

(*d*) **Series Reactor Control.** In this case a heavy duty coil wound on a core is placed in series with the furnace as and when designed. A part of applied voltage drop across the impedance produced by the coil and the rest appears across the furnace, thereby reducing the temperature inside the furnace. The circuit involved is shown in Fig. 2.9. The D.P.D.T. switch provides Low, High, two step mode of temperature control. Since the introduction of inductive reactance in series with the furnace can reduce the power factor, *a* power capacitor is simultaneously introduced in

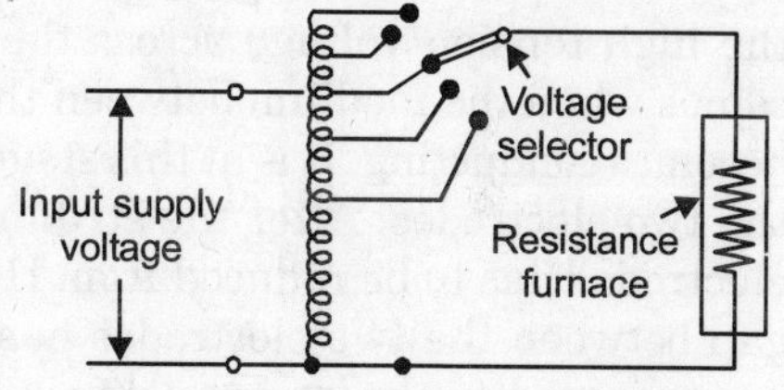

Fig. 2.9. Auto-transformer control.

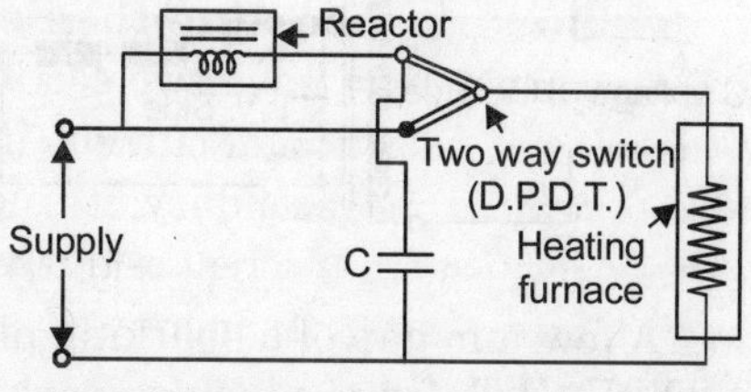

Fig. 2.10 Series Reactor Control.

the circuit for keeping the p.f. near unity. The inductor comes in series whereas the capacitor comes in parallel with the furnace load.

2.8.1. Arc Furnaces

The furnaces used for melting/extraction of ferrous and non-ferrous metals need a high temperature operation. One of the methods is through production of electric arc which gives an Arc temperature between 3000°C to 3500° C on L.T. operation. Arc is the flow of current through an air gap between two conducting bodies.

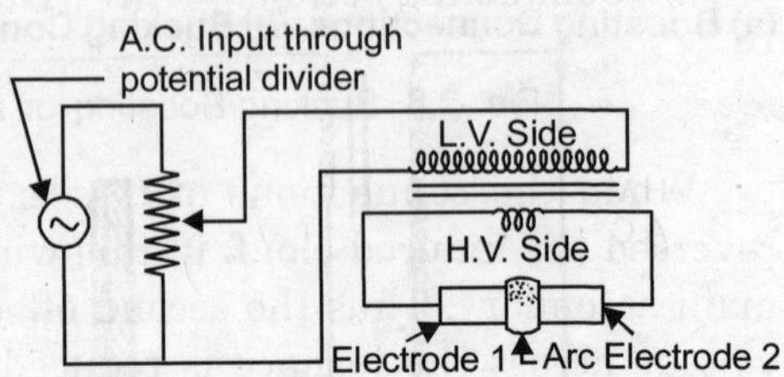

Fig. 2.11. H.T. method of striking an arc.

Methods of Striking Arc. There are two methods of striking an arc between two electrodes. On one of the methods is **H.T strike** whereas the second method is **L.T. strike method.**

In the first method a constant gap is maintained across a pair of electrodes say made of carbon. The electrodes are connected across the H.T. secondary of a step-up transformer. The primary is fed with variable A.C. voltage as shown in the circuit. To strike an arc the primary input voltage is gradually increased, thus increasing the high tension voltage across the secondary/electrodes. A stage comes when the medium between the two electrodes is ionised and becomes conducting. It is at this stage when an arc is struck between the two electrodes. After the striking of the arc the p.d. across the electrodes has to be reduced from H.T. to L.T. since the conducting gap between the two electrodes has now negligibly low resistance as compared to the gap resistance before striking of the arc. The H.T. required is dependent upon the size and nature of the gap*, but runs into thousands of volts.

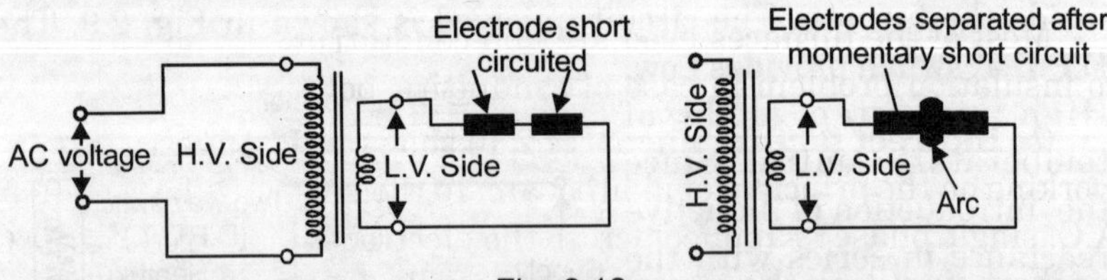

Fig. 2.12

* A few mm gap of a sparking plug used in auto-engines needs 10-20 kV, H.T. for producing spark, which is struck momentarily and not main-tained like an arc of the furnace.

Q.2.9 Explain the different types of electric arc furnaces giving sketch

Ans. (*a*) **Direct Arc furnace:** Electrically these furnaces can be further sub-divided into two categories *viz.*

(*i*) Conducting bottom (*ii*) Non conducting bottom

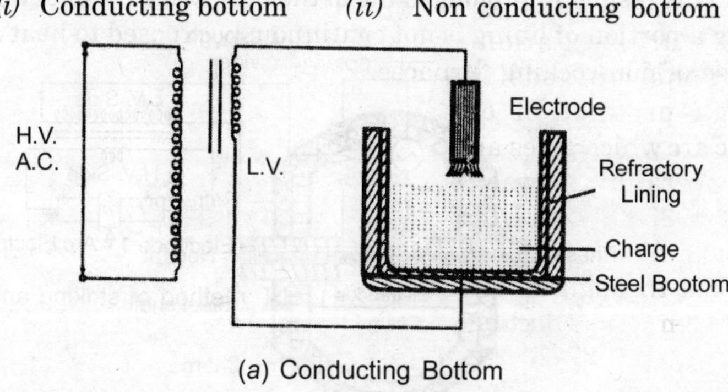

(*a*) Conducting Bottom

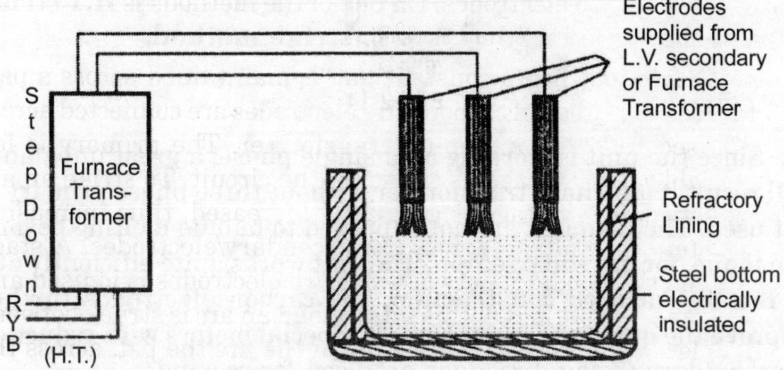

Fig. 2.13. (*b*) (Non-Conducting Bottom)

The former has the bottom of the furnace as a part of the electrical circuit whereas no current flows through the body of the furnace in the second case. It will be clear from the study of the Fig. 2.13.

Most of the furnaces used are non conducting bottom type due to insulation problem in case of conducting bottom.

(*b*) **Indirect fare furnace:** Fig 2.14 shows a rocking arc furnace working on the principal of indirect arc. In the given case low voltage A.C. single phase is applied across the electrodes. The arc is struck by short circuiting the electrodes manually or automatically for a moment and with drawing these apart. The heat from the arc is transferred to the charge to player and refractory lining, through radiation and from top layer of the charge to other portion of the charge through conduction. To distribute the heat properly, the furnace has to be rocked continuously, exposing further layers not only to the heat transferred from the arc, but from the exposed lining. This results in prolonged life of the lining as well since during rocking a portion of lining is not continuously exposed to heat as in the case of non-rocking furnaces.

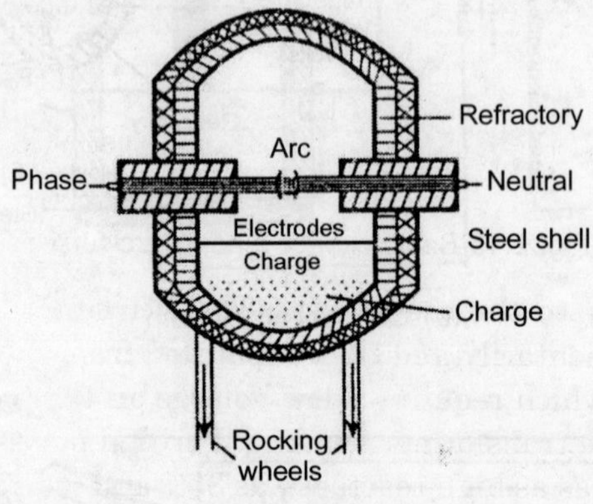

Fig. 2.14

Since the unit is working on a single phase, a great unbalance will result if a furnace transformer without three phase primary is not used. The furnaces are not supposed to handle melting beyond one tonne, for the same reason. The unit works quite efficiently due to totally enclosed construction, the carbon electrodes further improve the quality of ferrous melts. Special motors with reduction gears and reversing directions are used for rocking.

Q.2.10 Explain the basic ckt. for electric arc furnace showing thc arrangment of OCB control panels, CTS through relays, furnace transformer and arrangement of electrode movement.

Ans. The Fig 2.15 show the fundamental requirement of an electric arc funace. These requirements can be enumerated as :

(*i*) Supply to the electrodes

(*ii*) Arc stability

(*iii*) Over current protection

(*iv*) Isolation of the furnace from the supply

(*v*) Selection of electrode material.

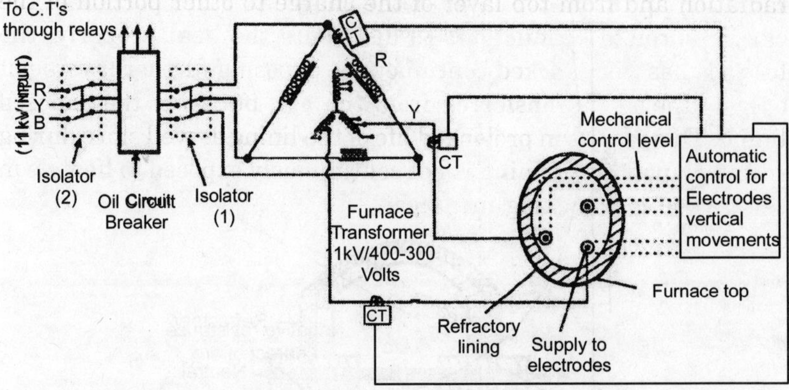

Fig. 2.15 Basic ckt. for electric arc furnace.

(*i*) **Supply to the electrodes:** The elctrodes have to be short circuited momentarily and then separated from the charge inside the furnace, which requires a low voltage on the secondary and a high reactance transformer winding. For this purpose the furnace transformer has as high reactance as 35% and multiple tapping on the secondary. The primary has a line voltage of 11KV whereas secondary varies between 340 to 400 volts on line side. The busbars supplied by this multiple tape secondary arc connected through flexible cables capable of carrying hundreds to thousands of amperes of current (depending upon melting capacity of furnace). The subsequent change of supply to electrodes results in temperature control of melt.

(*ii*) **Arc Stability:** After striking of the arc the electrodes burn out gradually thereby increasing the length of arc which will extinguish if gap is not maintained properly. Secondly the resistance of the metal undergoing melting, changes gradually with the rise of temp. Both these factors change the magnitude of arc current. The effect is transmitted with the help of current transformers (CT's) to an automatic regulator ckt. which adjust the height of the electrodes during the melting process. The stability of arc for a fixed given gap is due to high reactance of the furnace transformer.

(*iii*) **Over current/ground or leakage fault protection:** With the exception of momentary short ckt. the furnace is not supposed to draw an over-current (*i.e.* more than the furnace rating) continuously. For this purpose the current transformers operate a relay/relays ckt. tripping the oil ckt. breaker. The tripping can be due to 11 KV mains voltage fluctuation as well.

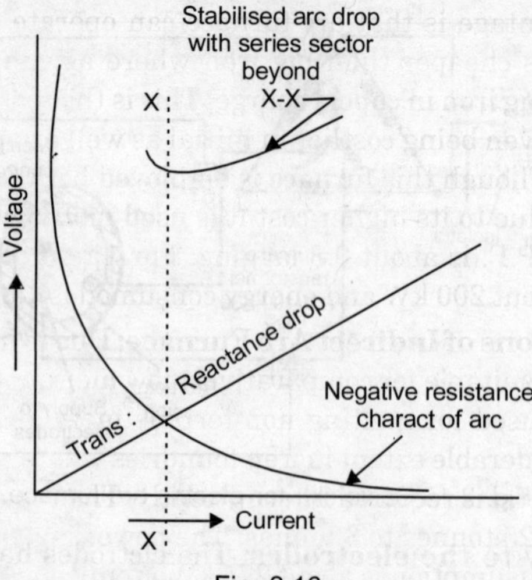

Fig. 2.16

(*iv*) **Isolation of the furnace from the supply:** When metal scrap has to be charged into the furnace or melt/slag removed from the furnace it has to be disconnected from the supply. For this purpose an isolator switch is connected between the oil circuit breaker and furnace transformer. The O.C.B. is tripped off and then line isolater opened.

A second isolator is provided between mains and the O.C.B., this is necessary to disconnect the O.C.B. from the line during maintainance and repairs. Frequent tripping carbonises the oil of the breaker and burns the contacts which have to be taken care of periodically.

(*v*) **Selection of electrode material:-** The electrodes used are either carbon or graphite. The later is a refined and specially treated form of carbon and has four times conductivity. The carbon electrodes are cheap and have a current density of about 8-15 A/cm^2 against 32 to 60 A/cm^2 for graphite. Thus carbon is suitable for low capacity arc furnaces only.

Q.2.11. Explain the application of direct and indirect-arc furnace in industry

Ans. Application of direct arc furnace. It is commonly used for production of steel. The usual size of such a furnace is between 5 and 10 tonnes. Though 50 and 100 tonne arc furnaces have also been developed. The main advantage of direct arc furnace over cupola method for production of steel is that purer production is obtained and the composition can be exactly controlled during refining process.

Another advantage is that arc furnace can operate on 100% steel scrap which is cheaper than pig iron where as cupola requires a proportion of pig iron in cupola charge. This is the reason, that direct arc furnaces even being costlier in initial as well as operating costs, is preferred. Though this furnace is employed both for melting and refining, but due to its higher cost it is used more for refining than melting. The P.F. is about 0.8 lagging. For 1 tonne furnace power required is about 200 kW and energy consumed is 1,000 kwh/tonne.

Applications of Indirect Arc Furnace: Due to indirect heating the furnace is suitable for comparatively low melting point. So these furnaces are used for melting non-ferrous metals. These are also used to a considerable extent in iron foundries where relatively small quantity of metal is required intermittently. The capacity of furnace varies from 0.25 tonnes to 3 tonnes. The power factor is about 0.85. The energy consumption is about 275 kWh/tonne for brass 325 kWh/ tonne for bronze and 400 kWh/tonne for copper.

Q.2.12. List the industrial applications of core type, coreless types and high frequency type furnaces.

Ans. Core type furnaces: This type of furnace is suitable only for continuous operation. These furnaces are widely used for melting and refining of brass and other heavy non-ferrous metals. Its efficiency is about 75%. Standard size of these furnaces range from 60 to 300 kW, all single phase, 50 Hz in this country for standard voltage upto 600 V.

Applications of Indirect Core Type Furnace: This furnace is directly in competition with resistance oven, but has comparatively poor power factor (0.8 approximately).

Applications of Coreless Induction Furnace: Coreless induction furnace is mainly employed as metal melting unit. An important application of this furnace is the production of carbon free ferrous alloy. Various special uses are vacuum melting, duplexing steel, heating of charges of non-conducting materials by use of conducting crucible etc. The energy consumption is 600 to 1000 kWh/ tonne of steel.

High Frequency Furnaces: Surface hardening, Anealing, Soldering, Welding, drying of paints, forging of bolt and rivet heads.

Q.2.13 Explain the principle of operations of induction furnaces with sketches (low and high frequency, Core type and Core less type).

Ans. Low Frequency Induction Heating

Induction heating processes make use of currents induced by electromagnetic action in the material to be heated. Induction heating is based on the principle of transformer. There is primary winding through which an A.C. current is passed. The coil is magnetically coupled with the metal to be heated. An electric current is induced in this metal when the A.C. current is passed through the primary coil.

For melting or refining metals various types of induction furnaces are available while for other processes like case hardening and soldering, the required current is induced in the article at the appropriate stage of its manufacture. The latter process always requires a high frequency and is commonly known as high frequency eddy current heating.

The following are different types of induction furnaces employed for melting purposes:

(1) **Core type**

(*a*) Direct core type,

(*b*) Vertical core type

(*c*) Indirect core type

(2) **Coreless type**

(*a*) **Direct Core type induction furnace:** It is like a transformer. The change forms the secondary winding and consists of one turn only formed by the metal to be melted. The current in the charge *is* very high, say of the order of several thousand amperes. The charge is magnetically coupled to the primary winding. Electromagnetic forces are set-up by the high current in the molten metal. When there is no molten metal, no current will flow in the secondary. To start the furnace molten metal is poured in the annular hearth or a sufficient quantity is left over from the previous charge. There are few drawbacks in the furnace which are given below:

(*i*) Leakage reactance is high, power factor is low due to poor magnetic coupling.

(*ii*) Low frequency supply is required and additional motor generator set or a frequency converter is required.

(*iii*) If current density increases about 5A/mm^2 the pinch effect is there and that causes a complete interruption of secondary.

(*iv*) Odd shape of crucibles are not convenient from metallurgical point of view.

(*v*) Furnace cannot function if the secondary circuit is open. It must be closed.

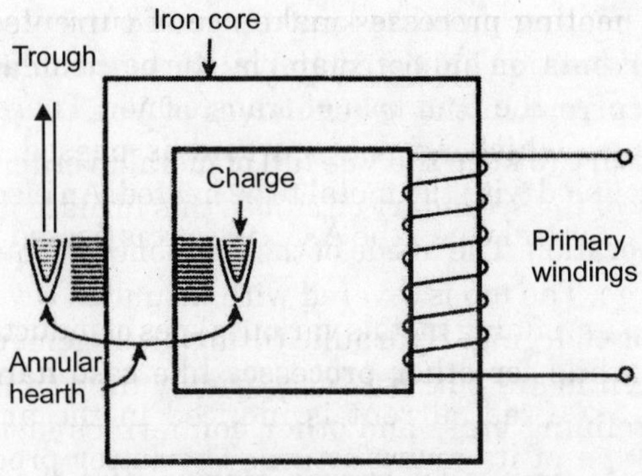

Fig. 2.17 Direct type induction furnace.

(b) **Vertical core-type induction furnace:** This furnace is an improved over core type furnace. It has vertical channel for the charge, thus the crucible used is also vertical which is suitable from metallurgical point of view. The modified form of this furnace is shown in Fig. 2.18 is known as **Ajax. Wyatt vertical core type furnace.**

The magnetic coupling in this furnace is better than core type. Hence leakage reactance is comparatively low and power factor is high, so it can operate from normal frequency supply. The tendency of the circuit to rupture due to the pinch effect, is counteracted by the weight of the charge in the main body of the crucible. The

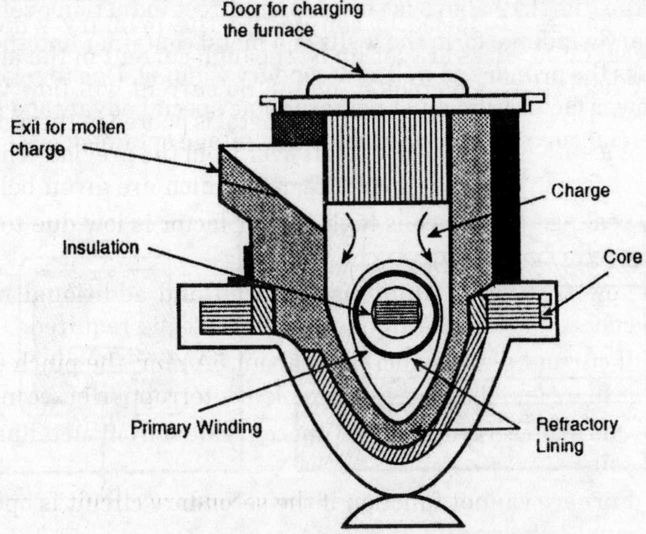

Fig. 2.18 Ajax Wyatt Vertical Core Furnace.

circulation of molten metal is kept up round the Vee portion by convection currents, as indicated and by the electromagnetic forces between the current in the lower halves of Vee.

It is necessary to keep the Vee full of metal in order to maintain the continuity of the secondary circuit so this furnace is suitable for continuous operation. The inside of the frequency is lined depending upon the charge. The top is covered with insulated cover which can be removed for charging. Hydraulic tilting arrangement is there to take the molten metal out. It is very widely used in foundries for melting and refining brass and other non ferrous metals.

Approximate figures for energy consumption for melting and refining of non-ferrous metals are :

Metal	Energy consumption/tonne
Copper and bronze	300-500 kWh
Brass	200-300 kWh
Zinc	100-150 kWh

(*c*) Indirect core type induction furnace. The induction principle can also be used for general heat treatment of metallic and other charges if an inductively heated element is employed to transmit the heat to charge by radiation. So far as the charge is concerned the conditions are similar to those as in the resistance oven.

In the Fig. 2.19 above is shown an indirect induction oven. The secondary windings form the walls of a metal container and the iron core links the primary as well as secondary winding. Due to relatively poor power factor (0.8) must possess some special advantage for its commercial success. The particular advantage obtained is a simple

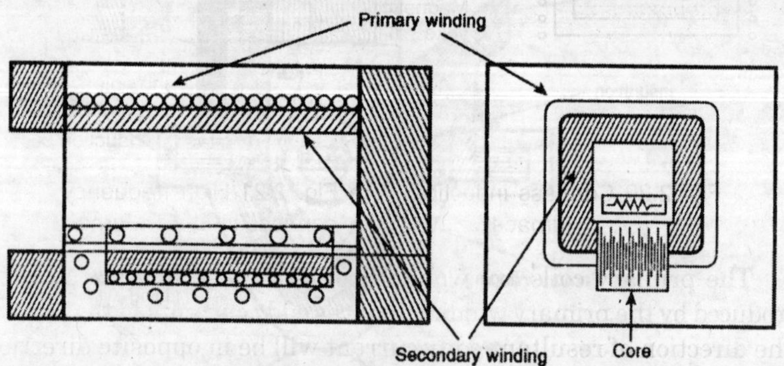

Fig. 2.19 Indirect core type Induction furnace.

and almost a fool proof method of temperature control. The part *AB* of the magnetic circuit is situated in the oven chamber and made from a special alloy which loses its magnetic properties at a particular temperature and regains them when cooled approximately to the same temperature. The oven temperature is thus limited to the critical value without the use of external control equipment. The bar AB is made detached so that it can be replaced by other having different critical temperature ranging between 400° *C* and 1000° C.

(*d*) **Coreless Induction Furnace.** The general construction of this type of furnace is shown in Fig. 2.20 which indicates that a convenient shape of crucible can be employed. The flux produced by the primary winding sets up eddy currents in the charge. The eddy currents developed in any magnetic current is given as :

Eddy current $\propto B^2 \times f^2$

where *B is* flux density, *f* is frequency. These eddy currents are sufficient to heat the metal to melting point and also sets up electromagnetic forces which produce a striking action. Construction of this type of furnace is shown in Fig. 2.21 below. The furnace has a refractory lining.

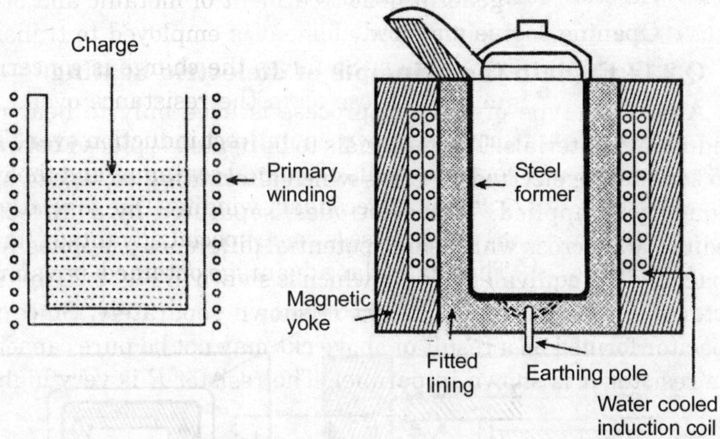

Fig. 2.20 Coreless induction furnace.

Fig. 2.21 High frequency coreless induction furnace.

The primary coils are wound around it. The alternating flux produced by the primary winding induces eddy currents in the charge. The direction of resultant eddy current will be in opposite direction to the current in primary coil. The artificial cooling of primary coil is necessary because of high amount of copper losses. The temperature of the primary winding increases considerably.

The coil is constructed in the form of hollow tube through which cold water is circulated.

In furnace a laminated yoke *is* built on the outside of induction coil to provide a definite path for the flux. This minimises the stray field and also reduces the exciting current necessary to produce the flux.

The following are the advantages of a coreless induction furnace over any other type:

(*i*) The time taken to reach the melting temperature is less in this furnace

(*ii*) Precise control of power into the charge can be employed

(*iii*) Crucible of any shape can be used

(*iv*) Intermittent operation possible without any loss in time for warming

(*v*) Charging and pouring is simple

(*vi*) There is no dust, smoke and noise etc.

(*vii*) Automatic stirring in the charge due to eddy currents

(*viii*) Erection cost of furnace is low

(*ix*) Opening cost is also low.

Q.2.14 Explain the principle of dielectric heating.

Ans. This type of heating process is used only to heat non conducting materials. The materials to be heated is placed between two conducting electrodes across which alternating voltage of high frequency is applied. Two electrodes, separated by *a* dielectric medium and across which some potential difference is applied for a capacitor. The equivalent ckt. of which is shown in Fig. 2.22 (b). The vector diagram for the same ckt is shown separately. Since the capacitor formed as a result of above ckt may not be pure capacitor so a resistor *R* is shown in parallel. The resistor *R* is very high so

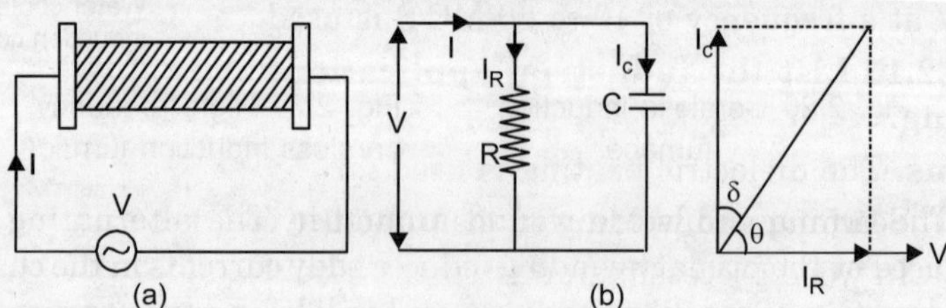

Fig. 2.22.

that current flowing through it is very small so much so that capacitor current I_C can be considered to be the same as the total current, I

$$P = \text{Power Consumed} = VI \cos\theta$$

$$I = I_c - I_R$$

but *III* is so small that it can be neglected

$$I = I_{C}\text{''}, \quad P = V.I_c \cos\theta$$

$$I_c = \frac{V}{X_c}, \quad X_C = \frac{1}{2\pi f c}$$

$$I_c = V.2\pi f c$$

$$P = V.(V.2\pi fc) \cos\theta = V^2 2\,\pi fc \cos\theta$$

$$\theta = (90° - \delta)$$

$$\cos\theta = \cos(90° - \delta) = \sin\delta = \tan\delta = \delta$$

where δ is small and it should be in radians

$$\therefore \quad P = 2\pi f V^2 C\,\delta \text{ watts}$$

This power consumed is converted into heat. The capacitance of parallel plate condensor is given by,

$$C = \frac{K°K°A}{t} \text{ farads}$$

$K°$ = Absolute permitivity of air

$$= \frac{1}{36 \times \pi \times 10^9} = 8.854 \times 10^{-12} \text{ farads/meter}$$

k = Relative permitivity of the material to be heated

A = Surface area of plates in m^2

t = thickness of the material to be heated in meters

Thus it is clear that heat produced/power consumed = $2\pi f V^2 C\delta$

Capacitance C and loss angle δ shall be constant for a particular material to be heated. Thus material produced is propotional to square of voltage and frequency. This is why high voltage at high frequency is used for dielectric heating. For sufficient heating voltage of about 20 KV at a frequency of 10 to 20 MC/S is used.

Q.2.15 List the industrial applications of the dielectric heating.

Ans. The dielectric heating is used for

(*i*) Seaming and welding in manufacture of synthetics

(*ii*) In wood processing industry

(*iii*) For baking foundary cores

(*iv*) For food processing

3. ELECTRIC WELDING

Q.3.1 Write the types and equipment of electric welding.

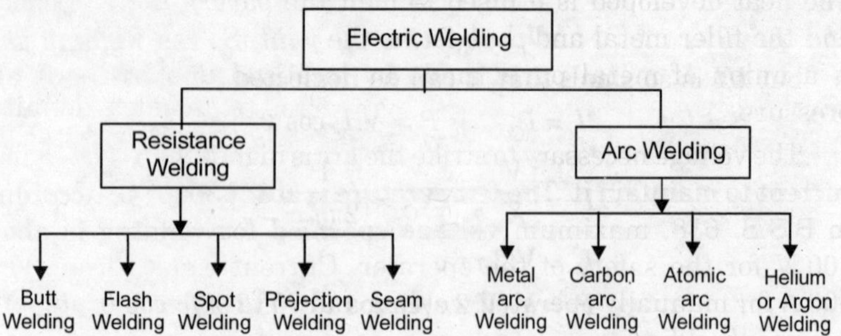

Ans. Electric arc welding equipment : Electric arc welding equipment falls into two natural sections:

(*a*) Those using D.C. at the arc or D.C. welding equipment

(*b*) Those using A.C. at the arc or A.C. welding equipment.

(*a*) **D.C. Welding Equipment:** It consists of mainly a motor generator set. The motor is a squirrel cage induction motor, and generator is differentially compound to give dropping character stics. In differentia] compound generator the terminal voltage falls automatically with the increase in load current.

If supply from existing system is used for welding then a ballast resistance is put in series with the equipment and control is obtained by the variation of the external resistance. This method is also suitable when a no. of operators are working on the same supply system. In such case each operator is provided with separate ballast.

(*b*) **A.C. Welding Equipment:** A transformer reduces the voltage from that of the supply drawn to about 100 V. To regulate the current and produce the dropping characteristics requires a resistance and reactance may be used. When a reactance is used it must be designed to operate well below the saturation point of its magnetic ckt. to prevent the introduction of harmonics which tend to lengthen the zero current period in each half cycle and allow cooling of the arc; a reactance with an air gap in the magnetic ckt. is generally employed.

Q.3.2 State different types of electric welding.

Ans. Arc welding: An electric arc is produced by bringing two conductors connected to a suitable source of electric current, momentarily in contact and then separating by small distance. The current continues to flow across small gap and give intense heat. The heat developed is utilised to melt the part of the work piece and the filler metal and thus forms the joint. So the welding joint is a union of metal parts made by localised heat without any pressure.

The voltage necessary to strike the arc is higher than the required current to maintain it. The temperature is about 3000° C. According to B.S.S. 638, maximum voltage specified for welding is about 100 V for the safety of the operator. Current ranges from 30 to 500 A for manually operated welding and 75 to 600 A for automatic operated/welding.

The electric welding is widely used for joining the metal parts to repair fractured castings and the fillings by the deposition of new metal on the worn parts. The heat produced by the arc is also used for cutting of metal. Arc welding is again sub-divided into four groups namely:

(*i*) Carbon arc welding
(*ii*) Metal arc welding
(*iii*) Atomic hydrogen arc welding
(*iv*) Helium or Argon arc welding

Q.3.3 Explain the principles and applications of (*a*) Spot (*b*) Seam and (*c*) Butt welding.

Ans. (*a*) **Spot welding:** This is the simplest and most universally adopted method of making lap welds in thin sheets upto a maximum thickness of 12.7 mm. In its simplest form spot welding machine consist of a transformer to produce high current at low voltage : electrodes are connected to the ends of the secondary winding for leading the current to the work and to apply the necessary mechanical pressure. In addition there must be a mechanical arrangement to bring the electrodes into contact with the work.

A diagrammatic representation is shown in fig. 3.1. In this are shown

(1) The transformer (2) The electrodes (3) The electrical control switch.

It is used to provide tappings on the primary winding of the transformer to regulate the secondary voltage & thus the welding current. A common arrangement employs tapping at both ends of this wdg., one end gives coarse adjustment and the other fine adjustment.

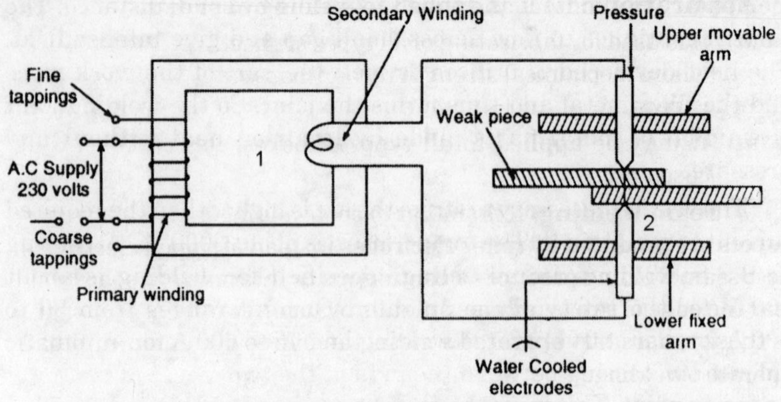

Fig. 3.1 Spot welding.

In this process the materials to be welded are overlapped and pressed between two water cooled electrodes and a impulse of current is passed through the assembly. The metals in the zone of pressure gets heated up to fusion and joint thus made gets cooled under pressure. For spot welding 5000 Amp. of current is required and voltage between the electrodes is usually less than 03 volts. The open ckt. voltage is less than 12 volts. The value of the welding current, the time for which the current flows and the pressure between the electrode tips (made of copper or copper alloys) are the main factors effecting the quality of spot weld. The time period of flow of current varies depending upon the thickness of sheets and kind of metal etc. For thin sheets this period is about 1/50 sec. for each 0.26 mm of total thickness of two sheets to be joined.

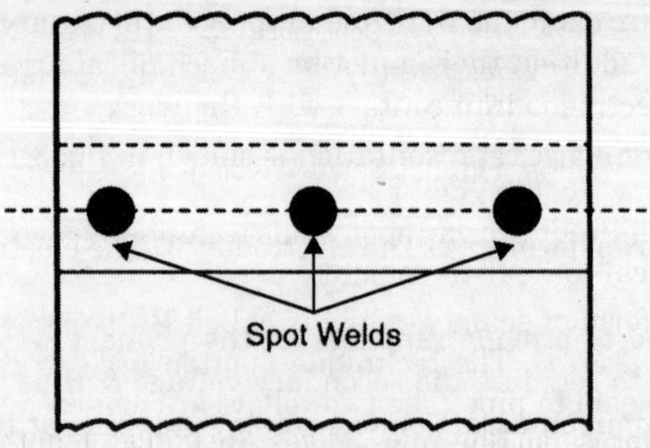

Fig. 3.2 Spot welding.

Applications: (*i*) It is applied to welding of sheets.

(*ii*) It is used for fabricating of all types of sheets metal structures where mechanical strength rather than water or air tightness is reqd.

(*iii*) It may be applied to all types of boxes, cores & enclosing cases etc.

(*b*) **Seam Welding:** It is similar to spot welding except that series of spots are produced by roller electrodes instead of tipped electrodes. Most seam welding produces a continuous or intermittent seam weld near the edge of two overlapped metals by using two roller electrodes. As these rollers travel over the metal the pieces are under pressure and current passing between them, heat the two pieces of metal to the fusion point. This might be called a seam spot weld method. Fig. 3.3 shows a schematic diagram of a typical seam welder in operation. The spaces can be timed to overlap or timed to have space between the spots. The main object of the overlapped spots is to produce gas and liquid leak-proof-lap joints. Two types of seam welds are shown in fig. 3.3.

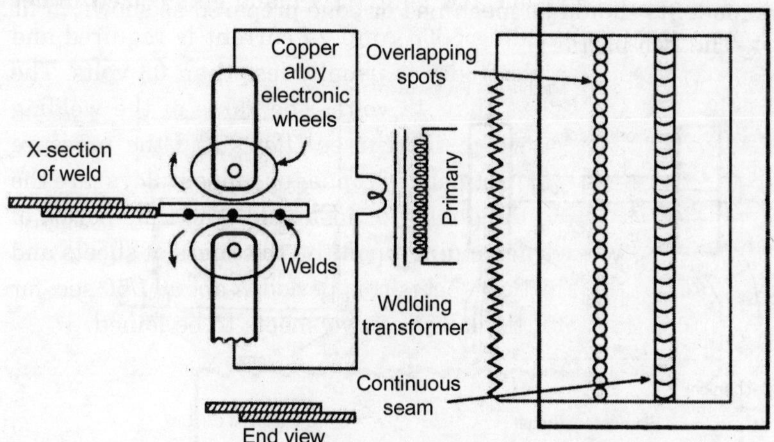

Fig. 3.3 Seam Welding.

The intermittent overlap spot welds technique is used for metals that are critical to heat treatment.

Another form of seam welding *is* called Butt-seam welding *as* shown in fig. 3:3 (*a*). The technique is often used to Butt weld a longitudinal seam in pipe. The two rollers are mounted at an angle and as they press on the pipe, edges are forced together and the electric current brings the metal to the plastic molten condition and fusion takes place.

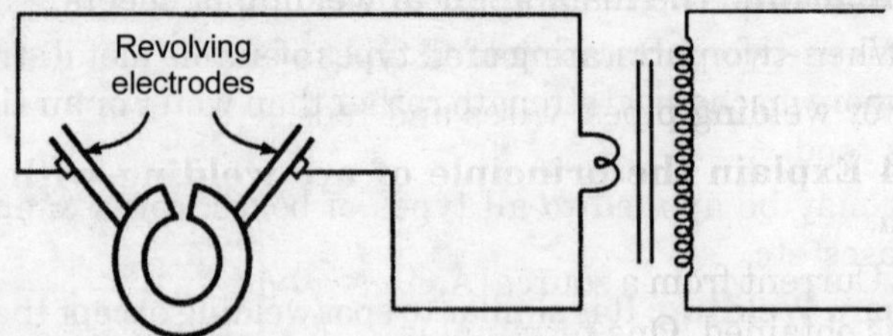

Fig. 3.3 (*a*) Principle of Butt seam Welding

Applications: (*i*) It is used for making lap and butt welds.

(*ii*) It is quicker than spot welding operation.

(*iii*) It is used on many types of pressure tight or leak proof tanks for various purposes and a no. of other products particularly of the relation type such as circular or rectangular containers, car body sections, transformer radiator units.

(*c*) **Butt welding:** In this process, heat is generated by the contact resistance between two components. The faces of the components should be machined or edge prepared as shown in fig. 3.4. The two parts

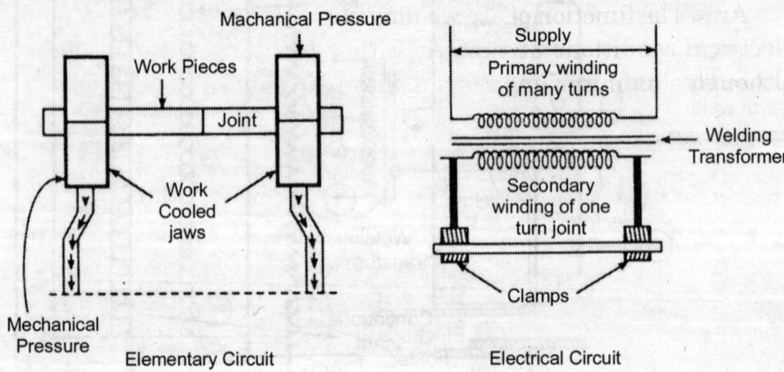

Fig. 3.4

of the components are brought together and the pressure is applied along the axial direction by a spring. A heavy current is passed from the welding T/F (which is having comparative large no. of turns on the primary and lesser no. of turns on the secondary) which creates necessary heat at the joints due to the comparatively high resistance of the contact area. The metal at the joint melts and the two parts fuse together producing a bulged joint.

Application: The main applications of butt welding are:

(*i*) When two parts are joined end to end or edge to edge.

(*ii*) For welding pipes, wires and rods.

Q.3.4 Explain the principle of arc welding with circuit diagram.

Ans. Current from a source (A.C. or D.C.) is obtained. One terminal is connected to the electrode and the other to the work piece and ckt. is completed through air gap as shown in fig. 3.5.

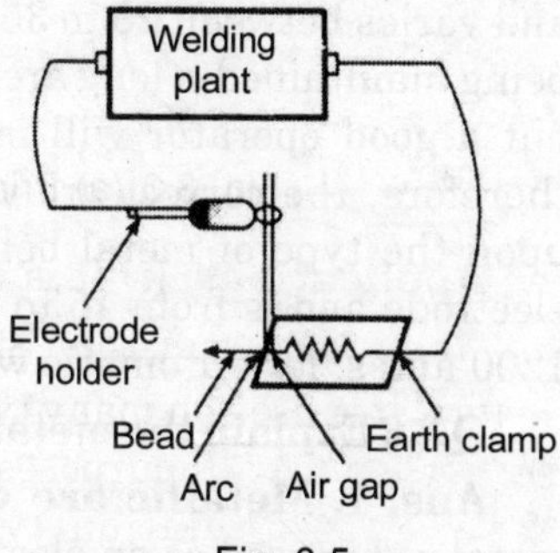

Fig. 3.5

The gap is provided between the tips of the electrode and the surface of workpiece by keeping the electrode at a distance of about 3-6 mm from the surface of the workpiece. Due to interruption by the air gap or gas, heat is produced and temperature attained varies from 3700° C to 4000° C. Electric energy is converted at the arc into heat energy

Q.3.5 List the conditions for successful welding.

Ans. The function of arc welding plant is to produce the required electrical conditions at the arc. Either D.C. or A.C. may be used, although aluminium and certain alloys can only be welded with D.C.

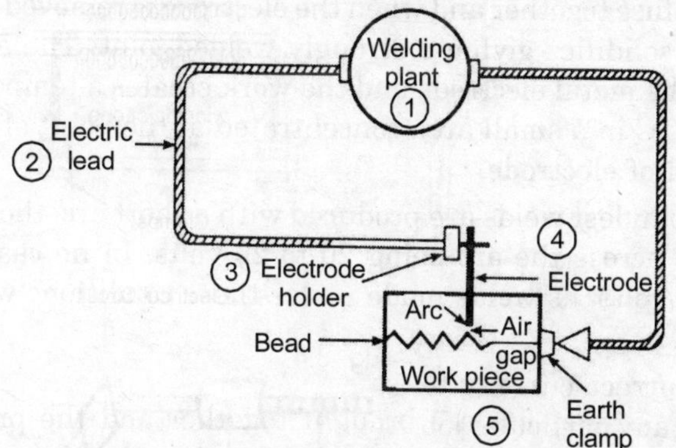

1. Welding plant 2. Two electric leads 3. Electric holder 4. Electrode 5. Workpiece

Fig. 3.6

The main requirements are a relatively high striking voltage to enable the arc to be struck and maintain stability a relatively low voltage to enable the arc to be maintained without being violent and a current of the correct value to melt the electrode and parent metal without burning.

The sparking voltage is usually in the region of 80 to 100 volts in the case of A.C. and 60 to 80 volts in the case of D.C. The arc voltage depends upon the type of metal and the flux of the electrode and varies between 20 to 35 volts. It is affected by the length of arc being maintained-a long arc requires a high voltage than a short arc but a good operator will maintain the correct length of arc and therefore, the correct arc voltage. The current is also dependent upon the type of metal being welded and the guage and type of electrode and is from 15 to 600 Amps, for hand welding and up to 1200 amps, for automatic welding.

Q.3.6 Explain the metal arc welding and carbon arc welding.

Ans. 1. Metallic arc welding: In the metallic arc system a metal rod is used as an electrode and the arc is struck between this electrode and work, which respectively form the two terminal A.C. supply from the secondary of the welding transformer or D.C. supply from the D.C. welding generator depending upon whether D.C. or A.C. is used, and is connected across the job and the electrode. The work is then suddenly touched by the electrode and then separated from it a little. This results in an arc between the job and the electrode. Due to the heat generated by the arc, a little portion of the work melts as also the tip of the electrode. The two pieces to be welded fuse together and when the electrode is removed the metal cools and solidifies giving a strongly welded joint. The arc drawn between the metal electrode and the work creates a temperature of over 3500° C in a small area concentrated at the point of weld and at the end of electrode.

The soundest welds are produced with a short arc, the potential difference across the arc being 20 to 25 volts. In no case must it exceed 30 volts, as welds made under these conditions will not be sound.

The correct current to be used in any particular job cannot be stated as, owing to cooling effect of the differential masses of the metal where the weld is made; A lap joint as shown in fig. 3.7 will require more current than the butt joint in the

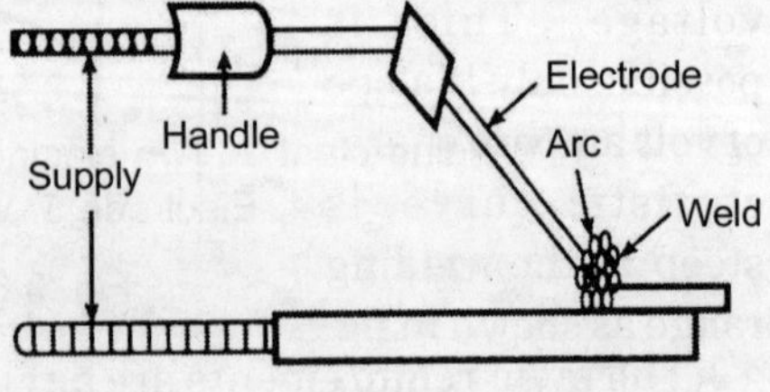

Fig. 3.7 Metal arc weld (Lap weld).

same material owing to the greater dissipation of heat by the greater thickness of metal under the arc where the sheet metal over-laps.

2. Carbon arc welding: The method is normally used for welding copper and its alloys. The carbon electrode which kept negative with respect to the work if D.C. is used. If the carbon electrode is made positive the carbon particles have a tendency to go into the welded joint and cause brittleness. The electrode should therefore, be kept negative and the work positive. For this type of welding only D.C. can be used. The heat from the arc forms a molten pool and the extra metal required to make the weld is supplied by a filler rod of the same composition as that of the molten metal. Two methods of carbon arc welding are used. In one method no flux is used and in the other method flux, either the form of powder or paste is used to prevent the weld from oxidation. Former method is limited to non-ferrous metals and latter method is usually applied for ferrous metals.

Q.3.7 Explain the characteristics of a welding generator.

Ans. Characteristic curve of a welding generator:

The high open ckt. or striking voltage and the low arc voltage is obtained by means of equipment designed to give a dropping characteristic. In this way, the high voltage is available to enable the arc to be struck but as a result of the current flowing the voltage drops to the arc voltage.

It is impossible to keep the arc length constant in manual arc welding. If there is any change in the length of arc correspondingly there is a change in the arc voltage. This is possible only if slope of volt ampere characteristic curve is steep in the welding range as shown in fig. 3.8. From above consideration and

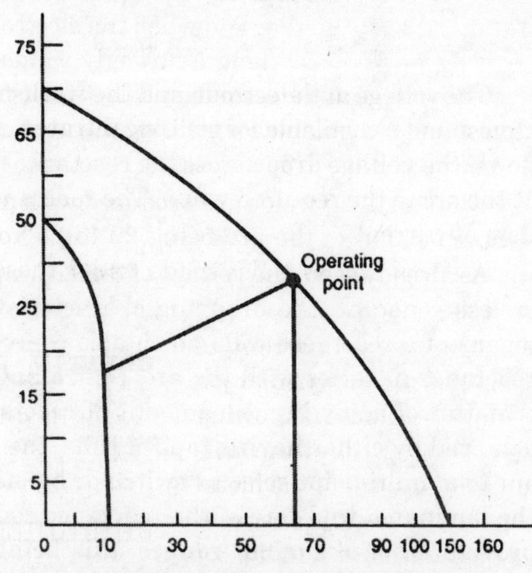

Fig. 3.8. Characterlstic curve of welding generator

also from arc stability point of view voltage source should have dropping characteristics.

This applies to both D.C. generators and to transfer equipment. The method of obtaining this characteristic, particularly in the case of generators, varies with different manufacturers.

Q.3.8 Explain with sketch the principle of operation of welding transformer used with a reactance coil.

Ans. The dropping characteristics is obtained in case of A.C. equipment by means of a reactance coil in series with the arc, as shown in fig. 3.9.

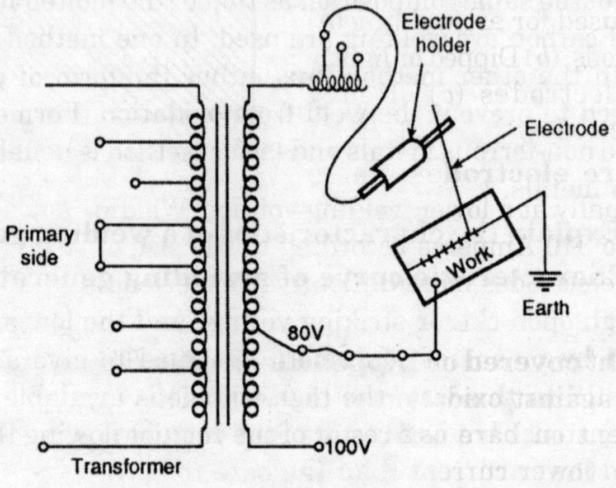

Fig. 3.9

The voltage at the secondary side of transformer remains constant almost and is available for striking the arc but as soon as the current flows, the voltage drop across the reactance coil reduces the voltage at the arc to the required value. The reactance coil also controls the flow of current.

As already seen, the voltage of the A.C. equipment remains more or less constant, the dropping characteristics being obtained by means of a reactance coil, which also controls the current. As the reactance in series with the arc, controlling the current is simply a matter of adjusting the amount of reactance in the ckt. This is achieved by either having tapped coils, the tapping being brought out to a multi-point selector switch or by mechanical movement of the laminated iron core of the reactance coil. Again it is normal to provide a no. of current ranges, this being done by tapping the reactance coil bringing the tapping out to plug points. In this way turns of the coil, can be cut out, thus changing the current range of the unit.

Q.3.9 Explain the different types of electrodes used for welding.

Ans. An electrode is made of metallic wire called core wire coated uniformly with flux. During fluxing the electrodes about 20 mm of its length bare at one end for inserting it into electrode holder. The coating of flux on the electrode is a bad conductor of electricity. There are three types of electrodes which are used for arc welding (*a*) Bare wire rods, (*b*) Dipped or lightly covered electrodes (*c*) Heavy coated electrodes.

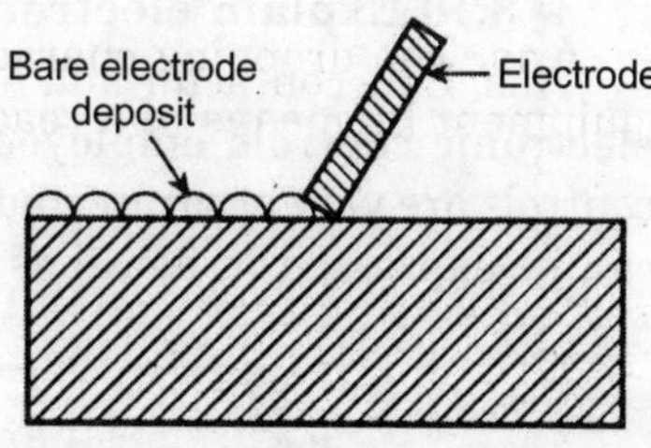

Fig. 3.10 Bare electrode welding.

(*a*) **Bare electrodes:** These are run usually at a lower welding voltage. Welding currents range from 160 to 310 Amperes according to the size of electrodes used. With bare electrodes more difficulty is experienced in starting the arc.

(*b*) **Light covered or Dipped Electrodes:** These provide better protection against oxidation of the weld. They are considered an improvement on bare conductors and require a higher welding voltage and lower current than the bare rods.

(*c*) **Heavy coated electrodes:** These are coated with a material of substantial thickness and solidity, calculated to provide protection to the electrode as shown in fig. 3.11 (*a*) & *(b)*

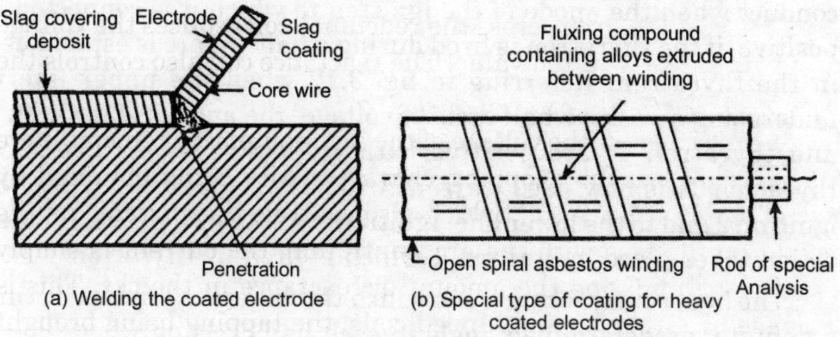

(a) Welding the coated electrode (b) Special type of coating for heavy coated electrodes

Fig. 3.11

Flux covered electrodes have in general a spiral wdg. of asbestos yarn as the basis of the covering. Asbestos is a valuable flux in itself, as it fuses and combine with iron oxide into a compound silicate which, being very light and mobile, is easily removed after solidification, leaving a clean surface—a ϕ important advantage for work which requires more than one run of welding. Asbestos yarn also makes an excellent vehicle and support for the other constituents of the covering and as a suitable binder makes tough surfaced sheath to the metal core. These electrodes, therefore, can be transported and handled without loss or damage.

Q.3.10 Explain electronic circuit used for welding.

Ans. Line contactor and heat control ckt are one of the types of electronic controls employed in resistance wdg. The electronic controls are very accurate and precise in operation. So, this method can be expressed in detail as:

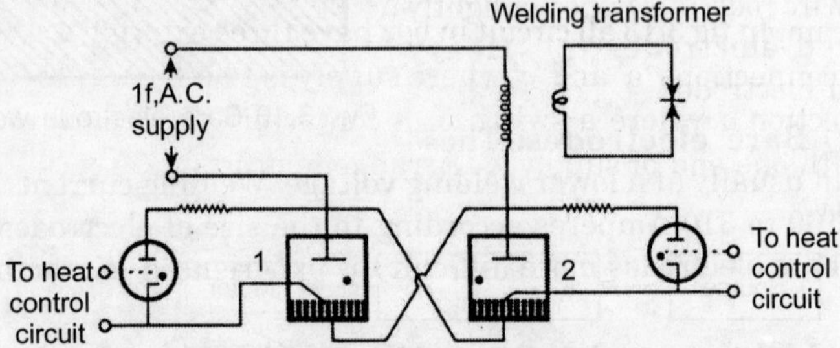

Fig. 3.12 Ignitorn contactor being fixed by thyratron

In this, the welding transformer is connected to the line when a control signal is given and the heat is achieved by accurate firing at ignitrons for a certain duration in each half cycle.

Working: When a thyratron conducts, the effect is to connect ignitor of the ignitron with its anode. Now, the thyratron can only conduct when the anode of the ignitron to which it is connected is positive, if the thyratron is fired during a half, an arc is established in the thyratron. Referring to fig. 3.12 when the upper line is undergoing a positive half cycle of voltage, the anodes of ignitron 2 and thyratron T_2 are positive with respect to the cathodes. If thyratron T_2 is now fired, current will pass through the ignitor of ignitron 2 and to the lower line. Ignitron 2 will therefore fire. Similar action takes place in ignitron 1 which is fired by thyratron 1.

The heat control ckt. is such which the ignitrons conduct during a definite, predetermined angle in each half cycle only.

Q.3.11 Explain the sequence of weld with a Block diagram.

Ans. In most of automatic welding processes, following operations are to be carried out and time required for completion of above operation has to be arranged:

1. During squeeze time, electrodes are brought close to the work and sufficient pressure is developed.
2. During weld time, current flows through the primary of welding T/F. It is during this time that welding heat is generated.
3. During hold time electrodes are held together after interruption of weld current and weld solidification.
4. During off time electrodes are opened, old work is removed and new work is moved to position ready for squeeze operation of next weld.

Before any block diagram of sequence weld timer is given, it is necessary to know how timer circuit will be denoted in the block diagram. In fig 3.13 all circuit in box have three external connections. Two connections 'a' and '*c*' where supply is to be connected and are connection *d* where a switch or a contactor is inserted to connect '*d*' with one end of supply 'c'. Arrangement is shown in fig 3.14 for clarity.

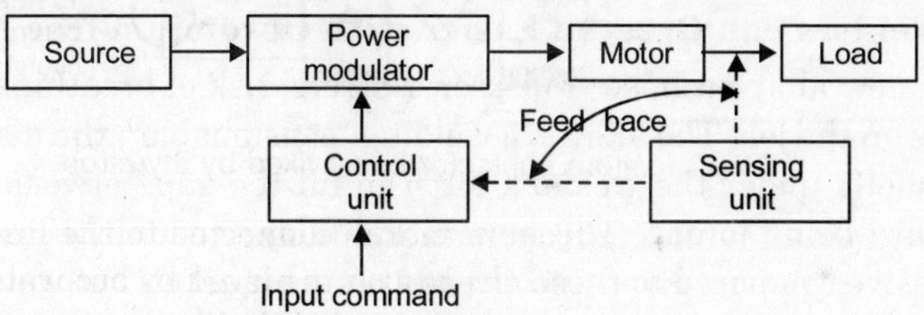

Fig. 3.13

Following symbols used in the circuit:

Contacts normally open

Contacts normally closed

Relay

Contacts operated by relay - C_R

Contacts operated by Timer - C_T

Short Note on Air Laser Technology:

Air Laser Technology to Detect Bombs from a Distance

American scientists claim to have developed *a* new laser sensing technology that will allow soldiers to detect hidden explosives from a distance. Researchers at Princeton University who were behind the invention said their technique involves generating a beam of laser light out of nothing but air. It will also help researchers measure airborne pollutants. "We are able to send a laser pulse out and get another pulse back from the air itself," Richard Miles, a professor of mechanical and aerospace engineering at Princeton, said. "The returning beam interacts with the molecules in the air and carries their fingerprints." Unlike past remote laser-sensing methods, in which the returning beam of light is just a reflection of the outgoing beam, the new "air laser" creates an entirely new laser beam generated by oxygen atoms whose electrons have been "excited" to high energy levels, the university said in a release. Using an ultraviolet laser pulse focused on a tiny patch of air, similar to the way a magnifying glass focuses sunlight into a hot spot, oxygen atoms in the hot spot become excited as their electrons get pumped up to high energy levels, eventually creating a coherent laser beam aimed straight back at the original laser, the researchers explained. *Source: The Times of India, February 1, 2011*

Short Note on use of chromium-free Alloy in welding:

Welders can Breathe Easier with Chromium-free Alloy

A new alloy promises to lessen welders' risk of breathing toxic fumes on the job. The alloy is a welding "consumable", the material that melts under the welder's torch to fill the gap between parts that are being joined. The new nickel alloy consumable is more expensive compared to those already on the market, but worth the cost in situations where adequate ventilation is a problem. That's why two Ohio State University engineers invented the alloy specifically to aid military and commercial welding personnel who work in tight spaces. In tests, welds made with the new consumable proved just as strong and corrosion-resistant as welds made with commercial stainless steel consumables. When melted, however, the new alloy does not produce fumes of hexavalent chromium, a toxic form of the element chromium, which has been linked to cancer. The consumable alloy that joins stainless steel parts together doesn't have to contain the metal as described by Gerald Frankel and John Lippold, Professors of Materials Science and Engineering at Ohio State University. Use of the new alloy essentially eliminates hexavalent chromium in the welding fumes. The engineers further estimated that it would cost five to ten times more than standard welding consumables, depending on metal *prices.Source : http://www.sciencedaily. com/releases/2011/02/110210111318.htm*

4. ELECTRIC DRIVES

Q.4.1. Give an introduction of the electric drives and mention the types of electric drives used in the industry. Also mention the choice of Electrical drives.

Ans. Introduction: By electric drive, we mean an arrangement which converts electrical energy in to mechanical energy output. **Drives employing electric motors are also known as electrical drives.**

Motion control is desired in large number of domestic and industrial applications like rolling mills, paper machines, transportation systems, machine tools, fans, pumps and washing machines etc. Systems employed for motion control are called **drives** and may employ any of the prime movers such as, Gas or steam turbines, Diesel or Petrol engines, steam engines, hydraulic motors and electric motors, for supplying mechanical energy for control of motion. Block diagram of an electrical drive is shown in fig 4.1.

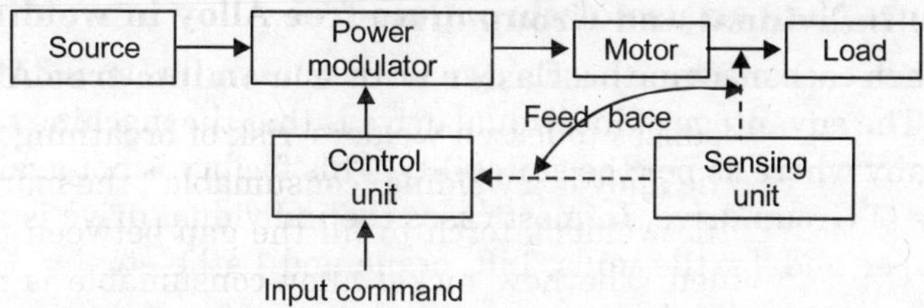

Fig. 4.1 Block Diagram of an Electrical Drive.

Normally load in a machinery is designed to accomplish a given task *e.g.* Pumps, fans, machine tools, trains and drills. Always load requirements, can be specified in terms of speed and torque demands.

A motor having speed—torque characteristics and capabilities comparable to the load requirements is chosen. Power modulator performs one or more of the following 4 functions:-

(i) It modulates flow of power from the source to the motor in such a way that motor is imparted speed-torque characteristics required by the load.

(ii) During transient operations, such as starting, braking and speed reversal it restricts source and motor currents within the permissible values; excessive current drawn from the source may overload it or may cause a voltage drop.

(iii) Converts electrical energy of the source in the form suitable to the motor, *e.g.* if the source is D.C. and an induction motor

is to be used then the power modulator is required to convert D.C. in to a variable frequency A.C.

(*iv*) It selects the mode of operation of the motor, *i.e.* motoring or braking.

When power modulator is employed to perform function *(ii) as* above, it is more correctly called **Converter** while (*iii*) is the main function, depending on its circuit, a converter can also perform other functions of power modulator.

Controls for power modulator are built in control unit which usually operate at much lower voltage and power levels. In addition to operating the power modulator if desired, it may also generate commands for the protection of power modulator and motor. Input command signal, which adjusts the operating point of the drive, forms an input to the control unit sensing of certain drive parameters, such as motor, current and speed, may be required either for protection or for closed loop operation.

(*a*) Individual and Group drive

When each machine has its own prime mover, it is an individual drive. The advantage of individual drive is that the machine can be fitted any where as per the suitability. This facility is not available in case of Group drive. In most cases the individual drive is must, *e.g.* Paper and Textile mills, Lift, crane and Lathe etc.

When several machines are connected with a common shaft through a belt and a single large motor drives all the machines, it is a **Group drive.** In many cases, Group drive is must such as flour mills, where sequence operation is required.

While selecting a drive, it should be first determined whether the drive should be individual or group drive. The advantages of individual drive are :

1. Disconnection of individual drive or its failure will not affect the smooth working of other machines
2. No necessity of a shaft for transmission of drive. But the cost of several machines is more than the cost of a single motor. But the failure of a single motor will bring the whole plant to stand-still.

The various factors can be incorporated in to the following broad based classification:

1. Type of Electric Supply : A.C. or D.C.
2. Nature of drive : Group drive or individual

3. Nature of load : Whether the load demands less or high starting torque.
4. Electrical characteristics of motor : (*i*) starting characteristics (*ii*) Running characteristics (*iii*) speed control (*iv*) Braking
5. Mechanical characteristics : (*i*) Type of enclosure (*ii*) Nature of bearings (*iii*) Transmission of drive and (*iv*) Noise
6. Size and rating of motors: Whether the motor is for continuous, intermittent or variable load cycle. Whether the over-load capacity or pull-out torque meet the requirements.
7. Economic Considerations : (*i*) Capital expenditure and (*ii*) Running expenditure.

Choice of an electrical drive depends on a number of factors. Some of the important factors are:

(*i*) *Steady state operation requirements:* Nature of speed torque characteristics, speed regulation, speed range, efficiency, duty cycle, quadrants of operation, speed fluctuations if any, ratings.

(*ii*) *Transient operation requirements:* Value of acceleration and deceleration, starting, braking and reversing performance.

(*iii*) *Requirements related to the source:* Type of source and its capacity, magnitude of voltage, voltage fluctuations, power factor, harmonics and their effect on other loads, ability to accept regenerated power.

(*iv*) Capital and running cost, maintainance needs, life.

(*v*) Space and weight restrictions if any.

(*vi*) Environment and location.

(*vii*) Reliability.

Q.4.2. State advantages of electric drive.

Ans. Advantages of electric drive :

(1) Cost is too low as compared to other systems of drive

(2) The system is more simple and clean.

(3) The control is very easy and smooth.

(4) Flexible in layout.

(5) Facility to remote control.

(6) Transmission of power from one place to other can be done with the help of cables instead of long shafts etc.

(7) Its maintenance cost is quite low

(8) It can be started at any time without delay.

Q.4.3 List the factors governing the selection of motors.

Ans. The selection of motor is based upon:

(*i*) Condition under which motor is supposed to work

(*ii*) Type of load the motor is supposed to handle. Accordingly the factors which are considered for selection of motor for a particular service are as under:

1. Electrical characteristics:
   (*a*) Running characteristics
   (*b*) Starting characteristics
   (c) Speed control
   (*d*) Braking characteristics
2. Mechanical characteristics:
   (a) Type of enclosure (*b*) Type of bearings
   (*c*) Transmission of drive (*d*) Noise level
3. Size and rating of motor:
   (a) Rating of motor
   (*b*) Whether the motor is to be used continuously, intermittently or for variable loads.
4. Cost
   (*a*) Capital cost (*b*) Running cost

Losses, power factor, maintenance and depreciation etc.

Q.4.4 Classify the electric drives and explain the difference between two drives.

Ans. Classification Of Electric Drives:

*According to developments that have taken place (Table 1.1), electric drives in industrial applications may be classified into three types: groups, individual and multi-motor drive systems.

Table 4.1 Classification of Electric Drive Systems

Loads and environment	Loads	Continuous or intermittent, positive, negative or varying sign, constant or variable in magnitude
	Environment	Temperature, humidity, dust and dirt
Mode of operation	Continuous duty Short time duty Intermittent duty	

Means of control	Manual	
	Semiautomatic	
	Automatic	
Number of machines and interrelations	Individual drive	
	Group drive	
	Multi-motor drive.	Independent, non-related control
		Interdependent, related control
Dynamic and transients	Uncontrolled transient period	
	Controlled transient period	
Methods of speed control	Reversible and nonreversible uncontrolled constant speed	
	Reversible and nonreversible step speed control	
	Reversible and nonreversible smooth speed control	
	Constant predetermined position control	
	Variable position control (servomechanism)	
	Composite control	

4.4.1 Group Electric Drive. The drive consists of a single motor, which drives one or more lines shafts supported on bearings. The line shafts may be fitted with either pulleys and belts or gears, by means of which a group of machines or mechanisms may be operated. It is also sometimes called the shaft drive. Group drives are seldom used now due to their low efficiency and other disadvantages, but such drives are more attractive economically

4.4.2 Individual Electric Drive. In this drive, each individual machine is driven by a separate motor. This motor also imparts motion to various other parts of the machine. Examples of such machines are single spindle drilling machines (universal motor is used) and lathes. In a lathe, the motor rotates the spindle, moves the feed and also with the help of gears, trasmits motion to lubricating and cooling pumps. A three-phase squirrel cage induction motor is used as the drive. In many such applications, the electric motor forms an integral part of the machine.

4.4.3 Multi-motor Electric Drive. In this drive system there are several drives, each of which serves to actuate one of the working parts of the driven machanism. Applications of such a drive are found incomplicated metal cutting machine tools, paper making machines, rolling mills, etc. The drives of a crane can also be considered as an example of a multi-motor drive system. This type of multi-motor drive incorporates three drives : first for vertical movement, second for side movement, and third for forward movement of the load. This type of multi-motor drive incorporates three drives: first for vertical movement, second for side movement and third for forward movement of the load. Each of these drives functions separately and the operator of the crane coordinates their functions. But there are some cases, when an automatic coordinated functioning of several drives is necessary. For instance, a tractiondrive of an electric locomotive consists of 4-8 motors. It is obvious that functioning of these motors should be coordinated in such a way that none loads the other.

A rolling mill incorporates several individual drives. To eliminate tearing and breakage of the product (rods, wires, plates), the speeds of individual drives should be coordinated to maintain continous uninterrupted flow of material at uniform speed and constant tension. Paper mill drives and cloth printing machines are other examples of multi-motor drives with coordinated functioning.

Q.4.5. Explain the classification of control schemes and methods of speed control.

4.5.1 Classification of control schemes. The operations of electric drives are selectively regulated and made partly or fully automatic to increase the productivity and efficiency of the industry. The controllers are defined under, three heads: (*a*) Manual, (*b*) Semi-automatic, and (c) Automatic.

The mode of operation *and* economics dictate the type of control. The load and its characteristics are the basic parameters for determining the motor rating. The load may be of continous or intermittent type. The magnitude of the load can be constant or variable. Environmental conditions should always be taken into account. External temperature, cooling method (natural or forced), humidity, density of dust and dirt, vibrations and shocks are also the important factors that determine the design and construction of the mechanical transmission and the type of motor. The types of control gear used may be electromechanical, electronic (solid state), and so on. From the point of view of the means, and consequently the ways of controlling the functioning of drives, three types of control schemes are identified (Table 4.1), namely manual, semi-automatic and automatic.

(*a*) **Manual Control.** The electric drives with manual control can be as simple as a room fan, incorporating one switch and a resistance for setting the required speed. It can be more complicated as in the case of a push button starter for a simple machine tool or a hoist drive, or drum or cam controllers for heavy duty crane drives and electric locomotives.

(*b*) **Semi-automatic Control.** The electric drives with semi-automatic control usually consist of a manual device for giving a certain command (starting, braking, reversing, change of speed) and an automatic device that, in response to a command, operates the drive in accordance with a preset sequence or order.

(*c*) **Automatic Control.** The electric drives with automatic control have a control gear, without manual devices (though as a security measure, such drives usually have a manual device for use, in case, the automatic system fails).

4.5.2 Classification of methods of speed control.

From the point of view of speed control, the electric drives are divided into five major groups (Table 4.1), although the outputs, modes of operation and means of automatic control of electric drives are all different depending upon the application.

(*i*) **The First Group.** These drives provide uncontrolled constant speed of rotation during normal operating conditions. The speed may be unidirectional or otherwise depending upon the requirement of a particular mechanism.

The magnitude of speed is not an important consideration and is permitted to fluctuate in response to variation in the magnitude and direction of the load torque. The drives for pumps, ventilators and compressors may serve a good example. For bidirectional rotation, the drives of a crane can be cited as examples.

(*ii*) **The Second Group.** These drives provide step changes in speed during normal operating conditions, while maintaining constant speed within the step. Rotation may be unidirectional or otherwise with or without speed control during transient periods. These drives are used in various machine tools. For example, in a drilling machine, an electric drive has to provide different speeds for different materials of the work-piece and diameters of the drill. Step control of speed is used in heavy duty cranes, lifts and mine winders in order to have appropriate speeds.

(*iii*) The **Third Group.** These drives provide smooth, or with narrow step, control of speed within certain ranges, during normal operating conditions. A drive of an electric locomotive can serve as an example, where smooth speed control of the traction motor is required. The main drive of a rolling mill can serve as another example of drives of this group, but compared to a locomotive traction drive, the requirements of speed control here are different.

The main drive of a blooming mill must provide minimum decrease of speed, when the billet is passing through the rolls and a fast reversal. The drives of a continuous rod and plate rolling mill, printing presses must rotate in one direction, but their speeds must be synchronized and accurately controlled so that the products (rods, plates, paper and linolium sheets) are uniformly rolled.

(*iv*) **The Fourth Group.** These drives have to provide control forfixed positions of the load. Drives for lifts and mine winders are examples of drives of this group. This type of drive is used for automatic production lines, skip lifting and loading installations in sugar and chemical industries.

(*v*) **The Fifth Group.** The drives are used for position control in response to a command. They are also called servomechanisms. They are widely used in recording and controlling instruments, copying devices for machine tools, drive systems for radio and radar antennas, etc.

Q.4.6 Give brief introduction of components of Electric Drives.

Ans. Components of electric drives

With the introduction of Thyristor or Silicon Controlled Rectified (SCR) as a power switching device in the sixties, and later the arrival of Gate Turn-Off (GTO) hyristor, bridge rectifiers (converters) and choppers fed from a.c. and d.c. supply respectively are used for speed control of d.c. motors. Choppers using GTOs are now preferred for high power d.c. drives as these devices can be turned off by proper signals applied to the gate. This is unlike thyristors, which can only be turned on by gate signal, *i.e.* cannot be turned off by gate signal. The inverters, fed from d.c. supply, are fabricated using GTOs or other devices from transistor family such as Bipolar Junction Transistors (BJTs). Insulated Gate Bipolar Transistors (IGBTs) and Metal Oxide Semiconductor Field Effect Transistor (MOSFETs). These are self-commutated devices, switched on or off by proper base drive, unlide thyristors, turned off only by forced commutation using circuits consisting of inductors and capacitors in conjunction with additional thyristors, Transistor chopper-fed d.c. drives are used for low and medium power applications. Two types of power electronic converters, namely invertors (mostly voltage source type) and cycloconverters, fabricated normally using thyristors, are used for speed control of a.c. motors, both induction and synchronous. Both types can provide variable voltage variable frequency (VVVF) control with constant voltage/frequency ratio to produce rated (constant) flux in the air gap of a.c. motors, at frequencies lower than base (rated) frequency. Sophisticated control strategies such as vector or field oriented control and other control laws developed using modern control theoretic techniques, can now be implemented using microprocessors, microcontrollers, Personal Com-puters (PCs), and/ or Digital Signal Processors (DSPs). The time needed for computation of control inputs has reduced with increase in clock frequency and higher capability for arithmetic and other functions, along with improved reliability due to reduced chip count in processor/PC-bases systems.

4.6.1 Motors. The motors are either d.c. or a.c., fed from d.c. and ac supply respectively. Various types of d.c. motors, such as separately excited or shunt, series and compound, are used to satisfy the speed and torque requirements of the drive. Normally, d.c. shunt, series and compound, are used to satisfy the speed and torque requirements of the drive. Normally, d.c.. shunt motors are used for applications, where precise speed control is needed, while series motors are preferred for traction and fan type loads, including compressors.

A.C. motors are of two types - induction and synchronous. Induction motors are fitted with wither squirrel cage or wound (slip ring) rotor. The first type is cheap, rugged and maintenance free. With the availability of inverters with VVVF control, variable speed a.c. drives using induction motors are emerging as a powerful alternative to d.c. drives. The use of Pulse Width Modulation (PWM) techniques in invertors has resulted in reduction of harmonics in output voltage waveforms and higher efficiency along with reduction of torque pulsations. In cases where high starting torque is required, the wound rotor (slip ring) motor is preferred, even though the cost is higher. The rotor resistance can be varied, by using a GTO/ transistor chopper in parallel with an additional resistance and changing its ON time, to control motor speed. Line commutated (phase controlled) rectifiers/inverters are used in the rotor circuit to run motors at sub-synchronous/super-synchronous speeds with slip power being either fed back or taken from the supply via a step-up transformer. For sub-synchronous speed operation, in which case slip power the rotor taken via a step-up transformer. For sub-synchronous speed operation, in which case slip power from the rotor taken via slip rings is fed to the supply, controlled (thyristor) bridge rectifier is changed to uncontrolled (diode) rectifier.

Synchronous motors operate at different power factor including leading, by varying field current (excitation), whereas induction motors take current at lagging power factor due to magnetizing component required to produce flux in the air gap. These motors are used for low speed, high power applications in cement and ball mills. Compared to conventional (wound field) type, variable reluctance (with no field winding) and permanent magnet (assumed as constant excitation) type motors are also coming into prominence for low power applications. Another development is the use of 'self-controlled' type, where the base drive to the triggering signals for the devices in inverters supplying power to the stator, are derived from the rotor position sensors. These motors run at different speeds, unlike the motors fed from variable frequency supply which cannot withstand large load torque variations due to stability problems.

4.6.2 Power Electronic Controllers. A power electronic controller consists of basic circuits termed power electronic converters, and other gate or base drive circuits needed to switch on/off the devices including analog/digital chips, but may not include any processor/PC-based system for control law implementation. The converter circuits are of the following types:

1. Bridge rectifiers (converters) and choppers for d.c. motors.
2. Inverters and cycloconverters for a.c.. motors.

The controlled rectifiers are fabricated using mostly thyristors, though for low voltage output as needed for low speed operation of d.c. motors, the power factor is low as delay angle (phase) is large. In certain cases, GTOs or transistors are used to obtain symmetrical output voltage as these devices can be switched off by gate/base drive, not needing additonal circuits for turning off like thyristors. This results in near unity power factor. If one controlled rectifier is used for motor operation in forward direction, it cannot be used for braking operation in forward direction as thyristors are unidirectional devices. However, the same rectifier can be used to brake the motor in reverse direction as needed for crane/hoist operation. To brake the motor in forward direction, either a reversing contactor with one rectifier or two rectifiers in back-to-back connection (dual converter) are needed. Such circuits are also suitable for four-quadrant operation of the drive. As stated, GTOs/transistors are used for chopper-fed d.c. series motor drives. Both motor and braking operations can be performed, but two choppers are used as current has to be reversed in armature circuit of the d.c. motor for braking.

Variable frequency invertor (voltage source) circuits using GTOs/ transistors have become simpler and occupy less space compared to thyristorized circuits needing bulky commutation circuits. The use of PWM technique in such inverters has resulted in their application to induction motor devices. Cycloconverters using thyristors are used to obtain low frequency needed to run synchronous motors at low speed.

Power electronic converter circuits require gate/drive signals to switch on/off the devices. These signals are obtained using different chips and transistors, which are needed at the last stage to obtain proper drive signals of required power (voltage/current) rating. These circuits perform various operations including sequencing and lock-out functions. Processor/PC-based systems are used to implement complex control algorithams to suit the requirements of d.c./a.c. drive systems for better performance.

Electric drive systems perform numerous operations such as frequent starting and stopping of motors including braking, changing connections as needed, running one or more motors in a particular sequence, interlocking to prevent malfunctioning and disconnecting the motor from supply when abnormal or faulty conditions are detected. Programmable Logic Controllers (PLCs) are used to reduce the hardware requirements, while at the same time software can be used to make changes in sequencing and other functions, if required, thus making the system flexible.

Q.4.7. State the advantages and disadvantages of two types of drives.

Ans. Advantages of Group drive:

Here, advantages of group drive will be disadvantages of individual drive

(*i*) **Initial Cost:** Initial cost of group drive is less as compared to that of individual drive.

(*ii*) **Sequence of operation:** Group drive system is useful because all the operations are stopped simultaneously.

(*iii*) **Space Requirement:** Less space is required in group drive as compared to individual drive.

(*iv*) **Low Maintainance Cost:-** It requires little maintenance as compared to individual drive.

Disadvantages of Group drive:

Although the initial cost of group drive is less yet the system is not used extensively because of the following disadvantages :

(*i*) **Power factor**: Group drive has low power factor

(*ii*) **Efficiency:** Group drive system when used and if all the machines are not working together the main motor shall work at very much reduced load.

(*iii*) **Reliability:** In group drive if the main motor fails whole industry will come to stand still.

(*iv*) **Flexibility:** Such arrangement is not possible in group drive *i.e.* this arrangement is not suitable for the place where flexibility is the prime factor.

(*v*) **Speed:** Group drives do not provide constant speed.

(*vi*) **Types of machines**: Group drive is not suitable for driving heavy machines such as cranes, lifts and hoists etc.

Q.4.8. Select the motors with given load

Ans. The following factors play an important part in selecting the size and suitable size of motor for a particular drive :

(i) **Load characteristics** It includes the load starting torque, speed requirements etc.

(*ii*) Surroundings

(*iii*) Thermal characteristics of the motor.

(*iv*) Noise

(*v*) Annual running charges of the motor.

Initially first four factors are thoroughly studied and the annual charges for the type of motors selected are assessed. The final selection made should be economical.

Load charac-teristics (1)	*Type of in-dustrial equip-ment* (2)	*Suitable D.C. motor* (3)	*Suitable A.C. motors* (4)
(*i*) Low torque nearly constant speed	Vacuum cleaner	Universal motor	(*i*) Single phase universal motor for constant load.

(*ii*) Frequent Start (*iii*) Load duration comparatively small	Portable tools Office appliances, Hair dryer, Kitchen appliances etc.		(*ii*) Capacitor start, 1ϕ I.M. for carrying loads
(*i*) Small load (*ii*) High torque (*iii*) High output	Small compressors Garage equipment Shoe machinery Wood working machinery Refrigerator		(*i*) Repulsion type 1ϕ motor (ii) Repulsion start I.M.
(*i*) Low load requirement between the range of 10-15 watts (*ii*) Intermittent on continuous load (*iii*) Constant speed (*iv*) Noise less requirement	Electric clocks, photography Instruments Hair dryers		Shaded pole single phase motor
(*i*) Constant speed (*ii*) Load requirement upto 1 KW (*iii*) Starting torque between 1.5 to 2 times full load torque.	Electric fans Blower pumps	Shunt motor	(*i*) Squirrel cage I.M. (*ii*) Synchronous motor with starting arrangement
(*i*) Unloaded or loaded start (*ii*) Nearly constant speed	Line shaft conveyors Grinders (Pulp)	Series motor	(*i*) Squirrel cage I.M. (*ii*) Slip ring I.M. (*iii*) Synchronous motor for line shaft
(*i*) Variable speed	Conveyors, shear	Compound motor	Squirrel cage I.M.

(*ii*) High starting torque	Floor mills		
(*i*) Constant speed (*ii*) High starting torque	Rolling mill, paper mill Rubber mill	Compound motor	(i) Squirrel cage I.M. (ii) Synchronous motor
(*i*) Frequently starting and reversing (*ii*) Starting at full load (*iii*) Intermittent service	Elevators, small cranes and hoists Large cranes & hoists	Series motor Series motor Series motor	Squirrel cage I.M. slip ring I.M. Squirrel cage I.M.
(*i*) High load (*ii*) High starting torque (*iii*) Motor should be capable to withstand heavy duty of rolling	Iron & steel works	Specially designed shunt motor	Specially designed I.M.
(*i*) Flammable gases surrounding the motor (*ii*) Variable load	Mining work	Compound motor	(i) For small load upto 10 H.P. squirrel cage I.M. (ii) For high load slip ring I.M.
(*i*) Variation in speed to a large extent (*ii*) High starting torque	Electric-trains	Series motor	(i) Single phase series motor (*ii*) Three phase I.M.
(*i*) Accuracy (*ii*) Constant speed (*iii*) Low load (*iv*) Low starting torque	Lathe Milling machines	Shunt motor	I.M. with pole changing arrangement

Q. 4.9. Sketch the typical curves of various types of industrial loads.

Ans. There are three different types of loads under which the motor is required to work, these are:

(*i*) Continuous loads,

(*ii*) Intermittent loads,

(*iii*) Variable or Fluctuating loads

Continuous load: Drives like pumps and fans etc. require a constant power as these keep on operating continuously and the choice of a suitable rating of motor is a simple affair. Manufacturers charts are referred.

It is essential to calculate the exact amount of power needed by the drive so as to make the correct choice of motor 'H.P'. If the H.P. of the motor is less than what is needed, the motor will overheat and subsequently burn out. On the other hand, if the H.P. of the motor is more than what it is demanded by the load, the motor will be operating at less than full load value and hence lower efficiency and power factor.

Intermittent loads: Intermittent loads may be of two types:

(*a*) The motor may be loaded for a short time and then shut off for a sufficiently long time allowing the motor to cool down. In such a case a motor with suitable short time rating is adopted such as in mixies etc.

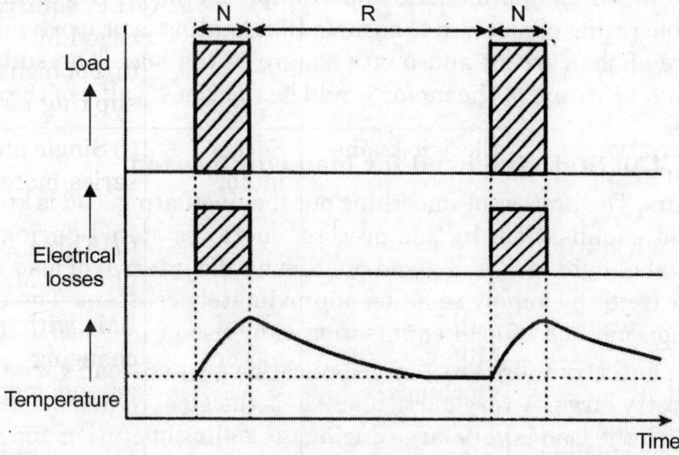

Fig. 4.2

(*b*) The motor may be loaded for some time and then shut off for some time. Before the motor can cool down completely the load cycle repeats. In such a case a suitable continuous or short time rated motor is chosen which when operating on the particular load cycle, will not exceed the specified temperature limits.

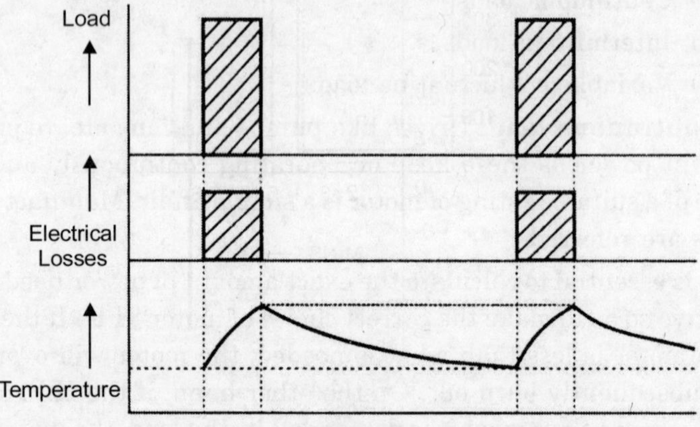

Fig. 4.3

Variable or fluctuating loads: In case of fluctuating loads the most accurate method of selection of a suitable rating of a motor is to draw the heating and cooling curves (as per load fluctuations) for a no. of motors. The smallest motor which does not exceed the permitted temperature rise, when operating on a particular load cycle is then chosen for the purpose.

However an approximate and a simple method of selection of a suitable rating of motor is to assume that heating is proportional to square of the current and hence square of the load. The suitable continuous rating of the motor would be the r.m.s. value of the load curve.

Q.4.10 State the need for load equilisation.

Ans. The process of smoothing out the fluctuating load is known as load equalisation. In this process, energy is stored during the interval of light load and given out during the interval of load thus power from the supply remains approximately constant. The most common method of load equilisation is by use of flywheel.

In industry there are many applications, where load variations are pretty large. A very particular *e.g.* is the steel rolling operation in which the load is very large during the rolling operation and very little as the charge passes by. A typical *e.g.* of the load cycle of this operation is shown in fig. 4.5 *a* & *b*.

It would require a motor of large output, though on the average, load on the motor is not large. It could well be a short time rated motor. As the load cycle is quite regular. It is quite possible to devise

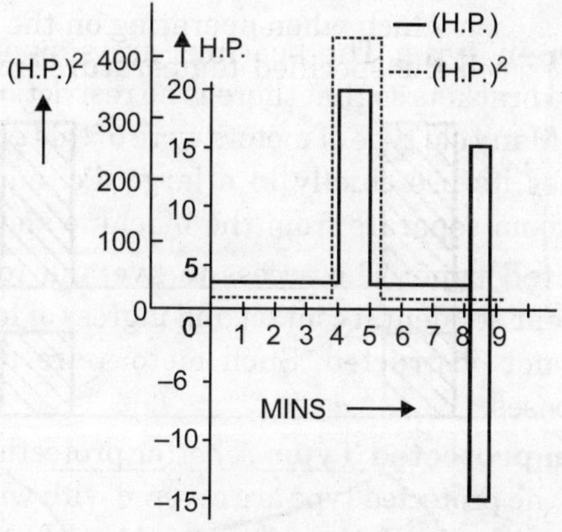

Fig. 4.4.

ways to cater to the large load torque to a shortwhile, thus effecting a considerable economy in motor size. The best way of doing this is to increase the moment of inertia of the motor by means of a flywheel which stores energy during the lean period of the cycle (when the load is small); and give it up during its peak period, when the load is large.

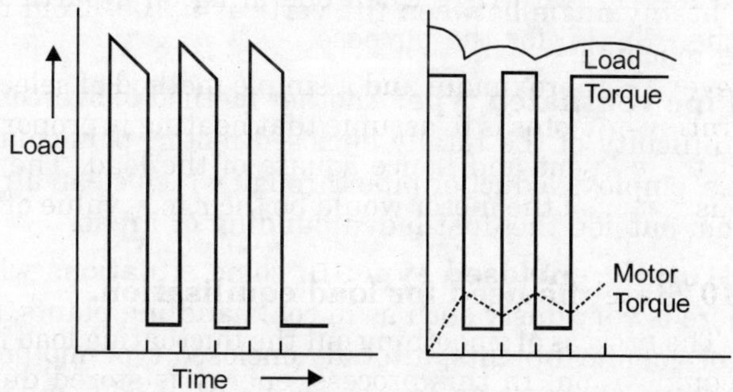

Fig. 4.5

Q.4.11 State the use of flywheels.

Ans. The flywheel stores energy when the load is small and the speed picks up; and gives it up to the system in time of demand, when the load is large. Thus, it brings about an equilization of load to a certain extent. It is thus practical to have a motor, whose torque is not as large as the load torque required, the gap being adjusted by flywheel.

Q.4.12 State the different types of enclosures with applications.

Ans. (*i*) **Open type:** The bearings are supported on simple pedestals or end brackets so that there is no restriction to ventilation and protection. Many old type of motors were of this type. But modern practice confines its use chiefly to a large d.c. motors which are installed in a room separate from the machine shop.

(*ii*) **Protected type:** The access to live and rotating parts is limited, to prevent accidental contact and ingress of foreign material. Ventilation is not obstructed. Such motors are used mostly for industrial purposes.

(*iii*) **Screen protected Type:** A better protection is afforded if the openings of the protected type are covered with wire mesh screen. The degree to which ventilation achieved depend on the fineness of the screen and the possibility of its becoming clogged with dirt.

(*iv*) **Drip proof type:** Where motors are to be installed in very damp situations, a drip proof enclosure can be used in which the openings are protected by overhanging cowls to prevent falling moisture from entering the machine.

(*v*) **Splash proof type:** In such machines ventilating openings are so constructed that liquid drops or dust particles reaching the machine at any angle between the vertical & 100° from it cannot enter the machine.

(*vi*) **Pipe ventilated type:** Another method of overcoming the heating difficulty of the totally enclosed motor, particularly with large *sizes,* employs a duct or pipe through which clean air is led to motor from outside the dust laden building or areas.

(*vii*) **Totally enclosed type:** In some situations where the atmosphere is very dusty such as in coal handling plants, saw mill or in oil or chemical plants, a totally enclosed type motor is used. Such machines have solid frames and end shields without any opening. These motors are cooled by the surface radiation only with the help of cooling fins. The output of such motors is about 50 to 75% of that obtainable from a protected motor of some dimensions. Motor of these types are of small ratings.

(*viii*) **Flame proof type:** For the use in coal mines or other situations where an explosive atmosphere may arise, a special form of totally enclosed motor must be used which is capable of withstanding any explosion of gas inside the motor without transmitting the flame to the outside. The ventilating opening in the machines are made long enough for the flame to be contained with in the frame.

Q.4.13. State the different types of bearings.

Ans. (*i*) Ball or roller bearing and

(*ii*) Sleeve or bush bearing

(*i*) **In ball bearing:** The shaft is supported on the walls or roller which are movable. Ball bearings are used for motor upto 100 H.P. It is expensive but lowers friction loss, requires less space and low maintenance. This bearing gives more noise so it is not recommended where noise is to be avoided.

(*ii*) **Sleeve or Brush bearings:** These are also known as plain bearings and friction in such cases is reduced by the use of lubricating film. For fractional horse power motor, it is in the form of self-aligning porous bronze bush known as sintered bush. For large horse powers it is in the form of journal bearings. When properly designed, the complete load is transferred to a hydrodynamic oil film developed in between the shaft and the sleeve. The advantage of this type of bearing is that it gives noiseless operation and has longer life.

Q.4.14. State the methods employed for reduction of noise.

Ans. To reduce the noise, journal bearing may be used in place of ball bearings. The motor should be mounted on a heavy concrete or cast iron block. The electrical connections should be made through the flexible conduits.

Q.4.15. Select the suitable motors for the following drives.

Ans. (*a*) **Domestic:** Small universal motor, usually series type is used for various domestic appliances such as domestic refrigerators, shavers, vacuum cleaners, mixi and cloth washing machines etc.

(*b*) **Refigeration and AC:** 1 – ϕ, 220 V capacitor type induction motors with DOL starters are usually employed.

(*c*) **Cranes:** D.C. series motors or D.C. compound motors are preferred on account of their high starting torque and easy speed control for lifts, hoists. Also for D.C. series motor and D.C. compound motors are used.

(*d*) **Lifts and hoists:** D.C. Compound wound or A.C. slip ring induction motors are used. To get speed control, shunt field regular method or Ward-Leonard method is used.

(*e*) **Lathe, drilling and grinding mills:** Lathes are usually driven by constant speed. Squirrel cage induction motor for milling and grinding machines, also. A.C. squirrel cage induction motors is used.

(*f*) **Pumpsets:** Drip proof, totally enclosed polyphase induction motor squirrel cage type in case of centrifugal pumps and slip ring type in case of reciprocating pumps are usually employed.

(*g*) **Punches, presses & shears:** Punches and shears are usually equipped with flywheel to supply momentary demand of power. High slip squirrel cage induction motor is used for punches & shears.

(*h*) **Planing machines:** D.C. shunt or compound motor is used if D.C. supply is available otherwise slip ring induction motor is used.

(*i*) **Wood working machine:** A D.C. shunt or compound motor may be used.

Q.4.16. State the types of loads for which drives are necessary.

Ans. 1. Lifting Load. This involves lifting of some weight by a crane or hoist etc. In lifting of loads, the torque requires independence of speed. Load-torque remains constant.

2. Air Fluid Friction. In these types of loads the torque changes with the square of speed such as in blowers, fans, water wheels etc. In this case, Torque $\propto$ (speed)^2.

3. Load due to friction. Whenever two surfaces move one over the other on a shaft, which rotates on a bearing or bush an opposite face always acts at the surface. This force is called the frictional force. However this force can be decreased by using a lubricant. When no lubricant is used friction is called dry friction and torque in that case is constant with speed. In fluid friction where a lubricant is used the torque is linearly proportional to speed.

4. Deformation Load. Such types of loads occur in crushing, grinding and metal drawing etc. In such cases torque is inversely proportional to the speed.

In most cases we have a combination of two or more types of loads.

Q.4.17. State the different types of mountings.

Ans. Motors are generally made in following mountings :

1. Foot mounting type. Where belt tension adjustment is required slide rail mounted motors are used.

2. Ceiling mounting type.

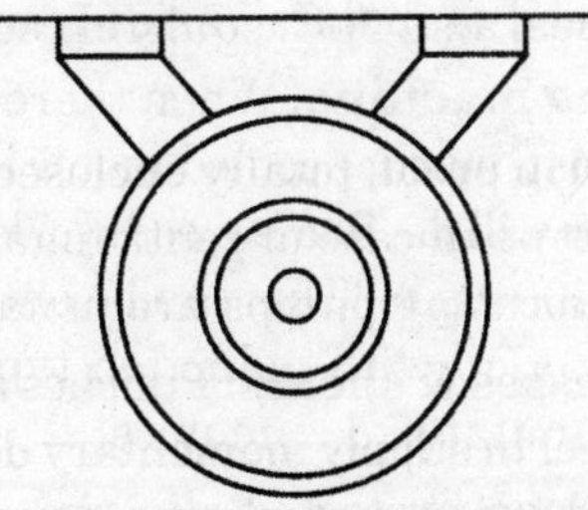

Fig. 4.6

3. Wall mounting type. With shaft horizontally or vertically down or vertically up.

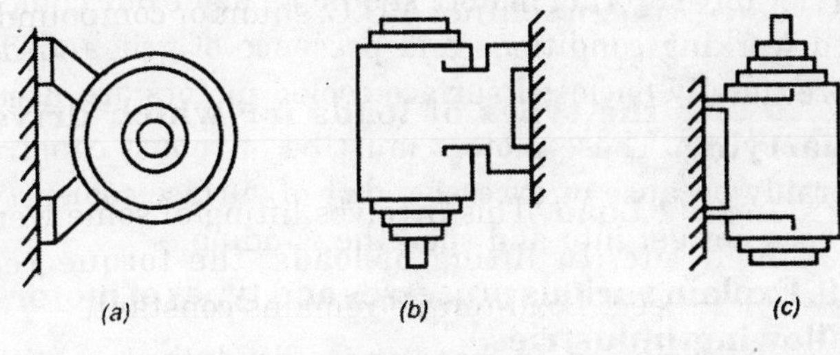

Fig. 4.7

4. Flange mounting type. Motors with flange is fixed directly on the driven machine. Hollow shaft deep well turbine pump motor is an example of flange mounting.

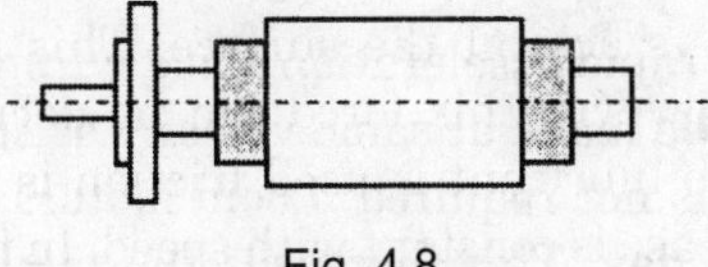

Fig. 4.8

5. Cradle mounting type. These motors are used as loom motors.

6. Body turned mounting type. For many of machine tools these motors are fitted into the body of the tool, this gives clean appearance.

Q.4.18. Select the suitable motors for following drives (*a*) Breweries, (*b*) Flour mills (*c*) Wood working machine (*d*) Printing (*e*) Belt conveyors (*f*) Quarrying.

(*a*) **Breweries.** Motors have to work mainly in damp atmosphere. Therefore totally enclosed surface cooled type motors are recommended.

(*b*) **Flour Mills.** Pipe ventilated or totally enclosed type motors should be used as speed of motors varies from 600—1000 r.p.m.

(*c*) **Wood working machine.** When there are chances of motor being buried in saw dust, surface cooled motor is preferred otherwise screen protected motor will do. Some wood working machines require speeds higher than 3000 r.p.m. This can be obtained with induction motors only in conjunction with induction type frequency changer or by the use of commutator d.c. motors.

(*d*) **Printing.** Screen protected slip ring motors where speed control is required. Otherwise squirrel cage motors are used.

(*e*) **Belt conveyors.** Belt conveyors are used for moving sand and gravel. Heavy loads have to be accelerated. High torque squirrel cage or preferably slip ring motors should be made our choice. Due to exposed working conditions and presence of grit and dust in atmosphere, totally enclosed surface cooled motors are used.

(*f*) **Quarrying.** These motors must be of robust construction and preferably because of excessive dirt of surface cooled type so that grit does not get into and spoil the winding.

Q.4.19. Explain various processes and types of motors used in the following industries.

(*a*) **Textile mills.** Loom motors operate in atmosphere which *is* laden with wool lint and moisture. Further the motor should have high starting torque of about 2 to $2\frac{1}{2}$ times full load torque. This is needed to ensure rapid acceleration of loom from rest to full speed as first pick has to be made at same velocity as subsequent picks if defective material is not required. Loom motors have to start and stop many times a day. This increases the motor heating during starting. Motors of low temp. rise under full load conditions are therefore selected so that under working condition of frequent starting and stopping, permissible temp. rise is not extended keeping above requirement in view totally enclosed, fan cooled, high torque squirrel cage motors are selected. Where only wool lint is present and no moist atmosphere, screen protected motors are used. Splash proof motors are used in wet location such as dye house. Brush shift adjustable speed motor is used in printing machines.

(*b*) **Paper mills.** Pulp stuff from storage tank after passing through strainers, comes to first section of paper making machine which consists of moving wire mesh. Subsequently this sheet of paper goes in between series of heated press rolls, drying rolls, Calenders, smoothing rolls, and finally to reelers.

Following requirements of drive must be fulfilled:

(*i*) To manufacture different types of paper we should be able to vary the speed of entire series of rolls.

(*ii*) Relative speeds of rolls should be constant otherwise tearing of paper may result.

(*iii*) We should be able to adjust the speed of any one group of rollers relative to others in order to allow the draw of the paper.

All the above requirements are fulfilled by using schrage motors. Master schrage motor drives a synchronous alternator from where variable frequency supply is obtained. This is supplied to the stator of all the sectional synchronous motors. Sectional synchronous motors are such that both stator and rotor are free to rotate. Stator of synchronous motor carries the brush gear of the sectional schrage motor. Rotor of the synchronous motor is driven mechanically by cone pulley drive from the shaft of sectional schrage motor. Normally the speeds of the rotor of synchronous motor and rotating magnetic field set up by stator currents are same. If due to any reason speed of section motor tends to fall, this will produce torque on the stator of the synchronous motor in such a manner as to increase the brush distance of sectional schrage motor, thereby condition (*ii*) is fulfilled. If we change the speed of master schrage motor frequency of supply to stator of all synchronous motors will change. This will again tend to rotate the stator of synchronous motors. This results in the movement of brush gear of all sectional schrage motor. This fulfills requirement (i) Speed of any group of rolls can be adjusted by means of sectional cone pulley drive. This fulfils requirement (*iii*).

(*c*) **Rolling Mills.** There are two types of mills and selection of the driving motor depend upon the type of mill. For instance in blooming and slabing mill, rolling operation requires production of heavy torque at low r.p.m. to effect heavy reduction of material from ingots to blooms or slabs. Blooming mill may have 2 rolls or 3 rolls. Two high blooming mill has arrangement of reversing the direction of rotation of the rolls in order to carry out rolling of material through successively reducing passes. In case of 3 high mill, direction of rotation need not be changed. By the use of tilt tables material can be passed between lower and middle rolls or between middle and top rolls. Manipulators are not only used to guide the blooms from groove to groove and to turn the bloom through 90° but they also straighten the pieces during their entry into passes. Blooms and slabs as obtained from blooming mills are further subjected to rolling to produce plates, structurals bars and billets. Sheet and tin bars are further rolled to sheets and tin plates. Billets when further rolled produce rounds, square, flat angles etc. Rolling subsequent to blooming, in most of the cases is carried out in continuous mills. Here rolling mill consist of a number of stands in series stock after coming out from first stand is turned through 90° by twist guides before being fed into the next stand. In order to minimize the formation of loops or tension in the material between stands, it become necessary to co-ordinate the speeds of rolls of various stands. If all the stands are driven by one motor this co-ordination of speeds is achieved by giving the drive from the shaft to various stands through level gears of diff. speed reduction ratio. If individual drive to each stand is adopted the speeds of diff. motors have to be synchronized. In order to achieve varied rolling programme say 6 mm to 25 mm diameter wire from 75 × 75 mm square billet, it requires reduction of cross-section area from $\frac{1}{11}$ to $\frac{1}{200}$ Taking into account this varied reduction of X-section area and the initial and final speeds of stock being rolled it becomes necessary to employ motors for continuous rolling mill having speed range in the ratio of 1:8. Depending upon the type of the mill, selection of motor has to satisfy following conditions also

(*i*) Motor should have robust construction so that it can with stand severest duty.

(*ii*) It should be capable of having wide speed variation.

(*iii*) It should be capable of developing short time torque to the extent of 2 to $2\frac{1}{2}$ times of rated torque.

(*iv*) It should be in a position to maintain preselected speed within close tolerances from no load to full load. Therefore it should have quick response of change.

(*a*) **Induction Motors.** We can change the speed of slip ring motor by introducing resistance in the rotor circuit. This method is wasteful as the reduction in speed is proportional to copper losses taking place in rotor ckt. As such slip ring motors are used for small re-rolling mills where wide speed range is not required. Besides wastage of energy taking place in rotor introduction of resistance in rotor circuit converts the speed torque characteristics of Induction motor from shunt to series. Therefore speed can not be maintained constant from no load to full load. This will therefore cause the motor to fly off to no load speed as soon as stock passes out of the stand. This reflects very much on the quality of rolling. As such slip ring motors are suitable for roughing stands of the rolling mill.

Wastage of energy in the rotor of slip ring motor which is proportional to the reduction of speed can be avoided by employing cascaded induction motor. Rotor power at slip frequency is rectified by means of silicon rectifier. In case of constant torque cascading rectified power is fed back to the a.c. system by means of mercury arc rectifier operating as inverter. In the first case speed of the main motor is adjusted by varying the excitation of the separately excited d.c. motor. In the second case, the amount of electrical power fed back to the system can be regulated by the grid control of the rectifier. Rating of the equipment required for cascading as compared to d.c. motor becomes cheaper where speed variation desired is not more than 30%.

Synchronous Motor. Synchronous motors are used for driving constant speed continuous mills where load is maintained for relatively long periods, flywheel can not be used and motor carries the entire load. For slow speed synchronous motor can be built more economically than induction motors and we at the same time, get P.F. correction.

Besides, main mill drive motors are required for auxiliary drive such as for driving the rolls of roller table, adjusting the gap between the rolls, driving the manipulators for chopping the tail ends cutting the material to the required length, winding the end product into coil etc.

In most of auxiliary drives, a large part of the energy converted is used for acceleration and braking of the flywheel masses. It is therefore necessary that the inherent moment of inertia of motor is kept low. It is also desirable that these motors should be compact as in most of the cases, space available for installation is restricted.

Ship Propulsion. Except for very small vessels, electric drive for propeller of ship is universal. It has the following advantages :

1. Steam turbine or diesel engine has most economical speed of about 3000 r.p.m. and 400 to 800 r.p.m. Propeller speed on the other hand is in between 100 to 200 r.p.m. We can run propeller and prime mover independently. This enables prime mover to run at its most economical speed.

2. At the time of slow speeds we can run only few prime mover sets to supply to all the propellers. Low speed cruising is now no longer economical.
3. Speed control and reversing in case of electric propulsion are easy and can be carried out from any convenient position instead of signalling to engine room.
4. Position of prime mover is not fixed by the propeller shaft. As such there is greater flexibility in the layout.

In case of very big ships, electric drive will be by 3 - ϕ induction or synchronous motors. Prime mover will be steam turbine, driving 3 - ϕ alternator. Speed control is done partly by variation of applied voltage and partly by frequency variation. Reversal is obtained by changing the phase sequence. Speed of induction motor can also be changed by pole changing method which gives two economical speeds. Disadvantage of induction motor is that its P.F. at low speed is low between 0.7 to 0.75. Power at the slip is generated by turbo alternator at 2.2 to 6.6 kV.

Small ships can be driven by d.c. motors which are controlled by means of Ward Leonard method. Power in the ship is generated by diesel electric set.

Instead of constant voltage system we can use constant current system of supply to feed armature of propeller motors whose fields are supplied from constant voltage supply. This system of supply is inherently against overloads, provides maximum safe torque without any risk of interruption of supply. Motor control is exercised through field control.

In case of very small vessels complications of using electrical generator and motor are not warranted. Propeller shafts are directly coupled to diesel engine through slip coupling. Speed of propeller shaft can be controlled by varying d.c. excitation of the slip coupling.

Excavators. Due to arduous and rugged work, totally enclosed motors are used.

(*f*) **Mines.** Motor used inside mines should have flame proof enclosure. For mine winders it is essential to have both speed control as well as braking capacity. Ward Leonard lligner system is most effective drive. Slip ring induction motor with d.c. dynamic braking is also used. We may also use d.c. motors supplied from controlled rectifier.

(*g*) **Cement works.** Owing to dusty atmosphere, pipe ventilated or totally enclosed surface cooled motors are recommended.

(*h*) **Woolen Mills.** Loom motors are totally enclosed type capable of developing 300% torque at starting. It is direct driven or alternatively started light and controlled through clutch.

Q.4.20. Classify the nature of supply as A.C. & D.C.

Ans. (*a*) **Sources:**

In India 1-phase and 3-phase 50 Hz ac supplies are readily available in most locations. Very low power drives are generally fed from 1-phase source. Rest of the drives are powered from 3-phase source; except in the case of traction drives where even at very high power levels, 1-phase supply is used because of economy. Most drives are powered from a.c. source either directly or through a converter link. When fed directly from 50 Hz, a.c. supply, maximum speeds of induction and synchronous motors are limited to 3000 rpm. For higher speeds, conversion to higher frequency supply becomes mandatory. Low and medium power motors (tens of kilowatts) are generally fed from 400 V supply; for high ratings, motors may be rated at 3.3 kV, 6.6 kV, 11 kV and higher.

In case of aircraft and space applications, 400 Hz a.c. supply is generally used to achieve high power to weight ratio for motors.

In main line traction a high voltage supply is preferred because of economy. In India 25 kV, 50 Hz supply is employed. In underground traction, major expenditure is cost of the tunnel which should be minimised by keeping its cross-section just enough for the train. Con-sequently, clearance between live conductor and the earth has also to be minimum. In view of this, underground traction systems employ a low voltage (500 to 750 V) d.c. supply. In Western India (Bombay to Igatpuri) 1500 V d.c. is used for mainline and suburban traction which is uneconomical, and therefore, future installations will not use it.

Some drives are powered from a battery, *e.g.* fork lift trucks and milk vans. Depending on size, battery voltage may have typical values of 6 V, 12 V, 24 V, 48 V and 110 V d.c. Another example of drives fed from a low voltage d.c. supply is solar powered drives used in space and water pumping applications. These drives, though presently very expensive have a great future for rural water pumping and low power transport applications.

Although choice of a motor does depend on the type of supply but there are many other factors which are even more important. Therefore, **a d.c. motor may be preferred over a.c. even when**

the a.c. supply is available and a.c. motors may be preferred over d.c. even when the supply is d.c.

(b) **Status Of D.C. And A.C. Drives:**

In the past, induction and synchronous motor drives were mainly used in fixed speed applications. Variable speed application were dominated by d.c. motor drives. Emergence of thyristors in 1957 lead to the development of variable speed induction motor drives in late sixties which were efficient and could match the performance of d.c. drives. Consequently, because of the advantages of squirrel-cage induction motors over d.c. motors, it was predicted that induction motor drives will replace d.c. drives in variable speed applications. However, following hurdles forbade the prediction to come true:

(i) Although squirrel-cage induction motor was cheaper than d.c. motor, the converter and control circuit of an induction motor drive was very expensive compared to those for a d.c. drive. Therefore, total cost of an induction motor drive was significantly higher than that of a d.c. drive.

(ii) While the technology of d.c. drives was well established, that of a.c. was new.

(iii) a.c. drives were not as reliable as d.c.

(iv) Developments in linear and digital ICs and VLSIs were helpful in improving the performance and reliability of a.c. drives. But then these developments also led to similar improvements in d.c. drives.

Improvement in thyristor capabilities, availability of power transistors in early seventies and that of GTOs and IGBTs in late seventies and late eighties respectively; reduction in cost of thyristors, power transistors and GTOs: developments of VLSIs and microprocessors; and improvement in control techniques of converters have resulted into reduction in cost, simple controllers and improvement in performance and reliability for a.c. drives. Although even now majority of variable speed applications employ d.c. drives, the a.c. are preferred over d.c. in a number of applications with the result, a.c. drive applications are growing. Induction motor drives find applications in low to high power applications and synchronous motor drives are employed in very high power (megawatts) drives. The brushless d.c. motor is being considered for replacing d.c. servo motors for fractional hp range. As the trend exists applications of A.C. drives will continue to grow. However, D.C. drives will also continue to be used for quite some time.

5. ELECTRIC TRACTION

Q.5.0 Explain the electric traction, tractive effort & write the requirements of electric traction.

Ans. (1) Electric Traction

Thousands of commuters all over the world get the benefit of EMU (Electric Multiple Unit) services regularly. In suburban railways, EMU service is very useful because of its advantages of quick start and stop enabling a high speed schedule. Electric traction system is without a parallel for underground as well as main line service. The system is very efficient, clean and pollution-free. But the major disadvantage lies in its capital cost, which is very high for a metro railway system. Power failure even for a short period can lead to a complete disruption of the running schedule of the trains.

(2) **Tractive Effort.**

Tractive effort is applied at the rim of the driving wheel through gears to propel the train. The gear ratio and wheel diameter are fixed. So, the tractive effort is proportional to motor torque. To provide a large tractive effort, more than one motor *is* normally used. The total tractive effort depends upon the number of traction motors. The effective force, necessary to propel the train, at the wheels of vehicle is called the **tractive effort.** It is tangential to the moving wheels & is measured in newtons. It is a vector quantity.

(3) **Requirements of Electric Traction**

(*a*) Traction equipment should be robust and sturdy enough to withstand continuous vibrations, dust and humid environemnt.

(*b*) Power to weight ratio of the traction motor should be high so that it occupies less space.

(*c*) High tractive effort at starting.

(*d*) It should be possible to over load the motor for a short period.

(*e*) Ability of traction motors to apply regenerative braking during descent.

(*f*) Coefficient of adhesion should be high.

(*g*) The traction motors must be capable of withstanding voltage fluctuations and interruptions of power supply.

(*h*) The motors should be amenable to simple speed control methods.

Q.5.1. State the requirement of supply voltages for the various types of electrical traction applications.

Supply Systems for Transportation. The following three alternatives are used for the various traction systems in various parts of the world.

1. Direct current system.

2. Single phase alternating current system at low frequency $16^{2/3}$ or 25 Hz.

3. Single phase alternating current system at industrial frequency (50 Hz in India, 60 Hz in USA)

4. Three phase a.c. system at 50 Hz.

In India, the following two supplies have been standardized.

1. 1500 V d.c. for suburban railways in Mumbai.

2. 25 kV, 50 Hz. 1 Ph. for main line service and suburban railways in Delhi and other places.

Q.5.1.1 List the type of services (Main Line, Suburban and Urban). Sketch the speed time curves for the above services.

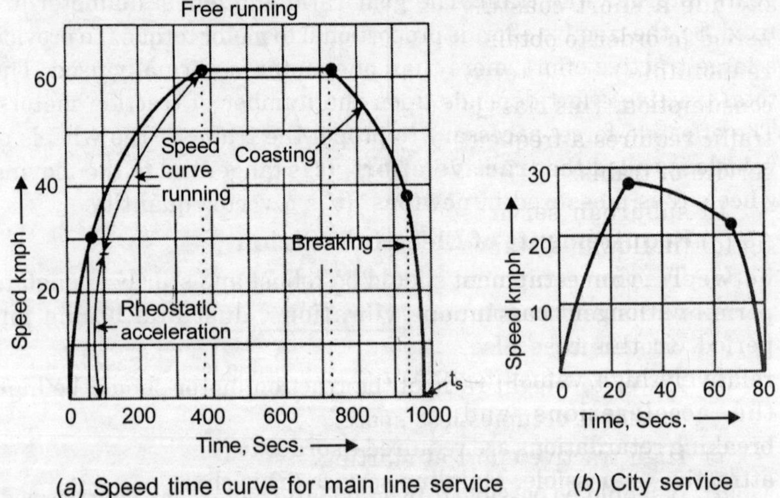

(*a*) Speed time curve for main line service (*b*) City service

Fig. 5.1.

Ans. Such type of curves for electric trains operating on passenger services are given in fig 5.1 a, 6 & c. Each curve corresponds to a particular class of traffic :

(1) Urban or city service. where the distance between the stop is of the order of say 1 km or so.

Suburban Service: Where the distance between the stops may average from 1.5 to 2 KM over a distance of 15 to 20 KM from the city terminous.

Main Line Service: Where the stops are infrequent.

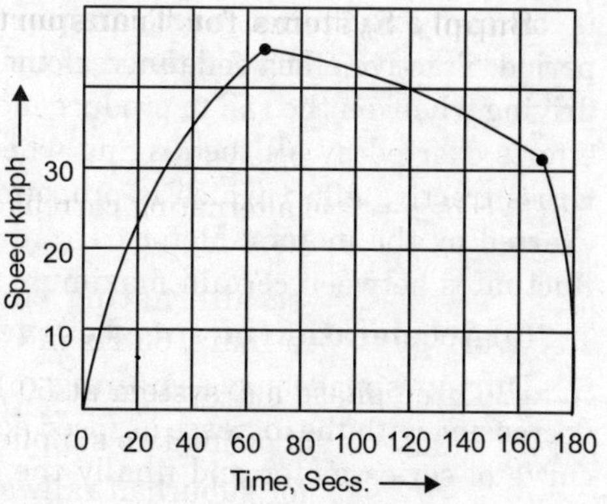

Fig. 5.1. (c) Suburban service

In fig. 5.1. (*b*) which is a representative of city service, relatively high values must be adopted for the acceleration and retardation in order to obtain a moderately high average speed between the stations. The short distance between the stations does not permit of a free running period since it is desirable to include a short coasting period in order to obtain a reasonable energy consumption. This class of traffic requires a frequent service of trains.

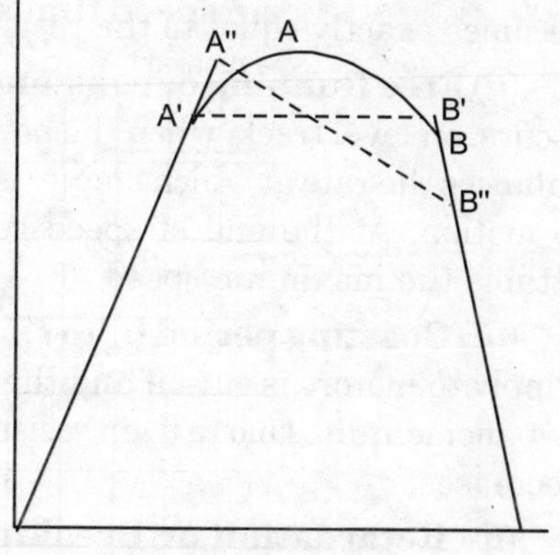

Fig. 5.1. (d)

In suburban services fig (*c*) the lower distance between the stations permits a longer coasting period. In this case also, relatively high values for the accelerations and breaking retardations are required in order to render the service as attractive as possible. Mainline service (*a*) is characterized by the long periods of free running at high speeds, the accelerating period being relatively un-important.

Q.5.2. State each stage of the speed-time curve with appropriate speeds used.

Ans. (*a*) **Acceleration while notching up or constant acceleration :**

During this period of run (o to t_1), starting resistance is gradually cut so that the motor current is limited to a certain value and the voltage across the motor is gradually increased. Thus tractive effort is constant and therefore acceleration remain, constant during this period. Tractive effort is defined as the force in newtons exerted by driving wheel on the rail to produce movement. Sum of the tractive efforts exerted by all the driving wheels of a locomotive is called Gross tractive effort which is propotional to the combined torque exerted by the motors. Motor current, during notching up period, fluctuates between certain maximum & minimum limits.

(*b*) **Acceleration on speed curve or speed curve running :**

During speed curve running (t_1 to t_2), the current starts decreasing with the increase in speed according to the characteristics curve of series motor and finally the current taken by the motor becomes constant. The voltage acting across the motor remains constant. During this period, though the train accelerates but acceleration decreases with the increase in speed & finally becomes zero at the speed at which the tractive effort developed by the motor becomes exactly equal to the resistance to the motion of the train.

(*ii*) **Free Running or constant speed run (t_2 to t_3):** This period occurs, on level track, when the power output from the driving axles balances the rate at which energy is expanded against the resistance to motion. At the end of speed curve running *i.e.* at t_2 the train attains the maximum speed.

(*iii*) **Coasting period (t_3 to t_4):** At the end of free running period, supply to motors is cut-off and the train is allowed to run under its own momentum. Due to train resistance, speed of the train gradually decreases.

(*iv*) **Retardation or Breaking period (t_4 to t_5):** At the end ofcoasting period brakes are applied to bring the train to stop.

Q.5.3 State the Importance of speed-time curves.

Ans. It is the curve drawn between instantaneous speed of train in kilometers per hour along Y-axis and time in seconds (or minutes) along X-axis. The speed-time curve gives complete information of the motion of the train. This curve gives the speed at various times after the start of run directly.

Further the distance travelled by the train during a given interval of time can be obtained by determining the area between the curve and the time axis corresponding to this interval.

Such speed time curves are of fundamental importance in electric traction, as when speed time curve is given and the resistance to the motion of train or vehicle are known, the energy required for propulsion can be calculated. Moreover when the characteristic curves of motor are known the energy input to the motor or vehicle can be calculated.

Q.5.4 Define Max. speed, average speed and scheduled speed.

Ans. (1) **Max. Speed:** The maximum speed attained by the train during the run is known as crest speed.

(2) **Average speed:** The mean speed from start to stop,

$$i.e. \ \frac{\text{Tolal distance (distance between stops)}}{\text{Running time (excluding stops)}}$$

(3) **Scheduled speed:** The ratio of distance covered between two stops and total time of run including time of stop is known as scheduled speed.

$$\text{Mathematically Scheduled speed} = \frac{\text{Total distance}}{\text{Total time including stops.}}$$

Q.5.5 List the factors affecting the scheduled speed.

Ans. Following are the factors which affect the scheduled speed of a train engaged in given service (*i.e.* with a given distance between stations)

(*i*) Acceleration and breaking retardation (*i.e.* value of α & β)

(*ii*) Maximum or crest speed, V_m

(*iii*) Duration of stop

For a given distance between the stations, greater the value of acceleration and retardation, less will be the actual time of run and more the schedule speed. Similarly for given distance between stops and fixed value of acceleration and retardation, greater the value of maximum speed reached, less will be the time of run and more the schedule speed. Average speed remaining same, increase in duration of stop decreases the scheduled speed. Station stoppage time has very much effect on the schedule speed in case of urban or suburban services where running time itself is small. It is due to this reason that station stops in the case of urban and suburban services are kept very small. When the distance between the stations is much and duration of actual run long, station stop has less conspicuous effect on the scheduled speed.

Q.5.6 Explain the simplified speed time curves and explain the practical importance of the above curves.

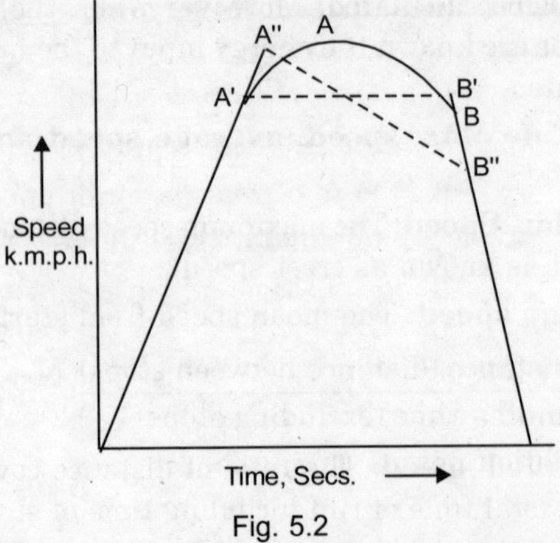

Fig. 5.2

Ans. Simplified speed time curves:- When comparative performances for a given service at various schedule speeds are required (for *e.g.*, in preliminary calculations for time tables etc.) the actual speed time curves of fig. 5.1 (a, *b, & c)* are replaced by simplified speed time curves, which do not involve a knowledge of the motor characteristics. These simplified speed time curves have simple geometric shapers and in consequence, the relationship between the acceleration, retardation, average speed and distance can be deducted by simple mathematics.

Trapezoidal speed time curve gives closer approximation of the conditions of main line service, where long distances are involved and quadrilateral speed time curve for urban and suburban services.

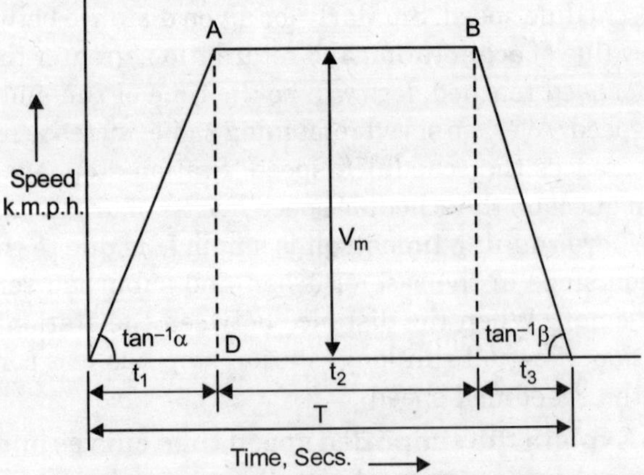

Fig. 5.3. Trapezoidal speed time curve

Q.5.7. Derive the expression for max. speed, acceleration and retardation.

Ans. S = Distance between stops in km

α = Rate of acceleration in km/hr/sec.

β = Rate of retardation in km/hr/sec.

V_m = maximum speed in km/hr.

T = Total time of run in ses.

$t_1,$ acceleration period or time in ses $= \dfrac{V_m}{\alpha}$

$t_3,$ retardation period or time in ses $= \dfrac{V_m}{\beta}$

$t_2,$ time for free running in ses $= T - (t_1 + t_3) = T - \left[\dfrac{V_m}{\alpha} - \dfrac{V_m}{\beta}\right]$

Total Distance of Run in km,

S = Distance travelled during acceleration + Distance travelled during free run + Distance travelled during braking.

Since the area under a speed-time curve represents the distance travelled

Hence, the total distance in km, S = Area$OABC$

= Area of Δ OAD + Area of rect. $ABED$ + Area of Δ BEC

$$= \frac{1}{2}\frac{V_m \cdot t_1}{3600} + \frac{V_m \cdot t_2}{3600} + \frac{1}{2}\frac{V_m \cdot t_3}{3600}$$

Substituting the value of t_1, t_2 and t_3 we have,

$$\text{S} = \frac{1}{2}V_{m,}\ \frac{V_m}{\alpha \cdot 3600} + \frac{V_{m.}}{3600},\ t_2 + \frac{1}{2}V_m \cdot \frac{V_m}{\beta 3600}$$

$$\therefore T - \left(\frac{V_m}{\alpha} + \frac{V_m}{\beta}\right) = t_2$$

$$= \frac{V_m^2}{7200\ \alpha} + \frac{V_m T}{3600} - \frac{V_m^2}{3600\alpha} - \frac{V_m^2}{3600\ \beta} + \frac{V_m^2}{7200\ \beta}$$

$$= \frac{V_m T}{3600} - \frac{V_m^2}{7200\ \alpha} - \frac{V_m^2}{7200\ \beta} \qquad \text{....(A)}$$

$$= \frac{V_m^2}{3600}\left[\frac{1}{2\alpha} + \frac{1}{2\beta}\right] - \frac{V_m T}{3600} + S = 0$$

$$= V_m^2\left[\frac{1}{2\alpha} + \frac{1}{2\beta}\right] - V_m T + 3600\ S = 0$$

This is a quadratic eqn. for V_m

Taking $\frac{1}{2\alpha}+\frac{1}{2\beta} = K$, we get,

$KV_m^2 - V_m T + 3600\,S = 0$

or $V_m = \text{T} \pm \sqrt{\frac{T^2 - 4K \times 3600\,S}{2K}}$

or $= \frac{T}{2k} \pm \sqrt{\frac{T^2}{4K^2} - \frac{3600S}{K}}$

The +ve sign gives very much high value of Vm which is not possible in practice

Hence -ve sign is adopted. ...[C.T.M.]

$$Vm = \frac{T}{2K} - \sqrt{\frac{T^2}{4k^2} - \frac{3600\,S}{k}}$$

$$S = \frac{V_m T}{3600} - \frac{V_m^2}{7200}\left[\frac{1}{\alpha}+\frac{1}{\beta}\right] \text{ considering eqn (A)}$$

Taking R.H.S. to the L.H.S., we get,

$$= \frac{V_m^2}{7200}\left[\frac{1}{\alpha}+\frac{1}{\beta}\right] - \frac{VmT}{3600} + S = 0$$

or $= V_m^2\left[\frac{1}{\alpha}+\frac{1}{\beta}\right] - 2\,VmT + 7200\,S = 0$

If $\left[\frac{1}{\alpha}+\frac{1}{\beta}\right] = K$ (Pl. note)

Then, $KV_m^2 - 2\,TVm + 7200\,S = 0$ ....(1)

This is quadratic eqn. for V_m,

$$V_m = \frac{2T \pm \sqrt{(2T)^2 - 4K \times 7200\,S}}{2K}$$

So, $V_m = \frac{T}{K} + - \sqrt{\frac{T^2}{K^2} - 7200\frac{S}{K}}$

Positive (+ve) sign gives much high value of V_m which is not possible in practice. Hence negative (-ve) sign is adopted

$\therefore$ $V_m = \frac{T}{K} - \sqrt{\frac{T^2}{K^2} - \frac{7200\,S}{K}}$

Eqn. (1) can be put as,

$$KV_m^2 - 2\left(\frac{3600S}{Vav}\right)V_m + 7200S = 0$$

$$V_{av} = \frac{3600\ S}{T}$$

$$\therefore \qquad T = \frac{3600S}{Vav}$$

$$\text{or} \qquad KV_m^2 = \frac{7200\ S.V_m}{V_{av.}} - 7200S = 0$$

$$\text{or} \qquad K = \frac{7200\ S.V_m}{V_{av.}V_m^2} - \frac{7200.\ S}{V_m^2} = \frac{7200.S}{V_{av.}V_m} - \frac{7200\ S}{V_m^2}$$

$$= \frac{7200\ S}{V_m^2}\left[\frac{V_m}{V_{av}} - 1\right]$$

$$\frac{1}{\alpha} + \frac{1}{\beta} = \frac{7200\ S}{V_m^3}\left[\frac{V_m}{V_{av}} - 1\right] \qquad \text{....(2)}$$

$$\left[\text{where } V_{av} = \text{Average velocity} - \frac{S}{T/3600} = \frac{3600\ S}{T}\right]$$

If V_{av}, V_m and S are given in a particular case, α & β can be adjusted by using eqn (2), to suit the traction motor characteristics and convenience of passengers respectively.

Eqn (2) can be put in the following formulae,

$$V_m = \sqrt{\frac{7200\ S}{K}\left(\frac{V_m}{V_{av}} - 1\right)}$$

If S and K are known, by choosing a suitable value for (V_m / V_{av}), we can predict the value of V_m which can be used for elementary calculations. The value (V_m / V_{av}) depends on the shape of trapezoid.

Q.5.8 Derive the expression for tractive effort for acceleration, to overcome gravity pull and train resistance and calculate the tractive effort under given conditions.

Ans. The total tractive effort required to run a train on track = Tractive effort during acceleration + Tractive effort to overcome the effect of gravity + Tractive effort used in overcoming frictional resistance

i.e. $$F_t = F_a + Fg + F_r$$

(*a*) **Tractive effort for acceleration** (F_a). Let the wt. of the train be *W* tonnes. If it is to be accelerated at a rate of ∝ km/hr/sec. then the accelerating force or tractive effort reqd. for linear acceleration, F_a will be given by,

$$F_a = m. \alpha \text{ (Mass × Acceleration)}$$

Where wt. of train = 1000 W kgf

Mass of the train = 1000 W kg

Acceleration, ∝ in km. p.h.p.s = $\dfrac{\alpha \times 1000}{3600}$ m/sec^2

Putting the values, we get

i.e. $F_a = m\,\alpha = 1000\,\omega \times 0.2778\,\alpha$

$= 277.8\,W\,\alpha$ Newtons.

When a train is accelerated in a linear direction, its rotating parts like the wheels and the armatures of motors have to be accelerated in an angular direction. Therefore, the accelerating mass of the train is greater than the head mass of the train by 8 to 15%. For most of the trains the effective or accelerating weight, W_e. is 1.1 times their dead wt., ω.

hence $F_a = 277.8\,W_e\,\alpha$ Newtons

(*b*) **Tractive effort to overcome the effect of gravity or to move the train up a gradient,** *(Fg):* The dead wt. of the train w tonnes can be assumed to be made up of two parts viz. ω cos θ and ω sin θ, where θ is the angle of inclination of the track to the horizontal (Fig. 5.4)

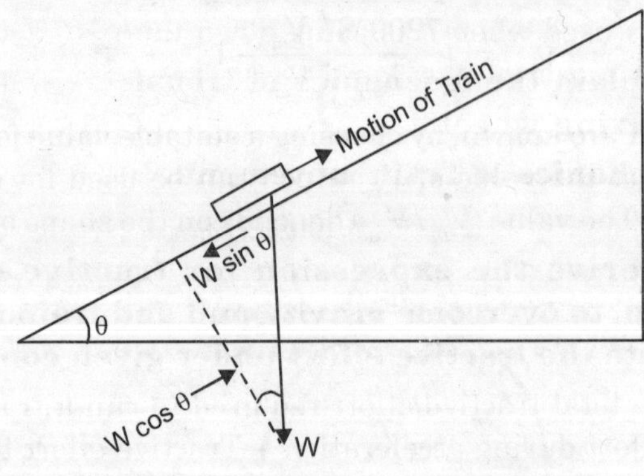

Fig. 5.4.

Since the component ω cos θ is perpendicular to the direction of motion of train, it has no effect on its motion but the other component ω sin θ tries to oppose the motion of train. Hence an additional tractive effort. F_g = ω sin θ tonnes must be provided for running the train on the inclined track.

If G = Percent gradient of the track

Then $G = 100 \sin \theta$

The tractive effort, F_g (in newtons) becomes

$$F_g = \pm 1000\ W \sin \theta \times 9.81 \text{ Newtons}$$
$$= \pm (100 \sin \theta) \times 98.1\ W \text{ Newtons}$$
$$= \pm G \times 98.1 \text{ W Newtons}$$
$$= 98.1\ WG \text{ Newtons.}$$

If the train is moving down the gradient then the component of dead wt. along the track *i.e.* $W \sin \theta$ will help the motion of the train.

(*c*) **Tractive effort to overcome the fricitional resistance (*F_r*):** If r is the specific train resistance in Newtons per tonne of the dead wt, then the tractive effort required to overcome this resistance,

i.e. $F_g = W \times r$ Newtons

where r is the function of velocity for a given train

For a normal train the value of specific resistance has been 40 to 70 newtons tonnes

Hence the total tractive effort required,

$$F_t = F_a + - F_f + F_r$$
$$= 277.8\ We\ \alpha + 98.1WG + W_r$$

The +ve sign is used when the train is going up the gradient and the -ve sign *is* used when it is going down the gradient.

Q.5.9 Explain the mechanics of transfer of power from motor to driving wheel.

Ans. Mechanics of Train movement:

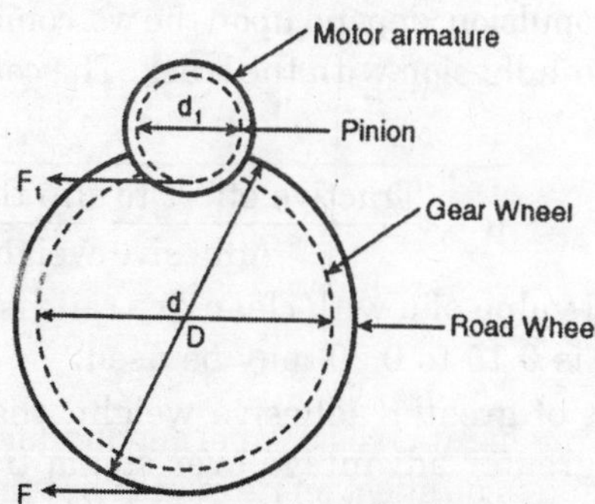

Fig. 5.5 Transmission of tractive effort.

Consider the essential driving mechanism of an electric vehicle. The armature of the driving motor has a pinion of diameter, d_1, meshed with the gear wheel on the driving wheel.

Let the driving motor exert a torque, T(in $N-m$) which will be transferred to the road wheel through the gear. F_1 *is* the tractive effort at the pinion.

Tractive effort at the edge of the pinion is given by the expression,

$$T = F_1 \times \frac{d_1}{2} \text{ and } F_1 = \frac{2T}{d_1}$$

Tractive effort transferred to the driving wheel,

$$F = \eta F_1 \left(\frac{d}{D}\right) = \eta. \frac{2T}{d_1}\left(\frac{d}{D_1}\right) = \eta.T\frac{2\gamma}{D}$$

$$\text{n.T.}\frac{2}{D}\left(\frac{d}{d_1}\right)$$

η = efficiency of the transmission

γ = is the gear ratio = d/d_1

d = dia. of gear wheel

D = dia. of driving wheel

The max. frictional force between the wheel and the track = μW, where μ is the coefficient of adhesion between the wheel and the track and ω is the wt. of the train on the driving axles. Slipping will not take place unless tractive effort, $F > \mu W$. For motion of trains without slipping, tractive effort, F should be less than or at the most equal to μW but in no case greater than μW.

Q5.10 Define Cofficient of adhesion.

Ans. The magnitude of tractive effort which can be usually employed for propulsion, depend upon the wt. coming over the driving wheels and their adhesion with the track. The cofficient of adhesion is the ratio,

i.e. $$\mu = \frac{\text{Tractive effort to slip the wheels}}{\text{Adhesive weight}}$$

The normal value of μ with clean dry rails is 0.25 and with wet or greasy rails is 0.15 to 0.20 may be used.

On account of greater adhesive weight, that electric traction possesses is a greater advantage over steam traction. In a motor coach 100 percent of the wt. is on the driving wheels, 70% in an electric locomotive and only 50% or less in a steam locomotive. The coefficient of adhesion is also greater in electric traction than in a steam traction for the following reasons:

(*a*) In electric traction the torque exerted is continuous whereas in steam traction the torque is pulsating which causes jolting and skidding.

(*b*) The driving wheels are distributed over much greater length in the case of electric traction whereas in steam traction they are very close to each other. The tractive effort in electric traction is much greater, therefore, electric train can be made to accelerate at a faster rate. This results in saving of time, especially when the distance between the stops is smalls *e.g.*, urban and suburban areas.

Q.5.11 Explain the term specific energy consumption & factors affecting specific energy consumption.

Ans. The total energy input to all the traction motors in a train is known as the "energy consumption" of the train. The specific energy consumption is the energy consumption in watt-hr., per tonne-km.

Hence, specific energy consumption,

$$= \frac{\text{Specifie energy output}}{\text{Efficiency of gear} \times \text{Efficiency of motor}}$$

$$= \frac{\text{Energey output in watt–hours}}{\eta\text{gear}\times\text{motor (wt. of train in tonnes}\times\text{Distance of run in km}}$$

Factors affecting specific energy consumption

The following are the factors to be considered for calculation of specific energy consumption:

(1) Distance between stops

(2) Acceleration

(3) Retardation

(4) Maximum speed

(5) Type of train and equipment used

(6) Nature of track

At a given scheduled speed, the specific energy consumption will be lower for higher acceleration and retardation, as this will increase the coasting period. Sufficient precaution should be taken to limit the weight of the equipment and its effect on the energy consumption. Also the overall efficiency will affect the specific energy consumption.

Q.5.12 State the Important requirements of traction motors.

Ans. (*i*) Maximum tractive effort should be exerted at starting in order that a rapid acceleration may be attained.

(*ii*) The equipment should be capable to bear overloads for short periods.

(*iii*) The wear caused on the track should be minimum.

(*iv*) The locomotive of the train unit should be self contained and able to run on any route.

(*v*) Braking should be possible without excessive wear on brake shoes and if possible the braking energy should be ' regenerated and returned to the supply.

Q.5.13 Explain with neat sketch the control of traction motor by auto-transformer method in single phase 25-KV system.

Ans. Tap changing on H.V. traction system: Tap changing on high voltage A.C. traction system at power frequency is generally through an auto-transformer connected to the traction supply as shown in fig 5.6.

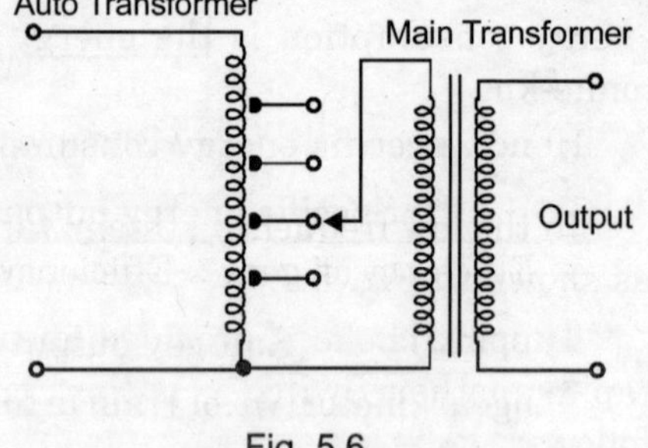

Fig. 5.6.

The odd no. taps. (1, 3, 5) are connected to one side of the tapping choke, while the even nos. to the other side

Full wave rectifiers are provided for D.C. supply to the traction motors. They are connected to the secondary side of the main transformer. Smoothing chokes are provided to reduce the ripple currents.

Voltage variation in A.C. systems: In A.C. system, the voltage is varied to control the starting current as well as for speed control. This is true for either system, whether at power frequency or at a lower frequency. A transformer is provided in the locomotive for stepping down the supply voltage. Tappings are provided on the secondary side for varying the voltage.

For high voltage tap changing, an additional transformer is generally employed. As the current in the primary wdg. is lower, more tappings can be provided on the H.V. side of the main transformer; because in that case to get a smaller voltage at start, the no. of turns on the primary winding will have to be large.

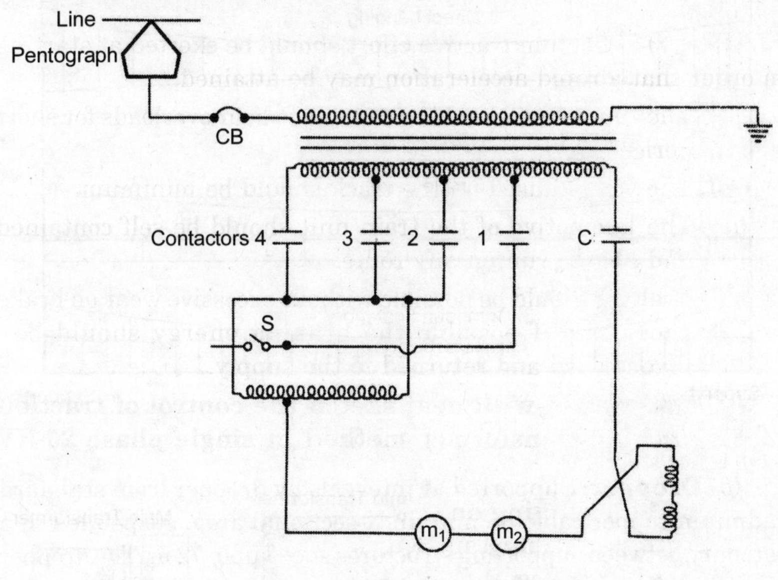

Fig. 5.7

In the low frequency system, taps are provided on the L.V. side as shown in Fig. 5.7.

Tapping choke is provided for a smooth change-over from one tap to another. Switch, *S* is short circuiting it. The operation is as follows:

On tap 1, contactor 1 & Switch, S is closed, 2, 3, 4 are open.

For change to tap 2, Open *S*, Close 2, open 1, Close *S*,

While the tap is changed, the tapping choke tend to stabilize the current.

Q.5.14. Explain the purposes and material used for:

Ans. (a) **Catenary:-** This is an overhead wire of special X-section. The catenary is supplied at 25 kV 50 c/s supply A.C. High voltage is tapped from catenary by pentograph.

Single Catenary Construction: Single Catenary construction is a polygonal system as employed on 25 kV single line as shown in Fig. 5.8 drawn below.

It consists of a solid grooved copper or cadmium copper wire 107 mm^2 in X-sectional area.

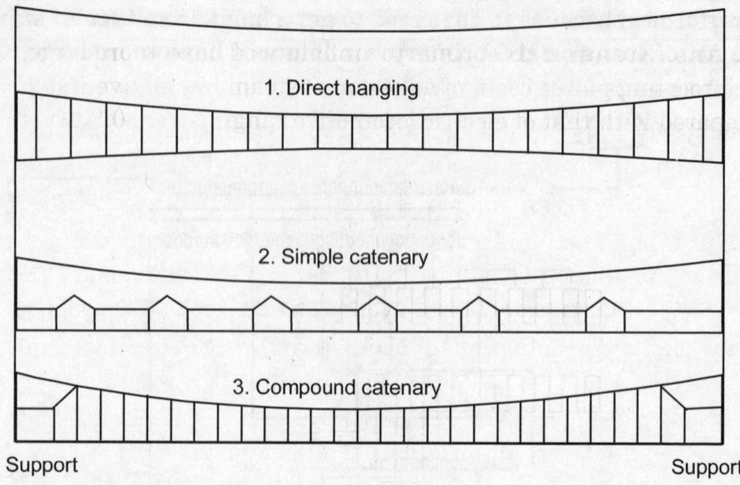

Fig. 5.8

(*b*) **Dropper:** Supported at intervals by dropper from standard cadmium copper cable 65 mm^2 in X-sectional area, suspended in a catenary between supporting structures speed upto 72 m. The dropper spacing is 9-m except the 1st and 2nd droppers which is 2.25 m and 6.75 m from support respectively.

(*c*) **Trolley wires:** The trolley wires must be suspended with the minimum of sag so that it remains practically horizontal. Small spans have therefore to be used and if they are not possible, same method is employed.

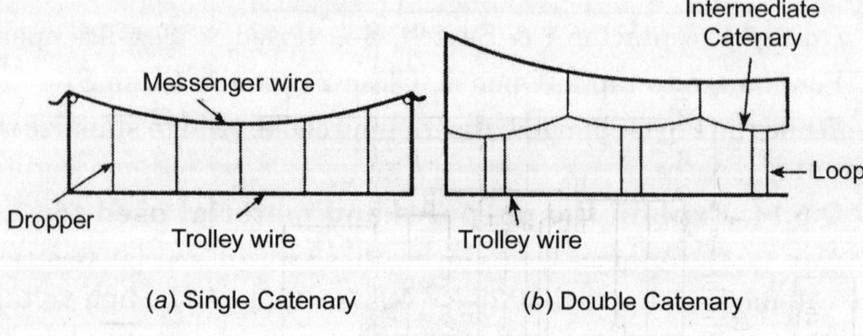

(*a*) Single Catenary (*b*) Double Catenary

Fig. 5.9

Q.5.15 Draw Electric wiring block diagram for an air conditioned coach.

Ans. Fig. 5.10

Q.5.16. State the typical values of coeff. of adhesion for steam, Diesel and electric drive.

Ans. Steam drive. Due to unbalanced forces produced by reciprocating masse coeff. of adhesion of steam locomotive is 0.25 as compared with that of electric locomotive ranging from 0.3 to 0.45.

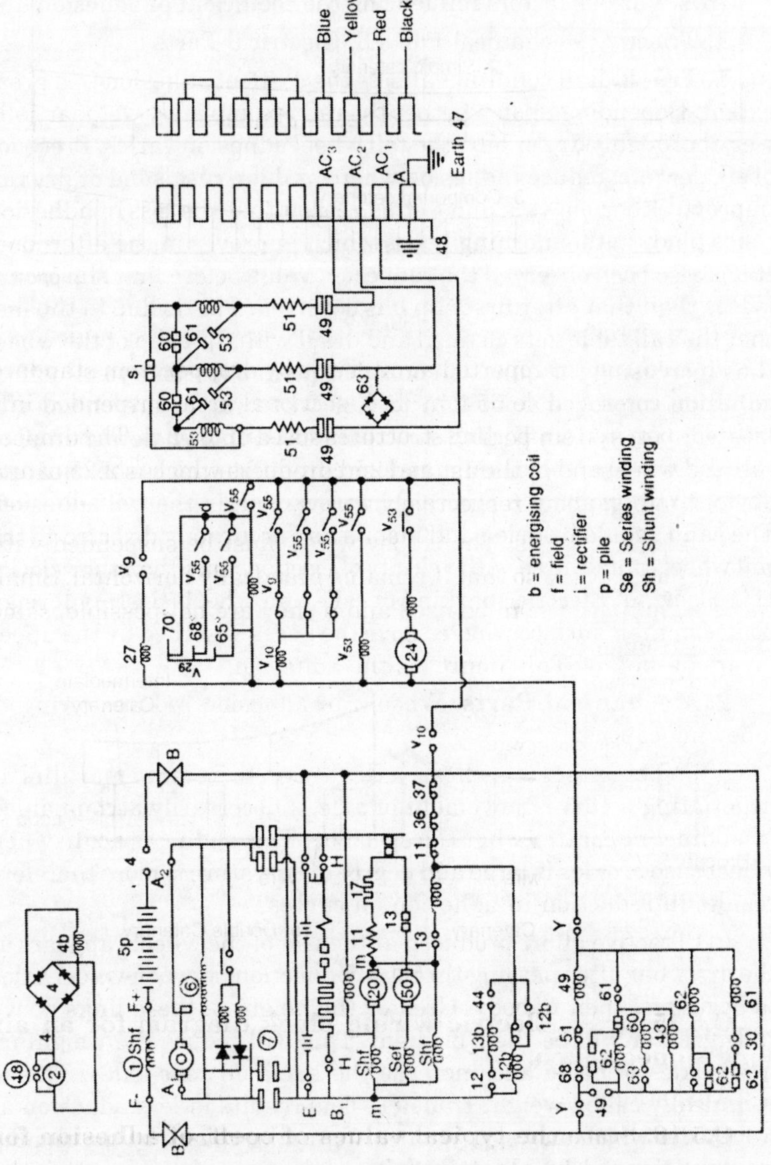

Fig. 10. Schematic diagram for air conditioning equipment

Electric drive. Due to absence of unbalanced forces produced by reciprocating masses coeff. of adhesion is more in case of electrici traction.

Q.5.17 List the factors affecting the coefficient of adhesion.

Ans. Various factors influencing the coefficient of adhesion are:

1. Track 2. Mechanical Parts 3. Electrical Parts

1. Track. Rail condition affects the coeff. of adhesion to a great extent. Depending upon whether the rail is completely wet, partially wet, thoroughly dry or oily, the tail wheel adhesion varies. Presence of oil, dew etc. reduces adhesion where as dust, rust, sand or dry rail improve adhesion. It is due to this reason that the loss of adhesion takes place in the morning hours which improves in the afternoon. It has also been observed that adhesion value before first slip occurs is less than that after first slip has occurred. This is due to the fact that the rail table gets cleaned and dried with first slip of the wheel thus increasing the adhesion on subsequent slips. In a test coeff. of adhesion improved from 0.25 to 0,33 when rails were chemically treated with system W_2O. Sand treatment although detrimental to rail and wheel and bringing track circuit difficulties is the cheapest device to bring about temporary improvement in the rail adhesion. The sand should be dried and mobile and accurately directed to the rail wheel contact area with velocity of ejection more than the velocity of cross wind. Electric sparking of rails and wheels destroy foreign bodies in their surfaces and remove the gases absorbed by the upper layers of metals. This improves the adhesion.

2. Mechanical Parts. Wheel rail adhesion reduces owing to following causes:

(i) When a bogie, whose axles have to remain parallel is negotiating a curve, movement of axle is necessarily accompanied by sliding, no matter what the speed be. This sliding is acute when wheel base of axles is large and degree of curvature is more therefore results in reduction in adhesion on curves.

(ii) Tractive effort produced at the rim of the wheels appears at the draw bar after passing through connection between wheel axles to bogie and then to body. Greater the height of these links above rails greater will be their moment. This will reduce the weight on front axles of bogie and increase the same on rear axles. This is technically called weight transfer. This results in less adhesion at front axles which may slip. Wt. transfer is also caused by the turning moment exerted by the motor. By employing what is technically called low traction the effect of weight transfer can be reduced.

(*iii*) Under the effect of vertical shock, extent of surface of contact between wheel and rail depends upon the elastic reaction of suspension. Better the elastic suspension and damping arrangement in bogie more will be the coeff. of adhesion.

(*iv*) Greater the moment of inertia of revolving masses achieved by coupling the axles by means of coupling rod or coupling gear wheels in monomotor bogie greater will be the adhesion and less will be the slip.

(*v*) For unequal static loading of axles, tendency of slipping or skidding is pronounced in the case of least loaded axles.

(*vi*) Bogie pitching oscillations cause appreciable variations of axle loads. This reduces adhesion.

3. Electrical Parts. It is due to the part played by electrical parts in affecting the adhesion that it has been possible to have light a.c. electric locomotive for the same tractive effort produced as compared to d.c. electric locomotive while d.c. locomotive is capable of hauling 1000 ton trailing load on a gradient of 10 in 1000. Its a.c. counterpart of equal weight can haul a trailing load of 1600 ton under same conditions.

1. Shape of tractive effort speed curve. In case of d.c. locomotives, motors are started by inserting resistance in series with the armature of motors. Insertion of resistance makes the torque speed curve flatter. In case of a.c. locomotives reduced voltage at starting as applied by step down T/F with the help of on-load tap changer and since no resistance is connected in series with armature speed torque curve is more steep as compared to that of d.c. locomotive slope of speed torque curve has an important effect on the slip and thereby on the adhesion. In case of a.c. locomotive on occurrence of slight slip torque developed is reduced by larger amount because of the steepness of the curve. This reduction of torque is such that new tractive effort is below the adhesion limit and this registers grip on the track. In case of d.c. locomotive since speed torque is more flat consequent to the occurrence of initial slip there will not be as much reduction of tractive effort now produced may still be above adhesion limit.

Q.5.18 State the method of improving co-efficient of adhesion.

Ans. 1. One motor in the leading bogie and one motor in the tracting bogie are permanently connected in series. The resulting 2 groups of motors are controlled in series parallel system. Tractive effort developed by the motor on the leading bogie is reduced to arrest the slipping tendency by reducing its field flux by providing shunting resistors across field.

2. In another design the armature of both the motors of the bogie are connected in series. The two motors combinations so formed are then operated in series or parallel during starting. Slipping of the motors of the leading bogie is prevented by shunting their armatures when the motors are in series. Current through all the motors is same but the current through the armature of motors of leading bogie is less than the torque produced by the leading bogie is less than the trailing bogie which prevents slip.

3. Yet in another method two motors of each bogie are connected in series and resulting combination group permanently in parallel. The starting of all the motors is done simultaneously by common resistance group. The field of the motors on trailing bogie are shunted by resistances. This increases the speed of motors on trailing bogie as compared to that of motors on leading bogie and prevents slipping tendency on the part of the latter.

Q.5.19. Drive the formulae for energy output of drive axles (*i*) to accelerate (*ii*) to over-come friction (*iii*) to over-come gradient.

(*a*) Energy output of driving axles to accelerate the train (E_a),

$$E_a = F_a \times \text{distance}$$

$$= 277.8\, W_e\, \alpha \left[\frac{1}{2}\frac{V_m \times 1000}{3600} \times \frac{V_m}{\alpha}\right] \text{ watt-sec}$$

$$= 277.8\, W_e\, \alpha \left[\frac{1}{2}\frac{V_m \times 1000}{3600} \times \frac{V_m}{\alpha}\right] \frac{1}{3600} \text{ watt-hours}$$

$$= 0.01072\, V_m^2\, W_e \text{ watt hour.}$$

(*b*) Energy output of driving axles to overcome friction,

$$E_r = F_r D_1\ 1000 \text{ Joules}$$

$$= \frac{W.r.D_1 1000}{3600} \text{Whr.} = 0.2778 \text{ W.r. } D_1 \text{ Whr.}$$

(*c*) Energy output of driving axles to overcome gradient,

$$E_g = F_g \times D_1$$

where D_1 is the distance over which power remains on and its maximum value is equal to area.

$$E_g = 9.81 \text{ WG} \times 1000\, D_1 \text{ Joules or watt sec.}$$

$$= \frac{9.81 \times 1000}{3600}\, WG\, D_1 \text{ Whrs.}$$

$$= 2.725\, WG\, D_1 \text{ Whrs.}$$

Q.5.20. Explain the suitability of different motors for traction.

Ans. Series motor:

1. $T \propto \frac{1}{N}$ where T is torque and N is speed. When speed is high torque is low or vice versa. This is exactly the requirement of traction units. High starting torque developed is needed to accelerate vehicle.
2. N-T and N-I characteristics of series motor are steep. Therefore for a given difference in rotational speed of various motors running in parallel caused by unequal wear of different driving wheels torque developed and current taken by each motor will not differ very much.
3. Commutating properties of series motor is also very good.
4. Series speed torque curve leads to stability in control.
5. Series motors of a given rating is capable of withstanding more heavy load torque.
6. $HP \propto T \times N$. But in series motor, $N \propto \frac{V}{I_a}$

$$\therefore HP \propto T.\frac{V}{I_a} \text{ but } T \propto I_a^2$$

$$HP \propto \sqrt{T}.V$$

7. Speed can be controlled by various methods.
8. These motors are not suitable for regenerative braking. D.C. series motors are better suited for urban and suburban services.

Shunt motors:

1. N-T characteristic does not match with the traction required.
2. N-T characteristic and N-I characteristics of shunt motors are flat so these are not suitable for parallel running.
3. With the increase in load speed decreases only very little. Motor therefore besides getting overload develops communication troubles.

4. Current drawn by the shunt motors is more as compared to that by series motor.
5. H.P. of shunt motor is directly proportional to torque. Therefore, for increase in torque, H.P. of motor will increase.

Q.5.21. Explain the following types of transmission of drive in traction.

(*a*) Direct drive (*b*) Nose-suspension gear drive.

(*a*) **Direct Drive.** In direct drive armature of the motor is directly mounted on the driving axle. Field poles, two in number along with yoke, are attached with the locomotive flame. In order that armature should not foul with pole faces due to vertical relative motion between the two, latter is made with flat faces. Disadvantages of this method of drive are : size of the armature of the motor is limited by the size of the driving wheel, more of unsprung weight of the motor and low centre of gravity, give rise to poor riding qualities besides large wear and tear of the track. In the absence of any reduction in speed, motor speed has to be low which makes the motor more bulky and costly. This method of drive is obsolete today and we have dealt with it for its historical interest only. There are however other examples to direct drive in which attempt has been made to do away with some of the disadvantages enumerated above.

(*b*) **Nose suspension gear drive.** By employing speed reduction gears, it is possible to select a high speed motor. This results in the reduction in physical size of the motor for the given H.P. The gear ratio depends upon the maximum safe speed of the motor and maximum designed speed of the vehicle. In this type of drive one end of the motor is supported on the axle by means of two plain bearings carried in the motor frame and the other end carries a nose or projection resiliently mounted on the bogie cross member. This arrangement maintain fixed centre to centre distance between armature shaft and axle and motor swings about the nose support as axle rises or falls in suspension on the bogie. Nearly 60% weight of the motor is spring borne. Remaining 40% unsprung weight on the axle gives rise to the hammer blow to the track. Motor itself is also subjected to direct shock from the track. Pinion on the motor engages with the gear wheel mounted on the axle. Resilient gear wheel is sometimes used to give some flexibility. In resilient gear wheel loose rim is connected to the fixed hub through rubber or spring mounting. In this method of drive, limitation is imposed on the overall size of the motor due to its one end being supported. This method of motor suspension is simple and robust and is used for low speed traffic such as in motor coaches and tramways. For high speed it is necessary that whole of the wt. of the motor should be spring borne for which more elaborate systems of motor mounting are employed.

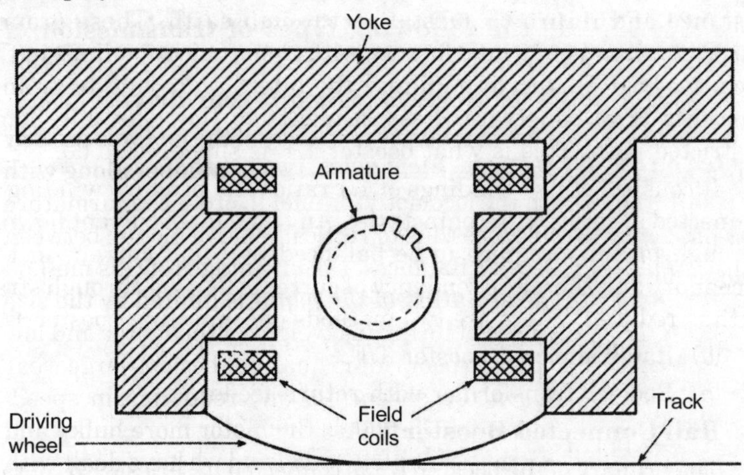

Fig. 5.11

Q.5.22. Materials required for following:

Bow collector, Pantograph collector.

Ans. Bow collector. Main advantage of bow collector is that it can be used for higher speeds. Bow collector consists of two trolley collector poles at the end of which is placed a light metal strip upto one meter long for current collection. Although provision of metal strip enables vehicle to run at higher speeds without running the risk of leaving the contacts of trolley wire. Its manoeuvrability is lost. It *is* as such not suitable for trolley buses. On tramways services trolley wire is hung at the centre line of the track with about 15 cm stagger to give uniform wear at the strips and prevent formation of groove on it. Collection strip is of soft material such as Cu, Al, so that it should wear instead of trolley wire,

Panto-graph collector. The collector strip material and contact pressure are such as to ensure minimum wear of contact wire.

Collector material must have long life itself and collecting high powers without overheating. Collector strip *is* made of such as Cu, Cu-steel ; silicon Mn steel, Al, steel or Bronze. This however requires some lubrication such as grease containing graphite etc. Difficulty of lubricating is overcome by the use of metallized carbon which is self lubricant.

Q.5.23. State the need of booster Transformers. Sketch the installation of the boosters by a line diagram.

Ans. In a.c. traction system return current which flows from the locomotive to the track soon leaks to the ground within short distance and return to substation through earth. These ground currents cause heavy interference with communication lines in the vicinity. Logical step to minimise the interference due to ground currents is to constrain these currents to flow through some restricted path. This is what booster T/F is supposed to do.

It consists of two windings of 1:1 ratio. The primary winding is connected in series with contact wire. Any amount of current flowing through primary require to be balanced by equal current in the secondary and therefore tendency of currents flowing through stray path is reduced. There are two methods of connecting booster T/F.

(*a*) Rail connected booster T/F.

(*b*) Booster transformer with return feeder.

Rail Connected Booster T/F.

The primary of the booster T/F *is* connected in series with contact wire and secondary in series with rails. The induced voltage into the secondary constrain the return current to flow through rails insulated from each other.

This method of connecting booster T/F suffers following drawbacks:

(i) All location of booster T/F we have to provide insulated rail joint with small neutral section in the OHE.

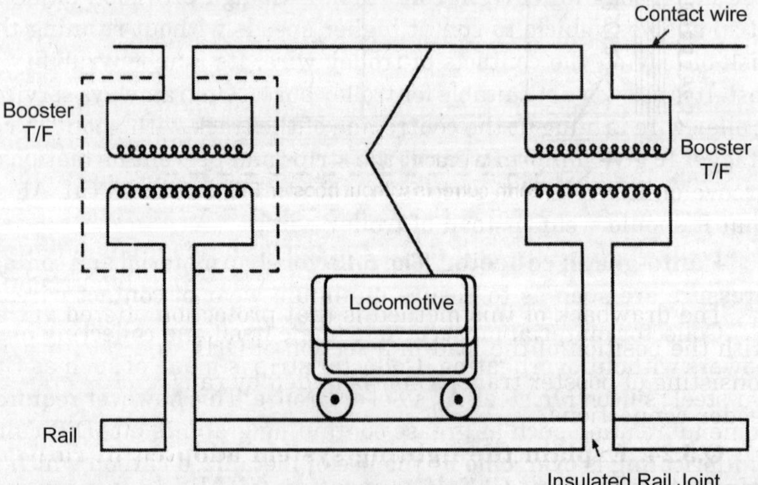

Fig. 5.12

(*ii*) Insulation punctured between rails of insulated joint may cause short circuiting the secondary and make booster T/F ineffective.

(*iii*) Voltage rise of 200 V between rail and 100 V between rail and earth are not uncommon. This may endanger life.

(*iv*) Protection offered being not more it requires closer spacing of booster T/F.

(*b*) **Booster T/F with return feeder**

The figure 5.13 shows midway between booster transformer rails are connected to return feeder.

By this arrangements rails are relieved of return current which now flows through the return feeder back to substation. Running the return feeder very close to the contact wire reduces the tendency of magnetic coupling between power line and telecommunication line. This method of connecting booster T/F is more effective for large space intervals.

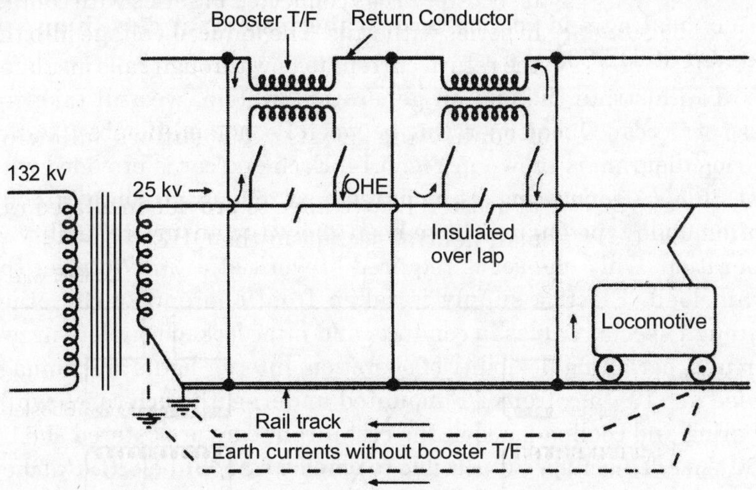

Fig, 5.13

The drawback of this method is that protection offered varies with the position of the load in a section of OHE and return path consisting of booster transformer bounded by rails and two return feeder connections.

Q.5.24. Explain the lighting system adopted in railway generation systems (i) Self generation *(ii)* Mid on generation and end on generation.

Ans. End on Generation. With the provision of more and more amenities to the passengers by the way of additional fans and lights the under frame generating system is becoming not only expensive at initial cost and maintenance but the provision of augmented capacity of dynamo itself may cause infringement with the moving dimensions. All the troubles can be overcome by employing mobile generator car which supplies the power requirement of the entire train. This is termed as end on generation as the generator car is attached at the end of the train. A second generator car at the other end of the train serves as a standby. This is the system used on deluxe trains.

In this system, electric power to the train is supplied from diesel generating sets at 400 V, 3 ϕ, 50 cycles a.c. generation set being mounted on each brake van at either end of the train. Coaches are provided with through wiring to transmit 400 V, 3 ϕ supply from end to end in duplicate through interlocked couplers. Step down T/Fs of (say) 5 kVA capacity are provided to feed the lights and fans at 100 V a.c. heating and cooling load in the restaurant cars, if any, is catered at 415/230 V a.c.

To illustrate the end on generation system, we will take the case of Deccan Queen operating between Poona and Bombay. Power wiring diagram is shown in Fig. 5.14. Each end car is provided with one 55 KVA generating set mounted next to luggage compartment. Automobile type engines have been chosen to ensure reliability in operation. Only one set is required to work at a time to meet the train load. Electric supply is taken from one end to the other through special Cables in conduits and inter locked couplers in two circuits, permiting flexibility of operation. Interior lights and fannage is fed at 110V a.c. from T/F mounted underneath each coach while heating and cooling loads in the restaurant cars are catered at 415/230 V a.c. This makes it possible to employ standard electric kitchen equipment. Control panel provided in the coaches consists of change over arrangement to select any one of the two feeders and also provisions for taking supply from external sources. Miniature circuit breaker are provided to afford necessary protection to various circuits. Fluorescent fittings are provided with condensers to improve P.F. of the load, which is not good due to inductive load of fans, static condensers of 25 kVA capacity are provided in the alternator room.

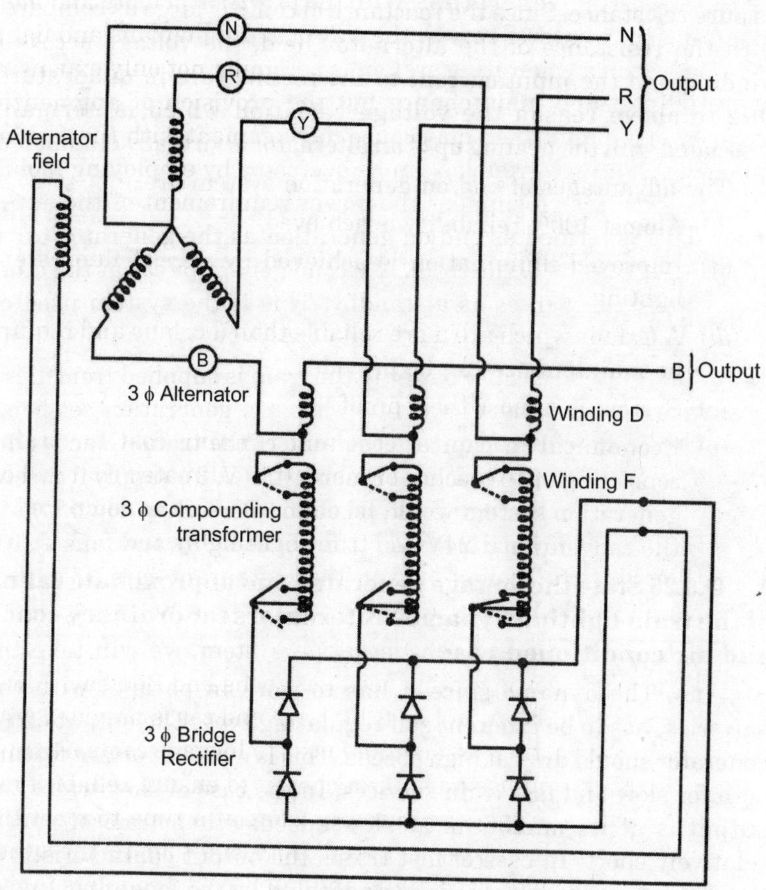

Fig. 5.14.

Alternator employs an extremely simple and reliable static excitation system making use of compounding T/F and a rectifier. The compounding T/F has an air gap in its magnetic circuit and is wound with two coils F and D. One no load winding F acts as a choke in series with the output of the alternator to supply the excitation via the rectifier. The compounding winding *D* which carries the load current of the alternator provides, by T/F action the additional excitation which is dependent on the magnitude and P.P. of the load. The phase relationship between these two components of the excitation is such that at zero P.F. lagging they are in phase and at unity P.F. they are in quadrature. This corresponds to the theoretical requirement of an ideal alternator. In case of 3 - ϕ alternator the rectifier and field winding behave in combination very approximately as pure resistance. Since the reactance of coil *F is* large as compared with the resistance of the alternator field, the voltage across *F* winding and the input current to the rectifier are in quadrature. Due to above reason the voltage variation which is normally associated with the heating up of an alternator is virtually eliminated.

The advantages of end on generation system are:

(*i*) Almost 100% reliability is achieved.

(*ii*) Improved illumination is achieved by way of fluorescent lighting.

(*iii*) A.C. fans which are more reliable than d.c. fans and require no maintenance can be provided.

(*iv*) System is almost theft proof.

(*v*) Economical in capital cost and running cost for trains consisting of 16 coaches or more. It is estimated that end on generation system would be of 66.67% cost as compared to the conventional 24V a.c. train lighting system.

Q.5.25 State the voltage generated and approximate rating of the train lighting dynamos/Alternators for ordinary coach and air conditioned coach.

Ans. The dynamo since it has to work in parallel with the batteries, has to be inherent self regulating type. The output of the generator should drop at high speeds. This is most important feature both for slow and fast train services. In the case of slow trains the output is at its maximum at slow speeds when time of run is relatively short. In case of fast trains the output characteristic of the dynamo drops but this is compensated by the longer periods of the run associated with the main line service. In this way we are in a position to have battery charge in ampere hours over a given period of time as same in case of either type of service.

The dynamo has two fields. The main field is connected across the +ve and -ve brushes and the aid field is connected across positive and third brush. The main field is designed to be relatively weaker so that armature reaction has strong influence on the pole flux distribution. Increase in load current will increase the armature reaction resulting in greater distortion of main field. Distortion of main field reduces the flux in the leading half of the pole which reduces the voltage between the dynamo +ve brush and third brush. This in turn reduces the aid field, current and thereby the total flux under the poles. This gives lower O/P voltage of the dynamo. As a direct result of reduced dynamo voltage load current will reduce providing, inherent regulation of the dynamo. The same control is exercised equally effectively for varying speeds of the dynamo. A constant field machine with the increase in speed would tend to generate voltage proportional to the speed. Its current output, which is proportional to the voltage will rise accordingly. This will cause increased armature reaction which reduces the potential across the aid field. The reduced aid field current lowers the output voltage of the dynamo and thus the load current. In this way automatic regulation of O/P is achieved irrespective of speed variation.

The third brush of the aid field is situated just opposite the main pole. As such it will be short circuiting the conductors which are under the maximum field string. The coil undergoing commutation over the aid brush will have dynamically induced emf. in addition to the reactances voltage developed due to the change in direction of the current. This will, therefore cause sparking at the aid brush. To minimize this the poles have to be cut away at the middle to appreciably reduce the flux density at this location.

To facilitate adjustment of the charging rate which varies during summer and winter months, a variable reactance known as 'output adjuster' is connected in series with the main shunt field. By increasing the resistance of shunt excitation is reduced thereby resulting in a lower dynamo voltage and consequently a lower output current. Usually 4 output settings of 100, 90, 80 and 70% are provided.

In the dynamo 3 resistances are connected in series with the aid field for obtaining flat, medium and drooping characteristics of the dynamo. These settings are quite independent of the O/P setting. Maximum resistance in aid field gives flat characteristics and minimum resistance gives drooping characteristics. With the increase in resistance the aid field current is reduced and as such it offers less control in the inherent regulation of the dynamo. With minimum resistance the aid field is comparatively strong and drooping characteristic required for inherent regulation of diagram.

Gauge	*Size of dynamo*
BG	100 Amp.
MG	60 Amp.
NG	40 Amp.

Q.5.26 (a) Know about obtaining unidirecting polarity.

Ans. Dynamo has rocker arm mounted on the shaft, friction tight. For the movement of armature clockwise, *X* terminal of the rocker arm will be +ve and Y terminal will be negative. Rocker arm being friction tight, will move in the direction of armature rotation. This will make X brush to touch brush P_1 and Y to touch brush N. This gives outside polarity of supply as shown in Fig. 5.15. If direction of rotation of shaft, terminal X will now *touch* N_1, and Y terminal will touch P. Once again outside polarity remains unchanged.

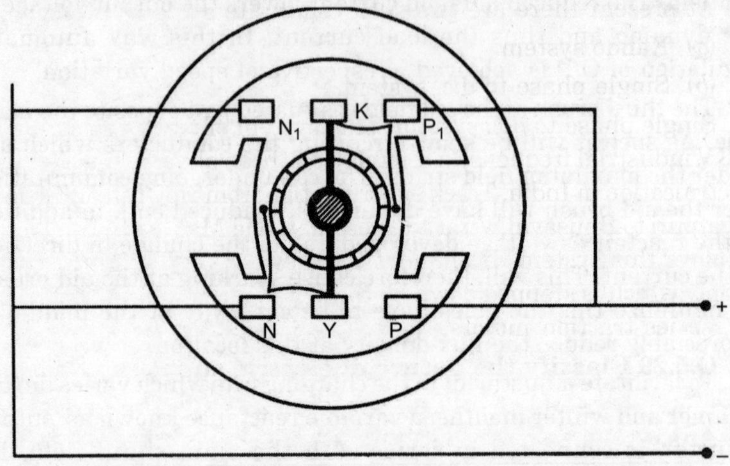

Fig. 5.15.

Q.5.27. Know about the method of obtaining constant output by using brushless alternator and their regulation.

Ans. As shown in Fig. 5.16 dynamo is hung from the carriage underframe and driven by means of belt from wheel axle pulley. Tensioning of the belt depends upon the component of weight of the dynamo and position of the tension screw. Any increase in the output makes armature heavy to rotate. This results in the belt to slip. This can be explained as. Increase in the drag on the armature results in the increase in belt tension which pulls the suspended dynamo towards pulley and belts get slack and slip.

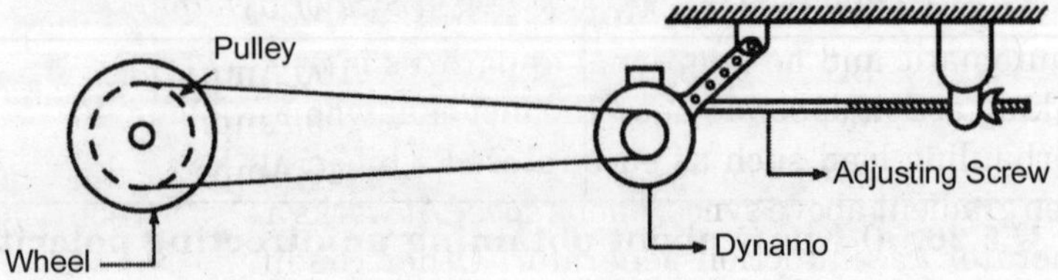

Fig. 5.16

Q.5.28 (*a*). Describe 1 - ϕ A.C. and composite system.

Ans. A.C. Systems: The A.C. system in India is being employed from Howrah to Tundla and from Madras to Tambram. The modern development in electric traction is to use 1ϕ A.C. supply.

Composite system.

At present there are two such composite systems in vogue :

(*a*) Kando system.

(*b*) Single phase to d.c. system.

Single phase to d.c. system. This system of traction employing 25KV industrial frequency a.c. supply has been adopted for all future electrification in India. Track electrification from Howrah and Tunia, Jagatpuri to Bhusaval, Virar to Sabarmati and Madras to Tambaram employs this system. It employs overhead 1 - ϕ, 25 kV, 50 cycles supply which is stepped down by the T/F to d.c. and used in driving d.c. series traction motors.

Q.5.29 Classify the nature of electric supply as D.C. and A.C.

Ans.

D.C.	**AC.**
It has constant value and does not vary with time.	It is of alternating nature and always vary with time.
Its wave shape is a straight line.	It has many wave shapes sinusoidal, square waveshape, triangle waveshape etc.
It can not step up or step down with the help of Transformers.	A.C. supply/voltage can step up or step down with the help of Transformers.

Q.5.30. Explain the suitability of different motors (A.C. and D.C.) for traction.

Ans. A.C Induction Motor. E.R.B. with 3-ϕ induction motor is automatic and no additional apparatus is needed. Regenerative braking action depends upon the fact that when motor is driven by overhauling load such as encountered while train is descending a steep gradient above synchronous speed, it works as non-synchronous generator *i.e.* induction generator. Other condition necessary for obtaining generation action, in view of induction machine being non-self exciting is that it needs supply system which supplies excitation and fixes frequency. It is therefore necessary that supply system should be at least connected with one synchronous alternator in addition to load dissipation arrangement speed-torque curves for both motor and generator operation are shown in Fig. 5.17. It is observed therefrom that if no resistance in the rotor circuit is inserted, speed of the train will be practically unaffected by this steepness of the gradient or weight of the train. If increased speeds are required with the light loads these can be obtained by inserting resistance in rotor circuit.

It will be further observed from Fig. 5.17 that maximum braking torque available is about 3 times the continuous full load torque.

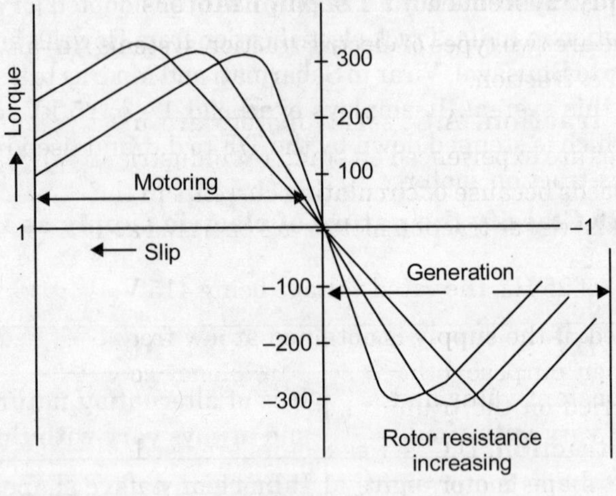

Fig. 5.17

Q.5.31 Briefly describe (i) **Suitability series motors** *(ii)* **Supply systems of traction motors.**

Ans. All the above requirements of electric traction are suitably met by d.c. series motors. A d.c. series motor is, therefore, the ideal-choice. The motor has to provide sufficient tractive effort (TE) to set the train in motion. The maximum TE is produced at starting when the motor is separately excited. The difference between the TE offered by the motor and the tractive resistance is the force available for accelerating the train. As the train gains speed, TE gradually decreases and eventually the balancing speed is reached. It is, therefore, seen that a series motor offers the best speed-torque characteristic for electric traction.

The heavy conductor sections used in the armature and field circuits make the series motor robust and reliable. Constant horse-power operation takes place. The motor current is restricted to a certain permissible limit. The motor can draw over-current for short periods. The motor efficiency through gears is normally high, above 80%. In traction systems, both electrical and mechanical forms of braking are applied. Electrical braking offers smoother and higher retardation than that obtained in mechanical braking. The only disadvantage with series motors for electric traction is that regenerative braking is not readily possible. A series motor has to be separately excited for regenerative braking. It may be noted that a d.c. series motor has a higher power to weight ratio than that of an a.c. series motor, which also gives commutation difficulties at low speeds.

Supply Systems and Traction Motors

There are two types of electric traction systems : (a) a.c. traction, and (b) d.c. traction.

A.C. traction. A.C. series motors are used. Commutation difficulties are experienced on standard industrial frequency supply at low speeds because of circulating currents in the short-circulated coil. That is why ac traction motors are operated at a lower frequency, *e.g.* $16\frac{2}{3}$ of 25 Hz, the rated voltage being 415 V. So, a transformer is required, if the supply is obtained at low frequency dedicated to the traction purpose, otherwise, a frequency converter is required to be carried on the train.

D.C. traction. D.C. series motors are used. The advantages of using d.c. series motors have already been stated. In this system, the d.c. operating voltage is 1500 V for urban and suburban services and 3000 V for main line service. The d.c. voltage is restricted to 3000 V to avoid commutation problems. Thyristor converters and choppers are used for driving d.c. traction motors resulting in high efficiency, reliability and smooth control. A transformer with tappings and on-load tap changers are also carried on the train to step down the voltage.

In both types of traction systems, electricity is collected from the overhead contact wire at 25 kV with the help of a pantograph. Adequate insulation has to be provided to protect the train from electrocution. The rail line serves as the return path for current.

Traction substations are set up enroute from one terminus to another terminus station. Assuming that 132 kV source is available, the substations house a 132/25 kV step-down transformer with proper switchgear and protection systems. In case of suburban lines, the substations are spaced 3 to 5 km apart, whereas for main line they are spaced 40 to 50 km apart. For low frequency operations, the substations may be spaced at a distance of 50 to 80 km apart. High voltage is preferred for overhead contact wire line so that its conductor section is reduced and the spacing between conductor sections increased. The a.c. system is preferable for overhead contact wire system from the point of view of current distribution. In metro railway systems, d.c. supply is given between the running rails.

Q.5.32 Explain the train movements in Electric Traction.

Ans. Train Movement: A train needs to be run optimally between the stops from the point of view of energy consumption, considering the weight carried and the distance traversed by the train. Speed-time and speed-distance curves are needed for the study of optimal train movement. In general, while moving from one station to another, the motion of the train may be of five types:

1. During rheostatic acceleration, the starting resistance is cut out step-by-step so as to maintain constant current and constant acceleration.
2. At the instant t_1, all starting resistance has been cut out. The train continues to accelerate up to the instant t_2, when balancing or steady speed is reached.
3. The train runs freely at steady speed in this section.
4. When the train approaches a stop, the power supply to the traction motors is switched off the speed falls, but the train continues to run due to momentum. The duration is called the *coasting period.*
5. First the electric brake and then the mechanical brake is applied to stop the train.

Q.5.33 Explain the Layout of Electric Drives and speed Control of traction motors.

Ans. Layout of Electric Drives :

The power for electric traction is provided by traction motors to the driving wheel through gears. The traction equipment, required for driving and controlling the train, is mounted in the underfloor bogies. A suburban or metro train consists of a number of units. Each unit consists of a motor coach, a trailer coach and a driven trailer coach. The traction motors are mounted under the trailer coach. The whole train can be operated by one driver from either end. Electric traction is more suitable for suburban trains where frequent start, stop and high speed schedules are required.

Electric locomotives are used for long distance trains. An underframe superstructure is mounted in two bogies of the electric locomotive. The traction motors and their control equipment are mounted on the superstructure. The roof of the locomotive is made strong to hold the pantograph. The driver's cab is also built in the locomotive with all controlling gadgets. The driver's control is for hand regulation. There is a master controller in the driver's cab to regulate train operations, *e.g.* starting, acceleration, speed control, coasting, braking, reversal of motors, etc. As the traction motors are mounted under the floor of the motor coach or the locomotive, they should be suitably protected from dust and humidity.

In adiesel electric locomotive, a motor-generator set is also carried in the locomotive to provide power to traction motors, which drives the wheels of the locomotive. Series generators and series motors are used. Diesel locomotives are suitable for hauling goods trains as they provide a high tractive effort to pull the large number of wagons involved. Two electric locomotives are sometimes deployed to run a goods train.

(*b*) **Speed Control**

The speed of traction motors is controlled by the combined series-parallel method and rheostatic method.

In electric traction, four or six identical motors are used. Constant horsepower operation takes place. There are separate notches of the master controller for different speeds. For long distance trains, acceleration and braking periods are negligible. On the other hand, free running is almost absent for suburban trains. There may be different arrangements of motor connections yielding different speeds. For example, a six motor locomotive having each motor rated for half the line voltage may undertake the following connections (A set comprises two motors connected in series):

(*a*) Three sets connected in parallel yielding full speed.

(*b*) Three sets connected in series yielding one-third of full speed.

If three motors are connected in series to form two sets across the line voltage, two-third of full speed is obtained. Similar interconnections may be made with four motors to have different speeds.

Accelerating resistance is used for adjustable voltage control of d.c. traction motors placed under the carriage. For getting various speeds, transformer output voltage are also adjusted with the help of on-load tap changers.

UTILISATION OF ELECTRICAL POWER AND ELECTRIC TRACTION

G.C. GARG

FEEDBACK FORM

1. Which topics from your examination point of view are inadequately or not covered in this book?

____

____

2. Is there any misprints/ mistakes/ factual inaccuracy in this book? If so, please specify.

____

____

3. What is your assessment of this book as regards the presentation of the subject-matter, expression, precision and price in relation to the other recommended books available on this subject?

____

____

4. Which competing books you regard as better than this? Please mention their author and publishers.

____

____

5. Any further suggestions/comments you would like to make for the improvement of the book?

____

____

PS: If needed, please attach a separate sheet.

Name : ____________________

College : ____________________

Address : ____________________

Email : ____________________ **Mobile** : ____________________

Please mail the feedback form at:

KHANNA BOOK PUBLISHING CO. (P) LTD.

4C/4344, Ansari Road, Daryaganj, New Delhi-110002

You can also email this feedback form at contact@khannabooks.com or Whatsapp at +91 – 99109 09320

BEST FEEDBACK WILL BE REWARDED SUITABLY